Advances in Geographic Information Science

The series aims to: present current and emerging innovations in GIScience; describe new and robust GIScience methods for use in transdisciplinary problem solving and decision making contexts, illustrate GIScience case studies for use in teaching and training situations, analyze GIScience tools at the forefront of scientific research, and examine the future of GIScience in an expanding knowledge-based economy. The scope of the series is broad and will encompass work from all subject disciplines that develop and use an explicitly spatial perspective for analysis and problem-solving. Advances in Geographic Information Science.

More information about this series at http://www.springer.com/series/7712

Advances in Geographic Information Science

The series aims to: present current and emerging innovations in GIScience; describe new and robust GIScience methods for use in transdisciplinary problem solving and decision making contexts, illustrate GIScience case studies for use in teaching and training situations, analyze GIScience tools at the forefront of scientific research, and examine the future of GIScience in an expanding knowledge-based economy. The scope of the series is broad and will encompass work from all subject disciplines that develop and use an explicitly spatial perspective for analysis and problem-solving. Advances in Geographic Information Science.

More information about this series at http://www.springer.com/series/7712

Dionissios T. Hristopulos

Random Fields for Spatial Data Modeling

A Primer for Scientists and Engineers

Springer

Dionissios T. Hristopulos
Technical University of Crete
Chania, Greece

ISSN 1867-2434 ISSN 1867-2442 (electronic)
Advances in Geographic Information Science
ISBN 978-94-024-1916-0 ISBN 978-94-024-1918-4 (eBook)
https://doi.org/10.1007/978-94-024-1918-4

This Springer imprint is published by the registered company Springer Nature B.V.
The registered company address is: Van Godewijckstraat 30, 3311 GX Dordrecht, The Netherlands

To my parents, Theodoros and Anna, for all that they have given me,

To Lisa for her love and support,

To Annalina and Thodoris for their random smiles, as well as the continuities and infinities that they represent,

To the Cretan seas for inspiration.

Preface

> It is through science that we prove, but through intuition that we discover.
> *Jules H. Poincaré*

Why Spatial Data Modeling?

Spatial data modeling encompasses an evolving landscape of concepts and methods that can be used to characterize, quantify, and exploit the spatial structure of extended variables. Spatial data analysis has significantly advanced over recent decades, partly due to the ongoing explosion in the availability of Earth observation data and faster computers. The field will most likely undergo a growth spurt in the coming years as a result of the continuously improving access of the research community to large and diverse spatial data sets. This development will drive methodological advances and generate tools suitable for the processing of massive spatial data.

Spatial data abound in the earth and environmental sciences—including climate science—epidemiology, social sciences, finance, astronomy, and many other research fields. The ability to extract information and patterns from large data sets will be of crucial importance in the years ahead. The development of methods, the computational performance of which scales favorably (e.g., linearly or sublinearly) with data size and is sufficiently flexible to handle different spatial features, will be instrumental in these efforts.

Spatial data analysis has applications in different scientific and engineering disciplines. Accordingly, quantitative methods of spatial analysis have been independently developed in fields such as statistics, applied mathematics, hydrology, geography, mining and mechanical engineering, geodesy, reservoir and surveying engineering, as well as machine learning. In addition, models and ideas that originate in statistical physics are very useful for characterizing correlations and spatial structure in extended systems.

Why Random Fields?

To date, spatial data modeling comprises an extensive body of methods and techniques. Hence, it is not surprising that this book covers only a small part of this evolving landscape. The book focuses on spatial random fields, their mathematical properties, and their applications in the analysis of spatially indexed scientific data. The overall goal is to present aspects of random field theory that are important for applications. Regarding statistical methods for estimation purposes, the focus is on intuitive understanding of selective methods rather than on an exhaustive presentation of the literature.

Random fields have become indispensable tools for the modeling and analysis of various natural and engineered processes that are characterized by complex variability and uncertainty. The term "complex" here loosely refers to variations that cannot be simply described by means of closed-form deterministic expressions.

Applications of random fields are widespread in fluid mechanics [587], computational and probabilistic engineering mechanics [286], materials science [636, 797], hydrological modeling [275, 459, 696], petroleum engineering [195, 241, 355], environmental monitoring [138, 425, 758], mining exploration and mineral reserves estimation [33, 210, 303], environmental health [141], geophysical signal processing [759], ecology [176], nanolithography [294, 295], image analysis [851, 852], astronomy [244], statistical cosmology [66], medical image registration [703], as well as the structural and functional mapping of the brain [115, 496, 751, 856].

Research that involves random fields and their applications is being pursued by applied mathematicians, statisticians, physicists, geophysicists, geostatisticians, epidemiologists, as well as mining, structural, mechanical, electrical, geotechnical, and reliability engineers.

Why should scientists care to read a book about random fields when a number of easy-to-use software tools that implement basic spatial data analysis are available? Here are some reasons:

1. Even practitioners of such black box software will benefit from an understanding of the concepts and assumptions involved in spatial data modeling. This will help them to recognize conditions under which the analysis can fail as well as possible signatures of trouble.
2. Further progress is needed in various research directions. These involve the processing of large data sets, advanced methods for coupled space and time correlations, as well as improved methods for the modeling, estimation, and simulation of nonstationary, non-Gaussian, and multivariate data as well as connections between random field theory and machine learning. This book discusses some of the recently proposed approaches for handling large data sets and briefly

comments on topics related to non-stationarity and multivariate dependence. Space-time methods are outside the scope of this book, but stochastic partial differential equations and local interaction models that are discussed herein provide flexible frameworks which can be used for developing statistical space-time models.

3. This book provides guidance to scientists and engineers who seek to analyze and design spatial patterns according to specific guidelines. For example, a topic of practical importance is the development of accurate and flexible methods for automated mapping and early warning systems that require minimal user intervention [215, 216]:

> Monitoring networks regularly collect and report local observations of a variable that need to be converted into information with spatial continuity, i.e., maps which are frequently essential for decision-making or further modeling. Ideally, these maps should be produced automatically in order to allow real-time assessments of the phenomenon monitored and to minimize human intervention in case of emergency. However, the task of automating the spatial interpolation step is everything but straightforward. First of all, no universally best mapping algorithm exists because each function has its advantages and drawbacks. Second, most functions require a number of parameters to be set, even arbitrarily.

Spatial Data Modeling and Statistical Physics

The approach adopted in this book emphasizes connections between spatial data modeling and concepts originating in statistical physics. There is a long history of successful transfer of ideas from statistical physics to statistics and spatial data analysis. Notable cases include Markov Chain Monte Carlo methods (e.g., Metropolis algorithm) that are widely used in simulation and Bayesian inference of spatial models [109] and simulated annealing [453] which is used in the conditional simulation of spatial data [624]. Statistical physics ideas are also important in machine learning, and this book highlights some of the connections between random fields and machine learning methods (mostly Gaussian process regression).

The motivation for writing this book is twofold: First, I hope that it will prove useful to researchers specializing in science and engineering applications of random fields and practitioners of spatial data analysis. Second, the book aims to appeal to statistical physicists involved with interdisciplinary research and the analysis of scientific spatial data arising, for example, in statistical seismology, hydrocarbon reservoir simulation, environmental physics, and climate modeling.

What Is a Random Field?

Random fields generalize the concept of random variables to spatially extended quantities (fields). An uncertain localized variable[1] that is independent of the spatial location and time is modeled as a *random variable*. For example, we can think of the wind speed at any given location as a random variable that takes different values in time. The random variable concept is thus linked with a number of possible (probable) states. In the *frequentist viewpoint*, these states are presumably observed with a frequency that is determined from the respective probability distribution.

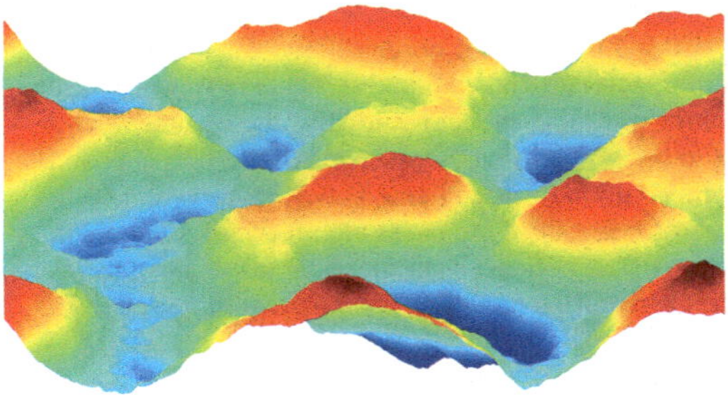

If we care about the evolution of a random variable in time, then the concept of *random processes* is used. For example, the wind speed can be sampled at different points in time and examined for possible patterns (regularities). This can be particularly interesting if the wind speed values at different times are not independent.

Furthermore, if the observables are spatially extended, we use the concept of *spatial random fields (SRFs)* or simply random fields. In reference to the wind speed example, if we freeze time and consider the values of the wind speed over a given area in space, the wind speed can be treated as an SRF. A random field can be viewed as a collection of random variables that are distributed in space. A key property of random fields is that the constituent random variables are *correlated*. Hence, the assumption of independent random variables which is common in statistics does not apply to random fields.

Random fields are also known as *random functions*. In certain scientific disciplines, e.g., in machine learning, the term *random process* is also used for random fields defined in general spaces. If the variable of interest exhibits dependence on both space and time, it can be modeled as a *space-time random field*, i.e., a random function whose values are correlated across space and time (space-time random fields are not covered in this book).

Word of caution: In statistical physics, the term "random field" specifically refers to nonuniform external electromagnetic fields, such as those generated by the presence of randomly distributed magnetic impurities in crystalline materials.

[1] The term "lumped variable" is also used for localized variables, which often represent coarse-grained quantities.

Target Audience

This book is an introduction to the theory of random fields with emphasis on applications in spatial data modeling. In addition, it aims to provide a bridge across different scientific disciplines that use random fields.

The *intended audience* involves advanced undergraduate students, graduate students, and practicing scientists and engineers who work with spatial data and would like to develop a better understanding of methods based on random fields. The book also aims to inspire the younger generation of scientists to investigate new frontiers at the boundaries between statistics, applied mathematics, machine learning, and statistical physics.

Geographic information systems (GIS) are routinely used nowadays for the visualization and preliminary analysis of spatial data. Modules that perform interpolation of spatial data are integrated in commercial GIS systems, such as the Geostatistical Analyst tool in ArcGIS [415]. GIS users who seek a deeper understanding of spatial interpolation and simulation will also benefit from this book.

A background in probability and statistics, linear ordinary differential equations, and Fourier transformations is required to read this book. Good knowledge of advanced calculus is also useful. Some parts of the book require at least some familiarity with partial differential equations as well as with stochastic ordinary and partial differential equations. More advanced mathematical background and familiarity with statistical physics will make certain passages of the book easier to read.

How to Read This Book

Chapter 1 provides notation and background information. The latter includes an overview of concepts (random fields, trends, fluctuations, noise) that are useful in spatial data modeling, as well as motivation for the use of random fields. The chapter also includes a discussion on the connections between random field theory and statistical mechanics on one hand and the theory of nonlinear systems on the other. The chapter ends with a personal selection of books related to spatial data modeling and random fields. This chapter could be skipped by impatient readers.

Chapter 2 focuses on the estimation of trend functions from data. Loosely speaking, the trend function is the deterministic component of random fields that describes large-scale spatial variability. We distinguish between statistical trend models and trend functions that are based on explicit solutions of physical laws. From the class of statistical models, we discuss linear regression and nonparametric smoothing approaches. Important concepts for the latter include moving window averaging, kernel regression, locally weighted regression, and polynomial (Savitzky-Golay) filters. A specific example of trend calculation for steady-state flow in porous media is presented in detail to illustrate the derivation of trend functions from physical laws. This chapter could be skipped by readers who are more interested in the stochastic aspects of random fields. On the other hand, the passages on regression and smoothing may be more useful to readers who are interested in coarse-scale, exploratory modeling of spatial data.

Chapter 3 offers a quick introduction to notions of probability theory that are used for the one-point and two-point descriptions of random fields including ensemble moments, moment and cumulant generating functions, and characteristic functions. This chapter introduces the covariance and variogram functions as well as the concept of statistical stationarity. Bochner's permissibility theorem is also discussed herein, and connections are drawn with the Wiener-Khinchin theorem in statistical physics. The relation between the covariance and the variogram as well as permissibility conditions for the latter are reviewed.

Chapter 4 continues the exploration of fundamental concepts and properties of random fields. It introduces the notions of ergodicity, isotropy, radial correlation functions and their properties, definitions of statistical anisotropy, as well as different types of correlation models and their properties. In addition, the mathematical framework for the study of multipoint (joint) properties of random fields is introduced including joint moments, cumulant, and moment-generating functions.

Chapter 5 investigates the geometric properties of random fields. These are founded on concepts of stochastic convergence and their applications to the continuity and differentiability of random fields. Spectral moments and their relation to differentiability, as well as different length scales that characterize spatial correlations in random fields, are also discussed. Then the focus shifts to fractal features, intrinsic random fields, and self-affine random fields that lack characteristic scales. The concepts of long-range dependence and roughness of random field surfaces are also discussed herein. Special attention is given to the fractional Brownian motion (fBm) model due to its central position in science and its broad range of applicability. Finally, the chapter closes with different random field classifications based on the joint probability distribution, the degree of non-stationarity, and the range of the correlations.

Chapter 6 begins by presenting mathematical properties of Gaussian random fields (GRFs). Following the approach used in statistical physics, we introduce GRFs by means of Boltzmann-Gibbs exponential joint density functions. This is followed by an introduction to the functional integral (field integral) formulation and its applications in evaluating moments and cumulants by means of respective generating functions. We also present the Wick-Isserlis and the Novikov-Furutsu-Donsker theorems that are quite useful for calculations with Gaussian random fields. We then address the use of perturbation expansions and the variational approximation for evaluating the moments of mildly non-Gaussian probability distributions. The perturbation analysis is also applied to the construction of nonstationary covariance functions. Finally, this chapter closes with a brief overview of methods used for generating nonstationary covariance models.

Chapter 7 presents the theory of Spartan spatial random fields (SSRFs). These are random fields with Boltzmann-Gibbs joint exponential density, defined by means of a quadratic energy function that is based on local interactions. SSRFs are a close cousin of the Gaussian model used in statistical field theory. The most important mathematical properties of SSRFs including their rational spectral density and their correlation functions in one, two, and three dimensions are presented. The chapter closes with a discussion of Gaussian random fields with Bessel-Lommel covariance functions which are motivated by SSRFs.

Chapter 8 continues the journey that begins in Chap. 7 by investigating lattice random field models that are obtained from discretized approximations of the SSRF continuum-space formulation. These include both isotropic and anisotropic spatial models. Connections with Gaussian Markov random fields and conditional autoregressive (CAR) models as well as basic ideas (e.g., conditional independence) and properties (e.g., conditional mean and variance) that pertain to the latter are discussed herein. We also investigate the question of whether it is possible to determine explicitly both the covariance and precision matrices for a lattice random field. The question is practically important, because a positive answer implies that the computationally costly operation of matrix inversion can be avoided.

Chapter 9 develops the connection between Spartan random fields and stochastic differential (Langevin) equations. It begins with an introduction to standard stochastic differential equations for the Brownian motion and the Ornstein-Uhlenbeck process. It then introduces the stochastic differential equation for the stochastic classical damped harmonic oscillator driven by white noise and establishes the equivalence between this model and the one-dimensional Spartan random field. The chapter then shifts to stochastic partial differential equations (SPDEs). Equations that involve linear combinations of partial derivatives, the fractional Whittle-Matérn SPDE, and SPDEs derived from polynomials of the diffusion operator are included. The SPDE associated with Spartan random fields in two and three spatial dimensions is derived, and its connections with the de Wijs process are discussed. There is also a discussion of the relation between covariance functions and the Green functions of associated partial differential equations. The chapter closes with a review of the linear time series ARMA models and highlights the connection between the Spartan random field in one dimension and the second-order autoregressive model.

Chapter 10 comprises an overview of spatial prediction methods for the interpolation of spatial data. It includes commonly used deterministic methods such as inverse distance weighting, minimum curvature, and natural neighbor interpolation. The main focus is on the stochastic interpolation method known as kriging, with emphasis on the two most common variants, i.e., simple and ordinary kriging. Various mathematical properties and practical issues related to the application of these methods and the validation of the results are discussed. The fundamental equations for the optimal kriging prediction and the kriging variance are presented for both methods. To spice up the discussion, kriging examples that employ Spartan covariance functions are solved. The connection between minimum curvature interpolation and kriging with a generalized covariance function is also discussed in this chapter.

Chapter 11 briefly reviews linear extensions of the basic kriging methods including ordinary kriging with intrinsic random fields, regression kriging, universal kriging, cokriging, and functional kriging. These are followed by brief expositions of nonlinear extensions of kriging, including indicator kriging, spin-based indicator models, and lognormal kriging. It also touches on the connection of kriging with the Bayesian framework in terms of Gaussian process regression and empirical Bayesian kriging. It then continues with linear prediction in the framework of Spartan spatial random fields and their discrete counterparts, i.e., the stochastic local interaction models. Connections between the latter and Gaussian Markov random fields are presented. The chapter concludes with a brief personal perspective on spatial prediction in the era of spatial data of continuously increasing size.

The focus of Chap. 12 is the estimation of the random field model from available data, which could be either irregularly spaced or distributed on regular lattices. This involves a vast area of research that cannot be adequately described herein. We opt to present certain classical concepts and methods which include desired statistical properties of estimators, estimation of the population mean from correlated data using ordinary kriging, and variogram estimation by means of the classical method of moments. The latter may be mathematically inferior to maximum likelihood estimation, but it has the advantages of visual clarity and intuitive appeal. The fitting of empirical variograms with suitable theoretical models is also discussed. Next, the basic elements of maximum likelihood estimation are presented and discussed. The final topic is the use of cross-validation for model parameter estimation.

Chapter 13 discusses random field estimation methods that are less popular or less tested. These involve the method of normalized correlations and maximum entropy. The former is analogous to the Yule-Walker method for estimating the coefficients of autoregressive models in time series analysis. Maximum entropy has a long history of applications in physics and has also been successfully used in geostatistical applications. As an example, the formulation of Spartan random fields is derived from the principle of maximum entropy and suitable spatial constraints. Next, parameter estimation and spatial prediction in the framework of stochastic local interaction models are explored. This section relies heavily on connections with Gaussian Markov random fields. The chapter closes with the formulation of an ergodic index; this is a heuristic measure that quantifies the suitability of the ergodic hypothesis.

Chapter 14 discusses spatial models with non-Gaussian probability distributions. It visits standard geostatistical topics such as trans-Gaussian random fields, Gaussian anamorphosis, and Box-Cox transformations. It also includes some newcomers to the literature such as Tukey g-and-h random fields and random fields based on the Kaniadakis exponential and logarithmic transforms. Student-t and log-Student-t random fields are also discussed in some detail, as well as the formulation of spatial prediction for such random fields. This is followed by the use of Hermite polynomials in nonlinear expansions of non-Gaussian random fields. A very brief introduction to copula models is also given herein. The chapter concludes with an introduction to the replica method, a tool that has been successfully applied to problems in statistical physics and machine learning but not in spatial data analysis.

Chapter 15 focuses on binary random fields such as the indicator field and the Ising spin model of statistical physics. The Ising model is the first example of a Markov random field introduced in statistics and has several applications in spatial data analysis. The mean-field theory of the Ising model is presented in some detail. This chapter also briefly discusses ongoing efforts for applying the Ising (and other discrete-valued spin models) to spatial data modeling. Mathematical results for binary random fields that were derived in porous media studies and are not well-known in the spatial statistics community are also presented. These include level cuts of Gaussian random fields, the leveled-wave model, and their applications in modeling porous media morphology. The chapter concludes with short references to generalized linear models, model-based geostatistics, and logistic regression.

Chapter 16 deals with the practically important topic of random field simulation. This involves a vast area of research that cannot be comprehensively presented in a single chapter. The material discussed herein involves standard methods for the unconditional and conditional simulation of random fields, including covariance matrix factorization, spectral approaches, and applications of low-discrepancy sequences. The spectral approaches include fast Fourier transform simulation and randomized spectral sampling with various sampling schemes (e.g., simple sampling, importance sampling, stratified sampling). The chapter also presents a brief review of Markov Chain Monte Carlo (MCMC) methods and especially the two most common variants based on the Metropolis-Hastings and the Gibbs sampling algorithms. Some of the simulation methods presented focus on Gaussian random fields, while MCMC methods have a broader scope. The simulation methods that focus on Gaussian models can be extended to non-Gaussian fields by means of the nonlinear transforms presented in Chap. 14. Examples that relate to the MCMC simulation of the binary Ising model, described in Chap. 15, are also given. An application of Gibbs sampling in sequential Gaussian simulation is discussed as well as the application of the Metropolis-Hastings scheme in simulated annealing. Finally, this chapter presents the Karhunen-Loève (K-L) optimal basis expansion. Optimal basis expansions are used in simulation as efficient tools for dimensionality reduction. In addition to the widely known Karhunen-Loève expansion of the Wiener process, this chapter also presents in detail the K-L expansion of one-dimensional Spartan random fields. This also represents the Karhunen-Loève expansion of a classical, linear, damped harmonic oscillator driven by white noise.

This book contains both classical results of random field theory and statistical analysis as well as less standard material. Certain parts of the book, e.g., sections in Chaps. 4, 6, 7, 9, 11, 14, and 16, are influenced by the authors' research. Some of the topics discussed in these chapters are motivated by connections between statistical field theory, applied mathematics, machine learning, and spatial random fields. One of the goals of this book is to emphasize such cross-disciplinary bridges and to encourage more interdisciplinary interaction, especially among young scientists.

Readers are welcome to send comments, suggestions, and corrections to the following electronic mail address: dionisi@mailaps.org.

Chania, Crete, Greece
September 2019

Dionissios Hristopulos

Acknowledgments

A career in science and writing of a book are continuous processes interspersed by a few critical events that are comparable to phase transitions in physical systems. By reason of continuity, I am indebted to great teachers for the gifts of inspiration and knowledge and to several colleagues for creating intellectually stimulating microcosms. The following paragraphs are an attempt to map the continuum.

To begin at the beginning, many teachers at the Ionidios High School of Pireas (Greece) played a crucial role in my education. Their inspired efforts to provide high-quality public education with modest means have shaped my choices in life.

At the National Technical University of Athens (NTUA), I benefited from the lectures of excellent science teachers, including my diploma thesis advisor, Alexandros Serafetinidis, as well as interactions with many bright and motivated classmates. The overall experience at NTUA, in spite of the politically charged climate of the times, convinced me to pursue graduate studies and a career in science.

At Princeton University, I was fortunate to learn from and find inspiration in the work of top-class physicists such as Philip W. Anderson (Nobel Prize in Physics, 1977), Sriram Shastry, Kirk MacDonald, James Peebles, Sol Gruner, Russ Gianetta, Bob Austin, and Duncan Haldane (Nobel Prize in Physics, 2016). During the Princeton years, I was also motivated and inspired by several fellow graduate students including Arif Babul, Camm Maguire, Nikos Nikopoulos, Jean Quashnock, Yong Ren, Charles Stafford, Greg Tucker, Hercules Vladimirou, Weimin Wang, and Alex Zaslavsky. Finally, the Center for Hellenic Studies, run by Dimitri Gondicas, provided a vital escape to a different universe, where I learned from great classical scholars such as Edmund Keeley and Robert Connor.

My first steps in science after Princeton were at the University of North Carolina at Chapel Hill, where I was introduced to theoretical and applied aspects of random fields by one of the pioneers of space-time random field modeling, George Christakos. During the Chapel Hill years, I had constructive collaborations with Marc Serre and Alexander Kolovos. I am also thankful for the encouragement that John Cushman (Purdue University) showed to my research during my first steps in a new field. Joshua Socolar at Duke University was a positive influence throughout this period.

After Chapel Hill, Tetsu Uesaka gave me the opportunity to apply stochastic methods in the study of physical properties of fiber networks at the Pulp and Paper Research Institute of Canada (Paprican). At Paprican, I learned the art of communicating research to broader audiences, enjoyed scientific conversations with Nick Provatas (currently at McGill University) and David Vidal (currently at Polytechnique, Montréal), and made several good friends.

At the Technical University of Crete (TUC), I am most thankful to Zach Agioutantis, Stelios Mertikas, and Nikos Varotsis for helping me adapt to the new environment. While at the TUC, I also had the opportunity for international collaborations with many colleagues including Gerard Heuvelink (Wageningen University), Edzer Pebesma (University of Muenster), Denis Allard (INRA), Mikhail Kanevski (Université de Lausanne), Juergen Pilz (Alpen-Adria Universität Klagenfurt), Dan Cornford (then at Aston University), Ricardo Olea (US Geological Survey), and Gregoire Dubois (ISPRA). Giorgio Kaniadakis (Politecnico di Torino) gave me the opportunity to cultivate the connection between spatial data and statistical physics by introducing me to the ΣΦ *International Statistical Physics* conference series and supporting the organization of special sessions on environmental applications of statistical physics over the years.

This book has been brewing in my mind over a decade, following the first paper on Spartan spatial random fields [362]. At the TUC, I benefited from collaborations with postdoctoral associates and graduate students who helped to develop, extend, and apply some of the ideas presented in this book. Samuel Elogne and Milan Žukovič significantly contributed to modeling and numerical implementations. Emmanouil Varouchakis applied methodologies to hydrological problems. Ersi Chorti, Manolis Petrakis, and Ioannis Spiliopoulos contributed to the development of anisotropy estimation methods. Vasiliki Agou, Andreas Pavlidis, and Panagiota Gkafa applied Spartan random fields in case studies that involve mineral resources data. Ivi Tsantili calculated the Karhunen-Loève expansion of Spartan random fields and contributed to the extension of local interaction concepts in the space-time domains. Aris Moustakas and Vasiliki Mouslopoulou proposed potential applications of random fields in ecology and seismology.

I am thankful for their hospitality during short visits to Denis Allard (Institut National de la Recherche Agronomique, France), Anastassia Baxevani (University of Cyprus, Nicosia, Cyprus), Patrick Bogaert (Université Catholique de Louvain, Belgium), Sujit K. Ghosh (North Carolina State University, USA), Markus Hilpert (Johns Hopkins University, USA), George Karniadakis (Brown University, USA), Kostas Konstantinou (National Central University, Taiwan), Valerie Monbet (Université de Rennes 1, France), and Emilio Porcu (University of Sassari, Italy, and University of Valparaiso, Chile).

Finally, I would like to thank the Springer editorial team, Aldo Rampioni and Kirsten Theunissen for their encouragement and gentle reminders during the long and arduous writing process, as well as Christopher Loughlin and Christoph Baumann for helping in the final stages of this project. My thanks also go to the Springer production editor, Clement Wilson, and his team for bringing this book to its final form.

Contents

Acronyms

1D	One-Dimensional
2D	Two-Dimensional
3D	Three-Dimensional
AIC	Akaike Information Criterion
ALC	Allowed Linear Combination
AMR	Adaptive Mesh Refinement
ARMA	Autoregressive Moving Average
ARIMA	Autoregressive Integrated Moving Average
BC	Box-Cox Transform
BIC	Bayesian Information Criterion
BG	Boltzmann-Gibbs
CLT	Central Limit Theorem
CDF	Cumulative Distribution Function
CGF	Cumulant Generating Function
CHI	Covariance Hessian Identity
DFA	Detrended Fluctuation Analysis
EOM	Equation of Motion
DFT	Discrete Fourier Transform
fBm	Fractional Brownian Motion
FBZ	First Brillouin Zone
fGn	Fractional Gaussian Noise
FT	Fourier Transform
FFT	Fast Fourier Transform
GCM	Global Climate Models
GDR	Gamma Dose Rate
GIS	Geographic Information Systems
GPR	Gaussian Process Regression
GRF	Gaussian Random Field
GMRF	Gaussian Markov Random Field
IDW	Inverse Distance Weighted
IFT	Inverse Fourier Transform

IFFT	Inverse Fast Fourier Transform
KL	Karhunen-Loève
LASSO	Least Absolute Shrinkage and Selection Operator
LMC	Linear Model of Coregionalization
LOESS	Locally Estimated Scatterplot Smoothing
LRD	Long-Range Dependence
LWR	Locally Weighted Regression
MC	Minimum Curvature
MCMC	Markov Chain Monte Carlo
MGF	Moment-Generating Functional
ME	Maximum Entropy
MLE	Maximum Likelihood Estimation
MSE	Mean Square Error
MMSE	Minimum Mean Square Error
MoM	Method of Moments
MoNC	Method of Normalized Correlations
MRF	Markov Random Field
NNI	Nearest Neighbor Interpolation
NaNI	Natural Neighbor Interpolation
NFD	Novikov-Furutsu-Donsker
NLL	Negative Log-Likelihood
OK	Ordinary Kriging
OLS	Ordinary Least Squares
PACF	Partial Autocorrelation Function
PDE	Partial Differential Equation
PDF	Probability Density Function
RF	Random Field
RG	Renormalization Group
RSSE	Root of Sum of Square Errors
SA	Simulated Annealing
SAR	Spatial Autoregressive
SDE	Stochastic Ordinary Differential Equation
SGF	Savitzky-Golay Filter
SK	Simple Kriging
SLI	Stochastic Local Interaction
SRF	Spatial Random Field
SVD	Singular Value Decomposition
S-T/RF	Space-Time Random Field
SSRF	Spartan Spatial Random Field
SOLP	Stochastic Optimal Linear Predictor
SPDE	Stochastic Partial Differential Equation
TGH	Tukey g-and-h Random Fields
WLS	Weighted Least Squares

Chapter 1
Introduction

> As you set out for Ithaca hope the voyage is a long one, full of adventure, full of discovery.
>
> *Ithaca, by C. P. Cavafy*

This chapter introduces various definitions and concepts that are useful in spatial data modeling: random fields, trends, fluctuations, spatial domain types, different spatial models, disorder and heterogeneity, noise and errors, inductive and empirical modeling, sampling and prediction, are among the topics discussed herein. There are also brief discussions of the connections between statistical mechanics and random fields as well as on stochastic versus nonlinear systems approaches.

1.1 Preliminary Remarks

System In the following, we will use the term "system" to refer to a physical entity that involves an ensemble of *interacting units* (e.g., collection of geological faults in the same area), or more simply a *complex structure* that may be static or dynamic, such as a composite material, a porous medium, a soil sample, or an ore deposit. The structure of static complex media can be represented in terms of random fields.

Process The term "process" will refer to a sequence of changes that occur over time in a system due to the interaction of the system elements between themselves or with external stimuli. We will also use the more specialized terms *stochastic process* and *random process* to denote a one-dimensional random field, as is often done in the literature.

D. T. Hristopulos, *Random Fields for Spatial Data Modeling*, Advances in Geographic Information Science, https://doi.org/10.1007/978-94-024-1918-4_1

Spatial data The term *spatial data* has different interpretations. In many cases, it implies samples of a measurable spatially extended variable, represented by a space function or field; for example, it could refer to measurements of an environmental pollutant at the locations of different monitoring stations. In this case, the spatial data represent a *partial sample* of the modeled process (cf. graph data below). The main goal of spatial analysis is to generate representations of the "non-measured" field values at the nodes of a user-specified mapping or simulation grid.

In the GIS literature the term *spatial data* refers to general features, both qualitative and quantitative, that are linked to specific locations in space (i.e., they are *geo-referenced*). The locations are determined in terms of an appropriate reference system. The latter could be a Cartesian coordinate system, if small areas are involved, a suitable cartographic projection for larger areas, or a spherical coordinate system for global processes that evolve on the surface of the Earth.

GIS data structures To represent general spatial features, Geographic Information Systems use *vector* and *raster* data models [509].

The vector model is *object oriented* and involves the definition of geographical features by means of geometric entities, such as lines, polygons, and points.

The raster model is *field oriented* and requires the values of the process of interest at the nodes of an underlying grid.

Each model has advantages and disadvantages that will not be considered in this book. These data types can fit within hierarchical data frameworks which provide multiresolution descriptions that can accommodate changing patterns of spatial variability. For more information on the application of hierarchical data structures in GIS we refer the reader to [880].

Graph data Another type of data structure involves *graphs* which are collections of nodes (vertices), connected by means of edges (lines). Each node carries a value that represents the observed process at the specific location. Hence, graphs can be used to model data that are partially sampled on irregular (unstructured) meshes. The graph structure is then simply imposed by the arbitrary choice of measurement locations (e.g., the nodes of earth-based observation networks are defined by the respective measuring station locations). The edges of these networks are not a priori different than any line segment in the domain of interest.

On the other hand, in certain cases the graph structure reflects an existing spatial organization (e.g., consider transportation and energy networks). In these cases, the graph represents an inherent part of the measured process, the graph structure is important as it impacts the studied process; typical examples include the spread of epidemics by the movement of people and cascading failures in electricity networks [106, 334]. In this case, the graph edges play a central role (e.g., vehicles are restricted in their motion by the road network, planes only fly on specific routes). The edges can be assigned values which determine the impact of each connection. Transportation, energy, biological, and social networks belong to this type of graph. The properties of graph processes are studied in the blossoming field of *graph signal processing* [633, 749].

Spatial random fields, on which this book focuses, can be used to model both *raster* and *graph* spatial data.

> As far as the laws of mathematics refer to reality, they are not certain; and as far as they are certain they do not refer to reality.
>
> *Albert Einstein*

1.2 Why Random Fields?

A deterministic description of a physical system requires complete knowledge of the following factors:

1. The natural laws and their mathematical representations (e.g., partial differential equations) that govern the evolution of the system.
2. Constitutive equations that determine the coefficients of the partial differential equations and the inputs of the system. In certain cases the coefficients are also derived from natural laws by means of coarse-graining fine-scale models, while in other cases they represent empirical models [171, 219, 401].
3. The initial and the boundary conditions of the system.

If these factors are known, a deterministic, albeit possibly complex, set of equations can be developed and solved for the response variables. The solution, however, is in most cases analytically intractable and requires numerical methods. To make things worse, the coefficients in the equations can be complicated functions of space or time.[1] This means that a deterministic description is undermined by the limited number of available observations and various uncertainties related to experimental measurements. Hence, a stochastic description becomes necessary both for the coefficients and the response variables of the system.

The resulting mathematical problem is similar to the problem of *statistical mechanics*, i.e., how to determine the motion of a large number of microscopic particles that obey known physical laws if the initial particle positions and velocities cannot be observed and the equations of motion that account for the interactions between the particles cannot be solved. In statistical mechanics this problem has been addressed using the concept of the *statistical ensemble*.

Lack of knowledge regarding the initial state and the individual particle trajectories is compensated by introducing a probability distribution (the *Boltzmann* distribution in classical statistical mechanics) for the energies of the particles. Whereas the trajectories of the individual particles in a chamber filled with gas

[1] I do not like using "and/or"; herein, "A or B" shall mean that A, or B, or both A and B are valid.

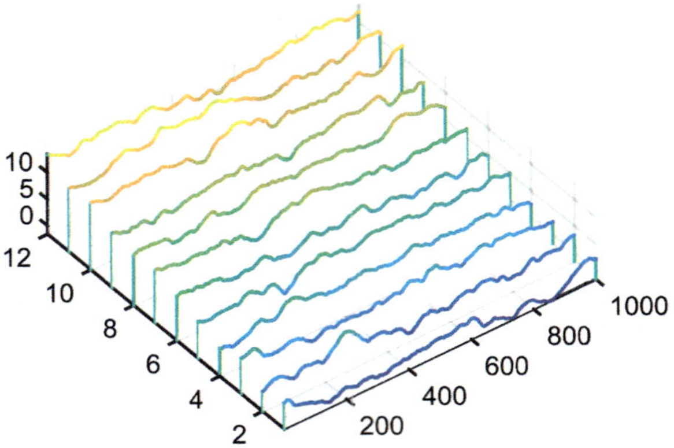

Fig. 1.1 Plot of twelve 1D realizations (states) of the same random field. In order to better discriminate the realizations visually a spatially uniform ascending trend (ranging from 1 to 12) has been added to the realizations

cannot be determined, the macroscopic behavior of the gas is expressed in terms of statistical moments (e.g., total energy, entropy) which are evaluated over the ensemble [480].

Random fields can be used to macroscopic quantities (e.g., atmospheric ozone concentration over an urban area, fluid permeability of a slab of porous medium, or solar radiation flux density over a desert valley) instead of single particles. Randomness is introduced into the system's structure and response due to uncertainty and complex spatial dependence. Probability distribution functions are used to account for the randomness. The random field thus involves an ensemble of *possible states or realizations* (see Fig. 1.1). Each state is presumed to appear with a frequency determined by the corresponding *probability density function (pdf)*. The statistical properties of the ensemble (e.g., the statistical moments) are determined by respective expectations that are calculated over all the states of the ensemble.

1.2.1 Random Fields, Trends, Fluctuations, Noise

Random fields, also known as *spatial random fields (SRFs)*, are mathematical entities that represent physical variables with complex dependence in space or time, that cannot be described by means of deterministic functions. Scalar random fields assign numerical values to each point within a spatial domain of interest. Scalar fields can be either real-valued or complex-valued. However, complex values are unnecessary for most static problems. To keep the notation simple, we will focus on real-valued random fields.

The crucial difference between a *spatial random field (SRF)* and a collection of spatially distributed random variables is that the fluctuations of the former are constrained by spatial correlations as shown in Fig. 1.2. The realization shown in Fig. 1.2b is completely random, whereas that shown in Fig. 1.2a displays a *disordered structure*.

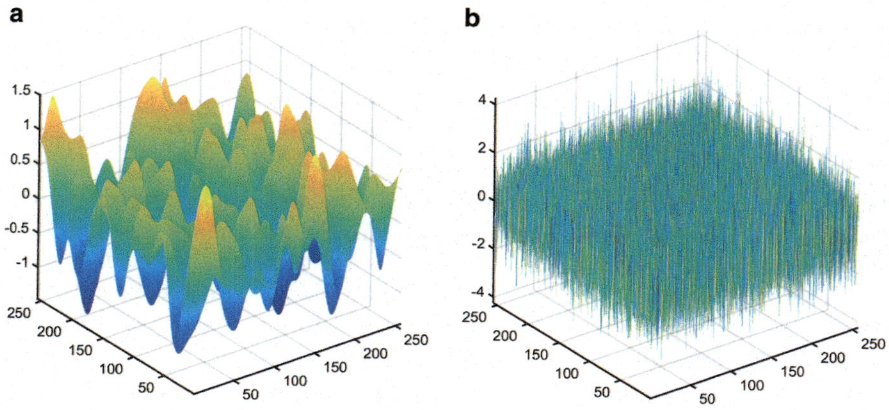

Fig. 1.2 (**a**) Realization of Gaussian random field with Gaussian covariance function ($m_{\mathrm{x}} = 0, \sigma_{\mathrm{x}} = 0.5, \xi = 16$) (**b**) Random Gaussian noise

Local correlations are evidenced in the relatively smooth (albeit irregular) variation of the "hills" and "valleys" of the landscape shown in Fig. 1.2a. *Disorder* implies that the hills and valleys do not follow a completely predictable, periodic pattern.

The first goal of spatial data analysis is to identify a suitable model for the local disordered structure and to infer the key statistical parameters of the model. The spatial model can then be exploited by making "probabilistic predictions" at points where measurements are not available.

Trend functions Random fields can incorporate both *deterministic* and *stochastic* components. Deterministic trends are expressed in terms of functions with closed-form expressions are represents large-scale variability. Stochastic trends typically represent slowly-changing random fields. For example, in time-series random walks represent stochastic trend models. In time series analysis such stochastic trends are removed by taking differences. Similar differencing operators can be used for random fields, as we discuss in Chap. 5 in relation to *intrinsic random fields*.

A trend is a systematic component that varies "regularly" in space and represents large-scale variations of the field. In the following, trends will represent *deterministic features*, such as a linear drift or a periodic modulation of the field.

We can view a random field as being composed of a *drift or trend* term that corresponds to the expectation (ensemble mean) of the field and a zero-mean stochastic fluctuation.

Trends are usually associated with large-scale features that extend over the domain of study (e.g., existence of a pressure gradient due to the imposed boundary conditions).

What is a trend? There are various perspectives but no unanimous agreement on what a trend is. For example, in the time series literature the trend can be a stochastic function that does not necessarily admit an explicit expression. In such cases, the trend is eliminated from the data by taking n-order differences; the higher n is, the more complex is the trend being subtracted. In spatial analysis, one can generalize this viewpoint by taking either derivatives or discrete differences (for lattice data) [138]. This approach is indeed used with *intrinsic random fields* (see Sect. 5.7). A comprehensive discussion of various trend definitions is given by Brillinger [103].

Fluctuations The stochastic component of the random field accounts for the *fluctuations* around the trend. Mathematically, the fluctuations are determined in terms of probability distributions. Fluctuations are attributed to the fine scale structure of physical processes (e.g., local pressure changes in a porous medium) that cannot be fully measured and characterized. This is a somewhat artificial distinction, which presumes that *separation of scales* is possible.

The hypothesis of scale separation may not always be true (for example in a porous medium) but is often a useful idealization. Physical problems that involve the interaction of multiple scales are of great interest in various fields of mathematical or applied research [219].

Some works in the geostatistical literature distinguish between trend and drift for historical reasons, pointing out that the "trend function" presumably carries all the useful information leading to a pure-noise residual, whereas in geological phenomena the drift is more suitable because it allows for correlated fluctuations in the residual [33]. Since this distinction is somewhat academic, the terms "trend" and "drift" are often used interchangeably [459]. A different perspective uses the term "drift" for the random field expectation and the term "trend" for the estimated drift based on the data. Ideally, the latter should be identical to the former, but this is not always the case.

It is helpful to distinguish between *complete randomness*, i.e., uncorrelated fluctuations and *correlated fluctuations*. The former is the spatial equivalent of *white noise* in the time domain, whereas the latter is similar to *colored noise*.[2] Measurements may include both sources of randomness. It is the latter case, however, which is of interest for random field applications: The existence of correlations implies that the measurement at one location contains information about the value of the field at neighboring locations. Hence, *prediction* at unobserved locations is possible, in a probabilistic sense, by taking advantage of correlations and the knowledge of the field's values at the measurement locations.

Remark In the following, the term noise is not applied to correlated fluctuations, since the latter contain information that can be exploited for prediction purposes.

Spatial models Spatial models aim to find and mathematically express patterns contained in the data that are possibly obscured by transformations imposed by the measurement process and the addition of noise. In spatially extended processes the useful information is contained in *latent random fields*.

---

[2]In technical jargon, different colors may be used for different types of noise. For example, pink noise, also known as flicker, has a power-law spectral density $\widetilde{C}_{xx}(\mathbf{k}) \propto \|\mathbf{k}\|^{-\alpha}$, where $\|\mathbf{k}\|$ is the wavenumber and $0 < \alpha < 2$. The spectral density of colored fluctuations does not necessarily follow a power law.

The vector of observations, $\mathbf{x}^* = (x_1^*, \ldots, x_N^*)^\top$ ($\top$ denotes the transpose), is assumed to be a sample of the observed random field $X^*(\mathbf{s}; \omega)$, where $\mathbf{s}$ is the location in space and ω the state index (see Sect. 1.3.2 for details). The observed field $X^*(\mathbf{s}; \omega)$ is linked to the intrinsic field $X(\mathbf{s}; \omega)$, which represents the physical process under observation by means of

$$X^*(\mathbf{s}; \omega) = \Phi[X(\mathbf{s}; \omega)] + \text{Noise}.$$

$\Phi(\cdot)$ is the *transfer functional* (in general nonlinear) that represents the effect of the measurements process on the intrinsic random field.

The noise term incorporates uncorrelated fluctuations. It may be comprise a component that reflects experimental errors and a second component due to micro-scale variations that are beyond the resolution of the available data set (this decomposition is further discussed in Sect. 10.5.3).

The ultimate goal of spatial data analysis is to identify a model and to predict the latent random field for $X(\mathbf{s}; \omega)$ at unmeasured points. Both the latent random field $X(\mathbf{s}; \omega)$ and the transfer functional $\Phi[\cdot]$ depend on a vector of parameters $\boldsymbol{\theta}$. These are not known a priori and should be estimated from the data, either by means of frequentist approaches (e.g., method of moments, maximum likelihood) or Bayesian methods that incorporate uncertain prior information (see Chap. 12). In the latter case, a prior probability distribution is assumed for the parameters. The prior is updated based on the data to derive a posterior probability distribution for the parameters.

Linear model with drift In the simplest case, the functional relation between the observed and the intrinsic random fields is linear, i.e.,

(A) Observed field = Drift + Fluctuations + Noise.

In this decomposition, the intrinsic random field comprises the drift term and stochastic correlated component (fluctuations), whereas the spatial noise an uncorrelated stochastic term (e.g., Gaussian white noise or heavy-tailed Lévy noise).

The construction of the spatial model involves determining suitable representations for the drift, the correlated fluctuations, and the noise. The resulting decomposition, however, is not necessarily unique. For example, if the decomposition is determined from the data without the foresight of a representative mathematical model, trend estimation is based on general knowledge of the process, the experience of the modeler, and exploratory analysis with various regression models. Thus, the drift function (i.e., the expectation of the random field) is often unknown a priori.

Linear model with trend A slightly different decomposition that corresponds more closely to the operational reality if (i) the drift is not known and (ii) the data involve a single realization of the observed process, is the following

(B) Observed field = Trend function + Residuals.

The *trend function* in the above relation is an estimate of the drift, whereas the residuals contain the *stochastic component*, i.e., both the correlated fluctuations and the noise. The analysis of spatial data is often based on one of the decomposition models (A) and (B) above.

Generalized linear model In certain cases, applying the linear model structure to a *nonlinear transformation* $G(\cdot)$ of the observed field than to the observed field itself gives more flexibility, i.e.,

(C) *G*(Observed field) = Trend function + Residuals.

This idea is at the heart of *generalized linear spatial models* which is investigated by Diggle and Ribeiro [201]. The use of generalized linear spatial models significantly increases the pool of spatial data that can be handled with geostatistical methods (see Chap. 15). Nonlinear transforms are also commonly used in geostatistical analysis to normalize non-Gaussian data (see Chap. 14).

Trend model selection As discussed above, rigorous estimates of the drift may not be known a priori. In such cases, the selection of the trend function is based on the fit of different candidate models to the data and physical insight or expert knowledge about the process. In the following, we assume for practical purposes that the trend function is a reasonable representation of the drift, and we do not further distinguish between the two terms.

In light of the models (A), (B), and (C), a variety of spatially extended variables can be modeled in terms of random fields. In fact, deterministic classical fields are idealizations obtained in the limit where the internal structure of the system is completely specified, and the local random forces acting on the system are negligible compared to applied external forcing.

Nonlinear regression model In some cases, especially for space-time processes that involve a time label t, the goal of the analysis is to determine the system's response to a set of excitations (control or input) variables. Then, the following general nonlinear regression model is used

$$\mathrm{Y}(\mathbf{s}, t; \omega) = \Phi[\mathbf{X}(\mathbf{s}, t; \omega), \epsilon(\mathbf{s}, t; \omega)] + \varepsilon(\mathbf{s}, t; \omega),$$

where $Y(\cdot)$ is the response variable (random field), $\mathbf{X}(\cdot)$ represents a vector of excitations (predictor variables), $\Phi[\cdot]$ is a transfer function specific to the observed physical system, and $\epsilon(\cdot)$, $\varepsilon(\cdot)$ are noise terms respectively added to the excitation and the response. We return to the regression model in Sect. 1.4.

1.2.2 Disorder and Heterogeneity

As discussed in the preface, random fields are used in many fields of science and engineering; this fact, however, is often obscured by the technical jargon of different fields of research. In Statistical and Condensed Matter Physics the terms *fluctuations* and *disorder* refer to spatial structures that are represented by random fields. In Solid State Physics, disorder refers to a departure of structural or other material properties from regularity, e.g., due to thermal excitations, impurity doping, or magnetic frustration [23, 297].

The disorder is called quenched if it is independent of time and annealed if it is dynamically determined by the evolution of the system.

Quenched disorder is an appropriate model for the structure of various engineered and naturally occurring porous media. Quenched disorder does not automatically imply that the response of the system is independent of time, since time dependence can develop via an externally applied stimulus that does not affect the disordered structure. For example, in models of diffusion in porous media the diffusion coefficient is characterized by quenched spatial disorder, but the concentration of the diffusing solute evolves over time. Most of the applications of spatial data analysis involve quenched disorder.

Annealed disorder, on the other hand, evolves dynamically due to interactions that take place in the system as the time evolves. If these interactions change the structure of the system, the disorder develops a time dependence. The principle of annealing is used in the combinatorial optimization method known as *simulated annealing (SA)*. The latter is employed in the conditional simulation of random fields (see Chap. 16).

Heterogeneity In engineering disciplines the term *heterogeneity* refers to the spatial variability generated by complex structures, such as those within porous media [8, 171, 797]. The heterogeneity of subsurface porous media is due to the natural complexity of the geological pore structures. Stochastic hydrology investigates models of groundwater flow and subsurface contaminant transport using the mathematical apparatus of random field models [174, 275, 459].

Random field estimation, interpolation, and simulation methods are also in the core of *geostatistics*. Such methods are instrumental in modeling the spatial distribution of mineral resources, concentrations of environmental interest, the potential of renewable energy resources, and the simulation of hydrocarbon reservoirs [195, 355, 419, 446].

1.2.3 Inductive Versus Empirical Modeling

Statistical mechanics and spatial random fields are both determined by underlying probability distributions. Hence, they share methods for calculating with such distributions. However, there are also significant differences between the two disciplines that are rooted in the fact that statistical mechanics is based on inductive modeling while spatial random fields are typically determined by empirical modeling.

Inductive models The *inductive* approach focuses on the formulation of physical models based on first principles. These models lead to a set of partial differential equations (PDEs) or logical rules that govern the evolution of the system. In this case, the physical laws are known and can be solved—at least in principle. The solutions then allow the construction of the respective pdf model. However, in some cases the physical laws involve coefficients with a complicated spatiotemporal dependence. Then, the coefficients (e.g., fluid permeability, diffusivity) are also modeled as random fields.

Example 1.1 An example of inductive modeling comes from stochastic subsurface hydrology [173, 275]. Flow in subsurface aquifers is governed by the *continuity equation* that conserves the fluid mass. The continuity equation for the steady state of an incompressible fluid in the linear Darcy regime is expressed in terms of the following partial differential equation (PDE) with random coefficients

$$\nabla \cdot \left[\mathsf{K}(\mathbf{s};\omega)\nabla \mathrm{H}(\mathbf{s};\omega)\right] = 0, \tag{1.1}$$

where $\mathsf{K}(\mathbf{s};\omega)$ is the *hydraulic conductivity tensor* and $\mathrm{H}(\mathbf{s};\omega)$ the hydraulic head in the aquifer. If we neglect variations in the depth of the aquifer, the hydraulic head is essentially the same as the pressure head.

The properties of the hydraulic conductivity random field $\mathsf{K}(\mathbf{s};\omega)$ are estimated from available measurements using empirical modeling (see below). If the hydraulic conductivity can be assumed to be uniform over the domain, there is no need for such empirical analysis. Assuming that the hydraulic conductivity is known, the pressure field $\mathrm{H}(\mathbf{s};\omega)$ is determined from the solution of the continuity PDE given the appropriate boundary conditions for the specific aquifer.

Empirical models The second approach is *empirical* and focuses on the development of statistical models based on the available data. Such models incorporate the spatial and temporal correlations inherent in the observations. In this case, the physical laws governing the system are not known, or they cannot be solved if they are known. Then, the spatiotemporal variability id described by random fields whose statistical properties are empirically defined [138].

For example, the spatial distribution of atmospheric concentrations of chemical pollutants (e.g., ozone) are determined by complicated chemical reactions and meteorological conditions that are difficult to model and solve. It is more feasible to model such concentrations as random fields based on observations (data) collected at a number of measuring stations [141].

Statistical mechanics In statistical mechanics, the experimental measurements (observable values) represent moments (expectations) calculated with respect to the microscopic probability distribution. For an equilibrium system, the pdf is determined from a suitable *Hamiltonian model*, whereas for the non-equilibrium systems by a *Fokker-Planck equation*. Observable quantities correspond to the moments of the underlying distributions. For example, the energy of a system at thermal equilibrium is given by the thermodynamic average of the energies of individual particle over the Maxwell-Boltzmann distribution.

Example The particles of an ideal gas at thermal equilibrium are in constant random motion, making it impossible to measure the positions and velocities of individual gas molecules. Due to the random motion of the molecules, it is reasonable to assume that the ensemble of probable microscopic states is adequately sampled. The microscopic probability distribution can also be expressed based on first principles. The observable quantities are coarse-grained averages that correspond to macroscopic variables such as the energy, pressure, et cetera. These quantities can be estimated based on the known microscopic probability distribution (inductive modeling).

Spatial data analysis In spatial data analysis, the observable quantities are usually obtained from single realizations (states) instead of random field moments. Less frequently, the observable variables may comprise several realizations of the field (e.g., time series of precipitation at different measurement stations). The ensemble moments are typically inferred from the available realizations. The pdf is also determined from the available data and not from first principles.

Example Let us consider data from drill holes that measure the thickness of lignite layers at different locations of a lignite deposit. We can view the variation of the thickness of lignite layers in space as a random field. Then, the drill-hole data represent a partial realization of a single state of the random field. Neither the probability distribution of the lignite thickness, nor its statistical moments (e.g., mean thickness) are observed. Both have to be empirically estimated from the data (empirical modeling). In fact, it takes a certain leap of faith to model the lignite layer thickness as a random field, since of all possible configurations only one is probed.

Inverse problem Modeling the pdf based on the available data is an *ill-defined inverse problem*, because a single state (realization) does not suffice to completely specify the random field model. Hence, an infinite number of random fields can be applied to any given data set. In view of this indeterminacy, ad hoc assumptions are used to constrain the possible outcomes. Such assumptions involve *statistical homogeneity* and *ergodicity*, and empirical knowledge about the behavior of the pdf in similar situations. These assumptions that eventually lead to a specific spatial model should be tested against the available information. The *cross validation* procedure achieves this goal by comparing model-based estimates (predictions) with a set of control data values (see Chap. 10).

1.2.4 Random Fields and Stochastic Systems

Simple models often provide a qualitative understanding of the mechanisms that govern the behavior of physical systems. However, they are usually inadequate for accurate estimation and prediction in complex environments.

Simple models are often deterministic, and the equations that describe them involve either constant coefficients or explicit expressions of spatial variability. In reality, many natural systems and technological media are *heterogeneous*, and their properties vary in space. At the same time, measurements are usually available only at a limited number of locations.

The combination of intrinsic variability and limited information implies *uncertainty* at the locations where data are not available. The impact of uncertainty and heterogeneity limits the practical scope of deterministic models. For example, in environmental engineering applications the inadequacy of simplified models is exemplified by the inaccurate estimates of groundwater remediation times at contaminated sites [461]. The erroneous estimates are obtained because deterministic models ignore the complexity of subsurface pore space that leads to significant fluid permeability variations.

In applied research, mathematical models that involve random fields, e.g., PDEs with random coefficients, are referred to as stochastic models. In other fields the term *stochastic partial differential equation* implies an equation of motion driven by an external force that involves an *irregular noise term* such as Wiener process [262, 621]. In such systems it may be necessary to use the formalism of *Itô calculus*, if the noise is a coefficient that multiplies the primary variable (multiplicative noise).

In the analysis of physical systems, both deterministic and random-field/stochastic approaches lead to a system of *partial differential equations* (PDEs) that describe the time evolution and spatial structure of the physical process. If the coefficients of the equations are random fields, the equations are called *partial differential equations with random coefficients*. The term *stochastic partial differential equation (SPDE)*, as discussed above, is reserved for equations that contain an irregular process as a driving term [757].

1.2.5 Connections with Nonlinear Systems

Random systems and deterministic nonlinear dynamical systems share a common property: *limited predictability*. In the case of random systems this property follows directly from the inherent uncertainty of the system. In the case of nonlinear systems, the reason for the lack of predictability is the more subtle sensitive dependence of the system on the initial conditions. This means that the evolution of the system can drastically change by small changes in the initial conditions. The sensitive dependence on the initial conditions implies that even deterministic systems can exhibit apparent randomness. The connections between randomness and determinism are explored without technical jargon in the wonderful book titled *Chance and Chaos* by David Ruelle [702].

Hence, nonlinear deterministic equations also possess solutions that exhibit “erratic behavior”, which in many respects appears to be random. The underlying dynamics, nevertheless, may only depend on a few degrees of freedom. In contrast, random fields in principle involve an infinite number of degrees of freedom.

For example, as shown in Chap. 16, to generate a realization of a Gaussian random field on a discretized $L \times L$ square domain, L^2 random numbers are used. Even for the discretized representation, as $L \to \infty$ so does the number of degrees of freedom.

How can one distinguish between fluctuations of random fields and the patterns generated by deterministic nonlinear equations?

Nonlinear systems exhibit characteristic structures that can be resolved in an appropriate phase space. For nonlinear time series, there exist methods for reconstructing a multivariate, embedding phase space with quantifiable geometry [2, 308, 432]. Mathematical tools of phase-space analysis can be used to determine a *low-dimensional attractor* and to estimate the number of dominant degrees of freedom. For a hydrological application of nonlinear systems analysis see [713]. Various tests have been proposed to determine whether a time series exhibits stochastic or nonlinear dependence.

Correlation dimension A commonly used measure of the attractor is the *correlation dimension* [308]: Small values of the correlation dimension imply *low-dimensional chaos* which is a signature of a deterministic system, whereas large values imply a stochastic system that involves many degrees of freedom. Typically, a large number of measurements is required to accurately estimate the value of the correlation dimension. However, this may not be possible for spatial data.

Stochastic or deterministic? The idea that the correlation dimension can distinguish between deterministic chaos and randomness has been challenged by the finding that *correlated randomness* has a small correlation dimension, just like a deterministic system [635]. This makes sense intuitively, since correlations imply a local dependence that, at short distances, partakes a deterministic flavor to random field realizations.

The question of how to determine whether a time series is deterministic or stochastic is a topic of ongoing research, e.g. [476]. The similarity of correlated random fields with low-dimensional chaotic systems may be even more pronounced for random fields that incorporate deterministic trends. Furthermore, in reality nonlinear dynamical systems also include stochastic components making the distinction between deterministic and stochastic models artificial. Finally, it should be mentioned that *low-dimensional approximations* of random fields are often used in practice for reasons of computational efficiency (see Sect. 16.11).

Quasi-randomness There is another issue that further blurs the distinction between randomness and nonlinearity: In reality, the computer simulation of stochastic systems is based on the use of "random numbers". The random number generators used, however, essentially employ deterministic, nonlinear algorithms that produce *quasirandom numbers* instead of purely random numbers [673]. These quasirandom numbers need to pass a number of numerical tests to ensure that their behavior is indistinguishable from that of truly random numbers. True random numbers are only generated by physical phenomena, such as radioactive decay and atmospheric noise. In contrast with quasirandom numbers that have a built-in period, truly random numbers are *aperiodic*.

Stochastic nonlinear systems The development of *stochastic nonlinear systems* that link nonlinearity with stochastic theories receives considerable attention in current research [219, 510]. Insight into the marriage between stochastic and deterministic approach, in particular as it applies to the modeling of atmospheric processes, can be found in the Introduction of the book by Lovejoy and Schertzer [512]. The importance of stochastic nonlinear systems in climate modeling is also emphasized in the recent review article [260].

Attractors of spatial systems To my knowledge, widely applicable methods for investigating the attractor and for estimating fractal dimensions of *spatially distributed systems* are not currently available. Estimates of fractal measures in spatially extended systems are hindered by the irregularity of sampling locations and the *dimensionality curse*: If the estimate of a fractal measure over q orders of magnitude requires 10^q measurements for a time series, it will require 10^{qd} for a spatially extended system in d dimensions.

In view of the above considerations, I believe that the theory of random fields is to date the most versatile mathematical and statistical framework for the study of spatially distributed processes and in particular for the modeling of spatial data. While nonlinearity is definitely an important factor in various physical systems, a flexible toolbox for spatial data analysis in the framework of dynamical systems theory is not at this point available. In contrast, for time-dependent processes there is a fertile interplay between the classical time series analysis and methods developed in the theory of dynamical systems [287].

1.3 Notation and Definitions

Some notation and definitions that will be used throughout the book to investigate SRFs and their properties are given below.

1.3.1 Notation

Spaces and Functions

Dimensionality	d denotes the number of spatial dimensions.
Real numbers	$\mathbb{R}$ denotes the set of real numbers.
Natural	$\mathbb{N}$ denotes the set of natural numbers, $1, 2, 3, \ldots$.
Integer	$\mathbb{Z}$ denotes the set of integers, i.e., $\ldots, -2, -1, 0, 1, 2, \ldots$.
Complex	$\mathbb{C}$ denotes the set of numbers on the complex plane.
Vectors	For any vector $\mathbf{A}$ or matrix $\mathbf{B}$, the symbols $\mathbf{A}^\top$ and $\mathbf{B}^\top$ denote, respectively, their transposes.
Coordinates	The vector $\mathbf{s} \in \mathbb{R}^d$ is used to denote the spatial location.

Location index	If we consider more than one location at the same time, we use the integer index $i = 1, \ldots, N$ to discriminate between different positions $\mathbf{s}_i$.
Position	In a Cartesian coordinate system position is denoted by $\mathbf{s}_i = (s_{i,1}, \ldots, s_{i,d})^\top$.
Spatial lag	The vector $\mathbf{r}_{i,j} = \mathbf{s}_i - \mathbf{s}_j$ denotes the lag between the locations $\mathbf{s}_i$ and $\mathbf{s}_j$. For conciseness we drop the indices i, j where possible.
Distance	The Euclidean norm of $\mathbf{r}$ will be denoted by $\|\mathbf{r}\| \in \mathbb{R}$.
Distance	The dimensionless lag is defined as $h = \|\mathbf{r}\|/\xi$.
Distance	Another normalized lag is $z = k_c \|\mathbf{r}\|$ (for Spartan spatial random fields).
Wavevector	The wavevector in reciprocal space is denoted by $\mathbf{k} \in \mathbb{R}^d$.
Wavenumber	The wavenumber $\|\mathbf{k}\| \in \mathbb{R}$ is the Euclidean norm of $\mathbf{k}$.
Unit Sphere	The surface area of the unit sphere in $\mathbb{R}^d$, denoted by $\mathcal{B}_d$, is given by $\mathcal{S}_d = 2\pi^{d/2}/\Gamma(d/2)$.
F.T.	$\tilde{x}(\mathbf{k})$ denotes the *Fourier transform* of the function $x(\mathbf{s})$, where $\mathbf{k}$ is the spatial frequency or wave-vector.
Domain	$\mathcal{D} \in \mathbb{R}^d$ denotes the sampling domain; $\partial\mathcal{D}$ is the boundary of the spatial domain.
Domain	$\mathcal{G} \in \mathbb{Z}^d$ denotes the sampling domain (grid); $\partial\mathcal{G}$ is the boundary of the grid.

Probability and Random Fields

Event space	The event space is denoted by Ω.
States	The state index is denoted by ω.
RF	We use the roman letter X to denote a *random variable* or a *random field* $\mathrm{X}(\mathbf{s};\omega)$.
PDF	We use the lowercase roman index x, e.g. $f_{\mathrm{x}}(\cdot)$ to denote the PDF of the random variable or random field $\mathrm{X}(\mathbf{s};\omega)$, and the lowercase italic x to denote specific point values, e.g., $f_{\mathrm{x}}(x)$.
Realization	We use $x(\mathbf{s},\omega)$ or simply $x(\mathbf{s})$ to denote the states (realizations) of a random field.
Distribution	$\mathrm{X}(\omega) \overset{d}{=} \mathrm{Y}(\omega)$ denotes that both variables follow the same probability distribution.
Mean	The expectation of a random field is denoted by $m_{\mathrm{x}}(\mathbf{s}) = \mathcal{E}[\mathrm{X}(\mathbf{s};\omega)]$.
Variance	The variance of a random variable $\mathrm{X}(\omega)$ or of a random field $\mathrm{X}(\mathbf{s};\omega)$ is denoted by σ_{x}^2.
Covariance	The auto-covariance function of a random field $\mathrm{X}(\mathbf{s};\omega)$ is denoted by $C_{\mathrm{xx}}(\mathbf{r})$.
Correlation	The auto-correlation function of a random field $\mathrm{X}(\mathbf{s};\omega)$ is denoted by $\rho_{\mathrm{xx}}(\mathbf{r})$.
Variogram	The variogram function of a random field $\mathrm{X}(\mathbf{s};\omega)$ is denoted by $\gamma_{\mathrm{xx}}(\mathbf{r})$.
Range	The integral range of an isotropic covariance is denoted by ℓ_c.

Radius	The correlation radius of an isotropic covariance is denoted by r_c.
Spectrum	A spectrum of random field length scales is denoted by $\lambda_c^{(\alpha)}$ where $0 \le \alpha \le 1$.
Gaussian	$X(\omega) \stackrel{d}{=} N(m_x, \sigma_x^2)$ means that the random variable $X(\omega)$ follows the Gaussian (normal) distribution with mean m_x and variance σ_x^2.
MultiGaussian	$\mathbf{X}(\omega) \stackrel{d}{=} N(\mathbf{m}_x, \mathbf{C}_{xx})$ means that the random vector $\mathbf{X}(\omega)$ follows the joint Gaussian (normal) distribution with mean $\mathbf{m}_x$ and covariance $\mathbf{C}_{xx}$.
MultiGaussian	$X(\mathbf{s}; \omega) \stackrel{d}{=} N(m_x(\cdot), C_{xx}(\cdot, \cdot))$—for short $X(\mathbf{s}; \omega) \stackrel{d}{=} N(m_x, C_{xx})$—means that the random field $X(\mathbf{s}; \omega)$ follows the joint Gaussian (normal) distribution with mean function $m_x(\mathbf{s})$ and covariance function $C_{xx}(\mathbf{s}, \mathbf{s}')$.
MultiStudent	$\mathbf{X}(\omega) \stackrel{d}{=} \mathcal{T}_\nu(\mathbf{m}_x, \boldsymbol{\Sigma})$ means that the random vector $\mathbf{X}(\omega)$ follows the multivariate Student's t-distribution with ν degrees of freedom, mean $\mathbf{m}_x$ and covariance $\boldsymbol{\Sigma}$.

Spartan Spatial Random Fields (SSRFs)

Amplitude	The SSRF amplitude coefficient is η_0.
Rigidity	The SSRF rigidity coefficient is η_1.
Length	The SSRF characteristic length is ξ.
Cutoff	The SSRF wavevector cutoff is k_c.
Cutoff	The dimensionless wavevector cutoff is $u_c = k_c\xi$.
Parameters	The vector of SSRF parameters is $\boldsymbol{\theta} = (\eta_0, \eta_1, \xi, k_c)^\top$.
Parameters	The vector of reduced SSRF parameters is $\boldsymbol{\theta}' = (\eta_1, \xi, k_c)^\top$.
Characteristic polynomial	The SSRF characteristic polynomial is $\Pi(u) = 1 + \eta_1 u^2 + u^4$.
Roots	The roots of the SSRF characteristic polynomial are $t_{\pm}^*$.
Discriminant	The discriminant of the characteristic polynomial is $\Delta = \sqrt{\eta_1{}^2 - 4}$.

Special Functions and Operators

$\Gamma(\cdot)$	The Gamma function is $\Gamma(\nu)$.
Indicator	$\mathbb{I}_A(x)$ is the indicator function of the set A, i.e., $\mathbb{I}_A(x) = 1,\ x \in A$ and $\mathbb{I}_A(x) = 0,\ x \notin A$.
$J_\nu(\cdot)$	The Bessel function of the first kind of order ν is $J_\nu(x)$.
$K_\nu(\cdot)$	The Modified Bessel function of the second kind of order ν is $K_\nu(x)$.
Lommel	The Lommel functions are denoted by $S_{\mu,\nu}(z)$.
Gradient	The gradient operator is defined by $\nabla = \left(\frac{\partial}{\partial s_1}, \ldots, \frac{\partial}{\partial s_d}\right)^\top$.
Laplacian	The Laplacian operator is defined by $\nabla^2 = \sum_{i=1}^{d} \left(\frac{\partial^2}{\partial s_1^2} + \ldots \frac{\partial^2}{\partial s_d^2}\right)$.
Biharmonic	The biharmonic (Bilaplacian) operator is defined by $\triangle^2 = \nabla^4$.
Expectation	The expectation operator over the ensemble of SRF states is denoted by $\mathcal{E}[\cdot]$.

Data and Estimates

Sampling set	$\mathbb{S}_{\mathrm{N}} = \{\mathbf{s}_1, \ldots, \mathbf{s}_N\}$ denotes the set of the sampling locations.
Prediction set	$\mathbb{G}_{\mathrm{P}} = \{\mathbf{z}_1, \ldots, \mathbf{z}_P\}$ denotes the set of "prediction" locations.
Sample values	$\mathbf{x}^* = \{x_1^*, \ldots, x_N^*\}$ set of data values at sampling locations $\mathbb{S}_{\mathrm{N}}$.
Predictions	$\hat{\mathbf{x}} = (\hat{x}_1, \ldots, \hat{x}_P)^\top$ denotes the vector of predictions at specified points.

Differential and finite difference operators

F.D.	$\widetilde{\delta}_i^m x(\mathbf{s})$ denotes the forward difference of order m in the lattice direction marked by $i = 1, \ldots, d$
C.D.	$\delta_i^m x(\mathbf{s})$ denotes the central difference of order m in the lattice direction marked by $i = 1, \ldots, d$
B.D.	$\breve{\delta}_i^m x(\mathbf{s})$ denotes the backward difference of order m in the lattice direction marked by $i = 1, \ldots, d$
Disc. Lapl.	The discrete Laplacian, $[\hat{\Delta}]_{m,n} x(\cdot)$, of a function $x(\cdot)$ is given by the sum $\sum_{i=1}^d \delta_i^2 x(\mathbf{s}_{m,n})$
Disc. Bilapl.	$\hat{\Delta}_{m,n}^2$ denotes the discrete Bilaplacian
4th-or. diff.	$\Delta^{(4)} x_{k,l}$ represents the fourth-order difference on grids
Differential	D_i denotes the differential operator in the lattice direction marked by $i = 1, \ldots, d$
F.D. operator	$\hat{\mathrm{D}}_i$ denotes the finite difference operator in the lattice direction marked by $i = 1, \ldots, d$
Latt. Lapl.	$\mathcal{L}$ denotes the lattice expansion of the Laplacian operator
Latt. Bilapl.	$\mathcal{B}$ denotes the lattice expansion of the Bilaplacian (Biharmonic) operator

1.3.2 Spatial Random Fields

Spatial Random Fields (SRF's) are mathematical abstractions that generalize the concept of *random processes* in d-dimensional spaces. A rigorous mathematical introduction is given in the classic text by Robert Adler [10]. Another classic reference that explains the main concepts in the context of random processes is the book by Cramér and Leadbetter [162].

The mathematical problems involved in spatial data modeling, such as parameter estimation, interpolation, simulation, and risk assessment can be formulated in the theoretical framework of random fields.

Definition 1.1 A spatial random field, also known as a *spatial random function*

$$\left\{\mathrm{X}(\mathbf{s};\omega) \in \mathbb{A}_{\mathrm{X}} \subseteq \mathbb{R};\ \mathbf{s} \in \mathcal{D} \subset \mathbb{R}^d;\ \omega \in \Omega\right\}$$

is defined as a mapping from the probability space $(\Omega, \mathcal{A}, P)$ into the space of real numbers.

1. For each fixed $\mathbf{s}$, $X(\mathbf{s}; \omega)$ is a measurable function of ω.
2. $\mathcal{D}$ is the spatial domain within which the SRF is defined.
3. $\mathbb{A}_{\mathrm{x}}$ is the domain (subset of real numbers) in which $X(\mathbf{s}; \omega)$ takes values.
4. The *sample space* Ω represents the set of all *possible states* of the random field and is assumed to be non-empty. The random field states, also known as *realizations*, are functions of space denoted by $x(\mathbf{s})$ or $x(\mathbf{s}, \omega)$.
5. The family $\mathcal{A}$ incorporates all *possible events*, and the function $P(A) \in [0, 1]$ assigns a probability to each event $A \in \mathcal{A}$.

Thus, an SRF involves by definition a multitude of probable *states* (*realizations*), which are indexed by ω [138, 863]. In the following, we suppress the dependence on ω where needed to keep the notation simple.

Notation We use roman uppercase letters, i.e, X, to denote random fields, which involve the ensemble of all probable states, and lowercase italic letters, i.e., x, to denote realizations.

The distinction between random fields and their realizations is often omitted. Instead, it is tacitly assumed that the difference between the field and the realizations is obvious based on the context.

Events versus sample space The difference between $\mathcal{A}$ and Ω is that events consist of subsets of all the possible states; $\mathcal{A}$ is *the space of all events*, whereas Ω is the *space of all states*.

States versus events An analogy with statistical mechanics helps to illustrate the difference between the *space of all possible states* and the *space of all possible events*: the set of microscopic energy states $\{\varepsilon_1, \ldots, \varepsilon_N\}$ of a system comprising N particles corresponds to the set of all possible states.

On the other hand, a system observable such as the *macroscopic energy* $E = \sum_{i=1}^{N} \varepsilon_i$ is a function of the individual energies of the states. Different values of the macroscopic energy E correspond to different events. Each event, i.e., each value of E, can be obtained by multiple combinations of microscopic states.

The formal mathematical definition of the probability space is given below for completeness. In the following, $A \in \mathcal{A}$ represents a unique event.

- A function $P(\cdot) : A \to [0, 1]$ is a *probability* if the following two conditions are met:

  1. $P(\Omega) = 1$.

2. For any family of disjoint sets $A_1, A_2, \ldots,$ the probability of their union is equal to the sum of the probabilities of the events:

$$P\left(\bigcup_{n=1}^{\infty} A_n\right) = \sum_{n=1}^{\infty} P(A_n).$$

The above are known as *Kolmogorov axioms* from the name of the founder of the theory of probability, Andrey Kolmogorov.

- The ordered triplet $(\Omega, \mathcal{A}, P)$ is called a *probability space* if the following conditions hold:

1. The space of all states Ω is non-empty.
2. The family of events $\mathcal{A}$ is a $\sigma-$algebra.
3. The function $P(A)$, for $A \in \mathcal{A}$, is a probability.

- A family $\mathcal{A}$ of subsets of Ω is an $\sigma-$algebra, also called a *Borel field*, if:

1. It contains the empty set, i.e., $\emptyset \in \mathcal{A}$.
2. It is closed under the operation of taking the complement with respect to Ω, that is, $A \in \mathcal{A} \Rightarrow A^c \in \mathcal{A}$.
3. It contains the union of countably many (either finite or enumerable) subsets, i.e., if:

$$A_1, A_2, \ldots, A_n \in \mathcal{A} \Rightarrow \bigcup_{i=1}^{n} A_i \in \mathcal{A} \text{ for all } n \in \mathbb{N}.$$

If the set Ω contains a finite or countable number of elements, then $\mathcal{A}$ includes the empty set, all possible subsets of Ω, and Ω itself. If Ω contains the real line $(-\infty, \infty)$, then $\mathcal{A}$ involves all open and closed intervals, as well as their countable unions.

Are all these technicalities concerning $\sigma-$algebras necessary? Essentially, they help to formalize the mathematical "objects" for which probabilities should be defined, in order to fully account for all the probable events. These concepts help to ensure that the "alphabet" used to describe random fields is complete. In applied studies, however, we do not typically think about these formalities any more than we think about the alphabet during our daily conversations.

Example 1.2 Consider the following simplistic example: A random field admits three different spatial configurations (states), which we will call for simplicity α, β, γ. Find the $\sigma-$algebra $\mathcal{A}$ which contains all possible events.

Answer According to the definition, the family of events $\mathcal{A}$ comprises the following elements: $\mathcal{A} = \Big\{\emptyset, \{\alpha\}, \{\beta\}, \{\gamma\}, \{\alpha, \beta\}, \{\alpha, \gamma\}, \{\beta, \gamma\}, \{\alpha, \beta, \gamma\}\Big\}$.

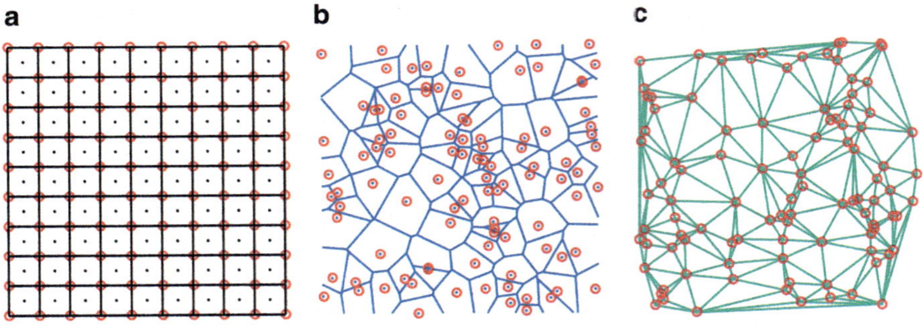

Fig. 1.3 (**a**) Regular lattice points (filled circles) and square grid (open circles and lines). (**b**) Irregular grid based on the Voronoi polygon tessellation of a random point set. (**c**) Network (graph) with links along the sides of the Delaunay triangles of a random point set

1.3.3 Spatial Domain

Continuum domains We consider random fields defined over spatial domains in d dimensions. *Continuum domains* are denoted by $\mathcal{D} \in \mathbb{R}^d$. The *domain volume* is denoted by $|\mathcal{D}|$ and the *domain boundary* by $\partial\mathcal{D}$. Unless otherwise stated, it is assumed that $\mathcal{D}$ is practically infinite. For practical purposes this means that the length of $\mathcal{D}$ in any direction significantly exceeds the random field integral range.[3]

Discrete domains In certain cases (e.g., digital images or simulations) the random field is not defined everywhere in a continuum domain but only at a point set that comprises the nodes of a *regular lattice* $\mathcal{G} \in \mathbb{Z}^d$. A lattice is called regular if it consists of uniform elements (cells); as a result, each node has exactly the same number of links as any other node.

In other cases, the point set that defines the lattice nodes consists of points irregularly scattered in space. Then, the corresponding lattice is called *irregular*. The volume $|\mathcal{D}|$ of the relevant domain in these cases is the volume enclosed by the convex hull of $\mathcal{D}$. Typical examples of irregular lattices are provided by the *Voronoi* (polygon) and the *Delaunay* (triangle) tessellations of an irregularly spaced point set.

The mess with meshes In the scientific literature several terms such as *lattice, mesh, grid, network, and graph* are used to refer to structures generated by point sets. These terms are often used interchangeably, although there are some differences.

The term *lattice* usually refers to a canonical (regular) discrete point set. In this case the lattice cells are uniform (see Fig. 1.3a). However, we can also use the terms

[3]The maximum integral range should be the reference scale, if the random field is anisotropic.

regular lattice and *irregular lattice* to differentiate between canonical and irregular (possibly random) point sets (see Fig. 1.3b, c).

The term *grid* typically includes the *connections (edges)* between the nearest nodes of the lattice in addition to the nodes. The term *mesh* implies the nodes and edges for irregular point sets. However, the terms "grid" and "mesh" are also used interchangeably [783, p. 308]. One also finds references to structured versus unstructured grids or meshes; the latter are used, for example, in finite element analysis.

The term *network* refers to a pair $G = (V, E)$ that consists of a set V of vertices (nodes) together with a set $E \in V \times V$ of edges (links) [590, 606]. The term *graph* is also used for networks, but the latter (network) is more often used in applications while the former (graph) is typically linked with the mathematical definition. Hence, grids and meshes can be viewed as networks (see Fig. 1.3c). However, networks are more general in the sense that the edges do not necessarily link near neighbors: certain neighboring sites may not be linked, while other, distant sites, could be connected on networks.

Is there any reason to differentiate between lattices and grids? For example, why do we refer to "lattice random fields" and not to "grid random fields"? The term "lattice" emphasizes the points where the random field values are defined. On the other hand, the term "grid" is used if the connections (edges) between nodes are important, e.g., if we need to calculate finite differences. This is probably why the term "grid" is more often used in applied mathematics, given the emphasis of this field on the solution of partial differential equations by means of finite differences.

1.3.4 Categories of Random Fields

Different classifications of random fields are possible, depending on the statistical properties (type of pdf, tail behavior, decay of correlation functions), the nature and range of their values, and the nature of the domain $\mathcal{D}$ over which they are defined.

The lists below classify random fields according to (i) the domain of values, $\mathbb{A}_{\mathrm{x}}$, assumed by their realizations, and (ii) the spatial domain $\mathcal{D}$ on which they are defined. In Chap. 5 other categorization schemes are discussed, based on (iii) the random field probability distribution (iv) the degree of homogeneity and (v) the properties of two-point correlations.

Quantitative random fields take numerical values which represent the outcome of a respective measurement process. These include the continuous and discrete random fields below, and they are the main focus of this book.

Categorical random fields take values which are non-numerical labels.

Value-based classification

Continuous SRF	The realizations of continuous SRFs take all possible values in a subset of the real numbers. For example, Gaussian random fields are defined for all real values, whereas lognormal random fields are only defined for positive real values.
Discrete SRF	Discrete SRFs take integer values in a countable set (e.g., integers). In statistical physics the *Ising model* and the *Potts model* are archetypical examples: in the Ising model (see Chap. 15) the field states take values ± 1, whereas in Potts models the values of the field states are integer numbers. In geostatistics the primary example is the binary *indicator function* which takes values zero and one (cf. Chap. 15).
Categorical SRF	Categorical random fields take values in a finite set. These values are non-numerical labels. *Ordinal random fields* are a sub-category of categorical SRFs, the values of which can be ordered. Categorical SRFs whose values cannot be ordered are called *nominal*.

Classification based on spatial domain of definition

Continuum	An SRF defined over a *continuum domain* $\mathcal{D} \subset \mathbb{R}^d$ is assigned values at every point in $\mathcal{D}$.
Regular lattice	A *lattice SRF* is defined at all the nodes (or the cell centers) of a regular lattice (structured mesh) $\mathcal{G}$. The domain of definition thus contains a countable set of points.
Irregular lattice	The random field is defined at the nodes of an *irregular lattice* (also known as *irregular network or unstructured mesh*). The nodes are a collection of irregularly spaced points so that the mesh cell size and geometry and the node connectivity vary.
Marked point process	A *marked point process* is an SRF that is defined over a discrete set of points. In contrast with lattice SRFs, the points are also defined by the process. A good example of a space-time marked point process is provided by earthquake sequences: the point process corresponds to the epicenters of the events and the marks to their magnitudes.

Different data types (combinations of field-value and spatial domain) that concern spatial analysis on both regular and irregular lattices are discussed by Besag [69].

Spatial sampling Regardless of the nature of the spatial domain where the SRF is defined, spatial data imply a sample of the field obtained over a *point set* $\mathbb{S}_N$. For example, a *sample* from an SRF defined in the continuum comprises values on a discrete point set, such as the nodes of an *irregular network (unstructured mesh)*. The data at these points may represent point measurements of the field or coarse-grained averages over some *measurement support* size that reflects the specifics of the measurement apparatus or setup. In addition, the spatial domain on which computer simulated SRFs are defined is discrete by construction.

Selective sampling The sampling points comprising $\mathbb{S}_N$ are often not completely random but selected by design that takes into account sampling costs, benefits, and experimental constraints (i.e., locations of sensors or measuring stations). Nevertheless, the configuration of all the points may resemble a random spatial pattern.

Spatial data are typically *spatially autocorrelated* and *heterogeneous* (i.e., their statistical properties vary in space, but the variations are correlated over some characteristic distance). Hence, spatial sampling does not comply with the assumption of *independent and identically distributed* (i.i.d.) data drawn from a population which is the assumption used in conventional sampling [828]. In the case of large data sets, *selective sampling* (random or stratified) may be used, to further reduce the data size that will be stored or processed [423].

Marked point processes In principle, the values of a continuous SRF can be measured at any point of $\mathcal{D}$. This is in contrast with marked point processes which take non-zero values only at certain points that are dynamically generated by the underlying process. We will not study marked point processes herein. The reason is twofold: First, there are excellent texts dedicated to marked point processes [40, 177]. Second, marked point processes can be viewed as the result of the interaction between dynamic processes and spatial random fields.

Earthquakes are a typical example of marked point process (see Example 1.3). However, earthquakes result from the competition between the dynamically evolving stress field of the seismic area and the fracture strength of the Earth's crust, both of which can be treated as random fields [375]. If the evolution of the joint probability distributions of these random fields in space and time were known, it would not be necessary to invoke the marked point process framework.

Example 1.3 An earthquake catalog can be viewed as a *marked point process*

$$C = \{\mathbf{s}_i, t_i, M_i\}_{i=1,\ldots,N},$$

where $\mathbf{s}_i$ is the location, t_i the time, and M_i the magnitude of each seismic event over a specified time period. Given a threshold magnitude M_c, a sequence of earthquake

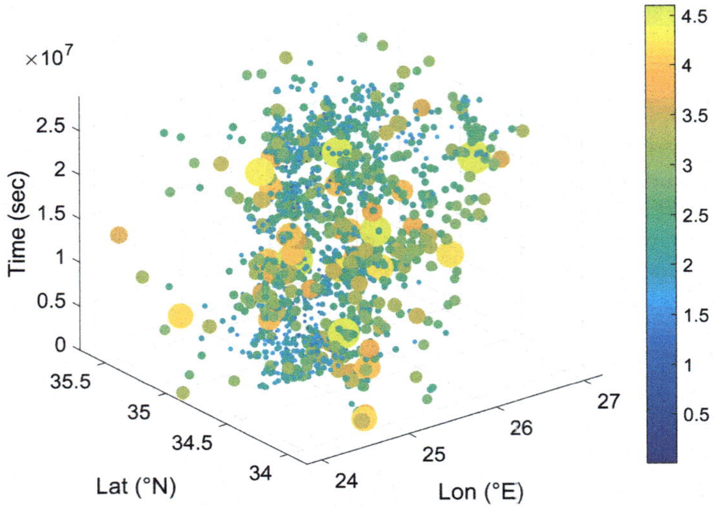

Fig. 1.4 Seismic sequence based on the data and analysis in [375]. The locations of the markers indicate the map positions of the epicenters and the times of the events. The size and color of the markers represents the magnitude of the events (local magnitude scale)

return times comprises the time intervals between events whose magnitude exceeds M_c, i.e.,

$$\{\tau_j = t_{j+k} - t_j \colon (M_j, M_{j+k} \geq M_c) \wedge (M_{j+1}, \ldots, M_{j+k-1} < M_c)\},$$

where $j = 1, \ldots, N_c - 1$, and N_c is the number of events with magnitude exceeding M_c. Figure 1.4 illustrates such a sequence of events; markers of different sizes are used to distinguish between the magnitudes of earthquakes.

1.3.5 Random Field Model of Spatial Data

In studies that involve real data, the variables of interest may involve a superposition of random fields with different statistical properties (e.g., characteristic scales and variability) as well as random noise components. As discussed in Sect. 1.2.1, it is helpful to analyze the data by means of a spatial model that involves a trend function, fluctuations, and noise.

We use the symbol $X^*(\mathbf{s}; \omega)$ for a random field with a noise component that is used to represent real data. In general, this random field is not necessarily the same as the random field $X(\mathbf{s}; \omega)$ due to the presence of noise.[4] The simplest possible spatial model for the data involves the following decomposition:

[4]Note that $X(\mathbf{s}; \omega)$ may represent a linear superposition of random fields with different properties.

$$\mathrm{X}^*(\mathbf{s};\omega) = \mathrm{X}(\mathbf{s};\omega) + \varepsilon(\mathbf{s};\omega) = m_{\mathrm{x}}(\mathbf{s}) + \mathrm{X}'(\mathbf{s};\omega) + \varepsilon(\mathbf{s};\omega). \quad (1.2)$$

In equation (1.2) the *trend* $m_{\mathrm{x}}(\mathbf{s})$ represents large-scale variability which is described by means of a deterministic function (e.g., a polynomial function of the coordinates). The trend is assumed to coincide with the *expectation* of the SRF, i.e.,

$$m_{\mathrm{x}}(\mathbf{s}) = \mathcal{E}[\mathrm{X}^*(\mathbf{s};\omega)] = \mathcal{E}[\mathrm{X}(\mathbf{s};\omega)]. \quad (1.3)$$

We will define the expectation more carefully in Chap. 3. For now, suffice it to say that it represents an average over all possible states of the random field. The result of such an average is that the fluctuations cancel out at every point, and the remainder represents spatial features that persist over all the realizations, i.e., trends.

The *fluctuation* $\mathrm{X}'(\mathbf{s};\omega)$ represents smaller-scaler variations around the trend, which can be considered as a *zero-mean, correlated SRF*. Hence,

$$\mathcal{E}[\mathrm{X}'(\mathbf{s};\omega)] = 0. \quad (1.4)$$

Finally, the *noise*, $\varepsilon(\mathbf{s};\omega)$, corresponds to uncorrelated fluctuations that are due to measurement errors or to random variability. The noise component contributes a discontinuity in the covariance and the variogram functions at the origin. In geostatistics, the discontinuity at the origin of the two-point functions is known as the *nugget term* [419]. The latter can be due either to measurement errors or to *microscale variability* that cannot be resolved based on the available spatial data.

As stated in Sect. 1.2.1, patterns of spatial variability that are based on single realizations can be modeled by partitioning the measurements between arbitrarily chosen trend function and fluctuations. Hence, unless constraining information is available from other sources, the trend-fluctuation-noise decomposition is not uniquely defined.

For visualization purposes, a synthetic 1D example illustrating the decomposition of the random field into trend-fluctuation-noise is shown in Fig. 1.5.

Example 1.4 We return to the subsurface fluid flow problem described in Example 1.1, inside an orthogonal parallelepiped domain.[5] The hydraulic head random field $\mathrm{H}(\mathbf{s};\omega)$ is typically decomposed into a trend $H_0(\mathbf{s}) = H_0 - \mathbf{J}\cdot\mathbf{s}$ and a fluctuation component $\mathrm{H}'(\mathbf{s};\omega)$.

H_0 is the boundary head value on one of the domain faces, and $\mathbf{J}$ is the average (macroscopic) hydraulic gradient; these are determined by the boundary conditions and the domain size. The fluctuation $\mathrm{H}'(\mathbf{s};\omega)$ is determined by $\mathbf{J}$ and the fluctuations of the hydraulic conductivity field [276].

---

[5] It is not presumed that readers will be able to solve the groundwater flow problem without training. This example serves more as an illustration and motivation for further reading.

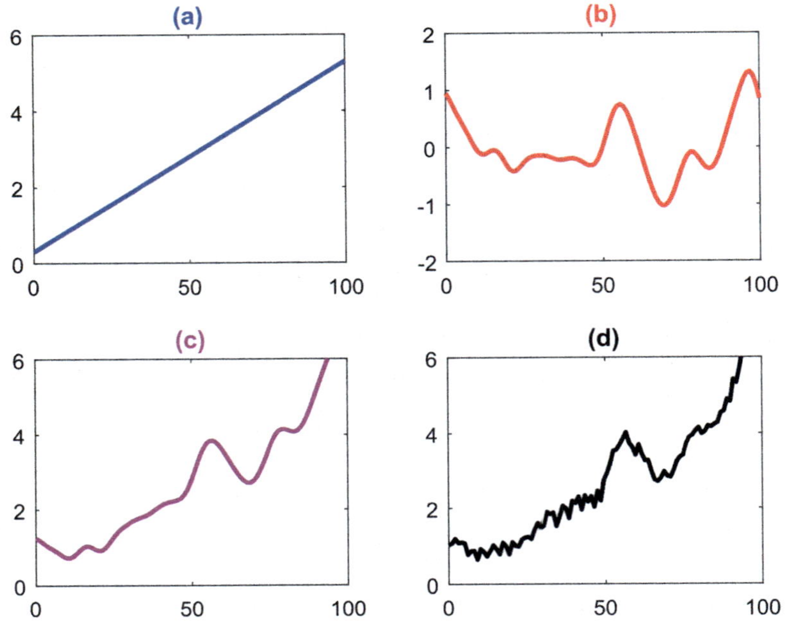

Fig. 1.5 Synthetic 1D example illustrating the trend-fluctuation-noise decomposition. Plot (**a**) shows the linear trend; plot (**b**) displays the correlated fluctuation random field; plot (**c**) shows the sum of the trend and the fluctuation, whereas plot (**d**) displays the composite random field after the addition of random noise

1.4 Noise and Errors

Noise and error are two closely linked concepts. There are different types of error and different types of noise. We further discuss the definition and the connections between noise and error below.

1.4.1 Noise

What is noise? The answer to this question is seemingly obvious to everyone; yet, it may mean different things to different people; to most, noise is an unpleasant audio disturbance. For scientists, noise is an erratic fluctuation that tends to obscure the measurement of the target variable. Noise is often confused with fluctuations. These statements, however, are not very precise for our purposes. Let us therefore consider more carefully the definition of noise.

Fluctuations versus noise Fluctuations represent noise if their correlation scale is much smaller than the lowest resolvable scale in the relevant problem, so that such fluctuations can be considered as statistically independent or at least uncorrelated.

The above definition means that stochastic fluctuations do not necessarily represent noise, if their correlation scale can be estimated by the available measurements and utilized to generate information, e.g., to "predict" missing values. This definition, however, is not universal. In many studies the term noise refers to both correlated and uncorrelated fluctuations if the target of investigation is the systematic trend.

Types of noise Noise enters in models of spatial and spatiotemporal data in various ways. We discuss the classification of noise depending on its interaction with the observable(s) following [510].[6] Therein, four categories of noise are defined based on the dependence of the observable on the noise: *observational, parametric, additive and multiplicative.*

Additive and multiplicative noise enters dynamic equations of motion that contain stochastic terms. Such equations are known as *Langevin equations*. In the case of pure time dependence they are ordinary stochastic differential equations, whereas in the case of joint space and time dependence they are stochastic partial differential equations.

- *Observational* noise does not affect the dynamic evolution of the system. It is inserted in the observations as additive noise. For example, assuming that the observations are given by a transform $\Phi[\cdot]$, possibly nonlinear, of the underlying random field $X(\mathbf{s}, t; \omega)$, the observational noise affects the system as follows:

$$X^{*}(\mathbf{s}, t; \omega) = \Phi\left[X(\mathbf{s}, t; \omega)\right] + \varepsilon(\mathbf{s}, t; \omega).$$

- *Parametric* noise affects some parameter of a dynamic system. This type of noise may have an impact on the system's state even under static conditions. For example, if the hydraulic conductivity field $\mathsf{K}(\mathbf{s}; \omega)$ contains a noise component, this will affect the steady-state hydraulic head random field as a result of the continuity equation $\nabla \cdot \left[\mathsf{K}(\mathbf{s}; \omega)\nabla H(\mathbf{s}; \omega)\right] = 0$.
- *Additive* noise is inserted in the equation of motion as a term that is independent of the state of the system. For example, if $\mathcal{L}[\cdot]$ is a linear differential operator (that may contain both time and space derivatives), additive noise enters the dynamic equation as follows

$$\frac{\partial X(\mathbf{s}, t; \omega)}{\partial t} = \mathcal{L}[X(\mathbf{s}, t; \omega)] + \varepsilon(\mathbf{s}, t; \omega).$$

[6] It will be useful to allow for dynamic changes of the noise; hence, in addition to space we include a time label for the following discussion.

- *Multiplicative* noise, on the other hand, depends on the state of the random field. An example of a stochastic partial differential equation with multiplicative noise is the following

$$\frac{\partial X(\mathbf{s}, t; \omega)}{\partial t} = \mathcal{L}[X(\mathbf{s}, t; \omega)] + g[X(\mathbf{s}, t; \omega)]\, \varepsilon(\mathbf{s}, t; \omega).$$

Spatial data analysis is mostly concerned with *observational noise.* In spatial modeling, however, additive noise is also useful. For example, it can be used in the framework of stochastic partial differential equations to derive spatial or space-time covariance models [379]. Parametric and multiplicative noise are mostly used in model construction by applied mathematicians.

1.4.2 Observational Noise

Observational noise may involve several components that are generated by different causes, some of which are listed below.

1. Random measurement errors due to the experimental apparatus.
2. Random ambient noise that contaminates the target process.
3. Sub-resolution variability, i.e., small-scale fluctuations that cannot be resolved with the available experimental setup.

From the modeling viewpoint, different sources of noise can be incorporated in a unified term $\varepsilon(\mathbf{s}; \omega)$, provided that they share the same statistical properties. Hence, for the purpose of estimation there is no need to discriminate between different sources of noise. It may be desirable to do so, nevertheless, if the aim is to determine the source of the noise and to implement procedures for improving the quality of the spatial data.

$1/f$ noise This type of noise is ubiquitous in physical, technological, economical, and biological processes even though its physical origins remain an open question [831]. The characteristic feature of $1/f$ noise is the power-law dependence of its spectral density, i.e., $\widetilde{C}_{xx}(f) \propto f^{-\alpha}$, where $\alpha > 0$ and f represents the frequency. In spatially extended processes f is replaced by the wavenumber $\|\mathbf{k}\|$. If the spectral density exponent is equal to one, then $1/f$ noise is also known as *pink noise*.

1.4.3 Gaussian White Noise

Gaussian white noise is the most widely used type of noise in geostatistics and other fields of science and engineering. The term "Gaussian" means that the probability distribution of the noise values is the normal (Gaussian) distribution. We denote this by writing

$$\varepsilon(\mathbf{s};\omega) \overset{d}{=} \mathcal{N}(0,\sigma_\varepsilon^2).$$

The symbol "$\overset{d}{=}$" denotes *equality in distribution*, and $\mathcal{N}(0,\sigma_\varepsilon^2)$ denotes the normal distribution with zero mean and variance equal to σ_ε^2. The main properties of Gaussian white noise are listed below.

1. The expectation of Gaussian white noise is zero

$$\mathcal{E}[\varepsilon(\mathbf{s};\omega)] = 0. \tag{1.5}$$

2. There are no spatial correlations between the values of the noise field at different locations. This statements is expressed mathematically as

$$\mathcal{E}[\varepsilon(\mathbf{s}_1;\omega)\,\varepsilon(\mathbf{s}_2;\omega)] = \sigma_\varepsilon^2\,\delta(\mathbf{s}_1-\mathbf{s}_2), \tag{1.6}$$

where $\delta(\cdot)$ is the Dirac delta function. This equation expresses the fact that the noise is uncorrelated and its variance is equal to σ_ε^2.

If the noise is Gaussian, the absence of two-point correlations implies that the field values at different locations are also statistically independent.

The term *white* emphasizes that the spectral density is flat, that is all possible frequencies contribute evenly to the spectrum,[7] just as white light contains a uniform mixture of all the frequencies in the visible range of the electromagnetic spectrum.

The realizations of white noise are very erratic functions, i.e., non-differentiable and discontinuous, due to the lack of correlations. A sample noise realization is shown in Fig. 1.6. Note that most of the values of the noise are in the interval $[-3, 3]$. This is due to the characteristic property of the normal distribution: about 99.73% of its values are included in the interval $[m-3\sigma, m+3\sigma]$.

1.4.4 Wiener Process (Brownian Motion)

The Wiener process is another archetypical noise model. In one dimension it can be viewed, non-rigorously, as the integral of a Gaussian white noise process (see Fig. 1.6). The Wiener process is continuous but non-differentiable everywhere. It is the prototypical *random walk* process that describes the position of a walker who successively takes completely random steps.

[7] To be more precise, for spatial noise case we should refer to wavenumbers instead of frequencies.

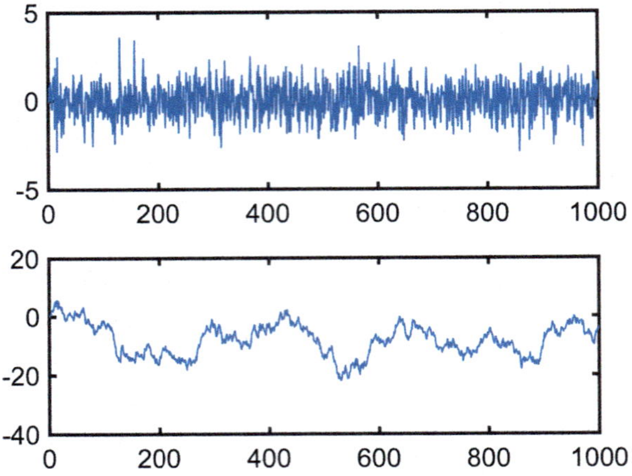

Fig. 1.6 (Top) Realization of Gaussian white noise with zero mean and unit variance. (Bottom) Realization of Brownian motion obtained by superposition of steps generated by a Gaussian white noise process

The Wiener process is denoted by $\mathrm{W}(t;\omega)$. The "steps" of the random walk are the increments of the Wiener process, which are denoted by $\{\mathrm{W}(t_n;\omega) - \mathrm{W}(t_{n-1};\omega)\}_{n=1}^{\infty}$. The main properties of the Wiener process are listed below.[8]

1. The initial value of the Wiener process is assumed to be zero, i.e., $\mathrm{W}(t_0;\omega) = 0$.
2. The increments of the Wiener process at n increasing times $t_n > t_{n-1} > \ldots t_2 > t_1$, i.e., the differences

$$\{\mathrm{W}(t_2;\omega) - \mathrm{W}(t_1;\omega), \mathrm{W}(t_3;\omega) - \mathrm{W}(t_2;\omega), \ldots, \mathrm{W}(t_n;\omega) - \mathrm{W}(t_{n-1};\omega)\},$$

   are independent for all $n > 1$.
3. The increment process is normally distributed, i.e., for all $n > 1$

$$\mathrm{W}(t_{n-1};\omega) - \mathrm{W}(t_{n-2};\omega) \stackrel{d}{=} \mathcal{N}(0, \sigma_W^2), \quad \text{where}\, \sigma_w^2 = |t_1 - t_2|\,.$$

4. The sample paths are continuous with probability one.[9]
5. The Wiener process $\mathrm{W}(t;\omega)$ for $t \in [0, 1]$ is the $n \to \infty$ limit of the following average

$$\mathrm{W}_n(t;\omega) = \frac{1}{\sqrt{n}} \sum_{i=1}^{\lfloor n\,t \rfloor} \chi_i(\omega),$$

   where $\chi_i(\omega)$ are *independent, identically distributed* random variables with zero mean and unit variance and $\lfloor n\,t \rfloor$ is the floor of nt, i.e., the largest positive integer that does not exceed nt [750]. The proof of the above proposition is based on Donsker's functional central limit theorem [206].

[8] We use the letter "W" to denote the Wiener process as is commonly done in the literature.

[9] For a precise definition of this statement see Sect. 5.1.

The Wiener process is a mathematical model for random motions observed in diffusion phenomena. This type of motion was first experimentally observed with particles of pollen particles immersed in fluid by the botanist Robert Brown [111]. In honor of his discovery, the phenomenon was named *Brownian motion*. The mathematical basis of Brownian motion was presented by Albert Einstein in a famous papers that laid the foundations for the physical theory of diffusion and provided a strong argument for the validity of the molecular theory of matter [226].

Extensions of the *Brownian motion* to multidimensional cases $\mathbf{W}(t;\omega)$, where $\mathbf{W}(\cdot)$ is a d-dimensional vector, are straightforward. In such cases, $\mathbf{W}(t;\omega)$ represents the random position vector at time t and defines the locus of the points visited by the Brownian motion at different times. This type of motion describes the random walk patterns observed in many natural phenomena that involve diffusion [92].

Extensions of the random walk to d-dimensional domains (i.e., $\mathbf{s} \in \mathbb{R}^d$) can be constructed using the formalism of *intrinsic random fields*—see Sect. 5.7.1 and [132]. A mathematical description of the Wiener process in terms of stochastic differential equations is given in Sect. 9.1.

1.4.5 Errors and Uncertainty

Broadly defined, the term "error" refers to any deviation between the actual value of a variable and its measurement or its estimate. Error analysis is an important component of data analysis, because it allows quantifying the confidence that can be attributed to our estimates.

There are various sources and types of errors. For example, errors can accumulate during the *observation phase*, or during the *registration and processing* of the data. Errors also result from *numerical operations* performed on digital computers. Finally, there are errors which are introduced by our *modeling assumptions*.

Errors lead to *uncertainty* regarding the actual value of a measured or modeled process. In recent decades the importance of characterizing and controlling uncertainty has been recognized by the scientific community. In response to this development, the research field of *Uncertainty Quantification (UQ)* has emerged, and dedicated research journals have recently been launched. Uncertainty quantification focuses on inter-disciplinary approaches to the characterization, estimation, and control of uncertainty.

The following is a brief and non-exhaustive review of different error sources.

1.4.5.1 Experimental Errors

Experimental errors occur during the phase of data generation and collection. They belong in two distinct categories: *systematic and random*.

Systematic errors accrue when the measurements procedure is affected by a deterministic perturbation. For example, inaccurate *calibration* may introduce a deviation of the reference level of an instrument from its intended value. A simple example of this type or error is a scale showing a non-zero reading in the absence of load. Another type of such error involves *homogenization errors*, i.e., the difference observed between measurements of the same variable obtained with different types of instruments. Homogenization is an important issue in the case of data sets compiled from several sources that possibly use different measurements standards and devices [755].

Outliers represent measurements that, under some definition of distance from the "bulk" of the probability distribution, are located at an abnormal distance from the rest of the data. This definition involves a degree of subjectivity, since the notion of abnormal distance needs to be specified. Often this involves some assumption regarding the form of the distribution that the data follow. For example, if the underlying data distribution is normal, various statistical tests can be used to determine if certain observations are outliers [352]. These tests are complemented by graphical statistical tools (e.g., box plots, histograms) that help to visually identify outliers.

Outliers may be due to errors caused by identifiable and localized perturbations of the measurement apparatus, sensor malfunctions, inadvertent human intervention, or extreme natural events (e.g., fires, earthquakes).

Extreme values Outliers sometimes are erroneously associated with *extreme values*. The latter represent infrequent events that are, however, allowed by the probability distribution of the observed processes. For example, earthquakes with magnitude seven or higher on the Richter scale are extreme events, that have a distinct, albeit small, probability of occurrence. On the other hand, a highly elevated radioactivity dose rate (e.g., ten times higher than the background) that is not corroborated by a similar reading at a neighboring station, is very likely due to a malfunctioning probe. The theory of *spatial extremes* is an interesting and current research topic [182, 183].

Spatial outliers The definition of outliers given above is based on the global behavior of the data. In spatial data analysis, however, one is often interested in *spatial outliers*. These refer to observations that are markedly different from their spatial neighbors. For example, such outliers could represent environmental pollution hot spots due to localized sources. The spatial outlier detection thus focuses on deviations that perturb the local continuity of the observed process.

Trick or treat? So, are outliers a pure nuisance, or do they contain useful information for the spatial analysis? Clearly, a scientist has to spend time and effort trying to detect and identify outliers. If they turn out to represent errors, they do not provide insight into the observed process, albeit they can be used as diagnostic tools to investigate the source of the errors. If, on the other hand, the outliers represent legitimate extreme values, they can reveal important aspects of the process that perhaps the traditional modeling approaches fail to capture.

Random errors ensue if the measurements are affected by random noise that might be due, among other causes, to uncorrelated fluctuations of the measured process, fluctuations of the measuring apparatus caused by random thermal variations, or electronic noise.

1.4.5.2 Numerical errors

The numerical processing of data on digital computers introduces two types of errors. The errors listed below are due to the limited memory size of digital computers.

Resolution errors are due to the fact that digital computers provide discretized representations of real numbers. Each computer is characterized by the minimum and maximum real numbers that can be represented in the computer's random access memory (RAM). For example, in MATLAB the corresponding real numbers are given by means of the commands `realmin` and `realmax` respectively. On the laptop computer with 4GB of RAM used to write parts of this book, the respective minimum and maximum real numbers are $2.2251\ 10^{-308}$ and $1.7977\ 10^{+308}$. These are very small (large), but still finite numbers. Values that are smaller (larger) than the above limits cannot be represented.

Relative accuracy refers to the computer's ability to resolve two real numbers with very similar values. It is defined as the minimum absolute difference that the computer can resolve and represents the computer's tolerance. The relative accuracy is typically a larger number than the minimum real number that can be represented on the computer. In MATLAB, the relative accuracy is given by means of the command `eps` which yields $2.22\,10^{-16}$ for the machine with the above characteristics. The accuracy of numerical calculations is affected by the relative accuracy of the computer.

Roundoff errors represent the loss of accuracy that is exclusively due to the finite memory size of computers. The memory limitations imply that numerical figures are truncated beyond a certain decimal place. In the course of numerical operations *roundoff errors* accumulate. Such accumulation can lead to significant error in the solutions of differential equations for large times.

Discretization errors result when we represent continuum systems and operations on inherently discrete systems such as the digital computers. The problem is that more than one discrete approximation of a any given continuum operator are possible, and each approximation has different accuracy. For example, the derivative can be approximated by forward or central finite differences; the former have an accuracy that increases linearly with the differencing step whereas the accuracy of the latter increases quadratically with the step. Hence, for small step values central differences provide a more accurate approximation. It is not possible to completely eliminate discretization errors, but it is desirable to characterize and control the errors introduced by different discretization schemes.

1.4.5.3 Modeling Errors

Spatial models aim to determine the optimal values—or even the predictive probability distribution—of observed spatial processes at unmeasured points. Spatial models, whether statistical or based on first principles, are subject to various errors.

Model error This type of error occurs if we use an inappropriate spatial model for the specific process that we are investigating. For example, we might use a diffusion model with a constant diffusion coefficient whereas the latter actually varies in space; or we may use a linear trend whereas the true trend is a second-order polynomial of the space coordinates. The model error is related to erroneous or insufficient assumptions regarding the underlying physical model of the process.

Accuracy and bias Let us assume that our target is to estimate the unknown value of the variable $\mathrm{X}(\mathbf{s};\omega)$ at some given point $\mathbf{s}$. Based on the available information, we construct statistical estimators $\hat{\mathrm{X}}(\mathbf{s};\omega)$ of $\mathrm{X}(\mathbf{s};\omega)$. The *accuracy* of the estimator characterizes how close $\hat{\mathrm{X}}(\mathbf{s};\omega)$ is to $\mathrm{X}(\mathbf{s};\omega)$. Of course, since $\mathrm{X}(\mathbf{s};\omega)$ and $\hat{\mathrm{X}}(\mathbf{s};\omega)$ are random variables, their proximity can be measured in several ways. Accuracy is a general, qualitative term used to assess the agreement between the estimates and the true values.

Bias is a specific measure of accuracy that represents the expected difference between the estimate and the true value, i.e.,

$$\text{Bias} = \mathcal{E}[\hat{\mathrm{X}}(\mathbf{s};\omega)] - \mathcal{E}[\mathrm{X}(\mathbf{s};\omega)]. \tag{1.7}$$

Lack of bias is often a requirement for statistical estimators.

If the estimation aims to determine the unknown value of a deterministic parameter θ (for example, if θ is a population parameter of the statistical distribution), the bias can be expressed as $\mathcal{E}[\hat{\theta}]-\theta$. The bias depends on the mathematical expressions or the algorithmic steps used in the estimator.

In experimental measurements, bias may reflect inadequate calibration of the measurement apparatus. Experimental bias is attributed to *systematic errors* (see Sect. 1.4.5) that can, at least in principle, be traced and eliminated.

Precision This term refers to the impact of randomness and uncertainty on the estimate. Precise estimates have a high degree of repeatability, whereas imprecise estimates exhibit significant dispersion. Lack of precision is caused by *random errors*. Hence, the precision of an estimator is often measured in terms of the *estimation variance*, i.e., by means of

$$\mathbb{V}\text{ar}\left\{\hat{\mathrm{X}}(\mathbf{s};\omega)\right\} = \mathcal{E}\left[\left\{\hat{\mathrm{X}}(\mathbf{s};\omega) - \mathcal{E}[\hat{\mathrm{X}}(\mathbf{s};\omega)]\right\}^2\right]. \tag{1.8}$$

Ideally, we would like our statistical estimators to be *infinitely precise*, which means that the estimation variance should vanish. Since this is not possible, in practice we settle for *minimum variance estimators* that minimize the variance

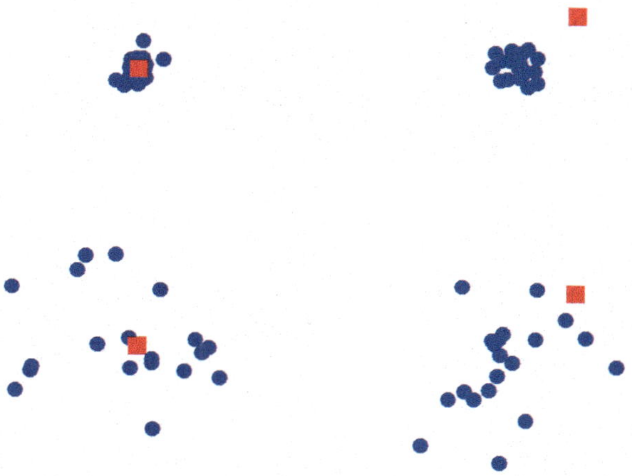

Fig. 1.7 Schematic diagram illustrating the difference between bias and precision. The red square represents the "target" value. The circles represent estimates of the target based on different sample. High precision (low variance) implies that the estimates have low dispersion. Low bias (high accuracy) implies that the center of the estimates is close to the target. An estimator with low bias and high precision (top left) is preferable. Top right: Low variance, high bias estimator. Bottom left: Low bias and high variance estimator. Bottom right: High bias and high variance estimator (this is the worst of the four cases shown)

given certain constraints. Typical examples are the kriging estimators discussed in Chap. 10.

Bias-variance tradeoff The difference between bias and precision is illustrated schematically in Fig. 1.7. The mean square error (MSE) of estimators involves the sum of the square of the bias and the estimator variance [330, p. 24]:

$$\text{MSE} = \text{Bias}^2 + \text{Variance}. \tag{1.9}$$

In general, complex statistical models are faced with the *bias-variance tradeoff*: as the model complexity increases, so does the variance of the predictions. The variance increase can be reduced by introducing some degree of bias in the prediction model [330, p. 37].

1.4.5.4 Heavy Tails and Outliers

In certain cases, spatial data include measurements that exhibit large excursions from the mean or from the values of the trend function at the respective locations. The number of such data can be quite small, but their presence raises questions about how they are generated and how they should be modeled. As discussed in Sect. 1.4.5, there are two distinct possibilities:

- These measurements are the signature of a heavy-tailed random field that has significant weight in its right tail (i.e., for high values). The probability distribution allows the appearance of extreme values with non-vanishing probability.

  Heavy-tailed processes are often described by *Lévy random fields* [709], although this is not the only possibility. The pdfs of Lévy random fields decay as power laws. This means that relatively high values have a much larger probability of occurrence than they would under an exponentially decaying probability distribution.
- The measurements represent *outliers*, i.e., values that are not captured on the basis of the underlying probability distribution for the studied process. Outliers are due to human errors, instrument malfunctions, registration errors, external stimuli generated by nuisance processes (e.g., an earthquake) that are unrelated to the measured process, or the "activation" of localized processes with different statistical properties than the background (e.g., accidental release of radioactivity in the environment).

If the spurious observations are an integral part of the system's response, the spatial model should be designed to account for them. If they can be classified with certainty as outliers due to errors or nuisance processes, they can be discarded from the subsequent analysis of the spatial model. On the other hand, outliers due to extrinsic localized processes should be further studied because they can carry significant information—if an elevated radioactivity level is measured, it is useful to know its source.

1.5 Spatial Data Preliminaries

Spatial data can be either scattered or distributed on a regular grid. The first case is common for drill-hole data and earth-based observation networks, whereas the latter for remote sensing digital images. In the first case we typically use a random field model defined over a continuum domain, whereas in the latter a discrete (lattice) point set is used (see Sect. 1.3.4 for a discussion on types of spatial domains).

1.5.1 Sampling Set

The set that comprises the sampling points will be denoted by $\mathbf{S}_{\mathrm{N}} = \{\mathbf{s}_1, \ldots \mathbf{s}_N\}$. The vector $\mathbf{x}^* = (x_1^*, \ldots, x_N^*)^\top$, comprising the measurements of the random field $\mathrm{X}(\mathbf{s}; \omega)$ represents a sample of the *observed random field* $\mathrm{X}^*(\mathbf{s}; \omega)$. The latter may contain deterministic trends, fluctuations, and noise. The sampling point pattern and the number of sampling points are to some extent determined by existing constraints (e.g., accessibility, economic costs, prevailing environmental conditions) and societal needs (e.g., denser monitoring near populated areas).

If sampling is conducted at the sites of a lattice $\mathcal{G} \in \mathbb{Z}^d$, the sampling domain includes all of the grid nodes. If the sampling geometry is irregular, the sampling domain $\mathcal{D}$ is typically defined as the locus points enclosed by the *convex hull* of the point set $\mathbb{S}_N$.

1.5.2 Prediction

We loosely use the term "prediction" to denote estimated field values based on the available information and a selected spatial model. In the above sense, both spatial interpolation and simulation are prediction problems: the former leads to a unique state, while the latter generates a number of probable alternatives.

Prediction points will be denoted by $\mathbf{z}_p$, where $p = 1, \ldots, P$. The *prediction set* $\mathbb{G}_P = \{\mathbf{z}_1, \ldots \mathbf{z}_P\}$ may comprise vertices of an unstructured grid or the nodes of a lattice $\mathcal{G}$. In most cases, prediction and sampling points belong in disjoint sets, i.e., $\mathbf{z}_p \notin \mathbb{S}_N$ and $\mathbf{s}_n \notin \mathbb{G}_P$, where $\mathbb{S}_N = \{\mathbf{s}_n\}_{n=1}^N$.

Prediction points may coincide with sampling points if (i) the processing goal is to smoothen the data or (ii) if the performance of the spatial model is assessed by means of a cross validation procedure.

Interpolation The interpolation problem refers to the estimation of *optimal values* of the random field at a number of locations $\{\mathbf{z}_p\}_{p=1}^P$ that are typically arranged on a regular grid $\mathcal{G}$ to allow for visualization and further processing of the optimal configuration. The interpolated field is usually estimated at each grid point independently from other grid points. Hence, we can focus on a single-point formulation of the interpolation problem. This also means that parallel implementations of the interpolation problem are possible. The exact definition of optimality is specified in Chap. 10, where we also discuss classic interpolation methods. More information on interpolation approaches is presented in Chap. 11.

Missing data problem Interpolation can also be used to derive optimal estimates of the field at a selected number of locations where the data are missing. The *missing data problem* is caused by various factors such as damaged or incomplete records, or, in the case of remote sensing images, sensor malfunctions and cloud coverage. For example, consider the ozone data shown in Fig. 1.8. The data are distributed on an orthogonal grid with $N_G = 180 \times 360$ nodes extending in latitude from 90S to 90N and in longitude from 180W to 180E [6]. The gaps (stripes in the North-South direction) on the left are mainly due to limited coverage on the particular day of the measurements. The graph on the right shows the image with the gaps filled by reconstructed data (for details about the reconstruction method see [900]).

Simulation In the *simulation problem* the goal is to generate different scenarios for the *states (realizations)* of the random field at the prediction sites that respect the joint probability distribution. Simulation has a significant advantage over interpolation, because it allows better visualization of the variability and uncertainty of the

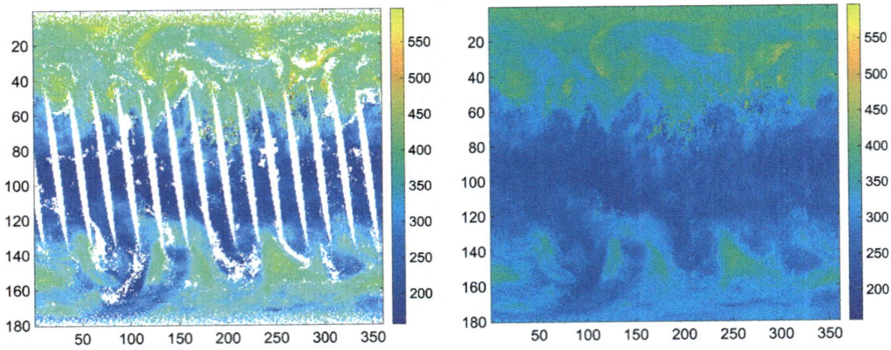

Fig. 1.8 Daily column ozone measurements on June 1, 2007 (Dobson units). The data set includes naturally missing (and therefore unknown) values. Directional Gradient-Curvature interpolation of the missing ozone data based on 10 simulations. (Figure taken from [900])

modeled process than the interpolated field. On the other hand, it is computationally more intensive, and it requires more complicated methodology than interpolation. We discuss different methods of simulation in Chap. 16.

1.5.3 Estimation

The general term "*estimation*" can have different meanings depending on the context. For example, if we aim to determine the parameters of a spatial model, "estimation" implies *parameter inference*.

The term "estimation" is also for a value that is obtained by applying a mathematical operation to the available data according to some specified constraints. A further distinction is between *filtering* that refers to smoothing the field value at a measured point, and *prediction* which refers to optimal estimates of the random field value at unmeasured points. Filtering aims to reduce or remove the noise and deliver a more accurate value for the monitored process. Prediction aims to provide information about the field at points where no measurements are available. If prediction aims to estimate the process value at a point inside the sampling domain, we speak of *interpolation*. On the other hand, if the prediction point is outside the sampling domain prediction refers to *extrapolation*.

1.6 A Personal Selection of Relevant Books

The theory and applications of random fields have been developed by many contributions over the years. The short list of manuscripts that follows is a personal selection rather than a comprehensive list of references or a list reflecting the historical development of the field.

For those interested in a deeper mathematical understanding of the theory of random fields, the books by Robert Adler [10], Robert Adler and Jonathan Taylor [11], and Akiva Yaglom [863] are very useful. A concise overview of the properties of Gaussian random fields can be found in the technical report written by Petter Abrahamsen [3]. These references focus on mathematical definitions and properties of random fields. The book by Michael Stein [774] gives a mathematical treatment of the problem of spatial interpolation using the framework of random fields. The simulation of random fields is investigated by Christian Lantuejoul in [487]. The level of mathematical sophistication required for the above texts should be within the grasp of physicists and engineers with graduate degrees.

Spatial statistics is concerned with the application of mathematical methods to spatial data. Hence, it involves the use of statistical approaches for the estimation of spatial models and their parameters. The spatial statistics perspective is summarized in the comprehensive handbook by Gelfand, Diggle, Fuentes and Guttorp [273] and in the books by Sherman [744], Schabenberger and Gotway [717].

Geostatistics is a branch of spatial statistics that uses random fields to analyze spatial data from the earth sciences. It is assumed that the data come from continuous underlying processes, and thus other types of spatial phenomena, e.g., point processes are excluded in the geostatistical analysis. Geostatistics was motivated by mining engineering problems, which is evident in the applications often discussed in geostatistical books [419, 624]. However, the domain of geostatistical applications is much broader to date. A simple introduction to geostatistics is given in the book of Margaret Armstrong [33]. Several other texts exist at different levels of mathematical sophistication including [165, 303, 823]. The comprehensive book by Jean-Paul Chilés and Pierrer Delfiner [132] skilfully combines theory and modeling aspects. A recent text by Noel Cressie and Christopher Wikle [169] presents a statistical approach to spatiotemporal data modeling from the perspective of Bayesian Hierarchical Modeling. The beautifully illustrated book by Gregoire Mariethoz and Jeff Caers in multiple-point geostatistics focuses on the use of training images for the simulation of geological patterns [543].

The engineering perspective with focus on applications of random fields in mechanics is given by Paul Spanos [286], Erik Vanmarcke [808] and Mircea Grigoriu [310]. Expositions of random field theory with applications in the earth sciences are presented by George Christakos [138] and Ricardo Olea [624], whereas applications in subsurface hydrology are described by Lynn Gelhar [275], Joram Rubin [696], and Peter Kitanidis [459]. Petroleum engineering applications are covered in the books of Mohan Kelkar and Godofredo Perez [446], Michael Hohn [355], and Clayton Deutsch [195]. Environmental applications are discussed by Christakos and Hristopulos [141], Olea [624] and Richard Webster and Margaret Oliver [837]. Applications of random fields in modeling the structure and physical properties of porous media are reviewed by Robert Adler [10], John Cushman [171], and Sol Torquato [797].

In machine learning, Gaussian random fields are known as Gaussian processes (GPs). GPs are employed in learning schemes that are based on the Bayesian framework [678]. Machine learning shares concepts and methods with statistics, and this cross fertilization has led to the field of *statistical learning* [330, 810]. A practical perspective on methods for scientific data mining that includes various types of spatial data is given by Chandrika Kamath [423]. The book titled "Information Field Theory" by Jörg Lemm combines ideas from statistical physics and machine learning using the Bayesian framework [495]. Applications of machine learning ideas to the analysis of spatial data are found in the book by Mikhail Kanevski and Michel Maignan [425].

In physics, random fields underlie the concept of classical field theories. Random fields are used to represent the fluctuations of spatially non-uniform system variables, and in the study of phase transitions where they represent order parameters of the system. There are many texts that include implicit references to random fields, such as those by Nigel Goldenfeld [297] and Mehran Kardar [434]. Useful ideas and methods can also be found in the classical field theory literature that includes monographs such as [21, 399, 593, 890].

Random fields are a key theoretical tool for understanding the statistical properties of turbulent velocity fields and tracer particles that are dispersed inside such fluids. Well-written references on statistical turbulence include the texts by Andrei Monin and Akiva Yaglom [587] and David McComb [557]. Finally, applications of data analysis methods, including geostatistical approaches, to astronomy are presented in the book by Jogesh Babu [244].

In various engineering applications, such as image analysis, random fields are defined on discrete domains such as unstructured grids (networks) and structured grids (lattices).[10] The mathematical tools used for such problems are usually based on Gibbs-Markov random fields. There are many references in this area including the texts by Leonhard Held and Håvard Rue [698], Gerhard Winkler [852], and Xavier Guyon [320].

[10]See the glossary for the definition of lattice used in this book.

Chapter 2
Trend Models and Estimation

> There is no subject so old that
> something new cannot be said about it.
> *Fyodor Dostoevsky*

In the preceding chapter we defined the trend as the component of a random field that represents the large-scale variations. In this chapter we will discuss different approaches for estimating the trend. Trend estimation is often the first step in the formulation of a spatial model.[1] There are many different approaches for trend estimation, and we will not be able to consider them all here. What follows is a collection of methods that provides enough flexibility to accommodate most practical problems. We will broadly distinguish between *empirical* and *systematic (process-based)* trend estimation methods.

In *empirical analysis*, the trend function is determined from general knowledge of the process or the exploratory analysis of the data. For example, it is expected that average temperatures are related to altitude (general knowledge). Data visualization may also reveal the existence of linear trends (exploratory analysis). In empirical trend estimation, the spatial models used are statistical, in the sense that they are not directly derived from natural laws.

On the other hand, *process-based* trend analysis attempts to infer the functional form of the trend, and possibly the coefficients of the trend function from physical, chemical, ecological or biological models. For example, let us consider steady-state flow in a homogeneous and isotropic porous medium which is saturated with a single-phase fluid. In this case, if we assume that the fluid permeability of the medium has a "coarse-grained" uniform scalar value (i.e., if we neglect local fluctuations and potential anisotropies of the permeability field), the solution of the Laplace equation with specified boundary conditions can give an estimate of the

[1] In some cases a nonlinear transformation of the data is performed before estimating the trend.

D. T. Hristopulos, *Random Fields for Spatial Data Modeling*, Advances in Geographic Information Science, https://doi.org/10.1007/978-94-024-1918-4_2

mean pressure field [173, p. 126]. In cases where the governing laws are known and admit analytical solutions for the coarse-grained behavior, the spatial models are based on knowledge of the underlying process(es).

What one needs to keep in mind about empirical trend estimation is that the *trend-fluctuation-noise decomposition* introduced in Sect. 1.2.1 is not uniquely specified. Thus, there are many (practically infinite) ways of "decomposing" the data into trend and fluctuation components.

2.1 Empirical Trend Estimation

In empirical trend estimation we can either assume a functional dependence for the trend function (parametric approach) or we can "construct" the trend from the data by means of a filtering (smoothing) operation (non-parametric approach). In both cases, it is important to provide physical motivation for the model selected. Very complicated trend models are not easily amenable to physical interpretation, and they have a high risk of being rejected by experts. In practice, modeling choices are usually guided by *Occam's razor* (also known as the *"law of parsimony"*). Occam's razor favors the simplest hypothesis (model) that is consistent with the sample data [82].

Empirical trend functions can be further classified into *global dependence* and *local dependence* models. In the case of global models, both the functional form and the coefficients are uniform throughout the spatial domain. In the case of local models, the coefficients or even the functional form of the trend function change over the domain.

In empirical trend modeling, *regression analysis* is the statistical method most often used to determine the coefficients of a particular functional form. Regression analysis accomplishes this by fitting the selected model function to the data. This procedure can be applied to different functional forms. If all the models have the same number of parameters, the "best" model is selected based on a measure-of-fit that quantifies the difference between the data and the model (e.g., the likelihood or the sum of the square errors). If the models tested involve different number of parameters, the "best model" is selected from all the trend functions considered using a *model selection criterion*, such as the Akaike Information Criterion (AIC), the Bayesian Information Criterion (BIC), et cetera [506].

2.2 Regression Analysis

Regression analysis considers two sets of variables: the *independent or predictor* variables and the *dependent or response* variables. The main goal of regression analysis is to determine a functional form that expresses the response variables in terms of the predictor variables. The predictor variables can be different monomials of the spatial coordinates or auxiliary physical variables that are related to the response (e.g., if precipitation is the response variable, it makes sense to use altitude and temperature as predictor variables).

This section is a brief introduction to regression analysis. To link regression analysis with trend estimation, we can loosely[2] think of the response variable as the trend function, and the predictor variables as various functions of the location vector $\mathbf{s} \in \mathcal{D} \subset \mathbb{R}^d$. This is further discussed in Sect. 2.3.

Warning on notation In this section we use $\mathbf{x}$ to represent the predictor variables and y for the response variable following the common notation in regression analysis.

Problem definition For simplicity, we focus on a single *response variable*, whose measurements are denoted by y_n, where $n = 1, \ldots, N$, and a vector of *predictor variables*, $\mathbf{x} = (x_1, \ldots, x_P)^\top$, $p = 1, \ldots, P$, where P is the dimensionality of the predictor set. The sample data involve the set of N response measurements, $\{y_n\}_{n=1}^N$ and $N \times P$ predictor measurements.

Let us assume that the response depends on the predictor variables through the following *regression model*

$$y_n = \Phi(\mathbf{x}_n; \boldsymbol{\beta}) + \upsilon_n, \quad \text{for } n = 1, \ldots, N,$$

where $\Phi(\cdot)$ is a nonlinear (in general) function, $\boldsymbol{\beta} = (\beta_1, \ldots, \beta_K)^\top$ is a vector of real-valued coefficients, $\beta_k \in \mathbb{R}$, where $k = 1, \ldots, K$, and $\upsilon_n = y_n^* - \Phi(\mathbf{x}_n; \boldsymbol{\beta})$, for all $n = 1, \ldots, N$ are the *regression residuals* which represent the difference between the data and the regression model.

Regression analysis aims to determine the optimal set of coefficients based on a number of measurements N of the predictor variables and the respective responses. The optimal coefficients are estimated by minimizing a *loss function* that involves the residuals.

Assumptions Typically, a number of modeling assumptions are implied in regression analysis. These assumptions (sufficient conditions) are used to ensure that the estimates of the regression coefficients have desired statistical properties. The assumptions involve the following:

- The residuals υ_n, $n = 1, \ldots, N$, represent realizations of the random variable $\Upsilon_n(\omega)$ whose mean is equal to zero, conditionally on the predictor variables.
- The sample is representative of the underlying population.
- The predictor variables are measured without errors.
- The predictor variables are linearly independent.
- The residuals are uncorrelated, i.e., $\mathcal{E}[\Upsilon_n(\omega)\, \Upsilon_m(\omega)] = \sigma_\upsilon^2\, \delta_{n,m}$, for $n, m = 1, \ldots, N$, where $\delta_{n,m}$ is the binary-valued Kronecker delta.
- The variance of the residuals is constant over the observed range of the predictor values, i.e., it does not depend on the predictor variables (this is known as the homoscedasticity assumption).

[2]In spatial data analysis, the residuals obtained by removing the trend from the data often are not uncorrelated, in contrast with the commonly used assumption in classical regression analysis.

In the following, we assume *linear regression models* of the form

$$y_n = \boldsymbol{\beta}^\top \mathbf{x}_n + \upsilon_n, \quad \text{for } n = 1, \ldots, N, \tag{2.1}$$

where $\boldsymbol{\beta} = (\beta_1, \ldots, \beta_K)^\top$ is the vector of linear *regression coefficients*. Note that in the linear case the dimensionality of $\boldsymbol{\beta}$ is the same as the number of prediction variables. To include a constant offset in the regression equation, we set $x_1 = 1$ so that β_1 represents the constant term in the regression equation (2.1). We will only present the most basic results of linear regression analysis below. Interested readers will find more information in the books by Seber and Lee [736] and Hastie, Tibshirani and Friedman [330].

Remark about the regression coefficients We have assumed that the residuals are realizations of the random variables $\{\Upsilon_n(\omega)\}_{n=1}^N$. The "true" value of the regression model coefficients, $\boldsymbol{\beta}^*$, is unknown to us. The regression coefficient vector $\boldsymbol{\beta}$ in (2.1) can take any values that we wish; some lead to a good fit with the data and some do not. The optimal coefficient vector is denoted by $\hat{\boldsymbol{\beta}}$, and it is based on the minimization of a specified loss function. The estimate $\hat{\boldsymbol{\beta}}$ is a random vector, since its values vary for different realizations of the residuals (usually we are given just one realization). The estimate $\hat{\boldsymbol{\beta}}$ is thus an approximation of $\boldsymbol{\beta}^*$. Different statistical measures can be used to assess the quality of the fit afforded by $\hat{\boldsymbol{\beta}}$ between the model and the data.

2.2.1 Ordinary Least Squares

The most commonly used loss function is the *residual sum of squares* which is given by the $\mathbb{L}_2$ norm of the residuals:[3]

$$\mathrm{RSS}(\boldsymbol{\beta}) = \sum_{n=1}^{N} \upsilon_n^2 = \mathbb{L}_2(\boldsymbol{\upsilon}). \tag{2.2}$$

The method of *ordinary least squares (OLS)* can then be used to estimate the optimal model parameters $\hat{\boldsymbol{\beta}}$ by minimizing the residual sum of squares, i.e.,

$$\hat{\boldsymbol{\beta}}_{\mathrm{OLS}} = \arg\min_{\boldsymbol{\beta}} \mathrm{RSS}(\boldsymbol{\beta}).$$

The minimization leads to a linear system of equations from which the OLS solution is obtained as follows:

$$\hat{\boldsymbol{\beta}}_{\mathrm{OLS}} = (\mathbf{X}^\top \mathbf{X})^{-1} \mathbf{X}^\top \mathbf{y}, \tag{2.3}$$

[3]The residuals $\boldsymbol{\upsilon}$ implicitly depend on the values of the regression coefficients $\boldsymbol{\beta}$.

where $\mathbf{X}$ is the $N \times P$ *design matrix* with elements $[\mathbf{X}]_{n,p} = x_{n,p}$, $n = 1, \ldots, N$ and $p = 1, \ldots, P$, and $\mathbf{y}$ is the vector of the response variable observations. So, the rows of $\mathbf{X}$ correspond to different sample points, while the columns of $\mathbf{X}$ represent different basis functions.

Finally, the estimate of the response variable is determined by the vector

$$\hat{\mathbf{Y}} = \mathbf{X}\,\hat{\boldsymbol{\beta}}_{\text{OLS}} = \mathbf{X}\,(\mathbf{X}^{\top}\,\mathbf{X})^{-1}\,\mathbf{X}^{\top}\,\mathbf{y} = \mathbf{H}\mathbf{y}, \tag{2.4}$$

where $\mathbf{H} = \mathbf{X}\,(\mathbf{X}^{\top}\,\mathbf{X})^{-1}\,\mathbf{X}^{\top}$ is the so-called *hat matrix*. The OLS estimate (2.4) is the orthogonal projection of $\mathbf{Y}$ onto the column space of the design matrix $\mathbf{X}$. It is straightforward to show the main properties of $\mathbf{H}$: (i) it is a symmetric $N \times N$ matrix, (ii) it is idempotent, meaning that $\mathbf{H}^2 = \mathbf{H}$, and (iii) $\mathbf{H}\mathbf{X} = \mathbf{X}$.

Problem 2.1 Based on the properties of the hat matrix $\mathbf{H}$ show that (i) The OLS estimate (2.4) is unbiased, (ii) the covariance of $\hat{\mathbf{Y}}$ is given by $\sigma_{\upsilon}^2\mathbf{H}$, where σ_{υ}^2 is the variance of the residuals and is the same as that of $\mathbf{Y}$ and (iii) the residuals are orthogonal to the column space of $\mathbf{X}$, i.e., $\boldsymbol{\upsilon}^{\top}\mathbf{X} = \mathbf{0}$.

Existence of OLS solution The existence of a solution for equation (2.3) assumes that the $P \times P$ matrix $\mathbf{X}^{\top}\,\mathbf{X}$ is invertible. This typically occurs if $N \gg P$. In addition, if $\text{rank}(\mathbf{X}) = p$, i.e., if the predictors are linearly independent, the solution $\hat{\boldsymbol{\beta}}_{\text{OLS}}$ is unique (for the definition of the matrix rank see the glossary and Appendix C). On the other hand, if $\text{rank}(\mathbf{X}) < p$ which occurs if $p > n$ (i.e., for high-dimensional data), the matrix $\mathbf{X}^{\top}\,\mathbf{X}$ is singular and the ordinary least squares problem admits multiple solutions. Problems admitting multiple solutions belong in the class of ill-posed problems (see glossary for the definition of ill-posed problems). In the case of ill-posed inverse problems, regularization techniques have been developed which help to achieve meaningful approximate solutions [602, 795].

Ordinary least squares is popular because it is possible to calculate the derivatives of the $\mathbb{L}_2$ norm with respect to the coefficients $\boldsymbol{\beta}$. In the case of a linear regression model, OLS leads to explicit equations for the optimal coefficients. Among other applications, the OLS method is used to determine the weights of the stochastic optimal linear predictor (kriging) as discussed in Chap. 10. Under the condition of independent and identically distributed samples, OLS can be shown to be equivalent to maximum likelihood estimation (see Chap. 12). We return to OLS in Sect. 2.3.4 where we discuss the application of the method in the estimation of spatial trend models.

2.2.2 *Weighted Least Squares*

The method of *weighted least squares* is a modification of OLS that assigns unequal weights to each of the residuals [165]. This method has an advantage over OLS if the variance of the residuals changes over the observation range, i.e., if the homoskedasticity assumption breaks down. In this case the weights are assigned values equal to the reciprocal of the variances.

Remark One has to be careful about choosing the size of the prediction vector. Large size p can improve the interpolation (in-sample) performance of the model. However, large values of p (especially $p \gg N$) can lead to large out-of-sample (extrapolation) errors.

2.2.3 Regularization

For ill-posed problems the norm of the unconstrained solution tends to infinity. The procedure known as *Tikhonov regularization* modifies the loss function by adding a penalty term $\lambda\, g(\boldsymbol{\beta})$, where λ is a real-valued parameter that controls the amount of regularization, and $g(\boldsymbol{\beta})$ is a function of the regression parameters. The penalty term controls the magnitude of the solution's norm [421].

The term Tikhonov regularization is used in numerical analysis to denote a procedure that renders ill-posed inverse problems tractable by introducing a constraint on the norm of the solution [602]. Linear regression can be viewed as an ill-posed problem because the data are often contaminated with noise. Hence, regularization methods can be used to obtain well-defined solutions. Another advantage of regularization is that it can reduce the number of essential predictor variables. This amounts to reducing the complexity of the regression model. The reduced complexity also allows better generalization (out-of-sample predictive performance), in the same way that a linear model (low complexity) of noisy data is better than a high-degree polynomial (high complexity) model.

2.2.4 Ridge Regression

In statistics and machine learning regularization is often referred to as ridge regression. In the case of *ridge regression* the function $g(\cdot)$ is given by the $\mathbb{L}_2$ norm of the coefficients $g(\boldsymbol{\beta}) = \|\boldsymbol{\beta}\|^2$ [353]. Then, the loss function becomes a *penalized residual sum of squares*

$$\text{PRSS}(\boldsymbol{\beta}) = \sum_{n=1}^{N} \upsilon_n^2 + \lambda\, \mathbb{L}_2(\boldsymbol{\beta}), \text{ where } \mathbb{L}_2(\boldsymbol{\beta}) = \sum_{k=1}^{K} \beta_k^2,$$

and λ represents the so-called *shrinkage parameter*. The optimal ridge regression coefficients are given by

$$\hat{\boldsymbol{\beta}}_{\text{ridge}} = \arg\min_{\boldsymbol{\beta}} \text{PRSS}(\boldsymbol{\beta}).$$

Assuming that the linear regression model (2.1) holds, the optimal ridge regression coefficients are obtained by the following equation

$$\hat{\boldsymbol{\beta}}_{\text{ridge}} = (\mathbf{X}^\top \mathbf{X} + \lambda \mathbf{I}_P)^{-1} \mathbf{X}^\top \mathbf{y}, \tag{2.5}$$

where $\mathbf{I}_P$ is the identity matrix of dimension $P \times P$; $[\mathbf{I}_P]_{n,m} = \delta_{n,m}$ for $n, m = 1, \ldots, P$, where $\delta_{n,m}$ is the Kronecker delta defined by $\delta_{n,m} = 1$ if $n = m$ and $\delta_{n,m} = 0$ if $n \neq m$. The ridge regression solution $\hat{\boldsymbol{\beta}}_{\text{ridge}}$ exists, even if the initial OLS problem is singular (i.e., if $(\mathbf{X}^\top \mathbf{X})$ cannot be inverted).

Example 2.1 Determine the ridge regression weight vector in the case of an orthogonal design matrix **X**.

Answer If **X** is orthogonal, then by definition $\mathbf{X}^\top \mathbf{X} = \mathbf{I}_P$. Then the ridge regression coefficients (2.5) become

$$\hat{\boldsymbol{\beta}}_{\text{ridge}} = \frac{\mathbf{X}^\top \mathbf{y}}{1+\lambda} = \frac{\hat{\boldsymbol{\beta}}_{\text{OLS}}}{1+\lambda}.$$

Hence, in this case the ridge coefficients are simply the OLS coefficients scaled by the factor $1 + \lambda$.

- The regularization term enforces desired smoothness properties on the regression solution by penalizing large values of the coefficients.
- In the case of *linear* ridge regression, an explicit system of linear equations can be derived for the optimal values of the regression coefficients.
- If $\lambda = 0$ the OLS solution is recovered.
- As $\lambda \to \infty$ the values of the regression coefficients tend to zero.[4]
- The ridge regression coefficients $\hat{\boldsymbol{\beta}}_{\text{ridge}}$ are biased estimators of the true $\boldsymbol{\beta}^*$, but their variance is reduced compared to $\hat{\boldsymbol{\beta}}_{\text{OLS}}$ (see also comments on the bias-variance tradeoff below). The variance reduction is due to the *shrinkage* of the coefficients and can lead to a reduction of the mean square error.

Bayesian formulation The problem of linear regression can also be formulated in the Bayesian framework. In the Bayesian setting, one assumes a prior distribution for the parameters (in this case the regression weights). The optimal weights can then be obtained by means of the *maximum a posteriori (MAP)* estimate which is based on the posterior distribution. The posterior distribution for the weights incorporates information from both the prior and the likelihood function (for more on the Bayesian framework see Chap. 11). The MAP estimate is given by the weights that maximize the logarithm of the posterior distribution. Assuming a

[4]The common convention in ridge and LASSO regression problems is that the predictor variables are standardized (i.e., their mean is equal to zero and their variance to one), and the response variable is centered, i.e., its mean is zero.

Gaussian prior for the weights, the MAP estimate yields the same result as ridge regression [561]. Hence, the latter can be viewed as the MAP estimate using the Gaussian prior.

2.2.5 LASSO

Ridge regression accomplishes coefficient shrinkage but not variable selection. However, variable selection is important in the case of complex regression models that involve many predictor variables, especially if $P \gg N$ (i.e., if the number of predictor variables greatly exceeds the number of observations).

To address this shortcoming, a regression method based on the *absolute value norm*, also known as $\mathbb{L}_1$ norm has been developed [794]. This approach is titled *least absolute shrinkage and selection operator (LASSO)*. The LASSO loss function is given by the following penalized residual sum of squares

$$\mathrm{PRSS}_{\mathbb{L}_1}(\boldsymbol{\beta}) = \sum_{n=1}^{N} \upsilon_n^2 + \lambda\, \mathbb{L}_1(\boldsymbol{\beta}), \ \text{where } \mathbb{L}_1(\boldsymbol{\beta}) = \sum_{k=1}^{K} \|\beta_k\| \,.$$

The optimal LASSO coefficients are given by

$$\hat{\boldsymbol{\beta}}_{\mathrm{lasso}} = \arg\min_{\boldsymbol{\beta}} \mathrm{PRSS}_{\mathbb{L}_1}(\boldsymbol{\beta}).$$

The $\mathbb{L}_1$-norm reduces the impact of potential outliers on the estimates of the optimal parameters compared with OLS. However, determining the optimal LASSO coefficients in general situations requires numerical methods of solution.

LASSO solution An explicit LASSO solution is only available if the design matrix $\mathbf{X}$ is orthonormal, i.e., if $\mathbf{X}^\top \mathbf{X} = \mathbf{I}_P$. The LASSO solution in this case is expressed in terms of the OLS coefficients and the regularization parameter as follows [794]

$$\hat{\beta}_{\mathrm{lasso},p} = \mathrm{sign}\left(\hat{\beta}_{\mathrm{OLS},p}\right) \mathrm{Pos}\left(\left|\hat{\beta}_{\mathrm{OLS},p}\right| - \lambda\right), \ \text{for } p = 1, \ldots, P, \tag{2.6}$$

where $\mathrm{sign}(x) = x/|x|$ is the sign function, and the function $\mathrm{Pos}(x) = x\,\Theta(x)$ returns the value of x if the latter is positive and zero otherwise. The function $\Theta(x)$ is the *unit step function or Heaviside function* defined by

$$\Theta(x) = \begin{cases} 1 & \text{if } x \geq 0 \\ 0 & \text{if } x < 0. \end{cases} \tag{2.7}$$

The LASSO allows for feature selection and can reduce the dimensionality of the prediction variable, which is important for complex regression problems in machine learning [330, 561, 794]. Feature selection implies that a number of the regression coefficients will be set exactly equal to zero, thus effectively eliminating the respective predictor variables from the regression equation. This property of LASSO regression is evidenced in the explicit solution (2.6): all LASSO coefficients which correspond to OLS coefficients with magnitude less than λ, i.e., for $\left|\hat{\beta}_{\text{OLS},p}\right| < \lambda$, are set to zero.

Bias-variance tradeoff Regression models are typically used for prediction at "points" $\mathbf{x} \in \mathbb{R}^P$ where no measurements are available. For example, given a specific predictor variable assignment $\mathbf{x}_0$, the unknown value of the response variable $\hat{y}_0$ is given by

$$\hat{y}_0 = \Phi(\mathbf{x}_0; \hat{\boldsymbol{\beta}}).$$

The prediction can be viewed as a random variable, since the coefficient $\hat{\boldsymbol{\beta}}$ is a random variable as discussed above. To assess the performance of this estimate, the mean square prediction error is typically used. The prediction error is defined as $y_0^* - \hat{y}_0$, where $y_0^* = \Phi(\mathbf{x}_0; \boldsymbol{\beta}^*)$ is the true value.

The *mean square error of the prediction* is then given by

$$\mathcal{E}[(\hat{y}_0 - y_0)^2] = \sigma_v^2 + \left[\text{Bias}(\hat{y}_0)\right]^2 + \mathbb{V}\text{ar}\left\{\hat{y}_0\right\}. \tag{2.8}$$

In the case of linear regression, the OLS estimate leads to unbiased estimates of the response. However, the variance of the response can be quite large, especially in complex regression models. Both ridge regression and LASSO can reduce the variance of the prediction by introducing some bias. This behavior is an example of the well-known *bias-variance tradeoff*.

The following toy example is often used to illustrate the bias-variance tradeoff.

Example 2.2 Consider a simple linear regression problem of the form $y_i = \beta x_i$. Furthermore, let us assume that the OLS estimate $\hat{\beta}$ follows the normal distribution $\mathcal{N}(1, 1)$; this implies that the true value of β is $\beta^* = 1$ and that the OLS estimate has a variance of one. What happens to the mean square error if the OLS estimator is replaced by $\hat{\beta}' = \beta/\alpha$, where $\alpha > 1$ is a shrinkage factor?

Answer The mean square error of the OLS estimator is given by

$$\text{MSE}(\hat{\beta}) = \mathcal{E}[(\hat{\beta} - 1)^2] = \mathbb{V}\text{ar}\left\{\hat{\beta}\right\}.$$

For the modified estimator $\hat{\beta}'$ the mean square error is respectively given by

$$\text{MSE}(\hat{\beta}') = \mathcal{E}[(\hat{\beta}' - 1)^2] = \mathbb{V}\text{ar}\left\{\hat{\beta}'\right\} + \left(\mathcal{E}[\hat{\beta}' - 1]\right)^2 = \frac{1}{\alpha^2} + \left(\frac{1}{\alpha} - 1\right)^2.$$

Due to shrinkage, the bias of $\hat{\beta}'$ is equal to $1/\alpha - 1 \neq 0$. In fact, the bias increases as $\alpha \to \infty$. On the other hand, the variance of $\hat{\beta}'$ declines as α increases. It is easy to determine that the stationary point of $\text{MSE}(\hat{\beta}')$ is $\alpha = 2$, and that the stationary point corresponds to a minimum for the mean square error.

The *elastic net* is a regularization and variable selection method that combines the LASSO with ridge regression [893]. The elastic net can presumably outperform the LASSO, while it shares with the latter the sparsity of representation afforded by variable selection.

2.2.6 *Goodness of Fit*

An important practical question is how to evaluate the performance of a linear regression model. Partly, this involves the question of how well the model fits the observed data. There exist various statistics for measuring the goodness of fit. One of them is the coefficient of determination.

Coefficient of determination Let us define the optimal model parameters by $\hat{\boldsymbol{\beta}}$ and the corresponding residuals by $\hat{\boldsymbol{\upsilon}}$. The quality of the fit between the response data $\hat{y}_n$ and the regression model $\Phi(\mathbf{x}_n; \hat{\boldsymbol{\beta}})$, $n = 1, \ldots, N$, is expressed by means of the *coefficient of determination*

$$R^2 = \frac{\|\hat{\mathbf{y}} - \overline{\mathbf{y}}\|^2}{\|\mathbf{y} - \overline{\mathbf{y}}\|^2}, \tag{2.9}$$

where (i) $\overline{\mathbf{y}}$ is an $N \times 1$ vector with all its elements equal to the average response value $\overline{y} = \frac{1}{N}\sum_{n=1}^{N} y_n$, (ii) $\hat{\mathbf{y}}$ is the response vector predicted by the model, and (iii) and $\mathbf{y} = (y_1, \ldots, y_N)^\top$ is the vector of observed responses.

In general the coefficient of determination takes positive values not exceeding one, i.e., $0 \leq R^2 \leq 1$. A value of R^2 close to one means that most of the variability in the response data is explained by the regression model. In contrast, a value near zero means that the predicted responses by the regression model are close to the sample mean. The coefficient of determination can be interpreted as the percent of the total variance of the response data that is explained by the regression model.

Other performance measures There are various statistical measures (in addition to goodness of fit) as well as graphical diagnostics that can be used to interpret the quality of the fit to the data afforded by the regression model. We will discuss some of these in Sect. 2.3.4 below and in Chap. 10.

A point to keep in mind is that in classical statistics a good regression model aims to achieve independent residuals. This means that the variability of the data is due to either the trend (and is explained by the model) or to independent fluctuations.

In linear time series analysis, the regression model involves both trend and fluctuation components, the latter being modeled typically by means of ARMA processes. In this case, the residuals are obtained after subtracting from the data estimates for both the trend and the fluctuations. Hence, if the model performs adequately, the residuals should represent uncorrelated white noise.

In spatial data modeling, on the other hand, the residuals obtained after the subtraction of trends typically contain correlated fluctuations. The latter are modeled in terms of random fields. The trend function, on the other hand, is used as a model of the expectation of the random field. Of course when both the trend function and the random field fluctuations are removed from the data, the resulting residuals should behave as white noise as well.

2.3 Global Trend Models

In the following, we will assume that the prediction variables $\mathbf{x}$ are replaced by a set of functions $\{f_p(\mathbf{s})\}_{p=1}^{P}$, where $f_p(\mathbf{s}) : \mathbb{R}^d \rightarrow R$, that are indexed by space variables and belong to a given basis. The trend is then formed as a superposition of such functions with linear coefficients that are determined by regression analysis. In general, the trend may be composed of a combination of such functions and prediction variables; e.g., the amount of rainfall on a mountain slope may depend not only on the location (map projection coordinates and altitude), but also on the orientation of the slope.

A global trend refers to functions which have the same functional form over the entire domain $\mathcal{D}$ of the data. The following global functions can be used either individually or in superposition to provide simple models of global trends.

2.3.1 *Linear Spatial Dependence*

Let us consider that $\mathbf{s} \in \mathbb{R}^d$ denotes a single point in the spatial domain $\mathcal{D}$. A linear trend function is given by the following equation

$$m_{\mathrm{x}}(\mathbf{s}) = m_0 + \mathbf{b} \cdot \mathbf{s}. \tag{2.10}$$

This trend has a uniform gradient given by $\nabla m_{\mathrm{x}}(\mathbf{s}) = \mathbf{b}$, where $\mathbf{b}^{\top} = (b_1, \ldots, b_d)$, and $b_j \in \mathbb{R}$, for all $j = 1, \ldots, d$. In this case the regression coefficient vector is $\boldsymbol{\beta} = (m_0, b_1, \ldots, b_d)^{\top}$.

2.3.2 Polynomial Spatial Dependence

A more flexible form of spatial dependence is the second degree polynomial in the spatial coordinates which is given by

$$m_{\mathrm{x}}(\mathbf{s}) = m_0 + \sum_{i=1}^{d} b_i\, s_i + \sum_{i=1}^{d}\sum_{j>i}^{d} c_{i,j}\, s_i\, s_j. \tag{2.11}$$

In (2.11), the coefficients satisfy $b_i \in \mathbb{R}$, for all $i = 1, \ldots, d$ and $c_{i,j} = c_{j,i} \in \mathbb{R}$, for all $i, j = 1, \ldots, d$. The linear trend (2.10) is a special case of the second degree polynomial obtained for $c_{i,j} = 0$, for all $i, j = 1, \ldots, d$. The regression coefficient vector $\boldsymbol{\beta}$ includes m_0, all the b_i and all the $c_{i,j}$.

Polynomials of higher degree than two are usually avoided. The reason is that they can lead to significant oscillations near the domain boundaries (*Runge phenomenon*) for equidistant grids [704], [673, p. 1090]. In addition, high-degree polynomials used as interpolating functions often perform very poorly if used for extrapolation outside the data domain.

2.3.3 Periodic Spatial Dependence

Periodic variation in space is expressed as a superposition of harmonic modes with different *wavelengths or spatial periods*

$$m_{\mathrm{x}}(\mathbf{s}) = m_0 + \sum\nolimits_{m=1}^{M} \left[A_m \cos(\mathbf{k}_m \cdot \mathbf{s}) + B_m \sin(\mathbf{k}_m \cdot \mathbf{s})\right], \tag{2.12}$$

where the *wavevectors* $\mathbf{k}_m$ correspond to *spatial frequencies*, and $A_m, B_m \in \mathbb{R}$ for $m = 1, \ldots, M$ are real-valued coefficients. The above is equivalent to the equation

$$m_{\mathrm{x}}(\mathbf{s}) = m_0 + \sum\nolimits_{m=1}^{M} C_m \cos(\mathbf{k}_m \cdot \mathbf{s} + \phi_m), \tag{2.13}$$

where the mode amplitudes

$$C_m = \sqrt{A_m^2 + B_m^2},\; m = 1, \ldots, M,$$

are real-valued coefficients and

$$\phi_m = \arctan\left(-\frac{B_m}{A_m}\right) \in [0, 2\pi),\; m = 1, \ldots, M$$

represent the phases of each mode. The dependence of the trend function (2.12) on the coefficients A_m and B_m is linear, whereas the function (2.13) depends nonlinearly on the phases. Hence, if the spatial frequencies $\mathbf{k}_m$ of the modes are known a priori, the coefficients A_m and B_m can be estimated by means of linear regression analysis.

2.3.4 *Multiple Linear Regression*

In certain cases the trend can be expressed as a linear superposition of known functions. For example, let the trend function be given by

$$m_{\mathrm{x}}(\mathbf{s}) = \sum_{p=1}^{P} \beta_p \, f_p(\mathbf{s}), \tag{2.14}$$

where $\beta_p \in \mathbb{R}$, for all $p = 1, \ldots, P$ are the regression model parameters, and the $f_p(\mathbf{s})$, $p = 1, \ldots, P$, with $f_1(\mathbf{s}) = 1$, comprise the set of *basis functions*. In this case, the trend involves known functions and the vector of undetermined coefficients $\boldsymbol{\beta}$.

To make the connection with the general formulation of regression analysis in Sect. 2.2, $\Phi(\mathbf{x}; \boldsymbol{\beta}) = m_{\mathrm{x}}(\mathbf{s})$. Since the dependence of the trend model on the coefficients is linear, *multiple linear regression* can be used to determine the optimal values of the coefficient vector $\boldsymbol{\beta}$.

The periodic model (2.12) does not comply with this representation, due to the nonlinear dependence on the wavevectors $\mathbf{k}_p$, unless the latter are assumed to be known. On the other hand, the linear (2.10) and polynomial (2.11) trend models conform with the expression (2.14).

Example 2.3 Linear model: Write the basis functions of the linear polynomial model in two spatial dimensions.

Answer In $d = 2$ the linear model is expressed in terms of three basis functions, $f_1(s_1) = 1$, $f_2(s_1) = s_1$ and $f_3(s_2) = s_2$.

Example 2.4 Second degree polynomial trend: Write the basis functions for a second degree polynomial trend in two dimensions.

Answer The trend function (2.14) is a second degree polynomial which includes the six basis functions shown in Table 2.1.

Table 2.1 Basis functions for a second degree polynomial trend. The location vector $\mathbf{s} = (s_1, s_2)$ is defined over a domain $\mathcal{D} \subset \mathbb{R}^2$

$f_1(\mathbf{s})$	$f_2(\mathbf{s})$	$f_3(\mathbf{s})$	$f_4(\mathbf{s})$	$f_5(\mathbf{s})$	$f_6(\mathbf{s})$
1	s_1	s_2	s_1^2	s_2^2	$s_1 s_2$

Matrix formulation of regression equations Let $\upsilon(\mathbf{s})$ denote the residual obtained by subtracting the trend from the data. The following linear system links the data with the trend and the residual

$$\begin{bmatrix} x_1^* \\ x_2^* \\ \vdots \\ x_N^* \end{bmatrix} = \begin{bmatrix} 1 & f_2(\mathbf{s}_1) & f_3(\mathbf{s}_1) & \cdots & f_P(\mathbf{s}_1) \\ 1 & f_2(\mathbf{s}_2) & f_3(\mathbf{s}_2) & \cdots & f_P(\mathbf{s}_2) \\ \vdots & \vdots & \vdots & \cdots & \vdots \\ 1 & f_2(\mathbf{s}_N) & f_3(\mathbf{s}_N) & \cdots & f_P(\mathbf{s}_N) \end{bmatrix} \begin{bmatrix} \beta_1 \\ \beta_2 \\ \beta_3 \\ \vdots \\ \beta_P \end{bmatrix} + \begin{bmatrix} \upsilon_1 \\ \upsilon_2 \\ \vdots \\ \upsilon_N \end{bmatrix}, \tag{2.15}$$

where $\upsilon_n = \upsilon(\mathbf{s}_n)$ for $n = 1, \ldots, N$. The above system of equations is expressed more concisely as follows

$$\mathbf{X}^* = \mathbf{S}\,\mathbf{B} + \mathbf{V}, \tag{2.16}$$

where the arrays $\mathbf{X}^*$, $\mathbf{S}$, $\mathbf{B}$ and $\mathbf{V}$ are defined as follows:

- $\mathbf{X}^*$ is the data vector of dimension $N \times 1$
- $\mathbf{S}$ is the *design matrix* of dimension $N \times P$, formed by the values of the P basis functions at the N sampling points
- $\mathbf{B}$ is the vector of the basis function coefficients of dimension $P \times 1$
- $\mathbf{V}$ is the vector of the residuals of dimension $N \times 1$.

The OLS solution of the system (2.16) minimizes the *sum of the squares of the residuals*, i.e., the inner product $\mathbf{V} \cdot \mathbf{V} = \mathbf{V}^\top \mathbf{V}$. If we assume that the matrix $\mathbf{S}^\top\,\mathbf{S}$ is invertible, the OLS solution is given by

$$\hat{\mathbf{B}} = \left(\mathbf{S}^\top\,\mathbf{S}\right)^{-1} \mathbf{S}^\top\,\mathbf{X}^*. \tag{2.17}$$

Trend prediction Consequently, the value of the trend at the sampling locations is given by

$$\hat{\mathbf{m}}_{\mathrm{x}} = \mathbf{S}\,\hat{\mathbf{B}} = \mathbf{S}\,\left(\mathbf{S}^\top\,\mathbf{S}\right)^{-1} \mathbf{S}^\top\,\mathbf{X}^*. \tag{2.18}$$

The residual vector which includes both the correlated fluctuations and potentially uncorrelated fluctuations is given by

$$\hat{\mathbf{V}} = \mathbf{X}^* - \hat{\mathbf{m}}_{\mathrm{x}} = \mathbf{X}^* - \mathbf{S}\,\left(\mathbf{S}^\top\,\mathbf{S}\right)^{-1} \mathbf{S}^\top\,\mathbf{X}^*. \tag{2.19}$$

Moreover, if we define the $N \times N$ *matrix of regression coefficients*

$$\mathbf{H} = \mathbf{S}\,\left(\mathbf{S}^\top\,\mathbf{S}\right)^{-1} \mathbf{S}^\top, \tag{2.20}$$

the equations for the trend and the residual are simplified as follows,

$$\hat{\mathbf{m}}_{\mathrm{x}} = \mathbf{H}\,\mathbf{X}^{*}, \tag{2.21}$$

$$\hat{\mathbf{V}} = (\mathbf{I}_N - \mathbf{H})\,\mathbf{X}^{*}, \tag{2.22}$$

where $\mathbf{I}_N$ is the identity matrix of dimension $N \times N$, such that $[\mathbf{I}_N]_{n,m} = \delta_{n,m}$, where $\delta_{n,m} = 1$ if $n = m$ and $\delta_{n,m} = 0$ if $n \neq m$ is the Kronecker delta. The above equations are standard for multiple linear regression, and their derivation can be found in any textbook on the subject.

Correlated residuals The OLS method is an efficient estimator of the model coefficients if the residuals are uncorrelated. However, in many problems of spatial data analysis the residuals are spatially correlated. Then, it is preferable to use the *generalized least squares (GLS)* estimator. In GLS, the sum of the squared residuals is replaced by $\mathbf{V}^{\top}\mathbf{C}^{-1}\,\mathbf{V}$ where $\mathbf{C}$ is the covariance matrix of the residuals. Then, the GLS solution for the coefficients is given by (e.g. [165, p. 21])

$$\hat{\mathbf{B}}_{\mathrm{GLS}} = \left(\mathbf{S}^{\top}\,\mathbf{C}^{-1}\mathbf{S}\right)^{-1}\mathbf{S}^{\top}\,\mathbf{C}^{-1}\,\mathbf{X}^{*}, \tag{2.23a}$$

while the variance of the GLS coefficients is given by the following $P \times P$ matrix

$$\mathbb{V}\mathrm{ar}\left\{\hat{\mathbf{B}}_{\mathrm{GLS}}\right\} = \left(\mathbf{S}^{\top}\,\mathbf{C}^{-1}\mathbf{S}\right)^{-1}. \tag{2.23b}$$

If the residuals follow the multivariate normal distribution, the GLS solution for the coefficients is equivalent to the *maximum likelihood* solution. If the residuals are uncorrelated their covariance matrix is diagonal and the OLS solution is recovered. A comparative analysis of the performance of OLS versus GLS for spatial data is presented in [54].

Example 2.5 Lignite[5] deposits are found in sedimentary basins. More sediment is typically deposited near the center of the basin giving rise to thicker and better-quality strata of lignite in these areas. Thus, the formation process is responsible for the appearance of global trends, such as the parabolic surface illustrated in Fig. 2.1. The data correspond to lignite content in ash water free (AWF).[6]

Answer Let $\mathbf{s} = (s_1, s_2)$ represent the position vector in the basin. We assume a second degree polynomial for the AWF trend that is expressed as follows

$$\hat{m}_{\mathrm{x}}(\mathbf{s}) = \beta_1 + \beta_2\,s_1 + \beta_3\,s_1^2 + \beta_4\,s_2 + \beta_5\,s_1\,s_2 + \beta_6\,s_2^2. \tag{2.24}$$

---

[5]Lignite is a form a low-quality coal used for electricity generation in various countries.

[6]Ash water free refers to the ash that remains after moisture is removed by burning. A low count of remaining ash is preferable.

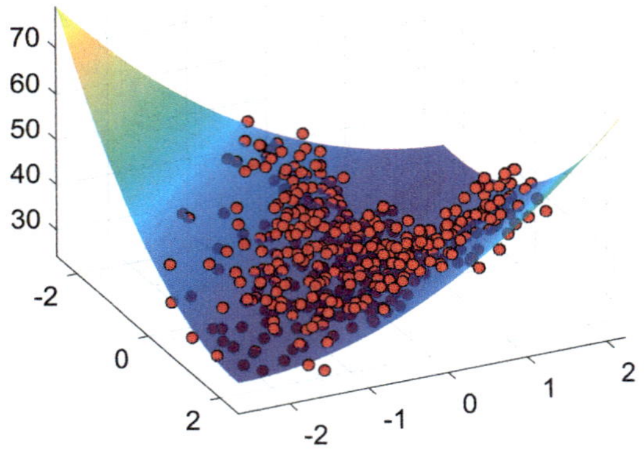

Fig. 2.1 Ash content (%) of lignite excavated from the South Field mine in Western Macedonia (Greece). Filled circles represent measurements of ash content whereas the continuum surface a parabolic trend (second degree polynomial) fitted to the data. Normalized spatial coordinates are used. See also [270]

Table 2.2 Coefficients of second degree polynomial trend (2.24) for the AWF data shown in Fig. 2.1

β_1	β_2	β_3	β_4	β_5	β_6
31.0840	0.2096	1.8712	0.3914	3.6114	1.8040

The optimal values for the polynomial coefficients are obtained using OLS. The regression problem is solved using the MATLAB function `regress` for multiple linear regression. The optimal polynomial coefficients thus obtained are given in Table 2.2.

The surface plot of the trend and the data is shown in Fig. 2.1. The *coefficient of determination* for the fit between the data and the trend model (2.24) is $R^2 \approx 0.5261$, which shows that considerable amount of the spatial variability is not captured by the trend model.

In linear regression the value of R^2 is equivalent to the square of the correlation coefficient between the data and the trend estimates. In addition to R^2, the function `regress` also returns an F value and a p value. The statistical F test is used to compare the regression model to the simplest possible alternative, i.e., a model that includes only a constant term β_1. The null hypothesis is that this simple model is no worse than the trend model (2.24). The alternative hypothesis is that the trend model with the optimal coefficients is an improved fit to the data.

The p value is related to the F value and represents the probability that the observed data would materialize—as a result of statistical fluctuations—if the null hypothesis were true. Hence, p values close to zero indicate that the null hypothesis is not very likely. If the p value is less than a specified significance level, which is typically set to 5% or 10%, the null hypothesis is rejected. In the present example, $p < 10^{-4}$. Hence, we deduce that the trend model is *statistically significant.*

It is also possible to evaluate the statistical significance of specific coefficients in the trend model. Coefficients that are not found to be statistically significant imply that the respective basis function does not improve the trend model in a statistically significant manner.

The R^2 value is a useful measure of model fit, but it suffers from the pitfall that it increases with the number of model parameters (predictor variables). This tendency is unsatisfactory, since an improvement of R^2 by including more predictor variables may just imply over-fitting. The *adjusted coefficient of determination* takes into account the number of prediction variables in the model. The R^2 value is the upper bound of the adjusted R^2, while the latter can also take negative values. These additional statistics can be calculated using the MATLAB function `fitlm`.

2.4 Local Trend Models

Trend models in this category obey simple functional forms with built-in local dependence. The coefficients of such models vary over the spatial domain. This additional flexibility, however, implies an increased computational cost. The main goal of such methods is to better adapt to local spatial patterns in the data.

The decision on whether to use a local or a global trend model can be guided by *model selection criteria*. Such criteria (e.g., AIC, BIC) weigh the fit between the model and the data against the model complexity.

Smoothing methods generate an approximating function that attempts to adapt to the most important, long-wavelength patterns in the data. When applied to spatial data, smoothing methods lead to local trend models. Such methods involve various digital filters (e.g., moving average, Savitzky-Golay filters), optimal filters, and locally weighted regression. The application of filtering methods is lucidly described in the book by Press et al. [673]. The idea of the optimal filter is very similar to the idea of the *stochastic optimal linear predictor* (SOLP), also known in geostatistics as *kriging*. Hence, we defer the discussion of optimal filtering until Chap. 10. Below, we discuss certain local trend models.

2.4.1 *Moving Average (MA)*

The method of moving average applies to the data a low-pass filter that is based on the calculation of local averages of the data values.

For illustration purposes, let us assume that the sample $\{x_1^*, \ldots, x_N^*\}$ is collected at the points $\{s_1, \ldots, s_N\}$ of a one-dimensional domain. The sampling step is assumed to be uniform, i.e., $s_n - s_{n-1} = a$ for all $n = 2, \ldots, N$. The equation for a *symmetric moving average filter of half width* L at an estimation point s_n, where $n = L + 1, \ldots N - L$, is given by

$$\hat{m}_{\mathrm{x}}(s_n) = \frac{1}{2L+1} \sum_{m=-L}^{L} x^*(s_n + ma). \qquad (2.25)$$

In the above equation, the *filter width* is equal to $2L + 1$, and the filter is symmetric with respect to the center point s_n. This filter suppresses frequencies that exceed the *cutoff frequency* $f_c = 1/L$.

At points that have fewer than L neighbors between them and the closest endpoint of the sample, the above equation is replaced by

$$\hat{m}_x(s_n) = \frac{1}{2l + 1} \sum_{m=-l}^{l} x^*(s_n + ma), \tag{2.26}$$

where $l = N - n$ if $n + L > N$ and $l = n - 1$ if $n - L < 1$.

An *asymmetric moving average filter* of width $2L+1$ involves a different number of neighbors on each side of the estimation point, i.e., $n_L \neq n_R$ where $n_L + n_R = 2L$.

Acausal filters If the filter is applied to a function of time and $n_R \neq 0$, the filter is called acausal, because its value at every instant depends on the values at "future" times. This is a distinct difference between time series that obey a definite time ordering, and spatial data for which there is no natural ordering. Spatial ordering emerges if the data are sampled along a linear subspace, e.g., drill-hole data can be ordered according to depth. However, even in this case there is no physical reason to apply only causal filters. More discussion on connections between time series and 1D spatial data is found in Chap. 7.

Example 2.6 Write the equations for a moving average filter of width equal to five for a one-dimensional sample that contains $N = 6$ values at equally spaced locations.

Answer

$$\hat{m}_x(s_1) = x^*(s_1)$$

$$\hat{m}_x(s_2) = \frac{1}{3}\left[x^*(s_1) + x^*(s_2) + x^*(s_3)\right],$$

$$\hat{m}_x(s_3) = \frac{1}{5}\left[x^*(s_1) + x^*(s_2) + x^*(s_3) + x^*(s_4) + x^*(s_5)\right],$$

$$\hat{m}_x(s_4) = \frac{1}{5}\left[x^*(s_2) + x^*(s_3) + x^*(s_4) + x^*(s_5) + x^*(s_6)\right],$$

$$\hat{m}_x(s_5) = \frac{1}{3}\left[x^*(s_4) + x^*(s_5) + x^*(s_6)\right],$$

$$\hat{m}_x(s_6) = x^*(s_6).$$

Grid data If the sample is distributed on a regular grid in higher than one dimensions, it is straightforward to extend the moving average filter equations using a stencil that matches the grid geometry (taking care to properly define the average near the boundary points of the grid).

Scattered data For spatially scattered samples, the filter width should be replaced by a *search radius* that defines a search neighborhood $B_\delta(\mathbf{s}_n)$ around each sampling point $\mathbf{s}_n$; for example, $B_\delta(\mathbf{s}_n) = \{\mathbf{s}_m : \|\mathbf{s}_n - \mathbf{s}_m\| \leq \delta\}_{m=1}^{N}$. The filter is then expressed in terms of the following equation

$$\hat{m}_{\mathrm{x}}(\mathbf{s}_n) = \frac{1}{n_\delta(\mathbf{s}_n)} \sum\nolimits_{\mathbf{s}_m \in B_\delta(\mathbf{s}_n)} x^*(\mathbf{s}_m), \tag{2.27}$$

where $n_\delta(\mathbf{s}_n)$ is the number of sampling points contained within the neighborhood of $\mathbf{s}_n$; in general, the "filter size" $n_\delta(\mathbf{s}_n)$ varies in space. In the following, the term "width" also refers to the radius of the search neighborhood, whereas the inverse of the width defines the *cutoff wavenumber* (or spatial frequency).

Properties of moving average filters The moving average filter is an efficient method for noise reduction. It is a good candidate for trend estimation, because it can eliminate fluctuations.

The application of the filter requires defining a "width" that has to be determined by experimentation. A larger width leads to a function that varies slowly, whereas a smaller width yields a function with more local fluctuations. Since we expect the trend to be a smoothly varying function of space, a larger width seems more appropriate; in the opposite case, the moving average filter starts to incorporate fluctuations into the trend.

In signal processing, the moving average filter is used for smoothing and noise suppression. In such cases, a smaller filter width is more appropriate than a larger filter width that may lead to excessive reduction of the fluctuations and possibly to suppression of high-frequency peaks. Nevertheless, this smoothing behavior is not a problem in trend estimation.

Local but not explicit One problem with the filtering methods, and other local trend methods as well, is that they *lack an explicit functional form for the trend.* Instead, the latter is expressed in terms of the data and respective weights. The lack of an explicit expression also implies that it is not possible to extrapolate the trend function beyond the spatial domain of the data (except for a "thin" area around the convex hull of the sampling points).

Moving median filtering Moving windows can be applied without reference to the mean. Hence, instead of calculating averages within the moving windows we can calculate the median of the data enclosed and use this statistic as the moving window estimate [165]. Median filtering, as this method is called, is less sensitive than the numerical average to errors and outliers. Thus, median filtering, if used with a window of appropriate size, can reduce errors while also preserving edges in digital images.

2.4.2 Savitzky-Golay Filters

Savitzky-Golay filters (SGF) are digital low-pass filters with adaptive capability that provide optimal (in the least squares sense) fitting to the data [516, 716]. They were introduced in 1964 by Savitzky and Golay in a short paper published in the journal *Analytical Chemistry* [716]. In 2000, the editors ranked it among the top five most influential papers ever published in this journal. SGFs are mostly used for smoothing noisy signals.

SGF in one dimension SGFs are also known as *least squares filters* and *digital smoothing polynomials*. To keep the presentation simple, we describe the SGF filter in $d = 1$ for regular sampling with uniform step $a > 0$. In higher spatial dimensions SGF becomes a special case of locally weighted regression, as we show below.

Assuming that the sampling points are given by $s_n = n\,a$, where $n = 1, \ldots, N$, the SGF equation in *one dimension* is

$$\hat{m}_{\mathrm{x}}(s_n) = \sum_{m=-n_L}^{n_R} c_m \, x^*(s_n + ma), \text{ for } N - n_R \geq n > n_L. \tag{2.28}$$

The filter coefficients $\{c_{-n_L}, c_{-n_L+1}, \ldots, c_0, c_1, \ldots, c_{n_R}\}$ are real numbers; their values are estimated from the data by minimizing the sum of the square errors. The filter width is equal to $n_L + n_R + 1$, i.e., to the number of data points that are included in the estimate at the target point s_n.

Filter construction: The main idea underlying SGFs is the following: A window of size equal to the filter width is defined at each estimation point s_n.

For example, for a symmetric filter with $n_L = n_R = L$, the window π_n around the point s_n contains the following values:

$$\left\{x^*(s_n - La),\; x^*(s_n - La + a),\; \ldots, x^*(s_n),\; \ldots, x^*(s_n + La)\right\}.$$

The data values contained within this window are then fitted to a polynomial model of degree p where $p \in \mathbb{N}$ such that $0 < p < n_R + n_L + 1$. *This polynomial is used only for the estimation of the trend at the point s_n.* The polynomial coefficients are assumed to be constant inside the window π_n around the point s_n. Hence, assuming that $n_R = n_L = L$ the following polynomial models the trend within the window π_n:

$$\hat{m}_{\mathrm{x}}(s_k) = \beta_{n,0} + \beta_{n,1}\, s_k + \beta_{n,2}\, s_k^2 + \ldots \beta_{n,p}\, s_k^p, \tag{2.29}$$

where $s_k = s_n + j\,a$, and $j = -L, \ldots, L$.

The error of the polynomial model at point s_k is defined as $x^*(s_k) - \hat{m}_{\mathrm{x}}(s_k)$. The polynomial coefficient vector $\boldsymbol{\beta}_n = (\beta_{n,0}, \ldots, \beta_{n,p})^\top$ is determined by minimizing the sum of the square "errors" (SSE) over all the points that lie within the window π_n, i.e.,

$$\hat{\boldsymbol{\beta}}_n = \arg\min_{\boldsymbol{\beta}} \left(\sum_{j=-L}^{L} \left[\hat{m}_{\mathrm{x}}(s_n + j\,a) - x^*(s_n + j\,a) \right]^2 \right). \tag{2.30}$$

The solution $\hat{\mathbf{B}} = \left(\hat{\beta}_{n,0}, \hat{\beta}_{n,1}, \ldots, \hat{\beta}_{n,p}\right)^\top$ that represents the optimal SGF coefficient vector is given by the following matrix product

$$\hat{\mathbf{B}} = \left(\mathbf{A}^\top \mathbf{A}\right)^{-1} \mathbf{A}^\top \mathbf{X}^*. \tag{2.31a}$$

The SGF *design matrix* $\mathbf{A}$ is given by [719]

$$A_{i,j} = [s_n + (i - L - 1)a]^j \,, \quad 1 \le i \le 2L + 1, \;\; j = 0, 1, \ldots, p. \tag{2.31b}$$

Note that the solution (2.31) for the SGF coefficients has the same form as the multiple linear regression solution (only the design matrices are different). This should not be surprising since both methods represent solutions to ordinary least squares problems.

The *smoothed signal* at s_n (i.e., our estimate of the trend), is given by $\hat{m}_{\mathrm{x}}(s_n)$. If the origin of the coordinate system is at s_n, then the trend value at s_n is given by

$$\hat{m}_{\mathrm{x}}(s_n) = \hat{\beta}_{n,0}, \quad \text{for all } N - n_R \ge n > n_L. \tag{2.32}$$

It is convenient to define local coordinate systems with origin at s_n, where s_n is the target point for trend estimation. Thus, the trend at every point is given by the constant term of the respective locally fitted polynomial.

Boundary points The edge points s_n such that $N \ge i > N - n_R$ and $1 \le i \le n_L$ cannot be smoothed with the SG filter (2.28). There are two options for dealing with such points: (i) They are either left unprocessed, or (ii) asymmetric filters with a structure specific to each point are constructed and applied to them.

Key points A local polynomial is defined at every point where the trend is estimated. Each local polynomial extends over a neighborhood of the target point and the data values in the neighborhood are used to determine the local polynomial at the target point. The trend at the neighbors of the target point, however, are described by their own local polynomials.

SGF in higher dimensions It is straightforward to extend the SG filter to regular lattices in two and three dimensions: the one dimensional window is replaced by a d-dimensional stencil and the 1-D local polynomial is replaced by a polynomial in d spatial dimensions.

For scattered data, the regular stencil is replaced with a neighborhood of specific radius as was done for the moving average filter in Sect. 2.4.1.

Applications It is somewhat surprising that SGFs are not well known in the signal processing community. A recent publication provides a readable exposition of SGFs and a discussion of their properties in the frequency domain [719]. Their use is also limited in spatial statistics, where a similar method, called locally weighted regression, is used (see Sect. 2.4.4 below).

The application of SG filters to spatial data generates a differentiable spatial function. Hence, the SG smoothing can be used to approximately estimate the parameters of geometric anisotropy based on the *covariance Hessian identity* [135] as discussed in Sect. 5.2. SGF is computationally efficient since the computations involved are local and involve only the data within each local neighborhood. On the other hand, the appropriate *neighborhood size* and the *degree of the local polynomials* are questions that need to be empirically resolved.

2.4.3 Kernel Smoothing

A different smoothing method employs weighted averages with the weights defined by means of *kernel functions* [150]. Pedagogical presentations of kernel smoothing methods are given in the review articles [38, 151]. These sources also include more information on the application of the method and in-depth analysis of relevant theoretical and computational issues.

Kernel smoothing has been developed by the research community of machine learning [731]. It has also found applications in the modeling of spatial data, mostly in the modeling of non-stationarity. The relevant papers are reviewed in [887]. Nevertheless, it is an intuitive and flexible method with potential for applications in spatial analysis.

The weights used in kernel smoothing are defined in terms of *weighting functions* [synonym: *kernel functions*] $K(r_{n,m})$, where $r_{n,m}$ is an appropriate measure of the distance between two points $\mathbf{s}_n$ and $\mathbf{s}_m$ for $n, m = 1, \ldots, N$.

Distance measures The distance between two points can be quantified by means of various mathematical relations. Some commonly used distance measures are listed below. A more complete list of proximity measures is given in [159].

Euclidean Distance: This is the most intuitive measure of distance between two points. It is *isotropic*, in the sense that all directions of space, i.e., $l = 1, \ldots, d$ contribute in the same way.

$$r_E(\mathbf{s}_n, \mathbf{s}_m) = \sqrt{\sum_{l=1}^{d} (s_{n,l} - s_{m,l})^2}. \tag{2.33}$$

Minkowski $\mathbb{L}^p$ Measure (where $p \in \mathbb{R}_{+,0}$): The Minkowski distance is based on the $\mathbb{L}^p$ norm

$$r_p(\mathbf{s}_n, \mathbf{s}_m) = \left(\sum_{l=1}^{d} \left| s_{n,l} - s_{m,l} \right|^p \right)^{1/p} . \tag{2.34}$$

The Minkowski distance is isotropic. For $0 < p < 2$ it emphasizes smaller distances, whereas for $p > 2$ the Minkowski distance emphasizes larger distances with respect to the Euclidean measure.

Diagonally Weighted Euclidean Distance: This is an anisotropic Euclidean distance measure that places different weights in different spatial directions.

$$r_a(\mathbf{s}_n, \mathbf{s}_m) = \sqrt{\sum_{l=1}^{d} a_l^2 \, (s_{n,l} - s_{m,l})^2}. \tag{2.35}$$

Fully Weighted Euclidean Distance: The so-called *Mahalanobis distance* is the Euclidean distance in a space that is generated by a linear transformation of the original space, i.e., $\mathcal{D} \ni \mathbf{s} \mapsto \mathbf{s}' = \mathbf{A}\mathbf{s} \in \mathcal{D}'$, where the $d \times d$ matrix $\mathbf{A}$ implements the linear transformation.

$$r_A(\mathbf{s}_n, \mathbf{s}_m) = \sqrt{(\mathbf{s}_n - \mathbf{s}_m)^\top \mathbf{A}^\top \mathbf{A} \, (\mathbf{s}_n - \mathbf{s}_m)} = r_E(\mathbf{A}\,\mathbf{s}_n, \mathbf{A}\,\mathbf{s}_m). \tag{2.36}$$

Fully Weighted Minkowski L^p Measure: This is defined using the Minkowski distance in the transformed space $\mathbf{s}'$, i.e., $\mathbf{s} \mapsto \mathbf{s}' = \mathbf{A}\,\mathbf{s}$, i.e.,

$$r_{p,A}(\mathbf{s}_n, \mathbf{s}_m) = r_p(\mathbf{A}\,\mathbf{s}_n, \mathbf{A}\,\mathbf{s}_m). \tag{2.37}$$

The use of weighted Euclidean distances and other distance transformations is motivated by the fact that the target variable may change with different rates in different directions. In such cases, an accurate spatial model needs to use some form of weighting to calculate representative measures of distance for a specific process.

For example, two locations at the center of a valley that are separated by 100 meters most likely receive approximately the same amount of annual rainfall. In contrast, two points with a difference of 100 meters in altitude may show considerable difference in precipitation.

Kernel functions Weighting functions satisfy certain properties that follow from logical and mathematical constraints.

Definition 2.1 A kernel (weighting) function $K(\mathbf{s}, \mathbf{s}')$ is a *non-negative, real-valued, integrable* function that represents a mapping $\mathbb{R}^d \times \mathbb{R}^d \rightarrow \mathbb{R}$. Let $\mathbf{s}_n$ and $\mathbf{s}_m$ be two points in the $\mathbb{R}^d$ Euclidean space and $u = r_{n,m}$ represent a measure of their distance. A kernel function satisfies the following properties:

1. *Non-negativity:* $K(u) \geq 0$, for all u.
2. *Symmetry:* $K(u) = K(-u)$ for all u.
3. *Normalization:* $\int_{-\infty}^{\infty} \mathrm{d}u\, K(u) = 1$.
4. *Finite second-order moment*: The integral $\int_{-\infty}^{\infty} \mathrm{d}u\, u^2\, K(u)$ exists.
5. *Mode at the origin:* $K(u)$ takes its maximum value at $u = 0$.
6. *Continuity:* $K(u)$ is a continuous function of u.
7. *Scaling:* If a function $K(u)$ is a kernel function and $h > 0$ is a non-negative number, the function $K_h(u) = \frac{1}{h} K(u/h)$ is also a kernel function.

Models of weighting functions Below, we list commonly used models of kernel functions. These functions depend on the normalized distance $u = r/h$, where $r > 0$ represents a measure of two-point distance (not necessarily the Euclidean measure) and $h > 0$ represents the *kernel bandwidth*.

The kernel models listed below are *radial functions* if $u = \|\mathbf{r}\|$. Then, the functions $K(u)$ depend only on the measure—but not the direction—of the Euclidean distance between two points.

1. Quadratic:

$$K(u) = \frac{3}{4}\,(1 - u^2)\,\mathbb{I}_{|u|\leq 1}(u).$$

2. Uniform:

$$K(u) = \frac{1}{2}\,\mathbb{I}_{|u|\leq 1}(u).$$

3. Triangular:

$$K(u) = (1 - |u|)\,\mathbb{I}_{|u|\leq 1}(u).$$

4. Quartic (biweight):

$$K(u) = \frac{15}{16}\left[\left(1 - |u|^2\right)^2\right]\mathbb{I}_{|u|\leq 1}(u).$$

5. Tricubic:

$$K(u) = \frac{70}{81} \left[\left(1 - |u|^3\right)^3 \right] \mathbb{I}_{|u|\leq 1}(u).$$

6. Epanechnikov:

$$K(u) = \frac{3}{4} \, (1-u)^2 \, \mathbb{I}_{|u|\leq 1}(u).$$

7. Spherical:

$$K(u) = \frac{4}{3} \left(1 - 1.5u + 0.5u^3\right) \mathbb{I}_{|u|\leq 1}(u).$$

8. Gaussian:

$$K(u) = \frac{1}{\sqrt{2\pi}} \exp(-u^2).$$

9. Exponential:

$$K(u) = \frac{1}{2} \exp(-|u|).$$

10. Cauchy:

$$K(u) = \frac{1}{\pi} \frac{1}{1+u^2}.$$

In the above, $\mathbb{I}_{|u|\leq 1}(u)$ denotes the *indicator function* that satisfies $\mathbb{I}_{|u|\leq 1}(u) = 1$ for $|u| \leq 1$ and $\mathbb{I}_{|u|\leq 1}(u) = 0$ for $|u| > 1$. The first seven kernel functions are compactly supported, while the last three have an unbounded support.

Distance weighted averaging In distance weighted averaging, the estimate at a single point is based on the linear superposition its neighboring values, i.e.,

$$\hat{m}_{\mathrm{x}}(\mathbf{s}_0) = \sum_{n=1}^{N} \lambda_n \, x^*(\mathbf{s}_n). \tag{2.38}$$

The above equation implements a *smoothing operation* if $\mathbf{s}_0 \in \mathbb{S}_N$. On the other hand, $\hat{m}_{\mathrm{x}}(\mathbf{s}_0)$ *interpolates* the data if $\mathbf{s}_0 \in \mathcal{D} \notin \mathbb{S}_N$.

Two different weighting approaches are used to determine the weights $\{\lambda_n\}_{n=1}^{N}$, that can be concisely characterized as *data weighting* and *error weighting*. The first approach weighs the data according to their distance from the estimation point.

The second approach weighs the model error according to the distance between its location and the estimation point.

Kernel regression In the method of data weighting, also known as *kernel regression*, the estimator is given by the following *weighted average* of the data:

$$\hat{m}_x(\mathbf{s}_0; h) = \frac{\sum_{n=1}^{N} x^*(\mathbf{s}_n)\, K_h(r_{0,n})}{\sum_{n=1}^{N} K_h(r_{0,n})}, \tag{2.39}$$

where $K_h(r) = K(r/h)$. Equation (2.39) is also known as the *Nadaraya-Watson kernel-weighted average* [594, 835]. Based on (2.39), the linear weights in (2.38) are given by the following kernel-based ratios

$$\lambda_n = \frac{K_h(r_{0,n})}{\sum_{n=1}^{N} K_h(r_{0,n})}, \quad \text{for all } n = 1, \ldots, N. \tag{2.40}$$

Error weighting In contrast with kernel regression, in the error weighting method the *cost function* associated with the weighted error criterion at the target point $\mathbf{s}_0$ is based on the following convex function of x_0

$$C(\mathbf{s}_0) = \sum_{n=1}^{N} \left[x^*(\mathbf{s}_n) - x_0\right]^2 K_h(r_{0,n}). \tag{2.41}$$

The optimal estimate of the process at $\mathbf{s}_0$ is obtained by minimizing the cost function with respect to x_0. The optimal estimate, $\hat{x}_0$, is determined by the *stationary point* of $C(\mathbf{s}_0)$, i.e.,

$$\left.\frac{\mathrm{d}C(\mathbf{s}_0)}{\mathrm{d}x_0}\right|_{x_0=\hat{x}_0} = 0. \tag{2.42}$$

Based on (2.42) and replacing $\hat{x}_0$ with $\hat{m}_x(\mathbf{s}_0; h)$, one can easily verify that the method of error weighting leads to the same estimator as data weighting.

Adaptive bandwidth selection The kernel regression estimates are based on kernel functions, the range of which is determined by the bandwidth h. The value of the bandwidth, however, is not known a priori. It makes sense to use the data to determine an optimal bandwidth. This can be achieved using the *leave-one-out cross validation* approach as described below.

1. First, we select a number of different bandwidths.
2. For each bandwidth we remove one sampling point at a time, repeating N times.

3. For each removed point $\mathbf{s}_n$, $n = 1, \ldots, N$ we estimate the missing value by means of the Nadaraya-Watson estimator (2.39) leading to $\hat{m}_{\mathrm{x}}(\mathbf{s}_n)$.
4. After all the sample points have been estimated, we calculate the sum of square errors (SSE) given by

$$\mathrm{SSE} = \sum_{n=1}^{N} \left[\hat{m}_{\mathrm{x}}(\mathbf{s}_n; h) - x^*(\mathbf{s}_n)\right]^2 .$$

Note that $\hat{m}_{\mathrm{x}}(\mathbf{s}_n; h)$ (and consequently the SSE as well) depends on the kernel function $K(\cdot)$ and the bandwidth h.
5. We select the optimal value of the bandwidth as the one that minimizes the SSE.

The above procedure is computationally intensive for large data sets because the Nadaraya-Watson estimator (2.39) involves a summation over all the sample points [594, 835]. The above approach uses a *global bandwidth*, i.e., it assumes that the bandwidth h does not depend on the site $\mathbf{s}_n$. This is sufficient if the sampling points are approximately uniformly distributed and the sampled function is spatially homogeneous. However, this approach is not adequate if the data are scattered or if the sampled function changes with different rates in different parts of the spatial domain $\mathcal{D}$. In such cases, better performance can be obtained by considering *local bandwidths* that adapt to the variations of the sampling density, e.g. [368].

2.4.4 Locally Weighted Regression (LWR)

Locally Weighted Regression (LWR), also known as locally weighted polynomial regression and LOESS, is a kernel-based smoothing method. LOESS defines a local statistical model (e.g., a polynomial function), the parameters of which vary at each estimation point [150]. Hence, a different statistical model is obtained at each estimation point. The statistical model is fitted to the data within a specified neighborhood around the target point. Thus, the model coefficients in general depend on the estimation point.

Locally weighted regression is reviewed in [38, 151]. The LWR weights are defined in terms of a *kernel function* $K(r_{n,m})$, where $r_{n,m}$ is a suitable distance measure between any two points $\mathbf{s}_n$ and $\mathbf{s}_m$.

If LWR is based on polynomial functions, it is similar to SGF. However, in contrast to SGF the LWR weights depend on the distance between the data points and the estimation points. On the other hand, if LWR is applied with a uniform kernel, then it is equivalent to a symmetric SGF.

Local trend representation Let $\mathbf{s}_0 \in \mathcal{D}$ denote the estimation point. We express the $\mathbf{s}_0$-centered *local trend* as follows in terms of a function basis

$$\hat{m}_{\mathrm{x}}(\mathbf{s}; \mathbf{s}_0) = \sum_{p=1}^{P} \beta_p^*(\mathbf{s}_0)\, f_p(\mathbf{s}), \tag{2.43}$$

where $f_1(\mathbf{s}) = 1$, and the set of functions $\{f_p(\mathbf{s})\}_{p=1}^{P}$ represents the basis. In the case of a local polynomial, the basis comprises monomial functions, i.e., $f_p(\mathbf{s}) = s_1^{\ell_1} \ldots s_d^{\ell_d}$, where $\ell_i \in \mathbb{N}$.

The coefficients $\{\beta_p^*(\mathbf{s}_0)\}_{p=1}^{P}$ have a dependence on $\mathbf{s}_0$ that emphasizes the local nature of the trend model. The trend function $\hat{m}_{\mathrm{x}}(\mathbf{s}; \mathbf{s}_0)$ has the same functional form for all estimation points, but the specific value of the coefficients is only valid for $\mathbf{s}_0$.

LWR solution To determine the locally dependent coefficients LOESS uses weighted data as explained in [38]. Each observation is assigned a weight that depends on the distance between the observation and the target point $\mathbf{s}_0$. The weights $\{w_{0,n}\}_{n-1}^{N}$, are determined by the kernel function $K(\cdot)$ as follows

$$w_{0,n} = \sqrt{K_h(\mathbf{s}_0 - \mathbf{s}_n)}, \text{ for all } n = 1, \ldots, N. \tag{2.44}$$

The optimal coefficients $\hat{\beta}_p(\mathbf{s}_0)$ for the specific bandwidth h are obtained by means of the ordinary least squares method, i.e., by minimizing the following sum of "weighted square errors"

$$\mathrm{SWSE}(\mathbf{s}_0; \beta_1, \ldots \beta_P) = \sum_{n=1}^{N} w_{0,n}^2 \left[x^*(\mathbf{s}_n) - \sum_{p=1}^{P} \beta_p\, f_p(\mathbf{s}_n) \right]^2 . \tag{2.45}$$

Thus, the LOESS coefficient vector $\hat{\mathbf{B}}_0 = \left(\hat{\beta}_1, \ldots, \hat{\beta}_P\right)^\top$ is given by

$$\hat{\mathbf{B}}_0 = \underset{\{\beta_1,\ldots,\beta_P\}}{\arg\min}\ \mathrm{SWSE}(\mathbf{s}_0; \beta_1, \ldots \beta_P)$$

The estimate of the local trend at $\mathbf{s}_0$ is then expressed based on (2.43) as follows

$$\hat{m}_{\mathrm{x}}(\mathbf{s}_0) = \sum_{p=1}^{P} \hat{\beta}_p(\mathbf{s}_0)\, f_p(\mathbf{s}_0), \tag{2.46}$$

We can set the origin of the coordinate system at $\mathbf{s}_0$ arbitrarily (as we did for SGF). Then, if the basis functions are monomials of the coordinates we obtain

$$\hat{m}_{\mathrm{x}}(\mathbf{s}_0) = \hat{\beta}_1. \tag{2.47}$$

The solution of the LOESS system that follows from the minimization of the SWSE can be concisely expressed using matrix notation. We define the following quantities:

- The $N \times 1$ *weighted data vector* $\mathbf{Y} = \left(w_{0,1}\, x^*(\mathbf{s}_1), \ldots, w_{0,N}\, x^*(\mathbf{s}_N)\right)^\top$.

- The diagonal $N \times N$ *weight matrix* with elements $W_{n,n} = w_{0,n}$ for $n = 1, \ldots, N$.
- The $N \times P$ *design matrix* $\mathbf{S}$ with elements $S_{n,p} = f_p(\mathbf{s}_n)$, where $n = 1, \ldots, N$ and $p = 1, \ldots, P$.
- The $N \times P$ *weighted design matrix* $\mathbf{Z} = \mathbf{W}\,\mathbf{S}$.
- The $P \times 1$ vector of LOESS coefficients $\hat{\mathbf{B}}_0 = \left(\hat{\beta}_1(\mathbf{s}_0), \ldots, \hat{\beta}_P(\mathbf{s}_0)\right)^\top$.

The vector of the LOESS coefficients $\hat{\mathbf{B}}_0$ is given by means of the matrix product

$$\hat{\mathbf{B}}_0 = \left(\mathbf{Z}^\top \mathbf{Z}\right)^{-1} \mathbf{Z}^\top \mathbf{Y}. \tag{2.48}$$

Bandwidth estimation In order to determine the optimal kernel bandwidth, we can apply the *leave-one-out cross validation* bandwidth selection method described in Sect. 2.4.3 above. The main difference is that the trend values in Step 3 are now estimated by means of (2.47) and (2.48).

Properties of LWR solution The properties of the kernel functions used for the weights determine the properties of the LWR generated local trend function. The following list summarizes some pertinent properties.

1. *Smoothness:* The smoothness of $K(r_{n,m})$ typically translates into a smooth variation of the LWR estimates.
2. *Exactitude:* For a kernel function that is singular at the origin, i.e., if

$$\lim_{r_{n,m} \to 0} K(r_{n,m}) = \infty,$$

   locally weighted regression leads to *exact interpolation*. This is due to the fact that the weight corresponding to the sample value at the estimation point overshadows the weights of any neighboring points.
3. *Smoothing:* In contrast with a singular kernel, finite values of the kernel at the origin lead to smoothing of the data.
4. *Computational cost:* LWR is computationally intensive for infinitely extended kernel functions, because the estimate at every target point involves the entire set of N sampling points. This implies a computational cost that scales as $O(N^2)$.
5. *Kernel compactness and efficiency:* For compactly supported kernel functions such that $K(r_{n,m}) = 0$ for $r_{n,m} > r_c$ where r_c is the range of the kernel, computationally efficient LWR implementations become possible. However, compactly supported kernel functions are less accurate in areas of low sampling density.
6. *Robustness to outliers:* LWR is less sensitive to outliers than the moving window filter, because the latter directly incorporates the data values in the estimate.
7. *Boundary point estimation:* The LWR estimates (2.46) can also be formulated for the boundary points.

Example 2.7 Consider a one-dimensional synthetic function which contains three periodic components with periods $T_1 = 2$, $T_2 = 1.24$, $T_3 = 5.43$, and a Gaussian white noise term $\sigma_\varepsilon \varepsilon(s; \omega)$, where $\sigma_\varepsilon > 0$ and $\varepsilon(s; \omega) \overset{d}{=} \mathcal{N}(0, 1)$. The periodic components are assumed to represent the local trend. The random function is thus defined by

$$\mathrm{X}(s; \omega) = 0.1 \cos\left(\frac{2\pi s}{2}\right) + 0.3 \cos\left(\frac{2\pi s}{1.24}\right) + 0.234 \sin\left(\frac{2\pi s}{5.43}\right) + \sigma_\varepsilon \varepsilon(s; \omega).$$

The goal is to fit *one realization* $x(s)$ sampled at 100 regularly spaced points inside the interval [0, 10] using locally weighted regression for two different levels of noise $\sigma_\varepsilon = 0.05$ and $\sigma_\varepsilon = 0.1$.

Answer We standardize the coordinates by means of $s'_n = (s_n - \bar{s})/\sigma_s$, for $n = 1, \ldots, 100$ where $\bar{s}$ and σ_s represent respectively the average and standard deviation of the sampling locations.[7]

We apply LWR with a second degree polynomial and the tricubic kernel. We test ten equidistant trial bandwidths in the interval $[h_{\max}/10, h_{\max}]$ where $h_{\max} = 0.35 r_{\max}$ and $r_{\max} \approx 3.41$ is the maximum distance between two sampling points (in normalized units). The range of bandwidth values was determined by experimentation based on the following guiding rules: (i) The bandwidth should be considerably smaller than the domain length to avoid over-smoothing and (ii) the bandwidth should not be smaller than a few multiples of the sampling step, in order for the local polynomial to include at least some neighbors at each point. Based on the above selections, $h_{\max} \approx 1.19$ and $h_{\min} \approx 0.12$, whereas the sampling step (in normalized units) is $a \approx 0.03$.

The optimal bandwidth selection based on leave-one-out cross validation is illustrated on the left-side plot in Fig. 2.2. The smoothed LWR approximation as well as the trend and the initial (noisy) data (with $\sigma_\varepsilon = 0.05$) are shown in the right-side plot of the same figure. It is possible to fine-tune the LWR fit by searching for the optimal bandwidth within a smaller neighborhood around the "optimal" value of $\approx$0.24 which is obtained from the initial discretization. In addition, it is possible to test more kernel functions and select the optimal among them. Nevertheless, even without these refining steps, Fig. 2.2 exhibits very good fitting of the data by means of the LWR model.

If we increase the noise level to $\sigma_\varepsilon = 0.1$, LWR still provides an accurate approximation of the trend as shown in Fig. 2.3. The pattern exhibited by the RSSE is unaltered with respect to that shown in Fig. 2.2 for $\sigma_\varepsilon = 0.05$, even though the RSEE values are higher due to the increased amplitude of the noise.

---

[7]The normalization of the sampling coordinates is not necessary in this example. It can improve numerical stability, however, if the coordinates involve very large numbers. The transformation does not alter the relative distances between pairs of sampling points, and thus it maintains existing correlations.

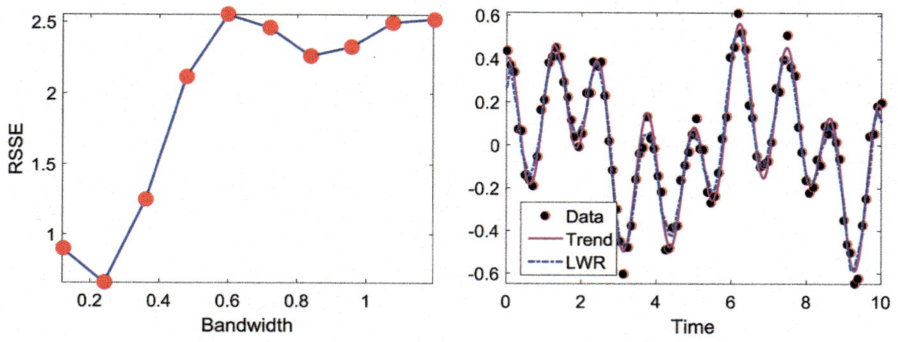

Fig. 2.2 Left: Root of the Sum of Square Errors (RSSE) obtained by cross validation using locally weighted regression with a second degree polynomial and the tricubic kernel. Equation (2.48) is used to determine the trend coefficients. Right: Sampled data with $\sigma_\varepsilon = 0.05$ (circles), periodic trend without noise (continuous line) and LWR curve ("-." blue line) using the optimal bandwidth

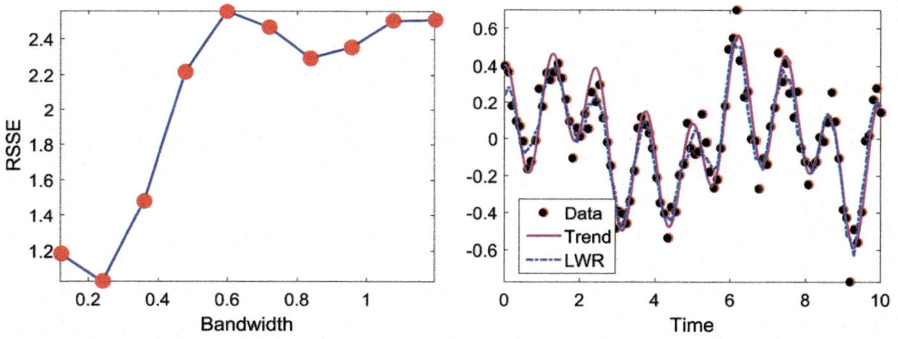

Fig. 2.3 Left: Root of the Sum of Square Errors (RSSE) obtained by cross validation using locally weighted regression with a second degree polynomial and tricubic kernel. Equation (2.48) is used to determine the trend coefficients. Right: Sampled data with $\sigma_\varepsilon = 0.1$ (circles), periodic trend without noise (continuous line) and LWR curve ("-." blue line) using the optimal bandwidth

2.5 Trend Estimation Based on Physical Information

If physical information is known about the process from which the spatial data are generated, it can be used to guide the trend estimation. This information could come in the form of *physical laws* or *empirical models* inspired by physical considerations. For example, if the physical laws are known, it may be possible to obtain solutions using average values of the spatially dependent coefficients. Assuming that the initial and boundary conditions are reasonably well specified, these coarse-grained solutions can be used to determine the functional form of the trend.

In other cases, the physical laws may not be exactly known or solvable, and the initial/boundary conditions may be highly uncertain. Then, a specific functional form cannot be determined from first principles. However, observations over several similar data sets may lead to empirical "laws." For example, the dependence of orographic precipitation on altitude and mountain ridge orientation is well documented [155]. If moist, warm air rising from the sea encounters a mountain range, it is lifted upwards gaining potential energy at the expense of kinetic energy, thus cooling off. At sufficiently high altitudes the temperature drop forces the water vapor in the uplifted air to condense and precipitate. On the lee side of the mountain, the descending air is dry and the precipitation significantly reduced. This phenomenon can lead to heavy precipitation upwind of a tall mountain range which is oriented across the prevailing direction of a wind from a warm ocean.

Both approaches for trend estimation are investigated in a study of the groundwater level of an aquifer on the Mediterranean island of Crete [815]. The hydraulic head measurements in this aquifer exhibit considerable variability, but they also display discernible large-scale patterns. Trend models of the hydraulic head based on topological variables were used. The auxiliary variables examined included the elevation above sea level and the distance from a seasonal river that traverses the basin. Combining these two factors in an empirical trend model improves the geostatistical estimation of the hydraulic head, compared to a spatial model with no explicit trend. Lower values of the hydraulic head were found along the river bed, where a large number of groundwater pumping stations are installed.

A different trend model for the hydraulic head was based on *Thiem's equation* for steady flow in confined, homogeneous aquifers with multiple wells [728]. This equation is based on physical principles but involves parameters that need to be estimated from the data. While the assumptions involved in Thiem's equations are not necessarily all satisfied in practice, the model is a useful approximation in confined aquifers. Using the Thiem-based trend model leads to a more accurate estimation of the groundwater level than the empirical trend model with auxiliary topological variables mentioned above. On the other hand, the variance of the geostatistical predictions (based on ordinary kriging of the fluctuations) is higher for Thiem's trend model than for the empirical model.

The advantage of physics-based trend models is that they can improve the performance of spatial interpolation. Their disadvantage is that it is not possible to formulate a suite of such models that can be applied to every imaginable problem. Actually, physics-based trend models are problem specific since they depend both on the physical processes involved and the respective boundary conditions (or the combination of initial and boundary conditions for dynamic problems).

Below, we will consider trend models derived from the Laplace equation with random coefficients. The Laplace equation is an elliptical partial differential equation that has a wide range of applications. The following sections require a higher level of mathematical sophistication than other parts of this book. The decision to skip the remainder of this section will not affect the reader's understanding of the following chapters.

2.6 Trend Model Based on the Laplace Equation

We return to the problem of saturated, single-phase, steady-state fluid flow in a random porous medium, i.e., equation (1.1). We assume that flow domain is a rectangular parallelepiped $\mathcal{D} \subset \mathbb{R}^d$ whose boundary $\partial\mathcal{D}$ involves six rectangular faces. We consider *Dirichlet* boundary conditions $H(\mathbf{s};\omega) = H_0$ and $H(\mathbf{s};\omega) = H_L$ on two parallel faces at distance L from each other and *Neumann* no-flow conditions at the four remaining faces.

The hydraulic conductivity at the local scale is modeled as a scalar random field $\mathrm{K}(\mathbf{s};\omega)$. We will denote its *realizations* by $K(\mathbf{s};\omega)$ and assume that they admit at least first-order partial derivatives in all directions.

The equation of steady-state fluid flow is expressed as follows

$$\nabla \cdot [K(\mathbf{s};\omega)\nabla H(\mathbf{s};\omega)] = 0, \quad \mathbf{s} \in \mathcal{D} \subset \mathbb{R}^d \tag{2.49a}$$

$$H(\mathbf{s};\omega) = H_0, \quad \mathbf{s} \in \partial\mathcal{D}_{\parallel}(1), \tag{2.49b}$$

$$H(\mathbf{s};\omega) = H_L, \quad \mathbf{s} \in \partial\mathcal{D}_{\parallel}(2), \tag{2.49c}$$

$$\mathbf{n} \cdot \nabla H(\mathbf{s};\omega) = 0, \quad \mathbf{s} \in \partial\mathcal{D}_{\perp}, \tag{2.49d}$$

where $\partial\mathcal{D}_{\parallel}(i)$, $i = 1, 2$ are the two faces along which a pressure gradient is applied, and $\partial\mathcal{D}_{\perp}$ denotes the faces with no flow condition. The symbol $\mathbf{n}$ denotes a unit vector that points perpendicular to the faces with no flow conditions. The boundary conditions are preserved in the following mathematical transformations of (2.49a).

Model of hydraulic conductivity Assuming that the hydraulic conductivity takes only positive values, we can express is as follows

$$K(\mathbf{s};\omega) = \mathrm{e}^{F(\mathbf{s};\omega)}, \tag{2.50}$$

where $F(\mathbf{s};\omega)$ is the *hydraulic log-conductivity* field. This representation is often used in studies of fluid flow in heterogeneous media. $F(\mathbf{s};\omega)$ is assumed to follow the *normal (Gaussian) joint distribution* . The probability distribution obeyed by the hydraulic conductivity field is the *lognormal distribution*.

The flow equation (2.49a) is expressed as

$$K(\mathbf{s};\omega)\nabla^2 H(\mathbf{s};\omega) + \nabla K(\mathbf{s};\omega) \cdot \nabla H(\mathbf{s};\omega) = 0. \tag{2.51}$$

Dividing both sides of the above equation by $K(\mathbf{s};\omega)$ and taking into account the transformation (2.50), the flow equation becomes

$$\nabla^2 H(\mathbf{s};\omega) + \nabla F(\mathbf{s};\omega) \cdot \nabla H(\mathbf{s};\omega) = 0.$$

If we assume that the log-conductivity field has a uniform mean, i.e., $F(\mathbf{s}; \omega) = \overline{F} + f(\mathbf{s}; \omega)$ where $\overline{F} = \mathcal{E}[F(\mathbf{s}; \omega)]$, the flow equation is transformed into

$$\nabla^2 H(\mathbf{s}; \omega) + \nabla f(\mathbf{s}; \omega) \cdot \nabla H(\mathbf{s}; \omega) = 0. \tag{2.52}$$

Equation (2.52) is an elliptical partial differential equation with a random coefficient given by $f(\mathbf{s}; \omega)$. A general, closed-form solution that is valid for all $f(\mathbf{s}; \omega)$ does not exist.

Integral equation solution of flow problem It is possible to rewrite the above equation as

$$\nabla^2 H(\mathbf{s}; \omega) = -\nabla f(\mathbf{s}; \omega) \cdot \nabla H(\mathbf{s}; \omega). \tag{2.53}$$

Let us view the right-hand side as a *source term*, i.e.,

$$S(\mathbf{s}; \omega) = -\nabla f(\mathbf{s}; \omega) \cdot \nabla H(\mathbf{s}; \omega). \tag{2.54}$$

We can do this if we pretend that $H(\mathbf{s}; \omega)$ on the right-hand side is known. Then, equation (2.53) becomes the *Poisson equation*, $\nabla^2 H(\mathbf{s}; \omega) = S(\mathbf{s}; \omega)$. It is possible to solve the Poisson equation, at least in principle. We show how this is done below.

Laplace equation First, we define the *homogeneous* partial differential equation

$$\nabla^2 H_0(\mathbf{s}) = 0, \text{ where } H_0(\mathbf{s}) = H_b(\mathbf{s}), \quad \mathbf{s} \in \partial\mathcal{D}, \tag{2.55}$$

where $H_b(\mathbf{s}),\ \mathbf{s} \in \partial\mathcal{D}$ is a shorthand for the boundary conditions specified in (2.49). Equation (2.55) describes steady-state flow in a homogeneous medium and is known as the *Laplace equation*. Solutions of the Laplace equation admit second-order continuous derivatives and are known as *harmonic functions*.

We also define the associated *Green function* problem

$$\nabla^2 G_0(\mathbf{s} - \mathbf{s}') = \delta(\mathbf{s} - \mathbf{s}'), \tag{2.56a}$$

$$G_0(\mathbf{s} - \mathbf{s}') = 0, \text{ if } \mathbf{s} \in \partial\mathcal{D} \text{ or } \mathbf{s}' \in \partial\mathcal{D}. \tag{2.56b}$$

The function $G_0(\mathbf{s} - \mathbf{s}')$ is known as the *fundamental solution or Green function* of the Laplace equation (2.56a).

Both the Laplace equation and its associated Green function equation can be solved analytically for different geometries, even if by means of series expansions [401]. The fundamental solution of the Laplace equation in infinite domains is given in Table 2.3.

Full solution of Poisson equation The full solution for $H(\mathbf{s}; \omega)$ is given by the superposition of the homogeneous solution which satisfies (2.55) and the specific

Table 2.3 Fundamental solution (Green function) $G_0(\mathbf{s}-\mathbf{s}')$ of the Laplace equation, $\nabla^2 G_0(\mathbf{s}-\mathbf{s}') = \delta(\mathbf{s}-\mathbf{s}')$, in an infinite domain for $d \in \mathbb{N}$. For $d > 3$, $S_d = 2\pi^{d/2}/\Gamma(d/2)$ represents the surface area of the unit sphere in d dimensions

Dimension	Green function
$d = 1$	$G_0(s-s') = \frac{1}{2}\, \lvert s - s' \rvert$
$d = 2$	$G_0(\mathbf{s}-\mathbf{s}') = \frac{1}{2\pi}\, \ln \lVert \mathbf{s}-\mathbf{s}' \rVert$
$d = 3$	$G_0(\mathbf{s}-\mathbf{s}') = -\frac{1}{4\pi}\, \lVert \mathbf{s}-\mathbf{s}' \rVert^{-1}$
$d > 3$	$G_0(\mathbf{s}-\mathbf{s}') = -\frac{1}{(d-2)S_d}\, \lVert \mathbf{s}-\mathbf{s}' \rVert^{2-d}$

solution which is obtained by the convolution of the Green function with the source term, i.e.,

$$H(\mathbf{s};\omega) = H_0(\mathbf{s}) + \int_{\mathcal{D}} d\mathbf{s}'\, G_0(\mathbf{s}-\mathbf{s}')\, S(\mathbf{s}';\omega). \tag{2.57}$$

To confirm that this expression solves the flow equation, apply the Laplacian operator ∇^2 to both sides of the above and add the term $\nabla f(\mathbf{s};\omega) \cdot \nabla H(\mathbf{s};\omega)$ to both sides. The left-hand side is equal to the left-hand side of (2.52). The right-hand side is then given by

$$\nabla^2 H_0(\mathbf{s}) + \nabla^2 \int_{\mathcal{D}} d\mathbf{s}'\, G_0(\mathbf{s}-\mathbf{s}')\, S(\mathbf{s}';\omega) + \nabla f(\mathbf{s};\omega) \cdot \nabla H(\mathbf{s};\omega).$$

The first term above is identically zero in light of the Laplace equation (2.55). In the second term, move the Laplacian ∇^2 inside the integral sign; it then operates onto the Green function, giving the delta function $\delta(\mathbf{s}-\mathbf{s}')$. The convolution of the delta function with the source term leads to $S(\mathbf{s};\omega)$. Recalling (2.54), the latter is equal to $-\nabla f(\mathbf{s};\omega) \cdot \nabla H(\mathbf{s};\omega)$ and thus cancels the third term. This verifies that the integral representation (2.57) of the head satisfies the flow equation (2.52).

So, we conclude that the head solution is given by the following *integral equation*

$$H(\mathbf{s};\omega) = H_0(\mathbf{s}) - \int_{\mathcal{D}} d\mathbf{s}'\, G_0(\mathbf{s}-\mathbf{s}')\, \nabla f(\mathbf{s}';\omega) \cdot \nabla H(\mathbf{s}';\omega). \tag{2.58}$$

Liouville-Neumann-Born series One might argue that we have not really made any progress by trading a partial differential equation for an integral equation, which is not any easier to solve. However, the integral equation (2.58) is the starting point for the *Liouville-Neumann-Born series expansion* [30, 779]. In particular, we can express the hydraulic head in terms of the following *perturbation series*

$$H(\mathbf{s};\omega) = H_0(\mathbf{s}) + \sum_{n=1}^{N} h_n(\mathbf{s};\omega).$$

By plugging in this expression in the left-hand side of (2.58) we obtain the following

$$\sum_{n=1}^{N} h_n(\mathbf{s};\omega) = -\int_{\mathcal{D}} d\mathbf{s}'\, G_0(\mathbf{s}-\mathbf{s}')\, \nabla f(\mathbf{s}';\omega) \cdot \nabla \left[H_0(\mathbf{s}') + \sum_{n=1}^{N} h_n(\mathbf{s}';\omega) \right].$$

The first-order term of the hydraulic head expansion is given by the integral

$$h_1(\mathbf{s};\omega) = -\int_{\mathcal{D}} d\mathbf{s}'\, G_0(\mathbf{s}-\mathbf{s}')\, \nabla f(\mathbf{s}';\omega) \cdot \nabla H_0(\mathbf{s}'). \tag{2.59}$$

Higher-order terms are given by the following recursive relation

$$h_n(\mathbf{s};\omega) = -\int_{\mathcal{D}} d\mathbf{s}'\, G_0(\mathbf{s}-\mathbf{s}')\, \nabla f(\mathbf{s}';\omega) \cdot \nabla h_{n-1}(\mathbf{s}';\omega), \text{ for } n = 2, 3, \ldots \tag{2.60}$$

Thus, based on (2.59) and (2.60) the second-order term is given by the following double integral

$$h_2(\mathbf{s};\omega) = \int_{\mathcal{D}} d\mathbf{s}_1 \int_{\mathcal{D}} d\mathbf{s}_2\, G_0(\mathbf{s}-\mathbf{s}_1)\, [\nabla f(\mathbf{s}_1;\omega) \cdot \nabla G_0(\mathbf{s}_1-\mathbf{s}_2)]\, [\nabla f(\mathbf{s}_2;\omega) \cdot \nabla H_0(\mathbf{s}_2)].$$

Higher-order terms ($n > 2$) are given, respectively, by the n-dimensional integrals

$$h_n(\mathbf{s};\omega) = \left(\prod_{i=1}^{n} \int_{\mathcal{D}} d\mathbf{s}_i \right) G_0(\mathbf{s}-\mathbf{s}_1)\, [\nabla_1 G_0(\mathbf{s}_1-\mathbf{s}_2) \ldots \nabla_{n-1} G_0(\mathbf{s}_{n-1}-\mathbf{s}_n) \nabla_n H_0(\mathbf{s}_n)]$$
$$\times (-1)^n\, [\nabla_1 f(\mathbf{s}_1;\omega) \ldots \nabla_n f(\mathbf{s}_n;\omega)].$$

In the right-hand side of the above equation, we use the shorthand notation

$$\nabla_m G_0(\mathbf{s}_m - \mathbf{s}_{m+1}) := \nabla_{\mathbf{s}_m} G_0(\mathbf{s}_m - \mathbf{s}_{m+1}).$$

We also imply that the inner product is taken between the Green function and log-conductivity gradients that have the same indices, i.e., for $m = 1, \ldots, n-1$

$$\nabla_m G_0(\mathbf{s}_m - \mathbf{s}_{m+1}) \nabla_m f(\mathbf{s}_m;\omega) := \sum_{l=1}^{d} [\nabla_m G_0(\mathbf{s}_m - \mathbf{s}_{m+1})]_l\, [\nabla_m f(\mathbf{s}_m;\omega)]_l\,,$$

and similarly for the inner product between the Green function and the head gradient, i.e., for $m = n$,

$$\nabla_n H_0(\mathbf{s}_n)\nabla_n f(\mathbf{s}_n;\omega) := \sum_{l=1}^{d} [\nabla_n H_0(\mathbf{s}_n)]_l \, [\nabla_n f(\mathbf{s}_n;\omega)]_l \, .$$

Normalized log-conductivity We introduce the *normalized fluctuation* $\chi(\mathbf{s};\omega)$ such that $f(\mathbf{s};\omega) = \sigma_f\,\chi(\mathbf{s};\omega)$ where σ_f is the standard deviation of the log-conductivity fluctuation and $\chi(\mathbf{s};\omega)$ is the normalized hydraulic log-conductivity field. If $f(\mathbf{s};\omega)$ follows the normal distribution, then $\chi(\mathbf{s};\omega)$ follows the $\mathcal{N}(0,1)$ distribution.

Using the *normalized fluctuation*, the Liouville-Neumann-Born series can be expressed as follows

$$h_n(\mathbf{s};\omega) = \sigma_f^n \left(\prod_{i=1}^{n}\int_{\mathcal{D}} d\mathbf{s}_i\right) G_0(\mathbf{s}-\mathbf{s}_1)\,\left[\nabla_1 G_0(\mathbf{s}_1-\mathbf{s}_2)\ldots\nabla_{n-1}G_0(\mathbf{s}_{n-1}-\mathbf{s}_n)\nabla_n H_0(\mathbf{s}_n)\right]$$
$$\times\,(-1)^n\left[\nabla_1\chi(\mathbf{s}_1;\omega)\ldots\,\nabla_n\chi(\mathbf{s}_{n-1};\omega)\right].$$

Hence, if the heterogeneity of the log-conductivity is mild, i.e., $\sigma_f \ll 1$, we can truncate the series at some low order in σ_f.

Trend estimation using low-order perturbation analysis The expected value of the hydraulic head is given by

$$\mathcal{E}[H(\mathbf{s};\omega)] = H_0(\mathbf{s}) + \sum_{n=1}^{\infty}\mathcal{E}[h_n(\mathbf{s};\omega)]\,.$$

Based on the normality of the log-conductivity fluctuations, the expectations of the odd-order terms $\mathcal{E}[h_{2m+1}(\mathbf{s};\omega)]$ vanish.[8] Hence, the expectation of the hydraulic head is given by the following infinite series

$$\mathcal{E}[H(\mathbf{s};\omega)] = H_0(\mathbf{s}) + \sum_{m=1}^{\infty}\mathcal{E}[h_{2m}(\mathbf{s};\omega)]\,.$$

The lowest non-zero correction to the homogeneous term $H_0(\mathbf{s})$ is $\mathcal{E}[h_2(\mathbf{s};\omega)]$. Even without calculating the corrections we can deduce the dependence of $\mathcal{E}[H(\mathbf{s};\omega)]$ on the heterogeneity parameter σ_f, i.e.,

$$\mathcal{E}[H(\mathbf{s};\omega)] = H_0(\mathbf{s}) + O(\sigma_f^2).$$

[8] We explain this in more detail in Chap. 6 which focuses on Gaussian random fields.

Hence, if $\sigma_f \ll 1$ it follows that $\mathcal{E}[H(\mathbf{s};\omega)] \approx H_0(\mathbf{s})$, even if it is not possible to explicitly calculate the perturbation terms.

Given the boundary conditions specified in (2.49), it follows that the homogeneous term $H_0(\mathbf{s})$ is given by

$$H_0(\mathbf{s}) = H_0 + \left(\frac{H_L - H_0}{L}\right) \hat{\mathbf{n}} \cdot \mathbf{s},$$

where $\hat{\mathbf{n}}$ is the unit vector in the direction of the imposed head gradient. Since the boundary conditions apply to the entire head solution, it follows that for all $m \in \mathbb{N}$ the fluctuations $h_{2m}(\mathbf{s};\omega)$ satisfy $h_{2m}(\mathbf{s};\omega) = 0$ for $\mathbf{s} \in \partial\mathcal{D}_{\parallel}$ and $\mathbf{n}\cdot\nabla h_{2m}(\mathbf{s};\omega) = 0$ for $\mathbf{s} \in \partial\mathcal{D}_{\perp}$.

Trend estimation based on statistical homogeneity Let us now try a different approach to obtaining the trend function. This requires digging a little deeper into the foundations of the steady-state flow equation (2.49a). The latter is based on two equations: one is the law of *conservation of mass*, also known as continuity equation, and the other is a phenomenological linear relation between the fluid flux (in units of volume per area per unit time) and the pressure difference, known as *Darcy's law* [171, 275, 696].

Darcy's law

$$\mathbf{Q}(\mathbf{s};\omega) = -K(\mathbf{s};\omega)\nabla H(\mathbf{s};\omega),$$

Continuity Equation

$$\nabla \cdot \mathbf{Q}(\mathbf{s};\omega) = 0.$$

The minus sign in Darcy's law denotes that the flux vector points in the direction of decreasing hydraulic head. The second equation ensures that the mass flowing through any elementary volume is conserved.

Let us now assume that the hydraulic conductivity is a *statistically homogeneous and isotropic random field*. We define these notions more carefully in Chap. 3. For now it suffices to say that the covariance function of $K(\mathbf{s};\omega)$ (and of $F(\mathbf{s};\omega)$ as well) depends only on the Euclidean measure of the spatial lag $\mathbf{s} - \mathbf{s}'$, but not separately on the locations $\mathbf{s}$ and $\mathbf{s}'$.

Then, the *effective Darcy's equation* links the expectation of the flux to the expectation of the pressure gradient by means of the following integral equation

$$\mathcal{E}[\mathbf{Q}(\mathbf{s};\omega)] = -\int_{\mathcal{D}} d\mathbf{s}'\, K^*(\mathbf{s}-\mathbf{s}')\,\mathcal{E}[\nabla H(\mathbf{s}';\omega)], \tag{2.61}$$

where $K^*(\mathbf{s} - \mathbf{s}')$ is a *hydraulic conductivity kernel* with units of hydraulic conductivity per volume. Unlike the hydraulic conductivity which is a random field, the hydraulic conductivity kernel is a deterministic function.

The connection between $K^*(\mathbf{s}-\mathbf{s}')$ and the moments of the hydraulic conductivity can be established based on the *Liouville-Neumann-Born expansion* [361, 370]. The dependence of $K^*(\mathbf{s} - \mathbf{s}')$ purely on the lag $\mathbf{s} - \mathbf{s}'$ is ensured by the statistical homogeneity of the hydraulic conductivity random field.

The *continuity equation* holds for the individual realizations, and therefore it is also valid for the expectation of the flux, i.e.,

$$\nabla \cdot \mathcal{E}[\mathbf{Q}(\mathbf{s};\omega)] = 0.$$

We can express this equation in reciprocal space (i.e., the space of wavevectors and spatial frequencies) using the *Fourier transform* $\widetilde{\mathbf{Q}}(\mathbf{k};\omega)$ of the flux vector $\mathbf{Q}(\mathbf{s};\omega)$.

We do not follow the mathematically rigorous spectral representation of random fields that is based on measure theory. For our purposes, we will assume—as is common in physics and engineering—that the realizations of the flux vector random field are absolutely integrable functions that possess a Fourier transform, which could involve generalized functions. A short introduction to Fourier transforms is given in Sect. 3.6.2.

The continuity equation in reciprocal space is then expressed as follows

$$\mathbf{k} \cdot \mathcal{E}[\widetilde{\mathbf{Q}}(\mathbf{k};\omega)] = 0.$$

In light of (2.61) the continuity equation is equivalent to

$$\widetilde{K}^*(\mathbf{k}) \left\{\mathbf{k} \cdot \mathcal{E}[\widetilde{\mathbf{J}}(\mathbf{k};\omega)]\right\} = 0, \text{ for all } \mathbf{k},$$

where $\mathbf{J}(\mathbf{s};\omega) = -\nabla H(\mathbf{s};\omega)$ is the hydraulic gradient field and $\widetilde{\mathbf{J}}(\mathbf{k};\omega)$ its Fourier transform. The above equation is trivially satisfied for $\mathbf{k} = \mathbf{0}$.

If the hydraulic conductivity is represented by a correlated random field, the Fourier transform of the kernel is non-zero at least within a finite "volume" of wavevectors $\mathbf{k}$. Hence, the continuity equation is satisfied if the following condition is true

$$\mathbf{k} \cdot \mathcal{E}[\widetilde{\mathbf{J}}(\mathbf{k};\omega)] = 0, \quad \text{for all } \mathbf{k}.$$

The vanishing of the above inner product is equivalent to the condition

$$\mathcal{E}[\widetilde{\mathbf{J}}(\mathbf{k};\omega)] = 0, \quad \text{for all } \mathbf{k} \neq \mathbf{0}.$$

Hence, only the zero-wavenumber component of $\widetilde{\mathbf{J}}(\mathbf{k};\omega)$ is allowed to have a finite, non-zero value. This means that $\mathcal{E}[\mathbf{J}(\mathbf{s};\omega)] = \mathbf{J}_0$ is a uniform vector. Based on the boundary conditions specified in (2.49) it follows that

$$\mathbf{J}_0 = -\left(\frac{H_L - H_0}{L}\right)\hat{\mathbf{n}}.$$

Finally, the mean hydraulic head is given by

$$\mathcal{E}[H(\mathbf{s};\omega)] = H_0 - \mathbf{J}_0 \cdot \mathbf{s}.$$

This analysis shows that the linear trend function is physically justifiable for the expectation of the hydraulic head, provided that suitable heterogeneity conditions apply and the boundary conditions have the relative simple form described at the beginning of the section. For more complicated geometries or boundary conditions, different trend functions may be necessary. In addition, if the flow domain includes sources or sinks, the trend function will also include respective terms that will be proportional to the Green function of the Laplace equation.

Given the simple answer for the head trend provided by the Fourier space analysis, one may wonder what is the use of the mathematically more complicated Liouville-Neumann-Born expansion. In fact, the latter is quite useful not so much for determining the head trend, but for deriving systematic perturbation approximations of the head covariance.

Effective hydraulic conductivity Given the linear hydraulic head trend, the nonlocal equation (2.61) is transformed as follows

$$\mathcal{E}[\mathbf{Q}(\mathbf{s};\omega)] = K_{\text{eff}}\,\mathbf{J}_0, \tag{2.62a}$$

$$K_{\text{eff}} = \int_{\mathcal{D}} d\mathbf{s}'\, K^*(\mathbf{s}-\mathbf{s}') = \widetilde{K}(\mathbf{k}=\mathbf{0}), \tag{2.62b}$$

where K_{eff} is the scalar *effective hydraulic conductivity*. Hence, the average flow is determined by a single effective conductivity measure that is based on the long-wavelength limit of the conductivity kernel.

The beautiful simplicity of (2.62a) is lost as soon as we relax various modeling assumptions. For example, if the hydraulic conductivity random field is homogeneous but statistically anisotropic, the effective hydraulic conductivity becomes a uniform tensor [276].

If we abolish the assumption of statistical homogeneity, the effective hydraulic conductivity is no longer uniform over the domain. In porous media near the percolation threshold, the continuum description of flow based on Darcy's equation with a random field hydraulic conductivity is not a suitable description [386, 387].

This section is a brief and incomplete description of the stochastic upscaling theory applied to steady-state, single phase flow in porous media. The description above is more of an appetizer than a full meal. The interested reader will find more extensive expositions of the subject in textbooks that specialize in stochastic subsurface hydrology [171, 173, 174, 275] and in review articles [172, 361].

The main motivation of this section is to show that the solutions of physical laws, if they can be obtained, even under simplifying assumptions, can guide the construction of insightful trend models. This possibility helps to build bridges between statistical spatial models and first-principles models of the physical processes involved.

Chapter 3
Basic Notions of Random Fields

> Science is the knowledge of consequences, and dependence of one fact upon another.
>
> *Thomas Hobbes*

3.1 Introduction

This chapter focuses on the mathematical properties of random fields. This may seem boring or superfluous to readers who are not mathematically oriented. However, as much as the ability to discern colors is necessary to appreciate paintings, it is also necessary to develop some degree of familiarity with random field properties in order to better understand their applications.

Probabilities are associated with events. The latter can be either simple or complex events that represent composites of simpler ones. The result of throwing a dice is a simple event.[1] A specific weather scenario for the next few days is a composite event that comprises several simpler events. As scientists, we tend to study and analyze complex events in terms of simpler ones. For example, the weather pattern of the next few days can be decomposed into different events that correspond to the weather pattern of each day, and the weather for each day can be further analyzed in terms of certain variables such as temperature, humidity, amount of precipitation, and wind speed. If we follow through with this reductionist approach, we end up working with measurable variables that take numerical values

[1] We can call it an *elementary event* because it cannot be analyzed into simpler events.

D. T. Hristopulos, *Random Fields for Spatial Data Modeling*, Advances in Geographic Information Science, https://doi.org/10.1007/978-94-024-1918-4_3

in the space $\mathbb{A}_x \subseteq \mathbb{R}$. Depending on the problem, $\mathbb{A}_x$ can be either continuum or discrete.

From a practical viewpoint, the random fields discussed herein can be used to model the variability of spatially distributed environmental variables, such as temperature and rainfall, elastic and transport coefficients of heterogeneous media, and various pollutant concentrations. They can also be used to represent the intensity per pixel of gray-level digital images acquired by means of various remote sensing sensors. Color images require the use of vector random fields.

3.2 Single-Point Description

This section focuses on the probabilistic description of random fields at single points. Hence, it is a review of basic probability for point-like random variables.

Assume that the variable of interest is represented by the spatial random field $X(\mathbf{s}; \omega)$ which takes values $x(\mathbf{s}) \in \mathbb{A}_x$ where $\mathbb{A}_x \subset \mathbb{R}$ is the field's domain of values (cf. Definition 1.1). Let us also focus at a single point $\mathbf{s}$ that can be anywhere in the domain $\mathcal{D}$.

Specific events A refer to different sets of values of $X(\mathbf{s}; \omega)$ at the point $\mathbf{s}$. For example, the event A could be defined as the set of values that fall inside a certain interval:

$$A := \{\, x(\mathbf{s}) = x, \ \text{ such that } x_1 \leq x \leq x_2\}, \ \text{ where } x_1 < x_2 \in \mathbb{R}.$$

The respective probability of the event A represents an interval probability defined by

$$P(A) = P(x_1 \leq x(\mathbf{s}) \leq x_2).$$

Cumulative distribution function We can calculate the probabilities of events such as the above if we know the *cumulative distribution function (cdf)*

$$F_x(x; \mathbf{s}) = P\left(X(\mathbf{s}; \omega) \leq x\right). \tag{3.1}$$

The right-hand side of the above equation represents the probability that the random field $X(\mathbf{s}; \omega)$ takes at $\mathbf{s}$ a value which is less than or equal to $x \in \mathbb{R}$. Hence, it follows from the fundamental axioms of probability that $F_x(x; \mathbf{s}) \in [0, 1]$. In terms of (3.1) it follows that the *interval probability* is given by

$$P\left(x_1 \leq x(\mathbf{s}) \leq x_2\right) = F_x(x_2; \mathbf{s}) - F_x(x_1; \mathbf{s}).$$

The probability that $X(\mathbf{s}; \omega)$ exceeds a specific value (*exceedance probability*) is given by the complement of the cdf, i.e.,

$$P\left(\mathrm{X}(\mathbf{s};\omega) > x\right) = 1 - F_{\mathrm{x}}(x;\mathbf{s}). \tag{3.2}$$

If $\mathrm{X}(\mathbf{s};\omega)$ is non-negative and the expectation $\mathcal{E}[\mathrm{X}(\mathbf{s};\omega)]$ is finite, a general upper bound for the exceedance probability is obtained by means of *Markov's inequality*, which states that for every $\alpha > 0$ the following holds:

$$P\left(\mathrm{X}(\mathbf{s};\omega) \geq \frac{x}{\alpha}\right) \leq \frac{\mathcal{E}[\mathrm{X}(\mathbf{s};\omega)]}{\alpha}. \tag{3.3}$$

If $\mathrm{X}(\mathbf{s};\omega)$ is not necessarily non-negative but both its expectation m_{x} and variance σ_{x}^2 are finite, an upper bound for the probability of fluctuations around the mean is given by *Chebychev's inequality* which states that for $\epsilon > 0$

$$P\left(|\mathrm{X}(\mathbf{s};\omega) - m_{\mathrm{x}}| \geq \epsilon\right) \leq \frac{\sigma_{\mathrm{x}}^2}{\epsilon^2}. \tag{3.4}$$

Probability density function If we assume that the function $F_{\mathrm{x}}(x;\mathbf{s})$ is differentiable in x, it is possible to define the *probability density function (pdf)* $f_{\mathrm{x}}(x;\mathbf{s})$ by means of the derivative of the cumulative probability, i.e.,

$$f_{\mathrm{x}}(x;\mathbf{s}) = \frac{\mathrm{d}F_{\mathrm{x}}(x;\mathbf{s})}{\mathrm{d}x}. \tag{3.5}$$

If the function $F_{\mathrm{x}}(x;\mathbf{s})$ is not differentiable but the points of non-differentiability compose at most a countable set, the $f_{\mathrm{x}}(x;\mathbf{s})$ can still be defined in terms of the Dirac delta function.

If the pdf exists the cdf can be obtained by means of the integral

$$F_{\mathrm{x}}(x;\mathbf{s}) = \int_{-\infty}^{x} \mathrm{d}x'\, f_{\mathrm{x}}(x';\mathbf{s}). \tag{3.6}$$

Example 3.1 A gray-level digital image can be viewed as an intensity random field defined at the nodes (or cell centers) of a square grid. Let us assume that the intensity field takes discrete values $x_k = 2^k$, where $k = 0, 1, \ldots, K$ with probabilities $0 \leq p_k \leq 1$ so that $\sum_{k=1}^{K} p_k = 1$. Find the single-point pdf and the cdf of the intensity field.

Answer Since the random field takes discrete values, the pdf is expressed as a superposition of Dirac delta functions located at $x = x_k$, i.e.,

$$f_{\mathrm{x}}(x;\mathbf{s}) = \sum_{k=1}^{K} p_k(\mathbf{s})\, \delta(x - x_k).$$

The cdf can be obtained by means of (3.6) which leads to

$$F_{\mathrm{x}}(x;\mathbf{s}) = \begin{cases} 0 & \text{if } x < 1, \\ \sum_{k=1}^{k_x} p_k(\mathbf{s}),\ k_x = \lfloor \log_2 x \rfloor, & \text{if } 1 \leq x \leq 2^K \\ 1, & \text{if } x > 2^K. \end{cases}$$

In the above, $\lfloor y \rfloor$, where $y > 0$, denotes the integer part of y, i.e, the largest non-negative integer which is no greater than y.

Median The *median* is a characteristic measure of the location of the pdf. The median is the value that divides the set area under the pdf into two equal areas. Hence, it is defined by

$$F_{\mathrm{x}}(x_{\mathrm{med}};\mathbf{s}) = 0.5 \Leftrightarrow \int_{-\infty}^{x_{\mathrm{med}}} \mathrm{d}x'\, f_{\mathrm{x}}(x';\mathbf{s}) = \int_{x_{\mathrm{med}}}^{\infty} \mathrm{d}x'\, f_{\mathrm{x}}(x';\mathbf{s}) = 0.5. \tag{3.7}$$

Gaussian (Normal) distribution The most common probability density function is the *Gaussian or Normal model*

$$f_{\mathrm{x}}(x;\mathbf{s}) = \frac{1}{\sqrt{2\pi}\,\sigma_{\mathrm{x}}(\mathbf{s})} \exp\left[-\frac{\{x - m_{\mathrm{x}}(\mathbf{s})\}^2}{2\sigma_{\mathrm{x}}^2(\mathbf{s})}\right]. \tag{3.8}$$

The Gaussian distribution is *symmetric* and *unimodal* (i.e., it has a single peak). All of its moments can be explicitly expressed in terms of the first and second order moments. The importance of the Gaussian is due not only to its modeling advantages, but also to the *Central Limit Theorem* (CLT).

We will study Gaussian random fields in detail in Chap. 6.

Other characteristic measures of the probability distribution are the ensemble moments (see below). The problem of reconstructing the probability distribution from the infinite sequence of moments is known as the *moment problem* [727].

The reconstruction of the probability distribution is not possible unless the moment sequence satisfies certain mathematical conditions that depend on the domain $\mathbb{A}_{\mathrm{x}}$ in which the field takes values, i.e., whether $\mathbb{A}_{\mathrm{x}} = (-\infty, \infty)$, $\mathbb{A}_{\mathrm{x}} = [0, \infty)$, or $\mathbb{A}_{\mathrm{x}}$ is a bounded interval. Suffice it to say here that the moments should not grow too fast as the moment order $n \to \infty$ for the reconstruction to be possible. The *lognormal distribution* is a characteristic example in which the moment sequence does not uniquely determine the probability distribution.

3.2.1 Ensemble Moments

The pdf $f_{\mathrm{x}}(x; \mathbf{s})$ contains the complete information about the behavior of the random field at the location $\mathbf{s}$. Partial information about the random field is included in the ensemble moments of the distribution. In some cases, it is helpful to work with the moments because they can be more easily or more accurately estimated from the data than the complete pdf.

The general ensemble average Let $\Phi\,[\mathrm{X}(\mathbf{s}; \omega)]$ represent a function, in general nonlinear, of the random field $\mathrm{X}(\mathbf{s}; \omega)$. The *ensemble average* of $\Phi\,[\mathrm{X}(\mathbf{s}; \omega)]$ is a weighted average over all probable random field states. The weights are provided by the respective values of the pdf for the specific state. Mathematically, this is expressed as follows:

$$\mathcal{E}[\,\Phi\,[\mathrm{X}(\mathbf{s}; \omega)]\,] = \int_{-\infty}^{\infty} \mathrm{d}x\, \Phi(x)\, f_{\mathrm{x}}(x; \mathbf{s}), \tag{3.9}$$

where $\Phi(x) = \Phi\,[\mathrm{X}(\mathbf{s}; \omega) = x]$.

Remarks

- In Eq. (3.9) the dependence on $\mathbf{s}$ derives from the pdf, not from the function $\Phi(x)$.
- The limits of integration are from $-\infty$ to ∞. This does not mean that $\mathrm{X}(\mathbf{s}; \omega)$ necessarily takes values in $(-\infty, \infty)$: If $\mathrm{X}(\mathbf{s}; \omega)$ takes values $x \in \mathbb{A}_{\mathrm{X}} \subseteq \mathbb{R}$, then $f_{\mathrm{x}}(x; \mathbf{s}) = 0$ for all $x \notin \mathbb{A}_{\mathrm{X}}$. This implies that the domain of integration is effectively restricted to $\mathbb{A}_{\mathrm{X}}$.
- Equation (3.9) is based on the premise that the integral of the product $\Phi(x) f_{\mathrm{x}}(x; \mathbf{s})$ exists for all $\mathbf{s} \in \mathcal{D}$. The convergence of the integral depends on the behavior of both $\Phi(x)$ and $f_{\mathrm{x}}(x; \mathbf{s})$. Convergence is not unconditionally ensured for all the moments of $\mathrm{X}(\mathbf{s}; \omega)$. For example, *Lévy random fields* have heavy tails and thus do not admit finite moments at all orders [709].
- Convergence problems are typically caused by the behavior of $f_{\mathrm{x}}(x; \mathbf{s})$ at low (near zero) or high ($x \to \infty$) values of x. In the former case we refer to *ultraviolet* and in the latter to *infrared* divergences.[2]

Digression on notation and terminology: The ensemble average is also known as *stochastic average* and *statistical average*. In engineering disciplines and often in physics the notation does not distinguish between the random field and its values. In such cases the significance of a symbol, i.e., if $x(\mathbf{s})$ refers to the random field or to a specific realization, is decided based on context. In physics different symbols for different types of ensemble average. For example:

[2]This terminology comes from physics: short (large) distances are affected by the spectral behavior at high (low) spatial frequencies; hence the terms ultraviolet (infrared) in analogy with the frequencies beyond the edges of the visible light spectrum.

- The average over *quenched disorder* is often denoted by the horizontal bar, i.e., $\overline{\Phi\,[\mathrm{X}(\mathbf{s};\,\omega)]}$. Quenched disorder refers to background randomness that is considered frozen in time with respect to fluctuating degrees of freedom in the system. For example, in the case of diffusive pollutant transport in groundwater the disorder of the permeability field is considered quenched if the porous medium is not modified on the time scale of observation.
- The average over an *external noise source* or a general probability distribution is often denoted by means of the angle brackets $\langle \Phi\,[\mathrm{X}(\mathbf{s};\,\omega)] \rangle$.

The above conventions are not universal since different notations appear in the literature. We use the symbol $\mathcal{E}[\cdot]$ to denote the *ensemble average* over a probability distribution, regardless of the origin of randomness. In addition, we use $\overline{\Phi\,[x(\mathbf{s})]}$ to denote the *spatial average* of the function $\Phi(\cdot)$ of a given realization over $\mathcal{D}$, i.e.,

$$\overline{\Phi\,[x(\mathbf{s})]} = \frac{1}{|\mathcal{D}|} \int_{\mathcal{D}} d\mathbf{s}\, \Phi\,[x(\mathbf{s})] \,. \tag{3.10}$$

We also use $\overline{\Phi\,[x(\mathbf{s});\,\omega]}$ to emphasize that the spatial average is a *random variable* that depends on the state ω.

3.2.1.1 Marginal Moments of Integer Order

Marginal moments of integer order n are ensemble averages of $\mathrm{X}^n(\mathbf{s};\,\omega)$ given by

$$\mathcal{E}[\mathrm{X}^n(\mathbf{s};\,\omega)] = \int_{-\infty}^{\infty} dx\, x^n\, f_{\mathrm{x}}(x;\,\mathbf{s}). \tag{3.11}$$

For $n = 1$ Eq. (3.11) returns the *expectation* or *ensemble mean* of $\mathrm{X}(\mathbf{s};\,\omega)$.

In practical applications of spatial data modeling, the expectation is modeled in terms of trend functions as described in Chap. 2.

In addition to (3.11) we can also define *centered moments* that are essentially the integer order moments of the *fluctuation field* $\mathrm{X}'(\mathbf{s};\,\omega) = \mathrm{X}(\mathbf{s};\,\omega) - m_{\mathrm{x}}(\mathbf{s})$:

$$\mathcal{E}_c[\mathrm{X}^n(\mathbf{s};\,\omega)] := \mathcal{E}[\mathrm{X}'^n(\mathbf{s};\,\omega)] = \int_{-\infty}^{\infty} dx\, [x - m_{\mathrm{x}}(\mathbf{s})]^n\, f_{\mathrm{x}}(x;\,\mathbf{s}), \tag{3.12}$$

where $\mathcal{E}_c[\cdot]$ denotes the *centered expectation*.

The moments defined above have spatial dependence through the pdf. If the random field is *translation invariant in the statistical sense* the moments are translation invariant functions. We will revisit this statement below, in the discussion of stationarity.

Some commonly used distributional measures that involve low-order moments are the following:

1. The *ensemble mean* or first-order moment

$$m_{\mathrm{x}}(\mathbf{s}) = \mathcal{E}[\mathrm{X}(\mathbf{s};\,\omega)]. \tag{3.13}$$

The ensemble mean determines the center of mass of the distribution.

2. The *variance* or centered second-order moment

$$\sigma_{\rm x}^2(\mathbf{s}) = \mathcal{E}\left[{\rm X}^2(\mathbf{s};\omega)\right] - \mathcal{E}\left[{\rm X}(\mathbf{s};\omega)\right]^2 = \mathcal{E}[\{{\rm X}'(\mathbf{s};\omega)\}^2]. \tag{3.14}$$

Remark The expression $\mathcal{E}[{\rm X}(\mathbf{s};\omega)]^n$ denotes the n-th power of $\mathcal{E}[{\rm X}(\mathbf{s};\omega)]$. The expectation of ${\rm X}^n(\mathbf{s};\omega)$ is denoted by $\mathcal{E}[{\rm X}^n(\mathbf{s};\omega)]$.

The variance measures the dispersion of the distribution around the mean. The *standard deviation* $\sigma_{\rm x}(\mathbf{s})$ has the same units as ${\rm X}(\mathbf{s};\omega)$.

3. The *coefficient of variation* is given by

$$c_{\rm v;x}(\mathbf{s}) = \frac{\sigma_{\rm x}(\mathbf{s})}{m_{\rm x}(\mathbf{s})}. \tag{3.15}$$

The coefficient of variation is a dimensionless quantity that measures the spread of the distribution normalized by the mean. It can be used as a perturbation parameter for expansions around the mean, provided that the latter is non-zero.

If the coefficient of variation is small, the fluctuations (disorder) can be treated using *low-order perturbation* expansions which facilitate the study of disordered systems. Examples of low-order perturbation analysis of flow in random porous media with application in subsurface hydrology are given in [276, 696].

The coefficient of variation is most effective if the distribution is symmetric. Nevertheless, even for asymmetric distributions (e.g., the Weibull distribution) it provides a useful measure of variability that allows comparing probability distributions with different mean values.

4. The *skewness coefficient* is the normalized, centered third-order moment which is given by

$$s_{\rm x}(\mathbf{s}) = \frac{\mathcal{E}\left[{\rm X}(\mathbf{s};\omega) - m_{\rm x}(\mathbf{s})\right]^3}{\sigma_{\rm x}^3(\mathbf{s})}. \tag{3.16}$$

The skewness coefficient is a measure of the distribution's *asymmetry*. It is equal to zero for symmetric probability distributions and takes nonzero values for asymmetric distributions.

5. The *kurtosis coefficient* is the normalized, centered fourth-order moment defined by

$$k_{\rm x}(\mathbf{s}) = \frac{\mathcal{E}\left[{\rm X}(\mathbf{s};\omega) - m_{\rm x}(\mathbf{s})\right]^4}{\sigma_{\rm x}^4(\mathbf{s})}. \tag{3.17}$$

The kurtosis coefficient is equal to three for the Gaussian distribution, and it can be used to identify deviations of symmetric distributions from the Gaussian.

If $k_{\mathrm{x}}(\mathbf{s}) > 3$ the distribution is called *leptokurtic*, whereas if $k_{\mathrm{x}}(\mathbf{s}) < 3$ it is called *platykurtic*. Leptokurtic distributions tend to be thinner at the center and have heavier tails than the Gaussian distribution (e.g., the Student's t-distribution with more than six degrees of freedom). In contrast, platykurtic distributions tend to have thinner tails than the Gaussian (e.g., the uniform distribution).

Remark The *excess kurtosis* is defined as the difference $k_{\mathrm{x}}(\mathbf{s}) - 3$. Sometimes the distinction between the two coefficients is not explicitly stated. If in doubt, calculate the kurtosis of a normally distributed random sample.

3.2.1.2 Useful Nonlinear Moments

In applications one encounters nonlinear marginal moments that are not of the form $\mathcal{E}[\mathrm{X}^n(\mathbf{s};\omega)]$. Two commonly used moments are the *geometric mean* and the *harmonic mean* [173] that are defined by means of

$$\ln X_{\mathrm{G}} = \mathcal{E}[\ln \mathrm{X}(\mathbf{s};\omega)], \tag{3.18}$$

$$\frac{1}{X_{\mathrm{H}}} = \mathcal{E}\left[\frac{1}{\mathrm{X}(\mathbf{s};\omega)}\right]. \tag{3.19}$$

These nonlinear expectations have application in stochastic subsurface hydrology. The *geometric mean* provides an estimate of the effective permeability of two-dimensional porous formations in which the permeability field follows the lognormal distribution. The *harmonic mean* gives the effective permeability in layered formations for transverse flow (in the direction perpendicular to the layers).

If the probability distribution of $\mathrm{X}(\mathbf{s};\omega)$ has non-zero weight for zero values, then both the geometric and the harmonic mean are equal to zero. For example, the harmonic mean vanishes for all Gaussian distributions, since the latter have a finite, even if tiny, probability for zero and negative values.

Divertissement To gain some insight into the behavior of the harmonic mean consider a grossly idealized example that involves the artificial porous medium shown in the schematic of Fig. 3.1. This artificial medium consists of N (where

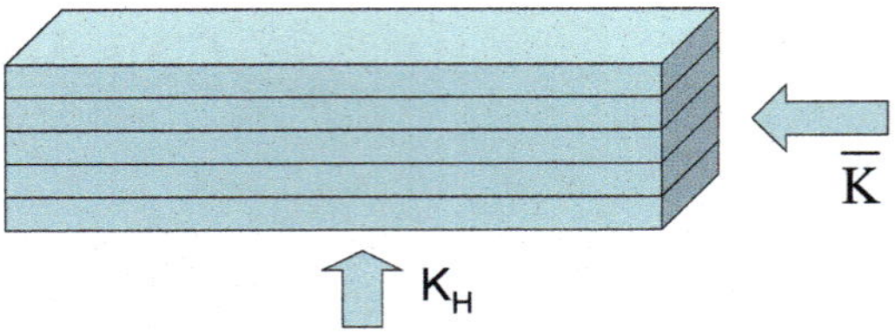

Fig. 3.1 Schematic of fluid flow in idealized layered formation. The effective permeability perpendicular to the layers is the harmonic mean, K_{H}, while parallel to the layers it is equal to $\overline{K}$

N is presumably large) parallel layers, with uniform permeability values in the layers given by $\{K_n\}_{n=1}^{N}$. Let us consider that a top-to-bottom pressure gradient J is applied leading to fluid flow in the perpendicular direction. The flux Q is constant throughout the medium and related to the gradient via

$$Q = -K_{\text{eff}}\, J.$$

The flux through each layer is also equal to Q, as a result of mass conservation; hence,

$$Q = -K_n J_n, \quad \text{for } n = 1, \ldots, N.$$

The average gradient over all the layers thus is equal to

$$J = \frac{1}{N} \sum_{n=1}^{N} J_n = -\frac{Q}{N} \sum_{n=1}^{N} \frac{1}{K_n}.$$

Comparing the above with the first equation, it follows that the effective permeability is given by the harmonic average, i.e.

$$K_{\text{eff}} = \left(\frac{1}{N} \sum_{n=1}^{N} \frac{1}{K_n} \right)^{-1}.$$

3.2.2 The Moment Generating Function

Let us assume that the marginal moments of $\mathrm{X}(\mathbf{s}; \omega)$ exist. We will see that the moments can be generated from the derivatives of a single function which is known as the moment generating function, provided that the latter exists.

Moment generating function The *moment generating function*, $\mathcal{M}_{\mathrm{x}}(\mathbf{s}; u)$, where $u \in \mathbb{R}$, is defined as follows

$$\mathcal{M}_{\mathrm{x}}(\mathbf{s}; u) = \mathcal{E}\left[\mathrm{e}^{u\,\mathrm{X}(\mathbf{s};\omega)} \right] = \sum_{n=0}^{\infty} \frac{u^n}{n!} \mathcal{E}\left[\mathrm{X}^n(\mathbf{s}; \omega) \right] \tag{3.20}$$

The moment generating function is well-defined in the moment of all orders n exist, and if the moment series in (3.20) converges.

The moments of order n are obtained from the moment generating function by means of the n-order derivatives with respect to u evaluated at $u = 0$, i.e.,

$$\mathcal{E}[\mathrm{X}^n(\mathbf{s};\omega)] = \left.\frac{\mathrm{d}^n \mathcal{M}_{\mathrm{x}}(\mathbf{s};u)}{\mathrm{d}u^n}\right|_{u=0}. \tag{3.21}$$

The existence of the pdf does not guarantee that the moments of all orders exist, i.e., that they are finite numbers. A standard example is the *Cauchy-Lorentz distribution* that has the following pdf:

$$f_{\mathrm{x}}(x;\mathbf{s}) = \frac{1}{\pi}\frac{\alpha}{(x-x_0)^2+\alpha^2} \tag{3.22}$$

This distribution does not have well-defined moments of order higher than one.

Lognormal Distribution It is possible that the moments $\mathcal{E}[\mathrm{X}^n(\mathbf{s};\omega)]$ exist, whereas the moment generating function does not; this happens if the series

$$\sum_{n=0}^{\infty}\frac{u^n}{n!}\,\mathcal{E}\left[\mathrm{X}^n(\mathbf{s};\omega)\right]$$

does not converge. For example, this lack of convergence is observed in the case of the *lognormal distribution*.

3.2.3 The Characteristic Function

Let us assume that the marginal pdf $f_{\mathrm{x}}(x;\mathbf{s})$ exists and is an *absolutely integrable function* of x for all $\mathbf{s} \in \mathcal{D}$. This is always true, because a pdf satisfies the inequality $f_{\mathrm{x}}(x;\mathbf{s}) \geq 0$ and the normalization constraint

$$\int_{-\infty}^{\infty} \mathrm{d}x\; f_{\mathrm{x}}(x;\mathbf{s}) = 1.$$

Since the function $f_{\mathrm{x}}(x;\mathbf{s})$ is absolutely integrable, its *Fourier Transform* $\phi_{\mathrm{x}}(\mathbf{s};u) := \mathcal{F}[f_{\mathrm{x}}(x;\mathbf{s})]$ exists [96]. It is given by

$$\phi_{\mathrm{x}}(\mathbf{s};u) \int_{-\infty}^{\infty} \mathrm{d}x\; f_{\mathrm{x}}(x;\mathbf{s})\,\mathrm{e}^{\mathrm{i}u\,x} = \mathcal{E}\left[\mathrm{e}^{\mathrm{i}u\,\mathrm{X}(\mathbf{s};\omega)}\right]. \tag{3.23}$$

The function $\phi_{\mathrm{x}}(\mathbf{s};u)$ is known as *the characteristic function* of the marginal pdf. The characteristic function has the advantage of being well-defined, even if the pdf is not (see Sect. 4.5.4).

If $\phi_{\mathrm{x}}(\mathbf{s};\,u)$ is known, the marginal pdf can be obtained by means of the *inverse Fourier transform*

$$f_{\mathrm{x}}(x;\,\mathbf{s}) = \mathcal{F}^{-1}[\phi_{\mathrm{x}}(\mathbf{s};\,u)] = \frac{1}{2\pi}\int_{-\infty}^{\infty} du\,\phi_{\mathrm{x}}(\mathbf{s};\,u)\,\mathrm{e}^{-\mathrm{i}u\,x}. \tag{3.24}$$

Connection with moment generating function If the moment generating function also exists, it is connected to the characteristic function via the equation

$$\phi_{\mathrm{x}}(\mathbf{s};\,u) = \mathcal{M}_{\mathrm{x}}(\mathbf{s};\,\mathrm{i}\,u). \tag{3.25}$$

Cauchy-Lorentz distribution While the characteristic function is always defined, it is possible that the moment generating function does not exist. A case in point is the Cauchy-Lorentz distribution, introduced above, with pdf (3.22). This probability distribution does not possess a moment generating function. Nevertheless, its characteristic function exists and is equal to

$$\phi_{\mathrm{x}}(\mathbf{s};\,u) = \mathrm{e}^{\mathrm{i}x_0 u - \alpha|u|}.$$

Equivalence of moments and pdf Let us assume that the moments of $\mathrm{X}(\mathbf{s};\,\omega)$ exist for all $n \in \mathbb{N}$. It is then straightforward to show that the *pdf is completely determined by the moments of all orders*. Specifically, based on (3.23) it follows that

$$\phi_{\mathrm{x}}(\mathbf{s};\,u) = \mathcal{E}\left[\mathrm{e}^{\mathrm{i}u\,\mathrm{X}(\mathbf{s};\omega)}\right].$$

The Taylor expansion of the exponential function leads to the following

$$\phi_{\mathrm{x}}(\mathbf{s};\,u) = \sum_{n=0}^{\infty}\frac{\mathrm{i}^n u^n \mathcal{E}\,[\mathrm{X}^n(\mathbf{s};\,\omega)]}{n!}. \tag{3.26}$$

Based on the above and (3.24) the correspondence between the moments $\mathcal{E}\,[\mathrm{X}^n(\mathbf{s};\,\omega)]$, $n \in \mathbb{N}$, and the pdf is established. In spite of this theoretical equivalence, the accurate estimation of higher-order moments from sample data becomes increasingly difficult with increasing order. This is due to the fact that the impact of errors in the data is amplified as the order n increases.

Moment extraction If the characteristic function is known, the marginal moments of $\mathrm{X}(\mathbf{s};\,\omega)$ can be evaluated by means of the nth-order derivatives of the characteristic function:

$$\mathcal{E}[\mathrm{X}^n(\mathbf{s};\,\omega)] = \frac{1}{\mathrm{i}^n}\left.\frac{\mathrm{d}^n\phi_{\mathrm{x}}(\mathbf{s};\,u)}{\mathrm{d}u^n}\right|_{u=0}. \tag{3.27}$$

3.2.4 The Cumulant Generating Function

If the marginal moments of $X(\mathbf{s};\omega)$ exist and the series in (3.20) converges, the moment generating functional can be defined as discussed in Sect. 3.2.2. We can then define the logarithm of the moment generating function. This is convenient since, as shown below, the *cumulant functions* can be derived from the derivatives of this logarithm.

Definition 3.1 The *cumulant generating function* is defined as follows

$$\mathcal{K}_x(\mathbf{s};u) := \ln \mathcal{M}_x(\mathbf{s};u) = \sum_{n=0}^{\infty} \frac{u^n}{n!}\,\kappa_{x;n}(\mathbf{s}), \tag{3.28}$$

where $\kappa_{x;n}(\mathbf{s})$ is the n-order marginal cumulant.

Connection with moment generating functional Cumulants of order n are related to moments of order $m \le n$. Vice versa, moments of order n are related to cumulants of order $m \le n$. In fact, a comparison of (3.28) with (3.20) shows that there must be a general relation between the moments and the cumulants, since the *moment and the cumulant generating functionals* are linked via

$$\mathcal{M}_x = \exp\left(\mathcal{K}_x\right). \tag{3.29}$$

To illustrate the moment-cumulant relations, let us denote the moment of order n by means of $\mu_{x;n}(\mathbf{s}) := \mathcal{E}[X^n(\mathbf{s};\omega)]$. Then, in light of (3.20) and (3.28) the cumulant expansion is linked to the moments as follows

$$\sum_{n=0}^{\infty} \frac{u^n}{n!}\,\kappa_{x;n}(\mathbf{s}) = \ln\left(1 + \sum_{n=1}^{\infty} \frac{u^n}{n!}\,\mu_{x;n}(\mathbf{s})\right).$$

The Taylor expansion of the natural logarithm $\ln(1+x)$ around $x=0$ is given by the following series

$$\ln(1+x) = \sum_{n=1}^{\infty} \frac{(-1)^{n+1}}{n}\,x^n.$$

By means of this series we can show that the expansions of the cumulant and the moment generating functions are related via the following equation

$$\sum_{n=0}^{\infty} \frac{u^n}{n!}\,\kappa_{x;n}(\mathbf{s}) = \sum_{l=1}^{\infty} \frac{(-1)^{l+1}}{l}\left(\sum_{m=1}^{\infty} \frac{u^m}{m!}\,\mu_{x;m}(\mathbf{s})\right)^l.$$

Relations between the cumulants and the moments are obtained by equating the coefficients of equal powers of u on both sides of the above equation.[3] Thus, we notice that the right-hand side of the equation does not involve a zero-order term; this leads to $\kappa_0 = 0$. More generally, to calculate the cumulant of order n we need to include terms from the right-hand side that satisfy $m\,l = n$. Hence, for the first five cumulants we obtain the expressions

$$\kappa_{x;1} = \mu_{x;1} \tag{3.30a}$$

$$\kappa_{x;2} = \mu_{x;2} - \mu_{x;1}^2 \tag{3.30b}$$

$$\kappa_{x;3} = \mu_{x;3} - 3\mu_{x;2}\mu_{x;1} + 2\mu_{x;1}^3 \tag{3.30c}$$

$$\kappa_{x;4} = \mu_{x;4} - 4\mu_{x;3}\mu_{x;1} + 12\mu_{x;2}\mu_{x;1}^2 - 3\mu_{x;2}^2 - 6\mu_{x;1}^4 \tag{3.30d}$$

$$\begin{aligned}\kappa_{x;5} = &\mu_{x;5} - 5\mu_{x;4}\mu_{x;1} - 10\mu_{x;3}\mu_{x;2} + 20\mu_{x;3}\mu_{x;1}^2 \\ &+ 30\mu_{x;2}^2\mu_{x;1} - 60\mu_{x;2}^2\mu_{x;1}^3 + 24\mu_{x;1}^5.\end{aligned} \tag{3.30e}$$

Moment-cumulant relations The above equations are low-order realizations of the general, recursive *moment-cumulant* relations

$$\mu_{x;n} = \sum_{p=0}^{n-1} \binom{n-1}{p} \mu_{x;p}\,\kappa_{x;n-p},$$

and

$$\kappa_{x;n} = \mu_{x;n} - \sum_{p=1}^{n-1} \binom{n-1}{n-p} \kappa_{x;p}\,\mu_{x;n-p}.$$

MGF and CGF for independent fields Assume that $X(\mathbf{s};\omega)$ and $Y(\mathbf{s};\omega)$ are two statistically independent random fields [the probability of $X(\mathbf{s};\omega) = x$ is independent of the probability that $Y(\mathbf{s};\omega) = y$ and vice versa].

Then, the MGF of the sum $X(\mathbf{s};\omega) + Y(\mathbf{s};\omega)$ is multiplicative and the respective CGF is additive, i.e.,

$$M_{x+y}(\mathbf{s};u) = M_x(\mathbf{s};u)\,M_y(\mathbf{s};u), \tag{3.31}$$

$$\kappa_{x+y}(\mathbf{s};u) = \kappa_x(\mathbf{s};u) + \kappa_y(\mathbf{s};u). \tag{3.32}$$

[3] In the following equations, we suppress the dependence of the moments and cumulants on $\mathbf{s}$ for the sake of brevity.

Properties of cumulants The cumulants contain the same information as the moments. They have, however, some advantages over the moments that motivate their use.

- For the Gaussian distribution cumulants of order ≥ 3 are zero. This is true both for the univariate and the multivariate Gaussian cases.
- The cumulants of the sum of two random variables are equal to the sum of the respective cumulants of the addends, i.e., (3.32). This implies that if many random variables are added, the cumulants of the sum tend to zero for all orders ≥ 3 by virtue of the Central Limit Theorem.[4]
- The cumulants are semi-invariants. This means that under the *affine transformation* $X' = \lambda X + c$, the cumulant of order p transforms as $\kappa_{X';p} = \lambda^p \kappa_{X;p}$.
- In physics, cumulants are known as *connected correlation functions* and *Ursell functions* [471]. Using Feynman diagrammatic methods, cumulants can be evaluated as sums of connected Feynman diagrams [593, 613].

Diagrammatic perturbation methods, introduced by Feynman, are quite powerful tools for handling perturbation expansions, since they allow replacing long mathematical expressions by insightful graphical terms. For an introduction to the diagrammatic perturbation theory applied in statistical field theory consider [297], [434, Chap. 5]. Applications of diagrammatic theory in turbulence studies are discussed in [557], while its use in the calculation of upscaled single-phase fluid permeability of random media can be found in [143, 370, 452].

3.3 Two-Point Properties of Random Fields

In the preceding section we focused on properties of the point-wise (marginal) probability distribution of random fields. The marginal distribution provides an incomplete perspective of spatial variation. In particular, it does not provide any information regarding the correlations between two or more points. In fact, if the random field is stationary (i.e., translation invariant in the statistical sense), the marginal distribution is completely independent of the position. In this section we focus on joint probabilities that characterize the dependence of random field values simultaneously at two points.

3.3.1 Joint Cumulative Distribution Function

To investigate two-point dependence we need to consider two-point joint probability functions. The joint cumulative distribution function at points $\mathbf{s}_1$ and $\mathbf{s}_2$ is the

[4] This assumes that the conditions of the central limit theorem hold; practically, this means that the tails of the addend probability density functions decline asymptotically at least as fast as a power law with exponent >3 and that the correlations are short-ranged.

probability that the values of the random field do not exceed x_1 at $\mathbf{s}_1$ and x_2 at $\mathbf{s}_2$. Formally, this is expressed as follows:

$$F_{\rm x}(x_1, x_2; \mathbf{s}_1, \mathbf{s}_2) = P\left(\mathrm{X}(\mathbf{s}_1; \omega) \leq x_1 \wedge \mathrm{X}(\mathbf{s}_2; \omega) \leq x_2\right), \tag{3.33}$$

where the symbol $\wedge$ denotes the logical "and" operator.

3.3.2 *Conditional Probability Function*

One of the most important theorems of probability is *Bayes' theorem*. It states that if $A, B \in \mathcal{A}$ are any two events, then the conditional probability, $P(B \mid A)$, of the event B given that the event A has been observed is related to the joint probability of the two events by means of

$$\mathrm{P(B \mid A)} = \frac{P(A \cap B)}{P(A)} = \frac{\mathrm{P(A \mid B)}\; P(B)}{P(A)}. \tag{3.34}$$

The two shaded quantities represent conditional probabilities. Note that in general $P(B \mid A)$ can be quite different than $P(A \mid B)$.

The equation of Bayes' theorem allows sequential updating of conditional probabilities if new information becomes available. In addition, two events are *independent* if the realization of one of them does not affect the probability that the other one will occur. Mathematically the independence condition is expressed as follows:

$$\text{If } A \text{ and } B \text{ are independent events, then } P(B \mid A) = P(B). \tag{3.35}$$

In the case of random fields, probabilities are measured by means of cumulative distribution functions. The conditional probability that the field does not exceed a certain value x_1 at point $\mathbf{s}_1$ given that the field at some other point $\mathbf{s}_2$ does not exceed the value x_2 is defined by

$$P\big(\underbrace{\mathrm{X}(\mathbf{s}_1; \omega) \leq x_1}_{B} \mid \underbrace{\mathrm{X}(\mathbf{s}_2; \omega) \leq x_2}_{A}\big) = F_{\rm x}(x_1; \mathbf{s}_1 \mid x_2; \mathbf{s}_2), \tag{3.36}$$

where $F_{\rm x}(x_1; \mathbf{s}_1 \mid x_2; \mathbf{s}_2)$ is the *conditional probability function*. Based on *Bayes' theorem*, the latter is related to the joint two-point probability function as follows

$$F_{\rm x}(x_1; \mathbf{s}_1 \mid x_2; \mathbf{s}_2) = \frac{F_{\rm x}(x_1, x_2; \mathbf{s}_1, \mathbf{s}_2)}{F_{\rm x}(x_2; \mathbf{s}_2)}. \tag{3.37}$$

If the value x_1 of $X(\mathbf{s};\omega)$ at $\mathbf{s}_1$ is *independent* of the value x_2 at $\mathbf{s}_2$, the conditional probability function is reduced to the marginal cumulative distribution function as a result of the independence, i.e.,

$$F_x(x_1;\mathbf{s}_1 \mid x_2;\mathbf{s}_2) = F_x(x_1;\mathbf{s}_1).$$

3.3.3 Joint Probability Density Function

In analogy with the single-point case, the two-point joint pdf can be defined by means of the second-order partial derivative of the joint cdf, i.e.,

$$f_x(x_1, x_2;\mathbf{s}_1,\mathbf{s}_2) = \frac{\partial^2 F_x(x_1, x_2;\mathbf{s}_1,\mathbf{s}_2)}{\partial x_1 \partial x_2}. \tag{3.38}$$

The *marginal pdf* at the position $\mathbf{s}_1$ is obtained by integrating the two-point joint pdf over all the possible values of the field at the other position $\mathbf{s}_2$, i.e., by means of the following integral

$$f_x(x_1;\mathbf{s}_1) = \int_{-\infty}^{\infty} \mathrm{d}x_2\, f_x(x_1, x_2;\mathbf{s}_1,\mathbf{s}_2). \tag{3.39}$$

3.3.4 Two-Point Correlation Functions

Various functions are used in the literature to measure the dependence of the random field values at two different points in space. Below we discuss the most commonly used ones. We will refer to such functions collectively as *two-point correlation functions*.[5]

Covariance function The covariance function is the expectation of the product of the field values at two different points minus the product of the respective mean functions at the same points, i.e.,

$$C_{xx}(\mathbf{s}_1,\mathbf{s}_2) = \mathcal{E}\left[X(\mathbf{s}_1;\omega)\, X(\mathbf{s}_2;\omega)\right] - \mathcal{E}\left[X(\mathbf{s}_1;\omega)\right]\, \mathcal{E}\left[X(\mathbf{s}_2;\omega)\right]. \tag{3.40}$$

It is straightforward to show (and hence left as a problem for the reader) that the above is equivalent to evaluating the expectation of the product of the field fluctuations at these locations, i.e.,

$$C_{xx}(\mathbf{s}_1,\mathbf{s}_2) = \mathcal{E}\left[X'(\mathbf{s}_1;\omega)\, X'(\mathbf{s}_2;\omega)\right]. \tag{3.41}$$

[5]This term is commonly used in physics.

Variance If both points coincide, i.e., $\mathbf{s}_2 \rightarrow \mathbf{s}_1$, the value of the covariance becomes equal to the variance of the random field at that point, i.e.,

$$C_{xx}(\mathbf{s}_1, \mathbf{s}_1) = \sigma_x^2(\mathbf{s}_1). \tag{3.42}$$

The covariance function declines as the distance between the two points increases and tends to zero as the distance tends to ∞. The decline is not necessarily monotonic. Indeed, there are covariance functions that exhibit an oscillatory behavior. In general, higher values of the covariance function indicate a higher degree of correlation between the field fluctuations at the two points than lower values (with the same sign) of the covariance function.

Correlation function The covariance function depends on the variance of the process and the spatial correlation between the field values at the selected points. In order to isolate the latter effect, the correlation function is defined as

$$\rho_{xx}(\mathbf{s}_1, \mathbf{s}_2) = \frac{C_{xx}(\mathbf{s}_1, \mathbf{s}_2)}{\sigma_x(\mathbf{s}_1)\, \sigma_x(\mathbf{s}_2)} \in [-1,\ 1]. \tag{3.43}$$

The correlation function takes values in the interval $[-1, 1]$, independently of the variance of the process. It measures how the linear dependence of the field values changes as a function of the distance between the two points. Hence, it can be used to compare the spatial extent of correlations in spatial processes with different variance.

Negative values of $\rho_{xx}(\mathbf{s}_1, \mathbf{s}_2)$ imply that, if all possible realizations are considered, higher values of the field at $\mathbf{s}_1$, tend to occur in synchrony with lower values at $\mathbf{s}_2$ and vice versa. In contrast, a positive value of $\rho_{xx}(\mathbf{s}_1, \mathbf{s}_2)$ implies that field values tend to behave similarly at $\mathbf{s}_1$ and $\mathbf{s}_2$ for all the realizations. Finally, if $\rho_{xx}(\mathbf{s}_1, \mathbf{s}_2) \approx 0$, the values of the field at $\mathbf{s}_1$ have little influence on the field values at $\mathbf{s}_2$ and vice versa.

In the physics literature, often the covariance and the correlation function are both referred to as *two-point correlation functions*. In geostatistics, the term *correlogram* is also used for the correlation function [132].

Variogram function The variogram function is proportional to the variance of the *field increment*, $X(\mathbf{s}_1; \omega) - X(\mathbf{s}_2; \omega)$, between two points $\mathbf{s}_1$ and $\mathbf{s}_2$ according to the equation

$$\gamma_{xx}(\mathbf{s}_1, \mathbf{s}_2) = \frac{1}{2} \operatorname{Var}\left\{X(\mathbf{s}_1; \omega) - X(\mathbf{s}_2; \omega)\right\}. \tag{3.44a}$$

If the expectation of the field increment is zero (i.e., if the random field has a constant mean), the variogram function is given by the following simpler expectation

$$\gamma_{xx}(\mathbf{s}_1, \mathbf{s}_2) = \frac{1}{2} \mathcal{E}\left\{[X(\mathbf{s}_1; \omega) - X(\mathbf{s}_2; \omega)]^2\right\}. \tag{3.44b}$$

The latter is valid for *stationary random fields* (discussed in Sect. 3.4) as well as for fields with zero-mean *stationary increments* (discussed in 5.7.1).

The variogram function is also used in the statistical theory of fluid turbulence where it is known as *structure function* [557], *delta variance* [396] and *difference correlation function* [305].

A theoretical treatment of the connections between the covariance function and the variogram is presented in [291]. The variogram has certain advantages over the covariance function. We discuss this topic in the next section, after we introduce the concept of stationarity.

There is a lingering issue related to the nomenclature of the variogram function. Historically, the function defined in (3.44) was named semivariogram due to the $1/2$ factor in front of the expectation (or the variance) operator. The term semivariogram is still in use, although not as widely accepted [132, p. 31]. Herein we use the term variogram for brevity and simplicity.

3.4 Stationarity and Statistical Homogeneity

> Simpler theories are preferable to complex ones.
>
> *Occam's razor*

Translation invariance is a mathematical property that is extremely useful in the study of physical systems. It implies that the physical laws governing the system and the properties of the system are independent of the position in space. This assumption is too drastic in the case of random fields, which vary over space by definition. Nevertheless, the concept of translation invariance is so useful that even a weaker version of it is desirable.

Stationarity Nomenclature Indeed, in the case of random fields, translation invariance is replaced by the concept of *stationarity or statistical homogeneity*. This implies that translation invariance is transferred to the moments (instead of the realizations) of the random field. The term stationarity was originally used in the time domain and is thus more common for random processes (i.e., random functions with temporal variation). However, its use has been extended to also encompass spatially distributed processes.

The term *statistical homogeneity* is used interchangeably with stationarity in the case of spatial random fields, and in some references it is preferred to stationarity. There are two types of stationarity (statistical homogeneity).

1. *Weak stationarity*, or *second-order stationarity*, or *wide-sense stationarity* is the type that we are concerned with in this section.
2. *Strong stationarity* implies stricter conditions that are more difficult to establish in practice. We will discuss it below in the Sect. 4.5.

Definition 3.2 A random field $X(\mathbf{s}; \omega)$ is called weakly stationary, or second-order stationary, or wide-sense stationary if the following two conditions hold:

1. $\mathcal{E}[X(\mathbf{s}; \omega)] = m_x$, for all $\mathbf{s} \in \mathcal{D}$.
2. $C_{xx}(\mathbf{s}_1, \mathbf{s}_2) = C_{xx}(\mathbf{s}_1 - \mathbf{s}_2)$, for all $\mathbf{s}_1, \mathbf{s}_2 \in \mathcal{D}$.

The vector $\mathbf{r} := \mathbf{s}_1 - \mathbf{s}_2$ is typically used to represent the spatial lag. If the random field is weakly stationary the mean and the covariance function are *translation invariant*. This means that the mean is constant and the covariance function a function of $\mathbf{r}$.

Consequences of stationarity The assumption of second-order stationarity has the following straightforward consequences:

1. Uniform variance:

$$\mathcal{E}\left[\{X'(\mathbf{s}; \omega)\}^2\right] = \sigma_x^2. \tag{3.45}$$

2. Translation-invariant correlation function:

$$\rho_{xx}(\mathbf{s}_1, \mathbf{s}_2) = \rho_{xx}(\mathbf{s}_1 - \mathbf{s}_2). \tag{3.46}$$

3. Variogram—Covariance relation:

$$\gamma_{xx}(\mathbf{s}_1 - \mathbf{s}_2) = \sigma_x^2 - C_{xx}(\mathbf{s}_1 - \mathbf{s}_2). \tag{3.47}$$

Variogram properties of stationary random fields In the case of stationarity, the variogram has certain characteristic properties that we review below.

1. The value of the variogram is zero at $\|\mathbf{r}\| = 0$.
2. In the presence of nugget, the variogram jumps discontinuously to the nugget variance at any lag $\|\mathbf{r}\| \to 0_+$ that is just slightly higher than zero.
3. The variogram tends to increase as the lag $\|\mathbf{r}\|$ increases. The increase may be non-monotonic for models with oscillatory dependence of the correlations.
4. The rate of the variogram's increase is determined by a *characteristic length*, possibly in combination with other parameters as well (e.g., as is the case for the rational quadratic, the Spartan, and the Matérn models).
5. The variogram tends to a *constant plateau* (sill) as $\|\mathbf{r}\|$ tends to infinity.
6. The value of the plateau depends on the direction of $\mathbf{r}$ if there is *zonal anisotropy* (see Sect. 4.3.5 below).
7. The rate at which the variogram sill is approached depends on the direction of $\mathbf{r}$ if geometric anisotropy (see Sect. 4.3.2) is present.

Stationary or not? Stationarity is a very convenient assumption that significantly simplifies the structure of spatial models. Simpler models are preferred to more complex ones based on *Occam's razor* (principle of parsimony), and thus stationarity is often used as a *working hypothesis* for modeling spatial data. If the stationary model's predictive accuracy and precision are deemed acceptable by some statistical measure(s) of fit (e.g., based on cross-validation metrics, maximum likelihood, information criteria), the stationary model provides a useful, albeit imperfect, approximation of reality.

If the properties of non-stationarity should be inductively inferred from the data rather than deduced from first principles, the inference process introduces additional uncertainties in the model [737]. In addition, non-stationary models should be cautiously used, since the introduction of non-stationarity in the wrong component of the spatial model may lead to adverse effects (such as excessive smoothing). Finally, the estimation of non-stationary models can be computationally intensive [266].

Reducible non-stationarity To model spatial data that exhibit clear spatial trends, it is not justifiable to use a stationary random field $X(\mathbf{s}; \omega)$. However, it may still be possible to model the residuals of the data after subtraction of a trend function by means of a stationary fluctuation $X'(\mathbf{s}; \omega)$ [815]. We shall refer to such non-stationarity as *reducible non-stationarity*.[6]

Another term used for non-stationarity that is caused by the presence of trends is *global non-stationarity*. In contrast, non-stationarity that is due to fluctuations is also referred to as *local non-stationarity* [266].

The simplicity of stationary models, however, is not always possible; a typical example of data that involve irreducible non-stationarity are geophysical signals (e.g., seismic signals), the variance of which changes—due to attenuation—with the distance from the source [51]. Another characteristic example is *fractional Brownian motion*, a ubiquitous non-stationary process with stationary increments (see Sect. 5.8).

3.5 Variogram Versus Covariance

Both the variogram and the covariance function describe two-point correlations. In principle, both functions contain the same information in the case of *stationary random fields*. However, important differences between the variogram and the covariance exist, especially in non-stationary cases. These differences are highlighted in the following list.

- If the covariance function of the random field $X(\mathbf{s}; \omega)$ depends only on the lag (i.e., if the covariance is translation invariant), the variogram is also translation invariant according to (3.47).

[6]This is not a generally accepted term. Nonetheless, it reflects the operational reality.

- The converse proposition, however, is not true: if the variogram of the random field $X(\mathbf{s}; \omega)$ is translation invariant, the covariance is not necessarily translation invariant.
- *Non-stationary random fields with stationary increments* exhibit translation invariance of the variogram but not of the covariance. This means that the variogram depends only on the lag vector, whereas the covariance function depends on the locations of both points.
- The variogram function can be well-defined for random fields with infinite variance, in contrast with the covariance [132].
- The estimation of the variogram does not require knowledge of the mean, and thus the variogram holds an advantage over the covariance with respect to estimation. If the mean is not known a priori, its estimate based on the sample average is typically used to determine the covariance function [132, p. 32–33]. However, the sample-average may provide a biased estimate of the mean for correlated random fields, depending on the spatial sampling configuration.
- The variogram definition (3.44) involves field increments. The difference operator acts as a filter that removes "stochastic trends". This filtering is analogous to the difference operator in time series analysis [750]. In covariance estimation, however, stochastic trends need to be explicitly subtracted in order to obtain an unbiased estimate. See also the discussion in [132, pp. 32–33].
- The variogram function (more generally, the geostatistical methodology) is not often used in the analysis of linear time series, where the prevalent modeling involves the covariance function. Recent studies investigate the advantages of the variogram function in the exploratory phase of time series analysis (e.g., for the detection of non-stationarity and periodicity) and in the prediction of missing data [184, 448]. A considerable advantage of the variogram function over the covariance is that variogram estimation does not require knowledge of the expectation of the random field. Inferring the latter from dependent data introduces a bias in the covariance function estimation.

Variograms of rough fields Non-stationary random fields are typical models of rough, *self-affine* surfaces.[7] The surface height is represented as a scalar random field $h(\mathbf{s}; \omega)$ where $\mathbf{s} = (s_1, s_2) \in \mathbb{R}^2$.

In the case of self-affine surfaces, the variogram of the surface height depends only on the lag $\mathbf{r} = \mathbf{s}_1 - \mathbf{s}_2$, so that $\gamma_h(\mathbf{r}) \propto \|\mathbf{r}\|^{2H}$, whereas the height-height covariance function also depends on the locations $\mathbf{s}_1$ and $\mathbf{s}_2$ [305].

In the theory of homogeneous and isotropic turbulence, the turbulent fluid velocity is modeled as a non-stationary random fields with stationary increments. In these models, the structure function is used to describe the correlations of the intermittent velocity variations [204]. The intrinsic model is also known as a *locally homogeneous random field* in the statistical theory of turbulence [557].

[7]For the definition of self-affine random fields see Definition 4.2.

3.6 Permissibility of Covariance Functions

In Sect. 4.2 we will examine some covariance functions commonly used in spatial data modeling. One may wonder if covariance functions obey any restrictions, or whether all real-valued functions or some subsets of them can be considered as covariance models.

In physics this question does not arise as often as in spatial data analysis, because the physical laws (system of differential equations) that describe the process of interest are usually known. The covariance functions are then obtained (by means of explicit approximations or numerical calculations if the full solutions are not analytically tractable) by solving the respective equations.

In spatial data modeling, however, covariance models do not typically follow from governing equations. In many cases, a model function is simply selected from the arsenal of well-known models. More creative souls, however, might prefer to design their own flavor of covariance function. For this purpose, mathematical tools are needed that help decide whether a specific candidate function is a permissible covariance. In case this seems pedantic, keep in mind that in the applied sciences "faulty" models of covariance (or correlation) functions have been proposed, due to ignorance of the mathematical constraints.

Let us see what intuition tells us about covariance functions. Given that a covariance function represents correlations between two locations $\mathbf{s}_1$ and $\mathbf{s}_2$ separated by a distance vector $\mathbf{r}$, one can reasonably expect that (i) the covariance function takes its maximum value at zero separation, $\|\mathbf{r}\| = 0$, and (ii) that its values decline with increasing distance. (iii) We could also guess that $C_{\mathrm{xx}}(\mathbf{r}) = C_{\mathrm{xx}}(-\mathbf{r})$, since $C_{\mathrm{xx}}(\mathbf{r})$ corresponds to correlations between points 1 and 2, whereas $C_{\mathrm{xx}}(-\mathbf{r})$ to correlations between points 2 and 1. Since in both cases the same pair of points is involved, there is no reason to expect different correlation values.

Now, let us see how this intuition compares with the mathematical constraints. The following three properties are *necessary* but *not sufficient* conditions that can be proved by simple arguments.

(C1). Non-negative variance: $$C_{\mathrm{xx}}(\mathbf{0}) \geq 0. \quad (3.48)$$

(C2). Symmetry: $$C_{\mathrm{xx}}(\mathbf{r}) = C_{\mathrm{xx}}(-\mathbf{r}). \quad (3.49)$$

(C3). Mode: $$|\, C_{\mathrm{xx}}(\mathbf{r})| \leq C_{\mathrm{xx}}(\mathbf{0}). \quad (3.50)$$

The variance lower bound (3.48) is a direct consequence of (3.45). The symmetry relation (3.49) reflects the fact that we can permute the order of appearance of $\mathbf{s}_1$ and $\mathbf{s}_2$ without any impact on the covariance function, i.e.,

$$\mathcal{E}[\mathrm{X}'(\mathbf{s}_1;\omega)\,\mathrm{X}'(\mathbf{s}_2;\omega)] = \mathcal{E}[\mathrm{X}'(\mathbf{s}_2;\omega)\,\mathrm{X}'(\mathbf{s}_1;\omega)].$$

The first two relations agree well with our intuition. The third relation is based on the *Cauchy-Schwartz* inequality. In the case of two random variables, the latter is expressed as follows:

Theorem 3.1 (Cauchy-Schwartz Inequality) *If* $\mathrm{X}(\omega)$ *and* $\mathrm{Y}(\omega)$ *are scalar random variables with finite first and second-order moments, the following relation holds:*

$$\sqrt{\mathcal{E}[\mathrm{X}^2(\omega)]\,\mathcal{E}[\mathrm{Y}^2(\omega)]} \geq \mathcal{E}[\mathrm{X}(\omega)\,\mathrm{Y}(\omega)]. \tag{3.51}$$

Setting $\mathrm{X}(\omega) = \mathrm{X}'(\mathbf{s}_1;\omega)$ and $\mathrm{Y}(\omega) = \mathrm{X}'(\mathbf{s}_2;\omega)$ we obtain (3.50).

The mode inequality (3.50) sets an upper bound on the absolute value of the covariance, but it does not otherwise constrain the covariance. In particular, it does not require monotonic decay of the correlations. As we will see below, monotonic decay is neither necessary nor always desired; in fact, there exist covariance functions which have an oscillatory dependence on the lag.

Are the above constraints sufficient to ensure permissibility? For example, is the "spherical box function" $C(\mathbf{r}) = C_0\,\mathbb{I}_{\|\mathbf{r}\|\leq R}(\mathbf{r})$ a valid covariance function? This function is constant for distances less than R and discontinuously drops to zero at slightly larger values R_+. It also satisfies the conditions (C1)–(C3) above. Nonetheless, as we show below, it is not a permissible covariance function.

3.6.1 Positive Definite Functions

Covariance functions are *positive definite functions*. This is the necessary and sufficient condition. It is based on the intuitive requirement that for any number $N \in \mathbb{N}$ of locations, and for any set of real-valued coefficients $\{c_n\}_{n=1}^N$, the following expectation of the weighted sum of squares is a non-negative number

$$\mathcal{E}\left[\left\{\sum_{n=1}^{N} c_n\,\mathrm{X}'(\mathbf{s}_n;\omega)\right\}^2\right] \geq 0. \tag{3.52}$$

The inequality (3.52) can also be expressed as follows

$$\sum_{n=1}^{N}\sum_{m=1}^{N} c_n\,c_m\,\mathcal{E}\left[\mathrm{X}'(\mathbf{s}_n;\omega)\,\mathrm{X}'(\mathbf{s}_m;\omega)\right] \geq 0 \Leftrightarrow \sum_{n=1}^{N}\sum_{m=1}^{N} c_n\,c_m\,C_{\mathrm{xx}}(\mathbf{s}_n - \mathbf{s}_m) \geq 0. \tag{3.53}$$

Definition 3.3 Functions $C_{\mathrm{xx}}(\cdot)$ are called *positive definite or non-negative definite* if they satisfy the following conditions:

1. They are continuous functions.
2. They are symmetric, i.e., $C_{\mathrm{xx}}(\mathbf{r}) = C_{\mathrm{xx}}(-\mathbf{r})$.
3. They satisfy the inequality (3.53) for any $\{\mathbf{s}_n\}_{n=1}^N$ and $\{c_n\}_{n=1}^N$ and for all $N \in \mathbb{N}$.

The term "non-negative definite" (sometimes used instead of positive definite) emphasizes that the double summation in (3.53) may be equal to zero, even if not all of the c_n are zero. Functions that satisfy (3.53) without the equality sign are known as *strictly positive definite*.

Common pitfalls Some points that often cause confusion regarding the meaning of positive definiteness are the following:

1. Functions $C_{xx}(\cdot)$ that take only positive values *are not necessarily positive definite*: Negative values of the coefficients $\{c_n\}_{n=1}^{N}$ can force the linear combination of covariances in (3.53) to negative values.
2. Functions $C_{xx}(\cdot)$ that admit *negative values* are not a priori excluded. We will encounter below covariance functions that take negative values over a range of lags. In such cases, one refers to covariance functions with *"negative hole(s)"* or *hole covariance functions*. A typical example is the cardinal sine covariance function (see Table 5.1).
3. Covariance functions can be either positive (non-negative) definite or strictly positive definite. A strictly positive definite function is also positive definite, but the converse is not true. A typical example of a covariance function that is positive definite but not strictly positive definite is the *cosine covariance function* by $C_{xx}(\mathbf{r}) = \cos(\mathbf{b} \cdot \mathbf{r})$ [132, p. 84].
4. A *strictly positive definite* covariance matrix has a unique lower-upper factorization $\mathbf{C} = \mathbf{L}\mathbf{L}^\top$ where $\mathbf{L}$ is a lower triangular matrix. This factorization is known as the *Cholesky decomposition* of the matrix.

It is practically impossible to prove comprehensively condition (3.53) for any given function, since it must be established for all possible real values of $\{c_n\}_{n=1}^{N}$, for all possible configurations of N points, and for all $N \in \mathbb{N}$. Fortunately, for stationary random fields *Bochner's theorem* reduces the constraint to a much simpler condition that can be easily tested [83].

The presentation of Bochner's theorem requires the mathematical toolbox of Fourier transforms. Below we review the definition of the Fourier transform pair for functions $C_{xx} : \mathbb{R}^d \to \mathbb{R}$. However, for a deeper understanding of Fourier transform and spectral methods in general, the reader is advised to consider the following sources: [83, 96, 296, 733] and [673, Chaps. 12–13].

Remark The non-negativity of the weighted sum of squares, i.e., the inequality (3.52), also holds for complex-valued coefficients $c_n \in \mathbb{C}$, $n = 1, \ldots, N$. In this case, the inequality is expressed as follows

$$\mathcal{E}\left[\left\| \sum_{n=1}^{N} c_n \, \mathrm{X}'(\mathbf{s}_n; \omega) \right\|^2\right] \geq 0,$$

where $\|z\|$ denotes the magnitude (absolute value) of the complex number z.

3.6.2 Fourier Transforms in a Nutshell

Let $C_{xx}(\mathbf{r})$ represent a function defined over the space of lag vectors $\mathbf{r} \in \mathbb{R}^d$. The Fourier transform, $\widetilde{C}_{xx}(\mathbf{k})$, maps $C_{xx}(\mathbf{r})$ to the function $\widetilde{C}_{xx}(\mathbf{k})$ that is defined over the *reciprocal (Fourier) space* of wavevectors $\mathbf{k} \in \mathbb{R}^d$. For a stationary SRF, let the *Fourier transform* of the covariance function be defined by means of the following improper[8] multiple integral:

$$\widetilde{C_{xx}}(\mathbf{k}) := \mathcal{F}[C_{xx}(\mathbf{r})] = \int_{\mathbb{R}^d} d\mathbf{r}\; e^{-i\mathbf{k}\cdot\mathbf{r}} C_{xx}(\mathbf{r}), \tag{3.54}$$

where the volume differential $d\mathbf{r}$ is given in Cartesian coordinates by means of

$$\int_{\mathbb{R}^d} d\mathbf{r} = \prod_{i=1}^{d} \int_{-\infty}^{\infty} dr_i .$$

By virtue of reflection symmetry, i.e., $C_{xx}(\mathbf{r}) = C_{xx}(-\mathbf{r})$, the integral (3.54) can also be expressed by means of the *cosine Fourier transform* as follows:

$$\widetilde{C_{xx}}(\mathbf{k}) = \int_{\mathbb{R}^d} d\mathbf{r}\; \cos(\mathbf{k}\cdot\mathbf{r})\, C_{xx}(\mathbf{r}). \tag{3.55}$$

The *inverse Fourier transform* is given by means of the improper multiple integral

$$C_{xx}(\mathbf{r}) := \mathcal{F}^{-1}[\widetilde{C_{xx}}(\mathbf{k})] = \frac{1}{(2\pi)^d} \int_{\mathbb{R}^d} d\mathbf{k}\; e^{i\mathbf{k}\cdot\mathbf{r}}\, \widetilde{C_{xx}}(\mathbf{k}). \tag{3.56}$$

In light of $C_{xx}(\mathbf{0}) = \sigma_x^2$, the *variance* of the random field is given by the integral of the spectral density, i.e.,

$$\sigma_x^2 = \frac{1}{(2\pi)^d} \int_{\mathbb{R}^d} d\mathbf{k}\; \widetilde{C_{xx}}(\mathbf{k}). \tag{3.57}$$

Existence The Fourier integral (3.54) exists if the covariance function $C_{xx}(\mathbf{r})$ is *absolutely integrable*. This requires that $C_{xx}(\mathbf{r})$ be *integrable* and *decay to zero sufficiently rapidly at infinity*, so that the spatial integral of $|C_{xx}(\mathbf{r})|$ exists. Then, the spectral density is a *continuous function* in $\mathbf{k} \in \mathbb{R}^d$ and satisfies the following upper bound [779]:

$$\widetilde{C}_{xx}(\mathbf{k}) \le \int_{\mathbb{R}^d} d\mathbf{r}\; |\, C_{xx}(\mathbf{r})\, | \,.$$

[8]The term "improper" denotes that the integration limits extend to infinity.

Riemann-Lebesgue lemma This lemma asserts that if $C_{xx}(\mathbf{r})$ is integrable and the integral of its absolute value also exists (is finite), then the spectral density tends to zero as the wavenumber tends to infinity [779, p. 785]:

$$\lim_{\|\mathbf{k}\|\to\infty} \widetilde{C}_{xx}(\mathbf{k}) = 0.$$

Similar conditions to the above also apply to the spectral density, $\widetilde{C_{xx}}(\mathbf{k})$, in order to ensure the existence of the inverse Fourier transform. The existence of the absolute value integral is *sufficient* but not *necessary*. Hence, it is possible for the Fourier transform to exist in cases where $C_{xx}(\mathbf{r})$ is integrable (in real space) while $|C_{xx}(\mathbf{r})|$ is not. In such cases the Riemann-Lebesgue lemma does not hold, and the Fourier transform does not necessarily decay to zero at infinity.

Symmetry Since $C_{xx}(\cdot)$ is a real-valued function, its symmetry in real space translates into a symmetric spectral density in reciprocal space. Hence, the integral (3.56) that defines the inverse Fourier transform can be replaced by the *inverse cosine Fourier transform*

$$C_{xx}(\mathbf{r}) = \frac{1}{(2\,\pi)^d} \int_{\mathbb{R}^d} d\mathbf{k}\, \cos(\mathbf{k}\cdot\mathbf{r})\, \widetilde{C_{xx}}(\mathbf{k}). \tag{3.58}$$

Differential operators in spectral domain One advantage of reciprocal space representations is that differential operators in real space transform into algebraic expressions in reciprocal space. Real-space partial derivatives transform in reciprocal space as follows

$$\frac{\partial}{\partial r_j} \xrightarrow{\mathcal{F}} \mathrm{i}\,k_j, \text{ for } j = 1, \ldots, d. \tag{3.59a}$$

Based on the above, the following relations are obtained for the gradient, the Laplacian, and the biharmonic derivatives

$$\nabla \xrightarrow{\mathcal{F}} \mathrm{i}\,\mathbf{k}, \tag{3.59b}$$

$$\nabla^2 \xrightarrow{\mathcal{F}} -\|\mathbf{k}\|^2, \tag{3.59c}$$

$$\nabla^4 \xrightarrow{\mathcal{F}} \|\mathbf{k}\|^4. \tag{3.59d}$$

How to calculate Fourier Transforms Fourier transforms involve multiple integrals that may look daunting at first. In certain cases these integrals can be analytically calculated. We then obtain explicit expressions for the spectral density that are very useful for testing the permissibility of covariance functions and for spectral simulation methods (see Chap. 16). Alternatively, we may first design a permissible spectral density and then determine its inverse Fourier transform to obtain the corresponding covariance function. A good place to learn about *multi-dimensional Fourier transforms* is the book by Laurent Schwartz [733].

In certain cases Fourier transform calculations can be simplified. To accomplish this task we often take advantage of symmetries respected by the integrand. Some examples are shown below.

Fourier Transform of separable functions A separable function can be expressed as a product of one-dimensional functions defined along orthogonal coordinate axes. For a *separable covariance* function of the form

$$C_{\mathrm{xx}}(\mathbf{r}) = \prod_{i=1}^{d} C_i(r_i),$$

the multiple Fourier integral (3.54) is decomposed into a product of one-dimensional Fourier integrals, i.e.,

$$\widetilde{C_{\mathrm{xx}}}(\mathbf{k}) = \prod_{i=1}^{d} \left[\int_{-\infty}^{\infty} \mathrm{d}r_i \, \mathrm{e}^{-\mathrm{i}k_i r_i} \, C_i(r_i) \right] = \prod_{i=1}^{d} \widetilde{C}_i(k_i).$$

Fourier Transform of radial functions Another case that yields to simplification involves radial functions that satisfy $C_{\mathrm{xx}}(\mathbf{r}) = C_{\mathrm{xx}}(\|\mathbf{r}\|)$. In the spherical coordinate system the integral (3.54) can be expressed as

$$\widetilde{C_{\mathrm{xx}}}(\mathbf{k}) = \int_0^{\infty} \mathrm{d}x \, x^{d-1} \, C_{\mathrm{xx}}(x) \int_{\mathcal{B}_d} \mathrm{d}\Omega_d(\hat{\mathbf{r}}) \, \mathrm{e}^{-\mathrm{i}\|\mathbf{k}\| \, x \cos\theta},$$

where $\mathcal{B}_d$ is the surface of the unit sphere in d dimensions, $\mathrm{d}\Omega_d(\hat{\mathbf{r}})$ is the differential of the solid angle subtended by the unit vector $\hat{\mathbf{r}}$, $x = \|\mathbf{r}\|$ is the spatial lag, $\|\mathbf{k}\|$ is the magnitude of the wavevector (wavenumber), and θ is the angle between $\hat{\mathbf{r}}$ and $\mathbf{k}$.

The *unit vector* $\hat{\mathbf{r}}$ in d dimensions is determined by $d-1$ *orientation angles*, $(\theta_1, \ldots, \theta_{d-1})$, where $0 \leq \theta_i \leq \pi$ for $i = 1, \ldots, d-2$, and $0 \leq \theta_{d-1} \leq 2\pi$. The integral over the solid angle in d dimensions is expressed in terms of the orientation angles as follows [593, p. 68]

$$\int_{\mathcal{B}_d} \mathrm{d}\Omega_d(\hat{\mathbf{r}}) = \int_0^{2\pi} \mathrm{d}\theta_1 \int_0^{\pi} \mathrm{d}\theta_2 \sin\theta_2 \ldots \int_0^{\pi} \mathrm{d}\theta_{d-2} \sin^{d-3}\theta_{d-2} \times \int_0^{\pi} \mathrm{d}\theta_{d-1} \sin^{d-2}\theta_{d-1}. \tag{3.60}$$

Thus, the integral of $\exp(-\mathrm{i}\|\mathbf{k}\|\, x \cos\theta)$ over the solid angle can be analytically evaluated [733, p. 200], leading to the expressions of the isotropic spectral representation (4.4a) and (4.4b) below.

Cartesian to spherical coordinates It is useful to know how the Cartesian coordinates of a *unit vector* $\hat{\mathbf{r}} = (\hat{r}_1, \ldots, \hat{r}_d)^\top$ are expressed in terms of the orientation angles $(\theta_1, \ldots, \theta_{d-1})$ in a spherical coordinate system. The transformation is accomplished by means of the following formulas [662]:

$$\begin{aligned}
\hat{r}_1 &= \cos\theta_1, \\
\hat{r}_2 &= \sin\theta_1 \cos\theta_2, \\
\hat{r}_3 &= \sin\theta_1 \sin\theta_2 \cos\theta_3, \\
&\vdots \\
\hat{r}_{d-1} &= \sin\theta_1 \sin\theta_2 \sin\theta_3 \ldots \sin\theta_{d-2} \cos\theta_{d-1}, \\
\hat{r}_d &= \sin\theta_1 \sin\theta_2 \sin\theta_3 \ldots \sin\theta_{d-2} \sin\theta_{d-1}.
\end{aligned} \tag{3.61}$$

In three dimensions ($d = 3$) it holds that $\theta_1 = \theta$, where θ is the *polar angle*, while $\theta_2 = \phi$, where ϕ is the *azimuthal angle*.

Volume integral of radial functions The d-dimensional volume integral of radial functions $f(\|\mathbf{k}\|)$ in reciprocal space, where $\mathbf{k} \in \mathbb{R}^d$, is calculated as follows:

$$\int_{\mathbb{R}^d} \mathrm{d}\mathbf{k}\, f(\|\mathbf{k}\|) = S_d \int_0^\infty \mathrm{d}k\, k^{d-1}\, f(k), \tag{3.62a}$$

where

$$S_d = \int_{\mathcal{B}_d} \mathrm{d}\Omega_d(\hat{\mathbf{r}}) = \frac{2\pi^{d/2}}{\Gamma(d/2)}, \tag{3.62b}$$

is the *surface area of the unit "hypersphere"* in d dimensions [733, p. 39], and $\Gamma(\cdot)$ is the *Gamma function*.

- In $d = 2$, equation (3.62b) yields $S_2 = 2\pi$, i.e., the circumference of the unit circle.
- In $d = 3$, it gives $S_3 = 4\pi$, i.e., the surface area of the unit sphere.
- Finally, in $d = 1$ it gives $S_1 = 2$ which is the number of endpoints of the real line.

Orthonormality of the Fourier basis The Fourier basis comprises plane-wave functions that are indexed (labeled) with respect to the wavevector $\mathbf{k}$ as follows

$$u_{\mathbf{k}}(\mathbf{r}) = \frac{e^{\mathrm{i}\mathbf{k}\cdot\mathbf{r}}}{(2\pi)^{d/2}}. \tag{3.63}$$

The Fourier basis is orthonormal with respect to the wavevectors, i.e.,

$$\int_{\mathbb{R}^d} d\mathbf{r}\, u_{\mathbf{k}}(\mathbf{r})\, u^{\dagger}_{\mathbf{k}'}(\mathbf{r}) = \frac{1}{(2\pi)^d} \int_{\mathbb{R}^d} d\mathbf{r}\, e^{i\,(\mathbf{k}-\mathbf{k}')\cdot\mathbf{r}} = \delta(\mathbf{k}-\mathbf{k}'). \tag{3.64}$$

The dagger $\dagger$ as superscript denotes the *complex conjugate.* The Dirac delta function in d-dimensional reciprocal space is the product of one-dimensional Dirac functions, i.e.,

$$\delta(\mathbf{k}-\mathbf{k}') = \prod_{i=1}^{d} \delta(k_i - k_i').$$

Remarks Fourier basis functions represent harmonic waves that extend throughout space with characteristic spatial frequencies determined by the wavevectors $\mathbf{k}$. Larger values of the wavenumber $\|\mathbf{k}\|$ correspond to basis functions that oscillate faster in space.

Orthonormality means that if we take two functions from the Fourier basis, replace one of them with its complex conjugate, and integrate their product over all space, the integral is non-zero only if the basis function labels are identical.

Closure of Fourier basis In addition to being orthonormal, the Fourier basis is also closed. The closure property is expressed in terms of the following integral

$$\int_{\mathbb{R}^d} d\mathbf{k}\, u_{\mathbf{k}}(\mathbf{r})\, u^{\dagger}_{\mathbf{k}}(\mathbf{r}') = \frac{1}{(2\pi)^d} \int_{\mathbb{R}^d} d\mathbf{k}\, e^{i\,(\mathbf{r}-\mathbf{r}')\cdot\mathbf{k}} = \delta(\mathbf{r}-\mathbf{r}'). \tag{3.65}$$

Remarks The integrals in (3.64) and (3.65) appear similar. However, the orthonormality condition involves a summation over space, while the closure involves a summation over the function labels (wavevectors). Both conditions are expressed as integrals because we consider an unbounded, continuum domain. In contrast, for functions defined on an infinite lattice $\mathcal{G}$, the spatial integral in (3.64) is replaced by a discrete sum over all lattice vectors.

Most environmental and energy resources data are defined on continuum spaces instead of regular lattices. Hence, it makes sense to focus in continuum representations. In addition, the domain size is usually significantly larger than the characteristic length of the fluctuations; this justifies the infinite domain assumption. On the other hand, crystalline materials and digital images are defined on regular lattice structures, and the discrete nature of the support should be accounted for in calculations of the Fourier transform [122].

3.6.3 Bochner's Theorem

Bochner's theorem is a cornerstone of random field modeling. The theorem provides conditions for testing the permissibility of candidate covariance functions in the spectral domain [83], [863, p.106].

Physicists are more familiar with the *Wiener-Khinchin* theorem that gives meaning to the spectral density of random processes [646, p. 416], [424]. The Wiener-Khinchin theorem embodies the idea that—for stationary random processes—the power spectrum (i.e., the spectral density) is equivalent to the covariance function [pp. 17–21] [471].

Theorem 3.2 *Bochner's permissibility theorem states that a real-valued, bounded and continuous function* $C_{xx}(\mathbf{r})$ *is a valid covariance function of a second-order stationary random field, if and only if it possesses a symmetric, non-negative, and integrable Fourier transform* $\widetilde{C_{xx}}(\|\mathbf{k}\|)$. *More specifically, the following conditions should hold:*

1. Symmetry:

$$\widetilde{C_{xx}}(\mathbf{k}) = \widetilde{C_{xx}}(-\mathbf{k}), \tag{3.66}$$

2. Non-negativity:

$$\widetilde{C_{xx}}(\mathbf{k}) \geq 0, \text{ for all } \mathbf{k} \in \mathbb{R}^d, \tag{3.67}$$

3. Bounded variation:

$$\int_{\mathbb{R}^d} \mathrm{d}\mathbf{k}\, \widetilde{C_{xx}}(\mathbf{k}) < \infty. \tag{3.68}$$

Sketch of the proof: The symmetry of the spectral density ensures real-space symmetry, i.e., $C_{xx}(\mathbf{r}) = C_{xx}(-\mathbf{r})$ which is a necessary condition for positive definiteness (see Definition 3.3). The bounded variation condition ensures that the SRF variance exists. Finally, the non-negativity condition enforces the inequality (3.53). It is easy to intuitively understand this if we express the weighed double summation of the covariance over the sites $\{\mathbf{s}_n\}_{n=1}^{N}$ as follows

$$\sum_{n=1}^{N}\sum_{m=1}^{N} c_n\, c_m\, C_{xx}(\mathbf{s}_n - \mathbf{s}_m) = \sum_{n=1}^{N}\sum_{m=1}^{N} c_n\, c_m \int_{\mathbb{R}^d} \frac{\mathrm{d}\mathbf{k}}{(2\pi)^d}\, \widetilde{C}_{xx}(\mathbf{k})\, \mathrm{e}^{\mathrm{i}\,\mathbf{k}\cdot(\mathbf{s}_n - \mathbf{s}_m)}$$

$$= \int_{\mathbb{R}^d} \frac{\mathrm{d}\mathbf{k}}{(2\pi)^d}\, \widetilde{C}_{xx}(\mathbf{k}) \sum_{n=1}^{N}\sum_{m=1}^{N} c_n\, c_m\, \mathrm{e}^{\mathrm{i}\,\mathbf{k}(\mathbf{s}_n - \mathbf{s}_m)}.$$

If we define the complex-valued function $z(\mathbf{k}) = \sum_{n=1}^{N} c_n\, \mathrm{e}^{\mathrm{i}\mathbf{k}\cdot\mathbf{s}_n}$, it follows that

$$\sum_{n=1}^{N}\sum_{m=1}^{N} c_n\, c_m\, \mathrm{e}^{\mathrm{i}\,\mathbf{k}(\mathbf{s}_n - \mathbf{s}_m)} = z(\mathbf{k})\, z^{\dagger}(\mathbf{k}),$$

where $z^{\dagger}(\mathbf{k})$ is the complex conjugate of $z(\mathbf{k})$. Hence, $z(\mathbf{k})\, z^{\dagger}(\mathbf{k}) = \|z(\mathbf{k})\|^2 \geq 0$. In fact, the equality sign is only valid if $c_n = 0$ for all $n = 1, \ldots, N$. Hence, we obtain

$$\sum_{n=1}^{N}\sum_{m=1}^{N} c_n\, c_m\, C_{xx}(\mathbf{s}_n - \mathbf{s}_m) = \int_{\mathbb{R}^d} \frac{\mathrm{d}\mathbf{k}}{(2\pi)^d}\, \widetilde{C}_{xx}(\mathbf{k})\, \|z(\mathbf{k})\|^2.$$

Using the triangle inequality, i.e., $\|\sum_{n=1}^{N} z_n\| \le \sum_{n=1}^{N} \|z_n\|$ for any $z_n \in \mathbb{C}$, and the fact that

$$\|c_n \exp(\mathrm{i}\,\mathbf{k}\cdot\mathbf{s}_n)\| = \|c_n\|\,\|\exp(\mathrm{i}\,\mathbf{k}\cdot\mathbf{s}_n)\| = \|c_n\|,$$

it follows that $\|z(\mathbf{k})\| \le \sum_{n=1}^{N} \|c_n\| = A$, where $A \ge 0$. Then, if $\widetilde{C}_{\mathrm{xx}}(\mathbf{k}) \ge 0$ for all $\mathbf{k}$ it holds that $\widetilde{C}_{\mathrm{xx}}(\mathbf{k})\,\|z(\mathbf{k})\|^2 \le A\,\widetilde{C}_{\mathrm{xx}}(\mathbf{k})$ for all $\mathbf{k}$. Thus, the spectral integral converges, because it is bounded from above by

$$A \int_{\mathbb{R}^d} \frac{\mathrm{d}\mathbf{k}}{(2\pi)^d}\,\widetilde{C}_{\mathrm{xx}}(\mathbf{k}).$$

The above integral corresponds to a finite value, due to the bounded variation of $\widetilde{C}_{\mathrm{xx}}(\mathbf{k})$. Thus, the function $C_{\mathrm{xx}}(\mathbf{r})$ is positive definite if $\widetilde{C}_{\mathrm{xx}}(\mathbf{k}) \ge 0$ for all $\mathbf{k}$ and strictly positive definite if $\widetilde{C}_{\mathrm{xx}}(\mathbf{k}) > 0$ for all $\mathbf{k}$.

Based on the above, it is sufficient to calculate the Fourier transform of a candidate covariance function in order to determine its permissibility as stationary covariance. The calculation of the Fourier transform can be performed by analytical or numerical means,[9] and it is definitely simpler than the task of checking the infinite combinations involved in (3.53).

Strict positive-definiteness If $\widetilde{C_{\mathrm{xx}}}(\mathbf{k}) > 0$ for all $\mathbf{k} \in \mathbb{R}^d$, the covariance function is *strictly positive definite*. If $\mathbf{A}$ is a covariance matrix constructed for a particular point set from a strictly positive definite covariance function, the inverse (i.e, the precision matrix) $\mathbf{A}^{-1}$ exists. As we shall see in Chap. 10, the existence of the precision matrix is necessary for the respective kriging system to have a unique solution.

The cosine covariance As a counterexample of a function that is positive definite but not strictly positive definite, consider the cosine function $\cos(\mathbf{b}\cdot\mathbf{s})$: its Fourier transform is given by the symmetric combination of two delta function peaks, i.e.,

$$\mathcal{F}[\cos(\mathbf{b}\cdot\mathbf{r})] = \frac{1}{2}\left[\delta(\mathbf{k}-\mathbf{b}) + \delta(\mathbf{k}+\mathbf{b})\right], \tag{3.69}$$

where $\delta(\cdot)$ is the Dirac delta function. The above FT satisfies the conditions of Theorem 3.2 and is thus a permissible covariance function. However, since the FT takes zero values for $\mathbf{k} \neq \pm\mathbf{b}$, the cosine function is not strictly positive definite.

Note that the cosine covariance cannot be used for log-Gaussian random fields [724]. This is due to the fact that permissibility implies that the function is an admissible covariance for some random field but not for every random field (i.e., for all probability distributions) [555]. For example, commonly used covariance models for Gaussian random fields are not necessarily admissible for indicator random fields [217].

[9] In the case of numerical calculation, special care is needed in the evaluation of the spectral density at high wavenumbers due to very fast oscillations of the integrand.

Caveat emptor The permissibility of a given covariance model is established with respect to a specific distance metric. In most applications that involve spatial data the Euclidean distance is used. However, there are cases in which the Euclidean distance is inappropriate.

For example, in spatial processes over networks, the Euclidean metric is not a suitable measure of distance between nodes. A more suitable measure of distance on networks is the *geodesic distance*, also known as the shortest distance. This is the length of the shortest path between two nodes of the network [606].

For spatial processes on the surface of the sphere, the great circle distance is the most suitable measure of distance [669]. For earth science applications there is also interest in formulating *non-Euclidean* distance metrics that incorporate information about landscape characteristics [518]. However, the fractional error incurred by using chordal distance as opposed to great circle distance to measure the distance of two points on the sphere is $1 - 2\sin(\theta/2)/\theta$, where θ is the angle formed by the radii passing through the respective locations. For most applications of geostatistical interest the error is small and does not warrant the use of great circle distances.

Permissibility of a specific covariance model based on the Euclidean distance does not ensure the permissibility of the model for a different distance metric [142]. Thus, the fact that the exponential function $\exp(-\|\mathbf{r}\|)$ is permissible if $\|\mathbf{r}\|$ is the Euclidean distance does not guarantee that the exponential is permissible if $\|\mathbf{r}\|$ is a different distance measure.

3.6.4 Wiener-Khinchin Theorem

We will now discuss how to define the spectral density of random processes and the Wiener-Khinchin theorem. First, we will recall some facts pertaining to the spectral content of deterministic functions in order to highlight the differences between deterministic and random functions.

Deterministic functions A "well-behaved" function $x(t)$ has a well-defined *Fourier transform* if the absolute value $|x(t)|$ is integrable. This condition implies that $x(t)$ decays faster than $1/|t|$ as $t \to \pm\infty$. The term "well-behaved" means that $x(t)$ has at most a finite number of extrema and discontinuities over any finite interval.

In contrast with the above, *fully extended, periodic* functions (e.g., sine and cosine) have a Fourier transform that is only defined in terms of *generalized functions* [96]. A typical example of a generalized function is the Dirac delta function $\delta(t - t')$. Generalized functions can be constructed from well-behaved functions $f_\sigma(t)$ indexed by a parameter σ at the limit $\sigma \to \infty$ or $\sigma \to 0$. For example, the delta function can be defined as the limit of a Gaussian function at the limit of extremely small variance, i.e.,

$$\delta(w) = \lim_{\sigma \to 0} \frac{1}{\sqrt{2\pi}\,\sigma} e^{-w^2/(2\sigma^2)}.$$

The delta function concentrates the density at a single point in the spectral domain. This makes sense for harmonic functions that oscillate with a single characteristic frequency—refer to the cosine function FT in (3.69).

Random processes Next, consider a random process $X(t; \omega)$: unlike "well-behaved" functions, the realizations $x(t; \omega)$ do not decay nicely at the limit $t \to \infty$. Hence, the existence of the Fourier transform of the realizations is questionable. Nonetheless, in non-rigorous treatments we pretend that the Fourier transform exists. After all, if the random process realizations do not contain singularities and more than a finite number of discontinuities, we can evaluate the respective Fourier integral over any bounded, however large, domain.

In fact, The Wiener-Khinchin theorem shows that the spectral density of the random process $X(t; \omega)$ can be obtained by first evaluating the Fourier integral over a finite-size domain and then taking the limit as the domain size goes to infinity. If the covariance function satisfies suitable conditions, this limiting process leads to a well-defined spectral density.

Theorem 3.3 *Wiener-Khinchin theorem: Let $C_{XX}(\tau)$ represent the auto-covariance function of a second-order stationary random process $X(t; \omega)$. Furthermore, let us assume that $C_{XX}(\tau)$ is absolutely integrable to ensure the existence of its Fourier transform.*

Let us then define the truncated spectral density

$$S_{XX;T}(w) = \frac{1}{T}\mathcal{E}\left[\left|\int_{-T/2}^{T/2} dt\, e^{-iwt}\, X(t; \omega)\right|^2\right],$$

where w is the angular frequency. The Wiener-Khinchin theorem states that:

1. *The limit $\lim_{T\to\infty} S_{XX;T}(w)$ exists and represents the spectral power density of the random process $X(t; \omega)$.*
2. *The spectral power density is the Fourier transform of the auto-covariance function $C_{XX}(\tau)$.*

Remarks on terminology (i) We use the term "auto-covariance" instead of the simpler term "covariance" to comply with the literature on random processes. (ii) The symbol ω is often used for the angular frequency. This notation, however, clashes with our use of ω as a state index. Thus, we opt to use w for the angular frequency.

The Wiener-Khinchin theorem enables the use of spectral analysis for random functions that are not square integrable, and therefore do not possess a Fourier transform. This is accomplished by restricting the Fourier transform over the compact support $[-T/2, T/2]$. The Wiener-Khinchin theorem forms the basis for the spectral representation of the covariance function of random fields [863].

3.6.5 Covariance Function Composition

It is also possible to build covariance functions based on simpler covariance models. The following three properties are particularly useful in this context.

Superposition: If the functions $\{C_n(\mathbf{r})\}_{n=1}^N$ are permissible covariance functions and $\{\alpha_n\}_{n=1}^N$ is a set of positive coefficients, then the superposition

$$C(\mathbf{r}) = \sum_{n=1}^{N} \alpha_n \, C_n(\mathbf{r})$$

is a permissible covariance function.

Product composition: If each function in the set $\{C_n(\mathbf{r})\}_{n=1}^N$ is a permissible covariance function, then the product

$$C(\mathbf{r}) = \prod_{n=1}^{N} C_n(\mathbf{r})$$

is also a permissible covariance function.

Subspace permissibility: If the function $C(\mathbf{r})$ is a permissible covariance function of a random field $X(\mathbf{s}; \omega)$ where $\mathbf{s} \in \mathbb{R}^d$, then $C(\mathbf{r})$ is a also a permissible covariance for a random field $X(\mathbf{s}'; \omega)$ where $\mathbf{s}' \in \mathbb{R}^{d'}$ and $d' < d$.

Subspace permissibility can also be proved by considering the definition of positive definite functions (3.52) and the permissibility condition (3.53). If the latter is satisfied for all possible sets of points in $\mathbb{R}^d$, then it is also satisfied by all subsets of these points which lie in lower-dimensionality subspaces; hence, the covariance function is permissible in the lower-dimensional subspace.

Vertical rescaling: This method leads to straightforward construction of non-stationary covariance functions from stationary ones. In particular, if $C_{xx}(\mathbf{s}, \mathbf{s}')$ is a valid covariance function (stationary or non-stationary) and $\alpha(\mathbf{s})$ is any real-valued function $\mathbb{R}^d \to \mathbb{R}$, then the function

$$K(\mathbf{s}, \mathbf{s}') = \alpha(\mathbf{s}) \, C_{xx}(\mathbf{s}, \mathbf{s}') \, \alpha(\mathbf{s}'), \tag{3.70}$$

is also a valid *non-stationary covariance* function [520]. The function $K(\mathbf{s}, \mathbf{s}')$ is non-stationary even if $C_{xx}(\mathbf{s}-\mathbf{s}')$ is a stationary covariance function. The variance of $K(\mathbf{s}, \mathbf{s}')$ is given by $\sigma_K^2 = \sigma_x^2 \, \alpha^2(\mathbf{s})$.

Warping (embedding): In this case, the physical space is embedded in a general vector space known as the feature space. Let $C(\mathbf{u}, \mathbf{u}') : \mathbb{R}^M \to \mathbb{R}$, where $\mathbf{u}$

is an M-dimensional vector in some suitably defined *feature space*, be a valid covariance function. In addition, assume that $\mathbf{s} \mapsto \mathbf{u}(\mathbf{s}) : \mathbb{R}^d \to \mathbb{R}^M$ represents a nonlinear transformation, not necessarily invertible, from the d-dimensional Euclidean space to the feature space. Then, the function

$$K(\mathbf{s}, \mathbf{s}') = C\left[\mathbf{u}(\mathbf{s}), \mathbf{u}(\mathbf{s}')\right], \tag{3.71}$$

is a valid covariance function in d dimensional space [520]. The dimensions d of the real space and M of the feature space do not need to be equal.
The covariance function $K(\mathbf{s}, \mathbf{s}')$ is *non-stationary*, even if the original covariance $C(\mathbf{u}, \mathbf{u}')$ is stationary. However, if the original covariance function $C(\mathbf{u}-\mathbf{u}')$ is *stationary*, the non-stationary covariance $K(\mathbf{s}, \mathbf{s}')$ obtained by warping has *uniform variance*, since $\sigma^2_{\mathrm{K}}(\mathbf{s}) = C(\mathbf{0})$.

Separability: If the real-valued functions $\{C_i(x)\}_{i=1}^d$ are permissible one-dimensional covariance functions, the product

$$C(\mathbf{r}) = \prod_{i=1}^{d} C_i(r_i)$$

is a permissible covariance function for a random field $X(\mathbf{s}; \omega)$ where $\mathbf{s} \in \mathbb{R}^d$. Moreover, since in one dimension it is possible to construct explicit *Karhunen-Loève expansions* (see Chap. 16) for certain covariance models [286, 379], separable covariance functions extend the benefits of explicit dimensional reduction to higher dimensional spaces as well. The assumption of separability, however, is a mathematical convenience that is not always justified by physical considerations.

The interested readers can prove the above statements by considering the Fourier transforms of the composite functions and by invoking Bochner's theorem. More properties of covariance functions can be found in [132, 138, 863].

Separability in Laplace equation In the case of solutions $X(\mathbf{s})$ of deterministic *partial differential equations* (PDE), the *method of separation of variables* can be applied, if the structure of the PDE allows decoupling the spatial dependence along orthogonal directions [240].

A case in point is the *Laplace equation*, $\nabla^2 X(\mathbf{s}) = 0$, which we discussed in Chap. 2. Note that in this problem $X(\mathbf{s})$ is a deterministic function, unless the boundary conditions include randomness. In an orthogonal Cartesian coordinate system the Laplace equation is expressed as follows:

$$\sum_{i=1}^{d} \frac{\partial^2 X(\mathbf{s})}{\partial s_i^2} = 0.$$

Assuming that $X(\mathbf{s}) \neq 0$ everywhere in the domain of interest, we can write the following trial solution

$$X(\mathbf{s}) = \prod_{i=1}^{d} X^{(i)}(s_i).$$

Then, if we insert the trial solution in the Laplace equation and divide all terms by $X(\mathbf{s})$, which is allowed since $X(\mathbf{s}) \neq 0$, for all $\mathbf{s}$, the Laplace PDE transforms into

$$\sum_{i=1}^{d} \frac{1}{X^{(i)}(s_i)} \frac{\partial^2 X^{(i)}(\mathbf{s})}{\partial s_i^2} = 0.$$

Each of the d terms in the above equation depends solely on a single coordinate s_i. Hence, the Laplace PDE becomes equivalent to the following system of one-dimensional ordinary differential equations (ODEs)

$$\frac{\partial^2 X^{(i)}(\mathbf{s})}{\partial s_i^2} = \begin{cases} c_i\, X^{(i)}(s_i), \; i = 1, \ldots, d-1 \\ -\left(\sum_{n=1}^{d-1} c_i\right) X^{(i)}(s_i), \; i = d, \end{cases}$$

where the c_i are *eigenvalues* determined by the boundary conditions. The solutions of the ODEs are *eigenfunctions* $X^{(i)}(s_i)$ given by harmonic or hyperbolic functions as determined by the sign of the respective eigenvalue, c_i (negative sign implies harmonic terms, while positive sign implies hyperbolic terms).

The ODEs typically admit a set of eigenvalues and associated eigenfunctions. The full solution consists of a superposition of the eigenfunctions, and is thus a non-separable function constructed by the superposition of separable functions. The interested reader can find solutions of the Laplace equation applied to electrostatic problems for different geometries and boundary conditions in the classic book by Jackson [401].

Practical use of permissibility properties The *superposition property* is used to construct covariance functions that have multiple correlation scales, or at least one long and one short correlation scales that correspond to different spatial structures. An uncorrelated (nugget) term is often also included to account for purely random (uncorrelated) fluctuations [303].

Covariance models that involve several spatial scales are referred to in the statistical literature as *nested models* [272, p. 499]. Typically, such models may comprise nugget, fine-scale dependence, and long-range components. For example, the following equation is a nested structure with nugget term, small-scale exponential variogram (length ξ_1) and large-scale exponential variogram (length ξ_2):

$$\gamma_{\mathrm{xx}}(\mathbf{r}) = c_0 \mathbb{I}_{(0,\infty)}(\|\mathbf{r}\|) + c_1 \left[1 - \exp\left(-\|\mathbf{r}\|/\xi_1\right)\right] + c_2 \left[1 - \exp\left(-\|\mathbf{r}\|/\xi_2\right)\right], \tag{3.72}$$

where c_0, c_1, c_2 are positive constants, while $\mathbb{I}_{(0,\infty)}(\|\mathbf{r}\|) = 1$ if $\|\mathbf{r}\| > 0$ and $\mathbb{I}_{(0,\infty)}(\|\mathbf{r}\|) = 0$ if $\|\mathbf{r}\| = 0$. The indicator function $\mathbb{I}_{(0,\infty)}(\|\mathbf{r}\|)$ implements the nugget discontinuity by adding to the variogram the term c_0 for lags that are infinitesimally larger than zero.

In nested models, different components can have different continuity and differentiability properties. For example, the nested variogram (3.72) contains a discontinuous term (the nugget), and two continuous but non-differentiable terms (the exponentials) with different correlation lengths.

Product composition is used to construct covariance functions with multiplicative components. This situation occurs if

$$X(\mathbf{s};\omega) = X_1(\mathbf{s};\omega)\, X_2(\mathbf{s};\omega), \tag{3.73}$$

where $X_1(\mathbf{s};\omega)$ and $X_2(\mathbf{s};\omega)$ are independent, zero-mean random fields.

For example, let us consider that the random field product (3.73) is a model of rainfall: The binary field $X_1(\mathbf{s};\omega)$ may be used to represent the *occurrence* of a rainfall event, while the continuum field $X_2(\mathbf{s};\omega)$ may be used to model the *intensity of the event* [464]. The composite field $X(\mathbf{s};\omega)$ is equal to zero if there is no precipitation at $\mathbf{s}$, and equal to the intensity of rainfall otherwise.

A different approach for modeling precipitation is based on the theory of point processes. For example, a *non-homogeneous Poisson process* can be used to model the occurrence of events. A locally varying intensity function $\lambda(\mathbf{s})$, typically modeled as a lognormal random field, can then be used to represent the intensity [202, 581].

Yet a different approach for modeling precipitation employs a single, latent Gaussian random field. The latent field is cut at some threshold level that is determined from the available data. If the value of the latent field at a certain point in space is below the threshold there is no precipitation. On the other hand, a value of the field that is above the threshold marks the presence of precipitation and also determines the intensity of the event [52].

Separability is a "cheap" method for constructing multidimensional covariance functions based on one-dimensional covariance functions. Separable covariance functions have certain computational advantages (e.g., in simulation studies). For example, the separability assumption is used to obtain explicit expressions for the Karhunen-Loève expansion in higher dimensional spaces. On the other hand, the resulting classes of covariance functions are rather restrictive and may lack certain desirable symmetries [808]. For example, in general separable covariances are not *radial functions*, with the notable exception of the *Gaussian covariance*

$$C_{\mathrm{xx}}(\mathbf{r}) = \sigma_{\mathrm{x}}^2\, \mathrm{e}^{-\|\mathbf{r}\|^2/\xi^2} = \sigma_{\mathrm{x}}^2 \prod_{i=1}^{d} \mathrm{e}^{-r_i^2/\xi^2},$$

that is both purely a function of $\|\mathbf{r}\|$ and naturally decomposes into a product of one-dimensional Gaussian covariance functions.

Subspace permissibility ensures that if a covariance function is permissible in three dimensions, e.g., the exponential function, it is also permissible in two and one dimensions.

Periodic covariance functions can be generated by *embedding*. Consider the harmonic embedding transformation $\mathbf{s} \in \mathbb{R}^d \mapsto \mathbf{u} \in \mathbb{R}^2$ defined by

$$u_1(\mathbf{s}) = \cos(\mathbf{k}\cdot\mathbf{s}),\;\; u_2(\mathbf{s}) = \sin(\mathbf{k}\cdot\mathbf{s}),$$

where $\mathbf{k} \in \mathbb{R}^d$ is a constant vector parameter. This embedding transformation maps the location vector $\mathbf{s}$ into the feature vector $\mathbf{u}$. Then $K(\mathbf{s}, \mathbf{s}') = C\left[\mathbf{u}(\mathbf{s}), \mathbf{u}(\mathbf{s}')\right]$ is a valid *periodic covariance function.*

Let us assume that $C(\ell_{\mathbf{u}})$, where $\ell_{\mathbf{u}} = \|\mathbf{u}(\mathbf{s}) - \mathbf{u}(\mathbf{s}')\|$, is the Euclidean distance in feature space, is a *stationary radial covariance*. The distance in feature space is given by

$$\begin{aligned}
\|\mathbf{u}(\mathbf{s}) - \mathbf{u}(\mathbf{s}')\| &= \left[u_1(\mathbf{s}) - u_1(\mathbf{s}')\right]^2 + \left[u_2(\mathbf{s}) - u_2(\mathbf{s}')\right]^2 \\
&= \left[\cos(\mathbf{k}\cdot\mathbf{s}) - \cos(\mathbf{k}\cdot\mathbf{s}')\right]^2 + \left[\sin(\mathbf{k}\cdot\mathbf{s}) - \sin(\mathbf{k}\cdot\mathbf{s}')\right]^2 \\
&= 4\sin^2\left(\frac{\mathbf{k}\cdot\left(\mathbf{s}-\mathbf{s}'\right)}{2}\right).
\end{aligned}$$

A different embedding option is offered by the mapping $f_i\colon \mathbb{R} \to \mathbb{R}^2, s_i \mapsto \mathbf{u}(s_i)$, for $i = 1, \ldots, d$. In contrast with the previous embedding, this transformation maps each coordinate of the real space location vector to a respective feature vector $\mathbf{u}$.

If the function $C\left[\mathbf{u}(x), \mathbf{u}(x')\right]$ is a valid covariance, it is possible to construct the separable covariance function

$$K(\mathbf{s}, \mathbf{s}') = \sigma_{\mathrm{x}}^{2(1-d)} \prod_{i=1}^{d} C\left[\mathbf{u}(s_i), \mathbf{u}(s_i')\right].$$

For example, let us assume that

$$C\left[\mathbf{u}(x), \mathbf{u}(x')\right] = \sigma_{\mathrm{x}}^2 \exp\left(-\frac{\|\mathbf{u}(x) - \mathbf{u}(x')\|^2}{\xi_x^2}\right).$$

Then, the *separable periodic covariance function* obtained from the above by means of the harmonic embedding is [520, 678]

$$K(\mathbf{s}, \mathbf{s}') = \sigma_{\mathrm{x}}^2 \prod_{i=1}^{d} \exp\left[-\frac{4}{\xi_i^2}\sin^2\left(\frac{k_i(s_i - s_i')}{2}\right)\right]. \tag{3.74}$$

Example 3.2 Show that a function obtained by means of vertical rescaling according to (3.70) is a valid covariance function.

Answer Let $C(\mathbf{s}, \mathbf{s}')$ denote the original covariance function. The rescaled function is $K(\mathbf{s}, \mathbf{s}') = \alpha(\mathbf{s})C(\mathbf{s}, \mathbf{s}')\alpha(\mathbf{s}')$, where $\alpha(\mathbf{s})\colon \mathbb{R}^d \to \mathbb{R}$. To show that $K(\mathbf{s}, \mathbf{s}')$ is a permissible covariance, we need to prove that for all $N \in \mathbb{N}$, all sampling sets $\{\mathbf{s}_n\}_{n=1}^N$, and all sets of real numbers $\{c_n\}_{n=1}^N$, the following inequality is true.

$$\sum_{n=1}^{N}\sum_{m=1}^{N} c_n\, c_m\, K(\mathbf{s}_n, \mathbf{s}_m) \geq 0.$$

The permissibility of $C(\mathbf{s}, \mathbf{s}')$ implies the inequality

$$\sum_{n=1}^{N}\sum_{m=1}^{N} \widetilde{c}_n\, \widetilde{c}_m\, C(\mathbf{s}_n, \mathbf{s}_m) \geq 0,$$

for all sets of real-valued coefficients $\{\widetilde{c}_i\}_{n=1}^{N}$. For every c_i we can define the mapping $c_n \mapsto \widetilde{c}_n = c_n\, \alpha(\mathbf{s}_n)$ for all $n = 1, \ldots, N$. If the above inequality is true for all possible real-valued c_n then it is also true for all the $\widetilde{c}_n$ obtained by the mapping. Hence, the required condition for $K(\cdot, \cdot)$ is shown to be true for all real-valued coefficients $\{c_n\}_{n=1}^{N}$ (since each c_n is uniquely mapped to a respective $\widetilde{c}_n$).

3.7 Permissibility of Variogram Functions

Bochner's theorem provides permissibility conditions for the covariance function but not for the variogram. This is not a problem for stationary random fields, since the variogram and the covariance function are connected via $C_{\mathrm{xx}}(\mathbf{r}) = \sigma_{\mathrm{x}}^2 - \gamma_{\mathrm{xx}}(\mathbf{r})$. However, for non-stationary random fields with stationary increments, it is convenient to have permissibility conditions for the variogram function, since the latter depends only on $\mathbf{r}$ unlike the covariance function that depends on both $\mathbf{s}$ and $\mathbf{s}'$.

Necessary conditions for the variogram can be easily obtained based on the definition (3.44). However, necessary conditions do not guarantee that a function is a permissible variogram, because permissibility involves both necessary and sufficient conditions. The permissibility conditions for the variogram are different from those for the covariance.

A permissible variogram is a *conditionally non-positive definite function*, i.e., it satisfies the inequality [136, 554]

$$\sum_{n=1}^{N}\sum_{m=1}^{N} \lambda_n \lambda_m\, \gamma_{\mathrm{xx}}(\mathbf{s}_n, \mathbf{s}_m) \leq 0 \tag{3.75a}$$

for any set of real or complex-valued coefficients $\{\lambda_n\}_{n=1}^{N}$ subject to the condition

$$\sum_{n=1}^{N} \lambda_n = 0. \tag{3.75b}$$

Remark To understand this condition intuitively we first consider a stationary random field for which $C_{\mathrm{xx}}(\mathbf{r}) = \sigma_{\mathrm{x}}^2 - \gamma_{\mathrm{xx}}(\mathbf{r})$. The permissibility condition for the covariance function requires the following inequality

$$\sum_{n=1}^{N}\sum_{m=1}^{N} \lambda_n \, \lambda_m \, C_{\mathrm{xx}}(\mathbf{s}_n, \mathbf{s}_m) = \sigma_{\mathrm{x}}^2 \left(\sum_{n=1}^{N} \lambda_n\right)^2 - \sum_{n=1}^{N}\sum_{m=1}^{N} \lambda_n \, \lambda_m \, \gamma_{\mathrm{xx}}(\mathbf{s}_n, \mathbf{s}_m) \geq 0.$$

The above must be true for all points $\mathbf{s}_n$, $n = 1, \ldots, N$. The first term at the right of the equality sign is non-negative and becomes equal to zero if $\sum_{n=1}^{N} \lambda_n = 0$. Hence, if the latter condition materializes, the second term must be positive to ensure the positive definiteness of the covariance. The variogram permissibility condition (3.75) can be proved more generally so that it also applies to non-stationary random fields [136].

The conditions (3.75) provide the are general definition of variogram permissibility. They can be used to disprove that a certain function is a valid variogram [537]: It suffices to find one combination of zero-sum coefficients for which (3.75a) does not hold in order to disprove the validity of a proposed variogram function.

On the other hand, the conditions (3.75) are not very practical for proving the validity of a given function as variogram model, because they cannot be verified for all the infinitely many combinations of points and coefficients—just like the general covariance permissibility condition. Hence, more practical conditions, analogous to Bochner's theorem, are needed. Such permissibility conditions are given in [136], [863, p. 434], [165, p. 87].

Theorem 3.4 *Conditionally non-positive definite functions:*
If $\gamma_{\mathrm{xx}}(\mathbf{r}) \colon \mathbb{R}^d \to \mathbb{R}$ *is a continuous function such that* $\gamma_{\mathrm{xx}}(\mathbf{0}) = 0$, *the following statements are equivalent [132, p. 69]:*

1. *The function* $\gamma_{\mathrm{xx}}(\mathbf{r})$ *is conditionally non-positive definite.*
2. *For all* $c > 0$ *the function* $\mathrm{e}^{-c\,\gamma_{\mathrm{xx}}(\mathbf{r})}$ *is positive definite (i.e., a permissible covariance function).*
3. *The function* $\gamma_{\mathrm{xx}}(\mathbf{r})$ *can be expressed in the following spectral representation*

$$\gamma_{\mathrm{xx}}(\mathbf{r}) = Q(\mathbf{r}) + \frac{1}{(2\pi)^d} \int_{\mathbb{R}^d} \mathrm{d}\mathbf{k} \frac{[1 - \cos(\mathbf{k} \cdot \mathbf{r})]}{\|\mathbf{k}\|^2} \, \widetilde{g}(\mathbf{k}), \tag{3.76}$$

where (i) $Q(\mathbf{r})$ *is a homogeneous polynomial of degree two (quadratic form), and (ii) the function* $\widetilde{g}(\mathbf{k})$ *is positive, symmetric, i.e.,* $\widetilde{g}(\mathbf{k}) = \widetilde{g}(-\mathbf{k})$, *continuous at the origin (it does not include a delta function), and its spectral integral defined below is finite:*

$$\int_{\mathbb{R}^d} \mathrm{d}\mathbf{k} \, \frac{\widetilde{g}(\mathbf{k})}{1 + \|\mathbf{k}\|^2} < \infty. \tag{3.77}$$

Comments (i) The third property uses the general concept of locally homogeneous or intrinsic random fields as defined in Sect. 5.7.

(ii) Most statistics textbooks express (3.76) in terms of a positive measure $\mathrm{d}F(\mathbf{k})$ which is equivalent to $\mathrm{d}\mathbf{k}\,\widetilde{g}(\mathbf{k})$, if the density $\widetilde{g}(\mathbf{k})$ exists. The measure theory formulation has formal advantages, but it can be avoided if we are willing to allow delta-function terms in $\widetilde{g}(\mathbf{k})$.

Let us now consider the origin of the quadratic form $Q(\mathbf{r})$ in (3.76). If we assume a spectral density $\widetilde{g}(\mathbf{k})$ that contains a Dirac delta function at the origin of the form $\widetilde{g}_0(\mathbf{k}) = A\delta(\mathbf{k})/\|\mathbf{k}\|^2$, where $A > 0$. Inserting this term in (3.76) and using the Taylor expansion

$$1 - \cos(\mathbf{k}\cdot\mathbf{r}) = \frac{(\mathbf{k}\cdot\mathbf{r})^2}{2} + O\left((\mathbf{k}\cdot\mathbf{r})^4\right),$$

we find that only the leading order term survives upon integration and thus $\gamma_{\mathrm{xx}}(\mathbf{r}) \propto A\,\|\mathbf{r}\|^2$. To gain more insight into how the integration of $(\mathbf{k}\cdot\mathbf{r})^2$ over $\mathbf{k}$ is performed, consider Sect. 5.8.2 below.

The definition of the variogram function as the variance of the increments places significant constraints on the type of functions that can be used as valid variograms. Below we investigate how such conditions can be obtained with simple arguments.

Example 3.3 Necessary variogram properties: Let $\gamma_{\mathrm{xx}}(\mathbf{r})$ be the variogram function of the random field $\mathrm{X}(\mathbf{s};\omega)\colon \mathbb{R}^d \to \mathbb{R}$ which has zero-mean increments, i.e.,

$$\mathcal{E}[\mathrm{X}(\mathbf{s}+\mathbf{r};\omega) - \mathrm{X}(\mathbf{s};\omega)] = 0.$$

Show that for all vectors $\mathbf{r}, \mathbf{r}_1, \mathbf{r}_2 \in \mathbb{R}^d$, $\gamma_{\mathrm{xx}}(\mathbf{r})$ satisfies the following inequalities:

$$2\gamma_{\mathrm{xx}}(\mathbf{r}_1) + 2\gamma_{\mathrm{xx}}(\mathbf{r}_2) \geq \gamma_{\mathrm{xx}}(\mathbf{r}_1 - \mathbf{r}_2), \qquad (3.78a)$$

$$4\,\gamma_{\mathrm{xx}}(\mathbf{r}) > \gamma_{\mathrm{xx}}(2\mathbf{r}). \qquad (3.78b)$$

If in addition $\gamma_{\mathrm{xx}}(\mathbf{r}) = \gamma_{\mathrm{xx}}(\|\mathbf{r}\|)$ is a radial function, then

$$4\,\gamma_{\mathrm{xx}}(2^{1/2}\|\mathbf{r}\|) > \gamma_{\mathrm{xx}}(\|\mathbf{r}\|), \qquad (3.78c)$$

Answer

(i) Consider the second-order increment process

$$\mathrm{Y}(\mathbf{s};\mathbf{r}_1,\mathbf{r}_2) = \mathrm{X}(\mathbf{s}+\mathbf{r}_1;\omega) + \mathrm{X}(\mathbf{s}+\mathbf{r}_2;\omega) - 2\mathrm{X}(\mathbf{s};\omega).$$

This can be expressed in terms of the *increment fields* $\Delta_{\mathrm{x}}(\mathbf{s},\mathbf{r}_i;\omega) = \mathrm{X}(\mathbf{s}+\mathbf{r}_i;\omega) - \mathrm{X}(\mathbf{s};\omega)$, where $i = 1, 2$ as follows:

$$\mathrm{Y}(\mathbf{s};\mathbf{r}_1,\mathbf{r}_2) = \Delta_{\mathrm{x}}(\mathbf{s},\mathbf{r}_1;\omega) + \Delta_{\mathrm{x}}(\mathbf{s},\mathbf{r}_2;\omega).$$

We can then use the identity $(u+v)^2 = 2(u^2+v^2) - (u-v)^2$, replace u with $\Delta_x(\mathbf{s}, \mathbf{r}_1; \omega)$ and v with $\Delta_x(\mathbf{s}, \mathbf{r}_2; \omega)$, and then evaluate

$$\mathcal{E}\Big[(u+v)^2\Big] = 2\mathcal{E}\Big[u^2+v^2\Big] - \mathcal{E}\Big[(u-v)^2\Big].$$

Hence, in light of the definitions for u and v the above is equivalent to

$$\mathcal{E}\Big[\mathrm{Y}^2(\mathbf{s}; \mathbf{r}_1, \mathbf{r}_2)\Big] = 4\gamma_{xx}(\mathbf{r}_1) + 4\gamma_{xx}(\mathbf{r}_2) - 2\gamma_{xx}(\mathbf{r}_1 - \mathbf{r}_2) \geq 0,$$

which proves the first inequality (3.78a).

(ii) The inequality (3.78b) follows from (3.78a) by setting $\mathbf{r}_1 = -\mathbf{r}_2 = \mathbf{r}$ and using the reflection symmetry of the variogram, i.e., $\gamma_{xx}(\mathbf{r}) = \gamma_{xx}(-\mathbf{r})$.

(iii) Let us define $\mathbf{r}_1 = \left(\frac{1}{2}, \frac{\sqrt{7}}{2}, 0\right)$ and $\mathbf{r}_2 = \left(-\frac{1}{2}, \frac{\sqrt{7}}{2}, 0\right)$, where $\mathbf{r}_1, \mathbf{r}_2 \in \mathbb{R}^3$. Then the inequality (3.78c) follows from (3.78a).

3.7.1 *Variogram Spectral Density*

In the case of second-order stationarity, the variogram and the covariance function are related via the equation (3.47), i.e.,

$$\gamma_{xx}(\mathbf{r}) = \sigma_x^2 - C_{xx}(\mathbf{r}).$$

This identity implies that the variogram function starts at zero and increases towards the asymptotic limit σ_x^2. Such a function is not absolutely integrable, and therefore its Fourier transform does not exist in the standard sense. However, it is possible to define the *variogram spectral density* $\widetilde{\gamma}_{xx}(\mathbf{k})$ as the Fourier transform of $\gamma_{xx}(\mathbf{r})$ in terms of a generalized function. This is accomplished by taking into account that the *Fourier transform* of the constant term σ_x^2 is given by the delta function, leading to

$$\widetilde{\gamma}_{xx}(\mathbf{k}) = (2\pi)^d \, \sigma_x^2 \, \delta(\mathbf{k}) - \widetilde{C}_{xx}(\mathbf{k}) \tag{3.79a}$$

or equivalently

$$\widetilde{C}_{xx}(\mathbf{k}) = (2\pi)^d \, \sigma_x^2 \, \delta(\mathbf{k}) - \widetilde{\gamma}_{xx}(\mathbf{k}), \tag{3.79b}$$

In deriving (3.79a) we used the orthonormality of the Fourier basis as expressed by the identity (3.64). Based on Bochner's theorem and (3.79a), the condition $\widetilde{C}_{xx}(\mathbf{k})$ for all $\mathbf{k}$ implies the following variogram permissibility conditions.

Condition 1. Sign of the variogram spectral density

$$\begin{cases} -\widetilde{\gamma}_{\mathrm{xx}}(\mathbf{k}) \geq 0, & \text{for all } \mathbf{k} \neq \mathbf{0} \\ (2\pi)^d \, \sigma_{\mathrm{x}}^2 \, \delta(\mathbf{k}) - \widetilde{\gamma}_{\mathrm{xx}}(\mathbf{k}) \geq 0, & \text{at } \mathbf{k} = \mathbf{0}. \end{cases}$$

Condition 2. Integral of the variogram spectral density The condition of Bochner theorem's on the integral of the spectral density, i.e.,

$$\frac{1}{(2\pi)^d} \int_{\mathbb{R}^d} \mathrm{d}\mathbf{k} \, \widetilde{C}_{\mathrm{xx}}(\mathbf{k}) = \sigma_{\mathrm{x}}^2,$$

implies that the variogram spectral density satisfies the following identity

$$\int_{\mathbb{R}^d} \mathrm{d}\mathbf{k} \, \widetilde{\gamma}_{\mathrm{xx}}(\mathbf{k}) = 0. \tag{3.80}$$

Condition 2 may seem to contradict the negative values of the variogram spectral density at every wavevector except for $\mathbf{k} \neq \mathbf{0}$. Nevertheless, if we recall (3.79a) Condition 2 follows naturally. The key is that $\widetilde{\gamma}_{\mathrm{xx}}(\mathbf{k})$ contains a Dirac delta function with positive strength at $\mathbf{k} = \mathbf{0}$ that cancels the contribution from the regular part of $\widetilde{\gamma}_{\mathrm{xx}}(\mathbf{k})$, which is derived from $-\widetilde{C}_{\mathrm{xx}}(\mathbf{k})$.

Chapter 4
Additional Topics of Random Field Modeling

Since we cannot change reality, let us
change the eyes which see reality.
Nikos Kazantzakis

We now turn our attention to specialized topics of random field modeling that include ergodicity, the concept of isotropy, the definition of different types of anisotropy, and the description of the joint dependence of random fields at more than two points. Ergodicity, isotropy and anisotropy are properties that have significant practical interest for the modeling of spatial data. On the other hand, the joint N-point dependence is a more advanced topic, primarily of modeling importance for non-Gaussian random fields. In the case of Gaussian random fields the N-point moments can be expressed in terms of the first and second-order moments.

4.1 Ergodicity

Ergodicity is a property that connects spatial and ensemble averages. It is thus useful for the estimation of random fields, because spatial averages can be evaluated from a single available sample while ensemble averages cannot. From a practitioner's perspective, ergodicity implies that the random field's statistical properties (i.e., the moments up to a specified order) can be estimated from a single sample.

In practical studies we are typically concerned with establishing second-order properties. To determine statistical properties from a single sample, it is necessary to assume *translational invariance*: without it, it is impossible to estimate the statistical properties at all points from a single snapshot of the field.[1]

[1] Exceptions are spatial data sets that exhibit slowly changing, quasi-stationary patterns.

D. T. Hristopulos, *Random Fields for Spatial Data Modeling*, Advances in Geographic Information Science, https://doi.org/10.1007/978-94-024-1918-4_4

Ergodicity The notion of ergodicity applies only to *stationary (statistically homogeneous)* random fields. Stationarity, however, is only a *necessary (but not sufficient)* condition for ergodicity.

If we restrict ourselves to second-order properties, we care about ergodicity in the mean and the covariance function.

Definition 4.1 The *sample average* of the random field $\mathrm{X}(\mathbf{s};\omega)$ over a spatial domain $\mathcal{D}$ is given by the following coarse-grained random variable

$$X(\mathcal{D};\omega) = \frac{1}{|\mathcal{D}|}\int_{\mathcal{D}} d\mathbf{s}\, \mathrm{X}(\mathbf{s};\omega). \tag{4.1}$$

A stationary random field $\mathrm{X}(\mathbf{s};\omega)$ with constant mean m_{x} and covariance function $C_{\mathrm{xx}}(\mathbf{r})$ that depends only on the lag, is ergodic in the mean if the coarse-grained average (4.1) tends asymptotically to the mean, i.e.,

$$\lim_{\mathcal{D}\to\infty} X(\mathcal{D};\omega) = m_{\mathrm{x}}.$$

Ergodicity implies that the spatial average $X(\mathcal{D};\omega)$ provides an accurate estimate of m_{x} as the domain $\mathcal{D}$ expands to infinity in every orthogonal direction.[2]

Theorem 4.1 *Slutsky's ergodic theorem: A stationary random field* $\mathrm{X}(\mathbf{s};\omega)$ *with constant mean* m_{x} *and covariance function* $C_{\mathrm{xx}}(\mathbf{r})$ *is ergodic in the mean if the following condition holds*

$$\lim_{V\to\infty} \frac{1}{|V|}\int_{V} d\mathbf{r}\, C_{\mathrm{xx}}(\mathbf{r}) = 0, \tag{4.2}$$

where V *is the vector space generated by the difference of all vectors* $\mathbf{s}\in\mathcal{D}$.

Proof It is easy to show that $\mathbb{E}[X(\mathcal{D};\omega)] = m_{\mathrm{x}}$. To complete the proof of ergodicity one needs to show that

$$\lim_{\mathcal{D}\to\infty} \mathrm{Var}\{X(\mathcal{D};\omega)\} = 0.$$

It is fairly straightforward and left as an exercise for the reader to show that the above is equivalent to Slutsky's condition.

---

[2]Expansion of the domain in all orthogonal directions is necessary in cases of stationary but anisotropic processes.

A *sufficient but not necessary* condition for Slutsky's theorem is that for all possible directions in space, the covariance function tends to zero as the magnitude of the lag tends to infinity.

Ergodicity in the covariance Without loss of generality we consider a zero-mean random field $\mathrm{X}(\mathbf{s};\omega)$. If the random field has a non-zero expectation $m_{\mathrm{x}}(\mathbf{s})$, we focus on the zero-mean fluctuations. To discuss ergodicity of a stationary random field with respect to the covariance, we need to construct the following *two-point coarse-grained random field*

$$Y(\mathcal{D};\mathbf{r},\omega)=\frac{1}{|\mathcal{D}|}\int_{\mathcal{D}}\mathrm{d}\mathbf{s}\,\mathrm{X}(\mathbf{s};\omega)\,\mathrm{X}(\mathbf{s}+\mathbf{r};\omega).$$

Ergodicity requires that at the limit of an infinite domain $Y(\mathcal{D};\mathbf{r},\omega)$ is an accurate estimate of the covariance function, i.e.,

$$\hat{Y}(\mathbf{r})=\lim_{\mathcal{D}\to\infty}Y(\mathcal{D};\mathbf{r},\omega)=C_{\mathrm{xx}}(\mathbf{r}).$$

Proof It is easy to show that $\hat{Y}(\mathbf{r})$ is an unbiased estimator of the covariance, since $\mathcal{E}[Y(\mathcal{D};\mathbf{r},\omega)]=C_{\mathrm{xx}}(\mathbf{r})$ for any $\mathcal{D}\subset\mathbb{R}^d$. However, to prove that the variance of $\hat{Y}(\mathbf{r})$ vanishes, i.e.,

$$\lim_{\mathcal{D}\to\infty}\mathbb{V}\mathrm{ar}\,\{Y(\mathcal{D};\mathbf{r},\omega)\}=0,$$

requires imposing conditions on the fourth-order moments of $\mathrm{X}(\mathbf{s};\omega)$. The situation is considerably simplified for *Gaussian random fields*, since the fourth-order moments can be decomposed in terms of second-order moments using the Wick-Isserlis theorem (see Chap. 6). Based on this decomposition, the condition for ergodicity in the covariance becomes [132, p. 21]

$$\lim_{\mathcal{D}\to\infty}\frac{1}{|\mathcal{D}|}\int_{\mathcal{D}}\mathrm{d}\mathbf{r}\,\left[C_{\mathrm{xx}}(\mathbf{r})\right]^2=0. \tag{4.3}$$

The above is satisfied if the absolute value of the covariance declines to zero sufficiently fast as the lag $\|\mathbf{r}\|$ tends to infinity.

Example 4.1 Assume that $C_{\mathrm{xx}}(\mathbf{r})$ is a continuous radial function that has a finite variance and decays asymptotically (for $\|\mathbf{r}\|\to\infty$) as a power law with exponent α, i.e., $C_{\mathrm{xx}}(\mathbf{r})\sim\|\mathbf{r}\|^{-\alpha}$. Derive the conditions on the tail exponent α for ergodicity in the mean and the covariance.

Answer Let us define the covariance integrals

$$I_1=|\mathcal{D}|^{-1}\int\mathrm{d}\mathbf{r}\,C_{\mathrm{xx}}(\mathbf{r}),\ \text{and}\ I_2=|\mathcal{D}|^{-1}\int\mathrm{d}\mathbf{r}\,|C_{\mathrm{xx}}(\mathbf{r})|^2.$$

Using the integration formula (3.62a) for radial functions, it follows that

$$I_1 = \frac{\mathcal{S}_d}{|\mathcal{D}|} \int_0^\infty \mathrm{d}t\, t^{d-1}\, C_{\mathrm{xx}}(t) = \frac{C_L}{|\mathcal{D}|} + \frac{\mathcal{S}_d}{|\mathcal{D}|} \int_L^\infty \mathrm{d}t\, t^{d-1-\alpha},$$

where $C_L = \mathcal{S}_d \int_0^L \mathrm{d}t\, t^{d-1}\, C_{\mathrm{xx}}(t)$ and $\mathcal{S}_d$ is the surface area of the unit sphere in d dimensions. The integral of the power law converges to a finite value if $d < \alpha$. Then, $\lim_{|\mathcal{D}|\to\infty} I_1 = 0$, and the field is ergodic in the mean according to (4.2).

Using similar arguments and based on (4.3), establishing ergodicity in the covariance requires that $d < 2\alpha$. On account of the exponent inequalities derived above, the random field is ergodic in both the mean and the covariance if $d < \alpha$. As we show in Sect. 5.6, this upper bound of the power-law tail exponent corresponds to the regime of short-range covariance functions.

Interested readers will find more information regarding ergodicity in the books of Cramér & Leadbetter [162], Adler [10], Yaglom [863], Christakos [138], Cressie [165], Lantuejoul [487], and Chilés & Delfiner [132]. We reprise the issue of ergodicity in Chap. 12, in connection with the estimation of random field moments from a single spatially distributed sample.

4.2 Statistical Isotropy

The stationarity assumption implies that the covariance function is translation invariant. To make this statement more precise consider that the position of a pair of points $\mathbf{s}_1$ and $\mathbf{s}_2$ in space is determined by two vectors: the position of the center of mass of the two points, $\mathbf{r}_{\mathrm{cm}} = (\mathbf{s}_1 + \mathbf{s}_2)/2$, and the spatial lag $\mathbf{r} = \mathbf{s}_1 - \mathbf{s}_2$.

Stationarity means that the degree of correlation between a pair of points separated by the distance vector $\mathbf{r} = \|\mathbf{r}\|\, \hat{\mathbf{r}}$ (where $\hat{\mathbf{r}}$ is the unit vector in the direction of $\mathbf{r}$) is the same, irrespectively of the pair's center of mass location. *Statistical isotropy* is a stricter condition: in addition to stationarity, it demands that the correlation function be independent of the *spatial direction* (determined by the unit vector $\hat{\mathbf{r}}$).

Radial functions Some definitions and properties of radial functions follow.

1. A stationary covariance $C_{\mathrm{xx}}(\mathbf{r})$ is a *radial function* if $C_{\mathrm{xx}}(\mathbf{r}) = C_{\mathrm{xx}}(\|\mathbf{r}\|)$, i.e., if it depends only on the Euclidean norm of the lag vector.
2. Similarly, a variogram is a radial function if $\gamma_{\mathrm{xx}}(\mathbf{r}) = \gamma_{\mathrm{xx}}(\|\mathbf{r}\|)$.
3. If a radial function in real space admits a Fourier transform, then the latter is a radial function in reciprocal space and vice versa.

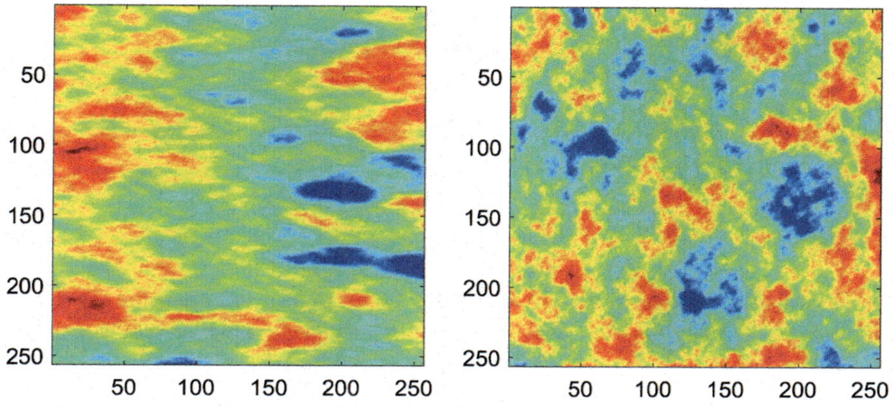

Fig. 4.1 Left: Realization obtained from stationary random field with directionally dependent covariance. Right: Realization of a stationary random field with radial covariance function

For the purpose of visual comparison, two realizations obtained from stationary Gaussian random fields, one with radial covariance function and the other with direction-dependent covariance function are shown in Fig. 4.1. The directionally dependent covariance leads to elongated features with a succinct orientation of the longer axis in the horizontal direction, whereas the radial covariance function leads to patterns with no characteristic orientation.

Remark For the sake of mathematical correctness we should write $C_{\rm xx}(\mathbf{r}) = C^{0}_{\rm xx}(\|\mathbf{r}\|)$, since $C_{\rm xx}(\cdot) : \mathbb{R}^d \to \mathbb{R}$, while $C^{(0)}_{\rm xx}(\cdot) : \mathbb{R} \to \mathbb{R}$. However, to keep the notation simple we use $C_{\rm xx}(\cdot)$ for both functions.

4.2.1 Spectral Representation

For *radial covariance* functions the pair of Fourier transforms (3.54) and (3.56) is expressed in terms of a pair of *Hankel transforms* as follows [863, p. 353], [132, p. 71]:

$$\widetilde{C_{\rm xx}}(\|\mathbf{k}\|) = \frac{(2\pi)^{d/2}}{\|\mathbf{k}\|^{\nu}} \int_0^{\infty} \mathrm{d}r\, r^{d/2} J_{\nu}(\|\mathbf{k}\| r)\, C_{\rm xx}(r), \tag{4.4a}$$

$$C_{\rm xx}(\|\mathbf{r}\|) = \frac{1}{(2\pi)^{d/2} \|\mathbf{r}\|^{\nu}} \int_0^{\infty} \mathrm{d}k\, k^{d/2} J_{\nu}(k \|\mathbf{r}\|)\, \widetilde{C_{\rm xx}}(k), \tag{4.4b}$$

where $J_{\nu}(x) \colon x \in \mathbb{R}$ is the *Bessel function of the first kind of order* $\nu = d/2 - 1$, and $\|\mathbf{k}\|$ is the Euclidean norm of the wavevector $\mathbf{k}$ in reciprocal space [730].

Bessel function in one dimension If $d = 1$ the Bessel function of the first kind of order $d/2 - 1$ that appears in (4.4a)–(4.4b) is given by the following expression

$$J_{-1/2}(x) = \sqrt{\frac{2}{\pi x}}\,\cos x, \; x \in \mathbb{R}.$$

Bessel function in three dimensions If $d = 3$ the respective Bessel function of the first kind of order $d/2 - 1$ is given by

$$J_{1/2}(x) = \sqrt{\frac{2}{\pi x}}\,\sin x.$$

The above simplifying expressions imply that for certain spectral densities in $d = 1$ and $d = 3$, the integrals in (4.4a)–(4.4b) can be evaluated using *Cauchy's theorem of residues* [194].

Bessel function in two dimensions In $d = 2$, the Bessel function $J_0(z)$ cannot be expressed in terms of simpler functions. Explicit calculation of the Hankel transform for spectral densities of the form

$$\widetilde{C}_{\mathrm{xx}}(\mathbf{k}) \propto \frac{1}{\left(\|\mathbf{k}\|^2 + z^2\right)^{\mu+1}}, \quad \text{where } \frac{d-5}{4} < \mu,$$

is made possible by means of the *Hankel-Nicholson integration* formula. The latter is given by [4, p. 488] and [306, Eq. 6.565.4]

$$\int_0^\infty du \frac{u^{\nu+1} J_\nu(h\,u)}{(u^2 + z^2)^{\mu+1}} = \frac{h^\mu z^{\nu-\mu}}{2^\mu \Gamma(\mu+1)} K_{\nu-\mu}(h\,z), \tag{4.5}$$

where $K_\lambda(x)$ is the *modified Bessel function of the second kind of order* λ, $h > 0$, in general z is a complex number with positive real part, and the real-valued indices ν and μ satisfy $-1 < \nu < 2\mu + \frac{3}{2}$.

Comment Note that in both (4.5) and (4.4), the difference between the power-law exponent in the numerator and the order of the Bessel function is equal to one.

The Hankel-Nicholson formula also allows evaluating the spectral densities of rational covariance functions such as the rational quadratic (see Table 4.4). The formula is not limited to calculations in one spatial dimension: it can also be used for radial functions in two and three dimensions with the corresponding value of ν.

The integrals that represent the *Hankel transform pairs* in $d = 1, 2, 3$ are given in Table 4.1.

Table 4.1 Expressions for the Hankel transform and its inverse for radial covariance functions in $d = 1, 2, 3$. In the following, $r = \|\mathbf{r}\|$ and $k = \|\mathbf{k}\|$ denote the Euclidean norms of the lag vector and the reciprocal-space wavevector respectively

d	Hankel transform	Inverse Hankel transform
1	$2\int_0^\infty dr\,\cos(k\,r)\,C_{\mathrm{xx}}(r)$	$\frac{1}{\pi}\int_0^\infty dk\,\cos(k\,r)\,\widetilde{C}_{\mathrm{xx}}(k)$
2	$2\pi\int_0^\infty dr\,r\,J_0(k\,r)\,C_{\mathrm{xx}}(r)$	$\frac{1}{2\pi}\int_0^\infty dk\,k\,J_0(k\,r)\,\widetilde{C}_{\mathrm{xx}}(k)$
3	$\frac{4\pi}{k}\int_0^\infty dr\,r\,\sin(k\,r)\,C_{\mathrm{xx}}(r)$	$\frac{1}{2\pi^2 r}\int_0^\infty dk\,k\,\sin(k\,r)\,\widetilde{C}_{\mathrm{xx}}(k)$

4.2.2 Isotropic Correlation Models

Certain commonly used isotropic correlation models are given below. The corresponding variogram models can be obtained from (3.47), while the covariance function is obtained by multiplying the correlation function with the respective variance. The list below is indicative and by no means exhaustive. For additional possibilities the reader is referred to [138], [487, Appendix] and [283]. A well-written discussion of the properties of several covariance functions is given in [678, Ch. 4].

Properties of isotropic correlation models Table 4.2 is a list of commonly used isotropic correlation models given by radial functions. The following comments clarify the notation used in the Table and provide some insight into the properties of the respective covariance functions. The local properties (continuity and differentiability) of these covariance functions are given in Table 5.1 of Sect. 5.1.4.

Table 4.2 List of commonly used isotropic correlation models. The dimensionless distance is $u = \|\mathbf{r}\|/\xi$. The covariance function is given by $C_{\mathrm{xx}}(\mathbf{r}) = \sigma_{\mathrm{x}}^2\, \rho_{\mathrm{xx}}(\mathbf{r})$

Model	Correlation
Nugget effect	$$\rho_{\mathrm{xx}}(\mathbf{r}) = \begin{cases} c_0 & \text{if } \|\mathbf{r}\| = 0, \\ 0 & \text{if } \|\mathbf{r}\| \neq 0 \end{cases} \quad (4.6)$$
Exponential	$$\rho_{\mathrm{xx}}(u) = \mathrm{e}^{-u} \quad (4.7)$$
Gaussian	$$\rho_{\mathrm{xx}}(u) = \mathrm{e}^{-u^2} \quad (4.8)$$
Spherical	$$\rho_{\mathrm{xx}}(u) = \left(1 - 1.5\,u + 0.5\,u^3\right) \mathbb{I}_{0 \leq u \leq 1}(u) \quad (4.9)$$
Cardinal Sine product	$$\rho_{\mathrm{xx}}(\mathbf{r}) = \prod_{i=1}^{d} \mathrm{sinc}(r_i/\xi) \quad (4.10)$$

(continued)

Table 4.2 (continued)

Generalized exponential	$$\rho_{xx}(u) = e^{-u^{\nu}}, \quad 0 < \nu < 2 \qquad (4.11)$$
Whittle-Matérn	$$\rho_{xx}(u) = \frac{\left(\sqrt{2\nu}\, u\right)^{\nu}}{2^{\nu-1}\,\Gamma(\nu)}\, K_{\nu}\left(\sqrt{2\nu}\, u\right), \quad \nu > 0 \qquad (4.12)$$
Bessel-J	$$\rho_{xx}(u) = 2^{\nu}\,\Gamma(\nu+1)\,\frac{J_{\nu}(u)}{u^{\nu}}, \quad \nu \geq d/2 - 1 \qquad (4.13)$$
Rational quadratic	$$\rho_{xx}(u) = \frac{1}{\left(1+u^2\right)^{b}}, \quad b > 0 \qquad (4.14)$$
Cauchy class	$$\rho_{xx}(u) = \frac{1}{(1+u^{\alpha})^{\beta/\alpha}}, \quad 0 < \alpha \leq 2,\ \beta \geq 0 \qquad (4.15)$$

Comments The following comments pertain to the notation and the properties of the correlation models presented in Table 4.2.

1. The correlation models defined in Table 4.2 are valid in any dimension d, provided that the Euclidean distance is used to measure the lag.
2. $\mathbb{I}_A(x)$ is the *indicator function* of the set A, i.e.,

$$\mathbb{I}_A(x) = \begin{cases} 1 & \text{if } x \in A \\ 0 & \text{if } x \notin A. \end{cases} \qquad (4.16)$$

   Hence, the correlations of the *spherical model* decay exactly to zero outside a spherical shell of radius $\|\mathbf{r}\| = \xi$ ($u = 1$, respectively).
3. $\Gamma(\nu)$ denotes the *Gamma function*, which is defined by means of the following convergent improper integral

$$\Gamma(\nu) \doteq \int_0^{\infty} dt\, t^{\nu-1}\, e^{-t}. \qquad (4.17)$$

In the above, ν can be a complex number with positive real part or a positive real number. If $\nu = n \in \mathbb{N}$, it can be shown using integration by parts that $\Gamma(n) = (n-1)!$.

4. The *cardinal sine* function is defined by $\text{sinc}(x) \doteq \sin(x)/x$.

The spectral densities of the correlation functions listed in Table 4.2, if they exist, are listed in Table 4.4. Below we discuss certain characteristic properties of the correlations models.

Nugget effect This correlation function that is non-zero only at zero lag describes spatially uncorrelated fluctuations. Hence, it can be used to model the following: (i) Independent measurement errors that are due to the measurement process. (ii) Purely random fluctuations endogenous to the system. (iii) Sub-resolution variability, i.e., fluctuations with characteristic length scales that are below the resolution limit of the observations. These fluctuations are also known as *microscale variability*.

With regard to a nugget term caused by sub-resolution variability, the nugget effect is similar to the *aliasing effect* in Fourier analysis [673, p. 605]. The term aliasing refers to the "contamination" of the spectrum with "phantom" peaks that are due to signal power at frequencies above the *Nyquist frequency*. These outlying frequencies are folded back inside the observable spectrum by the Fourier transform.

Continuum expression for the nugget effect The expression (4.6) for the nugget effect is easily understood if lag distances take discrete values. In the continuum case the nugget effect is expressed in terms of the Dirac delta function as $C_{xx}(\mathbf{r}) = c_0\,\delta(\mathbf{r})$, where c_0 is the magnitude of the discontinuity at the origin.

The nugget effect is typically used in the construction of *composite covariance* models. Such models are formed by the superposition of functions with finite correlation scales (e.g., corresponding to "slow" and "fast" spatial scales) and the nugget term that represents unresolved variability.

When to add a nugget term Adding a nugget effect to the covariance model is useful in the following situations:

1. If there are *closely spaced* but significantly different observations that cannot be smoothed (e.g., by averaging or by justifiable elimination of certain values). The only way to reconcile such fluctuations is to introduce a random error modeled by the nugget term.
2. In a linear combination with a *smooth covariance* function (such as the Gaussian model) to compensate for the inability of the smooth term to support abrupt changes over small distances.

In cases such as the above, the use of the nugget model is recommended to avoid numerical instabilities in the kriging system of equations (see Chap. 10) used in spatial interpolation [459].

Gaussian The Gaussian, $\rho_{xx}(\mathbf{r}) = \exp(-\|\mathbf{r}\|^2/\xi^2)$, is a special function, because it admits finite partial derivatives of all orders at $\|\mathbf{r}\| = 0$. As we discuss in Sect. 5.1.4 below, for Gaussian random fields the existence of the covariance derivatives at the origin ensures the existence of all the *sample path derivatives*. The implication is that the Gaussian covariance generates very smooth realizations.

The Gaussian's high degree of smoothness is desirable for modeling random coefficients of partial differential equations. However, it is often in disagreement with the observed patterns of spatial data (at least in the earth sciences). In addition, for $X(s; \omega)$ where $s \in \mathbb{R}$, knowledge of the partial derivatives of the field of all orders at zero distance, i.e., of the $X^{(n)}(0)$, $n = 0, 1, 2, \ldots$, fully determines the values of the random field at any location s (see [774, p. 30]).

The quasi-deterministic behavior of the Gaussian model can lead to ill-conditioned covariance matrices in the framework of spatial linear prediction (see Chap. 10). The problem of ill-conditioning translates into covariance matrices with very large condition numbers. The condition number of a positive-definite matrix is equal to the ratio of the largest to the smallest eigenvalue. The distribution of the eigenvalues is related to the spectral density of a covariance model. Since the Gaussian model has a spectral density that decays very fast (see Table 4.4) the respective condition numbers for Gaussian covariance matrices can be very large.[3] To compensate for these problems, the Gaussian covariance is often used in spatial models in combination with other covariance functions [132].

Cardinal sine product This correlation model is defined by the product of one-dimensional cardinal sine functions along d orthogonal principal directions, i.e.,

$$\rho_{xx}(\mathbf{r}) = \prod_{i=1}^{d} \text{sinc}(r_i/\xi).$$

Thus, each directional component takes negative values at distances r_i such that

$$(2k+1)\,\pi < \frac{r_i}{\xi} < (2k+2)\,\pi, \quad k \in \{0, 1, 2, \ldots\}, \; i = 1, \ldots, d.$$

This dependence implies that the cardinal sine product model exhibits "negative holes" which correspond to negative correlations.

Correlation models with negative correlation over some lag range(s) are mathematically allowed and supported by empirical (data based) covariance estimates. However, the depth of the negative troughs is limited by a lower bound that becomes shallower as the dimensionality increases.

---

[3] On this matter see also the comments by J. Kent in the Discussion of the paper [203].

Fig. 4.2 Isolevel contours of the cardinal sine product covariance on square lattice of size 301 × 301. The domain length is equal to $L = 8$ while the lattice step is 0.05. Isotropic correlation lengths $\xi_1 = \xi_2 = L/(2\pi)$ are used

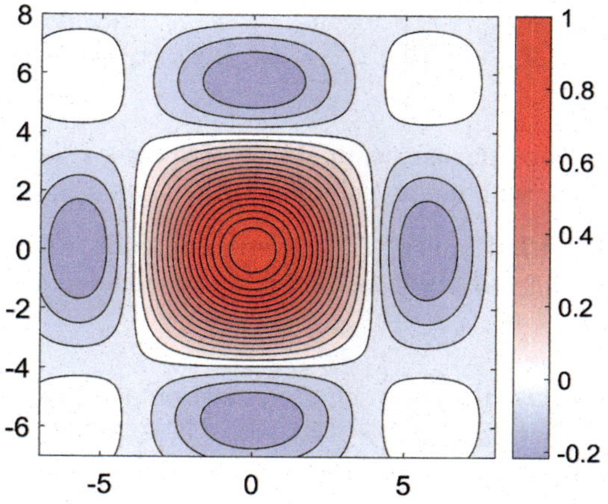

More specifically, all correlation functions of *isotropic random fields* satisfy the following inequality [10, p. 34]

$$\rho_{\mathrm{xx}}(\|\mathbf{r}\|) \geq -\frac{1}{d}, \quad \text{for all } \|\mathbf{r}\| \geq 0, \tag{4.18}$$

which defines $-1/d$ as the *lower bound* for negative correlations.

The *cardinal sine product* covariance (4.10) is not strictly isotropic, because it is not expressed in terms of the Euclidean distance $\|\mathbf{r}\|$. However, its dependence, on all the orthogonal projections of $\mathbf{r}$ on a Cartesian system of axes is the same; this means that the dependence of $C_{\mathrm{xx}}(\mathbf{r})$ on r_i is indistinguishable from the dependence on r_j for all $i = 1, \ldots, d$ and $j \neq i$. We will refer to such functions as *partially isotropic* functions.

Isolevel contours of the cardinal sine product covariance are plotted in Fig. 4.2. For small values of r_i/ξ $(i = 1, 2)$, the contours are very similar to circles. For higher lags, however, the contours deviate from the circular shape. The isolevel contour at a specific level $0 \leq c \leq 1$ is the locus of points $\mathbf{r} \in \mathbb{R}^d$ for which

$$\prod_{i=1}^{d} \operatorname{sinc}\left(\frac{r_i}{\xi}\right) = c, \quad 0 \leq c \leq 1.$$

For $r_i/\xi \ll 1$, we use the Taylor expansion of the cardinal sine around the origin, i.e., $\operatorname{sinc}(x) = 1 - x^2/6 + O(x^4)$. Inserting this series in the isolevel equation, the latter becomes

$$\prod_{i=1}^{d}\left[1-\sum_{i=1}^{d}\frac{r_i^2}{6\xi^2}+O(r_i^4)\right]=c\Rightarrow\sum_{i=1}^{d}r_i^2\approx 6\xi^2\,(1-c)\,.$$

In two dimensions, the locus of vectors $\mathbf{r}$ that satisfy the above equation forms a circle with radius $\xi\sqrt{6(1-c)}$.

Whittle-Matérn This covariance family is defined in terms of $K_\nu(\cdot)$, which is the *modified Bessel function* of the second kind of order ν.

Special case A characteristic property of the Whittle-Matérn correlation is that for $\nu = m + 1/2$, where $m \in \mathcal{N}$, the model is equivalent to the product of the exponential and a polynomial function, that is:

$$\rho_{\mathrm{xx}}(u)=\frac{\left(\sqrt{2\nu}\,u\right)^{\nu}}{2^{\nu-1}\,\Gamma(\nu)}K_\nu\left(\sqrt{2\nu}\,u\right)=\mathrm{e}^{-\sqrt{2\nu}\,u}\sum_{k=0}^{m}\frac{(m+k)!}{(2m)!}\binom{m}{k}\left(2\sqrt{2\nu}\,u\right)^{m-k}. \tag{4.19}$$

For $\nu = 1/2$ the Whittle-Matérn model coincides with the exponential correlation, whereas at the limit $\nu \to \infty$ it tends to the Gaussian.

Another commonly used form of the Whittle-Matérn correlation is given by

$$\rho_{\mathrm{xx}}(u)=\frac{u^{\nu}}{2^{\nu-1}\,\Gamma(\nu)}\,K_\nu(u). \tag{4.20}$$

This is identical to the expression (4.12), except that the normalized distance u is no longer multiplied with $\sqrt{2\nu}$. The equation (4.12) is preferable for estimation purposes, as we discuss in Chap. 12. On the other hand, the expression (4.20) is simpler and thus more often used in the literature. In Fig. 4.3 we plot the Whittle-Matérn correlation model (4.20) for different values of ν. If we plotted (4.12) instead, the graphs for different ν would show considerably less dispersion. This is due to the fact that the integral range of the Whittle-Matérn correlation model depends on both ν and ξ. The integral range is defined in Sect. 5.4.2. The dependence of the Whittle-Matérn covariance integral range on ν and ξ is discussed in [380].

The value of the Whittle-Matérn correlation at zero distance is determined by the following limit

$$\lim_{u\to 0}\left[K_\nu(u)\,u^{\nu}\right]=2^{\nu-1}\,\Gamma(\nu).$$

This ensures the proper normalization of the Whittle-Matérn correlation function, i.e., $\rho_{\mathrm{xx}}(0)=1$.

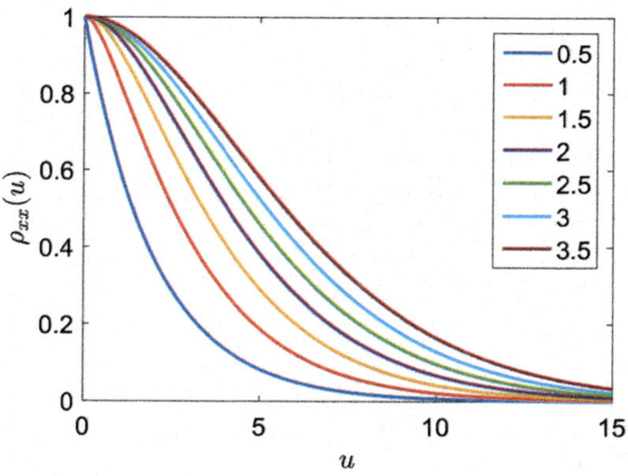

Fig. 4.3 Whittle-Matérn (W-M) correlation function (4.20) for smoothness index that ranges between 0.5 and 3.5 in steps of 0.5. The variance is equal to one and the characteristic length to $\xi = 2$. The smoothness index of the W-M functions increases from left to right. For the form of the W-M function with $\nu = 0.5, 1, 1.5, 2.5$ see also Table 4.3

Table 4.3 Special cases of the Whittle-Matérn isotropic correlation model. The dimensionless distance is $u = \|\mathbf{r}\|/\xi$. The formula (4.20) is used for the correlation function

ν	Expression	Common name
$\nu = 1/3$	$\left(2^{2/3}/\,\Gamma(1/3)\right)\, u^{1/3}\, K_{1/3}(u)$	von Kármán
$\nu = 1/2$	$\exp(-u)$	Exponential
$\nu = 1$	$u\, K_1(u)$	Whittle
$\nu = 3/2$	$(1+u)\exp(-u)$	Modified exponential
$\nu = 5/2$	$\left(1+u+\frac{1}{3}u^2\right)\exp(-u)$	Third-order autoregressive
$\nu = m + 1/2$	$\exp(-u)\, p_m(u)$	$p_m(u)$ is a polynomial of degree m

The asymptotic behavior of the Whittle-Matérn correlation is determined by the product of the modified Bessel function $K_\nu(\cdot)$ and the algebraic function, which decays asymptotically as [779, p. 291]

$$K_\nu(u)\, u^\nu \sim \sqrt{\frac{\pi}{2}}\, u^{\nu-1/2}\, \mathrm{e}^{-u}, \quad u \to \infty.$$

The notation $f(u) \sim g(u)$ means that the functions $f(u)$ and $g(u)$ are asymptotically equivalent in the sense that $\lim_{u\to\infty} \frac{f(u)}{g(u)} = 1$.

The Whittle-Matérn covariance function has been receiving considerable attention in the statistical literature [45, 290, 319, 327]. This popularity is partly due to its flexibility, and partly to its connection with a fractional partial differential equation which is the pivot for a powerful representation of spatial random fields [503, 846].

The Table 4.3, adapted from [319], presents special cases of the Whittle-Matérn covariance function. The formulation without the $\sqrt{2\nu}$ factor is used.

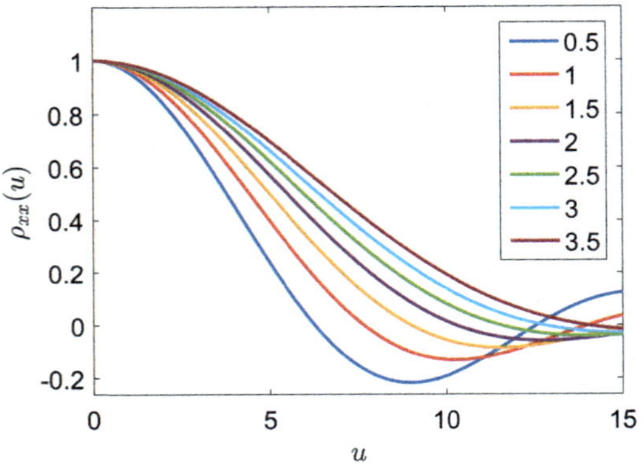

Fig. 4.4 Plots of the Bessel-J correlation function (4.13) for values of the index ν that range from 0.5 to 3.5 in steps of 0.5. The variance is set equal to one and the characteristic length equal to $\xi = 2$. The index ν increases as we move from left to right

J-Bessel The J-Bessel correlation model

$$\rho_{\mathrm{xx}}(u) = 2^{\nu}\,\Gamma(\nu+1)\,u^{-\nu}\,J_{\nu}(u), \tag{4.21}$$

where $\nu \geq d/2 - 1$, involves the *Bessel function* $J_\nu(\cdot)$ *of the first kind* of order ν [487]. The limiting behavior of the J-Bessel function as $u \to 0$ is given by [306, 8.402]

$$\lim_{u\to 0}\left[\frac{J_\nu(u)}{u^\nu}\right] = \frac{1}{2^\nu\,\Gamma(\nu+1)}. \tag{4.22}$$

The limit (4.22) ensures the proper normalization, $\rho_{\mathrm{xx}}(0) = 1$, of the J-Bessel correlation function.

In Fig. 4.4 we plot the Bessel-J correlation model (4.13) for different values of ν. The function $J_\nu(u)$ has an infinite number of real zeros for $\nu > -1$. Note that for $d = 1$ $\nu_{\min} = -1/2$, for $d = 2$ $\nu_{\min} = 0$, and for $d = 3$ $\nu_{\min} = 1/2$. Hence, all of the functions plotted are valid covariance models for $d \leq 3$.

Properties of $J_\nu(u)$ Below we discuss some useful mathematical properties of Bessel functions of the first kind. For more information, the reader can consult the classic text by Watson [834].

Series representation More generally, the Taylor series expansion of the J-Bessel correlation function around $u = 0$ is given by [306, 8.440]

$$\frac{J_\nu(u)}{u^\nu} = 2^{-\nu}\sum_{m=0}^{\infty}\frac{(-1)^m}{m!\,\Gamma(n+\nu+1)}\left(\frac{u}{2}\right)^{2m}. \tag{4.23}$$

Asymptotic expansion At large distances, the limit of the J-Bessel correlation (for real values of the argument) is given by the damped harmonic function [306, 8.451]

$$\frac{J_\nu(u)}{u^\nu} \sim \frac{\sqrt{2}}{\sqrt{\pi}\,u^{\nu+1/2}}\cos\left(u - \nu\frac{\pi}{2} - \frac{\pi}{4}\right),\quad u \to \infty.$$

Closure property The closure property of Bessel functions of the first kind is often useful in calculations. It is expressed as follows

$$\int_0^\infty dx\, J_\nu(a\,x)\, J_\nu(b\,x) = \frac{1}{a}\delta(a-b). \tag{4.24}$$

Orthogonality There are two useful orthogonality properties for Bessel functions of the first kind. The first relation applies to an infinite domain, while the second one to a bounded interval.

$$\int_0^\infty dx\, \frac{1}{x}\, J_\nu(x)\, J_\mu(x) = \frac{2}{\pi}\frac{\sin\left(\frac{\pi}{2}(\mu-\nu)\right)}{\mu^2-\nu^2}. \tag{4.25}$$

$$\int_0^1 dx\, x\, J_\nu(x\, u_{\alpha,m}) J_\nu(x\, u_{\alpha,n}) = \frac{\delta_{m,n}}{2}[J_{\nu+1}(u_{\alpha,m})]^2, \tag{4.26}$$

where the α_m and α_n $(m, n = 1, \ldots, \infty)$ represent, respectively, the m-th and n-th zeros of the Bessel function $J_\nu(\cdot)$.

Example 4.2 Show that the spectral density of the J-Bessel function with $\nu = d/2$ is given by

$$\widetilde{\rho}_{xx}(\kappa) = (2\xi)^d \pi^{d/2}\Gamma(d/2+1)\,\Theta(1-\kappa), \tag{4.27}$$

where $\Theta(\cdot)$ is the unit step function defined by (2.7).

The spectral density (4.2) is non-zero inside the unit-radius hyper-sphere in the reciprocal space $\mathbb{R}^d$. It represents an isotropic generalization of the one-dimensional boxcar spectral density which is the Fourier transform of the cardinal sine function.

Answer Since the spectral density above is a radial function, we will use the inverse Hankel transform (4.4b). This leads to the following integral for the correlation function (where $k_c = 1/\xi$)

$$\rho_{xx}(\|\mathbf{r}\|) = \frac{2^{d/2}\,\Gamma(d/2+1)}{k_c{}^d\, \|\mathbf{r}\|^{d/2-1}} \int_0^{k_c} dk\, k^{d/2} J_{d/2-1}(k\|\mathbf{r}\|).$$

By means of the transformation $k/k_c \mapsto x$ the integral is transformed as follows

$$\rho_{xx}(\|\mathbf{r}\|) = \frac{2^{d/2}\,\Gamma(d/2+1)}{(k_c\, \|\mathbf{r}\|)^{d/2-1}} \int_0^1 dx\, x^{d/2} J_{d/2-1}(x k_c\, \|\mathbf{r}\|).$$

The above can be evaluated using the *index-raising integral* of the Bessel function [306, Eq. 6.561.5, p. 676], i.e.,

$$\int_0^1 dx\, x^{d/2}\, J_{d/2-1}(xa) = \frac{J_{d/2}(a)}{a}, \quad \text{where } a = k_c\|\mathbf{r}\|.$$

Based on this result, the correct expression, according to (4.21), for the Bessel-J correlation with $\nu = d/2$ is obtained, i.e.,

$$\rho_{\mathrm{xx}}(\|\mathbf{r}\|) = 2^{d/2}\,\Gamma(1+d/2)\,\frac{J_{d/2}(k_c\,\|\mathbf{r}\|)}{(k_c\,\|\mathbf{r}\|)^{d/2}}.$$

Rational quadratic This correlation function, given by

$$\rho_{\mathrm{xx}}(u) = \frac{1}{(1+u^2)^b}, \quad \text{where } b > 0, \tag{4.28}$$

decays asymptotically (for $\|\mathbf{r}\| \to \infty$) algebraically with the lag distance, i.e. $\rho_{\mathrm{xx}}(\|\mathbf{r}\|) \sim \|\mathbf{r}\|^{-2b}$.

Construction The rational quadratic covariance can be viewed as a superposition of Gaussian covariance functions with distributed correlation lengths that represent different length scales. The *scale mixture* is constructed so that the inverse of the squared correlation lengths follows the gamma pdf [678, 764].

The gamma probability distribution has the following pdf

$$f_{\mathrm{x}}(x;\alpha,\beta) = \frac{1}{\beta^{\alpha}\Gamma(\alpha)}x^{\alpha-1}\exp\left(-x/\beta\right), \quad x > 0,\ \alpha,\ \beta > 0. \tag{4.29}$$

The *gamma distribution* is a commonly used model for the marginal distribution of skewed, positively-valued random variables. Special cases of the gamma distribution are the chi-square and Erlang distributions. The superposition of the squares of a number of zero-mean, independent and identically distributed Gaussian variables follows the gamma distribution [424].

A random field with rational quadratic correlation is called *long-ranged* if the d-dimensional volume integral of the covariance function diverges, i.e., if

$$\int_{\mathbb{R}^d} d\mathbf{r}\, C_{\mathrm{xx}}(\mathbf{r}) \to \infty.$$

The divergence occurs if $d \geq 2b$. Then, the necessary condition in *Slutsky's theorem* (Theorem 4.1) for ergodicity in the mean is not satisfied. Hence, the random field is non-ergodic if $b \leq d/2$. This implies that the slow decay of the correlations at large distances does not allow to sufficiently sample the mean based on a single realization.

For $b > d/2$ the covariance function is *short-ranged*, since the covariance volume integral converges [92]. For a non-negative covariance function, *Slutsky's condition* for ergodicity is equivalent to the sufficient condition for the Fourier transform of $C_{\mathrm{xx}}(\mathbf{r})$ to exist.

Cauchy class This family comprises correlation functions of the form

$$\rho_{\mathrm{xx}}(u) = \frac{1}{(1+u^{\alpha})^{\beta/\alpha}}, \text{ where } 0 < \alpha \leq 2, \ \beta \geq 0. \tag{4.30}$$

The Cauchy model exhibits power-law dependence with different exponents at small and large lags. In particular, the variogram function increases as a power law with exponent α at small lags, while the correlation function decays as a power-law with exponent β at large lags. In both limits, the correlation function exhibits *self-affine scaling*: $C_{\mathrm{xx}}(\lambda u) \sim \lambda^{-\beta} u^{-\beta}$ as $u \to \infty$ and $\gamma_{\mathrm{xx}}(\lambda u) \sim \lambda^{\alpha} u^{\alpha}$ as $u \to 0$.

Definition 4.2 *Self-affine scaling:* Self-affine scaling is a type of distorted self-similarity: a mathematical object is self-similar if it remains unchanged by a scaling transformation, i.e., by multiplying all length scales by the same factor. In self-affinity, for the object to remain invariant different scaling factors should be used in different directions. Hence, a random field $X(\mathbf{s};\omega)$ exhibits self-affine scaling if the statistical properties of the field remain invariant under the transformation $X(\mathbf{s};\omega) \mapsto \lambda^{\alpha} X(\lambda\mathbf{s};\omega)$.

In the case of a random field with Cauchy covariance, the length scaling transformation $\mathbf{s} \mapsto \lambda\mathbf{s}$ leaves the the long-distance correlation function invariant if the field is scaled as $X(\mathbf{s};\omega) \mapsto \lambda^{\beta/2} X(\lambda\mathbf{s};\omega)$.

The scaling behavior of the Cauchy model involves a decoupling of the *fractal dimension* (see Sect. 5.5), which characterizes self-affinity at short-distances, and the *Hurst exponent* that characterizes the long-distance self-affinity [292]. We will return to this topic below.

The *rational quadratic function* is a special case of the Cauchy class obtained for $\alpha = 2$ and $\beta = 2b$.

Other covariance models are presented in different parts of this book. For example, in Sect. 5.6 we discuss a covariance based on the incomplete gamma function that has a power-law tail. In Chap. 7 we present the classes of Spartan and Bessel-Lommel covariance functions.

4.2.3 Radon Transform

It has been shown that covariance functions in $\mathbb{R}^d$ can be constructed from one-dimensional projections [554]. Hence, it is possible to compose a d-dimensional covariance function by superposition of one-dimensional components that represent permissible covariance functions in $d = 1$. These constructions are simplified in the case of isotropic random fields. The mathematical tool that enables the generation of such covariance functions is the *Radon transform* [132, 187, 335].

Turning bands method Matheron showed that given a one-dimensional random process $\mathrm{Y}(t;\omega)$ with covariance function $C_1(\tau)$ and a unit random vector $\hat{\mathbf{p}}$, one can define the one-dimensional random field $\mathrm{X}_{\hat{\mathbf{p}}}(\mathbf{s};\omega) = \mathrm{Y}(\mathbf{s}\cdot\hat{\mathbf{p}};\omega)$ that has covariance function $C_1\left(\mathbf{r}\cdot\hat{\mathbf{p}}\right)$ [554]. The one-dimensional field $\mathrm{X}_{\hat{\mathbf{p}}}(\mathbf{s};\omega) = \mathrm{Y}(\mathbf{s}\cdot\hat{\mathbf{p}};\omega)$ can be viewed as the projection of a d-dimensional field $\mathrm{X}(\mathbf{s};\omega)$ along the direction $\hat{\mathbf{p}}$.

Then, the covariance of the original field $\mathrm{X}(\mathbf{s};\omega)$ is given by the following integral over all possible directions

$$C_{\mathrm{xx}}^{(d)}(\mathbf{r}) = \int_{\mathcal{B}_d} \mathrm{d}\Omega_d(\mathbf{u})\, C_1\left(\mathbf{r}\cdot\mathbf{u}\right)\, f_{\hat{\mathbf{p}}}(\mathbf{u}), \tag{4.31a}$$

where $\mathcal{B}_d$ denotes the surface of the unit sphere in d dimensions and $f_{\hat{\mathbf{p}}}(\mathbf{u})$ is a pdf that determines the distribution of the unit vector $\hat{\mathbf{p}}$ on the surface of the unit sphere. In the case of uniform distribution of the unit vectors on $\mathcal{B}_d$ there is no directional preference, and the integral (4.31) becomes

$$C_{\mathrm{xx}}^{(d)}(\|\mathbf{r}\|) = \frac{1}{S_d}\int_{\mathcal{B}_d} \mathrm{d}\Omega_d(\mathbf{u})\, C_{\mathrm{xx}}^{(1)}\left(\mathbf{r}\cdot\mathbf{u}\right). \tag{4.31b}$$

Dimension raising operator Equation (4.31b) leads to a direction-independent covariance function. By taking into account the radial symmetry, the covariance function in d dimensions is given by the following one-dimensional integral [554], [132, p. 505], [487, p. 195][4]

$$C_{\mathrm{xx}}^{(d)}(\|\mathbf{r}\|) = \frac{2S_{d-1}}{S_d}\int_0^1 \mathrm{d}z\,(1-z^2)^{(d-3)/2}\, C_{\mathrm{xx}}^{(1)}(z\|\mathbf{r}\|). \tag{4.32}$$

This integral expression can be used to derive covariance functions, $C_{\mathrm{xx}}^{(d)}(\|\mathbf{r}\|)$, in $d>1$ based on permissible covariance functions, $C_{\mathrm{xx}}^{(1)}(\|\mathbf{r}\|)$, in $d=1$. In particular, in $d=3$ the covariance functions $C_{\mathrm{xx}}^{(3)}(\|\mathbf{r}\|)$ and $C_{\mathrm{xx}}^{(1)}(\|\mathbf{r}\|)$ are related to each other by means of the following equations [233, 534]

$$C_{\mathrm{xx}}^{(3)}(\|\mathbf{r}\|) = \frac{1}{\|\mathbf{r}\|}\int_0^{\|\mathbf{r}\|} \mathrm{d}x\, C_{\mathrm{xx}}^{(1)}(x), \tag{4.33a}$$

$$C_{\mathrm{xx}}^{(1)}(\|\mathbf{r}\|) = \frac{\mathrm{d}}{\mathrm{d}\|\mathbf{r}\|}\left[\|\mathbf{r}\|\, C_{\mathrm{xx}}^{(3)}(\|\mathbf{r}\|)\right]. \tag{4.33b}$$

Equation (4.33a) is a straightforward application of (4.32), while (4.33a) simply follows from the inversion of (4.33a). Similar, albeit slightly more complicated relations are obtained for the two-dimensional covariances [487].

---

[4]In [487] this equation is expressed in terms of the volume V_d of the unit sphere which is related to the surface area by means of $S_d = d\,V_d$.

Various extensions of the exponential function (e.g., the modified exponential and the third-order autoregressive models in Table 4.3) are generated based on the dimension-raising operator. In addition, generalizations of the spherical correlation function (such as the cubic and the pentamodel) are also based on the above integral formula [132, pp. 87–89].

TBM realizations Equation (4.31) forms the basis of the so-called *turning bands method (TBM)* that allows the simulation of random fields based on the superposition of several synthetic one-dimensional processes [233, 534]. More specifically, if N_l different directions are used, the field $\mathrm{X}(\mathbf{s};\omega)$ where $\mathbf{s} \in \mathbb{R}^d$ is given by [487]

$$\mathrm{X}(\mathbf{s};\omega) = \frac{1}{N_l}\sum_{l=1}^{N_l} \mathrm{Y}(\mathbf{s}\cdot\hat{\mathbf{p}}_l;\omega). \tag{4.34}$$

In practice, the N_l directions are chosen randomly or based on quasi-random number sequences (see Chap. 16). The latter lead to a more homogeneous distribution of the unit vector endpoints on the surface of the unit sphere.

4.3 Anisotropy

A property is called anisotropic if its values depend on the space direction along which it is measured. Spatial data may exhibit various manifestations of anisotropy. For the purpose of this book, we are primarily interested in *statistical anisotropy* (defined below). We will make the distinction between statistical anisotropy and what we will call (for lack of a better term) *physical anisotropy*. Geostatisticians are usually more familiar with statistical anisotropy, while physicists and engineers are more likely familiar with the concept of physical anisotropy.

4.3.1 Physical Anisotropy

Coefficients that determine the response of physical systems to external excitations (e.g., hydraulic conductivity, dielectric permittivity) can take different values in different directions. This behavior implies a *tensorial* representation of the coefficients. For example, if $\mathsf{K}(\mathbf{s})$ is a spatially dependent function in three dimensions, it is expressed in matrix form as follows

$$\mathsf{K}(\mathbf{s}) = \begin{pmatrix} K_{1,1}(\mathbf{s}) & K_{1,2}(\mathbf{s}) & K_{1,3}(\mathbf{s}) \\ K_{2,1}(\mathbf{s}) & K_{2,2}(\mathbf{s}) & K_{2,3}(\mathbf{s}) \\ K_{3,1}(\mathbf{s}) & K_{3,2}(\mathbf{s}) & K_{3,3}(\mathbf{s}) \end{pmatrix},$$

where $\{K_{i,j}(\mathbf{s})\}_{i,j=1}^3$ are scalar functions of space (not necessarily random). $\mathsf{K}(\mathbf{s})$ may represent a deterministic, even uniform, quantity, or a *tensor random field* $\mathsf{K}(\mathbf{s};\omega)$.

In anisotropic systems and materials, tensor coefficients are used to relate a vector stimulus (e.g., electric field) to a vector response (e.g., polarization). This is the kind of anisotropy most commonly used in the study of physical and biological systems. For example, *diffusion tensor imaging* (DTI) is a magnetic resonance imaging method that exploits the tensor nature of the diffusion coefficient of water molecules in the brain in order to map the anisotropic distribution of white matter in the brain.

Tensor coefficients are in general represented by full matrices. There is, however, a system of principal coordinates in which the respective matrix becomes diagonal. The principal coordinate system depends on the symmetries of the observed system and is not necessarily aligned with the coordinate system.

4.3.2 Statistical Anisotropy

Statistical anisotropy refers to the dependence of statistical moments on the direction in space. In contrast with physical anisotropy that manifests in terms of tensor functions, statistical anisotropy is a property that characterizes the *statistical moments of scalar quantities*.

Trend function anisotropy Statistical anisotropy often emerges in terms of direction dependent coefficients in the trend function. For example, the linear trend function $m_{\mathrm{x}}(\mathbf{s}) = 10x + 3y + c$, where $\mathbf{s} = (x, y)$ changes faster along the x than along the y direction. The difference in the rate of change of $m_{\mathrm{x}}(\mathbf{s})$ along different directions is a mark of statistical anisotropy in the trend.

Anisotropy in correlation functions Statistical anisotropy can also be present in the correlation functions, where it is more difficult to estimate accurately. Statistical anisotropy may manifest itself by direction dependent variogram sill, range, and regularity at the origin [17].

Regularity parameter The *regularity parameter* determines how fast the variogram changes near the origin. For example, in the case of the fractional Brownian motion (see Chap. 5) the variogram function is $\gamma_{\mathrm{xx}}(\mathbf{r}) \propto \|\mathbf{r}\|^{2H}$, and the regularity parameter is determined by the Hurst exponent H. However, as shown in [179] and discussed in [17], if the regularity of a variogram *varies continuously* with the space direction, it must be uniform in all directions. Hence, it is not permissible to have continuously varying anisotropic dependence of the regularity parameter.

Range anisotropy is the most common type of statistical anisotropy. It refers to a covariance function (or equivalently to a variogram function) that has different characteristic lengths depending on direction. This type of anisotropy can apply to

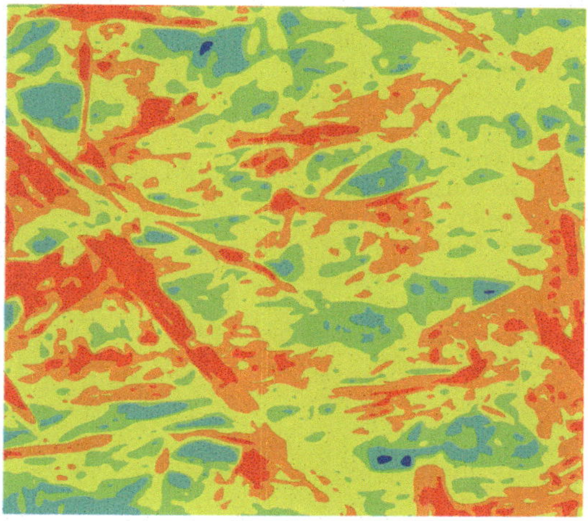

Fig. 4.5 Paper roughness (surface height above reference level) for a square paper sample of size $500 \times 500\,\mu\text{m}^2$ observed at resolution of $1\,\mu$m. A single fiber stretches nearly from the middle of the left boundary to the middle of the bottom boundary

scalar random field $X(\mathbf{s};\omega)$, vector random fields $\mathbf{X}(\mathbf{s};\omega)$, or tensor random fields $\mathsf{X}(\mathbf{s};\omega)$.

An implication of unequal characteristic lengths is that there exist at least two non-identical unit vectors $\hat{\mathbf{e}}_i$ and $\hat{\mathbf{e}}_j$ such that $C_{\mathrm{xx}}(\|\mathbf{r}\|\,\hat{\mathbf{e}}_i) \neq C_{\mathrm{xx}}(\|\mathbf{r}\|\,\hat{\mathbf{e}}_j)$. Nonetheless, in the case of range anisotropy the covariance function satisfies the relation

$$C_{\mathrm{xx}}(\mathbf{r}) = \sigma_{\mathrm{x}}^2\,\rho_{\mathrm{xx}}(\mathbf{r}),$$

which implies that (i) the variance is independent of direction and (ii) the anisotropy affects the correlation function $\rho_{\mathrm{xx}}(\mathbf{r})$ at distance lags $\mathbf{r} \neq \mathbf{0}$.

An example of a scalar random field with *range anisotropy* is illustrated in Fig. 4.5, which shows the surface height of a square piece of paper. The sample represents an anisotropic fiber network generated by the deposition of individual fibers (typical fiber length 4–5 mm).

Another example of *range anisotropy* is shown in Fig. 4.6, which displays the geological structure of layers in a lignite deposit. Essentially laminar structure is observed along the layers (i.e., horizontally), but the characteristic length scale that determines the variability in the vertical direction is much shorter, which is a clear sign of range anisotropy. The lignite structure is actually even more complicated, because in addition to anisotropy one can also observe geological discontinuities (slips) of the layers.

Transverse isotropy Geological media and environmental processes (e.g., precipitation) often exhibit a restricted type of anisotropy which is known as *transverse isotropy*. Transversely isotropic random fields have the same correlation lengths within the "horizontal" plane and a different correlation length in the "perpendicular" direction. The effect of transverse isotropy on the shape of the correlation function isolevel contours is illustrated in Fig. 4.7.

Fig. 4.6 Structure of lignite layers from a mine in the South Field of Western Macedonia (Greece). Range anisotropy is evidenced in the very different variability along the horizontal compared to the vertical direction. Note the discontinuity in the lignite layer caused by geological faulting near the top left corner of the figure

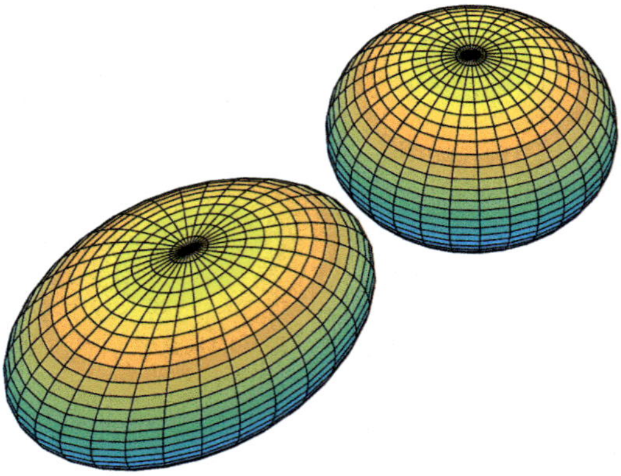

Fig. 4.7 Comparison of isolevel ellipsoidal surfaces for anisotropic (left) and transversely isotropic (right) correlation models

Range anisotropy From the practitioner's viewpoint, *range anisotropy* of random field $X(\mathbf{s}; \omega) : \mathbb{R}^d \rightarrow \mathbb{R}$, where $d \geq 2$, is evidenced in the presence of at least two non-identical characteristic lengths (along any two non-coinciding directions) in the covariance function and the variogram.

While the presence of anisotropic structures generates unequal correlation lengths, structural discontinuities caused by faults and other geological factors (cf. Fig. 4.6) can lead to points that are close in space having very different attribute values. Such effects can generate a downward jump of the covariance function (equivalently, an upward jump of the variogram) just after the origin (zero lag) that is commonly known as the *nugget effect*. In this perspective, the signature of discontinuities in the covariance (or the variogram) is indistinguishable from that of uncorrelated random fluctuations due to sampling errors.

We will next consider a specific type of range anisotropy, which is known as *elliptical anisotropy* and *geometric anisotropy*.

Principal anisotropy axes Let us assume that the correlation function $\rho_{xx}(\mathbf{r})$ that has range anisotropy can be expressed in terms of the one-dimensional function $\phi(h)\colon \mathbb{R} \to \mathbb{R}$ as follows:

$$\rho_{xx}(\mathbf{r}) = \phi(\mathbf{r}^{\top}\,\mathbf{A}\,\mathbf{r}), \tag{4.35}$$

where $\mathbf{A}$ is a real-valued, symmetric matrix with dimension $[\mathrm{L}]^{-2}$, so that the bilinear form $h = \left(\mathbf{r}^{\top}\,\mathbf{A}\,\mathbf{r}\right)^{1/2}$ represents a dimensionless distance.

An *isolevel contour* K of the correlation function at level $c \in \mathbb{R}$ such that $|c| \leq 1$, is the locus formed by all the points $\mathbf{r}$ such that[5]

$$\rho_{xx}(\mathbf{r}) = c.$$

Hence, the isolevel contour K_c is the locus of points $\mathbf{r}$ that satisfy the equation

$$K_c = \{\mathbf{r} \in \mathbb{R}^d :\ \text{such that } \mathbf{r}^{\top}\,\mathbf{A}\,\mathbf{r} = \phi^{-1}(c)\}. \tag{4.36}$$

Equation (4.36) defines an ellipsoid in three dimensions and an ellipse in two dimensions. This dependence of the isolevel contours justifies the term *elliptical anisotropy*. Note that the $\mathbf{r}$ represent lag vectors and the isolevel contours do not refer to the random field values but rather to the correlation function.

Isotropy restoring transformation Given a certain set of spatial data, both $\mathbf{r}$ and $\mathbf{A}$ depend on the selection of the coordinate system. Let us assume that the coordinate system is transformed by the action of an *orthogonal transformation matrix* $\mathbf{B}$.[6] The coordinates $\mathbf{r}'$ in the new (primed) system are related to the coordinates in the old system by means of

$$\mathbf{r}' = \mathbf{B}\,\mathbf{r}.$$

In the new coordinate system, the matrix $\mathbf{A}$ is transformed into $\mathbf{A}'$ by means of the following similarity transformation

$$\mathbf{A}' = \mathbf{B}\mathbf{A}\mathbf{B}^{-1}.$$

Then, it is straightforward to show that the correlation function in the transformed coordinate system satisfies the equation

---

[5]For simplicity we talk about isolevel contours, but in three dimensional problems these actually refer to surfaces.

[6]A matrix $\mathbf{B}$ is orthogonal if $\mathbf{B}^{\top} = \mathbf{B}^{-1}$.

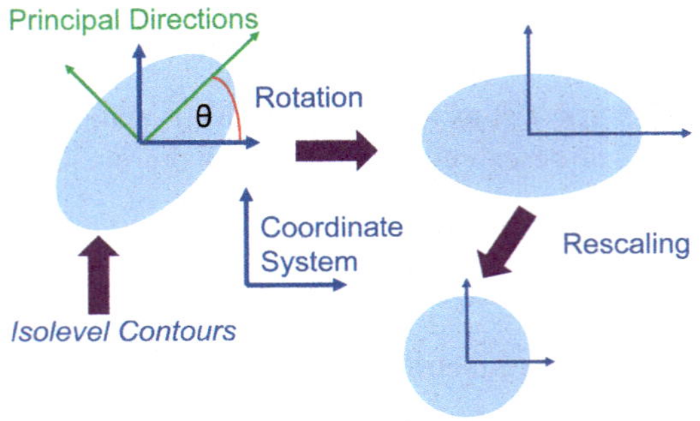

Fig. 4.8 Schematic of isotropy restoring transformation. Top left: Elliptical correlation isolevel contours in given coordinate system. Top right: Elliptical correlation isolevel contours in coordinate system aligned with the principal axes of correlation. Bottom right: Circular isolevel contours in rotated and rescaled coordinate system

$$\phi(\mathbf{r}^{\top}\,\mathbf{A}\,\mathbf{r}) = \phi(\mathbf{r}^{'\top}\,\mathbf{A}'\,\mathbf{r}').$$

The coordinate system in which the transformed matrix $\mathbf{A}'$ is diagonal is called the *system of principal axes of anisotropy*. The correlation lengths in this system are called *principal correlation lengths*. Furthermore, we can select one of the principal axes as reference and define $d-1$ aspect ratios of the principal lengths in the orthogonal (to the reference) directions over the reference principal length. A rescaling of the lengths along the $d-1$ axes by the respective aspect ratio generates a new coordinate system in which the random field appears statistically isotropic.

Isotropy restoring transformation in two dimensions In two space dimensions, the isolevel correlation contours are ellipses with principal axes that are possibly tilted with respect to the coordinate axes, as shown in Fig. 4.8.

Rotation angle We will refer to the angle $\theta \in (-\pi/2, \pi/2]$ between the horizontal axis of the coordinate system and the nearest anisotropy principal axis as the *rotation angle* (see Fig. 4.8.)

The angle is considered *positive* if the principal axis is rotated counter-clockwise by θ and *negative* if the principal axis is rotated clockwise with respect to the coordinate axis.

Remark In some research fields, the convention is to measure the rotation angle with respect to the vertical axis of the coordinate system; then, the rotation angle is $\theta' = \pi/2 - \theta$.

Rotation of the original coordinate system *counterclockwise* by θ aligns it with the principal axes of anisotropy. If θ is negative, the above operation is equivalent to a clockwise rotation. Furthermore, the isolevel contours transform into circles by a length transformation. This involves rescaling the lengths along one of the principal axes which is selected arbitrarily. For example, if we select the axis PA1, we divide all the coordinates in this direction by the *aspect ratio* $\lambda := \xi_1/\xi_2$, where ξ_1 is the correlation length along PA1 and ξ_2 is the correlation length along PA2. Following this transformation, the correlation lengths become identical in the rescaled system.

Degeneracy of isotropy restoring transformation The isotropy restoring transformation in two dimensions is determined by two parameters: the rotation angle θ and the aspect ratio λ. The aspect ratio can be defined with either characteristic length in the denominator. The characteristic length in the numerator is an overall scaling factor that does not affect the transformation.

The rotation angle is restricted to take values in the range between $-\pi/2$ and $\pi/2$. This is sufficient, because for angles $\pi > \theta > \pi/2$ the isolevel contours are invariant under the affine transformation $\theta \mapsto \pi - \theta$. Similarly, for $-\pi < \theta < -\pi/2$, the contours are invariant under the transformation $\theta \mapsto \pi + \theta$.

In addition, to the above, rotation angles $|\theta| > \pi/4$ lead to degenerate solutions, since the isolevel contours are invariant under the transformations

$$\lambda \mapsto \lambda = 1/\lambda, \quad \theta \mapsto \theta' = \begin{cases} \theta - \pi/2 : \theta > \pi/4, \\ \pi/2 + \theta : \theta < -\pi/4, \end{cases} \tag{4.37}$$

The above discussion shows that the isotropy restoring transformation is not uniquely defined, unless we restrict the range of the anisotropy parameters. The *principal (non-degenerate) branch* of the anisotropy parameters (θ, λ) can be defined in two equivalent ways as follows:

$$\theta \in \left[-\frac{\pi}{4}, \frac{\pi}{4}\right], \ \lambda \in [0, \infty) \ \vee \ \theta \in \left[-\frac{\pi}{2}, \frac{\pi}{2}\right], \ \lambda \in [0, 1].$$

Example 4.3 Let us consider the anisotropic Gaussian correlation function in $d = 2$. If $\mathbf{r} = (r_1, r_2)$, the anisotropic Gaussian is expressed as

$$\rho_{xx}(\mathbf{r}) = e^{-\left(a_1 r_1^2 + a_2 r_2^2 + a_{1,2} r_1 r_2\right)},$$

where a_1 and a_2 are positive coefficients and $a_{1,2}$ a real-valued coefficient.

Determine (i) the angle θ by which the coordinate system should be rotated in order to align with the principal axes of anisotropy and (ii) the characteristic lengths in the coordinate system of the principal axes.

Answer Let $\mathbf{r}' = (r_1', r_2')$ represent the lag vector in the *primed coordinate system*, which is rotated *counterclockwise* by θ with respect to the original system. The vector $\mathbf{r}'$ is then given by

$$\begin{pmatrix} r'_1 \\ r'_2 \end{pmatrix} = \begin{pmatrix} \cos\theta & -\sin\theta \\ \sin\theta & \cos\theta \end{pmatrix} \begin{pmatrix} r_1 \\ r_2 \end{pmatrix}. \tag{4.38}$$

Equivalently, the *original system* is obtained from the *primed system* through a *clockwise* rotation by θ, i.e.,

$$\begin{pmatrix} r_1 \\ r_2 \end{pmatrix} = \begin{pmatrix} \cos\theta & \sin\theta \\ -\sin\theta & \cos\theta \end{pmatrix} \begin{pmatrix} r'_1 \\ r'_2 \end{pmatrix}. \tag{4.39}$$

Let us now denote by $g(\mathbf{r}) := a_1 r_1^2 + a_2 r_2^2 + a_{1,2} r_1 r_2$ the exponent of the correlation function. Using the rotation transformation defined above, $g(\mathbf{r})$ is expressed in the primed system as follows:

$$\begin{aligned} g(\mathbf{r}) = & \, r'^2_1 \left(a_1 \cos^2\theta + a_2 \sin^2\theta - a_{1,2} \cos\theta \sin\theta \right) \\ & + r'^2_2 \left(a_1 \sin^2\theta + a_2 \cos^2\theta + a_{1,2} \cos\theta \sin\theta \right) \\ & + r'_1 r'_2 \left[(a_1 - a_2) \sin 2\theta + a_{1,2} (\cos^2\theta - \sin^2\theta) \right]. \end{aligned}$$

The cross term, which is proportional to $r'_1 r'_2$, should vanish in the principal system. This is accomplished if the rotation angle is given by

$$\theta = \frac{1}{2} \arctan\left(-\frac{a_{1,2}}{a_1 - a_2} \right).$$

Thus, the exponent of the anisotropic Gaussian correlation in the principal axes system becomes

$$\begin{aligned} g(\mathbf{r}) = & \, r'^2_1 \left(a_1 \cos^2\theta + a_2 \sin^2\theta - a_{1,2} \cos\theta \sin\theta \right) \\ & + r'^2_2 \left(a_1 \sin^2\theta + a_2 \cos^2\theta + a_{1,2} \cos\theta \sin\theta \right). \end{aligned}$$

Based on the above, the characteristic lengths in the system of the principal anisotropy axes are given by

$$\begin{aligned} \xi_1^2 = & \frac{1}{a_1 \cos^2\theta + a_2 \sin^2\theta - a_{1,2} \cos\theta \sin\theta}, \\ \xi_2^2 = & \frac{1}{a_1 \sin^2\theta + a_2 \cos^2\theta + a_{1,2} \cos\theta \sin\theta}. \end{aligned}$$

Let us now define the aspect ratios $\lambda_1 = a_2/a_1 > 0$ and $\lambda_2 = a_{1,2}/a_1$. Without loss of generality we can set $a_1 = 1$, so that a_1 is used as an overall scaling factor that is absorbed in the ξ_i, $i = 1, 2$. In light of the aspect ratios, the rotation angle becomes

$$\theta = \frac{1}{2} \arctan\left(-\frac{\lambda_2}{1 - \lambda_1} \right), \quad \text{where } -\pi/4 < \theta \leq \pi/4.$$

The characteristic lengths in the principal axes system are then expressed as

$$\xi_1^2 = \frac{1}{\cos^2\theta + \lambda_1 \sin^2\theta - \lambda_2 \cos\theta \sin\theta},$$

$$\xi_2^2 = \frac{1}{\sin^2\theta + \lambda_1 \cos^2\theta + \lambda_2 \cos\theta \sin\theta}.$$

Note that the coefficients a_1, a_2 and $a_{1,2}$ cannot take values completely independently. This will become better understood in terms of the following example.

Example 4.4 In the Example 4.3, let the anisotropic Gaussian correlation function be expressed as follows in the system of principal axes

$$\rho_{xx}(\mathbf{r}) = e^{-g(\mathbf{r}')}, \quad g(\mathbf{r}') = \frac{r_1'^2}{\xi_1^2} + \frac{r_2'^2}{\xi_2^2}.$$

(i) Determine the respective expression of the exponent $g(\mathbf{r})$ in the original coordinate system. (ii) Determine the values of $a_1, a_2, a_{1,2}$ that correspond to $\lambda = \xi_1/\xi_2$ and θ, where λ and θ take values in the principal branch of anisotropy.

Answer Based on (4.39) it follows that $g(\mathbf{r}') \mapsto g(\mathbf{r})$, where

$$g(\mathbf{r}) = r_1^2 \left(\frac{\cos^2\theta}{\xi_1^2} + \frac{\sin^2\theta}{\xi_2^2} \right) + r_2^2 \left(\frac{\sin^2\theta}{\xi_1^2} + \frac{\cos^2\theta}{\xi_2^2} \right) - 2 r_1 r_2 \cos\theta \sin\theta \left(\frac{1}{\xi_1^2} - \frac{1}{\xi_2^2} \right).$$

Hence, based on the above it follows that ($\lambda = \xi_1/\xi_2$):

$$a_1 = \frac{1}{\xi_1^2} \left(\cos^2\theta + \lambda^2 \sin^2\theta \right), \tag{4.40a}$$

$$a_2 = \frac{1}{\xi_1^2} \left(\sin^2\theta + \lambda^2 \cos^2\theta \right), \tag{4.40b}$$

$$a_{1,2} = -2 \frac{\cos\theta \sin\theta}{\xi_1^2} \left(1 - \lambda^2 \right). \tag{4.40c}$$

Without loss of generality we assume that $\xi_1 = 1$, since ξ_1 is just a multiplicative factor that equally affects all three coefficients $a_1, a_2, a_{1,2}$.

The dependence of the coefficients on θ and λ is shown in Fig. 4.9. It is evidenced in the above plots that the range of values taken by $a_1, a_2, a_{1,2}$ depends on θ and λ. The fact that only certain combinations of coefficient values are feasible is shown

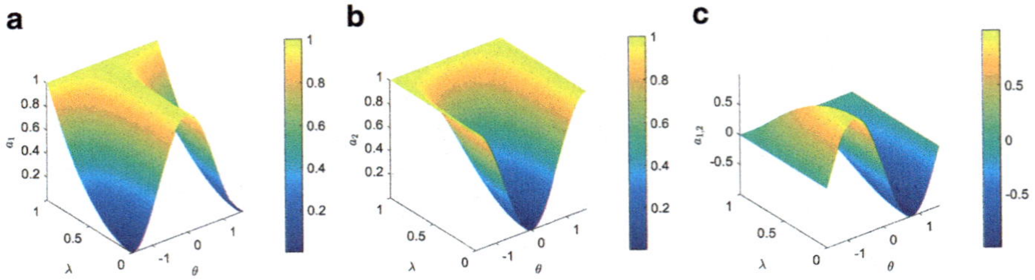

Fig. 4.9 Plot of the anisotropic coefficients $a_1, a_2, a_{1,2}$ based on (4.40a) versus the aspect ratio λ and the rotation angle θ. We use $\lambda \in [0, 1]$ and $\theta \in [-\pi/2, \pi/2]$ which is equivalent to $\lambda \in [0, \infty)$ and $\theta \in [-\pi/4, \pi/4]$ as discussed above

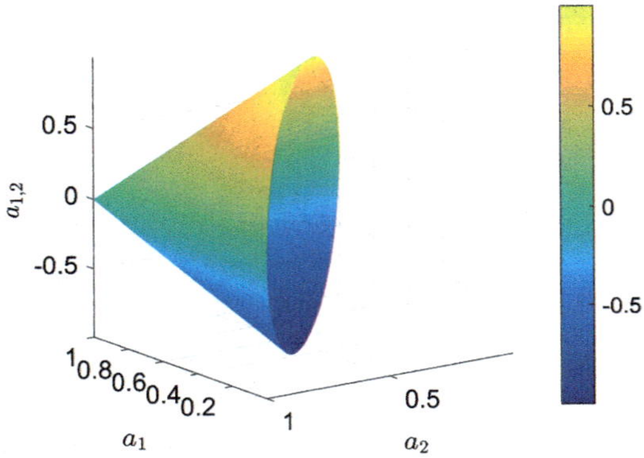

Fig. 4.10 Dependence of coefficient $a_{1,2}$ on the coefficients a_1 and a_2 for $\lambda \in [0, 1]$ and $\theta \in [-\pi/2, \pi/2]$

explicitly in Fig. 4.10. For example, consider the case $a_1 = a_2 = 1$; then, according to Fig. 4.10 the only value available for $a_{1,2}$ is zero. Why is that? Well, if $a_1 = a_2$ the characteristic lengths of the correlation function are equal; this means that the isolevel contours are circular. The equation of a circle, however, does not involve cross terms in any coordinate system, thus forcing $a_{1,2} = 0$.

4.3.3 *Anisotropy and Scale*

Above we distinguished between physical and statistical anisotropy. Whereas these are apparently distinct notions, there is a deeper connection between the two: *A scalar, space dependent local coefficient with anisotropic correlations can lead to an anisotropic (tensor) coefficient at larger scales.* We can appreciate this in terms of the diffusion tensor imaging example: The distribution of white matter in the brain is determined by a scalar density function with anisotropic spatial correlations. As a result of this structural anisotropy, the diffusion coefficient that determines the dispersion of water molecules in the brain becomes a tensor quantity.

In the geosciences, a characteristic example of anisotropy induced by upscaling is the hydraulic conductivity of random porous media [275, 276, 696]. Herein, we will be restricted to a bare-bones description of the problem.

The local-scale variability of the hydraulic conductivity is modeled as a scalar random field $\mathrm{K}(\mathbf{s};\omega)$ with anisotropic correlations that are presumed to represent the anisotropy of the geological medium. At the local scale, fluid flow within the medium is governed by Darcy's law, i.e.,

$$\mathbf{Q}(\mathbf{s};\omega) = -\mathrm{K}(\mathbf{s};\omega)\,\nabla\mathrm{H}(\mathbf{s};\omega), \tag{4.41}$$

where $\mathbf{Q}(\mathbf{s};\omega)$ is the local flux and $\nabla\mathrm{H}(\mathbf{s};\omega)$ is the gradient of the hydraulic head. At the field scale, under ergodic conditions,[7] the flow is presumably described by means of an *effective hydraulic conductivity* K_{eff}. The latter assumes the following general form

$$\mathcal{E}[\mathbf{Q}(\mathbf{s};\omega)] = -\mathsf{K}_{\text{eff}}\,\mathcal{E}[\nabla\mathrm{H}(\mathbf{s};\omega)], \tag{4.42}$$

where the ensemble average is calculated over the fluctuations of the local hydraulic conductivity field.

Evaluating the ensemble average is a technically challenging task that has attracted considerable attention in the porous media and hydrology literature (for a review see [360]). Let us assume that the average can be explicitly calculated, at least within the framework of low-order perturbation theory [276]. The main conclusions from the upscaling calculations are: (i) K_{eff} is a tensor. (ii) K_{eff} is diagonal tensor in the coordinate system that is aligned with the principal directions of anisotropy (these are determined by the principal axes of the local hydraulic conductivity correlations). (iii) The element of K_{eff} with the largest value is aligned with the principal direction that supports the largest correlation length.

4.3.4 Range Anisotropy Versus Elliptical Anisotropy

Range anisotropy is often referred to as elliptical anisotropy. Range anisotropy, however is more general than elliptical (geometric) anisotropy. Elliptical anisotropy assumes that the correlation function can be expressed in the functional form (4.35), i.e., as a function of the *quadratic form of the distance vector*, $\mathbf{r}^{\top}\mathbf{A}\,\mathbf{r}$. This mathematical expression indeed applies to many correlation functions.

---

[7]Ergodic conditions imply both translation invariance of the hydraulic conductivity correlations and infinite domain size (practically, much larger domain length in each orthogonal direction than the respective correlation length).

For example, consider the correlation functions listed in Sect. 4.2.2: All except for the first two can be extended to include elliptical anisotropy if the normalized distance u is replaced by $\left(\mathbf{r}^\top \mathbf{A}\,\mathbf{r}\right)^{1/2}$. Geometric anisotropy is not meaningful for the nugget effect which has a vanishing correlation range.

The cardinal sine model (4.10) is a *separable function* equal to the product of one-dimensional covariance functions; as a result, the cardinal sine covariance does not depend on the Euclidean distance, even in the "isotropic case" (i.e., when all the characteristic lengths are equal). Separable covariance functions (with the exception of the Gaussian) usually exhibit range anisotropy that is not reducible to elliptical.

Separability This mathematically desirable property allows decoupling the spatial dependence in terms of one-dimensional functions. A separable covariance model can in general be expressed as follows:

$$C_{\mathrm{xx}}(\mathbf{r}) = \sigma_{\mathrm{x}}^2 \prod_{i=1}^{d} \rho_{\mathrm{xx}}^{(i)}(r_i), \tag{4.43}$$

where the $\rho_{\mathrm{xx}}^{(i)}(r_i)$ are permissible one-dimensional correlation models. Separability has been used to construct covariance models in multi-dimensional spaces. With suitable choice of correlation lengths in each direction, it also generates anisotropic covariance models.

Super-ellipsoidal functions If we are willing to accept separable covariance functions (at least as components of a multi-scale superposition), there exist several possibilities for functions with *non-elliptical range anisotropy*, such as the super-ellipsoidal covariance functions introduced in [359]. An example is the family of *exponential super-ellipsoidal* covariance functions

$$\sigma_{\mathrm{x}}^2 \prod_{i=1}^{d} \exp\left[-\left(\frac{|r_i|}{\xi_i}\right)^{2/\alpha} \right], \quad \text{where } \alpha > 1. \tag{4.44}$$

The parameter α is the *non-ellipticity exponent*: If $\alpha \approx 1$, the isolevel contours are almost elliptical, while larger values of α lead to progressive deviations from the elliptical shape.

The functions (4.44) are permissible covariance functions, because their one-dimensional components belong in the class of the *generalized exponential* (4.11). The isolevel contours of the super-ellipsoidal covariance functions are non-elliptical as shown in Fig. 4.11, and the deviation from the elliptical shape becomes more pronounced with increasing α.

4.3.5 Zonal Anisotropy

Sometimes the analysis of spatial data exhibits dependence of the variogram sill on the direction in space. The signature of this phenomenon is the emergence of

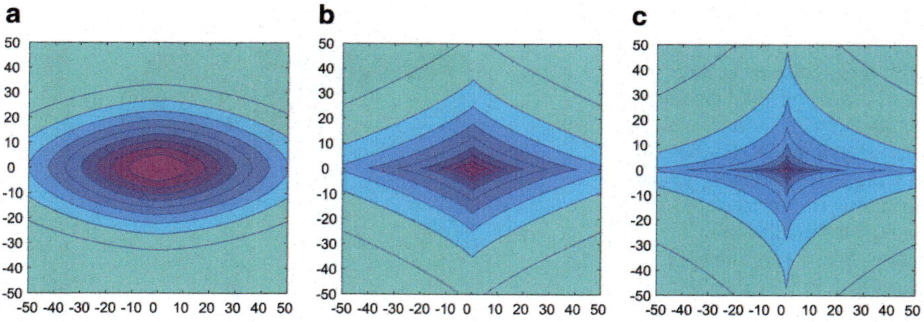

Fig. 4.11 Plots of isolevel contours for 2D superellipsoidal covariance functions (4.44) with $\xi_1 = 45$ and $\xi_2 = 20$ using three different values of the non-ellipticity exponent α. (**a**) $\alpha = 1.2$. (**b**) $\alpha = 2.4$. (**c**) $\alpha = 3.6$

Fig. 4.12 Low-dimensional wave pattern, caused by the motorboat's wake, embedded in two-dimensional space. Picture taken at Balos lagoon in Western Crete in the afternoon of a November day in 2012

unequal (direction dependent) sills in different directions. This property is known as *zonal anisotropy* [165, 303, 823].

From a modeling perspective, zonal anisotropy can be understood in terms of the superposition of two (or more) random fields, one of which has lower dimensionality than the embedding space. For example, this situation can occur if there is a practically linear pattern of variation superimposed on a two dimensional random field. In such cases, the overall variability is modeled as a composite random field that has higher variance in the direction(s) of the low-dimensional process. Figure 4.12 helps to visualize the concept of zonal anisotropy.

The variogram function of a process with zonal anisotropy in the direction $\hat{\mathbf{e}}_0$ is given by the equation

$$\gamma_{\mathrm{xx}}(\mathbf{r}) = \gamma^{(1)}(\mathbf{r}) + \gamma^{(2)}(\mathbf{r} \cdot \hat{\mathbf{e}}_0), \tag{4.45}$$

where $\gamma^{(1)}(\mathbf{r})$ is a variogram with isotropic dependence or range anisotropy, and $\gamma^{(2)}(\mathbf{r} \cdot \hat{\mathbf{e}}_0)$ is the variogram responsible for zonal anisotropy. Let c_1 be the sill of

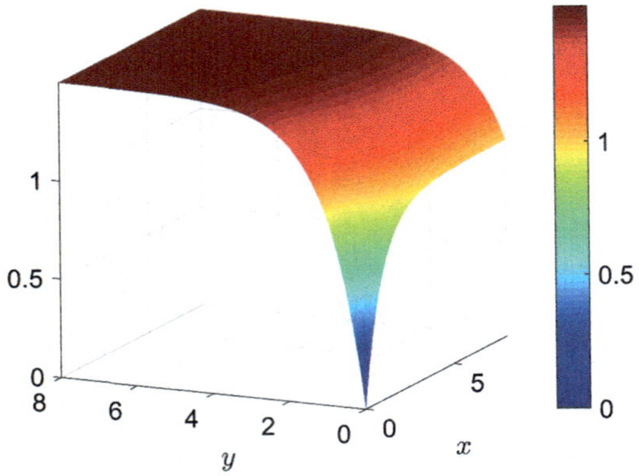

Fig. 4.13 Variogram isolevel contours for a random field with zonal anisotropy. An isotropic exponential model is combined with a directional exponential model. For both models $\xi = 1$, while the sill is equal to 1 for the isotropic and 0.5 for the directional model. The composite sill is equal to 1.5 in the direction y along $\hat{\mathbf{e}}_0 = (0, 1)$ and equal to one in the orthogonal direction x along $\hat{\mathbf{e}}_1 = (1, 0)$

$\gamma^{(1)}(\cdot)$ and c_2 be the sill of $\gamma^{(2)}(\cdot)$. For $\mathbf{r} = \|\mathbf{r}\|\hat{\mathbf{e}}_1$, the sill of $\gamma_{\mathrm{xx}}(\mathbf{r})$ depends on the direction of the unit vector $\hat{\mathbf{e}}_1$: If $\hat{\mathbf{e}}_1 \cdot \hat{\mathbf{e}}_0 = 1$, the sill of $\gamma_{\mathrm{xx}}(\mathbf{r})$ in the direction $\hat{\mathbf{e}}_1$ is equal to $c_1 + c_2$. If $\hat{\mathbf{e}}_1 \cdot \hat{\mathbf{e}}_0 = 0$, the sill of $\gamma_{\mathrm{xx}}(\mathbf{r})$ in the direction $\hat{\mathbf{e}}_1$ is equal to c_1.

More complex, composite variogram models that involve multiple directions of zonal anisotropy as well as components describing the variability at different length scales can be constructed. Zonal anisotropy is illustrated by means of the variogram of a two-dimensional field shown in Fig. 4.13. The covariance model displayed contains an isotropic exponential covariance $C^{(1)}_{\mathrm{xx}}(\mathbf{r}) = \exp\left(-\|\mathbf{r}\|/\xi\right)$ with unit variance and correlation length $\xi = 1$ combined with a directional exponential covariance $C^{(2)}_{\mathrm{xx}}(\mathbf{r}) = 0.5 \exp\left(-\mid \mathbf{r} \cdot \hat{\mathbf{e}}_0 \mid /\xi\right)$, where $\hat{\mathbf{e}}_0 = (0, 1)$.

Zonal anisotropy mixtures A necessary and sufficient characterization of range parameter anisotropy in terms of directional mixtures of zonal anisotropy variograms is given in [17]. This representation extends the classic models of zonal and range anisotropy by including anisotropy models with range parameters that vary continuously or discontinuously with direction.

4.4 Anisotropic Spectral Densities

Geometric anisotropy can be reduced to isotropy as shown in Sect. 4.3.2. Consequently, anisotropic spectral densities can be expressed in terms of their isotropic counterparts and the anisotropy parameters.

Let us define by $\mathbf{h}$ the dimensionless lag vector in the isotropic system. For example, the Gaussian correlation function is expressed as $\rho_{\mathrm{xx}}(\mathbf{h}) = \exp(-\|\mathbf{h}\|^2)$ in terms of $\mathbf{h}$.

4.4.1 Anisotropy in Planar Domains

The dimensionless vector $\mathbf{h}$ is related to the lag vector $\mathbf{r}$ in the initial anisotropic system by means of $\mathbf{h} = \mathbf{B}\mathbf{r}$. The transformation matrix $\mathbf{B} = \mathbf{L}\,\mathbf{U}$ consists of a rotation $\mathbf{U}$ followed by a rescaling $\mathbf{L}$.

In the transformed coordinate system, the *spectral density* of the correlation function is expressed as follows

$$\widetilde{\rho}_{xx}(\mathbf{k}) = \int d\mathbf{r}\, \rho_{xx}(\mathbf{r})\, e^{-i\mathbf{k}^\top\mathbf{r}} = \int d\mathbf{h}\, \det(\mathbb{J})\, \rho^*_{xx}(\mathbf{h})\, e^{-i\mathbf{k}^\top\mathbf{B}^{-1}\mathbf{h}}, \tag{4.46}$$

where $\rho^*_{xx}(\mathbf{h}) = \rho_{xx}(\mathbf{r})$ is the correlation function expressed in terms of the rotated and rescaled lag $\mathbf{h} = \mathbf{B}\,\mathbf{r}$, the matrix $\mathbb{J}$ is the *Jacobian* of the transformation $\mathbf{r} \mapsto \mathbf{h}$, and $\det(\mathbb{J})$ is the Jacobian's determinant. Conversely, the original lag vector is given by $\mathbf{r} = \mathbf{B}^{-1}\mathbf{h}$.

Thanks to the linearity of the rotation and recaling transformations, the elements of the Jacobian are given by the following partial derivatives

$$[\mathbb{J}]_{i,j} = \frac{\partial r_i}{\partial h_j} = \left[\mathbf{B}^{-1}\right]_{i,j}, \quad \text{for } i, j = 1, \ldots, d.$$

In *two spatial dimensions* $\mathbf{U}$ and $\mathbf{L}$ are given by the following:

$$\mathbf{B} = \underbrace{\begin{pmatrix} \frac{1}{\xi_1} & 0 \\ 0 & \frac{1}{\xi_2} \end{pmatrix}}_{\mathbf{L}} \underbrace{\begin{pmatrix} \cos\theta & \sin\theta \\ -\sin\theta & \cos\theta \end{pmatrix}}_{\mathbf{U}} = \begin{pmatrix} \frac{\cos\theta}{\xi_1} & \frac{\sin\theta}{\xi_1} \\ -\frac{\sin\theta}{\xi_2} & \frac{\cos\theta}{\xi_2}. \end{pmatrix} \tag{4.47a}$$

The inverse of the matrix $\mathbf{B}$ is given by

$$\mathbf{B}^{-1} = \begin{pmatrix} \xi_1 \cos\theta & -\xi_2 \sin\theta \\ \xi_1 \sin\theta & \xi_2 \cos\theta \end{pmatrix}. \tag{4.47b}$$

Based on (4.47b) the determinant of the transformation's Jacobian is given by

$$\det(\mathbb{J}) = \xi_1\, \xi_2.$$

In light of (4.46) and (4.47), the anisotropic spectral density $\widetilde{\rho}_{xx}(\mathbf{k})$ can be expressed as follows

$$\begin{aligned}\widetilde{\rho}_{xx}(\mathbf{k}) =& \xi_1\xi_2 \int d\mathbf{h}\, \rho^*_{xx}(\mathbf{h})\, e^{-i\mathbf{k}^\top\mathbf{B}^{-1}\mathbf{h}} = \xi_1\xi_2 \int d\mathbf{h}\, \rho^*_{xx}(\mathbf{h})\, e^{-i\mathbf{w}^\top\mathbf{h}} \\ =& \xi_1\xi_2\, \widetilde{\rho}^*_{xx}(\|\mathbf{w}\|),\end{aligned} \tag{4.48}$$

Table 4.4 Spectral densities of the isotropic correlation models in Table 4.2. The dimensionless distance is $u = \|\mathbf{r}\|/\xi$. The dimensionless wavenumber is $\kappa = \|\mathbf{k}\|\,\xi$. The spectral density expressions are listed for $\xi = 1$. Hence, all the densities (except for the nugget effect) should be multiplied by ξ^d for $\xi \neq 1$

Model	Covariance $\rho_{xx}(u)$	Spectral density $\widetilde{\rho}_{xx}(\kappa)$
Nugget effect	$\delta(\mathbf{r})$	1
Exponential [E]	$\exp(-u)$	$2^d \pi^{(d-1)/2}\,\Gamma(\frac{d+1}{2})(1+\kappa^2)^{-(d+1)/2}$
Gaussian [G]	$\exp(-u^2)$	$\sqrt{\pi}^d\,\exp(-\kappa^2/4)$
Spherical [S]	$\left(1-1.5\,u+0.5\,u^3\right)\,\mathbb{I}_{0\le u\le 1}(u)$	$3\pi^{d-1}\,2^{d-2}\,J^2_{3/2}(\kappa/2)$
Generalized exponential [GE]	$\exp(-u^\nu)$	$\sim c_\nu \kappa^{-d-\nu}$, for $0<\nu<2$
Matérn [M]	$\frac{u^\nu\,K_\nu(u)}{2^{\nu-1}\,\Gamma(\nu)}$	$2^d\,\pi^{d/2}\,\frac{\Gamma(\nu+d/2)}{\Gamma(\nu)}\,(1+\kappa^2)^{-(\nu+d/2)}$
Bessel-J [H]	$2^\nu\,\Gamma(\nu+1)\,\frac{J_\nu(u)}{u^\nu}$	$\frac{(4\pi)^{d/2}\,\Gamma(\nu+1)}{\Gamma(\nu+1-d/2)}\,\frac{\mathbb{I}_{0\le\kappa\le 1}(\kappa)}{(1-\lvert\kappa\rvert^2)^{\nu-d/2}}$
Rational quadratic [RQ]	$\left(1+u^2\right)^{-b}$	$\frac{(2\pi)^{d/2}}{2^{b-1}\,\Gamma(b)}\,K_{d/2-b}(\kappa)\,\kappa^{b-d/2}$
Cauchy class [C]	$(1+u^\alpha)^{-\beta/\alpha}$	N/A

[E,S] These spectral densities are obtained by multiplying the respective results in the book by Lantuejoul [487, p. 245] with $(2\pi\xi)^d$ to account for the different normalization. [G] See Ref. [678, p. 83]. [M] See Ref. [380]. [RQ] The result holds for $b > (d-1)/4$. See Ref. [380] and Ref. [306, 6.565.4]. [H] See Ref. [233]; the result holds for $\nu > d/2 - 1$. The spectral density is calculated using the integral [306, Eq. 6.567.1, p.711]. [GE;C] General explicit relations are not available for the generalized exponential and the Cauchy spectral density [502]. [GE] A closed-form expression for the spectral density exists for $\nu = 1$. The asymptotic expression for $0 < \nu < 2$ is valid for $\kappa \to \infty$

where $\mathbf{w}^\top = \mathbf{k}^\top \mathbf{B}^{-1}$, and $\mathbf{w}^\top \mathbf{h} = \mathbf{w}\cdot\mathbf{h}$. The spectral density $\widetilde{\rho}^*_{xx}(\cdot)\colon \mathbb{R} \to \mathbb{R}$ is thus a dimensionless radial function of $\|\mathbf{w}\|$.

A suitable isotropic spectral density $\widetilde{\rho}^*_{xx}(\cdot)$ can be selected from Table 4.4 which lists $\widetilde{\rho}_{xx}(\kappa)$ for various correlation models. If we take into account that $\widetilde{\rho}^*_{xx}(\mathbf{w})$ is dimensionless while the correlation models in Table 4.4 are proportional to ξ^d, we obtain

$$\widetilde{\rho}^*_{xx}(\mathbf{w}) = \frac{1}{\xi^2}\widetilde{\rho}_{xx}(\kappa = \|\mathbf{w}\|). \tag{4.49a}$$

The corresponding anisotropic spectral density can be obtained by expressing the wavevector $\mathbf{w}$ (which is in the transformed isotropic system) in terms of the wavevector $\mathbf{k}$ (which lives in the original coordinate system) using $\mathbf{w}^\top = \mathbf{k}^\top \mathbf{B}^{-1}$, i.e.,

$$\mathbf{w}^2 = w_1^2 + w_2^2 = g_{1,1}k_1^2 + g_{2,2}k_2^2 + 2g_{1,2}k_1\,k_2, \tag{4.49b}$$

where the *anisotropy coefficients* $\{g_{i,j}\}_{i,j=1}^2$ are given by

$$g_{1,1} = \xi_1^2 \cos^2\theta + \xi_2^2 \sin^2\theta, \tag{4.49c}$$

$$g_{2,2} = \xi_1^2 \sin^2\theta + \xi_2^2 \cos^2\theta, \tag{4.49d}$$

$$g_{1,2} = \sin\theta \cos\theta \left(\xi_1^2 - \xi_2^2\right). \tag{4.49e}$$

Finally, the anisotropic spectral density is obtained from (4.49a) after replacing $\|\mathbf{w}\|^2$ by means of (4.49b) and (4.49).

Example 4.5 Determine the anisotropic spectral densities for (i) the exponential, (ii) the Gaussian and (iii) the Whittle- Matérn correlation functions in two dimensions.

Answer We use the respective spectral densities from Table 4.4 for $d = 2$. We obtain the following expressions using (4.49) and

$$\tilde{\rho}_{\mathrm{xx}}(\mathbf{k}) = \frac{\xi_1\xi_2}{\xi^2}\tilde{\rho}_{\mathrm{xx}}(\kappa = \|\mathbf{w}\|).$$

1. Exponential

$$\tilde{\rho}_{\mathrm{xx}}(\mathbf{k}) = \frac{2\sqrt{\pi}\,\xi_1\xi_2}{(1 + g_{1,1}k_1^2 + g_{2,2}k_2^2 + 2g_{1,2}k_1\,k_2)^{3/2}}.$$

2. Gaussian

$$\tilde{\rho}_{\mathrm{xx}}(\mathbf{k}) = \pi\,\exp\left[-\frac{\left(g_{1,1}k_1^2 + g_{2,2}k_2^2 + 2g_{1,2}k_1\,k_2\right)}{4}\right].$$

3. Matérn

$$\tilde{\rho}_{\mathrm{xx}}(\mathbf{k}) = \frac{4\pi\,\nu\,\xi_1\xi_2}{(1 + g_{1,1}k_1^2 + g_{2,2}k_2^2 + 2g_{1,2}k_1\,k_2)^{\nu+1}}.$$

4.4.2 Anisotropy in Three-Dimensional Domains

The same approach can be used in three dimensions. The technical details are somewhat more complicated, because we need three *Euler angles* to rotate the initial coordinate system. The matrix $\mathbf{B}$ is the product of four matrices: the first matrix (on the right-hand side of the equality sign) performs the length rescaling, while the other three $\mathbf{U}_i$ matrices ($i = 1, 2, 3$) represent consecutive rotations by respective Euler angles [298]. Hence, the overall transformation is expressed as follows

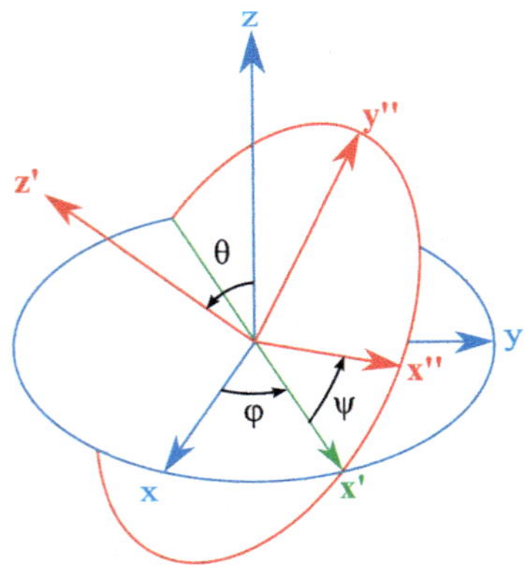

Fig. 4.14 Illustration of three-dimensional coordinate system rotation using the Euler angles (ϕ, θ, ψ). The schematic is based on [104]

$$\mathbf{B} = \underbrace{\begin{pmatrix} \frac{1}{\xi_1} & 0 & 0 \\ 0 & \frac{1}{\xi_2} & 0 \\ 0 & 0 & \frac{1}{\xi_3} \end{pmatrix}}_{\mathbf{L}} \underbrace{\begin{pmatrix} \cos\psi & \sin\psi & 0 \\ -\sin\psi & \cos\psi & 0 \\ 0 & 0 & 1 \end{pmatrix}}_{\mathbf{U}_3} \underbrace{\begin{pmatrix} 1 & 0 & 0 \\ 0 & \cos\theta & \sin\theta \\ 0 & -\sin\theta & \cos\theta \end{pmatrix}}_{\mathbf{U}_2} \underbrace{\begin{pmatrix} \cos\phi & \sin\phi & 0 \\ -\sin\phi & \cos\phi & 0 \\ 0 & 0 & 1 \end{pmatrix}}_{\mathbf{U}_1}. \tag{4.50}$$

The first rotation $\mathbf{U}_1$ is by an angle ϕ about the z axis. The second rotation $\mathbf{U}_2$ is by an angle θ about the axis x', which denotes the axis x after the first rotation. Finally, the third rotation $\mathbf{U}_3$ is by an angle ψ with respect to the axis z' which denotes the rotated (in the previous step) z axis (Fig. 4.14).

We can also rewrite the above as $\mathbf{B} = \mathbf{L}\mathbf{U}$, where $\mathbf{L}$ is the length rescaling matrix and $\mathbf{U} = \mathbf{U}_3\,\mathbf{U}_2\,\mathbf{U}_1$ is the composite rotation matrix. After performing all the matrix multiplications, we obtain the following expression for the rotation matrix $\mathbf{U}$

$$\mathbf{U} = \begin{pmatrix} \cos\psi\cos\phi - \cos\theta\sin\phi\sin\psi & \cos\psi\sin\phi + \cos\theta\cos\phi\sin\psi & \sin\psi\sin\theta \\ -\sin\psi\cos\phi - \cos\theta\sin\phi\cos\psi & -\sin\psi\sin\phi + \cos\theta\cos\phi\cos\psi & \cos\psi\sin\theta \\ \sin\theta\sin\phi & -\sin\theta\cos\theta & \cos\theta \end{pmatrix}. \tag{4.51}$$

The rotation matrix $\mathbf{U}$ is orthogonal, which implies $\mathbf{U}^{-1} = \mathbf{U}^\top$. Hence,

$$\mathbf{U}^{-1} = \begin{pmatrix} \cos\psi\cos\phi - \cos\theta\sin\phi\sin\psi & -\sin\psi\cos\phi - \cos\theta\sin\phi\cos\psi & \sin\theta\sin\phi \\ \cos\psi\sin\phi + \cos\theta\cos\phi\sin\psi & -\sin\psi\sin\phi + \cos\theta\cos\phi\cos\psi & -\sin\theta\cos\theta \\ \sin\psi\sin\theta & \cos\psi\sin\theta & \cos\theta \end{pmatrix}. \tag{4.52}$$

The anisotropic spectral density is then determined by the transformations

$$\widetilde{\rho}^{*}_{\mathrm{xx}}(\mathbf{w}) = \frac{1}{\xi^3}\widetilde{\rho}_{\mathrm{xx}}(\kappa = \|\mathbf{w}\|), \tag{4.53a}$$

where

$$\mathbf{w}^\top = \mathbf{k}^\top \mathbf{B}^{-1} = \mathbf{k}^\top \mathbf{U}^{-1} \mathbf{L}^{-1}, \tag{4.53b}$$

and the squared Euclidean norm of the dimensionless wavevector $\mathbf{w}$ is given by

$$\mathbf{w}^2 = \sum_{i=1}^{3} \sum_{j=1}^{3} g_{i,j} k_i k_j. \tag{4.53c}$$

To determine the anisotropy coefficients $\{g_{i,j}\}_{i=1}^{3}$, the square norm $\|\mathbf{w}\|^2$ is expressed as

$$\|\mathbf{w}\|^2 = \mathbf{w}^\top \mathbf{w} = \mathbf{k}^\top \mathbf{U}^{-1} \mathbf{L}^{-1} [\mathbf{L}^{-1}]^\top [\mathbf{U}^{-1}]^\top \mathbf{k}. \tag{4.54}$$

Taking into account that $\mathbf{L}^{-1}$ is a diagonal matrix and that $\mathbf{U}^{-1} = \mathbf{U}^\top$, we obtain

$$\|\mathbf{w}\|^2 = \mathbf{k}^\top \left(\mathbf{U}^{-1} \mathbf{L}^{-2} \mathbf{U} \right) \mathbf{k}, \tag{4.55}$$

where $\mathbf{L}^{-2}$ is given by the diagonal matrix of principal lengths

$$\mathbf{L}^{-2} = \begin{pmatrix} \xi_1^2 & 0 & 0 \\ 0 & \xi_2^2 & 0 \\ 0 & 0 & \xi_3^2 \end{pmatrix}. \tag{4.56}$$

From a comparison of (4.53b) with (4.55), it follows that the matrix of anisotropy coefficients $\mathbf{g}$ is related to the rotation and rescaling matrices via

$$\mathbf{g} = \mathbf{U}^{-1} \mathbf{L}^{-2} \mathbf{U}. \tag{4.57}$$

Problem The readers can confirm that the 2D coefficients $\{g_{i,j}\}_{i=1}^{2}$ given by (4.49) are also obtained from (4.57) using the respective 2D rotation and rescaling matrices.

Comment If the rotation matrix $\mathbf{U}$ is estimated from the data, it is possible to determine Euler angles that lead to the specific rotation matrix. However, the correspondence is not one-to-one, since there are different solutions for the Euler angles (i.e., different sequences of rotations) that lead to the same outcome.

Aligned coordinate system If the axes of the coordinate system coincide with the principal anisotropy axes, there is no need for the rotation transformations. In this case $\mathbf{B} = \mathbf{L}$. Furthermore, in the case of *transverse isotropy*, the rescaling matrix $\mathbf{L}$ involves only two parameters: ξ_3 and $\xi_1 = \xi_2$.

Planar misalignment of anisotropy Let us assume that the vertical axis of the coordinate system coincides with one of the principal axes of anisotropy, while the other two orthogonal coordinate axes are rotated with respect to the in-plane principal axes. Then $\mathbf{B} = \mathbf{L}\,\mathbf{U}_1$ and only the first rotation (by the angle ϕ) is necessary.

4.5 Multipoint Description of Random Fields

Up to this point we have presented several tools that can be used to characterize the behavior of random fields at isolated points in space and for configurations that involve pairs of points. Although quite informative, such descriptions are incomplete, because in most cases (Gaussian random fields being a notable exception), they cannot quantify the correlations of spatial patterns that involve three or four points. For example, if we consider the three vertices of a triangle, we might want to know how the correlations between these points depend on the lengths of the triangle's edges. In order to mathematically represent information of this type we need to consider probability functions that involve more than two points.

4.5.1 Joint Probability Density Functions

Joint probabilities determine the interdependence of the random field values at a set of points. It is easier to first focus on a countable set of points $\{\mathbf{s}_n\}_{n=1}^{N}$. The interdependence is incorporated in *joint probability density functions*, denoted by $f_{\mathrm{x}}(x_1, \ldots, x_N; \mathbf{s}_1, \ldots, \mathbf{s}_N)$.

A hierarchy of joint densities is obtained for $N \in \mathbb{N}$, up to arbitrarily large values. It is not possible to generate this hierarchy from the marginal pdf without additional information. It is also possible that the same $M < N$-point density can be obtained by integrating more than one N-point densities. In fact, joint densities that correspond to N and $M < N$ points respectively, are related via the following equation

$$f_{\mathrm{x}}(x_1, \ldots, x_M; \mathbf{s}_1, \ldots, \mathbf{s}_M) = \int_{-\infty}^{\infty} \mathrm{d}x_{M+1} \ldots \int_{-\infty}^{\infty} \mathrm{d}x_N \, f_{\mathrm{x}}(x_1, \ldots, x_N; \mathbf{s}_1, \ldots, \mathbf{s}_N). \tag{4.58}$$

4.5.2 Cumulative Joint Probability Function

The cumulative joint probability function is defined as follows:

$$F_{\mathrm{x}}(x_1, \ldots, x_N; \mathbf{s}_1, \ldots, \mathbf{s}_N) = \mathrm{Prob}\big(\mathrm{X}(\mathbf{s}_1; \omega) \le x_1 \ldots \mathrm{X}(\mathbf{s}_N; \omega) \le x_N\big). \tag{4.59}$$

The joint cdf can always be defined, in contrast with the joint pdf which requires that the joint cdf be differentiable. Assuming that the joint pdf exists, the joint cdf is given by means of the following integral equation

$$F_{\mathrm{x}}(x_1, \ldots, x_N; \mathbf{s}_1, \ldots, \mathbf{s}_N) = \left[\prod_{n=1}^{N} \int_{-\infty}^{x_n} \mathrm{d}x'_n\right] f_{\mathrm{x}}(x'_1, \ldots, x'_N; \mathbf{s}_1, \ldots, \mathbf{s}_N). \tag{4.60}$$

Reversely, the joint pdf is determined from the joint cdf by the partial derivatives of the latter according to

$$f_{\mathrm{x}}(x_1, \ldots, x_N; \mathbf{s}_1, \ldots, \mathbf{s}_N) = \frac{\partial^N F_{\mathrm{x}}(x_1, \ldots, x_N; \mathbf{s}_1, \ldots, \mathbf{s}_N)}{\partial x_1 \ldots \partial x_N}. \tag{4.61}$$

4.5.3 *Statistical Moments*

Joint statistical moments of random fields can be defined by means of straightforward extensions of the marginal expressions. Moments can be defined for multipoint expectations. It is possible to have different exponents, $(n_1, \ldots, n_N)$, at each point. The total order of the joint moment is equal to the sum of the point-wise exponents.

Non-centered moments of order $\mathbf{p} = (p_1, \ldots, p_N)$ are expectations of the respective monomials $\mathrm{X}^{p_1}(\mathbf{s}_1; \omega) \ldots \mathrm{X}^{p_N}(\mathbf{s}_N; \omega)$ of the random field, where p_n denotes the order of the moment at point $\mathbf{s}_n$:

$$\mu_{\mathrm{x};\mathbf{p}} := \mathcal{E}\left[\prod_{n=1}^{N} \mathrm{X}^{p_n}(\mathbf{s}_n; \omega)\right].$$

Centered moments will be denoted by

$$\mu^{(c)}_{\mathrm{x};\mathbf{p}} := \mathcal{E}_c\left[\prod_{n=1}^{N} \mathrm{X}^{p_n}(\mathbf{s}_n; \omega)\right] = \mathcal{E}\left[\prod_{n=1}^{N} \left\{\mathrm{X}(\mathbf{s}_n; \omega) - \mathcal{E}[\mathrm{X}(\mathbf{s}_n; \omega)]\right\}^{p_n}\right],$$

where $\mathcal{E}_c[\cdot]$ denotes the centered expectation operator.

If the joint pdf exists, the non-centered moments are given in terms of the joint pdf by means of the following multiple integral

$$\mu_{\mathrm{x};\mathbf{p}} = \left[\prod_{n=1}^{N} \int_{-\infty}^{\infty} \mathrm{d}x_n\right] f_{\mathrm{x}}(\{x_n\}; \{\mathbf{s}_n\})\, x_1^{p_1} \ldots x_N^{p_N}. \tag{4.62}$$

In the above, we replaced the joint pdf $f_{\mathrm{x}}(x_1, \ldots, x_N; \mathbf{s}_1, \ldots, \mathbf{s}_N)$ with $f_{\mathrm{x}}(\{x_n\}; \{\mathbf{s}_n\})$ for short.

A similar integral is defined for the centered moments:

$$\mu^{(c)}_{\mathrm{x};\mathbf{p}} = \left[\prod_{n=1}^{N} \int_{-\infty}^{\infty} \mathrm{d}x_n\right] f_{\mathrm{x}}(\{x_n\}; \{\mathbf{s}_n\}) \prod_{n=1}^{N} [x_n - m_{\mathrm{x}}(\mathbf{s}_n)]^{p_n} . \tag{4.63}$$

Practical limitations In practice, it is not an easy task to determine multipoint statistical dependence from spatial data. This in part due to the fact that estimates of higher-order moments are more sensitive to error than lower-order moments. Another factor is that the information required to estimate higher-order moments is limited at larger lag values.

To illustrate this limitation, consider the simple example of N points on a straight line, each point at distance a from its neighbor. If we consider correlations between pairs of points, the maximum possible lag allowed by this configuration is $(N-1)a$. There is, however, only one pair with this lag, which means that the statistical estimation of correlations is practically impossible at this lag. If we require, for example, at least 30 pairs to estimate correlation, the maximum lag that can be included is $(N-30)a$.

Next, let us concentrate on triplets and assume for simplicity of the argument that $N = 2M + 1$. If we consider triplets with their midpoint at equal distance from the endpoints, the maximum possible lag between the midpoint and the edges is $Ma = (N-1)a/2$, and there is only one triplet with such lag. If we aim for configurations that would give us at least 30 such triplets, the maximum possible lag is $(N-31)a/2$. So, we roughly have half as many triplets as pairs. The situation gets worse as we consider clusters that involve more than three points.

4.5.4 Characteristic Function

Let us now consider the finite-dimensional random vector

$$\mathbf{X}(\omega) = [\mathrm{X}(\mathbf{s}_1; \omega), \ldots, \mathrm{X}(\mathbf{s}_N; \omega)]^{\top}.$$

Assuming that $\mathbf{u} = (u_1, \ldots, u_N)^{\top}$, where $u_i \in \mathbb{R}$, plays the role of an *auxiliary field*. The characteristic function in this case is defined as follows, where

$$\phi_{\mathrm{x}}(\mathbf{u}) := \mathcal{E}\left[\mathrm{e}^{\mathrm{i}\,\mathbf{u}\cdot\mathbf{X}(\omega)}\right] = \mathcal{E}\left[\mathrm{e}^{\mathrm{i}\,\sum_{n=1}^{N} u_n\,\mathrm{X}(\mathbf{s}_n;\omega)}\right]$$
$$= \int_{-\infty}^{\infty} \mathrm{d}F_{\mathrm{x}}(\{x_n\}; \{\mathbf{s}_n\})\, \mathrm{e}^{\mathrm{i}\,\mathbf{u}\cdot\mathbf{x}}, \tag{4.64}$$

where $\mathrm{d}F_{\mathrm{x}}(\cdot;\cdot)$ denotes the Riemann-Stieltjes integral [863]. For notational brevity we only show the dependence of $\phi_{\mathrm{x}}(\cdot)$ on $\mathbf{u}$. However, the characteristic function also depends on the parameters of the probability distribution. If the random field is not statistically stationary, then $\phi_{\mathrm{x}}(\cdot)$ also depends implicitly on the locations through $\mathbf{u}$.

If the pdf is well-defined, the characteristic function is given by the integral

$$\mathcal{E}\left[e^{i\,\mathbf{u}\cdot\mathbf{X}(\omega)}\right] = \left[\prod_{n=1}^{N}\int_{-\infty}^{\infty} dx_n\right] f_{\mathrm{x}}(x_1,\ldots,x_N;\mathbf{s}_1,\ldots,\mathbf{s}_N)\, e^{i\mathbf{u}\cdot\mathbf{x}}, \tag{4.65}$$

which represents the *Fourier transform* of the pdf.

The *non-centered moments of order* $\mathbf{p}$, if they exist, can be obtained from the characteristic function. This is achieved by means of successive differentiations, followed by setting the auxiliary field to zero, i.e.,

$$\mu_{\mathrm{x};\mathbf{p}} = \frac{1}{i^n}\left.\frac{\partial^{p_1}\ldots\partial^{p_N}\phi_{\mathrm{x}}(\mathbf{u})}{\partial u_1^{p_1}\ldots\partial u_N^{p_N}}\right|_{\mathbf{u}=\mathbf{0}}. \tag{4.66}$$

Figuratively speaking, the auxiliary field acts as a flashlight beam: We turn the beam on to light up the field's properties at each point of interest. The more information (higher-order moments) we want to extract about any given point, the more times we turn the beam on. At the end of the calculation, we switch off the flashlight (set $\mathbf{u}$ to zero).

Normal distribution The characteristic function of a multivariate normal distribution an exponential function whose algorithm is given by

$$\ln\phi_{\mathrm{x}}(\mathbf{u}) = i\,\mathbf{u}^{\top}\mathbf{m}_{\mathrm{x}} - \frac{1}{2}\mathbf{u}^{\top}\mathbf{C}_{\mathrm{xx}}\,\mathbf{u}. \tag{4.67}$$

The characteristic function of the normal distribution depends only on the mean $\mathbf{m}_{\mathrm{x}}$ and the covariance $\mathbf{C}_{\mathrm{xx}}$. Then, in light of (4.67), the moment generating equation (4.66) confirms that the mean and the covariance suffice to determine all the higher-order moments in the Gaussian case.

Why characteristic functions? One may wonder about the practical use of characteristic functions. It is true that spatial data analysis mostly involves moments and probability density or cumulative probability functions. However, in certain cases the joint pdf or the moments do not exist, while the characteristic function is well-defined. Characteristic functions can be constructed using the respective *Bochner's permissibility Theorem* 4.2. The procedure is similar to the construction and testing of different covariance models.

Stable distributions A class of distribution functions that admits explicit forms for the characteristic function is the class of multivariate stable distributions. These are potentially asymmetric, heavy-tailed generalizations of the Gaussian distribution [709]. Loosely speaking, the concept of stability states that if the random variables $\mathrm{X}_n(\omega)$ follow a stable distribution, then any linear combination of translated and scaled random variables $\mathrm{Y}_n(\omega) = a\mathrm{X}_n(\omega) + b$ also follows the same probability distribution. The class of stable distributions includes the multivariate Gaussian and Cauchy distributions as special cases.

The characteristic function of the *multivariate Cauchy distribution* is given by [672]

$$\ln \phi_{\mathrm{x}}(\mathbf{u}) = \mathrm{i}\,\mathbf{m} \cdot \mathbf{u} - \left(\mathbf{u}^\top \mathbf{B}_k \mathbf{u}\right)^{1/2}. \tag{4.68}$$

Another family of stable distributions, known as *α-stable or Lévy distributions*, have a characteristic function given by [672]

$$\ln \phi_{\mathrm{x}}(\mathbf{u}) = \mathrm{i}\,\mathbf{m} \cdot \mathbf{u} - \frac{1}{2}\sum_{k=1}^{m} \left(\mathbf{u}^\top \mathbf{B}_k \mathbf{u}\right)^{\alpha/2}, \tag{4.69}$$

where m is the order of the distribution, $\mathbf{B}_k$, $k = 1, \ldots, m$ is a series of non-proportional matrices, and α is the width parameter. Note that both the Gaussian and Cauchy distributions can be viewed as special cases of Lévy distributions with $\alpha = 2$ and $\alpha = 1$ respectively.

Lévy distributions are used in various applications in statistical physics because they appear in the solution of anomalous random walk (also known as *Lévy flights*) problems [58, 227, 409, 566]. Lévy distributions have also been proposed as models of geological heterogeneity [316, 642–644]. However, reliable estimation of the tail dependence is difficult for geological data due to the limited size of the data sets. From the general class of stable distributions, only the normal, Cauchy, and Lévy distributions admit explicit pdf expressions.

Other distributions Characteristic functions are also available for the multivariate *Wishart distribution* [p. 258][26] and the *multivariate Student's t-distribution* [468, p. 36–43]. The latter is a family of heavy-tailed, symmetric distributions, which, unlike stable distributions, admit explicit expressions for the joint pdf.

In order to effectively design characteristic functions, we need to know their properties and the respective permissibility criteria.

Properties The following properties are easily obtained from the definition of the characteristic function. You can find more details in the monograph by Bochner [83].

- The characteristic function of a real-valued random vector $\mathbf{X}(\omega)$ can always be defined, because it is given by the expectation of the function $\exp\left[\mathrm{i}\mathbf{u} \cdot \mathbf{X}(\omega)\right]$ whose modulus is less than or equal to one.
- The maximum of the characteristic function's magnitude is obtained for zero external field, i.e., $\phi_{\mathrm{x}}(\mathbf{u} = \mathbf{0}) = 1$.
- Characteristic functions are bounded, i.e., $\|\phi_{\mathrm{x}}(\mathbf{u})\| \leq 1$.
- Characteristic functions are Hermitian, i.e., $\phi_{\mathrm{x}}(-\mathbf{u}) = \phi_{\mathrm{x}}^{\dagger}(\mathbf{u})$, where the dagger denotes the complex conjugate.

- Characteristic functions are uniformly continuous.

Permissibility criteria Various simple operations can generate permissible characteristic functions from other known characteristic functions.

- A *convex linear combination* $\sum_{n=1}^{N} a_n \phi_{\mathrm{x}_n}(\mathbf{u})$ of characteristic functions $\phi_{\mathrm{x}_n}(\mathbf{u})$ is also a characteristic function. A convex linear combination requires $a_n \geq 0$ and $\sum_{n=1}^{N} a_n = 1$.
- The product of a finite number of characteristic functions is also a characteristic function.
- If $\phi_{\mathrm{x}}(\mathbf{u})$ is a characteristic function and β is a real number, the functions

$$\phi_{\mathrm{x}}^{\dagger}(\mathbf{u}), \quad \Re[\phi_{\mathrm{x}}(\mathbf{u})], \quad \|\phi_{\mathrm{x}}(\mathbf{u})\|^2, \text{ and } \quad \phi_{\mathrm{x}}(\beta\mathbf{u})$$

are also characteristic functions for some random field:

In addition, it is possible to check if a function $\phi_{\mathrm{x}}(\mathbf{u})$ is a permissible characteristic function by means of the following theorems.

Theorem 4.2 *Assume that* $\mathbf{u} = (u_1, \ldots, u_N)^{\top}$ *is a real-valued* N*-dimensional vector. An arbitrary function* $\phi(\mathbf{u})\colon \mathbb{R}^N \mapsto \mathbb{C}$ *is the characteristic function of some random vector* $\mathbf{X}(\omega) = (\mathrm{X}_1(\omega), \ldots, \mathrm{X}_N(\omega))^{\top}$, *if and only if* $\phi(\mathbf{u})$ *is positive definite, continuous at the origin, and* $\phi(\mathbf{u} = \mathbf{0}) = 1$.

Based on the above, to test if a function $\phi(\mathbf{u})$ is positive definite, we need to calculate its Fourier transform. This involves an N-dimensional multiple integral. According to Bochner's Theorem 3.2, the function $\phi(\mathbf{u})$ is positive definite if its Fourier transform is a non-negative function.

4.5.5 Cumulant and Moment Generating Functions

In analogy with the univariate case, we can define the *multivariate moment generating function (mMGF)* in terms of the following expectation

$$\mathcal{M}_{\mathrm{x}}(\mathbf{u}) = \mathcal{E}\left[\mathrm{e}^{\mathbf{u}^{\top}\mathbf{X}(\omega)}\right] = \mathcal{E}\left[\mathrm{e}^{\mathbf{u}\cdot\mathbf{X}(\omega)}\right]. \tag{4.70}$$

In contrast with the characteristic function, the mMGF does not exist for all probability distributions. If the mMGF exists, the *multivariate cumulant generating function (mCGF)* is defined as follows[8]

$$\mathcal{K}_{\mathrm{x}}(\mathbf{u}) := \ln \mathcal{M}_{\mathrm{x}}(\mathbf{u}) = \ln \phi_{\mathrm{x}}(-\mathrm{i}\,\mathbf{u}). \tag{4.71}$$

[8] If the exponent is a functional of the random.

The cumulants of the random field $X(\mathbf{s};\omega)$ are the coefficients in the *multivariate Taylor expansion* (also known as *multipole expansion*) of $\mathcal{K}_x(\mathbf{u})$.

Multipole expansion Let $\mathbf{p} = (p_1, \ldots, p_N)^\top$, where $p_n \in \mathbb{N}$, be the vector of local moment orders, $p = p_1 + \ldots p_N$ be the sum of the local moment orders, and the multivariate derivative of order $\mathbf{p}$ be defined by

$$\mathrm{D}^{\mathbf{p}} K(\mathbf{u}) = \frac{\partial^p K(\mathbf{u})}{\partial u_1^{p_1} \ldots u_N^{p_N}}.$$

Assume that the *cumulant generating function* is infinitely differentiable with respect to $\mathbf{u}$. Then, the *multipole expansion* of the mCGF is given by

$$\mathcal{K}_x(\mathbf{u}) = \sum_{m=0}^{\infty} \sum_{p_1,p_2,\ldots p_N} \delta_{m,p} \frac{u_1^{p_1} u_2^{p_2} \ldots u_N^{p_N}}{p_1! p_2! \ldots p_N!} \kappa_{x;\mathbf{p}}, \tag{4.72}$$

where $\kappa_{x;\mathbf{p}} := \mathrm{D}^{\mathbf{p}} K(\mathbf{0})$ is the order-$\mathbf{p}$ cumulant, and p_n represents the local order of the cumulant with respect to the point $\mathbf{s}_n$, where $n = 1, \ldots, N$.

The inner sum in (4.72) runs over all possible combinations of integer values $0 \le p_n \le p$ such that $\sum_{n=1}^{N} p_n = p$. The Kronecker delta restricts the sum p of the local indices to be equal to m, where m is the order of the multipole expansion.

If the cumulant generating function is differentiable only up to order K, then the cumulant expansion is expressed as

$$\mathcal{K}_x(\mathbf{u}) = \sum_{m=0}^{K} \sum_{p_1,p_2,\ldots p_N} \delta_{m,p} \frac{u_1^{p_1} u_2^{p_2} \ldots u_N^{p_N}}{p_1! p_2! \ldots p_N!} \kappa_{x;\mathbf{p}} + R_K(\mathbf{u}),$$

where $R_K(\mathbf{u})$ is the residual that comprises the effect of cumulants of order $p > K$.

Based on the above, the order-$\mathbf{p}$ cumulant $\kappa_{x;\mathbf{p}}$ is given by the following partial derivative of the CGF with respect to $\mathbf{u}$:

$$\kappa_{x;\mathbf{p}} = \mathrm{D}^{\mathbf{p}} K(\mathbf{0}) = \left. \frac{\partial^{p_1+p_2+\ldots p_N} \mathcal{K}_x(\mathbf{u})}{\partial u_1^{p_1} \partial u_2^{p_2} \ldots \partial u_N^{p_N}} \right|_{\mathbf{u}=\mathbf{0}}, \tag{4.73}$$

where $\mathbf{u} = \mathbf{0}$ implies that $u_n = 0$, for all $n = 1, \ldots, N$.

To better understand the multipole expansion of the CGF, we write explicitly the low-order terms

$$\mathcal{K}_x(\mathbf{u}) = 1 + \nabla\mathcal{K}_x(\mathbf{0}) \cdot \mathbf{u} + \frac{1}{2} \sum_{n=1}^{N} \sum_{m=1}^{N} \frac{\partial^2 K(\mathbf{0})}{\partial u_n \partial u_{mj}} u_n u_m + \ldots$$

The second-order cumulant of $\mathrm{X}(\mathbf{s};\omega)$, i.e., $\mathrm{D}^{\mathbf{p}}K(\mathbf{0})$ with $\mathbf{n} = (p_1, \dots p_N)^\top$, $p_n = p_m = 1$ and $p_l = 0$ if $l \neq n, m$ is the *centered covariance* evaluated at points $\mathbf{s}_n$ and $\mathbf{s}_m$. Hence, based on (4.73) the covariance function is given by

$$C_{\mathrm{xx}}(\mathbf{s}_n, \mathbf{s}_m) = \left.\frac{\partial^2 \mathcal{K}_{\mathrm{x}}(\mathbf{u})}{\partial u_n \partial u_m}\right|_{\mathbf{u}=\mathbf{0}}. \tag{4.74}$$

Relations between moments and cumulants Similarly, the moments of $\mathrm{X}(\mathbf{s};\omega)$ are the coefficients in the respective multivariate expansion of the *moment generating function*, i.e.,

$$\mathcal{M}_{\mathrm{x}}(\mathbf{u}) = \sum_{m=0}^{\infty} \sum_{p_1, p_2, \dots p_N} \delta_{m,p} \frac{u_1^{p_1} u_2^{p_2} \dots u_N^{p_N}}{p_1! p_2! \dots p_N!} \mu_{\mathrm{x};\mathbf{p}}, \tag{4.75}$$

where the order-$\mathbf{p}$ moments are obtained from the partial derivatives of the moment generating functional evaluated at zero auxiliary field, i.e.,

$$\mu_{\mathrm{x};\mathbf{p}} = \mathrm{D}^{\mathbf{p}} \mathcal{M}_{\mathrm{x}}(\mathbf{0}) = \left.\frac{\partial^{p_1+p_2+\dots p_N} \mathcal{M}_{\mathrm{x}}(\mathbf{u})}{\partial u_1^{p_1} \partial u_2^2 \dots \partial u_N^{p_N}}\right|_{\mathbf{u}=\mathbf{0}}. \tag{4.76}$$

Relations between cumulants and moments are obtained from the expansions of the multivariate cumulant (4.72) and moment generating function (4.76) in combination with the Taylor expansion of the natural logarithm.

More specifically, the following relations are obtained for the low-order cumulants

$$\begin{aligned} \kappa_{\mathrm{x};\mathbf{p}_{n,m}} =& \mu_{\mathrm{x};\mathbf{p}_{n,m}} - \mu_{\mathrm{x};\mathbf{p}_n} \mu_{\mathrm{x};\mathbf{p}_m} \\ =& \mathcal{E}[\mathrm{X}(\mathbf{s}_n;\omega)\,\mathrm{X}(\mathbf{s}_m;\omega)] - \mathcal{E}[\mathrm{X}(\mathbf{s}_n;\omega)]\,\mathcal{E}[\mathrm{X}(\mathbf{s}_m;\omega)], \end{aligned} \tag{4.77}$$

$$\begin{aligned} \kappa_{\mathrm{x};\mathbf{p}_{n,m,k}} =& \mu_{\mathrm{x};\mathbf{p}_{n,m,k}} - \mu_{\mathrm{x};\mathbf{p}_{n,m}} \mu_{\mathrm{x};\mathbf{p}_k} - \mu_{\mathrm{x};\mathbf{p}_{n,k}} \mu_{\mathrm{x};\mathbf{p}_m} \\ & - \mu_{\mathrm{x};\mathbf{p}_{m,k}} \mu_{\mathrm{x};\mathbf{p}_n} + \mu_{\mathrm{x};\mathbf{p}_n} \mu_{\mathrm{x};\mathbf{p}_m} \mu_{\mathrm{x};\mathbf{p}_k}. \end{aligned} \tag{4.78}$$

In the above, we use the moment/cumulant index vectors $\mathbf{p}_{n,m,k}$, $\mathbf{p}_{n,m}$, and $\mathbf{p}_n$ to specify the location where the moment (cumulant) is evaluated. The components of these vectors are equal to one at location $\mathbf{s}_l$, if the location index l is included in the subscript set (e.g., if $l \in \{n, m, k\}$) and zero otherwise.

Chapter 5
Geometric Properties of Random Fields

> There is geometry in the humming of the strings, there is music in the spacing of the spheres.
>
> *Pythagoras*

This chapter deals with the main concepts and mathematical tools that help to describe and quantify the shapes of random fields. The geometry of Gaussian random functions is to a large extent determined by the mean and the two-point correlation functions. The classical text on the geometry of random fields is the book written by Robert Adler [10]. The basic elements of random field geometry are contained in the technical report by Abrahamsen [3]. A more recent and mathematically advanced book by Taylor and Adler exposes the geometry of random field using the language of manifolds [11].

5.1 Local Properties

The local properties of random fields are based on the concepts of *continuity* and *differentiability*. Both concepts are familiar from the calculus of deterministic functions. However, in the case of random fields these concepts are based on more complex mathematical notions of convergence. Before we address this topic, we will embark on a qualitative discussion to develop a practical perspective.

Random fields that are *continuous but non-differentiable* are often used to model geological variables that can exhibit irregular patterns of variability. *Smoother (differentiable) random fields* are used in atmospheric and oceanic processes. Random fields that support *discontinuous* changes are also of practical value, since smooth spatial fields that are contaminated by random noise become discontinuous.

D. T. Hristopulos, *Random Fields for Spatial Data Modeling*, Advances in Geographic Information Science, https://doi.org/10.1007/978-94-024-1918-4_5

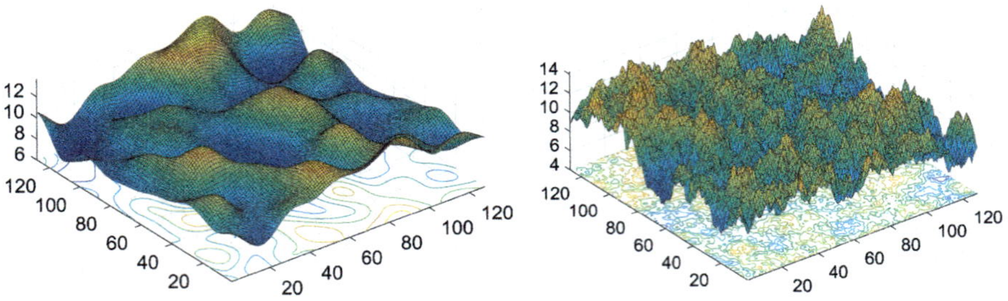

Fig. 5.1 Realizations of two Gaussian random fields with the same mean (equal to zero), standard deviation, and characteristic length. Left: Realization of random field with a differentiable Gaussian covariance function. Right: Realization of random field with non-differentiable, exponential covariance function

In addition, environmental variables also exhibit mathematical discontinuities as a result of *structural discontinuities*, e.g., geological faults.

We can visualize the difference between differentiable and continuous but non-differentiable random fields by means of the plots in Fig. 5.1.

In general, we can distinguish between at least three types of non-differentiability in random fields that are due to different causes.

(i) *Discontinuity of the covariance at the origin:* Additive white Gaussian noise, which is often present in spatial data sets, generates a covariance discontinuity at zero lag (known in geostatistics as the *nugget effect*). This effect is manifested by a discontinuous jump of the experimental variogram at lags infinitesimally larger than zero. The same effect can result from fast spatial variations that cannot be resolved with the available experimental resolution.

(ii) *Discontinuity of the covariance derivatives at the origin:* Several wide-sense stationary covariance models are non-differentiable at zero lag. Such models include the exponential, spherical, and Matérn models with $\nu < 1$. This type of discontinuity is due to a slope change of the covariance at zero lag and is milder than the discontinuity induced by uncorrelated noise.

(iii) *Non-stationarity:* Non-stationary random fields with stationary increments can have non-differentiable covariance functions, such as the celebrated *fractional Brownian motion (fBm)*, which we further discuss in Sect. 5.8. In this case, the lack of differentiability is due to the "roughness" of the increments. We illustrate this statement by means of a heuristic argument, applied for simplicity to a one-dimensional fBm. The variance of the zero-mean fBm increments satisfies

$$\mathcal{E}\left\{[\mathrm{X}_H(t;\omega) - \mathrm{X}_H(s;\omega)]^2\right\} = |t-s|^{2H}, \tag{5.1}$$

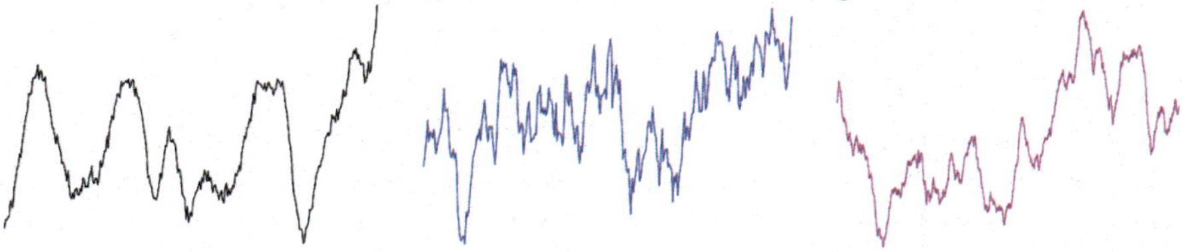

Fig. 5.2 Examples of the three non-differentiable classes of Gaussian random fields. (**a**) RF with Gaussian covariance function and superimposed uncorrelated noise. (**b**) RF with exponential covariance. (**c**) Fractional Brownian motion with Hurst exponent $H = 0.8$

where $0 < H < 1$ is the Hurst exponent. The fBm finite difference is defined by

$$\frac{\mathrm{X}_H(s + \delta t; \omega) - \mathrm{X}_H(s; \omega)}{\delta t},$$

where $\delta t > 0$. The variance of the fBm finite difference scales with δt as $(\delta t)^{2H-2}$, and therefore it blows up as $\delta t \to 0$. This divergence implies the lack of fBm differentiability.

The three types of non-differentiability are illustrated in Fig. 5.2 with one-dimensional examples. The fBm path shown corresponds to Hurst exponent $H = 0.8$. The respective path is considerably rougher for $H < 0.5$ (see Fig. 5.12), since such values correspond to *anti-persistent correlations* between the increments.

In agreement with the random field decomposition introduced in Sect. 1.3.5, we analyze random fields as a superposition of trend function and fluctuations. If a random field is continuous or differentiable, both the trend function and the fluctuation are continuous or differentiable, respectively. The trend is a deterministic function, and thus standard notions of continuity and differentiability apply. For the correlated fluctuations, however, more than one types of *stochastic convergence* exist, and they correspond to different mathematical conditions.

One may wonder why different modes of convergence are needed. Perhaps the following analogy would help to motivate the diversity of definitions: Think of the state index ω as a time label that indexes the position of a random variable in state space. In the deterministic case, the notion of convergence involves a sequence and a specific target point, which the sequence aims to approach. In the case of random variables, both the variable $\mathrm{X}(\omega)$ and the sequence $\mathrm{X}_n(\omega)$ are constantly moving around in state space. This additional degree of freedom implies that there are several ways to measure proximity, and thus different modes of convergence.

We keep the presentation to the bare minimum necessary to understand practical facts about continuity and differentiability. The readers who are eager for more mathematical details and rigorous analysis of continuity and differentiability are referred to [3, 10, 138, 162, 863].

5.1.1 Stochastic Convergence

The short discussion below follows Adler [10]. We begin by visiting the different modes of convergence that can be defined for random variables. The four modes are listed below, in order of decreasing strength. A visually oriented introduction to the concepts of stochastic convergence is given in [185].

Modes of convergence

Almost sure A random sequence $\{X_n(\omega)\}_{n=1}^{\infty}$ converges to the random variable $X(\omega)$ almost surely or with probability one, if

$$P\left(\lim_{n\to\infty} |X_n(\omega) - X(\omega)| < \epsilon\right) = 1, \text{ for all } \epsilon > 0.$$

In mean-square sense A random sequence $\{X_n(\omega)\}_{n=1}^{\infty}$ converges to the random variable $X(\omega)$ in the mean-square sense, if

$$\lim_{n\to\infty} \mathcal{E}\left[\{X_n(\omega) - X(\omega)\}^2\right] = 0.$$

In probability A random sequence $\{X_n(\omega)\}_{n=1}^{\infty}$ converges to the random variable $X(\omega)$ in probability, if

$$\lim_{n\to\infty} P\big(|X_n(\omega) - X(\omega)| < \epsilon\big) = 1, \text{ for all } \epsilon > 0.$$

In distribution A random sequence $\{X_n(\omega)\}_{n=1}^{\infty}$ with cdf $F_{X_n}(x)$ converges in distribution to the random variable $X(\omega)$ with cdf $F_X(x)$, if at every point x where $F_X(\cdot)$ is continuous

$$\lim_{n\to\infty} F_{X_n}(x) = F_X(x).$$

Convergence in distribution is the weakest of the four conditions. It is also known as (*weak convergence*), and it is denoted by $X_n(\omega) \stackrel{d}{=} X(\omega)$.

Convergence in probability implies that the distance between $X_n(\omega)$ and $X(\omega)$ does not necessarily vanish as $n \to \infty$. However, the *probability* that $|X_n(\omega) - X(\omega)|$ exceeds an arbitrarily small positive number ϵ tends to zero as $n \to \infty$. For example, it is possible that $|X_n(\omega) - X(\omega)| > \epsilon$ for a number $n_{\text{nc}} \propto 1/n$ of realizations. A sequence of random variables can converge in probability to either a random variable $X(\omega)$ or to a constant c_0.

An example Consider a dedicated student who practices math skills by repeatedly taking math tests. His/her score is measured by the random variable $X_n(\omega)$, where

n is the number of tests taken. The goal of the student is to minimize the distance between $X_n(\omega)$ and the perfect score per test, P; the perfect score could be a random variable, $P(\omega)$, that varies between tests or a constant (e.g., $P = 100$). After much practice the student is able to ace the tests most of the time. However, he/she might occasionally fail to achieve perfect score, even as $n \to \infty$. Nonetheless, if the frequency of such "failures" diminishes as n increases, the student's performance converges to perfection in probability.

A classical example of convergence in distribution is provided by the *Central Limit Theorem (CLT)*:

Theorem 5.1 *Let* $\{X_1(\omega), X_2(\omega), \ldots, X_n(\omega)\}$ *be a set of independent, identically distributed (i.i.d.) random variables with mean* m_X *and finite variance* σ_X^2. *The arithmetic average of the elementary variables,* $\overline{X}_n(\omega)$, *is defined by*

$$\overline{X}_n(\omega) = \frac{1}{n}\big[X_1(\omega) + X_2(\omega) + \ldots X_n(\omega)\big]. \tag{5.2}$$

The CLT states that the average $\overline{X}_n(\omega)$ *converges in distribution to a normally distributed random variable with mean equal to* m_X *and variance equal to* σ_X^2/n, *i.e.,*

$$\lim_{n\to\infty} \overline{X}_n(\omega) \overset{d}{=} \mathcal{N}\left(m_X, \frac{\sigma_X^2}{n}\right).$$

Weak law of large numbers Given the assumptions of the CLT 5.1, the average random variable $\overline{X}_n(\omega)$ converges in probability to the constant mean m_X of the elementary i.i.d. variables, i.e,

$$\lim_{n\to\infty} P\left(\left|\overline{X}_n(\omega) - m_X\right| < \epsilon\right) = 1, \text{ for all } \epsilon > 0.$$

Remark Note that in the discussion on the modes of convergence we use the index n to denote the sequence of random variables $X_n(\omega)$. In the CLT, however, the role of the random sequence $\{X_n(\omega)\}_{n=1}^{\infty}$ is played by the arithmetic average $\{\overline{X}_n(\omega)\}_{n=1}^{\infty}$.

Convergence in the mean square sense is probably the most commonly used mode of convergence in the analysis of random fields. It requires that the mean square of the difference $X_n(\omega) - X(\omega)$ tends to zero as $n \to \infty$.

Convergence in the mean square sense is denoted by the operator l. i. m. which stands for *limit in the mean*

$$\mathrm{X}(\omega) = \underset{n\to\infty}{\mathrm{l.\,i.\,m.}}\, \mathrm{X}_n(\omega) \quad \text{or} \quad \mathrm{X}_n(\omega) \xrightarrow{\mathrm{l.i.m.}} \mathrm{X}(\omega).$$

Almost sure convergence imposes the strictest convergence conditions. It can be defined either with respect to a random variable or with respect to a constant. Almost sure convergence is used to characterize the local properties of individual random field realizations (states).

This type of convergence implies that if n is sufficiently large, the distance between $\mathrm{X}_n(\omega)$ and $\mathrm{X}(\omega)$ is zero with probability one. Hence, in contrast with convergence in probability, almost sure convergence does not allow deviations $|\mathrm{X}_n(\omega) - \mathrm{X}(\omega)| > \epsilon$ as $n \to \infty$ even for a small number of states ω.

Almost sure convergence is denoted by $\mathrm{X}_n(\omega) \xrightarrow{\mathrm{a.s.}} \mathrm{X}(\omega)$.

Synthetic example We illustrate almost sure convergence in Fig. 5.3. The plot shows the difference between $\mathrm{X}_n(\omega)$ and $\mathrm{X}(\omega)$. The latter is a normally distributed random variable, $\mathrm{X}(\omega) \overset{d}{=} \mathcal{N}(m, \sigma^2)$, where $m = 10$ and $\sigma = 1$. The sequence $\mathrm{X}_n(\omega)$ is generated by the following affine transformation

$$\mathrm{X}_n(\omega) = m_n + \alpha_n\, [\mathrm{X}(\omega) - m]\,, \quad m_n = m + \frac{2}{n}, \quad \alpha_n = 1 + \frac{1}{\sqrt{n}}.$$

As shown in Fig. 5.3, the absolute-value norm of the difference between $\mathrm{X}_n(\omega)$ and $\mathrm{X}(\omega)$ converges to zero.

Next, consider the sequence

$$\mathrm{X}_n(\omega) = m_n + \alpha_n\, [\mathrm{Y}(\omega) - m]\,,$$

where $\mathrm{Y}(\omega)$ is a random variable, *identically distributed but independent* of $\mathrm{X}(\omega)$. In this case, $\mathrm{X}_n(\omega)$ converges to $\mathrm{X}(\omega)$ in distribution. However, $\mathrm{X}_n(\omega)$ does not converge to $\mathrm{X}(\omega)$ almost surely, since there exist $\epsilon > 0$ such that $|\mathrm{X}_n(\omega) - \mathrm{X}(\omega)| > \epsilon$ with finite probability as $n \to \infty$.

Solar fire Almost sure convergence is observed if a sequence of events inexorably leads to a "final or stationary state", either constant or random. Consider our sun, for example: it will eventually die with probability one. It may take billions of years,

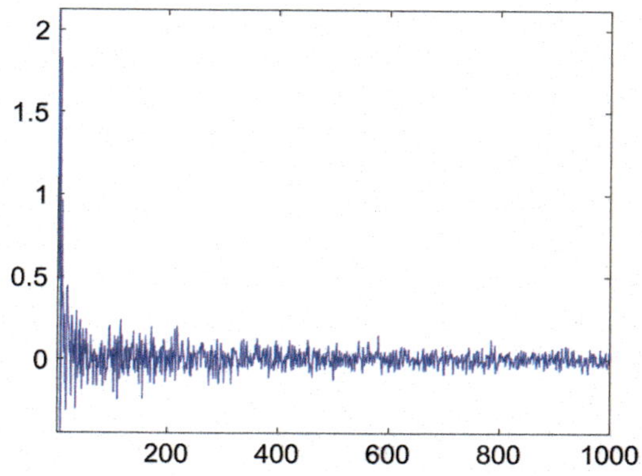

Fig. 5.3 Plot of the difference $X_n(\omega) - X(\omega)$ where $X(\omega)$ is a normally distributed random variable and $X_n(\omega)$ is a sequence generated by an affine transformation of $X(\omega)$ (see text for details). As $n \to \infty$ the absolute-value norm of the difference tends to be reduced. How might this plot differ in the case of convergence in probability?

but it will eventually exhaust its nuclear fuel and evolve first into a red giant and then into the final state of a white dwarf [732].

Strong law of large numbers Given the assumptions of the CLT 5.1, the arithmetic average $\overline{X}_n(\omega)$ of the i.i.d. elementary variables converges almost surely to the mean m_X of the common probability distribution of the elementary variables. In this case the a.s. convergence is to a constant and not to a random variable.

Relations between modes of convergence Almost sure convergence does not imply convergence in the mean square sense and vice versa. The following relations hold between the different modes of convergence:

- Almost sure convergence $\Rightarrow$ convergence in probability $\Rightarrow$ convergence in distribution
- Mean square convergence $\Rightarrow$ convergence in probability $\Rightarrow$ convergence in distribution

5.1.2 Random Field Continuity

Natura non facit saltus
Linnaeus, Philosophia Botanica

Many great scientists (including Gottfried Leibniz, one of the founders of calculus) firmly believed that natural processes do not "jump": *natura non facit saltus*. This Latin phrase vividly expresses the long-standing conviction among scientists that natural processes change smoothly (continuously).

Admittedly, phenomena of interest in spatial data analysis (e.g., atmospheric pollutant concentrations, coarse-grained fluid velocities in porous media) exhibit a certain degree of smoothness. Hence, we expect that even in the presence of randomness, the realizations of physical observables preserve continuity. Below, we discuss in more detail how continuity is expressed mathematically for random fields.

5.1.2.1 Modes of Continuity

The convergence modes of random variables can be extended to describe the continuity of random fields. To examine the continuity of an RF at $\mathbf{s}_0 \in \mathbb{R}^d$, we consider point sequences $\{\mathbf{s}_i\}_{i=1}^n$ such that $\|\mathbf{s}_n - \mathbf{s}_0\| \to 0$ as $n \to \infty$. We will focus on *almost sure* and *mean-square* convergence modes.

Almost sure: A random field $X(\mathbf{s}; \omega)$ is continuous almost surely at point $\mathbf{s}_0$, if

$$P\left(\lim_{\mathbf{s}_n \to \mathbf{s}_0} |X(\mathbf{s}_n; \omega) - X(\mathbf{s}_0; \omega)| < \epsilon\right) = 1, \text{ for all } \epsilon > 0.$$

An equivalent expression of the above is as follows:

$$P\Big(\omega \in \Omega : |X(\mathbf{s}_n; \omega) - X(\mathbf{s}_0; \omega)| \to 0 \text{ as } \mathbf{s}_n \to \mathbf{s}_0\Big) = 1.$$

Mean-square sense: A random field $X(\mathbf{s}; \omega)$ is continuous in the mean-square sense at point $\mathbf{s}_0$, if

$$\lim_{\mathbf{s}_n \to \mathbf{s}_0} \mathcal{E}\left[\left\{X(\mathbf{s}_n; \omega) - X(\mathbf{s}_0; \omega)\right\}^2\right] = 0.$$

Definition 5.1 A random field $X(\mathbf{s}; \omega)$ is continuous (almost surely or in the mean square sense) over $\mathcal{D} \subseteq \mathbb{R}^d$, if it is continuous at every point $\mathbf{s} \in \mathcal{D}$.

If a random field $X(\mathbf{s}; \omega)$ is *almost surely continuous* at every point $\mathbf{s}$, it is also known as a random field with *sample path continuity*. The term sample path is another way of referring to the realizations or states $x(\mathbf{s})$ of the random field.

5.1.2.2 Mean-Square Continuity

The concept of continuity in the mean-square sense requires less stringent conditions than almost sure continuity. In addition, it involves the first-order and second-order moments of the field instead of the full probability distribution. For many practical problems, the mathematical modeling does not go beyond second-order moments. For these reasons, the concept of mean square continuity is often preferred in engineering applications and spatial data analysis. Nonetheless, we should keep in mind that in many cases, such as for solutions of partial differential equations with random field coefficients, almost sure continuity (and differentiability) are more appropriate notions.

We present below the well-known theorems that determine conditions of mean-square continuity in terms of the random field covariance function. In the following we assume that the trend (i.e., the expectation of the random field) is a continuous function of the position $\mathbf{s}$.

Theorem 5.2 *A random field* $X(\mathbf{s};\omega)$ *with continuous expectation* $m_x(\mathbf{s})$ *is mean-square continuous at the point* $\mathbf{s}_0$, *if and only if the covariance function* $C_{xx}(\mathbf{s},\mathbf{s}')$ *is continuous as* $\mathbf{s} \to \mathbf{s}_0$ *and* $\mathbf{s}' \to \mathbf{s}_0$.

Hence, $X(\mathbf{s};\omega)$ is continuous in $\mathcal{D} \subseteq \mathbb{R}^d$ if the covariance function $C_{xx}(\mathbf{s},\mathbf{s})$ is continuous along the diagonal for every $\mathbf{s} \in \mathcal{D}$. This condition is equivalent to the variance $\sigma_x^2(\mathbf{s})$ being continuous $\forall \mathbf{s} \in \mathcal{D}$.

Theorem 5.3 *A wide-sense stationary or statistically homogeneous random field is continuous at every point* $\mathbf{s} \in \mathbb{R}^d$, *if* $C_{xx}(\mathbf{r})$ *is continuous at* $\mathbf{r} = \mathbf{0}$.

According to Theorem 5.3, for homogeneous random fields the continuity of the covariance function at the origin is a necessary and sufficient condition for the continuity of the random field (in mean square sense) everywhere in the domain.

The continuity conditions for homogeneous random fields are a straightforward consequence of the general continuity conditions and the translational invariance of the mean and the covariance imposed by statistical homogeneity.

Noise-induced discontinuity Let us consider a latent, stationary random field $X(\mathbf{s};\omega)$ with covariance function $C_{xx}(\mathbf{r})$. If the observed field $X^*(\mathbf{s};\omega)$ is contaminated by white noise, its variogram is given by

$$\gamma^*_{xx}(\mathbf{r}) = \gamma_{xx}(\mathbf{r}) + c_0\, \mathbb{I}_{(0,\infty)}(\|\mathbf{r}\|),$$

where $c_0\, \mathbb{I}_{(0,\infty)}(\|\mathbf{r}\|)$ is the nugget term. The indicator function $\mathbb{I}_{(0,\infty)}(\cdot)$ is equal to one if its argument is within the open interval $(0,\infty)$ and zero otherwise. The nugget introduces a step-function discontinuity at lags that are infinitesimally larger than zero. This discontinuity "obscures" the continuity of the latent random field.

5.1.3 Sample-Path Continuity

The conditions stated in Theorems 5.2 and 5.3 above are necessary and sufficient for mean square continuity, but only necessary for sample path continuity. Determining sufficient continuity conditions for random fields with general probability distributions requires more intricate technical steps, suitable for the mathematically minded (e.g., Chapter 3 in [10]). A general sufficient condition for sample path continuity is given by the following theorem:

Theorem 5.4 *A random field* $X(\mathbf{s}; \omega)$ *has continuous sample paths with probability one over any compact subset in* $\mathbb{R}^d$*, if for* $\alpha > 0$ *and* $\epsilon > \alpha$ *there is* $c > 0$ *such that*

$$\mathcal{E}\left[\,|\,X(\mathbf{s}; \omega) - X(\mathbf{s} + \mathbf{r}; \omega)|^{\alpha}\,\right] \le \frac{c\,\|\mathbf{r}\|^{2d}}{|\ln \|\mathbf{r}\|\,|^{1+\epsilon}}. \tag{5.3}$$

In the case of *Gaussian random fields*, the above theorem is simplified as follows:

Theorem 5.5 *A zero-mean, Gaussian random field* $X(\mathbf{s}; \omega)$ *with a continuous covariance function has continuous sample paths with probability one, if there is a finite-valued coefficient* $c > 0$ *and an exponent* $\epsilon > 0$ *such that*

$$\mathcal{E}\left[\,|\,X(\mathbf{s}; \omega) - X(\mathbf{s} + \mathbf{r}; \omega)|^{2}\,\right] \le \frac{c}{|\ln \|\mathbf{r}\|\,|^{1+\epsilon}}, \tag{5.4}$$

for all $\mathbf{s}$ *and* $\mathbf{r} \in \mathbb{R}^d$.

In the case of *stationary random fields*, the above sample path continuity condition is equivalent to

$$\gamma_{\mathrm{xx}}(\mathbf{r}) \le \frac{c}{2\,|\ln \|\mathbf{r}\|\,|^{1+\epsilon}}. \tag{5.5}$$

Abrahamsen [3] remarked that exceeding the upper bound specified by condition (5.5) requires a very sharp increase of the variogram for $\|\mathbf{r}\| \approx 0$. Thus, the above upper bound is satisfied by all the variogram functions commonly used in geostatistics. Motivated by this observation, he proposed the following conjecture:

Abrahamsen Conjecture: Gaussian random fields with continuous expectations and continuous covariance functions possess continuous sample paths with probability one [3].

5.1.4 Random Field Differentiability

Let $\partial_i X(\mathbf{s}; \omega)$ or equivalently, $X_i(\mathbf{s}; \omega)$, where $i = 1, \ldots, d$, denote the first-order partial derivative of the random field $X(\mathbf{s}; \omega)$ in the i-th direction. A random field $X(\mathbf{s}; \omega)$ is considered to be *almost surely differentiable* at point $\mathbf{s}$ if the partial derivative $\partial_i X(\mathbf{s}; \omega)$ exists almost surely, for all $i = 1, \ldots, d$.

More specifically, let $\hat{\mathbf{e}}_i$ denote the unit vector in the spatial direction i where $i = 1, \ldots, d$. In addition, define the finite difference

$$X_{i,a}(\mathbf{s}; \omega) = \frac{X(\mathbf{s} + a\, \hat{\mathbf{e}}_i; \omega) - X(\mathbf{s}; \omega)}{a}.$$

Then, the i-th first-order partial derivative for the state with index ω at the point $\mathbf{s}$ is defined by the following limit (if it exists):

$$\partial_i X(\mathbf{s}; \omega) \doteq \lim_{a \to 0} X_{i,a}(\mathbf{s}; \omega). \tag{5.6}$$

The existence of the limit can be investigated using either the *strong (almost sure)* or the *weak (with probability one)* notions of convergence.

Similarly, the partial derivative is said to exist in the *mean square sense* if the following limit exists

$$\partial_i X(\mathbf{s}; \omega) = \underset{a \to 0}{\mathrm{l.\,i.\,m.}}\, X_{i,a}(\mathbf{s}; \omega). \tag{5.7}$$

The above means that there exists a random field $\partial_i X(\mathbf{s}; \omega)$ with finite second-order moment such that

$$\lim_{a \to 0} \mathcal{E}\left[\left\{ X_{i,a}(\mathbf{s}; \omega) - \partial_i X(\mathbf{s}; \omega) \right\}^2 \right].$$

For stationary and Gaussian random fields, the existence of the partial derivative in the *mean square sense* is a *necessary condition* for the differentiability of the sample paths [3, p. 24]. For more general random fields, suitable conditions for sample-path differentiability are given in [670, 720, 721].

Comment We use the same notation, i.e., $\partial_i X(\mathbf{s}; \omega)$, for the partial derivatives whether they are defined in the mean square sense or almost surely.

5.1.5 Differentiability in the Mean-Square Sense

This section presents necessary and sufficient conditions for the existence of partial derivatives in the mean-square sense.

5.1.5.1 Non-homogeneous Random Fields

Theorem 5.6 *Let* $\mathbf{s} \in \mathcal{D} \subset \mathbb{R}^d$ *represent a point in the spatial domain of interest and the index* $i \in \{1, \ldots, d\}$ *denote the orthogonal directions in a Cartesian coordinate system. The partial derivative* $\partial_i \mathrm{X}(\mathbf{s}; \omega)$ *of random field* $\mathrm{X}(\mathbf{s}; \omega)$ *exists in the mean square sense, if the following conditions hold:*

- *The expectation* $m_{\mathrm{x}}(\mathbf{s})$ *admits the respective partial derivative* $\partial_i m_{\mathrm{x}}(\mathbf{s})$.
- *The second-order partial derivative of the covariance function,* $\partial^2 C_{\mathrm{xx}}(\mathbf{s}, \mathbf{s})/\partial s_i^2$ *exists at the point* $(\mathbf{s}, \mathbf{s})$ *and is finite.*

Hence, $\mathrm{X}(\mathbf{s}; \omega)$ is differentiable in $\mathcal{D} \subseteq \mathbb{R}^d$ if the partial derivatives $\partial^2 C_{\mathrm{xx}}(\mathbf{s}, \mathbf{s})/\partial s_i^2$, exist and are finite for every $\mathbf{s} \in \mathbb{R}^d$ and $i = 1, \ldots, d$. The above specify necessary regularity conditions. For *sufficient* conditions see [720] and references therein.

For a random field that admits n-th order partial derivatives in the ith direction in the mean-square sense, the expectation of the n-th derivative is given by the derivative of the expectation with the respective order, i.e.,

$$\mathcal{E}\left[\partial_i^n \mathrm{X}(\mathbf{s}; \omega)\right] = \frac{\partial^n m_{\mathrm{x}}(\mathbf{s})}{\partial s_i^n}, \text{ for every } \mathbf{s} \in \mathbb{R}^d, \text{ and } i = 1, \ldots, d. \tag{5.8}$$

5.1.5.2 Homogeneous Random Fields

The conditions of mean-square differentiability are simplified in the case of statistically homogeneous (wide-sense stationary) random fields. First of all, homogeneous random fields have a constant mean. This implies that the expectations of the first- and higher-order as well partial derivatives vanish, i.e.,

$$\mathcal{E}\left[\partial_i \mathrm{X}(\mathbf{s}; \omega)\right] = 0, \text{ for every } \mathbf{s} \in \mathbb{R}^d. \tag{5.9}$$

Theorem 5.7 *A statistically homogeneous (wide-sense stationary) random field* $\mathrm{X}(\mathbf{s}; \omega)$ *admits—in the mean square sense—the partial derivative* $\partial_i \mathrm{X}(\mathbf{s}; \omega)$ *where* $\mathbf{s} \in \mathcal{D}$ *and* $i \in \{1, \ldots, d\}$, *if*

$$\left. \frac{\partial^2 C_{\mathrm{xx}}(\mathbf{r})}{\partial r_i^2} \right|_{\mathbf{r}=\mathbf{0}},$$

exists and is a finite number.

If $X(\mathbf{s}; \omega)$ admits first-order partial derivatives in the mean square sense for all $i = 1, \ldots, d$, then the *gradient vector* $\nabla X(\mathbf{s}; \omega)$ is well-defined in the mean square sense, i.e.,

$$\nabla X(\mathbf{s}; \omega) = \left(\frac{\partial X(\mathbf{s}; \omega)}{\partial s_1}, \ldots, \frac{\partial X(\mathbf{s}; \omega)}{\partial s_d} \right)^\top . \tag{5.10}$$

It is straightforward to show that for *homogeneous random fields* the expectation of the gradient vector is zero, i.e.,

$$\mathcal{E}[\nabla X(\mathbf{s}; \omega)] = \mathbf{0}.$$

Hence, for *homogeneous random fields* the differentiability of the covariance function at the origin is a *necessary and sufficient* condition for the differentiability of the random field (in the mean square sense) everywhere in the domain $\mathcal{D} \subseteq \mathbb{R}^d$.

Covariance of the gradient vector The covariance of the gradient vector $\nabla X(\mathbf{s}; \omega)$ is a tensor whose elements are given by

$$\mathcal{E}\left[\frac{\partial X(\mathbf{s}; \omega)}{\partial s_i} \frac{\partial X(\mathbf{s}'; \omega)}{\partial s'_j} \right] = - \frac{\partial^2 C_{xx}(\mathbf{r})}{\partial r_i \partial r_j}, \quad \mathbf{r} = \mathbf{s} - \mathbf{s}', \; i, j = 1, \ldots, d. \tag{5.11}$$

The covariance of the gradient vector is equivalent to the expectation of the *gradient tensor* $\nabla X(\mathbf{s}; \omega) \nabla X(\mathbf{s}; \omega)$ with elements

$$[\nabla X(\mathbf{s}; \omega) \nabla X(\mathbf{s}; \omega)]_{ij} = \partial_i X(\mathbf{s}; \omega) \partial_j X(\mathbf{s}; \omega), \; \text{ for all } i, j = 1, \ldots, d.$$

We return to (5.11) and the gradient tensor in Sect. 5.2 in the discussion on anisotropy estimation. The topic of permissible covariance functions for random vector fields $\mathbf{V}(\mathbf{s}; \omega)$ that are either *divergence-free*, i.e.,

$$\operatorname{div} \mathbf{V}(\mathbf{s}; \omega) \equiv \nabla \cdot \mathbf{V}(\mathbf{s}; \omega) = 0$$

or *curl-free*, i.e.,

$$\operatorname{curl} \mathbf{V}(\mathbf{s}; \omega) \equiv \nabla \times \mathbf{V}(\mathbf{s}; \omega) = 0,$$

is investigated in [721]. Note that the gradient vector field $\mathbf{V}(\mathbf{s}; \omega) = \nabla X(\mathbf{s}; \omega)$ of a scalar random field $X(\mathbf{s}; \omega)$ is by construction curl-free due to the vector identity $\nabla \times \nabla x(\mathbf{s}) = 0$, for any continuously differentiable, real-valued scalar function $x(\mathbf{s})$.

Remark The divergence of a vector $\mathbf{v}(\mathbf{s})$ measures the density of the outward flux of $\mathbf{v}(\mathbf{s})$ emanating from a very small volume around the point $\mathbf{s}$. If $\mathbf{s} \in \mathbb{R}^d$, then

$$\nabla \cdot \mathbf{v}(\mathbf{s}) = \sum_{i=1}^{d} \frac{\partial v_i(\mathbf{s})}{\partial s_i}.$$

The curl on the other hand measures the circulation density (rotation) of the vector $\mathbf{v}(\mathbf{s})$ around an infinitesimal loop around the point $\mathbf{s}$. In three dimensions, the curl is given by

$$[\mathrm{curl}\mathbf{v}(\mathbf{s})]_1 = \frac{\partial v_3(\mathbf{s})}{\partial s_2} - \frac{\partial v_2(\mathbf{s})}{\partial s_3},$$

$$[\mathrm{curl}\mathbf{v}(\mathbf{s})]_2 = \frac{\partial v_1(\mathbf{s})}{\partial s_3} - \frac{\partial v_3(\mathbf{s})}{\partial s_1},$$

$$[\mathrm{curl}\mathbf{v}(\mathbf{s})]_3 = \frac{\partial v_2(\mathbf{s})}{\partial s_1} - \frac{\partial v_1(\mathbf{s})}{\partial s_2}.$$

Example 5.1 Show that a necessary condition for the first-order partial derivatives $\partial_i X(\mathbf{s}; \omega)$ of a wide-sense stationary random field $X(\mathbf{s}; \omega)$ to exist in the mean square sense for every $i, j = 1, \ldots, d$ is that

$$\left. \frac{\partial C_{\mathrm{xx}}(\mathbf{r})}{\partial r_i} \right|_{\mathbf{r}=\mathbf{0}} = 0, \text{ for all } i = 1, \ldots, d. \tag{5.12}$$

Answer Based on Theorem 5.7, the first-order mean-square differentiability is equivalent to the existence of the second-order partial derivatives of the covariance function at zero lag, i.e.,

$$\partial_i^2 C_{\mathrm{xx}}(0) \doteq \left. \frac{\partial^2 C_{\mathrm{xx}}(\mathbf{r})}{\partial r_i^2} \right|_{\mathbf{r}=\mathbf{0}}.$$

The existence of $\partial_i \partial_j C_{\mathrm{xx}}(0)$ presupposes the existence of the first-order derivatives of the covariance function at $\mathbf{r} = \mathbf{0}$.

At the same time, the covariance function is maximized at $\mathbf{r} = \mathbf{0}$. Since the function is differentiable, the maximum at $\mathbf{r} = \mathbf{0}$ is a critical point, and it must thus satisfy (5.12).

Covariance of higher-order derivatives Theorem 5.7 is easily generalized to partial derivatives of arbitrary order. More specifically, assuming that the field derivatives of order k and $m \in \mathbb{N}$ in the directions i and $j \in \{1, \ldots, d\}$ respectively, exist in the mean-square sense, it follows that the expectation of the product of partial derivatives is given by

$$\mathcal{E}\left[\frac{\partial^k X(\mathbf{s};\omega)}{\partial s_i^k}\frac{\partial^m X(\mathbf{s}';\omega)}{\partial s_j'^m}\right] = (-1)^m \frac{\partial^{k+m} C_{xx}(\mathbf{r})}{\partial r_i^k\,\partial r_j^m}, \tag{5.13}$$

for $\mathbf{r} = \mathbf{s} - \mathbf{s}'$, $i, j = 1, \ldots, d$, and $k, m \in \mathbb{Z}$.

Covariance of the gradient vector for isotropic random fields Let us now consider isotropic random fields $X(\mathbf{s};\omega)$ so that $m_x(\mathbf{s})$ is a constant and $C_{xx}(\mathbf{r})$ is a radial function of the Euclidean distance $\|\mathbf{r}\|$. The covariance of the gradient vector is given by (5.11).

Condition for gradient existence The existence of the first-order partial derivatives of $X(\mathbf{s};\omega)$ in the mean square sense is equivalent to the existence of the following second-order radial derivative of the covariance function

$$C_{xx}^{(2)}(0) \doteq \left.\frac{\mathrm{d}^2 C_{xx}(\|\mathbf{r}\|)}{\mathrm{d}\|\mathbf{r}\|^2}\right|_{\|\mathbf{r}\|=0}.$$

Proof Since the random field is assumed to be isotropic, it is also statistically homogeneous. Hence, based on the Theorem 5.7, the existence of $\partial_i X(\mathbf{s};\omega)$, $(i = 1, \ldots, d)$ requires the existence of the second-order partial derivatives of $C_{xx}(\mathbf{r})$.

The first-order partial derivatives of the covariance function are given in terms of the radial derivative by

$$\frac{\partial C_{xx}(\|\mathbf{r}\|)}{\partial r_i} = \frac{r_i}{\|\mathbf{r}\|}\frac{\mathrm{d}C_{xx}(\|\mathbf{r}\|)}{\mathrm{d}\|\mathbf{r}\|}, \quad \text{for all } i = 1, \ldots, d. \tag{5.14}$$

Based on the results of Example 5.1, a necessary condition for differentiability is the vanishing of the first-order radial derivative at zero lag, i.e.,

$$\left.\frac{\mathrm{d}C_{xx}(\|\mathbf{r}\|)}{\mathrm{d}\|\mathbf{r}\|} = 0\right|_{\|\mathbf{r}\|=0}. \tag{5.15}$$

The *second-order partial derivatives of* $C_{xx}(\|\mathbf{r}\|)$ are expressed in terms of the first- and second-order radial derivatives as follows:

$$\frac{\partial^2 C_{xx}(\|\mathbf{r}\|)}{\partial r_i \partial r_j} = \frac{1}{\|\mathbf{r}\|}\left[\left(\delta_{ij} - \frac{r_i r_j}{\|\mathbf{r}\|^2}\right)\frac{\mathrm{d}C_{xx}(\|\mathbf{r}\|)}{\mathrm{d}\|\mathbf{r}\|}\right] + \frac{r_i r_j}{\|\mathbf{r}\|^2}\frac{\mathrm{d}^2 C_{xx}(\|\mathbf{r}\|)}{\mathrm{d}\|\mathbf{r}\|^2}, \tag{5.16}$$

where $\delta_{i,j}$ is Kronecker's delta defined by $\delta_{i,j} = 1$, for $i = j$, and $\delta_{i,j} = 0$ for $i \neq j$.

Hence, combining (5.11) with (5.15) it follows that the gradient covariance tensor evaluated at zero lag is given by

$$-\partial_i \partial_j C_{xx}(0) = -\frac{r_i\, r_j}{\|\mathbf{r}\|^2}\, C_{xx}^{(2)}(0).$$

Thus, the existence of the second-order radial derivative of the covariance function at $\|\mathbf{r}\| = 0$ guarantees the existence of the field's gradient.

Local behavior A practical consequence of the above discussions on differentiability is that the local behavior of Gaussian random fields is determined by the properties of the covariance function at the origin. Differentiability also determines the coarseness of the field realizations. A non-differentiable random field has coarser realizations than a differentiable one, even if the main statistics (mean, variance, characteristic length) of the two fields are the same. This difference is evidenced in Fig. 5.4 which displays two realizations, one from a smooth differentiable Gaussian random field (left) and the other one from a non-differentiable random field which appears considerably "grainier" (right).

The difference in the covariance functions of the two fields which leads to distinctly different geometric features of the two realizations in Fig. 5.4 can be understood with the help of the simplified schematic in Fig. 5.5. The first curve is rounded at the peak, implying zero slope, whereas the second curve has a sharp peak. In the latter case, the value of the slope depends on the direction from which the peak is approached: the slope is positive on the left side of the peak but becomes negative on the right side. Thus, the slope at the peak depends on the direction of approach and the derivative is not well defined.

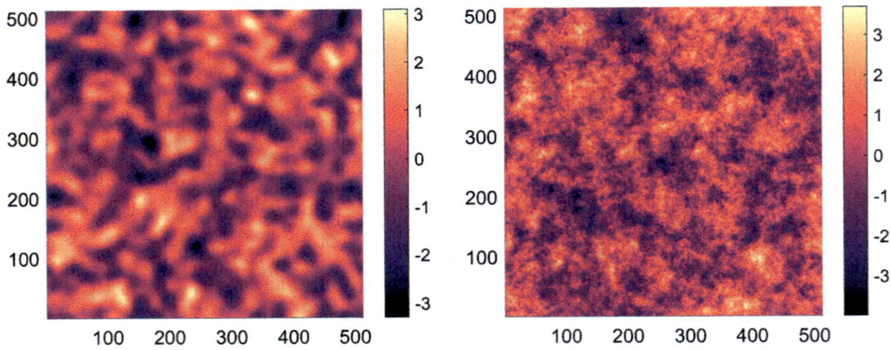

Fig. 5.4 Realizations of two Gaussian random fields with the same mean (equal to zero), standard deviation (equal to one), and characteristic length (equal to 20). The realization on the left is generated by a differentiable Gaussian covariance function, whereas the realization on the right is generated by a continuous but non-differentiable, exponential covariance function

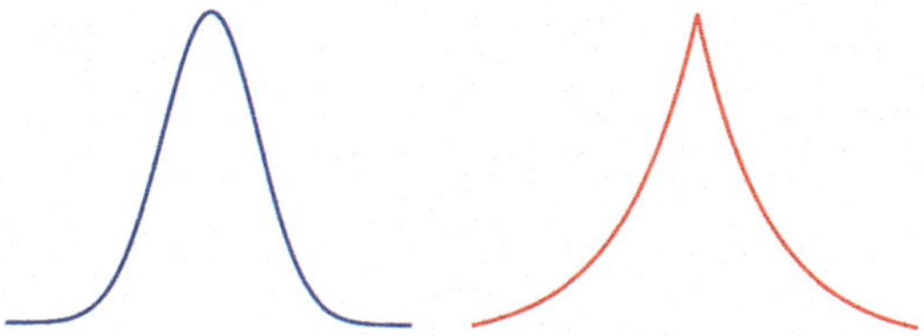

Fig. 5.5 Schematic of covariance functions corresponding to differentiable (Gaussian, left) and non-differentiable (exponential, right) random fields

Example 5.2 Investigate the first-order mean-square differentiability of two 2D random fields with (i) anisotropic Gaussian covariance and (ii) anisotropic exponential covariance. Assume that the coordinate system coincides with that of the principal anisotropy axes.

Answer The 2D Gaussian anisotropic covariance function is given by

$$C_{xx}(\mathbf{r}) = \exp\left(-\frac{r_1^2}{\xi_1^2} - \frac{r_2^2}{\xi_2^2}\right).$$

In the above we assumed, without loss of generality, that the variance is equal to one. The exponential anisotropic covariance function is given by

$$C_{xx}(\mathbf{r}) = \exp\left(-\sqrt{\frac{r_1^2}{\xi_1^2} + \frac{r_2^2}{\xi_2^2}}\right).$$

(i) The partial derivatives of the Gaussian covariance are given by the following equations

$$\frac{\partial C_{xx}(\mathbf{r})}{\partial r_1} = -\frac{2r_1}{\xi_1^2} \exp(-r_1^2/\xi_1^2 - r_2^2/\xi_2^2), \tag{5.17a}$$

$$\frac{\partial C_{xx}(\mathbf{r})}{\partial r_2} = -\frac{2r_2}{\xi_2^2} \exp(-r_1^2/\xi_1^2 - r_2^2/\xi_2^2). \tag{5.17b}$$

Based on the above, the first-order partial derivatives $\partial C_{xx}(\mathbf{r})/\partial r_i$, where $i = 1, 2$, vanish at $\|\mathbf{r}\| = 0$. Hence, the necessary condition (5.12) is satisfied. In addition, $\partial C_{xx}(\mathbf{r})/\partial r_i$ involves a product of two differentiable functions. Hence, the second-order partial derivatives exist at $\|\mathbf{r}\| = 0$, and thus the Gaussian covariance corresponds to realizations which are mean-square differentiable.[1]

(ii) To calculate the first-order partial derivatives of the exponential covariance we use the following notation: $C_{xx}(\mathbf{r}) = \exp[-\phi(u_1, u_2)]$, where $\phi(u_1, u_2) = \sqrt{u_1^2 + u_2^2}$ and $u_i = r_i/\xi_i$. Then,

[1]More generally, it can be shown that the Gaussian covariance function admits derivatives of all orders at $\|\mathbf{r}\| = 0$.

$$\frac{\partial C_{xx}(\mathbf{r})}{\partial r_i} = e^{-\phi(u_1,u_2)} \frac{\partial \phi(u_1,u_2)}{\partial u_i} \frac{\partial u_i}{\partial r_i} = -\frac{r_i\, e^{-\phi(u_1,u_2)}}{\xi_i^2 \sqrt{u_1^2+u_2^2}}.$$

Based on the above, the first-order derivatives are given by the following equations

$$\frac{\partial C_{xx}(\mathbf{r})}{\partial r_1} = -\frac{r_1}{\xi_1^2 \left(r_1^2/\xi_1^2 + r_2^2/\xi_2^2\right)^{1/2}} \exp\left(-\sqrt{r_1^2/\xi_1^2 + r_2^2/\xi_2^2}\right),$$

$$\frac{\partial C_{xx}(\mathbf{r})}{\partial r_2} = -\frac{r_2}{\xi_2^2 \left(r_1^2/\xi_1^2 + r_2^2/\xi_2^2\right)^{1/2}} \exp\left(-\sqrt{r_1^2/\xi_1^2 + r_2^2/\xi_2^2}\right).$$

We can show that the limit of the partial derivative near zero depends on the direction of approach. To calculate the derivative $\frac{\partial C_{xx}(\mathbf{r})}{\partial r_i}$ around the origin we first set $r_j = 0$ for $j \neq i$ in the equations above. Then

$$\left.\frac{\partial C_{xx}(\mathbf{r})}{\partial r_i}\right|_{r_i \approx 0} = -\frac{1}{\xi_i}\frac{r_i}{\mid r_i \mid} = -\frac{\operatorname{sign}(r_i)}{\xi_i}.$$

Hence, the value of the derivative around zero depends on which direction the point $r_i = 0$ is approached from, no matter how close we are to $r_i = 0$. This directional dependence implies that the derivative does not exist at $\|\mathbf{r}\| = 0$. Thus, the exponential covariance function is not differentiable even at first order.

Laplacian of radial functions The Laplacian of a second-order differentiable function $C_{xx}(\mathbf{r})$ is given by the following sum of partial derivatives

$$\nabla^2 C_{xx}(\mathbf{r}) = \sum_{i=1}^{d} \frac{\partial^2 C_{xx}(\|\mathbf{r}\|)}{\partial r_i^2}. \tag{5.18}$$

Based on the second-order partial derivatives (5.16) of radial covariance functions, the Laplacian is given by the following equation

$$\nabla^2 C_{xx}(\|\mathbf{r}\|) = \left(\frac{d-1}{\|\mathbf{r}\|}\right) \frac{\mathrm{d}C_{xx}(\|\mathbf{r}\|)}{\mathrm{d}\|\mathbf{r}\|} + \frac{\mathrm{d}^2 C_{xx}(\|\mathbf{r}\|)}{\mathrm{d}\|\mathbf{r}\|^2}. \tag{5.19}$$

Bilaplacian of radial function The *Bilaplacian operator* $\Delta^2 = \nabla^4$ represents the square of the Laplacian. The Bilaplacian is used in connection with the covariance function of Spartan random fields in Chaps. 8 and 9.

To evaluate the Bilaplacian of a radial function, we apply the Laplacian twice according to (5.19) to obtain

$$\nabla^2\left[\nabla^2 C_{xx}(\|\mathbf{r}\|)\right] = \nabla^2\left[\frac{d-1}{\|\mathbf{r}\|}\frac{\mathrm{d}C_{xx}(\|\mathbf{r}\|)}{\mathrm{d}\|\mathbf{r}\|}\right] + \nabla^2\left[\frac{\mathrm{d}^2 C_{xx}(\|\mathbf{r}\|)}{\mathrm{d}\|\mathbf{r}\|^2}\right].$$

The first term on the right-hand side of the above equation is of the form $\nabla^2 F_1(\|\mathbf{r}\|)$, where $F_1(\|\mathbf{r}\|) = \frac{d-1}{\|\mathbf{r}\|} C_{xx}^{(1)}(\|\mathbf{r}\|)$, and $C_{xx}^{(n)}(\|\mathbf{r}\|)$ denotes the n-th order derivative. We apply (5.19) to calculate the Laplacian of $F_1(\|\mathbf{r}\|)$, which leads to

Table 5.1 Local properties (mean square continuity and differentiability) of the isotropic correlation models listed in Table 4.2. The functions are expressed in terms of the dimensionless distance $u = \|\mathbf{r}\|/\xi$, where $\|\mathbf{r}\|$ is the Euclidean distance in $d \geq 2$, whereas $u = |r|$ in $d = 1$

Classes	Equation	Continuity	Differentiability[a]
Nugget effect	$\delta(\mathbf{r})$	No	No
Exponential	$\exp(-u)$	Yes	No
Gaussian	$\exp(-u^2)$	Yes	Yes (all orders)
Spherical	$\left(1 - 1.5\,u + 0.5\,u^3\right)\ \mathbb{I}_{0\leq u\leq 1}(u)$	Yes	No
Cardinal sine	$\mathrm{sinc}(u)$	Yes	Yes ($m \leq 2$)
Generalized exponential[b]	$\exp(-u^{\nu})$	Yes	No
Matérn[c]	$\dfrac{\left(\sqrt{2\nu}\,u\right)^{\nu}}{2^{\nu-1}\,\Gamma(\nu)}\ K_{\nu}\left(\sqrt{2\nu}\,u\right)$	Yes	Yes ($m < \nu$)
Bessel-J[d]	$2^{\nu}\,\Gamma(\nu+1)\left[\dfrac{J_{\nu}(u)}{u^{\nu}}\right]$	Yes	Yes[e]
Rational quadratic	$\dfrac{1}{(1+u^2)^{\beta}}$	Yes	Yes
Cauchy class[f]	$\dfrac{1}{(1+u^{\alpha})^{\beta/\alpha}}$	Yes	No

[a]The order of differentiability m refers to the random field, not to the covariance function
[b]$0 < \nu < 2$
[c,d]For ν such that $\nu > 0$
[e]The differentiability can be shown based on the J-Bessel correlation expansion (4.23)
[f]For α such that $0 < \alpha < 2$ and $\beta > 0$

$$\nabla^2 F_1(\|\mathbf{r}\|) = \frac{[(d-1)^2 - 2(d-1)]}{\|\mathbf{r}\|^2}\left[C^{(2)}_{\mathrm{xx}}(\|\mathbf{r}\|) + \frac{C^{(1)}_{\mathrm{xx}}(\|\mathbf{r}\|)}{\|\mathbf{r}\|}\right] + \frac{(d-1)}{\|\mathbf{r}\|}C^{(3)}_{\mathrm{xx}}(\|\mathbf{r}\|).$$

The second term on the right-hand side is of the form $\nabla^2 F_2(\|\mathbf{r}\|)$, where $F_2(\|\mathbf{r}\|) = C^{(2)}_{\mathrm{xx}}(\|\mathbf{r}\|)$. A further application of (5.19) leads to

$$\nabla^2 F_2(\|\mathbf{r}\|) = \frac{(d-1)}{\|\mathbf{r}\|}C^{(3)}_{\mathrm{xx}}(\|\mathbf{r}\|) + C^{(4)}_{\mathrm{xx}}(\|\mathbf{r}\|).$$

Adding the two terms, we obtain the following result for the *Bilaplacian of the radial function* $C_{\mathrm{xx}}(\|\mathbf{r}\|)$:

$$\nabla^4 C_{\mathrm{xx}}(\|\mathbf{r}\|) = C^{(4)}_{\mathrm{xx}}(\|\mathbf{r}\|) + \frac{2(d-1)}{\|\mathbf{r}\|}C^{(3)}_{\mathrm{xx}}(\|\mathbf{r}\|)$$
$$+ \frac{[(d-1)^2 - 2(d-1)]}{\|\mathbf{r}\|^2}\left[C^{(2)}_{\mathrm{xx}}(\|\mathbf{r}\|) + \frac{C^{(1)}_{\mathrm{xx}}(\|\mathbf{r}\|)}{\|\mathbf{r}\|}\right]. \tag{5.20}$$

5.2 Covariance Hessian Identity and Geometric Anisotropy

Let us consider a covariance function that admits second-order partial derivatives and define the *Covariance Hessian Matrix* **H** (CHM) which consists of elements $H_{ij}(\mathbf{r})$ that are defined as follows

$$H_{ij}(\mathbf{r}) = \frac{\partial^2 C_{xx}(\mathbf{r})}{\partial r_i \, \partial r_j}, \quad i, j = 1, \ldots, d. \tag{5.21}$$

Based on the above definition, the matrix $H_{ij}(\mathbf{r})$ is *symmetric*, i.e., $H_{ij}(\mathbf{r}) = H_{ji}(\mathbf{r})$.

The covariance equation (5.11) relating the field's gradient to the covariance tensor was introduced by Swerling in a paper on the statistics of random surfaces [786]. We will refer to (5.11) as the *Covariance Hessian Identity (CHI)*, since $\partial^2 C_{xx}(\mathbf{r})/\partial r_i \, \partial r_j$, where $i, j = 1, \ldots, d$ are the elements of the covariance Hessian matrix. In terms of the covariance Hessian matrix, CHI is expressed as follows:

Theorem 5.8 (Covariance Hessian Identity) *Let* $X(\mathbf{s}; \omega)$*, where* $\mathbf{s} \in \mathbb{R}^d$*, be a statistically stationary and anisotropic random field with a covariance function* $C_{xx}(\mathbf{r})$ *that (i) has finite principal correlation lengths and (ii) admits all second-order partial derivatives at the origin. Based on* (5.11)*, the* Covariance Hessian Identity *(CHI) is then expressed as follows* [786]*:*

$$\mathcal{E}\left[\partial_i X(\mathbf{s}; \omega)\, \partial_j X(\mathbf{s}; \omega)\right] = -\, H_{i,j}(\mathbf{r})\big|_{\mathbf{r}=\mathbf{0}}, \; \text{for } i, j = 1, \ldots, d, \tag{5.22}$$

where $H_{i,j}(\mathbf{r})$ *are the elements of the covariance Hessian defined by* (5.21).

For stationary random fields the existence of finite second-order derivatives of the covariance function at zero lag ensures the existence of first-order field derivatives in the mean-square sense according to Theorem 5.7 (see [162]). The covariance Hessian identity (5.11) allows estimating the parameters of *geometric anisotropy* for differentiable random fields without requiring the parametric assumption of a specific covariance form [135, 359].

5.2.1 CHI for Two-Dimensional Random Fields

The Covariance Hessian Identity (5.22) can be used in two dimensions to express the parameters of geometric anisotropy in terms of the elements of the covariance Hessian matrix [135, 359].

More specifically, let A_1 and A_2 represent the *principal axes of anisotropy*, so that in a coordinate system aligned with these axes $C_{xx}(\mathbf{u}) = \phi(\mathbf{u}^\top \mathbf{V} \mathbf{u})$, where $\mathbf{u} = (u_1, u_2)$ is the lag in the principal system, $\mathbf{V}$ is a diagonal 2×2 matrix, and $\phi(\cdot)$ is a positive definite function.

The *principal correlation lengths* of the random field $\mathrm{X}(\mathbf{s};\omega)$, where $\mathbf{s} \in \mathbb{R}^2$, are given by

$$\frac{1}{\xi_i^2} = -\frac{c_0}{\sigma_{\mathrm{x}}^2} \frac{\partial^2 C_{\mathrm{xx}}(\mathbf{u})}{\partial u_i^2}\bigg|_{\mathbf{u}=\mathbf{0}}, \quad i = 1, 2,$$

where $c_0 > 0$ is an $O(1)$ constant that depends on the functional form of the covariance.

We define the *anisotropy ratio* as $R = \xi_2/\xi_1$. We also define the *orientation (rotation) angle* θ as the angle between the horizontal axis of the reference frame and the nearest principal axis (we will assume that this is A_1). According to Sect. 4.3.2, we define $\theta > 0$ if the principal axis is rotated in the counterclockwise direction with respect to the horizontal axis. Let us also define for convenience the *covariance of the field's gradient* at zero lag as

$$Q_{i,j} \doteq \mathcal{E}\left[\partial_i \mathrm{X}(\mathbf{s};\omega)\, \partial_j \mathrm{X}(\mathbf{s};\omega)\right] = -\, H_{i,j}(\mathbf{r})\big|_{\mathbf{r}=\mathbf{0}}, \quad \text{for } i, j = 1, 2. \tag{5.23}$$

The diagonal elements $Q_{i,i}$ are non-negative, because they represent the *variance of the partial derivative* $\partial_i \mathrm{X}(\mathbf{s};\omega)$, for $i = 1, 2$. The quantities $Q_{i,i}$ take finite, non-zero values for finite correlation lengths. In addition, based on the equations above it is straightforward to show that the principal correlation lengths are related to the diagonal elements of the gradient covariance by means of

$$\xi_i^2 = -\frac{\sigma_{\mathrm{x}}^2}{c_0} \frac{1}{H_{i,i}(\mathbf{0})} = \frac{\sigma_{\mathrm{x}}^2}{c_0} \frac{1}{Q_{i,i}}, \quad \text{for } i = 1, 2. \tag{5.24}$$

The above equation shows that the longer principal correlation length is in the direction i of the smaller gradient variance $Q_{i,i}$. In addition, the above relation shows that $Q_{i,i} > 0$, for $i = 1, 2$.

Based on (5.22) it can be shown that the anisotropy parameters (R, θ) of a two-dimensional geometrically anisotropic, at least first-order differentiable random field, satisfy the following theorem [135][2]:

Theorem 5.9 (Covariance Hessian Identity in 2D) *Let* $\mathrm{X}(\mathbf{s};\omega)$, *where* $\mathbf{s} \in \mathbb{R}^2$, *be a statistically stationary and anisotropic random field whose covariance function (i) has finite principal correlation lengths and (ii) admits all second-order partial derivatives at the origin. In addition, let*

[2] In [135] $R_{2(1)} = \xi_1/\xi_2$ is used instead of R. The results from [135] are equivalent to those presented herein following the transformation $R_{2(1)} \mapsto 1/R$.

$$q_{\text{diag}} = \frac{Q_{22}}{Q_{11}}, \; q_{\text{off}} = \frac{Q_{12}}{Q_{11}} = \frac{Q_{21}}{Q_{11}},$$

represent ratios of the elements $\{Q_{i2}\}_{i=1}^{2}$ *of the gradient covariance tensor* (5.23) *normalized by* Q_{11}. *Then, the following relations hold between* q_{diag}, q_{off} *and the anisotropy parameters:*

$$q_{\text{diag}} \doteq \frac{Q_{22}}{Q_{11}} = \frac{1 + R^2 \tan^2\theta}{R^2 + \tan^2\theta}, \tag{5.25a}$$

$$q_{\text{off}} \doteq \frac{Q_{12}}{Q_{11}} = \frac{\tan\theta(R^2 - 1)}{R^2 + \tan^2\theta}. \tag{5.25b}$$

The anisotropy parameters are then obtained from the ratios q_{diag} *and* q_{off} *as follows:*

$$\theta = \frac{1}{2}\tan^{-1}\left(\frac{2q_{\text{off}}}{1 - q_{\text{diag}}}\right), \tag{5.26a}$$

$$R = \left[1 + \frac{1 - q_{\text{diag}}}{q_{\text{diag}} - (1 + q_{\text{diag}})\cos^2\theta}\right]^{-1/2}. \tag{5.26b}$$

The equations (5.26) are obtained by explicit inversion of the gradient covariance ratio equations (5.25).

The dependence of the gradient covariance ratios on the anisotropy parameters is displayed in Fig. 5.6. The diagonal ratio q_{diag} is symmetric around $\theta = 0$, whereas the off-diagonal ratio q_{off} is anti-symmetric.

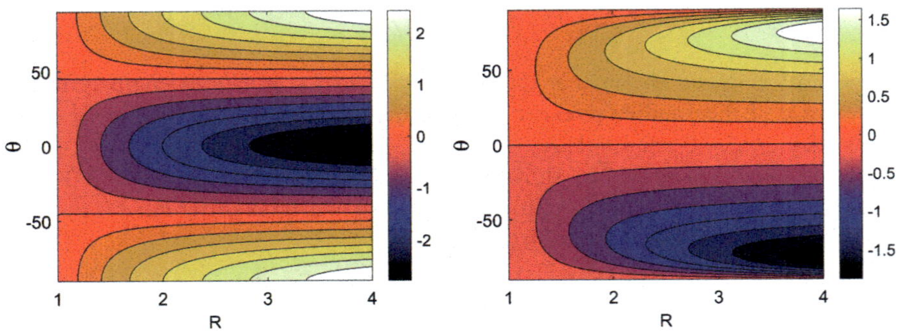

Fig. 5.6 Dependence of the gradient covariance ratios on the anisotropy parameters R and θ (in degrees). The natural logarithm, $\ln q_{\text{diag}}$, of the diagonal ratio q_{diag} is shown on the left, and the off-diagonal ratio q_{off} on the right

Positive definite **Q** We can easily show that the gradient covariance matrix **Q** is *positive definite.* First, note that $Q_{ii} = \sigma_i^2 > 0$ where $\sigma_i^2 = \mathbb{V}\text{ar}\{\partial_i X(\mathbf{s}; \omega)\}$. For $i \neq j$ it holds that $Q_{ij} = \sigma_i \sigma_j \rho_{i,j}$, where $|\rho_{i,j}| < 1$ is the correlation coefficient between the two components of the gradient. Hence, using the symmetry $Q_{12} = Q_{21}$, it follows that

$$Q_{11} Q_{22} - Q_{12} Q_{21} = Q_{11} Q_{22} - Q_{12}^2 = \sigma_1^2 \sigma_2^2 (1 - \rho_{1,2}^2) > 0.$$

Since $Q_{11} > 0$ the above inequality ensures that the determinant of the matrix **Q** is positive. Thus, the matrix **Q** is positive definite by means of *Sylvester's criterion*: a necessary and sufficient condition for a symmetric, real-valued matrix to be positive definite is that the leading principal minors be positive. If we divide both terms in the above inequality by Q_{11}^2, we obtain

$$q_{\text{diag}} > q_{\text{off}}^2, \text{ or equivalently } \sqrt{q_{\text{diag}}} > q_{\text{off}} > -\sqrt{q_{\text{diag}}}.$$

The solutions of equations (5.26) for R and θ are illustrated in Fig. 5.7. In light of the inequality satisfied by q_{diag} and q_{off}, not every combination $(q_{\text{diag}}, q_{\text{off}})$ is permissible. The white areas of the plots in Fig. 5.7 correspond to non-permissible combinations of q_{diag} and q_{off} values which fail to satisfy the condition of positive definiteness. Figure 5.7 (left plot) shows the natural logarithm of the anisotropy ratio as a function of q_{diag} and q_{off}. The larger values of $\ln R$ occur near the permissibility boundary. To understand why, consider the equations (5.25): if we assume that $R \gg 1$ and that θ is not close to $\pm\pi/2$ so that $R \gg \tan\theta$, it follows that $q_{\text{diag}} \approx \tan^2\theta$ and $q_{\text{off}} \approx \tan\theta$ which implies proximity of the permissibility boundary $q_{\text{diag}} = q_{\text{off}}^2$. On the other hand, based on (5.26a) $\theta \approx \pm\pi/2$ implies that $q_{\text{diag}} \approx 1$, which in light of (5.26b) leads to $R \approx 1$. Hence, the only regime where large values of R are possible is close to the permissibility boundary.

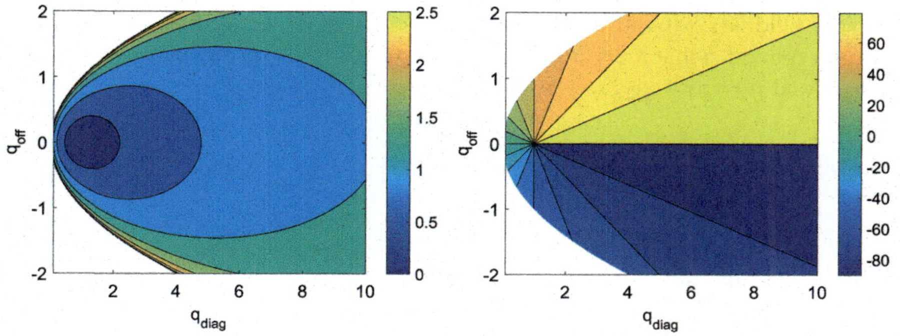

Fig. 5.7 Dependence of the anisotropy parameters R (left) and θ in degrees (right) on the mean gradient product ratios q_{diag} and q_{off}

Degeneracy of parameter space Equations (5.25) are invariant under the pair of transformations

$$\tan\theta \mapsto -(\tan\theta)^{-1}, \text{ that is, } \theta \mapsto \theta \pm \pi/2, \text{ and } R \mapsto 1/R.$$

By restricting the parameter space to $R \in [0, \infty)$ and $\theta \in [-\pi/4, \pi/4)$, or equivalently $R \in [1, \infty)$ and $\theta \in [-\pi/2, \pi/2)$, the pair (R, θ) that satisfies (5.25) for given $(q_{\mathrm{d}}, q_{\mathrm{o}})$ is unique, thus ensuring that the transformation $(q_{\mathrm{d}}, q_{\mathrm{o}}) \mapsto (R, \theta)$ is one-to-one (see also the discussion in Sect. 4.3.2).

Practical matters The CHI theorem permits estimating the anisotropy parameters if the slope tensor can be reliably estimated from the data. The advantage of the CHI approach is that the anisotropy parameters can be obtained in terms of simple equations without knowledge of the underlying covariance (or variogram) function, provided that the random field is at least first-order differentiable. Covariance functions that ensure differentiability of the random field in the mean square sense are listed in Table 5.1. The CHI approach is computationally more efficient than using a variogram-based estimation of anisotropy. However, one needs to be aware of the bias involved in estimating derivatives based on finite differences. The estimation of the gradient covariance from lattice and scattered data is investigated in [135, 359].

5.3 Spectral Moments

Spectral moments of a random field $X(\mathbf{s}; \omega)$ are defined with respect to integrals of the spectral density $\widetilde{C}_{\mathrm{xx}}(\mathbf{k})$. The usefulness of the spectral moments stems from their direct connection to different length scales of the random field and to the gradient covariance tensor.

Definition 5.2 A spectral moment $\Lambda_{\mathbf{p}}^{(p)}$ of order p of a wide-sense stationary random field $X(\mathbf{s}; \omega)$ with covariance function $C_{\mathrm{xx}}(\mathbf{r})$ is defined by the following integral (provided that it exists), where $\mathbf{p} = (p_1, \ldots, p_d)^\top$ is the vector of moment indices and $p = p_1 + p_2 \ldots + p_d$ [10, 809]

$$\Lambda_{\mathbf{p}}^{(p)} \doteq \int_{\mathbb{R}^d} d\mathbf{k}\, k_1^{p_1} \ldots k_d^{p_d}\, \widetilde{C}_{\mathrm{xx}}(\mathbf{k}), \;\; p_i \in \{0, 1, \ldots, p\}, \text{ such that } \sum_{i=1}^{d} p_i = p. \tag{5.27}$$

Note that multiple spectral moments of order p are possible, since the scalar value p does not uniquely specify the index vector $\mathbf{p}$.

5.3.1 Radial Spectral Densities

In the following, we assume that the covariance and the spectral density are radial functions that correspond to an isotropic random field.

Directional spectral moments with odd indices For a radial spectral density $\widetilde{C}_{xx}(\mathbf{k}) = \widetilde{C}_{xx}(\|\mathbf{k}\|)$, the spectral moments vanish if any of the moment indices p_i $(i = 1, \ldots, d)$ are odd integers. We show below, see (5.31), how this property is obtained for spectral second-order moments, i.e., $p_i = p_j = 1$, for $i \neq j = 1, \ldots, d$.

Radial spectral moments We will find useful to define the radial spectral moment of order 2α, where $0 \leq \alpha \leq 1$ as follows

$$\Lambda^{(2\alpha)} = \int_{\mathbb{R}^d} \mathrm{d}\mathbf{k}\, \|\mathbf{k}\|^{2\alpha}\, \widetilde{C}_{xx}(\mathbf{k}). \tag{5.28}$$

We use the radial spectral moments further in this chapter to define length scales for *non-differentiable* random fields that may admit fractional radial spectral moments of order less than two. For $\alpha = 1/2$ we obtain the first-order radial spectral moment. The latter does not vanish, in contrast with the directional odd-order spectral moments defined above, because $\|\mathbf{k}\|$ is an even function of $\mathbf{k}$.

Differentiability It can be shown that a zero-mean, isotropic random field $X(\mathbf{s}; \omega)$ with radial spectral density $\widetilde{C}_{xx}(\|\mathbf{k}\|)$ admits mean-square derivatives of order $p = p_1 + \ldots p_d$, if the spectral moment $\Lambda^{(2p)}$ exists. This requires the existence of the spectral integral

$$\int_0^\infty \mathrm{d}\|\mathbf{k}\|\; \|\mathbf{k}\|^{2p+d-1}\, \widetilde{C}_{xx}(\|\mathbf{k}\|).$$

Let us assume that the spectral density has no singularities and decays algebraically as $\|\mathbf{k}\| \to \infty$ according to $\widetilde{C}_{xx}(\|\mathbf{k}\|) \propto \|\mathbf{k}\|^{-\beta}$. Then, the above integral is well defined if $2p + d < \beta$. This condition ensures that the right tail of the spectral density decays sufficiently fast at large wavenumbers.

Hint in place of proof For the curious reader we give some clues regarding the connection between spectral moments and differentiability. The existence of the mean-square derivative of order p for an isotropic random field with a (radial) covariance function $C_{xx}(\mathbf{r})$ requires the existence of the following spectral integral

$$\int_{\mathbb{R}^d} \mathrm{d}\mathbf{k}\, k_1^{2p_1} \ldots k_d^{2p_d}\, \widetilde{C}_{xx}(\|\mathbf{k}\|).$$

This is obtained by expressing the covariance derivative of order $2p$ at the origin (for zero lag) using the Fourier transform of the covariance. If we express each wavevector component as $k_i = \|\mathbf{k}\| \cos\theta_i$, where $\cos\theta_i$, $i = 1, \ldots, d$ are the respective direction cosines of $\mathbf{k}$, the above multiple integral factorizes into a direction integral over the surface of the unit sphere $\mathcal{B}_d$ and a univariate integral over the wavenumber $\|\mathbf{k}\|$ as follows

$$\int_{\mathcal{B}_d} \mathrm{d}\hat{\mathbf{k}}\, \phi_{\mathbf{p}}(\hat{\mathbf{k}}) \int_0^\infty \mathrm{d}\|\mathbf{k}\|\, \|\mathbf{k}\|^{2p+d-1}\, \widetilde{C}_{\mathrm{xx}}(\|\mathbf{k}\|).$$

In the above expression $\hat{\mathbf{k}}$ is the unit direction vector in reciprocal space and $\phi_{\mathbf{p}}(\hat{\mathbf{k}})$ is a function that depends only on $\hat{\mathbf{k}}$. If

$$\widetilde{C}_{\mathrm{xx}}(\|\mathbf{k}\|) \propto \|\mathbf{k}\|^{-\beta}, \text{ as } \|\mathbf{k}\| \to \infty,$$

the integrand is asymptotically proportional to $\|\mathbf{k}\|^{2p+d-\beta-1}$. Thus, the integral over the wavenumbers is finite if the exponent satisfies $2p + d - \beta < 0$.

5.3.2 Second-Order Spectral Moments

Assume a mean-square differentiable random field that satisfies the conditions of Theorem 5.7. These conditions ensure the existence of first-order partial derivatives in the mean-square square. The *second-order moments* have integer indices that sum to two. This can be accomplished by means of the following two possibilities:

(i) $p_i = 2$, $p_j = 0$, for $i \in \{1, \ldots, d\}$ and $j \neq i$, and (ii) $p_i = p_j = 1$, for $i \neq j \in \{1, \ldots, d\}$ and $p_k = 0$ for all $k \neq i \neq j \in \{1, \ldots, d\}$.

Then, the second-order spectral moments are given by the following multiple integral[3]

$$\Lambda_{i,j} = \int_{\mathbb{R}^d} \mathrm{d}\mathbf{k}\, k_i\, k_j\, \widetilde{C}_{\mathrm{xx}}(\mathbf{k}), \text{ for } i, j = 1, \ldots, d. \tag{5.29}$$

Conversely, a spatial random field admits first-order partial derivatives if the spectral moments defined by (5.29) are well-defined, i.e., if the respective multidimensional integrals are finite.

In light of (3.59a), the moments $\Lambda_{i,j}$ can be expressed in terms of the second-order partial derivatives of the covariance function at zero lag as

$$(2\pi)^d\, \Lambda_{i,j} = -\left.\frac{\partial^2 C_{\mathrm{xx}}(\mathbf{r})}{\partial r_i \partial r_j}\right|_{\mathbf{r}=\mathbf{0}}, \text{ for } i, j = 1, \ldots, d.$$

Comparing the above with (5.22) and (5.23) it follows that the elements of the *gradient covariance matrix* $\mathbf{Q}$ are equal to the respective second-order spectral moments, i.e.,

$$Q_{i,j} = (2\pi)^d\, \Lambda_{i,j}, \text{ for } i, j = 1, \ldots, d. \tag{5.30}$$

[3] Note the slight difference of notation in $\Lambda_{i,j}$ compared to (5.27). For the second-order moments the indices i, j are sufficient instead of the vector $(p_1, \ldots, p_d)^\top$. We also drop the superscript (2) since the order of the moments is obvious.

Spectral moments of isotropic random fields In the case of isotropic random fields, the second-order spectral moments are given by the following one-dimensional integrals

$$\Lambda_{i,j} = \frac{\delta_{i,j}\, S_d}{d} \int_0^\infty du\, u^{d+1}\, \widetilde{C}_{xx}(u), \text{ for } i, j = 1, \ldots, d. \tag{5.31}$$

where $\delta_{i,j}$ is the Kronecker delta, and S_d is the surface area of a unit sphere in d dimensions given by $S_d = 2\pi^{d/2}/\Gamma(d/2)$. The matrix of spectral moments is a diagonal matrix of dimension $d \times d$.

Equation (5.31) is obtained directly by applying (3.62) to the volume integral of the radial function $\widetilde{C}_{xx}(\|\mathbf{k}\|)$:

$$\Lambda_{i,j} = \int_{\mathbb{R}^d} d\mathbf{k}\, k_i\, k_j\, \widetilde{C}_{xx}(\|\mathbf{k}\|) = \frac{S_d}{d}\, \delta_{i,j} \int_0^\infty d\|\mathbf{k}\|\, \|\mathbf{k}\|^{d-1}\, \widetilde{C}_{xx}(\|\mathbf{k}\|).$$

Let us see how this result is obtained in a more intuitive way.

(i) First, we consider the off-diagonal elements, and we assume $d = 2$ for simplicity (the argument is easily extended to $d > 2$). Then,

$$\Lambda_{1,2} = \int_{k_1=-\infty}^{\infty} dk_1 \int_{k_2=-\infty}^{\infty} dk_2\, k_1\, k_2\, \widetilde{C}_{xx}(\|\mathbf{k}\|).$$

We perform first the inner integral by splitting the real line into the $k_2 < 0$ and $k_2 \geq 0$ half lines. This leads to

$$k_1 \int_{k_2=-\infty}^{0} dk_2\, k_2\, \widetilde{C}_{xx}(\|\mathbf{k}\|) + k_1 \int_{k_2=0}^{\infty} dk_2\, k_2\, \widetilde{C}_{xx}(\|\mathbf{k}\|).$$

In the first integral we make the change of variables $k_2 \to -k_2$, leading to

$$k_1 \int_{k_2=\infty}^{0} dk_2\, k_2\, \widetilde{C}_{xx}(\|\mathbf{k}\|) + k_1 \int_{k_2=0}^{\infty} dk_2\, k_2\, \widetilde{C}_{xx}(\|\mathbf{k}\|) = 0.$$

In the above we used the fact that the $\mathbb{L}_2$ norm of the vector $\mathbf{k} = (k_1, k_2)^\top$ is the same as that of $\mathbf{k}' = (k_1, -k_2)^\top$; hence, the radial spectral density does not change under the transformation $(k_1, k_2) \mapsto (k_1, -k_2)$.

(ii) We next focus on the diagonal elements. Note that for a radial spectral density $\Lambda_{i,i}$ and $\Lambda_{j,j}$ for $i \neq j$ are identical. Hence, it follows that

$$\Lambda_{j,j} = \frac{1}{d} \sum_{i=1}^{d} \Lambda_{i,i}, \text{ for all } j = 1, \ldots, d.$$

Based on the definition of the spectral moments it follows that

$$\sum_{j=1}^{d} \Lambda_{j,j} = \int_{\mathbb{R}^d} d\mathbf{k}\, \|\mathbf{k}\|^2\, \widetilde{C}_{xx}(\mathbf{k}).$$

To simplify notation we use $u = \|\mathbf{k}\|$. Using equation (3.62), the above d-dimensional integral of the radial function $\|\mathbf{k}\|^2\, \widetilde{C}_{xx}(\mathbf{k})$ is reduced to the following one-dimensional integral

$$\int_{\mathbb{R}^d} d\mathbf{k}\, \|\mathbf{k}\|^2\, \widetilde{C}_{xx}(\mathbf{k}) = S_d \int_0^{\infty} du\, u^{d-1}\, u^2\, \widetilde{C}_{xx}(u),$$

which concludes the proof.

Spectral moments and spectral density tail Assume two isotropic random fields $X(\mathbf{s};\omega)$ and $Y(\mathbf{s};\omega)$ have equal variance and monotonically decreasing spectral densities. If $X(\mathbf{s};\omega)$ has larger diagonal spectral moments $\Lambda_{i,i}$ than $Y(\mathbf{s};\omega)$, equation (5.31) implies that the spectral density of the field $X(\mathbf{s};\omega)$ carries more weight in high spatial frequencies (wavenumbers) than the spectral density of $Y(\mathbf{s};\omega)$ which has smaller spectral moments.

5.3.3 Variance of Random Field Gradient and Curvature

If the parameters of the covariance function involve just the variance and a single characteristic length, then all the geometric properties of the zero-mean random field with the above covariance are determined by the characteristic length. This applies to single-component covariance functions such as the isotropic exponential, Gaussian, spherical, and other two-parameter models listed in Table 4.2. In contrast, if the covariance function involves additional shape parameters (e.g., degree of smoothness), a richer dependence of geometric patterns emerges. We illustrate these effects by focusing on the variance of the gradient and the curvature of second-order differentiable random fields.

For random fields that admit up to fourth-order radial derivatives of the covariance function, the *gradient*, $\nabla X(\mathbf{s};\omega)$, and the *linearized curvature*, $\nabla^2 X(\mathbf{s};\omega)$, exist in the mean square sense.

For stationary random fields, the mean squared gradient and curvature represent, respectively, the variances of the gradient and curvature fields. This is due to the fact that under stationarity the expectation of the gradient and the curvature are zero. Hence, the variance of the gradient (curvature) is equal to the expectation of the squared gradient (curvature).

The squared gradient based on (5.10) is given by

$$\left[\nabla X(\mathbf{s};\omega)\right]^2 = \sum_{i=1}^{d} \left(\frac{\partial_i X(\mathbf{s};\omega)}{\partial s_i} \right)^2,$$

while the squared curvature is given by following superposition of partial derivatives

$$\left[\nabla^2 \mathrm{X}(\mathbf{s};\omega)\right]^2 = \left(\sum_{i=1}^{d} \frac{\partial_i^2 \mathrm{X}(\mathbf{s};\omega)}{\partial s_i^2}\right)^2.$$

For isotropic random fields, the variances of the gradient and curvature are given, respectively, by the following expressions

$$\mathcal{E}\left\{[\nabla \mathrm{X}(\mathbf{s};\omega)]^2\right\} = d\, C_{\mathrm{xx}}^{(2)}(0), \tag{5.32}$$

$$\mathcal{E}\left\{[\nabla^2 \mathrm{X}(\mathbf{s};\omega)]^2\right\} = -\frac{d(d+2)}{3}\, C_{\mathrm{xx}}^{(4)}(0), \tag{5.33}$$

where the symbol $C_{\mathrm{xx}}^{(n)}(0)$ denotes the n-th order radial derivative of the covariance function $C_{\mathrm{xx}}(\|\mathbf{r}\|)$ at zero lag.

Equation (5.32) is obtained by means of (5.13) which expresses the covariance of the field gradient in terms of covariance derivatives, and equation (5.16) which expresses the second-order partial derivatives of radial covariance functions in terms of respective radial derivatives. Equation (5.33) also follows from (5.13). Equivalently, we can derive (5.32) and (5.33) from the lattice equations (8.20) and (8.21) at the limit of zero lattice step, i.e., $a \to 0$ (see Sect. 8.4).

Gradient to curvature variance ratio Let us define the dimensionless *gradient to curvature variance ratio*, g_c, which measures the relative strength of the gradient over the curvature variance. For isotropic random fields g_c is defined as follows:

$$g_c \doteq \frac{\mathcal{E}\left\{[\nabla \mathrm{X}(\mathbf{s};\omega)]^2\right\}}{\xi^2\, \mathcal{E}\left\{[\nabla^2 \mathrm{X}(\mathbf{s};\omega)]^2\right\}} = -\left(\frac{3}{d+2}\right) \frac{C_{\mathrm{xx}}^{(2)}(0)}{\xi^2\, C_{\mathrm{xx}}^{(4)}(0)}. \tag{5.34}$$

In the above, ξ is a typical length scale of the random field. The ratio g_c is dimensionless because the squared gradient has units $[L]^{-2}$, whereas the squared curvature has units $[L]^{-4}$, since in general $C_{\mathrm{xx}}^{(n)}(0) \propto \xi^{-n}$. Thus, for a covariance function whose parameters involve only the variance and ξ, the ratio g_c is a constant independent of σ_{x}^2 and ξ.

The ratio g_c exhibits richer behavior for covariance functions that involve more parameters as shown in Example 5.4 below.

Example 5.3 Calculate the gradient to curvature variance ratio g_c for the d-dimensional Gaussian correlation function $\rho_{\mathrm{xx}}(\mathbf{r}) = \exp(-\|\mathbf{r}\|^2/\xi^2)$.

Answer The radial derivatives of the Gaussian correlation function with respect to the Euclidean distance $r = \|\mathbf{r}\|$ are given by

Table 5.2 Hermite polynomials $\Pi_n(x)$ for $n = 1, 2, 3, 4$. These polynomials appear in the n-th order radial derivative, $\rho_{\mathrm{xx}}^{(n)}(x)$, of the Gaussian covariance function. The argument $x = r/\xi$ represents the normalized distance

n	1	2	3	4
$\Pi_n(x)$	$2x$	$4x^2 - 2$	$8x^3 - 12x$	$16x^4 - 48x^2 + 12$

$$\frac{\mathrm{d}^n \rho_{\mathrm{xx}}(\mathbf{r})}{\mathrm{d}r^n} = (-1)^n \frac{\mathrm{e}^{-r^2/\xi^2}}{\xi^n} \, \Pi_n(r/\xi), \tag{5.35}$$

where $\Pi_n(x)$ is the *Hermite polynomial* of degree n defined by

$$\Pi_n(x) = (-1)^n \exp(x^2) \frac{d^n \exp(-x^2)}{dx^n}. \tag{5.36}$$

The Hermite polynomials $\{\Pi_n(x)\}_{n=1}^4$ are listed in Table 5.2. Since the covariance and the correlation functions are linked by means of $C_{\mathrm{xx}}(\mathbf{r}) = \sigma_{\mathrm{x}}^2 \, \rho_{\mathrm{xx}}(\mathbf{r})$, the variance ratio (5.34) is equal to

$$g_c = -\frac{3}{d+2} \, \frac{\rho_{\mathrm{xx}}^{(2)}(0)}{\xi^2 \, \rho_{\mathrm{xx}}^{(4)}(0)}.$$

Hence, it follows from the above and from Table 5.2 that the variance ratio corresponding to a monoscale Gaussian covariance function is given by

$$g_c = \frac{1}{2(d+2)}.$$

Example 5.4 Calculate the coefficient g_c for the two-scale Gaussian correlation mixture with characteristic lengths $\xi_1 = \xi$ and $\xi_2 = b\,\xi$:

$$\rho_{\mathrm{xx}}(\mathbf{r}) = \alpha \, \mathrm{e}^{-\|\mathbf{r}\|^2/\xi^2} + (1-\alpha) \, \mathrm{e}^{-\|\mathbf{r}\|^2/b^2\,\xi^2}, \tag{5.37}$$

where $\alpha \in [0, 1]$ and $b > 0$ is a dimensionless scale factor.

Answer The same steps as in the previous example are followed. Based on (5.34) the following expression is obtained for the gradient-curvature variance ratio

$$g_c = \frac{1}{2(d+2)} \, \frac{\alpha + b^{-2}(1-\alpha)}{\alpha + b^{-4}(1-\alpha)}.$$

Hence, different values of g_c become possible for different combinations of a and b as illustrated in Fig. 5.8. The variance ratio of the two-scale Gaussian covariance

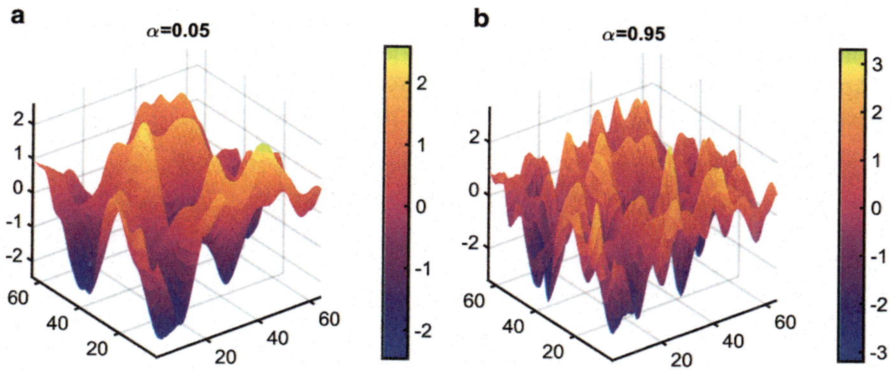

Fig. 5.8 Gaussian random field realizations on a square grid of length $L = 50$. The respective correlation functions are given by the Gaussian mixture (5.37) with $\xi = L/15$, $b = 2$ and $a = 0.05$ (left), $a = 0.95$ (right). The same set of random numbers is used for both realizations. The realization shown in (**b**) has lower ratio g_c than the realization shown in (**a**)

function depends both on the relative variance of the two components and the ratio of the respective correlation lengths. Similar behavior is displayed by the Spartan random fields (see Sect. 7.1), the covariance of which involves three parameters.

5.4 Length Scales of Random Fields

For random fields that are not scale-invariant, characteristic length scales are useful indicators of the spatial extent of correlations. Random fields with scale-invariant patterns, such as the fractional Brownian motion and the fractional Gaussian noise, have power-law correlation functions; thus, they are better classified in terms of characteristic exponents, as discussed in Sect. 5.6 below.

For statistically homogeneous random fields, various length scales can be used to measure the correlation patterns. Both local correlation measures such as the *correlation radius*, and global measures such as the *integral range* are used. The term *correlation length* is often used in the literature to denote either the integral range or the characteristic scale ξ [132, p. 99].

In the case of radial two-parameter correlations functions (e.g., Gaussian, exponential, cardinal sine, generalized exponential, and spherical models), the only parameter entering the definition of typical length scales is the *characteristic length* ξ. This length determines (except for "boring" constants) all the length scales of Gaussian random fields with such covariance functions. The variance determines the amplitude of the fluctuations but not their spatial extent, and therefore it does not affect the definition of length scales.

The dependence of characteristic length scales on ξ is easy to understand intuitively. Typical length scales are defined in terms of derivatives $\mathrm{d}\rho(\|\mathbf{r}\|/\xi)/\mathrm{d}\|\mathbf{r}\|$ at $\|\mathbf{r}\| = 0$ or integrals of the correlation function, such as $\int \mathrm{d}\|\mathbf{r}\|\, \rho(\|\mathbf{r}\|/\xi)$. In both cases, using the rescaling operation $\mathbf{r} \mapsto u = \|\mathbf{r}\|/\xi$ leads to

$$\int_0^\infty \mathrm{d}\|\mathbf{r}\|\, \rho(\|\mathbf{r}\|/\xi) = \xi \int_0^\infty \mathrm{d}u\, \rho(u),$$

and

$$\left. \frac{\mathrm{d}\rho(\|\mathbf{r}\|/\xi)}{\mathrm{d}\|\mathbf{r}\|} \right|_{\|\mathbf{r}\|=0} = \frac{1}{\xi} \left. \frac{\mathrm{d}\rho(u)}{\mathrm{d}u} \right|_{u=0}.$$

The covariance function models defined by (4.11)–(4.15) in Table 4.2, on the other hand, involve more parameters than the variance and ξ. Thus, we expect more interesting behavior of the various correlation length scales. This is further investigated in the following sections.

In the case of *non-homogeneous random fields*, the correlation function depends on both points and is not purely a function of the lag. Hence, representative correlation length scales also depend on the location $\mathbf{s}$. In addition, in the case of *self-affine* random fields, such as the fractional Brownian motion (fBm) the nature of the correlations is better described in terms of the Hurst exponent. This topic is also further discussed in Sect. 5.6. Finally, the correlations of *non-Gaussian random fields* involve higher-order moments. Consequently, different correlation length scales can be defined based on these moments.

5.4.1 Practical Range

A practical measure of the correlation range for statistical homogeneous random fields is the lag at which the correlation function has declined to 5% of its value at the origin [132].[4] In the case of damped oscillatory covariance functions that dip to negative values and then increase again, the practical range is the *smallest lag* at which the correlation function drops to 5% of its value at the origin. Equivalently, the *practical range* is defined as the smallest lag where the variogram function attains 95% of the sill value.

In the case of the Gaussian model, the practical range is $\approx 1.73\xi$, whereas in the case of the exponential model it is approximately equal to 3ξ. The practical range of both models is illustrated in Fig. 5.9 along with the Bessel-J correlation

[4]There is nothing magical about 5%; a different level, such as the distance at half-maximum could be selected instead.

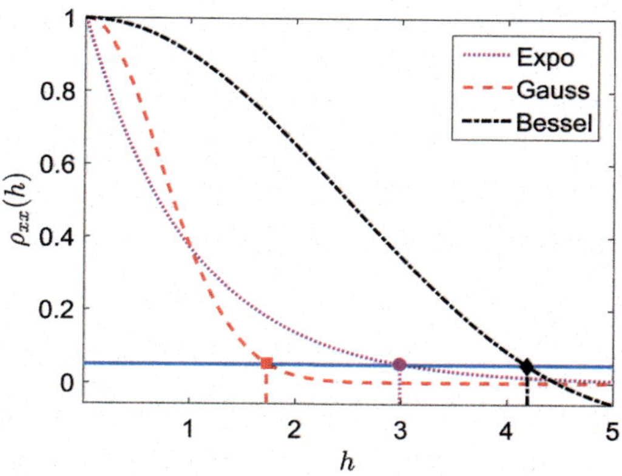

Fig. 5.9 Practical range for the exponential (Expo), Gaussian (Gauss) and Bessel-J (Bessel) with $\nu = 1.5$ correlation functions. The expressions for the correlation functions are given in Table 4.2. The horizontal axis measures the normalized lag $h = \|\mathbf{r}\|/\xi$. The horizontal solid line marks the 5% level. Markers denote the respective practical range for the Gaussian (square), exponential (circle), and Bessel-J (diamond) functions

(for $\nu = 1.5$), which displays an oscillatory behavior with increasing normalized lag h.

The practical range measures how fast the correlations decline with increasing lag. Hence, it can differ significantly from respective measures based on integrals (integral range and correlation radius) defined below. A case in point is provided by Spartan random fields in three dimensions (see Chap. 7): if the rigidity parameter is very large, the practical range of the respective random field is very small while the integral range remains large.

5.4.2 *Integral Range*

The *integral range* is an average measure of the distance over which two field values are correlated [380, 487]. It is estimated by integrating the correlation function over space. In the case of isotropic random fields, the integral range is defined using the following volume integral of the correlation function

$$\ell_c = \left[\int_{\mathbb{R}^d} d\mathbf{r}\, \rho_{xx}(\|\mathbf{r}\|) \right]^{1/d} = [\widetilde{\rho_{xx}}(\|\mathbf{k}\| = 0)]^{1/d} . \tag{5.38}$$

The expression in terms of the spectral density is based on the Fourier transform of the correlation function (see (3.54)). For random processes in $d = 1$, the integral range is equivalent to *correlation time* [646].

The integral range also incorporates negative contributions in the case of *oscillatory covariance functions* such as the cardinal sine and the J-Bessel functions. The presence of negative correlations tends to reduce the integral range.

Anisotropic integral range In the case of anisotropic correlations, we can define direction-dependent integral ranges by means of the following one-dimensional integrals

$$\ell_{c,i} = \int_{-\infty}^{\infty} dr_i \, \rho_{xx}(\mathbf{r} \cdot \hat{\mathbf{e}}_i), \text{ for } i = 1, \ldots, d, \tag{5.39}$$

where $\hat{\mathbf{e}}_i$ is the unit vector in the direction i.

In the case of isotropic random fields the anisotropic integral ranges become equal. However, the equal values of $\ell_{c,i}$ are not the same as ℓ_c obtained from the volume integral (5.38). Consider for example the case of the exponential covariance function. Based on (5.38) and the exponential's spectral density (listed in Table 4.2) we find that

$$\ell_c = 2\pi^{(d-1)/(2d)} \left[\Gamma\left(\frac{d+1}{2}\right) \right]^{1/d}.$$

On the other hand, the directional integral ranges defined by (5.39) are given by $\ell_{c,i} = 2$, for all $i = 1, \ldots, d$.

5.4.3 Correlation Radius

The range of correlations is also measured by means of the *correlation radius*. This is defined as the normalized second moment of the lag $\|\mathbf{r}\|$ with respect to the covariance function [21, p. 27]:

$$r_c^2 \doteq \frac{\int_{\mathbb{R}^d} d\mathbf{r} \, \|\mathbf{r}\|^2 \, C_{xx}(\|\mathbf{r}\|)}{\int_{\mathbb{R}^d} d\mathbf{r} \, C_{xx}(\|\mathbf{r}\|)} = -\frac{\nabla^2 \widetilde{C_{xx}}(0)}{\widetilde{C_{xx}}(0)}, \tag{5.40}$$

where $\nabla^2 \widetilde{C_{xx}}(0)$ denotes the value of the *Laplacian of the spectral density* $\widetilde{C_{xx}}(\|\mathbf{k}\|)$ evaluated at $\|\mathbf{k}\| = 0$. The correlation radius is thus well defined if the spectral density admits all the second-order partial derivatives at zero frequency.

Let us see how (5.40) is derived. First, the integral in the denominator

$$\int_{\mathbb{R}^d} d\mathbf{r} \, C_{xx}(\|\mathbf{r}\|) = \widetilde{C_{xx}}(0)$$

follows directly from the Fourier integral (3.54) evaluated at $\mathbf{k} = \mathbf{0}$. Second, since $\|\mathbf{r}\|^2 = \sum_{i=1}^{d} r_i^2$ we can express the numerator as the following sum

$$\int_{\mathbb{R}^d} \mathrm{d}\mathbf{r}\, \|\mathbf{r}\|^2\, C_{xx}(\|\mathbf{r}\|) = -\sum_{i=1}^{d} \left. \frac{\partial \widetilde{C_{xx}}(\|\mathbf{k}\|)}{\partial k_i^2} \right|_{\|\mathbf{k}\|=0} = -\left. \nabla^2 \widetilde{C_{xx}}(\|\mathbf{k}\|) \right|_{\|\mathbf{k}\|=0}.$$

Using the expression (5.16) for the second-order partial derivatives of a radial function, it follows that the *Laplacian of the radial spectral density* is given by

$$\nabla^2 \widetilde{C_{xx}}(\|\mathbf{k}\|) = \frac{1}{\|\mathbf{k}\|} \left[(d-1)\, \frac{\mathrm{d}\widetilde{C_{xx}}(\|\mathbf{k}\|)}{\mathrm{d}\|\mathbf{k}\|} \right] + \frac{\mathrm{d}^2 \widetilde{C_{xx}}(\|\mathbf{k}\|)}{\mathrm{d}\|\mathbf{k}\|^2}.$$

According to the above, for the second-order partial derivatives of the spectral density to exist at zero, the spectral density should admit the following Taylor expansion around zero

$$\widetilde{C_{xx}}(\|\mathbf{k}\|) = c_0 - c_1 \|\mathbf{k}\|^2 + O(\|\mathbf{k}\|^4),$$

i.e., the $O(\|\mathbf{k}\|)$ term should vanish. Then,

$$-\nabla^2 \widetilde{C_{xx}}(0) = 2(d-1)\, c_1 + 2\, c_1,$$

and finally the square of the correlation radius is given by

$$r_c^2 = \frac{2d\, c_1}{c_0}.$$

Based on the above, $r_c^2 > 0$ only if $c_1 > 0$, whereas if $c_1 \le 0$ the square of the correlation radius is a negative number. This is a direct consequence of the definition (5.40), which does not guarantee that the integral $\int \mathrm{d}\mathbf{r}\, \|\mathbf{r}\|^2\, C_{xx}(\|\mathbf{r}\|)$ is positive-valued. In fact, for covariance functions with negative hole(s) it is possible that (5.40) leads to zero or negative r_c^2. In such cases, it makes more sense to use the absolute value of the covariance function in the definition (5.40).

The correlation radius is determined by the spectral density's behavior at the origin, which corresponds to the long-wavelength limit of the correlation function. The correlation radius is equivalent to the correlation length used in statistical physics, in percolation theory [113, 769] and in statistical field theory [21].

Anisotropic radii For anisotropic random fields one can define correlation radii along specific directions using the following extension of (5.40)

$$r_{c;i}^2 = \frac{\int_{\mathbb{R}^d} \mathrm{d}\mathbf{r}\, r_i^2\, C_{xx}(\mathbf{r})}{\int_{\mathbb{R}^d} \mathrm{d}\mathbf{r}\, C_{xx}(\mathbf{r})} = -\frac{1}{\widetilde{C_{xx}}(0)} \left. \frac{\partial^2 \widetilde{C_{xx}}(\mathbf{k})}{\partial k_i^2} \right|_{\mathbf{k}=\mathbf{0}}, \quad \text{where } i = 1, \ldots, d. \tag{5.41}$$

5.4.4 Turbulence Microscale

The turbulence microscale [178], also known as the *Daley length scale*, is commonly used in oceanographic and meteorological data assimilation to characterise the spatial scale of a correlation function [576]. The Daley length scale is defined by

$$L_D = \sqrt{\frac{d}{-\nabla^2 \rho_{xx}(0)}}, \tag{5.42}$$

where $-\nabla^2 \rho_{xx}(0)$ is the Laplacian of the correlation function evaluated at zero lag, and d is as usual the spatial dimension. The Daley length scale defined by (5.42) is only valid for random fields that admit all first-order partial derivatives in the mean square sense.

In the case of the *Gaussian correlation* function, $\rho_{xx}(\mathbf{r}) = \exp\left(-\|\mathbf{r}\|^2/\xi^2\right)$, it follows from (5.42) that the Daley length scale is determined from ξ

$$L_D^{\text{Gauss}} = \frac{1}{\sqrt{2}}\,\xi.$$

On the other hand, for the *Whittle-Matérn correlation* function with smoothness index $\nu > 1$, the Daley length scale is given by the following expression that depends on both ξ and ν [576]

$$L_D^{\text{W-M}} = \sqrt{2(\nu - 1)}\,\xi.$$

In contrast with integral correlation measures, the Daley length scale is determined by means of the short-range behavior of the correlation function at the origin. Thus, it is closer in spirit to the practical range than to the integral measures.

5.4.5 Correlation Spectrum

Kolmogorov introduced the notion of the *dissipation length* in connection with the spectrum of homogeneous turbulence and in order to distinguish between large-scale eddies (inertial regime, in which viscosity is not important) and small-scale (dissipative) eddies for which viscosity plays a role.

The *smoothness microscale* is a generalization of the dissipation length. It is applicable to mean-square differentiable random fields with unimodal, radial spectral density [469]. The microscale λ denotes a length such that the SRF appears smooth and can be linearly interpolated over distances $r \ll \lambda$. The microscale is equivalent to the integral range of the gradient $\nabla X(\mathbf{s}; \omega)$ of the random field $X(\mathbf{s}; \omega)$.

The definition of the smoothness microscale was recently expanded to generate a continuum of length scales for stationary, but not necessarily mean-square differentiable random fields with unimodal, radially symmetric spectral density function [367].

Definition 5.3 Let the function $\widetilde{C_{xx}}(\|\mathbf{k}\|)$ be a permissible (according to Bochner's theorem 3.2 above) radial spectral density. In addition, let $\widetilde{C_{xx}}(\|\mathbf{k}\|)$ be a unimodal function[5] of the wavenumber $\|\mathbf{k}\|$.

The *correlation spectrum* $\lambda_c^{(\alpha)}$ indexed by the *scale index* α, where $0 \leq \alpha \leq 1$, is defined by means of the following ratio

$$\lambda_c^{(\alpha)} \doteq \left(\frac{\sup_{\mathbf{k}\in\mathbb{R}^d} \|\mathbf{k}\|^{2\alpha}\, \widetilde{C_{xx}}(\|\mathbf{k}\|)}{\int_{\mathbb{R}^d} d\mathbf{k}\, \|\mathbf{k}\|^{2\alpha}\, \widetilde{C_{xx}}(\|\mathbf{k}\|)} \right)^{1/d} = \left(\frac{\kappa_*^{2\alpha}\, \widetilde{C_{xx}}(\kappa_*)}{\Lambda^{(2\alpha)}} \right)^{1/d}, \tag{5.43}$$

where $\sup_{x\in A} f(x)$ corresponds to the *supremum* of $f(x)$, i.e., the smallest upper bound of the function $f(x)$ when $x \in A$ and κ_* is the wavenumber where the supremum is attained. The denominator in (5.43) is equal to the radial spectral moment of order 2α defined by (5.28).

Remarks

- The supremum $\sup(\cdot)$ operator can be replaced by the maximum $\max(\cdot)$, if the operand $\|\mathbf{k}\|^{2\alpha}\, \widetilde{C_{xx}}(\|\mathbf{k}\|)$ attains its maximum inside a closed subset of $\mathbb{R}^d$.
- The equation (5.43) returns a unique finite value even if $\widetilde{C_{xx}}(\|\mathbf{k}\|)$ is a bimodal, or in general multimodal function of $\|\mathbf{k}\|$. In such cases, however, an expression that can better resolve the length scales which correspond to the different peaks may be more desirable. For example, multimodal spectral densities are useful in the modeling of ocean waves [865].

Integral range ($\alpha = 0$) For $\alpha = 0$ the *integral range* (5.38) is recovered *if the peak of the spectral density occurs at* $\|\mathbf{k}\| = 0$. More precisely, for spectral densities that peak at zero wavenumber it holds that

$$\ell_c = 2\pi\, \lambda_c^{(0)}. \tag{5.44}$$

If the spectral density reaches a maximum at $k_{\max} > 0$, then $\lambda_c^{(0)}$ given by (5.43) is approximately equal to the width of the spectral density at maximum.

Smoothness microscale ($\alpha = 1$) For $\alpha = 1$ the equation (5.43) yields the *smoothness microscale*. The microscale expression involves in the denominator the Laplacian of the covariance function evaluated at zero lag, i.e., $[\nabla^2 C_{xx}(\|\mathbf{r}\|)]_{\|\mathbf{r}\|=0}$.

If the SRF is first-order differentiable in the mean square sense, the covariance Laplacian at zero lag is a finite number (see Sect. 5.1.4), and thus $\lambda_c^{(\alpha=1)} \in \mathbb{R}_+$.

In contrast, the Laplacian of the covariance diverges for non-differentiable SRFs, leading to $\lambda_c^{(\alpha=1)} = 0$. A zero value for the microscale denotes that the SRF appears rough at all length scales.

[5] A unimodal function is a function with a single peak.

Fractional scales Exponents $0 < \alpha < 1$ generate a spectrum of scales that emphasize different parts of the spectral density. In particular, correlation lengths obtained for $0 < \alpha < 1$ correspond to the integral range of the random field's *fractional derivative* of order α.

The length scales $\lambda_c^{(\alpha)}$ take finite, non-zero values if $\|\mathbf{k}\|^{2\alpha}\, \widetilde{C_{xx}}(\|\mathbf{k}\|)$ is integrable. A necessary condition for integrability is that

$$\widetilde{C_{xx}}(\|\mathbf{k}\|) \sim \|\mathbf{k}\|^{-q} \text{ as } \|\mathbf{k}\| \to \infty, \text{ where } q > 2\alpha + d.$$

For fixed q non-vanishing length scales are obtained for $\alpha < \alpha_{\max} = (q-d)/2$. The respective length scales $\lambda_c^{(\alpha)}$ can be used to quantify the "regularity" of continuous but non-differentiable random fields.

The fractional scales can be viewed in terms of the fractional Laplacian. The *fractional Laplacian* of a function $\phi(\mathbf{r})$ is denoted by $[-\nabla^2]^{\alpha}\, \phi(\mathbf{r})$ [708]. In Fourier domain, the fractional Laplacian operator transforms into $\|\mathbf{k}\|^{2\alpha}\widetilde{\phi}(\mathbf{k})$. Hence, the correlation spectrum for a fractional value of the scale index involves the fractional Laplacian of the covariance function at zero lag. A recent review of the fractional Laplacian is presented in [504].

5.5 Fractal Dimension

In this section we discuss concepts useful for describing random fields with complex geometries that involve multiple length scales. These concepts have been developed in the mathematical theory of fractals which was spearheaded by Benoit Mandelbrot [243, 530]. Fractals are geometric patterns that defy the Euclidean viewpoint which is based on smoothness and continuity. Fractal patterns are typically rough, discontinuous, and exhibit self-similar structure over a wide range of scales. Such structures are omnipresent in geography, materials, biological organisms, societies, and urban distributions as vividly described by the physicist Geoffrey West [844].

The classical notion of dimension (i.e., the topological dimension) refers to the number of coordinates that are required to specify the position of a point in space. This is an integer number that takes values equal to one for processes defined on a line (such as time series or measurements along a drillhole), two for planar processes (e.g., precipitation over a flat map area), three for fully three-dimensional functions (e.g., porosity of an aquifer), and four for space-time processes (e.g., dispersion of pollutants in the atmosphere); for space-time processes in linear or planar spaces the respective dimensions are equal to two and three. However, for geometric objects with complex internal structure, the topological dimension does not sufficiently specify dimensionality.

The *fractal or Hausdorff dimension* is a measure of the roughness (or smoothness) of geometric objects. Fractal dimensions allow measuring geometric properties (e.g., length, surface area, volume) of concrete or abstract objects as a function of the measurement scale (i.e., the yardstick) used. While classical geometrical

Fig. 5.10 Synthetic fractal mountains generated with the MATLAB code artificial_surf. Mona M. Kanafi (2016). Surface generator: artificial randomly rough surfaces. MATLAB Central File Exchange. (Retrieved August 15, 2019)

objects such as spheres and cubes have well defined length scales (i.e., radius and side length), fractal objects (snowflakes, coastlines, mountains) exhibit structure at a multitude of length scales, as evidenced in the synthetic fractal landscape shown in Fig. 5.10. The multi-scale features of fractal objects result from a construction process that involves the repeated use of the same geometric pattern at different length scales.

The Hausdorff dimension is formally defined using concepts from measure theory. More intuitive definitions of the fractal dimension are used in practice. These definitions may provide different estimates depending on the application and the targeted property. One of the commonly used fractal dimensions is the *box-counting dimension* d_{BC}. For a fractal object this is defined by means of the following limit

$$d_{BC} = \lim_{\epsilon \to 0} \frac{\ln N(\epsilon)}{\ln 1/\epsilon}, \tag{5.45}$$

where $\epsilon > 0$ denotes the side length of an elementary cube [6] and $N(\epsilon)$ is the smallest number of cubes of side length ϵ that are needed to cover the object.

The main assumption in (5.45) is that as $\epsilon \to 0$, the number of cubes of side length ϵ that is needed to completely cover the object increases as $N(\epsilon) = A\,\epsilon^{-d_{BC}}$. Then, the fractal dimension can be obtained by taking the logarithm on both sides leading to $\ln N(\epsilon) = \ln A - d_{BC} \ln \epsilon$. The constant factor $\ln A$ can be omitted, because $\ln A \ll |\ln \epsilon|$ as $\epsilon \to 0$.

Spatial (or temporal) data can also be viewed as geometric objects. Therefore, they are also characterized by fractal dimensions. Let us assume a random field $X(\mathbf{s}; \omega)$ where $\mathbf{s} \in \mathbb{R}^d$. A realization of this random field creates a "surface" indexed in $\mathbb{R}^d$. For example, the surface generated by the realization in Fig. 5.10 is indexed in $\mathbb{R}^2$.

The *graph* of a given random field realization $x(\mathbf{s})$ is defined by the set of points

$$\mathcal{G}_{x} = \Big\{ (\mathbf{s}, x(\mathbf{s})) \,,\ \mathbf{s} \in \mathbb{R}^d \Big\} \in \mathbb{R}^{d+1}. \tag{5.46}$$

[6] A "hyper-cube" is used in $d > 3$ dimensions; in $d = 1$ the "hyper-cubes" correspond to line segments and in $d = 2$ to squares.

$\mathcal{G}_x$ thus forms a geometric object in $\mathbb{R}^{d+1}$ which is characterized by a fractal dimension. The statistical properties of the random field depend on the fractal dimension. However, the fractal dimension does not uniquely specify all the properties. In spatial data analysis, the fractal dimension is not a priori known but needs to be inferred from the available data. Fractal dimension estimators for spatial data are compared and assessed in [293].

5.5.1 Fractal Dimension and Variogram Function

Random fields can exhibit irregular small-scale behavior which is typical of fractals. More precisely, there is a link between the fractional dimension and the small-scale dependence of the variogram [10, 293]. In particular, if the variogram function can be expressed for $\|\mathbf{r}\| \to 0$ as

$$\gamma_{xx}(\|\mathbf{r}\|) = c_2 \,\|\mathbf{r}\|^{\alpha} + O(\|\mathbf{r}\|^{\alpha+\epsilon}), \tag{5.47}$$

where $c_2, \epsilon > 0$ are positive constants and $\alpha \in (0, 2)$, it can then be shown that the "sample path" represented by the *graph* $\mathcal{G}_x$ has a fractal dimension that is almost surely given by

$$d_f = d + 1 - \frac{\alpha}{2}. \tag{5.48}$$

The fractal dimension is thus linked to the small-scale behavior of the variogram, and not to the long-distance decay of the covariance function. The fractal dimension can be estimated from the variogram function by means of the linear regression equation

$$\ln \gamma_{xx}(\|\mathbf{r}\|) = \ln c_2 + \alpha \,\ln \|\mathbf{r}\|.$$

The parameter c_2 is also known as *topothesy* [179], while the term $c_2 \,\|\mathbf{r}\|^{\alpha}$ is known as the *principal irregular term* of the variogram function [17].

Fractals and space filling If a random field is differentiable, the existence of the second-order derivative of the covariance at zero lag implies that $\alpha = 2$ in (5.47). Then, we obtain the relation $d_f = d$, which implies that the fractal dimension is equal to the topological space dimension. Hence, for $\alpha = 2$ the random field does not exhibit fractal behavior. Fractal patterns are obtained for $\alpha \in (0, 2)$. If the fractal exponent is in this range, the sample paths of the random field are continuous but non-differentiable. On the other hand, $\alpha = 2$ denotes the transition to differentiable random fields.

The relation between the *topological* and the *fractal dimension* is summarized in Table 5.3. Note that the fractal dimension d_f is equal to the topological dimension

Table 5.3 Topological and fractal dimension of graphs generated by random fields in d spatial dimensions in terms of the fractal exponent $\alpha \in (0, 2)$ where α is defined in (5.47). Rough surfaces are obtained for values of α near zero

Smoothness	Topological dimension	Fractal dimension
Smooth, differentiable surfaces	d	$d_f = d$
Rough, non-differentiable surfaces	d	$d_f = d + 1 - \alpha/2$

Table 5.4 Fractal exponent α for the relevant isotropic correlation functions of Table 4.2 (stationary models) as well as the fractional Brownian motion (non-stationary model)

Model	Fractal exponent α
Generalized exponential	$\alpha = \nu, \ \nu \in (0, 2)$
Matérn	$\alpha = 2\nu, \ \nu \in (0, 1)$
Cauchy	$\alpha \in (0, 2]$
Fractional Brownian motion	$2H, \ H \in (0, 1)$

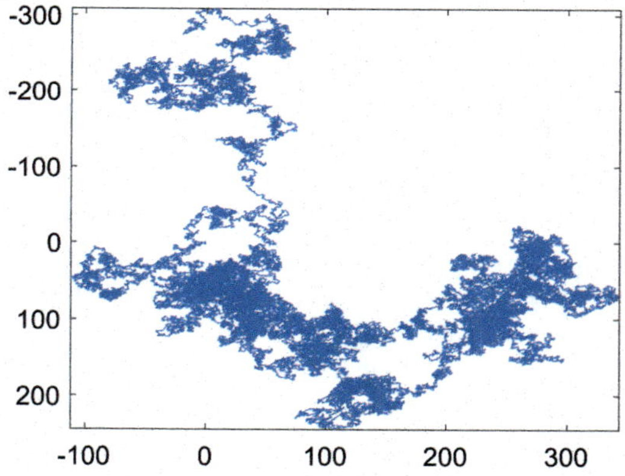

Fig. 5.11 Brownian motion path based on 10^5 steps drawn from the standard normal distribution

d if $\alpha = 2$ while $d_f \to d + 1$ as $\alpha \to 0$. A fractal dimension that exceeds the topological dimension by one indicates that fractals generate "space-filling" patterns due to the emergence of "wiggles" over a wide range of scales [844].

Table 5.4 lists the values of α for the covariance function models that are included in Table 4.2 and possess a fractal exponent α.

Note that in the case of Brownian motion $\alpha = 1$, which in light of (5.48) implies that $d_f = 1.5$. As shown in Fig. 5.11, a Brownian motion path tends to fill the two-dimensional space more fully than a typical "Euclidean" curve such as a circle or a triangle. With a fractal dimension of 1.5, the Brownian path exhibits a geometric pattern that seems like a hybrid between a curve and a surface.

5.6 Long-Range Dependence

Processes that exhibit long-range dependence (LRD) maintain significant correlations even between points that are separated by large distances. Such long-range correlations can provide insight into physical principles that involve the entire system or collective modes of the system's response. As an example of LRD, consider pressure fluctuations in an incompressible fluid: fluctuations propagate through the entire volume of the fluid due to the incompressibility constraint.

The mathematical signature of LRD is power-law dependence in the correlation functions.[7] Physical phenomena exhibit different types of power-law dependence. For example, power-law decay of the right tail of the probability density function is the hallmark of *heavy-tailed processes* [709, 766]. This type of power-law behavior is not linked to LRD, since it characterizes the marginal probability distribution but not the joint dependence of the field at different locations.

We focus on power-law dependence in two-point correlation functions which is a signature of LRD. In physical systems, power laws dependence of the correlation functions often emerges at the onset of thermodynamic phase transitions that mark the evolution of the system into a new state in response to changes in temperature. Typically one of the two states is disordered (at high temperatures) while the other state (at low temperatures) is ordered. The ordered state is mathematically characterized by a different renormalization group (RG) fixed point than the disordered state (for an introduction to RG and fixed points consider [297, 782]).

Long-range correlations also appear in systems with frozen fractal structure (e.g., porous media). LRD can have a significant impact on physical processes in such systems, because it tends to favor large-scale organization. For example, the role of LRD with respect to diffusion in random media is discussed at length in the review article by Bouchaud and Georges by means of RG analysis [92]. In addition to thermodynamic phase transitions, power laws are also observed in *complex systems* (e.g., non-equilibrium, driven, dissipative systems) that exhibit *self-organized criticality* [42, 544]. In this case, states with long-range organization emerge spontaneously as a result of the dynamic interactions in the system without tuning an external field.

Long-range dependence of covariance functions The definitions of the integral range and the correlation radius assume that the correlation volume integral $\int_{\mathbb{R}^d} d\mathbf{r}\, \rho_{xx}(\|\mathbf{r}\|)$ is a finite real number. For a radial function $\rho_{xx}(\|\mathbf{r}\|)$ this is equivalent to the existence of the univariate improper integral

[7]Condensed matter physicists call this less-than-perfect correlation quasi-long-range order, to distinguish it from "true" long-range order. The latter implies that all the system variables have the same value. A typical example is a ferromagnetic system in which all the magnetic moments are aligned in the same direction.

$$\int_0^\infty \mathrm{d}x\, x^{d-1}\, \rho_{\mathrm{xx}}(x). \tag{5.49}$$

In spatial data analysis, it is often tacitly assumed that this integral converges. After all, if this integral is well defined, the condition of Slutsky's ergodic theorem (4.2) is satisfied and the sample data can be used to infer distributional moments. In many spatially extended systems, however, the integral (5.49) diverges.

Common causes for the divergence of the above integral are the behavior of $\rho_{\mathrm{xx}}(x)$ near $x = 0$ and asymptotically as $x \to \infty$. We usually refer to the dependence of a function near $x = 0$ and at the limit $x \to \infty$ as *tail dependence*; the piece of the function near $x = 0$ represents the *left tail*, whereas the segment of the function $\rho_{\mathrm{xx}}(x)$ at $x > x'$, where $x' \to \infty$ is known as the *right tail*.

The left tail cannot have algebraic (power-law) dependence, because this would lead to a singularity at $x \to 0$, in conflict with $\rho_{\mathrm{xx}}(0) = 1$. Let us assume that the right tail has power-law dependence, i.e.,

$$\lim_{x\to\infty} \rho_{\mathrm{xx}}(x) \propto x^{-\mu},$$

where $\mu > 0$. The convergence of the integral (5.49) is controlled by the power-law behavior as $x \to \infty$. Then, the integrand in the integral is $\propto x^{\alpha}$ where $\alpha = d-1-\mu$. Convergence requires that the integrand decline faster than x^{-1} as $x \to \infty$. This implies the constraint $\alpha < -1$ which is equivalent to $d < \mu$.

Power-law tails If the right tail of the covariance function has power-law dependence, i.e., $\lim_{x\to\infty} \rho_{\mathrm{xx}}(x) \propto x^{-\mu}$, then the following holds:

- If $\mu > d$ the integral range and the correlation radius are finite. In this case, the correlations are characterized as *short ranged*. Short-range power laws with $\mu > d$ are thus no different than correlation functions with an exponentially decaying tail.
- If $\mu \le d$ the integral range and the correlation radius diverge. In this case, the correlations are characterized as *long range*.
- The case $\mu = d$ is special, since the integral range then has a logarithmic divergence, which is milder than the power-law divergence for $\mu < d$.

Long-range correlations are not extensively discussed in the spatial data literature. The reasons are threefold:

- Establishing the existence of long-range dependence requires data records that incorporate an adequate number of pairs at large lags to allow for precise estimates of long-range correlations. Often, this condition is not satisfied by spatial data sets.

- Identifying long-range dependence typically requires measurements over an extensive range of distances. Again, such records are not usually available in spatial studies—although they are more common in the analysis of time series [645]. Nevertheless, investigating long-range dependence in spatial data may become more feasible in the near future given the recent abundance of information from various environmental sensors.
- Finally, the interpolation of spatial processes relies more heavily on the information contained in the local neighborhood of each target point and less on long-range properties.

Long-range correlations are important in statistical physics, because they may indicate proximity to the critical point of a phase transition or the presence of states that exhibit *self-organized criticality* [42, 766]. In the presence of long-range dependence, small perturbations of the system's "equilibrium state" may lead to large-scale effects.

Covariance functions with power-law decay There are various covariance functions that exhibit LRD. From the models listed in Table 4.2 the *rational quadratic* and the *Cauchy* model exhibit power-law dependence with tail exponents respectively equal to $\mu = 2\beta$ and $\mu = \beta$.[8] Hence, LRD is obtained in the rational quadratic case for $\beta < d/2$ and in the Cauchy case for $\beta < d$.

The Whittle-Matérn correlation function does not exhibit LRD, since in the asymptotic regime it decays exponentially [501]:

$$\rho_{\mathrm{xx}}^{(WM)}(\|\mathbf{r}\|) \underset{\|\mathbf{r}\|\to\infty}{\sim} \|\mathbf{r}\|^{\nu-1/2}\, \mathrm{e}^{-\sqrt{2\nu}\|\mathbf{r}\|/\xi}.$$

A generalized version of the Whittle-Matérn correlation, proposed in [501], decays asymptotically as a power law. This model, however, does not admit an explicit relation for all possible lags $\|\mathbf{r}\|$, even though it possesses an analytical spectral density.

Incomplete gamma function covariance The incomplete gamma function covariance is given by [360]:

$$C_{\mathrm{xx}}(\|\mathbf{r}\|) = \|\mathbf{r}\|^{-2\beta}\, \gamma(2\beta, k_{\mathrm{c}}\|\mathbf{r}\|), \tag{5.50}$$

where $\beta > 0$ is the power-law exponent, $k_{\mathrm{c}} > 0$ is a critical wavenumber, and $\gamma(2\beta, x)$ is the *incomplete gamma function* defined by means of the integral [4, p. 260]

$$\gamma(2\beta, x) = \frac{1}{\Gamma(2\beta)} \int_0^x \mathrm{d}t\, \mathrm{e}^{-t}\, t^{2\beta-1}. \tag{5.51}$$

[8]The exponents β refer to the equations (4.14) and (4.15) respectively.

The incomplete gamma model has finite variance that depends on the power-law exponent and is given by the following expression

$$\sigma_{\mathrm{x}}^2 = \frac{k_{\mathrm{c}}{}^{2\beta}}{2\beta\,\Gamma(2\beta)}.$$

In addition, the covariance behaves as a power law with exponent $\mu = 2\beta$ for $\|\mathbf{r}\| \gg 1/k_{\mathrm{c}}$. Hence, LRD is obtained if $\beta < d/2$.

Rational quadratic covariance It is not always easy to obtain explicit expressions for the spectral density of LRD covariance functions such as the incomplete gamma covariance (5.50).

However, the *rational quadratic function* that decays as a power law for large distances, has the following spectral density (see [380] and Table 4.4)

$$\widetilde{\rho}_{\mathrm{xx}}(\mathbf{k}) = \frac{(2\pi)^{d/2}\,\xi^d}{2^{\beta-1}\,\Gamma(\beta)}\,\frac{K_{d/2-\beta}(\|\mathbf{k}\|\xi)}{(\|\mathbf{k}\|\xi)^{d/2-\beta}}. \tag{5.52}$$

The asymptotic expressions for the *modified Bessel function of the second kind* at the limits $x \to 0$ and $x \to \infty$ are given by

$$K_\nu(z) \underset{z\to 0}{\sim} \frac{\Gamma(\nu)}{2}\left(\frac{2}{z}\right)^{\nu}, \; \nu > 0, \tag{5.53a}$$

$$K_\nu(z) \underset{z\to\infty}{\sim} \sqrt{\frac{\pi}{2z}}\,\mathrm{e}^{-z}. \tag{5.53b}$$

In light of the above, the rational quadratic spectral density declines exponentially for large $\|\mathbf{k}\|$ independently of the value of β, i.e.,

$$\widetilde{\rho}_{\mathrm{xx}}(\mathbf{k}) \underset{\|\mathbf{k}\|\to\infty}{\sim} \mathrm{e}^{-\|\mathbf{k}\|\,\xi}\;(\|\mathbf{k}\|\xi)^{\beta-(d+1)/2}.$$

If the LRD condition $d > 2\beta$ holds, based on (5.53a) the spectral density develops a singularity as $\|\mathbf{k}\| \to 0$ according to

$$\widetilde{\rho}_{\mathrm{xx}}(\mathbf{k}) \underset{\|\mathbf{k}\|\to 0}{\sim} \frac{1}{\|\mathbf{k}\|^{\gamma}}, \quad \text{where } \gamma = d - 2\beta.$$

This singularity is related to the divergence of the integral range, since

$$\lim_{\|\mathbf{k}\|\to 0} \widetilde{\rho}_{\mathrm{xx}}(\mathbf{k}) = \int_{\mathbb{R}^d} d\mathbf{r}\, \rho_{\mathrm{xx}}(\mathbf{r}).$$

So, LRD in real space leads to power-law dependence of the covariance function for $\|\mathbf{k}\| \to 0$, but not necessarily for $\|\mathbf{k}\| \to \infty$.

Power-law tails The existence of power-law dependence in real space implies a respective power-law signature in reciprocal space and vice versa.

(i) If the left tail of the correlation has power-law dependence, i.e.,

$$\rho_{\mathrm{xx}}(x) \sim \|\mathbf{r}\|^{-\mu_l}, \quad \|\mathbf{r}\| \to 0$$

then the spectral density behaves asymptotically as a power law with exponent $\gamma_l = d - \mu_l$, i.e.,

$$\widetilde{\rho}_{\mathrm{xx}}(\mathbf{k}) \sim \|\mathbf{k}\|^{-\gamma_l}, \quad \|\mathbf{k}\| \to \infty.$$

(ii) Conversely, if the right tail of the correlation decays as a power-law, i.e.,

$$\rho_{\mathrm{xx}}(x) \sim \|\mathbf{r}\|^{-\mu_u}, \quad \|\mathbf{r}\| \to \infty,$$

then the spectral density near zero wavenumber behaves as a power law with $\gamma_u = d - \mu_u$, i.e.,

$$\widetilde{\rho}_{\mathrm{xx}}(\mathbf{k}) \sim \|\mathbf{k}\|^{-\gamma_u}, \quad \|\mathbf{k}\| \to 0.$$

Unbounded spectral density Let us now assume that the spectral density is given by the following unbounded function

$$\widetilde{C}_{\mathrm{xx}}(\mathbf{k}) = \frac{(2\pi)^d\, c_0}{\|\mathbf{k}\|^{\gamma}}, \quad \gamma > 0, \text{ for all } \|\mathbf{k}\| \geq 0.$$

It can be shown that the inverse Fourier transform of this function is given by [360]

$$C_{\mathrm{xx}}(\mathbf{r}) = \frac{c_0\, 2^{\mu}\, \pi^{d/2}\, \Gamma(\mu/2)}{\Gamma\left(\frac{\mu-d}{2}\right)} \frac{1}{\|\mathbf{r}\|^{\mu}}, \quad \text{where } \mu = d - \gamma \text{ and } 0 > \mu > -1/2.$$

Note that $\lim_{\|\mathbf{r}\|\to 0} C_{\mathrm{xx}}(\mathbf{r})$ does not exist. This is also guessed at by the singularity of the spectral density at $\|\mathbf{k}\| \to 0$. Therefore, according to Bochner's theorem this function is not a permissible covariance for a stationary random field since it does not possess a finite variance.

5.7 Intrinsic Random Fields

So far we have mostly focused on stationary random fields. In practice, however, spatial data may include non-stationary features. The presence of non-stationary patterns manifests itself through deterministic trend functions or more complex (stochastic) spatial dependencies. Readers familiar with time series analysis will recall that trend functions can be either deterministic or stochastic. In ARIMA time series models, stochastic trend functions are removed by taking differences [94].

Intrinsic random fields provide enhanced flexibility that can be used to model non-stationary stochastic dependencies. In this section we briefly discuss *generalized increments of order k* (GI-k) and *generalized covariance* functions. These concepts are intimately linked with the definition of intrinsic random fields. They are also closely connected with the concept of spatial filtering, which is motivated by the idea that the residual random field (obtained after removing local trends from the original field) is likely to be stationary.

GI-k generalize the k-differences that are used in ARIMA (auto-regressive integrated moving average) models while maintaining their ability to remove stochastic trends [588]. In fact, readers unfamiliar with generalized increments may benefit from the discussions in classical texts on time series modeling [94, 107]. Essentially, GI-k act as high-pass filters that remove large-scale variations while retaining the small-scale variability of the time series.

The main concepts of intrinsic random fields and generalized covariance functions were developed in [553, 862]. Applications in the earth sciences and in medical geography are given in [138, 141]. A detailed and accessible presentation of both subjects is given in the book by Chilès and Delfiner [132]. The engineering viewpoint is given by Kitanidis [459]. Several geostatistical articles address topics related to intrinsic random fields, e.g. [34, 136, 455, 456, 458, 775].

Definition 5.4 (Allowable measures) A set of real-valued coefficients $\{\lambda_n\}_{n=1}^{N}$, that correspond to N locations $\{\mathbf{s}_n\}_{n=1}^{N}$ defines an *allowable measure of order k*, if for all *polynomial functions* $p_k(\mathbf{s})$ *of degree less than or equal to k* the following condition holds

$$p_k(\lambda) := \sum_{n=1}^{N} \lambda_n \, p_m(\mathbf{s}_n) = 0, \quad \text{for all } m \leq k. \tag{5.54}$$

The polynomial $p_m(\mathbf{s}_n)$ where $\mathbf{s}_n = \left(s_{n,1}, \ldots, s_{n,d}\right)$, and $n = 1, \ldots, N$, stands for a superposition of monomials $s_{n,1}^{\ell_1} \ldots s_{n,d}^{\ell_d}$, such that the total (over all the coordinates) degree of each monomial is equal to m, i.e., $\ell_1 + \ldots \ell_d = m$.

The condition (5.54) is equivalent to requiring that the linear combinations (over all the sampling points) of all the monomials vanish, i.e.,

$$\sum_{n=1}^{N} \lambda_n \, s_{n,1}^{\ell_1} \dots s_{n,d}^{\ell_d} = 0, \text{ for } \ell_1 + \dots \ell_d = 0, \dots, k. \tag{5.55}$$

The minimum number of points required to enforce the conditions (5.54) as a function of k is given by the equation [132, p. 249–250]

$$N_{\min} = \frac{(k+d)!}{k!\,d!} + 1. \tag{5.56}$$

The allowable measure of order k is completely determined by the order of the polynomial and the spatial configuration of the points $\{\mathbf{s}_n\}_{n=1}^{N}$.

Generalized increments and intrinsic random fields Allowable measures provide a mathematical framework for removing polynomial trends from spatial data. For example, consider the following linear combination

$$\mathrm{X}(\lambda; \omega) = \sum_{n=1}^{N} \lambda_n \, \mathrm{X}(\mathbf{s}_n; \omega). \tag{5.57}$$

If the $\{\lambda_n\}_{n=1}^{N}$ define an allowable measure of order k, i.e., if they respect (5.54) or (5.55), then $\mathrm{X}(\lambda; \omega)$ is called a *generalized increment of order* k or allowable linear combination of order k (for short, ALC-k) [132, 775]. Typically, we would like to think of the point set $\{\mathbf{s}_n\}_{n=1}^{N}$ as consisting of points inside a local neighborhood. On a regular grid, such a neighborhood would correspond to a stencil that can be moved around the lattice, in order to implement locally derivative operations.

Polynomial trends of degree at most k are removed in the generalized increment $\mathrm{X}(\lambda; \omega)$. In addition, the filtering also removes stochastic trends that cannot be described by a deterministic equation. This is similar to removing piece-wise constant trends by means of finite differences in time series [750]. The filtering operation increases the likelihood that the generalized increment is a wide-sense stationary random field.

Definition 5.5 The random field $\mathrm{X}(\mathbf{s}; \omega)$ is an *intrinsic random field of order* k, also known as an IRF-k, if the generalized increment of order k, i.e., $\mathrm{X}(\lambda; \omega)$ defined by (5.57), is a zero-mean stationary random field [131].

Practical application It is straightforward to show (using the binomial theorem) that if the set $\{\lambda_n\}_{n=1}^N$ is an ALC-k in the sense of (5.55) for the point set $\{\mathbf{s}_n\}_{n=1}^N$, then it is also an ALC-k for the translated set of points $\{\mathbf{s}_0+\mathbf{s}_n\}_{n=1}^N$. This means that the N-point stencil defined by the set of coefficients $\{\lambda_n\}_{n=1}^N$ can be moved around to eliminate polynomial trends from all points in the domain leading to a stationary increment. For example, if the initial random field is sampled on a two-dimensional rectangular grid and contains a linear trend, application of the five-point Laplacian stencil removes the trend and leads to a potentially stationary residual.

Variance of generalized increment The generalized covariance $K(\mathbf{s}_n - \mathbf{s}_m)$ is a function that determines the variance of generalized increments. Assume that $X(\mathbf{s};\omega)$ is an *intrinsic random field of order k* (equivalently an IRF-k) and $X(\lambda;\omega)$ is an allowable measure of order k, as defined by (5.54) and (5.57). Then, there exists a *generalized covariance function* $K(\mathbf{s}_n - \mathbf{s}_m)$ such that the variance of the generalized increments of order k is given by [775]

$$\mathbb{V}\mathrm{ar}\,\{X(\lambda;\omega)\} = \sum_{n=1}^{N}\sum_{m=1}^{N} \lambda_n\,\lambda_m\,K(\mathbf{s}_n - \mathbf{s}_m). \tag{5.58}$$

The generalized covariance is a function that depends purely on the pair distance but it has a divergence at zero lag. This is linked to the lack of integrability of the spectral function of intrinsic random fields. The asymptotic behavior of the generalized covariance of an IRF-k under mild regularity conditions has been obtained by Matheron [554] (see also [165]):

$$\lim_{\|\mathbf{r}\|\to\infty} \frac{K(\mathbf{r})}{\|\mathbf{r}\|^{2k+2}} \to 0. \tag{5.59}$$

In addition, the generalized covariance of an IRF-k satisfies the following inequality

$$|K(0) - K(\|\mathbf{r}\|)| \le a + b\|\mathbf{r}\|^{2k}, \; a, b \ge 0. \tag{5.60}$$

The above inequality does not preclude a singularity of the generalized covariance at zero lag. If such a singularity exists, it can be removed by taking the difference $K(0) - K(\|\mathbf{r}\|)$. Generalized covariances with zero-lag singularities can be derived as solutions of partial differential equations, e.g. [460] and Sect. 5.7.2, as well as from certain Boltzmann-Gibbs probability density models (see Chap. 7).

Kitanidis argues that the generalized covariance is the essential part of the covariance function which contains the pertinent information for the purpose of kriging estimation [458]; see also [774, p. 39].

The generalized covariance function is estimated from the detrended data and is thus different from the covariance function of the original data. However, the original and the detrended data share the same generalized covariance function.

5.7.1 Random Fields with Stationary Increments

In contrast with stationary random fields, intrinsic random fields have unbounded variogram functions. If the spectral density[9] is non-integrable near the origin (i.e., near zero frequency), the corresponding random field is non-stationary, because it fails to satisfy Bochner's theorem. Intrinsic random fields have such non-integrable spectral densities [554, 774, p. 38].

Comment A stationary random field can also exhibit an apparently unbounded structure function, if it is sampled over a domain with characteristic length L, and the latter is not significantly larger than the integral range ℓ_c. Hence, a stationary random field appears as non-stationary if sampled over a finite, not sufficiently large domain. The apparent loss of stationarity is due to (i) the fact that in practice we often estimate the random field's statistical properties from a single realization, thus invoking the *ergodic hypothesis*, and (ii) that ergodicity is not an acceptable assumption if the condition $L \gg \ell_c$ is not satisfied.

A useful class of non-stationary fields are the *intrinsic random fields* of order zero (IRF-0), also known simply as intrinsic random fields and as *random fields with stationary increments* [138, 775, 863]. In studies of fluid turbulence, intrinsic random fields are used to model the velocity fluctuations. They are also known as *locally homogeneous random fields*, because their increments are stationary (statistically homogeneous) [557, 863, p. 101]. An intrinsic random field widely used in applications is *fractional Brownian motion* [529], which we further discuss in Sect. 5.8.

Definition 5.6 A random field $X(\mathbf{s};\omega)$ is called zero-order intrinsic (IRF-0) or simply intrinsic random field, if its increments are second-order stationary.

The stationarity of the increments is expressed in terms of the following conditions that apply to all position vectors $\mathbf{s}$ and lag vectors $\mathbf{r}$:

$$\mathcal{E}[X(\mathbf{s};\omega) - X(\mathbf{s}+\mathbf{r};\omega)] = 0, \tag{5.61}$$

$$\gamma_{xx}(\mathbf{r}) = \frac{1}{2}\,\mathbb{V}\mathrm{ar}\left\{X(\mathbf{s};\omega) - X(\mathbf{s}+\mathbf{r};\omega)\right\}. \tag{5.62}$$

In intrinsic IRF-0 random fields the expectation of $X(\mathbf{s};\omega)$ (if it exists) is at most constant, but the translation invariance of the covariance $C_{xx}(\mathbf{r})$ is lost. The variogram is purely a function of the space lag and increases without bound as

[9] In this case we should speak of a spectral function, since the spectral density is not well-defined.

$\|\mathbf{r}\| \to \infty$. One can incorporate anisotropy in intrinsic random fields by focusing on the increments, that are wide-sense stationary random fields.

Decomposition of intrinsic random fields IRF-0 can be viewed as comprising two components: a smooth but non-stationary random field generated by the singular part of the spectral density in a neighborhood near zero (i.e., the slowly varying frequencies) and a stationary but less smooth random field which is generated by the higher frequencies in the spectral density [554, 774, p. 39]. Obviously, such a decomposition can be constructed in more than one ways. Based on the decomposition principle, Stein opines [774, p. 39] that intrinsic random fields do not offer additional flexibility to stationary random fields regarding the local behavior. However, for many applications (e.g., spectral simulation) it is convenient to have a single spectral function that determines the behavior of the random field over all length scales.

IRF-0 and allowed linear combination Let us assume that $X(\mathbf{s};\omega)$ is an IRF-k with $k = 0$. Then based on (5.54) the ALC-0 conditions correspond to

$$\sum_{n=1}^{N} \lambda_n = 0. \tag{5.63}$$

According to (5.56), the minimum number of points required to enforce the condition (5.54) for $k = 0$ is $N_{\min} = 2$. Hence, $\lambda_1 = 1$ and $\lambda_2 = -1$ are admissible solutions. In this case, the generalized increment becomes the two-point difference used in Definition 5.6. The fact that the intrinsic order is zero implies that $X(\mathbf{s};\omega)$ has a constant mean (expectation), in agreement with Definition 5.6. This, however, does not imply that intrinsic random fields are second-order (wide-sense) stationary, since the covariance function is not required to depend only on the lag.

The variance of the generalized increment for zero-order intrinsic random fields is calculated from (5.58) using $\lambda_1 = 1$ and $\lambda_2 = -1$ and leads to

$$\mathbb{V}\mathrm{ar}\left\{X(\lambda;\omega)\right\} = -2\,K(\mathbf{s}_n - \mathbf{s}_m). \tag{5.64}$$

Since $\mathbb{V}\mathrm{ar}\left\{X(\lambda;\omega)\right\} = \mathbb{V}\mathrm{ar}\left\{[X(\mathbf{s}_n;\omega) - X(\mathbf{s}_m;\omega)]\right\}$, the above identity also implies that for intrinsic random fields the generalized covariance and the variogram are related through the equation

$$K(\mathbf{s}_n - \mathbf{s}_m) = -\gamma_{\mathrm{xx}}(\mathbf{s}_n - \mathbf{s}_m). \tag{5.65}$$

The following definition of intrinsic random fields is equivalent to Definition 5.6.

Stationarity of the increments As discussed above, an intrinsic random field $X(\mathbf{s};\omega)$ has wide-sense stationary increments. The stationarity of the increments implies that for any initial point $\mathbf{s}$ and lag vector $\mathbf{r}$, the *increment field*

$$\Delta_{\mathrm{x}}(\mathbf{s}, \mathbf{r}; \omega) = \mathrm{X}(\mathbf{s} + \mathbf{r}; \omega) - \mathrm{X}(\mathbf{s}; \omega) \tag{5.66}$$

is wide-sense stationary.

The following definition is more practical, but it assumes that the random field $\mathrm{X}(\mathbf{s}; \omega)$ admits the first-order partial derivatives in all directions [138].

Definition 5.7 A random field $\mathrm{X}(\mathbf{s}; \omega)$ that is at least once differentiable has wide-sense stationary increments if the partial derivatives

$$\mathrm{X}_i(\mathbf{s}; \omega) = \frac{\partial \mathrm{X}(\mathbf{s}; \omega)}{\partial s_i}, \quad \text{for all } i = 1, \ldots, d,$$

are wide-sense stationary random fields.

The Definition 5.7 does not apply, however, to non-differentiable intrinsic random fields such as the spatially extended fractional Brownian motion.

Random electric field If the vector random field $\mathbf{E}(\mathbf{s}; \omega)$ denotes the intensity of a randomly varying electric field, and $\Phi(\mathbf{s}; \omega)$ denotes the scalar electric potential field, the two are connected via $\mathbf{E}(\mathbf{s}; \omega) = -\nabla\Phi(\mathbf{s}; \omega)$. For example, such random electric fields are generated by the scattering of electromagnetic radiation from discrete random media [577]. Conversely, the potential difference between the point $\mathbf{s}$ and some arbitrary reference point $\mathbf{s}_0$ is given by the following line integral along any continuous path that connects the points $\mathbf{s}_0$ and $\mathbf{s}$, i.e.,

$$\Phi(\mathbf{s}; \omega) - \Phi(\mathbf{s}_0; \omega) = -\int_{\mathbf{s}_0}^{\mathbf{s}} \mathbf{dl}(\mathbf{s}') \cdot \mathbf{E}(\mathbf{s}'; \omega). \tag{5.67}$$

If the vector random field $\mathbf{E}(\mathbf{s}; \omega)$ has wide-sense stationary increments, then the potential field $\Phi(\mathbf{s}; \omega)$ is an IRF-0.

Variogram of intrinsic random fields Let us next define the covariance function $C_{\mathrm{yy}}(\mathbf{r})$ obtained by the summation of the covariance functions corresponding to the partial derivatives $\mathrm{X}_i(\mathbf{s}; \omega)$, i.e.,

$$C_{\mathrm{yy}}(\mathbf{r}) = \sum_{i=1}^{d} C_{\mathrm{X}_i\,\mathrm{X}_i}(\mathbf{r}) \doteq \sum_{i=1}^{d} \mathcal{E}[\mathrm{X}_i(\mathbf{s} + \mathbf{r}; \omega)\, \mathrm{X}_i(\mathbf{s}; \omega)]\,. \tag{5.68}$$

The following relation holds between the covariance function of the gradient fields and the variogram function of the initial scalar field $\mathrm{X}(\mathbf{s}; \omega)$

$$\nabla^2 \gamma_{\mathrm{xx}}(\mathbf{r}) = C_{\mathrm{yy}}(\mathbf{r}). \tag{5.69}$$

If we assume that the Fourier transform of the variogram function exists, at least in the sense of a generalized function [733], we obtain the following algebraic relation in the spectral domain

$$-\,\|\mathbf{k}\|^2\,\widetilde{\gamma}_{\mathrm{xx}}(\mathbf{k}) = \widetilde{C}_{\mathrm{yy}}(\mathbf{k}). \tag{5.70}$$

We can thus obtain the variogram function in real space by evaluating the inverse Fourier transform of $\widetilde{\gamma}_{\mathrm{xx}}(\mathbf{k})$

$$\begin{aligned}\gamma_{\mathrm{xx}}(\mathbf{r}) =& c_0 - \frac{1}{(2\pi)^d}\int_{\mathbb{R}^d} \mathrm{d}\mathbf{k}\,\frac{\widetilde{C}_{\mathrm{yy}}(\mathbf{k})}{\|\mathbf{k}\|^2}\,e^{i\mathbf{k}\cdot\mathbf{r}}\\ =& c_0 - \frac{1}{(2\pi)^d}\int_{\mathbb{R}^d} \mathrm{d}\mathbf{k}\,\frac{\widetilde{C}_{\mathrm{yy}}(\mathbf{k})}{\|\mathbf{k}\|^2}\,\cos(\mathbf{k}\cdot\mathbf{r}).\end{aligned} \tag{5.71}$$

The second line in the above is obtained by flipping the sign of $\mathbf{k}$ to $-\mathbf{k}$, and taking into account that $\widetilde{C}_{\mathrm{yy}}(\mathbf{k}) = \widetilde{C}_{\mathrm{yy}}(-\mathbf{k})$. The identity $\gamma_{\mathrm{xx}}(\mathbf{0}) = 0$ helps to determine the value of the integration constant c_0. Finally, we obtain

$$\gamma_{\mathrm{xx}}(\mathbf{r}) = \frac{2}{(2\pi)^d}\int_{\mathbb{R}^d} \mathrm{d}\mathbf{k}\,\frac{\widetilde{C}_{\mathrm{yy}}(\mathbf{k})\,\left[1-\cos(\mathbf{k}\cdot\mathbf{r})\right]}{\|\mathbf{k}\|^2}.$$

The term c_0 removes the zero-frequency (infrared divergence), which is due to the singular behavior of the term $\cos(\mathbf{k}\cdot\mathbf{r})\,\widetilde{C}_{\mathrm{yy}}(\mathbf{k})/\|\mathbf{k}\|^2$ at the limit $\|\mathbf{k}\| \to 0$. The divergence results from the term $\|\mathbf{k}\|^{-2}$, since $\widetilde{C}_{\mathrm{yy}}(\mathbf{k})$ tends to a constant as $\|\mathbf{k}\| \to 0$. In contrast,

$$1-\cos(\mathbf{k}\cdot\mathbf{r}) = \frac{1}{2}\|\mathbf{k}\|^2 + O(\|\mathbf{k}\|^4).$$

The $\|\mathbf{k}\|^2$ term in this expansion cancels the denominator of the variogram thus leading to a convergent integral.

Variogram properties of intrinsic random fields Intrinsic random fields that are not stationary can have variogram functions that increase with distance without reaching a sill. For example, the variogram function for fractional Brownian motion (fBm) increases as $\gamma_{\mathrm{xx}}(\mathbf{r}) \propto \|\mathbf{r}\|^{2H}$, where $0 < H < 1$ (see Sect. 5.8 below).

How fast can the variogram increase as a function of the spatial lag $\mathbf{r}$? It turns out that there are limits to the variogram's rapid ascent. This is thoroughly discussed by Yaglom [863, p. 396–402] and [864, p.136–137] as well as in [132, p. 61]. The answer involves two separate cases.

(i) If there is no discontinuity of the spectral density (i.e., delta function) at the origin, then

$$\lim_{\|\mathbf{r}\|\to\infty} \frac{\gamma_{xx}(\mathbf{r})}{\|\mathbf{r}\|^2} = 0. \tag{5.72}$$

This means that the variogram cannot increase asymptotically faster than $\|\mathbf{r}\|^2$.

(ii) If the spectral density contains a discontinuity at the origin, then for every $\mathbf{r}_0 \in \mathbb{R}^d$ there exists a positive constant $A_0 > 0$ such that

$$\gamma_{xx}(\mathbf{r}) \leq A_0 \|\mathbf{r}\|^2, \text{ for every } \mathbf{r} \in \mathbb{R}^d : \|\mathbf{r}\| > \|\mathbf{r}_0\|. \tag{5.73}$$

The derivation of these constraints is explained in Sect. 5.8.

Example 5.5 (Variogram asymptotic behavior) A characteristic example that leads to a variogram function $\gamma_{xx}(\mathbf{r}) = A_1 \|\mathbf{r}\|^2$ is given by Yaglom [863, p. 397–398]. Consider the random field

$$\mathrm{X}(\mathbf{s}; \omega) = X_0(\omega) + X_1(\omega)\, \mathbf{s},$$

where $X_0(\omega)$ and $X_1(\omega)$ are random variables. Determine the expectation and the variogram of $\mathrm{X}(\mathbf{s}; \omega)$.

Answer Let us define $m_i = \mathcal{E}[X_i(\omega)]$ and $A_i = \mathbb{V}\mathrm{ar}\{X_i(\omega)\}$, for $i = 0, 1$.

- The expectation is simply given by $\mathcal{E}[\mathrm{X}(\mathbf{s}; \omega)] = m_0 + m_1 \mathbf{s}$.
- The variogram is given by

$$\gamma_{xx}(\mathbf{r}) = \mathbb{V}\mathrm{ar}\{\mathrm{X}(\mathbf{s}+\mathbf{r}; \omega) - \mathrm{X}(\mathbf{s}; \omega)\} = \mathbb{V}\mathrm{ar}\{X_1(\omega)\}\ \|\mathbf{r}\|^2 = A_1\ \|\mathbf{r}\|^2.$$

Note that each realization of this random field is simply a straight line with random intercept and slope. The field is thus non-ergodic, since the statistical properties cannot be estimated from a single realization.

5.7.2 Higher-Order Stationary Increments

Fields with higher-order stationary increments (i.e., IRF-k where $k > 0$) can be defined in analogy with zero-order intrinsic random fields. Extension to higher intrinsic orders implies filtering operations that use higher-order derivatives.

Let us assume that $\mathrm{X}(\mathbf{s}; \omega)$ is a random field that admits all second-order partial derivatives [10]. For Gaussian random fields, this requires the existence of the fourth-order partial derivatives of the covariance function $C_{xx}(\mathbf{s}, \mathbf{s})$ at all $\mathbf{s} \in \mathcal{D}$.

In addition, let us assume that $X(\mathbf{s};\omega)$ satisfies the *stochastic Poisson equation*

$$-\triangle X(\mathbf{s}) := Y(\mathbf{s};\omega), \;\; \forall \mathbf{s} \in \mathcal{D}, \tag{5.74}$$

where $\triangle$ is the Laplacian operator, and the source term $Y(\mathbf{s};\omega)$ is a second-order stationary, zero-mean random field. The source term can then be viewed as the stationary increment of $X(\mathbf{s};\omega)$. For finite domains $\mathcal{D}$, the Poisson equation needs to be supplemented by boundary conditions on $\partial\mathcal{D}$.

Based on the above, $X(\mathbf{s};\omega)$ is a *first-order intrinsic random field*, also known as an IRF-1 [554, 775]. The intrinsic order is determined by the highest order of the polynomial trend that can be filtered by the differential operator. The intrinsic order is equal to one in this case, given that the order of the differential operator is equal to two. For the Poisson equation to be satisfied, the trend function (expectation) of $X(\mathbf{s};\omega)$ must be at most a linear polynomial in $\mathbf{s}$, i.e., $m_{\mathrm{x}}(\mathbf{s}) = a_0 + \mathbf{a}_1 \cdot \mathbf{s}$.

If we consider Poisson's equation (5.74) at two different points $\mathbf{s}$ and $\mathbf{s}'$, multiply the respective sides, and apply the expectation operator on both sides, it follows that the covariance $C_{\mathrm{xx}}(\mathbf{s},\mathbf{s}')$ obeys the following biharmonic PDE

$$\triangle_{\mathbf{s}} \triangle_{\mathbf{s}'} C_{\mathrm{xx}}(\mathbf{s},\mathbf{s}') = C_{\mathrm{yy}}(\mathbf{r}), \tag{5.75}$$

where (i) $\mathbf{r} = \mathbf{s}-\mathbf{s}'$, (ii) $C_{\mathrm{yy}}(\mathbf{r})$ is the covariance of the stationary increment, and (iii) $\triangle_{\mathbf{s}}(\cdot)$ implies that the Laplacian acts on the two-point function $C_{\mathrm{xx}}(\mathbf{s},\mathbf{s}')$ at point $\mathbf{s}$.

Solving a PDE is not always easy and it sometimes requires the use of numerical tools. Equation (5.75) is luckily a *linear elliptic* PDE with constant coefficients. Hence, an analytical solution can be derived. If you are unfamiliar with PDEs a good place to start is the classic text on electromagnetism by Jackson [401]; in spite of its focus on electromagnetic problems it contains the essential elements for the analytical solution of PDEs combined with clear physical interpretations.

The full solution of the covariance PDE comprises the homogeneous and particular solutions as defined below.

1. The *homogeneous solution* $C^{(0)}_{\mathrm{xx}}(\mathbf{s},\mathbf{s}')$ satisfies

$$\triangle_{\mathbf{s}} \triangle_{\mathbf{s}'} C^{(0)}_{\mathrm{xx}}(\mathbf{s},\mathbf{s}') = 0.$$

2. The *particular solution*, $k_{\mathrm{xx}}(\mathbf{s}-\mathbf{s}')$, satisfies the *biharmonic* (also known as *Bilaplacian*) PDE

$$\triangle^2 k_{\mathrm{xx}}(\mathbf{r}) := C_{\mathrm{yy}}(\mathbf{r}), \tag{5.76}$$

where $\triangle^2$ is the *biharmonic or Bilaplacian* operator.

The particular solution $k_{\mathrm{xx}}(\mathbf{r})$ can be obtained using the Green function method [401]. The latter is based on the *fundamental solution* $G(\mathbf{s},\mathbf{s}')$ of the biharmonic equation, i.e., $\triangle^2 G(\mathbf{s},\mathbf{s}') = \delta(\mathbf{s},\mathbf{s}')$. The particular solution $k_{\mathrm{xx}}(\mathbf{s}-\mathbf{s}')$ represents the *generalized covariance function*.

The *particular solution of the biharmonic* PDE (5.76) is obtained by the convolution of the biharmonic Green function (see Chap. 12) with the covariance of the stationary increment, namely by the convolution integral

$$k_{\mathrm{xx}}(\mathbf{r}) = \int_{\mathbb{R}^d} d\mathbf{r}' \, G(\mathbf{r} - \mathbf{r}') \, C_{\mathrm{yy}}(\mathbf{r}'). \tag{5.77}$$

The explicit equation for the biharmonic Green function over an infinite domain is given by (10.9) (see Chap. 10).

Finally, the *complete solution* for the covariance function is given by the superposition

$$C_{\mathrm{xx}}(\mathbf{s}, \mathbf{s}') = p_1(\mathbf{s}) \, p_1(\mathbf{s}') + k_{\mathrm{xx}}(\mathbf{s} - \mathbf{s}'),$$

where $p_1(\mathbf{s})$ and $p_1(\mathbf{s}')$ are the polynomials that respectively satisfy

$$\triangle_{\mathbf{s}} p_1(\mathbf{s}) = 0 \text{ and } \triangle_{\mathbf{s}'} p_1(\mathbf{s}') = 0.$$

If the covariance is calculated over a finite domain, respective boundary conditions should be respected by the complete solution. The boundary conditions will have an impact both on the functional form of $G(\mathbf{r} - \mathbf{r}')$ and on the solutions of the Laplace equation; hence, $p_1(\mathbf{s})$ will be replaced by a superposition of trigonometric and exponential functions that respect the boundary conditions (see [401] for details).

The Bilaplacian field Let us assume that the random field $\mathrm{X}(\mathbf{s}; \omega)$ satisfies the stochastic PDE

$$\Delta \mathrm{X}(\mathbf{s}; \omega) = -\mathrm{Y}(\mathbf{s}; \omega), \tag{5.78}$$

where $\mathrm{Y}(\mathbf{s}; \omega)$ is a white noise on $\mathbb{R}^d$ with covariance $C_{\mathrm{yy}}(\mathbf{r}') = \sigma_{\mathrm{y}}^2 \, \delta(\mathbf{r}')$.

The generalized covariance $k_{\mathrm{xx}}(\mathbf{r})$ of $\mathrm{X}(\mathbf{s}; \omega)$ is obtained by the convolution integral (5.77). For the white-noise covariance, the convolution integral leads to

$$k_{\mathrm{xx}}(\mathbf{r}) = \sigma_{\mathrm{y}}^2 \, G(\mathbf{r}). \tag{5.79}$$

Hence, the biharmonic Green function is the generalized covariance for a random field $\mathrm{X}(\mathbf{s}; \omega)$ with a white-noise increment obtained after Laplacian filtering. Generalized covariance functions carry sufficient information for determining the weights of kriging interpolation [132, 458].

The field $\mathrm{X}(\mathbf{s}; \omega)$ generated from the stochastic PDE (5.78) is also known as *Bilaplacian field* and belongs to the broader class of *fractional Gaussian fields* [507].

5.8 Fractional Brownian Motion

In this section we focus in a specific class of random fields with stationary increments. This class has a correlation structure that stems from the mathematical property of self-affinity and is known by the popular name of *fractional Brownian motion*.

Let us consider a random field $X(\mathbf{s}; \omega)$ defined over a spatial domain $\mathcal{D} \subset \mathbb{R}^2$. We can think of $X(\mathbf{s}; \omega)$ as the height of a flexible three-dimensional membrane. Let us also consider an arbitrary initial point $\mathbf{s}_0 \in \mathcal{D}$. Then the increments of the field with respect to the reference point, i.e., $X(\mathbf{s}; \omega) - X(\mathbf{s}_0; \omega)$, represent the difference of surface height between the points $\mathbf{s}$ and $\mathbf{s}_0$. If the increments are drawn from a stationary and uncorrelated random field the surface is completely random and corresponds to a two-dimensional extension of the *classical Brownian motion*.

The class of intrinsic random fields that we consider here have *self-affine* increments [863]. Self-affinity (see Definition 4.2) implies that the correlation function of the stationary increments decays as a power-law of the lag. Stationary random fields with power-law decay of the correlations are known as *fractional Gaussian noise (fGn)*. The British hydrologist Harold Edwin Hurst [388, 389] demonstrated that the increment variance for several natural processes grows with time as a power law with characteristic exponent H (where $0 < 2H < 2$). The parameter H was named *Hurst exponent* after him.

The mathematical properties of random fields with stationary fGn increments were investigated by Mandelbrot & Van Ness [529], who introduced the term *fractional Brownian motion (fBm)*. Given the fact that motion refers to one-dimensional random processes, the term *fractional Brownian field* is more suitable for random fields supported on $\mathbb{R}^d$. However, at this point the term fBm is widely used.

Definition 5.8 (Definition of fractional Brownian motion (fBm)) A random field $X(\mathbf{s}; \omega)$ belongs to the fBm class if and only if [230]:

1. $X(\mathbf{s}; \omega)$ is a zero-mean Gaussian random field.
2. The increments $\Delta_x(\mathbf{s}_0, \mathbf{s}; \omega) = X(\mathbf{s}; \omega) - X(\mathbf{s}_0; \omega)$ are stationary.
3. The random field $X(\mathbf{s}; \omega)$ is *self-affine* with index H, $0 < H < 1$, that is,

$$X(\mathbf{s}; \omega) \stackrel{d}{=} \lambda^{-H} X(\lambda \mathbf{s}; \omega), \tag{5.80}$$

where the symbol $\stackrel{d}{=}$ denotes equality in distribution.

The above definition establishes an equivalence between fBm and the class of all *Gaussian self-affine processes with stationary increments*.

5.8.1 Properties of fBm Fields

The following is a list of fBm properties that are important in spatial data modeling.

fBm sample paths Sample paths of fractional Brownian motion are (almost surely) (i) continuous, (ii) nowhere differentiable and (iii) of unbounded variation [230]. The properties of the fBm paths depend significantly on the value of the Hurst index H, as discussed in the following table and shown in Fig. 5.12.

The fractional Brownian motion does not belong in the same class as the random fields with stationary increments defined in Sect. 5.7 by taking partial derivatives. This procedure is not possible for fBm, which are non-differentiable as a result of self-affinity.

Second-order correlation functions The fBm covariance function is given by the following tripartite superposition with clear local dependence

$$C_{\mathrm{xx}}(\mathbf{s}, \mathbf{s}') = \frac{\sigma^2}{2} \left[\|\mathbf{s}\|^{2H} + \|\mathbf{s}'\|^{2H} - \|\mathbf{s} - \mathbf{s}'\|^{2H} \right]. \tag{5.81}$$

Based on (5.81) the non-uniform variance, $\sigma_{\mathrm{x}}^2(\mathbf{s}) = C_{\mathrm{xx}}(\mathbf{s}, \mathbf{s})$, is a power-law function of the Euclidean measure of $\mathbf{s}$

$$\sigma_{\mathrm{x}}^2(\mathbf{s}) = \sigma^2 \, \|\mathbf{s}\|^{2H}. \tag{5.82}$$

The increase of the fBm variance with the distance from the origin is reminiscent of the increased dispersion in the position of a random walker with the passing of time [566].

In contrast to $C_{\mathrm{xx}}(\mathbf{s}, \mathbf{s}')$, the *fBm variogram function* depends purely on the lag, i.e.,

$$\gamma_{\mathrm{xx}}(\mathbf{s}, \mathbf{s}') = \frac{\sigma^2}{2} \, \|\mathbf{s} - \mathbf{s}'\|^{2H}. \tag{5.83}$$

Classification based on Hurst exponent The geometry of the sample paths and the correlation properties of fBm depend crucially on the Hurst exponent H. To perceive this, let us calculate the covariance function of the increment $\mathrm{X}(\mathbf{s} + \mathbf{u}; \omega) - \mathrm{X}(\mathbf{s}; \omega)$ for a vector "step" $\mathbf{u}$, which we will denote by $C_{\Delta\mathrm{X}}(\mathbf{u}, \mathbf{r})$. Based on (5.81) and straightforward calculations, it can be shown that

$$C_{\Delta\mathrm{X}}(\mathbf{u}, \mathbf{r}) = \frac{\sigma^2}{2} \left[\|\mathbf{r} + \mathbf{u}\|^{2H} + \|\mathbf{r} - \mathbf{u}\|^{2H} - 2\|\mathbf{r}\|^{2H} \right]. \tag{5.84}$$

To gain insight in the correlation of the increments let us assume that $d = 1$ and $r > u > 0$. Then, the following properties hold:

1. For $H = 1/2$ the increments are uncorrelated and fBm defaults to the *classical Brownian motion*.
2. For $H < 1/2$ the increments are negatively correlated, and the fBm paths are rough as shown in Fig. 5.12 (top). In this case the correlations are called *anti-persistent*.
3. For $H > 1/2$ the increments have positive correlations, and the fBm paths exhibit a more regular variation as shown in Fig. 5.12 (bottom). In this case the correlations are called *persistent*.

Anisotropy Anisotropic fBm-like variogram functions can be constructed by means of different length scalings in different directions. For example, if $h = (\mathbf{r}^\top \mathbf{A}\, \mathbf{r})^{1/2}$ is the dimensionless distance as defined in Sect. 4.3.2, then it is possible to define the following anisotropic variogram

$$\gamma_{\mathrm{xx}}(\mathbf{r}) = c\, h^{2H}.$$

In the case of fBm random fields, the Hurst exponent determines the regularity parameter, i.e., the leading-order dependence of the variogram on distance near the origin. Then, as shown in [17], if the Hurst exponent varies continuously with the direction, it must be constant in all directions. Hence, it is not possible to construct fBm-like valid variogram functions with continuously varying Hurst exponents.

The definition, simulation and applications of *anisotropic fBm* fields in hydrology are investigated in [198, 514, 582], while applications in medical image analysis are proposed in [76].

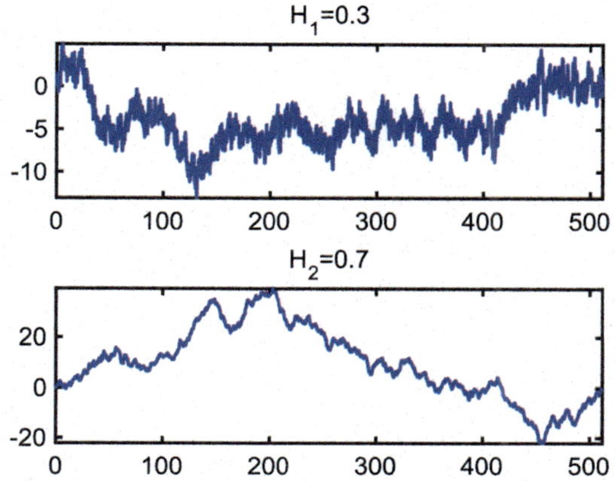

Fig. 5.12 Paths (realizations) of fractional Brownian motion with persistent correlations ($H = 0.7$) and anti-persistent correlations ($H = 0.3$). The fluctuations exhibit a higher degree of irregularity for $H = 0.3$ in contrast with the smoother pattern for $H = 0.7$

5.8.2 Spectral Representation

The power-law behavior of the fBm variogram function is reflected in a power-law spectral density of the form [396]

$$\tilde{g}(\mathbf{k}) = (2\pi)^d \, A \, \|\mathbf{k}\|^{-\delta}, \text{ where } A > 0, \; d+2 > \delta > d. \tag{5.85}$$

Note that (5.85) is linked with the variogram and does not represent the Fourier transform of the covariance function. The latter is a function of both the location and the lag, and thus it cannot possess a position-independent spectral density. In the case of non-stationary processes (and random fields), a more suitable spectral representation is based on the *Wigner-Ville spectrum* [548].

We will now take a closer look at how the spectral density (5.85) leads to the fBm variogram following [360]. According to (3.76), the variogram can be expressed using in terms of the following spectral integral

$$\gamma_{\mathrm{xx}}(\mathbf{r}) = A \int_{\mathbb{R}^d} \mathrm{d}\mathbf{k} \left(\frac{1 - \cos(\mathbf{k}\cdot\mathbf{r})}{\|\mathbf{k}\|^{\delta}} \right). \tag{5.86}$$

The integrability condition (3.77) is satisfied by $\tilde{g}(\mathbf{k}) \propto \|\mathbf{k}\|^{-\delta}$ for $\delta > 0$. Based on the integral transformation $1 - \cos u = \int_0^1 \mathrm{d}\chi \; \sin(\chi u)$, the variogram spectral integral is expressed as

$$\gamma_{\mathrm{xx}}(\mathbf{r}) = A \int_0^1 \mathrm{d}\chi \int_{\mathbb{R}^d} \mathrm{d}\mathbf{k} \, \frac{(\mathbf{k}\cdot\mathbf{r}) \, \sin(\chi \, \mathbf{k}\cdot\mathbf{r})}{\|\mathbf{k}\|^{\delta}}. \tag{5.87}$$

In order to evaluate the spectral integral, we define the function $\Phi(\cdot)$ by means of

$$\Phi(\mathbf{k}\cdot\mathbf{r}) = \frac{(\mathbf{k}\cdot\mathbf{r}) \, \sin(\chi \, \mathbf{k}\cdot\mathbf{r})}{\|\mathbf{k}\|^{\delta}}, \; d \geq 2,$$

and we use the following observations:

(i) $\mathbf{k}\cdot\mathbf{r} = \|\mathbf{k}\| \, \|\mathbf{r}\| \, \cos\theta$, where θ is the angle between the vectors $\mathbf{k}$ and $\mathbf{r}$. The vector $\mathbf{r}$ is considered constant during the integration over $\mathbf{k}$.
(ii) The volume integral is expressed in *spherical coordinates* as follows

$$\int_{\mathbb{R}^d} \mathrm{d}\mathbf{k} = \int_{\mathcal{B}_d} \mathrm{d}\hat{\mathbf{k}} \int_0^{\infty} \mathrm{d}\|\mathbf{k}\| \; \|\mathbf{k}\|^{d-1}, \tag{5.88}$$

where $\mathcal{B}_d$ is the surface of the "unit sphere" in $d > 1$ dimensions, $\hat{\mathbf{k}}$ is a unit vector in reciprocal space, $\|\mathbf{k}\|$ is the radius of the sphere, and $\mathrm{d}\hat{\mathbf{k}}$ denotes a differential element of $\mathcal{B}_d$.

The surface of the "unit sphere" implies the circumference of the circle in $d = 2$, the spherical surface in $d = 3$, and the surface of higher-dimensional hyperspheres in $d > 3$.

Then, the inner integral in the variogram spectral representation (5.87) becomes

$$\int_{\mathbb{R}^d} \mathrm{d}\mathbf{k}\, \Phi(\mathbf{k} \cdot \mathbf{r}) = \int_{\mathcal{B}_d} \mathrm{d}\hat{\mathbf{k}} \int_0^\infty \|\mathbf{k}\|^{d-1}\, \Phi\left(\|\mathbf{k}\|\,\|\mathbf{r}\| \cos\theta\right), \tag{5.89}$$

In the light of the above, we can express the variogram as follows

$$\gamma_{\mathrm{xx}}(\mathbf{r}) = A\, \|\mathbf{r}\| \int_0^1 \mathrm{d}\chi \int_0^\infty \mathrm{d}\|\mathbf{k}\|\, \|\mathbf{k}\|^{d-\delta}\, \Lambda_d(\chi\, \|\mathbf{k}\|\,\|\mathbf{r}\|), \tag{5.90a}$$

where $\Lambda(\cdot)$ is the spectral function defined by

$$\Lambda_d(x) = \int_{\mathcal{B}_d} \mathrm{d}\hat{\mathbf{k}}\, \cos\theta\, \sin\left(x \cos\theta\right), \text{ where } x = \chi\, \|\mathbf{k}\|\, \|\mathbf{r}\|. \tag{5.90b}$$

Closed-form expressions for the *spectral function* in 2D and 3D are as follows, where $J_1(\cdot)$ is the Bessel function of the first kind of order one:

$$\Lambda_d(x) = \begin{cases} 2\pi\, J_1(x), & d = 2 \\ \frac{4\pi}{x^2}\left(\sin x - x \cos x\right), & d = 3. \end{cases}$$

We now return to (5.90) and use the composite variable x defined in (5.90b) to express the variogram as follows

$$\gamma_{\mathrm{xx}}(\mathbf{r}) = \eta_d\, \|\mathbf{r}\|^{2H}, \text{ where } 0 < H < 1, \tag{5.91a}$$

$$\eta_d = A \int_0^1 \mathrm{d}\chi\, \chi^{2H-1} \int_0^\infty \mathrm{d}x\, x^{-2H}\, \Lambda_d(x), \tag{5.91b}$$

where $H = (\delta - d)/2$ is the *Hurst exponent.*

Exponent bounds The existence of the outer integral in (5.91b) requires that $H > 0$ for the outer integral to converge. A Taylor expansion of $\Lambda_d(x)$ around zero shows that $\Lambda_d(x) = a_d x + O(x^3)$, where $a_2 = \pi$ and $a_3 = 4\pi/3$. Hence, convergence of the inner integral at $x = 0$ requires that $H < 1$.

The bounds of the Hurst exponent imply that the spectral density exponent $\delta = 2H + d$ satisfies $d < \delta < d + 2$.

5.8.3 Long-Range Dependence

The Hurst exponent determines the persistence of fluctuations in fBm random fields. For $H > 1/2$, the fBm field exhibits long-range dependence. At the same time, the Hurst exponent also determines the fractal dimension, because the exponent of the *principal irregular term* is $\alpha = 2H$ (compare with (5.47)).

The estimation of the Hurst exponent from the data can be accomplished by means of various methods that include fitting the empirical variogram (see Sect. 12.3 for variogram estimation), maximum likelihood, and Mandelbrot's rescaled range analysis [292].

A popular method in statistical physics is *detrended fluctuation analysis (DFA)*. DFA was introduced by Peng et al. [659] as a tool for investigating long-range correlations in one-dimensional stochastic signals (in particular, nucleotide sequences).

For one-dimensional, uniformly-spaced data DFA is applied by means of the following steps:

1. The data sequence $\{x_i\}_{i=1}^N$ is integrated over time by calculating the cumulative sum of the data values.
2. The integrated series is divided into K segments (windows) of length n each.
3. Within each window of size n a trend function is fitted to the data—the trend function could be different in each window.
4. The fluctuations of the cumulative series are then evaluated by removing the respective trend function at each time instant.
5. The DFA variance is evaluated by averaging the squares of all the fluctuations.
6. The same procedure is repeated for different window sizes n and the DFA variance is plotted as a function of n.
7. If there is long-range dependence, the DFA variance $V(n)$ varies with the window size n as a power law with a characteristic DFA exponent, i.e., $V(n) \sim n^{2\alpha_{\text{DFA}}}$. Hence, the plot of the logarithm of the variance versus $\ln n$ is a straight line.
8. The slope of this line (i.e., the DFA exponent) is linearly related to the Hurst exponent, i.e., $\alpha_{\text{DFA}} = H + 1$, and thus it provides an estimate of the latter [292].

The effect of trends and non-stationarities on DFA has been studied in [129, 382]. The relation between DFA and the power spectral density for stochastic processes has been investigated in [337]. Empirical confidence intervals for the DFA exponent have been derived [843]. DFA has recently been extended to higher dimensions including two-dimensional multifractal fields [313, 827]. The higher-dimensional extensions are presently applicable only to data distributed on regular (e.g., square) grids. DFA has not been explored in spatial data analysis, probably because (i) the method has not yet been extended to scattered data, and (ii) it requires rather long sequences of data.

Note on Hurst exponent Two definitions of the Hurst exponent exist in the literature. The one adopted herein assumes that $2H$ is the exponent of the variogram's principal irregular term. In this context, the Hurst exponent takes values $0 < H < 1$ and is defined in connection with non-stationary processes with self-affine stationary increments. This definition implies that $\alpha_{\mathrm{DFA}} = H + 1$.

In the physics literature the Hurst exponent is sometimes taken as equal to α_{DFA}. If we denote this exponent by H', then $H' = H + 1$. The exponent H' is also well-defined for stationary processes. If $\mathrm{X}(\mathbf{s};\omega)$ is a stationary process with power-law correlations, then it holds that

$$\lim_{r\to\infty} \rho_{\mathrm{xx}}(r) \propto H'(2H'-1)\, r^{2H'-2},$$

where $0 < H' < 1$ [513]. In contrast, exponent values $1 < H' < 2$ are characteristic of non-stationary processes.

5.8.4 Random Walk Model

The random walk model is one of the best known models of non-stationary behavior with many applications in various branches of physics and finance [490, 766]. It is also a prototypical model of diffusion in both simple and complex environments [56]. The random walk models helps to develop some intuition into the generation of non-stationary processes.

We consider a simple one-dimensional example, in which a walker starts at $t = 0$ at $x_0 = 0$ and moves randomly along a straight line. The walker moves with constant time step $\delta t \ll 1$. The steps $h(t;\omega)$ are assumed to be realizations of a *correlated random process* with

$$\mathcal{E}[h(t;\omega)] = 0, \text{ and } C_{\mathrm{hh}}(\tau) = \mathcal{E}[h(t;\omega)\, h(t+\tau;\omega)] = \sigma^2 \exp(-|\tau|/\tau_0).$$

The position of the walker is a random process $\mathrm{X}(t;\omega)$. Our goal is to determine the expectation, the variance, and the variogram of the position process $\mathrm{X}(t;\omega)$.

The assumption $\mathcal{E}[h(t;\omega)] = 0$ means that backward steps are on average as likely as forward steps. The covariance function implies that the steps are exponentially correlated. The walker moves with a random velocity given by

$$V(t;\omega) \doteq h(t;\omega)/\delta t$$

where $\tau_0 \gg \delta t$. Since δt is small, we can treat the random walk as continuous in time. The distance of the walker from the starting point, i.e., the walker's displacement, is given by the integral of the velocity

$$\mathrm{X}(t;\omega) - x_0 = \int_0^t \mathrm{d}t'\, V(t';\omega).$$

Without any loss in generality, we will assume in the following that $x_0 = 0$. Then, the position coincides with the displacement.

Mean displacement In view of the velocity and the position integral, the expectation of the position is given by

$$\mathcal{E}\left[\mathrm{X}(t;\omega)\right] = \frac{1}{\delta t}\int_0^t \mathrm{d}t'\,\mathcal{E}\left[h(t';\omega)\right] = 0.$$

Covariance function Taking into account the zero mean displacement, the covariance function of the displacement is given by

$$C_{\mathrm{xx}}(t,\tau) = \mathcal{E}\left[\mathrm{X}(t;\omega)\,\mathrm{X}(t+\tau;\omega)\right] = \frac{1}{\delta t^2}\int_0^{t+\tau}\mathrm{d}t''\int_0^t \mathrm{d}t'\,\mathcal{E}\left[h(t';\omega)\,h(t'';\omega)\right]$$

$$= \frac{\sigma_{\mathrm{x}}^2}{\delta t^2}\int_0^{t+\tau}\mathrm{d}t''\int_0^t \mathrm{d}t'\,\mathrm{e}^{-|t'-t''|/\tau_0} \doteq \frac{\sigma_{\mathrm{x}}^2}{\delta t^2}\,I_C(t,\tau).$$

The double integral $I_C(t,\tau)$ is evaluated by expanding the absolute value in the exponent as follows

$$I_C(t,\tau) = \int_0^{t+\tau}\mathrm{d}t''\int_0^t \mathrm{d}t'\,\mathrm{e}^{-|t'-t''|/\tau_0} = \int_0^{t+\tau}\mathrm{d}t''\left[\int_0^{t''}\mathrm{d}t'\,\mathrm{e}^{(t'-t'')/\tau_0} + \int_{t''}^{t}\mathrm{d}t'\,\mathrm{e}^{-(t'-t'')/\tau_0}\right]$$

$$= \tau_0\int_0^{t+\tau}\mathrm{d}t''\left[2 - \mathrm{e}^{-t''/\tau_0} - \mathrm{e}^{-(t-t'')/\tau_0}\right]$$

$$= 2\tau_0(t+\tau) + \tau_0^2\left[\mathrm{e}^{-(t+\tau)/\tau_0} + \mathrm{e}^{-t/\tau_0} - \mathrm{e}^{-\tau/\tau_0} - 1\right].$$

This expression for $I_C(t,\tau)$ shows that $C_{\mathrm{xx}}(t,t')$ depends not only on the lag τ but also on the elapsed time t.

Position variance The position variance is given by $\sigma_{\mathrm{x}}^2(t) = C_{\mathrm{xx}}(t,\tau=0)$. Thus,

$$\sigma_{\mathrm{x}}^2(t) = \frac{2\tau_0\sigma^2}{\delta t^2}\left[t + \tau_0\left(\mathrm{e}^{-t/\tau_0} - 1\right)\right].$$

Hence, for $t \gg \tau_0$ the variance depends linearly on t. At $t < \tau_0$ the correlation of the steps introduces a nonlinear term in the variance. The impact of this term, however, is diminished as the time lag grows and the memory of the correlations is erased.

Position variogram To determine the position variogram, we first calculate the *displacement* over time τ, which is given by

$$X(t+\tau;\omega) - X(t;\omega) = \int_{t}^{t+\tau} dt'\, V(t';\omega) = \frac{1}{\delta t}\int_{t}^{t+\tau} dt'\, h(t';\omega).$$

Using the above expression for the displacement, the variogram function is given by

$$\begin{aligned}
\gamma_{xx}(\tau) &= \frac{1}{2}\mathcal{E}\Big[\,[X(t+\tau;\omega) - X(t;\omega)]^2\,\Big] \\
&= \frac{1}{2\delta t^2}\int_{t}^{t+\tau} dt' \int_{t}^{t+\tau} dt''\, \mathcal{E}\big[h(t';\omega)\, h(t'';\omega)\big] \\
&= \frac{\sigma^2}{2\delta t^2}\int_{t}^{t+\tau} dt' \int_{t}^{t+\tau} dt''\, e^{-|t'-t''|/\tau_0} \\
&= \frac{\sigma^2}{2\delta t^2}\int_{t}^{t+\tau} dt' \left[\int_{t}^{t'} dt''\, e^{-(t'-t'')/\tau_0} + \int_{t'}^{t+\tau} dt''\, e^{(t'-t'')/\tau_0}\right] \\
&= \frac{\tau_0\sigma^2}{2\delta t^2}\left[\int_{t}^{t+\tau} dt' \left[2 - e^{-(t'-t)/\tau_0} - e^{-(t+\tau-t')/\tau_0}\right]\right] \\
&= \frac{\tau_0\sigma^2}{\delta t^2}\left(\tau - \tau_0 + \tau_0\, e^{-\tau/\tau_0}\right).
\end{aligned}$$

For $\tau \gg \tau_0$, the exponential term is suppressed leading to a linear variogram function as is expected for the *classical Brownian motion.* At time lags up to $O(\tau_0)$ the step correlations modify the linear dependence through the exponential term.

Based on the above, the random walk with short-range correlated steps only shows deviations from the classical Brownian motion at finite time scales, unlike the fBm processes that differ from the classical model at all times. In the case of a real Brownian particle the correlation time is defined by the particle mass and the friction coefficient [903, pp. 3–12].

5.8.5 Applications of Fractional Brownian Motion

The random walk analogue helps to develop intuition regarding the fractional Brownian motion. The fBm model has also been applied to various spatial processes which are related to anomalous diffusion. However, fBm dependence has also been observed in data that are not related to diffusion in any obvious way. For example, some studies suggest that natural heterogeneities of geological media exhibit fractal correlation functions characterized by a Hurst exponent H [582].

In one such study, the fluid permeability of small sandstone samples (dimensions approximately $450 \times 250 \times 10\,\text{mm}^3$) was measured [522]. The integrated permeability along the two major axes revealed the characteristic dependence of fractional Brownian motion with Hurst coefficients $H \approx 0.82 - 0.90$. This result implies that the permeability random field is a stationary fractional Gaussian noise [529].

A different study [603] argues that the log-permeability of crystalline and sedimentary rocks has power-law variograms with anti-persistent correlations ($H \approx 0.25 - 0.35$) over a wide range of scales. This dependence is characteristic of fractional Brownian motion with stationary anti-correlated increments [529]. The two results are not necessarily contradictory, because they involve measurements at very different scales. Both studies are in contrast with the standard approach in stochastic subsurface hydrology, which treats the fluid permeability as a stationary random field with short-range correlations, e.g. [174, 275]. The short-range assumption is more easily justifiable for relatively homogeneous soils. Recent accounts of fBm applications in hydrogeology are given in [596, 604, 627].

At the same time, we should keep in mind that care is needed when "hunting" for power laws in empirical distributions, since other types of functional dependence can pose as power laws over a restricted range of scales. Clauset et al. emphasize the use of rigorous statistical analysis, including hypothesis testing, to investigate power laws and to accurately estimate power-law exponents [146]. A more recent study cautions that classical statistical analysis uses the hypothesis of statistical independence of the observations; for example, the log-likelihood is expressed as a sum of marginal log-likelihoods. However, if the observations are correlated, the independence assumption can lead to false rejections [285]. These issues are further discussed in Appendix D.

5.8.6 Roughness of Random Field Surfaces

Random fields $X(\mathbf{s}; \omega)$, $\mathbf{s} \in \mathcal{D} \subseteq \mathbb{R}^d$ are used to represent surfaces indexed in d-dimensional spaces. We can think of the random field values $x(\mathbf{s})$ as the local *height* of a surface at every point $\mathbf{s} \in \mathcal{D}$. Variations in surface height are responsible for the roughness of a surface. *Surface roughness* often plays a significant role in the engineering and technological properties of materials [491, 845]. Roughness is also a significant descriptive measure of landscape *surface topography* [312].

Even in the microscopic world, it is important to measure and characterize the roughness of *nanostructures* generated by nanolithography and plasma etching manufacturing processes (see Figs. 5.13 and 5.14). Such structures exhibit a considerable amount of surface disorder that can be modeled using random fields [294, 295].

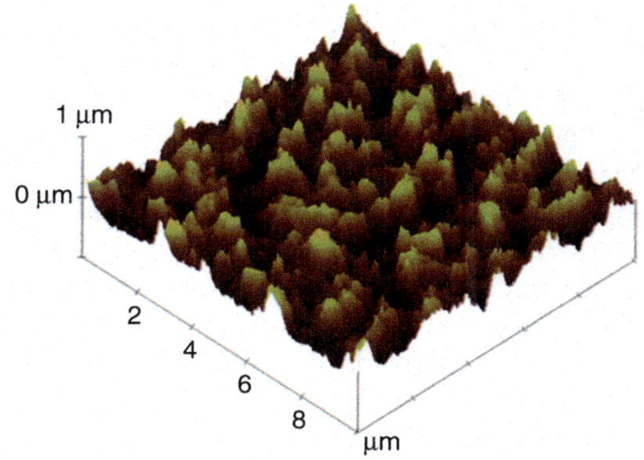

Fig. 5.13 Atomic Force Microscope image of the surface of a polymethyl methacrylate (PMMA) film (plexiglass) etched in oxygen plasma for five minutes [822]. (Courtesy of V. Constantoudis, NCSR Demokritos, Greece)

Fig. 5.14 Atomic Force Microscope image of the surface of a polydimethylsiloxane (PDMS) film etched in SF6 (sulfur hexafluoride) plasma for two minutes [294]. PDMS is a biostable synthetic polymer used for biomedical applications such as drug delivery. (Courtesy of V. Constantoudis, NCSR Demokritos, Greece)

Various measures are used to capture different aspects of surface roughness. These include summary statistical measures of the surface height distribution.

Even though humans have a good intuitive understanding of surface roughness, different statistical measures used to estimate roughness often lead to very different answers. Various roughness indices have been defined in terms of moments of the surface height distribution [845].

The *variance* is a commonly used measure of roughness. To a first approximation this makes sense, since the variance measures the fluctuation amplitude of spatial configurations. As we observe in Fig. 5.4, however, the apparent roughness of random fields with the same expectation, variance, and characteristic length can be quite different. The variance is useful in comparing stationary random fields with the same type and range of correlations. Nevertheless, a measure of roughness based on the variance has definite drawbacks. For example, the variance does not discriminate between two fields with different correlation structures. In addition, in the case of surfaces generated by *non-stationary random fields*, the variance depends on the spatial location. Hence, a global measure of variance is not a good descriptor of local roughness.

The *fractal dimension* was used to measure the roughness of metal surfaces and proposed as a universal measure of roughness for fractal objects by B. Mandelbrot [531, 532]. A similar definition of roughness involves the *DFA fractal exponent* [56]. A more detailed representation of roughness in terms of the *height variogram function* is discussed in [305]. This description, in addition to a fractal dimension uses three length scales to characterize roughness: (i) the mean surface height; (ii) the average surface thickness, which corresponds to the standard deviation of the height fluctuations; and (iii) the correlation length $\xi_{\|}$ parallel to the surface. Given these parameters, the height variogram of self-affine surfaces increases as a power law with Hurst exponent H for lags that are smaller than the characteristic correlation length $\xi_{\|}$, and saturates to a constant value equal to the variance of the height fluctuations for lags beyond $\xi_{\|}$.

On a more celestial topic, recent publications investigate the roughness of the lunar surface based on data from the Lunar Orbiter Laser Altimeter [691, 871]. The estimation of roughness is carried out over different baselines in order to investigate scale dependence, using geometric statistical measures such as median absolute slope, median differential slope, and the Hurst exponent.

5.9 Classification of Random Fields

Different classification schemes for random fields can be based on three criteria that involve (i) the joint probability density function (ii) the degree of statistical homogeneity and (iii) the range of spatial dependence as expressed through the two-point correlation functions.

5.9.1 Classification Based on Joint Probability Density Function

1. *Gaussian random fields.* The Gaussian distribution is widely used in spatial data modeling. According to the *Central Limit Theorem (CLT)* (see Theorem 5.1), the normal distribution is a strong attractor for the probability distribution of aggregate random fields that represent an average over smaller-scale fluctuations [92]. The CLT is often quoted as the main reason for the omnipresence of Gaussian random fields. In addition to that, Gaussian random fields have a central role in spatial data modeling, because at least certain non-Gaussian distributions can be normalized by application of suitable nonlinear transforms (e.g., see Chap. 14).
2. *Lognormal random fields.* These fields are derived from normally distributed ones, since the logarithm of the lognormal field $\mathrm{Y}(\mathbf{s};\omega)$ is the normally dis-

tributed field $X(\mathbf{s}; \omega) = \ln Y(\mathbf{s}; \omega)$. Lognormal fields are widely used models for data with asymmetric probability distributions that exhibit a heavier-than-Gaussian right tail. They are justified based on the Central Limit Theorem for products of random fields, which specifies (under mild conditions) that such products are attracted to the lognormal distribution. Physically motivated justifications of the lognormal distribution have also been proposed in the literature (for a review see [766]).

3. *Lévy random fields*. The probability density function of these fields decays slowly (as a power law) in the tails. Thus, Lévy random fields are used to model data with exceptionally high probabilities for large deviations from the mean. Large excursions from the mean state are characteristic of "extreme events". The impact of climate change on extreme events is a hot topic of current research [578, 872]. A consequence of the slow tail decay is that the statistical moments above a certain order may not be defined due to divergence of the respective integrals. In such cases, it is possible to define truncated Lévy distributions with well-defined moments [514, 627]. A disadvantage of Lévy random fields is that, except for special cases, they do not admit explicit expressions for the joint pdf.

   Heavy-tailed random field models with well-defined joint density functions include the Student-t (symmetric), the log-Student (asymmetric) random fields, and the κ-lognormal random fields (see Chap. 14 for more details). Furthermore, *max-stable random fields* are also used to model extreme events with spatial dependence [181, 686, 723].

5.9.2 Classification Based on Statistical Homogeneity

1. *Statistically homogeneous (stationary) random fields:* Weakly-stationary random fields are characterized by constant means and covariance functions that are translation invariant. Stationarity is a restrictive assumption, but it affords significant formalistic and computational simplifications in spatial modeling. In addition, if a stationary random field model is combined with the available data, the resulting conditional mean and covariance functions are non-stationary, thus relaxing the underlying stationarity assumption.

   A significant part of geostatistics is based on the stationarity assumption [165]. There are reasons in support of the stationarity assumption in the modeling of environmental extremes [737]. In addition, the usefulness of stationary models in the modeling of non-stationary data is discussed in [266].
2. *Random fields with statistically homogeneous increments:* These fields are statistically non-homogeneous, but their increments are statistically homogeneous (stationary). Such fields model processes that can be viewed as random walks over some physical or abstract space. Physical phenomena that lead to this type of behavior include *diffusion and growth* processes [56, 305].

   Common examples in this category are the classical and fractional Brownian motions. The spatial dependence of the increments is governed by the exponent

H. If $1/2 < H < 1$, the increments have *long-range (persistent)* statistical dependence that leads to smooth variations of the fBm paths. On the other hand, if $0 < H < 1/2$ the increments have negative correlations that lead to intense variability of the paths.

Other useful examples include the Lévy fractional motions, the increments of which follow heavy-tailed Lévy probability density functions [766].

3. *Random fields with non-homogeneous increments:* These are generalizations of random fields with homogeneous increments. Common examples involve *multifractal Brownian motions* [56, 512]. In this case, the statistical dependence between two points is characterized by power laws with characteristic exponents, but the value of the exponent depends on the distance between two points. Hence, the exponent has different values for small-range fluctuations than for large-range fluctuations. These fields allow abrupt changes and can be used to model *intermittent processes*.

5.9.3 Classification Based on the Type of Correlations

1. *Short-range correlations:* In this case the integral of the correlation function over $\mathbb{R}^d$ is finite. Characteristic examples comprise the correlation functions with a single length scale (e.g., exponential, Gaussian, and other), the linear superpositions of such functions, and certain correlation functions with multiple length scales which are linked to additional smoothness or rigidity parameters (such as the Matérn and Spartan models).
2. *Long-range correlations:* Long-range correlations are characterized by an increase of the volume integral of the correlation function as the integration volume increases. Typical examples include functions that decay as power laws at large distance with exponents smaller than the spatial dimension, i.e., $C_{xx}(\mathbf{r}) \propto \|\mathbf{r}\|^{-\alpha}$, where $0 < \alpha < d$.
3. *Fractal correlations:* Such correlations appear in spatial processes with fractal structure [530]. The fractal character manifests itself in the form of power law dependence of some suitable correlation function. Simple fractals are characterized by spatially uniform fractional exponents. Fractal dependence can be long-range depending on the exponent value.

   The archetypical example of fractal correlations in homogeneous random fields is *fractional Gaussian noise* (fGn). fGn has a power-law covariance function $C_{xx}(\mathbf{r}) \propto \|\mathbf{r}\|^{-\alpha}$. In the case of random fields with homogeneous increments the best known example is *fractional Brownian motion (fBm)*. The fBm correlations are fully determined by the Hurst exponent H. While the fBm covariance function depends on the position in space, its variogram function has the purely lag-based power-law dependence $\gamma_{xx}(\mathbf{r}) \sim \|\mathbf{r}\|^{2H}$.
4. *Multifractal correlations:* In this case an entire spectrum of exponents is necessary to describe the correlation structure: a different exponent characterizes the dependence of the field at short distances than at large distances. Multifractal

random fields can be used to describe spatial patterns with abrupt (discontinuous) changes as well as intermittency. Such models are used to describe the spatial structure of meteorological processes and climate patterns [512, 791].

5.9.4 Desired Properties of Random Field Models

Various types of random fields have been developed, either motivated by physical analogues or based on purely mathematical considerations. These random fields exhibit a wide range of mathematical properties. The following is a list of desirable features for continuously-valued RFs according to [316, 690].

1. *Fully specified probabilistic model:* This means that all the components (trend and fluctuations) of the random field are specified. This is necessary for the simulation of random fields. RF models that are based on the specification of the increments do not fully specify the trend, and thus they cannot be simulated without further assumptions. A characteristic example of non-fully specified models involve the Brownian and Lévy fractional motions [582, 642–644]. The practicality of such models for the representation of geological heterogeneity has thus been questioned [316].
2. *Symmetry (permutation invariance):* This property requires that for any set of n sites, and any set of n field values, the n-dimensional joint pdf is invariant under all permutations of the index set $\{1, 2, \ldots, N\}$ [863, p. 10]. The intuitive meaning of this property is that it should not matter how we choose to label the sites.
3. *Compatibility:* Let us consider two point sets $\mathbb{S}_1$ and $\mathbb{S}_2$ with cardinal numbers N_1 and N_2 respectively. Assume that $\mathbb{S}_1$ and $\mathbb{S}_2$ share a number of sites N_c but also contain some sites that are not commonly shared. The random field has a joint distribution for each set of sites. Then, the N_c-dimensional joint pdfs obtained respectively from the N_i-dimensional pdfs (where $i = 1, 2$) after integrating out the disjoint degrees of freedom must coincide [863].
4. *Marginal invariance:* It is best if all the finite-dimensional joint pdfs defined by the random field model belong to the same parametric class of probability density functions. This makes it easier to ensure the compatibility property and simplifies analytical calculations.
5. *Closure under additivity:* Spatial data often involve averages over different areas or volumes, which determine the *sample support*. Such data need to be harmonized (referred to the same spatial scale). Further, averaging may be required to obtain the estimates that will finally be reported. Thus, it is best for analytical calculations if the coarse-grained random field (obtained by spatial averaging) belongs to the same parametric class as the initial random field. This closure condition is satisfied by Gaussian random fields.

   A more general closure condition under *nonlinear transformations* can be applied in the vicinity of non-Gaussian fixed points in terms of the *renormalization group analysis* [417].

6. *Asymptotic independence:* It is required that the dependence between the field values at two separate locations diminish as the distance between them grows. The decline may be non-monotonic in the case of hole-type covariance functions. Then, the amplitude of the local covariance peaks (positive or negative) should diminish with increasing lag. Given the requirements for ergodicity specified in Sect. 4.1, asymptotic independence is necessary for the reliable estimation of model parameters from a single sample.
7. *Analytical tractability:* Spatial prediction is based on the conditional pdf at a target point, based on the available data at the observation locations. Deriving the conditional pdf from the general joint pdf requires integrating over a number of degrees of freedom. Since numerical integration is costly or even intractable in high dimensions, it is preferred that the joint pdf be amenable to analytical integration.
8. *Generality:* Ideally, a parametric random field model should define a broad family that is able to capture different types of spatial behavior, including different types of univariate marginal pdfs. In addition, it is preferable if certain combinations of parameters lead to known random field classes. Note that the parametrization can refer to the joint pdf or to the moment functions in the case of random fields with a specified joint pdf. For example, in the case of Gaussian random fields, the Matérn covariance family is a general covariance class that includes other known covariance models as specific cases.
9. *Model parsimony:* Guided by the general principle of parsimony as expressed in *Occam's razor*, it is preferable to define random field models that involve few parameters instead of models with many parameters [82]. This principle should be cautiously applied, keeping in mind the following maxim, attributed to Einstein:

   > Everything should be made as simple as possible, but not simpler.

   The most common application of Occam's razor is the use of linear regression for noisy data: a linear model often provides a more realistic representation of dependence than higher-order polynomial functions that fit every data point. The linear model involves a considerably smaller parameter set than higher-order models. This implies that the root mean square error of the linear model is worse than that of higher-order models. However, the linear model avoids over-fitting (adapting to the noise) and has more predictive power outside the range of the observations.

   Random field model parameters should ideally be interpretable in terms of intuitive quantities that have physical analogues (such as large-scale trends, characteristic lengths, smoothness indices, bending coefficients, et cetera.).

Chapter 6
Gaussian Random Fields

> The grand aim of all science is to cover the greatest number of empirical facts by logical deduction from the smallest number of hypotheses or axioms.
>
> *Albert Einstein*

Gaussian random fields have a long history in science that dates back to the research of Andrey Kolmogorov and his group. Their investigation remains an active field of research with many applications in physics and engineering. The widespread appeal of Gaussian random fields is due to convenient mathematical simplifications that they enable, such as the decomposition of many-point correlation functions into products of two-point correlation functions. The simplifications achieved by Gaussian random fields are based on fact that the joint Gaussian probability density function is fully determined by the mean and the covariance function.

Gaussian random fields are used to model a great variety of spatial processes. The omnipresence of the Gaussian distribution in nature is explained, to a large extent, by the *Central limit theorem (CLT)*. Under some not-too-strict mathematical conditions, the CLT states that the average of a large number of random variables tends to follow the Gaussian distribution. Multivariate extensions of the CLT are also available.

The generalization of the CLT to *vector random variables* is as follows [26]:

Theorem 6.1 (Multivariate CLT) *Assume a set of N independent and identically distributed D-dimensional vector random variables* $\{\mathbf{X}_k(\omega)\}_{k=1}^{N}$, *where*

$$\mathbf{X}_k(\omega) = \left(\mathrm{X}_{k;1}(\omega), \ldots, \mathrm{X}_{k;D}(\omega)\right)^{\top}, \quad k = 1, \ldots, N.$$

D. T. Hristopulos, *Random Fields for Spatial Data Modeling*, Advances in Geographic Information Science, https://doi.org/10.1007/978-94-024-1918-4_6

Let these vector random variables have a mean equal to $\mathbf{m} = (m_1, \ldots, m_D)^\top$ *and a* $D \times D$ *covariance matrix* $\mathbf{C}_{\mathrm{xx}}$. *Define the average random vector* $\overline{\mathbf{X}}(\omega)$ *according to*

$$\overline{\mathbf{X}}(\omega) \doteq \frac{1}{\sqrt{N}} \left[\mathbf{X}_1(\omega) + \ldots + \mathbf{X}_N(\omega)\right].$$

For $N \to \infty$ *the joint distribution of* $\overline{\mathbf{X}}(\omega)$ *converges to the multivariate normal with mean equal to* $\mathbf{m}$ *and covariance matrix* $\mathbf{C}_{\mathrm{xx}}$, *i.e.,*

$$\overline{\mathbf{X}}(\omega) \xrightarrow{d} \mathcal{N}\big(\mathbf{m}, \mathbf{C}_{\mathrm{xx}}\big). \tag{6.1}$$

Remark The proof of the multivariate CLT is based on the univariate CLT for the sequence of random variables $\{\mathrm{Y}_k(\omega)\}_{k=1}^N$ where $\mathrm{Y}_k(\omega) = \mathbf{X}_k(\omega) \cdot \mathbf{t}$, and $\mathbf{t} \in \mathbb{R}^D$. This means that the probability distribution of the random variables $\mathrm{Y}_k(\omega)$ should satisfy the condition for the validity of the univariate CLT for all possible $\mathbf{t}$. The condition in question involves the centered third-order moments of the absolute value of the fluctuations $\mathrm{Y}_k(\omega) - \mathbf{m} \cdot \mathbf{t}$, for $k = 1, \ldots, N$; the cubic root of the sum of these moments should be asymptotically (for $N \to \infty$) vanishingly small compared to the standard deviation of the $\mathrm{Y}_k(\omega)$ [161, p. 215].

The CLT (6.1) applies to finite-dimensional vectors, but it can easily be generalized to averages that involve infinite-dimensional spatial random fields with finite variance [89, 269, p. 260]. Loosely stated, if the pdf of a stationary random field lacks heavy tails and has finite-range correlations, then the random field average converges to the joint normal pdf [92].

To put the above comments in perspective, it should be noted that in many research disciplines there is strong interest in non-Gaussian distributions with long (heavy) tails. Processes that generate such non-Gaussian data are not adequately described by Gaussian random fields. Nonetheless, the latter retain a central position in spatial data modeling, especially since (i) there are types of non-Gaussian distributions that are easily reduced to the Gaussian model and (ii) mixtures of Gaussian distributions are effective at capturing certain types of non-Gaussianity that result from clustering processes and patchy spatial patterns. We return to non-Gaussian random fields in Chaps. 14 and 15.

6.1 Multivariate Normal Distribution

The multivariate normal distribution is relatively simple and admits explicit expressions for the moments of all orders. As discussed above, the multivariate CLT provides a strong argument for the applicability of the multivariate normal distribution in many cases.

Let the vector $\mathbf{x}$ consist of N values from the random field $\mathrm{X}(\mathbf{s}; \omega)$ which are sampled at the locations of the sampling set $\mathbb{S}_{\mathrm{N}} = \{\mathbf{s}_1, \ldots, \mathbf{s}_N\}$, i.e.,

$$\mathbf{x} \equiv \begin{bmatrix} x(\mathbf{s}_1) \\ \vdots \\ x(\mathbf{s}_N) \end{bmatrix}. \tag{6.2}$$

Accordingly, the symbol x_n corresponds to the value of the function $x(\mathbf{s}_n)$ at the location $\mathbf{s}_n$ for $n = 1, \ldots, N$. The vector $\mathbf{x}$ represents a realization of the random vector $\mathbf{X}(\omega)$ defined at the N sampling points in $\mathbb{S}_N$. These could represent a random set of points or a set of lattice nodes used for the simulation of the random field.

The *joint Gaussian pdf* of the vector $\mathbf{X}(\omega)$ is given by the following *multivariate normal density*

$$f_{\mathrm{x}}(\mathbf{x}) = \frac{1}{(2\pi)^{N/2}\,(\det \mathbf{C}_{\mathrm{xx}})^{1/2}} \exp\left[-\frac{1}{2}\,(\mathbf{x}-\mathbf{m}_{\mathrm{x}})^{\top}\,\mathbf{C}_{\mathrm{xx}}^{-1}\,(\mathbf{x}-\mathbf{m}_{\mathrm{x}})\right], \tag{6.3a}$$

$$\mathbf{m}_{\mathrm{x}} = \begin{bmatrix} \mathcal{E}[\mathrm{X}(\mathbf{s}_1;\omega)] \\ \vdots \\ \mathcal{E}[\mathrm{X}(\mathbf{s}_N;\omega)] \end{bmatrix}, \tag{6.3b}$$

$$\mathbf{C}_{\mathrm{xx}} = \begin{bmatrix} C_{\mathrm{xx}}(\mathbf{s}_1,\mathbf{s}_1) & \cdots & C_{\mathrm{xx}}(\mathbf{s}_1,\mathbf{s}_N) \\ \vdots & \vdots & \vdots \\ C_{\mathrm{xx}}(\mathbf{s}_N,\mathbf{s}_1) & \cdots & C_{\mathrm{xx}}(\mathbf{s}_N,\mathbf{s}_N) \end{bmatrix}. \tag{6.3c}$$

In the above equation $\mathbf{m}_{\mathrm{x}}$ is the *vector of expectations*, whereas $\mathbf{C}_{\mathrm{xx}}$ corresponds to the *covariance matrix* of the N sampling points. Hence, $\mathbf{C}_{\mathrm{xx}}$ is a *square and symmetric* matrix whose elements $C_{\mathrm{xx}}(n,m) = C_{\mathrm{xx}}(\mathbf{s}_n,\mathbf{s}_m)$ are evaluated from the covariance function of the random field $\mathrm{X}(\mathbf{s};\omega)$, i.e.,

$$C_{\mathrm{xx}}(\mathbf{s}_n,\mathbf{s}_m) := \mathcal{E}\left[\left(\mathrm{X}(\mathbf{s}_n;\omega) - m_{\mathrm{x}}(\mathbf{s}_n)\right)\left(\mathrm{X}(\mathbf{s}_m;\omega) - m_{\mathrm{x}}(\mathbf{s}_m)\right)\right].$$

The inverse covariance $\mathbf{C}_{\mathrm{xx}}^{-1}$ is also known as the *precision matrix* and satisfies

$$\mathbf{C}_{\mathrm{xx}}\,\mathbf{C}_{\mathrm{xx}}^{-1} = \mathbf{C}_{\mathrm{xx}}^{-1}\,\mathbf{C}_{\mathrm{xx}} = \mathbf{I}_N, \text{ or equivalently, } \sum_{k=1}^{N} C_{n,k}^{-1}\,C_{k,m} = \delta_{n,m}, \tag{6.4}$$

where $\mathbf{I}_N$ is the $N \times N$ identity matrix, i.e., $[\mathbf{I}_N]_{n,m} = \delta_{n,m}$. Finally, the symbol $\det \mathbf{C}_{\mathrm{xx}}$ denotes the *determinant* of the covariance matrix.

The multivariate normal distribution has certain useful mathematical properties.

1. *Subset normality:* If the components of the random vector $\mathbf{X}(\omega)$ are partitioned into two disjoint subsets, then each subset of values follows a multivariate normal

distribution as well. This property also implies that all the marginal distributions of a joint Gaussian pdf are univariate Gaussian.

2. *Superposition normality:* If a random vector $\mathbf{X}(\omega)$ follows the multivariate Gaussian distribution, then linear combinations of the vector's components also follow multivariate Gaussian distributions.
3. *Normality and independence:* In the case of a Gaussian-distributed vector $\mathbf{X}(\omega)$, the vector elements are independent from each other if and only if they are uncorrelated (i.e., independence and zero correlation are equivalent).
4. *Conditional normality:* The conditional probability distributions of subsets of a jointly Gaussian random vector $\mathbf{X}(\omega)$ follow respective multivariate Gaussian distributions. We return to Gaussian conditional distributions in Sect. 6.1.4.

6.1.1 Boltzmann-Gibbs Representation

A joint Gaussian pdf can be expressed in the following exponential form as a *Boltzmann-Gibbs* distribution

$$f_{\mathrm{x}}(\mathbf{x};\boldsymbol{\theta}) = \frac{\mathrm{e}^{-\mathcal{H}_0(\mathbf{x};\boldsymbol{\theta})}}{Z(\boldsymbol{\theta})}. \tag{6.5a}$$

In the Boltzmann-Gibbs expression, the following notation is used:

- $\boldsymbol{\theta}$ is the vector of random field parameters;
- $\mathcal{H}_0(\mathbf{x};\boldsymbol{\theta})$ is the random field's *energy function* given by

$$\mathcal{H}_0(\mathbf{x};\boldsymbol{\theta}) := \frac{1}{2}\left(\mathbf{x}-\mathbf{m}_{\mathrm{x}}\right)^{\top}\mathbf{C}_{\mathrm{xx}}^{-1}\left(\mathbf{x}-\mathbf{m}_{\mathrm{x}}\right); \tag{6.5b}$$

- $Z(\boldsymbol{\theta})$ is the *partition function* evaluated by summing the exponential factor over all possible realizations $\mathbf{x}$ of the random vector $\mathbf{X}(\omega)$. For a vector-valued $\mathbf{x}$ with N components, the partition function is given by the following multiple integral

$$Z(\boldsymbol{\theta}) = \int_{-\infty}^{\infty} \mathrm{d}x_1 \ldots \int_{-\infty}^{\infty} \mathrm{d}x_N\, \mathrm{e}^{-\mathcal{H}_0(\mathbf{x};\boldsymbol{\theta})}. \tag{6.5c}$$

Comparing the Boltzmann-Gibbs expression (6.5a) with the joint Gaussian pdf (6.3a), it follows that the Gaussian partition function is given by

$$Z(\boldsymbol{\theta}) = (2\pi)^{N/2}\left[\det\left(\mathbf{C}_{\mathrm{xx}}\right)\right]^{1/2}. \tag{6.6}$$

In the following, we will often drop the dependence on $\boldsymbol{\theta}$ to simplify notation.

6.1.2 Gaussian Second-Order Cumulant

Readers unfamiliar with the expression (6.5a) for the multivariate Gaussian pdf may wonder how this equation implies that the covariance matrix is given by $\mathbf{C}_{\mathrm{xx}}$. In the following we show that the covariance follows from the moment and the cumulant generating functionals, defined in (4.64) and (4.71) respectively. The derivation is a somewhat pedantic but instructive exercise.

The covariance function is given by the second-order cumulant of the joint pdf, i.e. (4.77). Thus, the covariance is generated from the CGF as follows

$$C_{\mathrm{xx}}(\mathbf{s}_1, \mathbf{s}_2) = \left.\frac{\partial^2 \mathcal{K}_{\mathrm{x}}(\mathbf{u})}{\partial u_1 \partial u_2}\right|_{\mathbf{u}=\mathbf{0}}. \tag{6.7}$$

In order to calculate the cumulant generating function, we first evaluate the moment generating function for the Gaussian pdf (6.5), which is given by the following expression

$$\mathcal{M}_{\mathrm{x}}(\mathbf{u}) = \frac{1}{Z}\int_{\mathbb{R}^N} \mathbf{dx}\; \mathrm{e}^{\mathbf{u}^\top \mathbf{x} - \frac{1}{2}(\mathbf{x}-\mathbf{m}_{\mathrm{x}})^\top \mathbf{C}_{\mathrm{xx}}^{-1}(\mathbf{x}-\mathbf{m}_{\mathrm{x}})},$$

where Z is given by (6.6) while the integration over $\mathbf{x}$ involves the following multiple integral

$$\int_{\mathbb{R}^N} \mathbf{dx} = \prod_{n=1}^{N}\int_{-\infty}^{\infty} \mathrm{d}x_n.$$

Shifting $\mathbf{x}$ by means of $\mathbf{x} - \mathbf{m}_{\mathrm{x}} \mapsto \mathbf{x}$, the moment generating function is transformed as follows

$$\mathcal{M}_{\mathrm{x}}(\mathbf{u}) = \frac{\mathrm{e}^{\mathbf{u}^\top \mathbf{m}_{\mathrm{x}}}}{Z}\int_{\mathbb{R}^N} \mathbf{dx}\; \mathrm{e}^{\mathbf{u}^\top \mathbf{x} - \frac{1}{2}\mathbf{x}^\top \mathbf{C}_{\mathrm{xx}}^{-1}\mathbf{x}}.$$

In order to evaluate the above multivariate integral, we add to and subtract from the exponent the term $\frac{1}{2}\mathbf{u}^\top \mathbf{C}_{\mathrm{xx}}\,\mathbf{u}$, where $\mathbf{u} \in \mathbb{R}^N$. We subsequently apply the transformation

$$\mathbf{x} \mapsto \mathbf{y} := \mathbf{x} - \mathbf{C}_{\mathrm{xx}}\,\mathbf{u}.$$

The moment generating function is then transformed into

$$\mathcal{M}_{\mathrm{x}}(\mathbf{u}) = \frac{\mathrm{e}^{\mathbf{u}^\top \mathbf{m}_{\mathrm{x}} + \frac{1}{2}\mathbf{u}^\top \mathbf{C}_{\mathrm{xx}}\,\mathbf{u}}}{Z}\int_{\mathbb{R}^N} \mathbf{dy}\; \mathrm{e}^{-\frac{1}{2}\mathbf{y}^\top \mathbf{C}_{\mathrm{xx}}^{-1}\mathbf{y}}.$$

The multiple integral over the vector $\mathbf{y}$ can be easily evaluated if the exponent is expressed as a sum of diagonal terms. This can be accomplished by diagonalizing the precision matrix $\mathbf{C}_{\mathrm{xx}}^{-1}$. Since the latter is a real-valued and symmetric matrix, it can be diagonalized by means of the orthogonal transformation

$$\mathbf{B} = \mathbf{S}\,\mathbf{C}_{\mathrm{xx}}^{-1}\,\mathbf{S}^{-1},$$

where $\mathbf{S}$ is a unitary matrix. The elements of the diagonal matrix $\mathbf{B}$ are $B_{n,m} = b_n \delta_{n,m}$. We also transform $\mathbf{y}$ by means of

$$\mathbf{y} \mapsto \mathbf{z} = \mathbf{S}^{-1}\,\mathbf{y}.$$

The *Jacobian determinant* $\det(\mathbb{J})$ of the above orthogonal transformation is given by

$$\det(\mathbb{J}) := \det\left[\mathbb{J}(\mathbf{y}, \mathbf{z})\right] = \det \begin{bmatrix} \frac{\partial y_1}{\partial z_1} & \cdots & \frac{\partial y_N}{\partial z_N} \\ \cdots & \cdots & \cdots \\ \frac{\partial y_N}{\partial z_1} & \cdots & \frac{\partial y_N}{\partial z_N} \end{bmatrix} = \det(\mathbf{S}) = 1.$$

The unit value of the Jacobian reflects the fact that $\mathbf{S}$ is a unitary matrix.

Next, we use the *multivariate transformation theorem*[1] (see Appendix A) to express the multiple integral in $\mathcal{M}_x(\mathbf{u})$ as follows

$$\begin{aligned} \int_{\mathbb{R}^N} \mathrm{d}\mathbf{y}\, \mathrm{e}^{-\frac{1}{2}\mathbf{y}^\top \mathbf{C}_{xx}^{-1}\mathbf{y}} &= \int_{\mathbb{R}^N} \mathrm{d}\mathbf{z}\, J(\mathbf{y}, \mathbf{z})\, \mathrm{e}^{-\frac{1}{2}\mathbf{z}^\top \mathbf{B}\,\mathbf{z}} \\ &= \prod_{n=1}^{N} \int_{-\infty}^{\infty} \mathrm{d}z_n\, \mathrm{e}^{-\frac{1}{2} b_n z_n^2} = \prod_{n=1}^{N} \left(\frac{2\pi}{b_n}\right)^{1/2} = \frac{(2\pi)^{N/2}}{\det(\mathbf{B})^{1/2}} \\ &= \frac{(2\pi)^{N/2}}{\det(\mathbf{C}_{xx}^{-1})^{1/2}} = (2\pi)^{N/2} \det(\mathbf{C}_{xx})^{1/2}. \end{aligned}$$

The above integral cancels out the partition function Z. Hence, the *moment generating function* is given by

$$\mathcal{M}_x(\mathbf{u}) = \exp\left(\mathbf{u}^\top \mathbf{m}_x + \frac{1}{2}\mathbf{u}^\top \mathbf{C}_{xx}\,\mathbf{u}\right). \tag{6.8}$$

Based on the above and (3.29) that links the moment and cumulant generating functions, the latter is given by

$$\mathcal{K}_x(\mathbf{u}) = \ln \mathcal{M}_x(\mathbf{u}) = \mathbf{u}^\top \mathbf{m}_x + \frac{1}{2}\mathbf{u}^\top \mathbf{C}_{xx}\,\mathbf{u}. \tag{6.9}$$

The first-order partial derivative of $\mathcal{K}_x(\mathbf{u})$ with respect to u_1 is given by

$$\frac{\partial \mathcal{K}_x(\mathbf{u})}{\partial u_1} = m_x(\mathbf{s}_1) + \frac{1}{2} C_{xx}(\mathbf{s}_1, \mathbf{s}_j)u_j + \frac{1}{2} u_j\, C_{xx}(\mathbf{s}_j, \mathbf{s}_1).$$

The above expression uses the *Einstein notation* which implies summation over the repeated indices. Using the symmetry of the covariance function, the following more compact expression is obtained

$$\frac{\partial \mathcal{K}_x(\mathbf{u})}{\partial u_1} = m_x(\mathbf{s}_1) + C_{xx}(\mathbf{s}_1, \mathbf{s}_j)u_j.$$

According to (4.74), the covariance function is given by the second-order derivative of the CGF, that is by

$$\frac{\partial^2 \mathcal{K}_x(\mathbf{u})}{\partial u_1 \partial u_2} = C_{xx}(\mathbf{s}_1, \mathbf{s}_2).$$

---

[1]The theorem essentially says that if we use a new set of integration variables, we need to take account of the Jacobian of the transformation from the old to the new variables.

Comment To obtain the covariance function of a Gaussian vector $\mathbf{X}(\omega)$ it is not necessary to set $\mathbf{u} = \mathbf{0}$ in the second-order derivative of the CGF, as in (4.74). This simplification occurs because the CGF is a quadratic function of $\mathbf{u}$, which guarantees that cumulants of higher than second order vanish.

6.1.3 Two-Dimensional Joint Gaussian pdf

It is difficult to visualize the joint pdf of N-dimensional random vectors, since it is an abstract mathematical object that lives in an N-dimensional space. To gain some insight, we investigate the two-dimensional (bivariate) joint Gaussian pdf of a *statistically homogeneous (stationary)* random field $X(\mathbf{s}; \omega)$.

Let us consider two points $\mathbf{s}_1, \mathbf{s}_2 \in \mathbb{R}^d$ and denote the respective field values by $x, y \in \mathbb{R}$. The two-dimensional joint Gaussian pdf is given by means of the following expression [396]

$$f_{\mathrm{x,y}}(x, y; \mathbf{s}_1, \mathbf{s}_2) = \frac{1}{2\pi\, \sigma_{\mathrm{x}}\sigma_{\mathrm{y}}\sqrt{1-\rho_{1,2}^2}} \exp\left\{ -\frac{1}{2} \frac{A(x, y)}{(1-\rho_{1,2}^2)} \right\}, \tag{6.10a}$$

$$A(x, y) = \left(\frac{x - m_{\mathrm{x}}}{\sigma_{\mathrm{x}}}\right)^2 + \left(\frac{y - m_{\mathrm{y}}}{\sigma_{\mathrm{y}}}\right)^2 - 2\rho_{1,2}\left(\frac{x - m_{\mathrm{x}}}{\sigma_{\mathrm{x}}}\right)\left(\frac{y - m_{\mathrm{y}}}{\sigma_{\mathrm{y}}}\right), \tag{6.10b}$$

where $\rho_{1,2} = \rho_{\mathrm{xx}}(\mathbf{s}_1, \mathbf{s}_2)$ is the value of random field's *correlation function* between the points $\mathbf{s}_1$ and $\mathbf{s}_2$.

Problem 6.1 Show that the expression (6.10) for the bivariate Gaussian pdf follows from the more general N-variate expression (6.5) for the Boltzmann-Gibbs pdf with Gaussian dependence.

Properties of the bivariate Gaussian pdf Using the standardized variables $x' = (x - m_{\mathrm{x}})/\sigma_{\mathrm{x}}$ and $y' = (y - m_{\mathrm{y}})/\sigma_{\mathrm{y}}$, the function $A(x, y)$ transforms into the quadratic form

$$A(x', y') = x'^2 + y'^2 - 2\rho_{1,2}\, x'\, y'.$$

The equation $A(x', y') = c$, where $c \in \mathbb{R}$ determines contours (*isolevel contours*) in the 2D phase space (x', y') along which the two-dimensional Gaussian pdf takes constant values. Due to the functional form of $A(x', y')$ the contours are elliptical. More specifically, the equation $A(x', y') = c$ leads to isolevel contours with the following shape:

1. If $\rho_{1,2} = 0$, the isolevel contours are circles (see Fig. 6.1).
2. For $\rho_{\mathrm{xy}} \neq 0$ the isolevel contours are elliptical. The major semiaxis of the ellipses is equal to $(1 - |\rho_{\mathrm{xy}}|)^{-1}$ whereas the length of the minor semiaxis is $(1 + |\rho_{\mathrm{xy}}|)^{-1}$.

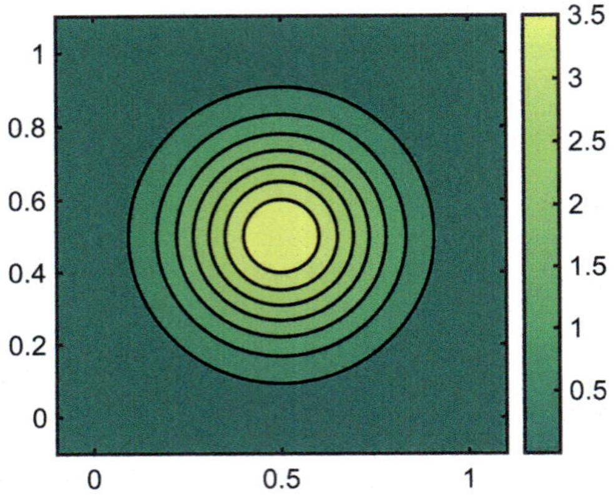

Fig. 6.1 Two-dimensional joint Gaussian pdf with expectation values $m_x = m_y = 0.5$, common standard deviation $\sigma_x = \sigma_y = 0.2$ and correlation coefficient $\rho_{1,2} = 0$

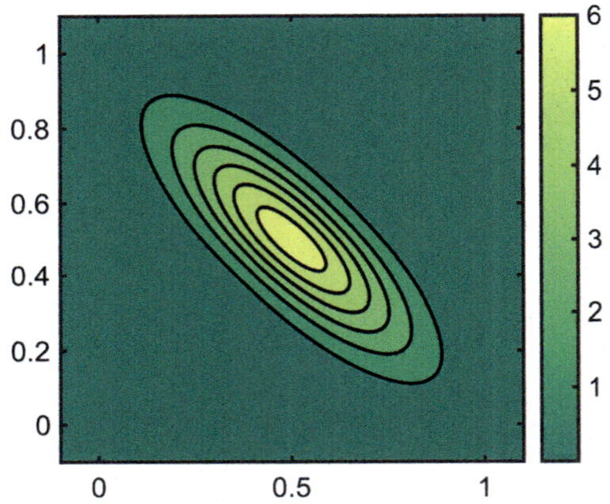

Fig. 6.2 Two-dimensional joint Gaussian pdf with expectation values $m_x = m_y = 0.5$, common standard deviation $\sigma_x = \sigma_y = 0.2$ and correlation coefficient $\rho_{1,2} = -0.8$

3. The angle of rotation of the greater semiaxis of the ellipse with respect to the horizontal is $-45°$ if $\rho_{xy} < 0$ (see Fig. 6.2) and $45°$ if $\rho_{xy} > 0$ (see Fig. 6.3).

Certain useful limits We now investigate the two-dimensional Gaussian pdf at the limits where (i) the correlation coefficient is equal to zero, (ii) it has a small magnitude (absolute value), or (iii) it has an absolute value close to one.

Zero correlation If $\rho_{1,2} = 0$ it follows from (6.10) that

$$f_{x,y}(x, y; \mathbf{s}_1, \mathbf{s}_2) = f_x(x; \mathbf{s}_1)\, f_y(y; \mathbf{s}_2).$$

The equation above implies independence of the field values at $\mathbf{s}_1$ and $\mathbf{s}_2$. This result confirms the well-known property that two jointly Gaussian, uncorrelated random variables are also statistically independent.

Weak correlations In the case of weak correlations $|\rho_{1,2}| \ll 1$. This situation can arise for a random field $X(\mathbf{s}; \omega)$ with finite-range correlations, if a pair of distant

Fig. 6.3 Two-dimensional joint Gaussian pdf with expectation values $m_x = m_y = 0.5$, common standard deviation $\sigma_x = \sigma_y = 0.2$ and correlation coefficient $\rho_{1,2} = 0.8$

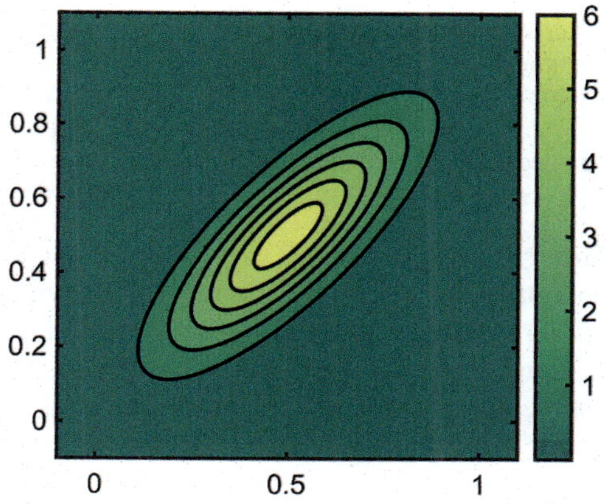

points is considered. We can then use $\rho_{1,2}$ as a small perturbation parameter and expand (6.10) in a Taylor series around $\rho_{1,2} = 0$. Using this assumption, the joint pdf is expressed as follows [396]:

$$f_{x,y}(x, y; \mathbf{s}_1, \mathbf{s}_2) = f_x(x; \mathbf{s}_1)\, f_y(y; \mathbf{s}_2) \left[1 + \rho_{1,2} \left(\frac{x - m_x}{\sigma_x}\right) \left(\frac{y - m_y}{\sigma_y}\right) + O(\rho_{1,2}^2)\right].$$

If we define the *standardized fluctuations*

$$\widetilde{X}(\omega) := \frac{X(\omega) - m_x}{\sigma_x} \quad \text{and} \quad \widetilde{Y}(\omega) := \frac{Y(\omega) - m_y}{\sigma_y},$$

and denote by $\widetilde{x}$, $\widetilde{y}$ their respective values, the *first-order approximation of the bivariate pdf* with respect to $\rho_{1,2}$ is given by

$$f_{x,y}^{(1)}(x, y; \mathbf{s}_1, \mathbf{s}_2) = f_x(x; \mathbf{s}_1)\, f_y(y; \mathbf{s}_2) \left(1 + \rho_{1,2}\widetilde{x}\,\widetilde{y}\right).$$

The approximation $f_{x,y}^{(1)}(x, y; \mathbf{s}_1, \mathbf{s}_2)$ is properly normalized. Based the fact that the expectation of the normalized Gaussian fluctuations is zero, it can be shown that

$$\int_{-\infty}^{\infty} dx \int_{-\infty}^{\infty} dy\, f_{x,y}^{(1)}(x, y; \mathbf{s}_1, \mathbf{s}_2) = 1 + \rho_{1,2}\, \mathcal{E}[\widetilde{X}(\omega)]\, \mathcal{E}[\widetilde{Y}(\omega)] = 1.$$

Warning In spite of the correct normalization, $f_{x,y}^{(1)}(x, y; \mathbf{s}_1, \mathbf{s}_2)$ is not a properly defined pdf, because it can take negative values if $\rho_{1,2}\widetilde{x}\,\widetilde{y} < -1$. If $|\rho_{1,2}| \ll 1$ this condition requires that $\widetilde{x}$ or $\widetilde{y}$ have very large magnitude. Such high deviations from zero imply that the absolute value of $f_{x,y}^{(1)}(x, y; \mathbf{s}_1, \mathbf{s}_2)$ is very low. This occurs because the product $\rho_{1,2}\widetilde{x}\,\widetilde{y}$ is multiplied by $f_x(x; \mathbf{s}_1)$ and $f_y(x; \mathbf{s}_2)$; the latter functions strongly suppress values $\widetilde{x}$ and $\widetilde{y}$ with large magnitudes. Thus, for practical purposes the weak violation of the non-negativity of $f^{(1)}_{x,y}(x, y; \mathbf{s}_1, \mathbf{s}_2)$ may not be important.

Alternatively, we may want to restrict the range of $\widetilde{x}$ and $\widetilde{y}$ such that (i) $\widetilde{x}\,\widetilde{y} \geq -1/\rho_{1,2}$ for $\rho_{1,2} > 0$ or (ii) $\widetilde{x}\,\widetilde{y} \leq 1/\left|\rho_{1,2}\right|$ for $\rho_{1,2} < 0$. This restriction ensures that $f^{(1)}_{x,y}(x, y; \mathbf{s}_1, \mathbf{s}_2)$ is non-negative at the expense of a small departure from normalization.

Strong correlations The strong correlation limit corresponds to $\left|\rho_{1,2}\right| \to 1$. The condition $\rho_{1,2} \to 1$ is realized if x and y are the values of a *correlated random field* at two nearby points. On the other hand, the condition $\rho_{1,2} \to -1$ can be realized if x and y are values of two *anti-correlated random variables*.[2] In both cases, the following approximation holds

$$\lim_{\left|\rho_{1,2}\right| \to 1} f_{x,y}(x, y; \mathbf{s}_1, \mathbf{s}_2) = f_x(x; \mathbf{s}_1)\,\delta\left(\frac{(x - m_x)\,\sigma_y\,\mathrm{sign}(\rho_{1,2})}{\sigma_x} - (y - m_y)\right), \tag{6.11}$$

where $\delta(\cdot)$ is the *Dirac delta function* and the *sign function* is defined as follows:

$$\mathrm{sign}(\rho_{1,2}) = \begin{cases} 1, & \rho_{1,2} > 0, \\ 0, & \rho_{1,2} = 0, \\ -1, & \rho_{1,2} < 0. \end{cases} \tag{6.12}$$

The presence of the delta function in (6.11) means that y is completely determined from x through the equation

$$y = m_y + \mathrm{sign}(\rho_{1,2})\left(\frac{x - m_x}{\sigma_x}\right)\sigma_y,$$

and that the joint pdf vanishes for all other values of y.

We briefly sketch how (6.11) is derived. Let us define for short $a := \sqrt{1 - \rho_{1,2}^2}$ and employ the standardized fluctuations defined in (6.10). By adding and subtracting the term $(x - m_x)^2/\sigma_x^2$ in the exponent, it follows that the joint pdf is given by

$$f_{x,y}(x, y; \mathbf{s}_1, \mathbf{s}_2) = f_x(x; \mathbf{s}_1)\frac{1}{2\sqrt{\pi}\,a\,\sigma_y}\exp\left\{-\frac{1}{2}\frac{(\widetilde{x}\,\rho_{1,2}\,\sigma_y - \widetilde{y})^2}{a^2\,\sigma_y^2}\right\}. \tag{6.13}$$

The limit $\left|\rho_{1,2}\right| \to 1$ implies that $a \to 0$, and thus the Gaussian density function

$$\frac{1}{2\sqrt{\pi}\,a\,\sigma_y}\exp\left\{-\frac{1}{2}\frac{(\widetilde{x}\,\rho_{1,2}\,\sigma_y - \widetilde{y})^2}{a^2\,\sigma_y^2}\right\}$$

[2] As we have seen in (4.18), the autocorrelation function is constrained by $\rho_{xx} \geq -1/d$.

which is centered at $\overline{x}\,\rho_{1,2}\,\sigma_y = \overline{y}$ becomes thinner and taller as $a\,\sigma_y \to 0$. In the limit of an infinitely narrow width ($a \to 0$) the Gaussian tends to the delta function $\delta(\overline{x}\,\rho_{1,2}\,\sigma_y - \overline{y})$.

Remark Note that (6.13) can even be used as an approximation if $a \approx 0$ (that is, for $\rho_{1,2} \approx 1$) but not exactly equal to zero. In addition, both (6.11) and (6.13) hold with x and y interchanged. This means that in the case of strong correlation we can express the joint pdf in terms of either the marginal pdf of $\mathrm{X}(\omega)$ or that of $\mathrm{Y}(\omega)$.

6.1.4 Conditional Probability Density Functions

The conditional pdf of a *Gaussian random field* $\mathrm{X}(\mathbf{s};\omega)$ is useful, because it incorporates information from the available data in the probability distribution of the random field at unmeasured locations. The conditional pdf is also a central concept in Bayesian methods, because it allows updating the probability of different outcomes based on the information made available by new measurements.

Let $\mathbf{x}^*$ represent the $N \times 1$ *data vector* at the set of sampling locations $\mathbb{S}_N = \{\mathbf{s}_1, \ldots, \mathbf{s}_N\}$, and $\mathbf{y}$ the $P \times 1$ *prediction vector* that represents the unknown field values at the set of "prediction" points $\mathbb{G}_P = \{\mathbf{z}_1, \ldots, \mathbf{z}_P\}$, which is typically disjoint from $\mathbb{S}_N$.

The *conditional probability density function* of the random vector $\mathbf{Y}(\omega)$ given the data $\mathbf{x}^*$ is denoted by

$$f_{\mathrm{y}|\mathrm{x}^*}(\mathbf{y}) = \mathcal{N}(\mathbf{m}_{\mathrm{y}|\mathrm{x}^*}, \mathbf{C}_{\mathrm{y}|\mathrm{x}^*}), \tag{6.14}$$

where the following notation is used:

1. $\mathcal{N}(\mathbf{m}_{\mathrm{y}|\mathrm{x}^*}, \mathbf{C}_{\mathrm{y}|\mathrm{x}^*})$ represents the P-dimensional Gaussian pdf with expectation given by the vector $\mathbf{m}_{\mathrm{y}|\mathrm{x}^*}$ and covariance matrix $\mathbf{C}_{\mathrm{y}|\mathrm{x}^*}$.
2. $\mathbf{m}_{\mathrm{y}|\mathrm{x}^*}$ is the *conditional expectation* of $\mathbf{Y}(\omega)$ given the data $\mathbf{x}^*$. It is determined by the equation

$$\underbrace{\mathbf{m}_{\mathrm{y}|\mathrm{x}^*}}_{P\times 1} = \underbrace{\mathbf{m}_{\mathrm{y}}}_{P\times 1} + \underbrace{\mathbf{C}_{\mathrm{y},\mathrm{x}^*}}_{P\times N}\,\underbrace{\mathbf{C}^{-1}_{\mathrm{x}^*,\mathrm{x}^*}}_{N\times N}\,\underbrace{(\mathbf{x}^* - \mathbf{m}_{\mathrm{x}^*})}_{N\times 1}. \tag{6.15}$$

The first term on the right-hand side, $\mathbf{m}_{\mathrm{y}}$, is the unconditional expectation of the random field $\mathrm{X}(\mathbf{s};\omega)$ at the prediction locations. The second term that involves the product of a covariance and a precision matrix and the fluctuation vector $\mathbf{x}^* - \mathbf{m}_{\mathrm{x}^*}$ represents the *conditioning correction* of the mean.
3. $\mathbf{C}_{\mathrm{y}|\mathrm{x}^*}$ is the *conditional covariance matrix*; it is determined by means of

$$\underbrace{\mathbf{C}_{\mathrm{y}|\mathrm{x}^*}}_{P\times P} = \underbrace{\mathbf{C}_{\mathrm{y,y}}}_{P\times P} - \underbrace{\mathbf{C}_{\mathrm{y,x}^*}}_{P\times N}\,\underbrace{\mathbf{C}^{-1}_{\mathrm{x}^*,\mathrm{x}^*}}_{N\times N}\,\underbrace{\mathbf{C}_{\mathrm{x}^*,\mathrm{y}}}_{N\times P}\,. \tag{6.16}$$

In (6.16), $\mathbf{C}_{\mathrm{y,x}^*} = \mathbf{C}^{\top}_{\mathrm{x}^*,\mathrm{y}}$ is the *cross-covariance matrix* between the data and the prediction locations. The matrix $\mathbf{C}_{\mathrm{y,y}}$ is the unconditional covariance of the prediction set. The second term in the right-hand side of (6.16) represents the conditioning correction of the covariance. Conditioning tends to reduce the prediction variance to levels below the unconditional variance.

Single estimation point It is instructive to consider the conditional mean and variance for a single estimation point. In this case, the equations (6.15) and (6.16) change as follows

$$m_{\mathrm{y}|\mathbf{x}^*} = m_{\mathrm{y}} + \mathbf{C}_{\mathrm{y},\mathbf{x}^*}\mathbf{C}^{-1}_{\mathbf{x}^*,\mathbf{x}^*}\,(\mathbf{x}^* - \mathbf{m}_{\mathbf{x}^*}) \tag{6.17a}$$

$$\sigma^2_{\mathrm{y}|\mathbf{x}^*} = \sigma^2_{\mathrm{y}} - \mathbf{C}_{\mathrm{y},\mathbf{x}^*}\,\mathbf{C}^{-1}_{\mathbf{x}^*,\mathbf{x}^*}\,\mathbf{C}_{\mathbf{x}^*,\mathrm{y}}. \tag{6.17b}$$

Example 6.1 Find the conditional joint pdf of the random variable $\mathrm{Y}(\omega)$ based on the measurement x^* of the dependent random variable $\mathrm{X}(\omega)$. Assume that $\mathrm{X}(\omega) \stackrel{d}{=} \mathcal{N}(m_{\mathrm{x}}, \sigma^2_{\mathrm{x}})$, $\mathrm{Y}(\omega) \stackrel{d}{=} \mathcal{N}(m_{\mathrm{y}}, \sigma^2_{\mathrm{y}})$, and that the correlation coefficient between $\mathrm{X}(\omega)$ and $\mathrm{Y}(\omega)$ is equal to ρ.

Answer According to (6.3a), the conditional pdf is given by

$$f_{\mathrm{y}|\mathrm{x}^*}(y) = \mathcal{N}(m_{\mathrm{y}|\mathrm{x}^*}, \sigma^2_{\mathrm{y}|\mathrm{x}^*}),$$

where the conditional expectation and variance are given by the following equations according to (6.17)

$$m_{\mathrm{y}|\mathrm{x}^*} = m_{\mathrm{y}} + \frac{\sigma_{\mathrm{y}}}{\sigma_{\mathrm{x}}}\,\rho(x^* - m_{\mathrm{x}}), \tag{6.18}$$

$$\sigma^2_{\mathrm{y}|\mathrm{x}^*} = \sigma^2_{\mathrm{y}}\,(1 - \rho^2). \tag{6.19}$$

Based on (6.18) the conditional expectation is equal to m_{y} if $\rho = 0$, while it is equal to x^* if $m_{\mathrm{x}} = m_{\mathrm{y}}$, $\sigma_{\mathrm{x}} = \sigma_{\mathrm{y}}$ and $\rho = 1$. Furthermore, according to (6.19), the conditional variance declines with increasing ρ as expected.

Example 6.2 Assume that ore grade follows the Gaussian distribution with mean $m = 4\%$ and standard deviation $\sigma = 0.5\%$. The grade at point A is measured at $c_A = 4.5\%$. In addition, assume that the correlation coefficient between grades at points A and B is $\rho_{A,B} = 0.75$. (i) Estimate the grade at point B. (ii) Determine the uncertainty of the estimate. (iii) Plot the estimate as a function of the ore grade at point A and the correlation coefficient.

Answer We will use the conditional pdf for two variables, i.e. (6.3a) with $x^* \mapsto c_A$ and $y^* \mapsto c_B$. Since both grades represent values from the same random field, $m_{\mathrm{x}} = m_{\mathrm{y}} = m$ and $\sigma_{\mathrm{x}} = \sigma_{\mathrm{y}} = \sigma$.

Given that the grade distribution is Gaussian, the mode of the distribution coincides with the mean value. Hence, the *optimal estimate* of the ore grade at B is equal to the conditional expectation of the Gaussian pdf. Based on (6.18) we obtain

$$\hat{c}_{B|A} = m + \rho_{A,B}\,(c_A - m) = m\,(1 - \rho_{A,B}) + c_A\,\rho_{A,B} = 4.375\%. \qquad (6.20)$$

The *uncertainty of the estimate* is determined by the conditional standard deviation, which in light of (6.19) is given by

$$\sigma_{B|A} = \sigma\,\sqrt{1 - \rho^2_{A,B}} = 0.5\,\sqrt{1 - 0.75^2}\% = 0.33\%. \qquad (6.21)$$

The conditional expectation at B exceeds the unconditional expectation as does the measurement at A. On the other hand, the standard deviation at B is less than the unconditional standard deviation. Both results are due to the positive correlation between A and B, which implies that the measurement at A "attracts" the estimate at B closer to its value, and at the same it reduces the uncertainty of the value at B. Note that the conditional standard deviation depends only on the absolute value of the correlation coefficient. Thus, the uncertainty would have the same value even if $\rho_{AB} = -0.75$.

The dependence of the ore grade estimate at point B as a function of c_A and $\rho_{A,B}$ is the saddle-shaped surface shown in Fig. 6.4. According to this graph, the concentration at B increases linearly with the concentration at A if $\rho_{A,B} > 0$. Reversely, the concentration at B decreases linearly with increasing concentration at A if $\rho_{A,B} < 0$.

Breaking stationarity by conditioning Conditioning a stationary random field on available data leads to breaking of the stationarity as evidenced in equations (6.17) for the conditional mean and variance. To illustrate this effect, let us consider a second-order stationary random field $\mathrm{X}(\mathbf{s};\omega)$ with mean m_{x} and covariance function $C_{\mathrm{xx}}(\mathbf{r})$. For the sake of simplicity we assume that $\mathrm{X}(\mathbf{s};\omega)$ is sampled at the point $\mathbf{s}$ where it takes the value x^*.

According to (6.18), the *conditional expectation* of the field at two "prediction" points $\mathbf{z}_1$ and $\mathbf{z}_2$ given the value at point $\mathbf{s}_0$ is given by

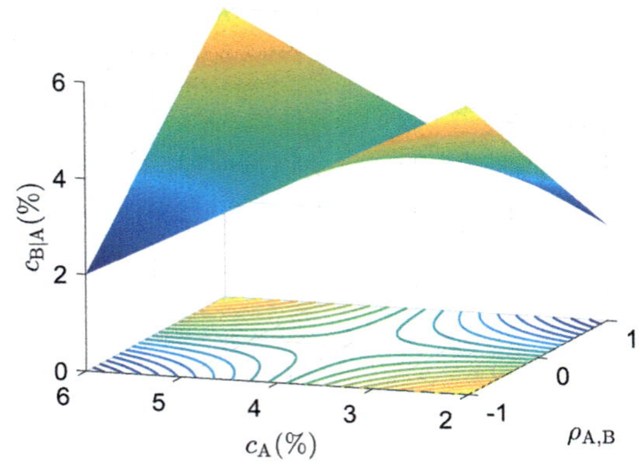

Fig. 6.4 Conditional expectation, $C_{B|A}$, at point B versus the value c_A at point A and the correlation coefficient $\rho_{A,B}$. The marginal expectation and the standard deviation at both points are respectively given by $m_x = m_y = 4\%$, and $\sigma_x = \sigma_y = 0.5\%$. The contour lines in the horizontal plane represent grade isolevel contours

$$m_{x|x^*}(\mathbf{z}_1) = m_x + \rho_{xx}(\mathbf{z}_1 - \mathbf{s})(x^* - m_x),$$

$$m_{x|x^*}(\mathbf{z}_2) = m_x + \rho_{xx}(\mathbf{z}_2 - \mathbf{s})(x^* - m_x).$$

Hence, it follows from the above that in general $m_{x|x^*}(\mathbf{z}_1) \neq m_{x|x^*}(\mathbf{z}_2)$ unless $\mathbf{z}_1 = \mathbf{z}_2$ or $\mathbf{z}_1 - \mathbf{s} = \mathbf{s} - \mathbf{z}_2$. Actually, if the field is isotropic the conditional means are equal for all $\mathbf{z}_1$ and $\mathbf{z}_2$ such that $\|\mathbf{z}_1 - \mathbf{s}\| = \|\mathbf{z}_2 - \mathbf{s}\|$.

Let us also apply (6.16) to calculate the *conditional covariance*. For short, we use $C_{xx}(1, 2)$ to denote $C_{xx}(\mathbf{z}_1 - \mathbf{z}_2)$ and $C_{xx}(i, 0)$ to denote $C_{xx}(\mathbf{s}_0 - \mathbf{z}_i)$, where $i = 1, 2$. Since our example involves a single conditioning point, the inverse data covariance is just the reciprocal of the variance, i.e.,

$$\mathbf{C}^{-1}_{x^*,x^*} = 1/\sigma_x^2,$$

while the *cross-covariance vector* between the data point and the prediction points is given by

$$\mathbf{C}^{\top}_{x^*,x} = [C_{xx}(1, 0) \quad C_{xx}(2, 0)].$$

Then, the *conditional covariance matrix* is given by

$$\begin{bmatrix} C_{x|x^*}(1,1) & C_{x|x^*}(1,2) \\ C_{x|x^*}(2,1) & C_{x|x^*}(2,2) \end{bmatrix} = \begin{bmatrix} \sigma_x^2 & C_{xx}(1,2) \\ C_{xx}(2,1) & \sigma_x^2 \end{bmatrix} - \begin{bmatrix} \rho_{xx}(1,0) \\ \rho_{xx}(2,0) \end{bmatrix} \begin{bmatrix} C_{xx}(1,0) \\ C_{xx}(2,0) \end{bmatrix}^{\top}$$

$$= \sigma_x^2 \begin{bmatrix} a_{1,1} & a_{1,2} \\ a_{2,1} & a_{2,2} \end{bmatrix}, \tag{6.22a}$$

where the matrix elements $a_{i,j}$ $(i, j = 1, 2)$ are given by

$$a_{i,i} = 1 - \rho_{xx}^2(i, 0), \quad \text{for } i = 1, 2, \tag{6.22b}$$

and

$$a_{1,2} = a_{2,1} = \rho_{xx}(1, 2) - \rho_{xx}(1, 0)\,\rho_{xx}(2, 0). \tag{6.22c}$$

Since the diagonal elements of the conditional covariance matrix $C_{x|x^*}(i, i)$ represent the variance at the points $\mathbf{z}_i$, it follows from (6.22b) that the conditional variance also depends on the location. This result also confirms that once measurements are taken into account, the stationarity is broken.

Broken symmetry In physics, the term "broken symmetry" refers to the appearance of a new state(s) with lower symmetry than the microscopic equations of the system [23, Chap. 2], [478, Chap. 26]. Symmetry breaking is associated with the following mechanisms:

- An external factor that generates a bias for a particular configuration (e.g., a magnetic field that polarizes all the spins of a magnetic material in one direction).
- A *phase transition* controlled by temperature such as the cooling of a liquid that leads to the emergence of a solid state. Solidification implies loss of the continuous translation symmetry of the liquid state (liquids look the same at every point and direction in space, while solids are characterized by more restrictive discrete crystal symmetries).
- Spontaneous organization in complex systems that leads to *emergent phenomena* (e.g., self-organized states). In such systems symmetry breaking often leads to the emergence of new phases [42].

The breaking of stationarity by conditioning does not require drastic changes of the system's state or the tuning of some control variable. Instead, it simply emerges as the result of measurements that destroy stationarity by "forcing" the values of the process observation points. Thus, the system loses its freedom to occupy an infinity of potential states at the sampling points.

Collapse of the probability density function Let us view stationarity breaking from a different perspective: For stationary random fields every point in space can take any of the values permitted by the pdf of the field. Once a measurement is made, however, that point is restricted to a unique value—within experimental error.

There is a close parallel to this effect in the theory of quantum mechanics, where the wavefunction[3] determines the probability that a particle occupies a given state. Typically, the wavefunction comprises a superposition of different eigenstates of the system. Nevertheless, once the particle is detected (i.e., registered by some measurement apparatus), its state is completely determined—at least within the

[3]More precisely, the real-valued amplitude of the complex-valued wavefunction.

bounds specified by *Heisenberg's uncertainty principle*. Hence, the other states originally included in the wavefunction are of no consequence any longer. Physicists refer to this phenomenon as the *wavefunction collapse*.

Similarly, for spatial random fields the act of measurement leads to a "collapse" of the pdf locally: At the measurement point the pdf is replaced by a delta function (assuming that the experimental uncertainty can be neglected.) The pdf continues to have non-zero width in the neighborhood of the measurement, but its dispersion is reduced. The increase of the information at the measurement point spreads in its vicinity, altering locally the statistical properties of the random field. Hence, the collapse of the pdf breaks the stationarity of the field.

The screening effect It has been observed in geostatistical applications that certain sample points may have almost no impact on the prediction (or even contribute with negative weight), compared to points that are closer to the prediction site $\mathbf{z}$. In general, if $\mathbf{s}_0$ is close to $\mathbf{z}$, its impact can overshadow the influence of other points $\mathbf{s}$, such that $\|\mathbf{s}_0 - \mathbf{z}\| < \|\mathbf{s} - \mathbf{z}\|$. This phenomenon is known as the screening effect [132, 776].

At the heart of the screening effect is the behavior of the *conditional covariance*. The main idea is that two spatially separated sites which are strongly correlated may become only weakly correlated after conditioning, if the conditioning is based on points neighboring one of the sites. Essentially, the measurements in the neighborhood of that site act as a shield for the correlations between the two sites. This effect is related to the concept of *conditional independence* which is discussed in Chap. 8.

Conditional covariance matrix Assume a weakly-stationary random field $X(\mathbf{s}; \omega)$ with covariance structure that is given by:

1. An exponential model $C_{xx}(\mathbf{r}) = \exp(-\|\mathbf{r}\|/\xi)$.
2. A Gaussian model $C_{xx}(\mathbf{r}) = \exp(-\|\mathbf{r}\|^2/\xi^2)$.

Let us consider a one-dimensional spatial configuration of three collinear points with coordinates s_1, s_2 and s_0 such that $s_1 \leq s_0 \leq s_2$. We assume that s_0 is the conditioning point.

We will calculate the *conditional covariance* $c^*_{1,2}$, between two points s_1 and s_2 based the value $x^*_0 = x^*(\mathbf{s}_0)$ at the point $\mathbf{s}_0$.

Using the conditional covariance matrix expression (6.22) (with the change of variables $\mathbf{z}_1 \mapsto s_1$, $\mathbf{z}_2 \mapsto s_2$, and $\mathbf{s} \mapsto s_0$), the *conditional covariance* $c^*_{1,2}$ between points s_1 and s_2 is given by

$$c^*_{1,2} = c_{1,2} - \frac{1}{\sigma_x^2} c_{1,0}\, c_{2,0},$$

where $c_{1,2} = C_{xx}(s_1, s_2)$ is the unconditional covariance between s_1 and s_2, and $c_{i,0} = C_{xx}(s_i, s_0)$ is the unconditional covariance between s_i $(i = 1, 2)$ and the conditioning point s_0.

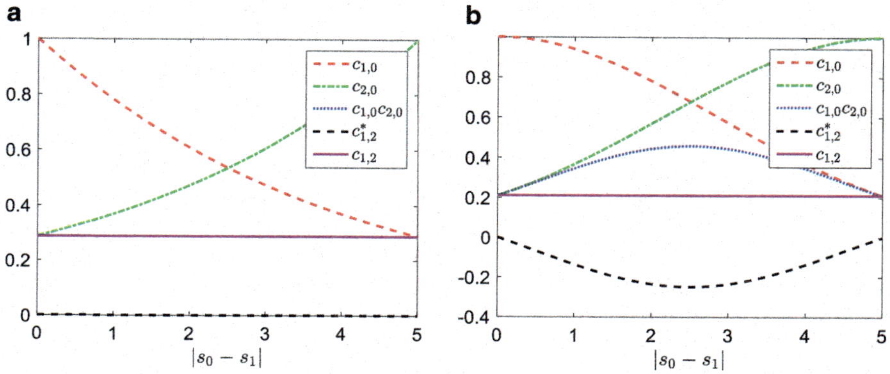

Fig. 6.5 Conditional covariance between the points $s_1 = 0$ and $s_2 = 5$ based on a measurement at the interior point s_0. (**a**) Exponential covariance model. (**b**) Gaussian covariance model. The interior point is allowed to move, and the horizontal axis represents the distance $|s_0 - s_1|$. The conditional covariance between s_1 and s_2 based on the measurement at s_0 is denoted by $c^*_{1,2}$

The conditional covariance $c^*_{1,2}$ is plotted in Fig. 6.5 versus the distance $|s_0 - s_1|$ for both covariance models. The screening effect leads to a vanishing of the conditional covariance for the *exponential model* ($c^*_{1,2}$ is a straight line coinciding with the horizontal axis). This is due to the fact that the two terms in the equation above cancel out. On the other hand, in the case of the *Gaussian model*, the screening effect leads to negative values of the conditional covariance $c^*_{1,2}$ even though $c_{1,2} > 0$.

Hint: The reader can easily verify that $|s_1 - s_0| + |s_0 - s_2| = |s_1 - s_2|$ if $s_1 \leq s_0 \leq s_2$. Hence, in the case of the exponential model $c_{1,2} = c_{1,0}\, c_{2,0}/\sigma_x^2$ thus leading to a vanishing conditional covariance. However, if the three points are not collinear, the identity involving the absolute values is replaced by the triangle inequality, i.e., $\|\mathbf{s}_1 - \mathbf{s}_0\| + \|\mathbf{s}_0 - \mathbf{s}_2\| \geq \|\mathbf{s}_1 - \mathbf{s}_2\|$.

An island metaphor for screening Anna and Stella live in Chora, on an island far away, and they both enjoy swimming. Every summer day the two young women walk to the sandy beach of Chora for a swim and socializing. Their summer daily activities are definitely highly correlated. Until one day the local paper publishes laboratory test results showing that the water in Chora beach has a rather high concentration of bacteria. Anna and Stella have to change their routine. Since they live in diametrically opposite ends of the town, the nearest clean beaches accessible to each of them are in opposite directions and farther away from their homes than Chora beach. Thus, Anna starts going to Milia beach while Stella starts frequenting Rodakino beach. As a result of the "conditioning" imposed by the water test results, their once highly correlated summer activities became decorrelated.

The role of the screening effect in kriging on regular lattices has been studied by Stein. He obtained sufficient conditions that should be satisfied by the spectral density for the screening effect to occur [774, 776] (see also Chap. 10). The screening effect has also been used to enable the efficient approximation of Gaussian processes [46, 440]. Iterative conditioning of Gaussian processes based on the screening effect is used to derive *sparse Cholesky decompositions* of covariance matrices [718].

6.2 Field Integral Formulation

The expression (6.3a) for the Gaussian joint pdf assumes that the number of points under consideration is either finite or at least countable (e.g., the nodes of an infinite regular grid). There are cases, however, in which continuum random field representations are natural. After all, many processes of interest take place in continuum spaces, even if we construct discrete models for modeling and computational convenience. A question of interest is how to calculate the sum, or the average, of a random field functional over all states (i.e., sample functions) in such spaces.

Making the transition to the continuum requires certain formalistic changes with respect to the discrete case. We know that at the continuum limit sums are replaced by integrals. In addition, the vector state $\mathbf{x}$ is replaced by the state function or field $x(\mathbf{s})$, which represents a vector in an infinite-dimensional *Hilbert space*. Hence, an integral over all possible values of a random vector used in (6.5c) should be replaced by a respective integral over all possible values of the function $x(\mathbf{s})$.

To make this discussion more concrete, consider the partition function given by (6.5c). We would like to calculate the partition function at the limit $(x_1, \ldots x_N)^\top \rightarrow x(\mathbf{s})$ and $\mathcal{H}_0(\mathbf{x}) \rightarrow \mathcal{H}_0\left[x(\mathbf{s})\right]$, where $\mathcal{H}_0\left[x(\mathbf{s})\right]$ is a functional of the field. The mathematical formalism used to handle this type of problems is the calculus of *field integrals* or *functional integrals*.

In physics, the related concept of path integrals was spearheaded by Richard Feynman in his formulation of quantum mechanics [246]. Applications of path integrals in quantum mechanical problems are described in several texts on quantum field theory, e.g. [5, 890, 892]. A more comprehensive view of path integral applications is given by Kleinert [465].

A functional integral formulation of classical statistical mechanics was presented in [665]. Lemm applies the functional integral formalism to signal processing and image reconstruction using a Bayesian perspective [495]. A concise introduction to path integrals and functional integrals is given by Zinn-Justin [891]. An application of path integrals to Bayesian inference for inverse problems is presented in [126]. Furthermore, another recent publication investigates the numerical approximation of functional integrals and functional differential equations [817].

Functional integral formulation of Boltzmann-Gibbs random fields Let us now revisit the Boltzmann-Gibbs representation of the Gaussian joint density (6.5). The energy function (6.5b) for a continuum field should be replaced by the following double integral over the space $\mathbb{R}^d \times \mathbb{R}^d$ [65, 396]

$$\mathcal{H}_0[x(\mathbf{s});\boldsymbol{\theta}] := \frac{1}{2}\int_{\mathcal{D}} d\mathbf{s} \int_{\mathcal{D}} d\mathbf{s}'\, x'(\mathbf{s})\, C_{\mathrm{xx}}^{-1}(\mathbf{s},\mathbf{s}')\, x'(\mathbf{s}'), \tag{6.23}$$

where $x'(\mathbf{s}) := x(\mathbf{s}) - m_{\mathrm{x}}(\mathbf{s})$ represents the fluctuation field. We use the notation $\mathcal{H}_0[x(\mathbf{s});\boldsymbol{\theta}]$ to denote that $\mathcal{H}_0$ is a functional of the state function $x(\mathbf{s})$, and thus it involves the values $x(\mathbf{s})$ at every $\mathbf{s} \in \mathcal{D}$. The vector $\boldsymbol{\theta}$ represents the parameters of the energy function. The dependence on $\boldsymbol{\theta}$ is not always shown for notational brevity.

Inverse covariance operator In equation (6.23) above, $C_{\mathrm{xx}}^{-1}(\mathbf{s},\mathbf{s}')$ is the *precision operator* also known as the inverse covariance operator. This satisfies the continuum extension of the precision matrix definition (6.4) which leads to the following integral equation

$$\int_{\mathcal{D}} d\mathbf{z}\, C_{\mathrm{xx}}^{-1}(\mathbf{z},\mathbf{s}')\, C_{\mathrm{xx}}(\mathbf{s},\mathbf{z}) = \delta(\mathbf{s}-\mathbf{s}'). \tag{6.24a}$$

If we assume that the random field $X(\mathbf{s};\omega)$ is statistically stationary, so that the covariance function depends only on the lag, (6.24) is replaced by

$$\int_{\mathcal{D}} d\mathbf{z}\, C_{\mathrm{xx}}^{-1}(\mathbf{z}-\mathbf{s}')\, C_{\mathrm{xx}}(\mathbf{s}-\mathbf{z}) = \delta(\mathbf{s}-\mathbf{s}'). \tag{6.24b}$$

Inverse spectral density At this point, we can use the Fourier transform of the covariance function, i.e., the spectral density, to put this equation in the spectral domain. It is left as an exercise for the reader to show that in terms of the Fourier transforms $\widetilde{C}_{\mathrm{xx}}(\mathbf{k}) := \mathcal{F}[C_{\mathrm{xx}}(\mathbf{r})]$ and $\widetilde{C_{\mathrm{xx}}^{-1}}(\mathbf{k}) := \mathcal{F}[C_{\mathrm{xx}}^{-1}(\mathbf{r})]$ the inverse spectral density satisfies the equation

$$\widetilde{C}_{\mathrm{xx}}(\mathbf{k})\, \widetilde{C_{\mathrm{xx}}^{-1}}(\mathbf{k}) = 1. \tag{6.25}$$

Note that $\widetilde{C_{\mathrm{xx}}^{-1}}(\mathbf{k})$ may represent the Fourier transform of a generalized function, since $\widetilde{C_{\mathrm{xx}}^{-1}}(\mathbf{k})$ is not always an absolutely integrable function. More details on the inverse of continuum covariance functions are given in [625].

The partition function So far so good. However, we still have to address the calculation of the partition function. Recall than for a countable set of points the partition function is given by the multiple integral (6.5c), i.e.,

$$Z = \prod_{n=1}^{N} \int_{-\infty}^{\infty} dx_n\, e^{-\mathcal{H}_0}.$$

What happens if we consider a continuum field $x(\mathbf{s})$ instead of the vector $\mathbf{x}$? To fully describe the field state $x(\mathbf{s})$ we need an infinity of points and, consequently, an infinite number of integrals. We could then envisage the following limiting procedure for the partition function at the continuum limit

$$Z = \lim_{N\to\infty} \left[\prod_{n=1}^{N} \int_{-\infty}^{\infty} \mathrm{d}x_n \right] \mathrm{e}^{-\mathcal{H}_0}. \tag{6.26}$$

We will denote the asymptotic limit of this multiple integral by means of the following notation

$$Z = \int \mathcal{D}x(\mathbf{s})\, \mathrm{e}^{-\mathcal{H}_0}, \tag{6.27}$$

where the symbol $\int \mathcal{D}x(\mathbf{s})$ stands for the *field integral.*

The continuum limit (in pursuit of infinity) The Gaussian partition function for finite N is given by (6.6). Note that as $N \to \infty$ the term $(2\pi)^{N/2}$ diverges. In addition, we need to calculate the limit of the covariance determinant, which is not in general amenable to explicit calculation.

To formally extend the result (6.6) for the Gaussian partition function to $N \to \infty$, we can use the *determinant trace identity*

$$\ln \det \mathbf{C}_{\mathrm{xx}} = \mathrm{Tr}(\ln \mathbf{C}_{\mathrm{xx}}). \tag{6.28}$$

If $\mathbf{A}$ is $N \times N$ square matrix, the symbol "Tr" refers to the *trace* of the matrix which is defined by $\mathrm{Tr}(\mathbf{A}) = \sum_{n=1}^{N} A_{n,n}$. Then, the partition function is expressed as follows

$$Z = (2\pi)^{N/2} \exp\left[\frac{1}{2}\, \mathrm{Tr}\,(\ln \mathbf{C}_{\mathrm{xx}}) \right]. \tag{6.29}$$

In general, it is preferred to calculate the logarithm of the partition function which is given by

$$\ln Z = \frac{N}{2}\, \ln(2\pi) + \frac{1}{2}\, \ln \det(\mathbf{C}_{\mathrm{xx}}) = \frac{N}{2}\, \ln(2\pi\, \sigma_{\mathrm{x}}^2) + \frac{1}{2} \sum_{n=1}^{N} \ln \lambda_n. \tag{6.30}$$

Equation (6.30) is derived by writing $\mathbf{C}_{\mathrm{xx}} = \sigma_{\mathrm{x}}^2\, \boldsymbol{\rho}$ where $\boldsymbol{\rho}$ is the *correlation matrix* and using the fact that $\det(\boldsymbol{\rho}) = \prod_{n=1}^{N} \lambda_n$, where λ_n, $n = 1, \ldots, N$ are the *eigenvalues of the correlation matrix.*

To gain some insight into the behavior of Z, let us consider a one-dimensional Gaussian random field with exponential covariance function $C_{\mathrm{xx}}(\mathbf{r}) = \sigma_{\mathrm{x}}^2 \exp(-|r|/\xi)$. Let us assume that the random field is defined over the line segment $[0, L]$, where $L = 300$ and sampled at N uniformly distributed locations over this interval, i.e., $L = N\alpha$, where α is the sampling step.

The elements of the respective correlation matrix are given by the equation

$$\rho_{\mathrm{xx}}(n, m) := \exp(-\left|s_n - s_m\right| / \xi) = \widetilde{\rho}_{\alpha}^{\,|n-m|},$$

where $\widetilde{\rho}_{\alpha} = \exp(-\alpha/\xi)$. The $N \times N$ matrix $\boldsymbol{\rho}$ is a *Toeplitz matrix* because $\rho_{n,m} = \rho_{|n-m|}$, for all $n, m = 1, \ldots, N$

In Fig. 6.6 we plot the dependence of $\ln Z$ on the number of sampling points N, where N increases from 100 to 1000 by 100. We consider three different ratios of L/ξ. As the three graphs in Fig. 6.6 show, the ratio L/ξ plays a key role in the dependence of $\ln Z$: for $L/\xi = 3000$ (i.e., for small correlation length) the dependence on N is linear; for $L/\xi = 20$ we observe a sublinear

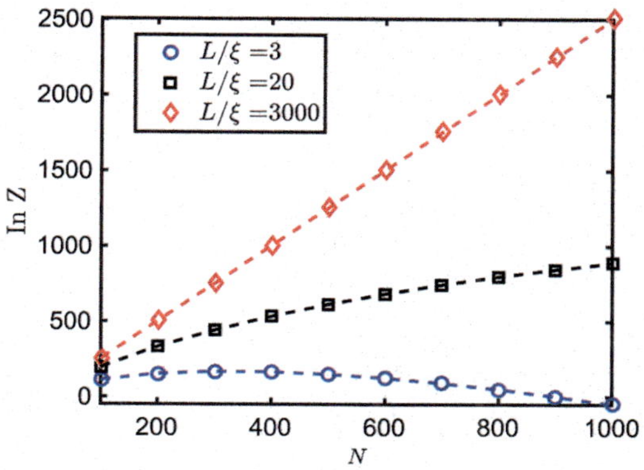

Fig. 6.6 Dependence of the logarithm of the partition function (6.30), on the number of sampling points N, for a one-dimensional Gaussian random field with exponential covariance function $C_{xx}(\mathbf{r}) = \sigma_x^2 \exp(-|r|/\xi)$ with $\sigma_x = 2$. The three graphs correspond to different ratios of domain to correlation length: $L/\xi = 3, 20, 3000$

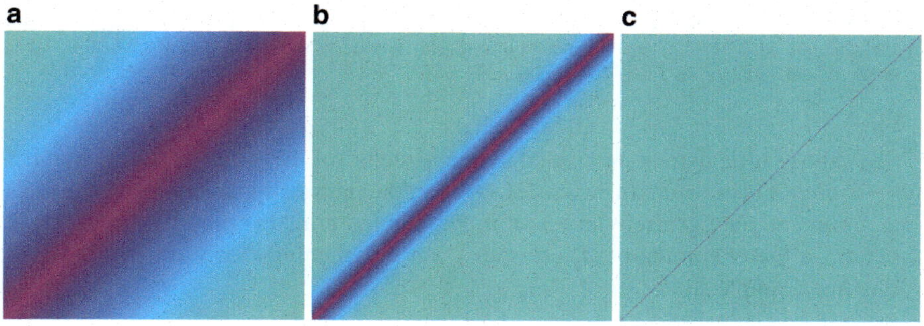

Fig. 6.7 Toeplitz correlation matrices obtained for 1000 uniformly spaced locations in the interval [0, 300]. The correlation matrices are generated by an exponential covariance function and correspond to three different values of the correlation scale ξ. The matrix entries decrease from one along the left diagonal to zero as we move away from the diagonal. (**a**) $L/\xi = 3$. (**b**) $L/\xi = 20$. (**c**) $L/\xi = 3000$

monotonic increase, whereas for $L/\xi = 3$, i.e., for $\xi \sim O(L)$, the initial increasing trend is even slower and then turns to a declining trend at higher N.

The respective covariance matrices for $N = 1000$ are shown in Fig. 6.7. The matrix in Fig. 6.7c in particular (i.e., for $\xi \ll L$) has a narrow diagonal band of nonzero elements. This reflects the fact that the random field is essentially a white noise process due to the very short correlation scale. In this case most of the eigenvalues are close to unity and their logarithms close to zero. Hence, the logarithm of the partition function is approximately given by

$$\ln Z \approx \frac{N}{2} \ln(2\pi\, \sigma_x^2),$$

which accounts for the linear increase of $\ln Z$ with N in Fig. 6.6. For the longer correlation lengths, the increase of $\ln Z$ with N is dampened or even reversed as N increases. To understand this behavior, we plot the logarithm of the eigenvalues of the respective correlation matrices in Fig. 6.8. These graphs show an increasing number of eigenvalues with negative logarithms as N increases, which slows down the increase of $\ln Z$ with N and then leads to a drop for $L/\xi = 3$.

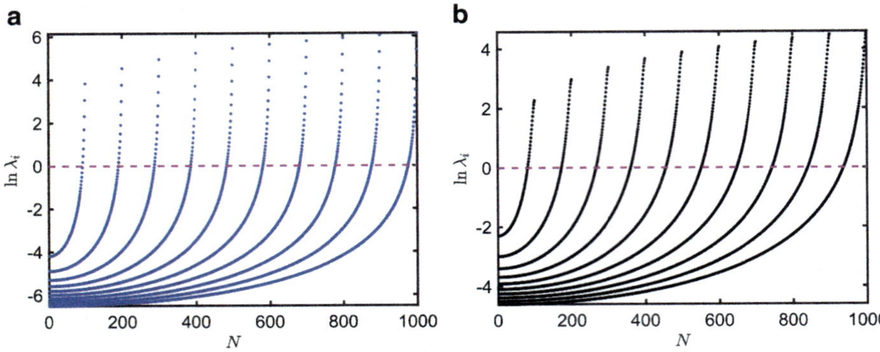

Fig. 6.8 Natural logarithms of the correlation matrix eigenvalues that correspond to the covariance matrices in plots (a) and (b) of Fig. 6.7. The points on each of the 10 curves represent the eigenvalues of an $N \times N$ exponential correlation matrix, where N takes values from 100 to 1000 in steps of 100. (**a**) $L/\xi = 3$. (**b**) $L/\xi = 20$

The above little detour shows that the dependence of the partition function in the $N \to \infty$ limit is not straightforward. However, the physically meaningful quantities (e.g., moments) are given by ratios of functional integrals that involve the partition function in the denominator. These ratios can be unambiguously evaluated as the following example shows.

Functional expectation Consider the expectation of the following exponential functional

$$\exp\left[\int d\mathbf{s}\, X(\mathbf{s};\omega)\, g(\mathbf{s})\right],$$

over the Boltzmann-Gibbs pdf with the Gaussian energy functional $\mathcal{H}_0$, as defined by (6.23). The expectation of the exponential function is given by the following ratio of functional integrals

$$F[g(\mathbf{s})] = \mathcal{E}\left[e^{\int d\mathbf{s}\, X(\mathbf{s};\omega)\, g(\mathbf{s})}\right] = \frac{\int \mathcal{D}x(\mathbf{s})\, e^{-\mathcal{H}_0 + \int d\mathbf{s}\, x(\mathbf{s})\, g(\mathbf{s})}}{\int \mathcal{D}x(\mathbf{s})\, e^{-\mathcal{H}_0}}.$$

The denominator represents the partition function of the Boltzmann-Gibbs pdf for the Gaussian random field $X(\mathbf{s};\omega)$. If an explicit expression is available for the numerator, the partition function can be obtained from the latter by setting $g(\mathbf{s}) = 0$.

To calculate the numerator we replace the functional integral with the following integral over an N-dimensional state vector $\mathbf{x}$:

$$\Phi(\mathbf{g}) := \int_{\mathbb{R}^N} d\mathbf{x}\, e^{-\frac{1}{2}\mathbf{x}^\top \mathbf{C}_{xx}^{-1}\mathbf{x} + \mathbf{g}\cdot\mathbf{x}} \tag{6.31a}$$

The above is a *multivariate Gaussian integral* with a linear term in the exponent. This integral can be evaluated by completing the square in the exponent leading to

$$\Phi(\mathbf{g}) = (2\pi)^{N/2} \det(\mathbf{C}_{xx})^{1/2} \, e^{\frac{1}{2}\mathbf{g}^\top \mathbf{C}_{xx}\, \mathbf{g}}. \tag{6.31b}$$

Based on the above, it follows that

$$F[\mathbf{g}] = \frac{\Phi(\mathbf{g})}{\Phi(0)} = e^{\frac{1}{2}\mathbf{g}^\top \mathbf{C}_{xx}\, \mathbf{g}}.$$

Hence, the value of the partition function $Z = \Phi(0)$ is not necessary to obtain the expectation, since Z cancels an equal factor in the numerator. At the limit $N \to \infty$, the above expression gives the ratio of the functional integrals as follows

$$F[g(\mathbf{s})] = e^{\frac{1}{2}\int \mathrm{d}\mathbf{s} \int \mathrm{d}\mathbf{s}'\, g(\mathbf{s})\, C_{xx}(\mathbf{s},\mathbf{s}')\, g(\mathbf{s}')}.$$

Orthogonal expansions Another approach for calculating functional integrals is based on orthogonal expansions of the random field in terms of a suitable basis that comprises a countable set of basis functions. One such possibility is the optimal bi-orthogonal Karhunen-Loève expansion that involves an infinite but countable number of terms (see Sect. 16.11). The Karhunen-Loève expansion can be truncated after a finite number of terms. The cutoff can be determined by the desired accuracy in the approximation of the variance.

Then, the functional integral over all states can be replaced by a finite number of integrals over the random coefficients of the orthogonal expansion. Other dimensionality reduction ideas can also be used to approximate the functional integral in terms of a finite set of integrals.

6.2.1 A Detour in Functional Derivatives

The following is a brief exposition of the formalism used for calculating statistical moments based on the functional integral formulation. The calculations are based on the concept of the *functional derivative*, also known as *variational derivative*. A pedagogic introduction to functional derivatives is given in [478].

The main idea Let us consider a function $\phi(\mathbf{s})$, where $\mathbf{s} \in \mathcal{D} \mapsto \mathbb{R}$. A *functional* is a mapping from the space of functions to the space of real numbers. For example, the functional $I[\phi]$ could represent the integral $I[\phi] = \int_a^b \mathrm{d}x\, \phi^2(x)$. Another possible functional is the value of the function itself at a given point, i.e., $\phi(\mathbf{s})$.

The functional derivative represents the rate of change of a given functional with respect to a function that is involved in the functional relation. For example, for the integral functional $I[\phi]$ defined above, $\delta I[\phi]/\delta\phi(x')$ represents how fast $I[\phi]$

changes with respect to a change of $\phi(\cdot)$ at the point x'. The functional derivative of $I[\phi]$ with respect to a function $g(\cdot)$ that is not involved in $I[\phi]$ is zero, since changes in $g(\cdot)$ do not affect $I[\phi]$.

Functional derivative of a function Let us consider the derivative of a function $\phi(\mathbf{s})$ with respect to $\phi(\mathbf{s}')$, where the points $\mathbf{s}$ and $\mathbf{s}'$ may (or not) be different. The functional derivative $\delta\phi(\mathbf{s})/\delta\phi(\mathbf{s}')$ is controlled by the function values inside a very local neighborhood around $\mathbf{s}'$, the radius of which tends to zero. Hence, the functional derivative $\delta\phi(\mathbf{s})/\delta\phi(\mathbf{s}')$ should be zero if $\mathbf{s} \neq \mathbf{s}'$, whereas it may take non-zero values only if $\mathbf{s} = \mathbf{s}'$. What should the value of the derivative be at $\mathbf{s} = \mathbf{s}'$?

Let us first assume that the function is only defined at the nodes of the lattice $\mathcal{G}$, so that $\mathbf{s} \to \mathbf{s}_n$ and $\mathbf{s}' \to \mathbf{s}_m$, where n, m are lattice indices. Then, the functional derivative becomes $\delta\phi_n/\delta\phi_m = \delta_{n,m}$, where $\delta_{n,m}$ is the *Kronecker delta*. Hence, the sum over the functional derivatives at all points $\mathbf{s}_m \in \mathcal{G}$ is equal to one. If we take the continuum limit, the sum is replaced by an integral, and the Kronecker delta $\delta_{n,m}$ is replaced by the *Dirac delta function* $\delta(\mathbf{s} - \mathbf{s}')$.[4]

Let us now put the above qualitative comments in a more formal framework. We assume that the functionals studied are sufficiently smooth that we can define the respective functional derivatives as needed. The interested reader can find more information in mathematical texts on functional analysis, while simplified presentations are given in physics textbooks [122, 297, 478].

Functional derivative definition If $\Phi[x(\mathbf{s})]$ is a functional of the function (possibly a random field realization) $x(\mathbf{s})$, the functional derivative of $\Phi[x(\mathbf{s})]$ with respect to $x(\mathbf{s})$ is given by

$$\frac{\delta\Phi[x(\mathbf{s})]}{\delta x(\mathbf{z})} = \lim_{\epsilon\to 0} \frac{\Phi\left[x(\mathbf{s}) + \epsilon\delta(\mathbf{s}-\mathbf{z})\right] - \Phi\left[x(\mathbf{s})\right]}{\epsilon} \tag{6.32}$$

The definition (6.32) can be iterated to obtain functional derivatives of higher orders. The definition (6.32) is similar to that of the *variational derivative*, the main difference being that in the latter the delta function is replaced with a function $\eta(\mathbf{s})$.

The functional derivative shares a number of properties with the classical derivatives, such as the chain rule and its application in series expansions.

Functional derivative of a function The derivative of the function $x(\mathbf{s})$ at point $\mathbf{s}$ with respect to a change in the value of the function at point $\mathbf{z}$ is given by

$$\frac{\delta x(\mathbf{s})}{\delta x(\mathbf{z})} = \delta(\mathbf{s} - \mathbf{z}). \tag{6.33}$$

[4] A clear explanation of the transformations from lattice space to the continuum is given in [122].

This follows directly from the definition (6.32) if we replace $\Phi[x(\mathbf{s})]$ with $x(\mathbf{s})$.

Chain rule This applies to the functional of a functional, i.e. $\Phi[G[x(\mathbf{s})]]$. The functional derivative of $\Phi[G[x(\mathbf{s})]]$ with respect to the change of the function $x(\cdot)$ at point $\mathbf{z}$ can be evaluated in terms of the changes in the functional $G[\cdot]$ with respect to the change at $\mathbf{z}$, i.e.,

$$\frac{\delta\Phi[G[x(\mathbf{s})]]}{\delta x(\mathbf{z})} = \int_{\mathbb{R}^d} d\mathbf{s}' \, \frac{\delta\Phi[G[x(\mathbf{s})]]}{\delta G[x(\mathbf{s}')]} \, \frac{\delta G[x(\mathbf{s}')]}{\delta x(\mathbf{z})}. \tag{6.34}$$

Product rule We can apply the functional derivative to products of two functionals. It comes as no surprise that the result is analogous to the product rule for the common derivatives, i.e.,

$$\frac{\delta\left(\Phi[x(\mathbf{s})]\, G[x(\mathbf{s})]\right)}{\delta x(\mathbf{z})} = \frac{\delta\Phi[x(\mathbf{s})]}{\delta x(\mathbf{z})}\, G[x(\mathbf{s})] + \frac{\delta G[x(\mathbf{s})]}{\delta x(\mathbf{z})}\, \Phi[x(\mathbf{s})], \tag{6.35}$$

where $\delta\Phi[x(\mathbf{s})]/\delta x(\mathbf{z})$ is defined in (6.34).

Functional Taylor expansion This expansion allows approximating a functional with respect to the function $x(\mathbf{s})$ in terms of the function $x^{(0)}(\mathbf{s})$ using the increments $\phi(\mathbf{s}) = x(\mathbf{s}) - x^{(0)}(\mathbf{s})$, i.e.,

$$\Phi[x(\mathbf{s})] = \sum_{n=0}^{\infty} \frac{1}{n!} \left[\prod_{l=1}^{n} \int_{\mathbb{R}^d} d\mathbf{s}_l\right] \left.\frac{\delta^n \Phi[x(\mathbf{s})]}{\delta x(\mathbf{s}_1)\ldots\delta x(\mathbf{s}_n)}\right|_{x^{(0)}(\mathbf{s})} \phi(\mathbf{s}_1)\ldots\phi(\mathbf{s}_n). \tag{6.36}$$

We use functional derivatives below in connection with the moment and cumulant generating functionals.

6.2.2 *Moment Generating Functional*

The field theory defined by means of the joint *Boltzmann-Gibbs probability density function*

$$f_{\mathrm{x}}[x(\mathbf{s});\boldsymbol{\theta}] = \frac{1}{Z(\boldsymbol{\theta})} e^{-\mathcal{H}_0[x(\mathbf{s});\boldsymbol{\theta}]}, \tag{6.37}$$

where $\mathcal{H}_0$ is given by (6.23) and $Z(\boldsymbol{\theta})$ by (6.27), provides a general starting point for modeling random fields in continuum spaces. We drop the dependence on $\boldsymbol{\theta}$ in the following for notational brevity.

In terms of the joint pdf (6.37), the expectation of the random field is given by the following functional integral

$$\mathcal{E}[\mathrm{X}(\mathbf{s}_1;\omega)] = \frac{1}{Z}\int \mathcal{D}x(\mathbf{s})\, x(\mathbf{s}_1)\, \mathrm{e}^{-\mathcal{H}_0}. \tag{6.38}$$

Similarly, higher-order moments can be defined by means of the following functional integral

$$\mathcal{E}[\mathrm{X}(\mathbf{s}_1;\omega)\ldots\mathrm{X}(\mathbf{s}_N;\omega)] = \frac{1}{Z}\int \mathcal{D}x(\mathbf{s})\, x(\mathbf{s}_1)\ldots x(\mathbf{s}_N)\, \mathrm{e}^{-\mathcal{H}_0}. \tag{6.39}$$

Next, we define the *moment generating functional* by extending the definition (4.71), which is valid for a countable set of points, to a non-denumerable set as follows

$$\mathcal{M}_{\mathrm{x}}[u(\mathbf{s})] = \mathcal{E}\left[\mathrm{e}^{\int \mathrm{d}\mathbf{s}\, u(\mathbf{s})\,\mathrm{X}(\mathbf{s};\omega)}\right] = \frac{1}{Z}\int \mathcal{D}x(\mathbf{s})\, \mathrm{e}^{-\mathcal{H}_0 + \int \mathrm{d}\mathbf{s}\, u(\mathbf{s})\, x(\mathbf{s})}. \tag{6.40}$$

Note that in the moment generating functional the exponent is a functional of the random field $\mathrm{X}(\mathbf{s};\omega)$.

It is straightforward to show based on the above and the definition (6.27) of the partition function that

$$\mathcal{M}_{\mathrm{x}}[u(\mathbf{s}) = 0] = 1. \tag{6.41}$$

Comment The term $\int \mathrm{d}\mathbf{s}\, u(\mathbf{s})\, x(\mathbf{s})$ does not appear in the pdf; the function $u(\mathbf{s})$ can be viewed as a flashlight that we turn on at will to illuminate moments of the random field at specific locations.

Returning to the moment generating functional, we can show using functional derivatives the following expressions for the mean and the second-order moments

$$\mathcal{E}[\mathrm{X}(\mathbf{s}_1;\omega)] = \left.\frac{\delta \mathcal{M}_{\mathrm{x}}[u(\mathbf{s})]}{\delta u(\mathbf{s}_1)}\right|_{u=0}, \tag{6.42}$$

$$\mathcal{E}[\mathrm{X}(\mathbf{s}_1;\omega)\,\mathrm{X}(\mathbf{s}_2;\omega)] = \left.\frac{\delta^2 \mathcal{M}_{\mathrm{x}}[u(\mathbf{s})]}{\delta u(\mathbf{s}_1)\delta u(\mathbf{s}_2)}\right|_{u=0}. \tag{6.43}$$

To verify these equations, we refer to the definition (6.40) of the moment generating functional. We assume that we can interchange the order of functional differentiation and integration. Based on (6.40), the first functional derivative of the moment generating functional with respect to $u(\mathbf{s}_1)$ operates on the exponent $\int d\mathbf{s}\, u(\mathbf{s})\, x(\mathbf{s})$. Using (6.33) we obtain

$$\frac{\delta \int d\mathbf{s}\, u(\mathbf{s})\, x(\mathbf{s})}{\delta u(\mathbf{s}_1)} = x(\mathbf{s}_1).$$

The function $u(\mathbf{s})$ is used to select the target point, $\mathbf{s}_1$. Putting together all the terms in the functional derivative $\delta \mathcal{M}_{\mathrm{x}}[u(\mathbf{s})]/\delta u(\mathbf{s}_1)$, it follows that

$$\frac{\delta \mathcal{M}_{\mathrm{x}}[u(\mathbf{s})]}{\delta u(\mathbf{s}_1)} = \frac{1}{Z} \int \mathcal{D}x(\mathbf{s})\, x(\mathbf{s}_1)\, \mathrm{e}^{-\mathcal{H}_0 + \int d\mathbf{s}\, u(\mathbf{s})\, x(\mathbf{s})}.$$

The above expression is very similar to (6.38), except for the extra term that involves $u(\mathbf{s})$. But we can fix this incongruence by setting $u = 0$ (i.e., $u(\mathbf{s}) = 0$ for all $\mathbf{s}$). So, essentially at the end of the calculation we turn off the auxiliary field $u(\mathbf{s})$, which we used to pinpoint the location of the moment. Similarly, we can verify the expression (6.43).

By iterative application of these operations we can show that moments of order N are given by the following functional derivative

$$\mathcal{E}[\mathrm{X}(\mathbf{s}_1;\omega) \ldots \mathrm{X}(\mathbf{s}_N;\omega)] = \left. \frac{\delta^N \mathcal{M}_{\mathrm{x}}[u(\mathbf{s})]}{\delta u(\mathbf{s}_1) \ldots \delta u(\mathbf{s}_N)} \right|_{u=0}. \tag{6.44}$$

Comment In some formulations, a minus sign precedes $\int d\mathbf{s}\, u(\mathbf{s})\, x(\mathbf{s})$ in the definition of $\mathcal{M}_{\mathrm{x}}[u(\mathbf{s})]$. Then, the equation that gives the N-order moment based on the moment generating functional becomes

$$\mathcal{E}[\mathrm{X}(\mathbf{s}_1;\omega) \ldots \mathrm{X}(\mathbf{s}_N;\omega)] = (-1)^N \left. \frac{\delta^N \mathcal{M}_{\mathrm{x}}[u(\mathbf{s})]}{\delta u(\mathbf{s}_1) \ldots \delta u(\mathbf{s}_N)} \right|_{u=0}. \tag{6.45}$$

6.2.3 Cumulant Generating Functional and Moments

The calculation of the random field cumulants is enabled by the following *cumulant generating functional*

$$\mathcal{K}_{\mathrm{x}}[u(\mathbf{s})] := \ln \left\{ \mathcal{E}\left[\mathrm{e}^{\int d\mathbf{s}\, u(\mathbf{s})\, \mathrm{X}(\mathbf{s};\omega)} \right] \right\}. \tag{6.46}$$

The cumulant generating functional can be expressed as the logarithm of the moment generating functional, i.e.,

$$\mathcal{K}_{\mathrm{x}}[u(\mathbf{s})] = \ln \mathcal{M}_{\mathrm{x}}[u(\mathbf{s})] = \ln \left(\frac{\int \mathcal{D}x(\mathbf{s})\, \mathrm{e}^{-\mathcal{H}_0 + \int d\mathbf{s}\, u(\mathbf{s})\, x(\mathbf{s})}}{\int \mathcal{D}x(\mathbf{s})\, \mathrm{e}^{-\mathcal{H}_0}} \right). \tag{6.47}$$

We can use this functional to generate cumulants of the random field. For example, the first-order cumulant, which is equal to the expectation of the field, is given by the first functional derivative of the cumulant generating functional, i.e.,

$$\mathcal{E}[X(\mathbf{s}_1;\omega)] = \left.\frac{\delta\mathcal{K}_x[u(\mathbf{s})]}{\delta u(\mathbf{s}_1)}\right|_{u(\mathbf{s})=0}. \tag{6.48}$$

Similarly, the second-order cumulant, i.e., the covariance function, is given by

$$\mathcal{E}[X'(\mathbf{s}_1;\omega)\,X'(\mathbf{s}_2;\omega)] = \left.\frac{\delta^2\mathcal{K}_x[u(\mathbf{s})]}{\delta u(\mathbf{s}_2)\,\delta u(\mathbf{s}_1)}\right|_{u(\mathbf{s})=0}. \tag{6.49}$$

Let us see how the differentiation of the cumulant generating functional works. The analysis below is based on the moment equations derived from the moment generating functional. Using the chain rule of functional differentiation and the definition (6.47) of the cumulant generating functional, it follows that

$$\frac{\delta\mathcal{K}_x[u(\mathbf{s})]}{\delta u(\mathbf{s}_1)} = \frac{1}{\mathcal{M}_x[u(\mathbf{s})]}\,\frac{\delta\mathcal{M}_x[u(\mathbf{s})]}{\delta u(\mathbf{s}_1)}.$$

At the limit $u(\mathbf{s}) \to 0$, the above functional derivative yields the expectation $\mathcal{E}[X(\mathbf{s}_1;\omega)]$ based on the moment generating functional equations (6.41) and (6.42).

By applying the second functional derivative, we obtain

$$\frac{\delta^2\mathcal{K}_x[u(\mathbf{s})]}{\delta u(\mathbf{s}_2)\delta u(\mathbf{s}_1)} = -\frac{1}{\{\mathcal{M}_x[u(\mathbf{s})]\}^2}\,\frac{\delta\mathcal{M}_x[u(\mathbf{s})]}{\delta u(\mathbf{s}_2)}\,\frac{\delta\mathcal{M}_x[u(\mathbf{s})]}{\delta u(\mathbf{s}_1)} + \frac{1}{\mathcal{M}_x[u(\mathbf{s})]}\,\frac{\delta^2\mathcal{M}_x[u(\mathbf{s})]}{\delta u(\mathbf{s}_1)\delta u(\mathbf{s}_2)}$$

Now, based on (6.42), (6.43), and (6.41), it follows that

$$\left.\frac{\delta^2\mathcal{K}_x[u(\mathbf{s})]}{\delta u(\mathbf{s}_2)\delta u(\mathbf{s}_1)}\right|_{u(\mathbf{s})=0} = \mathcal{E}\left[X(\mathbf{s}_1;\omega)\,X(\mathbf{s}_2;\omega)\right] - \mathcal{E}\left[X(\mathbf{s}_1;\omega)\right]\,\mathcal{E}\left[X(\mathbf{s}_2;\omega)\right].$$

Comment What can the functional integral formulation help us to accomplish? First of all, with the moment and cumulant generating functionals we escape the evaluation of the continuum limit of the partition function: the derivatives of both functionals yield well-defined moments. As we show below, moments and cumulants can be explicitly calculated in the Gaussian case from the energy functional $\mathcal{H}_0$. In addition, we will consider approaches that extend such calculations to random fields defined by means of non-Gaussian Boltzmann-Gibbs pdfs.

6.2.4 Non-Gaussian Densities and Perturbation Expansions

The definitions of the moment and cumulant generating functionals do not require quadratic energy functions $\mathcal{H}_0$, i.e., they are not limited to Gaussian random fields. In fact, the expressions for the moments and cumulants hold for non-Gaussian functionals as well. The Gaussian dependence, however, is amenable to explicit calculations. It is also possible to calculate moments of non-Gaussian pdfs, provided that the non-Gaussian term(s) can be treated as perturbations (see Sect. 6.4).

6.2.5 Related Topics

In astrophysics, researchers are concerned with inverse problems that include random field modeling (e.g., covariance estimation, denoising, spatial interpolation). The mathematical problems are similar to those faced in spatial statistics in the context of Earth-bound (e.g., environmental, energy, ecological, and natural resources) applications. Methods based on the moment and cumulant generating functionals, such as those described in this chapter, have been expressed in the framework of *information field theory* as developed in the book by Lemm [495] and elaborated for astrophysical applications in papers by Ensslin et al. [208, 234–236, 631, 632].

6.3 Useful Properties of Gaussian Random Fields

Gaussian random fields admit explicit expressions. This is a significant benefit that allows considerable simplifications in theoretical analysis and numerical calculations. Below we concentrate on zero-mean Gaussian random fields. Fields with nonzero expectation can be decomposed into a mean function and a zero-mean fluctuation.

6.3.1 Isserlis-Wick Theorem

A significant appeal of the Gaussian distribution is that all its moments of order higher than two can be expressed in terms of the expectation and the covariance function. This decomposition is accomplished by means of a theorem attributed to the mathematician L. Isserlis [398] and the physicist C. Wick [848].

Theorem 6.2 *Let* $\mathrm{X}(\mathbf{s};\omega)$ *be a zero-mean, Gaussian random field. The expectations of the form* $\mathcal{E}[\mathrm{X}(\mathbf{s}_1;\omega)\ldots\mathrm{X}(\mathbf{s}_N;\omega)]$, *are calculated as follows:*

1. *All odd-order Gaussian moments vanish, i.e.,*

$$\mathcal{E}[\mathrm{X}(\mathbf{s}_1;\omega)\ldots\mathrm{X}(\mathbf{s}_N;\omega)] = 0, \text{ if } N = 2m+1.$$

2. *For even-order moments* ($N = 2m$), *the Isserlis-Wick theorem states that*

$$\mathcal{E}[\mathrm{X}(\mathbf{s}_1;\omega)\ldots\mathrm{X}(\mathbf{s}_N;\omega)] = \sum_{\{i_1,\ldots i_N\}} \underbrace{C_{\mathrm{xx}}(\mathbf{s}_{i_1},\mathbf{s}_{i_2})\ldots C_{\mathrm{xx}}(\mathbf{s}_{i_{N-1}},\mathbf{s}_{i_N})}_{m} \tag{6.50}$$

The summation is over all possible pairings of the $N = 2m$ *points in* m *pairs, i.e., over* $(2m)!/(2^m m!)$ *terms.*

Comment The ordering of the points in pairs is significant, but the relative ordering of the pairs is irrelevant, i.e., the term $C_{\mathrm{xx}}(\mathbf{s}_1,\mathbf{s}_2)\,C_{\mathrm{xx}}(\mathbf{s}_3,\mathbf{s}_4)$ is identical to the term $C_{\mathrm{xx}}(\mathbf{s}_3,\mathbf{s}_4)\,C_{\mathrm{xx}}(\mathbf{s}_1,\mathbf{s}_2)$. Hence, equation (6.50) includes only one of any equivalent orderings.

Example 6.3 Evaluate the fourth-order moment

$$\mathcal{E}[\mathrm{X}(\mathbf{s}_1;\omega)\,\mathrm{X}(\mathbf{s}_2;\omega)\,\mathrm{X}(\mathbf{s}_3;\omega)\,\mathrm{X}(\mathbf{s}_4;\omega)]$$

of a zero-mean random field at the vertices of a square cell.

Answer Based on the Wick-Issserlis theorem the fourth-order moment is given by the sum over all possible combinations of the four points in pairs. Since we consider four points, $m = 2$. There are thus $4!\,/(2^2\,2!) = 3$ pairing combinations as shown in Fig. 6.9. The mathematical interpretation of this figure is

$$\begin{aligned}\mathcal{E}[\mathrm{X}(\mathbf{s}_1;\omega)\,\mathrm{X}(\mathbf{s}_2;\omega)\,\mathrm{X}(\mathbf{s}_3;\omega)\,\mathrm{X}(\mathbf{s}_4;\omega)] &= \mathcal{E}[\mathrm{X}(\mathbf{s}_1;\omega)\,\mathrm{X}(\mathbf{s}_3;\omega)]\,\mathcal{E}[\mathrm{X}(\mathbf{s}_2;\omega)\,\mathrm{X}(\mathbf{s}_4;\omega)]\\ &+ \mathcal{E}[\mathrm{X}(\mathbf{s}_1;\omega)\,\mathrm{X}(\mathbf{s}_2;\omega)]\,\mathcal{E}[\mathrm{X}(\mathbf{s}_3;\omega)\,\mathrm{X}(\mathbf{s}_4;\omega)]\\ &+ \mathcal{E}[\mathrm{X}(\mathbf{s}_1;\omega)\,\mathrm{X}(\mathbf{s}_4;\omega)]\,\mathcal{E}[\mathrm{X}(\mathbf{s}_2;\omega)\,\mathrm{X}(\mathbf{s}_3;\omega)].\end{aligned}$$

Example 6.4 Consider four points $\mathbf{s}_1, \mathbf{s}_2, \mathbf{s}_3, \mathbf{s}_4$ at the nodes of a square cell with side length equal to a as in the Example 6.3. Evaluate the fourth-order moment $\mathcal{E}\,[\mathrm{X}(\mathbf{s}_1;\omega)\,\mathrm{X}(\mathbf{s}_2;\omega)\,\mathrm{X}(\mathbf{s}_3;\omega)\,\mathrm{X}(\mathbf{s}_4;\omega)]$ of a zero-mean, statistically homogeneous and isotropic random field $\mathrm{X}(\mathbf{s};\omega)$ at the points $\{\mathbf{s}_i\}_{i=1}^4$.

Fig. 6.9 Schematic showing possible pairings of 4 vertices used in Wick-Isserlis theorem

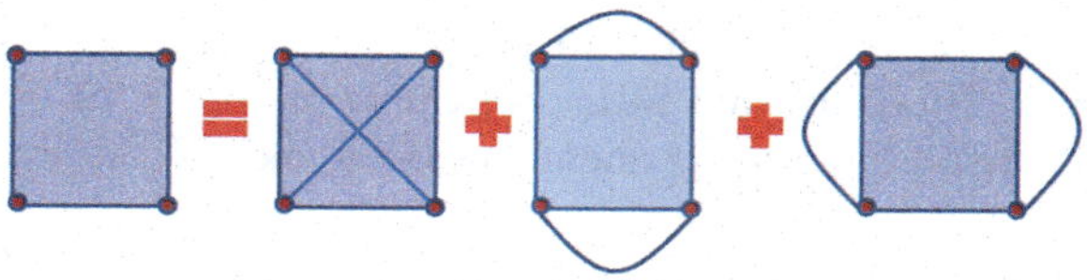

Answer Let $\mathbf{s}_1 = (0, 0), \mathbf{s}_2 = (0, a), \mathbf{s}_3 = (a, 0), \mathbf{s}_4 = (a, a)$. Using the Isserlis-Wick theorem we obtain the fourth-order moment decomposition as given in the Example 6.3.

Based on the statistical homogeneity and the zero mean property, it follows that

$$\mathcal{E}\left[\mathrm{X}(\mathbf{s}_n; \omega)\, \mathrm{X}(\mathbf{s}_m; \omega)\right] = C_{\mathrm{xx}}(\mathbf{s}_n - \mathbf{s}_m).$$

Based on the above, the fourth-order moment is given by

$$\begin{aligned}\mathcal{E}\left[\mathrm{X}(\mathbf{s}_1; \omega) \ldots \mathrm{X}(\mathbf{s}_4; \omega)\right] &= C_{\mathrm{xx}}(a)\, C_{\mathrm{xx}}(a) + C_{\mathrm{xx}}(a)\, C_{\mathrm{xx}}(a) + C_{\mathrm{xx}}(a\sqrt{2})\, C_{\mathrm{xx}}(a\sqrt{2}) \\ &= 2C_{\mathrm{xx}}^2(a) + C_{\mathrm{xx}}^2(a\sqrt{2}).\end{aligned}$$

Variance of the gradient tensor We encountered the gradient tensor in Sect. 5.2 in reference to geometric anisotropy and the Covariance Hessian Identity (CHI). The definition of the gradient tensor is recalled here

$$[\nabla \mathrm{X}(\mathbf{s}; \omega)\, \nabla \mathrm{X}(\mathbf{s}; \omega)]_{i,j} = \partial_i \mathrm{X}(\mathbf{s}; \omega)\, \partial_j \mathrm{X}(\mathbf{s}; \omega), \text{ for all } i, j = 1, \ldots, d.$$

The CHI theorem 5.9 gives the expectation of the gradient tensor according to (5.22). The width of the gradient tensor's pdf is measured by the square root of $\mathbb{V}\mathrm{ar}\left\{\partial_i \mathrm{X}(\mathbf{s}; \omega)\, \partial_j \mathrm{X}(\mathbf{s}; \omega)\right\}$, $i, j = 1, \ldots, d$. Let us then calculate the variance of the *gradient tensor* for a Gaussian, statistically homogeneous random field $\mathrm{X}(\mathbf{s}; \omega)$.

We define for short $X_{ij}(\mathbf{s}; \omega) = \partial_i \mathrm{X}(\mathbf{s}; \omega)\, \partial_j \mathrm{X}(\mathbf{s}; \omega)$. Based on the definition of the variance as the expectation of the square of the fluctuations, it follows that

$$\mathbb{V}\mathrm{ar}\left\{\partial_i \mathrm{X}(\mathbf{s}; \omega)\, \partial_j \mathrm{X}(\mathbf{s}; \omega)\right\} = \mathcal{E}\left[X_{ij}^2(\omega)\right] - \mathcal{E}\left[X_{ij}(\mathbf{s}; \omega)\right]^2.$$

The second term above is determined from the fact that $H_{ij}(\mathbf{0})$ is equal to the *expectation of the gradient tensor* (cf. Theorem 5.9)

$$\mathcal{E}\left[\partial_i \mathrm{X}(\mathbf{s}; \omega)\, \partial_j \mathrm{X}(\mathbf{s}; \omega)\right] = -H_{ij}(\mathbf{0}), \tag{6.51}$$

where $\mathbf{0} = (0, \ldots, 0)$ denotes the d-dimensional zero vector.

Since the gradient field components $\partial_i X(\mathbf{s};\omega)$, $i = 1, \ldots, d$ are zero-mean, normally distributed SRFs, we can apply the *Wick-Isserlis theorem* 6.2 to expand the first term of the gradient tensor variance as follows:

$$\mathcal{E}\left[X_{ij}^2(\mathbf{s};\omega)\right] = 2\,\mathcal{E}\left[X_{ij}(\mathbf{s};\omega)\right]^2 + \mathcal{E}\left[\{\partial_i X(\mathbf{s};\omega)\}^2\right]\mathcal{E}\left[\{\partial_j X(\mathbf{s};\omega)\}^2\right]. \tag{6.52}$$

The first term on the right-hand side of the above is equal to $2\,H_{ij}^2(\mathbf{0})$.

Finally, putting together equations (6.51) and (6.52), the variance of the gradient tensor is given by the following expression[5]:

$$\mathbb{V}\text{ar}\left\{\partial_i X(\mathbf{s};\omega)\,\partial_j X(\mathbf{s};\omega)\right\} = H_{ii}(\mathbf{0})\,H_{jj}(\mathbf{0}) + H_{ij}^2(\mathbf{0}), \quad i, j = 1, \ldots, d. \tag{6.53}$$

Let us assume that $H_{jj}(\mathbf{0}) \neq 0$. Based on the expectation of the gradient tensor (6.51) and the variance (6.53), the *coefficient of variation* for the non-zero gradient tensor elements becomes

$$\mathbb{C}\text{ov}\left\{X_{ij}(\mathbf{s};\omega)\right\} = \sqrt{\frac{H_{ii}(\mathbf{0})\,H_{jj}(\mathbf{0})}{H_{ij}^2(\mathbf{0})} + 1}. \tag{6.54}$$

Since $H_{ii}(\mathbf{0}) > 0$, for all $i = 1, \ldots, d$, it follows that $\mathbb{C}\text{ov}\left\{X_{ij}(\mathbf{s};\omega)\right\} > 1$, for all $i, j = 1, \ldots, d$. This value of the coefficient of variation emphasizes the high variability of the gradient tensor.

In addition, it is straightforward to show that the normalized diagonal elements $X_{ii}(\mathbf{s};\omega)/\sqrt{2}H_{ii}(\mathbf{0})$ of the gradient tensor random field, where $i = 1, \ldots, d$, follow the χ^2 (chi-squared) distribution with one degree of freedom.

Finally, the *covariance function* of the gradient tensor (for both the diagonal and non-diagonal elements) is calculated in [663].

6.3.2 Novikov-Furutsu-Donsker Theorem

The Novikov-Furutsu-Donsker (NFD) theorem allows calculating the correlation of a random field with a function $\Phi(\cdot)$ that depends implicitly on the random field. This differs from the Isserlis-Wick theorem 6.2 because the function $\Phi(\cdot)$ can have a more general form than the product of monomials of the field used in the Isserlis-Wick theorem.

Let us assume a *zero-mean, Gaussian random field* $X(\mathbf{s};\omega)$ that is discretely sampled so that the random variables $X_1(\omega), \ldots, X_N(\omega)$ represent random field samples at points $\mathbf{s}_n \in \mathcal{D}$, $n = 1, \ldots, N$.

[5] Summation is not implied over the indices i and j here.

6.3.2.1 Finite-Dimensional Formulation

Let $\Phi(x_1, \ldots, x_N)$ be a function that admits a multivariate Taylor series expansion around the "zero point" $(x_1, \ldots, x_N) = \mathbf{0}$. We will use the shorthand notation $\Phi\left(\mathrm{X}_1, \ldots, \mathrm{X}_N; \omega\right) := \Phi\left[\mathrm{X}_1(\omega), \ldots, \mathrm{X}_N(\omega)\right]$, where $\mathrm{X}_n(\omega) = \mathrm{X}(\mathbf{s}_n; \omega)$, for $n = 1, \ldots, N$.

The *Novikov-Furutsu-Donsker theorem* expresses the expectation of the product of $\mathrm{X}(\mathbf{s}; \omega)$ with a function $\Phi\left[\mathrm{X}_1(\omega), \ldots, \mathrm{X}_N(\omega)\right]$ in terms of the covariance function and the expectation of the partial derivatives of the random function $\Phi\left[\mathrm{X}_1(\omega), \ldots, \mathrm{X}_N(\omega)\right]$ [207, 268, 614].

Theorem 6.3 *The Novikov-Furutsu-Donsker (NFD) theorem states that the following identity holds for all* $\mathbf{s} \in \mathcal{D}$*:*

$$\mathcal{E}\left[\mathrm{X}(\mathbf{s}; \omega)\, \Phi(\mathrm{X}_1, \ldots, \mathrm{X}_N; \omega)\right] = \sum_{k=1}^{N} C_{\mathrm{xx}}(\mathbf{s}, \mathbf{s}_k)\, \mathcal{E}\left[\frac{\partial \Phi(\mathrm{X}_1, \ldots, \mathrm{X}_N; \omega)}{\partial \mathrm{X}_k(\omega)}\right], \tag{6.55}$$

where $C_{\mathrm{xx}}(\mathbf{s}, \mathbf{s}_k) = \mathcal{E}[\mathrm{X}(\mathbf{s}; \omega)\, \mathrm{X}(\mathbf{s}_k; \omega)]$ *is the covariance function between* $\mathbf{s}$ *and* $\mathbf{s}_k$.

Example 6.5 Evaluate the following expectation for the zero-mean Gaussian random field $\mathrm{X}(\mathbf{s}; \omega)$ using the NFD theorem:

$$\mathcal{E}\left[\mathrm{X}(\mathbf{s}_n; \omega)\, \mathrm{e}^{\mathrm{X}(\mathbf{s}_m; \omega)}\right].$$

Answer A straightforward application of (6.55), taking into account that the derivative of the exponential function is the function itself, leads to the following equation

$$\mathcal{E}\left[\mathrm{X}(\mathbf{s}_n; \omega)\, \mathrm{e}^{\mathrm{X}(\mathbf{s}_m; \omega)}\right] = C_{\mathrm{xx}}(\mathbf{s}_n, \mathbf{s}_m)\, \mathcal{E}\left[\mathrm{e}^{\mathrm{X}(\mathbf{s}_m; \omega)}\right].$$

The expectation $\mathcal{E}\left[\mathrm{e}^{\mathrm{X}(\mathbf{s}_m; \omega)}\right]$ is evaluated in Example 6.7.

6.3.2.2 Continuum-Space Formulation

The NFD theorem 6.3 also applies if $\Phi\left[\mathrm{X}(\mathbf{s}; \omega)\right]$ is a *functional* of the random field $\mathrm{X}(\mathbf{s}; \omega)$ defined over an infinite and denumerable set of points. Recall that a functional $\Phi\left[\mathrm{X}(\mathbf{s}; \omega)\right]$ of the random field $\mathrm{X}(\mathbf{s}; \omega)$ is a function of $\mathrm{X}(\mathbf{s}; \omega)$ that returns a numerical value, $\phi(\omega)$, for each realization of $\mathrm{X}(\mathbf{s}; \omega)$. For example, if we assume that the realizations of $\mathrm{X}(\mathbf{s}; \omega)$ are integrable functions, the volume integral $\Phi\left[\mathrm{X}(\mathbf{s}; \omega)\right] = \int_{\mathcal{D}} d\mathbf{s}\, \mathrm{X}(\mathbf{s}; \omega)$ is a possible functional.

The *continuum formulation of the NFD theorem* is then given by the following expression:

$$\mathcal{E}\big[\mathrm{X}(\mathbf{s};\omega)\,\Phi\,[\mathrm{X}(\mathbf{s};\omega)]\big] = \int_{\mathcal{D}} d\mathbf{s}'\, C_{\mathrm{xx}}(\mathbf{s},\mathbf{s}')\,\mathcal{E}\left[\frac{\delta\Phi\,[\mathrm{X}(\mathbf{s};\omega)]}{\delta\mathrm{X}(\mathbf{s}';\omega)}\right], \tag{6.56}$$

where $\delta\Phi\,[\mathrm{X}(\mathbf{s};\omega)]\,/\delta\mathrm{X}(\mathbf{s}';\omega)$ denotes the *functional derivative* of $\Phi\,[\mathrm{X}(\mathbf{s};\omega)]$ with respect to $\mathrm{X}(\mathbf{s}';\omega)$. A proof of the NFD theorem (based on the original proof for random processes by Novikov) is given in [735, pp. 252–254].

The functional derivative calculation involves the following steps: (i) Express $\Phi\,[\mathrm{X}(\mathbf{s};\omega)]$ in terms of a realization $x(\mathbf{s})$; (ii) evaluate the functional derivative $\delta\Phi\,[x(\mathbf{s})]\,/\delta x(\mathbf{s}')$; (iii) in the resulting expression, replace all occurrences of realizations $x(\cdot)$ with the respective random field $\mathrm{X}(\cdot;\omega)$.

Example 6.6 Consider the functional $\Phi\,[\mathrm{X}(\mathbf{s};\omega)] = \int_{\mathcal{D}} d\mathbf{s}\,\mathrm{X}(\mathbf{s};\omega)$. Use the NFD theorem to evaluate the expectation

$$\mathcal{E}\left[\mathrm{X}(\mathbf{s};\omega)\int_{\mathcal{D}} d\mathbf{s}\,\mathrm{X}(\mathbf{s};\omega)\right]$$

and confirm the result by direct evaluation (without using NFD).

Answer By applying the NFD theorem we obtain

$$\begin{aligned}\mathcal{E}\left[\mathrm{X}(\mathbf{s};\omega)\int_{\mathcal{D}} d\mathbf{s}'\,\mathrm{X}(\mathbf{s}';\omega)\right] &= \int_{\mathcal{D}} d\mathbf{z}\,C_{\mathrm{xx}}(\mathbf{s},\mathbf{z})\,\mathcal{E}\left[\int_{\mathcal{D}} d\mathbf{s}'\,\frac{\delta\mathrm{X}(\mathbf{s}';\omega)}{\delta\mathrm{X}(\mathbf{z};\omega)}\right] \\ &\int_{\mathcal{D}} d\mathbf{z}\,C_{\mathrm{xx}}(\mathbf{s},\mathbf{z})\int_{\mathcal{D}} d\mathbf{s}'\,\delta(\mathbf{s}'-\mathbf{z}) \\ &= \int_{\mathcal{D}} d\mathbf{s}'\,C_{\mathrm{xx}}(\mathbf{s},\mathbf{s}').\end{aligned}$$

The calculation of the functional derivative is the main step in the proof. A technical detail is that the position $\mathbf{s}$ inside the integral $\int_{\mathcal{D}} d\mathbf{s}\,\mathrm{X}(\mathbf{s};\omega)$ is an integration variable not to be confused with $\mathbf{s}$ in $\mathrm{X}(\mathbf{s};\omega)$ that appears outside the integral. Hence, we replace $\mathbf{s}$ inside the integral with $\mathbf{s}'$ to avoid confusion.

The NFD-based result is confirmed by straightforward calculation of the expectation, which is left as an exercise for the reader.

6.3.3 Gaussian Moments

The following properties are based on the Isserlis-Wick Theorem 6.2 and the Novikov-Furutsu-Donsker Theorem 6.3.

1. The *odd-order moments* of a zero-mean Gaussian field are zero, i.e.,

$$\mathcal{E}\left[X^{2m+1}(\mathbf{s};\omega)\right] = 0. \tag{6.57}$$

This property follows from the symmetry of the distribution and can be shown formally using the NFD theorem.

2. Based on the Isserlis-Wick theorem, the following relation holds for the marginal *even-order moments* of a zero-mean Gaussian field

$$\mathcal{E}\left[X^{2m}(\mathbf{s};\omega)\right] = \frac{\sigma_x^{2m}(2m)!}{2^m\, m!}. \tag{6.58}$$

3. The *kurtosis coefficient* of a zero-mean Gaussian random field is given by

$$k_x(\mathbf{s}) = 3. \tag{6.59}$$

The proof is straightforward using the Isserlis-Wick 6.2 theorem and the definition of the kurtosis coefficient

$$k_x(\mathbf{s}) = \frac{\mathcal{E}\left[X^4(\mathbf{s};\omega)\right]}{\sigma_x^4}. \tag{6.60}$$

Example 6.7 For a stationary Gaussian random field $X(\mathbf{s};\omega) \stackrel{d}{=} \mathcal{N}(m_x, \sigma_x^2)$ show that the expectation of the exponential $\exp\left[X(\mathbf{s};\omega)\right]$ is given by

$$\mathcal{E}\left[\,e^{X(\mathbf{s};\omega)}\right] = e^{m_x + \sigma_x^2/2}.$$

Answer We express the expectation using the mean-fluctuation decomposition of the random field

$$\mathcal{E}\left[\,e^{X(\mathbf{s};\omega)}\right] = \mathcal{E}\left[\,e^{m_x + X'(\mathbf{s};\omega)}\right] = e^{m_x}\,\mathcal{E}\left[\,e^{X'(\mathbf{s};\omega)}\right].$$

To evaluate $\mathcal{E}\left[\,e^{X'(\mathbf{s};\omega)}\right]$ we use the Taylor series expansion of the exponential, which leads to the following series for the expectation

$$\mathcal{E}[\,e^{X'(\mathbf{s};\omega)}] = \sum_{n=0}^{\infty} \frac{1}{n!} \mathcal{E}[\,X'^{n}(\mathbf{s};\omega)]$$

Based on properties (1) and (2) above, only the even-order moments contribute to the expectation. Finally, the following series expansion that is explicitly summable is obtained

$$\mathcal{E}[\,e^{X'(\mathbf{s};\omega)}] = \sum_{m=0}^{\infty} \frac{1}{(2m)!} \mathcal{E}[X'^{2m}(\mathbf{s};\omega)] = \sum_{m=0}^{\infty} \frac{1}{2^m\, m!}\, \sigma_{\mathrm{X}}^{2m} = e^{\sigma_{\mathrm{X}}^2/2}.$$

6.3.4 The Cumulant Expansion

Let $\mathrm{Y}(\mathbf{s};\omega)$ denote a random field. The *cumulant expansion* allows expressing the expectation of the exponential of $\mathrm{Y}(\mathbf{s};\omega)$ in terms of the exponential of its cumulants [170, 364]. More precisely, the cumulant expansion is given by

$$\mathcal{E}\left[e^{\mathrm{Y}(\mathbf{s};\omega)}\right] = \exp\left[\frac{\sum_{m=1}^{\infty} C_m[\mathrm{Y}(\mathbf{s};\omega)]}{m!}\right], \tag{6.61}$$

where $C_m[\mathrm{Y}(\mathbf{s};\omega)]$, $m \in \mathbb{N}$ is the cumulant of order m of the SRF $\mathrm{Y}(\mathbf{s};\omega)$.

If $\mathrm{Y}(\mathbf{s};\omega)$ is a Gaussian random field with non-zero mean m_{y}, only the first two cumulants are non-zero, i.e.

$$C_1[\mathrm{Y}(\mathbf{s};\omega)] = \mathcal{E}\,[\mathrm{Y}(\mathbf{s};\omega)] = m_{\mathrm{y}},$$

$$C_2[\mathrm{Y}(\mathbf{s};\omega)] = \mathcal{E}\left[\mathrm{Y}(\mathbf{s};\omega)^2\right] - \mathcal{E}\,[\mathrm{Y}(\mathbf{s};\omega)]^2 = \mathbb{V}\mathrm{ar}\,\{\mathrm{Y}(\mathbf{s};\omega)\}.$$

Thus, the *expectation of the exponential of a Gaussian random field* using the cumulant expansion is given by

$$\mathcal{E}\left[e^{\mathrm{Y}(\mathbf{s};\omega)}\right] = e^{\mathcal{E}[\mathrm{Y}(\mathbf{s};\omega)] + \frac{1}{2}\mathbb{V}\mathrm{ar}\{\mathrm{Y}(\mathbf{s};\omega)\}}. \tag{6.62}$$

6.3.5 The Lognormal Distribution

If the wide-sense stationary random field $Y(\mathbf{s}; \omega)$ follows the joint normal (multi-normal) distribution, then the exponentiated random field $X(\mathbf{s}; \omega) = \exp[Y(\mathbf{s}; \omega)]$ follows the lognormal distribution. The latter has the following marginal pdf

$$f_x(x) = \frac{1}{\sqrt{2\pi}\, x\, \sigma_y} \, e^{-(\ln x - m_y)^2/2\sigma_y^2}, \tag{6.63}$$

where m_y and σ_y are the mean and standard deviation of $Y(\mathbf{s}; \omega)$. Plots of lognormal density curves for three different values of σ_y are shown in Fig. 6.10.

The lognormal is a prime example of a skewed pdf with an extended right tail. The pdf becomes wider as the standard deviation σ_y of the logarithm increases. In addition, if $\sigma_y \gg 1$, the lognormal pdf plotted on a log-log scale looks very similar to a power law (i.e., it appears as a straight line on the log-log plot) over a wide range of values. Hence, it is not easy to distinguish between power laws and the lognormal density, especially if the available observations cover a limited range of scales [605].

The mean and the covariance function of the *lognormal random field* $X(\mathbf{s}; \omega)$ and the Gaussian random field $Y(\mathbf{s}; \omega) = \ln X(\mathbf{s}; \omega)$, are related by means of the following equations

$$m_x = e^{m_y + C_{yy}(0)/2}, \tag{6.64}$$

$$C_{xx}(\mathbf{r}) = m_x^2 \left[e^{C_{yy}(\mathbf{r})} - 1 \right], \tag{6.65}$$

(continued)

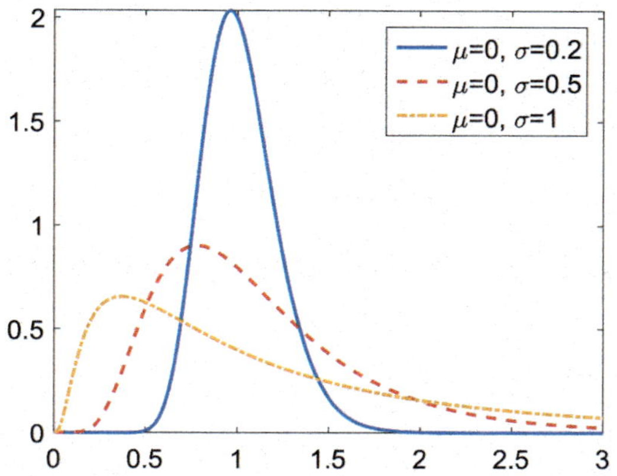

Fig. 6.10 Lognormal probability density functions for $\mu = m_y = 0$ and $\sigma = \sigma_y = 0.1, 0.5, 1$. Note the heavier right tail of the pdf as σ increases

$$m_{\mathrm{y}} = \mathcal{E}[\mathrm{Y}(\mathbf{s};\omega)], \text{ and } m_{\mathrm{x}} = \mathcal{E}[\mathrm{X}(\mathbf{s};\omega)] = \mathcal{E}\left[\mathrm{e}^{\mathrm{Y}(\mathbf{s};\omega)}\right]. \tag{6.66}$$

Equations (6.64) and (6.65) follow from the expression (6.62) for the expectation of the exponential of a random field. Equation (6.64) also employs the identity $\mathbb{V}\mathrm{ar}\{\mathrm{Y}(\mathbf{s};\omega)\} = C_{\mathrm{yy}}(0)$. Based on the definition of the covariance function, we obtain for $C_{\mathrm{xx}}(\mathbf{r})$ the following expression

$$C_{\mathrm{xx}}(\mathbf{r}) = \mathcal{E}[\mathrm{X}(\mathbf{s};\omega)\, X(\mathbf{s}+\mathbf{r})] - m_{\mathrm{x}}^2 = \mathcal{E}\left[\mathrm{e}^{\mathrm{Y}(\mathbf{s};\omega)+\mathrm{Y}(\mathbf{s}+\mathbf{r};\omega)}\right] - m_{\mathrm{x}}^2.$$

In order to evaluate the expectation of the exponential, we apply (6.62) to the sum $\mathrm{Y}(\mathbf{s};\omega) + \mathrm{Y}(\mathbf{s}+\mathbf{r};\omega)$. The respective cumulants are given by

$$C_1[\mathrm{Y}(\mathbf{s};\omega) + \mathrm{Y}(\mathbf{s}+\mathbf{r};\omega)] = 2m_{\mathrm{y}},$$

$$C_2[\mathrm{Y}(\mathbf{s};\omega) + \mathrm{Y}(\mathbf{s}+\mathbf{r};\omega)] = 2C_{\mathrm{yy}}(0) + 2C_{\mathrm{yy}}(\mathbf{r}),$$

thus leading to (6.65) by means of (6.62) and (6.64). Following the same approach, namely expressing $\mathrm{X}(\mathbf{s};\omega)$ as $\mathrm{Y}(\mathbf{s};\omega)$, we can evaluate higher-order moments and cumulants of the lognormal random field.

A characteristic property of the lognormal distribution is that its realizations are likely to contain a few extreme (very large) values, which are due to the extended tail of the lognormal pdf.

A recent study [657] shows that in high-resolution global ocean models the dissipation of kinetic energy by horizontal friction follows a log-normal distribution. This study emphasizes the impact of the lognormal's extreme values as follows:

> ...The present results have implications for a range of physicists and oceanographers. For example, the downscale cascade of energy through the ocean sub-mesoscale has been inferred through localized in situ observations [15] and regional numerical models [38–40], but we have shown that most of the dissipation at the mesoscale occurs in a small number of high-dissipation locations due to the log-normal distribution (90% of dissipation at a given depth occurs in about 10% of the world ocean). This distribution presents challenges when extrapolating regional turbulence observations to global or basin-wide statistics, where it is common to assume normal statistics.

6.4 Perturbation Theory for Non-Gaussian Probability Densities

While many data sets can be modeled using Gaussian random fields, given the diversity of spatial data it is desirable to have methods that can handle more general probability distributions. Below we discuss the calculation of stochastic moments for *symmetric but non-Gaussian probability densities*. These calculations rely on properties of Gaussian random fields, because they are based on perturbation expansions of the target pdf around a suitably chosen Gaussian pdf.

Since the expansions are developed with respect to a Gaussian pdf, it is logical that only symmetric densities yield reasonable results. In the case of non-symmetric densities, nonlinear transformations can symmetrize the pdf, which can then be approximated by means of the methods described below. Typical symmetrizing transforms include the logarithmic transformation and the more general power transform. The latter is also known as Box-Cox transform, and it is defined by [93]

$$Y = \begin{cases} \frac{(X+\lambda_2)^{\lambda_1}-1}{\lambda_1}, & \lambda_1 > 0 \\ \ln(X+\lambda_2), & \lambda_1 = 0. \end{cases} \tag{6.67}$$

Using perturbation methods, it is possible to construct multivariate distributions with local deviations from Gaussianity based on random fields with Boltzmann-Gibbs exponential densities. This approach has not been sufficiently exploited in geostatistics and can lead to novel models of spatial dependence [364].

In statistical physics there is an extensive literature on methods such as *variational approximations, Feynman diagrams, renormalization group, and replica averages*, that apply to non-Gaussian probability densities [122, 245, 297, 567]. These methods have wide-ranging applications in problems that admit perturbation series expansions. Typical examples include permeability fluctuations due to subsurface heterogeneity and non-Gaussian terms in a joint pdf [361, 369, 371] (and references therein). Further research on closed-form moment expressions and on the accuracy of such approximations in different regions of the parameter space is necessary in order to apply such methods in spatial data modeling.

Below, we briefly discuss *perturbation expansions*, *variational approximations* [50], [122, pp. 198–200], [245, pp. 71–77] and *cumulant expansions* [50, 245, 369]. In these approaches, the non-Gaussian pdf is expanded around an "optimal" Gaussian pdf. The latter is then used as a zero-point approximation for perturbation expansion of the moments [371, 567]. However, since the optimal pdf is Gaussian, the approximate odd-order moments will be equal to zero. Hence, such approximations are suitable for *symmetric* distributions—that nonetheless deviate from the Gaussian. The Box-Cox transform (6.67) can be applied first to reduce the asymmetry of non-symmetric probability distributions.

6.4.1 Perturbation Expansion

Let us assume a pdf in the *Boltzmann-Gibbs representation* defined by

$$f_{\mathrm{x}}(\mathbf{x};\boldsymbol{\theta}) = \frac{e^{-\mathcal{H}(\mathbf{x};\boldsymbol{\theta})}}{Z(\boldsymbol{\theta})}, \tag{6.68}$$

with an energy functional $\mathcal{H}(\mathbf{x}; \boldsymbol{\theta})$ of the following form

$$\mathcal{H} = \mathcal{H}_0 + \mathcal{H}_1. \tag{6.69}$$

In (6.69), the functional $\mathcal{H}_0$ is the quadratic, Gaussian component of the energy, and $\mathcal{H}_1$ is the non-Gaussian perturbation. We will denote the expectation with respect to the Gaussian Boltzmann-Gibbs pdf with energy $\mathcal{H}_0$. It will be convenient to assume that $\mathcal{H}_0$ is centered around zero so that $\mathcal{E}_0[X(\mathbf{s}; \omega)] = 0$. This is achieved by constructing $\mathcal{H}_0$ so that it involves interactions between fluctuations.

The expectation of a functional $\Phi(\cdot)$ of the random field $X(\mathbf{s}; \omega)$ is defined by means of functional integration as follows

$$\mathcal{E}[\Phi(\cdot)] = \frac{\int \mathcal{D}x(\mathbf{s})\, \Phi(\cdot)\, \mathrm{e}^{-\mathcal{H}}}{\int \mathcal{D}x(\mathbf{s})\, \mathrm{e}^{-\mathcal{H}}}. \tag{6.70}$$

Mapping to Gaussian expectation As stated above, $\mathcal{E}_0\,[\cdot]$ denotes the expectation over the trial Gaussian pdf with the Boltzmann-Gibbs expression

$$f_0[x(\mathbf{s}); \boldsymbol{\theta}] = \frac{\mathrm{e}^{-\mathcal{H}_0}}{\int \mathcal{D}x(\mathbf{s})\, \mathrm{e}^{-\mathcal{H}_0}}.$$

With the decomposition (6.69) and the above in mind, the numerator of (6.70) is proportional to the expectation of the term $\Phi[X(\mathbf{s}; \omega)] \exp\{-\mathcal{H}_1\,[X(\mathbf{s}; \omega)]\}$, while the denominator is proportional to the expectation of $\exp\{-\mathcal{H}_1\,[X(\mathbf{s}; \omega)]\}$. The factor of proportionality is equal to the inverse of the partition function Z_0 for both the numerator and the denominator.

Based on the above, the expectation (6.70) is expressed as follows

$$\mathcal{E}[\Phi(\cdot)] = \frac{\mathcal{E}_0\left[\Phi(\cdot)\,\mathrm{e}^{-\mathcal{H}_1}\right]}{\mathcal{E}_0\left[\mathrm{e}^{-\mathcal{H}_1}\right]} = \frac{N(\Phi; \mathcal{H}_1)}{D(\mathcal{H}_1)}. \tag{6.71a}$$

The numerator and denominator of the above can be expressed as Taylor series that follow from the Taylor expansion of the exponential $\mathrm{e}^{-\mathcal{H}_1}$, i.e.,

$$N(\Phi; \mathcal{H}_1) = \sum_{n=0}^{\infty} \frac{(-1)^n}{n!} \mathcal{E}_0\left[\Phi(\cdot)\,\mathcal{H}_1^n\right], \tag{6.71b}$$

$$D(\Phi; \mathcal{H}_1) = \sum_{n=0}^{\infty} \frac{(-1)^n}{n!} \mathcal{E}_0\left[\mathcal{H}_1^n\right]. \tag{6.71c}$$

Example 6.8 Assume that the random variable $X(\omega)$ satisfies a Boltzmann-Gibbs pdf with quartic energy function, defined by

$$f_X(x) = \frac{1}{Z} e^{-\mathcal{H}[x]}, \text{ where } \mathcal{H}[x] = \frac{\alpha^2}{2} x^2 + \beta^4 x^4. \tag{6.72}$$

Calculate the variance of $X(\omega)$ by means of the perturbation expansion around the Gaussian approximation. Compare the perturbation based result with the exact solution.

Answer The pdf (6.72) is centered around zero and thus symmetric. Hence, the linear expectation vanishes, i.e., $\mathcal{E}[X(\omega)] = 0$. We set $\mathcal{H}_0[x] = \frac{\alpha^2}{2} x^2$. Then, the variance is given by the expectation of the squared fluctuation, i.e.,

$$\sigma_X^2 = \mathcal{E}[X^2(\omega)] = \frac{\int_{-\infty}^{\infty} dx \, f_X(x) \, x^2}{\int_{-\infty}^{\infty} dx \, f_X(x)}. \tag{6.73}$$

Based on (6.71), the variance can be expressed as the following ratio of respective numerator and denominator series

$$\sigma_X^2 = \frac{\sum_{n=1}^{\infty} \frac{(-1)^n}{n!} \beta^{4n} \, \mathcal{E}_0 \left[X^{4n+2}(\omega) \right]}{\sum_{n=1}^{\infty} \frac{(-1)^n}{n!} \beta^{4n} \, \mathcal{E}_0 \left[X^{4n}(\omega) \right]}.$$

If we keep only the lowest-order terms in β in both the numerator and denominator we obtain the approximation

$$\sigma_X^2 \approx \frac{\mathcal{E}_0 \left[X^2(\omega) \right] - \beta^4 \mathcal{E}_0 \left[X^6(\omega) \right]}{1 - \beta^4 \mathcal{E}_0 \left[X^4(\omega) \right]}.$$

We use $\mathcal{E}_0 \left[X^2(\omega) \right] = \alpha^{-2}$ and the Isserlis-Wick relation (6.58) to evaluate the higher-order even moments; this leads to the equations

$$\mathcal{E}_0 \left[X^4(\omega) \right] = \frac{4!}{2! \, 2^2} \alpha^{-4} = 3 \, \alpha^{-4},$$

$$\mathcal{E}_0 \left[X^6(\omega) \right] = \frac{6!}{3! \, 2^3} \alpha^{-6} = 15 \, \alpha^{-6}.$$

Based on the above, it follows that the variance is approximated to the leading perturbation order by

$$\sigma_X^2 \approx \frac{1}{\alpha^2} \frac{1 - 15 \left(\frac{\beta}{\alpha} \right)^4}{1 - 3 \left(\frac{\beta}{\alpha} \right)^4}. \tag{6.74}$$

Similarly, if we include the $O(\beta^8)$ terms in both the numerator and denominator the following second-order approximation is derived

$$\sigma_{\mathrm{x}}^2 \approx \frac{\mathcal{E}_0\left[\mathrm{X}^2(\omega)\right] - \beta^4 \mathcal{E}_0\left[\mathrm{X}^6(\omega)\right] + \frac{\beta^8}{2}\mathcal{E}_0\left[\mathrm{X}^{10}(\omega)\right]}{1 - \beta^4 \mathcal{E}_0\left[\mathrm{X}^4(\omega)\right] + \frac{\beta^8}{2}\mathcal{E}_0\left[\mathrm{X}^8(\omega)\right]}$$

$$= \frac{1}{\alpha^2}\,\frac{1 - 15\left(\frac{\beta}{\alpha}\right)^4 + \frac{945}{2}\left(\frac{\beta}{\alpha}\right)^8}{1 - 3\left(\frac{\beta}{\alpha}\right)^4 + 105\left(\frac{\beta}{\alpha}\right)^8}. \tag{6.75}$$

In Fig. 6.11 the variance of the quartic model is plotted as a function of the *aspect ratio* β/α and compared with the approximations derived by first- and second-order expansions of both the numerator and the denominator in $(\beta/\alpha)^4$. For $\beta/\alpha < 0.2$, both approximations accurately track the exact variance. The first-order approximation starts to diverge toward negative values, whereas the second-order approximation starts to diverge, at higher ratios β/α than the first-order, toward positive values.

Note that to obtain the leading-order in β/α variance correction, it is necessary to expand the denominator, which is of the form $1-x$, where $x = 3(\beta/\alpha)^4$, in a Taylor series around $x = 0$. The following is the leading order expansion $(1-x)^{-1} = 1 + x + O(x^2)$. Based on the above and (6.74) it is easy to show that the leading-order variance correction is

$$\sigma_{\mathrm{x}}^2 = \frac{1}{\alpha^2}\left[1 - 12\left(\frac{\beta}{\alpha}\right)^4\right] + O\left((\beta/\alpha)^8\right). \tag{6.76}$$

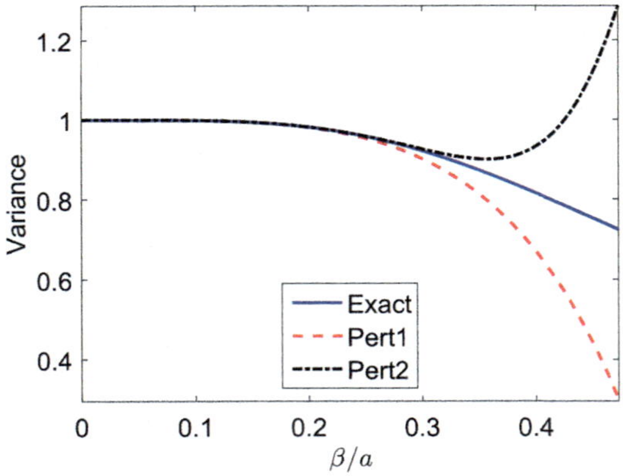

Fig. 6.11 Variance of the Boltzmann-Gibbs pdf with quartic Hamiltonian as defined in (6.72). The "exact" result is based on numerical integration of the density. "Pert-1" represents the leading-order perturbation approximation given by (6.74), and "Pert-2" represents the second-order perturbation approximation given by (6.75)

Connection with RG analysis A multivariate version of the energy function (6.72) for an N-dimensional vector $\mathbf{x}$ is

$$\mathcal{H}(\mathbf{x}) = \frac{1}{2}\sum_{n=1}^{N}\sum_{m=1}^{N} x_n J_{n,m} x_m - \frac{g}{4!}\sum_{n=1}^{N} x_n^4. \tag{6.77}$$

This energy function comprises a Gaussian component which is given by the quadratic part with couplings $J_{n,m}$ between two variables x_n and x_m and a quartic part. The model (6.77) can be justified based on the principle of maximum entropy (see Chap. 13) if the available information (statistical constraints) involves the covariance matrix and the mean kurtosis of all the variables. For systems with many degrees of freedom, such as (6.77), it is often useful to seek for simplified (coarse-grained) descriptions as discussed in [97, 371]. Renormalization group (RG) analysis is a coarse-graining procedure developed in statistical physics based on seminal contributions by Leo Kadanoff and Kenneth G. Wilson. RG analysis determines if the coarse-grained system obtained after eliminating the finer-scale modes will tend towards the Gaussian distribution or some other non-Gaussian form [92, 97, 417].

6.4.2 *The Cumulant Expansion*

The preceding section demonstrates how to obtain perturbation expansions for statistical moments of non-Gaussian random fields by means of the perturbation expansion (6.71). A difficulty of the perturbation expansion is that the expressions for the statistical moments involve the ratio of two infinite series. This implies some further manipulations and rearrangement of terms, if the result needs to be computed to a specified order in the perturbation parameter. For example, the variance (6.74) is based on numerator and denominator expansions that are $O(\beta^4)$, but the combined result is not $O(\beta^4)$.

In this section we recover the results of the perturbation expansion using a more systematic approach which is based on the cumulant generating functional defined in Sects. 4.5.5 and 6.2.3. The cumulants of $X(\mathbf{s};\omega)$ are then given by respective derivatives of the CGF with respect to the auxiliary field $u(\mathbf{s})$. This allows for the efficient calculation of cumulants at the specified order of approximation.

Based on (6.47) the *cumulant generating functional* of the SRF $X(\mathbf{s};\omega)$ is given by the following logarithm of the ratio of two functional integrals

$$\mathcal{K}_{\mathrm{x}}[u(\mathbf{s})] = \ln\left(\frac{\int \mathcal{D}x(\mathbf{s})\, \mathrm{e}^{-\mathcal{H}+\int \mathrm{d}\mathbf{s}\, u(\mathbf{s})\, x(\mathbf{s})}}{\int \mathcal{D}x(\mathbf{s})\, \mathrm{e}^{-\mathcal{H}}}\right), \tag{6.78}$$

where according to (6.69) the energy function $\mathcal{H}$ can be separated into a Gaussian component $\mathcal{H}_0$ and a non-Gaussian perturbation $\mathcal{H}_1$ as in (6.69), i.e.,

$$\mathcal{H} = \mathcal{H}_0 + \mathcal{H}_1.$$

The relation between the cumulant generating functional and the *cumulants of order n* of the random field X(**s**; ω) is given by extending (4.73) to the continuum. This is achieved by means of the transformation $\mathbf{u} \rightarrow u(\mathbf{s})$ and the following functional derivative

$$\kappa_{\mathrm{x};\mathbf{p}}(\mathbf{s}_1, \ldots, \mathbf{s}_N) = \left. \frac{\delta^{p_1+p_2+\ldots p_N} \mathcal{K}_{\mathrm{x}}[u(\mathbf{s})]}{\delta u_1^{p_1} \delta u_2^{p_2} \ldots \delta u_N^{p_N}} \right|_{u(\mathbf{s})=0}. \tag{6.79}$$

The index vector $\mathbf{p}^\top = (p_1, p_2, \ldots, p_N)$ denotes the *order of differentiation* at each point and $u_n = u(\mathbf{s}_n)$, for $n = 1, \ldots, N$. The vector **p** also denotes the order of the cumulant per location.

Remark The CGF defined in (6.78) can be expressed as the difference between the logarithms of two partition functions, i.e.,

$$\mathcal{K}_{\mathrm{x}}[u(\mathbf{s})] = \ln Z[u(\mathbf{s})] - \ln Z[u(\mathbf{s}) = 0], \tag{6.80a}$$

where the partition function $Z[u(\mathbf{s})]$ is defined by the following functional integral

$$Z[u(\mathbf{s})] = \int \mathcal{D}x(\mathbf{s})\, \mathrm{e}^{-\mathcal{H}[x(\mathbf{s})] + \int \mathrm{d}\mathbf{s}\, u(\mathbf{s})\, x(\mathbf{s})}. \tag{6.80b}$$

These expressions will be useful in the discussion of the replica method in Chap. 14.

Let us continue with the derivation of the cumulant expansion. We define the *perturbation potential function* which involves both the interaction of the auxiliary field with $x(\mathbf{s})$ and the non-Gaussian energy $\mathcal{H}_1$ as follows

$$V = -\mathcal{H}_1 + \int_{\mathbb{R}^d} \mathrm{d}\mathbf{s}\, u(\mathbf{s})\, x(\mathbf{s}). \tag{6.81}$$

Next, we express the cumulant generating functional using (6.78) and the well-known property of logarithms $\ln(a/b) = \ln a - \ln b$ as follows

$$\mathcal{K}_{\mathrm{x}}[u(\mathbf{s})] = \ln \left[\int \mathcal{D}x(\mathbf{s})\, \mathrm{e}^{-\mathcal{H}_0 + V} \right] - \ln K_0,$$

where K_0 is given by the following functional integral

$$K_0 = \int \mathcal{D}x(\mathbf{s})\, \mathrm{e}^{-\mathcal{H}_0-\mathcal{H}_1}.$$

The term K_0 does not involve the auxiliary field $u(\mathbf{s})$. Hence, it is irrelevant for the evaluation of the cumulants, because the latter involve functional derivatives with respect to $u(\mathbf{s})$. Without regret, we drop the irrelevant term $\ln K_0$ in the following. The relevant term involves the functional integral of the exponential with energy $\mathcal{H}_0 - V$. This term can be rearranged by multiplying and dividing the functional integral inside the logarithm with the Gaussian partition function

$$Z_0 = \int \mathcal{D}x(\mathbf{s})\, \mathrm{e}^{-\mathcal{H}_0}.$$

The key observation is that the ratio of the two functional integrals that is formed represents a Gaussian expectation, i.e.,

$$\frac{\int \mathcal{D}x(\mathbf{s})\, \mathrm{e}^{-\mathcal{H}_0+V}}{\int \mathcal{D}x(\mathbf{s})\, \mathrm{e}^{-\mathcal{H}_0}} = \mathcal{E}_0\left[\mathrm{e}^{V}\right],$$

where $\mathcal{E}_0[\cdot]$ is the expectation with respect to the Gaussian pdf.[6] In light of the above, the non-Gaussian cumulant generating functional is expressed as

$$\mathcal{K}_{\mathrm{x}}[u(\mathbf{s})] = \ln\left\{Z_0\, \mathcal{E}_0\left[\mathrm{e}^{V}\right]\right\} = \ln Z_0 + \ln\left\{\mathcal{E}_0\left[\mathrm{e}^{V}\right]\right\}.$$

The term Z_0 is also independent of the auxiliary field and thus irrelevant for the calculation of the cumulants; hence it can be dropped without affecting the rest of the calculation.

The *cumulant expansion* is based on the following equation

$$\mathcal{K}'_{\mathrm{x}}[u(\mathbf{s})] = \ln \mathcal{E}_0\left[\mathrm{e}^{V}\right] = \sum_{n=1}^{\infty} \frac{1}{n!}\, C_n[V], \tag{6.82}$$

where $C_n[V]$ is the order-n cumulant of the potential function V defined by (6.81) and $\mathcal{E}_0[\cdot]$ denotes the expectation with respect to the Gaussian pdf.

The cumulant expansion is obtained from (6.61) (see also [399, p. 77], [170, p. 364]).

[6] The expectation operator acts over the degrees of freedom of the potential function $V[\mathrm{X}(\mathbf{s};\omega)]$ which herein is denoted by V for short.

Remark The left-hand side of (6.82) is the cumulant generating functional of $\mathrm{X}(\mathbf{s};\omega)$, whereas the right-hand side involves the cumulants of the potential function V with respect to the Gaussian distribution.

The logarithmic term of (6.82) can be expanded using the Taylor series of the exponential as follows

$$\ln\left\{\mathcal{E}_0\left[1+\sum_{n=1}^{\infty}\frac{V^n}{n!}\right]\right\}.$$

Furthermore, we can use the Taylor series expansion of the logarithm, i.e.,

$$\ln(1+x)=x-\frac{x^2}{2}+\frac{x^3}{3}-\frac{x^4}{4}+\ldots.$$

The left-hand side of (6.82) can thus be expanded using the Taylor series expansions of the exponential and the logarithm. The cumulants of order n on the right-hand side of (6.82) correspond to the terms that involve same-order moments[7] of the potential function V. This leads to the following *cumulant equation hierarchy* for the random field V, which is defined by means of (6.81):

$$C_1[V]=\mathcal{E}_0\left[V\right], \tag{6.83a}$$

$$C_2[V]=\mathcal{E}_0\left[V^2\right]-\mathcal{E}_0\left[V\right]^2, \tag{6.83b}$$

$$C_3[V]=\mathcal{E}_0\left[V^3\right]-3\,\mathcal{E}_0\left[V^2\right]\mathcal{E}_0\left[V\right]+2\,\mathcal{E}_0\left[V\right]^3, \tag{6.83c}$$

$$\begin{aligned}C_4[V]=&\mathcal{E}_0\left[V^4\right]-4\,\mathcal{E}_0\left[V^3\right]\mathcal{E}_0\left[V\right]-3\,\mathcal{E}_0\left[V^2\right]^2\\&+12\,\mathcal{E}_0\left[V^2\right]\mathcal{E}_0\left[V\right]^2-6\,\mathcal{E}_0\left[V\right]^4,\end{aligned} \tag{6.83d}$$

$$\begin{aligned}C_5[V]=&\mathcal{E}_0\left[V^5\right]+24\,\mathcal{E}_0\left[V\right]^5-5\,\mathcal{E}_0\left[V^4\right]\mathcal{E}_0\left[V\right]\\&-10\,\mathcal{E}_0\left[V^3\right]\mathcal{E}_0\left[V^2\right]+20\,\mathcal{E}_0\left[V^3\right]\mathcal{E}_0\left[V\right]^2\\&+30\,\mathcal{E}_0\left[V^2\right]^2\mathcal{E}_0\left[V\right]-60\,\mathcal{E}_0\left[V^2\right]\mathcal{E}_0\left[V\right]^3.\end{aligned} \tag{6.83e}$$

Calculations with Cumulants Note that the hierarchy of equations (6.83) applies to the cumulants of the potential function V. The calculation of the cumulants $\kappa_{\mathrm{x};\mathbf{n}}(\cdot)$

[7] This includes products of $k=1,\ldots,K$ moments of order $m_k<n$ such that $\sum_{k=1}^{K}=n$.

of $\mathrm{X}(\mathbf{s};\omega)$ is based on the cumulant generating functional (6.79) and the cumulant expansion (6.82).

Using the *binomial expansion* of the potential function (6.81) in terms of the two summands, $-\mathcal{H}_1$ and $\int d\mathbf{s} u(\mathbf{s})x(\mathbf{s})$, it can be shown that the terms in the n-th order cumulant $C_n[V]$ involve expectations of quantities such as

$$\frac{n!}{m!\,(n-m)!}(-1)^{n-m}\,\mathcal{H}_1^{n-m}\left(\int d\mathbf{s}\, u(\mathbf{s})\, \mathrm{X}(\mathbf{s};\omega)\right)^m, \quad 1 \le k \le n.$$

The calculation of a cumulant $\kappa_{\mathrm{x};\mathbf{n}}(\cdot)$ of order m involves the m-th order derivative of the cumulant generating functional with respect to $u(\mathbf{s})$ according to (6.79); thus, only the terms of order $k = n - m \ge 0$ in the non-Gaussian perturbation $\mathcal{H}_1$ are involved. For leading-order approximations, only the terms with $k = 0, 1$ are necessary.

The following identities, which are valid for even integer m, are useful for calculations with the cumulant expansion

$$\left.\frac{\delta^m \mathcal{E}_0\left[\left\{\int d\mathbf{s}\, u(\mathbf{s})\, \mathrm{X}(\mathbf{s};\omega)\right\}^m\right]}{\delta u(\mathbf{s}_1)\, \delta u(\mathbf{s}_2)\, \ldots\, \delta u(\mathbf{s}_m)}\right|_{u(\mathbf{s})=0} = m!\, \mathcal{E}_0\left[\mathrm{X}(\mathbf{s}_1;\omega)\mathrm{X}(\mathbf{s}_2;\omega)\ldots\mathrm{X}(\mathbf{s}_m;\omega)\right]. \tag{6.84}$$

$$\left.\frac{\delta^m \mathcal{E}_0\left[\left\{\int d\mathbf{s}\, u(\mathbf{s})\, \mathrm{X}(\mathbf{s};\omega)\right\}^{2l}\right]^k}{\delta u(\mathbf{s}_1)\, \delta u(\mathbf{s}_2)\, \ldots\, \delta u(\mathbf{s}_m)}\right|_{u(\mathbf{s})=0} = m!\, \mathcal{E}_0\left[\mathrm{X}(\mathbf{s}_1;\omega)\mathrm{X}(\mathbf{s}_2;\omega)\ldots\mathrm{X}(\mathbf{s}_{2l};\omega)\right], \quad \text{for } m = 2lk. \tag{6.85}$$

The m-order derivatives appear in the calculation of the cumulants of order m.

We illustrate the application of the cumulant expansion in the calculation of the covariance function and the kurtosis below.

Covariance function approximation The zero-order approximation of the covariance function is obtained by ignoring the perturbation $\mathcal{H}_1$. Then, the covariance is simply given by

$$C_{\mathrm{xx}}^{(0)}(\mathbf{s}_1, \mathbf{s}_2) = C_0(\mathbf{s}_1, \mathbf{s}_2),$$

where $C_0(\mathbf{s}_1, \mathbf{s}_2)$ is the covariance function corresponding to $\mathcal{H}_0$.

Let us now confirm that this result is also obtained from the cumulant expansion and calculate non-Gaussian corrections. Based on (6.82) the first term of the cumulant generating functional is obtained for $n = 1$ and involves $C_1[V]$.

- First cumulant $n = 1$

Based on (6.83a) the leading-order (in $\mathcal{H}_1$) covariance approximation is

$$C^{(0)}_{\mathrm{xx}}(\mathbf{s}_1, \mathbf{s}_2) = \left. \frac{\delta^2 C_1[V]}{\delta u_1 \delta u_2} \right|_{u(\mathbf{s})=0}. \tag{6.86}$$

From (6.83a) it follows that $C_1[V]$ contains two terms:

$$-\mathcal{E}_0\left[\mathcal{H}_1\right], \quad \text{and} \quad \mathcal{E}_0\left[\int \mathrm{d}\mathbf{s}\, u(\mathbf{s})\, \mathrm{X}(\mathbf{s};\omega)\right].$$

The first term is independent of $u(\mathbf{s})$ and thus does not contribute to the $\mathrm{X}(\mathbf{s};\omega)$ cumulants. The second term also vanishes because $\mathcal{E}_0\,[\mathrm{X}(\mathbf{s};\omega)] = 0$ for a zero-mean SRF. Thus, non-zero contributions will follow from higher orders in V. From the above calculation we keep the result

$$\mathcal{E}_0\,[V] = -\mathcal{E}_0\left[\mathcal{H}_1\right]. \tag{6.87}$$

- Second cumulant $n = 2$

  Based on (6.83b), $C_2[V]$ contains two terms: $\mathcal{E}_0\left[V^2\right]$ and $\mathcal{E}_0\,[V]^2$. The second term is independent of the auxiliary field and does not contribute to the covariance. The first term is expanded as follows

$$\mathcal{E}_0\left[V^2\right] = \mathcal{E}_0\left[\mathcal{H}_1^2\right] + \mathcal{E}_0\left[\left\{\int \mathrm{d}\mathbf{s}\, u(\mathbf{s})\, \mathrm{X}(\mathbf{s};\omega)\right\}^2\right] - 2\,\mathcal{E}_0\left[\mathcal{H}_1 \int \mathrm{d}\mathbf{s}\, u(\mathbf{s})\, \mathrm{X}(\mathbf{s};\omega)\right]. \tag{6.88a}$$

  Only the second of the three terms on the right-hand side has a non-vanishing second derivative with respect to $u(\mathbf{s})$: The first term does not contain $u(\mathbf{s})$, while the third term includes a vanishing expectation $\mathcal{E}_0\,[\mathrm{X}(\mathbf{s};\omega)\,]$. In particular, based on (6.84), the second-order derivative of the second term leads to

$$\frac{\delta^2 \mathcal{E}_0\left[V^2\right]}{\delta u(\mathbf{s}_1)\delta u(\mathbf{s}_2)} = 2\,\mathcal{E}_0\,[\mathrm{X}(\mathbf{s}_1;\omega)\mathrm{X}(\mathbf{s}_2;\omega)] = 2\,C_0(\mathbf{s}_1, \mathbf{s}_2). \tag{6.88b}$$

  Combining the above result with (6.79) and (6.82) it follows that indeed the leading-order covariance approximation is $C_0(\mathbf{s}_1, \mathbf{s}_2)$.

  So, after considerable grinding using the $n = 2$ order of the cumulant expansion we recovered the intuitive zero-order result! At least this exercise confirms that the cumulant expansion agrees with our expectations. To generate perturbation corrections it is necessary to consider $n = 3$.
- Third cumulant $n = 3$

  $C_3[V]$ contains three terms according to (6.83c). The third one, $\mathcal{E}_0[V]^3$ does not contribute because it is independent of $u(\mathbf{s})$. The other two terms are:

1. $C_{3,a}[V] = \mathcal{E}_0\left[V^3\right]$
2. $C_{3,b}[V] - 3\,\mathcal{E}_0\left[V^2\right]\mathcal{E}_0\left[V\right]$.

In light of (6.82), let us denote the *covariance contributions* that are derived from each term as follows

$$C_{n=3;\mathrm{a}}(\mathbf{s}_1, \mathbf{s}_2) = \left.\frac{1}{6}\frac{\delta^2\mathcal{E}_0\left[V^3\right]}{\delta u(\mathbf{s}_1)\delta u(\mathbf{s}_2)}\right|_{u(\mathbf{s})=0}, \tag{6.89}$$

$$C_{n=3;\mathrm{b}}(\mathbf{s}_1, \mathbf{s}_2) = -\left.\frac{1}{2}\frac{\delta^2\mathcal{E}_0\left[V^2\right]\mathcal{E}_0\left[V\right]}{\delta u(\mathbf{s}_1)\delta u(\mathbf{s}_2)}\right|_{u(\mathbf{s})=0}. \tag{6.90}$$

It is straightforward to evaluate $C_{n=3;\mathrm{b}}(\mathbf{s}_1, \mathbf{s}_2)$ using (6.87) for the expectation $\mathcal{E}_0\left[V\right]$ and (6.88b) for the second-order derivative of $\mathcal{E}_0[V^2]$ which lead to

$$C_{n=3;\mathrm{b}}(\mathbf{s}_1, \mathbf{s}_2) = \mathcal{E}_0\left[\mathcal{H}_1\right] C_0(\mathbf{s}_1, \mathbf{s}_2).$$

On the other hand, the term $\mathcal{E}_0[V^3]$ in $C_{n=3;\mathrm{a}}(\mathbf{s}_1, \mathbf{s}_2)$ is given by

$$\begin{aligned}\mathcal{E}_0\left[V^3\right] = & -\mathcal{E}_0\left[\mathcal{H}_1^3\right] + \mathcal{E}_0\left[\left\{\int \mathrm{d}\mathbf{s}\, u(\mathbf{s})\, \mathrm{X}(\mathbf{s};\omega)\right\}^3\right] \\ & - 3\,\mathcal{E}_0\left[\mathcal{H}_1\left\{\int \mathrm{d}\mathbf{s}\, u(\mathbf{s})\, \mathrm{X}(\mathbf{s};\omega)\right\}^2\right] \\ & + 3\,\mathcal{E}_0\left[\mathcal{H}_1^2 \int \mathrm{d}\mathbf{s}\, u(\mathbf{s})\, \mathrm{X}(\mathbf{s};\omega)\right].\end{aligned}$$

The fact that $C_{n=3;a}(\mathbf{s}_1, \mathbf{s}_2)$ involves the second-order functional derivative with respect to $u(\mathbf{s})$, evaluated at $u(\mathbf{s}) = 0$, implies that only terms proportional to $u^2(\mathbf{s})$ contribute; hence, only the third term on the right-hand side of the above makes a non-zero contribution that can be calculated using (6.84). This term has a multiplicity factor equal to six—due to the pre-factor of three and the doubling that results from the functional derivative (6.84). This cancels out the $1/6$ prefactor in (6.89) and finally leads to the following covariance correction

$$C_{n=3;a}(\mathbf{s}_1, \mathbf{s}_2) = -\mathcal{E}_0\left[\mathcal{H}_1 \mathrm{X}(\mathbf{s}_1;\omega)\, \mathrm{X}(\mathbf{s}_2;\omega)\right].$$

Finally, collecting all the terms derived from $C_n(V)$ where $n = 2, 3$ we find that the *covariance function* with the *leading-order perturbation corrections* is given by

$$C_{xx}(\mathbf{s}_1, \mathbf{s}_2) = C_0(\mathbf{s}_1, \mathbf{s}_2) - \mathcal{E}_0\left[\mathcal{H}_1\, X(\mathbf{s}_1;\omega)\, X(\mathbf{s}_2;\omega)\right] + \mathcal{E}_0\left[\mathcal{H}_1\right] C_0(\mathbf{s}_1, \mathbf{s}_2). \quad (6.91)$$

The cumulant expansion avoids separate series expansions for the numerator and denominator and the collection of equal-order terms, which is necessary within the framework of the "naive" perturbation approach. We use the cumulant expansion (6.91) in the following section to derive non-stationary covariance functions.

Variance Based on (6.91), the variance including the leading-order perturbation corrections is given by means of the equation

$$\sigma_x^2(\mathbf{s}) = \sigma_0^2(\mathbf{s}) - \mathcal{E}_0\left[\mathcal{H}_1 X^2(\mathbf{s};\omega)\right] + \sigma_0^2(\mathbf{s})\, \mathcal{E}_0\left[\mathcal{H}_1\right]. \quad (6.92)$$

We leave it as an exercise for the reader to verify that in the case of marginal non-Gaussian pdf of the Example 6.8 the expression (6.92) coincides with the leading-order perturbation expansion (6.76).

Example 6.9 Consider the point-like random variable $X(\omega)$ with a marginal pdf defined by the quartic energy function $\mathcal{H}[x] = \frac{\alpha^2}{2} x^2 + \beta^4 x^4$ defined in the Example 6.8. Confirm that the $O(\beta^4)$ variance correction calculated with the "naive" perturbation expansion in the Example 6.8 agrees with (6.92) which is derived using the cumulant expansion.

6.4.3 Non-stationarity Caused by Non-uniform Perturbations

Non-stationarity is an important research topic in spatial data modeling. The detection of *non-stationarity* may indicate the presence of processes such as attenuation, point-like sources, time-variant amplitudes, and so on. The development of non-stationary covariance models is an evolving research field. This section focuses solely on some possibilities resulting from the cumulant expansion. The readers who would like a spherical perspective on the current literature are referred to the review articles [256, 887] (see also Sect. 6.2.5).

In the following, we show how the covariance cumulant expansion can be used to construct *non-stationary covariance functions* in which the non-stationarity is linked to non-uniform perturbations of the energy functional.

As in the above, we adopt the Boltzmann-Gibbs representation of the joint pdf (6.68). Let us assume that $\mathcal{H}_0$ is a centered quadratic functional with known stationary covariance function $C_0(\mathbf{r})$, and that the perturbation energy is given by

$$\mathcal{H}_1 = \int d\mathbf{s}\, \lambda(\mathbf{s})\, x^2(\mathbf{s}), \tag{6.93}$$

where the *perturbation strength* $\lambda(\mathbf{s})$ is a known, real-valued function of space. This function introduces non-uniformity in the energy model parameters. For leading-order corrections to the covariance function to be adequate, it should be ensured that the term $\mathcal{H}_1$ can be treated as a small perturbation. The condition $\left|\mathcal{E}_0\left[\mathcal{H}_1\right]\right| \ll \mathcal{E}_0\left[\mathcal{H}_0\right]$ establishes the smallness of $\mathcal{H}_1$, at least as an expectation. A stricter condition would require that $|\mathcal{H}_1| \ll \mathcal{H}_0$ for all functions $x(\mathbf{s})$.

First, we calculate the covariance function using the leading-order cumulant expansion result (6.91). The covariance function can be obtained within the framework of low-order cumulant expansions. Thus, using (6.91) and letting $\mathbf{r} = \mathbf{s}_1 - \mathbf{s}_2$ denote the spatial lag we obtain

$$\begin{aligned} C_{xx}(\mathbf{s}_1, \mathbf{s}_2) =& C_0(\mathbf{r}) - \int d\mathbf{s}\, \lambda(\mathbf{s})\, \mathcal{E}_0\left[\mathrm{X}^2(\mathbf{s};\omega)\, \mathrm{X}(\mathbf{s}_1;\omega)\, \mathrm{X}(\mathbf{s}_2;\omega)\right] \\ &+ \int d\mathbf{s}\, \lambda(\mathbf{s})\, \mathcal{E}_0\left[\mathrm{X}^2(\mathbf{s};\omega)\right] C_0(\mathbf{r}). \end{aligned}$$

We use the *Wick-Isserlis theorem* 6.2 to express the second term by means of products of second-order moments. After some rearrangement and the cancellation of the third term (proportional to σ_0^2) by an equal but opposite-sign contribution generated from the second term, we obtain the following expression for the covariance

$$C_{xx}(\mathbf{s}_1, \mathbf{s}_2) = C_0(\mathbf{r}) - 2 \int d\mathbf{s}\, \lambda(\mathbf{s})\, C_0(\mathbf{s} - \mathbf{s}_1)\, C_0(\mathbf{s} - \mathbf{s}_2). \tag{6.94}$$

Permissibility There are two ways to approach the question of permissibility for covariances obtained by means of perturbations. First, we can consider conditions that ensure a non-negative energy function $\mathcal{H}_0 + \mathcal{H}_1$ for all $x(\mathbf{s})$. Such conditions imply that the covariance function of the joint Boltzmann-Gibbs pdf (6.68) with the specific energy is permissible. However, this covariance function would not be given by the leading cumulant correction (6.94) if the perturbation is not small.

Second, we can consider permissibility conditions for covariance functions determined by the perturbation expansion (6.94). This expression is obtained by the leading-order cumulant expansion. Therefore, it represents the covariance of the joint pdf with energy $\mathcal{H}_0 + \mathcal{H}_1$, where $\mathcal{H}_1$ is given by (6.93), only if the latter term can be considered as a small perturbation. However, the function (6.94) can be investigated for permissibility outside the approximation framework of the Boltzmann-Gibbs pdf. General permissibility conditions thus do not assume that (6.94) is derived from the leading-order cumulant expansion of the energy functional $\mathcal{H}_0 + \mathcal{H}_1$.

Permissibility of energy functional Let us assume that $\mathcal{H}_0$ is given by the quadratic functional (6.23). The perturbed energy functional is then given by the expression

$$\mathcal{H} = \frac{1}{2} \int_{\mathcal{D}} \mathrm{d}\mathbf{s} \int_{\mathcal{D}} \mathrm{d}\mathbf{s}'\, x'(\mathbf{s})\, A(\mathbf{s}, \mathbf{s}')\, x'(\mathbf{s}'), \tag{6.95}$$

where $A(\mathbf{s}, \mathbf{s}') = C_0^{-1}(\mathbf{s} - \mathbf{s}') + \lambda(\mathbf{s})\, \delta(\mathbf{s} - \mathbf{s}')$. Since the above energy functional is still quadratic, the joint pdf remains Gaussian, and $A(\mathbf{s}, \mathbf{s}')$ represents the *precision operator*. However, $A(\mathbf{s}, \mathbf{s}')$ does not depend purely on the lag distance $\mathbf{s} - \mathbf{s}'$ due to the presence of the term $\lambda(\mathbf{s})$. Hence, the inverse covariance (and consequently the covariance function) is not translation invariant. If $A(\mathbf{s}, \mathbf{s}')$ is strictly positive definite, then it is invertible, and its inverse $A^{-1}(\mathbf{s}, \mathbf{s}')$ is a permissible covariance function.

A sufficient condition for $A(\mathbf{s}, \mathbf{s}')$ to be strictly positive definite is that $\mathcal{H} > 0$ for all $x(\mathbf{s}) : \mathbb{R}^d \to \mathbb{R}$ except for $x(\mathbf{s}) = 0$. Since $\mathcal{H}_0 > 0$ for all $x(\mathbf{s})$, it is sufficient to show that $\mathcal{H}_1 > 0$. In light of (6.93), the positivity of the energy perturbation is ensured by $\lambda(\mathbf{s}) > 0$.

6.4.4 Non-stationarity Caused by Localized Perturbations

In this section, we determine the covariance function obtained from (6.94) if $\lambda(\mathbf{s})$ is given by a superposition of *localized perturbations*, represented by a series of delta functions that are centered at the points $\mathbf{z}_k$, $k = 1, \ldots, K$, i.e.,

$$\lambda(\mathbf{s}) = \sum_{k=1}^{K} \lambda_k \delta(\mathbf{s} - \mathbf{z}_k).$$

The energy function that is obtained by means of this perturbation is given by

$$\mathcal{H} = \mathcal{H}_0 + \sum_{k=1}^{K} \lambda_k x^2(\mathbf{s}_k).$$

This energy function is non-negative for all realizations $x(\mathbf{s})$, for all $K \in \mathbb{N}$, and $\{\lambda_k\}_{k=1}^{K}$, if $\lambda_k > 0$ for all $k = 1, \ldots, K$.

By substituting the expression for the strength $\lambda(\mathbf{s})$ in the convolution integral (6.94), the covariance function becomes

$$C_{\mathrm{xx}}(\mathbf{s}_1, \mathbf{s}_2) = C_0(\mathbf{r}) - 2 \sum_{k=1}^{K} \lambda_k\, C_0(\mathbf{z}_k - \mathbf{s}_1)\, C_0(\mathbf{z}_k - \mathbf{s}_2). \tag{6.96}$$

The covariance function obtained by means of the localized perturbations is *non-stationary*, even though the initial covariance is stationary. This expression cannot be an accurate covariance representation if the condition $\lambda_k \ll 1/\sigma_0^2$ is not satisfied.

The *non-stationary variance* is obtained from (6.96) by setting $\mathbf{s}_1 = \mathbf{s}_2 = \mathbf{s}$, which leads to

$$\sigma_{\mathrm{x}}^2(\mathbf{s}) = \sigma_0^2 - 2 \sum_{k=1}^{K} \lambda_k \, C_0^2(\mathbf{z}_k - \mathbf{s}). \tag{6.97}$$

The procedure described above provides a simple approach for constructing non-stationary covariance functions that are physically motivated. For example, in studies of environmental pollutant distributions, local perturbations could arise due to the presence of "point-like sources" of pollution such as factories and sewage treatment plants. A critical point, however, is the permissibility of these covariance functions.

Permissibility A complete investigation requires that permissibility conditions on $\{\lambda_k\}_{k=1}^{K}$ be specified. Bochner's theorem cannot be used for this purpose because of the non-stationarity of the covariance function.

Let us consider the permissibility of (6.96) without reference to the origin of this equation. A *necessary condition* for the function $C_{\mathrm{xx}}(\mathbf{s}_1, \mathbf{s}_2)$ to be positive definite is that $C_{\mathrm{xx}}(\mathbf{s}_1, \mathbf{s}_1) \geq 0$. This is equivalent to

$$C_{\mathrm{xx}}(\mathbf{s}_1, \mathbf{s}_1) = \sigma_0^2 - 2 \sum_{k=1}^{K} \lambda_k \, C_0^2(\mathbf{z}_k - \mathbf{s}_1) \geq 0.$$

Based on the upper bound $C_0^2(\mathbf{z}_k - \mathbf{s}_1) \leq \sigma_0^4$, the above inequality holds if

$$1 - \sigma_0^2 \sum_{k=1}^{K} \lambda_k \geq 0. \tag{6.98}$$

Non-stationary 1D covariances We illustrate non-stationary covariance functions for one-dimensional fields (random processes). We consider a stationary, zero-mean Gaussian random process with *exponential covariance*

$$C_{0;n,m} = \exp(-\left| s_n - s_m \right| / \xi),$$

with correlation length $\xi = 20$. The process is sampled at $L = 1000$ sites $s_n = n\,a$ where $n = 1, \ldots, L$, that are uniformly spaced with unit step a. Non-stationary covariance functions are obtained by means of (6.96) that expresses the impact of a superposition of localized, stationarity breaking, sources.

Random source allocation First, we generate sources at $N_s = 60$ random sample locations with strength coefficients λ_k, $k = 1, \ldots, N_s$, randomly distributed in the interval $[-1/2, 0]$. The non-stationary covariance matrix calculated using (6.96) is displayed in Fig. 6.12a. A non-stationary realization is compared against a stationary realization obtained by means of $C_{0;n,m}$, where $n, m = 1, \ldots, L$, in Fig. 6.12c.

Both realizations are generated using the *covariance matrix factorization* approach (see Sect. 16.2) with the same set of random numbers, so that the differences observed are exclusively due to the covariance models. As shown in Fig. 6.12c, the non-stationary realization exhibits overall larger fluctuations than the original stationary one. The variance of the non-stationary process, obtained from (6.97), is plotted in Fig. 6.12e and supports this observation. For comparison purposes, note that the variance of the stationary process is equal to one.

Deterministic source allocation We repeat the experiment with non-stationary sources located at every sampling site with coefficients $\lambda(s_n) = -0.1\, s_n/L$, for $n = 1, \ldots, L$. The respective plots are shown on the right column of Fig. 6.12. The non-stationary realization, shown in Fig. 6.12d, exhibits a more definite departure from the stationary realization due to the continuous distribution of the sources. In addition, the non-stationary variance, shown in Fig. 6.12f, shows a linear increasing trend. This trend is cut-off near the domain's right boundary due to the smaller number of sources that affect sites. Based on (6.97) the number of source points that impact the non-stationary variance is determined from the correlation length of the stationary covariance C_0. The same effect is also present at the left domain boundary, but its impact is not as noticeable there due to the lower magnitude of the perturbation source coefficients near the left boundary.

6.4.5 The Variational Method

If the CGF can be explicitly determined, all the information regarding the moments of the pdf can be obtained in terms of cumulants. However, the CGF in most cases cannot be exactly evaluated for non-Gaussian joint pdfs. In such cases, accurate approximations of the CGF are useful. A method that is widely used both in statistical physics and machine learning is the so-called *variational approximation* [245, 400, 561, 630, 805]. The applications of the variational method in statistical inference are reviewed in [80].

The main idea underlying the variational method is that the statistical moments of the true Boltzmann-Gibbs distribution can be approximated using a simpler probability distribution that is more amenable to explicit calculations. This approximative distribution contains one or more "variational parameters". The term "variational" implies that these parameters are not known a priori; instead, they are varied so as to optimize a suitable measure of proximity between the true and the variational distribution.

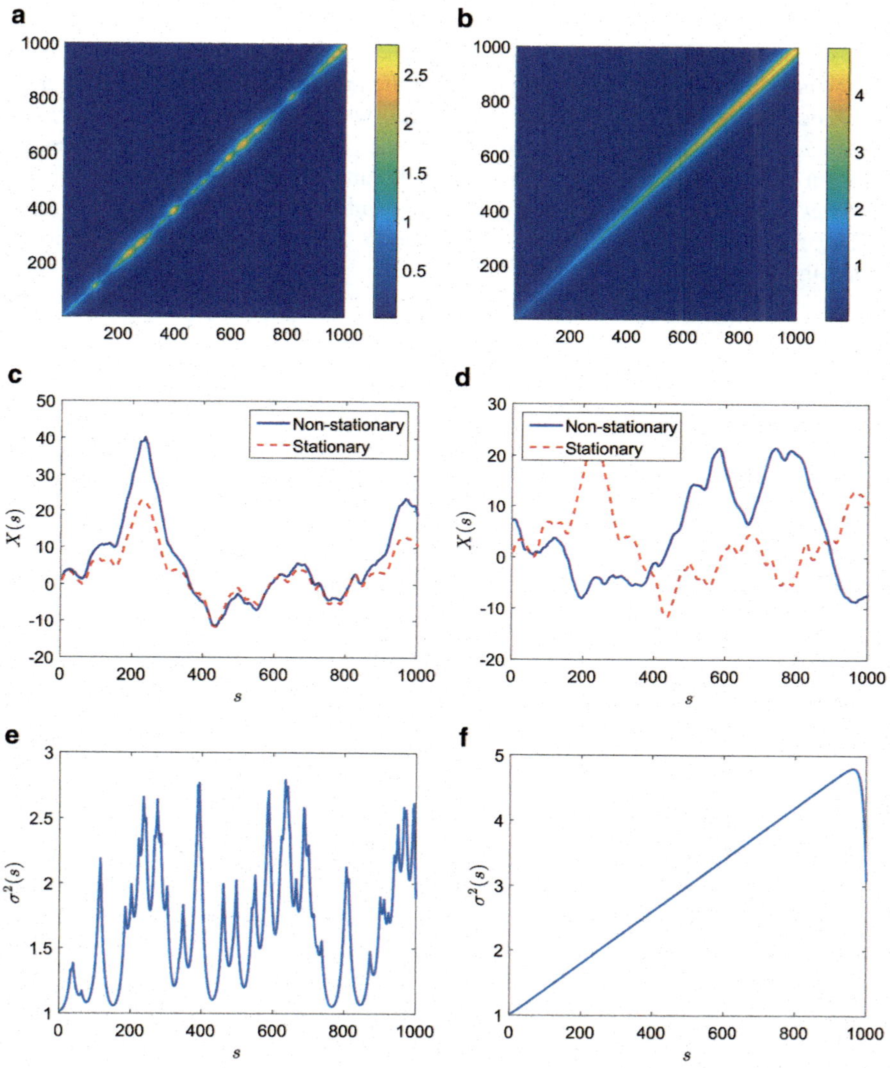

Fig. 6.12 Diagrams illustrating non-stationary covariance functions generated by (6.96) using random sources (left column plots) and sources with linearly increasing magnitude (right column plots). Top row: covariance matrices. Middle row: realizations for the unperturbed stationary covariance (dotted lines) and for the non-stationary covariance with inclusion of sources (solid lines). Bottom row: non-stationary variance based on (6.97). (**a**) Covariance—random sources. (**b**) Covariance—linear sources. (**c**) Realization—random sources. (**d**) Realization—linear sources. (**e**) Variance—random sources. (**f**) Variance—linear sources

In statistical physics, a proximity measure between two probability distributions is based on the *Gibbs free energy*.[8] For the purpose of spatial data analysis, the free energy is defined in terms of the logarithm of the partition function as $F = -\ln Z$. Let F denote the free energy of the true model (which is based on $\mathcal{H}$) and F_G the free energy of the variational approximation (which is based on $\mathcal{H}_G$). The "distance" between the true and the approximate distributions is measured by means of the *Kullback-Leibler divergence* (synonym: relative entropy).

The relative entropy of the true distribution with respect to the variational approximation is given by (cf. Sect. 16.7):

$$D_{KL}(f_{\mathcal{H}_G} \parallel f_{\mathcal{H}}) = -\int \mathcal{D}x(\mathbf{s})\, f_{\mathcal{H}_G}[x(\mathbf{s})]\, \ln\left(\frac{f_{\mathcal{H}}[x(\mathbf{s})]}{f_{\mathcal{H}_G}[x(\mathbf{s})]}\right). \tag{6.99}$$

The relative entropy is non-negative and asymmetric, i.e.,

$$D_{KL}(f_{\mathcal{H}_G} \parallel f_{\mathcal{H}}) \neq D_{KL}(f_{\mathcal{H}} \parallel f_{\mathcal{H}_G}).$$

The following general relation holds between the free energies of the true and the variational distributions [521]

$$F_G = F + D_{KL}(f_{\mathcal{H}_G} \parallel f_{\mathcal{H}}). \tag{6.100}$$

If we take into account the non-negativity of the relative entropy, the identity (6.100) leads to the inequality

$$F_G \geq F, \tag{6.101}$$

where the equality sign holds if $\mathcal{H}_G$ and $\mathcal{H}$ are identical. Hence, the variational free energy is an upper bound of the true free energy. This means that an *optimal variational approximation* can be obtained by minimizing F_G (or the Kullback-Leibler divergence) with respect to the variational parameter(s).

A typical choice for the variational distribution is the *Gaussian model* due to its nice analytical properties. A different choice is a model based on the product of marginal pdfs. This approach is used in the context of mean-field theories (cf. Sect. 15.2.3). In this section, we focus on the variational Gaussian approximation.

Variational Gaussian distribution In the following we consider an energy function $\mathcal{H}$ and a *variational Gaussian* energy functional $\mathcal{H}_G$. If $\mathcal{H}$ has a Gaussian component $\mathcal{H}_0$, the optimal $\mathcal{H}_G$ is not necessarily equal to $\mathcal{H}_0$. For simplicity, we will consider a joint pdf for a zero-mean vector $\mathbf{X} \in \mathbb{R}^N$. The results can be extended

[8] The statistical physics definition of the free energy involves the temperature T and the Boltzmann constant k_B, i.e., $F = -k_B T \ln Z$. These factors do not play a role in determining the approximation of the partition function. They are also irrelevant if Z represents the partition function of a spatial random field instead of a system of particles at thermal equilibrium.

to a random field that involves an infinite number of points and has a general mean (trend) function. Let us assume that

$$\mathcal{H}_G = \frac{1}{2}\mathbf{X}^\top \mathbf{C}_G{}^{-1}\,\mathbf{X},$$

where $\mathbf{C}_G$ is the (unknown initially) *variational covariance* matrix. The free energy corresponding to the variational Gaussian is denoted by F_G and the corresponding partition function is denoted by Z_G.

The expectation of a functional $\Phi\,[\mathbf{X}(\omega)]$ with respect to the variational pdf is then defined as follows:

$$\mathcal{E}_G\,[\Phi\,[\mathbf{X}(\omega)]] = \frac{\left[\prod_{n=1}^{N}\int_{-\infty}^{\infty} \mathrm{d}x_n\right]\,\Phi\,(\mathbf{x})\;\mathrm{e}^{-\mathcal{H}_G(\mathbf{x})}}{\left[\prod_{n=1}^{N}\int_{-\infty}^{\infty} \mathrm{d}x_n\right]\,\mathrm{e}^{-\mathcal{H}_G(\mathbf{x})}}. \tag{6.102}$$

Gibbs-Bogoliubov inequality The non-Gaussian free energy F (that corresponds to the energy $\mathcal{H}$) is bounded from above and from below. For any $\mathcal{H}_G$, the lower and upper bound are provided by means of the *Gibbs-Bogoliubov inequality* [397], [245, p. 71],

$$F_G + \mathcal{E}\left[\mathcal{H} - \mathcal{H}_G\right] \le F \le F_G + \mathcal{E}_G\left[\mathcal{H} - \mathcal{H}_G\right]. \tag{6.103}$$

The rightmost part of the inequality (6.103) is also known as the *Jensen-Feynman* inequality [641]. Let us briefly sketch how (6.103) is derived. The simplest way to prove this upper bound of the free energy is by means of *Jensen's inequality* [408] as shown in [641]. Jensen's inequality states that for any *continuous* and *convex* function $\phi(\cdot)$ the following holds

$$\mathcal{E}\left\{\phi\,[\mathrm{X}(\mathbf{s};\,\omega)]\right\} \ge \phi\,(\mathcal{E}[\mathrm{X}(\mathbf{s};\,\omega)])\,. \tag{6.104}$$

The partition function Z corresponding to the original $\mathcal{H}$ is given by the integral

$$Z = \left[\prod_{n=1}^{N}\int_{-\infty}^{\infty} \mathrm{d}x_n\right]\,\mathrm{e}^{-\mathcal{H}(\mathbf{x})} = Z_G\,\mathcal{E}\left[\mathrm{e}^{-(\mathcal{H}-\mathcal{H}_G)}\right].$$

Using Jensen's inequality, the right-hand side of the above leads to

$$Z_G\,\mathcal{E}_G\left[\mathrm{e}^{-(\mathcal{H}-\mathcal{H}_G)}\right] \ge Z_G\,\mathrm{e}^{-\mathcal{E}_G[\mathcal{H}-\mathcal{H}_G]}.$$

In light of the two relations above, taking logarithms and introducing the negative sign leads to the variational inequality

$$-\ln Z \le -\ln Z_G + \mathcal{E}_G\left[\mathcal{H} - \mathcal{H}_G\right].$$

The right-hand side of inequality (6.103) provides the *Gibbs-Bogoliubov upper bound* of the original free energy in terms of the Gaussian free energy and the Gaussian average of the perturbation $\mathcal{H} - \mathcal{H}_G$. This bound is easier to work with because it involves the expectation with respect to the variational Gaussian, unlike the lower bound which involves an expectation over the true distribution.

Variational solution The variational functional $\mathcal{H}_{G^*}$ that yields the *optimal approximation* to F is obtained by minimizing the upper variational bound with respect to the parameters of $\mathcal{H}_G$.

The minimization can be accomplished by the following system of equations that determines a stationary solution

$$\frac{\partial}{\partial [C_G]_{n,m}} \left(F_{\mathrm{G}} + \mathcal{E}_{\mathrm{G}} \left[\mathcal{H} - \mathcal{H}_G \right] \right) = 0, \quad (n, m = 1, \ldots, N).$$

To ensure that the solution corresponds to a maximum, it must be verified that the corresponding Hessian matrix is positive definite.[9] This approach, however, is not very efficient, since it requires calculating multiple partial derivatives and solving a large linear system. In addition, it must be checked that the resulting solution $[C_G]_{n,m}$, for $n, m = 1, \ldots, N$ is indeed a permissible covariance function.

Closed-form covariance expression The situation is greatly simplified if the covariance matrix $\mathbf{C}_{\mathrm{G}}$ is assumed to follow a specific spatial dependence, e.g., the exponential form $C_{\mathrm{G};n,m} = \sigma_0^2 \exp\left(-\|\mathbf{s}_n - \mathbf{s}_m\|/\xi\right)$, with unknown parameters σ_0^2 and ξ. Then, the minimization problem is reduced to solving the following set of equations

$$\frac{\partial F_{\mathrm{G}} + \mathcal{E}_{\mathrm{G}} \left[\mathcal{H} - \mathcal{H}_G \right]}{\partial \sigma_0^2} = 0, \tag{6.105a}$$

$$\frac{\partial F_{\mathrm{G}} + \mathcal{E}_{\mathrm{G}} \left[\mathcal{H} - \mathcal{H}_G \right]}{\partial \xi} = 0. \tag{6.105b}$$

If the mean of the variational Gaussian is non-zero, then in addition to the above the following partial derivative should vanish

$$\frac{\partial F_{\mathrm{G}} + \mathcal{E}_{\mathrm{G}} \left[\mathcal{H} - \mathcal{H}_G \right]}{\partial m_{\mathrm{x}}} = 0. \tag{6.105c}$$

[9]A symmetric $N \times N$ real matrix $\mathbf{H}$ is said to be positive definite if $\mathbf{x}^\top \mathbf{H} \mathbf{x}$ is non-negative for every non-zero column vector $\mathbf{x}$ of dimension N.

The equations (6.105) are not necessarily linear and may require numerical solution. If the covariance function involves additional parameters (e.g., as in the case of Matérn or Spartan models), the set of equations is accordingly augmented.

The optimal *variational approximation of the CGF* is then given by means of

$$F \approx F_{\mathrm{up}} = F_{\mathrm{G}^*} + \mathcal{E}_{\mathrm{G}^*}\left[\mathcal{H} - \mathcal{H}_G\right],$$

where $\mathcal{E}_{\mathrm{G}^*}$ denotes the expectation with respect to the *Gaussian pdf* with the optimal variational parameters, as determined from the above system.

It is then possible to obtain approximate estimates of the non-Gaussian covariance function and other moments using the *optimal Gaussian CGF* $\mathcal{K}_{\mathrm{G}^*}(\mathbf{u})$ for cumulant generation. In particular, the cumulants of the original distribution are approximately given by means of respective derivatives of the optimal variational CGF $\mathcal{K}_{\mathrm{G}^*}(\mathbf{u})$ (compare with (4.74)).

Optimal variational CGF Since the variational approximation is Gaussian, the respective CGF is given according to (6.9) by

$$\mathcal{K}_{\mathrm{G}^*}(\mathbf{u}) = \mathbf{u}^\top \mathbf{m}_{\mathrm{G}^*} + \frac{1}{2}\mathbf{u}^\top \mathbf{C}_{\mathrm{G}^*}\,\mathbf{u}.$$

Hence, the *variational estimate of the covariance function* is given by means of the following equation

$$C^*_{\mathrm{xx}}(\mathbf{s}_n, \mathbf{s}_m) = \left.\frac{\partial^2 \mathcal{K}_{\mathrm{G}^*}(\mathbf{u})}{\partial u_n \partial u_m}\right|_{\mathbf{u}=\mathbf{0}} = C_{\mathrm{G}^*}(\mathbf{s}_n, \mathbf{s}_m). \tag{6.106}$$

Note that the approximations of odd-order moments vanish, because the variational approximation is based on a Gaussian pdf. Even-order moments are obtained in terms of products of covariance functions. However, this implies that within the variational approximation the kurtosis coefficient is equal to the Gaussian value of three, regardless of the actual kurtosis value in the original distribution.

Example 6.10 Let us recall the Example 6.8. We will now calculate the variance of the Boltzmann-Gibbs pdf with the quartic energy function $\mathcal{H}[x] = \frac{\alpha^2}{2}x^2 + \beta^4 x^4$ using the variational approximation. The goal is to find a Gaussian functional $\mathcal{H}_{G^*}$ that provides a better expansion point than the functional $\mathcal{H}_0[x] = \frac{\alpha^2}{2}x^2$.

Answer The following Gaussian variational expression will be used as an approximation of $f_{\mathrm{x}}(x)$:

$$f_{\mathrm{G}}(x) = \left(\sqrt{2\pi}\sigma_0\right)^{-1} \exp\left(-x^2/2\sigma_0^2\right). \tag{6.107}$$

The variational energy functional is $\mathcal{H}_G = x^2/2\sigma_0^2$ and σ_0 is the *variational parameter*. The partition function that corresponds to $\mathcal{H}_G$ is $Z_G = \sqrt{2\pi}\sigma_0$. The corresponding free energy is

$$F_G = -\ln(\sqrt{2\pi}\sigma_0). \tag{6.108}$$

To calculate the expectation of the energy difference $\mathcal{H} - \mathcal{H}_G$ we take into account that $\mathcal{E}_G[x^2] = \sigma_0^2$, and we use the Isserlis-Wick theorem 6.2 for $\mathcal{E}_G[x^4]$. Thus, the following expression is obtained

$$\begin{aligned}\mathcal{E}_G\left[\mathcal{H} - \mathcal{H}_G\right] &= \frac{\alpha^2}{2}\,\mathcal{E}_G\left[x^2\right] + \beta^4\mathcal{E}_G\left[x^4\right] - \frac{1}{2\sigma_0^2}\,\mathcal{E}_G\left[x^2\right] \\ &= \frac{\alpha^2}{2}\,\sigma_0^2 + 3\,\beta^4\,\sigma_0^4 - \frac{1}{2}.\end{aligned} \tag{6.109}$$

Finally, we obtain the following variational upper bound of the free energy

$$F_{\text{up}} = -\ln(\sqrt{2\pi}\sigma_0) + \frac{\alpha^2}{2}\,\sigma_0^2 + 3\,\beta^4\,\sigma_0^4 - \frac{1}{2}.$$

The upper bound is minimized for the value of σ_0 that renders the first derivative of the free energy equal to zero, i.e., if the following condition is satisfied

$$\frac{\partial F_{\text{up}}}{\partial \sigma_0} = -\frac{1}{\sigma_0} + \alpha^2\sigma_0 + 12\beta^4\,\sigma_0^3 = 0.$$

For $\beta = 0$ the above leads to $\sigma_0^* = 1/\alpha$. For $\beta \neq 0$ the above is equivalent to a quadratic equation that admits the following non-negative solution

$$\alpha\sigma_0^* = \frac{1}{2\sqrt{6}\rho^2}\left[\sqrt{1 + 48\,\rho^4} - 1\right]^{1/2}. \tag{6.110}$$

The variational upper bound F_{up} is a convex function of σ_0 as can be shown by evaluating the second derivative, i.e.,

$$\frac{\partial^2 F_{\text{up}}}{\partial \sigma_0^2} = \frac{1}{\sigma_0^2} + \alpha^2 + 36\beta^4\,\sigma_0^2,$$

which is obviously positive. Hence, the stationary point defined in (6.110) corresponds to a minimum of the free energy upper bound F_{up}.

The dependence of the variational variance on the aspect ratio is shown in Fig. 6.13. The agreement between the exact calculation and the variational approximation is excellent even for values of β/α greater than one.

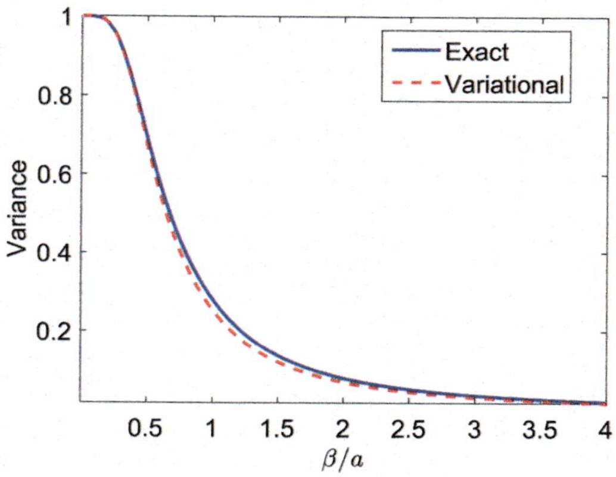

Fig. 6.13 Variance of the Boltzmann-Gibbs pdf with quartic energy function. The "exact" result is based on numerical integration over the probability density function. The variational approximation is based on (6.110)

6.5 Non-stationary Covariance Functions

The development of *non-stationary (non-homogeneous)* covariance models is a field of continuing research interest. In Sects. 6.4.3 and 6.4.4 we showed that non-stationary covariance functions can be obtained using the cumulant expansion and localized energy perturbations. There are several other approaches in the literature.

Convolution approach This approach focuses on constructing non-stationary covariance functions by convolving kernel functions with known stationary covariance models [342]. The convolution approach has been generalized by means of a non-stationary, anisotropic distance measure, leading to non-stationary extensions of the Whittle-Matérn covariance models [639, 640].

Space deformation approach In this framework, non-stationary covariance functions are built using models that are stationary in a deformed space [710, 726].

Related to this is the *warping approach* [520] that we discussed in Sect. 3.6.5. Warping is based on the use of a suitable feature space (see also (3.71)) in which the physical space is embedded. A recently proposed spectral approach uses symmetrized dual-wavevector spectral densities that are numerically integrated by means of Monte Carlo integration methods [796].

Stochastic weighting A different model for non-stationary covariance functions is based on the convolution of an orthogonal random measure with a spatially varying stochastic weighting function. With a suitable choice of the stochastic weighting functions, this approach retrieves classes of closed-form non-stationary covariance functions with locally varying geometric anisotropy [258].

A recent review of the available approaches for non-stationary covariance modeling is given by Fouedjio [256]. Extensions of such approaches in the framework of the *linear coregionalization model* for applications in *multivariate random fields* are presented in [257].

Vertical rescaling This method was discussed in Sect. 3.6.5 (see equation (3.70)). Vertical rescaling leads to straightforward construction of non-stationary covariance functions from stationary ones. The non-stationary covariance is built from a stationary covariance model by multiplying on both sides (i.e., both at $\mathbf{s}$ and $\mathbf{s}'$) with a space-dependent function $\alpha(\mathbf{s})$. This implies that the variance is given by $\sigma^2(\mathbf{s}) = \sigma_x^2\, \alpha^2(\mathbf{s})$, where σ_x^2 is the variance of the stationary component.

Vertical rescaling can be used to model the so-called *proportional effect* which is marked by spatial dependence of the variance that track the variations of the mean. This behavior leads to fluctuations with larger amplitude at locations where the mean is higher. The proportional effect is often observed in geo-scientific data [132, Chap. 2],[394]. A recent review of the approaches that are used in geosciences to model the proportional effect is presented in [528].

Neural network covariance Non-stationary covariance functions are also constructed for machine learning applications [520, 678]. A particular covariance function that has its origin in the field of Gaussian processes is the so called "neural network covariance" that is given by [678, 850]

$$C(\mathbf{s}_n, \mathbf{s}_m) = \frac{2}{\pi} \sin^{-1}\left(\frac{2\tilde{\mathbf{s}}_n^\top\, \boldsymbol{\Sigma}\, \tilde{\mathbf{s}}_m}{\sqrt{\left(1 + 2\tilde{\mathbf{s}}_n^\top\, \boldsymbol{\Sigma}\, \tilde{\mathbf{s}}_n\right)\left(1 + 2\tilde{\mathbf{s}}_m^\top\, \boldsymbol{\Sigma}\, \tilde{\mathbf{s}}_m\right)}} \right), \tag{6.111}$$

where $\tilde{\mathbf{s}}_n = (1, s_{n,1}, \ldots, s_{n,d})^\top$ is an augmented position vector and $\boldsymbol{\Sigma}$ is a $(d+1) \times (d+1)$ diagonal weight matrix with elements $W_{0,0} = \sigma_0^2$, and the square of inverse characteristic distances $W_{i,i} = 1/\xi_i^2$, for $i = 1, \ldots, d$.

This covariance function is based on a result by Radford Neal which states that *single-layer, feed-forward Bayesian neural networks with infinite number of hidden units* are equivalent to Gaussian processes [598]. The architecture of such a neural network is shown in Fig. 6.14. The priors used for the network parameters are assumed to follow the i.i.d. assumption.

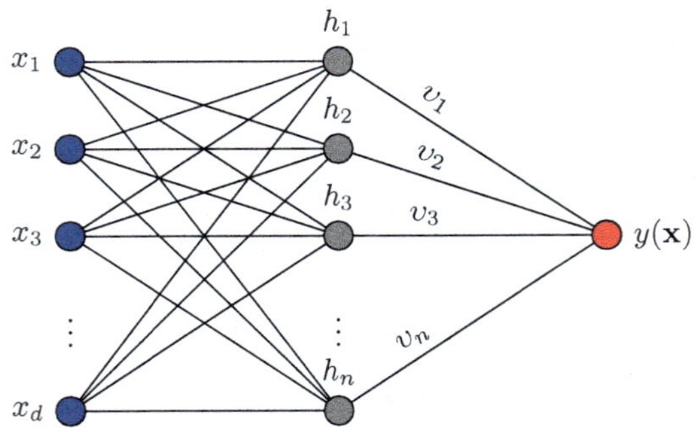

Fig. 6.14 Architecture of single-layer, feed-forward Bayesian neural networks with d-dimensional input $\mathbf{x} = (x_1, \ldots, x_d)^\top$, transfer function $h(\cdot)$, and infinite number ($n \to \infty$) of hidden units. The hidden-unit weights $\{u_i\}_{i=1}^d$ (not shown), and the output weights $\{\upsilon_k\}_{k=1}^n$ are zero-mean, Gaussian, independent random variables

Building on this result, Williams developed the covariance function (6.111) assuming that the transfer function of the hidden units is the sigmoidal function, i.e., $h(\mathbf{s}; \mathbf{u}) = \Phi(\mathbf{u} \cdot \tilde{\mathbf{s}})$ where $\Phi(\cdot)$ is the error function, and that the weights $\mathbf{u}$ follow the zero-mean Gaussian distribution with covariance matrix $\boldsymbol{\Sigma}$ [850]. In the initial formulation, $\tilde{\mathbf{s}}$ is replaced by $\mathbf{x}$ which represents a general feature vector that is not necessarily limited to the spatial location.

Networks with an infinite number of hidden units are also known as infinitely wide. Recently, Lee and co-workers derived an exact equivalence between *infinitely wide deep networks and Gaussian processes* [492, 493]. They also developed a computationally efficient recipe for computing the respective Gaussian process covariance function for wide neural networks with a finite number of layers. They used the resulting Gaussian processes to perform Bayesian inference for wide deep neural networks. They found that the accuracy of the trained neural network approaches that of the corresponding Gaussian process as the layer width increases. In addition, they determined that the Gaussian process uncertainty is strongly correlated with the prediction error of the trained network. Finally, they established that the performance of Gaussian process predictions is typically superior to that of finite-width trained networks, and that the agreement between the Gaussian process and the respective neural network increases with increasing layer width.

The time seems ripe for developing new bridges between the theory of random fields (Gaussian processes) and deep learning methods. The emergence of such connections will benefit both machine learning and spatial statistics.

Chapter 7
Random Fields Based on Local Interactions

Taking a new step, uttering a new word, is what people fear most
Fyodor Dostoyevsky

In this chapter we look at random fields from a perspective that is common in statistical physics but not so much in spatial data analysis. This perspective is useful, because it can lead to computationally efficient methods for spatial prediction, while it is also related with Markovian random fields. In addition, it enables the calculation of new forms of covariance functions and provides a link with *stochastic partial differential equations*.

In spatial studies, the random field models are motivated by the available data. Hence, they are defined in terms of the moments (typically the mean and the covariance) that can be obtained from the data and some plausible assumption (typically Gaussian) for the pdf. Statistical physicists, on the other hand, are more accustomed to thinking in terms of *Boltzmann-Gibbs (BG)* distributions. The latter follow the exponential form

$$f_{\mathrm{x}}(\mathbf{x};\boldsymbol{\theta}) \propto \mathrm{e}^{-\mathcal{H}_0(\mathbf{x};\boldsymbol{\theta})},$$

where $\mathcal{H}_0$ is an energy function, as in (6.5a).

The difference may seem trivial at first, since the geostatistical approach—at least for jointly normal distributions—can be recast in the BG framework by expressing the energy in terms of the precision matrix $\mathbf{C}_{\mathrm{xx}}^{-1}$ and the field values $\mathbf{x}$ as follows

$$\mathcal{H}_0 = \frac{1}{2}\mathbf{x}^{\top}\,\mathbf{C}_{\mathrm{xx}}^{-1}\,\mathbf{x}. \tag{7.1}$$

There is, however, a deeper difference: in statistical physics the "energy" $\mathcal{H}_0$ is defined in terms of field interactions and respective coupling strengths that

D. T. Hristopulos, *Random Fields for Spatial Data Modeling*, Advances in Geographic Information Science, https://doi.org/10.1007/978-94-024-1918-4_7

correspond to model parameters. The covariance function is a priori unknown and needs to be calculated, if possible, from the BG model. This means that the pdf model is specified in terms of a few parameters (the coupling strengths) instead of the entire covariance matrix.

One can reasonably argue that in geostatistics the problem is not very different, since the covariance matrix is determined from a covariance model that also involves a small number of parameters which are inferred from the data (see Chap. 12).

The one unequivocal difference between the two approaches is that in statistical physics $\mathcal{H}_0(\mathbf{x})$ is *explicitly* expressed in terms of the inverse covariance matrix, i.e., the *precision matrix*.[1] In geostatistics, the covariance matrix needs to be inverted in order to evaluate the joint pdf and to make predictions by interpolation (see Chap. 10). The numerical inversion of large, non-sparse matrices is a computationally intensive task, since its algorithmic complexity scales as $O(N^3)$ where $N \times N$ is the size of the covariance matrix. This poor scaling with data size becomes especially costly if the inversion is repeated several times, e.g., as in maximum likelihood parameter estimation [784].

7.1 Spartan Spatial Random Fields (SSRFs)

We live by selected fictions
Lawrence Durrell, in Balthazar

7.1.1 Overture

Sparta was a famous city in ancient Greece, known for the disciplined lifestyle and the prowess of its soldiers in combat. In what sense then is the term "Spartan" appropriate for random fields? The name seemed suitable for describing a more compact (i.e., local) representation of random fields than the classical geostatistical formulation based on the covariance matrix. In addition, the compact representation of spatial correlations provided by the *sparse precision matrix* of Spartan random fields invokes the compactness of the Spartan army's battle formation; hence the name "*Spartan spatial random fields*".

Spartan spatial random fields are based on the Boltzmann-Gibbs exponential expression for the joint density (6.68). The central idea on which Spartan random fields are based is that *local interactions* can lead to *sparse precision matrices*, which offer computational advantages compared to the classical covariance models.

[1] In continuum random fields the precision matrix is replaced by a respective precision operator.

In statistical physics, mathematical constructions similar to Spartan random fields are used (among other things) in the study of phase transitions. In this case, the exponent of the joint distribution involves non-Gaussian terms (i.e., contributions that involve interactions between triplets or quartets of field values at different points).

In contrast with statistical physics, in spatial data modeling the random fields of interest do not exhibit phase transitions between disordered and fully ordered phases. Hence, the focus herein is on Gaussian random field models that share many similarities with the classical Gaussian field theory [297, 434, 593]. In *Gaussian field theory*, the joint pdf of the field is exponential and the exponent involves an energy function that is quadratic in the field values. More precisely, the exponent of the Boltzmann-Gibbs pdf can be expressed as follows

$$\mathcal{H}_0 = \int \mathrm{ds} \left[\{\nabla x(\mathbf{s})\}^2 + m^2\, x^2(\mathbf{s}) \right]. \tag{7.2}$$

In the above, $x(\mathbf{s})$ represents a realization of a scalar field, m is a free parameter, and $\nabla x(\mathbf{s})$ is the gradient of the field [890]. The energy $\mathcal{H}_0$ penalizes large fluctuations of the field and large gradients since they both lead to higher values of $\mathcal{H}_0$, which according to (7.2) have lower probability of occurrence than smaller values of $\mathcal{H}_0$.

The two-point correlation function corresponding to the above functional can be evaluated explicitly based on (6.49) in domains with $d = 1, 2, 3$ dimensions [593, p. 227]. However, note that in $d \geq 2$ this correlation function corresponds to a *generalized covariance* function, since it diverges at $\|\mathbf{r}\| = 0$.

7.1.2 The Fluctuation-Gradient-Curvature Energy Function

SSRFs correspond to the *Gaussian field theory* generated by the following energy function

$$f_{\mathrm{x}}[\{x(\mathbf{s})\}; \boldsymbol{\theta}] = \frac{1}{Z(\boldsymbol{\theta})} \mathrm{e}^{-\mathcal{H}_0[x(\mathbf{s})\}; \boldsymbol{\theta}]}, \tag{7.3}$$

where the energy function $\mathcal{H}_0[\cdot]$ given by

$$\mathcal{H}_0 = \frac{1}{2\eta_0 \xi^d} \int_{\mathcal{D}} \mathrm{ds} \left\{ x(\mathbf{s})^2 + \eta_1\, \xi^2\, [\nabla x(\mathbf{s})]^2 + \xi^4\, [\Delta x(\mathbf{s})]^2 \right\}, \tag{7.4}$$

is more flexible than (7.2), while the parameter vector $\boldsymbol{\theta}$ comprises the SSRF coefficients, i.e., $\boldsymbol{\theta} = (\eta_0, \eta_1, \xi)^\top$.

In the energy expression (7.4), the quantity $\nabla x(\mathbf{s})$ represents the gradient of the realization $x(\mathbf{s})$, which is defined by means of

$$\nabla x(\mathbf{s}) = \left(\frac{\partial x(\mathbf{s})}{\partial s_1}, \ldots, \frac{\partial x(\mathbf{s})}{\partial s_d} \right)^{\top},$$

while the square of the gradient is defined by means of the inner product

$$[\nabla x(\mathbf{s})]^2 = \nabla x(\mathbf{s}) \cdot \nabla x(\mathbf{s}) = \sum_{i=1}^{d} \left(\frac{\partial x(\mathbf{s})}{\partial s_i} \right)^2,$$

and $\Delta x(\mathbf{s})$ is the *Laplacian* of the realization $x(\mathbf{s})$ given by the sum of the second-order derivatives in all directions, i.e.

$$\Delta x(\mathbf{s}) = \sum_{i=1}^{d} \frac{\partial^2 x(\mathbf{s})}{\partial s_i^2}.$$

On the role of curvature The Laplacian provides a linearized approximation of the *local curvature* of the field (see also Sect. 10.2.3). The mean square Laplacian, in particular, is a measure of the global curvature of the random field [679].

Statistical field theories focus on phase transitions generated by long-wavelength (low-frequency) modes that drastically change the macroscopic state of the system (e.g., in magnetic systems the transition is from a disordered state of zero magnetization to an ordered state with finite magnetization). In the long-wavelength regime the curvature term in (7.4) is irrelevant, since the gradient term dominates. The curvature term is more significant for short-distance variations.[2]

In contrast, in spatial data modeling higher-order derivatives than the gradient (such as the curvature) contribute to local structure and are important. In fact, a popular interpolation method in geophysics is based on the principle of *minimum curvature* (see Sect. 10.2.3 below).

Curvature also plays a critical role in the statistical mechanics of flexible membranes and surfaces [600]. The Ginzburg-Landau theory of ternary amphiphilic systems includes a free energy very similar to that of Spartan random fields [301, 302]. Amphiphiles are substances that exhibit both hydrophilic and lipophilic properties (e.g., surfactants or soaps). Adding an amphiphile to a water-oil system reduces the tension between the two phases due to the creation of an amphiphilic monolayer at the water-oil interface. Small scalar displacements of the interface from a flat reference state are then described by a Spartan-like free-energy given by equation (11) in [302].

[2]As shown below, in the spectral domain the gradient term is $\propto \|\mathbf{k}\|^2$, while the curvature term is $\propto \|\mathbf{k}\|^4$ where $\|\mathbf{k}\|$ is the wavenumber.

SSRF properties The random fields with joint pdf determined by the energy functional (7.4) have *zero expectation*, and they are *statistically isotropic* and *jointly Gaussian*.

1. *Zero expectation:* The zero-mean property follows from the fact that the energy functional, and thus the pdf, include only terms even in $x(\mathbf{s})$. It can be easily relaxed by replacing $x^2(\mathbf{s})$ in the SSRF energy (7.4) with $\{x(\mathbf{s}) - m_{\mathrm{x}}\}^2$, where m_{x} is a real-valued constant that represents the expectation.
2. *Isotropy:* This property first implies that the SSRF model is statistically homogeneous (stationary). The stationarity is guaranteed because the model coefficients, η_0, η_1, ξ are *translation invariant*. In addition, the model does not have directional dependence since the contributions of the partial field derivatives in (7.4) are the same in every direction, thus implying isotropy. This constraint is easily relaxed by introducing different characteristic lengths along respective principal anisotropy directions.
3. *Gaussianity:* The Gaussian character of the joint pdf follows from the *quadratic dependence* of the energy on the field. We return to the Gaussian nature of the random field in Sect. 7.1.5.
4. *Scaling invariance:* The energy functional (7.4) remains invariant under the scaling transformations $\mathbf{s} \rightarrow \lambda\,\mathbf{s}$ and $\xi \rightarrow \lambda\,\xi$. This means that the SSRF energy remains unaffected by a change in the units used to measure length.

The mathematical development and some applications of Spartan random fields are described in [228, 229, 362, 363, 366–368, 372, 373, 378].

7.1.3 SSRF Model Parameters

The SSRF model is determined by the parameter vector $\boldsymbol{\theta} = (\eta_0, \eta_1, \xi, k_{\mathrm{c}})^{\top}$, which contains the following elements.

- *The amplitude coefficient* η_0 controls the overall magnitude of the fluctuations: higher values of η_0 imply larger fluctuations. Its role is analogous to temperature in statistical mechanics.
- *The characteristic length* ξ controls, in conjunction with η_1, the relative strength of the squared Laplacian versus the squared gradient term. It is also proportional to all the length scales (integral range, correlation radius, et cetera.) that characterize the random field
- *The rigidity coefficient* η_1 controls the resistance of the field to changes of its gradient. Hence, larger values of η_1 imply that the SSRF realizations are less "floppy" than the realizations that correspond to lower η_1. Negative values of η_1 in particular are linked with oscillatory correlation functions.
- *The spectral cutoff,* k_{c} represents an implicit upper bound on the wavevector magnitude in reciprocal (wavevector) space. It is necessary to enforce the

differentiability of SSRFs in $\mathcal{D} \subset \mathbb{R}^d$ where $d > 1$. In practice, however, explicit expressions for the covariance function can only be obtained at the limit $k_c \to \infty$. We discuss the issue of cutoff in more detail below. However, unless otherwise noted, we will assume that the SSRF is determined by the three parameters η_0, ξ, η_1.

- *The spatial dimension d* enters the expressions of SSRF moments in real space. In contrast with most of the classical covariance models studied in Chap. 3, the expressions for SSRF covariance functions depend on d, while the spectral density is independent of d.

SSRF alternate parametrization One may wonder why not to use a simpler parametrization of the SSRF energy functional (7.4) such as

$$\mathcal{H}_0 = \frac{\lambda}{2} \int_{\mathcal{D}} \mathrm{d}\mathbf{s} \left\{ x(\mathbf{s})^2 + c_1 \, [\nabla x(\mathbf{s})]^2 + c_2 \, [\Delta x(\mathbf{s})]^2 \right\}.$$

The above also contains three parameters, λ, c_1, c_2 with linear dependence of the energy on them. This is advantageous in model estimation for calculating the derivatives of the log-likelihood with respect to the parameter vector. On the other hand, the coefficients λ, c_1, c_2 depend on the dimensionality d of the spatial domain $\mathcal{D}$ and the measure of length. In addition, they do not admit a straightforward physical interpretation. In contrast, the formulation (7.4) is based on parameters η_0, η_1, ξ that have clear physical interpretations and are independent of the dimension of space.

7.1.4 More on Nomenclature

The term *FGC Spartan functional* denotes the energy function (7.4). The acronym "FGC" refers to Fluctuation-Gradient-Curvature. FGC specifies the type of the interactions involved in $\mathcal{H}_0$ and distinguishes (7.4) from different energy functions with local structure. For example, it is possible to define an SSRF that involves the square of third-order partial field derivatives. Henceforth we use the abbreviation SSRF to denote FGC Spartan Spatial Random Fields.

Local random fields Covariance models of essentially the same form as those obtained by SSRFs were rediscovered in the meteorological literature by studying polynomials of the diffusion operator [836, 865, 866]. A few years after the first papers on Spartan spatial random fields, Richard Farmer, working independently, introduced very similar expressions to the energy functional (7.4). He named the respective RFs *local Gaussian fields* [242].

The term *local Gaussian fields* is also appropriate for Spartan random fields, since it emphasizes the *emergence of correlations from localized operations.* SSRFs are local in the sense that the field value at any point $\mathbf{s}$ is determined by interactions that only involve points in a local neighborhood of $\mathbf{s}$. Such local interactions are typically expressed in terms of low-order partial derivatives of the field. SSRFs are not the only possible model for local random fields. Other types of random fields with local dependence have been developed by Annika Lang and coworkers [482, 483].

The meaning of locality The dependence imposed in (7.4) through partial derivatives is local compared to the extended couplings realized by a generally dense precision matrix in (7.1). On the other hand, readers familiar with statistical physics will recognize the presence of partial derivatives as *nonlocal dependence*, in the sense that the field values at any given point are related to the neighbors, e.g. [74]. Hence, *locality* may mean different things in different fields.

Herein, locality denotes the dependence of the field value at any point $\mathbf{s}$ on the state of the field within a compact neighborhood of $\mathbf{s}$. In the case of discretized models of local random fields, locality is connected with the concept of *conditional independence* (see Chap. 8).

7.1.5 Are SSRFs Gaussian Random Fields?

Field theory practitioners, being familiar with the free field theory (7.2), have no doubt that the energy functional (7.4) is associated with a Gaussian random field. Readers with statistics background who have in mind the standard bilinear form $\mathcal{H}_0(\mathbf{x}) = \frac{1}{2}\mathbf{x}^\top \mathbf{C}_{\text{xx}}^{-1}\,\mathbf{x}$ for the exponent of the joint Gaussian density, may wonder whether SSRF energy function terms such as $\int d\mathbf{s}\; [\nabla x(\mathbf{s})]^2$ can be expressed in the more familiar form.

Let us then investigate the two terms in (7.4) that involve the squared gradient and curvature. The main idea is to express these terms using the *quadratic form* $\sum_{n,m=1}^{N} x_n A^{(l)}_{n,m} x_m$, where $l = 0, 1, 2$, $\mathbf{A}^{(l)}$ is an $N \times N$ real-valued, symmetric matrix, and the indices $n, m \in \{1, \ldots, N\}$ denote spatial locations.

The above quadratic form applies to finite-dimensional vectors $\mathbf{x}$. In the case of infinite-dimensional vectors, i.e., functions $x(\mathbf{s})$, the quadratic form is given by the following double integral, where the $A^{(l)}(\mathbf{s}, \mathbf{s}')$ $(l = 0, 1, 2)$ involve real-valued, generalized functions

$$\int_{\mathcal{D}} d\mathbf{s} \int_{\mathcal{D}} d\mathbf{s}'\, x(\mathbf{s})\, A^{(l)}(\mathbf{s}, \mathbf{s}')\, x(\mathbf{s}'). \tag{7.5}$$

The function $A^{(0)}(\mathbf{s}, \mathbf{s}')$ is such that the double integral (7.5) leads to $\int d\mathbf{s}\, x^2(\mathbf{s})$. Hence, it follows that $A^{(0)}(\mathbf{s}, \mathbf{s}') = \delta(\mathbf{s} - \mathbf{s}')$.

For the terms that involve derivatives, the main idea is to use extensions of integration by parts in order to recast these terms in the form of (7.5). The key mathematical tool is *Green's first identity*, which replaces a "volume" integral over an $\mathbb{R}^d$ domain with a "surface" integral over $\partial\mathcal{D} \subset \mathbb{R}^{d-1}$.

Theorem 7.1 *Green's first identity: If $\phi(\mathbf{s})$ and $\psi(\mathbf{s})$ are once and twice continuously differentiable functions, respectively, then [401, p. 41]*

$$\int_{\mathcal{D}} d\mathbf{s} \left[\phi(\mathbf{s}) \nabla^2 \psi(\mathbf{s}) + \nabla \phi(\mathbf{s}) \cdot \nabla \psi(\mathbf{s}) \right] = \int_{\partial \mathcal{D}} d\mathbf{a} \cdot \nabla \psi(\mathbf{s})\, \phi(\mathbf{s}). \tag{7.6}$$

In the above, $\partial \mathcal{D}$ is the boundary of the domain $\mathcal{D}$ and $d\mathbf{a} = \mathbf{n}\, da$, where $\mathbf{n}$ is the outward pointing normal unit vector at the boundary, and da is the surface area differential on $\mathcal{D}$. Hence, it follows that $d\mathbf{a} \cdot \nabla \psi(\mathbf{s}) = [\mathbf{n} \cdot \nabla \psi(\mathbf{s})]\, da$.

The squared gradient term We can now use Green's first identity to evaluate the integral of the squared gradient term in the SSRF energy function. If we set $\psi(\mathbf{s}) = \phi(\mathbf{s}) = x(\mathbf{s})$ in the identity (7.6) and assume that the realizations $x(\mathbf{s})$ admit continuous second-order partial derivatives, it follows that the integral of the squared gradient term is given by

$$\int_{\mathcal{D}} d\mathbf{s}\, [\nabla x(\mathbf{s})]^2 = - \int_{\mathcal{D}} d\mathbf{s}\, x(\mathbf{s})\, \nabla^2 x(\mathbf{s}) + \int_{\partial \mathcal{D}} d\mathbf{a} \cdot \nabla x(\mathbf{s})\, x(\mathbf{s}). \tag{7.7}$$

We will assume that the boundary integral in (7.7) vanishes. This is usually justified by the fact that for large domains the boundary is much smaller than the interior of the domain of interest. Alternatively, we can think of the boundary as being very far away where the field values can be safely assumed to be very small and hence negligible. If we omit the boundary term, the remaining integral on the right-hand side is expressed as

$$- \int_{\mathcal{D}} d\mathbf{s} \int_{\mathcal{D}} d\mathbf{s}'\, x(\mathbf{s})\, \nabla^2 \delta(\mathbf{s} - \mathbf{s}')\, x(\mathbf{s}').$$

Hence, according to the above, the function $A^{(1)}(\mathbf{s}, \mathbf{s}')$ is given by

$$A^{(1)}(\mathbf{s}, \mathbf{s}') = -\nabla^2 \delta(\mathbf{s} - \mathbf{s}').$$

This may seem a bit strange, since it involves the second derivative of a singular (generalized) function. We discuss such derivatives below.

The square curvature term To calculate the integral that involves the square of the Laplacian $[\Delta x(\mathbf{s})]^2$ we apply Green's first identity twice, using $\psi(\mathbf{s}) = x(\mathbf{s})$ and $\phi(\mathbf{s}) = \nabla^2 x(\mathbf{s})$ the first time, and $\psi(\mathbf{s}) = \nabla^2 x(\mathbf{s})$, $\phi(\mathbf{s}) = x(\mathbf{s})$ the second time. To perform these operations "legitimately", it is assumed that $x(\mathbf{s})$ admits at least continuous fourth-order partial derivatives. The following equation is obtained for the curvature term [364]:

$$\int_{\mathcal{D}} d\mathbf{s} \left[\nabla^2 x(\mathbf{s}) \right]^2 = \int_{\mathcal{D}} d\mathbf{s}\, x(\mathbf{s})\, \nabla^4 x(\mathbf{s}) + \int_{\partial \mathcal{D}} d\mathbf{a} \cdot \nabla x(\mathbf{s})\, \nabla^2 x(\mathbf{s})$$

$$- \int_{\partial \mathcal{D}} \mathbf{da} \cdot \nabla \left[\nabla^2 x(\mathbf{s}) \right] x(\mathbf{s}), \tag{7.8}$$

Neglecting the two boundary terms, as we did for the squared gradient term, it follows that the operator $A^{(1)}(\mathbf{s}, \mathbf{s}')$ is given by

$$A^{(2)}(\mathbf{s}, \mathbf{s}') = \nabla^4 \delta(\mathbf{s} - \mathbf{s}').$$

Gaussian form of the energy If we collect all the terms and take account of the expressions for $A^{(1)}(\mathbf{s}, \mathbf{s}')$ and $A^{(2)}(\mathbf{s}, \mathbf{s}')$, the energy functional (7.4) is expressed as

$$\mathcal{H}_0 = \frac{1}{2\eta_0 \xi^d} \int_{\mathcal{D}} \mathbf{ds} \left\{ x(\mathbf{s})^2 - \eta_1 \xi^2 x(\mathbf{s}) \nabla^2 x(\mathbf{s}) + \xi^4 x(\mathbf{s}) \, \Delta^2 x(\mathbf{s}) \right\}. \tag{7.9}$$

In light of the above and the field theory representation presented in Sect. 6.2, the SSRF energy functional can be expressed as follows [396]

$$\mathcal{H}_0 = \frac{1}{2} \int_{\mathcal{D}} \mathbf{ds} \int_{\mathcal{D}} \mathbf{ds}' \, x(\mathbf{s}) \, C_{\mathrm{xx}}^{-1}(\mathbf{s} - \mathbf{s}') \, x(\mathbf{s}'). \tag{7.10}$$

The above is the continuum limit of the discrete energy functional (6.5b).

The energy function given by (7.10) is identical to (7.9), if the *inverse SSRF covariance kernel* $C_{\mathrm{xx}}^{-1}(\mathbf{s} - \mathbf{s}')$ is given by the following symmetric[3] and translation invariant generalized function:

Inverse SSRF covariance kernel

$$C_{\mathrm{xx}}^{-1}(\mathbf{s} - \mathbf{s}') = \frac{1}{\eta_0 \xi^d} \left(1 - \eta_1 \xi^2 \Delta_{\mathbf{s}} + \xi^4 \Delta_{\mathbf{s}}^2 \right) \delta(\mathbf{s} - \mathbf{s}'). \tag{7.11}$$

Derivatives of the Dirac function Equation (7.11) involves the second-order and fourth-order partial derivatives of the generalized Dirac function $\delta(\mathbf{s} - \mathbf{s}')$, namely,

$$\frac{\partial^{2m} \delta(\mathbf{s} - \mathbf{s}')}{\partial s_i^{2m}}, \quad i = 1, \ldots, d, \text{ and } m = 1, 2.$$

[3]The precision operator contains differential operators, e.g., $\Delta_{\mathbf{s}}$, which seem asymmetric since they act on one of the positions involved. However, the value of the quadratic form does not change if we use $\Delta_{\mathbf{s}'}$ instead, as required by symmetry.

The derivatives of the Dirac function are defined in the distributional sense [745]. The main idea is that they appear inside integrals as products with differentiable functions, e.g., $\phi(\mathbf{s})$. Through integration by parts the action of the derivatives is transferred to the differentiable function $\phi(\mathbf{s})$; the integral then involves the product of an ordinary function and the Dirac delta function, which can be evaluated using the properties of the latter.

The action of the *Laplacian*, $\Delta_{\mathbf{s}}$, and *biharmonic*, $\Delta_{\mathbf{s}}^2$, operators on the Dirac delta function at $\mathbf{s}$ is also defined in terms of partial derivatives, i.e.,

$$\Delta_{\mathbf{s}}\, \delta(\mathbf{s}-\mathbf{s}') = \sum_{i=1}^{d} \frac{\partial^2 \delta(\mathbf{s}-\mathbf{s}')}{\partial s_i^2}, \tag{7.12a}$$

$$\Delta_{\mathbf{s}}^2\, \delta(\mathbf{s}-\mathbf{s}') = \Delta_{\mathbf{s}}\left[\Delta_{\mathbf{s}}\, \delta(\mathbf{s}-\mathbf{s}')\right] = \sum_{i=1}^{d}\sum_{j=1}^{d} \frac{\partial^4 \delta(\mathbf{s}-\mathbf{s}')}{\partial s_i^2 \partial s_j^2}. \tag{7.12b}$$

In order to evaluate the integral over $d\mathbf{s}'$ in (7.10) we will use the properties of the derivatives of the delta function, and more specifically

$$\int_{\mathcal{D}} d\mathbf{s}' \, \frac{\partial^2 \delta(\mathbf{s}-\mathbf{s}')}{\partial s_i^2} x(\mathbf{s}') = \frac{\partial^2 x(\mathbf{s})}{\partial s_i^2}. \tag{7.13}$$

In the above, we neglect the boundary terms that arise from integration by parts; these terms can be dropped because the delta function vanishes away from $\mathbf{s}' = \mathbf{s}$.

Based on the above, it is straightforward to show that the SSRF energy functional (7.9) is recovered by inserting the inverse SSRF covariance kernel $C_{\mathrm{xx}}^{-1}(\mathbf{s}-\mathbf{s}')$ from (7.11) into the general equation for the Gaussian energy functional (7.10).

Example 7.1 Evaluate the following integral that represents the convolution of the Gaussian function with the second-order derivative of the Dirac function

$$I = \int_{-\infty}^{\infty} du\, e^{-u^2/\xi^2} \frac{d^2 \delta(s-u)}{d^2 u}.$$

Answer The integral I can equivalently be expressed as follows:

$$I = \int_{-\infty}^{\infty} du\, e^{-u^2/\xi^2} \frac{d}{du}\left[\frac{d\delta(s-u)}{du}\right].$$

Using integration by parts and the fact that the Gaussian function vanishes at the boundaries (i.e. for $u \to -\infty, \infty$), the integral can be expressed as follows

$$I = -\int_{-\infty}^{\infty} du\, \frac{d}{du} e^{-u^2/\xi^2} \frac{d\delta(s-u)}{du} = \frac{2}{\xi^2}\int_{-\infty}^{\infty} du\, u\, e^{-u^2/\xi^2} \frac{d\delta(s-u)}{du}.$$

A second integration by parts leads to

$$I = -\frac{2}{\xi^2}\int_{-\infty}^{\infty} \mathrm{d}u\, \mathrm{e}^{-u^2/\xi^2}\,\delta(s-u) + \frac{4}{\xi^4}\int_{-\infty}^{\infty} \mathrm{d}u\, u^2\, \mathrm{e}^{-u^2/\xi^2}\,\delta(s-u)$$

$$= \left(-\frac{2}{\xi^2} + \frac{4\,s^2}{\xi^4}\right)\mathrm{e}^{-s^2/\xi^2}.$$

It is easy to verify that the above is equivalent to the second derivative of the Gaussian function at s in agreement with (7.13).

7.1.6 Spectral Representation

The easiest way to determine the SSRF spectral density is based on (6.25) that links the spectral density with its inverse. The latter follows directly from the inverse spectral kernel (7.11), if we take into account the following *Fourier transforms* of the generalized delta function and its derivatives

$$\mathcal{F}\left[\delta(\mathbf{s}-\mathbf{s}')\right] = 1, \tag{7.14a}$$

$$\mathcal{F}\left[\nabla\delta(\mathbf{s}-\mathbf{s}')\right] = \mathrm{i}\,\mathbf{k}, \tag{7.14b}$$

$$\mathcal{F}\left[\Delta\delta(\mathbf{s}-\mathbf{s}')\right] = -\,\|\mathbf{k}\|^2, \tag{7.14c}$$

$$\mathcal{F}\left[\Delta^2\delta(\mathbf{s}-\mathbf{s}')\right] = \|\mathbf{k}\|^4. \tag{7.14d}$$

Based on the above and (7.11), the *SSRF inverse spectral density* is given by

$$\widetilde{C}^{-1}_{\mathrm{xx}}(\mathbf{k}) = \frac{1+\eta_1\,\xi^2\,\|\mathbf{k}\|^2 + \xi^4\,\|\mathbf{k}\|^4}{\eta_0\,\xi^d}. \tag{7.15}$$

Thus the *SSRF spectral density* is given by

$$\widetilde{C}_{\mathrm{xx}}(\mathbf{k}) = \frac{\eta_0\,\xi^d}{1+\eta_1\,\xi^2\,\|\mathbf{k}\|^2 + \xi^4\,\|\mathbf{k}\|^4}. \tag{7.16}$$

The spectral density declines monotonically for $\eta_1 > 0$ (see Fig. 7.1), whereas for $\eta_1 < 0$ it exhibits a peak at the stationary point $\mathrm{d}\widetilde{C}_{\mathrm{xx}}(\mathbf{k})/\mathrm{d}\|\mathbf{k}\| = 0$, i.e., at the wavenumber (spatial frequency)

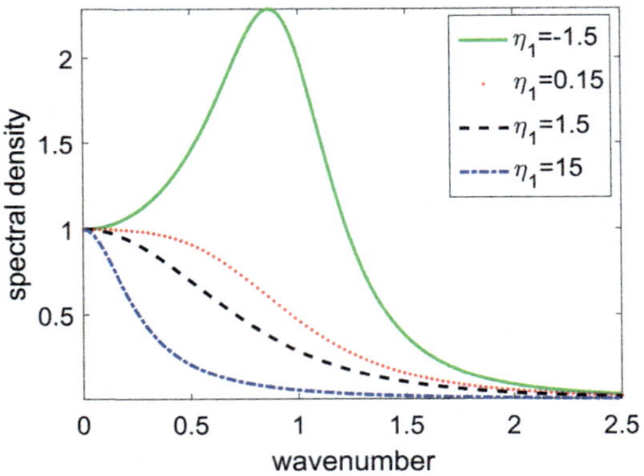

Fig. 7.1 SSRF isotropic spectral density for $\eta_0 = 1$, $\xi = 1$, and different values of η_1. For $\eta_1 < 0$ the dependence of the spectral density on the wavenumber is non-monotonic and develops a peak at $k^* \approx 0.866$. The spectral density declines monotonically with $\|\mathbf{k}\|$ for $\eta_1 > 0$

$$k^* = \xi^{-1}\sqrt{-\eta_1/2}. \tag{7.17}$$

Hole effect The peak of the spectral density implies that the covariance function exhibits damped oscillations in real space, cf. Figs. 7.4, 7.13, and 7.17. This damped oscillatory behavior leads to covariance functions with the characteristic "*negative hole effect*" , which is often encountered in geostatistical applications.

Typical models of hole-effect covariance functions include the damped cosine function [823]

$$C_{\mathrm{xx}}(\mathbf{r}) = \sigma_{\mathrm{x}}^2\, \mathrm{e}^{-h}\, \cos(\lambda\, h), \;\; \text{where } \lambda > 1/\sqrt{3},$$

and the cardinal sine function [624]

$$C_{\mathrm{xx}}(\mathbf{r}) = \sigma_{\mathrm{x}}^2\, \mathrm{sinc}(h).$$

Oscillatory covariance functions have been observed in the analysis of real data. For example, an oscillatory covariance was obtained for the time series of sea surface elevation during a rogue wave event, that was recorded in 1995 at an offshore oil platform [77]. More "hole effect" covariance models and information on their applications are given in [132, 367, 487].

The SSRF covariance in real space is a radial function that is obtained by means of the *spectral representation* (4.4b), i.e., by the following univariate integral

$$C_{\mathrm{xx}}(\mathbf{r}) = \frac{\eta_0 \xi^d\, \|\mathbf{r}\|^{1-d/2}}{(2\pi)^{d/2}} \int_0^\infty \mathrm{d}k\, \frac{k^{d/2}\, J_{d/2-1}(\|\mathbf{r}\|k)}{1 + \eta_1\,(k\xi)^2 + (k\xi)^4}, \tag{7.18}$$

where $J_\nu(\cdot)$ is the Bessel function of the first kind of order ν. The variance $C(0)$ is proportional to η_0 and depends nonlinearly on η_1 [372]. Explicit expressions for $C_{\mathrm{xx}}(\mathbf{r})$ in $d = 1, 2, 3$ at the limit $k_c \to \infty$ are given in [367, 372].

SSRF Permissibility The spectral density $\widetilde{C}_{xx}(\mathbf{k})$ given by (7.16) needs to satisfy the conditions of Bochner's Theorem 3.2, i.e., (i) $\widetilde{C}_{xx}(\mathbf{k}) \geq 0, \forall \mathbf{k} \in \mathbb{R}^d$ and (ii) $\int d\mathbf{k}\, \widetilde{C}_{xx}(\mathbf{k}) < \infty$. For the rational function (7.16) the permissibility conditions are determined by the SSRF *characteristic polynomial*

$$\Pi(x) = 1 + \eta_1 x^2 + x^4, \quad x = \|\mathbf{k}\|\, \xi. \tag{7.19a}$$

Since $\Pi(x)$ contains only even orders of x, it can also be expressed as follows

$$\Pi_1(u) = 1 + \eta_1 u + u^2, \quad u = (\|\mathbf{k}\|\, \xi)^2. \tag{7.19b}$$

The *discriminant of the characteristic polynomial* $\Pi_1(u)$ is given by

$$\Delta = \left(\eta_1{}^2 - 4\right)^{\frac{1}{2}}, \tag{7.20}$$

and the two roots of $\Pi_1(u)$ are given by

$$u_{\pm} = \frac{-\eta_1 \pm \Delta}{2}. \tag{7.21}$$

The value of the SSRF discriminant determines the type of the roots and consequently the behavior of the covariance function (as shown explicitly in the following sections). For $\Delta = 0$ the characteristic polynomial $\Pi_1(u)$ admits a double real root, for $\Delta > 0$ it develops two real roots, while for imaginary Δ the polynomial admits two complex roots that are conjugate of each other.

The non-negativity of $\widetilde{C}_{xx}(\mathbf{k})$ is ensured if $\Pi_1(u) > 0$ for all u. The integrability condition is satisfied if $\Pi_1(u) > 0$, since the long-wavenumber (infrared) limit of the integrand is $\propto \|\mathbf{k}\|^{d-5}$ which is integrable for $d < 4$. For $d \geq 4$ a finite cutoff k_c should be used to ensure the integrability of the spectral density.

SSRF permissibility conditions The function $\widetilde{C}_{xx}(\mathbf{k})$ given by (7.16) is a permissible SSRF spectral density if the following conditions are satisfied:

1. *No cutoff* ($k_c \to \infty$) *case:* The function $\widetilde{C}_{xx}(\mathbf{k})$ is positive for all $\|\mathbf{k}\|$ if the roots of the characteristic polynomial $\Pi_1(u)$ given by (7.19b) are complex numbers and the variance is well-defined [362]. This leads to the following conditions:

$$d < 4, \; \eta_0 > 0, \; \xi > 0, \; \eta_1 > -2.$$

(continued)

2. *Hard cutoff ($k_c > 0$) case:* In this case we arbitrarily cut off $\widetilde{C}_{xx}(\mathbf{k})$ at the wavenumber k_c. Two possibilities are discussed in [372]:

   a. For $\eta_1 > -2$ the conditions are as in the infinite k_c case, i.e.,

   $$d < 4, \ \eta_0 > 0, \ \xi > 0.$$

   b. The above conditions can be relaxed for $k_c > 0$, because the characteristic polynomial $\Pi_1(u)$ given by (7.19b) can then admit real roots $u_r > k_c\xi$. This allows rigidity coefficients $\eta_1 \leq -2$ if the following conditions are met:

   $$k_c\xi < \frac{1}{\sqrt{2}}\sqrt{|\eta_1| - |\Delta|}, \text{ where } \Delta = \sqrt{\eta_1{}^2 - 4},$$

   or equivalently

   $$0 < k_c\xi < 1 \text{ and } \eta_1 > -(k_c\xi)^2 - (k_c\xi)^{-2}.$$

Cutoff and differentiability The rational spectral density (7.16) does not include a cutoff for high wavenumbers. The absence of such a cutoff leads to non-differentiability of the SSRF covariance function, at least for $d > 1$. The same problem is encountered in statistical field theory [890], and is known as the *ultraviolet divergence* [399], [21, Chap. 2].

In principle, there are several ways of implementing a cutoff.

- We can discretize the energy functional (7.4) on a lattice of step $a > 0$. Then, the covariance function is given by the inverse Fourier transform of the spectral density over the first Brillouin zone, which extends between $(-\pi/a, \pi/a]$ in every direction of the reciprocal space [222]. This leads to $C_{xx}(\mathbf{r})$ that is differentiable everywhere, but explicit expressions for general $\mathbf{r}$ are not available [420].
- We can set the spectral density equal to zero for wavenumbers that exceed k_c, or we can multiply the rational density (7.16) by a function that decays exponentially for large $\|\mathbf{k}\|$ above a characteristic k_c, which is known as the *ultraviolet cutoff*. More details are given in [362]. The main shortcoming of these approaches, however, is that they do not lead to explicit expressions for the covariance function in real space.
- We can implement a *soft cutoff* by multiplying the SSRF spectral density with a non-negative function $\widetilde{\phi}(\mathbf{k})$ that decays to zero as $\|\mathbf{k}\| \to \infty$ so as to render the SSRF spectral moments of any desired order finite. As discussed below, this approach is equivalent to using a colored noise field in the Langevin equation that defines SSRFs (see Chap. 9).

The SSRF covariance functions evaluated without a cutoff are non-differentiable at the origin, $\|\mathbf{r}\| = 0$, for $d > 1$; hence, strictly speaking they do not correspond to the covariance function of the random field $X(\mathbf{s}; \omega)$ with the energy functional (7.4), since the definition of the latter involves derivatives of the field realizations.

Nonetheless, the functions derived with no spectral cutoff are valid covariances, because their spectral density satisfies the conditions of *Bochner's theorem* [83]. They also provide excellent approximations of the differentiable, finite-k_c covariance functions—obtained by numerical integration of the spectral density—for all lags $\|\mathbf{r}\| \gg 1/k_c$. From this latter perspective they are analogous to the asymptotic *Ornstein-Zernicke approximations* used in statistical field theory [467, Chap. 2]. We revisit the issue of the spectral cutoff in Sect. 7.2.6.

Killing the curvature If we omit the curvature term of SSRFs, the spectral function is given by $\widetilde{C}_{xx}(\mathbf{k}) = \eta_0\, \xi^d \left(1 + \eta_1\, \xi^2\, \|\mathbf{k}\|^2\right)^{-1}$ and corresponds to the Gaussian field theory model (7.2). In this case the spectral function is not a well-defined spectral density in $d \geq 2$ (since its integral over the reciprocal space diverges and thus the variance does not exist). For example, in $d = 2$ the inverse Fourier transform of the spectral function has a logarithmic divergence. This implies that the inverse Fourier transform is proportional to the modified Bessel function of the second kind of order zero. This function diverges at zero lag, and thus it represents a generalized covariance function. The Gaussian model (7.2) defines an *intrinsic random field*.

The singularity of the covariance function at zero lag can create problems in calculations, e.g., in kriging interpolation. Peter Kitanidis suggested an approach for taming singularities which is based on introducing a cutoff by means of a so-called microstructural parameter [460]. Essentially, the main idea is to replace the distance $\|\mathbf{r}\|$ in the covariance function by means of $\sqrt{\|\mathbf{r}\|^2 + a^2}/a$, where a is the microstructural parameter. The latter is presumably taken equal to a fraction of the lattice spacing. This is a practical engineering approach to dealing with singularities in the case of discrete sums (in continuum problems integrable singularities can be handled analytically). However, we should keep in mind that altering the distance metric affects the spectral density. Ideally, we should ensure that the modification of the distance does not destroy the positive definiteness of the covariance.

7.2 Two-Point Functions and Realizations

In the following, we consider the SSRF two-point functions at the limit $k_c \to \infty$. We use the normalized (dimensionless) lag $h = \|\mathbf{r}\|/\xi$. The parameter Δ, defined by (7.20), is the discriminant of the characteristic polynomial (7.19). In addition, the following coefficients appear in the explicit expressions

$$\beta_{1,2} = \frac{|2 \mp \eta_1|^{1/2}}{2}, \tag{7.22a}$$

$$w_{1,2} = \left(\frac{|\eta_1 \mp \Delta|}{2}\right)^{1/2}. \tag{7.22b}$$

In (7.22), the coefficient β_1 represents a wavenumber that characterizes periodic variations of the covariance, whereas β_2 as well as $w_{1,2}$ represent reciprocal lengths that characterize the relaxation of the fluctuations.

The evaluation of the spectral integral (7.18) depends on the dimensionality d. The Bessel function $J_{d/2-1}(k\,\|\mathbf{r}\|)$ is given by the following expressions for $1 \leq d \leq 3$

$$J_{d/2-1}(x) = \begin{cases} \sqrt{\frac{2}{\pi x}} \cos x, & d = 1, \\ J_0(x), & d = 2, \\ \sqrt{\frac{2}{\pi x}} \sin x, & d = 3. \end{cases} \tag{7.23}$$

Given the above expressions of $J_{-1/2}(x)$ and $J_{1/2}(x)$, the spectral integral is evaluated in $d = 1$ and $d = 3$ by means of *Cauchy's theorem of residues* [779, p. 621]. The details of the calculations are given in [372]. In $d = 2$ the Bessel function is not reducible to a simpler expression. In this case the integration can be performed using the *Hankel-Nicholson* integration formula [4, p. 488]. The relevant calculations are described in [367].

Comment SSRF covariance functions, unlike standard covariance models such as exponential, Gaussian, et cetera, have the same spectral density in all dimensions $d < 4$. However, the real-space covariance functions depend on the spatial dimensionality.

7.2.1 SSRF Length Scales

As mentioned above, the SSRF two-point functions in real space depend on the dimensionality of the embedding space. On the other hand, the SSRF spectral representation is independent of the spatial dimension. Below we focus on characteristic SSRF length scales that do not require real-space covariance expressions. The correlation spectrum is also in this category, but we discuss it in detail in Sect. 7.3.4.

Correlation radius The correlation radius is based on the second-order moment of the spatial lag weighted by the covariance. As defined by (5.40), the SSRF correlation radius is given by

$$r_c = \sqrt{2d\eta_1}\, \xi. \tag{7.24}$$

This result is meaningful for $\eta_1 > 0$. For $\eta_1 < 0$ the equation (5.40) leads to $r_c^2 < 0$, due to the negative hole in the covariance function. For negative rigidity values, it makes more sense to define the correlation radius in terms of the absolute value of the covariance function.

Integral range Based on (5.38), the SSRF integral range (based on the volume integral of the correlation function) is expressed in terms of the covariance function as

$$\ell_c = \left(\frac{1}{C_{xx}(0)} \int d\mathbf{r}\, C_{xx}(\mathbf{r}) \right)^{1/d} = \left(\frac{\widetilde{C}_{xx}(0)}{\sigma_x^2} \right)^{1/d}.$$

Taking into account the SSRF spectral density (7.16), it follows that $\widetilde{C}_{\mathrm{xx}}(0) = \eta_0\,\xi^d$, and the integral range is expressed as

$$\ell_c = \xi \left(\frac{\eta_0}{\sigma_{\mathrm{x}}^2}\right)^{1/d}. \tag{7.25}$$

The SSRF variance, however, depends on the spatial dimension and on η_1. Hence, we will give explicit expressions for the integral range specific to $d = 1, 2, 3$ below.

7.2.2 One-Dimensional SSRFs

Random fields defined over $\mathcal{D} \subset \mathbb{R}$ are essentially random processes that can be used to model time series or drillhole data. Applications of SSRFs in the modeling of financial and environmental pollution time series are given in [894, 895]. It is also possible to develop a time series model analogous to autoregressive models using SSRFs. We elaborate on this connection in Sect. 9.7.

One-dimensional random fields can also be used to study the variability of core samples obtained at different depths along a drill hole. In the case of hydrocarbon reservoir exploitation, information collected from drill holes at oil wells (e.g., torque and weight of drillstring, fluid pressure and flow) can be represented in the form of one-dimensional random fields or time series. Analysis of such 1D data is important for detecting the presence of subsurface structures and discontinuities (faults) [616]. Such a spatial model, however, provides a partial view of the variability restricted along the vertical direction.

In addition, 1D random fields find applications in geotechnical engineering. One-dimensional models include vertical seepage in soils, where the hydraulic conductivity is modeled as a stationary lognormal random field, and shallow landslide models in which the vertical variation of strength follows a stationary normal random field. Recent research focuses on the impact of correlation parameters other than the characteristic length, such as the smoothness index ν of the Whittle-Matérn model, on the probability of failure in geotechnical applications [133].

Covariance and variogram functions In the following, we use the *normalized lag* $h = |r|/\xi$. The covariance function is determined from the spectral representation (7.18) using the expression (7.23) for the Bessel function of the first kind of order $-1/2$ ($d = 1$) and (7.19a) for the SSRF characteristic polynomial. These lead to the following integral in terms of the dimensionless variable $x = \|\mathbf{k}\|\xi$ (see also Eq. (36) in [372]):

$$C_{\mathrm{xx}}(h;\boldsymbol{\theta}) = \frac{\eta_0}{\pi}\int_0^{\infty} \mathrm{d}x\, \frac{\cos(x\,h)}{\Pi(x)}. \tag{7.26}$$

The SSRF covariance function in one dimension has three different branches that are determined by the value of η_1. The covariance in each branch is given by the following equations

$$C_{xx}(h;\boldsymbol{\theta}) = \eta_0\, e^{-h\beta_2}\left[\frac{\cos(h\beta_1)}{4\,\beta_2} + \frac{\sin(h\beta_1)}{4\,\beta_1}\right],\ |\eta_1| < 2, \tag{7.27a}$$

$$C_{xx}(h;\boldsymbol{\theta}) = \eta_0\,\frac{(1+h)}{4\,e^{h}},\ \eta_1 = 2 \tag{7.27b}$$

$$C_{xx}(h;\boldsymbol{\theta}) = \frac{\eta_0}{\sqrt{\eta_1{}^2-4}}\left(\frac{e^{-h\,w_1}}{2w_1} - \frac{e^{-h\,w_2}}{2w_2}\right),\ \eta_1 > 2. \tag{7.27c}$$

The coefficients β_i, w_i $(i = 1,2)$ are determined by (7.22), i.e., $\beta_{1,2} = \frac{1}{2}|2 \mp \eta_1|^{1/2}$ and $w_{1,2} = [(\eta_1 \mp \Delta)/2]^{1/2}$, where $\Delta = \left(\eta_1{}^2 - 4\right)^{\frac{1}{2}}$. The coefficients $w_{1,2}$ are only defined for $\eta_1 > 2$; then Δ is a real number less than two, thus ensuring that the $w_{1,2}$ are also real numbers. The normalized lag is given by $h = |r|/\xi$.

Example 7.2 As we show in Sect. 7.2.6 the SSRF covariance function in one dimension admits second-order derivatives for all lags. Evaluate the second-order derivative of the covariance function, i.e., $d^2C_{xx}(r)/d^2r$, for $\eta_1 = 2$ at $r = 0$.

Answer We focus on $\eta_1 = 2$, but similar arguments hold for the other branches as well. The SSRF covariance for $\eta_1 = 2$ is also known as the *modified exponential.*

Based on (7.27b) the covariance function is $C_{xx}(r) \propto (1+h)\,e^{-h}$. The function e^{-h} is not differentiable at $h = 0$ due to the absolute value in $h = |r|/\xi$. Nevertheless, we can expand $C_{xx}(r)$ in Taylor series around $r = 0_+$ and $r = 0_-$ (i.e., for lags slightly to the right and to the left of zero, respectively). The two series can then be combined into the following

$$\begin{aligned} C_{xx}(r) &= \left(1+\frac{|r|}{\xi}\right)\left[1-\frac{|r|}{\xi}+\frac{r^2}{2\xi^2}+O\left(|r|^3\right)\right] \\ &= 1-\frac{r^2}{2\xi^2}+O\left(|r|^3\right). \end{aligned} \tag{7.28}$$

The equation above shows that even though the exponential is non-differentiable at $r = 0$, the terms proportional to $|r|/\xi$ that cause the non-differentiability cancel out. As a result, we can show that

$$\left.\frac{dC_{xx}(r)}{dr}\right|_{r=0} = 0, \quad \text{and}$$

$$\left.\frac{d^2 C_{xx}(r)}{dr^2}\right|_{r=0} = -\frac{1}{\xi^2}.$$

The covariance function is plotted for different η_1 in Fig. 7.2, and the respective variogram functions are shown in Fig. 7.3. The graphs demonstrate that the SSRF variance depends on η_1 in addition to η_0. In particular, for fixed η_0 and ξ the variance declines with increasing η_1. This dependence reflects the fact that higher η_1 imply higher gradient "cost" in the energy functional (7.4), thus leading to relative suppression of large variations. More specifically, the variance is given by the following equations [372]

$$\sigma_x^2 = \frac{\eta_0}{2\sqrt{2+\eta_1}}, \qquad |\eta_1| < 2, \tag{7.29a}$$

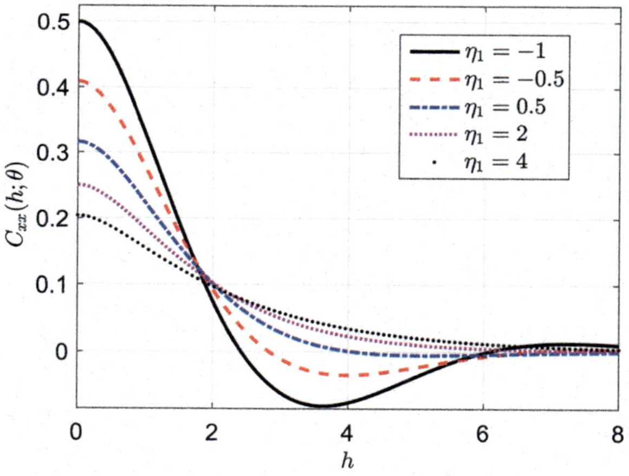

Fig. 7.2 SSRF covariance function in one dimension for $\eta_0 = 1$ and different values of η_1. For $\eta_1 > 2$ the covariance is obtained from Eq. (7.27c), for $\eta_1 = 2$, it is obtained from Eq. (7.27b), whereas for $|\eta_1| < 2$, it is obtained from Eq. (7.27a)

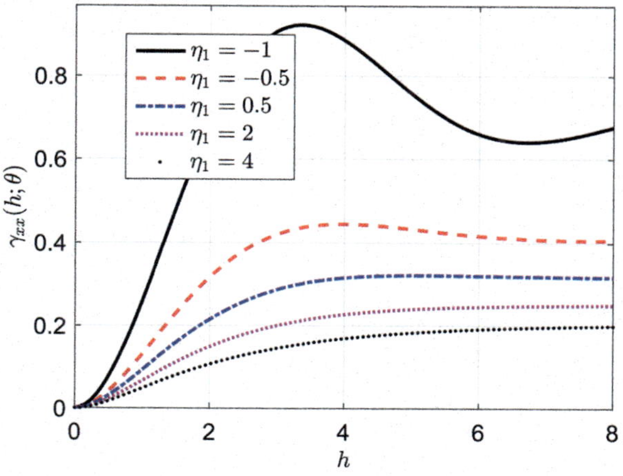

Fig. 7.3 SSRF variogram function in one dimension for $\eta_0 = 1$ and different values of η_1. The variogram is obtained from the covariance by means of $\gamma_{xx}(\mathbf{r}) = \sigma_x^2 - C_{xx}(\mathbf{r})$

$$\sigma_{\mathrm{x}}^2 = \frac{\eta_0}{4}, \quad \eta_1 = 2, \tag{7.29b}$$

$$\sigma_{\mathrm{x}}^2 = \frac{\eta_0}{2\sqrt{\eta_1{}^2 - 4}} \left(\frac{1}{w_1} - \frac{1}{w_2} \right), \quad \eta_1 > 2. \tag{7.29c}$$

In the above, the coefficients β_i, w_i ($i = 1, 2$) are determined by (7.22), i.e., $\beta_{1,2} = \frac{1}{2}|2 \mp \eta_1|^{1/2}$ and $w_{1,2} = [\,(\eta_1 \mp \Delta)\,/2]^{1/2}$, where $\Delta = \left(\eta_1{}^2 - 4\right)^{\frac{1}{2}}$.

Correlation function The correlation function is given by means of the following expressions:

$$\rho_{\mathrm{xx}}(h; \boldsymbol{\theta}) = \mathrm{e}^{-h\beta_2} \left[\cos(h\beta_1) + \frac{\beta_2}{\beta_1} \sin(h\beta_1) \right], \quad |\eta_1| < 2, \tag{7.30a}$$

$$\rho_{\mathrm{xx}}(h; \boldsymbol{\theta}) = (1 + h)\,\mathrm{e}^{-h}, \quad \eta_1 = 2, \tag{7.30b}$$

$$\rho_{\mathrm{xx}}(h; \boldsymbol{\theta}) = \frac{w_2\,\mathrm{e}^{-h\,w_1} - w_1\,\mathrm{e}^{-h\,w_2}}{w_2 - w_1}, \quad \eta_1 > 2. \tag{7.30c}$$

In the above equations, the coefficients β_i, w_i ($i = 1, 2$) are determined by (7.22), i.e., $\beta_{1,2} = \frac{1}{2}|2 \mp \eta_1|^{1/2}$ and $w_{1,2} = [\,(\eta_1 \mp \Delta)\,/2]^{1/2}$, where $\Delta = \sqrt{\eta_1{}^2 - 4}$ is the discriminant of the SSRF polynomial. Plots of the correlation function for different values of η_1 are shown in Fig. 7.4.

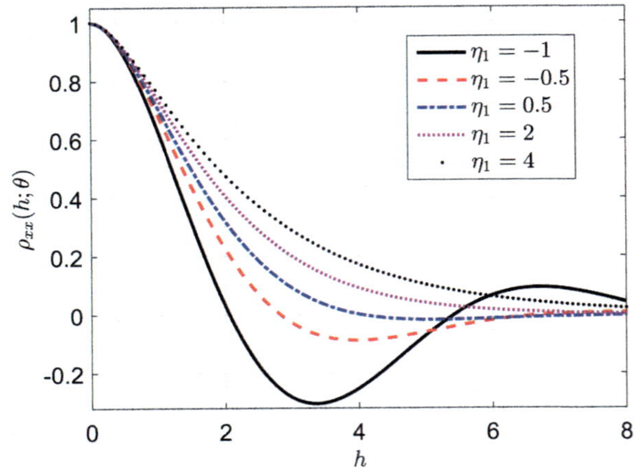

Fig. 7.4 SSRF correlation function in one dimension for different values of η_1. The correlation function is obtained from the covariance by means of $\rho_{\mathrm{xx}}(\mathbf{r}) = C_{\mathrm{xx}}(\mathbf{r})/C_{\mathrm{xx}}(0)$. The equations for the correlation function are given by (7.30)

The SSRF correlation function for $\eta_1 = 2$ and $d = 1$ is identical to the Whittle-Matérn covariance function for $\nu = 3/2$, as it can be seen by comparing (7.30b) with (4.12). This function is also known as the *modified exponential* [767].

Integral range The integral range is evaluated according to (5.38), and it is given by the following equations as shown in [372]

$$\ell_c = 2\xi \sqrt{|2 + \eta_1|}, \quad |\eta_1| < 2, \tag{7.31a}$$

$$\ell_c = 4\xi, \quad \eta_1 = 2, \tag{7.31b}$$

$$\ell_c = \sqrt{2}\xi \left(\sqrt{\eta_1 + \Delta} + \sqrt{\eta_1 - \Delta} \right), \quad \eta_1 > 2. \tag{7.31c}$$

As it follows from (7.31), the integral range increases with η_1, due to the increased rigidity of the field that opposes large gradients. Moreover, as $\eta_1 \to \infty$ then $\ell_c \to \infty$ signaling the onset of long-range dependence. The increase of ℓ_c with η_1 is reflected in the behavior of the correlation function $\rho_{xx}(\mathbf{r})$ shown in Fig. 7.4. Note also that $\ell_c \to 0$ as $\eta_1 \to -2$. Does this make sense? At the limit $\eta_1 \to -2$, the coefficient β_2 in (7.30a) tends to zero, implying that $\rho_{xx}(\mathbf{r})$ tends to a pure cosine term. The latter has zero integral range due to its oscillatory behavior. Hence, the limiting value of ℓ_c based on (7.31) agrees with the insight gained from the correlation function.

Realizations We compare the SSRF realizations for different values of η_1 in Fig. 7.5. First, we note that the states obtained for $\eta_1 = -1.999$ are quasi-periodic with an almost uniform wavelength. This quasi-periodicity is caused by the spectral density peak (7.17) at $k^*\xi = \sqrt{-\eta_1/2}$.

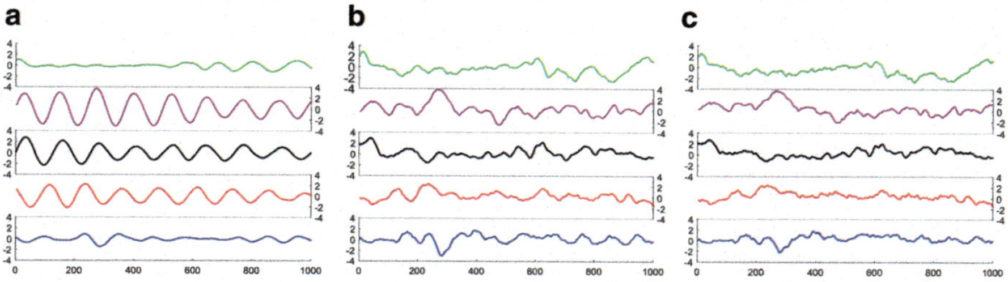

Fig. 7.5 Five realizations of one-dimensional SSRFs with $\eta_1 = -1.999$, 0.15 and 15 respectively. The domain contains 1000 points sampled with unit step. The characteristic length is $\xi = 20$ and the scale factor is $\eta_0 = 1$. The same random number generator seed is used for all three cases to ensure that the differences between the respective states (from 1 to 5) are only due to differences in η_1. (**a**) $\eta_1 = -1.999$. (**b**) $\eta_1 = 0.15$. (**c**) $\eta_1 = 15$

The wavelength of the periodic variation is determined by $\xi\beta_1 = \xi\sqrt{2-\eta_1}/2$ according to (7.30a). For $\eta_1 = -1.999$—close to the permissibility boundary—this implies a wavelength $\lambda^* \approx 2\pi\,\xi \approx 125.67$, in visual agreement with the respective graphs in Fig. 7.5. On the other hand, the integral range according to (7.31) is approximately equal to 0.6325 due to the oscillations of the covariance function.

One may wonder if such periodic variations of the correlations are justified on physical grounds. For time series, it is well established that periodic variations correspond to identifiable cyclical behavior, such as seasonal patterns. There is also a close connection between the one-dimensional SSRF and the classical damped harmonic oscillator—further elaborated in Sect. 9.2—which provides a physical mechanism for the oscillations.

In two and three dimensions, which are more relevant for spatial data analysis, periodic variations are observed in phenomena that involve waves [865, 866] and in spatial patterns that result from dynamic processes that are subject to periodic forcing during their temporal evolution [132, 56–57].

The respective realizations for $\eta_1 = 0.15$ and $\eta_1 = 15$ shown in Fig. 7.5 look very similar; e.g., the peaks and valleys appear approximately at the same locations. This is due to the fact that the same random numbers are used for all the realizations in each row. Nevertheless, there are subtle differences between $\eta_1 = 0.15$ and $\eta_1 = 15$.

(i) The peaks and valleys are overall more pronounced for $\eta_1 = 0.15$ than for $\eta_1 = 15$; this is due to the suppression of the SSRF variance at higher η_1 (see Fig. 7.7b).
(ii) The realizations for $\eta_1 = 15$ exhibit more fine-scale fluctuations than those for $\eta_1 = 0.15$. These correlated fluctuations contribute to a slower decline of the two-point correlation function for $\eta_1 = 15$, in agreement with the trend shown in Fig. 7.4. The role of η_1 is further analyzed below.

7.2.3 *The Role of Rigidity*

Rigidity coefficient values $|\eta_1| < 2$ lead to a "negative hole" in the covariance function which implies that the field values are anti-correlated at the respective lags. The hole becomes deeper as η_1 approaches the permissibility boundary of -2, whereas the damped oscillations are less pronounced for $0 < \eta_1 < 2$.

In the following we define a *correlation length* λ and a *pseudo-period* (or wavelength if the 1D process is spatially supported) T in terms of η_1. It should be kept in mind, however, that since we use normalized lags h, the actual values of λ and T measured in un-normalized coordinates should be multiplied with ξ.

In Chap. 9 we will demonstrate a correspondence between SSRFs and the harmonic damped oscillator, in light of which λ and T are connected with the oscillator *damping time* and *angular frequency*.

$|\eta_1| < 2$ In this case we can recast the covariance expression (7.27a) as

$$C_{xx}(h;\boldsymbol{\theta}) = \eta_0\, e^{-h\beta_2}\, g(h;\eta_1),$$

where the function $g(h;\eta_1)$, defined by

$$g(h;\eta_1) = \frac{\cos(h\beta_1)}{4\,\beta_2} + \frac{\sin(h\beta_1)}{4\,\beta_1},$$

provides an oscillatory modulation of the exponential decline. For $|\eta_1| < 2$ the roots of the SSRF characteristic polynomial are complex conjugate.

The oscillations of the SSRF covariance are due to the oscillations of $g(h;\eta_1)$. For $\eta_1 < 0$ the oscillations of the covariance are expected as a result of the peak of the spectral density, as discussed in Sect. 7.1.6. The behavior of the modulating function for $\eta_1 > 0$ is illustrated in Fig. 7.6. Somewhat surprisingly, the fluctuations of $g(h;\eta_1)$ seem to increase as $\eta_1 \to 2_-$ (the subscript indicates that the limit is approached from the left on the real axis).

To better understand the behavior near $\eta_1 = 2$, we express $g(h;\eta_1)$ in terms of the period T, i.e.,

$$g(h;\eta_1) = \frac{1}{4}\left[\frac{2\cos(h/T)}{\sqrt{2+\eta_1}} + h\,\mathrm{sinc}\left(\frac{h}{T}\right)\right], \quad \text{where } T = \frac{2}{\sqrt{2-\eta_1}}.$$

$\eta_1 = 2$ Based on the above equation, as $\eta_1 \to 2$ it follows that $T \to \infty$, while both $\cos(h/T)$ and $\mathrm{sinc}(h/T) \to 1$ for $h \ll T$. Hence, even though the amplitude of $g(h;\eta_1)$ increases with η_1, the period T also increases towards infinity as $\eta_1 \to 2$. Thus, at $\eta_1 = 2$ the periodic dependence is replaced by the monotonic increase $1 + h$.

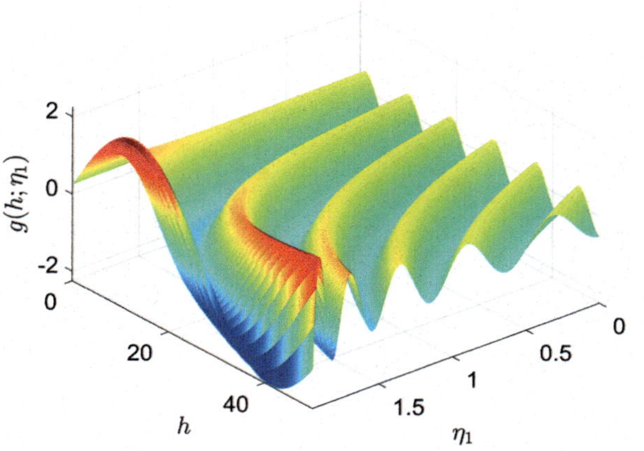

Fig. 7.6 Modulating function $g(h;\eta_1)$ for the one-dimensional SSRF covariance function for different values of η_1 in the interval [0, 1.95]

The first-degree polynomial, i.e., $1 + h$, in the SSRF covariance for $\eta_1 = 2$ is due to the presence of a double root in the SSRF characteristic polynomial which is equivalent to a double pole in the spectral density.

$\eta_1 > 2$ In this case, the roots of the SSRF characteristic polynomial (7.19b) separate again leading to a combination of two exponentially damped functions. The covariance function (7.27c) is then expressed—using (7.22b) for $w_{1,2}$ and (7.29a) for the variance—as follows

$$C_{xx}(h) = \frac{\eta_0}{2\sqrt{\eta_1{}^2 - 4}} \left(\lambda_1 \, e^{-h/\lambda_1} - \lambda_2 \, e^{-h/\lambda_2} \right), \tag{7.32}$$

where the dimensionless *damping times* λ_1 and λ_2 are given respectively by

$$\lambda_1 = \sqrt{\frac{2}{\eta_1 - \sqrt{\eta_1{}^2 - 4}}}, \text{ and } \lambda_2 = \sqrt{\frac{2}{\eta_1 + \sqrt{\eta_1{}^2 - 4}}} < \lambda_1 .$$

The damping times $\lambda_1 > \lambda_2$ represent, respectively, the characteristic scales of "slow" and "fast" (more heavily damped) exponential modes. As η_1 increases, so does the separation between the scales and their relative intensities: at the limit $\eta_1 \to \infty$ it follows that $\lambda_1 \to \infty$ while $\lambda_2 \to 0$. The dependence of the two damping times and the variance on η_1 is illustrated in Fig. 7.7.

The "fast" mode enters in (7.32) with a negative sign which implies that the fine-scale fluctuations are anti-correlated. We believe that these negative correlations are responsible for the "fine-scale" variability evidenced for $\eta_1 = 15$ in Fig. 7.5.

Narrow-band noise Let us consider the covariance for $|\eta_1| < 2$. We define the period $T = 1/\beta_1 = 2/\sqrt{2 - \eta_1}$ and the damping constant $\lambda = 1/\beta_2 = 2/\sqrt{2 + \eta_1}$. The covariance function (7.27a) can also be expressed as

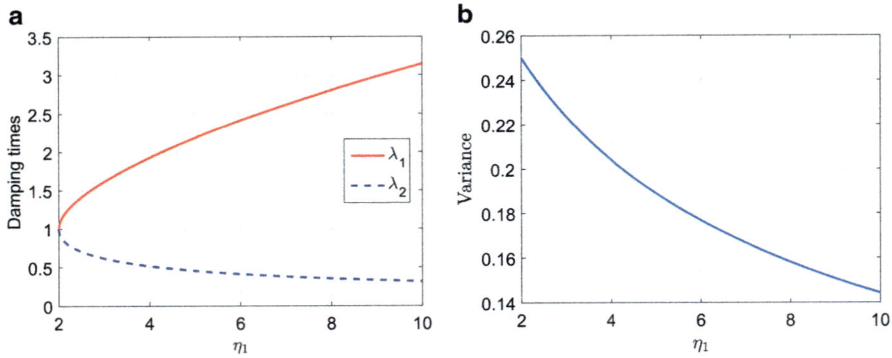

Fig. 7.7 (**a**) Characteristic damping times $\lambda_1 = \sqrt{2/(\eta_1 - \Delta)}$ and $\lambda_2 = \sqrt{2/(\eta_1 + \Delta)}$ versus $\eta_1 > 2$, where $\Delta = \sqrt{\eta_1{}^2 - 4}$. (**b**) Variance based on $\sigma_x^2 = \eta_0\,(\lambda_1 - \lambda_2)/(2\Delta)$ for $\eta_0 = 1$ and different values of η_1

$$C_{xx}(h;\boldsymbol{\theta}) = \frac{\eta_0\,e^{-h/\lambda}}{\sqrt{4-\eta_1{}^2}}\cos\left(\frac{h}{T}+\phi\right), \tag{7.33}$$

$$\phi = \arctan\left(\frac{T}{\lambda}\right) = \arctan\left(\sqrt{\frac{2+\eta_1}{2-\eta_1}}\right).$$

This expression is similar to the covariance function of *narrow-band noise* that is obtained for $\phi = 0$ [780, 821]. The covariance (7.33) has a characteristic *quasi-period* T that depends on η_1, and a *correlation length (time)*, λ, that also depends on η_1. This means that the realizations oscillate with a wavelength T, while their amplitude and phase persist over a distance (time) determined by the correlation length.

The variance of the narrow-band noise depends on both η_1 and η_0 via

$$\sigma_x^2 = \frac{\eta_0}{\sqrt{4-\eta_1{}^2}}\cos\phi = \frac{\eta_0}{2\sqrt{\eta_1+2}}. \tag{7.34}$$

The correlation function of narrow-band noise for different values of η_1 is shown in Fig. 7.8. It exhibits a transition from the oscillatory decay regime for $\eta_1 < 0$ to slow, seemingly monotonic decline for $\eta_1 > 0$.

This behavior is better understood in terms of Fig. 7.9a: As $\eta_1 \to 2$, the wavelength (T) tends to infinity, marking the transition to the modified exponential dependence. Since λ tends to decrease as $\eta_1 \to 2$, it follows that $\lambda/T \ll 1$, and the oscillations are much slower than the decay of the correlations. On the other hand, the correlation length tends to infinity as $\eta_1 \to -2$, tends to a minimum value, thus establishing the almost periodic behavior shown in Fig. 7.5a.

Based on (7.34), the variance increases towards infinity at the limit $\eta_1 \to -2$ as shown in Fig. 7.9b. The variance decreases monotonically as η_1 increases, tending to $1/4$ as $\eta_1 \to 2$.

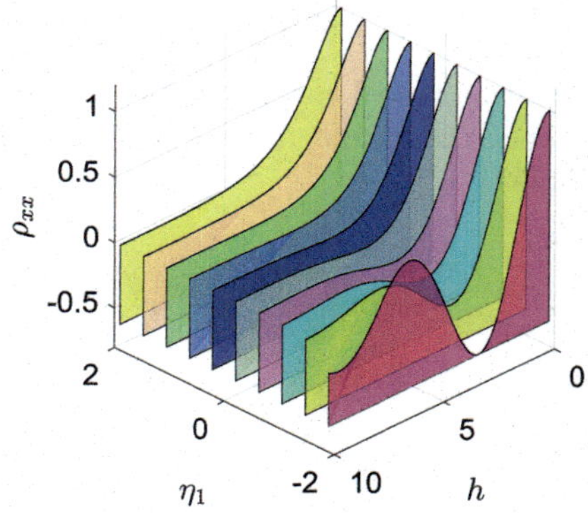

Fig. 7.8 SSRF correlation function $\rho_{xx}(h) = C_{xx}(h)/\sigma_x^2$ in one dimension for different values of η_1 in the interval $[-1.9, 1.9]$. The covariance function is given by (7.33) and the variance by (7.34)

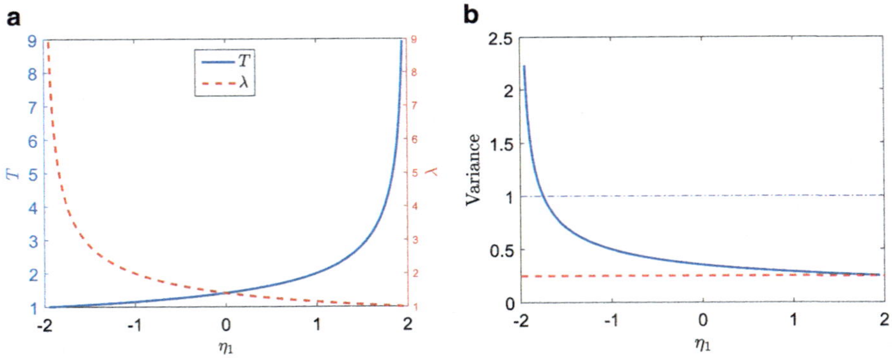

Fig. 7.9 (**a**) Dependence of the characteristic wavelength (period) $T = 2/\sqrt{2-\eta_1}$ and the correlation length $\lambda = 2/\sqrt{2+\eta_1}$ on the rigidity coefficient η_1. (**b**) Narrow-band noise variance based on (7.34) for $\eta_0 = 1$ versus $\eta_1 \in [-1.9, 1.9]$. The red dashed line marks the value 1/4 (equal to the variance at $\eta_1 = 2$), while the blue dash-dot line marks the value 1

Note For the plot in Fig. 7.9b the variance was numerically calculated using the expression $\eta_0/2\sqrt{\eta_1+2}$ instead of the equivalent $\eta_0 \cos\phi/\sqrt{4-\eta_1{}^2}$. The latter is prone to numerical errors due to the inversion of $\tan\phi = \sqrt{(2+\eta_1)/(2-\eta_1)}$, because the tangent tends to infinity as $\eta_1 \to 2$.

7.2.4 Two-Dimensional SSRFs

In two and three dimensions, the same general features hold for the two-point functions and the field realizations as in one dimension, even though the expressions for the covariance functions are different. A significant difference is that in two and three dimensions the Spartan random fields (without spectral cutoff) are non-differentiable in the mean square sense, unlike their one-dimensional counterparts.

In the following, we use the normalized lag distance $h = \|\mathbf{r}\|/\xi$.

Covariance and variogram functions The covariance function in two dimensions is obtained by the following *Hankel transform of zero order* (see equation (9) in [367]):

$$C_{xx}(h;\boldsymbol{\theta}) = \frac{\eta_0}{2\pi}\int_0^\infty dx\, \frac{x\, J_0(x\,h)}{\Pi(x)}, \tag{7.35}$$

where $x = \|\mathbf{k}\|\xi$ is a dimensionless variable, $\Pi(x)$ is the characteristic polynomial defined by (7.19), and $J_0(\cdot)$ is the Bessel function of the first kind of order zero. The above integral can be evaluated by means of the Hankel-Nicholson integration formula [4, p. 488].

Roots of characteristic polynomial In the following, if $\phi(\cdot): \mathbb{R} \to \mathbb{C}$ is a complex-valued function, $\Re[\phi(\cdot)]$ denotes the real part, while $\Im[\phi(\cdot)]$ denotes its imaginary part. In addition, $\phi^{\dagger}(\cdot)$ denotes the complex conjugate of $\phi(\cdot)$ so that $\Re[\phi^{\dagger}(\cdot)] = \Re[\phi(\cdot)]$ and $\Im[\phi^{\dagger}(\cdot)] = -\Im[\phi(\cdot)]$.

Let $z_{\pm}$ denote the double roots of the SSRF characteristic polynomial (7.19a), which is expressed as $\Pi(u) = (z^2 - z_+^2)(z^2 - z_-^2)$. The roots $z_{\pm}$, which determine the spatial dependence of the covariance function, are given by the following expression

$$z_{\pm}^2 = \frac{\eta_1 \mp \Delta}{2}, \text{ where } \Delta = \left(\eta_1{}^2 - 4\right)^{\frac{1}{2}}. \tag{7.36}$$

The dependence of the roots $z_{\pm}$ on η_1 is shown in Fig. 7.10 and reveals the following behavior in each of the three rigidity branches:

- For $\eta_1 > 2$, both z_+ and z_- are real numbers.
- For $\eta_1 = 2$ there is a double root, $z_{\pm} = \sqrt{\eta_1/2}$.
- For $|\eta_1| < 2$, the roots are complex conjugates of each other, i.e., $z_+ = z_-^{\dagger}$, or equivalently $\Re(z_+) = \Re(z_-)$ and $\Im(z_+) = -\Im(z_-)$.

Upon integration of (7.35), the three rigidity branches of the SSRF covariance function are given by the following equations that involve the *modified Bessel function of the second kind*, $K_{\nu}(\cdot)$, of orders $\nu = 0, -1$ [367].

$$C_{xx}(h;\boldsymbol{\theta}) = \frac{\eta_0 \, \Im\left[K_0(h\, z_+)\right]}{\pi\sqrt{4 - \eta_1{}^2}}, \quad |\eta_1| < 2, \tag{7.37a}$$

$$C_{xx}(h;\boldsymbol{\theta}) = \left(\frac{\eta_0\, h}{4\pi}\right) K_{-1}(h), \quad \eta_1 = 2, \tag{7.37b}$$

(continued)

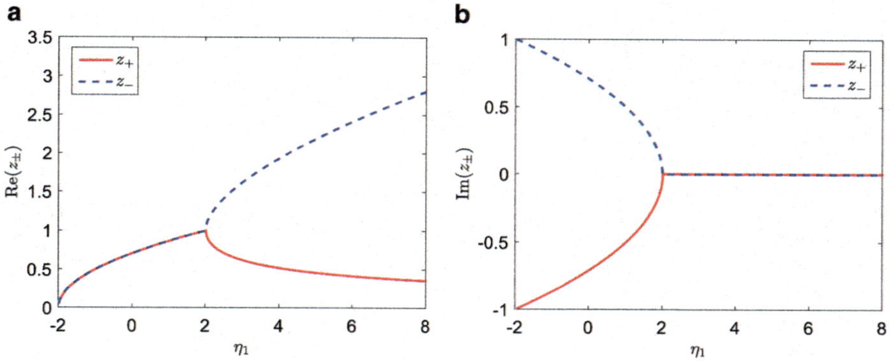

Fig. 7.10 Real (**a**) and imaginary (**b**) parts of the roots $z_+ = \sqrt{-t_+^*}$ and $z_- = \sqrt{-t_-^*}$ of the characteristic polynomial $\Pi(u) = 1 + \eta_1 u^2 + u^4$ according to (7.36)

$$C_{xx}(h;\boldsymbol{\theta}) = \frac{\eta_0\,[K_0(h\,z_+) - K_0(h\,z_-)]}{2\pi\sqrt{\eta_1{}^2 - 4}}, \quad \eta_1 > 2. \tag{7.37c}$$

The covariance expression for $|\eta_1| < 2$ can also be derived by means of (7.37c). However, in this case the two modified Bessel functions have complex-valued arguments due to $z_\pm$. In order to render the covariance expression for $\eta_1 < 2$ explicitly real-valued, we use an analytic continuation property of the modified Bessel function as given in [4, p. 377]. According to this property, for any $z \in \mathbb{C}$, it holds that $K_0(z^\dagger) = K_0^\dagger(z)$. Then, $K_0(h\,z_+) - K_0(h\,z_-) = K_0(h\,z_+) - K_0^\dagger(h\,z_+)$ which leads to the explicitly real-valued expression (7.37a).

Remark In (7.37b) we can replace $K_{-1}(h)$ with $K_1(h)$ taking into account the even symmetry of the modified Bessel function of the second kind $K_\nu(\cdot)$ with respect to the order ν. The latter follows from the integral representation of $K_\nu(h)$ [4, p. 376]

$$K_\nu(h) = \int_0^\infty dx\, \exp(-h\cosh x)\,\cosh(\nu x).$$

The dependence of $C_{xx}(h;\boldsymbol{\theta})$ on h for different η_1 is shown in Fig. 7.11, and it is qualitatively similar to the one-dimensional case. The values $C_{xx}(h = 0;\boldsymbol{\theta})$ agree with the following variance expressions

$$\sigma_x^2 = \frac{\eta_0}{2\pi\,|\Delta|}\left[\frac{\pi}{2} - \arctan\left(\frac{\eta_1}{|\Delta|}\right)\right], \quad \text{for } |\eta_1| < 2, \tag{7.38a}$$

$$\sigma_x^2 = \frac{\eta_0}{4\pi}, \quad \text{for } \eta_1 = 2, \tag{7.38b}$$

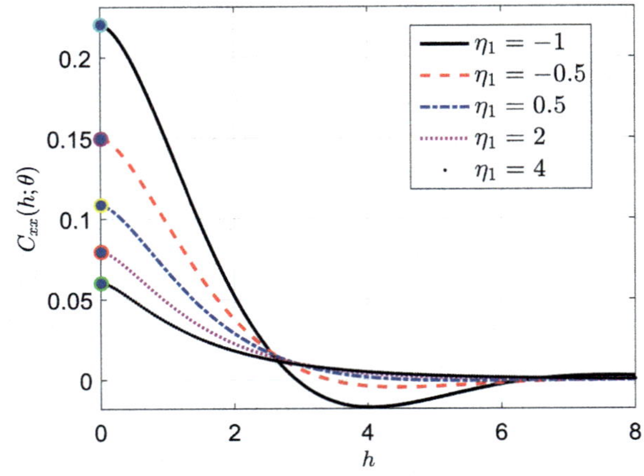

Fig. 7.11 SSRF covariance in two dimensions for different values of η_1. For $\eta_1 > 2$ the covariance is calculated from Eq. (7.37c), for $\eta_1 = 2$, it is calculated from Eq. (7.37b), whereas for $|\eta_1| < 2$, it is calculated from Eq. (7.37a). For all the graphs shown $\eta_0 = 1$. Filled circles at the origin ($h = 0$) represent the variance calculated independently from (7.38)

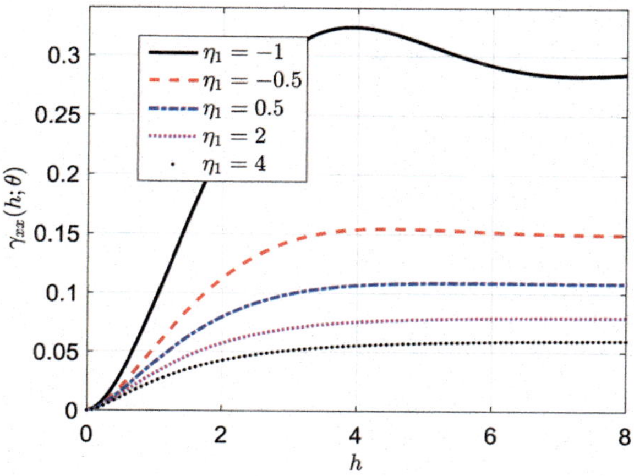

Fig. 7.12 SSRF variogram function in two dimensions for $\eta_0 = 1$ and different values of η_1. The variogram is obtained from the covariance expressions (7.37) by means of $\gamma_{\mathrm{xx}}(\mathbf{r}) = \sigma_{\mathrm{x}}^2 - C_{\mathrm{xx}}(\mathbf{r})$

$$\sigma_{\mathrm{x}}^2 = \frac{\eta_0}{4\pi\,\Delta} \ln\left(\frac{\eta_1 + \Delta}{\eta_1 - \Delta}\right), \quad \text{for } \eta_1 > 2, \tag{7.38c}$$

which are independently obtained [362]. In general, higher values of η_1 reduce the variance and increase the spatial coherence as evidenced in the slower decay of the tail of $C_{\mathrm{xx}}(h;\boldsymbol{\theta})$. This behavior reflects the higher stiffness of SSRF realizations for higher η_1. The dependence of the variogram on the lag is shown in Fig. 7.12.

Correlation function The two-dimensional correlation function is given by means of the following expressions [367]

$$\rho_{\mathrm{xx}}(h;\boldsymbol{\theta}) = \frac{\Im\left[K_0(h\,z_+)\right]}{\frac{\pi}{2} - \arctan\left(\frac{\eta_1}{|\Delta|}\right)}, \quad |\eta_1| < 2, \tag{7.39a}$$

$$\rho_{\mathrm{xx}}(h;\boldsymbol{\theta}) = h\,K_{-1}(h), \quad \eta_1 = 2, \tag{7.39b}$$

$$\rho_{\mathrm{xx}}(h;\boldsymbol{\theta}) = \frac{2\left[K_0(h\,z_+) - K_0(h\,z_-)\right]}{\ln\left(\frac{\eta_1+\Delta}{\eta_1-\Delta}\right)}, \quad \eta_1 > 2. \tag{7.39c}$$

Plots of the correlation function for different values of the rigidity coefficient are shown in Fig. 7.13. Sample realizations are shown in Fig. 7.14.

Integral range The integral range is evaluated according to (5.38). It is given by the following equations [380]:

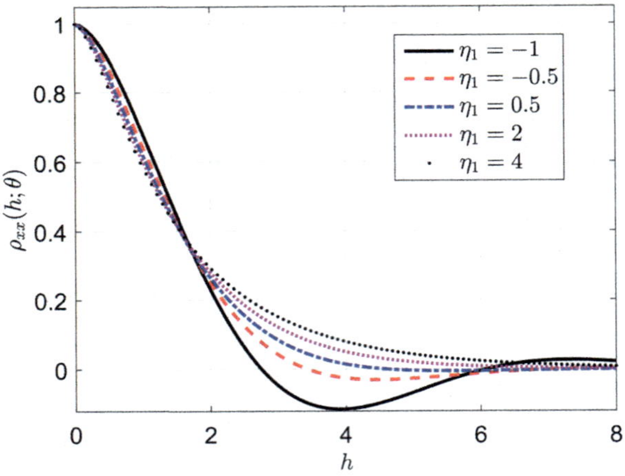

Fig. 7.13 SSRF correlation function in two dimensions for different values of η_1. The correlation function is obtained from the covariance by means of $\rho_{xx}(\mathbf{r}) = C_{xx}(\mathbf{r})/C_{xx}(0)$. The equations that determine the correlation function are given by (7.39)

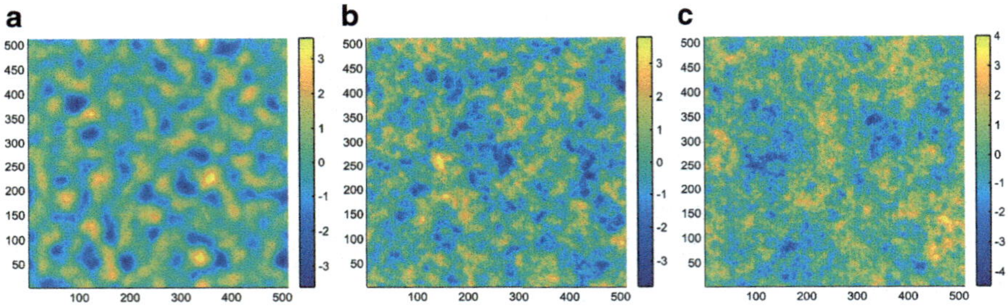

Fig. 7.14 Realizations of two-dimensional SSRFs with $\eta_1 = -1.9$ (**a**), $\eta_1 = 1.9$ (**b**) and $\eta_1 = 10$ (**c**) respectively, on a square domain with unit step and 512 nodes per side. The characteristic length is $\xi = 10$ and η_0 is selected so that $\sigma_x^2(\eta_1) = 1$. Different random number generator seeds are used for each realization

$$\ell_c = \xi \sqrt{\frac{4\,|\Delta|}{1 - \frac{2}{\pi}\arctan\left(\frac{\eta_1}{|\Delta|}\right)}}, \quad |\eta_1| < 2, \tag{7.40a}$$

$$\ell_c = 2\sqrt{\pi}\,\xi, \quad \eta_1 = 2, \tag{7.40b}$$

$$\ell_c = 2\xi \sqrt{\frac{\pi\,\Delta}{\ln(\eta_1 + \Delta) - \ln(\eta_1 - \Delta)}}, \quad \eta_1 > 2. \tag{7.40c}$$

It follows from (7.40) that the integral range increases with η_1, because increased rigidity opposes large gradients. Moreover, as $\eta_1 \to \infty$ then $\ell_c \to \infty$ which reflects the long-range dependence. The increase of ℓ_c with η_1 is reflected in the behavior of the correlation function $\rho_{xx}(\mathbf{r})$ shown in Figs. 7.13.

7.2.5 Three-Dimensional SSRFs

Covariance and variogram functions The covariance function in three dimensions is obtained by the following spectral integral (see Sect. 7.2.2 and Eq. (47) in [372]):

$$C_{xx}(h; \boldsymbol{\theta}) = \frac{\eta_0}{2\pi^2 h} \int_0^{\infty} dx \, \frac{x \sin(x\, h)}{\Pi(x)}. \tag{7.41}$$

The covariance function is plotted for different η_1 in Fig. 7.15, whereas the respective variogram functions are shown in Fig. 7.16. The graphs exhibit the same trends as in one and two dimensions. However, the negative holes of the covariance for $\eta_1 < 2$ are less pronounced in three dimensions compared to one and two dimensions.

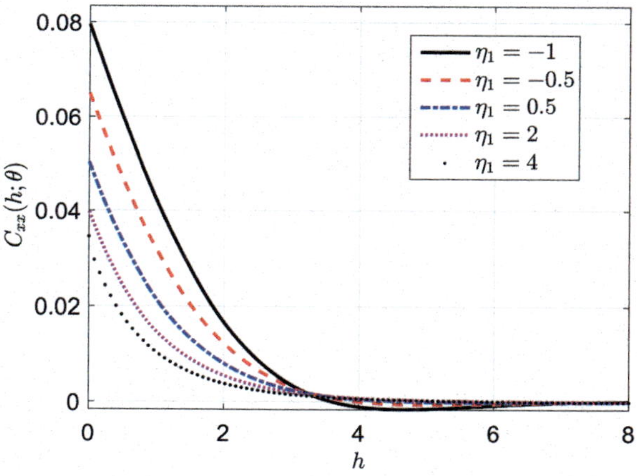

Fig. 7.15 SSRF covariance function in three dimensions for $\eta_0 = 1$ and different values of η_1. For $|\eta_1| < 2$, it is calculated from Eq. (7.42a); for $\eta_1 = 2$, it is calculated from Eq. (7.42b); for $\eta_1 > 2$ the covariance is calculated from Eq. (7.42c)

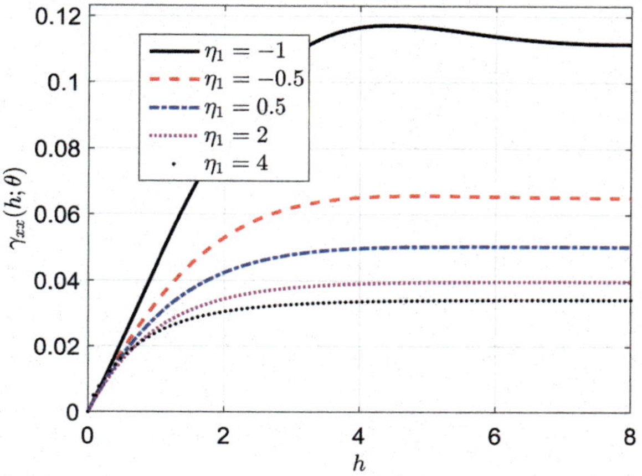

Fig. 7.16 SSRF variogram function in three dimensions for $\eta_0 = 1$ and different values of η_1. The variogram is obtained from the covariance (7.42) by means of $\gamma_{xx}(\mathbf{r}) = \sigma_x^2 - C_{xx}(\mathbf{r})$

$$C_{xx}(h;\boldsymbol{\theta}) = \eta_0 \, \frac{e^{-h\beta_2}}{2\pi\,|\Delta|} \left[\frac{\sin(h\beta_1)}{h} \right], \quad |\eta_1| < 2, \tag{7.42a}$$

$$C_{xx}(h;\boldsymbol{\theta}) = \frac{\eta_0}{8\pi}\, e^{-h}, \quad \eta_1 = 2, \tag{7.42b}$$

$$C_{xx}(h;\boldsymbol{\theta}) = \frac{\eta_0}{4\pi\,\Delta} \left(\frac{e^{-h\omega_1} - e^{-h\omega_2}}{h} \right), \quad \eta_1 > 2. \tag{7.42c}$$

As evidenced in the above equations, the covariance for $\eta_1 = 2$ is equivalent to the *exponential model*. For $|\eta_1| < 2$, the covariance is equal to the product of two permissible covariance models, the cardinal sine and the exponential. For $\eta_1 > 2$ the covariance involves a superposition of two exponential functions divided by the lag distance. The combination is permissible in spite of the negative sign in front of $e^{-h\omega_2}$ in (7.42c), because the coefficients are obtained by integration of the non-negative SSRF spectral density (7.16).

The SSRF variance is given by the following equations [372]

$$\sigma_x^2 = \frac{\eta_0}{4\pi\,\sqrt{2+\eta_1}}, \quad |\eta_1| < 2, \tag{7.43a}$$

$$\sigma_x^2 = \frac{\eta_0}{8\pi}, \quad \eta_1 = 2, \tag{7.43b}$$

$$\sigma_x^2 = \frac{\eta_0}{4\pi\,\Delta}\,(w_2 - w_1), \quad \eta_1 > 2. \tag{7.43c}$$

The coefficients β_i, w_i ($i = 1, 2$) are determined by (7.22). We recall that their values are given by $\beta_{1,2} = \frac{1}{2}|2 \mp \eta_1|^{1/2}$ and $w_{1,2} = [\,(\eta_1 \mp \Delta)\,/2]^{1/2}$, where $\Delta = \sqrt{\eta_1{}^2 - 4}$. Based on (7.43a) the variance diverges as the permissibility boundary is approached, i.e., $\eta_1 \to -2$ (in agreement with the one- and two-dimensional cases.)

Comment The variance for $\eta_1 > 2$ can be obtained from (7.42c) as $\sigma_x^2 = \lim_{h\to 0_+} C_{xx}(h;\boldsymbol{\theta})$. This is the limit of an indeterminate form, because both the numerator and denominator tend to zero at $h \to 0$; the limit, however, can be evaluated using L'Hôspital's rule.

Correlation function The three-dimensional correlation function is given by means of the following expressions [372]

$$\rho_{xx}(h;\boldsymbol{\theta}) = e^{-h\beta_1}\, \frac{\sin(h\beta_2)}{h\beta_2}, \quad |\eta_1| < 2, \tag{7.44a}$$

$$\rho_{xx}(h;\boldsymbol{\theta}) = e^{-h}, \quad \eta_1 = 2, \tag{7.44b}$$

$$\rho_{xx}(h;\boldsymbol{\theta}) = \frac{e^{-h\,w_1} - e^{-h\,w_2}}{h\,(w_2 - w_1)}, \quad \eta_1 > 2. \tag{7.44c}$$

Representative plots of the correlation function are shown in Fig. 7.17. The correlation hole for $\eta_1 < 0$ is less pronounced than in the one-dimensional case. This reflects the fact that the lower bound (4.18) for negative correlations becomes shallower as the space dimensionality increases.

Integral range The integral range is evaluated according to (5.38). It is given by the following equations as shown in [372]

$$\ell_c = 2\xi \, (\pi \, \beta_2)^{1/3}, \quad |\eta_1| < 2, \tag{7.45a}$$

$$\ell_c = 2\xi \, \pi^{1/3}, \quad \eta_1 = 2, \tag{7.45b}$$

$$\ell_c = 2\xi \left[\frac{\pi (w_1 + w_2)}{2} \right]^{1/3}, \quad \eta_1 > 2. \tag{7.45c}$$

Based on (7.45) it follows that the integral range increases with η_1, due to the increased rigidity of the field that opposes large gradients. Moreover, as $\eta_1 \to \infty$ then $\ell_c \to \infty$ which reflects long-range dependence. The increase of ℓ_c with η_1 is reflected in the behavior of the correlation function $\rho_{xx}(\mathbf{r})$ shown in Fig. 7.17.

Subspace permissibility The three-dimensional SSRF covariance can also be used in $d = 1, 2$ based on the subspace permissibility property. In lower dimensions, however, the spectral density does not coincide with the SSRF density. In addition,

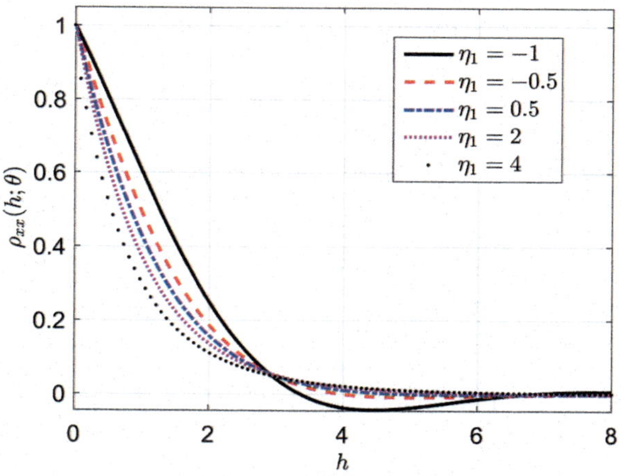

Fig. 7.17 SSRF correlation function in three dimensions for different values of η_1. The correlation function is obtained from the covariance by means of $\rho_{xx}(\mathbf{r}) = C_{xx}(\mathbf{r})/C_{xx}(0)$. The equations for the correlation function are given by (7.44)

in $d = 1, 2$ the three-dimensional covariance, albeit permissible, is not associated with the SSRF energy functional of the respective dimensionality.

> I have committed the ultimate sin, I have predicted the existence of a particle that can never be observed.
> *Wolfgang Pauli*

7.2.6 On Mean-Square Differentiability

SSRFs have a curious mathematical property that may even seem to undermine the self-consistency of the model. Physicists are familiar with it, but do not consider it a problem—for the reasons explained below. Engineers are practically minded, and will therefore not care about a mathematical oddity that does not have practical consequences in most cases. Mathematicians and statisticians on the other hand will, almost certainly, frown.

Existence? of first-order derivatives As discussed in Sect. 5.3.2, a necessary condition for the first-order partial derivatives of the field to exist in the mean square sense is the existence of the spectral moments $\Lambda_{i,j}$ given by (5.29). For SSRFs, this requires the convergence of the integral

$$\int_0^\infty \mathrm{d}k \, \frac{k^{d+1}}{1 + \eta_1 k^2 + k^4}.$$

The convergence of this integral depends on the behavior of the integrand at the limit $k \to \infty$, where the dominant behavior is k^{d-3}. Hence, the integral converges only for $d = 1$. Therefore, the mean-square-sense first-order derivative exists in one dimension (i.e., for random processes), but the partial first-order derivatives do not exist for $d \geq 2$.[4]

The non-differentiability leads to the seeming "paradox" that the SSRF covariance function is derived from an energy functional that involves the derivatives of a non-differentiable field. This paradox arises only at the limit of an infinite spectral cutoff, $k_c \to \infty$, which is what we used to calculate the SSRF covariance functions. The non-differentiability is due to the fact that, as we show in Chap. 9, the SSRF realizations are solutions of spatial Langevin equations driven by white noise [i.e., equation (9.36a)]. Hence, the discontinuity of the noise source naturally leads to the SSRF non-differentiability.

[4]In $d = 2$ the singularity is only marginal, i.e., logarithmic.

For finite k_c, however large, the covariance derivatives at zero distance are well-defined at all orders; hence, the sample function derivatives of all orders also exist in the mean-square sense. There are three different approaches for implementing a *finite frequency cutoff* as discussed in Sect. 7.1.6.

(i) The integral of the spectral density is cut off at some finite, arbitrary wavenumber k_c. This can be achieved either by a sharp cutoff or by multiplying the spectral density with a function that smoothly decays to zero around k_c. To our knowledge, this approach has not yet yielded explicit results for the covariance function. In addition, a sharp cutoff at k_c is likely to introduce oscillatory behavior in the real-space covariance function at lags $\propto 1/k_c$.

(ii) The SSRF is defined on a regular lattice with finite step size (e.g., a cubic grid with step a). The SSRF spectral density on cubic grids is given by [360]

$$\widetilde{C}_{xx}(\mathbf{k}) = \frac{\eta_0\, \xi^d}{1 + 4\eta_1 \frac{\xi^2}{a^2} \sum_{i=1}^{d} \sin^2\left(\frac{a\,k_i}{2}\right) + \frac{16\xi^4}{a^4} \sum_{i=1}^{d} \sin^4\left(\frac{a\,k_i}{2}\right)}.$$

The above spectral density at the limit $a \to 0$ tends to the SSRF spectral density (7.16). This can be shown by means of the Taylor expansion of the sine function around zero, i.e., $\sin x = x + O(x^3)$.

Inverting the spectral density implies calculating the spectral integral over the wavevectors in the "first Brillouin zone", i.e., within a hypercubic volume that spans the interval $[-\pi/a, \pi/a)$ in each orthogonal direction. This ensures that divergences caused by large wavevectors are eliminated. Integrals of this type also occur in the calculation of *lattice Green's functions* used in solid state physics problems. Explicit expressions however, cannot be derived except for specific values of the lag distance [318].

(iii) The SSRF spectral density $\widetilde{C}_{xx}(\mathbf{k})$ is transformed by means of $\widetilde{C}_{xx}(\mathbf{k}) \mapsto \widetilde{C}_{xx}(\mathbf{k})\, \widetilde{\phi}(\mathbf{k})$, where $\widetilde{\phi}(\mathbf{k})$ is a non-negative function that decays at infinity *at least as fast* as $\|\mathbf{k}\|^{-\alpha}$, where $\alpha > d + 2m - 4$, and m is the desired order of differentiation. For example, in $d = 2$ in order for the first-order derivatives to be well defined in the mean-square sense, it is required that $\alpha > 0$.

The introduction of $\widetilde{\phi}(\mathbf{k})$ is equivalent to driving the Langevin equation that defines SSRFs by means of *colored noise* with spectral density $\widetilde{\phi}(\mathbf{k})$ instead of white noise (see Chap. 9). However, explicit expressions for real-space covariance functions obtained by means of this approach are not available to date.

The asymptotic limit We presented SSRF spatial covariance functions obtained from the spectral representation at the asymptotic limit of the cutoff $k_c \to \infty$. How are we to interpret the lack of differentiability of these covariance functions in $d > 1$ in connection with the SSRF energy functional that involves spatial derivatives? This dilemma is well-known in statistical field theory as well [890]. There are two approaches out of this hiatus:

(i) First, the covariance functions derived by integrating the SSRF spectral density are valid covariance functions based on the permissibility of the spectral density, independently of the SSRF energy functional.
(ii) Secondly, we can view the SSRF as a random function defined on a lattice with step a where $k_c \gg 1/a$. Then, the derivatives in the SSRF energy functional are replaced by finite differences and differentiability is not an issue. We are, however, left with the task of reconciling a very large but finite k_c with the covariance expressions obtained by assuming an infinite k_c. However, since $k_c \gg 1/a$, ignoring the cutoff only affects the covariance function at distances $r < a$.

The careful reader will notice that the spectral integral used to derive the covariance functions assumes a radial spectral density, while the spectral density of an SSRF on a square grid is not a radial function. Nonetheless, the isotropy of lattice SSRFs is broken only for large wavenumbers and thus only distances smaller than a (that do not interest us) are significantly affected.[5]

In conclusion, we can use the SSRF covariance functions derived above as valid covariance models. In two and three dimensions, they correspond to non-differentiable (in the mean-square sense) random fields. Such covariance functions are useful models for spatial processes of limited smoothness (e.g., models of geological media, processes driven by white noise).

We return to the issue of non-differentiability in Chap. 9, where we trace its origin to the structure of the *Langevin equation* associated with SSRFs. In one dimension, as we discuss in detail in Chap. 9, the SSRF is equivalent to a *classical, damped harmonic oscillator* driven by white noise.

With respect to the SSRF energy functional, we can view the continuum SSRF covariance functions as approximate covariances for the lattice-based SSRFs described in Chap. 8. The approximation is accurate for $r \gg a$, while its accuracy declines at distances $\approx a$.

7.3 Statistical and Geometric Properties

A characteristic feature of SSRF covariance functions is that the spectral density (7.16) is independent of the number of spatial dimensions d. This, however, implies that the covariance in real space depends on d. We have already encountered in Sect. 7.2.6 how the dimensionality dependence impacts mean-square differentiability. Herein we focus on the variance, various characteristic length scales, and the behavior of the variogram in some interesting limits. Some properties of SSRFs are summarized in the Appendix B.

---

[5] A hexagonal (honeycomb) grid is a more isotropic structure than a square lattice. Hexagonal lattices have been explored in statistical physics models but not much in spatial statistics.

7.3.1 SSRF Variance

The variance in one, two, and three dimensions is plotted in Fig. 7.18 based on equations (7.29), (7.38), and (7.43). The main feature is the decline of the variance with (i) η_1 (for fixed d) and (ii) with d (for fixed η_1).

The decline of the variance for higher η_1 implies that increased rigidity makes the field less flexible due to the increased energy cost of large gradients. This tends to reduce the amplitude of the fluctuations leading to variance reduction.

The decline of the variance with d implies that a given level of rigidity generates more pronounced fluctuations in lower dimensions. This behavior is related to the fact that the SSRF is mean-square differentiable, hence more flexible, in one dimension.

As we have shown in [362], the SSRF variance is given by

$$\sigma_{\mathrm{x}}^2 = \frac{\eta_0 \xi^d}{2^{(d-2)/2}\, \Gamma\left(\frac{d}{2}\right)\, (2\pi)^{d/2}} \int_0^\infty \frac{d\|\mathbf{k}\|\; \|\mathbf{k}\|^{d-1}}{1 + \eta_1\, (\|\mathbf{k}\|\xi)^2 + (\|\mathbf{k}\|\xi)^4}.$$

After the variable transformation $\|\mathbf{k}\|\xi \mapsto x$, the variance is given by the following integral

$$\sigma_{\mathrm{x}}^2 = \frac{\eta_0}{2^{(d-2)/2}\, \Gamma\left(\frac{d}{2}\right)\, (2\pi)^{d/2}} \int_0^\infty \frac{\mathrm{d}x\, x^{d-1}}{1 + \eta_1 x^2 + x^4}. \tag{7.46}$$

The above shows that the denominator of the integral increases with d. In addition, as d increases, the numerator is reduced for values of $x \approx 0$, where the spectral density has more weight than for larger x. The combined action of these two factors leads to the decline of the variance with d.

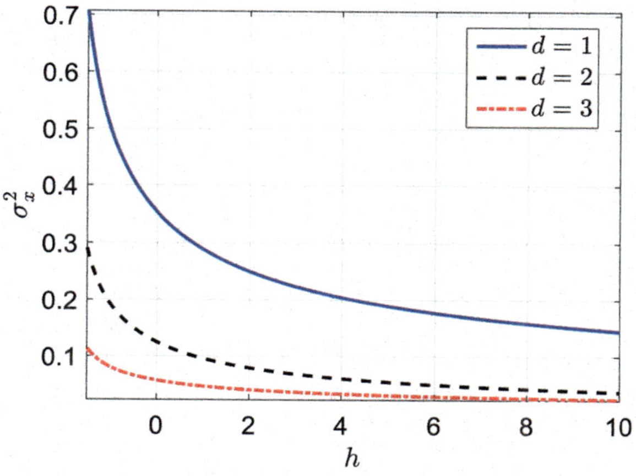

Fig. 7.18 SSRF variance versus η_1 in one, two, and three dimensions based on (7.29) ($d = 1$), (7.38) ($d = 2$) and (7.43) ($d = 3$), respectively. All plots correspond to $\eta_0 = 1$

7.3.2 Integral Range

The integral range is determined by the equations (7.31), (7.40), and (7.45). The dependence of ℓ_c on η_1 for $d = 1, 2, 3$ is shown in Fig. 7.19. For $\eta_1 < 0$ the integral range is larger for higher d. This is due to the suppression of the negative hole with increasing dimensionality. On the other hand, for $\eta_1 > 0$ a given rigidity level gives lower integral ranges in higher dimensions. This is due to the increased roughness of the SSRF in higher dimensions.

7.3.3 Large Rigidity

Let us now consider SSRFs with large rigidity, i.e., $\eta_1 \gg 1$. The Taylor expansion of the SSRF wavenumber and damping coefficients (7.22) in $x = 1/\eta_1$ around the *infinite rigidity* point, $x = 0$, shows that the SSRF coefficients satisfy the following approximate relations

$$w_1 \approx 1/\sqrt{\eta_1}, \quad w_2 \approx \sqrt{\eta_1}, \quad \Delta \approx \eta_1,$$

$$z_+ \approx 1/\sqrt{\eta_1}, \quad z_- \approx \sqrt{\eta_1}.$$

In the following, we examine the behavior of the variance, the covariance function and the variogram at the large rigidity limit.

One dimension Based on the above and (7.29), the variance is given by

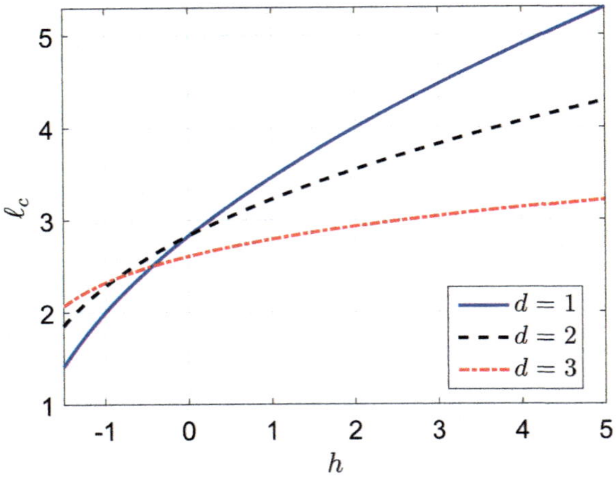

Fig. 7.19 SSRF integral range versus η_1 in one, two, and three dimensions based on (7.31) ($d = 1$), (7.40) ($d = 2$) and (7.45) ($d = 3$), respectively. All plots correspond to $\eta_0 = 1$

$$\sigma_x^2 \approx \frac{\eta_0}{2\sqrt{\eta_1}}.$$

The SSRF covariance function (7.27c) involves "slow" and "fast" exponential components; the former has a characteristic radius of $\sqrt{\eta_1}$, whereas the latter is $\propto 1/\sqrt{\eta_1}$. For $\eta_1 \gg 1$ the first component dominates and the covariance function is given by

$$C_{xx}(h) \approx \frac{\eta_0}{2\sqrt{\eta_1}}\, e^{-h/\sqrt{\eta_1}}. \tag{7.47}$$

Based on the above and the relation $\gamma_{xx}(h) = \sigma_x^2 - C_{xx}(h)$ it follows that for a wide range of lags such that $\|\mathbf{r}\| \ll \xi\sqrt{\eta_1}$ the variogram is approximately equal to the linear function[6]

$$\gamma_{xx}(h) \approx \frac{\eta_0}{2\,\eta_1\,\xi}\,\|\mathbf{r}\|. \tag{7.48}$$

Hence, the SSRF variogram is identical to that of *classical Brownian motion* except for very large lags. This behavior is illustrated in Fig. 7.20 which shows that the linear regime extends to larger lags as η_1 grows.

Two dimensions Based on (7.38c) the variance is given by

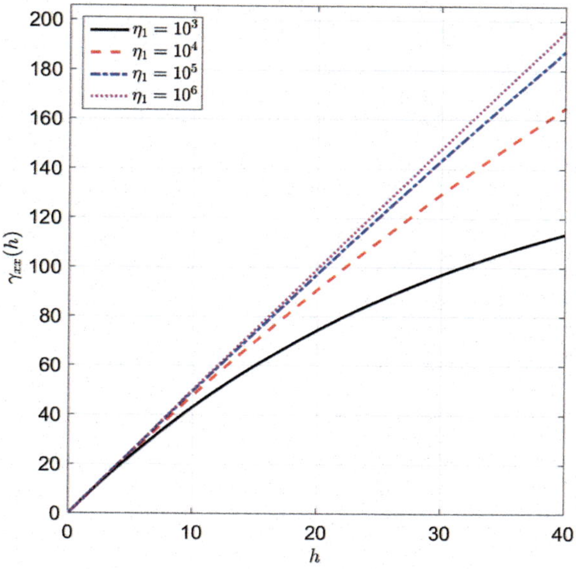

Fig. 7.20 One-dimensional SSRF variogram based on (7.27c). All plots correspond to $\xi = 1$. The values of η_1 are given by 10^m, and the respective values of η_0 are given by $\eta_0 = 10^{m-3}$, for $m = 3, 4, 5, 6$

[6]Remember that $h = \|\mathbf{r}\|/\xi$.

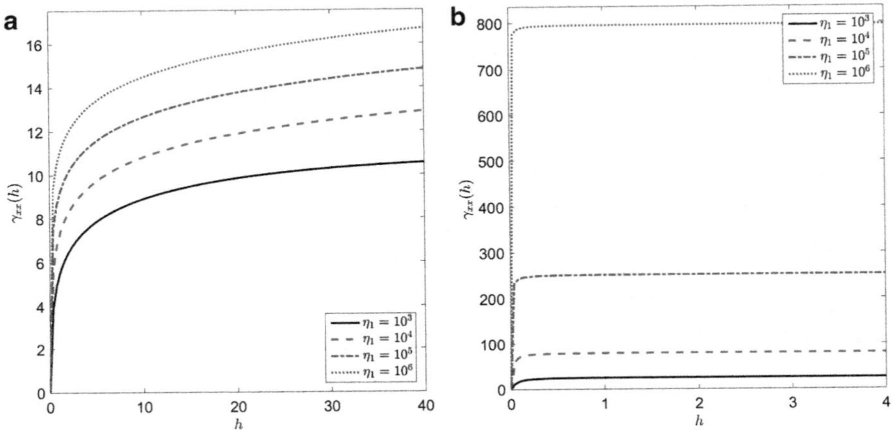

Fig. 7.21 Two- and three-dimensional SSRF variograms based on (7.37c) and (7.42c). All plots correspond to $\xi = 1$. The values of η_1 are given by 10^m, and the respective values of η_0 are given by $\eta_0 = 10^{m-3}$, for $m = 3, 4, 5, 6$. (**a**) 2D SSRF Variogram. (**b**) 3D SSRF Variogram

$$\sigma_{\mathrm{x}}^2 \approx \eta_0 \, \frac{\ln(\eta_1)}{2\pi \, \eta_1}.$$

The behavior of the covariance function is not as simple near $h = 0$ as in $d = 1$, because both the fast and the slow term have singularities that cancel each other out. As shown in Fig. 7.21a, in contrast with the one-dimensional case, the variogram is a nonlinear function of the lag.

Three dimensions In three dimensions, the variance (7.43), is given by

$$\sigma_{\mathrm{x}}^2 \approx \frac{\eta_0}{4\pi \sqrt{\eta_1}}.$$

The variogram function is plotted in Fig. 7.21b and resembles, except at very small lags, the nugget variogram that is characteristic of white noise. The quick rise of the variogram is determined by the fast length scale $\propto \xi/\sqrt{\eta_1}$. In spite of the fast equilibration of the variogram function (and the covariance), the integral range is approximately given, according to (7.45), by $\xi \left(4\pi \sqrt{\eta_1}\right)^{1/3}$. Hence, the integral range is determined by the slow length scale. The different length scales that correspond to the slow and fast scales are captured by different values of the *correlation spectrum index* as shown in Sect. 7.3.4.

At this point, readers could recall the discussion related to the difference between the practical range and the integral scales (integral range and correlation radius) in Sect. 5.4. An extreme example is provided by the delta function covariance which corresponds to a spatial white noise field: the practical range of this covariance is zero, while its integral range is a finite number since $\int \mathrm{d}\mathbf{r}\delta(\mathbf{r}) = 1$.

7.3.4 SSRF Correlation Spectrum

This section focuses on the calculation of the SSRF correlation spectrum. Based on (5.43), the correlation spectrum that corresponds to SSRF covariance functions in d dimensions is given by the following equation

$$\lambda_c^{(\alpha)} = \xi \left[\frac{(\widetilde{\kappa}_1\, \xi)^{2\alpha}}{\Lambda_{2\alpha}(\eta_1; d)\left(1 + \eta_1 (\widetilde{\kappa}_1 \xi)^2 + (\widetilde{\kappa}_1 \xi)^4\right)} \right]^{1/d}, \tag{7.49}$$

where $\widetilde{\kappa}_1$ is the wavenumber that maximizes the function $\|\mathbf{k}\|^{2\alpha}\, \widetilde{C}_{\mathrm{xx}}(\mathbf{k})$, and $\Lambda_{2\alpha}(\eta_1; d)$ is the radial spectral moment of order 2α defined by (5.28).

According to the discussion in Sect. 5.4.5, the $\alpha = 0$ spectrum is equivalent to the integral range if the spectral density peaks at $\|\mathbf{k}\| = 0$; in the case of SSRFs this occurs for $\eta_1 > 0$. In particular,

$$\ell_c = 2\pi\, \lambda_c^{(0)}, \text{ for } \eta_1 \geq 0. \tag{7.50}$$

On the other hand, for $\eta_1 < 0$, the zero-index ($\alpha = 0$) correlation spectrum differs essentially from the integral range, because the former takes into account the width of the spectral density's peak.

The equation (7.49) involves the characteristic wavenumber $\widetilde{\kappa}_1$ at which the product $\|\mathbf{k}\|^{2\alpha}\widetilde{C}_{\mathrm{xx}}(\mathbf{k})$ peaks and the spectral integral $\Lambda_{2\alpha}(\eta_1; d)$.

Spectral peak location Since the SSRF spectral density is independent of d, the wavenumber $\widetilde{\kappa}_1$ is given by the same expression for all d, i.e.,

$$\widetilde{\kappa}_1 = \frac{1}{\xi}\sqrt{\frac{\sqrt{{\eta_1}^2\,(1-\alpha)^2 + 4\alpha(2-\alpha)} - \eta_1\,(1-\alpha)}{2(2-\alpha)}}. \tag{7.51a}$$

The dependence of the peak wavenumber $\widetilde{\kappa}_1$ on η_1 and α is shown in Fig. 7.22. For $\alpha = 0$ and $\eta_1 > 0$ it holds that $\widetilde{\kappa}_1 = 0$, while for $\eta_1 < 0$ it follows that $\widetilde{\kappa}_1 > 0$. The wavenumber $\widetilde{\kappa}_1$ increases with α.

The radial spectral moment The radial spectral moment $\Lambda_{2\alpha}(\eta_1; d)$ is given by the following multiple integral, where S_d is the surface area of the unit sphere given by (3.62b),

$$\Lambda_{2\alpha}(\eta_1; d) = \int d\mathbf{k}\, \frac{\|\mathbf{k}\|^{2\alpha}}{1 + \eta_1\,(\|\mathbf{k}\|\,\xi)^2 + (\|\mathbf{k}\|\,\xi)^4}$$

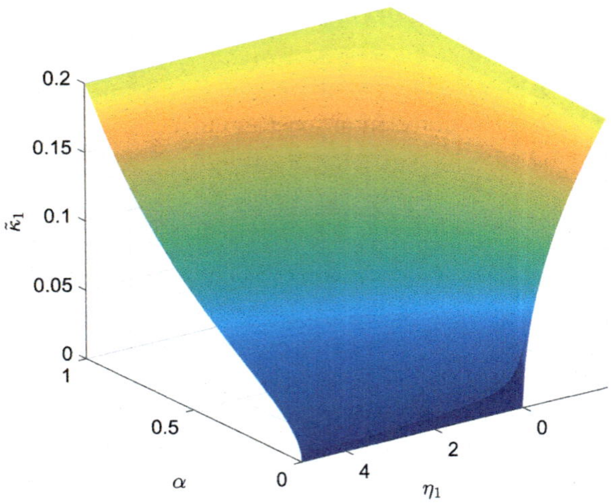

Fig. 7.22 Dependence of the wavenumber $\tilde{\kappa}_1$, where the maximum of the spectral function $\|\mathbf{k}\|^{2\alpha}\,\widetilde{C}_{xx}(\|\mathbf{k}\|)$ is attained, on α and η_1. The plot is based on (7.51a) for α from zero to one and η_1 from -1.9 to 5 assuming $\xi = 5$. Zero values of $\tilde{\kappa}_1$ are obtained for $\alpha = 0$ and $\eta_1 \geq 0$ reflecting a maximum at the origin of the frequency band. Higher values of $\tilde{\kappa}_1$ are obtained for higher α and for negative η_1

$$=S_d \int_0^\infty \mathrm{d}x \, \frac{x^{2\alpha+d-1}}{1+\eta_1 x^2 + x^4}.$$

The spectral moment $\Lambda_{2\alpha}(\eta_1; d)$ is evaluated using the integration formula [306, 3.264.2, p. 330] which leads to the following result[7]

$$\Lambda_{2\alpha}(\eta_1; d) = \frac{2^{-\nu}\,\pi^{1+d/2}}{\Gamma(d/2)} \, \frac{\left(\eta_1 - \sqrt{\eta_1{}^2-4}\right)^\nu - \left(\eta_1 + \sqrt{\eta_1{}^2-4}\right)^\nu}{\sin\left[(\nu+1)\pi\right]\,\sqrt{\eta_1{}^2-4}}, \tag{7.51b}$$

where $\nu = \alpha + d/2 - 1$.

The function $\Lambda_{2\alpha}(\eta_1; d)$ contains the factor $\sqrt{\eta_1{}^2-4}$ which becomes imaginary for $\eta_1 < 2$. Hence, one wonders if $\Lambda_{2\alpha}(\eta_1; d)$ is real-valued for $|\eta_1| < 2$. Let us use the composite exponent $\beta = d/2 + \alpha - 1$. According to (7.51b), $\Lambda_{2\alpha}(\eta_1; d)$ is given by

$$\Lambda_{2\alpha}(\eta_1; d) = \frac{2^{-\beta}\,\pi^{1+d/2}}{\Gamma(d/2)\,\sin\left[(\beta+1)\pi\right]} \, \frac{(\eta_1-\Delta)^\beta - (\eta_1+\Delta)^\beta}{\Delta}.$$

The term

$$\frac{2^{-\beta}\,\pi^{1+d/2}}{\sin(\beta\pi+\pi)\,\Gamma(d/2)}$$

is clearly real-valued. To calculate the remaining factor, we use the notation $\sqrt{\eta_1{}^2-4} = \mathrm{i}\,|\Delta|$, where $|\Delta| = \sqrt{4-\eta_1{}^2}$. This leads to

[7] In three dimensions the above is valid for $\alpha < 1/2$.

$$\frac{(\eta_1-\Delta)^\beta-(\eta_1+\Delta)^\beta}{\Delta}=\frac{(\eta_1-\mathrm{i}\,|\Delta|)^\beta-(\eta_1+\mathrm{i}\,|\Delta|)^\beta}{\mathrm{i}\,|\Delta|}$$

Using the polar form for the complex variables in the numerator, it follows that $\eta_1-\mathrm{i}\,|\Delta| = 2\,\mathrm{e}^{\mathrm{i}\theta}$ where $\tan\theta=-|\Delta|/\eta_1$. Hence,

$$\frac{(\eta_1-\Delta)^\beta-(\eta_1+\Delta)^\beta}{\Delta}=\frac{(\eta_1-\mathrm{i}\,|\Delta|)^\beta-(\eta_1+\mathrm{i}\,|\Delta|)^\beta}{\mathrm{i}\,|\Delta|}=\frac{2^{\beta+1}\,\sin(\beta\,\theta)}{|\Delta|}. \tag{7.52}$$

This calculation confirms that $\Lambda_{2\alpha}(\eta_1;d)$ is a real-valued function of η_1 and α. It can also be shown that $\Lambda_{2\alpha}(\eta_1;d)\in\mathbb{R}$ for all $\eta_1>-2$ and $d=1,2$, while for $d=3$ the function $\Lambda_{2\alpha}(\eta_1;d)$ is well defined only for $0\le\alpha<0.5$.

Large rigidity in three dimensions In Sect. 7.3.3 we noticed that in three dimensions an SSRF with large rigidity ($\eta_1\gg 1$) has a high integral range, while its variogram increases very fast towards the sill. The difference between the two scales is captured by different indices α of the correlation spectrum. In Fig. 7.23 we plot $\lambda_c^{(\alpha)}$ as a function of the index α for values of $\alpha\in[0,0.5)$ ($\lambda_c^{(\alpha)}$ is not defined in three dimensions for $\alpha\ge 0.5$) for two different values of $\eta_1=2.5,10^4$. As shown by the graphs, the difference between $\lambda_c^{(0)}$ and $\lambda_c^{(0.5)}$ is more pronounced for $\eta_1=10^4$ than for $\eta_1=2.5$.

Spectrum for $\alpha=0$ The zero-index correlation spectrum ($\alpha=0$) is of particular interest: for $\eta_1>0$ it is equivalent to the integral range (except for a constant factor) as expressed by (7.50), while for $\eta_1<0$ it yields a meaningful (positive-valued) length scale. The calculation of the zero-index correlation spectrum is based on (7.49). At $\alpha=0$, the position of the spectral peak is

$$\widetilde{\kappa}_1(\alpha=0)=\frac{1}{2\xi}\sqrt{|\eta_1|-\eta_1}=\begin{cases}0, & \text{for } \eta_1\ge 0\\ \frac{\sqrt{|\eta_1|}}{\sqrt{2}\xi}, & \text{for } \eta_1<0.\end{cases} \tag{7.53}$$

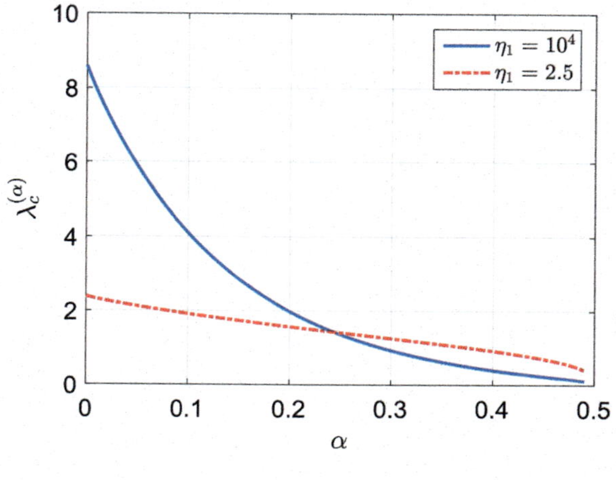

Fig. 7.23 Dependence of λ_c^{α} on α for $d=3$, $\xi=5$ based on (7.49). Two rigidity values equal to $\eta_1=10^4$ and $\eta_1=2.5$ are used. The wavenumber $\tilde{\kappa}_1$ is calculated based on (7.51a) and the function $\Lambda_{2\alpha}(\eta_1;d)$ based on (7.51b)

Using the result (7.53) for $\widetilde{\kappa}_1$ at $\alpha = 0$, and the result (7.51b) for the radial spectral moment, the zero-index correlation spectrum is given by the following expressions where $\nu = d/2 - 1$, $\eta_1 > -2$, and $\Delta = \sqrt{\eta_1^2 - 4}$:

$$\lambda_c^{(0)} = \begin{cases} \xi\ [g_d(\eta_1)]^{1/d}, & \text{for } \eta_1 \geq 0, \\ \xi\ \left(\frac{g_d(\eta_1)}{1-\eta_1^2/4}\right)^{1/d}, & \text{for } \eta_1 < 0, \end{cases} \tag{7.54a}$$

where

$$g_d(\eta_1) = \frac{2^{\nu}\ \Delta\, \Gamma(\nu+1)\ \sin(\pi + \nu\,\pi)}{\pi^{\nu+2}\left[(\eta_1 - \Delta)^{\nu} - (\eta_1 + \Delta)^{\nu}\right]}, \quad \text{where } \nu = d/2 - 1. \tag{7.54b}$$

Warning The equations (7.54) are valid in any $d = 1, 2, 3$ and for any η_1. However, they cannot be used in this form for numerical calculations. The reason is that they lead to indefinite, e.g., $0/0$ operations for certain d and η_1 values. It is thus necessary to analytically derive the limits for the cases in question.

1. The limit $\eta_1 = 2$

   Note that both the numerator and the denominator become zero in (7.54b) for $\Delta = 0$, i.e., for $\eta_1 = 2$. Thus, we recast equation (7.54b) as follows

$$g_d(\eta_1 = 2) = \frac{2^{\nu}\ \Gamma(\nu+1)\ \sin(\pi + \nu\,\pi)}{\pi^{\nu+2}} \lim_{\Delta \to 0} g_d^*(\Delta),$$

$$g_d^*(\Delta) = \frac{\Delta}{(\eta_1 - \Delta)^{\nu} - (\eta_1 + \Delta)^{\nu}}.$$

   To evaluate $g_d^*(\Delta)$ at the limit $\Delta = 0$, we use *l'Hospital's rule*[8] to obtain

$$\lim_{\Delta \to 0} g_d^*(\Delta) = \frac{-1}{\lim_{\Delta \to 0} \nu\left[(\eta_1 - \Delta)^{\nu-1} + (\eta_1 + \Delta)^{\nu-1}\right]} = \frac{-2^{-\nu}}{\nu}.$$

   Based on this limit and (7.54) it follows that $\lambda_c^{(0)}$ for $\eta_1 = 2$ is given by

$$\lambda_c^{(0)} = \xi\left[-\frac{\Gamma(\nu+1)\ \sin(\pi + \nu\pi)}{\nu\,\pi^{\nu+2}}\right]^{1/d}, \quad \text{for } \nu \neq 0. \tag{7.55a}$$

[8] Let us assume that the limit $\lim_{x\to c} f(x)/g(x)$ is undetermined, because both the numerator and the denominator tend to either zero or infinity at the limit $x \to c$. L'Hospital's rule states that $\lim_{x\to c} f(x)/g(x) = \lim_{x\to c} f'(x)/\lim_{x\to c} g'(x)$, where the prime denotes the first derivative.

The above expression, however, is undetermined for $\nu = 0$, that is, for $d = 2$. A second application of l'Hospital's rule to the expression inside the radical leads to

$$\lambda_c^{(0)} = \frac{\xi}{\sqrt{\pi}}, \quad \text{for } \nu = 0. \tag{7.55b}$$

2. The limit $d = 1$ ($\nu = -1/2$)

In one dimension the correlation spectrum for $\alpha = 0$ is given by $\lambda_c^{(0)} = \xi\, g_1(\eta_1)$ for $\eta_1 > 0$ and by $\lambda_c^{(0)} = \xi\, g_1(\eta_1)/(1 - {\eta_1}^2/4)$, where, based on (7.54), $g_1(\eta_1)$ is given by

$$g_1(\eta_1) = \frac{2^{-1/2}\,\Delta}{\pi\,\left[(\eta_1 - \Delta)^{-1/2} - (\eta_1 + \Delta)^{-1/2}\right]}.$$

The above is well defined for $\eta_1 > 2$. At $\eta_1 = 2$ we need to evaluate the limit using l'Hospital's rule, and for $\eta_1 < 0$ we need to account for the complex-valued Δ.

1. Case $\eta_1 = 2$

   We need to calculate the limit of $g_1(\eta_1)$ as $\Delta \to 0$.

$$\lim_{\eta_1 \to 2} g_1(\eta_1) = \frac{2^{-1/2}}{\pi\, \lim_{\Delta \to 0}\left[\frac{1}{2}\,(2 - \Delta)^{-3/2} + \frac{1}{2}\,(2 + \Delta)^{-3/2}\right]} = \frac{2}{\pi}.$$

   The result for $\eta_1 = 2$ agrees with (7.55a) that was derived by taking the limit $\eta_1 \to 2$ for any d.

2. Case $-2 < \eta_1 < 2$

   We follow the same approach as in the proof that $\Lambda_{2\alpha}(\eta_1; d)$ is real-valued (given above). We thus obtain

$$g_1(\eta_1) = \frac{2^{-1/2}\,|\Delta|}{2^{1/2}\,\pi\, \sin(\theta/2)} = \frac{|\Delta|}{2\pi\, \sin(\theta/2)},$$

   where $\tan\theta = |\Delta|/\eta_1$.

   We use the identity $\sin(\theta/2) = \sqrt{(1 - \cos\theta)/2}$ and the cosine dependence $\cos\theta = \eta_1/2$, to obtain $\sin(\theta/2) = \sqrt{(1 - \eta_1/2)/2}$. Then, $g_1(\eta_1)$ is given by

$$g_1(\eta_1) = \frac{\sqrt{2 + \eta_1}}{\pi}.$$

In light of the above, the *index* $\alpha = 0$ *correlation spectrum in one dimension* is given by

$$\lambda_c^{(0)} = \xi \times \begin{cases} \dfrac{2^{-1/2}\,\Delta}{\pi\left[(\eta_1-\Delta)^{-1/2}-(\eta_1+\Delta)^{-1/2}\right]}, & \text{for } \eta_1 > 2, \\ \dfrac{2}{\pi}, & \text{for } \eta_1 = 2, \\ \dfrac{\sqrt{2+\eta_1}}{\pi}, & \text{for } 0 \le \eta_1 < 2, \\ \dfrac{\sqrt{2+\eta_1}}{\pi\left(1-\frac{\eta_1{}^2}{4}\right)}, & \text{for } -2 < \eta_1 < 0. \end{cases} \quad (7.56)$$

The first equation is a straightforward application of (7.54) for $\eta_1 > 0$ and $\nu = -1/2$. The second relation follows from (7.55a) for $\nu = -1/2$ using $\Gamma(1/2) = \sqrt{\pi}$. The third row follows from (7.54) for $\eta_1 > 0$ and $\nu = -1/2$ and from (7.52). Finally, the fourth equation follows from (7.54) for $\eta_1 < 0$ and $\nu = -1/2$ and from (7.52).

The three first formulas for $\lambda_c^{(0)}$ given above are in agreement with the relation $\ell_c = 2\pi\,\lambda_c^{(0)}$ between the integral range and the correlation spectrum that holds for $\eta_1 > 0$, and the SSRF integral ranges for $d = 1$ that are listed in Table B.2.

The limit $d = 2\,(\nu = 0)$ In this case, both the numerator and denominator of the right-hand side formulas in (7.54) tend to zero. We can, however, apply the by now familiar l'Hospital's rule to the ratio of the "dangerous terms", treating separately the cases $\eta_1 > 2$, $\eta_1 = 2$, and $-2 \le \eta_1 < 2$. The case $\eta_1 = 2$ was actually addressed above, and the result is given by (7.55b). Hence, we focus on the other two cases. The prime in the following denotes the derivative with respect to ν.

Let us write express the function $g_d(\eta_1)$ defined by (7.54b) as follows

$$g_2(\eta_1) = \frac{\Delta}{\pi^2} \lim_{\nu\to 0} \frac{\sin(\pi + \nu\,\pi)}{\left[(\eta_1 - \Delta)^{\nu} - (\eta_1 + \Delta)^{\nu}\right]}.$$

The limit is evaluated as follows depending on the value of η_1:

1. **Case $\eta_1 > 2$**

$$g_2(\eta_1) = \frac{-\pi}{\left(e^{\nu\,\ln(\eta_1-\Delta)}\right)' - \left(e^{\nu\,\ln(\eta_1+\Delta)}\right)'\Big|_{\nu=0}} = \pi\left[\ln\left(\frac{\eta_1+\Delta}{\eta_1-\Delta}\right)\right]^{-1}.$$

2. **Case $-2 \le \eta_1 < 2$**

   The definition (7.54b) is valid in this case as well, but $\Delta = \mathrm{i}\,|\Delta|$ is now an imaginary number. Hence, we use the polar representation of complex numbers in the denominator of $g_2(\eta_1)$. This leads to $\eta_1 \mp \mathrm{i}\,|\Delta| = 2\,e^{\pm \mathrm{i}\theta}$ where $\tan\theta = -|\Delta|/\eta_1$. Based on this transformation, the function $g_2(\eta_1)$ is calculated as

$$g_2(\eta_1) = \frac{\Delta}{\pi}\left[\ln\left(\frac{\eta_1+\Delta}{\eta_1-\Delta}\right)\right]^{-1} = -\frac{|\Delta|}{2\pi\,\theta}.$$

Using this expression for $g_2(\eta_1)$ and (7.54a), the zero-index correlation spectrum $\lambda_c^{(0)}$ is given by

$$\lambda_c^{(0)} = \begin{cases} \xi \left(\frac{\Delta}{\pi \ln\left(\frac{\eta_1+\Delta}{\eta_1-\Delta}\right)} \right)^{1/2}, & \eta_1 > 2, \\ \xi \frac{1}{\sqrt{\pi}}, & \eta_1 = 2, \\ \xi \left(\frac{|\Delta|}{\pi \arctan(|\Delta|/\eta_1)} \right)^{1/2}, & 0 \leq \eta_1 < 2, \\ \xi \left(\frac{|\Delta|}{\pi \left(1-\frac{\eta_1{}^2}{4}\right) \arctan(|\Delta|/\eta_1)} \right)^{1/2}, & -2 < \eta_1 < 0. \end{cases} \tag{7.57}$$

We can confirm that (i) the first and the third of the above expressions yield the second expression at the limit $\eta_1 = 2$, and (ii) the above formulas for $\lambda_c^{(0)}$ are in agreement with the integral ranges listed in Table B.2 for $d = 2$ and for $\eta_1 > 0$.

The limit $d = 3$ $(\nu = 1/2)$ In three dimensions the correlation spectrum for $\alpha = 0$ is given by $\lambda_c^{(0)} = \xi\, g_3(\eta_1)$ for $\eta_1 > 0$ and by $\lambda_c^{(0)} = \xi\, g_3(\eta_1)/(1 - \eta_1{}^2/4)$. Based on (7.54), $g_3(\eta_1)$ is given by

$$\begin{aligned} g_3(\eta_1) &= \frac{\Delta}{\sqrt{2}\,\pi^2 \left[(\eta_1 + \Delta)^{-1/2} - (\eta_1 - \Delta)^{-1/2}\right]} \\ &= \frac{1}{2^{3/2}\,\pi^2} \left(\sqrt{\eta_1 + \Delta} + \sqrt{\eta_1 - \Delta} \right). \end{aligned}$$

The above is well defined for $\eta_1 > 2$. For $\eta_1 \leq 2$ we distinguish two separate cases.

1. Case $\eta_1 = 2$ $(\Delta = 0)$

   Based on (7.55a), which is valid for all dimensions, and using $\Delta = 0$ we obtain

$$g_3(\eta_1) = \frac{1}{\pi^2}.$$

2. Case $-2 < \eta_1 < 2$

   We follow the same steps as in the second case for $d = 1$. Taking into account the different constant coefficients, we obtain

$$g_3(\eta_1) = \frac{\sqrt{2 + \eta_1}}{2\pi^2}.$$

In light of the above, the $\alpha = 0$ correlation spectrum in $d = 3$ is given by

$$\lambda_c^{(0)} = \xi \times \begin{cases} \frac{\left[(\eta_1-\Delta)^{-1/2}+(\eta_1+\Delta)^{-1/2}\right]^{1/3}}{\sqrt{2}\,\pi^{2/3}}, & \text{for } \eta_1 > 2, \\ 2\pi^{1/3} & \text{for } \eta_1 = 2, \\ (4\,\pi)^{1/3}\,(2+\eta_1)^{1/6} & \text{for } 0 \le \eta_1 < 2, \\ (4\,\pi)^{1/3}\,\frac{(2+\eta_1)^{1/6}}{\left(1-\frac{\eta_1^2}{4}\right)^{1/3}}, & \text{for } -2 < \eta_1 < 0. \end{cases} \tag{7.58}$$

7.4 Random Fields with Bessel-Lommel Covariance

As evidenced in the preceding sections, SSRFs lead to covariance functions that include negative values, if the rigidity coefficient is in the range $-2 < \eta_1 < 2$. Covariance functions with negative values are used in stochastic hydrology [43, 250] and in ocean wave models [865]. Simple covariance functions with oscillatory behavior include the *cosine*, the *cardinal sine*, the *modified exponential* that involves the product of harmonic and exponential functions [487], and the *separable periodic covariance* [678]. These models are also known as *hole covariance functions*, because they exhibit negative correlations for a range of lags (see also Sect. 3.6.1).

In one-dimensional stochastic models of groundwater flow, a Gaussian log-permeability field with a hole covariance function leads to a stationary hydraulic head field, in contrast with the exponential covariance that leads to a non-stationary head field [43]. Hole covariance functions have also been associated with spherical inclusions in porous media [250]. Covariance models with a hole effect but without oscillatory dependence can also be constructed by means of linear combinations of non-orthogonal components that correspond to different length scales [811].

In [367] covariance functions are constructed based on spectral density functions that are proportional to the SSRF characteristic polynomial $\Pi(u) = 1 + \eta_1\, u + u^2$ given by (7.19) where $u = \|\mathbf{k}\|^2\,\xi^2$. Since $\Pi(u)$ is an increasing function of u, it is necessary to cut off the spectral density at some maximum wavenumber k_{c} in order to ensure its integrability. The explicit expressions for the covariance function in real space involve products of *Bessel functions of the first kind* and *Lommel functions*, thus prompting the name *Bessel-Lommel covariance functions*.

The *Bessel-Lommel spectral density* is given by

$$\widetilde{C_{\mathrm{xx}}^{BL}}(\|\mathbf{k}\|;\boldsymbol{\theta}) = \begin{cases} c_0 + c_1\|\mathbf{k}\|^2 + c_2\|\mathbf{k}\|^4, & \|\mathbf{k}\| \le k_{\mathrm{c}}, \\ 0, & \|\mathbf{k}\| > k_{\mathrm{c}}, \end{cases} \tag{7.59}$$

where $\boldsymbol{\theta} = (c_0, c_1, c_2)^\top$ is the parameter vector. In order to connect with the SSRF spectral density, the Bessel-Lommel coefficients are given by

$$c_0 = \lambda_0 \xi^d, \quad c_1 = \eta_1 \xi^2 c_0, \quad c_2 = \xi^4 c_0,$$

where the constant coefficient c_0 has units $[X]^2$ in order for the Bessel-Lommel spectral density to be correctly dimensioned. Conversely, the parameters (λ_0, η_1, ξ) are determined from (c_0, c_1, c_2) as follows

$$\eta_1 = \frac{c_1}{\sqrt{c_0 c_2}}, \quad \xi = \left(\frac{c_2}{c_0}\right)^{1/4}, \quad \lambda_0 = \frac{c_0^{1+d/4}}{c_2^{d/4}}.$$

The *permissibility conditions* for the Bessel-Lommel random fields are $\eta_1 > -2$ and $k_c > 0$. Typical plots of the Bessel-Lommel spectral density are shown in Fig. 7.24.

Since $\widetilde{C_{\rm xx}^{BL}}(\|\mathbf{k}\|; \boldsymbol{\theta})$ is a radial spectral density, the inverse Fourier transform is given by the *spectral integral* (4.4b). For the Bessel-Lommel spectral density this is expressed as follows (where $z = \|\mathbf{k}\|$):

$$C_{\rm xx}^{BL}(\mathbf{r}; \boldsymbol{\theta}) = \frac{\|\mathbf{r}\|^{1-d/2}}{(2\pi)^{d/2}} \int_0^{k_c} dz\, z^{d/2} J_{d/2-1}(z\|\mathbf{r}\|)\, \widetilde{C_{\rm xx}^{BL}}(z; \boldsymbol{\theta}). \tag{7.60}$$

On differentiability The Bessel-Lommel spectral density has no singularities for $\|\mathbf{k}\| \in [0, k_c]$ and is cut off at $\|\mathbf{k}\| = k_c$. Hence, the spectral moments of all orders are well-defined. As shown in Sect. 5.3.1, the existence of the spectral moments of all orders implies that the Bessel-Lommel covariance function is infinitely differentiable at zero lag. Consequently, the *Bessel-Lommel covariance*

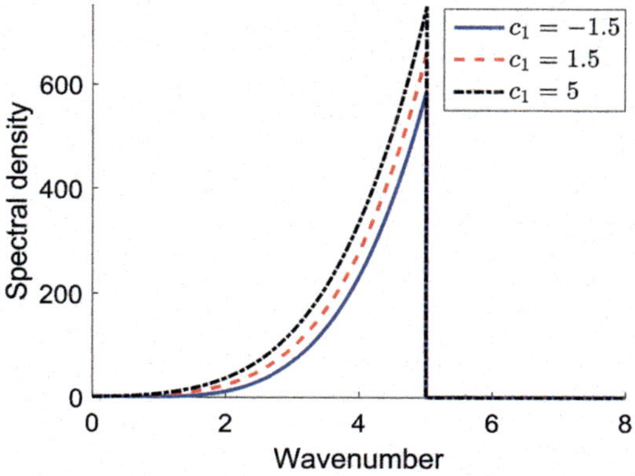

Fig. 7.24 Dependence of Bessel-Lommel spectral density (7.59) on the wavenumber. The parametrization of the density is determined by $\lambda_0 = 1$, $c_2 = 1$, and $k_c = 5$. The densities for three different values of c_1 are shown

class involves *positive definite, infinitely differentiable* functions. It is thus suitable for smooth spatial processes defined in Euclidean $\mathbb{R}^d$ spaces.

Bessel-Lommel covariance The spectral integral (7.60) is evaluated in [367]. The resulting Bessel-Lommel covariance functions are given by means of the following tripartite sum, where $z = k_c \|\mathbf{r}\|$, $d \geq 2$, and $\nu = d/2 - 1$:

$$C_{xx}^{BL}(z; \boldsymbol{\theta}) = \sum_{l=0,1,2} \frac{g_l(\boldsymbol{\theta})}{z^{2\nu+2l+1}} \left[(2\nu + 2l) J_\nu(z)\, S_{\nu+2l,\nu-1}(z) - J_{\nu-1}(z)\, S_{\nu+2l+1,\nu}(z) \right], \tag{7.61a}$$

$$g_0(\boldsymbol{\theta}) = \frac{\lambda_0 (\xi k_c)^d}{(2\pi)^{d/2}}, \; g_1(\boldsymbol{\theta}) = \eta_1 (k_c \xi)^2 g_0(\boldsymbol{\theta}), \; g_2(\boldsymbol{\theta}) = (k_c \xi)^4 g_0(\boldsymbol{\theta}). \tag{7.61b}$$

The *Lommel functions* $S_{\mu,\nu}(z)$ that appear in (7.61a) are shown in Table 7.1. For $0 < k_c < \infty$ and $\eta_1 > -2$, the equations (7.61a) and (7.61b) define positive definite, infinitely differentiable, radial covariance functions. A brief introduction to the properties of Lommel functions is given below.

Lommel Functions The *Lommel functions* $S_{\mu,\nu}(z)$ are solutions of the *inhomogeneous Bessel equation*

$$z^2 \frac{d^2 w(z)}{dz^2} + z \frac{dw(z)}{dz} + (z^2 - \nu^2)\, w(z) = z^{\mu+1}.$$

If either $\mu + \nu$ or $\mu - \nu$ are an odd positive integer, the respective Lommel functions are expressed as a terminating series. In descending order in the powers of z, the series is given by the following equation [834, p.347]

Table 7.1 Lommel functions $S_{\nu+2l,\nu-1}(z)$ and $S_{\nu+2l+1,\nu}(z)$ for $l = 0, 1, 2$ used in the Bessel-Lommel covariance function $C_{xx}^{BL}(z; \boldsymbol{\theta})$ defined in (7.61). The expressions are based on (7.62), where $d \geq 2$ is the space dimension and $\nu = d/2 - 1$

Notation	Expression
$S_{\nu,\nu-1}(z)$	$z^{\nu-1}$
$S_{\nu+2,\nu-1}(z)$	$z^{\nu+1}\left(1 - \frac{4\nu}{z^2}\right)$
$S_{\nu+4,\nu-1}(z)$	$z^{\nu+3}\left[1 - \frac{8(1+\nu)}{z^2} + \frac{32\nu(1+\nu)}{z^4}\right]$
$S_{\nu+1,\nu}(z)$	z^{ν}
$S_{\nu+3,\nu}(z)$	$z^{\nu+2}\left[1 - \frac{4(1+\nu)}{z^2}\right]$
$S_{\nu+5,\nu}(z)$	$z^{\nu+4}\left[1 - \frac{8(\nu+2)}{z^2} + \frac{32(\nu+1)(\nu+2)}{z^4}\right]$

$$S_{\mu,\nu}(z) = z^{\mu-1}\left[1 - \frac{(\mu-1)^2-\nu^2}{z^2} + \frac{[(\mu-1)^2-\nu^2][(\mu-3)^2-\nu^2]}{z^4} - \ldots\right]. \tag{7.62}$$

If $\mu - \nu = 2l + 1$, where $l \in \mathbb{Z}_{+,0}$ the series (7.62) terminates after $l + 1$ terms. The respective Lommel functions are given by

$$S_{\nu+2l+1,\nu}(z) = z^{\nu+2l}\left(1 + \mathbb{I}_{l>0}(l)\sum_{k=1}^{l}(-1)^k \frac{\prod_{j=0}^{k-1}\left[(\nu+2(k-j))^2-\nu^2\right]}{z^{2k}}\right), \tag{7.63}$$

where the indicator function $\mathbb{I}_{l>0}(l)$ limits the sum to only positive values of l.

Bessel-Lommel variance Based on (3.56) and (7.59), the variance $C_{\mathrm{xx}}^{BL}(0;\boldsymbol{\theta})$ is given by the following spectral integral over the d-dimensional sphere $\mathcal{B}_d(k_{\mathrm{c}})$ with radius equal to k_{c}

$$\begin{aligned} C_{\mathrm{xx}}^{BL}(0;\boldsymbol{\theta}) &= \lambda_0 \xi^d \int_{\mathcal{B}_d(k_{\mathrm{c}})} \frac{\mathrm{d}\mathbf{k}}{(2\pi)^d}\left(1 + \eta_1\,\xi^2\,\|\mathbf{k}\|^2 + \xi^4\,\|\mathbf{k}\|^4\right) \\ &= \lambda_0 \frac{2^{1-d}\,(\xi k_{\mathrm{c}})^d}{\pi^{d/2}\,\Gamma(d/2)}\left[\frac{1}{d} + \frac{\eta_1\,(k_{\mathrm{c}}\xi)^2}{d+2} + \frac{(k_{\mathrm{c}}\xi)^4}{d+4}\right]. \end{aligned} \tag{7.64}$$

Equation (7.64) is obtained using the expressions (3.62) for the d-dimensional volume integral of radial functions and $\mathcal{S}_d = 2\pi^{d/2}/\,\Gamma(d/2)$ for the surface of the unit sphere in d dimensions [733, p. 39]. $\Gamma(\cdot)$ is the *Gamma function*.

Bessel-Lommel correlation function The spatial dependence of the Bessel-Lommel covariance is better interpreted in terms of the *correlation function* $\rho_{\mathrm{xx}}^{BL}(z;\boldsymbol{\theta}')$, where $\boldsymbol{\theta}' = (\eta_1, \xi, k_{\mathrm{c}})^T$. The correlation function is given by the ratio of the covariance over the variance, i.e.,

$$\rho_{\mathrm{xx}}^{BL}(z;\boldsymbol{\theta}') = \frac{C_{\mathrm{xx}}^{BL}(z;\boldsymbol{\theta})}{C_{\mathrm{xx}}^{BL}(0;\boldsymbol{\theta})}. \tag{7.65}$$

The dependence of $\rho_{\mathrm{xx}}^{BL}(z;\boldsymbol{\theta}')$ on d is shown in Fig. 7.25, where the characteristic oscillations of the Bessel-Lommel covariance functions are clearly shown. The amplitude of the negative peak of $C_{\mathrm{xx}}^{BL}(\|\mathbf{r}\|;\boldsymbol{\theta})$ diminishes with increasing d, whereas the distance at which $C_{\mathrm{xx}}^{BL}(\|\mathbf{r}\|;\boldsymbol{\theta})$ first crosses zero moves to higher values. The "flattening" of the negative holes with increasing d is in agreement with the $-1/d$ lower bound of the correlation function, as defined in (4.18). The Bessel-Lommel correlation function in $d = 3$ has a more pronounced oscillatory behavior than that of SSRF covariance functions. This type of correlations is encountered in the analysis of digital images from porous media [64].

Bessel-Lommel integral range The *integral range* is defined by (5.38). For the Bessel-Lommel covariance function $C_{\mathrm{xx}}^{BL}(z;\boldsymbol{\theta})$ the integral range is given by the following expression

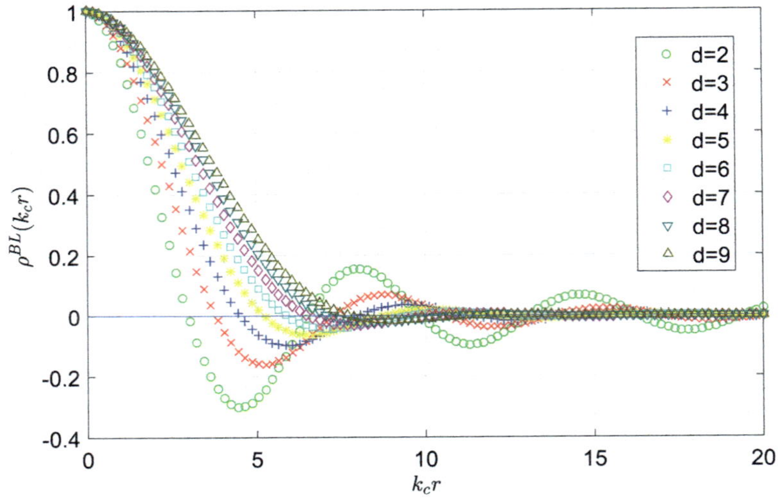

Fig. 7.25 Dependence of the Bessel-Lommel autocorrelation function (7.65) on the spatial lag and d. The parameter values used are $\lambda_0 = 1$, $\xi = 1$, $\eta_1 = 2$, and $k_c = 2$

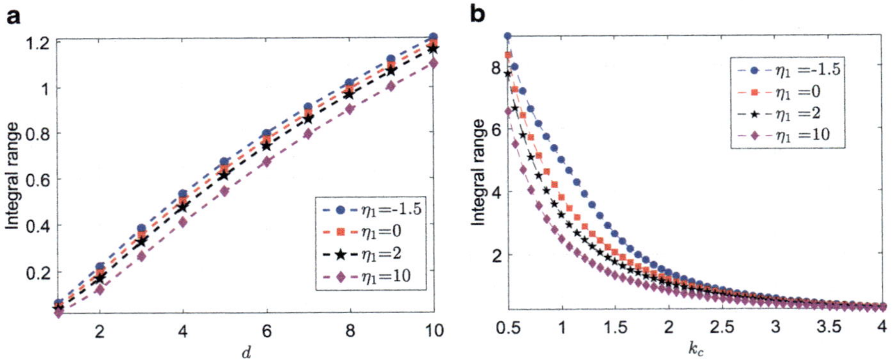

Fig. 7.26 (**a**) Bessel-Lommel integral range ℓ_c versus d based on (7.66). (**b**) Integral range ℓ_c versus k_c for different η_1 based on (7.66); in both cases $\xi = 2$

$$\ell_c = \left(\frac{\widetilde{C^{BL}_{xx}}(0;\boldsymbol{\theta})}{C^{BL}_{xx}(0;\boldsymbol{\theta})} \right)^{1/d} = \frac{\pi^{1/2}\, 2^{1-1/d}}{k_c} \left(\frac{\Gamma(d/2)}{\frac{1}{d} + \eta_1 \frac{(k_c\xi)^2}{d+2} + \frac{(k_c\xi)^4}{d+4}} \right)^{1/d} . \tag{7.66}$$

Equation (7.66) is derived using (7.59) for the zero-wavenumber spectral density $\widetilde{C^{BL}_{xx}}(\|\mathbf{k}\| = 0; \boldsymbol{\theta}) = \lambda_0\, \xi^d$ and (7.64) for $C^{BL}_{xx}(z = 0; \boldsymbol{\theta})$.

The dependence of the integral range is investigated in the parametric curves of Fig. 7.26a, where it is shown that ℓ_c increases with the dimensionality d. This is due to the suppression of negative peaks as d increases (cf. Fig. 7.25). In addition, the integral range decreases with increasing ξ and η_1. The reason is that the Bessel-Lommel spectral density (7.59) increases faster with $\|\mathbf{k}\|$ for higher ξ or η_1, thus shifting more spectral weight to higher wavenumbers and reducing the spatial coherence of the field. This tendency of the Bessel-Lommel integral range is opposite to that of the SSRF integral range.

The dependence of the integral range on ξ is through the dimensionless product $\kappa_c = k_c\xi$. Hence, for fixed κ_c, d and η_1, equation (7.66) implies that $\ell_c \propto 1/k_c$ as shown in Fig. 7.26b. This dependence reflects that higher k_c signifies a wider spectral band with more weight in higher wavenumbers that reduces spatial coherence. The inverse relation between ℓ_c and k_c is rooted in the fact that the Bessel-Lommel covariance is a function of $z = \kappa_c h = k_c\|\mathbf{r}\|$, implying that the characteristic distance scale is ${k_c}^{-1}$ instead of ξ.

Bessel-Lommel correlation spectrum The correlation spectrum of the Bessel-Lommel covariance function with spectral density (7.59) is determined by the following equation (the proof is given in [367]):

$$\lambda_c^{(\alpha)} = \begin{cases} \frac{g_d}{k_c}\left(1+\eta_1\kappa_c^2+\kappa_c^4\right)^{1/d}, & \eta_1 > \eta_{1,c} \text{ or } k_c < k_- \\ \frac{g_d}{k_c}\left(1+\eta_1\kappa_-^2+\kappa_-^4\right)^{1/d}, & \eta_{1,c} \geq \eta_1 > -2 \text{ and } k_+ \geq k_c > k_- \\ \frac{g_d}{k_c}\left(1+\eta_1\kappa^{*2}+\kappa^{*4}\right)^{1/d}, & \eta_{1,c} \geq \eta_1 > -2 \text{ and } k_c > k_+. \end{cases} \tag{7.67a}$$

- g_d is a dimensionless function of d and the index α given by

$$g_d = \left[\frac{\Gamma(d/2)}{2\pi^{d/2}\left(\frac{1}{d+2\alpha}+\frac{\eta_1 {k_c}^2\xi^2}{d+2\alpha+2}+\frac{{k_c}^4\xi^4}{d+2\alpha+4}\right)}\right]^{1/d}, \tag{7.67b}$$

- $\eta_{1,c}$ is a critical rigidity coefficient that depends on the index α via

$$\eta_{1,c} = -\frac{2\sqrt{\alpha(\alpha+2)}}{(\alpha+1)}, \tag{7.67c}$$

- $k_\pm$ and k^* are characteristic wavenumbers ($\kappa_\pm = k_\pm\xi$, $\kappa^* = k^*\xi$ are their dimensionless counterparts) of the Bessel-Lommel spectrum and α, that are defined as follows

$$k_\pm = \frac{1}{\xi}\sqrt{\frac{-(\alpha+1)\eta_1 \pm \sqrt{(\alpha+1)^2{\eta_1}^2 - 4\alpha(\alpha+2)}}{2(\alpha+2)}}, \tag{7.67d}$$

$$k^* = \arg\max\left(k_-^{2\alpha}\,\Pi(k_-\xi),\ {k_c}^{2\alpha}\,\Pi(k_c\xi)\right), \tag{7.67e}$$

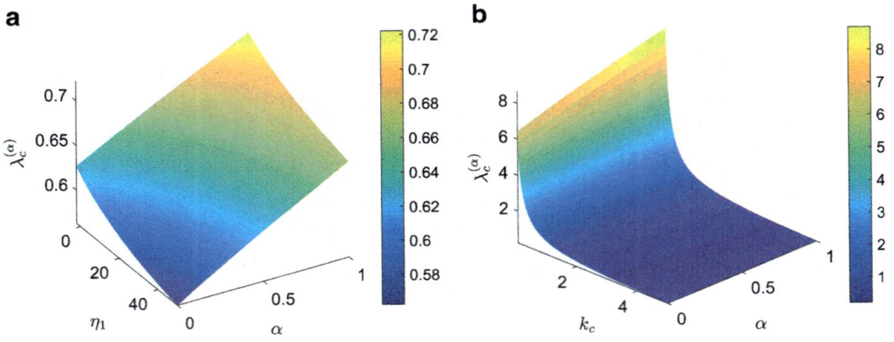

Fig. 7.27 Spectrum of Bessel-Lommel correlation scales in $d = 2$. (**a**) $\lambda_c^{(\alpha)}$ for $\eta_1 \in [0, 50]$ and $\alpha \in [0, 1]$ with $\xi = 5$, $k_c = \pi/2$. (**b**) $\lambda_c^{(\alpha)}$ for $k_c \in [0.1, 5]$ and $\alpha \in [0, 1]$ with $\eta_1 = 3$, $\xi = 5$

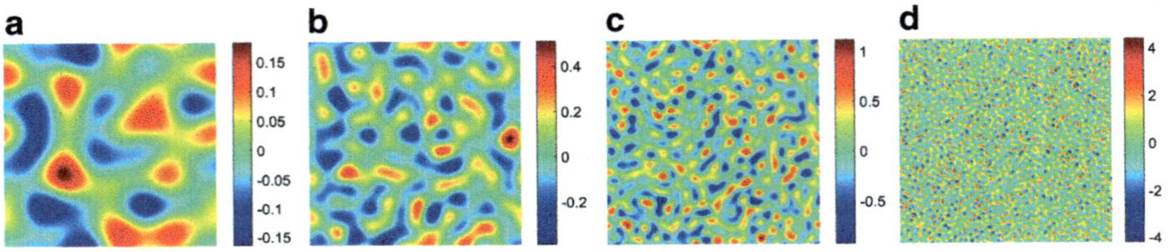

Fig. 7.28 Bessel-Lommel SRF realizations with covariance function (7.37) on square 512×512 grid. For all realizations $\lambda_0 = 0.01$, $\eta_1 = 1$, $\xi = 2$. The FFT spectral simulation method with identical random generator seed is used. (**a**) $k_c = 0.05$. (**b**) $k_c = 0.1$. (**c**) $k_c = 0.2$. (**d**) $k_c = 0.5$

- and $\Pi(x)$ is the characteristic polynomial (7.19a).

The dependence of the Bessel-Lommel correlation spectrum on η_1, α, k_c is illustrated in Fig. 7.27. Overall, higher η_1 and k_c tend to reduce the correlation scale. The k_c dependence is more pronounced and follows asymptotically $\lambda_c^{(\alpha)} \sim 1/{k_c}^3$. We can intuitively understand this effect as the result of increased spectral weight in higher wavenumbers that tends to reduce the spatial extent of correlations.

More surprising is the increase of $\lambda_c^{(\alpha)}$ with α, which signifies that the *smoothness microscale* ($\alpha = 1$) exceeds the *integral range* ($\alpha = 0$). This effect is due to two underlying causes:

1. The differentiability of the Bessel-Lommel random field implies smoother fluctuations.
2. The oscillations in the Bessel-Lommel covariance function reduce the integral range.

Realizations of Bessel-Lommel random fields Four different realizations of Bessel-Lommel SRFs are shown in Fig. 7.28. The realizations are generated using the spectral *Fast Fourier Transform* simulation method (see Sect. 16.4) using the same seed for the random number generator in all cases. The realizations correspond to identical λ_0, η_1 and ξ but different k_c. Two patterns of dependence are obvious:

1. The variance of the fluctuations increases with k_c.
2. The spatial extent of characteristic spatial patterns (i.e., areas that predominantly contain values above or below a certain threshold) is reduced as k_c increases.

The first pattern is due to increasing spectral weight at higher wavenumbers that results from the increase of k_c. This extra "high-frequency" contribution leads to higher variability. The second pattern reflects the decline of the integral range with increasing k_c according to (7.66).

Random fields with Bessel-Lommel covariance functions can be loosely viewed as spatial analogues of moving average (MA) time series models. The analogy is based on the polynomial dependence of the Bessel-Lommel spectral density (7.59) on $\|\mathbf{k}\|^2$ (up to the cutoff wavenumber), which is reminiscent of the MA(q) spectral density (9.58). Essentially, we can think of the Bessel-Lommel random fields as being generated by taking suitably modified "derivatives" of the white noise random field. The similarity with MA processes is not complete: The latter have covariance functions with a sharp cutoff (the covariance function vanishes after a lag distance equal to the order of the model), while the Bessel-Lommel random fields feature spectral densities with a sharp cutoff.

Chapter 8
Lattice Representations of Spartan Random Fields

> The most fruitful areas for the growth of the sciences were those which had been neglected as a no-man's land between the various established fields.
>
> *Norbert Wiener*

Our discussion of SSRFs has so far assumed that the values of the field are defined on continuum domains $\mathcal{D} \subset \mathbb{R}^d$. This assumption, which is inherent in the energy functional (7.4), reflects the fact that geophysical and environmental processes take place in a spatial continuum.[1]

On the other hand, spatial systems such as hydrocarbon reservoirs simulated on a digital computer involve sampled field values at the nodes of rectangular grids (lattices). In digital image analysis [852], the "observed" fields are a priori defined on discrete domains, due to the coarse-graining operations imposed by the respective generative processes (e.g., photographic capturing device). Earth-based observations of spatially distributed processes are typically collected on irregular sampling networks due to socioeconomic and technical feasibility constraints. Thus, spatial random fields on discrete domains are motivated by practical and computational reasons.

In the spirit of SSRFs, it is possible to define computationally efficient *local interaction models* on regular grids. Such models can be constructed so as to approximate SSRFs. This can be accomplished by replacing the derivatives used in the SSRF energy functional (7.4) with finite differences. The resulting models only involve nearby lattice neighbors (e.g., k-nearest neighbors where $k \in \mathbb{N}$ is a specified integer neighbor order).

[1] Solid state physics focuses on the microscopic properties of materials with periodic crystal lattice structure. However, the latter does not explicitly appear in studies of macroscopic material properties and response to external stimuli for which continuum theories are employed.

D. T. Hristopulos, *Random Fields for Spatial Data Modeling*, Advances in Geographic Information Science, https://doi.org/10.1007/978-94-024-1918-4_8

For these local interaction models, however, it is not possible to obtain closed-form expressions of the real-space covariance function as in the continuum case. In order to construct discretized lattice models with covariance functions that accurately approximate the SSRF continuum covariance functions, higher-order discretization schemes should be used (see Sect. 8.5).

Discretized versions of SSRFs on regular lattices generate *Gaussian Markov Random Fields (GMRFs)* with energy functional structure inherited from the specific local interactions.

8.1 Introduction to Gauss-Markov Random Fields (GMRFs)

This section is a brief and incomplete introduction to Gauss-Markov random fields. For a more rigorous and complete treatment consider [698]. Shorter introductions to Markov random fields (Gaussian and non-Gaussian) are given in [699, 700].

For our purposes, a GMRF is a random field defined on a lattice that (i) follows the Gaussian joint pdf and (ii) satisfies *Markov properties* of *conditional independence* [698, 852]. The hallmark of GMRFs is the Markov property of *local dependence* which specifies that the field value at each lattice point is determined from its local neighborhood. This also implies that the joint pdf of the field is fully determined by the conditional distributions at every site. In addition, the conditional distributions exclusively interact with the lattice sites inside a finite neighborhood around the target point.

Markov property in random processes The *Markov property* for random processes specifies that the state (realization) at any time t_n is fully determined by the state at the previous time t_{n-1}, where $n = 1, 2, \ldots$.

Let $f_{\mathrm{x}}(x_n \mid x_{n-1})$ represent the *conditional pdf* of the random variable $\mathrm{X}(t_n; \omega)$ at t_n given the value x_{n-1} at the preceding time t_{n-1}.

The Markov property means that $f_{\mathrm{x}}(x_n \mid x_{n-1})$ contains the same information for $\mathrm{X}(t_n; \omega)$ as the joint pdf of the random vector $\left(\mathrm{X}(t_1; \omega), \ldots, \mathrm{X}(t_n; \omega)\right)^{\top}$ that comprises observations of the process at all times t_i, $i = 1, \ldots, n$.

An equivalent way to express the Markov property given a sequence of times $t_1, \ldots, t_n$ and the vector of values $\mathbf{x} = (x_1, \ldots, x_n)^{\top}$ is the following factorization of the joint pdf

$$f_{\mathrm{x}}(\mathbf{x}) = f_{\mathrm{x}}(x_1)\, f_{\mathrm{x}}(x_2 \mid x_1)\, f_{\mathrm{x}}(x_3 \mid x_2) \ldots\, f_{\mathrm{x}}(x_n \mid x_{n-1}).$$

GMRFs are used in spatial statistics to model various types of lattice data [69, 72, 682], [165, Chap. 6]. For the interpolation of the field at the locations of missing lattice data, both GMRFs and geostatistical methods can be used. A comparison between GMRFs and classical geostatistical models is given in [765]. In addition, GMRFs have been explored in statistical physics with respect to various data-oriented applications such as image restoration and seismic tomography [439, 473, 595, 706].

Extension of the Markov property to spatial models provides the mathematical foundation of locality in GMRFs. In temporal problems the arrow of time imposes a natural ordering on the data. In contrast, in spatial data the ordering of different sites (even on regular lattices) is not uniquely defined. However, the main idea embodied in the Markov property is that the behavior of the process at any given point depends only on the field values in its vicinity. Hence, we need to introduce concepts that allow us to define neighborhood relations.

Definition 8.1 The following notation will be used in the discussion of the local properties of Markov random fields. The point $\mathbf{s}_n$, $n = 1, \ldots, N$, denotes any node of the underlying grid.[2] Our main interest herein are *regular grids* that allow straightforward definition of local interactions. However, the local properties that will be discussed also apply to irregular grids. In the following, $\mathbb{S}_\mathrm{N} = \{\mathbf{s}_1, \ldots, \mathbf{s}_N\}$ is the set of sampling points that includes the grid nodes.

1. $B(\mathbf{s}_n)$: neighborhood of the point $\mathbf{s}_n$ excluding $\mathbf{s}_n$.
2. $\mathbf{x}_{-n}$: values of the random field $\mathrm{X}(\mathbf{s};\omega)$ at every grid point $\mathbb{S}_\mathrm{N} \setminus \{\mathbf{s}_n\}$ except $\mathbf{s}_n$.
3. $\mathcal{B}(\mathbf{x}_{-n})$: set of random field values inside the local neighborhood $B(\mathbf{s}_n)$.
4. $\perp$: symbol of *conditional independence*.
5. $\mathrm{X}_n(\omega)$: random variable that corresponds to the random field at the location $\mathbf{s}_n$.
6. $\mathbf{X}_{\{-n,B(\mathbf{s}_n)\}}(\omega)$: random vector of field values at all lattice sites excluding $\mathbf{s}_n$ and its "connected" neighbors $B(\mathbf{s}_n)$.

The Markov property implies that given a grid $\mathcal{G}$ and the values of the field on a subset of sites $\mathcal{G}_1$ the following is true: The remaining sites $\mathcal{G}\setminus\mathcal{G}_1$ can be partitioned into two subsets $\mathcal{G}_2$ and $\mathcal{G}_3$, disjoint with each other and $\mathcal{G}_1$, such that the values of the field on $\mathcal{G}_2$ are *conditionally independent* of those on $\mathcal{G}_3$ given the values on $\mathcal{G}_1$.

Gaussian Markov random fields are equivalent to *Gaussian conditional autoregression (CAR)* models with a Markov property [699]. Hence, not all CAR models are equivalent to Markov random fields, since CAR models need to exhibit the Markov property in order to be equivalent to GMRFs. CAR models extend the framework of autoregressive (AR) time series models to the space domain (see Chap. 9).

[2] Alternatively, the centers of the grid cells can be viewed as the field locations.

8.1.1 Conditional Independence

The concept of *conditional independence* has its roots in a simple logical proposition: it states that if two events A and B are correlated, and their correlation is due to a common cause, e.g., to a third event C, then A and B are independent if their conditional probabilities given C are calculated. The principle of conditional independence is illustrated in Fig. 8.1.

The principle of *conditional independence* can also be expressed more formally. If the correlation between events A and B are caused by a third event C, then the following statements are true:

$$P(A \cap B) > P(A)\, P(B) \tag{8.1a}$$

$$P(A \cap B \mid C) = P(A \mid C)\, P(B \mid C). \tag{8.1b}$$

A clear physical interpretation of the principle of conditional independence in the context of tsunami incidence in the Pacific is given in [666]:

> "For example, the incidence of tsunamis in Chile is statistically correlated with that of tsunamis in Japan. In statistical terms, the combined probability for two tsunamis is greater than the product of the separate probabilities for tsunamis in Chile and Japan. But neither event is a cause of the other. If we condition the tsunamis' probabilities on the knowledge that an earthquake has occurred in the Pacific basin, then we should find that the events are independent: the combined (conditional) probability of the two is equal to the product of the separate (conditional) probabilities. In other words, the correlation disappears. Given our knowledge of an earthquake, the news that a tsunami occurred in Chile no longer gives us

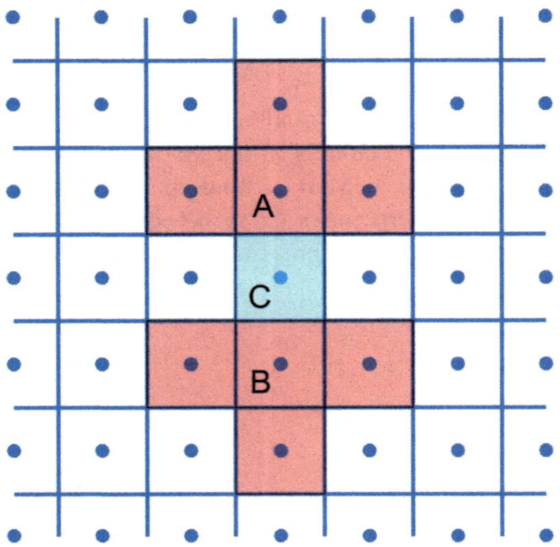

Fig. 8.1 Illustration of *conditional independence*. It is assumed that GMRF interactions between sites (cell centers) are restricted only between nearest neighbors. Then, the values at the sites A and B are conditionally independent given the values of the field at the site C, since the latter is the only grid point that interacts with both A and B

any extra information about the probability that a tsunami occurred in Japan. Reichenbach's conditional independence suggests that the earthquake might be the common cause of the tsunamis in the two regions."

The Markov property can be expressed mathematically in two equivalent ways that are known as the *local and global Markov* properties.

Local Markov property The local Markov property means that the random field value at each location, given the values of its neighbors, is conditionally independent of the field values at all other lattice points (see Fig. 8.2).

Using the *full conditional pdf* $f_{\mathrm{x}}(x_n \mid \mathbf{x}_{-n})$, the *local Markov property* is expressed as follows

$$f_{\mathrm{x}}(x_n \mid \mathbf{x}_{-n}) = f_{\mathrm{x}}\left[x_n \mid \mathcal{B}(\mathbf{x}_{-n})\right], \; n = 1, \ldots, N. \tag{8.2}$$

The local Markov property is also denoted in the following way:

$$\mathrm{X}_n(\omega) \perp \mathbf{X}_{\{-n, B(\mathbf{s}_n)\}}(\omega) \mid \mathcal{B}(\mathbf{x}_{-n}),$$

where the notation used above is introduced in Definition 8.1.

Global Markov property The global Markov property states that if A, B and C are disjoint, non-empty sets of lattice sites such that C separates A and B, then the random vector $\mathbf{X}_A(\omega)$ is conditionally independent of $\mathbf{X}_B(\omega)$ given the values on the separating set $\mathbf{X}_C(\omega) = \mathbf{x}_C$, i.e.,

$$\mathbf{X}_A(\omega) \perp \mathbf{X}_B(\omega) \mid \mathbf{x}_C. \tag{8.3}$$

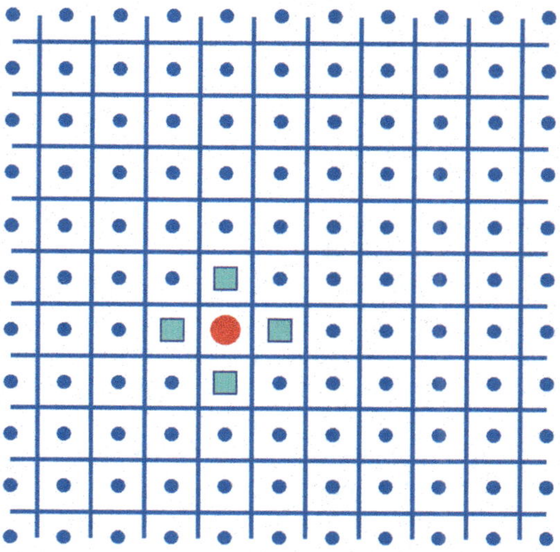

Fig. 8.2 Illustration of the *local Markov property*. The sampling sites (grid cell centers) are indicated by dots. The field values are determined by the Gibbs pdf with energy function (8.11). The field value at the target location (red circle) is determined by its four neighbors (cyan squares). If the values of these neighbors are known, the target value is *conditionally independent* of the field values at the remaining locations

8.1.2 GMRF Conditional Probability Distribution

The connection between the conditional Gauss-Markov distribution at a specific location (given the values at all the other locations) and the joint distribution can be expressed in precise mathematical terms [69, 765].

Let us assume that the conditional distribution at the point $\mathbf{s}_n$, given the values x_k at the locations $\mathbf{s}_k$ for $n \neq k$, is the following Gaussian distribution

$$X(\mathbf{s}_n; \omega) \mid \mathbf{x}_{-n} \stackrel{d}{=} \mathcal{N}\left(m_n + \sum_{k=1}^{N} b_{n,k}(x_k - m_k),\ \sigma_n^2 \right), \tag{8.4}$$

where m_n is the mean at the site $\mathbf{s}_n$, σ_n^2 is the *conditional variance* of $X(\mathbf{s}_n; \omega)$ given the $x_k \in \mathcal{B}(\mathbf{x}_{-n})$, and $b_{n,k}$ are *coupling parameters*.

The coupling parameters and the local variance satisfy the following conditions:

1. $b_{n,n} = 0$, for all $n = 1, \ldots, N$. This condition reflects the fact that the self-interaction of x_n is embodied in m_n.
2. $b_{n,k} = 0$, unless $\mathbf{s}_k \in B(\mathbf{s}_n)$. This second condition expresses the local property, i.e., the dependence of x_n exclusively on the surrounding local neighborhood.
3. $b_{n,k}\,\sigma_k^2 = b_{k,n}\,\sigma_n^2$, for all $k, n = 1, \ldots, N$. The local interactions in (8.4) are expressed in terms of a *precision matrix* that should be symmetric for the interaction between $\mathbf{s}_n$ and $\mathbf{s}_k$ to be identical to the interaction between $\mathbf{s}_k$ and $\mathbf{s}_n$. This third condition yields a symmetric precision matrix, as shown below in (8.6).

8.1.3 GMRF Joint Probability Distribution

If the joint distribution is known, the full conditional distributions can be constructed from it. The converse is not so obvious. *Brook's lemma* shows how to construct the unique joint pdf from the full conditional densities $f_{\mathrm{x}}(x_n \mid \mathbf{x}_{-n})$, for $n = 1, \ldots, N$ [44, 108, 700]. The full conditional densities need to be compatible for the joint pdf to be well-defined. In addition, the joint pdf may be improper (i.e., its integral over all states may not be equal to one), even if the conditional densities are proper [44, p. 78-79]

Brook's lemma specifies how the joint pdf can be constructed from the full conditionals. The *Hammersley-Clifford* theorem [69, 152, 165] gives necessary and sufficient conditions for a probability distribution to define a Markov random field and establishes the equivalence between Gibbs and Markov random fields. In the case of Markov random fields, the full conditional distributions can be locally determined as in (8.4). For a Gaussian conditional distribution of the form (8.4), the condition that enforces the precision matrix symmetry is necessary and sufficient for the conditional densities to be compatible.

It can be shown by Brook's lemma and Hammersley-Clifford's theorem that the *GMRF joint probability distribution* for the random vector

$$\mathbf{X}(\omega) = (\mathrm{X}_1(\omega), \ldots, \mathrm{X}_N(\omega))^\top$$

is given by

$$\mathbf{X}(\omega) \stackrel{d}{=} \mathcal{N}(\mathbf{m}, \mathbf{C}), \text{ where } \mathbf{C} = \boldsymbol{\Sigma}\,(\mathbf{I}_N - \mathcal{B})^{-1}. \tag{8.5}$$

In the above equation, $\mathbf{m}$ is the vector of the mean values, $\mathbf{C}$ is the covariance matrix, $\mathcal{B}_{n,k} = b_{n,k}$, and $\boldsymbol{\Sigma}$ is the diagonal matrix of conditional variances, $\Sigma_{n,k} = \sigma_n^2\,\delta_{n,k}$, where $n, k = 1, \ldots, N$ [69].

The elements of the *GMRF precision matrix* $\mathbf{J} = \mathbf{C}^{-1}$ are given by

$$J_{n,n} = \frac{1}{\sigma_n^2}, \quad J_{n,k} = -\frac{\beta_{n,k}}{\sigma_n^2}, \; n \neq k, \; n, k = 1, \ldots, N, \tag{8.6}$$

provided that $\mathbf{J}$ is a *symmetric and positive-definite* matrix.

Based on the above, if the precision matrix $\mathbf{J}$ is known, the *conditional variance* at any site is equal to the inverse of the respective diagonal precision matrix element.

The coupling parameters must obey conditions which ensure that the precision matrix $\mathbf{J}$ is not only symmetric but also positive definite. Such conditions are in general difficult to establish. Instead, it is easier to establish conditions for *diagonal dominance*.

Definition 8.2 A square $N \times N$ matrix $\mathbf{J}$ is called diagonally dominant if

$$\left|J_{n,n}\right| > \sum_{k=1,\neq n}^{N} \left| J_{n,k} \right|, \text{ for all } n = 1, \ldots, N.$$

A matrix that is diagonally dominant is also *positive definite*.

Given the dependence (8.6) of the precision matrix elements on the coupling parameters, diagonal dominance requires that $\sum_{k=1,\neq n}^{N} \left|\beta_{n,k}\right| < 1$. Since diagonal dominance is a stricter (sufficient but not necessary) condition than positive definiteness, the above constraint may be too restrictive [698].

Conditional moments The precision matrix can be used to obtain expressions for the conditional moments of GMRFs. More precisely, the conditional mean, the conditional variance, and the *partial autocorrelation function (PACF)* are given by

$$\mathcal{E}[X(\mathbf{s}_n;\omega) \mid \mathbf{x}_{-n}] = m_n - \frac{1}{J_{n,n}} \sum_{\mathbf{s}_k \in B(\mathbf{s}_n)} J_{n,k}\,(x_k - m_k)\,, \tag{8.7}$$

$$\mathbb{V}\text{ar}\,\{X(\mathbf{s}_n;\omega) \mid \mathbf{x}_{-n}\} = \frac{1}{J_{n,n}}, \tag{8.8}$$

$$\mathbb{C}\text{or}\,\left\{X(\mathbf{s}_n;\omega), X(\mathbf{s}_k;\omega) \mid \mathbf{x}_{-\{n,k\}}\right\} = -\frac{J_{n,k}}{\sqrt{J_{n,n}J_{k,k}}},\quad n \neq k. \tag{8.9}$$

Partial correlations In (8.9) the symbol $\mathbb{C}\text{or}\left\{X(\mathbf{s}_n;\omega), X(\mathbf{s}_k;\omega) \mid \mathbf{x}_{-\{n,k\}}\right\}$ denotes the *partial correlation* between the sites $\mathbf{s}_n$ and $\mathbf{s}_k$ conditioned on the values $\mathbf{x}_{-\{n,k\}}$ at the remaining sites of the network. The proof of equation (8.9) can be found in [489, pp. 129–130].

Partial correlations are an important concept in the analysis and estimation of network data, because they represent pairwise interactions *after the indirect correlations are removed.* Owing to this property, partial correlations are a better tool for uncovering connectivity/interactivity than classical correlations. For example, if two locations A and C interact with location B but not directly with each other, the correlation coefficient between A and C has a finite value; on the other hand, the partial correlation between A and C is zero. Partial correlations are used as connectivity measures in network analysis since they are more suitable than correlation coefficients for determining *causal connections*.

8.2 From SSRFs to Gauss-Markov Random Fields

After the brief introduction to Gauss-Markov random fields we turn our attention to the definition of Spartan random fields on regular lattices. As we show below, SSRFs are transformed to GMRFs by replacing the derivatives in the energy functional (7.4) with respective lattice-based finite differences. The characteristic feature of the SSRF-GMRFs is the specific energy structure that is derived from the discretization of the continuum SSRF functional (7.4).

The computational memory and time requirements of SSRF-GMRFs scale *linearly* with the data size, due to the fact that each lattice site only interacts with a small set of neighbors in its vicinity. This is a significant computational advantage shared by all Gauss-Markov models. More details on algorithms for the estimation and simulation of GMRFs can be found in [698–700].

8.2.1 Lattice SSRF Model

Next, we describe the transformation from continuum SSRFs to random fields that are defined on regular lattices but maintain similar spatial structure. In terms of notation, we will use the index "i" to denote the orthogonal directions of the lattice, while the lattice positions will be denoted by the index "n".

- We assume "hypercubic" grids with lattice step a and unit lattice vector $\hat{\mathbf{e}}_i$ in the i-th direction, where $i = 1, \ldots, d$. (More generally, we can consider direction-dependent step sizes a_i.)
- For a lattice with L_i nodes per direction, the total number of nodes is $N = \prod_{i=1}^{d} L_i$.
- The vector $\mathbf{s}_n$, $n = 1, \ldots, N$ marks the position of the lattice nodes on the grid.
- The value of the field at any site $\mathbf{s}_n$, where $n = 1, \ldots, N$, is denoted by $x(\mathbf{s}_n)$.

Gradient and Laplacian The simplest discrete approximations of the gradient (vector) and the Laplacian (scalar) are given by the following finite differences

$$\nabla_a x(\mathbf{s}_n) = \left(\frac{x(\mathbf{s}_n + a\hat{\mathbf{e}}_1) - x(\mathbf{s}_n)}{a}, \ldots, \frac{x(\mathbf{s}_n + a\hat{\mathbf{e}}_d) - x(\mathbf{s}_n)}{a} \right)^\top, \tag{8.10a}$$

$$\nabla_a^2 x(\mathbf{s}_n) = \sum_{i=1}^{d} \frac{x(\mathbf{s}_n + a\hat{\mathbf{e}}_i) + x(\mathbf{s}_n - a\hat{\mathbf{e}}_i) - 2x(\mathbf{s}_n)}{a^2}. \tag{8.10b}$$

The lattice functions $\nabla_a x(\mathbf{s}_n)$ and $\nabla_a^2 x(\mathbf{s}_n)$ represent realizations of the *discrete gradient* and the *discrete Laplacian* random fields $\nabla_a \mathrm{X}(\mathbf{s};\omega)$ and $\nabla_a^2 \mathrm{X}(\mathbf{s};\omega)$, respectively.

Equation (8.10a) employs the forward finite difference approximation for first-order partial derivatives, whereas (8.10b) is the central difference approximation of the *Laplacian*.

8.2.2 Lattice SSRF with Isotropic Structure

Based on (8.10) above, the lattice approximation of the isotropic SSRF energy functional (7.4) on a cubic grid becomes

$$\mathcal{H}_0(\mathbf{x}) = \frac{1}{2\eta_0 \xi^d} \sum_{n=1}^{N} \left\{ \{x(\mathbf{s}_n) - m_{\mathrm{x}}\}^2 + \eta_1 \xi^2 \sum_{i=1}^{d} \left[\frac{x(\mathbf{s}_n + a\hat{\mathbf{e}}_i) - x(\mathbf{s}_n)}{a} \right]^2 \right.$$
$$\left. + \xi^4 \sum_{i=1}^{d} \left[\frac{x(\mathbf{s}_n + a\hat{\mathbf{e}}_i) - 2x(\mathbf{s}_n) + x(\mathbf{s}_n - a\hat{\mathbf{e}}_i)}{a^2} \right]^2 \right\}, \tag{8.11}$$

where $\mathbf{x} = (x_1, \ldots, x_N)^\top$ is the vector comprising the values $x(\mathbf{s}_n)$ at all sites $\mathbf{s}_i$, where $n = 1, \ldots, N$.[3]

Positive definite SSRF precision matrix Since $\mathcal{H}_0(\mathbf{x})$ is a symmetric bilinear form, it can be expressed as a quadratic function of $\mathbf{x}$, i.e.,

$$\mathcal{H}_0(\mathbf{x}) = \frac{1}{2}\mathbf{x}^\top \mathbf{J}\,\mathbf{x}.$$

The *SSRF precision matrix* $\mathbf{J}$ is positive definite if

$$\mathbf{x}^\top \mathbf{J}\,\mathbf{x} > 0,$$

for all non-zero $\mathbf{x} \in \mathbb{R}^N$ and for all $N \in \mathbb{N}$. This condition is satisfied if $\mathcal{H}_0(\mathbf{x}) > 0$.

Hence, a *sufficient* (but not necessary) condition for positive definiteness is that $\eta_1 > 0$. While this condition suffices to obtain a well-defined precision matrix, it does not allow the oscillatory covariance behavior that is possible in SSRFs with negative values of η_1. On the other hand, as shown in Sect. 8.3 below, the permissible boundary of η_1 moves closer to zero as the lattice dimensionality $d \to \infty$.

Estimation The energy functional (8.11) contains four parameters, i.e., $\eta_0, \eta_1, \xi, m_{\mathrm{x}}$. The estimation of these parameters from the available data can be accomplished by means of various methods such as maximum likelihood and Besag's pseudo-likelihood method [70], [269, Chap. 5], [697], [682] expectation-maximization [193, 320], Markov Chain Monte Carlo, and normalized correlations (see Chap. 13).

The expression (8.11) does not discriminate between spatial directions, since the same coefficients apply to all $\hat{\mathbf{e}}_i$, $i = 1, \ldots, d$. However, the resulting random field is not strictly isotropic, since the covariance is not a radial function, at least for length scales of $O(a)$. This is due to the lattice structure that is generated by the translation of unit lattice vectors in d orthogonal directions.

The *joint pdf* of the lattice SSRF is given by

$$f_{\mathrm{x}}(\mathbf{x}) = \frac{\mathrm{e}^{-\mathcal{H}_0(\mathbf{x})}}{Z}, \tag{8.12}$$

where Z is the partition function that normalizes the pdf. For lattice random fields the latter is given by the multiple integral (6.5c).

[3]For notational economy we use $\mathcal{H}_0(\mathbf{x})$ for the discrete energy functional as for the continuum SSRF.

Definition 8.3 The joint Gibbs pdf (8.12) equipped with the SSRF energy functional (8.11) satisfies the two defining properties of *Markov random fields*:

- $f_{\mathrm{x}}(\mathbf{x})$ is positive for all $\mathbf{x} \in \mathbb{R}^N$ and for all $N \in \mathbb{N}$.
- The conditional probability that $\mathrm{X}(\mathbf{s}_n; \omega) = x_n$ given the values $\mathbf{x}_{-n}$, for $n = 1, \ldots, N$, is completely determined from the subset $\mathcal{B}(\mathbf{x}_{-n})$ of $\mathbf{x}_{-n}$ that comprises the sites in the local neighborhood $B(\mathbf{s}_n)$.

The strictly positive joint pdf (8.12) also defines a *Gibbs random field* [69]. According to the *Hammersley-Clifford* theorem, a random field is a *Markov random field* if and only if the joint probability distribution can be expressed as a Gibbs distribution.

8.2.3 Lattice SSRF with Anisotropic Structure

It is straightforward to extend the above energy functional (8.11) to anisotropic SSRFs. We assume for simplicity that the SSRF principal axes are aligned with the grid axes. Furthermore, instead of η_0, η_1, ξ we use a complementary parametrization in which the above parameters are replaced by linear *gradient coefficients* $c_{1,i}$, *Laplacian coefficients*, $c_{2,i}$, where $i = 1, \ldots, d$, and an overall *scale factor* λ.

$$\mathcal{H}_0 = \frac{1}{2\lambda} \sum_{n=1}^{N} \left\{ [x(\mathbf{s}_n) - m_{\mathrm{x}}]^2 + \sum_{i=1}^{d} c_{1,i} \left[\frac{x(\mathbf{s}_n + a_i \hat{\mathbf{e}}_i) - x(\mathbf{s}_n)}{a_i} \right]^2 \right.$$
$$\left. + \sum_{i=1}^{d} c_{2,i} \left[\frac{x(\mathbf{s}_n + a_i \hat{\mathbf{e}}_i) - 2x(\mathbf{s}_n) + x(\mathbf{s}_n - a_i \hat{\mathbf{e}}_i)}{a_i^2} \right]^2 \right\}. \tag{8.13}$$

The anisotropic energy functional (8.13) comprises $2(d+1)$ parameters, i.e., the mean m_{x}, the scale factor λ, and the dimensionless directional coefficients $\{c_{1,i}\}_{i=1}^{d}$, $\{c_{2,i}\}_{i=1}^{d}$. The lattice steps, a_i, are also allowed to vary with the direction.

8.3 Lattice Spectral Density

We assume that the Fourier transforms of individual field states (realizations) exist. This assumption may not be rigorous mathematically, but it can be made meaningful with proper definition. After all, the discrete Fourier transform of sampled realizations can be numerically evaluated. Anyhow, the ultimate goal is the spectral density that is a well-defined quantity.

Continuum Fourier transform In the continuum case, the Fourier transform of a function $x(\mathbf{s})$ is given by

$$\widetilde{x}(\mathbf{k}) = \int_{\mathbb{R}^d} \mathrm{d}\mathbf{s}\, x(\mathbf{s})\, \mathrm{e}^{-\mathrm{i}\mathbf{k}\cdot\mathbf{s}}. \tag{8.14}$$

Lattice Fourier transform The finite lattice step implies that integrals of functions over real space should be replaced by respective sums over the lattice sites (centres). This change also impacts the Fourier transform. The integral (8.14) can be expressed in terms of a discrete sum that involves the lattice sites $\mathbf{s}_n$ as follows

$$\widetilde{x}_{\mathrm{latt}}(\mathbf{k}) = \upsilon_c \sum_{n=1}^{N} x(\mathbf{s}_n)\mathrm{e}^{-\mathrm{i}\mathbf{k}\cdot\mathbf{s}_n},$$

where $\upsilon_c = V_d/N$ is the volume of the unit lattice cell (V_d is the lattice volume in d dimensions).

It would be handy to have a continuum (integral) formulation that smoothly transfers to the discrete case. To accomplish this goal, we replace the function $x(\mathbf{s})$ with the *lattice function* $x_{\mathrm{latt}}(\mathbf{s}) = \upsilon_c\, x(\mathbf{s})\rho(\mathbf{s})$, where $\rho(\mathbf{s})$ is a *lattice density operator* given by a superposition of Dirac functions placed at the lattice sites

$$\rho(\mathbf{s}) = \sum_{n=1}^{N} \delta(\mathbf{s} - \mathbf{s}_n).$$

In light of the above, the Fourier transform on the lattice function $x_{\mathrm{latt}}(\mathbf{s})$ becomes [122, p. 101]

$$\begin{aligned}\widetilde{x}_{\mathrm{latt}}(\mathbf{k}) &= \int \mathrm{d}\mathbf{s}\, x_{\mathrm{latt}}(\mathbf{s})\, \mathrm{e}^{-\mathrm{i}\mathbf{k}\cdot\mathbf{s}} = \upsilon_c \int \mathrm{d}\mathbf{s}\, x(\mathbf{s})\, \rho(\mathbf{s})\mathrm{e}^{-\mathrm{i}\mathbf{k}\cdot\mathbf{s}} \\ &= \upsilon_c \sum_{n=1}^{N} x(\mathbf{s}_n)\, \mathrm{e}^{-\mathrm{i}\mathbf{k}\cdot\mathbf{s}_n}.\end{aligned} \tag{8.15a}$$

Lattice inverse Fourier transform The discretization of the lattice Fourier transform is due to the minimum length scale, i.e., the lattice step. In the case of an infinite spatial domain, there is no minimum scale in reciprocal space (the wavenumber step is set by the inverse of the domain size). Hence, the inverse Fourier transform is given by an integral over the reciprocal space that involves wavevectors $\mathbf{k} \subset \mathbb{R}^d$.

The integration in reciprocal space does not extend to infinity, since the finite lattice steps in direct space imply that spatial frequencies higher than $1/a_i$, $i =$

$1, \ldots, d$ cannot be resolved. Hence, the relevant part of the reciprocal space is bounded within a "box" that extends from $-\pi/a_i$ to π/a_i in each lattice direction.[4] This hyper-rectangular box is defined by the *Cartesian product* of intervals

$$\left[-\frac{\pi}{a_1}, \frac{\pi}{a_1}\right] \times \left[-\frac{\pi}{a_2}, \frac{\pi}{a_2}\right] \ldots \times \left[-\frac{\pi}{a_d}, \frac{\pi}{a_d}\right],$$

and is known in solid state physics as the *first Brillouin zone (FBZ)*. If $L_i \gg a_i$ for all $i = 1, \ldots, d$, we can ignore the discretization of the reciprocal space and treat $\mathbf{k}$ as a continuous vector variable.

We thus obtain the following inverse Fourier transform

$$x_{\text{latt}}(\mathbf{s}_n) = \frac{1}{(2\pi)^d} \int_{\text{FBZ}} d\mathbf{k}\, \widetilde{x}_{\text{latt}}(\mathbf{k})\, e^{i\,\mathbf{k}\cdot\mathbf{s}_n}, \tag{8.15b}$$

where the subscript "FBZ" denotes that the integration domain is the first Brillouin zone. More information about lattice Fourier transforms is found in [122, App. 2A].

Lattice SSRF spectral density Using the pair of Fourier transforms (8.15), and the SSRF energy functional (7.4), the *lattice SSRF spectral density* is given by [362]

$$\widetilde{C}_{\text{xx}}^{(\text{latt})}(\mathbf{k}) = \frac{\eta_0\, \xi^d}{1 + 4\eta_1 \frac{\xi^2}{a_i^2} \sum_{i=1}^{d} \sin^2\left(\frac{a_i\, k_i}{2}\right) + \frac{16\xi^4}{a_i^4} \sum_{i=1}^{d} \sin^4\left(\frac{a_i\, k_i}{2}\right)}. \tag{8.16}$$

Permissibility The function (8.16) is a permissible spectral density if

$$\widetilde{C}_{\text{xx}}^{(\text{latt})}(\mathbf{k}) > 0, \text{ for all } \mathbf{k} \in \mathbb{R}^d.$$

This condition is satisfied for $\eta_1 \geq 0$. If $\eta_1 < 0$ we need to ensure that the denominator of $\widetilde{C}_{\text{xx}}^{(\text{latt})}(\mathbf{k})$ is positive for all $\mathbf{k} \in \mathbb{R}^d$. As we show below, this condition is satisfied if and only if

$$\eta_1 > -\frac{2}{\sqrt{d}}. \tag{8.17}$$

For $d > 1$ the constraint (8.17) is more restrictive than the respective $\eta_1 < -2$ for continuous SSRFs. The reason for this difference is that the fourth power of the sine function—third term in the denominator of (8.16)—cannot compensate for negative

[4]The reason for the presence of π is that the reciprocal space is spanned by wavevectors which correspond to cyclical frequencies $w = 2\pi f$ in the reciprocal time domain.

values coming from the term proportional to η_1—second term in the denominator of (8.16)—to the same extent that $\|\mathbf{k}\|^4$ in the continuum spectral density (7.16) can offset negative values due to $\eta_1 \, \|\mathbf{k}\|^2$.

Proof of permissibility condition Let us define the following ratios

$$\psi_i = \frac{4\xi^2}{a_i^2} \sin^2\left(\frac{a_i\, k_i}{2}\right), \text{ so that } \psi_i \in \left[0, \frac{4\xi^2}{a_i^2}\right], \text{ for } i = 1, \ldots, d.$$

Then, the permissibility condition is expressed as

$$\phi(\boldsymbol{\psi}) = 1 + \eta_1 \sum_{i=1}^{d} \psi_i + \sum_{i=1}^{d} \psi_i^2 > 0, \text{ where } \boldsymbol{\psi} = (\psi_1, \ldots \psi_d)^\top .$$

Since $\psi_i \geq 0$, it holds that $\phi(\boldsymbol{\psi}) > 0$ if $\eta_1 \geq 0$. In the case of $\eta_1 < 0$, for $\phi(\boldsymbol{\psi}) > 0$ to hold it is necessary and sufficient that

$$\phi(\boldsymbol{\psi}^*) > 0, \text{ where } \boldsymbol{\psi}^* = \arg\min_{\boldsymbol{\psi}} \phi(\boldsymbol{\psi}).$$

In order to find the stationary points of $\phi(\boldsymbol{\psi})$ we need to solve the equations $\partial\phi(\boldsymbol{\psi})/\partial\psi_i = 0$, for all $i = 1, \ldots, d$. These equations lead by straightforward calculations to $\psi_i = -\eta_1/2$.

It is easy to verify that the stationary point $\boldsymbol{\psi}^* = -\eta_1/2\,(1, \ldots 1)^\top$ corresponds to a minimum since $\partial^2\phi(\boldsymbol{\psi})/\partial\psi_i\partial\psi_j = 2\delta_{i,j}$ and thus the Hessian is a positive definite matrix.

The value $\phi(\boldsymbol{\psi}^*)$ at the minimum is given by

$$\phi(\boldsymbol{\psi}^*) = 1 - \frac{d\,{\eta_1}^2}{4}.$$

Hence, it follows that $\phi(\boldsymbol{\psi}^*) > 0$ if $|\eta_1| < 2/\sqrt{d}$, which, given that $\eta_1 < 0$, leads to (8.17).

The spectral density (8.16) is a *partially isotropic function*, in the sense that the dependence of $\widetilde{C}_{\mathrm{xx}}^{(\mathrm{latt})}(\mathbf{k})$ is the same on all k_i, $i = 1, \ldots, d$. However, $\widetilde{C}_{\mathrm{xx}}^{(\mathrm{latt})}(\mathbf{k})$ is not a radial function of the wavenumber $\|\mathbf{k}\|$. We have discussed partial isotropy in Chap. 3 in connection with the cardinal sine covariance function.

For $k_i a_i \ll 1$ (for $i = 1, \ldots, d$) it holds that

$$4\,\sin^2\left(\frac{a_i\, k_i}{2}\right) \approx (a_i k_i)^2, \text{ and } 16\,\sin^4\left(\frac{a_i\, k_i}{2}\right) \approx (a_i k_i)^4.$$

Hence, in the small wavevector limit, the spectral density can be approximated by the isotropic SSRF spectral density (7.16)

$$\widetilde{C}_{\mathrm{xx}}^{(\mathrm{latt})}(\mathbf{k}) \approx \widetilde{C}_{\mathrm{xx}}(\mathbf{k}) = \frac{\eta_0\, \xi^d}{1 + \eta_1\, (\|\mathbf{k}\|\, \xi)^2 + (\|\mathbf{k}\|\, \xi)^4}. \tag{8.18}$$

To gain insight into how the spectral density (8.16) is obtained from the energy functional (8.11), consider the following example.

Example 8.1 Find an equivalent expression in the spectral domain for the sum of the squared increments

$$S_1 = \sum_{n=1}^{N} \sum_{j=1}^{d} \left[x(\mathbf{s}_n + a_j \hat{\mathbf{e}}_j) - x(\mathbf{s}_n) \right]^2,$$

that appears in the energy functional (8.11).

Answer Using the inverse lattice Fourier transform (8.15b), and replacing $\widetilde{x}_{\mathrm{latt}}(\mathbf{k})$ with $\widetilde{x}(\mathbf{k})$ for brevity, the squared field increment is expressed as follows

$$S_1 = \frac{1}{(2\pi)^{2d}} \sum_{n=1}^{N} \sum_{j=1}^{d} \int_{\mathrm{FBZ}} \mathrm{d}\mathbf{k} \int_{\mathrm{FBZ}} \mathrm{d}\mathbf{k}' \, \widetilde{x}(\mathbf{k}) \, \widetilde{x}(\mathbf{k}') \, \mathrm{e}^{\mathrm{i}\,(\mathbf{k}+\mathbf{k}')\cdot \mathbf{s}_n}$$
$$\times \left(\mathrm{e}^{\mathrm{i}\, a_j \, \mathbf{k}\cdot\hat{\mathbf{e}}_j} - 1 \right) \left(\mathrm{e}^{\mathrm{i}\, a_j \, \mathbf{k}'\cdot\hat{\mathbf{e}}_j} - 1 \right).$$

The summation over the lattice sites affects the exponential terms $\exp(\mathrm{i}\,(\mathbf{k}+\mathbf{k}')\cdot\mathbf{s}_n)$. If $\mathbf{k} = -\mathbf{k}'$ the N terms add coherently leading to a sum equal to N; however, if $\mathbf{k} \neq -\mathbf{k}'$ the sum is incoherent and tends to zero as $N \to \infty$. Therefore, the sum of exponentials at the continuum limit is given by [122, p. 101]

$$\sum_{n=1}^{N} \mathrm{e}^{\mathrm{i}\,(\mathbf{k}+\mathbf{k}')\cdot\mathbf{s}_n} = N \, \delta_{\mathbf{k}+\mathbf{k}',\mathbf{0}} \underset{N\to\infty}{\rightarrow} \frac{(2\pi)^d}{\upsilon_c} \, \delta(\mathbf{k}+\mathbf{k}'),$$

where $\upsilon_c = V_d/N$ is the volume of the unit lattice cell and $\delta_{\mathbf{k}+\mathbf{k}',\mathbf{0}}$ is the Kronecker delta that is equal to one if $\mathbf{k}+\mathbf{k}' = \mathbf{0}$ and zero otherwise. Thus, the sum of the squared increments is given by

$$S_1 = \frac{1}{\upsilon_c (2\pi)^d} \int_{\mathrm{FBZ}} \mathrm{d}\mathbf{k} \, \widetilde{x}(-\mathbf{k}) \sum_{j=1}^{d} \left(2 - \mathrm{e}^{\mathrm{i}\, a_j \, \mathbf{k}\cdot\hat{\mathbf{e}}_j} - \mathrm{e}^{-\mathrm{i}\, a_j \, \mathbf{k}\cdot\hat{\mathbf{e}}_j} \right) \widetilde{x}(\mathbf{k})$$
$$= \frac{2}{\upsilon_c (2\pi)^d} \int_{\mathrm{FBZ}} \mathrm{d}\mathbf{k} \, \widetilde{x}^{\dagger}(\mathbf{k}) \sum_{j=1}^{d} \left[1 - \cos(k_j a_j) \right] \widetilde{x}(\mathbf{k}),$$
$$= \frac{4}{\upsilon_c (2\pi)^d} \int_{\mathrm{FBZ}} \mathrm{d}\mathbf{k} \, \widetilde{x}^{\dagger}(\mathbf{k}) \sum_{j=1}^{d} \sin^2\left(\frac{k_j a_j}{2} \right) \widetilde{x}(\mathbf{k}).$$

We can similarly work out the other terms in (8.11). The spectral representation of the energy function is thus given by

$$\mathcal{H}_0 = \frac{1}{2\upsilon_c(2\pi)^d} \int_{\text{FBZ}} \mathrm{d}\mathbf{k}\, \widetilde{x}^{\dagger}(\mathbf{k}) \left[\widetilde{C}_{\text{xx}}^{(\text{latt})}(\mathbf{k})\right]^{-1} \widetilde{x}(\mathbf{k}). \tag{8.19}$$

The above equation is the reciprocal-space analogue of the real-space equation (7.10); the spectral density $\widetilde{C}_{\text{xx}}^{(\text{latt})}(\mathbf{k})$ is given by (8.16). In the above equations, the fact that $x(\mathbf{s})$ is real-valued implies that $\widetilde{x}^{\dagger}(\mathbf{k}) = \widetilde{x}(-\mathbf{k})$.

8.4 SSRF Lattice Moments

The estimation of the SSRF parameters can be accomplished by means of the methods mentioned in Sect. 8.2.2. Alternatively, the *method of moments* and *maximum entropy* (see Chap. 13) can also be used. These approaches require the calculation of expectations over the ensemble of SSRF states [137, 141, 406]. The expectations (used to construct the joint pdf in the maximum entropy framework), include the squared gradient and curvature.

We will next calculate the expected squared gradient and curvature calculations for lattice SSRFs. We will assume that the field $X(\mathbf{s};\omega)$ is sampled on a hypercubic lattice $\mathcal{G}$ in d dimensions with uniform lattice step a. Both the gradient and the curvature can be expressed in terms of random field increments, which will be assumed *statistically isotropic*. The assumption of stationary increments allows us to also consider *intrinsic random fields*. Extensions of these expectations to anisotropic dependence are also possible and left as an exercise for the reader.

In the following, $\gamma_{\text{xx}}(\cdot)$ denotes the radial variogram function. The lowest-order *discretized gradient* and *Laplacian* are respectively given by (8.10a) and (8.10b). The expectations of the respective squares are given by means of the following expressions

$$\mathcal{E}\left\{[\nabla_a X(\mathbf{s};\omega)]^2\right\} = \frac{2d}{a^2}\, \gamma_{\text{xx}}(a), \tag{8.20}$$

$$\mathcal{E}\left\{[\nabla_a^2 X(\mathbf{s};\omega)]^2\right\} = \frac{1}{a^4}\left[8d^2\, \gamma_{\text{xx}}(a) - 4d(d-1)\, \gamma_{\text{xx}}(\sqrt{2}a) - 2d\, \gamma_{\text{xx}}(2a)\right]. \tag{8.21}$$

The second term in (8.21) results from the coupling of the field values in orthogonal directions. Hence, this term should vanish in $d = 1$, which is indeed ensured by the coefficient $d - 1$.

Lattice SSRF variogram An exact expression is not available for the lattice SSRF variogram, in contrast with the continuum SSRF variograms. For lags $\gg a$ the lattice variogram is practically the same as the continuum variogram, because the lattice structure mainly affects the short-distance behavior. This argument is also supported by the near-isotropy of the lattice SSRF spectral density (8.16) in the short-wavevector (long spatial lag) regime.

The value of the lattice SSRF variogram at lags equal to a is influenced by (i) the cutoff of the spectral density at high wavenumbers due to the finite lattice step and (ii) the departure from full isotropy due to the lattice structure that breaks radial dependence for short-distance correlations. For stationary random fields with the lattice SSRF energy function, the covariance (or equivalently, the variogram function) can be obtained by numerical evaluation of the respective spectral integral using the lattice spectral density (8.16) [362].

Mean squared gradient and curvature expressions We demonstrate the derivation of (8.20) and (8.21). Equation (8.20) is straightforward considering that

$$\mathcal{E}\left\{[\nabla_a X(\mathbf{s};\omega)]^2\right\} = \sum_{i=1}^{d} \frac{\mathcal{E}\left[X(\mathbf{s}+a\hat{\mathbf{e}}_i;\omega) - X(\mathbf{s};\omega)\right]^2}{a^2}.$$

Given the partial isotropy of the lattice variogram, the expectation of the squared increment in the numerator is equal to $2\gamma_{\mathrm{xx}}(a)$ for the orthogonal lattice directions $\{\hat{\mathbf{e}}_i\}_{i=1}^{d}$. This leads to the constant factor d in (8.20).

The proof of (8.21) is more complicated. The three terms involved arise as follows: The first term represents the expectation of squared collinear increments, the second term the expectation of orthogonal increment products, and the third term the expectation of products of collinear increments in opposite directions. The main steps of the calculations are outlined below.

Definition 8.4 Let $x(\mathbf{s})\colon \mathbb{R}^d \to \mathbb{R}$ denote a real-valued function and $m \in \mathbb{N}$ a positive integer. We denote by $\overline{\delta}_i^n x(\mathbf{s})$, $\delta_i^n x(\mathbf{s})$ and $\breve{\delta}_i^n x(\mathbf{s})$ respectively the mth-order *forward, central and backward* finite differences of $x(\mathbf{s})$ in the direction $\hat{\mathbf{e}}_i$ $(i = 1, \ldots, d)$. These differences are defined as follows

$$\overline{\delta}_i^m x(\mathbf{s}) = \sum_{k=0}^{m} (-1)^k \binom{m}{k} x\big(\mathbf{s} + (m-k)a\,\hat{\mathbf{e}}_i\big), \tag{8.22a}$$

$$\delta_i^m x(\mathbf{s}) = \sum_{k=0}^{m} (-1)^k \binom{m}{k} x\big(\mathbf{s} + (m/2-k)a\,\hat{\mathbf{e}}_i\big), \tag{8.22b}$$

$$\breve{\delta}_i^m x(\mathbf{s}) = \sum_{k=0}^{m} (-1)^k \binom{m}{k} x\big(\mathbf{s} - (m-k)a\,\hat{\mathbf{e}}_i\big). \tag{8.22c}$$

We assume a lattice random field $X(\mathbf{s}; \omega)$ where $\mathbf{s} \in \mathcal{G}$. For the sake of simplicity we also assume that $X(\mathbf{s}; \omega)$ is a *statistically isotropic random field* or an *intrinsic random field with isotropic increments*. We then define the following translation invariant expectation

$$\Phi(a) = \sum_{i=1}^{d} \sum_{j=1}^{d} \mathcal{E}[\, \overline{\delta}_i^1 X(\mathbf{s}; \omega) + \breve{\delta}_i^1 X(\mathbf{s}; \omega) \,] \, [\, \overline{\delta}_j^1 X(\mathbf{s}; \omega) + \breve{\delta}_j^1 X(\mathbf{s}; \omega) \,]$$

$$= \sum_{i=1}^{d} \sum_{j=1}^{d} \mathcal{E}[\, \overline{\delta}_i^1 X(\mathbf{s}; \omega) \, \overline{\delta}_j^1 X(\mathbf{s}; \omega)] + \mathcal{E}[\, \breve{\delta}_i^1 X(\mathbf{s}; \omega) \, \breve{\delta}_j^1 X(\mathbf{s}; \omega) \,]$$

$$+ 2\mathcal{E}[\, \overline{\delta}_i^1 X(\mathbf{s}; \omega) \, \breve{\delta}_j^1 X(\mathbf{s}; \omega) \,]. \tag{8.23}$$

Based on (8.22a) and (8.22c) the first-order forward and backward finite differences are given by

$$\overline{\delta}_i^1 X(\mathbf{s}; \omega) = X\left(\mathbf{s} + a\, \hat{\mathbf{e}}_i\right) - X(\mathbf{s}; \omega), \quad \breve{\delta}_i^1 X(\mathbf{s}; \omega) = X\left(\mathbf{s} - a\, \hat{\mathbf{e}}_i\right) - X(\mathbf{s}; \omega).$$

Then, using the definition of the discretized Laplacian (8.10b) and the definition of second-order differences according to (8.22), it follows that

$$\mathcal{E}[\{\nabla_a^2 X(\mathbf{s}; \omega)\}^2] = \frac{1}{a^4} \Phi(a).$$

The proof of (8.21) is completed using the identities in Table 8.1. The first three rows of the table involve relations for collinear increments, i.e., $i = j$, while the remaining three refer to increments in orthogonal directions, i.e., ($i \neq j$). The first two relations follow trivially from the variogram definition (3.44) and the isotropy of the increments. The remaining four relations follow from the *binomial identity*. The latter applies to the expectation of the product of two random variables $A(\omega)$ and $B(\omega)$

$$\mathcal{E}\,[A(\omega)\, B(\omega)] = \frac{1}{2} \left\{ \mathcal{E}[A^2(\omega)] + \mathcal{E}[B^2(\omega)] - \mathcal{E}\left[\{A(\omega) - B(\omega)\}^2 \right] \right\}.$$

Table 8.1 Expectations of increment products used in the calculation of $\Phi(a)$. The third column lists the multiplicity of each term, i.e., the number of times that each term appears in the expression (8.23) for $\Phi(a)$. For collinear increments it holds that $i = j$ while for increments in orthogonal directions $i \neq j$

Expectation	Result	Collinearity	Multiplicity
$\mathcal{E}\left[\overline{\delta}_i^1 X(\mathbf{s}; \omega)\, \overline{\delta}_i^1 X(\mathbf{s}; \omega)\right]$	$2\gamma_{xx}(a)$	$i = j$	d
$\mathcal{E}\left[\breve{\delta}_i^1 X(\mathbf{s}; \omega)\, \breve{\delta}_i^1 X(\mathbf{s}; \omega)\right]$	$2\gamma_{xx}(a)$	$i = j$	d
$\mathcal{E}\left[\overline{\delta}_i^1 X(\mathbf{s}; \omega)\, \breve{\delta}_i^1 X(\mathbf{s}; \omega)\right]$	$2\gamma_{xx}(a) - \gamma_{xx}(2a)$	$i = j$	$2d$
$\mathcal{E}\left[\overline{\delta}_i^1 X(\mathbf{s}; \omega)\, \overline{\delta}_j^1 X(\mathbf{s}; \omega)\right]$	$2\gamma_{xx}(a) - \gamma_{xx}(a\sqrt{2})$	$i \neq j$	$d(d-1)$
$\mathcal{E}\left[\breve{\delta}_i^1 X(\mathbf{s}; \omega)\, \breve{\delta}_j^1 X(\mathbf{s}; \omega)\right]$	$2\gamma_{xx}(a) - \gamma_{xx}(a\sqrt{2})$	$i \neq j$	$d(d-1)$
$\mathcal{E}\left[\overline{\delta}_i^1 X(\mathbf{s}; \omega)\, \breve{\delta}_j^1 X(\mathbf{s}; \omega)\right]$	$2\gamma_{xx}(a) - \gamma_{xx}(a\sqrt{2})$	$i \neq j$	$2d(d-1)$

In Chap. 5 we presented continuum expressions for the mean squared gradient and curvature of differentiable random fields. The aforementioned equations, given by (5.32) and (5.33), are equivalent to the variance of the gradient and the curvature. The derivation of (5.32) and (5.33) is based, respectively, on (8.20) and (8.21) and the *Taylor expansion of the variogram* around zero, i.e.,

$$\gamma_{xx}(a) = \frac{a^2}{2}\,\gamma_{xx}^{(2)}(0) + \frac{a^4}{4!}\,\gamma_{xx}^{(4)}(0) + O(a^6). \tag{8.24}$$

In the Taylor expansion of the variogram, we denote by $\gamma_{xx}^{(n)}(0)$ the n-th order radial derivative at zero lag. We also use (i) $\gamma_{xx}(0) = 0$ and (ii) the fact that odd-order radial derivatives of the variogram at zero lag vanish.

8.5 SSRF Inverse Covariance Operator on Lattices

In the preceding section, we focused on discrete versions of SSRF functionals based on a particular discretization of the gradient and curvature terms. We can, however, derive more than one lattice models from the same continuum SSRF model, since the derivatives in the energy can be approximated as an infinite series of finite differences that can be truncated at some specified order. We discuss this issue in more detail using tools from numerical analysis.

There is an ulterior motive in this exercise. In the continuum, we have derived explicit forms for the SSRF covariance functions in one, two, and three dimensions. We can use these expressions to calculate the SSRF covariance matrix of lattice data. However, on lattices we can also obtain explicit expressions for the precision kernel using *discrete approximations* of the continuum SSRF model (7.11). An intriguing question is whether and under what conditions explicit expressions are possible for both the covariance (derived from the continuum SSRF expressions) and the precision matrix (derived from the discrete approximations).

In both the isotropic (8.11) and the anisotropic (8.13) cases the energy functional $\mathcal{H}_0$ is a quadratic function of the field realizations $\mathbf{x}$. Hence, $\mathcal{H}_0$ can also be expressed in terms of the *precision matrix* $\mathbf{J}$ so that

$$\mathcal{H}_0(\mathbf{x}) = \frac{1}{2}\mathbf{x}^\top\,\mathbf{J}\,\mathbf{x}. \tag{8.25}$$

The precision matrix is necessary in parameter estimation using the method of maximum likelihood and for optimal (kriging) prediction at unmeasured points (see Chap. 10). Knowing both the covariance matrix and its inverse (the precision matrix) has significant practical implications: in such cases the computational cost associated with the numerical inversion of large, dense covariance matrices can be avoided. The numerical inversion of dense covariance matrices scales poorly with size as $O(N^3)$ [784]. Methods for approximating the covariance inverse without full inversion were investigated in [625].

Let us begin with the inverse SSRF covariance operator (7.11) in continuum space. This expression is repeated below for easy reference:

$$J(\mathbf{s}-\mathbf{s}') = \frac{1}{\eta_0 \xi^d}\left[1 - \eta_1 \xi^2 \Delta_{\mathbf{s}} + \xi^4 \, \Delta_{\mathbf{s}}^2\right] \delta(\mathbf{s}-\mathbf{s}'), \tag{8.26}$$

where $\Delta_{\mathbf{s}}$ and $\Delta_{\mathbf{s}}^2$ are, respectively, the Laplace and biharmonic operators acting at $\mathbf{s}$. On regular lattices, the above local operators are expressed as *sparse matrices* using finite-difference approximations.

8.5.1 Low-Order Discretization Schemes

A low-order discretization of the SSRF precision operator implies a low-order approximation of the derivatives in the energy functional in terms of finite differences, such as those used in (8.11) and the anisotropic (8.13).

A discretized approximation of the precision operator $J(\mathbf{s}-\mathbf{s}')$ (8.26) on a regular grid is obtained by means of the $N \times N$ precision matrix $\hat{\mathbf{J}}$ as follows

$$\hat{J}_{m,n} = \frac{\upsilon}{\eta_0 \xi^2}\left[\delta_{m,n} - \eta_1 \xi^2 [\hat{\Delta}]_{m,n} + \xi^4 \, [\hat{\Delta}^2]_{m,n}\right], \; m,n = 1,\ldots,N, \tag{8.27}$$

where υ is a normalizing constant, while $[\hat{\Delta}]_{m,n}$ and $[\hat{\Delta}^2]_{m,n}$ are, respectively, discrete (matrix) approximations of the Laplace and biharmonic operators. The indices m, n refer to the lattice sites $\mathbf{s}_n$ and $\mathbf{s}_m$. The discretized Laplace and biharmonic operators on square grids with $N = L \times L$ nodes correspond to $N \times N$ matrices.

Labeling of grid nodes For *2D square grids* $\mathcal{G}$ with step a and nodes $\mathbf{s}_n$, $n = 1,\ldots,N$, we can also use the *row-column* notation, in which $n \to (k_n, l_n)$, where $k_n, l_n = 1,\ldots,L$ are, respectively, the row and column indices of the site $\mathbf{s}_n$. To simplify notation, we use (k,l) instead of (k_n, l_n), and we denote by $x_{k,l}$ the field value $x(\mathbf{s}_{k,l})$ at the point $\mathbf{s}_{k,l}$.

2D Discrete Laplacian The lowest-order discretization is the *five-point discretization of the Laplacian* which is given by the following expression [4, p. 885]:

$$\hat{\Delta} x_{k,l} = \frac{x_{k+1,l} + x_{k,l+1} + x_{k-1,l} + x_{k-1,l} - 4x_{k,l}}{a^2}. \tag{8.28}$$

The approximation above is of $O(a^2)$, i.e., $\Delta x(\mathbf{s}_{k,l}) = \hat{\Delta} x_{k,l} + O(a^2)$. An $O(a^4)$ nine-point approximation of the Laplacian is also given in [4, p. 885].

The discretization (8.28) is expressed by means of the *five-point stencil* shown schematically in Fig. 8.3. A stencil is a group of nodes surrounding the target node $\mathbf{s}_{k,l}$ that lies at the center of the stencil. Each node is assigned a coefficient that determines the linear weight with which the specific node contributes in the Laplacian matrix.

The five-point scheme is valid at the *interior* lattice points $\mathbf{s}_{k,l}$ that do not lie along the lattice boundary. The expression for the discrete Laplacian at the boundary points depends on the boundary conditions used. If we assume *open boundary conditions*, the Laplacian at the boundary lattice nodes is calculated from (8.28) by setting equal to zero points $x_{k\pm1,l\pm1}$ that lie outside the lattice. If *periodic* (toroidal) boundary conditions are used, the lattice is supposed to close on itself like a doughnut. Hence, the left-hand (downward) nearest neighbors of the nodes on the left (bottom) boundary are the nodes on the same row (column) of the right (top) boundary.

Example of Laplacian matrix The non-zero elements of the Laplacian matrix for a 4×4 lattice are shown in the schematic of Fig. 8.4. All the entries along the main diagonal are equal to -4, while all non-zero entries along the superdiagonal, the subdiagonal, as well as the $(L + 1)$ diagonal, $-(L + 1)$ diagonal are equal to 1. The diagonals right above and below the main diagonal contain contributions from adjacent nodes along the vertical direction, while the two $(L + 1)$ and $-(L + 1)$

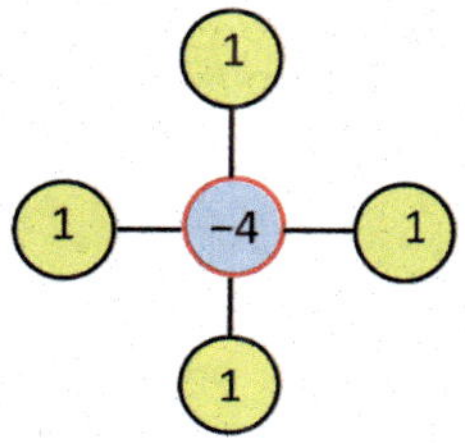

Fig. 8.3 Five-point stencil for the two-dimensional Laplacian based on the leading (second-order) central difference approximation given by (8.28)

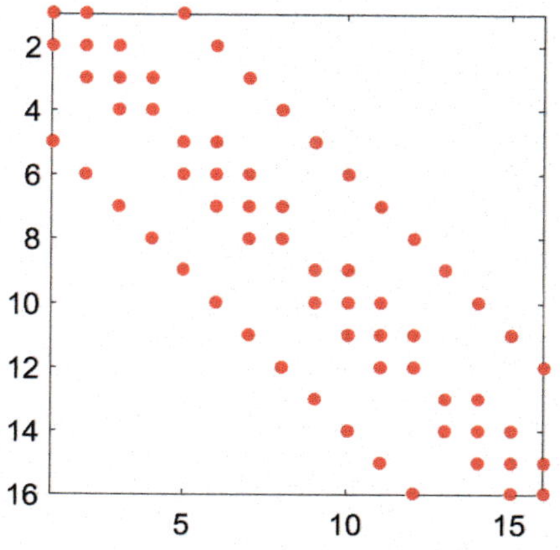

Fig. 8.4 Schematic representation of non-zero elements (dots) of the discretized Laplace operator (five-point stencil) for a 4×4 square lattice with open boundary conditions. The axes labels correspond to the column and row indices. The elements of the main diagonal are equal to -4 while the entries of the other four diagonals are equal to one

diagonals correspond to adjacent nodes along the horizontal direction. Two nodes are considered to be *adjacent* if they are at opposite ends of the same edge.

2D Discrete Bilaplacian The discretized biharmonic operator can be obtained by applying the Laplacian operator twice, i.e.,

$$\hat{\Delta}^2 x_{k,l} = \frac{1}{a^4}\,\Delta^{(4)} x_{k,l}. \tag{8.29a}$$

The fourth-order difference operator $\Delta^{(4)} x_{k,l}$ is given to leading order approximation by the following *thirteen-point* stencil [4, p. 885], [395, p. 165]

$$\begin{aligned}\Delta^{(4)} x_{k,l} = & \, x_{k,l+2} + x_{k,l-2} + x_{k-2,l} + x_{k+2,l} \\ & - 8\left(x_{k,l+1} + x_{k,l-1} + x_{k+1,l} + x_{k-1,l}\right) \\ & + 2\left(x_{k+1,l+1} + x_{k-1,l+1} + x_{k+1,l-1} + x_{k-1,l-1}\right) + 20 x_{k,l}.\end{aligned} \tag{8.29b}$$

The thirteen-point stencil with the weights defined in (8.29b) is shown in Fig. 8.5. The Bilaplacian $\hat{\Delta}^2 x_{k,l}$ obtained by means of $\Delta^{(4)} x_{k,l}$ is an $O(a^2)$ discretization scheme in the sense that $\Delta^2 x_{k,l} = \hat{\Delta}^2 x_{k,l} + O(a^2)$. An $O(a^4)$ discretization scheme can be obtained using a twenty-five-point stencil (not shown here).

Example 8.2 Prove (8.29b) using the definition of the discrete Laplacian (8.28) and the three-point discretization of second-order partial derivatives given by

$$\begin{aligned}\delta_1^2 x(\mathbf{s}_{k,l}) =& x(\mathbf{s}_{k,l+1}) + x(\mathbf{s}_{k,l-1}) - 2x(\mathbf{s}_{k,l}) \\ \delta_2^2 x(\mathbf{s}_{k,l}) =& x(\mathbf{s}_{k+1,l}) + x(\mathbf{s}_{k-1,l}) - 2x(\mathbf{s}_{k,l}).\end{aligned}$$

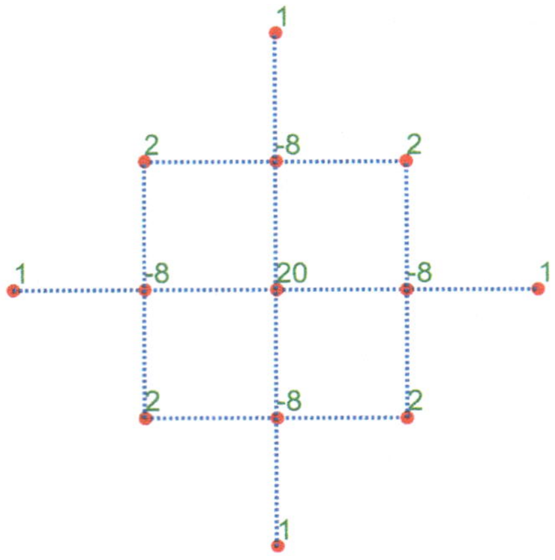

Fig. 8.5 Thirteen-point stencil for the two-dimensional biharmonic operator based on the leading (second-order) central difference approximation of the Laplacian. The nodal weights are as defined in (8.29b)

Answer: The discretized Laplace operator in two dimensions is expressed as $\hat{\Delta} = \left(\delta_1^2 x(\mathbf{s}) + \delta_2^2 x(\mathbf{s})\right)/a^2$. The Bilaplacian is obtained by successive application of the Laplacian operator leading to

$$\hat{\Delta}^2 = \hat{\Delta}\hat{\Delta} = \frac{1}{a^4}\left(\delta_1^4 + \delta_2^4 + 2\delta_1^2\delta_2^2\right).$$

The second-order partial differences are defined above. The fourth-order partial differences and the product of the second-order partial differences are given by

$$\begin{aligned}
\delta_1^4 x(\mathbf{s}_{k,l}) =& x(\mathbf{s}_{k,l+2}) + x(\mathbf{s}_{k,l-2}) + 6x(\mathbf{s}_{k,l}) \\
& - 4x(\mathbf{s}_{k,l+1}) - 4x(\mathbf{s}_{k,l-1}), \\
\delta_2^4 x(\mathbf{s}_{k,l}) =& x(\mathbf{s}_{k+2,l}) + x(\mathbf{s}_{k-2,l}) + 6x(\mathbf{s}_{k,l}) \\
& - 4x(\mathbf{s}_{k+1,l}) - 4x(\mathbf{s}_{k-1,l}), \\
\delta_1^2\delta_2^2 x(\mathbf{s}_{k,l}) =& x(\mathbf{s}_{k+1,l+1}) + x(\mathbf{s}_{k-1,l+1}) + 4x(\mathbf{s}_{k,l}) \\
& + x(\mathbf{s}_{k+1,l-1}) + x(\mathbf{s}_{k-1,l-1}) - 2x(\mathbf{s}_{k,l+1}) \\
& - 2x(\mathbf{s}_{k,l-1}) - 2x(\mathbf{s}_{k+1,l}) - 2x(\mathbf{s}_{k-1,l}).
\end{aligned}$$

Adding the first two terms to the third term (the latter is multiplied by two) we obtain (8.29b).

We have presented above low-order discretization schemes for the Laplacian and the biharmonic operators on two-dimensional lattices. It is also possible to define *higher-order schemes*, as well as isotropic schemes in which the discretization error does not have preferred directions in space as shown in [655].

8.5.2 Higher-Order Discretization Schemes

In this section we use formal expansions of partial differential operators in terms of finite differences. This approach leads to higher-order discretization of the Laplacian and the Bilaplacian operators on d-dimensional lattices. The ultimate goal is to construct higher-order lattice-based models that closely approximate continuum SSRFs. More precisely, we would like to construct finite-difference-based approximations of the Laplacian and the Biharmonic operators such that the resulting precision matrix closely approximates the directly inverted SSRF covariance matrix (where the latter is derived from the continuum SSRF covariance function.)

Lattice partial differential operators First, we define the dimensionless lattice partial differential operators

$$\mathrm{D}_i x(\mathbf{s}) = a_i \, \frac{\partial x(\mathbf{s})}{\partial s_i},$$

where a_i, $i = 1, \ldots, d$, is the step in the i-th orthogonal direction of the lattice. The *unit vector* in the i-th direction is denoted by $\hat{\mathbf{e}}_i$.

The SSRF precision operator (8.26) is expressed in terms of the dimensionless partial differential operators as follows

$$J(\mathbf{s}-\mathbf{s}') = \frac{1}{\eta_0 \, \xi^d} \left[1 - \eta_1 \left(\sum_{i=1}^{d} \frac{\xi^2}{a_i^2} \mathrm{D}_i^2 \right) + \left(\sum_{i=1}^{d} \sum_{j=1}^{d} \frac{\xi^4}{a_i^2 \, a_j^2} \mathrm{D}_i^2 \, \mathrm{D}_j^2 \right) \right] \delta(\mathbf{s}-\mathbf{s}'). \tag{8.30}$$

On rectangular grids, the delta functions are replaced by Kronecker deltas according to the mapping $\delta(\mathbf{s}_i - \mathbf{s}_j) \mapsto \delta_{i,j}/\upsilon_c$, where $\upsilon_c = \prod_{i=1}^{d} a_i$ is the grid cell volume.

To obtain a lattice-based model that corresponds as closely as possible to continuum SSRFs, it is necessary to develop accurate lattice approximations for the differential operators D_i, $i = 1, \ldots, d$. We show investigate such approximations below. First, we introduce convenient notation for the representation of finite difference approximations. This generalizes the discussion on the labeling of the nodes on two-dimensional grids in Sect. 8.5.1.

Vector index notation For d-dimensional orthogonal lattices $\mathcal{G}$, the position *vector index* $\mathbf{n} = (n_1, \ldots, n_d)^{\top}$, can be used to label the lattice sites. The scalar indices n_i take integer values in the respective sets $\{1, 2, \ldots, L_i\}$, where $i = 1, \ldots, d$. In solid state physics this notation is known as the *Miller index* notation. The lattice positions $\mathbf{s}_n$, where $n = 1, \ldots, N$, are then determined by means of the vector

$$\mathbf{s_n} = (n_1 a_1, \; n_2 a_2, \; \ldots, n_d a_d)^{\top} = \sum_{i=1}^{d} n_i \, a_i \, \hat{\mathbf{e}}_i \,.$$

The value of the random field's realization at the grid node $\mathbf{s}_n$ is denoted by $x_{\mathbf{n}}$.

First-order central differences According to (8.22b) the central finite difference approximation of the first-order partial derivative is given by

$$\delta_i^1 \, x_{\mathbf{n}} = x_{\mathbf{n}+\hat{\mathbf{e}}_i/2} - x_{\mathbf{n}-\hat{\mathbf{e}}_i/2}, \tag{8.31a}$$

where $i = 1, \ldots, d$ and $x_{\mathbf{n}\pm\hat{\mathbf{e}}_i/2} = x(\mathbf{s_n} \pm a_i\hat{\mathbf{e}}_i/2)$ represent the values of the field at points located midway between neighboring nodes.

Second-order differences The second-order central difference $\delta_i^2 x_{\mathbf{n}} = \delta_i^1\{\delta_i^1 x_{\mathbf{n}}\}$ is similarly given by the following equation

$$\delta_i^2 x_{\mathbf{n}} = x_{\mathbf{n}+\hat{\mathbf{e}}_i} + x_{\mathbf{n}-\hat{\mathbf{e}}_i} - 2x_{\mathbf{n}}. \tag{8.31b}$$

Higher-order differences More generally, the difference of order $2p$ is given by the following iterative equation

$$\delta_i^{2p+2} x_{\mathbf{n}} = \delta_i^2\{\delta_i^{2p} x_{\mathbf{n}}\}, \tag{8.31c}$$

for all integer values of p. The central differences δ_i^{2p} of order $2p$ for the lattice function $x_{\mathbf{n}}$ can be evaluated by means of the following expansion [4, p. 877]

$$\delta_i^{2p} x_{\mathbf{n}} = \sum_{l=0}^{2p}(-1)^l \binom{2p}{l} x_{\mathbf{n}+(p-l)\,\hat{\mathbf{e}}_i} = \sum_{l=0}^{2p}(-1)^l \frac{(2p)!}{l!\,(2p-l)!} x_{\mathbf{n}+(p-l)\,\hat{\mathbf{e}}_i}\,. \tag{8.31d}$$

The respective equations for the differences of order $2p$, where $p = 1, 2, \ldots, 6$ are given in Table 8.2.

Formal expansion of differential operators in terms of finite differences The differential operator D_i has a discrete analogue in the *finite difference operator* $\hat{\mathrm{D}}_i$ such that $\hat{\mathrm{D}}_i \to \mathrm{D}_i$ as $a_i \to 0$. The discrete operator $\hat{\mathrm{D}}_i$ is linked to the central difference operator δ_i^1 by means of the following relation [347]

Table 8.2 Central finite differences of order $2p$ ($p = 1, 2, \ldots, 6$) along the orthogonal lattice direction i ($i = 1, \ldots, d$), of a hypercubic grid with uniform lattice step $a = 1$. The symbol $x_{\mathbf{n}} = x(\mathbf{s}_n)$ denotes random field values at the locations specified by the Miller vector index $\mathbf{n}$. F. D. stands for finite differences

F. D.	Lattice terms involved
$\delta_i^2 x_{\mathbf{n}} =$	$x_{\mathbf{n}+\hat{\mathbf{e}}_i} + x_{\mathbf{n}-\hat{\mathbf{e}}_i} - 2x_{\mathbf{n}}$
$\delta_i^4 x_{\mathbf{n}} =$	$x_{\mathbf{n}+2\hat{\mathbf{e}}_i} + x_{\mathbf{n}-2\hat{\mathbf{e}}_i} - 4x_{\mathbf{n}+\hat{\mathbf{e}}_i} - 4x_{\mathbf{n}-\hat{\mathbf{e}}_i} + 6x_{\mathbf{n}}$
$\delta_i^6 x_{\mathbf{n}} =$	$x_{\mathbf{n}+3\hat{\mathbf{e}}_i} + x_{\mathbf{n}-3\hat{\mathbf{e}}_i} - 6x_{\mathbf{n}+2\hat{\mathbf{e}}_i} - 6x_{\mathbf{n}-2\hat{\mathbf{e}}_i} + 15x_{\mathbf{n}+\hat{\mathbf{e}}_i} + 15x_{\mathbf{n}-\hat{\mathbf{e}}_i} - 20x_{\mathbf{n}}$
$\delta_i^8 x_{\mathbf{n}} =$	$x_{\mathbf{n}+4\hat{\mathbf{e}}_i} + x_{\mathbf{n}-4\hat{\mathbf{e}}_i} - 8x_{\mathbf{n}+3\hat{\mathbf{e}}_i} - 8x_{\mathbf{n}-3\hat{\mathbf{e}}_i} + 28x_{\mathbf{n}+2\hat{\mathbf{e}}_i} + 28x_{\mathbf{n}-2\hat{\mathbf{e}}_i} - 56x_{\mathbf{n}+\hat{\mathbf{e}}_i}$ $-56x_{\mathbf{n}-\hat{\mathbf{e}}_i} + 70x_{\mathbf{n}}$
$\delta_i^{10} x_{\mathbf{n}} =$	$x_{\mathbf{n}+5\hat{\mathbf{e}}_i} + x_{\mathbf{n}-5\hat{\mathbf{e}}_i} - 10x_{\mathbf{n}+4\hat{\mathbf{e}}_i} - 10x_{\mathbf{n}-4\hat{\mathbf{e}}_i} + 45x_{\mathbf{n}+3\hat{\mathbf{e}}_i} + 45x_{\mathbf{n}-3\hat{\mathbf{e}}_i} - 120x_{\mathbf{n}+2\hat{\mathbf{e}}_i}$ $-120x_{\mathbf{n}-2\hat{\mathbf{e}}_i} + 210x_{\mathbf{n}+\hat{\mathbf{e}}_i} + 210x_{\mathbf{n}-\hat{\mathbf{e}}_i} - 252x_{\mathbf{n}}$
$\delta_i^{12} x_{\mathbf{n}} =$	$x_{\mathbf{n}+6\hat{\mathbf{e}}_i} + x_{\mathbf{n}-6\hat{\mathbf{e}}_i} - 12x_{n_i+5,n_j} - 12x_{n_i-5,n_j} + 66x_{\mathbf{n}+4\hat{\mathbf{e}}_i} + 66x_{\mathbf{n}-4\hat{\mathbf{e}}_i} - 220x_{\mathbf{n}+3\hat{\mathbf{e}}_i}$ $-220x_{\mathbf{n}-3\hat{\mathbf{e}}_i} + 495x_{\mathbf{n}+2\hat{\mathbf{e}}_i} + 495x_{\mathbf{n}-2\hat{\mathbf{e}}_i} - 792x_{\mathbf{n}+\hat{\mathbf{e}}_i} - 792x_{\mathbf{n}-\hat{\mathbf{e}}_i} + 924x_{\mathbf{n}}$

$$\mathrm{D}_i \to \hat{\mathrm{D}}_i \equiv \frac{2}{a_i} \sinh^{-1}\left(\frac{\delta_i}{2}\right). \tag{8.32}$$

Equation (8.32) is a symbolic expression that leads to Taylor series expansions for partial finite differences (e.g., the second-order, $\hat{\mathrm{D}}_i^2$, and the fourth-order, $\hat{\mathrm{D}}_i^4$), in terms of δ_i^1. These series expansions are given by the following equations

$$a_i^2\, \hat{\mathrm{D}}_i^2 = \delta_i^2 - \frac{\delta_i^4}{12} + \frac{\delta_i^6}{90} - \frac{\delta_i^8}{560} + \frac{\delta_i^{10}}{3150} - \frac{\delta_i^{12}}{16632} + O\left(\delta_i^{14}\right), \tag{8.33a}$$

$$a_i^4\, \hat{\mathrm{D}}_i^4 = \delta_i^4 - \frac{\delta_i^6}{6} + \frac{7\delta_i^8}{240} - \frac{41\delta_i^{10}}{7560} + \frac{479\delta_i^{12}}{453600} + O\left(\delta_i^{14}\right). \tag{8.33b}$$

Lattice expansions of the Laplacian and Bilaplacian operators Based on (8.30), the SSRF precision operator on a hypercubic grid of unit length is given by

$$\mathbf{J} = \frac{1}{\eta_0\, \xi^d} \left(\mathbf{I} - \eta_1\, \xi^2\, \mathcal{L} + \xi^4\, \mathcal{B}\right), \tag{8.34}$$

where $\mathbf{I}$ is the identity matrix, $\mathcal{L}$ and $\mathcal{B}$ represent the lattice expansions of the Laplacian and the biharmonic operators respectively. These lattice expansions in terms of the partial difference operators are given by

$$\mathcal{L} = \sum_{i=1}^{d} \hat{\mathrm{D}}_i^2, \tag{8.35}$$

$$\mathcal{B} = \sum_{i=1}^{d} \hat{\mathrm{D}}_i^4 + \sum_{i=1}^{d} \sum_{j=1,\neq i}^{d} \hat{\mathrm{D}}_i^2\, \hat{\mathrm{D}}_j^2. \tag{8.36}$$

Using the identity (8.32) that connects the differential and the partial difference operators, $\mathcal{L}$ and $\mathcal{B}$ can be expanded as infinite series in terms of the central different operator δ_i^1. By truncating these series at some arbitrary order, e.g., $2p = 12$, we obtain respective lattice expansions of the *Laplacian* and *Bilaplacian* operators

$$\mathcal{L}^{(12)} = \sum_{i=1}^{d} \left[\delta_i^2 - \frac{\delta_i^4}{12} + \frac{\delta_i^6}{90} - \frac{\delta_i^8}{560} + \frac{\delta_i^{10}}{3150} - \frac{\delta_i^{12}}{16632}\right], \tag{8.37}$$

$$\mathcal{B}^{(12)} = \sum_{i=1}^{d} \left(\delta_i^4 - \frac{\delta_i^6}{6} + \frac{7\,\delta_i^8}{240} - \frac{41\,\delta_i^{10}}{7560} + \frac{479\,\delta_i^{12}}{453600} \right)$$
$$+ \sum_{i=1}^{d} \sum_{j=1,\neq i}^{d} \left(\delta_i^2 \delta_j^2 - \frac{\delta_i^2\,\delta_j^4}{6} + \frac{\delta_i^2 \delta_j^6}{45} - \frac{\delta_i^2 \delta_j^8}{280} + \frac{\delta_i^2 \delta_j^{10}}{1575} \right.$$
$$\left. + \frac{\delta_i^4\,\delta_j^4}{144} - \frac{\delta_i^4\,\delta_j^6}{540} + \frac{\delta_i^4\,\delta_j^8}{3360} + \frac{\delta_i^6\,\delta_j^6}{8100} \right). \tag{8.38}$$

The $(2d+1)$-point discretization of the *Laplacian operator* $\triangle$ on a d-dimensional grid is given by the first term of the series (8.37). By taking into account Table 8.2, this leads to

$$\mathcal{L}^{(2)} = \hat{\triangle} = \sum_{i=1}^{d} \delta_i^2. \tag{8.39}$$

Based on (8.31b), the *leading-order* expansion of the *Laplace operator* is given by the following $N \times N$ symmetric matrix [395]

$$\hat{\triangle}_{n,m} = \begin{cases} -2d, \text{ if } n = m, \\ 1, \text{ if } \mathbf{s}_n \neq \mathbf{s}_m \text{ but } \mathbf{s}_n \text{ and } \mathbf{s}_m \text{ are adjacent} \\ 0, \text{ otherwise.} \end{cases} \tag{8.40}$$

The above relations can also be expressed in terms of a suitably sized stencil. In two dimensions this is illustrated by means of the five-point stencil shown in Fig. 8.3.

Similarly, the leading-order expansion of the *biharmonic operator* $\mathcal{B}$ is given by the following expression

$$\mathcal{B}^{(4)} = \sum_{i=1}^{d} \delta_i^4 + \sum_{i=1}^{d} \sum_{j=1,\neq i}^{d} \delta_i^2\,\delta_j^2. \tag{8.41}$$

In two dimensions, this leads to the thirteen-point biharmonic stencil shown in Fig. 8.5.

Comparison of precision matrix with finite-difference approximations We compare the inverse SSRF covariance kernel as obtained by direct numerical inversion of the continuum SSRF covariance function with the lattice approximation of the precision kernel obtained by means of the central finite difference scheme (as explained above).

For simplicity we use the one-dimensional SSRF covariance function for $\eta_1 = 2$ according to (7.27b), i.e., the modified exponential function

$$C_{xx}(h) = \frac{\eta_0}{4}(1 + h)\, e^{-h}.$$

We assume a $d = 1$ SSRF process $X(s;\omega)$ observed at L sites s_l, $l = 1, \ldots, L$. The covariance matrix is an $L \times L$ symmetric, positive-definite matrix whose values are obtained from the above modified exponential for the respective lags $h_{l,k} = |s_l - s_k|/\xi$. The *inverse covariance (precision)* matrix is evaluated by numerical inversion of the covariance matrix obtained from $C_{xx}(h)$.

Figure 8.6 compares the vector $C_{xx}^{-1}(L/2, k)$, for $k = 1, \ldots, L$, obtained by inverting the SSRF covariance matrix, with the *finite-difference-based approximations* $J^{(2p)}(L/2, k)$ $(p = 2, 4, 5, 6)$ of the inverse SSRF covariance kernel obtained from (8.34), (8.37), and (8.38). The expressions for the finite differences used in (8.37) and (8.38) are taken from Table 8.2.

The plots demonstrate the convergence of the finite-difference-based approximation of the inverse covariance to the precision matrix values (obtained by the numerical inversion of the covariance matrix) as the order of the approximation increases from six to twelve.

Based on the above, discrete lattice approximations of the continuum SSRF model seem feasible. However, further analysis is required to establish the accuracy of the approximations and the impact of the boundary conditions. In summary, achieving good agreement between the explicit precision matrix of the discrete (approximate) model and that obtained by numerical inversion of the continuum SSRF model's covariance requires including a high-order discrete model. Hence, eating the cake and having it at the same time (explicit expressions for both the SSRF covariance function and the precision matrix) is only approximately feasible.

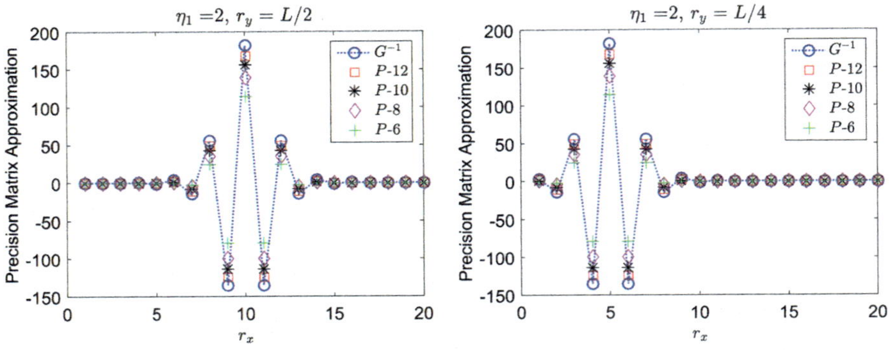

Fig. 8.6 Central finite difference approximations of the SSRF inverse kernel of order six (crosses), eight (diamonds), ten (triangles), and twelve (squares) versus the SSRF precision values (circles). The latter are evaluated by inversion of the $L \times L$ covariance matrix based on the $d = 1$ SSRF covariance function. A one-dimensional series of length $L = 20$ and unit step $a = 1$ is assumed. The SSRF parameters are $\eta_0 = 10$, $\eta_1 = 2$, and $\xi = 10$. The covariance function is given by (7.27b). The horizontal axis represents the precision matrix elements with row and column indices $(r_x, L/2)$ (left) and with indices $(r_x, L/4)$ (right)

Chapter 9
Spartan Random Fields and Langevin Equations

I don't know anything, but I do know that everything is interesting if you go into it deeply enough.

Richard Feynman

In this chapter we look at Spartan random fields from a different perspective. Our first goal is to show that solutions of linear stochastic partial differential (Langevin) equations are random fields with rational spectral densities [694]. In addition, the respective covariance function is the Green's function of a suitable (i.e., derivable from Langevin equation) partial differential equation. Finally, the joint dependence of random fields that satisfy Langevin equations driven by a Gaussian white noise process can be expressed in terms of an exponential Gibbs-Boltzmann pdf; the latter has a quadratic energy function that involves local (i.e., based on low-order field derivatives) terms.

First, we identify connections with seemingly unrelated topics, such as the *classical harmonic oscillator*, which is an archetypical model in physics. More precisely, we establish an equivalence between the one-dimensional SSRF and the classical damped harmonic oscillator driven by white noise. The resulting motion of the oscillator is represented by a stochastic differential equation (Langevin equation). Hence, we begin this chapter with a brief introduction to stochastic differential equations.

Following up on the equivalence of one-dimensional SSRFs with the harmonic oscillator, we demonstrate the connection between Spartan random fields in two and three dimensions and a specific *linear stochastic partial differential equation* (SPDE). This is analogous to the link between the Whittle-Matérn covariance function and the associated fractional Langevin equation described by Whittle in the

D. T. Hristopulos, *Random Fields for Spatial Data Modeling*, Advances in Geographic Information Science, https://doi.org/10.1007/978-94-024-1918-4_9

classic paper [846]. The connection between random fields and the stochastic partial differential equations in the Whittle-Matérn family is pursued in recent research by Håvard Rue, Finn Lindgren and coworkers [87, 503, 701]. Linear SPDEs that correspond to SSRFs [242, 360, 367, 372] and to polynomials of the diffusion operator [866] are also promising research directions.

We also highlight the equivalence between the SSRF covariance functions and the *Green functions* of associated deterministic partial differential equations. This equivalence is a property of covariance functions of random fields that are defined by means of SPDEs. However, not every Green function is a valid covariance function for some stationary random field. For example, the Green function of the Laplace equation is a generalized covariance. The correspondence between covariance functions of SPDEs and Green functions of associated PDEs has been noted in the physics literature [205, 289].

Another interesting connection is that random fields which are obtained as solutions of linear SPDEs driven by white Gaussian noise can also be expressed in terms of Boltzmann-Gibbs joint probability density functions. This means that the Spartan random fields presented in Chap. 7 can also be expressed by means of an associated SPDE.

The chapter closes with an excursion in time series modeling. This gives us the opportunity to explore the connection between one-dimensional Spartan random fields and *second-order autoregressive processes*. We have commented in Chap. 8 on the connections between lattice SSRFs in higher dimensions and Gaussian Markov random fields, as well as between the latter and conditional autoregressive models.

9.1 Introduction to Stochastic Differential Equations (SPDEs)

Stochastic Differential Equations (SDEs) describe the dynamic evolution of a scalar, $X(t; \omega)$, or vector, $\mathbf{X}(t; \omega)$, observable in time under the action of forces that involve a stochastic (random) component. Hence, $X(t; \omega)$ is mathematically described by a random process. The stochastic forces can be either internal or external (e.g., ambient noise) to the system described by the random observable. The SDEs are also known as Langevin equations in honor of the French physicist who introduced them to describe Brownian motion. Physics-based introductions to SDEs are given in [402, 471, 903], while a more mathematical introduction is given in [621].

The best known stochastic differential equations represent *Brownian motion* (see Sect. 1.4.4) and the *Ornstein-Uhlenbeck process*. They are both defined by means of first-order differential equations:

Brownian motion: $$\frac{\mathrm{dX}(t;\omega)}{\mathrm{d}t} = \varepsilon(t;\omega), \tag{9.1}$$

Ornstein-Uhlenbeck equation: $$\frac{\mathrm{dX}(t;\omega)}{\mathrm{d}t} + \gamma\left[\mathrm{X}(t;\omega) - m_{\mathrm{x}}\right] = \varepsilon(t;\omega), \tag{9.2}$$

where $\varepsilon(t;\omega)$ is a *zero-mean, unit-variance independent Gaussian noise*. The unit variance constraint can be easily relaxed by multiplying $\varepsilon(t;\omega)$ with the square root of the desired variance.

The notation used above follows a common convention in physics and engineering, which pretends that the white-noise process is a well-defined function and that the derivative of the process $\mathrm{X}(t;\omega)$ exists for every realization of $\varepsilon(t;\omega)$. This is acceptable if we assume that $\varepsilon(t;\omega)$ in reality represents some well-behaved approximation of white noise (i.e., filtered noise with the high-frequency tail of the spectrum removed).

Mathematical notation In mathematics, the preferred notation uses the random process increments to emphasize the discontinuity of the noise term. In terms of the process $\mathrm{dX}(t;\omega)$ differential over the small time increment δt, the above equations are expressed as

Brownian motion: $$\mathrm{dX}(t;\omega) = \mathrm{d}W(t;\omega), \tag{9.3}$$

Ornstein-Uhlenbeck: $$\mathrm{dX}(t;\omega) + \gamma\left[\mathrm{X}(t;\omega) - m_{\mathrm{x}}\right]\delta t = \mathrm{d}W(t;\omega), \tag{9.4}$$

where the differential $\mathrm{d}W(t;\omega)$ is a Gaussian random process with zero mean and variance equal to δt, i.e.,

$$\mathcal{E}[\mathrm{d}W(t;\omega)] = 0, \text{ and } \mathbb{V}\mathrm{ar}\,\{\mathrm{d}W(t;\omega)\} = \delta t.$$

Numerical integration Various numerical methods exist for integrating SDEs, including the simple *Euler-Maruyama* scheme, and the higher-order Milstein and Runge-Kutta methods. A nicely written introduction to numerical integration of SDEs is given by Higham [344], while a physicist's perspective on stochastic processes is given by Jacobs [402].

Brownian motion Using the Euler-Maruyama discretization scheme, we replace the differential equation (9.1) with the difference equation

$$\mathrm{W}(t+\delta t;\omega) - \mathrm{W}(t;\omega) = \sqrt{\delta t}\,\varepsilon(t;\omega),$$

where δt is the time step. The random process $W(t; \omega)$ can be viewed as the linear displacement of a "walker" (e.g., a moving particle) from an initial position w_0 at time t—for the specific realization denoted by ω. This random walk is named Brownian motion after the botanist Robert Brown who noticed the random motion of pollen particles in water. It was later theoretically shown, in a beautiful paper by Einstein, that the observed motion is due to the random collisions of the pollen particles with the water molecules as a result of the thermal motion of the latter [226]. Brownian motion thus provided convincing evidence for the validity of the atomic theory that was not universally accepted at the time.

Continuous time limit As the time step δt between successive moves of the walker tends to zero (i.e., at the continuous time limit), the random walk is determined by the differential equation (9.1). The stochastic process $W(t; \omega)$ that satisfies (9.1) is known in mathematics as the *Wiener process* after the mathematician Norbert Wiener. It is easy to show that the Brownian motion's expectation is equal to w_0, while its variance grows linearly with time.

Diffusion modeling It is straightforward to extend the model of Brownian motion to higher dimensions by replacing $W(t; \omega)$ with a displacement random field $\mathbf{R}(t; \omega)$ driven by a white noise vector with uncorrelated components. It can then be shown that the mean square displacement is given by $\mathbb{V}\text{ar}\{\mathbf{R}(t; \omega)\} \propto D\,t$, where D is the *diffusion coefficient*. Brownian motion is thus the archetypical model for *classical diffusion*, exhibiting a mean square displacement that grows linearly with time. Anomalous diffusion models can be obtained by introducing correlations in the driving noise term, as in the case of fractional Brownian motion [92, 566].

Note that unlike the solutions of linear ordinary differential equations that are in general smooth functions, the solutions of SDEs exhibit rough trajectories due to the impact of the random driving term. This is evidenced in the ten random walk realizations shown in Fig. 9.1.

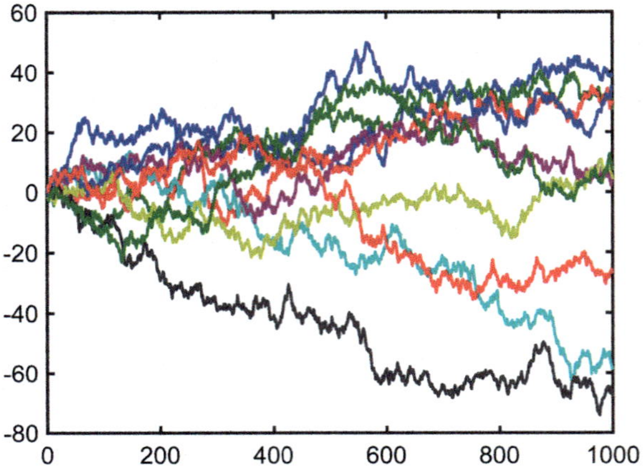

Fig. 9.1 Ten random walk realizations with the same initial position $x_0 = 0$. The horizontal axis corresponds to time, while the vertical axis displays the position of the walkers. Each realization is obtained with a different set of random numbers from the Gaussian distribution that correspond to the steps

Ornstein-Uhlenbeck process The Ornstein-Uhlenbeck (O-U) process is defined by (9.2) or (9.4). It describes a random walk with a *restoring force* which is equal to

$$F_{\text{resto}} = -\gamma\left[\mathrm{X}(t;\omega) - m_{\mathrm{x}}\right].$$

The restoring force opposes deviations from the equilibrium position m_{x}.

The initial value of the O-U process is in general arbitrary. However, as the time tends to infinity, the process tends to return to the equilibrium position, m_{x}, regardless of the initial condition (starting point). Because of this property, the O-U process is called a *mean-reverting process*.

The O-U process is a prototypical model for *stochastic relaxation*: It describes a system that is perturbed from its equilibrium position m_{x} due to the initial displacement and the random noise forcing. At the same time, the system is pulled towards the equilibrium position by the restoring force. The approach to equilibrium follows a random trajectory due to the impact of the random force. In discrete-time systems, the O-U process becomes equivalent to the autoregressive random process of order one, i.e., to an AR(1) process (see Sect. 9.7).

9.2 Classical Harmonic Oscillator

This section focuses on a slightly more complex SDE that corresponds to the *classical damped harmonic oscillator* driven by white noise. The complication is that the classical harmonic oscillator equation involves both the first- and the second-order time derivatives. The classical harmonic oscillator is an important and well-studied paradigm in physics [298]. In astronomy, the stochastically driven damped harmonic oscillator has been used to model the spectral features in the solar oscillation power spectrum [22] and asteroseismic oscillations [253].

The simple harmonic oscillator can be extended to coupled, spatially extended oscillator systems that model wave propagation [163], and to nonlinear, driven, dissipative models that can simulate earthquakes [116]. Its universality is underlined by the fact that it is used to model mechanical, electrical, biological, chemical systems, while its quantum counterpart is used in theories of atomic systems.

We demonstrate that the SSRF in one dimension is equivalent with the classical damped harmonic oscillator driven by white noise. This correspondence connects the abstract random field with the concrete mechanical example of an oscillatory system. The analogy helps to intuitively understand the "negative holes" in the SSRF covariance function, which are attributed to oscillatory motion.

The emergence of connections between seemingly unrelated scientific fields is elegantly expressed in the *Feynman Lectures on Physics* as follows [247]:

...But a strange thing occurs again and again: the equations which appear in different fields of physics, and even in other sciences, are often almost exactly the same, so that many phenomena have analogs in these different fields.

Equation of motion of free oscillator If a damped harmonic oscillator of mass $m \neq 0$ is disturbed from its equilibrium position, its displacement evolves according to Newton's second law of motion $F = m\,a$, where F is the force, m is the oscillator's mass, and a is its acceleration. If the oscillator moves only under the influence of the restoring spring force and friction, its equation of motion is given by

$$m\,\ddot{x}(t) = F = -\gamma\,\dot{x}(t) - k\,x(t),$$

where t measures time, $x(t)$ is the displacement of the oscillator from the equilibrium position, $\dot{x}(t)$ is the velocity, $\ddot{x}(t)$ is the acceleration, F is the applied force, $\gamma > 0$ is the friction coefficient, $-\gamma\,\dot{x}(t)$ is the friction force, k is a positive constant, and $-k\,x(t)$ is the restoring force.

Upon dividing both sides of the above equation by the oscillator mass m, the following is obtained

$$\ddot{x}(t) + \Gamma\dot{x}(t) + w_0^2 x(t) = 0,$$

where w_0 is the *resonant angular frequency*, and $\Gamma = \gamma/m$ is the *damping constant* that has units of inverse time. Thus, the damping constant Γ defines a characteristic *damping time* $\tau_c = 1/\Gamma$.

If the oscillator is displaced from its equilibrium position and is left to move subject only to the restoring and friction forces, it swings back and forth with progressively decreasing maximum displacement—due to energy dissipation caused by damping—and eventually returns to the equilibrium position.

Oscillator driven by white noise Let us now assume that the oscillator is placed in contact with a *heat bath*. This term is used in physics to denote a thermal reservoir that does not change its temperature when it comes in contact with the oscillator. The system of the heat bath and the oscillator is thus in thermal equilibrium.

For macroscopic processes of geostatistical interest, large bodies of water that maintain their temperature unaffected by environmental perturbations (e.g., by potential injections of pollutants) can be modeled as heat baths.

The heat bath exerts a random force $f(t)$ on the oscillator. As a result, the oscillator displacement becomes a random process $X(t;\omega)$. The realizations $x(t)$ of the oscillator's displacement are governed by the following equation of motion (EOM)[1]

[1]To simplify the notation and avoid confusing ω with the angular frequency w, we replace $X(t;\omega)$ with $x(t)$, where the latter represents the trajectory for a specific realization.

$$\ddot{x}(t) + \Gamma \dot{x}(t) + w_0^2 x(t) = \frac{f(t)}{m}. \tag{9.5}$$

For simplicity we assume that the heat-bath force $f(t)$ is *Gaussian white noise*. Since the force is random, the EOM is a second-order stochastic differential equation. At the limit $w_0 = 0$, if we set $y(t) = \dot{x}(t)$, the *Ornstein-Uhlenbeck* process which describes Brownian motion with inertia is obtained from (9.5).

The thermal random force has zero mean, i.e., $\mathcal{E}[f(t)] = 0$. In addition, the white-noise correlations imply the following expression for the covariance function

$$\mathcal{E}\left[f(t)\, f(t')\right] = \sigma_{ff}^2 \delta(t - t').$$

The expectation and variance of the transient response of the damped harmonic oscillator to random forcing (not necessarily white noise) were studied in [121].

Spectral domain analysis The dynamic equation of motion becomes an algebraic relation in the Fourier domain. Assuming that $\widetilde{x}(w)$ is the Fourier transform of $x(t)$, and $\widetilde{f}(w)$ the respective transform of the force, it follows that

$$\widetilde{x}(w) = \widetilde{G}(w)\widetilde{f}(w) = \frac{\widetilde{f}(w)}{m\left(w_0^2 - w^2 + \mathrm{i}w\Gamma\right)},$$

where w represents the *circular frequency*.[2] Strictly speaking, the Fourier transform $\widetilde{f}(w)$ does not exist, because the Fourier integral of white noise does not converge. However, we are interested in spectral densities, which are well defined by virtue of the *Wiener-Khinchin Theorem* 3.3.

The oscillator's power spectral density is given by $\widetilde{C}_{\mathrm{xx}}(w) = |\widetilde{x}(w)|^2$. In view of the expression for $\widetilde{x}(w)$, the spectral density is

$$\widetilde{C}_{\mathrm{xx}}(w) = \frac{\sigma_{ff}^2}{m^2 \left[\left(w_0^2 - w^2\right)^2 + (w\,\Gamma)^2\right]}, \tag{9.6}$$

where the power spectral density of the random force is $\widetilde{C}_{ff}(w) = \sigma_{ff}^2$ for all w due to the delta function correlations of the white noise. If we can estimate σ_{ff}^2, the power spectral density is fully determined. To accomplish this, the position variance is evaluated using two different approaches: the first one involves the spectral integral and the second one the equipartition theorem.

(i) The integral of the power spectral density is equal to the position variance. Assuming that the equilibrium position is $x_0 = 0$, the position variance is given by

[2] w corresponds to the wavenumber k which represents the "spatial" frequency in reciprocal space.

$$\mathcal{E}\left[x^2(t)\right] = \int_{-\infty}^{\infty} \mathrm{d}w\, \widetilde{C}_{\mathrm{xx}}(w) = \frac{\sigma_{ff}^2}{2m^2 w_0^2 \Gamma}. \tag{9.7}$$

(ii) The energy involves a kinetic term which is proportional to the square of the velocity, and a potential term which is proportional to the square of the displacement. The energy oscillates between these two degrees of freedom. The heat bath also transfers energy to the oscillator, and the latter dissipates energy due to damping.

The *equipartition theorem* states that for a system in thermal equilibrium the average energy stored in each quadratic degree of freedom is equal to

$$\mathcal{E}[E] = \frac{1}{2} k_{\mathrm{B}}\, T,$$

where k_{B} is the Boltzmann constant and T is the temperature. Hence, the average potential energy of the oscillator is given by the following statistical average

$$\frac{1}{2} m\, w_0^2\, \mathcal{E}\left[x^2(t)\right] = \frac{1}{2} k_{\mathrm{B}}\, T. \tag{9.8}$$

In light of (9.7) and (9.8), it follows that

$$\frac{k_{\mathrm{B}}\, T}{m\, w_0^2} = \frac{\sigma_{ff}^2}{2m^2 w_0^2 \Gamma} \Rightarrow \sigma_{ff}^2 = 2m\, k_{\mathrm{B}}\, T\, \Gamma.$$

Based on the above and assuming that the resonant frequency w_0 is finite, the *spectral density* of the thermally driven classical harmonic oscillator is given by [331]

$$\widetilde{C}_{\mathrm{xx}}(w) = \frac{2k_{\mathrm{B}} T\, \Gamma / m w_0^4}{\left(\frac{w}{w_0}\right)^4 + \left(\frac{\Gamma^2}{w_0^2} - 2\right)\left(\frac{w}{w_0}\right)^2 + 1}. \tag{9.9}$$

Comparison with SSRFs Compare the spectral density (9.9) of the harmonic oscillator with the one-dimensional SSRF spectral density (7.16), which is given by

$$\widetilde{C}_{\mathrm{xx}}(k) = \frac{\eta_0 \xi}{(k\xi)^4 + \eta_1\, (k\xi)^2 + 1}.$$

The SSRF spectral density is thus equivalent to the harmonic oscillator spectral density upon performing the following variable transformations:

$$k \mapsto w, \quad \xi \mapsto w_0^{-1}, \quad \eta_1 \mapsto \Gamma^2 w_0^{-2} - 2, \quad \eta_0 \mapsto \frac{2k_B T \Gamma}{m w_0^3}.$$

Conversely, the damping coefficient and the temperature are related to the SSRF parameters as follows

$$\Gamma^2 \mapsto w_0^2 (\eta_1 + 2) = \frac{\eta_1 + 2}{\xi^2}, \quad 2k_B T \mapsto \frac{m \eta_0}{\xi^2 \sqrt{\eta_1 + 2}}$$

In light of the above relations, the SSRF permissibility condition $\eta_1 > -2$ follows naturally from the physical constraint $\Gamma^2/w_0^2 > 0$. The limiting case $\eta_1 = -2$ ($\Gamma = 0$) corresponds to a non-damped (free) oscillator. The three different regimes of the harmonic oscillator correspond to the SSRF covariance regimes as follows:

Damped Harmonic Oscillator Regimes

- *Underdamping* is observed for $-2 < \eta_1 < 2$ (the upper bound implies that $4w_0^2 > \Gamma^2$). For these values of η_1 the oscillations of the covariance function persist in spite of the damping. In this case the oscillator's response is determined by the *modified angular frequency* w_d, which is given by

  $$w_d = \sqrt{w_0^2 - \frac{\Gamma^2}{4}}.$$

  The decline of the correlations with increasing time lag is determined from the damping time $\tau_c = 1/\Gamma$.
- *Critical damping* occurs for $\Gamma^2 = 4w_0^2$, i.e., for $\eta_1 = 2$. For fixed w_0, the critically damped oscillator returns to its equilibrium position faster than an oscillator with any other value of Γ. In this case, the system's response is determined by the damping time $\tau_c = 1/\Gamma$.
- *Overdamping* occurs for large η_1 which corresponds to $\Gamma \gg w_0$; this also agrees with the exponential decay of the SSRF covariance for $\eta_1 > 2$ with two different characteristic times that correspond to slow and fast decay.

Harmonic oscillator covariance The harmonic oscillator covariance functions are given by the equations (7.27) with the modified parametrization.

$$C_{xx}(\tau;\boldsymbol{\theta}) = \frac{\eta_0}{4}\,e^{-\tau/2\tau_c}\left[2w_0\tau_c\cos(\tau w_d) + \frac{w_0}{w_d}\sin(\tau w_d)\right], \quad \Gamma/w_0 < 2, \tag{9.10a}$$

$$C_{xx}(\tau;\boldsymbol{\theta}) = \eta_0\,\frac{(1+\tau/2\tau_c)}{4\,e^{\tau/2\tau_c}}, \quad \Gamma/w_0 = 2, \tag{9.10b}$$

$$C_{xx}(\tau;\boldsymbol{\theta}) = \frac{\eta_0\,w_0^2}{8\,|w_d|}\left(\frac{e^{-\tau/\tau_-}}{w_0^2\,\tau_+} - \frac{e^{-\tau/\tau_+}}{w_0^2\,\tau_-}\right), \quad \Gamma/w_0 > 2, \tag{9.10c}$$

where the fast and slow decay times, τ_+ and τ_- respectively, are given by

$$\tau_\pm = \frac{2\tau_c}{1 \pm 2\tau_c\,|w_d|}. \tag{9.10d}$$

The parameter vector is $\boldsymbol{\theta} = (\eta_0, w_0, \Gamma)^\top$ or equivalently $\boldsymbol{\theta} = (\eta_0, w_0, \tau_c)^\top$. The normalized spatial lag h used in SSRFs is replaced with the dimensionless time distance $w_0\tau$; this is equivalent to $h = \|\mathbf{r}\|/\xi$ taking into account that for time-dependent processes $\|\mathbf{r}\|$ is replaced by τ and ξ by $1/w_0$.

The covariance expressions (9.10) are in agreement with equations that were recently obtained in a study focusing on the stochastic dynamics of damped harmonic oscillators in heat baths [612].

Mean square displacement Following the above discussion, we can view the plots in Fig. 7.5 as representing the position of a harmonic oscillator in contact with the heat bath.[3] We should also mention that the variogram is proportional to the mean square displacement of the oscillator from the equilibrium position at time τ, since

$$2\gamma_{xx}(\tau) = \mathcal{E}\left\{[x(\tau) - x(0)]^2\right\}.$$

9.3 Stochastic Partial Differential Equations

Stochastic partial differential equations (SPDE) are dynamic equations that model the evolution (in space and time) of a spatially extended variable in the presence of stochastic forces (external or internal), e.g., Gaussian white noise. Hence, SPDEs enlarge the scope of stochastic differential equations from the description of lumped, point-like variables to fields that are extended in space. While SPDEs that involve both space and time derivatives are possible, herein we focus on purely spatial SPDEs.

[3]The horizontal axis corresponds to time, while the vertical axis denotes the position.

A recent exposition of numerical methods for the solution of SPDEs driven by white noise is given by Zhongqiang Zhang and George Karniadakis [881]. The simulation of SPDEs is also investigated in the dissertation of Annika Lang [481]. Finally, a general mathematical framework for random field models using SPDEs and distribution theory is presented in the recent dissertation of Ricardo Vergara [818, 819].

9.3.1 Linear SPDEs with Constant Coefficients

Let us consider an SPDE defined by the following equation over the domain $\mathcal{D} \subset \mathbb{R}^d$ subject to specified boundary conditions at the domain boundary $\partial\mathcal{D}$:

$$\mathcal{L}_{\mathbf{s}}^{*}\, \mathrm{X}(\mathbf{s};\omega) = \sigma_0\, \varepsilon(\mathbf{s};\omega). \tag{9.11}$$

$\mathcal{L}_{\mathbf{s}}^{*}$ is a linear differential operator that comprises spatial partial derivatives and acts on $\mathrm{X}(\mathbf{s};\omega)$ at the location identified by the subscript $\mathbf{s}$ and $\sigma_0 > 0$ measures the amplitude of the driving noise term.

Typically, the SPDE forcing term $\varepsilon(\mathbf{s};\omega)$ is a *spatial Gaussian white noise field* that follows a standard normal marginal distribution (zero mean and unit variance) and has localized correlations expressed by the delta-function covariance, that is,

$$\mathcal{E}[\varepsilon(\mathbf{s};\omega)\,\varepsilon(\mathbf{s}';\omega)] = \delta(\mathbf{s}-\mathbf{s}'). \tag{9.12}$$

We will assume that $\mathcal{L}_{\mathbf{s}}^{*}$ is defined by a multivariate polynomial of degree M in the partial derivatives

$$\mathcal{L}_{\mathbf{s}}^{*} = \sum_{m=0}^{M}\sum_{i=1}^{d} \alpha_{m,i}\, \frac{\partial^m}{\partial s_i^m}, \tag{9.13}$$

where $\{\alpha_{m,i}\}_{i=1,m=0}^{i=d,m=M}$ are real-valued, constant coefficients.

Based on (9.11), the expectation $m_{\mathrm{x}}(\mathbf{s})$ of the random field satisfies the following PDE (with the specified boundary conditions)

$$\mathcal{L}_{\mathbf{s}}^{*} m_{\mathrm{x}}(\mathbf{s}) = 0.$$

For simplicity, we assume that $\mathcal{D}$ coincides with $\mathbb{R}^d$. Then, the solution of the above PDE is satisfied by a multivariate polynomial function $m_{\mathrm{x}}(\mathbf{s})$ of degree $M - 1$. In the following, we will assume that the polynomial trend has been removed, so that we can set $m_{\mathrm{x}} = 0$ without loss of generality.

9.3.2 Covariance Equation in Real Space

Under the conditions specified above, the covariance equation of motion is obtained by (i) replicating (9.11) with a different spatial label $\mathbf{s}'$, (ii) multiplying the respective sides of both equations (at $\mathbf{s}$ and $\mathbf{s}'$), and (iii) evaluating the stochastic expectation. These steps lead to the following PDE

$$\mathcal{L}^*_{\mathbf{s}}\, \mathcal{L}^*_{\mathbf{s}'}\, \mathcal{E}[\mathrm{X}(\mathbf{s};\omega)\mathrm{X}(\mathbf{s}';\omega)] = \sigma_0^2\, \mathcal{E}[\varepsilon(\mathbf{s};\omega)\, \varepsilon(\mathbf{s}';\omega)] = \sigma_0^2\, \delta(\mathbf{s}-\mathbf{s}'), \tag{9.14a}$$

where $\mathcal{L} := \mathcal{L}^*_{\mathbf{s}}\, \mathcal{L}^*_{\mathbf{s}'}$ is the *linear differential operator* given by the following polynomial of degree $2m$ in the partial derivatives

$$\mathcal{L} = \sum_{m=0}^{M}\sum_{l=0}^{M}\sum_{i=1}^{d}\sum_{j=1}^{d} \alpha_{m,i}\,\alpha_{l,j}\, \frac{\partial^m}{\partial s_i^m}\,\frac{\partial^l}{\partial s'^l_i}. \tag{9.14b}$$

The covariance PDE (9.14a) is equivalently expressed as

$$\mathcal{L}C_{\mathrm{xx}}(\mathbf{r}) = \sigma_0^2\, \delta(\mathbf{r}), \ \text{ where } \mathbf{r} = \mathbf{s} - \mathbf{s}'. \tag{9.15}$$

In deriving the above, we used the translation invariance of the SPDE (imparted by the constant linear coefficients) in order to express the covariance PDE in terms of the spatial lag $\mathbf{r}$. The differential operator $\mathcal{L}$ is expressed in terms of $\mathbf{r}$ using the following transformations of the partial derivatives

$$\frac{\partial}{\partial s_i} = \frac{\partial}{\partial r_i}, \quad \frac{\partial}{\partial s'_j} = -\frac{\partial}{\partial r_j}, \quad \frac{\partial^m}{\partial s_i^m}\,\frac{\partial^l}{\partial s'^l_i} = (-1)^l\, \frac{\partial^{m+l}}{\partial r_i^m\, \partial r_j^l}.$$

Precision operator Let us rewrite the covariance PDE (9.15) as $\mathcal{L}C_{\mathrm{xx}}(\mathbf{r})/\sigma_0^2 = \delta(\mathbf{r})$. The operator $\mathcal{L}/\sigma_0^2$ acting on the covariance function returns the delta function, which is the continuum-space generalization of the identity matrix. Thus, it is intuitive to identify $\mathcal{L}/\sigma_0^2$ with the *precision operator*.

Non-directional linear coefficients Let us assume that the coefficients of the SPDE are direction independent, i.e., $\alpha_{m,i} = \alpha_m$ for all $i = 1, \ldots, d$ and $m = 1, \ldots, M$. Then, the covariance PDE is expressed as

$$\sum_{m=0}^{M}\sum_{l=0}^{M}(-1)^l\,\alpha_m\,\alpha_l\,\frac{\partial^{m+l}C_{xx}(\mathbf{r})}{\partial r_i^m\,\partial r_j^l}=\sigma_0^2\,\delta(\mathbf{r}). \tag{9.16}$$

Covariance from non-directional SPDE For a concrete example of a covariance PDE derived from a respective SPDE, let us consider the second-order SPDE ($M = 2$), with $\alpha_0 = 1$, $\alpha_1 = \alpha \in \mathbb{R}$, and $\alpha_2 = \beta \in \mathbb{R}$. The SPDE is then expressed as

$$1+\alpha\sum_{i=1}^{d}\frac{\partial \mathrm{X}(\mathbf{s};\omega)}{\partial s_i}+\beta\sum_{i=1}^{d}\frac{\partial^2 \mathrm{X}(\mathbf{s};\omega)}{\partial s_i^2}=\sigma_0\varepsilon(\mathbf{s};\omega). \tag{9.17}$$

Based on the general formulation (9.16) the covariance function PDE is given by[4]

$$1+\beta^2\sum\frac{\partial^4 C_{xx}(\mathbf{r})}{\partial r_i^2\partial r_j^2}+\alpha\beta\sum\left(\frac{\partial^3 C_{xx}(\mathbf{r})}{\partial r_i^2\partial r_j}-\frac{\partial^3 C_{xx}(\mathbf{r})}{\partial r_i\partial r_j^2}\right)-\alpha^2\sum\frac{\partial^2 C_{xx}(\mathbf{r})}{\partial r_i\partial r_j}$$
$$+\beta\sum\left(\frac{\partial^2 C_{xx}(\mathbf{r})}{\partial r_i^2}+\frac{\partial^2 C_{xx}(\mathbf{r})}{\partial r_j^2}\right)+\alpha\sum\left(\frac{\partial C_{xx}(\mathbf{r})}{\partial r_i}-\frac{\partial C_{xx}(\mathbf{r})}{\partial r_j}\right)=\sigma_0\delta(\mathbf{r}),$$

where the summation symbol in the above represents the double sum over the spatial directions. Upon calculating these sums, the terms proportional to α and $\alpha\beta$ vanish. Thus, the covariance PDE is expressed as follows:

$$1+\beta^2\nabla^4 C_{xx}(\mathbf{r})+\left(2\beta-\alpha^2\right)\nabla^2 C_{xx}(\mathbf{r})-\alpha^2\sum_{i=1}^{d}\sum_{j=1,\neq i}^{d}\frac{\partial^2 C_{xx}(\mathbf{r})}{\partial r_i\partial r_j}=\sigma_0\,\delta(\mathbf{r}), \tag{9.18}$$

where ∇^2 denotes the *Laplacian* and ∇^4 the *biharmonic* derivative operators.

Covariance for second-order SDE If $d = 1$ the last term on the left-hand side of (9.18) that contains the off-diagonal partial derivatives drops out. In this case the covariance PDE is replaced by the following ODE

$$1+\beta^2\frac{\mathrm{d}^4 C_{xx}(r)}{\mathrm{d}r^4}+\left(2\beta-\alpha^2\right)\frac{\mathrm{d}^2 C_{xx}(r)}{\mathrm{d}r^2}=\sigma_0\,\delta(r),$$

in which the *precision operator* is given by

$$\mathcal{L}=1+\beta^2\frac{\mathrm{d}^4}{\mathrm{d}r^4}+\left(2\beta-\alpha^2\right)\frac{\mathrm{d}^2}{\mathrm{d}r^2}.$$

[4]For the sake of space economy we use the abbreviation $\sum$ for $\sum_{i=1}^{d}\sum_{j=1}^{d}$.

Remark Other than requiring α and β to be real numbers, we have not yet constrained their values. However, not all real values are acceptable, since the solution of (9.18) must satisfy Bochner's theorem in order to provide a valid covariance function. The constraints on the coefficients can be more efficiently expressed in terms of the spectral density, as shown below.

9.3.3 Spectral Density from Linear SPDEs

Our aim is to derive an expression for the *spectral density* of the random field $\mathrm{X}(\mathbf{s};\omega)$ that is defined by means of the general SPDE (9.11). We will follow the approach used for the harmonic oscillator, boldly assuming that the Fourier transforms of $\mathrm{X}(\mathbf{s};\omega)$ and $\varepsilon(\mathbf{s};\omega)$ exist and are given respectively by $\widetilde{\mathrm{X}}(\mathbf{k};\omega)$ and $\widetilde{\varepsilon}(\mathbf{k};\omega)$. Then, the SPDE (9.11) is transformed in the spectral domain into

$$\widetilde{\mathcal{L}}^{*}(\mathbf{k})\,\widetilde{\mathrm{X}}(\mathbf{k};\omega) = \sigma_0\,\widetilde{\varepsilon}(\mathbf{k};\omega), \tag{9.19}$$

where $\widetilde{\mathcal{L}}^{*}(\mathbf{k})$ is a *multivariate polynomial* in $\mathbf{k} = (k_1,\ldots,k_d)^\top$. This polynomial is derived using the transformation of the partial space derivatives, i.e., $\partial/\partial s_i \mapsto \mathrm{i}\,k_i$. The polynomial $\widetilde{\mathcal{L}}^{*}(\mathbf{k})$ includes *complex-valued coefficients*, if $\mathcal{L}^{*}$ includes odd-order partial derivatives.

The *spectral density* of the random field $\mathrm{X}(\mathbf{s};\omega)$ is defined by means of the expectation

$$\widetilde{C}_{\mathrm{xx}}(\mathbf{k}) = \mathcal{E}\left[\widetilde{\mathrm{X}}(\mathbf{k};\omega)\,\widetilde{X}^{\dagger}(\mathbf{k};\omega)\right]. \tag{9.20}$$

Based on (9.19) and (9.20) the spectral density is given by the following expression

$$\widetilde{C}_{\mathrm{xx}}(\mathbf{k}) = \frac{\sigma_0^2\,\mathcal{E}\left[\widetilde{\varepsilon}(\mathbf{k};\omega)\,\widetilde{\varepsilon}^{\dagger}(\mathbf{k};\omega)\right]}{\widetilde{\mathcal{L}}^{*}(\mathbf{k})\,\widetilde{\mathcal{L}}^{*\dagger}(\mathbf{k})} = \frac{\sigma_0^2}{\|\widetilde{\mathcal{L}}^{*}(\mathbf{k})\|^2}. \tag{9.21}$$

Alternatively, we can apply the Fourier transform directly to (9.15) to obtain

$$\widetilde{C}_{\mathrm{xx}}(\mathbf{k}) = \frac{\sigma_0^2}{\widetilde{\mathcal{L}}(\mathbf{k})}, \tag{9.22}$$

where $\widetilde{\mathcal{L}}(\mathbf{k})$ is the multivariate polynomial that corresponds to the Fourier transform of the covariance linear differential operator $\mathcal{L}$.[5] We return to $\widetilde{\mathcal{L}}(\mathbf{k})$ in the discussion on polynomials of the diffusion operator below.

[5]Note that $\mathcal{L}$ is the operator acting on the covariance, while $\mathcal{L}^{*}$ is the operator acting on the random field.

Spectral density for second-order SPDEs To be more specific, let us calculate the spectral density of the random field that satisfies the linear, second-order SPDE (9.17) with constant, non-directional coefficients. The respective equation satisfied by the covariance function is (9.18). The spectral expressions of the Laplacian and the biharmonic operators are $-\|\mathbf{k}\|^2$ and $\|\mathbf{k}\|^4$ respectively according to (3.59). Hence, the spectral density that is obtained from the second-order SPDE is given by

$$\widetilde{C}_{\rm xx}(\mathbf{k}) = \frac{\sigma_0^2}{1 + \beta^2 \|\mathbf{k}\|^4 - (2\beta - \alpha^2) \|\mathbf{k}\|^2 + \alpha^2 \sum_{i=1}^{d} \sum_{j=1,\neq i}^{d} k_i k_j}. \tag{9.23}$$

The denominator of the spectral density (9.23) is a fourth-degree d-variate polynomial, similar to the characteristic polynomial of the SSRF spectral density (7.16). Bochner's permissibility theorem requires that

$$\alpha^2 - 2\beta \geq 0 \tag{9.24}$$

in order for (9.23) to represent a valid spectral density.

A crucial difference between SSRFs and the spectral density (9.23) is that the latter contains the non-diagonal terms $\sum_{i=1}^{d} \sum_{j=1,\neq i}^{d} k_i k_j$. The presence of such terms complicates the evaluation of the covariance function in real space, since the inverse Fourier transform requires the calculation of a truly d-dimensional integral. Off-diagonal terms are in general present in the spectral density, if the linear SPDE involves odd-order spatial derivatives.

The off-diagonal term is eliminated if $\alpha = 0$; in this case the permissibility condition (9.24) requires that $\beta < 0$. The respective spectral density becomes

$$\widetilde{C}_{\rm xx}(\mathbf{k}) = \frac{\sigma_0^2}{\left(1 + |\beta| \, \|\mathbf{k}\|^2\right)^2}.$$

This density is equivalent to the $\eta_1 = 2$ SSRF spectral density provided that $|\beta| = \xi^2$.

9.3.4 Polynomials of the Diffusion Operator

In order to avoid the nuisance of off-diagonal terms in the spectral density, the precision operator can be constructed as a superposition of powers of the Laplacian, i.e.,

$$\mathcal{L} = 1 + \sum_{l=1}^{L} c_l \nabla^{2l}. \tag{9.25a}$$

Hence, in light of $\mathcal{F}[\nabla^{2n}] = (-1)^n \|\mathbf{k}\|^{2n}$, $\widetilde{\mathcal{L}}(\mathbf{k})$ is given by the expression

$$\widetilde{\mathcal{L}}(\mathbf{k}) = \sum_{l=0}^{L} c_l (-1)^l \|\mathbf{k}\|^{2l}, \; c_0 = 1. \tag{9.25b}$$

The Laplacian is also known as the *diffusion operator*, because of its role in the diffusion equation

$$\frac{\partial x(\mathbf{s}, t)}{\partial t} = D\, \nabla^2 x(\mathbf{s}, t),$$

which governs the dispersion of a non-uniform initial concentration $x(\mathbf{s}, t_0)$ towards a spatially uniform final state.

The spectral density corresponding to the diffusion polynomial operator (9.25) is given by

$$\widetilde{C}_{\mathrm{xx}}(\mathbf{k}) = \frac{\sigma_0^2}{1 + \sum_{l=1}^{L} (-1)^l \, c_l \, \|\mathbf{k}\|^{2l}}. \tag{9.26}$$

In analogy with the SSRF characteristic polynomial (7.19a), the *characteristic polynomial* of (9.26) is obtained from the inverse of the spectral density and is given by

$$\Pi(x) = 1 + \sum_{l=1}^{L} (-1)^l \, c_l \, x^{2l}. \tag{9.27}$$

Equation (9.26) is a rational function. Hence, it follows that rational spectral densities result from precision operators that are polynomials of the diffusion (Laplacian) operator.

Note that for $L = 2$ the characteristic polynomial (9.27) is of the same form as the SSRF characteristic polynomial (7.19a).

Multiscale covariance functions The approach based on polynomials of the diffusion operator was followed by Yaremchuk & Smith who obtained covariance functions similar to the SSRF expressions [866]. In fact, they showed that if the degree L of the characteristic polynomial is even, i.e., $L = 2M$, it is possible to construct the spectral density using complex-valued coefficients $z_m = a_m + \mathrm{i} b_m$, where $a_m, b_m \in \mathbb{R}$, $m = 1, \ldots, M$ and their complex conjugates $z_m^{\dagger}$. Each coefficient z_m corresponds to a different physical scale. The spectral density is then given by the following expression, which is by construction *positive definite*

$$\widetilde{C}_{\mathrm{xx}}(\mathbf{k}) = \frac{\sigma_0^2 \prod_{m=1}^{M} |z_m^2|^2}{\prod_{m=1}^{M} \left(\|\mathbf{k}\|^2 + z_m^2\right) \left(\|\mathbf{k}\|^2 + z_m^{\dagger 2}\right)}. \tag{9.28}$$

Consequently, the covariance function in d dimensions is obtained by direct integration of the spectral representation for radial functions (4.4). The result is given by the following mathematical expression

$$C_{\mathrm{xx}}(\mathbf{r}) = \sigma_0^2 \frac{2\|\mathbf{r}\|^{2-d}}{(2\pi)^{d/2}} \sum_{m=1}^{M} \Re\left[q_m \left(z_m \|\mathbf{r}\|\right)^{\nu} K_\nu(z_m \|\mathbf{r}\|)\right], \tag{9.29a}$$

$$q_m = \frac{\prod_{m=1}^{M} \left|z_m^2\right|^2}{\left(z_m^{\dagger 2} - z_m^2\right) \prod_{m \neq j} \left(z_m^2 - z_j^2\right)\left(z_m^2 - z_j^{\dagger 2}\right)}, \tag{9.29b}$$

where $\nu = d/2 - 1$ is the *dimensionality index*, $K_\nu(\cdot)$ is the modified Bessel function of the second kind of order ν, and $\Re[\cdot]$ denotes the real part of the expression inside the brackets.

The *multiscale covariance function* (9.29) involves $L = 2M$ parameters, which can be used to capture different interacting length scales. The multiscale covariances maintain the central advantage of Spartan covariance functions: their inverses admit *sparse representations*, because they involve local derivative operators.

Regarding the *differentiability* of random fields that have covariance functions given by the expression (9.29), note that the spectral densities (9.28) decay algebraically as $\widetilde{C}_{\mathrm{xx}}(\|\mathbf{k}\|) \propto \|\mathbf{k}\|^{-2L}$ at the limit $\|\mathbf{k}\| \to \infty$. As discussed in Sect. 5.3.1, the existence of the radial spectral moment of order n requires that the tail exponent satisfies the inequality $2n < 2L - d$. Hence, as L increases so does the order of differentiability (smoothness) of the respective random field.

9.4 Spartan Random Fields and SPDEs

In Chap. 7 we showed that the joint pdf of SSRFs is defined in terms of the energy functional (7.4). The realizations of SSRFs can also be viewed as solutions of an associated SPDE that is linked to the SSRF precision operator. Subsequently, the covariance function is determined by a (deterministic) partial differential equation. We demonstrated in Sect. 9.2 this equivalence in the case of the stochastic classical damped harmonic oscillator—which is equivalent to a one-dimensional SSRF. In the following, we study the PDE obeyed by the SSRF covariance function and the SPDE that governs the spatial variability of SSRFs.

9.4.1 Partial Differential Equation for SSRF Covariance Functions

In general, the precision operator associated with an invertible covariance function is defined by means of the identity (6.24a), namely,

$$\int_{\mathcal{D}} d\mathbf{s}''\, C_{\mathrm{xx}}^{-1}(\mathbf{s}-\mathbf{s}'')\, C_{\mathrm{xx}}(\mathbf{s}''-\mathbf{s}') = \delta(\mathbf{s}-\mathbf{s}'). \tag{9.30a}$$

Let us recall equation (7.11) which defines the inverse SSRF covariance kernel. We repeat it below for easy reference:

$$C_{\mathrm{xx}}^{-1}(\mathbf{s}-\mathbf{s}') = \frac{1}{\eta_0 \xi^d}\left[1 - \eta_1\, \xi^2\, \Delta_{\mathbf{s}} + \xi^4\, \Delta_{\mathbf{s}}^2\right] \delta(\mathbf{s}-\mathbf{s}'). \tag{9.30b}$$

The above expression for $C_{\mathrm{xx}}^{-1}(\mathbf{s}-\mathbf{s}')$ involves generalized delta functions. We therefore perform the convolution integral in (9.30a) using the method of integration by parts as described below the equation (7.11). These calculations (which the readers can verify for their amusement) lead to the following PDE:

PDE associated with SSRF covariance functions

$$\mathcal{L}\left[C_{\mathrm{xx}}(\mathbf{s}-\mathbf{s}')\right] = \delta(\mathbf{s}-\mathbf{s}'), \tag{9.31a}$$

where $\mathcal{L}$ is the *self-adjoint SSRF precision* operator

$$\mathcal{L} = \frac{1}{\eta_0 \xi^d}\left[1 - \eta_1\, \xi^2\, \Delta_{\mathbf{s}} + \xi^4\, \Delta_{\mathbf{s}}^2\right]. \tag{9.31b}$$

To derive the expression (9.31b) we assumed that the covariance function vanishes for all $\mathbf{s}$, $\mathbf{s}' \in \partial\mathcal{D}$, so that the boundary terms resulting from the application of Green's first identity are eliminated (see Sect. 7.1.5).

On differentiability The non-differentiability of SSRF covariance functions (in $d \geq 2$) discussed in Sect. 7.2.6 has its origin in (9.31): The delta function in the SSRF covariance PDE leads to a zero-lag discontinuity in the Bilaplacian and a slope discontinuity in the Laplacian of the field.

The delta function is associated with the correlations of the white noise process that drives the respective Langevin equation (see Sect. 9.3.2 and the following section). Hence, smoother random fields can be obtained from SSRF Langevin equations that are driven by *colored* (correlated) spatial noise. Then, the PDE (9.31a) is replaced by

$$\mathcal{L}\left[C_{xx}(\mathbf{s}-\mathbf{s}')\right] = c_{no}(\mathbf{s}-\mathbf{s}'), \tag{9.32}$$

where $c_{no}(\mathbf{s}-\mathbf{s}')$ is the short-ranged noise covariance with characteristic length scale ξ_{no} and spectral density $\widetilde{c}_{no}(\mathbf{k})$.

9.4.2 *Spatial Harmonic Oscillators*

As demonstrated in connection with the classical harmonic oscillator in Sect. 9.2, an associated Langevin equation can be constructed for SSRF realizations $x(s)$ in one-dimensional space ($s \in \mathbb{R}$). In this case, the Langevin equation involves a differential operator $\mathcal{L}^*$ that comprises first-order and second-order derivatives with respect to s (or with respect to the time t).

However, as shown in Sect. 9.3, if the differential operator $\mathcal{L}^*$ involves first-order partial derivatives of the field, for $d > 1$ the precision operator involves terms with cross derivatives in addition to the Laplacian and the biharmonic operator. Hence, it becomes obvious that the Langevin equation that leads to SSRF spectral densities cannot be constructed using an operator $\mathcal{L}^*$ composed of linear combinations of first- and second-order partial derivatives.

Since the SSRF covariance PDE (9.31) involves the Laplacian, it is intuitively expected that the associated Langevin equation should involve a term proportional to the square root of the Laplacian (more precisely, of the negative Laplacian, as we discuss below). In one dimension the first-order derivative, $\mathrm{d}/\mathrm{d}t$, in the Langevin equation directly leads to the second-order time derivative in the covariance ODE. In higher dimensions it is necessary to adopt a similar, meaningful notion of the half-Laplacian.

Fractional Laplacian The *half Laplacian*, denoted by $-(-\Delta)^{1/2}$, is a special case of the *fractional Laplacian* operator, $-(-\Delta)^s$, where $s \in (0, 1)$ [504].

The *fractional Laplacian* is the most basic linear, elliptic, integro-differential operator. It is possible to define it in different ways. As shown recently several of the definitions are equivalent [475]. The fractional derivative of order $s = 1/2$ in (9.36b) can be viewed as a shifted Riesz fractional derivative that admits a formal series representation [634, 708].

To intuitively understand the reason for the negative sign inside the square root in $-(-\Delta)^{1/2}$, recall that the Fourier transform of the Laplacian is $-\|\mathbf{k}\|^2$. Hence, to obtain real-valued fractional powers of the Laplacian operator, it is necessary to multiply $-\|\mathbf{k}\|^2$ with -1. This is best appreciated in the spectral domain, where the fractional Laplacian is defined by means of

$$\mathcal{F}[(-\Delta)^s\, x(\mathbf{s})] = \|\mathbf{k}\|^{2s}\, \widetilde{x}(\mathbf{k}). \tag{9.33}$$

If we lift the restriction that the Fourier transform be real-valued, the above equation can be used to define the Fourier transform of Δ^s as follows

$$\mathcal{F}[\Delta^s x(\mathbf{s})] = \mathrm{i}^{2s} \, \|\mathbf{k}\|^{2s} \, \tilde{x}(\mathbf{k}). \tag{9.34}$$

Both of the above are consistent with the Fourier transform of the Laplacian at the limit $s = 1$.

De Wijs process In $d = 2$ the half-Laplacian appears in the SPDE

$$(-\Delta)^{1/2} \mathrm{X}(\mathbf{s}; \omega) = \varepsilon(\mathbf{s}; \omega). \tag{9.35}$$

The SPDE (9.35) defines a generalized random field known as the *de Wijs process* [73, 586]. The de Wijs process has a generalized covariance function that satisfies the same equation as the Green's function of the Laplace equation, i.e., $\Delta G(\mathbf{r}) = -\delta(\mathbf{r})$. Thus, the two-dimensional de Wijs process admits the logarithmic variogram function $c_0 \ln(\|\mathbf{r}\|/r_s)$. The logarithmic variogram is used for variables regularized over a finite support. Then, the logarithmic dependence holds only for distances exceeding the support size r_s. Due to the nice properties of the logarithmic function, the de Wijs variogram has been amply used in geostatistical applications (see [132, chapter 2.5] for a review).

Incidentally, the SPDE of the de Wijs process defines a random field with Boltzmann-Gibbs joint pdf (6.5a) and energy functional

$$\mathcal{H}_0[x(\mathbf{s})] = \frac{1}{2} \int d\mathbf{s} \, [\nabla x(\mathbf{s})]^2 .$$

The energy above corresponds to a *massless Gaussian field theory*. A mathematical treatment of this theory is given in [741]. The interested reader will find in [741] more information about the definition of the fractional Laplacian for $s \neq 1/2$.

The model is called massless because it does not involve a term $\propto x^2(\mathbf{s})$. The absence of such a term implies that there is no penalty in the energy for large values of $x(\mathbf{s})$. Hence, the massless model does not have a bounded variance. Actually, the fact that the energy depends only on the squared gradient $[\nabla x(\mathbf{s})]^2$ implies that the corresponding random field model is an *intrinsically stationary of order zero* (or locally homogeneous) random field.

Spartan random fields After the brief detour into fractional Laplacians, we return to the SPDE representation of SSRFs in d-dimensional spaces ($d = 2, 3$). The SSRF realizations can be viewed as solutions of the following *stochastic partial differential equation*

$$\mathcal{L}^*[\mathrm{X}(\mathbf{s}; \omega)] = \varepsilon(\mathbf{s}; \omega), \tag{9.36a}$$

where $\varepsilon(\mathbf{s}; \omega)$ is a Gaussian white noise field, and the *associated partial differential operator* $\mathcal{L}^*$ is expressed in terms of the Laplacian Δ as follows

$$\mathcal{L}^{*} = \frac{1}{\sigma_0}\left[1 - \left(\sqrt{2+\eta_1}\right)\,\xi\,(-\Delta)^{1/2} + \xi^2\Delta\right], \tag{9.36b}$$

and the coefficient $\sigma_0 = \sqrt{\eta_0\,\xi^d}$ can be thought as the amplitude of the noise field $\varepsilon(\mathbf{s};\omega)$. Hence, the SPDE (9.36a) generates the spatial variability of SSRFs.

Connection with the Boltzmann-Gibbs joint density The Langevin equation (9.36a) leads directly to the Boltzmann-Gibbs joint density representation of Spartan random fields given by (7.9). This can be easily shown using the definition of the characteristic function (4.64) and the expression (4.67) for the Gaussian characteristic function. Assuming that the formal solution for $X(\mathbf{s};\omega)$ is

$$X(\mathbf{s};\omega) = \Phi(\mathbf{s},\cdot) * \varepsilon(\cdot;\omega),$$

where $\Phi(\cdot)$ is the inverse of the partial differential operator $\mathcal{L}$ and $*$ denotes the convolution operator, it follows from (4.67) that the covariance function satisfies

$$C_{xx}(\mathbf{s},\mathbf{s}') = \Phi(\mathbf{s},\cdot) * \Phi(\cdot,\mathbf{s}') = \int d\mathbf{z}\,\Phi(\mathbf{s},\mathbf{z})\,\Phi(\mathbf{z},\mathbf{s}').$$

Spectral domain Based on the representation of $\mathcal{L}$ given above, the SSRF operator $\widetilde{\mathcal{L}}^{*}$ in the spectral domain is given by the following second-degree polynomial[6]

$$\widetilde{\mathcal{L}}^{*}(\mathbf{k}) = \frac{1}{\sigma_0}\left[1 + \mathrm{i}\left(\sqrt{2+\eta_1}\right)\,\|\mathbf{k}\|\,\xi - \xi^2\|\mathbf{k}\|^2\right]. \tag{9.37}$$

This model is associated with the precision operator (9.31b). It is straightforward to show that the spectral SSRF operator is linked to the spectral density as follows [cf. (9.6)]

$$\widetilde{C}_{xx}(\mathbf{k}) = \frac{1}{\widetilde{\mathcal{L}}(\mathbf{k})} = \frac{1}{\left|\widetilde{\mathcal{L}}^{*}(\mathbf{k})\right|^2} = \frac{\sigma_0^2}{1+\eta_1\|\mathbf{k}\|^2\xi^2 + \|\mathbf{k}\|^4\xi^4}.$$

Harmonic oscillator analogy We can view the SPDE (9.36b) as a model for a "spatial harmonic oscillator", in which the first- and second-order time derivatives are replaced respectively by the half-Laplacian and the Laplacian. The SSRF model is different from a classical harmonic oscillator in two or three dimensions: in the latter case, the oscillator displacement $x(t)$ is replaced by $\mathbf{x}(t)$, but the equation of motion is still expressed in terms of first- and second-order time derivatives. In

[6] In contrast with Sect. 9.3, herein we absorb the coefficient σ_0 in $\widetilde{\mathcal{L}}^{*}(\mathbf{k})$ and consider forcing with unit-variance white noise.

contrast, we can consider the SPDE (9.36b) that defines SSRFs as an harmonic oscillator equation of motion in d-dimensional-time.

9.5 Covariances and Green's Functions

There is a formal connection between covariance functions of random fields governed by linear SPDEs driven by white noise and the Green's functions of associated (with the SPDE) partial differential equations. This relation was expressed by Dolph and Woodbury as follows [205]:

> ...our method is based on the intrinsically interesting relationship that exists between covariance functions of random processes generated by driving n-th order linear differential equations by so-called "pure noise" and the Green's function of a suitably defined self-adjoint equation of order $2n$.

Indeed, based on (9.31), the SSRF covariance function $C_{xx}(\mathbf{s} - \mathbf{s}')$ is also a Green's function that satisfies the equation[7]

$$\mathcal{L}[G(\mathbf{s} - \mathbf{s}')] = \delta(\mathbf{s} - \mathbf{s}'). \tag{9.38}$$

A Green's function is known as the *fundamental solution* or *impulse response function* associated with the deterministic linear PDE

$$\mathcal{L}[\phi(\mathbf{s})] = 0.$$

The deterministic function $\phi(\mathbf{s})$ should not be confused with the random field realizations $x(\mathbf{s})$ that are governed by the operator $\mathcal{L}^*$ subject to white noise forcing.

The relation between covariance functions and Green's functions is not one to one: There are spatial covariance functions that are not associated with a linear PDE, e.g., the Gaussian covariance model $C_{xx}(\mathbf{r}) = \sigma_x^2 \exp(-\|\mathbf{r}\|^2/\xi^2)$.

(i) A covariance function is equivalent to a Green function only if the random field with the above covariance admits an SPDE representation such as (9.36a).
(ii) Conversely, a Green function that satisfies a linear PDE such as (9.38) represents the covariance of the stationary random field that obeys the associated SPDE

$$\mathcal{L}^* X(\mathbf{s}; \omega) = \varepsilon(\mathbf{s}; \omega),$$

provided that the inverse of the generalized Fourier transform $\widetilde{\mathcal{L}}(\mathbf{k})$, i.e.,

[7] The Green's function equation is sometimes expressed as $\mathcal{L}[G(\mathbf{s} - \mathbf{s}')] = -\delta(\mathbf{s} - \mathbf{s}')$, i.e., with a negative sign in front of the delta function.

$$\frac{1}{\widetilde{\mathcal{L}}(\mathbf{k})} = \frac{1}{\|\widetilde{\mathcal{L}}^{*}(\mathbf{k})\|^2}$$

yields an admissible spectral density. This condition is satisfied for SSRFs, because according to (9.31b) the inverse covariance operator is

$$\mathcal{L} = \frac{1}{\eta_0 \xi^d}\left[1 - \eta_1\,\xi^2\,\Delta_{\mathbf{s}} + \xi^4\,\Delta_{\mathbf{s}}^2\right],$$

and its spectral counterpart is given by the polynomial

$$\mathcal{F}[\mathcal{L}] = \widetilde{\mathcal{L}}(\mathbf{k}) = \frac{1}{\eta_0 \xi^d}\left[1 + \eta_1\,\xi^2\,\|\mathbf{k}\|^2 + \xi^4\,\|\mathbf{k}\|^4\right].$$

There exist, however, Green's functions of linear PDEs that are *not admissible stationary covariance functions.* One such example is the Green function of the Laplace equation over an infinite domain, which is not a valid covariance; nonetheless, it provides a generalized covariance for the de Wijs process discussed above. Another example is the Green function of the biharmonic equation [see (10.9)] that is also not an admissible stationary covariance but usable as a generalized covariance.

9.6 Whittle-Matérn Stochastic Partial Differential Equation

The connection between SPDEs and random fields was noted by Whittle [846, 847] who realized that the so-called *Whittle-Matérn* random fields are solutions of the following family of fractional stochastic partial differential equations

$$\left(\kappa^2 - \Delta\right)^{\alpha/2} \mathrm{X}(\mathbf{s};\omega) = \sigma_0\,\varepsilon(\mathbf{s};\omega), \tag{9.39}$$

where the parameters and the source term are defined as follows:

- $\mathbf{s} \in \mathbb{R}^d$ is the position vector in d-dimensional space,
- $\varepsilon(\mathbf{s};\omega)$ is a Gaussian white noise with unit variance,
- $\sigma_0 > 0$ is a coefficient that modifies the noise standard deviation,
- $\kappa > 0$ is an inverse length, i.e., $\xi = 1/\kappa$ is a characteristic length,
- Δ is the d-dimensional Laplacian,
- $\alpha = \nu + d/2$, is a dimension-dependent power-law exponent,
- $\nu > 0$ is the *smoothness index* that determines the differentiability of the covariance function, and subsequently of the realizations (the Whittle-Matérn field is mean-square differentiable m times, where $m = \lceil \nu \rceil - 1$).

The SPDE (9.39) generates random fields with Whittle-Matérn covariance functions. Adopting the "simple" expression for the Whittle-Matérn correlation given by (4.20), the respective covariance is given by

$$C_{\mathrm{xx}}(u) = \frac{\sigma_{\mathrm{x}}^2}{2^{\nu-1}\,\Gamma(\nu)}\, u^{\nu}\, K_{\nu}(u). \tag{9.40}$$

As shown in [503], if $\nu + d/2$ is an integer, the Gaussian random field with the respective Whittle-Matérn covariance can be accurately approximated by means of a Gaussian Markov random field (GMRF) with sparse precision matrix. Precision matrix examples for specific values of ν are shown in [538].

The variance of the Whittle-Matérn field is related to the coefficient σ_0^2 and the model parameters $\nu,\ \alpha,\ \xi$ by means of

$$\sigma_{\mathrm{x}}^2 = \frac{\sigma_0^2\,\Gamma(\nu)\,\xi^{2\nu}}{(4\pi)^{d/2}\,\Gamma(\alpha)}. \tag{9.41}$$

The spectral density of the Whittle-Matérn model is given by

$$\widetilde{C}_{\mathrm{xx}}(\mathbf{k}) = \frac{1}{(2\pi)^d}\,\frac{1}{(1+\kappa^2)^{(\nu+d/2)}}. \tag{9.42}$$

The Whittle-Matérn model has been recently generalized to include an additional fractional exponent. The second exponent emerges by replacing the Laplacian operator in (9.39) with the fractional Laplacian, thus introducing a new parameter $\in (0, 1]$. This new parameter is used to include spatial memory effects in the covariance that determine the asymptotic power-law dependence of the covariance tail. The generalized model has been used to model wind speed data [501]. However, it may be more relevant for processes that exhibit stronger spatial dependence.

SPDE formalism Recently, Linden et al. [503] proposed to generate random fields by solving the SPDE (9.39) in an approximation space, where the solution becomes a superposition of basis functions weighted with stochastic coefficients that represent the random dimension of the field. This representation leads to computationally efficient implementations, if the basis functions are locally supported. The SPDE approach extends the theory of Markov random fields to scattered data, and it also provides a computational framework for non-stationary covariance models [87, 503].

Nonlinear models In both the SSRF and the Whittle-Matérn cases, the associated SPDEs are linear with respect to the field $\mathrm{X}(\mathbf{s};\omega)$. In statistical physics there are SPDE models that describe the evolution of non-equilibrium systems and involve nonlinear terms. Typical examples are the Kardar-Parisi-Zhang (KPZ) equation that models surface growth phenomena [435] and the Swift-Hohenberg equation that provides a prototype for pattern formation [787, 788].

SPDEs with nonlinear terms lead to more complex structures and long-range connectivity than their linear counterparts, but they are not usually amenable to exact analytic solutions. The theoretical analysis of nonlinear SPDEs involves the formalism of field theory, perturbation expansions, and Feynman diagrams [351, 547]. To our knowledge such approaches have not yet been explored in problems of spatial statistics with the exception of [364, 371].

9.7 Diversion in Time Series

In one dimension, SSRFs are equivalent to random processes $X(s; \omega)$, $s \in \mathbb{R}$, defined over continuous time domains $\mathcal{D}$. In data analysis, one is often interested in random processes that are sampled at discrete times $\{s_i\}_{i=1}^{N}$. The term *time series* is used for such processes. If the sampling occurs with uniform sampling step δt, then $s_i = i\,\delta t$. Time series applications of the SSRF model were investigated in [894, 895].

The SSRF model is defined in continuum time, while the time series models assume that the samples are taken at discrete times. In the case of the classical harmonic oscillator, the effect of sampling the positions at discrete times was investigated in [612]. It is therein shown that the covariance function and the mean square displacement of the oscillator position are not affected by the finite sampling step δt. This result also applies to the one-dimensional SSRF model.

Below we investigate a discrete version of the one-dimensional SSRF model and its connections with the continuum SSRF model. As a result of the discretization, the stochastic differential equation that describes the SSRF dynamics is replaced by a respective difference equation that represents the evolution of the respective time series. The discrete SSRF formulation is shown to correspond to an autoregressive process of order two.

9.7.1 Brief Overview of Linear Time Series Modeling

In this section we briefly review the theory of *autoregressive moving average (ARMA)* processes and draw connections with SSRFs. There are several textbooks on time series analysis that the interested reader can consult for further information on ARMA processes [325, 588, 750].

We assume that the time series is a sample of the random process $X(t; \omega)$. The sampling times can be expressed as $t_m = m\,\delta t$, where m is a positive integer or zero and δt is a constant time step. The value of the random process at time t_m is denoted by $x_m \doteq x(t_m)$, and the time series comprises the set of values $\{x_1, \ldots, x_M\}$. The term time series is often used to denote both the discrete random process $\{X_1(\omega), \ldots, X_M(\omega)\}$ and its realizations.

In the following we consider stationary time series with zero mean. Time series with constant mean m_x can be treated by replacing x_t with $x_t - m_x$.

ARMA processes are compactly formulated in terms of the *lag or backshift operator* $B(\cdot)$ which is defined by

$$B(x_t) = x_{t-\delta t}.$$

Higher-order powers of the backshift operator can also be defined by

$$B^k(x_t) = \underbrace{B \ldots B(x_t)}_{k} = x_{t-k\delta t}, \; k \in \mathbb{N}.$$

Then, an *ARMA process of autoregressive order p and moving average q* is expressed in terms of the following difference equation

$$\phi(B)\, x_t = \theta(B)\, \varepsilon_t, \tag{9.43}$$

where $\phi(B)$ is the AR (autoregressive) backshift operator polynomial

$$\phi(B) = 1 - \sum_{i=1}^{p} \phi_i\, B^i, \tag{9.44}$$

$\theta(B)$ is the MA (moving average) operator polynomial,

$$\theta(B) = 1 + \sum_{j=1}^{q} \theta_j\, B^j, \tag{9.45}$$

and ε_t is a white noise with variance σ_ε^2. If the noise is Gaussian, then the time series defined by (9.43) is also a Gaussian random process.

The ARMA model expresses the time series at time t in terms of its values at earlier times through the autoregressive operator polynomial, and a linear combination of the noise values at the current and previous times via the moving average polynomial.

The key idea of autoregressive time-series models, codified by means of (9.44), can be extended to the spatial domain leading to *spatial autoregressive models (SAR)* defined on lattices [682]. In these models, the random field at a location $\mathbf{s}_n$ is determined as a function of the values of its neighbors and a spatial noise (uncorrelated) process. If the noise is Gaussian and the autoregression employs a linear superposition of the neighboring values, the resulting SRF is Gaussian [165].

9.7.2 Properties of ARMA Models

Below we review the fundamental properties of ARMA models.

Linear processes A time series is characterized as a *linear process*, if it can be defined as a linear combination of white noise values (not necessarily Gaussian) at different times, i.e.,

$$x_t - m_x = \sum_{j=-\infty}^{\infty} \psi_j \, \varepsilon_{t-j\delta t}, \text{ where } \psi_j \in \mathbb{R} \text{ and } \sum_{j=-\infty}^{\infty} |\psi_j| < \infty. \tag{9.46}$$

The above equation defines x_t in terms of both its past ($j > 0$) and future ($j < 0$) values. A *causal linear process* is obtained if $\psi_j = 0$ for all $j < 0$, because then the process depends only on the past and the present. Hence, causal linear processes have a built-in time arrow: the past influences the future but not vice versa.

The theorem known as *Wold decomposition* states that a stationary (non-deterministic) time series can be expressed as a causal linear process, i.e.,

$$x_t - m_x = \sum_{j=0}^{\infty} \psi_j \, \varepsilon_{t-j\delta t}, \text{ where } \sum_{j=0}^{\infty} |\psi_j| < \infty. \tag{9.47}$$

Stationarity An ARMA(p,q) model leads to a *stationary time series* if the AR polynomial $\phi(z)$, obtained by replacing in (9.44) the backshift operator B with a complex-valued variable z, does not have roots on the unit circle.

Causality An ARMA(p,q) process is called *causal*, if all the roots $\{z_r\}_{r=1}^{p}$ of the AR polynomial $\phi(z)$ lie outside the unit circle on the complex plane, i.e., $|z_r| > 1$. Then, the time series can be expressed in terms of the *causal form*, i.e.,

$$x_t = \phi(B)^{-1} \, \theta(B) \, \varepsilon_t = \psi(B) \, \varepsilon_t, \tag{9.48a}$$

where the polynomial $\psi(\cdot)$ is defined by

$$\psi(z) = \sum_{j=0}^{\infty} \psi_j z^j = \frac{\theta(z)}{\phi(z)}, \quad \text{for } |z| \leq 1. \tag{9.48b}$$

AR(1) process As an example of the stationarity condition, consider the autoregressive process defined by the following difference equation

$$x_t = \phi_1 \, x_{t-\delta t} + \varepsilon_t. \tag{9.49}$$

If $|\phi_1| < 1$, the root of the AR polynomial $\phi(z) = 1 - \phi_1 z$ lies outside the unit circle, and the AR(1) process is causal and stationary according to the above. We can easily verify that

$$\mathcal{E}[X_t(\omega)] = \phi_1\, \mathcal{E}[X_{t-\delta t}(\omega)] + \mathcal{E}[\varepsilon_t(\omega)],$$

and since $\mathcal{E}[\varepsilon_t(\omega)] = 0$, it follows that $\mathcal{E}[X_t(\omega)] = 0$ as well.

The variance of the AR(1) process is obtained from the following equation

$$\mathcal{E}[X_t^2(\omega)] = \phi_1^2\, \mathcal{E}[X_{t-\delta t}^2(\omega)] + \sigma_\varepsilon^2.$$

Taking into account the stationarity condition $\mathbb{V}\mathrm{ar}\,\{X_t(\omega)\} = \mathbb{V}\mathrm{ar}\,\{X_{t-\delta t}(\omega)\}$, the above equation admits a positive solution for $\mathcal{E}[X_t^2(\omega)]$ if $|\phi_1| < 1$. Thus, we obtain

$$\mathbb{V}\mathrm{ar}\,\{X_t(\omega)\} = \frac{\sigma_\varepsilon^2}{1-\phi_1^2}. \tag{9.50}$$

Similarly, the covariance function for $\tau > 0$ is given by the following recursive equation[8]

$$\begin{aligned}\mathcal{E}[X_{t+\tau}(\omega)X_t(\omega)] =& \phi_1\, \mathcal{E}[X_{t+\tau-\delta t(\omega)}\, X_t(\omega)] + \mathcal{E}[X_t(\omega)\, \varepsilon_{t+\tau}(\omega)]\\ \Rightarrow& C_{\mathrm{xx}}(\tau) = \phi_1\, C_{\mathrm{xx}}(\tau - \delta t).\end{aligned}$$

In the above we used the AR(1) equation (9.49) to express the random process at $t+\tau$ and took advantage of the fact that the future noise values at $t+\tau$ are uncorrelated with the random process at t. We can iteratively use the above equation setting $\tau = m\,\delta t$ to show that the AR(1) covariance function declines exponentially, i.e.,

$$C_{\mathrm{xx}}(m) = \phi_1^m\, C_{\mathrm{xx}}(0) = \sigma_\varepsilon^2\, \frac{\phi_1^m}{1-\phi_1^2} = \frac{\sigma_\varepsilon^2}{1-\phi_1^2}\, \exp\left(-m \mid \ln \phi_1 |\right),\ m \in \mathbb{N}. \tag{9.51}$$

Based on the expectation and the covariance, the AR(1) process is second-order stationary if $|\phi_1| < 1$; this condition also ensures that the variance is positive. For $\phi_1 = 1$ the variance tends to infinity, which marks the departure from stationarity. Applying the same steps to $\tau = -m\delta t$ yields the same result as (9.51) if m is replaced by $-m$. This confirms the time reflection symmetry of the covariance.

Random walk as an AR(1) process Let us now revisit (9.49) setting $\phi_1 = 1$; we then find that the difference equation is expressed as

$$x_t - x_{t-\delta t} = \varepsilon_t,$$

[8] In the time series literature the autocovariance function is usually denoted by the symbol γ, which we have reserved for the variogram as is common in spatial statistics.

which defines a random walk process. As we have discussed in Chap. 1, the random walk has stationary random increments (as defined by the noise process), but it is a non-stationary random process. This is confirmed by the fact that $\phi_1 = 1$ in the respective AR(1) polynomial.

Yule-Walker equations Equation (9.51) is a simple example of the Yule-Walker equations that determine the covariance function of autoregressive processes. For an AR(p) process they are given by

$$C_{xx}(m) = \sum_{k=1}^{p} \phi_k \, C_{xx}(m-k) + \sigma_\varepsilon^2 \, \delta_{m,0}, \text{ for } m = 0, \ldots, p. \tag{9.52a}$$

The above defines a linear system of $p+1$ equations that can be used to estimate (i) the covariance function for the first p lags, if the AR model (i.e., its coefficients ϕ_k) is known; and (ii) the coefficients ϕ_k of the AR(p) model from sample estimates of the covariance function for the first m lags. To solve the Yule-Walker equations the reflection symmetry of the covariance, i.e., $C_{xx}(-m) = C_{xx}(m)$ should be taken into account. A moment-based method for estimating the parameters of SSRF time series models [894] and random fields [229, 897] employs estimates of linear combinations of the variogram function at small lags, in the same spirit as the Yule-Walker based approach (see also Chap. 13).

The general solution for the AR correlation function for times $m > p$ is given by

$$\rho_{xx}(m) = \sum_{i=1}^{p'} z_i^{-m} \, P_i(m). \tag{9.52b}$$

In the above, $\{z_i\}_{i=1}^{p'}$ where $p' \leq p$ are the *distinct roots* roots of the *AR(p) characteristic polynomial*

$$\phi(z) = 1 - \sum_{l=1}^{p} \phi_l \, z^l. \tag{9.52c}$$

The distinct roots $\{z_i\}_{i=1}^{p'}$ have multiplicities $\{r_i\}_{i=1}^{p'}$. The sum of the multiplicities is equal to the order of the model, i.e., $\sum_{i=1}^{p'} r_i = p$. Finally, $P_i(m) = P_i(z = m)$ where the $P_i(z)$ are polynomials of degree $r_i - 1$ for $i = 1, \ldots, p' \leq p$ [750].

Invertibility A time series model is called *invertible* if and only if white noise can be obtained by a time-lag polynomial that operates on the time series, i.e., if

$$\varepsilon_t = \theta(B)^{-1} \, \phi(B) = \pi(B) \, x_t = \sum_{j=0}^{\infty} \pi_j x_{t-j\delta t},$$

where $\pi_0 = 1$. An ARMA(p,q) model is invertible if the roots of the polynomial $\theta(z)$ are outside the unit circle in the complex plane. Then, the inversion polynomial $\pi(z)$ is obtained from

$$\psi(z) = \sum_{j=0}^{\infty} \pi_j \, z^j = \frac{\phi(z)}{\theta(z)}, \qquad \text{for } |z| \leq 1.$$

Spectral analysis As discussed in the context of Bochner's Theorem 3.2, a given function represents the covariance of a random process, if and only if it possesses a valid (non-negative and integrable) spectral density.

If $C_{\mathrm{xx}}(\tau)$ is the covariance of a stationary time series and if $C_{\mathrm{xx}}(\tau)$ is absolutely summable, i.e.,

$$\sum_{m=-\infty}^{\infty} |C_{\mathrm{xx}}(\tau_m)| < \infty, \text{ where } \tau_m = m\,\delta t,$$

then the covariance function admits the following representation

$$C_{\mathrm{xx}}(\tau) = \int_{-\pi/\delta t}^{\pi/\delta t} \frac{\mathrm{d}w}{2\pi} \, \mathrm{e}^{\mathrm{i}w\tau} \, \widetilde{C}'_{\mathrm{xx}}(w), \tag{9.53}$$

where w is the *circular frequency* and $\widetilde{C}'_{\mathrm{xx}}(w)$ is the *spectral density* defined by

$$\widetilde{C}'_{\mathrm{xx}}(w) = \sum_{m=-\infty}^{\infty} C_{\mathrm{xx}}(m\,\delta t)\, \mathrm{e}^{-\mathrm{i}wm\delta t}. \tag{9.54}$$

Remark Comparing the above with the inverse Fourier transform (3.56) and the direct Fourier transform (3.54) of random fields in continuum domains, we notice two main differences: (i) The integral of the spectral density is over the bounded domain $[-\pi/\delta t,\ \pi/\delta t]$ in (9.53). (ii) The integral over the covariance function in (3.54) is replaced with a summation in (9.54). Both differences originate in the definition of the time series at a discrete set of points, multiples of the time step δt. It is straightforward to check that $\widetilde{C}_{\mathrm{xx}}(w) = \delta t \, \widetilde{C}'_{\mathrm{xx}}(w)$.

ARMA Spectral Density The spectral density of ARMA processes is easily obtained using the autoregressive and moving average polynomials, as well as the fact that the lag operator is elegantly expressed in the spectral domain as

$$\mathcal{F}\,[B(x_t)] = \mathrm{e}^{-\mathrm{i}\,w\,t}\, \widetilde{x}(w).$$

Based on the above, the following expression is derived for the ARMA spectral density

$$\widetilde{C}'_{\mathrm{xx}}(w) = \sigma_{\varepsilon}^{2}\, \frac{\left|\theta(\mathrm{e}^{-\mathrm{i}\,w\,\delta t})\right|^{2}}{\left|\phi(\mathrm{e}^{-\mathrm{i}\,w\,\delta t})\right|^{2}}. \tag{9.55}$$

The expression (9.55) is also known as the *rational spectrum.*

Based on (9.55) the spectral density of the AR(1) process is given by the following equation

$$\widetilde{C}'_{\mathrm{xx}}(w) = \sum_{\tau=-\infty}^{\infty} C_{\mathrm{xx}}(\tau)\, \mathrm{e}^{-\mathrm{i}\,w\,\tau} = \frac{\sigma_{\varepsilon}^{2}}{1 - 2\,\phi_1 \cos(w\delta t) + \phi_1^{2}}.$$

The spectral density of the MA(1) process is given by

$$\widetilde{C}'_{\mathrm{xx}}(w) = \sum_{\tau=-\infty}^{\infty} C_{\mathrm{xx}}(\tau)\, \mathrm{e}^{-\mathrm{i}\,w\,\tau} = \sigma_{\varepsilon}^{2}\left(1 + 2\,\theta_1 \cos(w\delta t) + \theta_1^{2}\right).$$

AR Spectral Approximation In practical applications, more than one ARMA models may give similar results for any given data set. Then, an "optimal model" should be selected based on the available knowledge about the physical process studied or a *model selection criterion* such as the Akaike information criterion (AIC) or the Bayesian information criterion (BIC). The close performance of different models is a hint that the more complex models cannot be distinguished from simpler models based on the data. Based on *Occam's razor*, the simpler models are preferable.

The converse question is also of interest: Given a stationary process with covariance function $C_{\mathrm{xx}}(\tau)$, is it always possible to find an ARMA model with this covariance function? The AR spectral approximation theorem states that there is an AR(p) process with (i) all the roots of the AR polynomial outside the unit circle and (ii) a covariance function $C_p(\tau)$ such that the absolute difference between $C_p(\tau)$ and $C_{\mathrm{xx}}(\tau)$ is arbitrarily small for all $-\pi/\delta t < w \leq \pi/\delta t$ [750], [325, p.157].

The Spectral Representation Theorem Wold's decomposition (9.47) allows expressing a stationary random process as a superposition of white noise variates in the time domain. The counterpart of this theorem in the spectral domain is the spectral representation theorem, which expresses a stationary random process as a superposition of harmonic functions. The rigorous expression of the spectral representation theorem involves the Fourier-Stieltjes transform. However, for our purposes the simplified representation given by Hamilton [325, p. 157] is sufficient.

Theorem 9.1 (Spectral representation) *Every second-order stationary, mean-square continuous process* $\mathrm{X}_t(\omega)$ *that admits a spectral density can be analyzed into a superposition of sine and cosine functions as follows*

$$\mathrm{X}_t(\omega) = m_{\mathrm{x}} + \int_0^{\pi/\delta t} dw\big[a(w;\omega)\, \cos(wt) + b(w;\omega)\, \sin(wt)\big], \tag{9.56}$$

where $a(w;\omega)$, $b(w;\omega)$ are zero-mean spectral random processes with variance

$$\mathcal{E}[a^2(w;\omega)] = \mathcal{E}[b^2(w;\omega)] = \frac{\widetilde{C}_{xx}(w)}{2\pi},$$

and uncorrelated (orthogonal) variations. The variations refer to the integrated spectral functions $a(w;\omega)$ and $b(w;\omega)$ over a range of frequencies. The orthogonality of the variations implies that for the ordered set of frequencies

$$0 < w_1 < w_2 < w_3 < w_4 < \pi,$$

the correlations of the integrated spectral processes over the non-overlapping frequency intervals $[w_1, w_2]$ and $[w_3, w_4]$ vanish, i.e.,

$$\mathcal{E}\left[\int_{w_1}^{w_2} \mathrm{d}w\, a(w;\omega) \int_{w_3}^{w_4} \mathrm{d}w\, a(w;\omega)\right] = 0,$$

while the same equation also holds for $b(w;\omega)$. In addition, the cross-correlation of the variations over any two, potentially overlapping frequency intervals $[w_1, w_2]$ and $[w_3, w_4]$ also vanishes, i.e.,

$$\mathcal{E}\left[\int_{w_1}^{w_2} \mathrm{d}w\, a(w;\omega) \int_{w_3}^{w_4} \mathrm{d}w\, b(w;\omega)\right] = 0.$$

Example 9.1 Derive the spectral density of (i) an AR(p) process with $\phi(B) = 1 - \sum_{i=1}^{p} \phi_i B^i$ and (ii) of an MA(q) process $\theta(B) = 1 + \sum_{j=1}^{q} \theta_j B^j$ where $p, q > 1$.

Answer In both cases the results are derived from (9.55). For the AR(p) process it follows that $\widetilde{C}'_{xx}(w) = \sigma_\varepsilon^2/\Phi(w)$. The term $\Phi(w)$ is evaluated using the identity $|z|^2 = zz^\dagger$, where $z^\dagger$ is the conjugate of the complex variable z. This leads to the following expression

$$\Phi(w) = \left|1 - \sum_{n=1}^{p} \phi_n \mathrm{e}^{-\mathrm{i}\, n\, w\, \delta t}\right|^2 = \left(1 - \sum_{n=1}^{p} \phi_n \mathrm{e}^{-\mathrm{i}\, n\, w\, \delta t}\right)\left(1 - \sum_{m=1}^{p} \phi_m \mathrm{e}^{\mathrm{i}\, m\, w\, \delta t}\right)$$
$$= 1 + \sum_{n=1}^{p} \phi_n^2 - 2\sum_{n=1}^{p} \phi_n \cos(nw\delta t) + 2\sum_{n=1}^{p}\sum_{m>n}^{p} \phi_n\phi_m \cos\left[w(n-m)\delta t\right]. \quad (9.57)$$

Similarly, for an MA(q) process the spectral density is $\widetilde{C}'_{xx}(w) = \sigma_\varepsilon^2\, \Theta(w)$, where the function $\Theta(w)$ is given by

$$\Theta(w) = 1 + \sum_{n=1}^{q} \theta_n^2 + 2\sum_{n=1}^{q} \theta_n \cos(nw\delta t) + 2\sum_{n=1}^{q}\sum_{m>n}^{q} \theta_n \theta_m \cos\left[w(n-m)\delta t\right]. \quad (9.58)$$

9.7.3 Autoregressive Formulation of SSRFs

In this section we highlight the connection between SSRFs and the formalism of autoregressive models. Autoregressive random field models have been studied among others by Whittle [846] and Vanmarcke [808]. In models of autoregressive time series the value at any target time is determined by a linear combination of the time series values at past times and a random component, typically a white noise process. Autoregressive time series models, expressed in terms of difference equations, are applicable if the observation times are discrete. If we consider continuous-time processes, the AR difference equations are replaced by stochastic differential equations (SDEs). One application of such SDEs is to model the *trajectories* of moving objects in space [102].

In analogy with the Langevin equation (9.5) of the classical harmonic oscillator, the one-dimensional SSRF is governed by the following second-order, linear, stochastic ordinary differential equation

$$\left(1 + \sqrt{2+\eta_1}\,\xi\,\frac{\mathrm{d}}{\mathrm{d}t} + \xi^2\,\frac{\mathrm{d}^2}{\mathrm{d}t^2}\right) x(t) = \sqrt{\eta_0 \xi}\;\varepsilon(t), \quad (9.59)$$

where $\varepsilon(t) \overset{d}{=} \mathcal{N}(0, 1)$ is a standard Gaussian white noise, and the SSRF coefficients satisfy $\eta_1 > -2$ and $\xi > 0$. The parameter ξ corresponds in this case to a characteristic time. The spectral counterpart of the random process $x(t)$ is given by the complex-valued function

$$\widetilde{x}(w) = \frac{\sqrt{\eta_0 \xi}\,\widetilde{\varepsilon}(w)}{1 + \mathrm{i}\sqrt{2+\eta_1}\,(\xi w) - (\xi w)^2}. \quad (9.60)$$

In the above, we assume that the Fourier transforms of the process $x(t)$ and the white noise realizations exist without delving into the intricacies of the spectral representation theorem. This does not cause any harm, however, since the covariance spectral density is well defined by means of

$$\widetilde{C}_{\mathrm{xx}}(w) = |\widetilde{x}(w)|^2 = \frac{\eta_0 \xi}{1 + \eta_1 (\xi w)^2 + (\xi w)^4}. \quad (9.61)$$

The above is the one-dimensional analogue of the SSRF spectral density (7.16) if the wavenumber $\|\mathbf{k}\|$ is replaced by the angular (circular) frequency w.

The SSRF-AR(2) model Let us focus on a random process sampled at discrete times $t_m = m\delta t$, where δt is the time step. The characteristic time ξ must be larger than the time step δt, otherwise the sampled state x_t will essentially look like noise. Hence, it makes sense to focus on $\xi/\delta t > 1$. The time step δt can be absorbed in ξ by replacing the latter with the dimensionless *correlation scale* $\tau_c = \xi/\delta t$.

If we replace the derivatives in (9.59) with finite differences, we obtain the following *SSRF difference equation*

$$x_t + \alpha_1\,(x_{t+1} - x_t) + \alpha_2\,(x_{t+1} + x_{t-1} - 2x_t) = \alpha_3\,\varepsilon_t, \tag{9.62}$$

where the time index t corresponds to discrete times $t_m = m\delta t$, and the $\{\alpha_i\}_{i=1}^3$ are dimensionless, non-negative coefficients given by

$$\alpha_1 = \tau_c\sqrt{2+\eta_1}, \quad \alpha_2 = {\tau_c}^2, \quad \alpha_3 = \sqrt{\eta_0\tau_c}, \tag{9.63}$$

Since $\tau_c > 1$, it also follows from the above that $\alpha_2 > 1$. This SSRF-based model has been applied to financial and environmental time series [894, 895].

Equation (9.62) is a *bilateral autoregression model*, meaning that the process at the current time t depends not only on the past time $t-1$ but also on the future time $t+1$ [846]. The dependence of the process at the current time on future values implies that the autoregressive model is *acausal*, i.e., it does not respect causality.

The SSRF difference equation (9.62) can also be expressed by grouping together terms that refer to the same time instant, i.e.,

$$a\,x_{t-1} - x_t + b\,x_{t+1} = c\,\varepsilon_t, \tag{9.64a}$$

where the positive coefficients a, b, c are given by[9]

$$a = \frac{\alpha_2}{\alpha_1 + 2\alpha_2 - 1}, \quad b = \frac{\alpha_1 + \alpha_2}{\alpha_1 + 2\alpha_2 - 1}, \quad c = \frac{\alpha_3}{\alpha_1 + 2\alpha_2 - 1}. \tag{9.64b}$$

Equation (9.64a) can be expressed as an equivalent unilateral process according to Whittle [846]. The main idea is to apply the transformation $t+1 \mapsto t$ and then recast (9.64a) in the form of the following causal model

$$x_t - b^{-1}\,x_{t-1} + a\,b^{-1}\,x_{t-2} = c\,b^{-1}\varepsilon'_t, \tag{9.65}$$

where $\varepsilon'_t = \varepsilon_{t-1}$ is the shifted standard Gaussian white noise.

Thus, the bilateral autoregression has the same covariance function as the following causal AR(2) model

$$x_t = \phi_1\,x_{t-1} + \phi_2\,x_{t-2} + \sigma_\epsilon\,\varepsilon_t, \tag{9.66a}$$

[9]The positivity of the coefficients is ensured by the fact that $\alpha_1, \alpha_3 > 0$ and $\alpha_2 > 1$.

where the coefficients of the AR polynomial and the standard deviation of the noise are given by

$$\phi_1 = \frac{1}{b} = \frac{\alpha_1 + 2\alpha_2 - 1}{\alpha_1 + \alpha_2}, \tag{9.66b}$$

$$\phi_2 = -\frac{a}{b} = -\frac{\alpha_2}{\alpha_1 + \alpha_2}, \tag{9.66c}$$

$$\sigma_\epsilon = \frac{c}{b} = \frac{\alpha_3}{\alpha_1 + \alpha_2}. \tag{9.66d}$$

Based on the positivity of the coefficients a, b, c it follows that $\phi_1 > 0$, $\phi_2 < 0$, and $\sigma_\epsilon > 0$. We will refer to the model defined by (9.66a) as the *SSRF-AR(2) model.*

AR(2) autocorrelation function The autocorrelation function of the AR(2) process (9.66a) admits an explicit expression that involves the roots of the associated polynomial [750]:

$$\phi(z) = 1 - \phi_1 z - \phi_2 z^2.$$

The roots of the polynomial, denoted by z_1 and z_2, are given by

$$z_{1,2} = -\frac{\phi_1 \pm \sqrt{\phi_1^2 + 4\phi_2}}{2\phi_2} = \frac{\phi_1 \pm \sqrt{\phi_1^2 - 4\,|\phi_2|}}{2|\phi_2|}. \tag{9.67}$$

The causality condition requires that $|z_1|, |z_2| > 1$. In terms of the AR(2) coefficients, this is expressed as

$$\phi_1 + \phi_2 < 1, \;\; \phi_2 - \phi_1 < 1, \;\; \text{and } |\phi_2| < 1. \tag{9.68}$$

Assuming that $\eta_1 > -2$ so that $\alpha_1 > 0$ and $\alpha_2 > 1$, it can be easily shown based on (9.66) that the above causality conditions are satisfied.

The first two values of the correlation function are given by [750, p. 102]:

$$\rho_{xx}(0) = 1, \tag{9.69a}$$

$$\rho_{xx}(1) = \frac{\phi_1}{1 - \phi_2}. \tag{9.69b}$$

The first equation follows from the definition of the correlation function. For the second equation see Example 9.2 below. The first causality condition (9.68) guarantees that $\phi_1 < 1 - \phi_2$ and thus $\rho_{xx}(1) < 1$. In addition, since $\phi_1 > 0$ and $\phi_2 < 0$, the correlation at $\tau = 1$ is positive, i.e., $\rho_{xx}(1) > 0$. The equations (9.69) provide initial conditions for the Yule-Walker equations (9.52) of the AR(2) correlation function.

Example 9.2 (i) Prove the expression for the autocorrelation function of the SSRF AR(2) process at lag equal to one [equation (9.69b)]. (ii) Calculate the variance of the model.

Answer (i) Based on (9.66a) we obtain the expression for $X_{t+1}(\omega)$ as follows

$$X_{t+1}(\omega) = \phi_1 X_t(\omega) + \phi_2 X_{t-1}(\omega) + \sigma_\epsilon\, \varepsilon_{t+1}(\omega).$$

Then, the product of the process values at two consecutive times is given by

$$X_{t+1}(\omega)\, X_t(\omega) = \phi_1 X_t^2(\omega) + \phi_2 X_t(\omega)\, X_{t-1}(\omega) + \sigma_\epsilon X_t(\omega)\, \varepsilon_{t+1}(\omega).$$

Evaluating the expectation on both sides of the above and taking into account the symmetry of the covariance function, i.e., $C_{XX}(1) = C_{XX}(-1)$, we obtain the following equation

$$C_{XX}(1)(1 - \phi_2) = \phi_1\, \sigma_X^2.$$

Note that $\mathcal{E}[X_t(\omega)\, \varepsilon_{t+1}(\omega)]$ vanishes because the process is not correlated with future values of the noise. Since $\rho_{XX}(1) = C_{XX}(1)/\sigma_X^2$, the result (9.69b) follows easily from the equation above. The same result can also be directly obtained from the Yule-Walker equations (9.52).

(ii) Next, we calculate the variance of $X_t(\omega)$ by means of $\mathcal{E}[X_t^2(\omega)]$ using (9.66a) to express the random process $X_t(\omega)$, and we then multiply both sides by $X_t(\omega)$.[10] This leads to the following expression

$$\sigma_X^2 = \phi_1 C_{XX}(1) + \phi_2 C_{XX}(2) + \sigma_\epsilon^2.$$

The last term is obtained from the expectation $\sigma_\epsilon\, \mathcal{E}[X_t(\omega)\, \varepsilon_t(\omega)]$. We can evaluate $C_{XX}(2)$ from the Yule-Walker equations (9.52) which lead to

$$C_{XX}(2) = \phi_1 C_{XX}(1) + \phi_2 C_{XX}(0).$$

Using the expression above for $C_{XX}(2)$ in combination with the expression for $C_{XX}(1)$, it follows that the variance is given by

$$\sigma_X^2 = \frac{\sigma_\epsilon^2}{1 - \phi_2^2 - \frac{\phi_1^2\,(1+\phi_2)}{1-\phi_2}}. \tag{9.70}$$

The *SSRF AR(2) autocorrelation* function at any lag $\tau > 0$ is given by different expressions in each branch. These expressions for an AR(2) process are given

[10]The equation for the realizations x_t holds for $X_t(\omega)$ as well.

in [750, p. 104–105]. If $\tau < 0$ the same equations apply upon replacing τ with $|\tau|$.

1. *Underdamped case:* If $\widetilde{\Delta} = \phi_1^2 + 4\phi_2 < 0$, then the roots of the AR(2) polynomial are a pair of complex conjugate numbers $z_1 = z_2^{\dagger}$. Using the dimensionless time constant $\widetilde{\tau} = 1/\ln|z_1|$, the correlation function is given by an exponentially damped harmonic function

$$\rho_{xx}(\tau) = c_1 e^{-\tau/\widetilde{\tau}} \cos(w_0\,\tau + c_2), \tag{9.71a}$$

$$w_0 = \arctan\left(\frac{\Im(z_1)}{\Re(z_1)}\right) = \arctan\left(\sqrt{\frac{4\,|\phi_2|}{\phi_1^2} - 1}\right),$$

   where the cyclical frequency w_0 determines the quasi-periodicity of the damped harmonic motion. Equation (9.71a) has the same form whether $\tau = m \in \mathbb{N}$ or $\tau = m\,\delta t$. In the latter case $\widetilde{\tau} \mapsto \widetilde{\tau}\delta t$.
2. *Critical damping:* If $\widetilde{\Delta} = 0$, then the AR(2) polynomial has a double real root, i.e., $z_1 = z_2 > 0$. If we define the characteristic time

$$\tau_0 = \frac{1}{\ln z_1} = \frac{1}{\ln\left(\frac{\phi_1}{2\,|\phi_2|}\right)}, \tag{9.71b}$$

   the correlation function is given by the modified exponential function

$$\rho_{xx}(\tau) = e^{-\tau/\tau_0}\,(c_1 + c_2\,\tau)\,. \tag{9.71c}$$

3. *Underdamping:* If $\widetilde{\Delta} > 0$, the roots z_1 and z_2 are real and distinct. Based on the equation (9.67) that determines the roots in terms of the AR(2) coefficients, $z_1 > z_2$; we can thus define $\tau_i = 1/\ln z_i$, where $i = 1, 2$, and $\tau_1 < \tau_2$. The correlation function is then given by the sum of two exponentials, i.e.,

$$\rho_{xx}(\tau) = c_1\,e^{-\tau/\tau_1} + c_2\,e^{-\tau/\tau_2}. \tag{9.71d}$$

The stationarity condition ensures an exponential decay of the correlation function with increasing lag in all three cases. The coefficients c_1 and c_2 are determined by imposing the initial conditions (9.69) on the correlation function.

The equations (9.71) for the SSRF AR(2) correlation functions correspond to the three regimes of the one-dimensional SSRF correlation function given by (7.30) and to the respective harmonic oscillator regimes defined in Sect. 9.2. However, the discretization of the process in the time domain affects the geometry of the regimes in the space of the SSRF coefficients (η_1, ξ) as illustrated in Fig. 9.2.

Example 9.3 Plot the correlation function of the SSRF-AR(2) model with parameters $\eta_0 = 1$, $\eta_1 = 2$, $\xi = 2$ for time lags between zero and ten, using two different

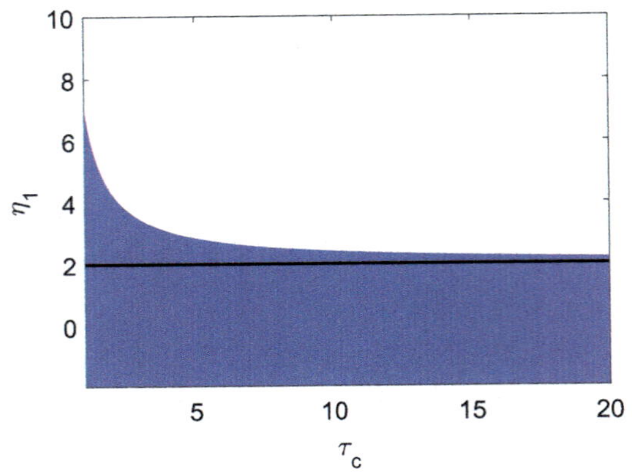

Fig. 9.2 The regime of critical damping is obtained for $\tilde{\Delta} = 0$. It coincides with the boundary line between the light (overdamped, $\tilde{\Delta} > 0$) and the dark (underdamped, $\tilde{\Delta} < 0$) areas. In the continuum, the regime boundary becomes the solid horizontal line at $\eta_1 = 2$. For finite δt, the continuum boundary is asymptotically approached as $\tau_c = \xi/\delta t \to \infty$, i.e., as $\xi \gg \delta t$. For finite δt the underdamped regime for low τ_c expands into areas with $\eta_1 > 2$

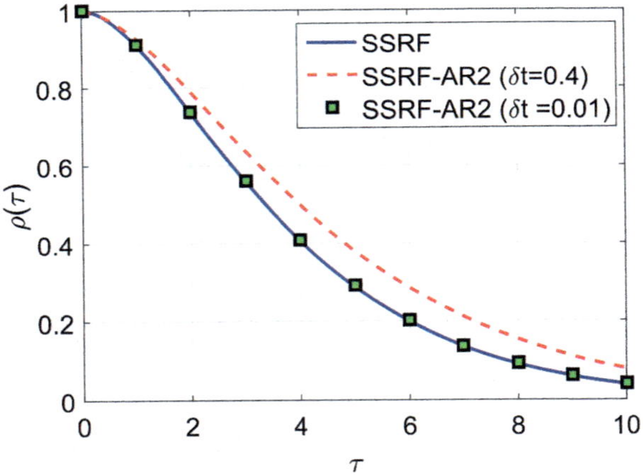

Fig. 9.3 Plots of the correlation function of the SSRF-AR(2) model with $\eta_0 = 1$, $\eta_1 = 2$, $\xi = 2$ for times steps equal to $\delta t = 0.4$ (broken line) and $\delta t = 0.01$ (circles). The SSRF-AR(2) correlation functions are compared with the continuum SSRF correlation function obtained for the same parameters. The SSRF-AR(2) model with the smaller time step is an accurate approximation of the continuum SSRF covariance

time discretization schemes (i) $\delta t = 0.4$ and (ii) $\delta t = 0.01$. Compare with the SSRF model with the same values of η_0, η_1, ξ.

Answer The plots are shown in Fig. 9.3. The continuum SSRF correlation function is obtained from (7.30b) using the respective values for the parameters.

The horizontal axis in Fig. 9.3 is the time lag τ (equal to $\xi\, h$ in terms of the dimensionless lag). The SSRF-AR(2) correlation function follows from (9.71c), and the coefficients c_1, c_2 follow from the initial conditions (9.66a) which lead to

$$c_1 = 1, \quad c_2 = \frac{\phi_1\, e^{1/\tau_0}}{1 - \phi_2} - 1.$$

The characteristic time τ_0 is evaluated using (9.71b), the parameters AR-2 coefficients ϕ_1, ϕ_2 are based on (9.66b) and (9.66c) respectively, while the coefficients α_1, α_2 are obtained from (9.63) using $\tau_c = \xi/\delta t$. The plots shown in Fig. 9.3 demonstrate that the SSRF-AR(2) correlation function is a very good approximation of the continuum SSRF correlation for $\delta t = 0.01$, i.e., for $\tau_c = 200$, while for $\delta t = 0.4$, i.e., for $\tau_c = 5$ the SSRF-AR(2) correlation deviates from the SSRF curve.

Power spectral density We calculate the SSRF-AR(2) spectral density using the ARMA spectral density expression (9.55). We use $\widetilde{C}_{\mathrm{xx}}(w) = \delta t\, \widetilde{C}'_{\mathrm{xx}}(w)$, where $\widetilde{C}'_{\mathrm{xx}}(w)$ is given by (9.55). We also use equation (9.66d) for the noise standard deviation and equations (9.63) for the coefficients $\{\alpha_i\}_{i=1}^{3}$ which lead to

$$\widetilde{C}_{\mathrm{xx}}(w) = \frac{\alpha_3^2}{(\alpha_1+\alpha_2)^2}\, \frac{\delta t}{\left|1-\phi_1 \mathrm{e}^{-\mathrm{i}\, w\, \delta t} - \phi_2\, \mathrm{e}^{-2\mathrm{i}\, w\, \delta t}\right|^2} = \tag{9.72}$$

$$\frac{\eta_0 \tau_c^{-1}}{\left(\tau_c + \sqrt{\eta_1+2}\right)^2}\, \frac{\delta t}{1+\phi_1^2+\phi_2^2 - 2\phi_1(1-\phi_2)\cos(w\delta t) - 2\phi_2\cos(2w\delta t)}.$$

This expression looks more complicated than the SSRF spectral density (9.61). The difference is due to the finite sampling step δt which implies a *Nyquist frequency* of $1/2\delta t$ and consequently an upper bound on the angular frequency $\pi/\delta t$. Hence, the spectral density of the SSRF-AR(2) model is modified with respect to the continuum density at least in the vicinity of the cutoff. This is also true for the spectral density of SSRF lattice models in higher dimensions, e.g., see (8.16) and [362].

The relation between the AR(2) coefficients ϕ_1, ϕ_2 and the SSRF parameters η_1, η_0, and τ_c is established by means of (9.64) and (9.66). In Fig. 9.4 we compare the continuum SSRF spectral density (9.61) with the SSRF-AR(2) spectral

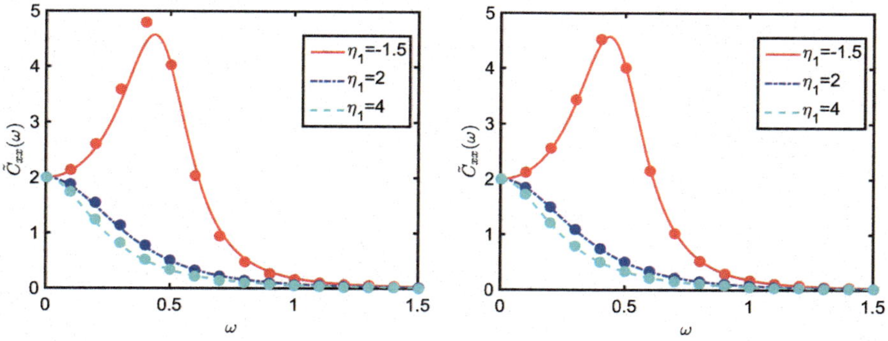

Fig. 9.4 Spectral density plots for the SSRF continuum model (continuous lines) and the SSRF-AR(2) model (circles) for three different values of $\eta_1 = (-1.5, 2, 4)$. For all three models $\eta_0 = 1$ and $\xi = 2$. The SSRF-AR(2) models use a time step equal to $\delta t = 0.4$ (left) and $\delta t = 0.1$ (right)

density (9.72) for three different values of η_1 and two time steps ($\delta t = 0.4$ and $\delta t = 0.1$). The spectral densities of the continuum SSRF and the SSRF-AR(2) models exhibit small differences for $\delta t = 0.4$, while for $\delta t = 0.1$ the respective spectral densities practically coincide. A difference, not evidenced in these figures, is that the SSRF-AR(2) spectral density cannot be observed beyond the edge of the Brillouin band (for $w > \pi/\delta t$), while the SSRF spectral density decays asymptotically to zero as $w \to \infty$.

In summary, we have established the close connection between the continuum SSRF model in one dimension and a respective second-order autoregressive time series model. We have also shown that the main statistical properties of the two models are quite similar if the SSRF characteristic time ξ is significantly larger than the sampling step δt.

Chapter 10
Spatial Prediction Fundamentals

> To boldly go where no data have gone before, to seek out plausible trends and fluctuations.

The analysis of spatial data involves different procedures that typically include *model estimation, spatial prediction, and simulation*. Model estimation or model inference refers to determining a suitable spatial model and the "best" values for the parameters of the model. Parameter estimation is not necessary for certain simple deterministic models (e.g., nearest neighbor method), since such models do not involve any free parameters. *Model selection* is then used to choose the "optimal model" (based on some specified statistical criterion) among a suite of candidates.

Due to various practical reasons, we often need to analyze *incomplete data*. *Spatial prediction* refers to the estimation of the unknown (missing) values of the spatial process. If the prediction points are "inside" the sampling area, the spatial prediction problem is referred to as *interpolation*. If the spatial prediction is conducted at points "outside" the sample area, the procedure is known as *extrapolation*. The distinction between interpolation and extrapolation requires defining a notion of proximity between the prediction point and the data points. We return to this issue below. Finally, *simulations* involve the generation of multiple probable scenarios of the spatial process, possibly under a set of constraints.

Practical motivation In the case of terrestrial observations, e.g., environmental monitoring and mineral resources exploration, the sampling locations can be viewed as the nodes of an irregular (non-uniform, random) grid. In these cases the geometry of the sampling network is dictated by financial restrictions, accessibility and environmental constraints, and the focus of the investigation (e.g., interest in urban centers, areas with pollution sources, domains of high mineralization, et cetera). In the case of remote sensing images the sampling is spatially uniform, but the image

D. T. Hristopulos, *Random Fields for Spatial Data Modeling*, Advances in Geographic Information Science, https://doi.org/10.1007/978-94-024-1918-4_10

is potentially contaminated with patches of missing data. The causes of incomplete sampling include sensor malfunctions and obscured sectors of the image (e.g., due to cloud coverage).

More recently, spatial prediction methods are used in system engineering in *multi-fidelity* modeling frameworks. Multi-fidelity approaches aim to efficiently combine high-fidelity (high accuracy and precision) data (e.g., computationally expensive numerical solutions of partial differential equations) with lower-fidelity, computationally cheaper, surrogate models [254, 447, 660]. In astrophysical research, estimates of velocity fields in distant galaxy clusters have been derived using stochastic interpolation methods [244, 875].

Interpolation and visualization To visualize the spatial process of interest, it is necessary to construct maps that are based on estimates of the field at the nodes of a regular *map grid*. The use of a regular grid structure allows visualizing the underlying process by means of isolevel contours, surface plots (in two dimensions) or volume plots (in three dimensions). Maps also allow resolving details at specified scales, such as the exploitable mineral reserves within grid blocks of desired size (selective mining units) [210]. In addition, if the estimated field represents a spatially variable coefficient in some partial differential equation (PDE), obtaining gridded values allows the application of numerical discretization methods for the solution of the PDE.

Downscaling In the case of digital images, it may be desirable to artificially "enhance" the spatial resolution, i.e., to generate data on finer grids—a procedure known as downscaling [880]. This operation is useful for the fusion of data with different resolutions. *Downscaling* is also used to obtain fine-grained estimates of meteorological variables from the output of Global climate models (GCMs). The downscaling of GCM outputs can be accomplished either by means of statistical methods or dynamical methods that use the GCM outputs to drive regional scale models [855]. Downscaling has hydrological applications as well, since it is used to obtain fine-grain representations of the subsurface physical properties [336]. The mathematical procedures employed to estimate the field at missing-value locations employ spatial prediction methods. A schematic illustrating a spatial prediction (more precisely, interpolation) problem in two dimensions is shown in Fig. 10.2.

Trends and fluctuations As we have discussed in Chaps. 1 and 2, we can decompose spatial processes in terms of trends and fluctuations. Trends are assumed to represent large-scale variations. Thus, their behavior can be "predicted" at distant points, provided that the conditions that establish the trend are valid over the entire area. For example, a spatial trend that characterizes the groundwater level within a certain hydrological basin is not necessarily accurate in a different, albeit neighboring, basin. The spatial prediction of fluctuations, on the other hand, works best if the estimation points are close to at least some of the samples. How close is close enough depends on the spatial persistence of the fluctuations in space, which can be quantified by means of the various correlation measures discussed in Chap. 5.

Interpolation and convex hull In light of the short range of the fluctuations, a conservative approach is to restrict prediction only to points that are contained inside the *convex hull* of the sample set. The convex hull is the smallest *convex set* that includes the points in question. A convex set of points defines a region such that any two points inside it can be connected with a straight line that lies completely within the region.

An illustration of the convex hull for a random set of points is given in Fig. 10.1. Computational geometry provides efficient methods for constructing the convex hull of random point sets. There is no a priori guarantee that every point inside the convex hull, in particular a point near the boundary, has a sufficient number of neighbors in the sample set to ensure accurate and precise estimation. Nonetheless, by focusing inside the convex hull, we avoid estimating at locations that often have fewer sample neighbors in their vicinity than the points inside the hull.

Extrapolation, i.e., estimation of the process outside the boundary of the convex hull is possible in principle, at least with some methods, and especially if there is

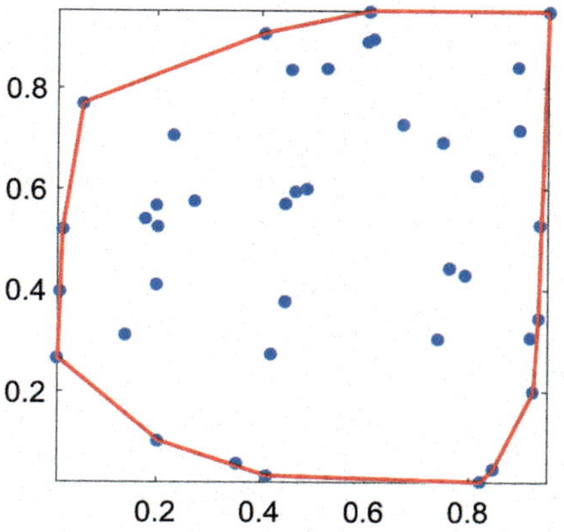

Fig. 10.1 Convex hull of a set of forty random points uniformly distributed over the unit square. The convex hull was constructed using the MATLAB function *convhull*

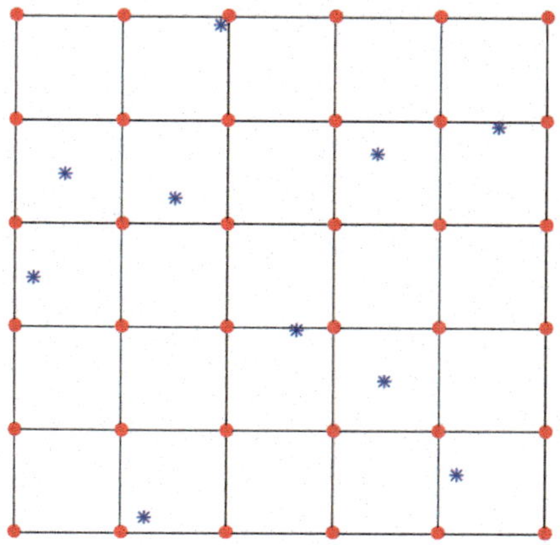

Fig. 10.2 Schematic of idealized sampling pattern depicting ten sample points (stars) and interpolation grid points (filled circles). The number of prediction points is $N = L^2$, where $L = 6$ is the number of grid nodes per side. Geostatistical methods typically require the inversion of dense $N \times N$ covariance matrices, which is an operation with computational complexity $O(N^3)$

a dominant trend. However, the range of validity of extrapolated values should be carefully considered. In general, the accuracy and the precision of the predictions deteriorate moving away from the boundary of the convex hull, especially if the estimates are mostly based on correlated fluctuations. For example, if we are trying to estimate a random field with a certain correlation range ξ, extrapolation of the field at distances $r \gg \xi$ from the nearest sampling point is not very informative.[1] On the other hand, long-range trends or first-principles governing equations—if they are available—can be used for extrapolation purposes.

In the following, we focus on *interpolation*, so that the predictions can be reliably based on the information provided by neighboring data.

Motivation for research Why is spatial prediction a topic of continuing research interest? In part, it is due to the ever increasing amount of available spatial information. The era of big data is already having an impact in many fields, including the geosciences. Increased interest is spurred by the abundance of various remote sensing products and earth-based observations [7, 770]. A similar data explosion is under way in other scientific and engineering fields [13, 857] and the social sciences as well [451, 533]. In computer science, there is also considerable interest in the mining of big spatial data and its applications to pattern recognition, climate change, social-media-based geographic information and mobility applications. Some of the relevant spatial methods and the computational challenges facing them are reviewed in [816]. In addition, most spatial prediction methods are based on some set of idealized assumptions that are not necessarily met by real-life spatial data. Hence, the persisting goal is to develop new, more flexible, and faster spatial prediction methods.

Speed The advent of low-cost spatially distributed sensors [91] and the explosive growth of remote sensing data products drive the need for the design of accurate and efficient algorithms for the analysis of large data sets. Among other technical requirements, these trends require methods with a computational time that scales favorably (i.e., preferably linearly) with data size. In addition, some popular simulation algorithms also involve an interpolation step (e.g., the polarization-based method of conditional simulation discussed in Chap. 16). The role of simulations is expanding in spatial studies, because they provide versatile tools for assessing uncertainty and extreme-case scenarios. Thus, simulation methods will also profit from advances that result in faster spatial prediction algorithms.

Continuous versus discrete random fields We will mostly consider the spatial prediction of *continuously-valued fields*. We will shortly discuss the spatial prediction of *discretely-valued fields*. The latter are useful for modeling geological structures, discrete phases (e.g., the pore space and solid matrix in porous media), and various *categorical variables*. The spatial prediction of categorical variables cannot be adequately accomplished by straightforward adaptations of methods

[1] This statement is clarified in the section on kriging.

designed for continuous data [280]. In the case of continuous fields that follow non-Gaussian probability distributions there are advantages to working with discretized representations [896, 899, 900]. The spatial prediction of discrete-valued random fields can also be viewed as a *classification problem*, since the field at the estimation points is assigned to one of several possible classes. Machine learning includes several methods that can be used for classification such as k-nearest neighbors, support vector machines, and random forests [609].

Continuum versus discrete supports Most of the spatial processes that we consider herein are assumed to take place in the continuum (e.g., we do not examine processes that evolve on networks). On the other hand, the representation of spatial processes often employs a spatial grid. For example, this could be a numerical grid as in Fig. 10.2 used for computational and visualization purposes. In classical geostatistics, random field models defined over continuum supports are used. On the other hand, Gauss-Markov random fields are defined over discrete lattices. If we are not interested in sub-grid scale correlations, there is no a priori preference for one type of model over another. Continuum-support models have the advantage that they are easily applicable to scattered spatial data and adjustable to variable spatial resolution in different areas, while discrete-support models can benefit from local representations which lead to improved computational complexity.

Deterministic versus stochastic methods There are many spatial prediction methods in the scientific literature, and different academic disciplines have favorite prediction tools. On a conceptual level one can distinguish between deterministic and stochastic interpolation methods. *Deterministic* interpolation methods are simpler to understand and implement. On the other hand, *stochastic* methods have distinct advantages as we discuss below. A review of interpolation methods used for environmental sciences applications is given in [499].

In the deterministic approach, the data are treated as a partial sample of a well-defined mathematical function. Hence, deterministic interpolation leads to a single value at each domain point. The stochastic approach views the data as a—possibly noisy—sample of a random field. This perspective implies that the predictions at the missing value locations are actually described by respective probability distributions.

We return to stochastic interpolation in Sect. 10.3. For reasons of space and personal preference we will not present methods based on neural networks and machine learning [425] or Bayesian field theory [495]. We note, however, that there is a strong connection between the geostatistical approach presented herein and *Gaussian processes* [661, 678]. This connection, as well as a Bayesian extension of stochastic interpolation, are briefly discussed in Chap. 11.

10.1 General Principles of Linear Prediction

In *linear interpolation methods* the prediction of a random field $X(\mathbf{s}; \omega)$ at some unmeasured site $\mathbf{z}$ is formulated as a linear superposition of the sample data. Every sample value is weighted by means of a coefficient (weight) λ_n, for $n = 1, \ldots, N$. For example, the prediction $\hat{x}(\mathbf{z})$ at the point $\mathbf{z}$ is given by

$$\hat{x}(\mathbf{z}) = \sum_{n=1}^{N} \lambda_n \, x_n^*, \tag{10.1}$$

where $\{x_n^*\}_{n=1}^N$ are the sample values (data) at the *nodes* of the sampling network.

To generate an interpolation map of the random field $X(\mathbf{s}; \omega)$, the prediction point $\mathbf{z}$ traces sequentially every site $\{\mathbf{z}_p\}_{p=1}^P$ of the map grid. The "hat" (circumflex) over $\hat{x}(\mathbf{z})$ denotes an estimate of the true value of $X(\mathbf{s}; \omega)$ (for the specific, partially sampled realization) which is unknown.

Two useful properties that help to classify the performance of different interpolators are exactitude and convexity.

Definition 10.1 An interpolator is *exact* if it reproduces the data at every sampling point. Exactitude is tested by keeping the interpolation point in the sample set and comparing the interpolated value at this point with the datum.

Exactitude implies that if $\mathbf{z}$ coincides with the sampling point $\mathbf{s}_k$, where $k \in \{1, \ldots, N\}$, then $\lambda_k = 1$ and $\lambda_m = 0$, for all $m \neq k$.

Definition 10.2 An interpolation method is called *convex*, if the predictions are within the range of the data., i.e., for all $\mathbf{z} \in \mathcal{D}$ it holds that

$$\min\{x_1^*, \ldots, x_N^*\} \le \hat{x}(\mathbf{z}) \le \max\{x_1^*, \ldots, x_N^*\}.$$

10.2 Deterministic Interpolation

Deterministic interpolation methods view the data as finite-dimensional samples of a deterministic function $x(\mathbf{s})$. There are numerous deterministic interpolation methods, including various splines techniques [825], radial basis functions [842], natural neighbor interpolation [84, 85], inverse distance weighting [743], and minimum curvature [101]. Since our focus is on stochastic approaches we just give a brief taste of four commonly used deterministic methods.

Remark Deterministic methods do not generate estimates of *prediction uncertainty*. Hence, they do not provide any information regarding the probability distribution that characterizes the interpolated field values.

10.2.1 The Linearity Hypothesis

Deterministic linear interpolation is based on a weighted combination of the measured values at near neighbors of the prediction point. In *one dimension*, this simply means that if s_1 and s_2 are the nearest left and right neighbors of s, so that $s_1 < s < s_2$, then $\hat{x}(s) = a_0 + a_1 s$, where the coefficients a_0 and a_1 are determined by setting $\hat{x}(s_n) = x^*(s_n)$, for $n = 1, 2$.

In two dimensions, linear interpolation is extended to *bilinear interpolation*. Bilinear interpolation also involves products of linear functions in each of the orthogonal directions, and thus terms that are proportional to $s_1\, s_2$. On rectangular grids, bilinear interpolation involves the four nearest neighbors of the target point, i.e., the nodes of a 2×2 grid cell. In this case, one uses for spatial prediction at the point $\mathbf{s} = (s_1, s_2)$ the function

$$\hat{x}(\mathbf{s}) = a_0 + a_1\, s_1 + a_2\, s_2 + a_3\, s_1\, s_2 .$$

The coefficients a_0, a_1, a_2, a_3 are determined by solving the consistency equations $\hat{x}(\mathbf{s}_n) = x_n^*$, where $n = 1, \ldots, 4$ and the x_n^* represent the measurements of the field at the four grid neighbors.

For scattered data, linear interpolation is based on the *Delaunay triangulation* of the sampling network. Prediction is formulated as a weighted averages of the sample values at the vertices of the Delaunay triangle that encloses the target point.

10.2.2 Inverse Distance Weighting

A conceptually simple deterministic methods is the so-called *inverse distance weighting (IDW)*. It is also known as *Shepard's method* from the name of its inventor [743]. In spite of its simplicity, the method is recommended in the Hydrology Handbook [28], and it is commonly used for the estimation of missing data in the Earth sciences.

The method assumes that the unknown value at some target point $\mathbf{z}$ (e.g., a node of the map grid as the one shown in Fig. 10.2) is given by a linear combination of the sample values (e.g., points marked with stars in Fig. 10.2). IDW weighs each sample point with a coefficient λ_n which is inversely proportional to a power of the distance between the target, $\mathbf{z}$, and the sample points $\{\mathbf{s}_n\}_{n=1}^N$. The interpolation equation is given by (10.1), and the weights are determined by

$$\lambda_n = \frac{\|\mathbf{z} - \mathbf{s}_n\|^{-p}}{\sum_{n=1}^{N} \|\mathbf{z} - \mathbf{s}_n\|^{-p}}, \quad p > 0, \; n = 1, \ldots, N. \tag{10.2}$$

The exponent p is empirically estimated from the data by finding the value that maximizes a selected distance measure between the data and the predictions (see Sect. 12.5 on cross validation). More commonly, the exponent is arbitrarily set to a fixed value (e.g., $p = 2$ is often used).

Properties The properties described in items 1–6 below are derived from the equations (10.1) and (10.2). The property in item (7) is due to the fact that IDW has a computational complexity that scales linearly with the sample size N.

1. Larger weights are assigned to data points that are closer to $\mathbf{z}$ than to more distant points.
2. Higher values of p increase the relative impact of sample values near $\mathbf{z}$, whereas lower p values imply more uniform weights.
3. An "optimal" p value can be obtained using cross validation approaches.
4. If $p = 0$ the IDW estimate reduces to the sample mean.
5. IDW weights are positive, i.e., $\lambda_n > 0$, and normalized so that their sum equals one, i.e., $\sum_{n=1}^{N} \lambda_n = 1$.
6. IDW is an *exact* and *convex* interpolation method.
7. The computational cost of IDW scales linearly with data size. A cutoff radius may be required for larger data sets, i.e., for $N > 10^3$ [338].
8. The method's shortcomings involve (i) the arbitrary choice of the weighting function (ii) relatively low accuracy (iii) the lack of an associated uncertainty measure and (iv) directionally independent weight functions.

Unlike IDW, the stochastic kriging methods incorporate anisotropic correlations. The non-directionality of IDW may lead to less realistic isolevel contours compared to those obtained with kriging [124].

Extensions It is possible to extend the IDW method using different notions of distance than the Euclidean norm. For example, the following transformation can be used to provide a smoother IDW interpolant

$$\frac{1}{\|\mathbf{z} - \mathbf{s}_n\|^p} \rightarrow \frac{1}{\left(\|\mathbf{z} - \mathbf{s}_n\|^2 + \sigma^2\right)^{p/2}}, \quad \text{for } n = 1, \ldots, N.$$

In principle, it is possible to include *anisotropic dependence* in IDW by means of an anisotropic distance measure, such as the ones presented in Sect. 2.4.3.

A variant of IDW is the *fixed radius IDW* [790]. In this version of IDW the weights λ_n in (10.2) are non-zero only if $\|\mathbf{z} - \mathbf{s}_n\| < R$, where R is the user-selected fixed radius. Fixed radius IDW focuses on data values that are not too distant from the target point $\mathbf{z}$, while it ignores data at distances larger than R from $\mathbf{z}$.

Example 10.1 Verify that IDW is (i) exact and (ii) convex.

Answer Let us assume that $\mathbf{z} \to \mathbf{s}_k$, where $k \in \{1, \ldots, N\}$. (i) Then, $\lambda_k \to \infty$ and $\hat{x}(\mathbf{z}) \to x(\mathbf{s}_k)$ which proves exactness. (ii) Since the weights satisfy the constraints $0 \le \lambda_n \le 1$ and $\sum_{n=1}^{N} \lambda_n = 1$, IDW is a convex interpolation method.

A physical analogy: Critics of IDW argue that it is arbitrary and devoid of physical meaning. Nevertheless, we can motivate IDW (at least for $p = 1$) by an analogy with the electric field that is generated by point charges. Remember that the electric potential due to an isolated point charge q_n at distance $\|\mathbf{r}_n\|$ from q_n is given by the inverse distance law, i.e.,

$$V_n = \frac{1}{4\pi\epsilon_0} \frac{q_n}{\|\mathbf{r}_n\|},$$

where ϵ_0 is the dielectric constant of the vacuum [324]. For a system of charges $\{q_n\}_{n=1}^{N}$, the electric potential is given by the superposition of the elementary potentials

$$V_{\text{tot}}(\mathbf{z}) = \frac{1}{4\pi\epsilon_0} \sum_{n=1}^{N} \frac{q_n}{\|\mathbf{r}_n\|}.$$

If we replace $\mathbf{r}_n$ with $\mathbf{z} - \mathbf{s}_n$ and the charges $q_n/4\pi\epsilon_0$ with x_n^*, the total potential at $\mathbf{z}$ becomes

$$V_{\text{tot}}(\mathbf{z}) = \sum_{n=1}^{N} \frac{x_n^*}{\|\mathbf{z} - \mathbf{s}_n\|}.$$

Using equations (10.1) and (10.2), the latter for $p = 1$, we can express the potential field as

$$V_{\text{tot}}(\mathbf{z}) = \hat{x}(\mathbf{z}) \sum_{n=1}^{N} \frac{1}{\|\mathbf{z} - \mathbf{s}_n\|},$$

where $\hat{x}(\mathbf{z})$ is the IDW predictor. In light of the above, we can identify $\hat{x}(\mathbf{z})$ as a "uniform effective charge" which, if placed at each sampling point, generates the same electric potential at $\mathbf{z}$ as the configuration of the $\{q_n\}_{n=1}^{N}$ point charges. The same reasoning can be extended to more general potential functions for $p \neq 1$. The above argument simply provides a physical analogy for IDW, but it does not imply its efficacy in interpolation.

10.2.3 *Minimum Curvature Interpolation*

Minimum curvature interpolation (MCI) has been introduced and applied to geophysical data by Briggs [101] and Sandwell [711]. The principle on which the method is based is intuitively simple.

Let us view a continuum process $x(\mathbf{s})$, where $\mathbf{s} \in \mathcal{D} \subset \mathbb{R}^d$, as the surface of a stretched membrane in $d+1$-dimensional space. The membrane is anchored at the sampling points, so that its height at these points corresponds to the respective

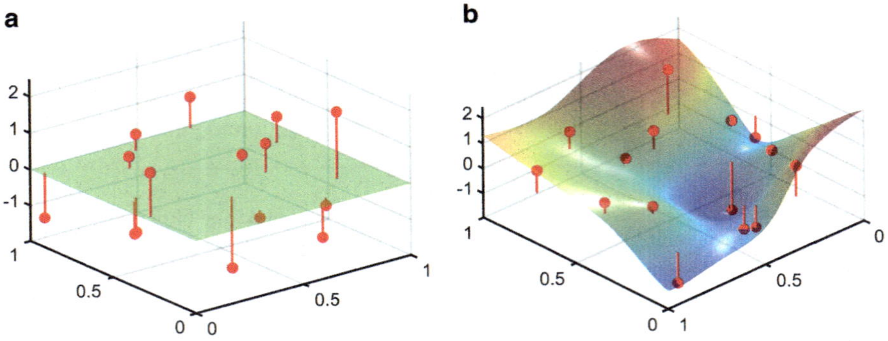

Fig. 10.3 Schematic illustrating minimum curvature interpolation in two-dimensional space. The stem plot on the left represents the sampling points (the stem traces on the semi-transparent plane through the middle of the cube) and the respective sample values (given by the stem heights). The interpolated surface on the right is pinned at the sample points by the stem values and depicts the minimum curvature function. (**a**) Sampling points. (**b**) Interpolated surface

sample values. To estimate the function at other points $\mathbf{s} \in \mathcal{D}$, we search for the membrane configuration that globally minimizes the square of the linearized curvature. The linearized curvature is given by the Laplacian of $x(\mathbf{s})$. The global linearized curvature is defined by the integral of the Laplacian over the domain. The main idea is illustrated schematically in Fig. 10.3.

Derivation of the MCI equation Next, let use see how the minimum curvature interpolation equation can be derived. The main assumptions are that (i) the function $x(\mathbf{s})$ is smooth and admits at least all second-order partial derivatives so that the *Laplace operator* $\triangle = \nabla^2$ is well-defined and (ii) the integral $\int_{\mathcal{D}} d\mathbf{s}\, [\triangle x(\mathbf{s})]^2$ over the domain $\mathcal{D}$ exists. Then, the following expression

$$C_x := \frac{1}{|\mathcal{D}|} \int_{\mathcal{D}} d\mathbf{s}\, [\triangle x(\mathbf{s})]^2, \tag{10.3}$$

defines the average of the squared linearized curvature over $\mathcal{D}$. Minimum curvature interpolation focuses on minimizing C_x, conditioned on the available data $\mathbf{x}^*$.

Since the minimization of the average square curvature is a constrained optimization problem, its solution is obtained by minimizing the following objective function

$$C'_x = C_x - \sum_{n=1}^{N} \int_{\mathcal{D}} d\mathbf{s}\, w_n \left[x(\mathbf{s}) - x_n^*\right] \delta(\mathbf{s} - \mathbf{s}_n), \tag{10.4}$$

where the coefficients w_n represent the *Lagrange multipliers* for the data constraints, and the delta functions enforce the constraints at the sampling points.

The curvature term in (10.4) involves a volume integral. Its evaluation involves *Green's theorem* which extends integration by parts to volume integrals [733]. Using Green's theorem, the curvature term in (10.3) is expressed as follows [364]:

$$\int_{\mathcal{D}} \mathrm{d}\mathbf{s} \left[\nabla^2 x(\mathbf{s})\right]^2 = \int_{\mathcal{D}} \mathrm{d}\mathbf{s}\, x(\mathbf{s})\, \nabla^4 x(\mathbf{s}) + \int_{\partial\mathcal{D}} \mathbf{da} \cdot \nabla x(\mathbf{s}) \nabla^2 x(\mathbf{s})$$
$$- \int_{\partial\mathcal{D}} \mathbf{da} \cdot \nabla \left[\nabla^2 x(\mathbf{s})\right] x(\mathbf{s}), \tag{10.5}$$

where $\int_{\partial\mathcal{D}} \mathbf{da}$ denotes the surface integral along the boundary of the domain of interest, and $\mathbf{da} \cdot \mathbf{b}(\mathbf{s})$ denotes the *inner product* of the vectors $\mathbf{da}$ and $\mathbf{b}(\mathbf{s})$. The vector $\mathbf{da}$ has a norm that is equal to the differential of the surface element and points perpendicular to the surface in the outward direction.

Let $\delta[.]/\delta x(\mathbf{z})$ denote the *functional derivative* of the curvature functional (10.4) with respect to the field value at the estimation point $\mathbf{z}$. The minimization of the objective functional (10.4) occurs at the stationary point

$$\frac{\delta C'\left(x(\mathbf{s}); \{w_n\}_{n=1}^N\right)}{\delta x(\mathbf{z})} = 0.$$

The minimization leads to the following PDE

$$\nabla^4 x(\mathbf{z}) - \sum_{n=1}^{N} w_n\, \delta(\mathbf{s}_n - \mathbf{z}) = 0. \tag{10.6}$$

The PDE (10.6) represents the *biharmonic equation* with a source term that involves a weighted superposition of delta functions. To find the solution of (10.6) in the presence of sources, we need the fundamental solution, i.e., the *Green function* of the biharmonic equation. The latter is defined as the solution of the biharmonic equation in the presence of a delta function source, i.e.,

$$\nabla^4 G_0(\mathbf{s} - \mathbf{s}') = \delta(\mathbf{s} - \mathbf{s}'). \tag{10.7}$$

In light of the linearity of the PDE (10.6), the *specific solution* can be expressed as the convolution of the fundamental solution, $G_0(\mathbf{s}-\mathbf{s}')$, with the source term (see Sect. 2.6), i.e.,

$$x(\mathbf{z}) = \int \mathrm{d}\mathbf{s}' G_0(\mathbf{z} - \mathbf{s}') \sum_{n=1}^{N} w_n\, \delta(\mathbf{s}_n - \mathbf{s}'). \tag{10.8}$$

Fortunately, explicit solutions for the fundamental solution of the biharmonic equation are available [14, 711].[2]

Biharmonic Green function The Green function $G_0(\cdot)$ of the biharmonic equation over an unbounded spatial domain in $\mathbb{R}^d$ is given by the following equation, where S_d is the surface area of the unit sphere in d dimensions [14]

$$G_0(\mathbf{r}) = \begin{cases} \frac{1}{8\pi} \|\mathbf{r}\|^2 \left(\ln \|\mathbf{r}\| - 1\right), & d = 2 \\ -\frac{1}{8\pi} \|\mathbf{r}\|, & d = 3 \\ -\frac{1}{8\pi^2} \ln \|\mathbf{r}\|, & d = 4 \\ \frac{1}{2(d-2)(d-4)S_d} \|\mathbf{r}\|^{4-d}, & d \geq 5. \end{cases} \tag{10.9}$$

In light of the Green function definition (10.7) and (10.8), $x(\mathbf{z})$ is given by the following linear superposition of Green's functions

$$x(\mathbf{z}) = \sum_{n=1}^{N} w_n \, G_0(\mathbf{s}_n - \mathbf{z}). \tag{10.10}$$

The above equation is valid for any $\mathbf{z} \in \mathcal{D}$. Thus, the Lagrange multipliers $\{w_n\}_{n=1}^{N}$ represent interpolation coefficients that are independent of the interpolation point. The coefficients are therefore determined by forcing the interpolated function (10.10) to agree with the data at the sampling points $\mathbf{s}_k$, for $k = 1, \ldots, N$. This constraint requires solving the $N \times N$ system of linear equations

$$x_k^* = \sum_{n=1}^{N} w_n \, G_0\left(\mathbf{s}_n - \mathbf{s}_k\right), \; k = 1, \ldots, N. \tag{10.11}$$

Note that the linear system (10.11) is well defined in $d = 2, 3$ where $G_0(\mathbf{0}) = 0$. In higher dimensionality, the biharmonic Green function has a singularity at the origin.

The minimum curvature method has some attractive conceptual features. First, it expresses interpolation as an optimization problem that involves a geometric quantity (i.e., global squared curvature). Second, if the optimization problem is cast in the form of a PDE, the data constraints can be handled as source/sink terms in that PDE. The first idea is closely related with the formulation of Spartan spatial random fields that are discussed in Chap. 7.

[2] The reference [711] does not include the constant, dimension-dependent coefficients of the biharmonic Green functions given in (10.9). This is not an issue for MCI, since constant factors are automatically compensated by the weights w_n.

Multiple-point interpolation Let us now consider interpolation of the data at $P \in \mathbb{Z}_+$ points. As discussed in Sect. 1.5.2, the data points may be scattered or the nodes of a regular grid. The same spatial arrangements are also possible for the prediction sites.

To use a unified notation, we assume that the prediction values are contained in a vector denoted by $\hat{\mathbf{x}}$. In the case of a regular grid, this means that the vector index corresponds to the row and column indices of the respective grid point.

Using vector matrix notation, we can express the prediction as follows

Minimum curvature interpolation equation

$$\hat{\mathbf{x}} = \mathbf{G}_{p,s}\, \mathbf{G}_{s,s}^{-1}\, \mathbf{x}^{*}. \tag{10.12}$$

The equation (10.12) uses the following notation:

1. $\hat{\mathbf{x}}$ is the $P \times 1$ vector of estimates at the prediction points
2. $[\mathbf{G}_{s,s}]_{n,m} = G_0(\mathbf{s}_n, \mathbf{s}_m)$ is the $N \times N$ Green function matrix between data points
3. $[\mathbf{G}_{p,s}]_{k,l} = G_0(\mathbf{z}_k, \mathbf{s}_l)$ is the $P \times N$ Green function matrix between prediction and data points.
4. $\mathbf{x}^{*}$ is the $N \times 1$ vector of the data values.

Using the general notation of the linear interpolation equation (10.1), the MCI prediction is expressed as follows

$$\begin{pmatrix} \hat{x}_1 \\ \hat{x}_2 \\ \vdots \\ \hat{x}_P \end{pmatrix} = \begin{pmatrix} \lambda_{1,1} & \lambda_{1,2} & \ldots & \lambda_{1,N} \\ \lambda_{2,1} & \lambda_{2,2} & \ldots & \lambda_{2,N} \\ \vdots & \ldots & \ldots & \vdots \\ \lambda_{P,1} & \lambda_{P,2} & \ldots & \lambda_{P,N} \end{pmatrix} \begin{pmatrix} x_1^{*} \\ x_2^{*} \\ \vdots \\ x_N^{*} \end{pmatrix}, \tag{10.13}$$

where

$$\boldsymbol{\lambda} = \mathbf{G}_{p,s}\, \mathbf{G}_{s,s}^{-1}$$

is the matrix of the *interpolation weights*. The p-th row of $\boldsymbol{\lambda}$ comprises the weights used to obtain the estimate for the prediction point $\mathbf{z}_p$, where $p = 1, \ldots, P$.

MCI properties

1. MCI is suitable for *smooth (differentiable)* surfaces.
2. The MC method is an *exact* interpolator.
3. As a result of the exactitude property, MCI may lead to *unrealistic oscillations*, especially if the sample contains neighboring points with significantly different values. For example, this condition may arise if the data are contaminated by noise.

4. To account for this shortcoming, the MC method has been modified by including a *regularization term* proportional to the square of the gradient in the cost function.

   The *biharmonic operator* ∇^4 is then replaced by the following linear combination of the biharmonic and the Laplacian operators

$$(1-g)\,\nabla^4 + g\,\nabla^2,$$

   where $g \in [0, 1]$ is an empirical *tension factor* that determines the balance between the two terms [760], and ∇^2 represents the "cost" attributed to gradients. Note the similarity between this formulation of MCI and the differential operators in the PDE of the SSRF covariance functions (9.31b).
5. The main numerical cost of minimum curvature interpolation comes from the inversion of the Green function matrix $\mathbf{G}_{s,s}$, which is an operation with an $O(N^3)$ computational complexity.
6. The solution (10.12) of the MCI system has the same form as the *simple kriging* solution (10.32) if the covariance function in the latter is replaced with the Green function $G_0(\mathbf{r})$ given by (10.9). Hence, we can view MCI as a stochastic interpolation method with a generalized covariance function given by the Green function of the biharmonic equation.

Example 10.2 Confirm (10.9) for the fundamental solution of the biharmonic equation.

Answer Let $G_{\mathcal{L}}(\mathbf{r})$ denote the fundamental solution of the Laplace equation $\nabla^2 G_{\mathcal{L}}(\mathbf{r}) = \delta(\mathbf{r})$. The respective functional forms are given in Table 2.3. Consequently, the fundamental solution of the biharmonic equation satisfies $\nabla^2 G_0(\mathbf{r}) = G_{\mathcal{L}}(\mathbf{r})$, as it can easily be verified that $\nabla^4 G_0(\mathbf{r}) = \delta(\mathbf{r})$.

Next, we recall that the fundamental solutions of the Laplace and biharmonic equations over unbounded domains are radial functions of $r = \|\mathbf{r}\|$. This means that the Laplacian operator, expressed in spherical coordinates, depends only on r. Consequently the Laplacian of $G_0(r)$ is given by:

$$\nabla^2 G_0(r) = \frac{1}{r^{d-1}} \frac{\mathrm{d}}{\mathrm{d}r}\left[\left(r^{d-1}\frac{\mathrm{d}G_0(r)}{\mathrm{d}r}\right)\right]$$

Hence, the following ODE is obtained for the fundamental solution of the biharmonic equation [14]

$$\frac{1}{r^{d-1}} \frac{\mathrm{d}}{\mathrm{d}r}\left[\left(r^{d-1}\frac{\mathrm{d}G_0(r)}{\mathrm{d}r}\right)\right] = G_{\mathcal{L}}(r).$$

It is straightforward to solve this ODE by integration leading to

$$G_0(r) = \int_{-\infty}^{r} \mathrm{d}y \frac{1}{y^{d-1}} \int_{-\infty}^{y} \mathrm{d}x\, x^{d-1}\, G_{\mathcal{L}}(x).$$

The double integral can be evaluated by taking into account the expressions in Table 2.3 for the Laplace Green function, and it yields (10.9).

10.2.4 Natural Neighbor Interpolation

One of the simplest deterministic interpolants is *nearest neighbor interpolation* (NNI). This method assigns to each node of the interpolation grid a value that is equal to the value of its nearest neighbor *in the sample*. By definition, NNI interpolated values are drawn exclusively from the sample set. It is very simple to generate NNI-based maps, even in high-dimensional spaces. However, such maps suffer from lack of continuity that gives them a grainy texture.

The *natural neighbor interpolation* (NANI) method is an extension of NNI that addresses some of the former's shortcomings. NANI is based on the Voronoi tessellation of a discrete set of spatial points [84, 85]. It expresses the prediction at the target point as a weighted sum of the sample values at the natural neighbors, i.e., the centers of the neighboring Voronoi polygons.

Both the selection of the neighbors and the respective weights are based on the *Voronoi tessellation* of the sampling point set and the set that results after inserting the interpolation point (see Fig. 10.4 for an illustration).

1. First, the Voronoi tessellation of the sampling network is determined.

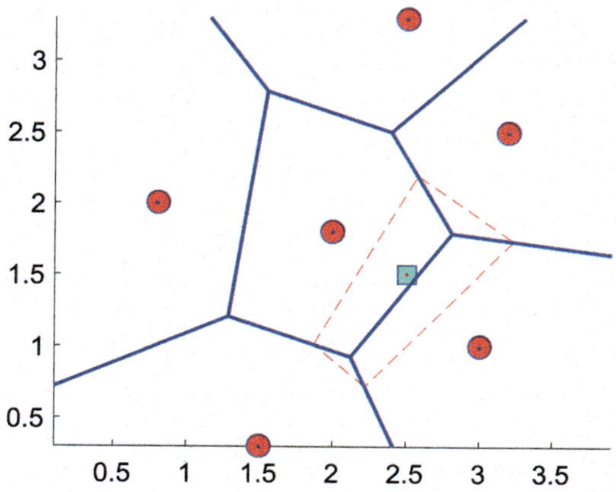

Fig. 10.4 Example of Voronoi construction for natural neighbor interpolation. The continuous lines represent the Voronoi tessellation of the sampling point set which is marked by circles. The prediction point is marked by a square. The Voronoi cell attached to the prediction point is marked by the broken line. The natural neighbor set, $S_{\mathbf{z}}$, of $\mathbf{z}$ comprises four points. The Voronoi diagrams were constructed using the MATLAB function *voronoi*

2. Then, the prediction point $\mathbf{z}$ is added. This inserts a new Voronoi polygon centered at the prediction point and at the same time it reduces the area "owned" by the polygons of the neighboring sampling points.
3. The new polygon, centered at $\mathbf{z}$, intersects Voronoi cells generated by the sampling points $\mathbf{s}_n$ that are in the neighborhood of the prediction point. These sampling points are the *natural neighbors* of the prediction point.
4. The NANI prediction is then formed by the following weighted sum over the set $\mathcal{S}_{\mathbf{z}}$ of the natural neighbors of point $\mathbf{z}$

$$x(\mathbf{z}) = \sum_{\mathbf{s}_n \in \mathcal{S}_{\mathbf{z}}} w_n x_n^*. \tag{10.14}$$

5. The weight w_n associated with each natural neighbor of $\mathbf{z}$ is proportional to the percentage of the area of the new Voronoi polygon (centered at the prediction point) that overlaps the polygon of the respective natural neighbor (as determined before the insertion of the prediction point).

Efficient procedures for performing natural neighbor interpolation so far exist only in two dimensions.

NANI properties

1. NANI can handle data with different probability distributions, because it does not involve a distributional assumption.
2. It does not require any user-defined parameters and is thus suitable for automatic spatial prediction.
3. It is an exact interpolator (see Definition 10.1).
4. It ensures continuous first and second derivatives of the interpolated function everywhere except at the sampling points. However, in practice these discontinuities are not an issue for scattered data interpolation, since the spatial prediction grid is unlikely to include sampling points.

10.3 Stochastic Methods

The deterministic spatial prediction methods presented above (IDW, MCI, NNI, NANI) are conceptually simple and permit a preliminary exploration of the data. These methods are either parameter-free or they involve only a small number of parameters that can be determined by trial and error or by optimizing a selected fitting criterion.

Stochastic methods, on the other hand, seem rather mysterious by comparison. The standard approach relies on defining the "prediction error" as the difference between the true (unknown) value of the random field and its prediction. Even though the prediction error is unknown, the intrepid modeler charges ahead, requiring that the elusive error satisfy constraints that render the prediction optimal in some specified sense. If these constraints are followed to their logical conclusion, explicit expressions for the prediction "magically" emerge in terms of parameters that can be inferred from the available spatial data.

Stochastic methods are conceptually more complex and typically involve several parameters that are determined by optimizing the "fit" of the model to the spatial data. This task can be accomplished by means of statistical methods such as maximum likelihood and the method of moments. The stochastic spatial prediction methods are also known as *kriging methods*. Since stochastic methods provide tools for characterizing uncertainty, they have gained acceptance in the earth sciences where uncertainties abound. A non-technical introduction to the application of these methods in petroleum engineering is given in [123, 124].

Stochastic linear prediction In stochastic methods the linear spatial prediction equation (10.1) has a dual significance. First, as in the deterministic sense, it expresses the prediction in terms of the sample values. In the stochastic framework, however, the sample values are assumed to represent realizations of a random field. Hence, more generally it can be claimed that the following relation holds between the random field at the unknown location and the sampling sites

$$\hat{X}(\mathbf{z};\,\omega) = \sum_{n=1}^{N} \lambda_n\, X(\mathbf{s}_n;\,\omega). \tag{10.15}$$

The *prediction error* of this estimator with respect to the true, albeit unknown, random variable $X(\mathbf{z};\,\omega)$, is given by

$$\epsilon(\mathbf{z};\,\omega) = X(\mathbf{z};\,\omega) - \hat{X}(\mathbf{z};\,\omega). \tag{10.16}$$

Stochastic estimators should ideally satisfy certain statistical properties that determine their quality. For example, it is desired that they are *unbiased* and have *minimum variance*. It is often required that they are also *exact* in the sense explained in Definition 10.1. The condition of exactitude typically assumes that the data are reliable, i.e., that we have a large degree of confidence in their values.

Definition 10.3 A spatial prediction method is *unbiased* if the expectation of the estimate at points $\mathbf{z} \in \mathcal{D}$ is equal to the true expectation, i.e.,

$$\mathcal{E}\left[\hat{X}(\mathbf{z};\,\omega)\right] = \mathcal{E}\left[X(\mathbf{z};\,\omega)\right]. \tag{10.17}$$

The zero-bias property should hold at every point of the map grid. An interesting observation is that often we do not know a priori neither the left nor the right-hand side of the zero-bias equation (10.17). Nevertheless, the equality (10.17) suffices to provide constraints that render the estimator unbiased. We shall revisit this idea below.

Definition 10.4 A spatial prediction method is called a *minimum-mean-square error* estimator if it minimizes the *mean square error* of the estimate, i.e., the following expectation

$$\mathrm{MSE} = \mathcal{E}\left[\left\{\hat{\mathrm{X}}(\mathbf{z};\omega) - \mathrm{X}(\mathbf{z};\omega)\right\}^2\right]. \tag{10.18}$$

Optimal weights For a linear estimator defined by (10.15), the minimization is carried out with respect to the weights $\{\lambda_n\}_{n=1}^N$. Note that based on (10.15), the estimator $\hat{\mathrm{X}}(\mathbf{z})$ is a function of the weights, even though this dependence is not explicitly shown. Hence, the mathematical problem of the estimation is a *constrained minimization* problem. It is expressed as follows

$$\hat{\boldsymbol{\lambda}} = \underset{\lambda_1,\lambda_2,\ldots,\lambda_N}{\arg\min} \left(\mathcal{E}\left[\epsilon^2(\mathbf{z};\omega)\right] \mid \mathcal{E}[\epsilon(\mathbf{z};\omega)] = 0\right), \tag{10.19}$$

where $\hat{\boldsymbol{\lambda}} = (\hat{\lambda}_1, \ldots, \hat{\lambda}_N)^\top$ is the vector of optimal weights and $\epsilon(\mathbf{z};\omega)$ is the prediction error defined in (10.16). The vertical bar indicates that the minimum MSE should respect the zero-bias condition.

The concepts of zero bias and variance minimization are essential in the stochastic framework, because they characterize the performance of the interpolator and they are used to derive the optimal values of the linear weights. The stochastic predictor that minimizes the mean square error and satisfies the above condition of zero bias is known as *stochastic optimal linear predictor* (SOLP) and as *best linear unbiased estimator* (BLUE) .

Finally, the optimal linear prediction at the point $\mathbf{z}$ based on the data vector $\mathbf{x}^*$ is given by

$$\hat{x}(\mathbf{z}) = \sum_{n=1}^{N} \hat{\lambda}_n\, x_n^*. \tag{10.20}$$

Notational convention Often the optimal weights $\hat{\lambda}_n$ are not distinguished in notation from the "free" weights λ_n. To simplify the notation, we will also drop the "hat" from the optimal weights in the following.

Prediction error distribution The stochastic estimator $\mathrm{X}(\mathbf{z};\omega)$ is actually a random variable as evidenced in (10.15) and so is the prediction error (10.16). This means that the error follows a respective probability distribution. Our "optimal estimator" (10.19) controls the mean and the variance of the error: the mean is zero and the variance is minimized. These constraints are appropriate for symmetric error distributions, and they completely determine the error distribution if the latter is

Gaussian. If the error, however, follows an asymmetric probability distribution, e.g., a pdf with heavy tails, the minimum mean square estimator may not be adequate to control the error.

Historical note Since the ideas on which the optimal linear estimator are based are quite general, "BLUE methods" have been developed independently in different research fields. We will skip the historical account of the development of kriging which involves an international cast of researchers from the Soviet Union (Andrey Kolmogorov, Lev Gandin), South Africa (Danie Krige), and France (Georges Matheron). Interested readers, however, can find more information in [132] and in [164]. In the geosciences, BLUE methods are known as *kriging* in honor of Danie Krige, a mining engineer who first used them in applications focusing in mineral resources estimation.

10.4 Simple Kriging (SK)

The simplest kriging method is appropriately known by the name of *simple kriging* (SK). *The main assumptions used in SK are (i) that the random field* $\mathrm{X}(\mathbf{s};\omega)$ *is wide-sense stationary and (ii) that the constant mean* $\mathcal{E}[\mathrm{X}(\mathbf{s};\omega)] = m_{\mathrm{x}}$ *is known.* In light of the stationarity condition, the variance $\mathbb{V}\mathrm{ar}\,\{\mathrm{X}(\mathbf{s};\omega)\} = \sigma_{\mathrm{x}}^2$ is constant everywhere, while the covariance depends only on the lag vector between locations, i.e., $C_{\mathrm{xx}}(\mathbf{s}_n, \mathbf{s}_m) = C_{\mathrm{xx}}(\mathbf{s}_n - \mathbf{s}_m)$.

The assumption of known constant mean implies that the mean can be derived from theoretical analysis or can be accurately estimated from the data. The constant mean assumption is relaxed in the related method of ordinary kriging.

Simple Kriging equation
In the case of simple kriging, the general SOLP equation (10.20) is expressed by taking advantage of the constant mean as follows

$$\hat{x}(\mathbf{z}) := m_{\mathrm{x}} + \hat{x}'(\mathbf{z}) = m_{\mathrm{x}} + \sum_{n=1}^{N} \lambda_n \, x'^{*}_{n}, \tag{10.21}$$

where x'^{*}_{n} is the fluctuation of the data around the mean at the sampling location $\mathbf{s}_n$, given by

$$x'^{*}_{n} = x^{*}_{n} - m_{\mathrm{x}}, \quad \text{for } n = 1, \ldots, N.$$

More generally, the random field fluctuation at $\mathbf{z}$ can be expressed as a linear combination of the fluctuations at the sampling locations, i.e.,

$$\hat{X}(\mathbf{z};\omega) = m_x + \sum_{n=1}^{N} \lambda_n \, X'(\mathbf{s}_n;\omega). \tag{10.22}$$

In (10.22) the weights are real numbers. Their optimal values are obtained by enforcing the minimum mean square error condition (see below).

SK is an unbiased estimator The SK estimator (10.22) is unbiased in the sense of (10.17). This is a straightforward conclusion from (10.22) by taking the expectation on both sides and recalling that the expectation of the fluctuations vanishes.

The optimality criterion used to determine the optimal weights is the minimization of the mean square error, $\mathcal{E}[\epsilon^2(\mathbf{z};\omega)]$. Simple kriging is thus an MMSE (minimum mean square error) estimator .

The main steps in the derivation of the SK predictor are outlined below.

1. Let us denote by $\sigma^2_{\rm sk}(\mathbf{z}) = \mathcal{E}[\epsilon^2(\mathbf{z};\omega)]$ the variance of the prediction error (10.16). Using the interpolation equation (10.22), the error variance is expressed as follows

$$\begin{aligned}\sigma^2_{\rm sk}(\mathbf{z}) &= \mathcal{E}\left[\left(\sum_{n=1}^{N} \lambda_n \, X'(\mathbf{s}_n;\omega) - X'(\mathbf{z};\omega)\right)^2\right] \\ &= \sum_{\alpha=1}^{N}\sum_{\beta=1}^{N} \lambda_\alpha \lambda_\beta \, \mathcal{E}\left[X'(\mathbf{s}_\alpha;\omega) X'(\mathbf{s}_\beta;\omega)\right] \\ &\quad - 2\sum_{\alpha=1}^{N} \lambda_\alpha \mathcal{E}\left[X'(\mathbf{s}_\alpha;\omega)\, X'(\mathbf{z};\omega)\right] + \mathcal{E}\left[X'(\mathbf{z};\omega)^2\right].\end{aligned}$$

   The last term is equal to $\sigma^2_{\rm x}$ due to the stationarity of the random field.

2. By taking into account the definition of the covariance function (3.40), the *mean square error of the prediction* is expressed as follows

$$\sigma^2_{\rm sk}(\mathbf{z}) = \sigma^2_{\rm x} + \sum_{\alpha=1}^{N}\sum_{\beta=1}^{N} \lambda_\alpha \lambda_\beta \, C_{\rm xx}(\mathbf{s}_\alpha - \mathbf{s}_\beta) - 2\sum_{\alpha=1}^{N} \lambda_\alpha C_{\rm xx}(\mathbf{s}_\alpha - \mathbf{z}). \tag{10.23}$$

3. The minimum mean square error conditions are determined by solving the linear system that follows from the *first-order optimality criterion*; the latter requires that the partial derivatives of the error variance with respect to the linear weights vanish, leading to

$$\frac{\partial \sigma^2_{\rm sk}(\mathbf{z})}{\partial \lambda_\alpha} = 0, \quad \text{for } \alpha = 1, \ldots, N. \tag{10.24}$$

4. The equations (10.24) lead to the following linear system for the optimal weights

$$\sum_{\beta=1}^{N} \lambda_\beta \, C_{xx}(\mathbf{s}_\alpha - \mathbf{s}_\beta) = C_{xx}(\mathbf{s}_\alpha - \mathbf{z}), \quad \text{for } \alpha = 1, \ldots, N, \tag{10.25}$$

5. To ensure that the weights obtained by the solution of (10.25) determine the minimum of the error variance, we evaluate the *Hessian matrix* of the error variance with respect to the weights. The Hessian matrix involves the second-order partial derivatives which are given by

$$H_{\alpha,\beta} = \frac{\partial^2 \sigma^2_{\text{sk}}(\mathbf{z})}{\partial \lambda_\alpha \partial \lambda_\beta}, \text{ for } \alpha = 1, \ldots, N, \text{ and } \beta = 1, \ldots, N.$$

If the Hessian matrix is *positive definite*, the stationary point $\boldsymbol{\lambda}$ defined by (10.25) minimizes the variance. Calculating the second-order derivatives of (10.23) with respect to the weights leads to

$$H_{\alpha,\beta} = C_{xx}(\mathbf{s}_\alpha - \mathbf{s}_\beta), \text{ for } \alpha = 1, \ldots, N, \text{ and } \beta = 1, \ldots, N. \tag{10.26}$$

Thus, the Hessian of the error variance coincides with the covariance matrix of $X(\mathbf{s}; \omega)$. Based on Bochner's theorem, the covariance is positive definite. Hence, the nature (minimum) of the stationary point is confirmed.

Based on the above, the linear system of simple kriging equations given by (10.25) minimizes the mean square error of the SK predictor.

Using the expression for the kriging weights from (10.25) in the mean square error expression (10.23), we obtain the following equation for the *kriging variance*

$$\sigma^2_{\text{sk}}(\mathbf{z}) = \sigma^2_{\text{x}} - \sum_{\alpha=1}^{N} \lambda_\alpha C_{xx}(\mathbf{s}_\alpha - \mathbf{z}). \tag{10.27}$$

The system of kriging equations (10.25) is expressed in matrix form as follows

Simple Kriging System—Covariance formulation

$$\begin{bmatrix} \sigma^2_{\text{x}} & C_{xx}(\mathbf{s}_1 - \mathbf{s}_2) & \ldots & C_{xx}(\mathbf{s}_1 - \mathbf{s}_N) \\ C_{xx}(\mathbf{s}_2 - \mathbf{s}_1) & \sigma^2_{\text{x}} & \ldots & C_{xx}(\mathbf{s}_2 - \mathbf{s}_N) \\ \vdots & \vdots & \vdots & \vdots \\ C_{xx}(\mathbf{s}_N - \mathbf{s}_1) & C_{xx}(\mathbf{s}_N - \mathbf{s}_2) & \vdots & \sigma^2_{\text{x}} \end{bmatrix} \begin{bmatrix} \lambda_1 \\ \lambda_2 \\ \vdots \\ \lambda_N \end{bmatrix} = \begin{bmatrix} C_{xx}(\mathbf{s}_1 - \mathbf{z}) \\ C_{xx}(\mathbf{s}_2 - \mathbf{z}) \\ \vdots \\ C_{xx}(\mathbf{s}_N - \mathbf{z}) \end{bmatrix}. \tag{10.28}$$

The above can also be expressed in terms of the correlation function. This is achieved by dividing both sides of the equation (10.28) with σ_x^2. The correlation-based linear system is thus given by

Simple Kriging System—Correlation formulation

$$\begin{bmatrix} 1 & \rho_{xx}(\mathbf{s}_1 - \mathbf{s}_2) & \dots & \rho_{xx}(\mathbf{s}_1 - \mathbf{s}_N) \\ \rho_{xx}(\mathbf{s}_2 - \mathbf{s}_1) & 1 & \dots & \rho_{xx}(\mathbf{s}_2 - \mathbf{s}_N) \\ \vdots & \vdots & \vdots & \vdots \\ \rho_{xx}(\mathbf{s}_N - \mathbf{s}_1) & \rho_{xx}(\mathbf{s}_N - \mathbf{s}_2) & \vdots & 1 \end{bmatrix} \begin{bmatrix} \lambda_1 \\ \lambda_2 \\ \vdots \\ \lambda_N \end{bmatrix} = \begin{bmatrix} \rho_{xx}(\mathbf{s}_1 - \mathbf{z}) \\ \rho_{xx}(\mathbf{s}_2 - \mathbf{z}) \\ \vdots \\ \rho_{xx}(\mathbf{s}_N - \mathbf{z}) \end{bmatrix}. \tag{10.29}$$

10.4.1 *Compact Form of SK Equations*

It is easier to recall the kriging equations using the more compact matrix notation. The covariance-based formulation of kriging, i.e., (10.28), is expressed as follows

$$\mathbf{C}_{d,d}\,\boldsymbol{\lambda} = \mathbf{C}_{d,p}, \tag{10.30}$$

where $\mathbf{C}_{d,d}$ represents the $N \times N$ covariance matrix of the observation (data) points,

$$[\mathbf{C}_{d,d}]_{n,m} = C_{xx}(\mathbf{s}_n - \mathbf{s}_m), \text{ for all } n, m = 1, \dots, N,$$

$\boldsymbol{\lambda}$ stands for the vector of kriging weights

$$\boldsymbol{\lambda} = (\lambda_1, \dots, \lambda_N)^\top,$$

and $\mathbf{C}_{d,p}$ is the $N \times 1$ vector of the covariance function evaluated for all pairs that involve the prediction and the sampling points, i.e.,

$$[\mathbf{C}_{d,p}]_n = C_{xx}(\mathbf{z} - \mathbf{s}_n), \text{ for } n = 1, \dots, N.$$

A unique solution of (10.30) exists if $\mathbf{C}_{d,d}$ is invertible. The precision matrix $\left[\mathbf{C}_{d,d}\right]^{-1}$ exists if $\mathbf{C}_{d,d}$ is a positive definite matrix. The matrix is positive definite since the generating covariance function $C_{xx}(\mathbf{r})$ is *positive definite* [75].

Then, the solution for the vector of the SK weights is given by

$$\boldsymbol{\lambda} = \left[\mathbf{C}_{\mathrm{d,d}}\right]^{-1} \mathbf{C}_{\mathrm{d,p}}. \qquad (10.31)$$

Based on the above and (10.22), the *SK point prediction* is given by the equation

$$\hat{x}(\mathbf{z}) = m_{\mathrm{x}} + \left[\mathbf{C}_{\mathrm{d,d}}\right]^{-1} \mathbf{C}_{\mathrm{d,p}}(\mathbf{x}^{*} - \mathbf{m}_{\mathrm{x}}). \qquad (10.32)$$

In addition, based on (10.31), the *kriging variance* (10.27) is given by

$$\sigma^2_{\mathrm{sk}}(\mathbf{z}) = \sigma^2_{\mathrm{x}} - \mathbf{C}^{\top}_{\mathrm{d,p}} \left[\mathbf{C}_{\mathrm{d,d}}\right]^{-1} \mathbf{C}_{\mathrm{d,p}}. \qquad (10.33)$$

The 95% prediction interval for SK is given by

$$\left[\hat{x}(\mathbf{z}) - 1.96\sigma_{\mathrm{sk}}, \quad \hat{x}(\mathbf{z}) + 1.96\sigma_{\mathrm{sk}}\right]. \qquad (10.34)$$

General prediction intervals can be defined by replacing 1.96 with the desired quantile z_p of the normal distribution. The prediction interval defined above assumes that the random field follows the Gaussian distribution. It neglects the possibility of *model error*, e.g., misspecification of the covariance function. In addition, it also neglects possible uncertainty due to the estimation of the covariance model parameters. Hence, the SK prediction intervals tend to underestimate the uncertainty of the modeled process.

10.4.2 Properties of the SK Predictor

Variance bounds *The kriging variance does not exceed the variance of the random field.* This useful result can be simply derived from (10.33) as shown below.

- The right-hand side of (10.33) is non-negative because it represents the mean square error.
- We will demonstrate that the second term on the right-hand side is non-negative.
- Since $\mathbf{C}_{\mathrm{d,d}}$ is a positive definite matrix, the same is true for its inverse matrix $\mathbf{J} = [\mathbf{C}]^{-1}$.
- The vector $\mathbf{C}_{\mathrm{d,p}}$ of covariance elements may contain negative terms, if the covariance function has negative holes. Let us denote this vector by $\boldsymbol{\chi}$.
- The second term in (10.33) is then equal to $\sum_{\alpha,\beta} \chi_\alpha \, J_{\alpha,\beta} \, \chi_\beta \geq 0$. The inequality is true because the matrix $\mathbf{J}$ is positive definite.
- Based on the above, it is clear that $0 \leq \sigma^2_{\mathrm{sk}}(\mathbf{z}) \leq \sigma^2_{\mathrm{x}}$.

Differentiability of the kriging prediction surface The prediction points $\mathbf{z}_p \in \mathcal{D}$, (10.32) define a "hyper-surface" in $(d+1)$-dimensional space (i.e., a surface if $d = 2$ and a curve if $d = 1$).[3] The differentiability of the prediction surface is determined by the covariance function.

Of the three terms involved in the prediction equation (10.32), only $\mathbf{C}_{\mathrm{d,p}}$ depends on the prediction points $\mathbf{z}$. Hence, the differentiability of the prediction surface is controlled by the differentiability of the covariance function. More precisely, the prediction surface is differentiable at the sampling points only if the covariance function is differentiable at zero lag, since if $\mathbf{z}$ coincides with a sampling point $\mathbf{s}_n$ the derivative of the covariance function at zero lag must exist. According to the above, covariance functions that correspond to non-differentiable random fields, e.g., the exponential covariance, generate prediction surfaces that are non-differentiable at the sampling points. We illustrate this effect in the Examples 10.4 and 10.5.

Connection of kriging with conditional mean and variance According to (10.22) the SK prediction is a linear combination of the field values at the sampling sites. If the observed values are drawn from a Gaussian random field, the prediction also follows the Gaussian distribution. In addition, the SK prediction given by (10.32) coincides with the Gaussian *conditional mean* given by (6.17a), while the *kriging variance* (10.33) is identical to the *conditional variance* (6.17b) .

Hence, within the Gaussian framework the kriging predictor suffices to determine the predictive distribution by conditioning on the available data.

Computational issues The covariance matrix that is inverted in (10.32) to obtain the optimal weights is of size $N \times N$. The covariance matrix is typically dense (most of its elements are nonzero), albeit there are elements that are almost zero for lags that significantly exceed the correlation length. The computational complexity for inverting a dense matrix scales as $O(N^3)$, and this scaling creates a computational bottleneck for large N. Various approaches have been developed to address the computational cost (see Sect. 11.7).

Kriging neighborhood To alleviate the computational costs, kriging is often applied using a finite *search neighborhood* also known as *kriging neighborhood* around each estimation point. This approach restricts the size of the data covariance matrix that needs to be inverted. On the other hand, the data covariance matrix is different for each prediction point. This is in contrast with the *global neighborhood* approach (where the search neighborhood encompasses the entire domain) that generates a single data covariance matrix for all the prediction points. Nevertheless, the computational gains imparted by the inversion of small covariance matrices usually offset the cost of repeating the local data covariance matrix inversion at every prediction point.

The kriging neighborhood approach is empirically justified by invoking the *screening effect* (see Sect. 10.7.1). The main drawback of using finite kriging neigh-

[3]A hyper-surface is a manifold of dimension $d - 1$ embedded in an ambient space of dimension d.

borhoods is the appearance of artifacts (e.g., spurious discontinuities) in the contour maps and the potential for numerical instabilities due to ill-conditioned covariance matrices in the kriging system. Actually, the problem of *ill-conditioned covariance matrices* (with or without search neighborhood) often plagues kriging and Gaussian process regression applications if there are closely-spaced observations. Unstable covariance matrices have a large condition number. The impact of the covariance function on the condition number is studied in [1]. Different regularization methods can be used to address the issue of matrix stability [580].

Recommendations for selecting a suitable kriging neighborhood are given in [132, p. 204]. Empirical guidance for the definition of the neighborhood includes the following: (i) In case of geometric anisotropy the neighborhood should be an ellipse (or ellipsoid) with principal axes matching those of the anisotropy. (ii) The radii of the ellipsoid should be approximately equal to the correlation lengths in the respective directions. (iii) Each neighborhood should include a sufficient number of sample points to avoid directional bias. This implies an adequate number of lag vectors between the target point and sample points inside the neighborhood along each of the orthogonal directions of anisotropy.

10.4.3 Examples of Simple Kriging

Example 10.3 Consider a prediction point $\mathbf{z}$ the search neighborhood of which contains a single data point $\mathbf{s}_1$ with value x_1^*. Determine the SK weight, the optimal prediction at $\mathbf{z}$, and the prediction variance.

Answer We will use the correlation function formulation (10.51). In the case of a single data point, the system of equations (10.30) becomes

$$\lambda = \rho_{\mathrm{xx}}(\mathbf{z} - \mathbf{s}_1).$$

Then, the optimal estimate based on (10.21) is

$$\hat{x}(\mathbf{z}) = m_{\mathrm{x}} + \lambda \,(x_1^* - m_{\mathrm{x}}) = m_{\mathrm{x}} + \rho_{\mathrm{xx}}(\mathbf{z} - \mathbf{s}_1)\,(x_1^* - m_{\mathrm{x}}). \tag{10.35}$$

The SK prediction tends to x_1^* as $\mathbf{z} \to \mathbf{s}_1$, whereas $\hat{x}(\mathbf{z}) \to m_{\mathrm{x}}$ if $\|\mathbf{z} - \mathbf{s}_1\| \to \infty$. Hence, if there are no sampling points in the neighborhood of the prediction point, the optimal estimate of the SK equation is the constant mean.

In addition, using (10.27) the kriging variance is given by

$$\sigma_{\mathrm{sk}}^2 = \sigma_{\mathrm{x}}^2 \left[1 - \rho_{\mathrm{xx}}^2(\mathbf{z} - \mathbf{s}_1)\right].$$

The expression for the kriging variance shows that $\sigma_{\mathrm{sk}}^2 \to 0$ if $\mathbf{z} \to \mathbf{s}_1$, whereas $\sigma_{\mathrm{sk}}^2 \to \sigma_{\mathrm{x}}^2$ if $\|\mathbf{z} - \mathbf{s}_1\| \to \infty$. These limits make sense intuitively, since the error variance tends to zero close to the sampling location where the outcome is known, while if the prediction point is very far from the sampling point, the error variance is determined by the unconstrained random field variance σ_{x}^2.

Example 10.4 Next, consider that the search neighborhood of the prediction point $\mathbf{z}$ contains two non-identical data points $\mathbf{s}_1$ and $\mathbf{s}_2$ with values x_1^* and x_2^* respectively. Determine (i) the optimal weights (ii) the simple kriging prediction, and (iii) the prediction variance.

Answer For notational convenience let $\rho_{1,2} = \rho_{\mathrm{xx}}(\mathbf{s}_1 - \mathbf{s}_2)$, $\rho_{n,0} = \rho_{\mathrm{xx}}(\mathbf{s}_n - \mathbf{z})$, where $n = 1, 2$. The simple kriging equations based on the correlation function formulation (10.29) lead to the following linear system

$$\begin{bmatrix} 1 & \rho_{1,2} \\ \rho_{1,2} & 1 \end{bmatrix} \begin{bmatrix} \lambda_1 \\ \lambda_2 \end{bmatrix} = \begin{bmatrix} \rho_{1,0} \\ \rho_{2,0} \end{bmatrix}.$$

(i) The solution of the above system for the weights is given by

$$\lambda_1 = \frac{\rho_{1,0} - \rho_{2,0}\rho_{1,2}}{1 - \rho_{1,2}^2},$$

$$\lambda_2 = \frac{\rho_{2,0} - \rho_{1,0}\rho_{1,2}}{1 - \rho_{1,2}^2}.$$

Based on the above, the difference of the weights is given by

$$\lambda_1 - \lambda_2 = \frac{\rho_{1,0} - \rho_{2,0}}{1 - \rho_{1,2}}.$$

The denominator is positive because $|\rho_{1,2}| < 1$. If the correlation is a monotonically decreasing function of the lag (this is not true for damped oscillatory covariance functions), then $\lambda_1 > \lambda_2$ if $\mathbf{s}_1$ is closer to $\mathbf{z}$ than $\mathbf{s}_2$, and $\lambda_1 < \lambda_2$ in the opposite case. If $\mathbf{z}$ is at the same distance form both $\mathbf{s}_1$ and $\mathbf{s}_2$, then $\lambda_1 = \lambda_2$.

(ii) Based on (10.22), the SK optimal prediction for the two-point sample is

$$\hat{\mathrm{x}}(\mathbf{z}) = m_{\mathrm{x}} + \frac{\left(\rho_{1,0} - \rho_{2,0}\rho_{1,2}\right)(x_1^* - m_{\mathrm{x}}) + \left(\rho_{2,0} - \rho_{1,0}\rho_{1,2}\right)(x_2^* - m_{\mathrm{x}})}{1 - \rho_{1,2}^2}. \tag{10.36}$$

The following observations can be made by analyzing the SK prediction (10.36):

1. If $x_1^* = x_2^* = x^*$, the SK prediction is given by

$$\hat{x}(\mathbf{z}) = m_x + \left(\frac{\rho_{1,0} + \rho_{2,0}}{1 + \rho_{1,2}}\right)(x^* - m_x).$$

2. If the prediction point $\mathbf{z}$ is equidistant from $\mathbf{s}_1$ and $\mathbf{s}_2$, and $\rho_{xx}(\cdot)$ is a radial function, then $\rho_{1,0} = \rho_{2,0} = \rho$ and the SK prediction is given by

$$\hat{x}(\mathbf{z}) = m_x + \frac{\rho}{1 + \rho_{1,2}}(x_1^* + x_2^* - 2m_x).$$

3. If the prediction point $\mathbf{z}$ is very far from both $\mathbf{s}_1$ and $\mathbf{s}_2$ (meaning that the maximum of the distances $\|\mathbf{s}_1 - \mathbf{z}\|$ and $\|\mathbf{s}_2 - \mathbf{z}\|$ is much larger than the larger correlation length), then the SK prediction is equal to the mean.
4. If only of the sampling points $\mathbf{s}_n$, e.g., $\mathbf{s}_2$ is far from both $\mathbf{z}$ and $\mathbf{s}_1$, the SK prediction becomes

$$\hat{x}(\mathbf{z}) = m_x + \rho_{1,0}(x_1^* - m_x).$$

This equation is identical to the single-point SK prediction (10.35).

(iii) According to (10.27), the SK kriging variance is given by

$$\sigma_{sk}^2(\mathbf{z}) = \sigma_x^2\left(1 - \frac{\rho_{1,0}^2 + \rho_{2,0}^2 - 2\rho_{1,0}\rho_{2,0}\rho_{1,2}}{1 - \rho_{1,2}^2}\right). \tag{10.37}$$

If the points $\mathbf{s}_1$ and $\mathbf{s}_2$ are equidistant from $\mathbf{z}$, so that $\rho_{1,0} = \rho_{2,0} := \rho_0$, then the kriging variance is given by

$$\sigma_{sk}^2(\mathbf{z}) = \sigma_x^2\left(1 - \frac{2\rho_0^2}{\rho_{1,2} + 1}\right).$$

Further, if $\mathbf{s}_1 = \mathbf{s}_2$ so that $\rho_{1,2} = 1$, the above variance equation becomes equivalent to the SK variance obtained for one sampling point in Example 10.3.

Example 10.5 Returning to the preceding example, assume that the three points are collinear. Consider that $s_1 = 2$, $s_2 = 4$ are the positions of the sampling points with values $x_1^* = 5$, $x_2^* = 2$ respectively. Further, let us assume that the random field has $m_x = 3$ and $\sigma_x^2 = 0.2$. We investigate the SK prediction based on the SSRF covariance functions in $d = 1, 2, 3$.

Answer In one dimension the SSRF covariance functions for $d = 2$ and $d = 3$ are also permissible, and we use them for comparison purposes. The SK predictions for $s \in [0, 8]$ are shown in Fig. 10.5. The sample values are marked by circles. The one-dimensional covariance functions are differentiable. This leads to smooth

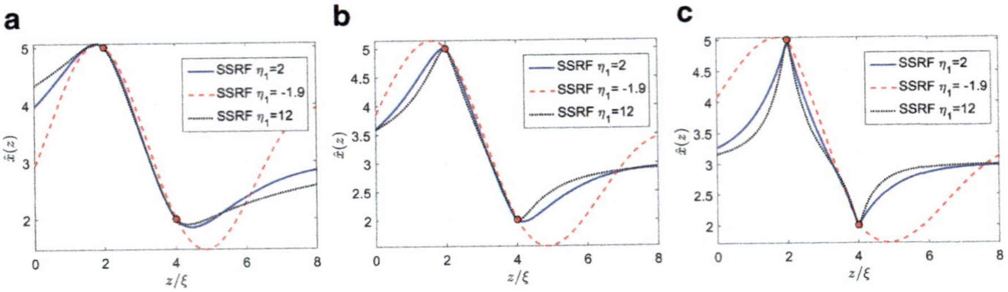

Fig. 10.5 Simple kriging predictions based on SSRF covariance functions in $d = 1$ (**a**), $d = 2$ (**b**), $d = 3$ (**c**), with $\eta_1 = 2, -1.9, 12$. The other SSRF parameters are $\xi = 1$ and $\eta_0 = 1$ while the mean value of the field is $m_x = 3$. The sample values are denoted by red circles. The horizontal axis marks the position of the prediction point normalized by the characteristic length ξ

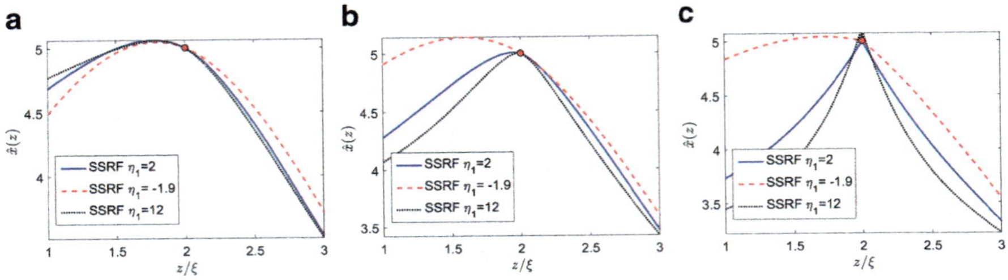

Fig. 10.6 Details of the simple kriging predictions shown in Fig. 10.5. The sample value is denoted by red circles. The horizontal axis marks the position of the prediction point normalized by the characteristic length ξ. (**a**) $d = 1$. (**b**) $d = 2$. (**c**) $d = 3$

prediction curves around the sample points. The prediction curves show overall larger variability for $\eta_1 = -1.9$ than for $\eta_1 > 0$ and smaller for $\eta_1 = 12$.

The same trends persist for the covariance functions in higher dimensions. For $d = 2$ the covariance function is barely non-differentiable — due to the logarithmic singularity of the integral $\int d\mathbf{k}\, \|\mathbf{k}\|^2\, \widetilde{C}_{xx}(\mathbf{k})$ caused by the limit $\|\mathbf{k}\| \to \infty$. As a result, the prediction curve still looks visually smooth around the sample points. In contrast, the three-dimensional covariance functions are non-differentiable, leading to prediction curves with discontinuous derivatives at the sample points. The derivative discontinuities are especially visible for $\eta_1 = 2$ and $\eta_1 = 12$.

To better visualize the impact of covariance differentiability on the SK predictions around the sampling points, consider Fig. 10.6 which shows in more detail the SK prediction around $s_1 = 2$. As evidenced in Fig. 10.6a, the prediction curve is smooth around $s_1 = 2$ for the $d = 1$ SSRF covariance. In $d = 2$, the prediction shown in Fig. 10.6b looks smooth, but a closer inspection reveals that the derivative of the prediction curve is not continuous at $s_1 = 2$.

The signature of non-differentiability is obvious for $d = 3$: for $\eta_1 = -1.9$ the magnitude of the prediction curve's slope at $s_1 = 2$ changes, while for $\eta_1 = 2$ the

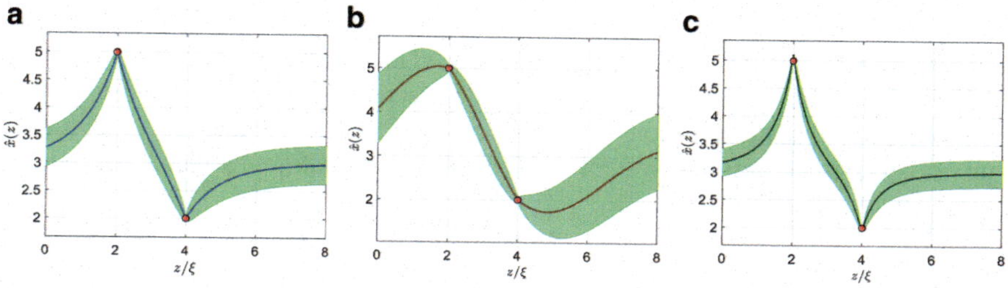

Fig. 10.7 Simple kriging predictions with prediction intervals (shaded regions) based on one kriging standard deviation. The sample values are those used in Example 10.5. An SSRF $d = 3$ covariance is assumed with $\eta_1 = 2, -1.9, 12$. The other SSRF parameters are $\xi = 1$ and $\eta_0 = 3$. (**a**) $\eta_1 = 2$. (**b**) $\eta_1 = -1.9$. (**c**) $\eta_1 = 12$

sign of the slope changes in addition to the magnitude. For $\eta_1 = 12$ the prediction curve exceeds the sample value on both sides of $s_1 = 2$. This overshooting is an example of the non-convexity of the kriging estimator. An intuitive interpretation is that higher values of the rigidity coefficient enforce sharper local bending of the prediction curve near the sampling points in order to respect the constraints.

Prediction variance Next, we calculate the uncertainty of the SK predictions for the example 10.4. The uncertainty estimation is based on the kriging variance given by (10.37). For each of the three values of η_1 we plot the prediction curves obtained by means of the $d = 3$ SSRF covariance function in Fig. 10.7.

We also plot shaded zones around the prediction curves which demarcate the 68% prediction intervals. The lower and upper envelopes of these zones are the curves defined by $\hat{x}(z) \mp \sigma_{sk}(z)$, for $z \in [0, 8]$. The width of the prediction intervals decreases with increasing η_1. This is expected based on the discussion in Sect. 7.3.1, since the increased rigidity reduces the variance of the field.

10.4.4 Impact of the Nugget Term

A nugget term is often detected in the empirical variograms of real spatial data. The nugget appears as a discontinuity of the variogram near the origin. The nugget term also implies a discontinuity of the covariance, the value of which drops from $C_{xx}(0) = \sigma_x^2 + c_0$ at the origin to σ_x^2 at lags infinitesimally larger than zero.

Often the kriging system obtained with a Gaussian covariance (or variogram) model is ill-posed, which implies that the condition number of the covariance matrix is high. A small nugget term is then added to the Gaussian variogram (and covariance) in order to obtain a stable kriging system. This procedure is called *regularization*. An ill-posed problem does not have a unique solution that changes continuously with the data [345]. The term regularization refers to finding an approximate but stable solution of an ill-posed problem.

Various regularization approaches with application to kriging systems are discussed in [580]. The addition of a nugget term in the covariance to obtain a numerically stable kriging system is equivalent to *Tikhnonov regularization* in regression analysis (see Chap. 2).

The disadvantage of a variogram with a nugget term is that the kriging estimates for *exact interpolation* tend to the data values discontinuously as the estimation point approaches a sampling point [624]. However, it is also possible to use a *smoothing formulation of kriging* which leads to continuous kriging estimates (see [165] and Sect. 10.5.3 herein).

Example 10.6 Consider a prediction point $\mathbf{z}$ whose search neighborhood contains a single data point $\mathbf{s}_1$ with value x_1^*. Determine the SK weight, the optimal prediction, and the prediction variance if the covariance function includes a nugget term c_0.

Answer We follow the same approach as in Example 10.3. The kriging weight is then given by

$$\lambda = \frac{\sigma_{\mathrm{x}}^2 \rho_{\mathrm{xx}}(\mathbf{z} - \mathbf{s}_1) + c_0\, \delta_{\mathbf{z},\mathbf{s}_1}}{\sigma_{\mathrm{x}}^2 + c_0}.$$

Then, the optimal estimate based on (10.21) is

$$\hat{x}(\mathbf{z}) = m_{\mathrm{x}} + \frac{\rho_{\mathrm{xx}}(\mathbf{z} - \mathbf{s}_1) + w\, \delta_{\mathbf{z},\mathbf{s}_1}}{1 + w}\,(x_1^* - m_{\mathrm{x}}), \tag{10.38}$$

where $w = c_0/\sigma_{\mathrm{x}}^2$ is the *noise to signal ratio*, and $\delta_{\mathbf{z},\mathbf{s}_1}$ is equal to one if $\mathbf{z} = \mathbf{s}_1$ and zero otherwise. The SK prediction becomes equal to x_1^* if $\mathbf{z} = \mathbf{s}_1$ but changes discontinuously around $\mathbf{s}_1$. The magnitude of the jump discontinuity increases with w. In addition, the linear weight depends on w.

We illustrate this behavior for the one-dimensional SSRF covariance function (7.27b) in Fig. 10.8. For $w = 0$ the prediction changes smoothly around the sampling point. As w increases, a discontinuity with increasing magnitude develops at $s = 0$. At $s \approx 5\xi$, the SK prediction tends to the mean m_{x}.

10.4.5 Properties of the Kriging Error

Useful properties of the kriging error follow from the theory of *Hilbert spaces* [188]. More specifically, the orthogonality of the kriging prediction and the error, and the smoothing effect of kriging can be understood in this framework.

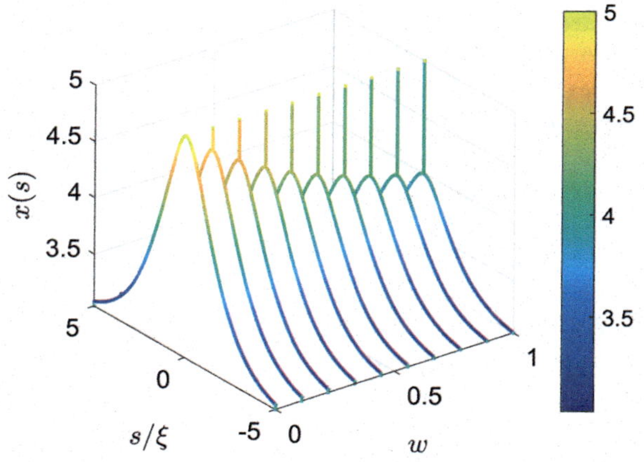

Fig. 10.8 Simple Kriging prediction based on one sample with value $x_1^* = 5$ located at $s = 0$. The prediction is plotted as a function of s along a line extending from -5ξ to 5ξ for noise to signal ratio, w, between zero and one. The sample is assumed to come from a random field with $m_x = 3$ and SSRF covariance function with $\xi = 3$, $\eta_0 = 1$, and $\eta_1 = 2$

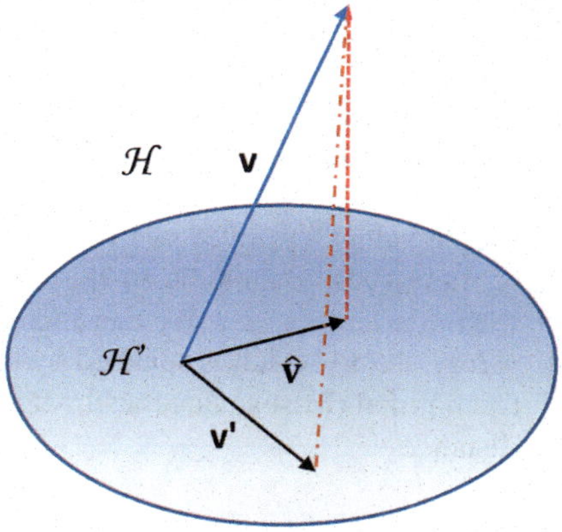

Fig. 10.9 Orthogonal projection $\hat{\mathbf{v}}$ of vector $\mathbf{v} \in \mathcal{H}$ onto the Hilbert subspace $\mathcal{H}'$. The distance between $\mathbf{v}$ and $\hat{\mathbf{v}}$ is smaller than the distance between $\mathbf{v}$ and any other vector $\mathbf{v}'$ in the subspace $\mathcal{H}'$

Definition 10.5 A Hilbert space $\mathcal{H}$ is a complete space equipped with an inner product. A space is complete, if every convergent (Cauchy) sequence $\{x_n\}_{n=1}^{\infty}$ converges in norm to some element $x \in \mathcal{H}$.[4]

Hilbert spaces are abstract generalizations of d-dimensional Euclidean spaces. They include the definition of an inner product $< \mathbf{v}_1, \mathbf{v}_2 >$ between two elements $\mathbf{v}_1, \mathbf{v}_2 \in \mathcal{H}$, which enables defining the concepts of distance and angle between elements of the Hilbert space. Thus, classical notions of geometry can be applied to these more general spaces. Gaussian random fields can be studied in the framework of Hilbert spaces [405, 774].

An important tool in the theory of Hilbert spaces is the *projection theorem* which defines the projection of vector in $\mathcal{H}$ onto some subspace $\mathcal{H}'$. The main idea of the theorem is illustrated in Fig. 10.9.

[4]In mathematics, the term "space" refers to a set with some specified rules that impose structure.

Theorem 10.1 *Projection theorem: Let $\mathcal{H}$ be a Hilbert space, and $\mathcal{H}' \subset \mathcal{H}$ be a subspace of $\mathcal{H}$ which is closed under the operation of linear combination. The closure property means that a linear combination of any two vectors in $\mathcal{H}'$ is also a vector in $\mathcal{H}'$.*

If $\mathbf{v} \in \mathcal{H}$ is a vector in the Hilbert space $\mathcal{H}$, then, there exists a unique element $\hat{\mathbf{v}} \in \mathcal{H}'$ such that

$$\|\mathbf{v} - \hat{\mathbf{v}}\| \leq \|\mathbf{v} - \mathbf{v}'\|, \text{ for all } \mathbf{v}' \in \mathcal{H}'.$$

The vector $\hat{\mathbf{v}}$ is the orthogonal projection of $\mathbf{v}$ in the subspace $\mathcal{H}'$.

In addition, it can be shown that the above is true if and only if $\hat{\mathbf{v}} \in \mathcal{H}'$ and $\mathbf{v} - \hat{\mathbf{v}} \in \mathcal{H}'^{\perp}$, where $\mathcal{H}'^{\perp}$ is the complement of $\mathcal{H}'$ in $\mathcal{H}$.

Least squares estimation Linear regression based on the minimization of the square errors can be viewed as a projection problem in Hilbert space: Let us consider two variables (x, y) measured at a number of N points so that the sample consists of the pairs $\{(x_n, y_n)\}_{n=1}^{N}$. Let the x_n represent the independent variable and the y_n the dependent variable. The goal of linear regression is to find real-valued coefficients a and b such that $y = ax + b$ is an "optimal" model, in the sense of minimum mean square error.

The N values $(y_1, \ldots, y_N)$ define a vector $\mathbf{y}$ in the N-dimensional Euclidean space. Let us also denote by $\mathbf{v}_1$ the vector in $\mathbb{R}^N$ with all entries equal to one and by $\mathbf{v}_2 = (x_1, \ldots, x_N)^\top$. We can then generate the subspace $W \subset \mathbb{R}^N$ such that $W = b\,\mathbf{v}_1 + a\,\mathbf{v}_2$ which is spanned by the vectors $\mathbf{v}_1$ and $\mathbf{v}_2$.

The optimal (least squares) estimate $\hat{\mathbf{y}} = \hat{b}\,\mathbf{v}_1 + \hat{a}\,\mathbf{v}_2$ is the vector that minimizes the distance

$$D = \|\hat{\mathbf{y}} - \mathbf{y}\|^2.$$

Hence, $\hat{\mathbf{y}}$ can be considered as the orthogonal projection of $\mathbf{y}$ on the subspace W that is spanned by $\mathbf{v}_1$ and $\mathbf{v}_2$.

In the geostatistical literature, Journel formulated kriging prediction as a Hilbert space projection [418]. The basic idea is that random fields can be viewed as Hilbert spaces in which the inner product is defined in terms of the expectation, i.e.,

$$\langle \mathrm{X}(\mathbf{s};\omega)\, \mathrm{X}(\mathbf{s}';\omega)\rangle = \mathcal{E}\left[\mathrm{X}(\mathbf{s};\omega)\, \mathrm{X}(\mathbf{s}';\omega)\right].$$

We can consider the kriging prediction $\hat{\mathrm{X}}(\mathbf{z};\omega)$, defined by means of (10.16), as the projection of the unknown true value $\mathrm{X}(\mathbf{z};\omega)$ in the sampling subspace $\{\mathrm{X}(\mathbf{s}_1;\omega), \ldots, \mathrm{X}(\mathbf{s}_N;\omega)\}$. The projection theorem then ensures that the prediction error is orthogonal to the elements of the sampling subspace, i.e.,

Orthogonality of Kriging error and sampled field

$$\mathcal{E}[\epsilon(\mathbf{z};\omega)\,\mathrm{X}(\mathbf{s}_n;\omega)] = 0, \text{ for all } n = 1, \ldots, N. \tag{10.39}$$

Since the kriging prediction is a linear combination of the sampled values and the sampling subspace is closed under the operation of linear superposition, it follows that the prediction error is orthogonal to the prediction, i.e.,

Orthogonality of Kriging error and prediction

$$\mathcal{E}\left[\left\{\mathrm{X}(\mathbf{z};\omega) - \hat{\mathrm{X}}(\mathbf{z};\omega)\right\}\hat{\mathrm{X}}(\mathbf{z};\omega)\right] = \mathcal{E}\left[\epsilon(\mathbf{z};\omega)\,\hat{\mathrm{X}}(\mathbf{z};\omega)\right] = 0. \tag{10.40}$$

The above orthogonality expression leads to the following equation for the kriging variance

$$\begin{aligned}\mathcal{E}\left[\epsilon^2(\mathbf{z};\omega)\right] =& \mathcal{E}\left[\epsilon(\mathbf{z};\omega)\,\hat{\mathrm{X}}(\mathbf{z};\omega)\right] - \mathcal{E}[\epsilon(\mathbf{z};\omega)\,\mathrm{X}(\mathbf{z};\omega)] \\ =& \mathcal{E}\left[\mathrm{X}^2(\mathbf{z};\omega)\right] - \mathcal{E}\left[\hat{\mathrm{X}}(\mathbf{z};\omega)\,\mathrm{X}(\mathbf{z};\omega)\right].\end{aligned} \tag{10.41}$$

Smoothing effect In general, we can express the true value of the field in terms of the prediction and the error as follows

$$\mathrm{X}(\mathbf{z};\omega) = \hat{\mathrm{X}}(\mathbf{z};\omega) + \epsilon(\mathbf{z};\omega).$$

By taking the squares on both sides of the above equation and using the orthogonality of the error with the prediction we obtain the following identity

$$\mathcal{E}[\mathrm{X}^2(\mathbf{z};\omega)] = \mathcal{E}[\hat{\mathrm{X}}^2(\mathbf{z};\omega)] + \mathcal{E}[\epsilon^2(\mathbf{z};\omega)].$$

If we combine the above with the lack of bias of the kriging prediction, i.e., $\mathcal{E}[\mathrm{X}(\mathbf{z};\omega)] = \mathcal{E}[\hat{\mathrm{X}}(\mathbf{z};\omega)] = m_{\mathrm{x}}$ and $\mathcal{E}[\epsilon(\mathbf{z};\omega)] = 0$, the following equation is obtained for the *kriging variance* $\mathbb{V}\mathrm{ar}\,\{\epsilon(\mathbf{z};\omega)\}$:

Kriging smoothing effect

$$\mathbb{V}\mathrm{ar}\,\{\mathrm{X}(\mathbf{z};\omega)\} = \mathbb{V}\mathrm{ar}\left\{\hat{\mathrm{X}}(\mathbf{z};\omega)\right\} + \mathbb{V}\mathrm{ar}\,\{\epsilon(\mathbf{z};\omega)\}, \tag{10.42}$$

The above equation expresses the *smoothing effect* of the kriging predictor. First, it shows that the kriging variance is smaller than the true variance of the field. This is in agreement with the conclusion drawn in Sect. 10.4.1. The current relation, however, is based on general principles and does not use the covariance function. Second, the variance of the kriging prediction—the first term on the right-hand side—increases near the data points, where the kriging variance tends to be small. Third, the prediction variance tends to zero as the prediction point moves away from the data points and the kriging variance tends to the true variance. This reflects the fact that away from the data points the kriging prediction tends to the mean.

Practical implications of smoothing The smoothing effect of kriging is an advantage if our purpose is to reduce the impact of noise in the data. On the other hand, it is a disadvantage if the goal is to investigate the variability of the spatial configurations allowed by the random field model. Kriging maps, for example, may excessively smoothen the natural variability of geological media and natural processes. For this reason, the current trend is for spatial analysis to rely more on simulations than on kriging interpolation.

Comment The smoothing effect of kriging does not imply that the response surfaces generated by kriging are necessarily smooth (i.e., differentiable). In fact, as we discussed in Sects. 10.4.2 and 10.4.3, if the covariance function is non-differentiable at the origin, the prediction surface (or curve in $d = 1$) is also non-differentiable at the sampling points.

10.5 Ordinary Kriging (OK)

This section focuses on the most commonly used kriging method which is known as *ordinary kriging* (OK).

The main assumptions used in OK are: *(i) that the random field* $\mathrm{X}(\mathbf{s};\omega)$ *is wide-sense stationary and (ii) the constant mean* $\mathcal{E}[\mathrm{X}(\mathbf{s};\omega)]$ *is not known a priori.*

Ordinary kriging is more flexible than simple kriging, because the former does not require estimating the mean of the random field from the data. Recall that the sample average is not an accurate estimate of the mean in the presence of correlations and non-uniform sampling conditions. In practice, the sampling may preferentially target areas of low or high values, leading to biased estimates of the mean.

As we discuss in Sect. 11.1.1, condition (i) can be relaxed. The variogram-based formulation of OK is valid also for *intrinsic random fields* with constant mean. Thus, OK has a wider scope than SK.

The OK prediction is based on the linear superposition (10.15), i.e.,

$$\hat{\mathrm{X}}(\mathbf{z};\omega) = \sum_{n=1}^{N} \lambda_n \, \mathrm{X}(\mathbf{s}_n;\omega).$$

The above equation is equivalent to the SK prediction equation (10.22), if the latter is expressed as

$$\hat{X}(\mathbf{z};\omega) = m_{x}\left(1-\sum_{n=1}^{N}\lambda_{n}\right) + \sum_{n=1}^{N}\lambda_{n}\,X(\mathbf{s}_{n};\omega), \tag{10.43}$$

and provided that the weights are forced to satisfy the constraint $\sum_{n=1}^{N}\lambda_{n}=1$.

We will illustrate below the main steps of the calculation that determines the optimal weights of the OK predictor. We will use the stationarity assumption for $X(\mathbf{s};\omega)$, i.e., that the mean is constant and the covariance depends only on the lag. It is also possible to derive the OK equations (in terms of the variogram function) for intrinsic random fields with constant mean. This is further discussed in Sect. 11.1.1.

In light of the stationarity condition, the variance, σ_{x}^{2}, of the random field $X(\mathbf{s};\omega)$ is constant everywhere in space, while the covariance depends only on the lag vector between two locations.

1. Let us denote by $\sigma_{\text{ok}}^{2}(\mathbf{z}) = \mathcal{E}[\epsilon^{2}(\mathbf{z};\omega)]$ the *variance of the prediction error* (10.16), which is also known as the estimation variance. Using the interpolation equation (10.15), the error variance is expressed as follows

$$\sigma_{\text{ok}}^{2}(\mathbf{z}) = \sigma_{x}^{2} + \sum_{\alpha=1}^{N}\sum_{\beta=1}^{N}\lambda_{\alpha}\lambda_{\beta}\,\mathcal{E}\left[X(\mathbf{s}_{\alpha};\omega)X(\mathbf{s}_{\beta};\omega)\right] \tag{10.44}$$

$$-\,2\sum_{\alpha=1}^{N}\lambda_{\alpha}\mathcal{E}[X(\mathbf{s}_{\alpha};\omega)\,X(\mathbf{z};\omega)] + 2\mu\left(\sum_{\beta=1}^{N}\lambda_{\alpha}-1\right). \tag{10.45}$$

The constant 2μ is the Lagrange multiplier used to enforce the zero-bias condition.

2. By taking into account the definition of the covariance function (3.40), the above equation is expressed as follows

$$\sigma_{\text{ok}}^{2}(\mathbf{z}) = \sigma_{x}^{2} + \sum_{\alpha=1}^{N}\sum_{\beta=1}^{N}\lambda_{\alpha}\lambda_{\beta}\,C_{xx}(\mathbf{s}_{\alpha}-\mathbf{s}_{\beta}) - 2\sum_{\alpha=1}^{N}\lambda_{\alpha}C_{xx}(\mathbf{s}_{\alpha}-\mathbf{z}) + 2\mu\left(\sum_{\beta=1}^{N}\lambda_{\alpha}-1\right). \tag{10.46}$$

The terms involving the unknown constant mean from the second and third summands cancel out in the above equation. It is a good exercise to confirm this cancellation.

3. The minimum mean square estimation error is determined by solving the linear equation system which follows from the *first-order optimality criterion*. This requires that partial derivatives of the error variance with respect to the linear weights and the Lagrange multiplier vanish, i.e.,

$$\frac{\partial\sigma_{\text{ok}}^{2}(\mathbf{z})}{\partial\lambda_{\alpha}} = 0, \quad \text{for } \alpha = 1,\ldots,N, \tag{10.47a}$$

$$\frac{\partial\sigma_{\text{ok}}^{2}(\mathbf{z})}{\partial\mu} = 0. \tag{10.47b}$$

4. The above conditions lead to the following system of linear equations for the kriging weights

$$\sum_{\beta=1}^{N} \lambda_\beta\, C_{xx}(\mathbf{s}_\alpha - \mathbf{s}_\beta) + \mu = C_{xx}(\mathbf{s}_\alpha - \mathbf{z}), \quad \text{for } \alpha = 1, \ldots, N, \tag{10.48a}$$

$$\sum_{\alpha=1}^{N} \lambda_\alpha = 1. \tag{10.48b}$$

5. To ensure that the stationary point obtained above corresponds to a minimum (and not to a maximum) of the error variance, we need to consider the *Hessian matrix* determined by the partial derivatives of the error variance with respect to the weights. This is given by

$$H_{\alpha,\beta} = \frac{\partial^2 \sigma^2_{\text{ok}}(\mathbf{z})}{\partial \lambda_\alpha \partial \lambda_\beta}, \quad \text{for } \alpha = 1, \ldots, N, \text{ and } \beta = 1, \ldots, N.$$

If the stationary point defined by (10.48a) corresponds to a minimum of the variance, the Hessian matrix should be positive definite. Differentiating (10.47a) with respect to the weights for a second time leads to

$$H_{\alpha,\beta} = C_{xx}(\mathbf{s}_\alpha - \mathbf{s}_\beta), \text{ for } \alpha, \beta = 1, \ldots, N. \tag{10.49}$$

Thus, the Hessian matrix is given by the covariance matrix for the sampling points. The latter is always positive definite for permissible covariance functions based on Bochner's theorem.

The system of the kriging equations derived in (10.48) is compactly expressed in matrix form as follows

Ordinary Kriging System—Covariance formulation

$$\begin{bmatrix} \sigma_x^2 & C_{xx}(\mathbf{s}_1 - \mathbf{s}_2) & \ldots & C_{xx}(\mathbf{s}_1 - \mathbf{s}_N) & 1 \\ C_{xx}(\mathbf{s}_2 - \mathbf{s}_1) & \sigma_x^2 & \ldots & C_{xx}(\mathbf{s}_2 - \mathbf{s}_N) & 1 \\ \vdots & \vdots & \vdots & \vdots & \vdots \\ C_{xx}(\mathbf{s}_N - \mathbf{s}_1) & C_{xx}(\mathbf{s}_N - \mathbf{s}_2) & \vdots & \sigma_x^2 & 1 \\ 1 & 1 & \ldots & 1 & 0 \end{bmatrix} \begin{bmatrix} \lambda_1 \\ \lambda_2 \\ \vdots \\ \lambda_N \\ \mu \end{bmatrix} = \begin{bmatrix} C_{xx}(\mathbf{s}_1 - \mathbf{z}) \\ C_{xx}(\mathbf{s}_2 - \mathbf{z}) \\ \vdots \\ C_{xx}(\mathbf{s}_N - \mathbf{z}) \\ 1 \end{bmatrix}. \tag{10.50}$$

In a completely equivalent formulation, the ordinary kriging predictor can be expressed in terms of the correlation function as follows

Ordinary Kriging System—Correlation formulation

$$\begin{bmatrix} 1 & \rho_{xx}(\mathbf{s}_1-\mathbf{s}_2) & \dots & \rho_{xx}(\mathbf{s}_1-\mathbf{s}_N) & 1 \\ \rho_{xx}(\mathbf{s}_2-\mathbf{s}_1) & 1 & \dots & \rho_{xx}(\mathbf{s}_2-\mathbf{s}_N) & 1 \\ \vdots & \vdots & \vdots & \vdots & \vdots \\ \rho_{xx}(\mathbf{s}_N-\mathbf{s}_1) & \rho_{xx}(\mathbf{s}_N-\mathbf{s}_2) & \vdots & 1 & 1 \\ 1 & 1 & \dots & 1 & 0 \end{bmatrix} \begin{bmatrix} \lambda_1 \\ \lambda_2 \\ \vdots \\ \lambda_N \\ \mu' \end{bmatrix} = \begin{bmatrix} \rho_{xx}(\mathbf{s}_1-\mathbf{z}) \\ \rho_{xx}(\mathbf{s}_2-\mathbf{z}) \\ \vdots \\ \rho_{xx}(\mathbf{s}_N-\mathbf{z}) \\ 1 \end{bmatrix}. \tag{10.51}$$

The above follows from (10.50) by dividing both sides with σ_x^2 and defining μ' by means of $\mu = \mu' \sigma_x^2$.

The ordinary kriging predictor can also be expressed in terms of the variogram function as follows

Ordinary Kriging System—Variogram formulation

$$\begin{bmatrix} \gamma_{xx}(\mathbf{0}) & \gamma_{xx}(\mathbf{s}_1-\mathbf{s}_2) & \dots & \gamma_{xx}(\mathbf{s}_1-\mathbf{s}_N) & 1 \\ \gamma_{xx}(\mathbf{s}_2-\mathbf{s}_1) & \gamma_{xx}(\mathbf{0}) & \dots & \gamma_{xx}(\mathbf{s}_2-\mathbf{s}_N) & 1 \\ \vdots & \vdots & \vdots & \vdots & \vdots \\ \gamma_{xx}(\mathbf{s}_N-\mathbf{s}_1) & \gamma_{xx}(\mathbf{s}_N-\mathbf{s}_2) & \vdots & \gamma_{xx}(\mathbf{0}) & 1 \\ 1 & 1 & \dots & 1 & 0 \end{bmatrix} \begin{bmatrix} \lambda_1 \\ \lambda_2 \\ \vdots \\ \lambda_N \\ -\mu \end{bmatrix} = \begin{bmatrix} \gamma_{xx}(\mathbf{s}_1-\mathbf{z}) \\ \gamma_{xx}(\mathbf{s}_2-\mathbf{z}) \\ \vdots \\ \gamma_{xx}(\mathbf{s}_N-\mathbf{z}) \\ 1 \end{bmatrix} \tag{10.52}$$

Problem 10.1 Confirm that the two formulations above (i) are obtained from each other by means of the substitution $\gamma_{xx}(\cdot) \mapsto -C_{xx}(\cdot)$ and (ii) that they are indeed equivalent based on the stationarity property (3.47) and the fact that the kriging weights sum to one.

10.5.1 Compact Form of Ordinary Kriging Equations

The kriging equations can be easier recalled by introducing a more compact matrix notation. The covariance-based formulation of kriging, i.e., (10.50), is expressed as follows:

$$\left[\begin{array}{c|c} \mathbf{C}_{\mathrm{d,d}} & \mathbf{1} \\ \hline \mathbf{1}^\top & 0 \end{array}\right] \left[\begin{array}{c} \boldsymbol{\lambda} \\ \hline \mu \end{array}\right] = \left[\begin{array}{c} \mathbf{C}_{\mathrm{d,p}} \\ \hline 1 \end{array}\right]. \tag{10.53}$$

Similarly, the variogram-based formulation of kriging, i.e., (10.52), is expressed as follows:

$$\left[\begin{array}{c|c} \boldsymbol{\Gamma}_{\mathrm{d,d}} & \mathbf{1} \\ \hline \mathbf{1}^\top & 0 \end{array}\right] \left[\begin{array}{c} \boldsymbol{\lambda} \\ \hline -\mu \end{array}\right] = \left[\begin{array}{c} \boldsymbol{\Gamma}_{\mathrm{d,p}} \\ \hline 1 \end{array}\right]. \tag{10.54}$$

The above matrix formulations of the kriging equations use the following notation:

1. $\boldsymbol{\lambda}$ is the $N \times 1$ vector of the kriging weights
2. $\mathbf{1}$ is the $N \times 1$ vector of ones
3. $\mathbf{C}_{\mathrm{d,d}}$ is the $N \times N$ covariance matrix between the sampling points
4. $\mathbf{C}_{\mathrm{d,p}}$ is the $N \times 1$ covariance vector between the prediction point and the sampling points
5. $\boldsymbol{\Gamma}_{\mathrm{d,d}}$ is the $N \times N$ variogram matrix between the sampling points
6. $\boldsymbol{\Gamma}_{\mathrm{d,p}}$ is the $N \times 1$ variogram vector between the prediction point and the sampling points.

10.5.2 *Kriging Variance*

The kriging variance measures the precision of the kriging estimate, i.e., the deviation of the kriging prediction from the true value of the field. The kriging variance is defined as the *minimum mean square error* of the kriging predictor and is given by means of the equation

$$\sigma_{\mathrm{ok}}^2(\mathbf{z}) = \sigma_{\mathrm{x}}^2 - \sum_{\alpha=1}^{N} \lambda_\alpha C_{\mathrm{xx}}(\mathbf{z} - \mathbf{s}_\alpha) - \mu. \tag{10.55}$$

This can be shown by inserting the equation for the kriging weights (10.48a)

$$\sum_{\beta=1}^{N} \lambda_\beta \, C_{\mathrm{xx}}(\mathbf{s}_\alpha - \mathbf{s}_\beta) = C_{\mathrm{xx}}(\mathbf{s}_\alpha - \mathbf{z}) - \mu,$$

in the mean square error equation (10.46) to obtain (10.55).

The three terms on the right-hand side of the kriging variance (10.55) have straightforward interpretations.

- The first term represents the random field variance; it is the unconstrained variability expected in the absence of correlations (e.g., at prediction points far from the sampling sites).
- The second term, which is a weighted combination of the covariance function values between the prediction and the sampling locations, reduces the variance due to the presence of the conditioning data and reflects the impact of spatial correlations.
- Finally, the Lagrange multiplier μ is a negative real number. Hence, the term $-\mu$ tends to increase the kriging variance. This effect represents the additional uncertainty which is due to the indeterminacy of the mean.

Prediction intervals for ordinary kriging can be formulated as in (10.34) for simple kriging.

10.5.3 How the Nugget Term Affects the Kriging Equations

The general form of the *simple and ordinary kriging* equations does not change in the presence of a nugget term. However, the presence of nugget affects the expressions that will be used for the covariance (or variogram) function in the kriging equations [165].

The nugget term can be modeled as a sum of two components. The first term involves *microscale variations* which correspond to spatial features of the process that cannot be resolved by the data (e.g., sub-meter variability cannot be detected by means of data that are spaced hundreds of meters apart). The second term incorporates random variability that is due to *measurement error*. If both terms are present, the nugget variance is expressed as follows:

$$c_0 = \sigma^2_{\text{micro}} + \sigma^2_{\varepsilon}.$$

An empirically observed nugget term could be due to either microscale variability or measurement errors or a combination of both. The following spatial model distinguishes between the two sources of nugget

$$\mathrm{X}^*(\mathbf{s};\omega) = \mathrm{X}(\mathbf{s};\omega) + W(\mathbf{s};\omega) + \varepsilon(\mathbf{s};\omega). \tag{10.56}$$

The random field $W(\mathbf{s};\omega)$ represents the microscale variations and $\varepsilon(\mathbf{s};\omega)$ the measurement errors. Neither term is predictable, but whether we treat the nugget as measurement error or microscale variation has an impact on the expression of the kriging system and the solutions for the spatial prediction and its variance.

An additional consideration is whether we aim to predict the *observed process* $X^*(\mathbf{s};\omega)$ or the *smoothed process* $X(\mathbf{s};\omega)$. Hence, four different cases encompass all the combinations of nugget origin and prediction purpose: microscale-observed, microscale-smoothed, noise-observed, and noise-smoothed, where the term "noise" refers to measurement error. In addition, we need to consider the possibility that the prediction point coincides with one of the sampling points.

The simple kriging equations are given by (10.31)–(10.33); these equations are repeated here for easy reference:

$$\boldsymbol{\lambda} = \left[\mathbf{C}_{\mathrm{d,d}}\right]^{-1} \mathbf{C}_{\mathrm{d,p}},$$

$$\hat{x}(\mathbf{z}) = m_{\mathrm{x}} + \left[\mathbf{C}_{\mathrm{d,d}}\right]^{-1} \mathbf{C}_{\mathrm{d,p}} \left(\mathbf{x}^* - \mathbf{m}_{\mathrm{x}}\right),$$

$$\sigma^2_{\mathrm{sk}}(\mathbf{z}) = \sigma^2_{\mathrm{x}} - \mathbf{C}^{\top}_{\mathrm{d,p}} \left[\mathbf{C}_{\mathrm{d,d}}\right]^{-1} \mathbf{C}_{\mathrm{d,p}}.$$

The respective equations for ordinary kriging will not be repeated, but the following statements apply to them as well. In the presence of a nugget term, the above equations hold with minor modifications which follow from a set of intuitive rules [638]:

- The nugget term appears on the diagonal of the data-data covariance matrix $\mathbf{C}_{\mathrm{d,d}}$ in all cases (i.e., microscale-observed, microscale-smoothed, noise-observed, and noise-smoothed).
- If the prediction point is different from the sampling points, the data-data covariance $\mathbf{C}_{\mathrm{d,d}}$ is the only matrix of the kriging system where the nugget variance appears. The *prediction equations* have the same form for all cases.
- If the prediction point is different from the sampling points, the kriging variance equation is the same regardless of the origin of the nugget. In the case of predicting the *observed process*, the nugget appears in the kriging variance equation where the field variance at the prediction point is replaced by $\sigma^2_{\mathrm{x}} + c_0$. However, in the case of *smoothing* the field variance does not include the nugget term.
- If the prediction point coincides with a sampling point, exact interpolation holds only in the *microscale-observed* case. In the other three cases the same kriging prediction equation is used, which does not involve the nugget variance in the data-prediction covariance vector $\mathbf{C}_{\mathrm{d,p}}$.
- If the prediction point coincides with a sampling point, the kriging variance is zero in the *microscale-observed* case. In the case of *smoothing* the kriging variance does not involve the nugget term at the prediction point regardless of the nugget's origin. On the other hand, in the *noise-observed* case the field variance at the prediction point includes the nugget variance.

Exact interpolation: If the nugget term is attributed to *microscale variations* and the goal is *exact interpolation*, i.e., prediction of the observed process, the nugget term is included in both the data-data matrices, i.e., $\mathbf{C}_{\mathrm{d,d}}$ (or $\boldsymbol{\Gamma}_{\mathrm{d,d}}$ if the variogram-based formulation is used) and the data-prediction matrices, i.e., $\mathbf{C}_{p,d}$ (or $\boldsymbol{\Gamma}_{p,d}$). In addition, the first term of the conditional variance at the prediction point includes the nugget term and is therefore equal to $\sigma_x^2 + c_0$. The nugget term appears in the data-prediction matrices only if the prediction point coincides with one of the sampling points.

Smoothing: If the nugget term is attributed to *microscale variations* and the goal is *smoothing* of the observed process, the nugget term is considered only in the data-data matrices, i.e., $\mathbf{C}_{d,d}$ (or $\boldsymbol{\Gamma}_{d,d}$), but not in the data-prediction matrices $\mathbf{C}_{p,d}$ (or $\boldsymbol{\Gamma}_{p,d}$). In this case, the first term of the conditional variance at the prediction point is σ_x^2 and does not include the nugget term.

If the nugget term is due to *measurement errors*, the smoothing formulation is often preferred. In this case, the nugget is included only in the data-data matrices. If the smoothing formulation is used, kriging is not an exact interpolator since the predictions are not constrained to satisfy the data at the sampling points.

10.5.4 Ordinary Kriging Examples

The following examples aim to provide a better understanding of the application of ordinary kriging and the impact of different factors on the prediction and the kriging variance.

Example 10.7 Assume that the search neighborhood of the prediction point $\mathbf{z}$ contains a single data point $\mathbf{s}_1$ with value x_1^*. Determine the optimal weights, the ordinary kriging prediction, and the respective prediction variance.

Answer Let us use the correlation function formulation (10.51). For a single data point, the linear system of equations becomes

$$\begin{bmatrix} 1 & 1 \\ 1 & 0 \end{bmatrix} \begin{bmatrix} \lambda_1 \\ \mu' \end{bmatrix} = \begin{bmatrix} \rho_{xx}(\mathbf{s}_1 - \mathbf{z}) \\ 1 \end{bmatrix}.$$

The solution of the above linear system of equations is given by

$$\lambda_1 = 1, \quad \mu' = \rho_{xx}(\mathbf{s}_1 - \mathbf{z}) - 1.$$

The solution $\lambda_1 = 1$ may seem overly simplistic, but it is necessary to ensure that the estimate of the field at $\mathbf{z}$ is unbiased.

In light of the above, the ordinary kriging prediction is given by

$$\hat{x}(\mathbf{z}) = x_1^*.$$

In addition, using (10.55) and $\mu = \sigma_x^2 \, \mu'$, it follows that the kriging variance is given by the following equation

$$\sigma_{\mathrm{ok}}^2 = 2\left[\sigma_{\mathrm{x}}^2 - C_{\mathrm{xx}}(\mathbf{s}_1 - \mathbf{z})\right].$$

It follows from the above that

$$\sigma_{\mathrm{ok}}^2 \to 0, \text{ if } \mathbf{z} \to \mathbf{s}_1, \text{ while } \sigma_{\mathrm{ok}}^2 \to 2\sigma_{\mathrm{x}}^2, \text{ if } \|\mathbf{z} - \mathbf{s}_1\| \to \infty.$$

It is somewhat surprising that the uncertainty of the kriging predictor far from the sampling point is equal to twice the random field variance. This result reflects the fact that the prediction variance accounts both for the unknown mean and the fluctuation around the mean, and each of these terms contributes a term equal to σ_{x}^2.

Example 10.8 Consider that the search neighborhood of the prediction point $\mathbf{z}$ contains two data points $\mathbf{s}_1$ and $\mathbf{s}_2$ with values x_1^* and x_2^* respectively. Determine (i) the optimal weights and the Lagrange multiplier, (ii) the ordinary kriging prediction, and (iii) the prediction variance.

Answer Let us denote $\rho_{1,2} = \rho_{\mathrm{xx}}(\mathbf{s}_1 - \mathbf{s}_2)$ and $\rho_{n,0} = \rho_{\mathrm{xx}}(\mathbf{s}_n - \mathbf{z})$ for $n = 1, 2$. Using the correlation function formulation (10.51), the ordinary kriging equations are given by the following 3×3 linear system

$$\begin{bmatrix} 1 & \rho_{1,2} & 1 \\ \rho_{1,2} & 1 & 1 \\ 1 & 1 & 0 \end{bmatrix} \begin{bmatrix} \lambda_1 \\ \lambda_2 \\ \mu' \end{bmatrix} = \begin{bmatrix} \rho_{1,0} \\ \rho_{2,0} \\ 1 \end{bmatrix}.$$

(i) The solution of the above system for the weights λ_1, λ_2 and the Lagrange multiplier μ' is

$$\lambda_1 = \frac{\begin{vmatrix} \rho_{1,0} & \rho_{1,2} & 1 \\ \rho_{2,0} & 1 & 1 \\ 1 & 1 & 0 \end{vmatrix}}{\begin{vmatrix} 1 & \rho_{1,2} & 1 \\ \rho_{1,2} & 1 & 1 \\ 1 & 1 & 0 \end{vmatrix}} = \frac{1}{2}\left(1 + \frac{\rho_{1,0} - \rho_{2,0}}{1 - \rho_{1,2}}\right),$$

$$\lambda_2 = \frac{\begin{vmatrix} 1 & \rho_{1,0} & 1 \\ \rho_{1,2} & \rho_{2,0} & 1 \\ 1 & 1 & 0 \end{vmatrix}}{\begin{vmatrix} 1 & \rho_{1,2} & 1 \\ \rho_{1,2} & 1 & 1 \\ 1 & 1 & 0 \end{vmatrix}} = \frac{1}{2}\left(1 - \frac{\rho_{1,0} - \rho_{2,0}}{1 - \rho_{1,2}}\right),$$

$$\mu' = \frac{\begin{vmatrix} 1 & \rho_{1,2} & \rho_{1,0} \\ \rho_{1,2} & 1 & \rho_{2,0} \\ 1 & 1 & 1 \end{vmatrix}}{\begin{vmatrix} 1 & \rho_{1,2} & 1 \\ \rho_{1,2} & 1 & 1 \\ 1 & 1 & 0 \end{vmatrix}} = \frac{1}{2}\left(\rho_{1,0} + \rho_{2,0} - \rho_{1,2} - 1\right).$$

Hence, the difference of the two weights is given by the following equation

$$\lambda_1 - \lambda_2 = \frac{\rho_{1,0} - \rho_{2,0}}{1 - \rho_{1,2}}.$$

If the sampling points do not coincide, the denominator is positive because $|\rho_{1,2}| < 1$. If the correlation is a monotonically decreasing function of the lag, the numerator is positive if $\mathbf{s}_1$ is closer to $\mathbf{z}$ than $\mathbf{s}_1$, and negative in the opposite case. If $\mathbf{z}$ is at the same distance from both $\mathbf{s}_1$ and $\mathbf{s}_2$, then both weights are equal.

(ii) The ordinary kriging prediction at the point $\mathbf{z}$ is given by

$$\hat{x}(\mathbf{z}) = \lambda_1 x_1^* + \lambda_2 x_2^* = \frac{x_1^* + x_2^*}{2} + \frac{\rho_{1,0} - \rho_{2,0}}{2(1 - \rho_{1,2})}\left(x_1^* - x_2^*\right). \tag{10.57}$$

The following remarks can be made regarding the OK prediction.

1. If $x_1^* = x_2^*$ the prediction is equal to the common value of the sample points, irrespectively of the spatial configuration of the three points.
2. If the prediction point $\mathbf{z}$ is at the same distance from $\mathbf{s}_1$ and $\mathbf{s}_2$ and the correlation function $\rho_{xx}(\cdot)$ is isotropic, the prediction is equal to the average of the sample values.
3. If the prediction point $\mathbf{z}$ is very far from both $\mathbf{s}_1$ and $\mathbf{s}_2$ (i.e., if the maximum of $\|\mathbf{s}_1 - \mathbf{z}\|$ and $\|\mathbf{s}_2 - \mathbf{z}\|$ is much larger than the maximum correlation length), the prediction is equal to the average of the sample values. This property holds regardless of whether the correlation function is isotropic or anisotropic.
4. If one of the sampling points, e.g., $\mathbf{s}_2$ is far from both $\mathbf{z}$ and $\mathbf{s}_1$, the ordinary kriging prediction becomes

$$\hat{x}(\mathbf{z}) = \frac{x_1^* + x_2^*}{2} + \frac{\rho_{1,0}}{2}\left(x_1^* - x_2^*\right). \tag{10.58}$$

Note that this equation differs from the single-point prediction of the preceding example, in spite of the large distance between $\mathbf{s}_2$ and $\mathbf{z}$.

(iii) According to (10.55) and taking into account that $\mu = \sigma_x^2\,\mu'$, the kriging variance is given by

$$\sigma_{\text{ok}}^2(\mathbf{z}) = \sigma_{\text{x}}^2 \left[1 - (\rho_{1,0} + \rho_{2,0}) - \frac{(\rho_{1,0} - \rho_{2,0})^2}{2(1 - \rho_{1,2})} + \frac{\rho_{1,2} + 1}{2} \right]. \tag{10.59}$$

If the points $\mathbf{s}_1$ and $\mathbf{s}_2$ are equidistant from $\mathbf{z}$, so that $\rho_{1,0} = \rho_{2,0} := \rho_0$, then the kriging variance is given by

$$\sigma_{\text{ok}}^2(\mathbf{z}) = \sigma_{\text{x}}^2 \left(1 - 2\rho_0 + \frac{\rho_{1,2} + 1}{2} \right).$$

Further, if $\mathbf{s}_1 = \mathbf{s}_2$ so that $\rho_{1,2} = 1$, the above variance equation leads to the kriging variance obtained for a single sampling point (see Example 10.7).

Example 10.9 Let us revisit the Example 10.8, assuming that the covariance function contains a nugget term with variance equal to c_0 such that the ratio of the nugget variance to the correlated variance is $q = c_0/\sigma_{\text{x}}^2$. Determine (i) the optimal weights and the Lagrange multiplier, (ii) the optimal prediction, and (iii) the prediction variance.

Answer Let us define $c_{1,2} = C_{\text{xx}}(\mathbf{s}_1 - \mathbf{s}_2)$, $c_{n,0} = C_{\text{xx}}(\mathbf{s}_n - \mathbf{z})$, for $n = 1, 2$. The ordinary kriging equations based on the covariance function formulation (10.50) are given by the following 3×3 linear system of equations

$$\begin{bmatrix} \sigma_{\text{x}}^2 + c_0 & c_{1,2} & 1 \\ c_{1,2} & \sigma_{\text{x}}^2 + c_0 & 1 \\ 1 & 1 & 0 \end{bmatrix} \begin{bmatrix} \lambda_1 \\ \lambda_2 \\ \mu \end{bmatrix} = \begin{bmatrix} c_{1,0} \\ c_{2,0} \\ 1 \end{bmatrix}.$$

(i) The solution of the above linear system for the weights and the Lagrange multiplier is given by

$$\lambda_1 = \frac{\begin{vmatrix} c_{1,0} & c_{1,2} & 1 \\ c_{2,0} & \sigma_{\text{x}}^2 + c_0 & 1 \\ 1 & 1 & 0 \end{vmatrix}}{\begin{vmatrix} \sigma_{\text{x}}^2 + c_0 & c_{1,2} & 1 \\ c_{1,2} & \sigma_{\text{x}}^2 + c_0 & 1 \\ 1 & 1 & 0 \end{vmatrix}} = \frac{1}{2} \left(1 + \frac{c_{1,0} - c_{2,0}}{\sigma_{\text{x}}^2 + c_0 - c_{1,2}} \right),$$

$$\lambda_2 = \frac{\begin{vmatrix} \sigma_{\text{x}}^2 + c_0 & c_{1,0} & 1 \\ c_{1,2} & c_{2,0} & 1 \\ 1 & 1 & 0 \end{vmatrix}}{\begin{vmatrix} \sigma_{\text{x}}^2 + c_0 & c_{1,2} & 1 \\ c_{1,2} & \sigma_{\text{x}}^2 + c_0 & 1 \\ 1 & 1 & 0 \end{vmatrix}} = \frac{1}{2} \left(1 - \frac{c_{1,0} - c_{2,0}}{\sigma_{\text{x}}^2 + c_0 - c_{1,2}} \right),$$

$$\mu = \frac{\begin{vmatrix} 1 & c_{1,2} & \rho_{1,0} \\ \rho_{1,2} & 1 & \rho_{2,0} \\ 1 & 1 & 1 \end{vmatrix}}{\begin{vmatrix} 1 & \rho_{1,2} & 1 \\ \rho_{1,2} & 1 & 1 \\ 1 & 1 & 0 \end{vmatrix}} = \frac{1}{2}\left(c_{1,0} + c_{2,0} - c_{1,2} - \sigma_{\mathrm{x}}^2 - c_0\right).$$

(ii) The optimal prediction in this case is given by

$$\hat{\mathrm{x}}(\mathbf{z}) = \frac{x_1^* + x_2^*}{2} + \frac{c_{1,0} - c_{2,0}}{2(\sigma_{\mathrm{x}}^2 + c_0 - c_{1,2})}\left(x_1^* - x_2^*\right). \tag{10.60}$$

(iii) Based on (10.55), the kriging variance is given by means of

$$\sigma_{\mathrm{ok}}^2(\mathbf{z}) = \sigma_{\mathrm{x}}^2\left[1 - (\rho_{1,0} + \rho_{2,0}) - \frac{(\rho_{1,0} - \rho_{2,0})^2}{2(1 + q - \rho_{1,2})} + \frac{1 + \rho_{1,2} + q}{2}\right]. \tag{10.61}$$

If we compare the above with the nugget-free kriging variance (10.59) it follows that (i) the kriging variance in the presence of nugget is higher and (ii) the nugget-free expression is obtained from (10.61) at the limit $q \to 0$.

10.6 Properties of the Kriging Predictor

The following is list of common properties of the simple and ordinary kriging predictor in the absence of a nugget term.

- *Independence of prediction from the variance:* The kriging prediction is independent of the field variance σ_{x}^2. This is most easily seen in terms of (10.51) which involves only the (normalized) correlation function. As a result the linear weights $\{\lambda_n\}_{n=1}^N$ are independent of the variance.
- *Kriging variance:* In contrast with the kriging prediction, the kriging variance is proportional to σ_{x}^2. For example, for OK this follows from (10.55) by taking into account that the weights $\{\lambda_n\}_{n=1}^N$ as well as μ' are independent of the variance and $\mu = \mu' \sigma_{\mathrm{x}}^2$.
- *The meaning of optimality:* The ordinary kriging prediction is *optimal* under two assumptions: (i) the random field from which the data are sampled is Gaussian and (ii) the predictor is formulated in terms of the "true" covariance (or variogram) function. In practice none of these conditions are exactly satisfied. If the data represent a realization of a non-Gaussian random field, kriging is the minimum mean square *linear* predictor (assuming that the covariance or

variogram functions are known). However, the linear predictor is not necessarily optimal in this case.

- *Conditional mean and kriging:* The connection between the kriging prediction and the normality of the probability distribution can be further appreciated by noting that the *simple kriging* prediction is identical to the *conditional mean* of the joint Gaussian distribution given by (6.15).
- *Exactitude:* Simple and ordinary kriging are *exact interpolators* according to the definition 10.1, because the predictions at the sampling points coincide with the sampling values. The exactitude property is lost if there is a nugget term and the smoothing formulation of kriging is used.
- *Non-convexity:* If an interpolation method generates a convex function, the predicted values are constrained to lie within the range of the data values. Kriging is not necessarily a convex predictor [303, 499]. In light of (10.1) convexity is guaranteed if the kriging weights satisfy $0 \leq \lambda_n \leq 1$. In practice, however, the kriging weights can be larger than one or even negative in the presence of the screening effect. In such cases, the estimates are not constrained to lie inside the range of the data values. For a specific example see [132, p. 189].
- *Robustness:* Kriging is a robust method, in the sense that small changes in the model parameters lead to small changes in the predictions. This is attributed to the linearity of the kriging predictor—notwithstanding the fact that the weights depend nonlinearly on parameters such as the correlation length.
- *Smoothing:* Kriging methods tend to generate overly smooth contours. This is a common problem for all interpolation methods. In addition, the small scale variability is not sufficiently represented by kriging, and extreme values tend to be underestimated.
- *Gaussian Process Regression:* There is a close connection between Gaussian random fields and Gaussian processes (GPs) [678]. In particular, the simple kriging predictor and the kriging variance are identical to the posterior mean and variance in Gaussian process regression.

  A nice introduction to Gaussian processes from the signal processing viewpoint is given in [661]. This review touches on a number of interesting topics such as *sparse GPs*, non-stationarity and *adaptive GPs*, and *warped (nonlinearly transformed)* GPs for non-Gaussian distributions (see also Sect. 11.3 below).

10.7 Topics Related to the Application of Kriging

Practical applications of the kriging methods to spatial data typically rely on a number of assumptions as discussed below. These assumptions should be validated in terms of statistical tests or based on general knowledge of the spatial processes involved.

- It is assumed that the data are derived from (i) a random field which is statistically homogeneous or (ii) a random field that has homogeneous increments or (iii) a random field that can be analyzed into a trend function and a statistically homogeneous residual, as explained in Sect. 1.3.5.

- Spatial predictions based on the *minimum mean square error* (MMSE) criterion are appropriate if the joint distribution of the data follows a symmetric, preferably Gaussian, probability density function.
- After removing a drift function, it is assumed that the residuals are represented by a *statistically homogeneous random field* or a *random field with homogeneous increments*. Then, the covariance function, or at least the variogram, depends on the spatial lag but not the location.
- *Variogram or covariance?* The ordinary kriging predictor is expressed in terms of both the covariance and the variogram. Is there a practical difference between the two formulations? If we use ordinary kriging, the mean is assumed to be unknown; hence, the covariance cannot be estimated using the method of moments estimator—although a maximum likelihood estimate is possible. Hence, the variogram is usually estimated from the data instead of the covariance. Once the variogram is determined, the covariance function can also be obtained if the field is statistically homogeneous by means of (3.47). If the field has homogeneous increments, using the variogram is the best option.
- *Optimality:* As stated earlier, kriging is the optimal predictor if the data is Gaussian and the true variogram is known. Since the variogram is estimated from the data, in practice kriging is at best approximately optimal. For non-Gaussian data, kriging is at best the optimal linear predictor.
- Kriging is an MMSE predictor and thus subject to the impact of outliers. *Robust methods* for estimating the variogram and minimizing the impact of outliers in prediction (robust kriging) have been developed [165, 166, 332]. The main idea is that the influence of large (non-Gaussian) errors and heavy tails is reduced by means of suitable transformations of the respective values.
- *Incorporating the trend in the kriging system:* It is possible to include in the kriging system a trend function constructed as a superposition of known basis functions. The coefficients of the superposition are determined from the solution of the kriging system. Respective Lagrange multipliers must be included to ensure an unbiased estimate in the presence of the undetermined trend coefficients. This formulation is known as *Universal Kriging* [132, 552]. In practice, universal kriging may not perform better than ordinary or regression kriging [813].
- Generalizations of kriging can be used to model non-Gaussian and vector data. Some of these extensions are briefly discussed in Chaps. 11 and 14.
- *Prediction intervals* can be constructed based on the appropriate kriging prediction and the respective kriging standard deviation [cf. (10.34)]. In practice, however, such intervals tend to be too narrow, since they do not account for the possibility of model error or for the uncertainty in the estimates of the model parameters. The latter can be addressed by means of Bayesian extensions of kriging.

10.7.1 The Screening Effect

It has been noticed in geostatistical studies that predictions primarily depend on the values of the closest observations [33, 165, 623]. It seems that the points in a local neighborhood around the prediction point effectively shield the prediction point from the impact of more distant observations. This phenomenon is known as the *screening effect* or as screen effect (see also Chap. 6).

The screening effect is discussed in detail with illustrative examples in [132]. On the practical side, screening motivates the construction of kriging predictors based on a subset of observations located in the neighborhood of the prediction point. This is particularly useful for problems including many observations, since it allows reducing the computational burden of kriging. Hence, the screening effect is often offered as a justification for using kriging neighborhoods in large spatial data sets.

A mathematically rigorous treatment of the screening effect is given by Stein [774, 776]. He shows that for data located on a regular grid, linear predictors are subject to the screening effect as the grid becomes progressively denser. The main factors that control the appearance of the screening effect are (i) the lack of singularities in the RF spectral density and (ii) the decay rate of the spectral density's tail at large wavenumbers. The simplest form of the tail condition requires that as $\|\mathbf{k}\| \to \infty$ the spectral density decay no faster than a power law $\propto \|\mathbf{k}\|^{-\alpha}$ where $\alpha > d$.[5]

Based on Table 4.4, the screening conditions are satisfied for (i) the exponential (ii) the Matérn models. Based on (7.16) the condition also holds for SSRF models. In addition, according to (5.85) the condition is satisfied for *fractional Brownian motion* whose spectral density is proportional to $\|\mathbf{k}\|^{-\alpha}$, where $\alpha \in (d, d+2)$.

Stein also examines the common assertion that the presence of a nugget term in the variogram cancels the screening effect. He shows that a nugget term can indeed reduce the screening effect, but he also finds a non-monotonic relation between the nugget variance and the screening effect [776].

10.7.2 Impact of Anisotropy on Kriging Weights

Often the simplest choice for modeling spatial data is to use a radial variogram function to capture spatial correlations. However, using an isotropic variogram to model anisotropic data can have significant impact on the kriging prediction and variance. The effect is stronger if only a small number of observations are available near the prediction point. This effect is investigated in [744, pp. 89-90], by focusing on a single prediction point and four equidistant observations in a spatial configuration as the one shown in Fig. 10.10.

[5] In full detail, the condition on the spectral density's tail allows for a product between an algebraic function $\|\mathbf{k}\|^{\alpha}$ and a slowly varying function of $\mathbf{k}$, i.e., combinations of the form $\|\mathbf{k}\|^{-\alpha} \ln \|\mathbf{k}\|$. However, such spectral densities are not typically used.

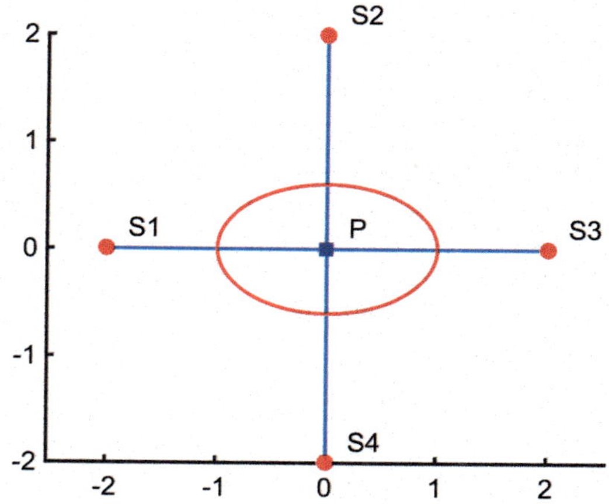

Fig. 10.10 Point configuration used to investigate the impact of anisotropy on kriging weights. The four points $S1$, $S2$, $S3$, and $S4$, marked by circles, denote observations. The point at the center, denoted by the square marker, is the prediction point. The elliptical contour illustrates the anisotropy of the variogram

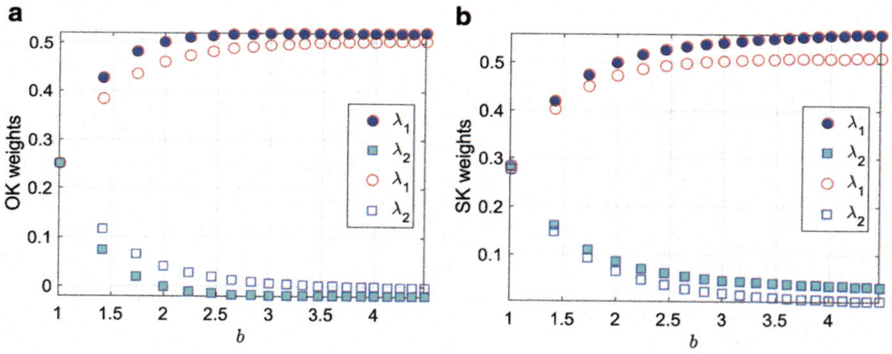

Fig. 10.11 Kriging weights λ_1 (associated with points S1 and S3 in Fig. 10.10) and λ_2 (associated with S2 and S4 in Fig. 10.10) for ordinary kriging (**a**) and simple kriging weights (**b**) versus the anisotropic length ratio $b = \xi_{\|}/\xi_{\perp}$. The distance between points P and S2 is equal to 2/3. A two-dimensional SSRF covariance with rigidity coefficient $\eta_1 = -1.9$ (solid markers) and 2 (open markers) is assumed. The lateral SSRF characteristic length is $\xi_{\|} = 1$ and the scale factor $\eta_0 = 10$. The vertical characteristic length takes values $\xi_{\perp} = \xi_{\|}/b$

Toy example We assume a Spartan variogram with principal axes of anisotropy aligned with the coordinate system. The longer correlation length, $\xi_{\|}$, is along the horizontal axis, while the shorter correlation length $\xi_{\perp}$, along the vertical axis. We control the anisotropy by maintaining $\xi_{\|}$ constant while reducing $\xi_{\perp}$ by the anisotropy length ratio b. This operation implies that the normalized horizontal lag, $r_{\|}/\xi_{\|}$ remains fixed, while the normalized vertical lag, $r_{\perp}/\xi_{\perp}$ increases.

As shown in Fig. 10.11, the kriging weights associated with points S1 and S3 increase as $\xi_{\perp}$ is reduced. In contrast, the kriging weights associated with points S2 and S4 decrease. This tendency is exhibited by the weights of both simple and ordinary kriging.

The ordinary kriging weights associated with S2 and S4 become negative for $\eta_1 = -1.9$ as the anisotropy ratio increases. The same behavior is observed for the Gaussian model [744, p. 90]. The simple kriging weights at S2 and S4, however, tend to zero but remain non-negative. The emergence of negative weights is reminiscent of the *screening effect*. Due to the anisotropy, the normalized distance $r_{\parallel}/\xi_{\parallel}$ between points S1 and S3 is smaller than the normalized distance $r_{\perp}/\xi_{\perp}$ between S2 and S4. The vertical normalized distance increases relatively to the horizontal normalized distance as b increases. Thus, the more "distant" points are "shielded" by the presence of the "closer" observations, leading to negative values of the weights.

Real data The impact of anisotropy on the interpolation of real data is not as easily analyzed. The anisotropy parameters are not known a priori as in the example above; instead, they have to be inferred from the data. Anisotropy estimates can be obtained by means of fitting directional variograms [303, 823], the maximum likelihood method [884], the Bayesian approach [220, 221], or the Covariance Hessian identity [135, 663]. All methods involve considerable uncertainties, and the anisotropy estimates are influenced by the sampling density and the geometry of the sampling network among other things.

Nonetheless, there is empirical evidence that the use of anisotropic variograms in kriging can improve interpolation estimates. For example, anisotropic kriging has been shown to improve the interpolation of wind speeds over geographically large areas compared to isotropic kriging [261]. A recent study reviews non-parametric methods for testing the hypothesis of isotropy in spatial data [839]. Such tests can help researchers decide whether to include anisotropy in the spatial model.

10.8 Evaluating Model Performance

In the above, we have tacitly assumed that it is possible to build a suitable spatial model, which we then use to predict the unknown values of the process at prescribed target points. It is fair to ask how the performance of the spatial model can be measured. There are various possible answers to this question, which rely on statistical measures that probe the agreement between the model and the data. These measures (i) quantify the agreement between the model predictions and the data, and (ii) test the consistence between the initial assumptions regarding the spatial model structure and the data (set out in Sect. 1.2.1). Such tasks belong in the domain of *model validation*. Established procedures for validation of spatial models are available in the geostatistical literature [132, 165, 459].

Model residuals A mathematical formulation of model validation is based on the notion of *model residuals*. They represent the differences between the measured values of the process and the respective model estimates, i.e.,

$$\widetilde{\epsilon}(\mathbf{s}_n) = x_n^* - \hat{x}_{-n}(\mathbf{s}_n), \; n = 1, \ldots, N, \tag{10.62}$$

where $\hat{x}_{-n}(\mathbf{s}_n)$ is the prediction obtained at $\mathbf{s}_n$ based on the sampling set minus the target point $\mathbf{s}_n$. Thus, we use the notion of *leave-one-out cross validation* which assumes that the estimate $\hat{x}_{-n}(\mathbf{s}_n)$ is based on the $N-1$ points in the set $\mathbf{S}_\mathrm{N} \setminus \{\mathbf{s}_n\}$. Note that if $\mathbf{s}_n$ were not excluded, the above equation would not be very helpful: kriging being an exact interpolator, it would have implied $\widetilde{\epsilon}(\mathbf{s}_n) = 0$.

It is also straightforward to define the *residual field*

$$\widetilde{\varepsilon}(\mathbf{s}_n; \omega) = \mathrm{X}(\mathbf{s}_n; \omega) - \hat{\mathrm{X}}_{-\mathrm{n}}(\mathbf{s}_n; \omega).$$

Cross validation The evaluation of the model performance by means of cross-validation is based on various statistical measures (e.g., mean absolute value, root mean square error) of the residuals and on correlation coefficients between the data and the model estimates (e.g., linear Pearson, Spearman and Kendall rank coefficients). We further discuss cross-validation measures in Chap. 12.

Standardized residuals Typically, one defines the *standardized residuals* as follows

$$\breve{\epsilon}(\mathbf{s}_n) = \frac{\widetilde{\epsilon}(\mathbf{s}_n)}{\sigma_{\mathrm{kr}}(\mathbf{s}_n)}, \quad \breve{\varepsilon}(\mathbf{s}_n; \omega) = \frac{\widetilde{\varepsilon}(\mathbf{s}_n; \omega)}{\sigma_{\mathrm{kr}}(\mathbf{s}_n)}, \tag{10.63}$$

where $\sigma_{\mathrm{kr}}(\mathbf{s}_n)$ is the standard deviation of the kriging method used and $\breve{\epsilon}(\mathbf{s}_n)$ is the sample value of the standardized residual. The standardized residual is generalized to the respective residual random field $\breve{\varepsilon}(\mathbf{s}_n; \omega)$.

Statistics of residuals The performance of the interpolation method can then be evaluated by means of the following *residual statistics* [165]

$$S_1(\omega) = \frac{1}{N}\sum_{n=1}^{N} \breve{\epsilon}(\mathbf{s}_n; \omega), \quad S_2(\omega) = \frac{1}{N}\sum_{n=1}^{N} \breve{\epsilon}^2(\mathbf{s}_n; \omega),$$

The expectation of the averaged residuals and the expectation of the averaged squared residuals are given respectively by $\mathcal{E}[S_1(\omega)] = 0$ and $\mathcal{E}[S_2(\omega)] = 1$. Thus, it is expected that the sample values S_1 and S_2 should be close to zero and one respectively, especially for large N. The proximity of the sample values to the respective expectations also depends on the probability distributions followed by the statistics $S_i(\omega)$.

Normality of residual statistics Under the assumption that the random field $\mathrm{X}(\mathbf{s}; \omega)$ follows the joint Gaussian distribution, $S_1(\omega)$ follows the univariate standard normal distribution. Appealing to the Central Limit Theorem under weak dependence (e.g., [92]), it can also be argued that $S_2(\omega)$ also follows the normal distribution asymptotically (for $N \gg 1$).

Model hypothesis tests The standard statistical theory can be used to test the *null hypothesis* that any given spatial model adequately describes the data. Given the normal distribution of the residual statistics, acceptable values for the latter should not deviate from the respective expectations by more than a few times the standard deviation. For example, the spatial model at confidence level ≈5% is rejected if [457]

$$|S_i - \mathcal{E}[S_i(\omega)]\,| > \ell\,\sqrt{\mathbb{V}\mathrm{ar}\,\{S_i(\omega)\}},\ \text{ for } i = 1, 2,$$

where a value of $\ell = 2$ would imply that there is a ≈ 5% chance that the model would be erroneously rejected.

Orthonormal residuals The problem with the above procedures is that analytical calculation of the variance of the residual statistics $S_i(\omega)$, $i = 1, 2$, involves complicated expressions that include the variogram function, the sampling locations, and the kriging coefficients. This complexity is due to the presence of correlations between the standardized residuals at different locations [165, 457, p. 104]. Kitanidis addressed this issue by defining uncorrelated *orthonormal residuals* that are used to build respective hypothesis tests [457, 459].

Information theory measures Cross-validation measures that are based on causality and information theoretic methods are used for the identification, estimation, and prediction of nonlinear and complex dynamical systems from available data [830]. Such methods are not commonly used in spatial statistics. However, for spatial data that represent responses of nonlinear and complex systems to various excitations, it would be worthwhile to further explore the application of such measures.

Chapter 11
More on Spatial Prediction

> More is different.
>
> *Philip Warren Anderson*

This chapter begins with linear extensions of kriging that provide higher flexibility and allow relaxing the underlying assumptions on the method. Such generalizations include the application of ordinary kriging to *intrinsic random fields* that can handle non-stationary data, as well as the methods of *regression kriging* and *universal kriging* that incorporate deterministic trends in the linear prediction equation [338]. *Cokriging* allows combining multivariate information in the prediction equations. Various *nonlinear extensions* of kriging have also been proposed (indicator kriging, disjunctive kriging) that aim to handle non-Gaussian data.

All of the kriging extensions are treated herein briefly and superficially. There are two reasons motivating this choice. Firstly, residual kriging and the application of ordinary kriging to intrinsic random fields do not involve substantial new ideas. Secondly, based on personal experience and the information available in the scientific literature, I feel that the "cost-benefit" balance does not warrant, at the current stage, more emphasis on these generalizations. Interested readers can find more information on these methods in standard geostatistical textbooks [132, 165, 684, 823].

Next, the chapter focuses on the formulation of kriging in the Bayesian framework. This is introduced by means of *Gaussian Process Regression*—a method that has its origins in machine learning but is also linked with kriging—and *Bayesian Kriging*.

The short promenade in the Bayesian woods is followed by a discussion of the continuum formulation of linear prediction and its connection with Spartan spatial random fields. Finally, *linear prediction* is discussed in the framework of the *local interaction* approach that is derived from the discretization of the SSRF precision kernel. The chapter closes with a short comment on computational challenges that confront the analysis big spatial data sets.

D. T. Hristopulos, *Random Fields for Spatial Data Modeling*, Advances in Geographic Information Science, https://doi.org/10.1007/978-94-024-1918-4_11

11.1 Linear Generalizations of Kriging

This section is concerned with linear prediction methods that generalize the simple and ordinary kriging framework. Such methods include ordinary kriging with intrinsic random fields, regression kriging, universal kriging, cokriging and functional kriging.

11.1.1 Ordinary Kriging with Intrinsic Random Fields

Unlike simple kriging, ordinary kriging can also be used with intrinsic random fields, thus allowing application to data sets that display more complex types of spatial dependence. The mathematical operations on which this extension is based are described in detail in [624]. The main result is that for intrinsic random fields the formulation (10.52)—as well as its abbreviated form (10.54)—are valid.

We outline the main mathematical steps involved in the proof. At the heart of the proof is the following identity for the covariance of the increments

$$\begin{aligned}\mathbb{C}\text{ov}\left\{X(\mathbf{s}_1;\omega) - X(\mathbf{s};\omega), X(\mathbf{s}_2;\omega) - X(\mathbf{s};\omega)\right\} =& \gamma_{xx}(\mathbf{s}_1-\mathbf{s}) + \gamma_{xx}(\mathbf{s}_2-\mathbf{s}) \\ &- \gamma_{xx}(\mathbf{s}_1-\mathbf{s}_2). \end{aligned} \tag{11.1}$$

This identity is valid for intrinsic random fields (it does not require the second-order stationarity assumption). It is obtained from the definition of the variogram $\gamma_{xx}(\mathbf{s}_1 - \mathbf{s}_2)$ by adding and subtracting the term $X(\mathbf{s};\omega)$ inside the expectation. Based on this identity, the mean square prediction error (kriging variance) is given by

$$\sigma^2_{\text{ok}}(\mathbf{z}) = 2\sum_{\alpha=1}^{N}\lambda_\alpha\,\gamma_{xx}\left(\mathbf{z}-\mathbf{s}_\alpha\right) - \sum_{\alpha=1}^{N}\sum_{\beta=1}^{N}\lambda_\alpha\lambda_\beta\,\gamma_{xx}\left(\mathbf{s}_\alpha-\mathbf{s}_\beta\right). \tag{11.2}$$

The objective function used in the minimization includes, in addition to $\sigma^2_{\text{ok}}(\mathbf{z})$, the term $2\mu\left(\sum_{\alpha=1}^{n}\lambda_n - 1\right)$ that involves the Lagrange multiplier μ. It is then straightforward to show that the *first-order optimality* condition yields the variogram-based formulation of the ordinary kriging system (10.52).

An equivalent expression for the kriging variance in terms of the variogram function is given by the following expression, in which $\lambda_0 = -1$:

$$\sigma^2_{\text{ok}}(\mathbf{z}) = -\sum_{\alpha=0}^{N}\sum_{\beta=0}^{N}\lambda_\alpha\lambda_\beta\,\gamma_{xx}\left(\mathbf{s}_\alpha-\mathbf{s}_\beta\right). \tag{11.3}$$

fBm interpolation Since the OK equations are valid for intrinsic random fields they can be used to interpolate *fractional Brownian fields* (i.e., fractional Brownian motion). A simple example of such an application is given in [132, pp. 165–167].

11.1.2 Regression Kriging

In regression kriging (also known as residual kriging) the target variable is decomposed into a deterministic trend function and a residual field. The trend function is typically expressed in terms of a superposition of suitable selected basis functions or auxiliary variables. For example, if the target variable is the amount of monthly rainfall at a specific geographic location, the altitude of the location is an auxiliary variable that affects rainfall and should be incorporated in the trend function. The spatial model is thus defined by

$$X(\mathbf{s};\omega) = \sum_{k=1}^{K} \beta_k \, f_k(\mathbf{s}) + X'(\mathbf{s};\omega), \tag{11.4}$$

where $f_k(\mathbf{s}), k = 1, \ldots, K$ are the basis functions, β_k are the respective coefficients, and $X'(\mathbf{s};\omega)$ is the fluctuation field.

Regression kriging allows incorporating in the spatial model information from secondary variables in the form of a trend function; it is then known as *kriging with external drift* [340, 692]. Regression kriging can also incorporate deterministic trends derived from the solution of physical laws or other auxiliary information (see Chap. 2 and [815]).

Once the trend function is determined, the residuals are estimated by subtracting from the data the trend values at the specific locations. Then, a random field model is estimated for the residuals. The residual field is interpolated by means of ordinary kriging to derive estimates at unmeasured locations. Finally, the interpolated residual field is added to the trend function.

A brief review of spatial data analyses that are based on regression kriging are given in [814]. In addition, this study presents an application of regression kriging to groundwater level modeling, using a trend function that is based on physical law.

11.1.3 Universal Kriging

Universal kriging (UK) is another version of kriging that aims to model spatial data with apparent trends. In UK, the random field is assumed to comprise drift (assumed to represent the expectation of the random field) and fluctuation components.

The drift is constructed by means of a superposition of known *basis functions* as described in Chap. 2 by (2.14), i.e.,

$$m_{x}(\mathbf{s}) = \sum_{p=1}^{P} \beta_p \, f_p(\mathbf{s}),$$

and it is typically assumed that $f_1(\mathbf{s}) = 1$ represents the constant term.

The fluctuations are assumed to be either a stationary or an intrinsic random field. The spatial model has thus the same form as (11.4). However, the coefficients of the trend function are not determined separately but instead through the solution of the kriging system. This approach has the advantage over regression kriging that the uncertainty in the trend coefficients is taken into account.

The UK predictor is given by the following linear superposition

$$\hat{X}(\mathbf{z};\omega)=\sum_{n=1}^{N}\lambda_n\, X(\mathbf{s}_n;\omega).$$

The kriging equations are derived, as usual, by satisfying the conditions for zero bias and minimum variance. This requirement leads to an $(N+K)\times(N+K)$ linear system of equations that involves either the covariance function (stationary case) or the variogram (intrinsic case). The N unknown coefficients correspond to the weights $\{\lambda_n\}_{n=1}^{N}$, while the remaining K coefficients correspond to Lagrange multipliers $\{\mu_k\}_{k=1}^{K}$ that enforce the *zero-bias constraints*, i.e.,

$$\sum_{n=1}^{N}\lambda_n f_k(\mathbf{s}_n)=f_k(\mathbf{z}),\ \text{ for } k=1,\ldots,K.$$

The kriging system is supplemented by the equations that result from the minimum variance constraint.

In the *stationary case* the UK system is given by

$$\sum_{m=1}^{N}C_{\mathrm{xx}}(\mathbf{s}_n-\mathbf{s}_m)\lambda_m+\sum_{k=1}^{K}\mu_k f_k(\mathbf{s}_n)=C_{\mathrm{xx}}(\mathbf{z}-\mathbf{s}_n),\ n=1,\ldots,N. \tag{11.5}$$

In the *intrinsic case* the UK system is expressed in terms of the variogram function as follows

$$\sum_{m=1}^{N}\gamma_{\mathrm{xx}}(\mathbf{s}_n-\mathbf{s}_m)\lambda_m-\sum_{k=1}^{K}\mu_k f_k(\mathbf{s}_n)=\gamma_{\mathrm{xx}}(\mathbf{z}-\mathbf{s}_n),\ n=1,\ldots,N. \tag{11.6}$$

For the kriging system to have a unique solution, the vectors obtained by the values of the basis functions at the sampling points need to be linearly independent.

The *UK variance* is given by the following equation

$$\sigma^2_{\mathrm{uk}}=\sum_{m=1}^{N}\lambda_m\,\gamma_{\mathrm{xx}}(\mathbf{z}-\mathbf{s}_m)-\sum_{k=1}^{K}\mu_k f_k(\mathbf{z}). \tag{11.7}$$

Hence, the UK variance also takes into account, by means of the Lagrange multipliers, the uncertainty that results from the estimation of the trend function.

The UK linear system assumes that the covariance (or variogram) function is known or can be estimated from the data. However, this assumption implies that the drift function is known—which is not typically the case. Hence, one needs to estimate both the drift and the variogram at the same time, not a simple task.

One possible solution is to use an iterative approach of drift-variogram estimation [32]. A different answer to this problem is provided by using the Bayesian framework to address the model uncertainties [628].

11.1.4 Cokriging

Cokriging is a multivariate extension of kriging that allows incorporating in the prediction equations information from secondary variables that are strongly (either positively or negatively) correlated with the primary (target) variable.

The use of cokriging can be beneficial if only sparse measurements of the primary variable are available. Then, the sparse observations can be complemented by including secondary data (which may be more easily available) that are strongly correlated with the primary variable. This approach is in contrast with regression kriging, in which the auxiliary variables are used to define trend functions.

One of the main difficulties with the application of cokriging is the *coregionalization* problem. This refers to the definition of physically meaningful and mathematically permissible *cross-covariance models* for the correlations between the primary and the secondary variables [823]. We consider coregionalization in a little more detail, given its central role in the application of cokriging.

Definition 11.1 Let us denote by $\mathbf{X}(\mathbf{s};\omega) = (X_1(\mathbf{s};\omega), \ldots, X_D(\mathbf{s};\omega))^\top$ a zero-mean, stationary multivariate RF with D components $X_p(\mathbf{s};\omega)$, $p = 1, \ldots D$.

The *matrix covariance function* of the above RF is defined by

$$C_{p,q}(\mathbf{s}_1 - \mathbf{s}_2) := \mathcal{E}[X_p(\mathbf{s}_1;\omega)\, X_q(\mathbf{s}_2;\omega)], \quad p, q = 1, \ldots, D.$$

A generalization of Bochner's theorem for the multivariate case was developed by Cramer. His theorem allows testing the permissibility of proposed matrix covariance functions for multivariate fields [160, 823]. In the following, we use $\mathbf{C}(\mathbf{r})$ to denote the $D \times D$ matrix covariance function, $\widetilde{\mathbf{C}}(\mathbf{k})$ to denote the *matrix spectral density*, and $\widetilde{C}_{p,q}(\mathbf{k})$ (where $p, q = 1, \ldots, D$) the scalar spectral density elements of $\widetilde{\mathbf{C}}(\mathbf{k})$.

Theorem 11.1 (Cramer's Theorem) *The matrix function* $\mathbf{C}(\mathbf{r})$ *is a valid matrix covariance of a continuous, stationary, multivariate RF if its Fourier transform* $\widetilde{\mathbf{C}}(\mathbf{k})$ *exists, and if the following conditions hold:*

1. *The marginal (auto-spectral) densities exist, i.e., the functions* $\widetilde{C}_{p,p}(\mathbf{k})$ *are positive definite functions of* $\mathbf{k}$, *which implies*

$$\widetilde{C}_{p,p}(\mathbf{k}) \geq 0, \; \textit{for all } \mathbf{k} \in \mathbb{R}^d, \quad \textit{for all } p = 1, \ldots, D.$$

2. *The integrals of the auto-spectral densities are finite and equal to the respective variances, that is,*

$$\int_{\mathbb{R}^d} \frac{\mathrm{d}\mathbf{k}}{(2\pi)^d} \, \widetilde{C}_{p,p}(\mathbf{k}) = \sigma_p^2, \; \textit{for all } p = 1, \ldots, D.$$

3. *The cross-spectral densities (i.e., the off-diagonal elements of the matrix spectral density) have bounded variation, i.e.,*

$$\int_{\mathbb{R}^d} \mathrm{d}\mathbf{k} \, \left| \widetilde{C}_{p,q}(\mathbf{k}) \right| < \infty, \; \textit{for all } p \neq q = 1, \ldots, D.$$

4. *The matrix* $\widetilde{\mathbf{C}}(\mathbf{k})$ *(obtained for fixed* $\mathbf{k} \in \mathbb{R}^d$*) is positive definite for all* $\mathbf{k} \in \mathbb{R}^d$*.*

The first two conditions of Cramer's theorem ensure that the diagonal spectral elements $\widetilde{C}_{p,p}(\mathbf{k})$ correspond to permissible spectral densities for the respective univariate processes in agreement with Bochner's theorem. These conditions are straightforward to establish as is usually done for scalar random fields. The third condition ensures the integrability of the off-diagonal spectral densities.

The main difficulty with Cramer's Theorem stems from the fourth condition. To appreciate this consider that a real-valued, $D \times D$ symmetric matrix $\boldsymbol{A}$ is positive definite if for any real vector $\mathbf{z} \in \mathbb{R}^D$ it holds that $\mathbf{z}^\top \boldsymbol{A}\mathbf{z} \geq 0$. To prove positive definiteness, one needs to show that all the eigenvalues for all the principal minors of $\boldsymbol{A}$ are nonnegative. The fourth condition of Cramer's theorem for $\widetilde{\mathbf{C}}$ requires demonstrating positive definiteness for all $\mathbf{k} \in \mathbb{R}^d$. However, there are no general methods available for this task.

Separable model The simplest mathematical construction that allows establishing the permissibility of a matrix covariance function is based on the *separability hypothesis*. According to this hypothesis, we can write

$$C_{p,q}(\mathbf{r}) := C(\mathbf{r}) \, a_{p,q}, \; \text{for } p, q = 1, \ldots, D,$$

where $C(\mathbf{r})$ is a marginal covariance function and $\boldsymbol{A} = [a_{p,q}]_{p,q=1}^{D}$ a *positive definite coefficient matrix*. This *separable coregionalization* model is straightforward but inflexible. In addition, it is not supported by physical models.

Linear coregionalization model The popular *linear model of coregionalization* (LMC) assumes that the vector components are linear combinations of latent, independent, univariate spatial processes [304, 823]. More precisely, a multivariate model that involves L latent fields is defined by

$$\mathrm{X}_p(\mathbf{s};\omega) = \sum_{\alpha=1}^{L} \mathrm{X}_p^{(\alpha)}(\mathbf{s};\omega), \text{ for } p = 1, \ldots, D. \tag{11.8}$$

The latent fields $\mathrm{X}_p^{(\alpha)}(\mathbf{s};\omega)$ have zero expectation, i.e., $\mathcal{E}[\mathrm{X}_p^{(\alpha)}(\mathbf{s};\omega)] = 0$ for all $\alpha = 1, \ldots, L$, $p = 1, \ldots, D$. In addition, the covariance functions of the latent fields is given by

$$\mathcal{E}\left[\mathrm{X}_p^{(\alpha)}(\mathbf{s};\omega)\,\mathrm{X}_q^{(\beta)}(\mathbf{s}+\mathbf{r};\omega)\right] = \begin{cases} C_{p,q}^{(\alpha)}(\mathbf{r}), & \alpha = \beta, \\ 0, & \alpha \neq \beta. \end{cases} \tag{11.9}$$

In the coregionalization model the covariance functions of the latent fields are assumed to be given by

$$C_{p,q}^{(\alpha)}(\mathbf{r}) = a_{p,q}^{(\alpha)}\,\rho^{(\alpha)}(\mathbf{r}),$$

where $\rho^{(\alpha)}(\mathbf{r})$ is a common correlation function for all the vector components that correspond to the nested model specified by α. Then, the *matrix covariance function* is given by

$$\mathbf{C}(\mathbf{r}) = \sum_{\alpha=1}^{L} \boldsymbol{A}^{(\alpha)}\,\rho^{(\alpha)}(\mathbf{r}),$$

where the matrices $\boldsymbol{A}^{(\alpha)}$ for $\alpha = 1, \ldots, L$ are positive definite.

The LMC construction is inadequate in many situations, since the smoothness properties are dominated by the roughest of the latent components [290]. Beyond the LMC framework, a multivariate model in which both the marginal and the cross-covariance functions are of Matérn type is introduced in [290]. For $D > 2$ the proposed model uses a parsimonious assumption that is limited to common scale factors (correlation lengths) and restricted smoothness parameters for the cross covariances. A class of valid Matérn cross-covariance functions with different smoothness properties and correlation lengths for each component is presented in [29].

Factorial kriging or kriging analysis aims at estimating and mapping different sources of spatial variability in the experimental variogram [823]. In particular, factorial kriging determines different spatial correlation scales and uses them to construct the coregionalization model. In a sense, factorial kriging is similar to spectral analysis, because it allows resolving different length scales (i.e., corresponding to different spatial frequencies or different frequency bands).

Multivariate Spartan random fields A multivariate extension of Spartan spatial random fields that is based on the concept of *diagonal dominance* is presented in [378]. Diagonal dominance is mathematically a stricter condition than positive definiteness.[1] However, diagonal dominance can provide tractable, sufficient conditions for positive definiteness. In particular, the diagonal dominance condition requires that the elements of the matrix spectral density satisfy the inequalities

$$\widetilde{C}_{p,p}(\mathbf{k}) > \sum_{q=1,\neq p}^{D} \left| \widetilde{C}_{p,q}(\mathbf{k}) \right|, \text{ for all } p = 1, \ldots, D, \text{ for all } \mathbf{k} \in \mathbb{R}^d.$$

The spectral density of multivariate Spartan random fields is given by

$$\widetilde{C}_{p,q}(\mathbf{k}) = \frac{\eta_{0;p,q}\xi_{p,q}^{d}}{1 + \eta_{1;p,q}\xi_{p,q}^{2} \|\mathbf{k}\|^2 + \xi_{p,q}^{4} \|\mathbf{k}\|^4}, \quad (11.10)$$

where the coefficients $\eta_{0;p,q}$ represent scale factors with dimensions $[X]^2$, the $\eta_{1;p,q}$ represents rigidity parameters, and $\xi_{p,q}$ characteristic length scales. The permissibility conditions involve the following inequalities

1. C1: $\eta_{0;p,q} > 0$, for all $p, q = 1, \ldots, D$.
2. C2: $\xi_{p,p} > 0$, for all $p, q = 1, \ldots, D$.
3. C3: $\eta_{1;p,p} > -2$, for all $p = 1, \ldots, D$.
4. C4: $\eta_{0;p,q}\, \xi_{p,p}^{d} > \sum_{p\neq q} \left| \eta_{0;p,q}\, \xi_{p,q}^{d} \right|$, for all $p = 1, \ldots, D$.
5. C5: $\xi_{p,q} > \xi_{p,p}$, for all $p, q = 1, \ldots, D$.
6. C6: $\eta_{1;p,q}\xi_{p,q}^{2} > \eta_{1;p,p}\xi_{p,p}^{2}$, for all $p, q = 1, \ldots, D$.

The conditions (C1–C3) ensure that the first condition of Cramer's theorem is valid, i.e., that the diagonal spectral (marginal) densities are non-negative. The second condition of Cramer's theorem, i.e., the integrability of the marginal spectral densities is ensured for $\mathbf{k} \in \mathbb{R}^d, d \leq 3$, by the dependence of the spectral density (11.10). The third condition of Cramer's theorem, i.e., the integrability of the absolute values of the cross spectral densities is also ensured by the form of the spectral density. The fourth condition of Cramer's theorem, i.e., the positive definiteness of the spectral density matrix for all $\mathbf{k} \in \mathbb{R}^d$ is satisfied by means of the diagonal dominance conditions (C4–C6).

The Spartan multivariate covariance is expressed in terms of the following dimensionless (normalized) parameters:

$$\Delta_{p,q} = \left| \eta_{1;p,q}{}^2 - 4 \right|^{1/2}, \quad \text{discriminant of spectral density polynomial}$$

$$h = \frac{\|\mathbf{r}\|}{\xi_{p,q}}, \quad \text{normalized lag}$$

[1] A square $N \times N$ matrix $\mathbf{A}$ is diagonally dominant if $\left|A_{i,i}\right| > \sum_{j=1,\neq i}^{N} \left| A_{i,j} \right|$ for all $i = 1, \ldots, N$.

$$u_{p,q} = \frac{1}{2} \left|2 - \eta_{1;p,q}\right|^{1/2}, \quad \text{wavenumber for } |\eta_{1;p,q}| < 2,\ d = 1, 3$$

$$\beta_{p,q} = \frac{1}{2} \left|2 + \eta_{1;p,q}\right|^{1/2}, \quad \text{inverse relaxation length for } |\eta_{1;p,q}| < 2,\ d = 1, 3$$

$$w_{1;p,q} = \frac{1}{2} \left|2 - \Delta_{p,q}\right|^{1/2}, \quad \text{inverse relaxation length for } \eta_{1;p,q} > 2,\ d = 1, 3$$

$$w_{2;p,q} = \frac{1}{2} \left|2 + \Delta_{p,q}\right|^{1/2}, \quad \text{inverse relaxation length for } \eta_{1;p,q} > 2,\ d = 1, 3$$

$$\lambda_{1;p,q} = \left(\frac{\eta_{1;p,q} - \Delta_{p,q}}{2}\right)^{1/2} \quad \text{inverse relaxation length, } d = 2$$

$$\lambda_{2;p,q} = \left(\frac{\eta_{1;p,q} + \Delta_{p,q}}{2}\right)^{1/2} \quad \text{inverse relaxation length, } d = 2.$$

The multivariate Spartan covariance function $C_{p,q}(\|\mathbf{r}\|)$ $(p, q = 1, \ldots, D)$ obtained from the matrix spectral density (11.10) is given by the following expressions that depend on the dimensionality d of the space $\mathbb{R}^d$ in which the position vectors are defined.

The multivariate Spartan covariance function in $d = 1$ is given by

$$C_{p,q}(h) = \eta_{0;p,q}\, \mathrm{e}^{-h\,\beta_{p,q}} \left[\frac{\sin(h\,u_{p,q})}{4\,u_{p,q}} + \frac{\cos(h\,u_{p,q})}{4\,\beta_{p,q}}\right], \qquad \left|\eta_{1;p,q}\right| < 2 \tag{11.11a}$$

$$C_{p,q}(h) = \frac{\eta_{0;p,q}\,(1+h)\,\mathrm{e}^{-h}}{4}, \qquad \eta_{1;p,q} = 2 \tag{11.11b}$$

$$C_{p,q}(h) = \frac{\eta_{0;p,q}}{2\,\Delta_{1;p,q}} \left(\frac{\mathrm{e}^{-h\,w_{1;p,q}}}{w_{1;p,q}} - \frac{\mathrm{e}^{-h\,w_{2;p,q}}}{w_{2;p,q}}\right), \qquad \eta_{1;p,q} > 2. \tag{11.11c}$$

In light of the discussion in Sect. 9.2, the multivariate Spartan covariance function in one dimension represents the covariance for a system of D coupled, damped, classical harmonic oscillators driven by white noise.

The multivariate Spartan covariance function in $d = 2$ is given by

$$C_{p,q}(h) = \frac{\eta_{0;p,q}\,\Im\left[K_0(h\lambda_{2;p,q})\right]}{\pi\,\Delta_{p,q}}, \qquad \left|\eta_{1;p,q}\right| < 2, \tag{11.12a}$$

$$C_{p,q}(h) = \frac{\eta_{0;p,q}\, h}{4\pi}\, K_{-1}(h), \qquad \eta_{1;p,q} = 2, \tag{11.12b}$$

$$C_{p,q}(h) = \frac{\eta_{0;p,q}\left[K_0(h\,\lambda_{2;p,q}) - K_0(h\lambda_{1;p,q})\right]}{2\pi\,\Delta_{p,q}}, \qquad \eta_{1;p,q} > 2 \tag{11.12c}$$

where $K_\nu(\cdot)$ is the modified Bessel function of the second kind and order ν, and $\Im\,[\cdot]$ denotes the imaginary part.

Finally, the equations that define the multivariate Spartan covariance function in $d = 3$ are

$$C_{p,q}(h) = \frac{\eta_{0;p,q}\,\mathrm{e}^{-h\,\beta_{p,q}}}{2\pi\,\Delta_{p,q}}\left[\frac{\sin(h\,u_{p,q})}{h\,u_{p,q}}\right], \qquad \left|\eta_{1;p,q}\right| < 2 \tag{11.13a}$$

$$C_{p,q}(h) = \frac{\eta_{0;p,q}\,\mathrm{e}^{-h}}{8\pi}, \qquad \eta_{1;p,q} = 2 \tag{11.13b}$$

$$C_{p,q}(h) = \frac{\eta_{0;p,q}\left(\mathrm{e}^{-h\,w_{1;p,q}} - \mathrm{e}^{-h\,w_{2;p,q}}\right)}{4\pi\,\Delta_{p,q}\,h}, \qquad \eta_{1;p,q} > 2. \tag{11.13c}$$

In the above equations, keep in mind that h is actually a shorthand for $h_{p,q}$.

11.1.5 Functional Kriging

Functional data analysis is a rapidly developing field of statistics [356, 676]. The main premise of functional data theory is that a stream of data at a given location in space can be viewed as a continuous function (or stochastic process). A spatially distributed collection of such data generates an ensemble of curves, the properties of which are correlated in space. This framework is quite suitable for the analysis of spatiotemporal data, where time series at specific locations can be viewed as samples of continuous functions. In certain cases, the functional data can represent probability density functions.

Aiming to take advantage of the spatial continuity of functional data, several linear generalizations of kriging, known as *functional kriging*, have been proposed in the literature [191, 225, 562, 563, 601]. These involve extensions of both ordinary and universal kriging methods to functional data. Functional kriging methods can thus be applied to interpolate both probability density functions (if the marginal pdf of a physical attribute depends on space) and spatiotemporal data.

The data used in functional kriging are spatially extended, possibly contaminated by noise, and represent mathematical objects (curves) of infinite dimensionality. Hence, both smoothing and dimensionality reduction methods are often applied to functional data [829].

In the case of purely spatial (as opposed to space-time) data, it is assumed that every point in space represents an elementary volume, within which different classes (states) of the process are observed (for example, different percentages of grain diameters in soil). The observed classes allow the definition of suitable probability density functions at the observation locations. The goal of functional kriging is to interpolate these probability density functions to other locations where measurements are not available.

11.2 Nonlinear Extensions of Kriging

Several generalizations of kriging have been proposed that aim to extend the scope of optimal linear estimation to non-Gaussian data. These generalizations are based on various *nonlinear methods*. We briefly consider these topics, referring the reader to the classical geostatistical texts for further information [132, 165, 823].

A pertinent question is whether nonlinear methods perform better than the simpler, linear kriging methods. Numerous studies in the scientific literature debate the relative benefits and shortcomings of different kriging methods on different spatial data sets. Other studies more generally compare different kriging methods with other interpolation methods [214, 215, 591, 814]. A search on Google Scholar with the keywords "kriging comparison" yields more than 70,000 results. Since the results of such comparisons usually depend on various user-specified modeling assumptions, computational implementation platforms, and the specific data sets involved, it is not possible to derive general conclusions.

An empirical comparative study that involves some of the methods discussed finds that for marginally skewed data the nonlinear methods and linear ordinary kriging perform similarly in terms of the estimates' precision at validation points. On the other hand, nonlinear methods better represent the conditional distribution than linear kriging, albeit the differences are not significant [591]. However, the study also found that if the data skewness is significant (e.g., larger than two), even the nonlinear methods fail to perform well.

Information content Currently used approaches for comparison of spatial interpolation methods often involve different data sets that are represented by different underlying probability distributions. Intuitively we expect that a data set of size N provides more information for the construction of the spatial model, and thus bestows higher predictive capability, if the underlying probability law is Gaussian rather than a long-tailed distribution.

Measures of information content, based on *entropic concepts*, could be used to set standards for the comparison of different probability distributions conditioned on available data. Then, one could focus on comparisons between scenarios with similar information content. Such notions have not been formalized, to my knowledge, in a quantitative framework for spatial data. In the theory of complex systems, entropy-based measures of information content and effective complexity have been proposed by Gell-Man and Loyd [277].

11.2.1 Indicator Kriging

The indicator function is often defined with reference to a spatial random field $X(\mathbf{s};\omega)$ and an arbitrary threshold x_c. The indicator is a binary random function that becomes equal to one at points $\mathbf{s}$ where $X(\mathbf{s};\omega) > x_c$ and is equal to zero where $X(\mathbf{s};\omega) \le x_c$, i.e.,

$$I_x(\mathbf{s}, x_c;\omega) = \Theta\left(X(\mathbf{s};\omega) - x_c\right), \tag{11.14}$$

where $\Theta(\cdot)$ is the unit step function defined in (2.7). Note that the above is not the only possible definition of the indicator; the reverse definition $I_x(\mathbf{s}, x_c;\omega) = \Theta\left[x_c - X(\mathbf{s};\omega)\right]$ is also encountered.

One can choose a number of different thresholds, typically of $O(10)$, estimate variograms for each indicator field, and finally predict the indicator field at unmeasured locations. Based on the indicator predictions, the conditional cumulative distribution function can be estimated at the target points.

Indicator kriging is often used by practitioners who are confronted with the analysis of strongly non-Gaussian data, because it does not assume a specific model for the probability distribution of the data. However, the method is subject to mathematical shortcomings: It ignores the cross-correlations between indicators that correspond to different thresholds, and it models inherently non-Gaussian (binary) random fields by means of two-point (variogram) functions. These inconsistencies often lead to unstable estimates (e.g., non-monotonic behavior) of the conditional probability distribution at the estimation points [591]. Therefore, the method is not in favor with statisticians [273].

Disjunctive kriging is a related nonlinear method that is based on nonlinear transformations of the original field. If the nonlinear transformation used is the indicator transform, disjunctive kriging is equivalent to the cokriging of indicators [684].

11.2.2 Spin-Based Indicator Models

Discretely-valued spin models (e.g., Ising and Potts) that originated in statistical physics do not suffer from the inconsistencies of indicator kriging [897, 898]. There also exist continuously-valued spin models such as the *rotator model* in which spins take values continuously in the interval $[-1, 1]$ [902].

The correlations of Ising and Potts spin fields are determined from Gibbs pdfs that involve well-defined energy functions. Hence, spin-based methods do not require variogram estimation, in contrast with indicator kriging. Another difference between indicator kriging and the Ising spin model is that the latter employs a bottom-up propagation of information (from the lower to the higher thresholds), which ensures that the conditional distribution function is well-defined (i.e., it has monotonic behavior). Ising spin models are further discussed in Chap. 15.

The shortcoming of spin-based models for spatial data analysis is that they have not yet been adapted to scattered data. However, this is in principle possible using kernel functions to spread the local interactions involved in the energy function of spin models in the same spirit as in [368].

11.2.3 Lognormal Kriging

The lognormal kriging model is often used for the interpolation of *positively-valued, asymmetric* data that follow probability distributions with positive skewness. Lognormal kriging applies the kriging framework to the logarithm of the data values. The variogram model is estimated for the logarithmized data, and the predictions are then evaluated using ordinary kriging. Eventually, the predictions in the space of the original data are obtained by inverting the logarithmic transform.

If the inversion is implemented by simple exponentiation of the kriging predictions of the logarithms, the resulting estimates are biased. Then, a *bias correction factor* needs to be included in the exponent [165, p. 135]. The bias correction accounts for the nonlinearity of the logarithmic and exponential transformations.

In order to gain some insight into the bias correction, consider a Gaussian random field $\mathrm{Y}(\mathbf{s};\omega)$. Assume that the optimal value of the field is estimated at point $\mathbf{z}$ based on the available sample $\mathbf{y}^* = (y_1, \ldots y_N)^\top$ at the locations $\{\mathbf{s}_1, \ldots, \mathbf{s}_N\}$. The estimate $\hat{y}(\mathbf{z})$ coincides with the mean (and thus also the median) of the conditional distribution since the latter is Gaussian.

Bias and bias correction Let $\mathrm{Y}(\cdot;\omega) \mapsto \mathrm{X}(\cdot;\omega) = G[\mathrm{Y}(\cdot;\omega)]$ represent a nonlinear, monotonic transformation from $\mathrm{Y}(\cdot;\omega)$ to a non-Gaussian random field $\mathrm{X}(\cdot;\omega)$, both referring to point $\mathbf{z}$. The median of the distribution remains invariant under this transformation, i.e., $x_{\text{med}} = G(y_{\text{med}})$. This is based on the principle of *quantile invariance* under monotonic transformations (see Sect. 14.1). However, the mean is not invariant under $G(\cdot)$, so that in general $\mathcal{E}[\mathrm{X}(\mathbf{z};\omega)] \neq G[\mathcal{E}[\mathrm{X}(\mathbf{z};\omega)]]$.

Assuming that the observed field $\mathrm{X}(\mathbf{s};\omega)$ is lognormal so that the transformed field $\mathrm{Y}(\mathbf{s};\omega) = \ln \mathrm{X}(\mathbf{s};\omega)$ is Gaussian, the *lognormal kriging prediction* at point $\mathbf{z}$ is given by

$$\hat{\mathrm{x}}(\mathbf{z}) = C_0 \exp\left[\hat{\mathrm{y}}_{\text{ok}}(\mathbf{z})\right], \text{ where } C_0 = \exp\left[\frac{1}{2}\left(\sigma_{\mathrm{y}}^2 - \sigma_{\text{ok};\mathbf{y}^*}^2(\mathbf{z})\right)\right], \tag{11.15a}$$

$$\hat{\mathrm{y}}_{\text{ok}}(\mathbf{z}) = \boldsymbol{\lambda}^\top \mathbf{y}^*. \tag{11.15b}$$

The following symbols are used in the prediction equations (11.15a):

- $\mathbf{y}^*$ is the vector of the log-transformed data ($y_n^* = \ln x_n^*$, for $n = 1, \ldots, N$),
- $\boldsymbol{\lambda}$ is the vector of the kriging weights,
- σ_{y}^2 is the variance of the Gaussian field,
- $\hat{\mathrm{y}}_{\mathrm{ok}}(\mathbf{z})$ is the ordinary kriging prediction of the optimal y-value at $\mathbf{z}$,
- $\sigma_{\mathrm{ok};\mathbf{y}^*}^2(\mathbf{z})$ is the ordinary kriging variance of the logarithmized variable at the target point, and
- C_0 is the *bias-correction* factor.

Lognormal kriging is a special case of nonlinear kriging, also known as *trans-Gaussian kriging*, which is further discussed in Chap. 14. In trans-Gaussian kriging a nonlinear transformation is applied to the original data. The goal of the transformation is to normalize the marginal distribution of the transformed data. Usually, the so-called *Box-Cox* nonlinear transformation is employed (see Chap. 14). The logarithm is included as a special case of the Box-Cox transform.

11.3 Connections with Gaussian Process Regression

In the field of machine learning, Gaussian random fields are known as *Gaussian processes*, e.g. [521, 678]. Gaussian processes are a more general concept than Gaussian RFs, because the former define mappings from a general feature vector $\mathbf{s} \in \mathbb{R}^d$ into $\mathbb{R}$. The feature vector $\mathbf{s}$ is not necessarily limited to spatial locations and could involve more features than just the coordinates.

Gaussian processes are integrated in the Bayesian framework, which means that *prior* (initial) distributions are assumed and a posteriori (posterior) distributions that incorporate the data are formulated for the process parameters (e.g., the variance and the correlation length). Predictions (corresponding to either interpolation or extrapolation of the data) are formulated in terms of *predictive probability distributions* obtained by Gaussian Process Regression (GPR) analysis.

As Gelfand points out, geostatistical modeling can be cast in a *hierarchical framework* (see Sect. 14.6.5) in which Gaussian processes appear at different levels of the hierarchy [271, 274]. This perspective implies a Bayesian framework for inference and *Markov Chain Monte Carlo* methods (see Sect. 16.8.2) for the estimation of model parameters.

A substantial difference of GPR and geostatistics is the following: In geostatistics, the random field $\mathrm{X}(\mathbf{s};\omega)$ is assumed to link the value of the variable X to the spatial position. In GPR input variable is not necessarily the coordinate vector, since $\mathbf{x}$ could represent a general feature vector [661]. The *input* variable $\mathbf{s} \in \mathbb{R}^d$, where d is the dimension of the feature space, is linked to the *output* variable $x(\mathbf{s})$ in terms of the Gaussian process $G(\mathbf{s})$, i.e., $x(\mathbf{s}) = G(\mathbf{s})$.

Predictive probability distribution The predictive probability distribution at some point $\mathbf{z} \in \mathbb{R}^d$ is given by the normal distribution with mean equal to the conditional mean and variance given by the conditional variance, i.e.,

$$X(\mathbf{z}; \omega \mid \mathbf{x}^*) \overset{d}{=} \mathcal{N}\left(\hat{x}(\mathbf{z}), \sigma^2_{\text{sk}}(\mathbf{z});\ \boldsymbol{\theta}\right), \tag{11.16}$$

where the mean $\hat{x}_{\mathbf{z}}$ is given by the simple kriging mean (10.22), $\sigma^2_{\text{sk}}(\mathbf{z})$ is the kriging variance (10.33), $\mathbf{x}^*$ denotes the data set, and $\overset{d}{=}$ denotes equality in distribution.

We can also express the above equation by stating that the prediction at point $\mathbf{z}$ follows the *predictive distribution* with pdf given by

$$f_{\text{pred}}\left(x_{\mathbf{z}} \mid \mathbf{x}^*, \boldsymbol{\theta}\right) = \mathcal{N}\left(\hat{x}_{\mathbf{z}}, \sigma^2_{\text{sk}}(\mathbf{z});\ \boldsymbol{\theta}\right). \tag{11.17}$$

Equation (11.16) establishes the connection between GPR and Kriging. Expressions that involve the Ordinary Kriging mean and variance are obtained if the GPR mean is assumed to be unknown.

Posterior distribution In the case of GPR, the parameter $\boldsymbol{\theta}$ is allowed to vary as a random vector that follows its own distribution. The so-called *posterior distribution* follows from a pdf that is given by [678]

$$f_{\text{post}}(\boldsymbol{\theta} \mid \mathbf{x}^*) = \frac{\mathcal{L}(\boldsymbol{\theta}; \mathbf{x}^*)\, f_{\text{prior}}(\boldsymbol{\theta})}{\int \mathrm{d}\boldsymbol{\theta}\, \mathcal{L}(\boldsymbol{\theta}; \mathbf{x}^*)\, f_{\text{prior}}(\boldsymbol{\theta})}. \tag{11.18}$$

The equation for the posterior density (11.18) involves the following quantities:

- $f_{\text{post}}(\boldsymbol{\theta} \mid \mathbf{x}^*)$ is the *posterior* pdf of the parameter vector given the data
- $f_{\text{prior}}(\boldsymbol{\theta})$ is the *prior* pdf of the parameters
- $\mathcal{L}(\boldsymbol{\theta}; \mathbf{x}^*)$ is the *likelihood* of the data given the specific GP model
- $\int \mathrm{d}\boldsymbol{\theta}\, \mathcal{L}(\boldsymbol{\theta}; \mathbf{x}^*)$ is the *marginal likelihood* (where the marginalization takes place over the parameter values).

After averaging over the model parameters $\boldsymbol{\theta}$, the *posterior predictive density* is given by the following expression

$$f_{\text{pred}}\left(x_{\mathbf{z}} \mid \mathbf{x}^*\right) = \int \mathrm{d}\boldsymbol{\theta}\, f_{\text{post}}(\boldsymbol{\theta} \mid \mathbf{x}^*)\, f_{\text{pred}}\left(x_{\mathbf{z}} \mid \mathbf{x}^*, \boldsymbol{\theta}\right), \tag{11.19}$$

where the posterior pdf of the model parameter vector $\boldsymbol{\theta}$ is obtained from (11.18).

The calculation of the above integral, however, is not analytically feasible, while its numerical evaluation is expensive. Approximate methods are based on the optimal parameters $\hat{\boldsymbol{\theta}}$ that maximize the posterior density (11.18).

Flat prior In addition, if the prior density is rather flat (i.e., it does not favor specific values or ranges of values), the optimal parameters are approximately equal to those that maximize the likelihood $\mathcal{L}(\boldsymbol{\theta}; \mathbf{x}^*)$, i.e.,

$$\hat{\boldsymbol{\theta}} = \arg\max_{\boldsymbol{\theta}} \mathcal{L}(\boldsymbol{\theta}; \mathbf{x}^*). \tag{11.20}$$

Then, the posterior predictive density is approximated by a delta function centered at $\hat{\boldsymbol{\theta}}$, and the approximate predictive pdf becomes [589]

$$f_{\text{pred}}\left(x_{\mathbf{z}} \mid \mathbf{x}^*\right) \approx f_{\text{pred}}\left(x_{\mathbf{z}} \mid \mathbf{x}^*, \hat{\boldsymbol{\theta}}\right). \tag{11.21}$$

The approximate predictive density (11.21) agrees with the predictive density obtained by kriging, in which the optimal value of the parameter vector is plugged in.

For large spatial data sets, Gaussian processes are beset by the same computational problems as kriging. These problems stem from the cubic dependence $O(N^3)$ of the numerical complexity for the inversion of large and dense covariance matrices. Remedies for this problem are discussed in Sect. 11.7.

11.4 Bayesian Kriging

Gaussian process regression incorporates plausible assumptions or empirical knowledge regarding parameter uncertainty in terms of the prior probability distribution. The method of Bayesian kriging, which was developed in statistics and geostatistics, is similar to GPR. The main contribution is that the uncertainty due to the estimation of the model parameters is also taken into account in the predictions.

Kriging is the optimal linear predictor if the following conditions are met:

1. The data follow the joint normal (Gaussian) distribution.
2. The mean (expectation) and the covariance function of the random field are known.

In geostatistical studies, it has been recognized that prediction uncertainty may be significantly underestimated if deviations of the data from normality are ignored. For example, this situation is observed if kriging is applied using the model parameters estimated from the data by means of normality assumptions. The underestimate of uncertainty affects all kriging method (e.g., simple, ordinary, universal or disjunctive).

Researchers responded to the above challenges by (i) developing Bayesian kriging methods that can better handle the uncertainty [327, 327, 456] and (ii) studying non-linear transformations that can—at least to some extent— normalize the non-Gaussian fluctuations, leading to the so-called trans-Gaussian kriging approach [186, 667]. A historical account of the development of Bayesian methods in geostatistics can be found in [667].

Empirical Bayesian kriging Bayesian kriging shares with GPR the problem of integration over the prior parameter distribution. This step is necessary to obtain the posterior and the posterior predictive distribution. A heuristic approach to overcome this difficulty is *empirical Bayesian kriging* [667].

Bootstrap The empirical approach employs the "Bayesian bootstrap" idea, which allows generating the posterior distribution of the model parameters from simulated (bootstrapped) samples of the random field [695]. The bootstrap method involves a resampling procedure which generates realizations of the random field with the parameters that are estimated from the data. The variogram model is then calculated for each of the replicated fields, and the estimated variograms create a "variogram cloud". The cloud helps to estimate the uncertainty in the estimation of the variogram parameters. The application of the bootstrap method in the estimation of variogram uncertainty is explored in [650, 651].

In its simplest form, the empirical Bayesian kriging procedure includes the following steps:

1. Estimate the "optimal" model parameters from the data by means of your favorite method, e.g., maximum likelihood (see Chap. 12).
2. Generate a set of N_{sim} realizations from the random field with the "optimal" parameters (see Chap. 16 for simulation methods). These realizations consist of simulated field values at the sampling locations.
3. For each realization, estimate new "optimal" spatial model parameters. These parameters will be different, in general, than the parameters estimated from the initial sample.
4. Construct the *empirical posterior parameter distribution* based on the estimates of the "optimal" parameter sets obtained for each of the N_{sim} simulated samples. Each parameter set can be weighed in the predictive distribution with a weight that is proportional to its likelihood given the initial sample.
5. Determine the *empirical predictive distribution* by averaging over the predictive distributions obtained for each bootstrapped sample.

Non-Gaussian data Mild deviations of the data distribution from the Gaussian law are typically treated by means of the so-called *Box-Cox transform*, which is discussed in Chap. 14. In trans-Gaussian Bayesian kriging, the Box-Cox transform is first used to normalize the data. The transformed data are then processed according to the empirical Bayesian kriging approach described above. At the end of the calculation, the results are back-transformed by inverting the Box-Cox transform.

Bagging The empirical Bayesian approach to kriging is similar to the process of bagging (short for bootstrap aggregation) which is employed in random forest modeling [100]. Random forests is a stochastic method that allows growing classification and regression trees. Bagging refers to a procedure in which a bootstrap sample is taken from the training set, and a predictor corresponding to a specific tree is constructed for this sample. The above step is repeated many times leading to a "bag" of several predictors, one per bootstrap sample. The final prediction is then based on an average or a majority vote over all the predictors in the bag. Bagging belongs to the class of *ensemble methods*. Random ensembles can be used to reduce the estimation variance [561].

11.5 Continuum Formulation of Linear Prediction

Let us now consider linear prediction in the case of a second-order stationary random field $X(\mathbf{s};\omega)$ defined over a continuum domain $\in \mathbb{R}^d$. We assume that the estimate at the prediction point $\mathbf{z}$ for a given realization $x(\mathbf{s})$ is given by means of the following convolution integral

$$\hat{x}(\mathbf{z}) = \int_{\mathbb{R}^d} d\mathbf{s}\lambda(\mathbf{z},\mathbf{s})\, x(\mathbf{s}), \tag{11.22}$$

where $\lambda(\mathbf{z},\mathbf{s})$ is a *linear weight kernel* analogous to the set of discrete kriging weights $\{\lambda_n\}_{n=1}^N$. The weight kernel is obtained by the continuum version of the simple kriging weight matrix equation (10.31), which is given by the following convolution integral

$$\lambda(\mathbf{z},\mathbf{s}) = \int_{\mathbb{R}^d} d\mathbf{s}'\, C_{\mathrm{xx}}^{-1}(\mathbf{s}-\mathbf{s}')\, C_{\mathrm{xx}}(\mathbf{s}'-\mathbf{z}). \tag{11.23}$$

The *precision operator* $C_{\mathrm{xx}}^{-1}(\cdot,\cdot)$ is obtained by extending the definition of the precision matrix, i.e., $\mathbf{C}^{-1}\,\mathbf{C} = \mathbf{I}$ to the continuum by means of the following integral equation

$$\int_{\mathbb{R}^d} d\mathbf{u}\, C_{\mathrm{xx}}^{-1}(\mathbf{s}-\mathbf{u})\, C_{\mathrm{xx}}(\mathbf{u}-\mathbf{s}') = \delta(\mathbf{s}-\mathbf{s}'). \tag{11.24}$$

In the Fourier domain, the convolution integral is replaced by the product of the respective Fourier transforms. Thus, the definition of the inverse kernel in the Fourier domain leads to

$$\widetilde{C}_{\mathrm{xx}}(\mathbf{k})\, \widetilde{C_{\mathrm{xx}}^{-1}}(\mathbf{k}) = 1. \tag{11.25}$$

Exactitude In light of the definition (11.24) of the precision kernel, the weight kernel is given by $\lambda(\mathbf{z},\mathbf{s}) = \delta(\mathbf{z}-\mathbf{s})$. Then, the predictive equation (11.22) leads to $\hat{x}(\mathbf{z}) = x(\mathbf{z})$. At first this result may seem surprising. However, if the realization $x(\mathbf{s})$ is known for every $\mathbf{s} \in \mathcal{D}$ as we assumed, it makes sense that the best estimate at $\mathbf{z}$ coincide with the value $x(\mathbf{z})$. This result confirms the *exactitude property* of the linear MMSE estimator: the estimate coincides with the value of the field at the target point, if the latter is included in the sample.

11.5.1 Continuum Prediction for Spartan Random Fields

In light of the SSRF spectral density (7.16) and the definition (11.25) of the precision kernel, the precision operator in the Fourier domain is given by

$$\widetilde{C_{xx}^{-1}}(\mathbf{k}) = \frac{1}{\eta_0 \xi^d} \left(1 + \eta_1 \xi^2 \|\mathbf{k}\|^2 + \xi^4 \|\mathbf{k}\|^4 \right). \tag{11.26}$$

Equation (11.26) can be inverted to derive the precision kernel in real space using generalized functions as described in [625]. In real space, this leads to an expression that involves partial derivatives of the Dirac delta function, i.e.,

$$C_{xx}^{-1}(\mathbf{s}-\mathbf{s}') = \frac{1}{\eta_0 \xi^d} \left[\delta(\mathbf{s}-\mathbf{s}') - \eta_1 \xi^2 \nabla^2 \delta(\mathbf{s}-\mathbf{s}') + \xi^4 \nabla^4 \delta(\mathbf{s}-\mathbf{s}') \right]. \tag{11.27}$$

Since the Dirac delta is a generalized function, its derivatives are defined with respect to test functions [188, p. 294]. To make sense of the partial derivatives of the delta function we assume that $f(\cdot)\colon \mathbb{R}^d \to \mathbb{R}$ is a smooth, compactly supported or rapidly decreasing at infinity function. Then, the following identity is obtained

$$\int_{\mathbb{R}^d} d\mathbf{s}'\, \nabla\delta(\mathbf{s}-\mathbf{s}')\, f(\mathbf{s}'-\mathbf{z}) = -\nabla_{\mathbf{s}} f(\mathbf{s}-\mathbf{z}).$$

The above is based on integration by parts and the fact that the boundary terms (at infinity) due to the extreme localization of the delta functions.

Based on (11.23) and (11.27), it follows that for an FGC-SSRF the interpolating weights are given by the following explicit expression

$$\lambda(\mathbf{z},\mathbf{s}) = \frac{1}{\eta_0 \xi^d} \left[C_{xx}(\mathbf{s}-\mathbf{z}) - \eta_1 \xi^2 \nabla_{\mathbf{s}}^2\, C_{xx}(\mathbf{s}-\mathbf{z}) + \xi^4 \nabla_{\mathbf{s}}^4\, C_{xx}(\mathbf{s}-\mathbf{z}) \right]. \tag{11.28}$$

Comment In the case of SSRFs we could have arrived at the linear prediction equation (11.22) with the kernel function (11.27) by requiring that the prediction $\hat{x}(\mathbf{z})$ be a stationary point of the SSRF energy function (7.4) with respect to changes of the field at the location $\mathbf{z}$. The stationary point is obtained by requiring that the first-order functional derivative of the energy vanish, i.e.,

$$\left. \frac{\delta \mathcal{H}_0[x(\mathbf{s})]}{\delta x(\mathbf{z})} \right|_{x(\mathbf{z})=\hat{x}(\mathbf{z})} = 0.$$

11.5.2 From Continuum Space to Discrete Grids

Let us now assume that observations are available at the N nodes of a sampling network, and that the prediction points coincide with the nodes of a regular grid that covers the domain of interest. Then, for any prediction point it should be possible to obtain from (11.28) a set of linear weights $\{\lambda_n\}_{n=1}^{N}$ that can be used to interpolate the measurements.

If we want to use the SSRF model, a central question is how to discretize the Laplacian operator in terms of the nodes $\{\mathbf{s}_n\}_{n=1}^{N}$ of an unstructured sampling grid. One possible approach is to consider the sampling network as a graph. Then, we can replace the Laplacian operator with the *graph Laplacian*, which we will denote by $\mathcal{L}_G$ [606]. On graphs the Laplacian operator is expressed as a matrix.

The biharmonic operator can be expressed as the product of the Laplacian with itself, i.e., $\mathcal{B}_G = \mathcal{L}_G\, \mathcal{L}_G$. Hence, the biharmonic operator can be expressed in terms of the approximation of the Laplacian operator with the graph Laplacian.

However, if the operators $\mathcal{L}_G$ and $\mathcal{B}_G$ are discretized, the solution (11.28) for the SSRF linear weights at the point $\mathbf{z}$ is given by the following discretized expression

$$\lambda_n^{*} = \frac{1}{\eta_0 \xi^d} \left[C_{\mathrm{xx}}(\mathbf{s}_n - \mathbf{z}) - \eta_1\, \mathcal{L}_G\, \xi^2\, C_{\mathrm{xx}}(\mathbf{s}_n - \mathbf{z}) + \xi^4 \mathcal{B}_G\, C_{\mathrm{xx}}(\mathbf{s}_n - \mathbf{z}) \right],$$

which, in contrast with kriging methods, can be evaluated without matrix inversion calculations.

11.6 The "Local-Interaction" Approach

In this section we focus on lattice random field models that are motivated by the SSRF energy functional (7.4). The advantage of these models is that their joint pdf is expressed in terms of local interactions between field values at neighboring points. The local nature of the interactions implies a neighborhood structure for each point. This is a distinctive property of *Markov random fields*, the statistical properties of which were spearheaded by Besag and coworkers, e.g. [69, 72].

An explicit form of the interactions leads to an *explicit and sparse* form for the *precision matrix*. The sparsity implies that a large fraction of pairs of points do not interact and thus have zero couplings. These crucial insights have been introduced in early papers on spatial modeling, e.g. [192].

The value of the field at an unmeasured location is determined through its interactions with its neighbors in the sampling set. The optimal value of the field at the target point can be selected as the one that minimizes the overall energy of the system, thereby maximizing the conditional pdf. Hence, spatial prediction in local lattice models can be formulated as a *minimum energy* problem.

The minimization of the energy can be accomplished in a semi-analytic form that does not require the inversion of the covariance matrix. Thus, the local interaction approach is attractive for the processing of large data sets. On the negative side, the covariance function of lattice spatial models is not known a priori. This means that estimates of the covariance require the numerical inversion of the sparse precision matrix. The required matrix inversion, however, is less demanding computationally for sparse matrices than for full covariance matrices.

In the following sections we elaborate on the definition of the energy function and the precision matrix for local interaction models defined on discrete grids (both lattices and unstructured grids).

11.6.1 Lattice Site Indexing

Let us now return to the Miller index notation for the labeling of lattice sites on a grid $\mathcal{G} \in \mathbb{Z}^d$ that we introduced in Sect. 8.5.2. We assume that the lattice has L_i nodes per side for $i = 1, \ldots, d$. Each lattice location $\mathbf{s}_n$, $n = 1, \ldots, N$ where $N = \prod_{i=1}^{d} L_i$ is uniquely specified by the *Miller index vector*

$$\mathbf{n} = (n_1, \ldots, n_d)^\top, \text{ where } 1 \leq n_i \leq L_i, \text{ for } 1 \leq i \leq d.$$

It is often preferable to refer to sites with a scalar index such as the index n in $\mathbf{s}_n$ above. The scalar index is useful for data on irregular grids, but it is also convenient for lattice data in cases where it is beneficial to treat the lattice data as a vector. For example, we use the expression (see below) $x_n\, J_{n,m}\, x_m$ where each of the scalar indices n, m refers to a specific lattice location and $\mathbf{x} = (x_1, \ldots, x_N)^\top$.

Each vector index $\mathbf{n}$ corresponds to a respective *scalar matrix index* $n_{\mathbf{n}}$ (the dependence of $n_{\mathbf{n}}$ on $\mathbf{n}$ can be omitted for brevity).

The mapping from the Miller index to the scalar index $\mathbf{n} \mapsto n_{\mathbf{n}}$ is accomplished as follows[2]

$$n_{\mathbf{n}} = n_1 + \sum_{k=2}^{d} (n_k - 1) \prod_{l=1}^{k-1} L_l. \tag{11.29}$$

Example 11.1 For example, if $d = 2$, so that $\mathbf{n} = (n_1, n_2)^\top$, and $n_i = 1, \ldots, L_i$ for $i = 1, 2$, then the scalar index is given by $i_{\mathbf{n}} = (n_2 - 1)\, L_1 + n_1$. Hence, if n_1 is the row index and n_2 is the column index of each matrix entry, the scalar index defined by (11.29) first increases along the rows and then along the columns.

[2]The transformation from a vector to a scalar index is uniquely defined. Equation (11.29) employs the ordering used in MATLAB linear indexing.

Example 11.2 Next, we illustrate the two labeling schemes for a 3×4 matrix (i.e., $L_1 = 3$ and $L_2 = 4$). The entries in the matrix on the left define the vector index, while the scalar indices are the entries in the matrix on the right (neither matrix contains the actual values of the variable represented by the matrix).

$$\begin{pmatrix} 1,1 & 1,2 & 1,3 & 1,4 \\ 2,1 & 2,2 & 2,3 & 2,4 \\ 3,1 & 3,2 & 3,3 & 3,4 \end{pmatrix} \rightarrow \begin{pmatrix} 1 & 4 & 7 & 10 \\ 2 & 5 & 8 & 11 \\ 3 & 6 & 9 & 12 \end{pmatrix} \tag{11.30}$$

Nearest-neighbor labels The nearest neighbors of the site $\mathbf{s}_n$ are the sites $\mathbf{s}_n \pm a_i\, \hat{\mathbf{e}}_i$ where $i = 1, \ldots, d$. The vector index $\mathbf{n}(i\pm)$ of the nearest-neighbor sites for sites $\mathbf{s}_n$ that are not on the domain boundary $\partial \mathcal{G}$ is given by

$$\mathbf{n}(i\pm) = (n_1, \ldots, n_i \pm 1, \ldots n_d)^\top, \text{ for } i = 1, \ldots, d, \tag{11.31}$$

where $n_i = 2, 3, \ldots, L_i - 1$. The respective scalar indices are then determined from (11.29).

Special care must be used for determining the nearest neighbors of sites that are on the domain boundary (see Fig. 11.1). The treatment of these sites depends on the boundary conditions.

If *open boundary conditions* are used, the nearest neighbors of the boundary sites are only those for which $L_i \geq n_i \pm 1 \geq 1$ according to (11.31). This means that point on the right boundary do not have right-hand neighbors, while points on the left boundary lack left-hand neighbors (similarly for points on the top and bottom boundaries).

Another option is *periodic boundary conditions*, which assumes that the lattice is wrapped around at its ends as a torus. Hence, the right (left) side nearest neighbors of the points on the right (left) boundary are the same-row points on the left (right) boundary. The nearest neighbors are similarly determined for the top and bottom boundaries.

Assuming that the point $\mathbf{s}_n$ (where n is the scalar index of the site) is not on the grid boundary, the scalar indices $n^{(i\pm)}$ of its nearest lattice neighbors in the direction $i = 1, \ldots, d$ are given by

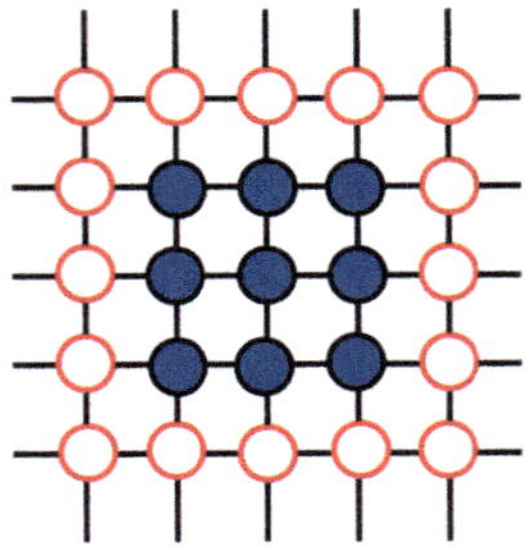

Fig. 11.1 Schematic of 2D grid. Filled circles represent sites in the bulk of the lattice, while open circles correspond to boundary sites

$$n^{(i\pm)} = n \pm \delta_{i,1} \pm L_1 \delta_{i,2} + \ldots \pm \delta_{i,d} \prod_{l=1}^{d-1} L_l. \tag{11.32}$$

Example 11.3 Consider the element with scalar index 8 (shaded yellow) in the 4×3 matrix (11.30); the corresponding vector index is (2, 3).

The left and right nearest neighbors of this site have vector indices (2, 2) and (2, 4) respectively. The relevant orthogonal direction is $i = 2$, since both neighbors are obtained by changing the second component of the vector index by ∓ 1 respectively. Thus, scalar indices of the nearest neighbors according to (11.32) are $8 - 3 = 5$ and $8 + 3 = 11$ respectively (as shown by the horizontally aligned cyan shaded entries above).

The top and bottom nearest neighbors, on the other hand, have vector indices (1, 3) and (3, 3) respectively. For these neighbors the relevant orthogonal direction is $i = 1$, and thus the scalar indices for the top and bottom neighbors are $8 - 1 = 7$ and $8 + 1 = 9$ respectively—as shown by the vertically aligned, shaded cyan entries in the right-hand side matrix in (11.30).

11.6.2 Local Interactions on Regular Grids

We have discussed the discretization of SSRFs on regular grids in Chap. 7. Therein we expressed the energy functional of an anisotropic SSRF on a d-dimensional rectangular grid $\mathcal{G}$ by means of (8.13), which we repeat below:

$$\mathcal{H}_0(\mathbf{x}) = \frac{1}{2\lambda} \sum_{n=1}^{N} \left\{ [x(\mathbf{s}_n) - m_{\mathrm{x}}]^2 + \sum_{i=1}^{d} c_{1,i} \left[\frac{x(\mathbf{s}_n + a_i \hat{\mathbf{e}}_i) - x(\mathbf{s}_n)}{a_i} \right]^2 + \sum_{i=1}^{d} c_{2,i} \left[\frac{x(\mathbf{s}_n + a_i \hat{\mathbf{e}}_i) - 2x(\mathbf{s}_n) + x(\mathbf{s}_n - a_i \hat{\mathbf{e}}_i)}{a_i^2} \right]^2 \right\}, \tag{11.33}$$

where $\{\mathbf{s}_n\}_{n=1}^{N} \in \mathcal{G}$. The energy functional (11.33) implies that each site interacts only with its nearest neighbors. This property ensures that the SSRF is a *Markov random field* [69, 279, 698, 700]. The symmetry (i.e., the fact that the coefficient of the terms $x_n x_m$ and $x_m x_n$ are equal for $m \neq n$) and bilinearity (i.e., the dependence on products of two values) of the interactions further implies that the energy function (11.33) defines a *Gaussian Markov SRF*.

Single point predictor Let us assume that only one grid value is missing at the point $\mathbf{s}_p$. The optimal prediction of $\hat{x}(\mathbf{s}_p)$ is the value that maximizes the conditional pdf at this point given the remaining points $\mathbf{s}_n \in \mathcal{G} \setminus \{\mathbf{s}_p\}$. Thus, the optimal value $\hat{x}_p := \hat{x}(\mathbf{s}_p)$ minimizes the energy, i.e.,

$$\hat{x}_p = \arg\min_{x_p} \mathcal{H}_0(x_p \mid \mathbf{x}_{-p}),$$

where $\mathbf{x}_{-p}$ is the set of the values at all grid points excluding $\mathbf{s}_p$.

The energy $\mathcal{H}_0(x_p \mid \mathbf{x}_{-p})$ is a convex, non-negative function of x_p with a unique *stationary point*. The latter corresponds to the value of x_p that renders the derivative of the energy—with respect to x_p—equal to zero, i.e.

$$\left. \frac{\partial \mathcal{H}_0(x_p \mid \mathbf{x}_{-p})}{\partial x_p} \right|_{x_p = \hat{x}_p} = 0. \tag{11.34}$$

The solution of the stationary point equation (11.34) is easily derived and involves only the nearest neighbors of the point $\mathbf{s}_p$ [362]

$$\hat{x}_p = \frac{m_{\mathrm{x}} + \sum_{i=1}^{d} \frac{c_{1,i}}{a_i^2}\, x(\mathbf{s}_p + a_i \hat{\mathbf{e}}_i) + \sum_{i=1}^{d} \frac{c_{2,i}}{a_i^4} \left[x(\mathbf{s}_p + a_i \hat{\mathbf{e}}_i) + x(\mathbf{s}_p - a_i \hat{\mathbf{e}}_i) \right]}{1 + \sum_{i=1}^{d} \frac{c_{1,i}}{a_i^2} + 2 \sum_{i=1}^{d} \frac{c_{2,i}}{a_i^4}}. \tag{11.35}$$

The equation (11.35) can be used for fast interpolation of missing values. In order to predict a missing value, at least one value of the field should be available within the target point's neighborhood.

Non-uniform mean In the above we have assumed a uniform mean m_{x}. This assumption can be relaxed to a variable trend $m_{\mathrm{x}}(\mathbf{s})$, so long as the latter does not significantly change over distances equal to the lattice step. Thus, the lattice SSRF model can also be used with non-stationary data provided that the non-stationarity is adequately captured by the trend function.

Precision matrix formulation The SSRF energy (11.33) can also be expressed in terms of the following quadratic function[3]

$$\mathcal{H}_0(\mathbf{x}) = \frac{1}{2} \left(\mathbf{x} - \mathbf{m}_{\mathrm{x}}\right)^{\top} \mathbf{J} \left(\mathbf{x} - \mathbf{m}_{\mathrm{x}}\right), \tag{11.36}$$

where $\mathbf{J}$ is the *SSRF precision matrix*. Let us assume that $\mathbf{J}$ is a known, symmetric, positive-definite matrix. This representation is also known as the *conditional autoregressive* formulation [72]. Then, the value of the energy function at the prediction point, conditionally on the values of the neighbors, becomes

[3]We could normalize the first term by dividing with N as in [368]. In this case, the first component of the precision matrix (11.42a) should also be divided by N.

$$\mathcal{H}_0(x_p \mid \mathbf{x}_{-p}) = \mathcal{H}_0(\mathbf{x}_{-p}) + J_{p,p}(x_p - m_x)^2 + 2\sum_{n=1}^{N}(x_p - m_x)\, J_{p,n}\,(x_n - m_x)\,\left(1 - \delta_{n,p}\right). \tag{11.37}$$

The factor $1 - \delta_{n,p}$ removes the point $\mathbf{s}_p$ from the summation over all the grid nodes. The stationarity condition (11.34) is then expressed as

$$J_{p,p}(\hat{x}_p - m_x) + \sum_{n=1}^{N} J_{p,n}\,(x_n - m_x)\,\left(1 - \delta_{n,p}\right) = 0.$$

Thus, the stationary point of the SSRF energy function is determined by

$$\hat{x}_p = m_x - \sum_{n=1}^{N} \frac{J_{p,n}}{J_{p,p}}\,(x_n - m_x)\,\left(1 - \delta_{n,p}\right). \tag{11.38}$$

Evaluation of the second derivative of the energy with respect to x_p leads to

$$\left.\frac{\partial^2 \mathcal{H}_0(x_p \mid \mathbf{x}_{-p})}{\partial^2 x_p}\right|_{x_p = \hat{x}_p} = J_{p,p}. \tag{11.39}$$

Hence, the prediction (11.38) corresponds to a minimum of the energy if $J_{p,p} > 0$. An explicit form of $\mathbf{J}$ can be derived from the energy functional (11.33) as we show below.

Note that the predictive equation (11.38) that accounts for the local (sparse) nature of the precision matrix $\mathbf{J}$, is equivalent to the expression (8.7) for the conditional means of Gaussian Markov random fields.

Conditional prediction pdf More generally, based on the Gaussian nature of the pdf corresponding to the energy function (11.33), it can be shown that the *conditional pdf at the prediction point* is given by the following normal distribution [72]

$$X_p(\omega) \mid \mathbf{x}_p \overset{d}{=} N(\hat{x}_p, \hat{\sigma}_p^2), \text{ where } \hat{\sigma}_p = \sqrt{1/J_{p,p}}. \tag{11.40}$$

Precision matrix structure For a d-dimensional rectangular grid with $N = L_1 L_2 \ldots L_d$ sites, the precision matrix $\mathbf{J}$ is of dimension $N \times N$. The structure of the precision matrix depends on the neighbor structure of the grid. Hence, we will use the lattice site indexing developed in Sect. 11.6.1 and in particular (11.29).

The anisotropic SSRF energy function (11.33) contains three distinct terms. This energy can be expressed in terms of the precision matrix $\mathbf{J}$ as in (11.36). In direct analogy with the three terms of the energy, we decompose the precision matrix into three sub-matrices, the diagonal sub-matrix, $\mathbf{J}_0$, the Laplacian sub-matrix, $\mathbf{J}_1$ and the Bilaplacian sub-matrix, $\mathbf{J}_2$, as follows:

$$\mathbf{J}(\boldsymbol{\theta}) = \mathbf{J}_0 + \mathbf{J}_1 + \mathbf{J}_2, \tag{11.41}$$

$$\mathbf{J}_0 = \frac{\mathbf{I}_N}{\lambda} \quad \text{where} \quad [\mathbf{I}_N]_{n,m} = \delta_{n,m}, \text{ for all } n, m = 1, \ldots, N, \tag{11.42a}$$

$$J_{1;\mathbf{n},\mathbf{m}} = \sum_{i=1}^{d} \frac{c_{1,i}}{\lambda\, a_i^2} \left(\delta_{\mathbf{n},\mathbf{m}(i+)} + \delta_{\mathbf{n},\mathbf{m}(i-)} - 2\,\delta_{\mathbf{n},\mathbf{m}}\right), \tag{11.42b}$$

$$\begin{aligned} J_{2;\mathbf{n},\mathbf{m}} = \sum_{i=1}^{d} \frac{c_{2,i}}{\lambda\, a_i^4} \Big[& 6\,\delta_{\mathbf{n},\mathbf{m}} - 4\,\delta_{\mathbf{n},\mathbf{m}(i+)} - 4\,\delta_{\mathbf{n},\mathbf{m}(i-)} + \delta_{\mathbf{n},\mathbf{m}(i++)} + \delta_{\mathbf{n},\mathbf{m}(i--)} \\ & + \sum_{j=1,\neq i}^{d} \big(\delta_{\mathbf{n},\mathbf{m}(i+j+)} + \delta_{\mathbf{n},\mathbf{m}(i+j-)} + \delta_{\mathbf{n},\mathbf{m}(i-j+)} + \delta_{\mathbf{n},\mathbf{m}(i-j-)} \\ & -2\delta_{\mathbf{n},\mathbf{m}(i+)} - 2\delta_{\mathbf{n},\mathbf{m}(i-)} - 2\delta_{\mathbf{n},\mathbf{m}(j+)} - 2\delta_{\mathbf{n},i_{\mathbf{m}(j-)}} + 4\delta_{\mathbf{n},\mathbf{m}}\big)\Big], \end{aligned} \tag{11.42c}$$

where $\mathbf{n}$, $\mathbf{m}$ are the Miller indices of the sites $\mathbf{s}_n$ and $\mathbf{s}_m$. The indices $\mathbf{m}(i\pm)$ denote the nearest-neighbors of the site $\mathbf{s}_m$ in the orthogonal lattice directions $\hat{\mathbf{e}}_i$, where $i = 1, \ldots, d$, as described in Sect. 11.6.1. The Kronecker delta $\delta_{\mathbf{n},\mathbf{m}(i+)}$ is equal to one if the lattice point with index $\mathbf{n}$ coincides with the point with index $\mathbf{m}(i+)$, i.e., the nearest neighbor $\mathbf{s}_m + \hat{\mathbf{e}}_i$ of $\mathbf{s}_m$, and is equal to zero otherwise. Similarly, the indices $\mathbf{m}(i++)$ and $\mathbf{m}(i--)$ represent the *next-nearest neighbors* of $\mathbf{s}_m$ in the orthogonal directions $\hat{\mathbf{e}}_i$.

We next show in more detail the steps used to obtain the above expressions. First, the energy function is written explicitly in terms of the Laplacian (8.39) and the Bilaplacian (8.41) operators, i.e.,[4]

[4]The components of the Laplacian and the Bilaplacian in the orthogonal directions are multiplied by directional coefficients; hence, strictly speaking, the sums over the orthogonal directions are not, strictly speaking, equal to the Laplacian and the Bilaplacian.

$$\mathcal{H}_0 = \frac{1}{2}\mathbf{x}'^{\top}\mathbf{J}\mathbf{x}' = \frac{1}{2}\mathbf{x}'^{\top}\left[\mathbf{J}_0 + \sum_{i=1}^{d}\frac{c_{1,i}}{\lambda\, a_i^2}\delta_i^2 + \sum_{i=1}^{d}\frac{c_{2,i}}{\lambda\, a_i^4}\left(\delta_i^4 + \sum_{j=1,\neq i}^{d}\delta_i^2\delta_j^2\right)\right]\mathbf{x}',$$

where $\mathbf{x}' = \mathbf{x} - \mathbf{m}$ is the fluctuation data vector, while $\boldsymbol{\delta}_i^{2p}$, where $p = 1, 2$, are the matrix representations of the central finite difference operators of order $2p$ as given in the Table 8.2. For example

$$\delta_i^2 x_{\mathbf{m}} = x_{\mathbf{m}(i+)} + x_{\mathbf{m}(i-)} - 2x_{\mathbf{m}}.$$

The above justifies (11.42b) for the Laplacian precision sub-matrix. Similarly we obtain the following expression for the fourth-order difference

$$\delta_i^4 x_{\mathbf{m}} = \delta_i^2\delta_i^2 x_{\mathbf{m}} = x_{\mathbf{m}(i++)} + x_{\mathbf{m}(i--)} + 6x_{\mathbf{m}} - 4x_{\mathbf{m}(i+)} - 4x_{\mathbf{m}(i-)},$$

and for the cross terms $(i \neq j)$

$$\begin{aligned}\delta_i^2\delta_j^2 x_{\mathbf{m}} =& x_{\mathbf{m}(i+j+)} + x_{\mathbf{m}(i+j-)} + x_{\mathbf{m}(i-j+)} + x_{\mathbf{m}(i-j-)}\\ &+ 4x_{\mathbf{m}} - 2x_{\mathbf{m}(i+)} - 2x_{\mathbf{m}(i-)} - 2x_{\mathbf{m}(j+)} - 2x_{\mathbf{m}(j-)},\end{aligned}$$

where $\mathbf{m}(i \pm j\pm)$ denotes the vector $\mathbf{s_m} \pm \hat{\mathbf{e}}_i \pm \hat{\mathbf{e}}_j$. The above proves (11.42c).

Precision matrix in one dimension In the case of a one-dimensional random field or time series, the precision sub-matrices defined by (11.42) are given by the following relations [894]:

$$\mathbf{J}_1 = \frac{c_1}{\lambda\, a^2}\begin{pmatrix} 1 & -1 & 0 & \cdots & 0 \\ -1 & 2 & -1 & \cdots & 0 \\ & \ddots & \ddots & \ddots & \\ 0 & \cdots & -1 & 2 & -1 \\ 0 & \cdots & 0 & -1 & 1 \end{pmatrix}, \tag{11.43a}$$

$$\mathbf{J}_2 = \frac{c_2}{\lambda\, a^4}\begin{pmatrix} 1 & -2 & 1 & 0 & \cdots & 0 & 0 \\ -2 & 5 & -4 & 1 & \cdots & 0 & 0 \\ 1 & -4 & 6 & -4 & 1 & \cdots & 0 \\ \ddots & \ddots & \ddots & \ddots & \ddots & \ddots & \ddots \\ 0 & \cdots & 1 & -4 & 6 & -4 & 1 \\ 0 & 0 & \cdots & 1 & -4 & 5 & -2 \\ 0 & 0 & \cdots & 0 & 1 & -2 & 1 \end{pmatrix}. \tag{11.43b}$$

The first equation for the Laplacian sub-matrix $\mathbf{J}_1$ follows from the lowest-order discretization of the Laplacian, given by (8.40) for $d = 1$. Similarly, the second equation for the Bilaplacian sub-matrix $\mathbf{J}_2$ follows from (8.41), taking into account that in $d = 1$ the cross terms $\delta_i^2 \delta_j^2$ (with $i \neq j$) vanish and after inserting the expression for δ_i^4 from Table 8.2.

Bilaplacian in two dimensions Under isotropic conditions (non-directional coefficients c_1 and c_2) the equation (11.42c) for the Bilaplacian in d dimensions gives the weights that are shown in Fig. 11.2.

There are four types of terms in the 2D Bilaplacian:

1. Self-interaction terms that couple each point to itself (weight equal to 20).
   The self-interaction weight comes from the first and the last terms in (11.42c), taking into account that in 2D the sum over $j \neq i$ involves a single term for each i (i.e., if $i = 1$ then $j = 2$ and if $i = 2$ then $j = 1$), and that the summation over the orthogonal directions contributes a multiplicative factor of 2.
2. Terms that couple each point and its nearest neighbors along the orthogonal lattice directions (weights equal to -8).
   The nearest-neighbor couplings are derived from the second and third terms on the first line of (11.42c) and the first four terms on the last line of the same equation.
3. Terms that couple each point and its diagonal neighbors (weights equal to -4).
   The diagonal neighbor coupling is derived from the terms in the second line of (11.42c); the weight (equal to 2) is obtained by summing over the orthogonal directions.
4. Terms that couple each point with its next-nearest neighbors in the orthogonal lattice directions (weights equal to 1).
   The next-nearest neighbor coupling involves the last two terms on the first line of (11.42c); the summation over i ensures that all the orthogonal directions are accounted.

Multiple missing values In typical applications, there are several locations with missing values as illustrated in Fig. 11.3. In such cases, the energy can be simul-

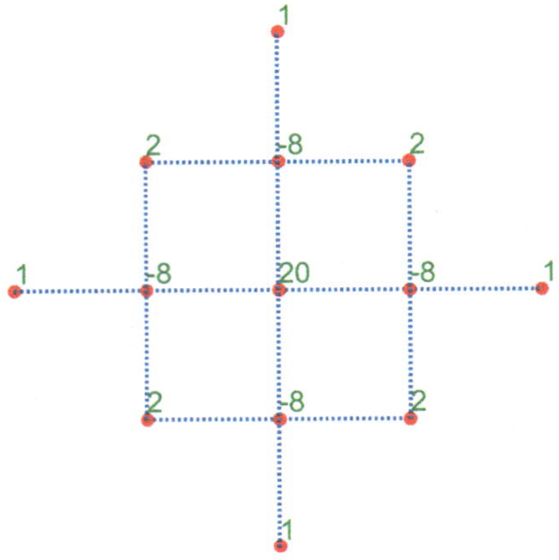

Fig. 11.2 Thirteen-point stencil for the two-dimensional biharmonic operator based on the leading (second-order) central difference approximation of the Laplacian. The nodal weights are as defined in (8.29b). Note the negative sign in front of the weights corresponding to the nearest neighbors

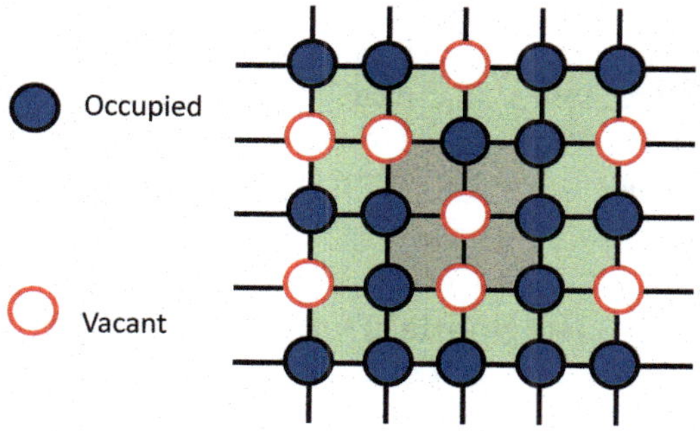

Fig. 11.3 Schematic of 2D grid with multiple missing values. Filled circles represent sites with existing sample values, while open circles correspond to sites with missing values (vacant sites). All the vacant sites are connected with at least one sample site

taneously minimized with respect to all of the missing values leading to a linear system of dimension $M \times M$, using the multiple point predictor equation given below.

Occasionally there may be locations that do not have any nearest neighbors in the sample. We will call such vacant sites "disconnected", while we will call vacant sites with at least one neighbor in the sample connected. If there exist *disconnected sites*, it is best to first sequentially assign values to the vacant sites and include the new estimates to the sample. This approach increases the chances that initially disconnected sites become connected (albeit the connection is with estimated—not sampled—values).

The sequential prediction approach resembles time series forecasting with low-order autoregressive models. For example, if an AR(1) model is used, the value of the time series at the next time step, $\hat{x}_{t+1}$, is estimated based on the last value of the sample, i.e., x_t; for more distant times, the estimates $\hat{x}_{t+n}$, where $n > 1$, are based on the forecasts $\hat{x}_{t+n-1}$.

Multiple point predictor The conditional mean predictor (11.38) can be generalized to P prediction points as follows

$$\hat{\mathbf{x}}_P = \mathbf{m}_x - \widetilde{J}_{P,S}\,(\mathbf{x} - \mathbf{m}_x), \tag{11.44a}$$

where $\mathbf{x}$ is the $N \times 1$ vector of known values,[5] $\mathbf{m}_x$ is an $N \times 1$ vector of constant values equal to m_x and $\widetilde{J}_{P,S}$ is a $P \times N$ *transfer matrix* defined by

$$\left[\widetilde{J}_{P,S}\right]_{p,n} = \frac{J_{p,n}}{J_{p,p}}, \text{ for all } n = 1, \ldots, N, \; p = 1, \ldots, P. \tag{11.44b}$$

Prediction properties (i) The SSRF prediction (11.44) is *unbiased* in view of the vanishing *row sum property* satisfied by the precision matrix—as shown by (13.36).

[5]If these values represents real data, then $\mathbf{x}$ is replaced by $\mathbf{x}^*$.

(ii) The SSRF prediction (11.44) is independent of the parameter λ which sets the amplitude of the fluctuations, because the transfer matrix $\widetilde{J}_{P,S}$ involves the ratio of precision matrix elements. This property is analogous to the independence of the kriging predictor from the random field variance. Hence, leave-one-out cross validation does not determine the optimal value of λ, which is obtained from (13.46).

11.7 Big Spatial Data

In the case of large spatial data sets, most stochastic methods of analysis suffer from the poor scalability of the computational resources required to store and invert large covariance matrices and to calculate their determinants. This scaling problem essentially plagues both geostatistics and machine learning applications. For dense covariance matrices, the memory storage requirements scales as $O(N^2)$ while the computational time for the inversion of the covariance matrix as $O(N^3)$. Hence, both disciplines have explored approaches to reduce the computational complexity of the operations involved. A recent study compares a variety of methods constructed to handle large spatial datasets based on their interpolation performance with simulated and real (satellite) data [333]. Some possible remedies to the scaling problem are briefly reviewed below.

Sparsity A possible approach for reducing the computational cost is to ensure that the covariance matrix, or its inverse (i.e., the precision matrix) are sparse, that is, a significant portion of their entries are equal to zero. Sparse matrices require less computer memory for storage and can be inverted more efficiently than dense matrices. However, most of the classical covariance models lead to dense covariance matrices. This means that the computational cost for inverting such matrices scales as $O(N^3)$ for an $N \times N$ matrix. Hence, processing a digital image with $L = 1024$ pixels per side implies that the sample size is $N \approx 10^6$ and respectively $N^3 \approx 10^{18}$.

Covariance cutoff A straightforward approach for increasing the sparsity of the covariance matrix is to set all the matrix elements with magnitude below a given threshold to zero, thus generating to a sparse covariance matrix. While this sounds appealing, one needs to be careful since the truncated covariance matrix is not necessarily positive definite. The perils and the rigorous mathematical treatment of thresholds are exposed in [314].

Local and approximate methods Improved scaling of spatial models with data size can also be achieved by means of *local approximations*, *dimensionality reduction* techniques, *determinant approximations* and *parallel algorithms*. Spatial statistics methods for large data applications are reviewed in [784]. Such approaches employ clever approximations to reduce the computational complexity. They involve local approximations of the likelihood such as composite likelihood which is defined on smaller data subsets [812], pseudo-likelihood [807], and approximate likelihood [264, 777]. A different approach involves covariance tapering which leads to non-essential correlations outside a specified range [213, 267, 441].

Dimensionality reduction This approach leads to spatial models with a reduced number of degrees of freedom. Dimensionality reduction includes methods such as *fixed rank kriging*, a method that models the precision (inverse covariance) matrix by means of a matrix with a fixed rank $r \ll N$ [167, 617].

Gaussian Markov random fields (GMRFs) also take advantage of local representations of the precision matrix using factorizable joint densities. The local feature of the precision matrix leads to sparse matrix representations that can be handled more efficiently than full matrices. The application of MRFs in spatial data analysis was initially limited to structured grids [69, 698, 852]. However, recently a framework that links Gaussian random fields and GMRFs via stochastic partial differential equations (SPDEs) has been developed. This framework has extended the scope of GMRFs to scattered data as well [503].

Spartan spatial random fields also employ a local approximation of the precision matrix (or the precision operator in continuum space) that leads to a specific GMRF energy structure on regular grids. The sparse precision matrix of SSRFs can also be extended to scattered data [368, 373].

Determinant approximations aim to efficiently calculate the covariance determinant by means of the sparse inverse approximation [680], or stochastic approximations that are applicable to large computational problems including Bayesian inference [208].

Recent advances in linear algebra also give hope that the cost of inverting large covariance matrices and calculating their determinants can be reduced by taking advantage of the internal structure of covariance matrices [19]. These efforts are based on hierarchical factorization of the covariance matrix into a product of block low-rank updates of the identity matrix that can be calculated in $O(N \log_2 N)$ computational time.

Chapter 12
Basic Concepts and Methods of Estimation

> No amount of experimentation can ever prove me right; a single experiment can prove me wrong.
>
> *Albert Einstein*

In previous chapters we have tacitly assumed that the parameters of the random field model are known. An exception is Chap. 2 where we discussed regression analysis for estimating the coefficients of trend functions. However, the parameters of spatial models constructed for spatial data sets are typically not known a priori. In simulations we use spatial models with specified parameters, which have been derived from available data using *model estimation* procedures. This chapter examines some of the methods that can be used to estimate the model parameters from available spatial data. The focus will be on the estimation of parameters for stochastic models, instead of the simpler deterministic models examined in Chap. 10.

In the following, we will assume that the expectation of the random field is determined by the trend function. The parameters of the latter can be estimated by means of the methods discussed in Chap. 2. Hence, this chapter focuses on estimating the parameters of *zero-mean SRF models*.

Estimation of the trend function is followed by detrending the data and estimation of the spatial model for the residuals (fluctuations). Subsequently, spatial prediction is performed on the residuals, and the final result is composed by adding the trend function to the interpolated residuals. This approach is followed in the framework of *regression kriging* [340]. Alternatively, the trend function can be incorporated in the prediction problem by means of *universal kriging*, which accommodates the uncertainty due to the trend model as well.

Finally, regardless of what approach we use to estimate the spatial model, we must select an optimal model between several candidates (e.g., different trend or

D. T. Hristopulos, *Random Fields for Spatial Data Modeling*, Advances in Geographic Information Science, https://doi.org/10.1007/978-94-024-1918-4_12

variogram functions). This can be achieved by means of a suitable *model selection criterion*.

Warning A mathematically more rigorous approach is to estimate both the trend function and the fluctuations in a unified maximum likelihood framework. For more details on this approach consider the references [201, p. 54], [885, 886]. In addition, *model selection* for spatial data is a delicate matter, since spatial correlations, if not properly treated, can have a confounding effect on the selection of predictor variables used in the trend function; for more information on this topic consider [354].

The basic steps in the estimation of the spatial model for SRF residuals can be described in terms of the following algorithm:

Algorithm 12.1 Basic steps of spatial model estimation. The "model pool" contains different types of spatial models (e.g., exponential and spherical variogram) for the fluctuation SRF. Selection criteria are used to determine the "optimal" model. Model selection can be based on cross-validation or information criteria that penalize models with more parameters according to the *principle of parsimony* [145]

Detrend the data (i.e., determine and subtract the trend function)
Determine the "model pool" for the fluctuation SRF
while model pool is not exhausted **do**
    Select SRF model for the residual fluctuations
    Fit model to the data $\rightarrow$ optimal parameters
    Calculate model fit according to selection criteria
end while
Select optimal model based on selection criteria
For prediction, add the optimal residual SRF model to the trend function

---

This chapter includes a brief review of estimator properties, the estimation of uniform SRF mean using Ordinary Kriging, methods for estimating the variogram function (moment estimator and the non-parametric kernel approach), the method of maximum likelihood for covariance function estimation, and cross validation as a tool for parameter estimation and model selection.

12.1 Estimator Properties

Estimators aim to generate informed guesses for the values of unknown model parameters based on the available data. The parameters of the investigated model will be referred to as the *target or population parameters*. If the data set changes, so will the estimates of specific parameters (see Fig. 12.1). Hence, the estimates are uncertain quantities that are best represented as random variables.

Different types of problems require different degrees of precision in determining the model parameters. If the goal is to estimate the height of a mountain, an error of

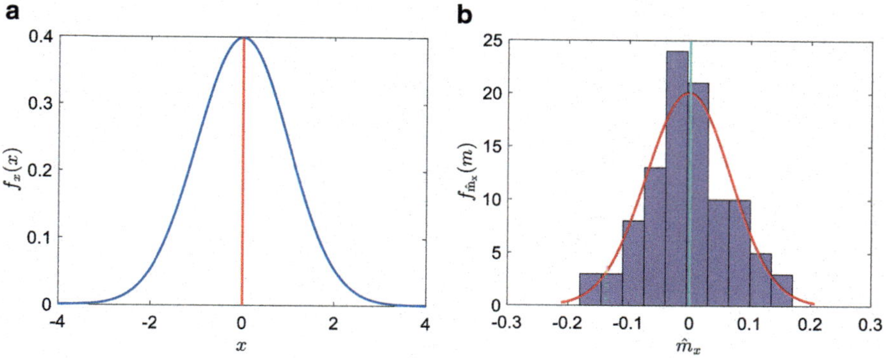

Fig. 12.1 Normal pdf, $\mathcal{N}(0, 1)$, with a mean value equal to zero (continuous vertical line) and a standard deviation equal to one (left). Histogram of average values of 200 samples comprising each 100 random numbers from the $N(0, 1)$ distribution (right). The latter plot illustrates that estimates of the mean derived from a single sample are realizations of a random variable. On the other hand, the mean of the average sample values (marked by the light blue vertical line) deviates from the population mean only in the third decimal place

a few centimeters may be insignificant. However, the same error could be significant in other applications, such as measuring the displacement of high-rise buildings due to intense wind loading [374].

More than one estimator can be designed for each target parameter. For example, the mean of a random field can be estimated from a single sample of spatially scattered data using the simple average or a weighted average of the measurements. The weights could be designed to emphasize certain data values at the expense of other ones. Weighting may be necessary if *preferential sampling* of low or high values is used, or if the data are collected at clustered locations. To compensate for such effects, various declustering methods have been developed. For example, in the case of mining data gathered at irregularly spaced drillholes, such methods help to obtain more accurate estimates of mineral grades [303].

Since different estimators can be constructed for any particular property, objective criteria are needed in order to evaluate the performance of such estimators (see also the discussion in Sect. 1.4.5.3).

An introduction to the mathematical theory of statistical estimation is presented in the classic book of Cramér [161, Chap. 32]. More applied perspectives are given in the references [90, 556].

Notation Let us recall that $\mathbf{x}^* = (x_1^*, \ldots, x_N^*)^\top$ denotes the data vector. The symbol $\boldsymbol{\theta}$ denotes a parameter vector for a specific statistical model. In general, we treat $\boldsymbol{\theta}$ as a vector variable. The "hat" symbol will be used to denote estimates of the respective quantity based on the data. The estimates depend on the particular realization from which the data are drawn; therefore, we will use $\hat{\boldsymbol{\theta}}(\omega)$ to denote estimated parameter vectors. Furthermore, we will use $\hat{\boldsymbol{\theta}}^*(\omega)$ to refer to the optimal estimate according to some specified optimality criterion.

Table 12.1 List of symbols used for the parameter vector, its true (population) value, the sample estimate and the optimal sample estimate. Often there is no explicit distinction between $\hat{\boldsymbol{\theta}}(\omega)$ and $\hat{\boldsymbol{\theta}}^*(\omega)$

Symbol	Meaning
$\boldsymbol{\theta}$	Parameter vector
$\boldsymbol{\theta}^*$	Population value
$\hat{\boldsymbol{\theta}}(\omega)$	Estimate based on single realization
$\hat{\boldsymbol{\theta}}^*(\omega)$	"Optimal" estimate based on single realization

The true value of the parameter vector will be denoted by $\boldsymbol{\theta}^*$.[1] For a univariate normal probability distribution with known population parameters $\boldsymbol{\theta}^* = (m_{\mathrm{x}}, \sigma_{\mathrm{x}}^2)^\top$. In the case of an isotropic exponential covariance model $\boldsymbol{\theta}^* = (\sigma_{\mathrm{x}}^2, \xi)^\top$. Table 12.1 summarizes the symbols used for the parameter vector.

We also use the hat symbol to denote the estimate of a random field at some unmeasured location, e.g., $\hat{\mathrm{x}}(\mathbf{z})$ denotes the estimate of $\mathrm{X}(\mathbf{s};\omega)$ at some location $\mathbf{z}$. The remarks below also apply to estimators of random field values at unmeasured points, such as the Kriging estimators reviewed in Chap. 10.

Desired estimator properties Good estimators should satisfy certain mathematical properties which aim to ensure that the estimates are close to the true values. Since there are several ways of measuring the proximity of random variables to a fixed target (the true values), it is not surprising that there are also several estimator properties. These are reviewed below, using the parameter vector $\boldsymbol{\theta}$ to illustrate them.

1. An estimator $\hat{\boldsymbol{\theta}}$ is considered *unbiased* if its expectation is equal to the true (population) value of the parameter, i.e.,

$$\mathcal{E}\left[\hat{\boldsymbol{\theta}}(\omega)\right] = \boldsymbol{\theta}^*.$$

The *bias* of an estimator is thus defined as

$$B\left(\hat{\boldsymbol{\theta}}(\omega)\right) = \mathcal{E}\left[\hat{\boldsymbol{\theta}}(\omega)\right] - \boldsymbol{\theta}^*. \tag{12.1}$$

2. An estimator $\hat{\boldsymbol{\theta}}(\omega)$ is called *consistent* if it converges to the true value of the parameter, $\boldsymbol{\theta}^*$, for large N (more precisely, as $N \to \infty$). Formally, there are two definitions of consistency. The *strong definition of consistency* requires that

$$\lim_{N\to\infty} P(\hat{\boldsymbol{\theta}}(\omega) = \boldsymbol{\theta}^*) = 1. \tag{12.2}$$

[1] The symbol $*$ that distinguishes the population parameters $\boldsymbol{\theta}^*$ from the estimates should not be confused with the symbol used in $\mathbf{x}^*$ to denote the data vector.

The *weak definition of consistency* requires that

$$\lim_{N\to\infty} P(\|\hat{\boldsymbol{\theta}}(\omega) - \boldsymbol{\theta}^*\| \geq \epsilon) = 0, \text{ where } \epsilon > 0. \tag{12.3}$$

Consistency implies that the estimator is asymptotically unbiased and its variance tends asymptotically to zero.

3. The estimator with the smallest possible variance is called *efficient*.

The "best" estimator is unbiased, consistent, and efficient.

Mean square error To better understand the impact of the estimator's bias and variance on the estimate, consider the *mean square error (MSE) of the estimator* defined by

$$\begin{aligned}\mathcal{E}\left[\left(\hat{\boldsymbol{\theta}}(\omega) - \boldsymbol{\theta}^*\right)^2\right] =& \mathcal{E}\left[\left(\hat{\boldsymbol{\theta}}(\omega) - \mathcal{E}[\hat{\boldsymbol{\theta}}(\omega)]\right)^2\right] + \left(\mathcal{E}[\hat{\boldsymbol{\theta}}(\omega)] - \boldsymbol{\theta}^*\right)^2 \\ &= \text{Variance}\left(\hat{\boldsymbol{\theta}}(\omega)\right) + \left[\text{Bias}\left(\hat{\boldsymbol{\theta}}(\omega)\right)\right]^2. \end{aligned} \tag{12.4}$$

If the estimator involves an additive noise term (with zero mean and variance σ_ε^2), the estimator MSE also includes the irreducible variance term σ_ε^2. The irreducible variance places a lower bound on the MSE, since it cannot be extinguished by optimizing the estimator. In the case of the kriging predictors, the irreducible variance corresponds to the nugget variance.

It is straightforward to prove the above identity by adding to and subtracting from the estimation error the expectation of the estimator, i.e., by writing the identity

$$\hat{\boldsymbol{\theta}}(\omega) - \boldsymbol{\theta}^* = \left(\hat{\boldsymbol{\theta}}(\omega) - \mathcal{E}[\hat{\boldsymbol{\theta}}(\omega)]\right) + \left(\mathcal{E}[\hat{\boldsymbol{\theta}}(\omega)] - \boldsymbol{\theta}^*\right).$$

The square of the estimator error, $\hat{\boldsymbol{\theta}}(\omega) - \boldsymbol{\theta}^*$, is calculated by taking into account that the bias $\mathcal{E}[\hat{\boldsymbol{\theta}}(\omega)] - \boldsymbol{\theta}^*$ is a constant (non-random) vector and that the expectation of the second term on the right-hand side of the above vanishes, i.e., $\mathcal{E}\left[\hat{\boldsymbol{\theta}}(\omega) - \mathcal{E}[\hat{\boldsymbol{\theta}}(\omega)]\right] = \mathbf{0}$, thus leading to (12.4).

The mean square error thus contains contributions from both the variance of the estimator and the bias. Based on (10.18) an *unbiased* and *efficient* estimator has the smallest possible mean square error. Keep in mind, however, that the *bias-variance tradeoff*, discussed in Chaps. 1 and 2 in connection with regression methods, is present in general estimation problems.

12.2 Estimating the Mean with Ordinary Kriging

The most basic parameter of a random field that one would like to infer from a spatial data set is the mean of the underlying probability distribution (population).[2] Under conditions of second-order stationarity the mean is constant.

Let us assume a sample comprising N values of a second-order stationary random field at different locations. The simplest estimator of the population mean is the sample average. If the data represent independent identically distributed random variables, the sample average is demonstrably an efficient estimator of the true mean (i.e., it is the unbiased, minimum variance estimator).

The nice efficiency property, however, is not necessarily true if the sample contains a finite number of correlated values. Especially if the sample locations are clustered in certain areas, *declustering techniques* may be necessary in order to accurately estimate the population mean. For example, it is possible to weigh each sample value proportionally to the volume of the *Voronoi cell* that is centered at the respective location [303]. This method produces smaller weights in areas of dense sampling than in areas where the sampling is sparse.

A more systematic declustering approach is based on Ordinary Kriging (OK). The OK declustering effect is due to the fact the the kriging estimates take into account the covariance dependence and thus incorporate the impact of spatial correlations. We describe below how OK can be used to estimate the unknown population mean from scattered data over the domain $\mathcal{D}$.

The estimate of the mean, $\hat{m}_{\mathrm{x}}$, is given by the linear combination of the data, i.e.,

$$\hat{m}_{\mathrm{x}} = \sum_{n=1}^{N} \lambda_n \, x_n^*, \tag{12.5}$$

where the linear weights $\{\lambda_n^*\}_{n=1}^N$ are to be determined by ordinary kriging.

Remark In agreement with the notation used in this chapter, we use $*$ to distinguish the optimal weights, λ_n^*, from the λ_n that are not necessarily optimized. Note that in Chap. 10 this distinction was not made for reasons of notational simplicity.

The zero-bias condition for the mean estimate, i.e., $\mathcal{E}[\hat{m}_{\mathrm{x}}] = m_{\mathrm{x}}$, implies that the linear weights $\{\lambda_n\}_{n=1}^N$ should satisfy the condition

$$\sum_{n=1}^{N} \lambda_n = 1. \tag{12.6}$$

[2]We assume that the population mean is a well-defined parameter. There are probability distributions, such as the Cauchy distribution, for which the mean is not well defined.

The optimal weights, $\{\lambda_n^*\}_{n=1}^N$, are determined by minimizing the mean square error of $\hat{m}_x$, i.e., $\mathcal{E}[(\hat{m}_x - m_x)^2]$, under the zero-bias constraint (12.6). It is left as an exercise for the reader to show that the optimal weights satisfy the following $(N+1) \times (N+1)$ linear system of equations

$$\sum_{\beta=1}^{N} \lambda_\beta^* \, C_{xx}(\mathbf{s}_\alpha - \mathbf{s}_\beta) = \mu_m, \quad \text{for } \alpha = 1, \ldots, N, \tag{12.7a}$$

$$\sum_{\alpha=1}^{N} \lambda_\alpha^* = 1, \tag{12.7b}$$

where μ_m is the Lagrange multiplier used to enforce the zero-bias constraint. The Lagrange multiplier determines the variance of the estimator $\hat{m}_x$ according to [33]

$$\mathbb{V}\text{ar}\left\{\hat{m}_x\right\} = \mu_m. \tag{12.8}$$

In matrix form, the solution to the kriging estimate of the mean is given by

Ordinary Kriging System for Estimation of the Mean

$$\begin{bmatrix} \sigma_x^2 & C_{xx}(\mathbf{s}_1 - \mathbf{s}_2) & \ldots & C_{xx}(\mathbf{s}_1 - \mathbf{s}_N) & -1 \\ C_{xx}(\mathbf{s}_2 - \mathbf{s}_1) & \sigma_x^2 & \ldots & C_{xx}(\mathbf{s}_2 - \mathbf{s}_N) & -1 \\ \vdots & \vdots & \vdots & \vdots & \vdots \\ C_{xx}(\mathbf{s}_N - \mathbf{s}_1) & C_{xx}(\mathbf{s}_N - \mathbf{s}_2) & \vdots & \sigma_x^2 & -1 \\ 1 & 1 & \ldots & 1 & 0 \end{bmatrix} \begin{bmatrix} \lambda_1^* \\ \lambda_2^* \\ \vdots \\ \lambda_N^* \\ \mu_m \end{bmatrix} = \begin{bmatrix} 0 \\ 0 \\ \vdots \\ 0 \\ 1 \end{bmatrix}. \tag{12.9}$$

In contrast with the sample average, the ordinary kriging estimator takes into account correlations between sampling locations. This feature implies that kriging can better handle clustered observations, a scenario which is often encountered in environmental monitoring and natural resources estimation.

12.2.1 A Synthetic Example

We study the kriging estimate of the population mean using a synthetic data set simulated from a Gaussian random field that is sampled over a square grid. The stationary random field has zero mean and an SSRF covariance function given by (7.37).

We consider two different sampling scenarios: In the first case a *random sampling* design is employed. This involves 100 points that are randomly distributed over the square grid as shown in Fig. 12.2a. In the second case, a *clustered sampling* design is used (see Fig. 12.3a). The grid is divided into four equal-size quadrants; eighty points are randomly distributed in the bottom left quadrant, each of the top two quadrants contains ten points, while the bottom right quadrant is left empty.

We generate five hundred realizations from the SSRF with covariance function (7.37) and parameters $\eta_0 = 1.1501, \eta_1 = 1.5$, and $\xi = 10$. We use the method of covariance matrix factorization (based on the square root factorization) to simulate the random field (see Chap. 16).

For each realization we estimate the *sample average*, $\overline{\mu}^{(k)}$, and the *kriging estimate of the mean*, $\hat{m}_{\mathrm{x}}^{(k)}$, based on the solution of the ordinary kriging system (12.9), where $k = 1, \ldots, 500$ is the realization index. We construct the ordinary kriging (OK) system based on the covariance function used to generate the realizations. The means of these estimates over the 500 realizations and their standard deviations are reported in Table 12.2.

The results listed in Table 12.2 and the histograms shown in Figs. 12.2a and 12.3b exhibit the following properties:

1. For random sampling, the estimates of the population mean obtained by means of the sample average and those obtained by the OK-based estimates are similar and both close to the true mean (zero).

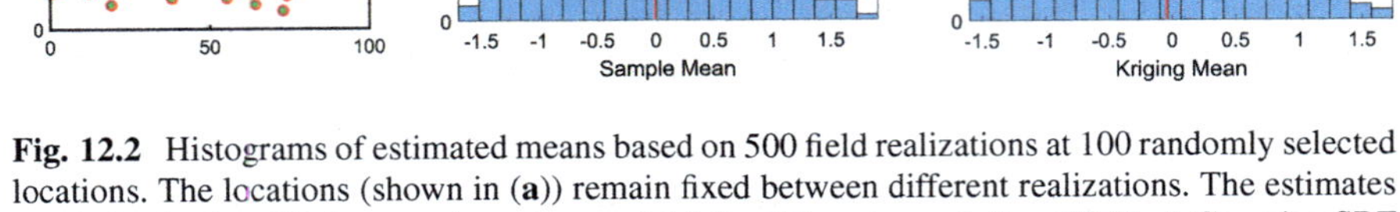

Fig. 12.2 Histograms of estimated means based on 500 field realizations at 100 randomly selected locations. The locations (shown in (**a**)) remain fixed between different realizations. The estimates are derived using (**b**) the sample average and (**c**) the OK system solution (12.9). A Gaussian SRF with SSRF covariance function is sampled at 100 randomly distributed locations inside a square grid with side length $L = 100$. The SSRF parameters are $\eta_0 = 1.15, \eta_1 = 1.5, \xi = 10$. The locations of the ensemble averages of the estimates are marked by vertical solid red lines on the histogram plots

Table 12.2 Estimates of the mean of a stationary random field. A zero-mean Gaussian SSRF with covariance parameters $\eta_0 = 1.1501$, $\eta_1 = 1.5$, and $\xi = 10$ is used to generate 500 simulations. Two sampling designs (random and clustered) are employed. The table entries report the mean and standard deviations of the estimators based on the respective averages over the simulation ensemble

	Sample average estimator		OK-based estimator	
	Mean	Standard deviation	Mean	Standard deviation
Random sampling	−0.0077	0.6454	−0.0134	0.5264
Clustered sampling	−0.0797	1.0464	−0.0406	0.5943

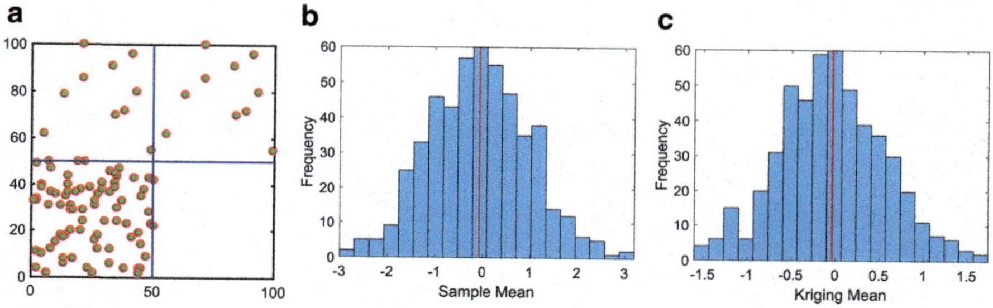

Fig. 12.3 Histograms of estimated means based on 500 simulations using a clustered sampling design. (**a**) The estimates are derived using (**b**) the sample average and (**c**) the OK system solution (12.9). A Gaussian SRF with SSRF covariance function is sampled at 100 locations distributed over a square grid with side length $L = 100$. The grid is sub-divided in four equal-size quadrants. Eighty locations are randomly distributed in the lower left quadrant, ten locations are randomly distributed in each of the top two quadrants, while the bottom right quadrant is left empty. The SSRF parameters are $\eta_0 = 1.15$, $\eta_1 = 1.5$ and $\xi = 10$. The ensemble averages of the estimates are marked by vertical solid red lines on the histogram plots

2. For random sampling, the standard deviation of the sample average is slightly higher than that of the OK-based estimates ($\approx$0.65 versus $\approx$0.53).
3. The magnitude of the bias for both methods is higher for clustered than for random sampling; the magnitude of the bias for the simple sample average is almost twice that of the OK-based estimate ($\approx$0.08 versus $\approx$0.04).
4. For clustered sampling, the OK-based estimate is more precise (standard deviation $\approx$0.59) than the sample average (standard deviation $\approx$1.05).

Remark The careful reader may have noticed that the OK-based estimate of the mean is cyclical: we use the covariance function to estimate the mean, but we need an estimate of the latter to determine the covariance function. In the synthetic example studied above we bypassed this problem by using the "exact" covariance function that generated the realizations; this is not possible with real data. In the case of stationary random fields this problem can be circumvented, because the covariance function is directly linked to the variogram through (3.47). As we have discussed in Sect. 10.7 and further demonstrate below, the estimation of the variogram of stationary random fields does not require knowledge of the mean.

12.3 Variogram Estimation

The variogram function is the best known measure of two-point correlations in spatial statistics. The variogram is defined by means of the variance the field increment, i.e. via equation (3.44). As we have discussed in Sect. 3.3.4, for stationary random fields the variogram and the covariance function contain the same information about the correlations. However, the variogram is preferred for reasons already mentioned—see text below (3.44). Since the variogram function has a central role in spatial data analysis, we will briefly survey some of the approaches proposed in the literature for its estimation.

12.3.1 Types of Estimation Methods

Variogram estimation methods can be classified as (i) *indirect* if they require the estimation of an experimental (empirical) variogram function and (ii) *direct* if the variogram model is obtained by optimizing a selected objective function [650].

Direct methods (e.g., maximum likelihood) estimate the variogram model without the intermediate step of empirical variogram estimation. In direct methods, the optimal variogram model is estimated by directly fitting the model to the data by means of some appropriate fitness measure.

Indirect methods involve an additional step: first, an *empirical variogram* is estimated from the data, and it is then fitted to a *model variogram function*. The model fitting step is necessary, because the empirical variogram is calculated at a discrete set of lags, and it does not necessarily satisfy the conditional non-negativity constraint. Indirect methods allow a visual comparison between the data-based estimator (empirical variogram) and the model function.

For example, in Fig. 12.4 we show the empirical variogram (circles) calculated with the method of moments and its fit to the SSRF variogram model (continuous line). The data involve topsoil chromium concentration (in units of mg/kg) measured at 259 locations in the Swiss Jura mountains.

About the data: The Swiss Jura data set used to generate the variogram in Fig. 12.4 is available with the `R` language package `gstat`. The package can be downloaded from https://cran.r-project.org/web/packages/gstat/.

Parametric variogram models employ specific functional forms and thus involve some assumptions (usually subjective) regarding the nature of the underlying process.

Both direct and indirect estimation approaches, including maximum likelihood (direct) [650], the classical method of moments (indirect) [550, 551], and the robust variogram estimator (indirect) [166] can be used to estimate parametric variogram models. The method of moments is often preferred by practitioners due to its simplicity and visual appeal (see Fig. 12.4). Robust methods attempt to reduce the impact of outliers on the estimation of the empirical variogram.

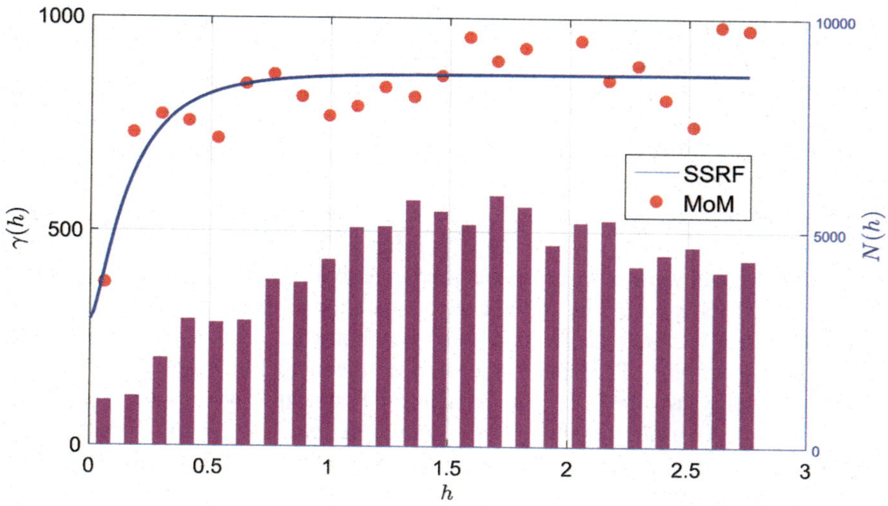

Fig. 12.4 Variogram estimation using the method of moments. The horizontal axis corresponds to lag distances (measured in km), while the left vertical axis represents the variogram values. Filled circle markers (•) denote empirical variogram values. The continuous line represents the best-fit SSRF variogram. The bar plot shows the number of pairs per lag distance (values displayed on the right vertical axis) used in variogram estimation

Non-parametric variogram methods do not rely on variogram model functions that contain a small number of parameters. Instead, they construct the variogram function as a superposition of known functions with suitably selected weights.

12.3.2 *Method of Moments (MoM)*

The sample-based variogram is also known as the *empirical variogram*. The classical method of variogram estimation is based on the *method of moments (MoM)* estimator developed by Matheron [419, 550]:

$$\hat{\gamma}_{\mathrm{xx}}(\mathbf{r}_k) = \frac{1}{2\,N_p(\mathbf{r}_k)} \sum_{n,m=1}^{N} \left\{ \left[x^*(\mathbf{s}_n) - x^*(\mathbf{s}_m)\right]^2 \right\} \vartheta_{n,m}(\mathbf{r}_k), \quad k = 1, \ldots, N_c,$$

$$\vartheta_{n,m}(\mathbf{r}_k) = \begin{cases} 1, & \text{if } \mathbf{s}_n - \mathbf{s}_m \in B(\mathbf{r}_k) \quad \text{for } n, m = 1, \ldots, N. \\ 0, & \text{if } \mathbf{s}_n - \mathbf{s}_m \notin B(\mathbf{r}_k) \quad \text{for } n, m = 1, \ldots, N. \end{cases} \tag{12.10}$$

The following is a list of factors that influence the estimation of the empirical variogram.

1. *Set of lags:* The sample variogram is estimated for a finite set of discrete lags $\{\mathbf{r}_k\}_{k=1}^{N_c}$. The cardinality of the lag set is equal to the total *number of classes* N_c.
2. *Maximum lag:* Typically, we restrict our attention to lags with magnitude $\|\mathbf{r}_k\| \leq r_{\max}/2$, where $r_{\max}$ is the maximum distance between sampling points. The rationale for this choice is (i) to emphasize shorter distances, due to their importance in interpolation and (ii) to avoid the calculation of the variogram at large lags that may only involve a small number of pairs.
3. The *class membership function* $\vartheta_{n,m}(\mathbf{r}_k)$ defines different lag groups (classes), by selecting lag vectors within a specified neighborhood $B(\mathbf{r}_k)$ around the target vector $\mathbf{r}_k$. This is necessary in the case of scattered data, because pairs of points with distance vectors exactly equal to $\mathbf{r}_k$ may not exist.
4. *Lag neighborhood:* The shape and size of the neighborhood depend on the dimensionality of the sampling domain $\mathcal{D}$ and the potential presence of anisotropy. Assuming isotropy, the *omnidirectional* empirical variogram is calculated; the latter does not distinguish between lag vectors with the same magnitude $\|\mathbf{r}_k\|$ but different direction. If anisotropy is present, *directional* empirical variograms are calculated in specified directions. Schematics that illustrate the neighborhoods used in these two cases are shown in Fig. 12.5.
5. *Number of pairs:* The parameter $N_p(\mathbf{r}_k)$ is the cardinality of the set containing pairs of points that are separated by lags that are approximately equal to $\mathbf{r}_k$, i.e., the set of lags inside $B(\mathbf{r}_k)$.
6. *Neighborhood selection:* A schematic illustrating the neighborhood selection for omnidirectional and directional variograms is shown in Fig. 12.5. In the case of *omnidirectional variogram* estimation, the only parameter required for neighborhood selection is the *radial tolerance* δr. In the case of *directional variogram* estimation, the *radial tolerance* is complemented by respective

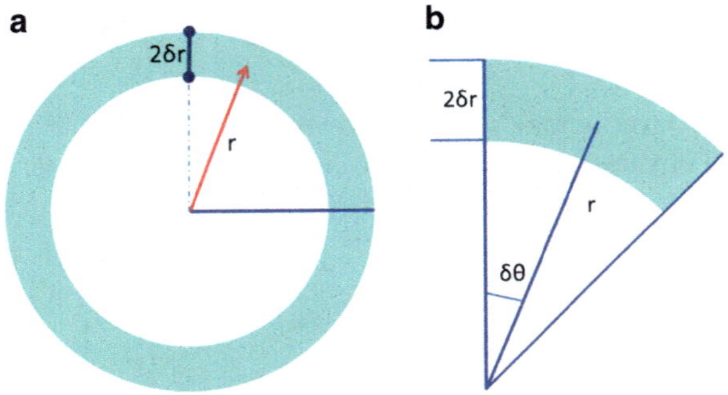

Fig. 12.5 Schematic of lag class neighborhoods $B(\mathbf{r})$ around the lag vector $\mathbf{r}$ in two dimensions. The drawing in (**a**) represents the neighborhood (annulus-shaped shaded area) for the calculation of the omnidirectional variogram at a lag distance equal to $r = \|\mathbf{r}\|$. The tolerance $2\delta r$ defines the width of the annulus. The drawing in (**b**) represents a neighborhood (shaded area shaped as an annulus segment) corresponding to a lag vector with magnitude r and angle θ (measured with respect to the horizontal axis). In this case the neighborhood is defined using a radial tolerance δr and an angular tolerance $\delta\theta$

angular tolerances. In two dimensions only one angle tolerance is necessary (see Fig. 12.5b).

7. *Bandwidth and large lags:* For large radial distances, in addition to the radial and angular tolerances, a *bandwidth* parameter can be specified to constrain the average over the pairs that lie inside a cylinder with diameter equal to the bandwidth; for more information see [195, 623]. The bandwidth parameter restricts the maximum deviation from the target vector.
8. *Unbiased estimator:* The MoM estimator is unbiased if the random field from which the data are derived is stationary or intrinsically stationary. This means that the expectation of the increment field is zero and the variogram depends only on the lag.

Based on the above description of available options and the respective parametrization that they entail, it follows that the MoM sample variogram is a non-uniquely defined discrete function. In particular, the empirical variogram depends on several user assumptions regarding stationarity and anisotropy, the number N_c of lag classes, their tolerances as specified by $B(\mathbf{r}_k)$, the maximum lag considered, the use of overlapping versus non-overlapping lag neighborhoods, and other modeling preferences of the MoM estimator.

Practical guides for the calculation and interpretation of the variogram are given in the articles [311, 622]. More details regarding variogram estimation are found in the textbooks [165, 624]. Comparisons between the MoM and other methods are investigated in [488, 897]. Thanks to its simplicity and in spite of the assumptions required, the MoM remains a widely used estimator of the variogram function.

12.3.3 Properties of MoM Estimator

For the function $\hat{\gamma}_{xx}(\mathbf{r}_k)$ defined by (12.10) to be a "good" estimator of $\gamma_{xx}(\mathbf{r}_k; \boldsymbol{\theta}^*)$ two key properties are necessary:

1. The average of the squared increments at lag $\approx \mathbf{r}_k$ should accurately approximate the respective expectation $\mathcal{E}\left[\{X(\mathbf{s};\omega) - X(\mathbf{s}+\mathbf{r}_k;\omega)\}^2\right]$.
2. The average square increment should be precise, i.e., it should not fluctuate significantly between different sample realizations. A necessary (but not sufficient) condition for the above property to hold is the validity of the *ergodic hypothesis*, which allows interchanging the expectation and the spatial average.

For the ergodic hypothesis to hold, the following conditions should be fulfilled:

1. The *increment field* $\Delta_x(\mathbf{s}+\mathbf{r}_k, \mathbf{s}; \omega) = X(\mathbf{s};\omega) - X(\mathbf{s}+\mathbf{r}_k;\omega)$ must be *statistically homogeneous*. In certain cases, meeting this condition requires dividing the study domain into smaller units that can be considered as approximately homogeneous.
2. The *number of pairs* in each class must be sufficiently large to increase the *reliability of the MoM estimator*, i.e., to ensure that the sample average of the square increments tends to the respective expectation. The random error of the

variogram estimates declines roughly in proportion to the inverse of the number of pairs contained in each class.

3. The *number of classes* used in the discrete variogram estimator must be large to ensure *adequate resolution* and to provide a dense approximation of the theoretical variogram function.
4. The resolution and the reliability requirements are in conflict, because the maximum number of pairs that can be used by the MoM estimator is finite, i.e.,

$$\sum_{k=1}^{N_c} N_p(\mathbf{r}_k) \leq N(N-1)/2.$$

In practice, a compromise is made between resolution and reliability. As a rule of thumb, the minimum number of pairs per class should be at least thirty.

12.3.4 Method of Moments on Regular Grids

The MoM estimator defined in (12.10) is suitable for scattered data. The computational complexity of the estimator is $O(N^2)$, and thus the calculation can become quite cumbersome for large data sets. In such cases, more efficient implementations of the variogram estimation based on kd-tree structures can be used [309].

The computational complexity, however, can be reduced for data sampled on regular grids. For example, remote sensing images can be quite large (e.g., they could easily involve 512×512 grids). For such grid data it is possible to estimate *marginal variograms* along the orthogonal directions of the grid. If the spatial correlations of the image are isotropic, the marginal variograms should be identical (within statistical precision).

Digital image on rectangular grid Consider a gray-level digital image defined on a two-dimensional rectangular grid with L_1 rows, L_2 columns, and sampling step equal to a. We use the labeling of the grid sites introduced in Sect. 11.6.2: each site is determined by the *row index* n_1 (which measures the position along the y axis) and the *column index* n_2 (which measures the position along the x axis). A data point is then denoted by $x^*(n_2, n_1)$ where $n_i = 1, \ldots, L_i$, for $i = 1, 2$.

Marginal MoM variogram estimators The marginal variograms in the two orthogonal directions can be estimated as follows (where p_x, p_y determine the lags in the orthogonal directions)

$$\hat{\gamma}_{\mathrm{xx}}(p_x a, 0) = \frac{1}{L_1} \sum_{n_1=1}^{L_1} \frac{1}{L_2 - p_x} \sum_{n_2=1}^{L_2-p_x} \left[x^*(n_2 + p_x, n_1) - x^*(n_2, n_1)\right]^2,$$

$$p_x = 0, \ldots, \frac{L_2}{2}, \tag{12.11a}$$

$$\hat{\gamma}_{xx}(0, p_y a) = \frac{1}{L_2} \sum_{n_2=1}^{L_2} \frac{1}{L_1 - p_y} \sum_{n_1=1}^{L_1 - p_y} \left[x^*(n_2, n_1 + p_y) - x^*(n_2, n_1)\right]^2,$$

$$p_y = 0, \ldots, \frac{L_1}{2}. \tag{12.11b}$$

The above marginal variogram estimators do not require the definition of neighborhood structures. Thus, they have a computational complexity that *scales linearly* as $O(N)$, where $N = L_1 L_2$, for each lag evaluated. The full variogram can be obtained from the marginal variograms, if (i) the random field is isotropic or (ii) the random field has geometric anisotropy and the principal directions coincide with the grid axes.

In the case of *geometric anisotropy* that is not aligned with the grid axes, the following grid variogram estimator can be used

$$\hat{\gamma}_{xx}(p_x a, p_y a) = \frac{1}{L_1 - p_y} \sum_{n_1=1}^{L_1 - p_y} \frac{1}{L_2 - p_x} \sum_{n_2=1}^{L_2 - p_x} \left[x^*(n_2 + p_x, n_1 + p_y) - x^*(n_2, n_1)\right]^2, \tag{12.12}$$

where $p_x = 0, \ldots, L_2/2$, and $p_y = 0, \ldots, L_1/2$.

If the grid includes missing data, a suitable mask can be used to remove from the summation pairs of points for which the data are missing at one or both points.

12.3.5 Fit to Theoretical Variogram Model

To guarantee the property of conditional negative definiteness, the sample (experimental) variograms obtained with the MoM should be fitted to valid mathematical variogram models.

Various non-linear least-squares methods have been used to fit experimental variograms to theoretical functions [165]. *Ordinary least squares* and *weighted least squares* methods of variogram fitting have a lower computational cost than the direct likelihood-based approaches, while they are competitive in terms of estimation accuracy. Other estimators are discussed in [269, Chap. 5].

Weighted least squares The *weighted least squares* (hereafter WLS) estimator is based on minimizing the weighted sum of the squared variogram residuals. The WLS objective function is defined as follows:

$$\text{RSS}(\boldsymbol{\theta}) = \sum_{k=1}^{N_c} N_p(\mathbf{r}_k) \left[\frac{\hat{\gamma}_{xx}(\mathbf{r}_k) - \gamma_{xx}(\mathbf{r}_k; \boldsymbol{\theta})}{\gamma_{xx}(\mathbf{r}_k; \boldsymbol{\theta})}\right]^2, \tag{12.13}$$

where $\hat{\gamma}_{xx}(\mathbf{r}_k)$, $k = 1, \ldots, N_c$ is the experimental variogram and $\gamma_{xx}(\mathbf{r}_k; \boldsymbol{\theta})$ is respectively the model variogram calculated for the parameter vector $\boldsymbol{\theta}$. The summation is evaluated over the N_c lag classes, each of which contains $N_p(\mathbf{r}_k)$ pairs of points.

Then, the optimal parameter vector, $\hat{\boldsymbol{\theta}}^*$ is given by minimizing the residual sum of squares (RSS):

$$\hat{\boldsymbol{\theta}}^* = \arg\min_{\boldsymbol{\theta}} \mathrm{RSS}(\boldsymbol{\theta}). \tag{12.14}$$

The WLS objective function emphasizes (i) lags with a higher number of pairs over lags with fewer pairs and (ii) shorter over longer lags due to the denominator $\gamma_{xx}(\mathbf{r}_k; \boldsymbol{\theta})$ which is close to zero (or to the nugget variance if a nugget effect is present) for small lags.

The WLS estimator has been found to perform well with different types of data [888]. A discussion of the WLS variogram estimator and its relation with maximum likelihood (see below), is given in [886]. The fit between the omnidirectional variogram of the Jura data and the SSRF variogram model in Fig. 12.4 is obtained with the WLS method.

12.3.6 Non-parametric Variogram Estimation

Various *non-parametric methods* have been proposed for variogram estimation. These methods do not rely on explicit variogram model functions. Instead, they express the unknown variogram using a linear combination of known, simple, valid variogram functions. The coefficients of this expansion are determined by using an optimal fit criterion between the empirical variogram and the estimated fitting function [710, 740].

Non-parametric estimator Non-parametric estimates of the covariance function and the variogram can be constructed based on *modified kernel regression*. The modifications of the regression are needed in order to ensure the positive-definiteness (for covariances) and conditional positive definiteness (for variograms) [322]. In particular, if $K(\mathbf{s}, \mathbf{s}')$ is a kernel function (see Definition 2.1), a non-parametric, regression-based estimate of the variogram function for isotropic SRFs is given by

$$\widehat{\gamma_{xx}}(\mathbf{r}) = \sum_{n=1}^{N} \sum_{m=1}^{N} \frac{\hat{Y}_{n,m}\, K\left(\frac{\|\mathbf{r}-\mathbf{r}_{n,m}\|}{h}\right)}{K\left(\frac{\|\mathbf{r}-\mathbf{r}_{n,m}\|}{h}\right)}, \quad h > 0, \tag{12.15a}$$

$$\hat{Y}_{n,m} = \left[x^*(\mathbf{s}_n) - x^*(\mathbf{s}_m)\right]^2, \tag{12.15b}$$

$$\mathbf{r}_{n,m} = \mathbf{s}_n - \mathbf{s}_m, \quad \text{for } n, m = 1, \ldots, N, \tag{12.15c}$$

where the bandwidth h is determined by minimizing the asymptotic mean square error of the variogram estimator [564, 565].

The function defined in (12.15a) is not necessarily conditionally negative definite. The covariance estimator, $\widehat{C_{xx}}(\mathbf{r})$, can be defined by replacing $Y_{n,m}$ with $\left(x_n^* - \overline{x^*}\right)\left(x_m^* - \overline{x^*}\right)$. Similarly, the covariance thus derived is not necessarily a positive definite function.

Restoring permissibility The solution proposed by Hall et al. in [322] for the non-permissibility problem of the kernel-based covariance estimator is as follows: (i) calculate the spectral density of $\widehat{C_{xx}}(\mathbf{r})$, (ii) truncate the density in order to remove negative tails and possibly to smoothen it, and (iii) evaluate the inverse Fourier transform of the "filtered" density that will give the permissible covariance estimate.

Other nonparametric methods are also based on the isotropic spectral representation of positive definite functions [130, 740]. A more recent proposal is based an optimal orthogonal discretization of the spectral representation that employs Fourier-Bessel matrices. This method provides smooth and positive definite non-parametric estimators in the continuum [283].

Generalizing the method of moments Alternatively, the function $\widehat{\gamma_{xx}}(\mathbf{r})$ can be viewed as a *kernel-based generalization of the method-of-moments estimator.* Actually, the method-of-moments estimator is obtained from (12.15a) if the uniform kernel $K(\cdot)$ is used. Then, $\widehat{\gamma_{xx}}(\mathbf{r})$ still needs to be fitted to a parametric model.

Parametric or non-parametric? Comparisons between the parametric and non-parametric approaches based on different data sets find that both approaches have similar performance [130]. Non-parametric methods are proclaimed to have some useful properties in terms of speed and ease of use [130]. More recently, the kernel-based methods have been extended to non-stationary, anisotropic covariance functions [258].

12.3.7 Practical Issues of Variogram Estimation

Variogram estimation methods typically involve selecting between various options and identifying certain patterns in the spatial data that affect the estimation. The following list of topics is based on the study [650].

- If the sample data contain *outliers*, robust variogram estimation methods can be used to reduce the impact of outliers [166, 282]. Outliers can affect the variogram estimate in the same way that they affect the estimation of marginal statistical moments (e.g., variance and kurtosis).
- The use of *non-parametric variogram* estimation methods has not been widely explored. With non-parametric methods, users need not worry about fitting an empirical variogram to a model function, since the non-parametric approach provides a function that is defined at every lag (but it is not necessarily permissible without further processing). It is, however, necessary to optimize

parameters in the non-parametric approaches as well. For example, estimating an isotropic variogram with kernel functions requires determining an *optimal kernel bandwidth*.

- In the case of spatial data with *long-range non-homogeneities*, it is necessary to estimate the trend or drift before estimating the variogram. If a trend is detected, it is best to estimate the variogram of the data residuals. Methods for trend estimation were discussed in Chap. 2.
- If the data set can be augmented to include more points, principles of *optimal sampling design* could be used to increase the usefulness of additional points in variogram estimation [592]. The goal of sampling design is to determine the spatial locations where data should be collected in order to obtain efficient estimation of the model parameters or prediction of the field values at non-sampled locations. Spatial sampling design is based on the optimization of some design criterion, such as the average kriging prediction variance over all the prediction sites [883].
- For a given sampling configuration, it is possible that some of the spatial model's parameters cannot be "correctly" estimated. In this respect the concept of *microergodic parameters* (introduced by Georges Matheron) refers to parameters that can be correctly estimated from a single realizations. For precise definitions of the notions of "correctly estimated" and "micro-ergodicity" we refer the reader to [774, Chap. 6]. Micro-ergodicity is not an inherent property of the spatial model, since the classification of a parameter as microergodic depends on the spatial distribution of the data. As shown by Hao Zhang, spatial sampling design affects which variogram parameters (or combinations of parameters) can be reliably estimated [878]. For example, in the case of an exponential variogram function $\gamma_{xx}(\mathbf{r}) = \sigma_x^2 \left[1 - \exp(-\|\mathbf{r}\|/\xi)\right]$ only the ratio σ_x^2/ξ can be consistently estimated under *infill asymptotics*.[3]
- For data sets of *small size*, *Bayesian estimation* can incorporate a priori knowledge about the variogram parameters in terms of prior densities [647]. However, in such cases the estimate will be influenced by the assumptions on the prior. In this framework, it is possible to obtain the posterior distribution of the parameters by means of (11.18). The Bayesian approach to parameter estimation is also used in *empirical Bayesian kriging* as described in Sect. 11.4.
- The *optimal variogram* model function may significantly depend on the method of variogram estimation. The impact of specific modeling choices on the kriging variance, and consequently on the kriging prediction intervals, can be significant. These topics are investigated in [888].
- In the case of data that follow *asymmetric (skewed)* probability distributions, *nonlinear transformations* (e.g., logarithmic, square root) can help to restore

[3]In asymptotic studies the domain size is assumed to scale as $\propto N^{\delta}$, where N is the number of points and $0 \leq \delta \leq 1$. Infill asymptotics corresponds to $\delta = 0$ and implies fixed domain size with decreasing distance between observations. Cases with $\delta > 1$ correspond to expanding domain asymptotics.

normality of the marginal probability distribution. We further discuss this topic in Chap. 14.

- *Trend removal* may lead to residuals that are closer to the normal distribution than the original data. However, if the residuals contain negative values, certain nonlinear transforms that assume positivity of the data (e.g., the square root and logarithmic transforms) cannot be applied to them. In addition, the nonlinear transforms are typically constructed to normalize the marginal data distribution. Thus, they do not ensure that the joint pdf of N points in an N-dimensional Gaussian, as required by geostatistical methods.
- Various variogram models may provide reasonable fits to the empirical variogram or show similar performance in terms of cross-validation measures. In such cases, it may be desirable to apply *information criteria for model selection* [867].
- Different sets of parameter vectors can provide very similar fits to the experimental variogram. This *quasi-degeneracy* occurs in particular if the variogram model involves more than two parameters (e.g., Spartan and Matérn models). This effect implies that some parameters are inter-dependent [380, 800, 897]. In such cases nonlinear parameter combinations that lead to a new vector of orthogonal parameters with respect to estimation can help to identify optimal parameters [201].

12.4 Maximum Likelihood Estimation (MLE)

This section is an overview of maximum likelihood estimation as applied to spatial data analysis. The spatial data are denoted by the vector $\mathbf{x}^*$. A classical study on the application of maximum likelihood in spatial data problems is [541]. A clear exposition of the application of the method to soil data is given in [575]. This paper also examines the application of *restricted maximum likelihood (REML)*, an approach developed by Kitanidis to reduce potential bias in the simultaneous maximum likelihood estimation of trend and variogram parameters [455, 456, 462]. REML is based on stationary data increments.

Definition of the likelihood Given a data set $\mathbf{x}^*$ and a candidate spatial model with parameter vector $\boldsymbol{\theta}$, the *likelihood* $\mathcal{L}(\boldsymbol{\theta}; \mathbf{x}^*) = f_{\mathrm{x}}(\mathbf{x}^* \mid \boldsymbol{\theta})$ is the probability that the specific data is observed if the spatial model with the specified parameter vector is true. The likelihood can be expressed in terms of the joint pdf of the spatial model. Hence, for Gaussian data the likelihood can be fully specified in terms of the mean and the covariance function.

Once we have decided on the structure of the spatial model that will be fitted to the data (e.g., Gaussian random field with linear trend function and Spartan covariance), the optimal parameter vector $\boldsymbol{\theta}^*$ can be estimated by maximizing the likelihood of the data set, i.e.,

$$\boldsymbol{\theta}^* = \arg\max_{\boldsymbol{\theta}} \mathcal{L}(\boldsymbol{\theta}; \mathbf{x}^*). \tag{12.16}$$

This method of parameter estimation is known as *maximum likelihood estimation (MLE)*.

MLE properties Maximum likelihood estimation is the method of choice for many statisticians due to its desirable *asymptotic properties* and the fact that it is based on a joint pdf model. In particular, for N independent and identically distributed samples it can be shown that ML estimators are consistent, asymptotically normal, and efficient (i.e., their variance attains the Cramer-Rao lower bound) [161]. Moreover, in the case of *linear regression* problems, maximum likelihood estimation is equivalent to the least squares solution provided that the error distribution is Gaussian.

MLE for spatial models In the case of *spatial models*, the likelihood dependence on the covariance model parameters (e.g., the correlation length) is nonlinear. As a result, the maximization of the likelihood requires the use of computational algorithms. MLE is also used in the analysis of high energy physics and astrophysical data [244, 263]. A concise introduction to the application of MLE in spatial problems is presented in [885]. Conditions for the asymptotic normality and consistency of MLE in spatial data were provided by Mardia [541].

One of the most celebrated methods for determining the maximum likelihood in the case of problems with *latent variables* is the *Expectation-Maximization (EM)* algorithm developed by Dempster et al. [193]. Problems with latent variables typically appear if the observed spatial process values drawn from a multimodal probability density function. Such multimodal distributions can be obtained from the superposition of different pdfs that represent different classes. For example, in the case of atmospheric pollution, the observed quantities represent the concentrations of different pollutants. However, the observed concentrations may result as a superposition of the contributions from different sources, the locations and emission characteristics of which are unknown. In this case, the information about the sources is incorporated in latent (unobserved) variables. If we denote the latent variables by the vector $\mathbf{u}$, the *marginal likelihood* of the observed data is $\mathcal{L}(\boldsymbol{\theta}; \mathbf{x}^*) = \int d\mathbf{u} f_{\mathrm{x}}(\mathbf{x}^*, \mathbf{u} \mid \boldsymbol{\theta})$, where $f_{\mathrm{x}}(\mathbf{x}^*, \mathbf{u} \mid \boldsymbol{\theta})$ is the joint pdf of both the observations and the latent variables given the parameter vector. Maximization of the observed likelihood requires the evaluation of the integral over the latent variables for different values of the parameter vector, which is not easy to compute. Instead, the EM algorithm uses an iterative approach that converges to the optimum solution. This approach involves two steps as follows:

1. *Expectation step:* Let us assume that an estimate, $\boldsymbol{\theta}^{(n)}$, of the parameter vector is available. The E-step involves the calculation of the expectation of the log-likelihood function $\ln \mathcal{L}(\boldsymbol{\theta}; \mathbf{u}, \mathbf{x}^*)$ over the latent variables, i.e.

$$Q(\boldsymbol{\theta} \mid \boldsymbol{\theta}^{(n)}) = \int d\mathbf{u} f(\mathbf{u} \mid \mathbf{x}^*, \boldsymbol{\theta}^{(n)}) \ln \mathcal{L}(\boldsymbol{\theta}; \mathbf{u}, \mathbf{x}^*).$$

This integral involves the conditional pdf of the latent variables given the data and the current estimate of the parameter vector. In case the latent variables are discrete, the integral is replaced by a summation.

2. The M-step maximizes the expected log-likelihood with respect to the parameter vector $\boldsymbol{\theta}$, thus leading to a new estimate $\boldsymbol{\theta}^{(n+1)}$, i.e.,

$$\boldsymbol{\theta}^{(n+1)} = \arg\max_{\boldsymbol{\theta}} Q(\boldsymbol{\theta} \mid \boldsymbol{\theta}^{(n)}).$$

The above sequence is repeated for a number of times until a specified convergence criterion is reached. Most of the computational cost is typically due to the E-step.

Model selection If the functional form of the true spatial model is not a priori known, several different options can be investigated. The candidate models may differ with respect to the number of parameters. This issue makes the selection if the optimal model problematic, since models with a higher number of parameters tend to perform better. On the other hand, is the additional gain in performance justified by the increase in the complexity of the model? In certain cases in which the governing laws are known we can argue in favor of more complicated models based on physical grounds. However, in other cases all the available information is the data. Clearly there is a need for methods that allow selecting an optimal model among several possibilities.

The selection procedure typically employs *penalized likelihood* criteria such as the *Akaike information criterion (AIC)* and *Schwartz's Bayesian information criterion (BIC)* which combine the likelihood with a penalty term that favors parsimonious models (i.e., models with fewer parameters). These criteria favor models with fewer parameters over models with a higher number of parameters, unless the latter significantly improve the likelihood of the data. Hence, the application of model selection criteria conforms with the principle of parsimony embodied in *Occam's razor*.

The method of maximum likelihood can thus be used to derive "optimal" trend and covariance models among different candidates. However, if the covariance model and the trend are known, the variogram can also be obtained. Thus, the maximum likelihood solution also provides as an estimator of the variogram [535, 647, 652].

12.4.1 Basic Steps of MLE

This section briefly reviews the MLE steps as applied in the analysis of spatial data. More details can be found in spatial statistics reference books [165, 774, pp. 169–175]. First, we introduce some notation to facilitate the discussion.

1. Let $\mathcal{L}(\boldsymbol{\theta}; \mathbf{x}^*)$ define the *likelihood* which is the probability of observing the data $\mathbf{x}^* = (x_1^*, \ldots, x_N^*)^\top$ given a specific spatial model characterized by the parameter vector $\boldsymbol{\theta} = (\theta_1, \ldots, \theta_{N_\theta})^\top$.
2. In the following we focus on the residuals after the trend has been removed from the data. This is justifiable if the trend model is determined through a physically inspired regression procedure as described in Chap. 2. It is also possible, however, to perform MLE without removing the trend function [541]. In this case, the parameter vector $\boldsymbol{\theta}$ also includes the trend parameters.
3. The residual fluctuations (after trend removal) are assumed to follow the joint normal pdf. Then, the respective *likelihood* of the model is expressed as

$$\mathcal{L}(\boldsymbol{\theta}; \mathbf{x}^*) = (2\pi)^{-N/2} \det[\mathbf{C}_{\mathrm{xx}}(\boldsymbol{\theta})]^{-1/2} \, \mathrm{e}^{-\frac{1}{2}\mathbf{x}^{*\top} \mathbf{C}_{\mathrm{xx}}^{-1}(\boldsymbol{\theta})\, \mathbf{x}^*}, \tag{12.17}$$

where $\mathbf{C}_{\mathrm{xx}}(\boldsymbol{\theta})$ is the covariance matrix and $\det(\cdot)$ denotes the matrix determinant.

The maximization of the likelihood entails the following steps.

1. Given that the logarithm is a monotonically increasing function, maximizing the likelihood is equivalent to maximizing its logarithm (the log-likelihood). The latter is *concave*, easier to work with, and numerically more stable (fast changes of the likelihood are dampened by taking the logarithm). Maximizing the log-likelihood is equivalent to minimizing the *negative log-likelihood (NLL)*, which is defined by the following convex function

$$\mathrm{NLL}(\boldsymbol{\theta}; \mathbf{x}^*) = -\ln \mathcal{L}(\boldsymbol{\theta}; \mathbf{x}^*). \tag{12.18}$$

In light of the NLL definition (12.18) and the Gaussian likelihood definition (12.17) we obtain the following expression for the Gaussian NLL

$$\mathrm{NLL}(\boldsymbol{\theta}; \mathbf{x}^*) = \frac{1}{2}\mathbf{x}^{*\top} \mathbf{C}_{\mathrm{xx}}^{-1}(\boldsymbol{\theta})\, \mathbf{x}^* + \frac{1}{2} \ln \det[\mathbf{C}_{\mathrm{xx}}(\boldsymbol{\theta})] + \frac{N}{2} \ln(2\pi). \tag{12.19}$$

Hint If $\mathbf{C}_{\mathrm{xx}}$ is an $N \times N$ positive definite matrix, the logarithm of its determinant is given by

$$\ln \det[\mathbf{C}_{\mathrm{xx}}(\boldsymbol{\theta})] = \sum_{n=1}^{N} \ln \lambda_n,$$

where the λ_n are the non-negative eigenvalues of the covariance matrix $\mathbf{C}_{\mathrm{xx}}$.

2. The minimization of NLL involves solving the system of nonlinear equations that determines the NLL *stationary point*, i.e.,

$$\left. \frac{\partial\, \mathrm{NLL}(\boldsymbol{\theta}; \mathbf{x}^*)}{\partial \theta_i} \right|_{\boldsymbol{\theta}=\hat{\boldsymbol{\theta}}} = 0, \quad \text{for} \quad i = 1, \ldots, N_\theta. \tag{12.20}$$

Remark To ensure that the stationary point corresponds to a minimum, we also need to consider if the *NLL Hessian matrix* $\nabla\nabla\, \mathrm{NLL}(\boldsymbol{\theta}; \mathbf{x}^*)$ defined by

$$\left[\nabla\nabla\,\mathrm{NLL}(\boldsymbol{\theta};\mathbf{x}^*)\right]_{i,j} = \frac{\partial^2\,\mathrm{NLL}(\boldsymbol{\theta};\mathbf{x}^*)}{\partial\hat{\theta}_i\,\partial\hat{\theta}_j},\quad i,j = 1,\ldots,N_\theta$$

is a *positive definite* matrix. A matrix $\mathbf{M}$ is positive definite if for every real-valued vector $\mathbf{x} = (x_1,\ldots,x_N)^\top$, the product $\mathbf{x}^\top\mathbf{M}\,\mathbf{x}$ is a non-negative number.

3. In most realistic problems, it is impossible to find the NLL minimum analytically. However, the NLL can be analytically maximised over the covariance scale parameter, thus reducing the number of parameters that need to be numerically optimized by one.

   The *optimal value of the variance* can be analytically obtained by expressing the covariance matrix as $\mathbf{C}_{\mathrm{XX}}(\boldsymbol{\theta}) = \sigma_{\mathrm{X}}^2\,\boldsymbol{\rho}(\boldsymbol{\theta}')$, where $\boldsymbol{\rho}(\boldsymbol{\theta}')$ is the *correlation matrix* and $\boldsymbol{\theta}' = \boldsymbol{\theta}\setminus\{\sigma_{\mathrm{X}}^2\}$ is the *reduced parameter vector* that excludes σ_{X}^2.

   Substituting the above expression for the covariance in the NLL (12.19) we obtain

$$\mathrm{NLL}(\boldsymbol{\theta};\mathbf{x}^*) = \frac{1}{2}\left\{\frac{\mathbf{x}^{*\top}\left[\boldsymbol{\rho}(\boldsymbol{\theta}')\right]^{-1}\mathbf{x}^*}{\sigma_{\mathrm{X}}^2} + N\ln\left(2\pi\,\sigma_{\mathrm{X}}^2\right) + \ln\det\left[\boldsymbol{\rho}(\boldsymbol{\theta}')\right]\right\}. \tag{12.21}$$

   Hint In deriving the above we used $\ln(\det\mathbf{C}_{\mathrm{XX}}) = N\,\ln(\sigma_{\mathrm{X}}^2) + \ln\det\left[\boldsymbol{\rho}(\boldsymbol{\theta}')\right]$.

   The above expression can then be numerically minimized with respect to $\boldsymbol{\theta}$. However, it helps to evaluate, when possible, part of the solution analytically. This can be done for the optimal variance which is expressed as a function of the remaining parameters as shown below. The analytical calculation of the variance reduces the number of parameters by one and thus makes the numerical problem easier to handle.

4. Based on the above expression for NLL, setting the first partial derivative with respect to the variance equal to zero leads to

$$\frac{\partial\,\mathrm{NLL}(\boldsymbol{\theta}';\mathbf{x}^*)}{\partial\sigma_{\mathrm{X}}^2} = -\frac{\mathbf{x}^{*\top}\left[\boldsymbol{\rho}(\boldsymbol{\theta}')\right]^{-1}\mathbf{x}^*}{2(\sigma_{\mathrm{X}}^2)^2} + \frac{N}{2\sigma_{\mathrm{X}}^2} = 0.$$

   Thus, the *optimal variance* is given by

$$\hat{\sigma}_{\mathrm{X}}^2(\boldsymbol{\theta}') = \frac{1}{N}\,\mathbf{x}^{*\top}\left[\boldsymbol{\rho}(\boldsymbol{\theta}')\right]^{-1}\mathbf{x}^*. \tag{12.22}$$

   The above "solution" is a function of the reduced model parameter vector $\boldsymbol{\theta}'$ which is unknown at this stage.

5. If we now replace the optimal MLE variance (12.22) in the NLL expression (12.21) we obtain the *modified NLL*

$$\mathrm{NLL}^*(\boldsymbol{\theta}';\mathbf{x}^*) = c_N + \frac{N}{2}\,\ln\left(\mathbf{x}^{*\top}\left[\boldsymbol{\rho}(\boldsymbol{\theta}')\right]^{-1}\mathbf{x}^*\right) + \frac{1}{2}\,\ln\det\left[\boldsymbol{\rho}(\boldsymbol{\theta}')\right], \tag{12.23}$$

where the coefficient $c_N = \frac{N}{2}[1 + \ln(2\pi) - \ln N]$ is independent of $\boldsymbol{\theta}'$ and does not affect the minimization with respect to $\boldsymbol{\theta}'$.

6. The minimization of NLL* with respect to $\boldsymbol{\theta}'$ is equivalent to determining the stationary point that solves the system of equations

$$\left. \frac{\partial \mathrm{NLL}^*(\boldsymbol{\theta}'; \mathbf{x}^*)}{\partial \theta_i'} \right|_{\boldsymbol{\theta}'=\hat{\boldsymbol{\theta}}'} = 0, \quad i = 1, \ldots, N_\theta - 1,$$

7. The optimal solution $\hat{\boldsymbol{\theta}}'$ is then inserted in (12.22) to determine the optimal variance $\hat{\sigma}_{\mathrm{x}}^2(\hat{\boldsymbol{\theta}}')$. If the minimization is successful, the vector $\left(\hat{\sigma}_{\mathrm{x}}^2(\hat{\boldsymbol{\theta}}'), \hat{\boldsymbol{\theta}}'\right)$ accurately estimates the *true stationary point* $\boldsymbol{\theta}^*$.

In the presence of a trend model, $\mathbf{x}^*$ in the likelihood equations is replaced by $\mathbf{x}^* - \mathbf{f}\boldsymbol{\beta}$, where $\mathbf{f}$ is an $N \times L$ matrix of L basis functions and $\boldsymbol{\beta}$ is a respective vector of L coefficients. In addition, $\mathbf{x}^*$ in (12.22) is replaced by $\mathbf{x}^* - \mathbf{f}\hat{\boldsymbol{\beta}}$, where $\hat{\boldsymbol{\beta}}$ is determined from

$$\hat{\boldsymbol{\beta}} = \left(\mathbf{f}^\top \rho^{-1}(\boldsymbol{\theta}')\,\mathbf{f}\right)^{-1} \mathbf{f}^\top \rho^{-1}(\boldsymbol{\theta}')\,\mathbf{x}^*.$$

The rate-determining calculation in likelihood estimation is the computation of the *precision matrix* $\mathbf{C}_{\mathrm{xx}}^{-1}$. If the covariance matrix is *dense*, the calculation of the inverse covariance (precision matrix) is an $O(N^3)$ operation. The other steps in the NLL equation (12.19) scale as $O(N^2)$ or faster.

The minimization of the NLL* is usually conducted numerically. If a *gradient-based optimization* algorithm, such as the *steepest descent* or *conjugate gradient method* [561, 673] is used, the *Hessian matrix of the likelihood* (with respect to the parameters) should also be provided in addition to the gradient. The Hessian captures the curvature of the likelihood surface (more specifically, the singular values of the Hessian are inversely proportional to the squares of the likelihood's local curvatures). Hence, gradient-based methods can use the Hessian to automatically adjust the step size, taking larger steps in small-curvature (flat) directions and smaller steps in steep directions with large curvature [561]. Note, however, that for some commonly used covariance functions, such as the spherical model (see Table 4.2), the second derivative of the likelihood with respect to the range parameter does not exist [542].

Log-likelihood gradient The covariance matrix $\mathbf{C}_{\mathrm{xx}}$ is the only factor in the log-likelihood (12.19) that depends on $\boldsymbol{\theta}$. In the modified negative log-likelihood $\mathrm{NLL}^*(\boldsymbol{\theta}'; \mathbf{x}^*)$ (12.23), the respective factor is the correlation matrix $\boldsymbol{\rho}(\boldsymbol{\theta}')$. To calculate the gradient of the log-likelihood, the partial derivatives of the precision matrix and the covariance matrix determinant with respect to $\{\theta_i\}_{i=1}^{N_\theta}$ are needed. These partial derivatives are given by the following equations [678, App. A]

$$\partial_{\theta_i} \mathbf{C}_{\mathrm{xx}}^{-1} = -\mathbf{C}_{\mathrm{xx}}^{-1} \left(\partial_{\theta_i} \mathbf{C}_{\mathrm{xx}}\right) \mathbf{C}_{\mathrm{xx}}^{-1} \tag{12.24a}$$

$$\partial_{\theta_i} \left(\det \mathbf{C}_{\mathrm{xx}}\right) = \left(\det \mathbf{C}_{\mathrm{xx}}\right) \mathrm{Tr}(\mathbf{C}_{\mathrm{xx}}^{-1}\, \partial_{\theta_i} \mathbf{C}_{\mathrm{xx}}). \tag{12.24b}$$

Hint To prove the first equation take the derivative on both sides of the equation $\mathbf{C}_{\mathrm{xx}}\,\mathbf{C}_{\mathrm{xx}}^{-1} = \mathbf{I}$. To show the second equation, express the determinant as $\det \mathbf{C}_{\mathrm{xx}} = \prod_{n=1}^{N} \lambda_n$, where λ_n are the eigenvalues of the covariance matrix.

Based on the above partial derivatives, the components of the log-likelihood gradient are given by [678]

$$\partial_{\theta_i} \ln \mathcal{L}(\boldsymbol{\theta}; \mathbf{x}^*) = \frac{1}{2} \mathbf{x}^{*\top}\, \mathbf{C}_{\mathrm{xx}}^{-1} \left(\partial_{\theta_i} \mathbf{C}_{\mathrm{xx}}\right) \mathbf{C}_{\mathrm{xx}}^{-1}\, \mathbf{x}^* - \frac{1}{2} \operatorname{Tr}\left(\mathbf{C}_{\mathrm{xx}}^{-1}\, \partial_{\theta_i} \mathbf{C}_{\mathrm{xx}}\right), \; i = 1, \ldots, N_\theta. \tag{12.25}$$

If the variance is analytically estimated, the log-likelihood gradient takes the form [589]

$$\nabla_{\boldsymbol{\theta}'} \ln \mathcal{L}(\boldsymbol{\theta}; \mathbf{x}^*) = \frac{1}{2\hat{\sigma}_{\mathrm{x}}^2} \mathbf{x}^{*\top}\, \boldsymbol{\rho}^{-1} \left(\nabla_{\boldsymbol{\theta}'} \boldsymbol{\rho}\right) \boldsymbol{\rho}^{-1}\, \mathbf{x}^* - \frac{1}{2} \operatorname{Tr}\left(\boldsymbol{\rho}^{-1}\, \nabla_{\boldsymbol{\theta}'} \boldsymbol{\rho}\right), \tag{12.26}$$

where $\hat{\sigma}_{\mathrm{x}}^2(\boldsymbol{\theta}')$ is the variance evaluated by means of (12.22).

Computational hint It is not recommended to evaluate the matrix–matrix products in (12.25) directly, because matrix multiplication is an $O(N^3)$ operation. It is more efficient to evaluate the first term using matrix–vector products: instead of performing three matrix multiplications to calculate $\mathbf{C}_{\mathrm{xx}}^{-1} \left(\partial_{\theta_i} \mathbf{C}_{\mathrm{xx}}\right) \mathbf{C}_{\mathrm{xx}}^{-1}$, first multiply each of the $\mathbf{C}_{\mathrm{xx}}^{-1}$ with the respective data vector; thus, only matrix-vector multiplications are used. In the second term, only the diagonal elements that contribute to the calculation of the trace need to be evaluated. Both of these operations (i.e., the matrix-vector multiplications and the trace) scale as $O(N^2)$. Thus, once the precision matrix is calculated, the log-likelihood gradient in (12.25) can be evaluated in $O(N^2)$. Hence, for large samples the additional computational cost of the log-likelihood gradient is relatively small compared to the cost of calculating the log-likelihood [589, 678].

Hessian of negative log-likelihood Differentiating the gradient in (12.25) to evaluate the second-order partial derivatives at the stationary point $\boldsymbol{\theta} = \boldsymbol{\theta}^*$ (maximum of the likelihood), gives the NLL Hessian. The *NLL Hessian* $\mathbf{H}_{\mathcal{L}}$ is defined as follows

$$\left[\mathbf{H}_{\mathcal{L}}(\boldsymbol{\theta}^*; \mathbf{x}^*)\right]_{i,j} = -\, \partial_{\theta_i} \partial_{\theta_j} \ln \mathcal{L}(\boldsymbol{\theta}; \mathbf{x}^*)\big|_{\boldsymbol{\theta}=\boldsymbol{\theta}^*}, \; i, j = 1, \ldots, N_\theta. \tag{12.27}$$

In light of the expression (12.25) for the log-likelihood gradient, the NLL Hessian is given by

$$\begin{aligned}\left[\mathbf{H}_{\mathcal{L}}(\boldsymbol{\theta}^*; \mathbf{x}^*)\right]_{i,j} =& \frac{1}{2} \mathbf{x}^{*\top} \left[2\mathbf{C}_{\mathrm{xx}}^{-1}\, \partial_{\theta_i} \mathbf{C}_{\mathrm{xx}}\, \mathbf{C}_{\mathrm{xx}}^{-1} \partial_{\theta_j} \mathbf{C}_{\mathrm{xx}}\, \mathbf{C}_{\mathrm{xx}}^{-1} - \mathbf{C}_{\mathrm{xx}}^{-1}\, \partial_{\theta_i} \partial_{\theta_j} \mathbf{C}_{\mathrm{xx}}\, \mathbf{C}_{\mathrm{xx}}^{-1}\right] \mathbf{x}^* \\ & - \frac{1}{2} \operatorname{Tr}\left(\mathbf{C}_{\mathrm{xx}}^{-1}\, \partial_{\theta_i} \mathbf{C}_{\mathrm{xx}}\, \mathbf{C}_{\mathrm{xx}}^{-1}\, \partial_{\theta_j} \mathbf{C}_{\mathrm{xx}} - \mathbf{C}_{\mathrm{xx}}^{-1}\, \partial_{\theta_i} \partial_{\theta_j} \mathbf{C}_{\mathrm{xx}}\right).\end{aligned} \tag{12.28}$$

For the NLL Hessian in the case of an analytically calculated variance consult [589]. After the precision matrix is computed, the Hessian can be evaluated at relatively small cost in the same manner as the log-likelihood gradient.

12.4.2 Fisher Information Matrix and Parameter Uncertainty

A matter of practical interest is to characterize the amount of information (respectively, the uncertainty) that the measurements $\mathbf{x}^*$ carry about the parameter vector $\boldsymbol{\theta}$. This task can be accomplished by means of the second-order derivatives of the likelihood function which define the Fisher information matrix [251]. The logarithm of the determinant of the Fisher information matrix is also used as a sampling design criterion for the estimation of the covariance function parameters, e.g. [883].

The likelihood can be expanded around the stationary point $\boldsymbol{\theta}^*$ by means of the *Gaussian approximation* [752]

$$\ln \mathcal{L}(\boldsymbol{\theta};\mathbf{x}^*) \approx \ln \mathcal{L}(\boldsymbol{\theta}^*;\mathbf{x}^*) - \frac{1}{2}\delta\boldsymbol{\theta}^\top \mathbf{H}_{\mathcal{L}}(\boldsymbol{\theta}^*;\mathbf{x}^*)\,\delta\boldsymbol{\theta}, \tag{12.29}$$

where $\delta\boldsymbol{\theta} = \boldsymbol{\theta} - \boldsymbol{\theta}^*$. This expansion ignores terms of $O\left((\delta\boldsymbol{\theta})^3\right)$. Note that the first-order term involving the gradient of the likelihood is absent because the expansion takes place around the stationary point.

Thus, the NLL Hessian can provide error estimates for the parameter vector $\boldsymbol{\theta}^*$, since the inverse of the Hessian is the *covariance matrix* of the maximum likelihood estimator of the parameters. More specifically, based on (12.29) it follows that

$$\mathcal{L}(\boldsymbol{\theta};\mathbf{x}^*) \approx \mathcal{L}(\boldsymbol{\theta}^*;\mathbf{x}^*)\,\mathrm{e}^{-\frac{1}{2}\delta\boldsymbol{\theta}^\top \mathbf{H}_{\mathcal{L}}\,\delta\boldsymbol{\theta}}. \tag{12.30}$$

The NLL Hessian, defined by (12.27), is a function of the data values. The *Fisher information matrix* is defined as the expectation[4] of the NLL Hessian, i.e.,

$$F_{i,j}(\boldsymbol{\theta}^*) = \mathcal{E}\left[\mathbf{H}_{\mathcal{L}}(\boldsymbol{\theta}^*;\mathbf{x}^*)\right]_{i,j} = -\mathcal{E}\left[\partial_{\theta_i}\partial_{\theta_j} \ln \mathcal{L}(\boldsymbol{\theta};\mathbf{x}^*)\big|_{\boldsymbol{\theta}=\boldsymbol{\theta}^*}\right]. \tag{12.31}$$

Hence, the Fisher information matrix removes the explicit dependence on the specific data set (field realization). On the other hand, since $\boldsymbol{\theta}^*$ is determined by optimizing the NLL given the specific data, there is an implicit dependence on the data through $\boldsymbol{\theta}^*$. Furthermore, assuming that there is no dependence of the data vector $\mathbf{x}^*$ on the model parameters, it can be shown that the Fisher matrix is given by the equation [455, 541]

$$F_{i,j}(\boldsymbol{\theta}^*) = \frac{1}{2}\,\mathrm{Tr}\left[\mathbf{C}_{\mathrm{xx}}^{-1}\,\frac{\partial \mathbf{C}_{\mathrm{xx}}}{\partial\theta_i}\,\mathbf{C}_{\mathrm{xx}}^{-1}\,\frac{\partial \mathbf{C}_{\mathrm{xx}}}{\partial\theta_j}\right]\Bigg|_{\boldsymbol{\theta}=\boldsymbol{\theta}^*}. \tag{12.32}$$

Remark If the spatial model includes a parametric trend $\mathbf{m}_{\mathrm{x}}$, the Fisher information matrix contains the additional term $\partial_{\theta_i}\mathbf{m}_{\mathrm{x}}^\top\,\mathbf{C}_{\mathrm{xx}}^{-1}\,\partial_{\theta_j}\mathbf{m}_{\mathrm{x}}$. This term will be nonzero only for the components θ_i and θ_j that enter $\mathbf{m}_{\mathrm{x}}$.

---

[4]The expectation is evaluated over the ensemble of states of the random field $X(\mathbf{s};\omega)$.

In light of the Gaussian likelihood approximation (12.30), the Fisher information matrix is an approximation of the parameter precision matrix. The *parameter covariance matrix* is then given by the inverse of the Fisher matrix

$$\mathbf{C}_{\theta} = \mathbf{F}^{-1}.$$

The Fisher matrix characterizes the Gaussian uncertainty of inter-dependent parameters. It is expected that MLE is an asymptotically normal estimator such that (i) the asymptotic mean of the ML estimates coincides with the true values of the parameters while (ii) the asymptotic covariance matrix of the parameters is given by the inverse of the Fisher information matrix [773].

The inverse Fisher information matrix might contain elements with very different scales. In this case, large diagonal elements correspond to parameters that are not microergodic while small diagonal elements correspond to microergodic parameters that can be consistently estimated [883].

Sloppiness is a term coined by the physicist James Sethna whose group has been investigating the estimation of multi-parameter complex systems. Sloppy models may involve many parameters, but their behavior mostly depends on a few stiff parameter combinations. A parameter (or parameter combination) is called stiff if a small change in its value has a large effect on the system. In sloppy models the majority of the parameters are sloppy, i.e., their values can change considerably without significantly affecting the response of the system. Hence, sloppy parameters are practically unimportant for model predictions which means that sloppy models are amenable to *dimensionality reduction*. Many such models have been identified in systems biology, physics and mathematics, e.g. [519, 799, 800]. It turns out that sloppiness can be quantified by the spectrum of eigenvalues of the Fisher information matrix. In particular, in the case of sloppy models the eigenvalues of the Fisher information matrix follow the exponential distribution.

The concept of sloppiness is linked to the concept of micro-ergodicity in spatial statistics: Micro-ergodic parameters (or combinations of parameters) can be consistently estimated from the data for a given spatial sampling configuration [774, 878, 883]. Hence, the concept of stiff parameters in sloppy models seems related to the concept of microergodic parameters. On the other hand, the dimension of the parameter space of spatial statistics models is typically much smaller than that of sloppy models describing complex systems.

12.4.3 MLE Properties

The popularity of the MLE is due to its nice statistical properties that include *consistency, asymptotic unbiasedness, and asymptotic normality*. The term "asymptotic" here refers to a large sample size N, which in theory should tend to infinity. In addition, ML is an *efficient estimator* (i.e., it achieves the minimum possible variance) among all estimators with these asymptotic properties.

A mathematical treatment of asymptotic MLE properties for spatial data is given in [541]. Information regarding the statistical properties of ML estimates and the asymptotic behavior of MLE is given in [165, Chap. 2 and Chap. 7] and [885].

Small samples For small N, however, ML estimates are not necessarily either unbiased or efficient [81, 789]. On the other hand, since there is no general method for finding an unbiased, efficient estimator, MLE is often the method of choice. The application of the method in the analysis of spatial data has been investigated in several studies [462, 572, 652].

Computational complexity The main drawback of MLE is its computational complexity (see also the discussion in Sects. 11.7 and 12.4.1). If the covariance matrix of the data is *dense*, the memory storage requirements scale as $O(N^2)$, where N is the sample size.

For every set of parameters visited by the iterative likelihood optimization algorithm, the covariance matrix is calculated and inverted. The computational time required for the inversion of *dense covariance matrices* scales as $O(N^3)$. Hence, good initial estimates of the parameters can accelerate the convergence of MLE by minimizing the number of the required matrix inversions.

Another potential difficulty is the purported multimodality of the likelihood surface [542, 832]. However, the existence of likelihood multimodality is a controversial issue [774, p. 173]. The interested reader can find more information in the references given in [540].

The "big data" problem For large sample sizes ($N \gg 1$), approximation methods help to reduce the computational burden of maximum likelihood estimation [264, 777]. Improved scaling with N can be achieved by means of *local approximations, dimensionality reduction techniques, and parallel algorithms*. Available methods for large spatial data problems have been recently reviewed in [784].

Local approximations focus on computing the likelihood over sub-domains of smaller area than the entire sample domain $\mathcal{D}$. These approximations involve ideas such as *composite likelihood* [812] and *pseudo-likelihood* [70, 682, 682, 807]. The pseudo-likelihood is the product of the conditional densities of single observations given the values of their neighbours. The pseudo-likelihood concept has evolved in the idea of composite likelihood methods in which the likelihood is approximated by means of the product of smaller component likelihoods.

Covariance tapering neglects correlations outside an empirically specified range [213, 267, 441]. The main idea is to taper the covariance function to zero beyond a certain range. One way that this can be achieved is by multiplying a permissible covariance model with a positive definite, compactly supported function. The product is then guaranteed to be a permissible covariance function. Tapering functions include the spherical covariance and compact-support Wendland functions [841]. A recent proposal adaptively modifies the tapering range according to the local sampling density [88].

Dimensionality reduction methods lower the dimensionality of the problem by constructing low-dimensional approximations. For example, in *fixed-rank kriging* the precision matrix is modeled as a fixed rank $r \ll N$ matrix [167, 617].

Covariance factorization: A different approach that aims to address the large data problem involves the hierarchical factorization of the covariance matrix into a product of block low-rank updates of the identity matrix. The factorization of the covariance matrix allows algorithms with computational complexity $O(N \log^2 N)$ for covariance inversion and $O(N \log N)$ for covariance determinant calculation [19, 705].

12.5 Cross Validation

Cross validation is a statistical approach that can be used to estimate model parameters, to evaluate the performance of specific spatial models, and to perform *model selection*. An introduction to cross validation and its connections with the jackknife and bootstrap statistical approaches is given in [224]. Herein we focus on the application of cross validation as a tool for evaluating model performance.

Training and validation sets To apply cross validation, it is necessary to split the data into two sets: a "*training set*" that involves the data $\mathbf{x}^*_{\text{tr}}$ at the locations $\mathbb{S}_{\text{tr}} \subset \mathbb{S}_N$ and a "*validation set*" that involves the data $\mathbf{x}^*_{\text{pr}}$ at the locations $\mathbb{S}_{\text{pr}} = \mathbb{S}_N \setminus \mathbb{S}_{\text{tr}}$.[5] The training data are used for the estimation of the model parameters, while the validation data are used to compare with the model predictions.

> The main idea of *model performance evaluation* by means of cross validation is that the "quality" of a given spatial model determined from the training data set (e.g., with parameters estimated by means of maximum likelihood or method of moments) can be evaluated by comparing its predictive performance with the process values in the validation data set.

Various statistical measures of predictive performance can be used in cross validation; a non-exhaustive list is given below in Sect. 12.5.2. In classification problems, cross validation is shown to be comparable with the bootstrap and the Akaike selection criterion in terms of model selection performance [330].

[5] In the machine learning literature the term "test set" is used for the data that are used to check the performance of a model, while the term "validation set" applies to data which are used to make some choices in the learning process.

12.5.1 Splitting the Data

Three different methodologies can be used to split the data into training and validation data sets for the purpose of cross validation: k-fold cross validation, leave-P-out cross validation, and leave-one-out cross validation [330]. Regardless of the approach used to split the data into training and validation sets, the essential steps of cross validation are the same: use the training set to determine the spatial model and the validation set to test the performance of the model against the observations.

K-fold cross validation In K-fold cross validation, the sample locations are split into K disjoint subsets $\{\mathbb{S}_k\}_{k=1}^{K}$ such that $\mathbb{S}_k \subset \mathbb{S}_\mathrm{N}$ for $k \in \{1, \ldots, K\}$, and $\mathbb{S}_k \cap \mathbb{S}_l = \emptyset$ for $l \in \{1, \ldots, K\}_{-k}$. We choose the *training set* $\mathbb{S}_\mathrm{tr}$ as the union of any $K-1$ subsets, i.e.,

$$\mathbb{S}_\mathrm{tr} = \begin{cases} \mathbb{S}_2 \cap \mathbb{S}_3 \ldots \cap \mathbb{S}_K, & l = 1, \\ \mathbb{S}_1 \cap \ldots \mathbb{S}_{l-1} \cap \mathbb{S}_{l+1} \ldots \cap \mathbb{S}_K, & l \in \{2, \ldots, K-1\}, \\ \mathbb{S}_1 \cap \mathbb{S}_2 \cap \ldots \cap \mathbb{S}_{K-1}, & l = K. \end{cases}$$

The remaining subset $\mathbb{S}_l$, where $l = 1, \ldots, K$ then becomes the validation set. This splitting procedure is repeated K times, using every time a different subset as the validation set. The validation measure (or measures) is obtained as an average over the K different configurations of the validation set. Typical values for the number of subsets used in K-fold cross validation are $K = 4$ and $K = 10$.

Leave-P-out cross validation In leave-P-out cross validation, we partition the data set into a training set which contains $N - P$ sample points, chosen at random, and a validation set consisting of the remaining P sample points. The spatial model is constructed based on the $N - P$ points of the training set, and the predictions are compared with the observations at the P points of the validation set.

This process should be repeated as many times as the possible partitions of the set of N points into two sets with P and $N-P$ points. However, the number of such combinations is $N!/\big(P!\,(N-P)!\big)$, which can be very large even for moderate N and P. Hence, this approach can be computationally expensive.

The *poor man's approach:* To avoid the computational cost of conducting cross validation over all possible partitions, in practical studies one may apply leave-P-out cross validation with a single training and a single prediction set of dimensions $N - P$ and P respectively [215].

Leave-one-out cross validation Leave-one-out cross validation is a specific case of K-fold cross validation when $K = N$. In this case, the training set at every iteration contains $N - 1$ points, while the prediction set contains a single point. The procedure is repeated N times, and the cross-validation measure is calculated as an average over the N partitions.

12.5.2 Cross Validation Measures

The following statistical measures are typically used to evaluate the performance of the predictions in a cross validation study. To simplify the presentation, we assume that leave-one-out cross validation is performed. It is straightforward to modify the definitions of the measures for the other cross-validation schemes. It is also assumed that the sample values x_n^*, where $n = 1, \ldots, N$ are viewed as the "true" values of the measured process. Hence, the potential presence of measurements errors is not taken into account.

The values $\{\hat{x}_n\}_{n=1}^N$ below represent model estimates obtained from the N data vectors $\mathbf{x}_{-n}^*$, that exclude x_n^* from the full set $\mathbf{x}^*$ for $n = 1, \ldots, N$.

- *Mean error (bias)*

$$\text{ME} = \frac{1}{N} \sum_{n=1}^{N} \left(\hat{x}_n - x_n^* \right). \tag{12.33}$$

  Small bias is a necessary but not sufficient condition for a good estimator. The condition is not sufficient since large positive errors may be compensated by large negative errors still leading to small bias.
- *Mean absolute error (MAE)*

$$\text{MAE} = \frac{1}{N} \sum_{n=1}^{N} \left| x_n^* - \hat{x}_n \right|. \tag{12.34}$$

  The MAE quantifies the magnitude (absolute value) of the deviations between the estimates and the true values. Hence, unlike the bias, it is not subject to the compensation of negative and positive deviations. Ideally, the estimator should have both small bias and small MAE.
- *Root mean square error (RMSE)*

$$\text{RMSE} = \sqrt{\frac{1}{N} \sum_{n=1}^{N} \left(x_n^* - \hat{x}_n \right)^2}. \tag{12.35}$$

  The root mean square error also measures the magnitude of deviations between the estimates and the true values. However, it weighs large deviations more heavily than the MAE, due to the squaring of the errors.
- *Mean absolute relative error (MARE)*

$$\text{MARE} = \frac{1}{N} \sum_{n=1}^{N} \left| \frac{x_n^* - \hat{x}_n}{x_n^*} \right|. \tag{12.36}$$

The MARE is a normalized error measure that expresses the difference between the estimates and the true values in terms of a dimensionless number. If the MARE is multiplied by 100 it yields the percent absolute error of the estimator, which is easier to interpret than the MAE.

However, if the observations include values equal to or close to zero, the MARE may not be defined or takes large values. In such cases, the MARE is not a reliable measure of performance.

- *Root mean square relative error (RMSRE)*

$$\text{RMSRE} = \sqrt{\frac{1}{N}\left(\frac{x_n^* - \hat{x}_n}{x_n^*}\right)^2}. \tag{12.37}$$

The RMSRE combines the advantages of the MARE with higher penalties for large deviations. It is equally unreliable if the measured process takes values close to zero.

- *Pearson correlation coefficient* (ρ)

$$\rho = \frac{\sum_{n=1}^{N}\left(x_n^* - \overline{x_n^*}\right)\left(\hat{x}_n - \overline{\hat{x}_n}\right)}{\sqrt{\sum_{n=1}^{N}\left(x_n^* - \overline{x_n^*}\right)^2}\sqrt{\sum_{n=1}^{N}\left(\hat{x}_n - \overline{\hat{x}_n}\right)^2}}. \tag{12.38}$$

The ρ value measures the statistical correlation between the data and the estimates. As it can be shown based on the Cauchy-Schwartz inequality (3.51), it holds that $-1 \leq \rho \leq 1$.[6]

The Pearson ρ is sensitive to linear correlations between the data and the estimates, but it is not reliable in the case of nonlinear relations. In addition, the Pearson ρ cannot distinguish between differences in magnitude, because the random variables $\hat{\mathrm{X}}(\omega)$ and $\mathrm{X}^*(\omega)$ have the same correlation coefficient as $\mathrm{X}^*(\omega)$ and $a\,\hat{\mathrm{X}}(\omega)$, where a is a positive constant.

- *Spearman rank correlation coefficient* (r_S)[7]

$$r_S = 1 - \frac{6\sum_{n=1}^{N}\left[R(x_n^*) - R(\hat{x}_n)\right]^2}{N\,(N^2 - 1)}, \tag{12.39}$$

where $R(x_n)$ is the *rank (order)* of the value x_n, $n = 1, \ldots, N$, and $R(\hat{x}_n)$ the rank of the respective estimate.

The highest value of each set (measurements or estimates) is assigned a rank equal to one, with smaller values assigned correspondingly increasing ranks. If

[6] In the current use of the inequality the expectation operator $\mathcal{E}[\cdot]$ is replaced by the averaging operator $\mathcal{A} = \frac{1}{N}\sum_{n=1}^{N}(\cdot)$.

[7] This formula can only be used for integer ranks (i.e., no tied ranks). Otherwise, the formula for the linear correlation of the ranks is used.

one value appears multiple times in the set, then all appearances are assigned the same rank, which is equal to the average of the ranks of the individual appearances.

The Spearman rank correlation also takes values between -1 and 1 as the Pearson correlation coefficient. However, the Spearman rank correlation is a non-parametric measure, because it quantifies if the two variables are monotonically related but not if their relation assumes a specific (e.g., linear) form.

The above cross validation measures are based on univariate statistics. It is possible to examine validation measures that are based on higher-order statistics that also quantify the reproduction of spatial dependence, such as the *variogram function*. In addition, it is possible to formulate validation measures that focus on *spatial extremes*, that is on the ability of the estimator to reproduce both the locations and the magnitudes of extreme values. Furthermore, it is possible to define various performance measures based on *information theory* [278, 840].

12.5.3 Parameter Estimation via Cross-validation

It is possible to formulate a *cross validation cost function* $\Phi(\mathbf{x}^*; \boldsymbol{\theta})$ that measures the deviation between the validation set values and the respective model predictions based on the training set values. The cross validation cost function is typically based on some indicator of "predictive quality" such as the measures described above (e.g., correlation coefficient, root mean square error or mean absolute error).

A different approach relies on the *standardized residuals* that were defined in (10.63) as follows:

$$\breve{\epsilon}(\mathbf{z}_p) = \frac{\hat{x}(\mathbf{z}_p) - x^*(\mathbf{z}_p)}{\sigma_{\text{ok}}(\mathbf{z}_p)}, \quad p = 1, \ldots, P,$$

where $\sigma_{\text{ok}}(\mathbf{z}_p)$ is the OK standard deviation given by (10.46), $\{\mathbf{z}_p\}_{p=1}^{P}$ the points of the cross validation set, $x^*(\mathbf{z}_p)$ the data value at point $\mathbf{z}_p$ and $\hat{x}(\mathbf{z}_p)$ the cross-validation estimate at this point.

Then, the following cross validation cost function can be defined

$$\Phi(\mathbf{x}^*; \boldsymbol{\theta}) = \left(\frac{1}{P} \sum_{p=1}^{P} \breve{\epsilon}^2(\mathbf{z}_p) - 1 \right)^2 . \tag{12.40}$$

The above cost function can be further refined by using the orthonormal residuals instead of the standardized residuals [457].

A suitable cost function $\Phi(\mathbf{x}^*; \boldsymbol{\theta})$ can also be used to determine the *optimal parameter vector* via optimization

$$\hat{\boldsymbol{\theta}}^* = \arg\min_{\boldsymbol{\theta}} \Phi(\mathbf{x}^*; \boldsymbol{\theta}),$$

for a specific spatial model. The "optimal parameter vector" will in general depend on the specific form of the cost function.

The cross validation approach for parameter estimation represents an alternative option to MLE. Cross validation is computationally more efficient than MLE for spatial models that are based on local interactions, and thus allow the formulation of computationally fast linear predictors based on the precision operator. In contrast with MLE, cross validation avoids the calculation of the covariance matrix determinant [368]. However, the MLE computational efficiency could be improved by using sparse matrix methods to calculate the determinant.

The performance of cross validation for variogram parameter inference was empirically tested against other estimation methods for small data sets in [650]. Cross-validation has also been used as a means of determining relative regression and geostatistical components in a combined spatial model for infilling incomplete records of precipitation data [413]. A theoretical analysis of the performance of cross validation for *parameter selection* is reviewed in the recent survey [31], where an attempt is made to distinguish empirical statements from rigorous results. More information regarding the application of cross validation methods to spatial data is given in [39, 190, 678, 879].

Chapter 13
More on Estimation

> In desperation I asked Fermi whether he was not impressed by the agreement between our calculated numbers and his measured numbers.
> He replied, "How many arbitrary parameters did you use for your calculations?"
> I thought for a moment about our cut-off procedures and said, "Four."
> He said, "I remember my friend Johnny von Neumann used to say, with four parameters I can fit an elephant, and with five I can make him wiggle his trunk."
>
> *Freeman Dyson*

This chapter discusses estimation methods that are less established and not as commonly used as those presented in the preceding chapter. For example, the *method of normalized correlations* is relatively new, and its statistical properties have not been fully explored. The method of *maximum entropy* was used by Edwin Thompson Jaynes to derive statistical mechanics based on information theory [406, 407]. Following the work of Jaynes, maximum entropy has found several applications in physics [674, 753], image processing [315, 754], and machine learning [521, 561].

We first present the method of *normalized correlations* and discuss its connections with the Yule-Walker method used in time series analysis. The method of normalized correlations is attractive because it focuses on local correlation properties. This feature renders it computationally efficient and potentially suitable for big and densely sampled data sets.

D. T. Hristopulos, *Random Fields for Spatial Data Modeling*, Advances in Geographic Information Science, https://doi.org/10.1007/978-94-024-1918-4_13

Then, we review the main ideas of the *maximum entropy principle* and demonstrate that the *Spartan random field model* can be derived from the maximum entropy principle. We also discuss the topic of parameter estimation for *stochastic local interaction* models.

The section that follows focuses on the stochastic local interaction (SLI) model. We discuss the definition, properties, parameter estimation, and spatial prediction equations for the SLI model. Due to an explicit precision matrix expression, the SLI model can be used for computationally efficient interpolation of spatially scattered data. In addition, the construction of the precision matrix in terms of kernel functions ensures that the model is valid (positive-definite) even on manifolds with non-Euclidean distances.

Finally, this chapter closes with the presentation of an *empirical ergodic index* that can be used to assess the suitability of the ergodic hypothesis for specific random field realizations on bounded (finite-size) domains.

13.1 Method of Normalized Correlations (MoNC)

The idea motivating the *method of normalized correlations* (MoNC) is that short-range (local) spatial constraints are often more important for spatial interpolation and can be more accurately estimated than long-range constraints. The local constraints can be matched with corresponding stochastic counterparts [228, 373, 897]. The latter are evaluated in terms of the assumed joint pdf of the model random field. For Gaussian random fields, the stochastic constraints are expressed in terms of the first two moments (the mean and the covariance function). Thus, they depend on the unknown parameters of the random field. MoNC is based on the premise that the stochastic and sample constraints can be matched, similarly with moment-based and maximum entropy methods. The constraint matching leads to a nonlinear optimization problem, the solution of which provides estimates of the model parameters.

In order to equate the sample-based constraints with their stochastic counterparts, certain conditions should be satisfied. The conditions include that (i) the model SRF is an adequate representation of the spatial process, and that (ii) the sample-based constraints are accurate and precise estimates of the stochastic constraints. The condition (i) above is a premise for all spatial models and thus not specific to MoNC. Condition (ii) presumes that second-order *ergodic conditions* hold. Based on the above description, MoNC is essentially a method of moments. However, it focuses only on short-range correlations instead of the entire covariance (or variogram) function.

13.1.1 MoNC Constraints for Data on Regular Grids

This section defines "local constraints" that can be used in MoNC applications to the estimation of *stationary random fields* $\mathrm{X}(\mathbf{s};\omega)$ sampled on regular grids. It is assumed that the SRF mean is equal to m_{x} and the SRF covariance function is given by $C_{\mathrm{xx}}(\mathbf{r})$.

Let $\hat{\mathbf{e}}_i$ denote the unit vector in the ith direction on a *hypercubic grid* $\mathcal{G}_d \in \mathbb{R}^d$ with uniform step a. The *central second-order difference* of the field $\mathrm{X}(\mathbf{s};\omega)$ along $\hat{\mathbf{e}}_i$ is defined as follows according to (8.31b):

$$\delta_i^2[\mathrm{X}(\mathbf{s}_n;\omega)] = \frac{1}{a^2}\Big[\mathrm{X}(\mathbf{s}_n + a\hat{\mathbf{e}}_i;\omega) + \mathrm{X}(\mathbf{s}_n - a\hat{\mathbf{e}}_i;\omega) - 2\mathrm{X}(\mathbf{s}_n;\omega)\Big].$$

The following local functions of the field are featured in the MoNC constraints

$$S_0(\mathbf{s}_n;\omega) = [\mathrm{X}(\mathbf{s}_n;\omega) - m_{\mathrm{x}}]^2\,, \tag{13.1a}$$

$$S_1(\mathbf{s}_n;\omega) = \frac{1}{a^2}\sum_{i=1}^{d}\Big[\mathrm{X}(\mathbf{s}_n + a\hat{\mathbf{e}}_i;\omega) - \mathrm{X}(\mathbf{s}_n;\omega)\Big]^2, \tag{13.1b}$$

$$S_2(\mathbf{s}_n;\omega) = \sum_{i,j=1}^{d} \delta_i^2[\mathrm{X}(\mathbf{s}_n;\omega)]\,\delta_j^2[\mathrm{X}(\mathbf{s}_n;\omega)]. \tag{13.1c}$$

The equations (13.1) define respectively the squared fluctuation, the discrete squared gradient, and the discrete squared Laplacian (curvature) random fields.

Model constraints The ensemble moments that represent the model constraints $\mathcal{E}[S_0(\mathbf{s}_n;\omega)]$, $\mathcal{E}[S_1(\mathbf{s}_n;\omega)]$, and $\mathcal{E}[S_2(\mathbf{s}_n;\omega)]$ are expressed in terms of the correlation function $C_{\mathrm{xx}}(\mathbf{r})$ as follows [897]:

$$\mathcal{E}[S_0(\mathbf{s}_n;\omega)] = C_{\mathrm{xx}}(0) = \sigma_{\mathrm{x}}^2, \tag{13.2a}$$

$$\mathcal{E}[S_1(\mathbf{s}_n;\omega)] = \frac{2}{a^2}\sum_{i=1}^{d}\big[C_{\mathrm{xx}}(0) - C_{\mathrm{xx}}(a\hat{\mathbf{e}}_i)\big], \tag{13.2b}$$

$$\begin{aligned}\mathcal{E}[S_2(\mathbf{s}_n;\omega)] =& \frac{2}{a^4}\sum_{i,j=1}^{d}\Big[2C_{\mathrm{xx}}(0) - 2C_{\mathrm{xx}}(a\hat{\mathbf{e}}_i) - 2C_{\mathrm{xx}}(a\hat{\mathbf{e}}_j) + \\ & C_{\mathrm{xx}}(a\hat{\mathbf{e}}_i - a\hat{\mathbf{e}}_j) + C_{\mathrm{xx}}(a\hat{\mathbf{e}}_i + a\hat{\mathbf{e}}_j)\Big].\end{aligned} \tag{13.2c}$$

Remark If $\mathrm{X}(\mathbf{s};\omega)$ is a differentiable random field with characteristic length ξ, the constraints are expressed at the limit $a \to 0$ (practically for $a \ll \xi$) as partial derivatives of $C_{\mathrm{xx}}(\mathbf{r})$ (see Sect. 5.3.3).

Sample constraints For data on $\mathcal{G}_d$, the sample constraints $\overline{S_0(\mathbf{x}^*)}$, $\overline{S_1(\mathbf{x}^*)}$ and $\overline{S_2(\mathbf{x}^*)}$ are obtained as grid averages of the respective counterparts defined in (13.1). In the following, $\overline{\mu}_{\mathrm{x}}$ denotes the sample mean, i.e., $\overline{\mu}_{\mathrm{x}} = \sum_{n=1}^{N} x^*(\mathbf{s}_n)/N$.

$$\overline{S_0(\mathbf{x}^*)} = \frac{1}{N} \sum_{n=1}^{N} \left[x^*(\mathbf{s}_n) - \overline{\mu}_{\mathrm{x}} \right]^2, \tag{13.3a}$$

$$\overline{S_1(\mathbf{x}^*)} = \frac{1}{N} \sum_{n=1}^{N} \sum_{i=1}^{d} \left[\frac{x^*(\mathbf{s}_n + a\hat{\mathbf{e}}_i) - x^*(\mathbf{s}_n)}{a} \right]^2, \tag{13.3b}$$

$$\overline{S_2(\mathbf{x}^*)} = \frac{1}{N} \sum_{n=1}^{N} \sum_{i=1}^{d} \left[\frac{x^*(\mathbf{s}_n + a\hat{\mathbf{e}}_i) - 2x^*(\mathbf{s}_n) + x^*(\mathbf{s}_n - a\hat{\mathbf{e}}_i)}{a^2} \right]^2. \tag{13.3c}$$

The above definitions use the same length scale (the lattice step a) in both the discrete gradient and curvature terms. In general it is possible to use different length scales, a_1 and a_2, not necessarily equal to a, for the gradient and curvature terms. The use of different length scales shifts the emphasis from the smallest possible lag to larger distances.

13.1.2 MoNC Constraints for Data on Irregular Grids

Calculating the sample constraints in the case of irregularly spaced (scattered) data is not as straightforward as in the case of gridded data for two reasons. (i) There is no well-defined characteristic length such as the grid step. (ii) It is not obvious how to compute the equivalent quantities of the squared gradient and the Laplacian.

Sample gradient constraint For random sampling grids we use non-negative, compactly supported or exponentially decaying *kernel functions* to compensate for the lack of a well-defined length scale. Sample constraints can then defined by means of *Watson-Nadaraya kernel weighted averages* (2.39) as shown in [229, 373]. For example, the following sample average can be used for the squared gradient

$$\overline{S_1(\mathbf{x}^*, \hat{a})} = \frac{1}{\hat{a}^2} \frac{\sum_{n=1}^{N} \sum_{m=1,\neq n}^{N} K\left(\frac{\|\mathbf{s}_n - \mathbf{s}_m\|}{h_n}\right) \left(x_n^* - x_m^*\right)^2}{\sum_{n=1}^{N} \sum_{m=1,\neq n}^{N} K\left(\frac{\|\mathbf{s}_n - \mathbf{s}_m\|}{h_n}\right)}, \tag{13.4}$$

where $\hat{a}$ is a *characteristic grid length*, h_n are *local kernel bandwidths*, and $\sum_{m=1,\neq n}$ denotes a summation over all points $\mathbf{s}_m \in \mathbf{S}_{\mathrm{N}} \setminus \mathbf{s}_n$.

The corresponding constraint expressed in terms of the respective random field is defined as follows:

$$S_1(\mathbb{S}_N;\omega) = \frac{1}{\hat{a}^2}\frac{\sum_{n=1}^{N}\sum_{m=1,\neq n}^{N} K\left(\frac{\|\mathbf{s}_n-\mathbf{s}_m\|}{h_n}\right)[\mathrm{X}(\mathbf{s}_n;\omega)-\mathrm{X}(\mathbf{s}_m;\omega)]^2}{\sum_{n=1}^{N}\sum_{m=1,\neq n}^{N} K\left(\frac{\|\mathbf{s}_n-\mathbf{s}_m\|}{h_n}\right)}.$$

Stochastic gradient constraint In light of the above definition, the stochastic gradient constraint can be expressed as follows in terms of the kernel and the variogram functions

$$\mathcal{E}[S_1(\mathbb{S}_N;\omega)] = \frac{2}{\hat{a}^2}\frac{\sum_{n=1}^{N}\sum_{n\neq m} K\left(\frac{\|\mathbf{s}_n-\mathbf{s}_m\|}{h_n}\right)\gamma_{\mathrm{xx}}(\mathbf{s}_n-\mathbf{s}_m;\boldsymbol{\theta})}{\sum_{n=1}^{N}\sum_{n\neq m} K\left(\frac{\|\mathbf{s}_n-\mathbf{s}_m\|}{h_n}\right)}. \quad (13.5)$$

Characteristic grid length The expressions for the sample and the stochastic gradient constraints involve the characteristic grid length $\hat{a}$. This length aims to replace the constant lattice step by providing some average over the shortest distances in the sampling network.

Let $\mathrm{DT}(\mathbb{S}_N)$ denote the *Delaunay triangulation* of the sampling point set $\mathbb{S}_N$. In two dimensions, $\mathrm{DT}(\mathbb{S}_N)$ comprises triangles whose vertices coincide with the points in $\mathbb{S}_N$. The triangles share the following property: If the circumscribed circle is drawn (this circle passes through all the vertices of the triangle), then the only points from $\mathbb{S}_N$ inside the circle are the triangle vertices. Delaunay triangulation is extended in higher dimensions than $d = 2$ by means of the geometric concept of *simplices* which generalize triangles and tetrahedra.

$\mathrm{DT}(\mathbb{S}_N)$ contains a set of triangles and the respective set of the triangle edges. Let $N_0 = \#(\text{Delaunay triangle edges})$ and $D_p = \|\mathbf{s}_n - \mathbf{s}_m\|$, where $p = 1, \ldots, N_0$ denote the set of near neighbor distances in the Delaunay triangulation. An example of Delaunay triangulation is given in Fig. 13.1.

A *characteristic grid length* $\hat{a}$ that reflects the connectivity of the network at short distances can be defined by means of the following *Minkowski average*

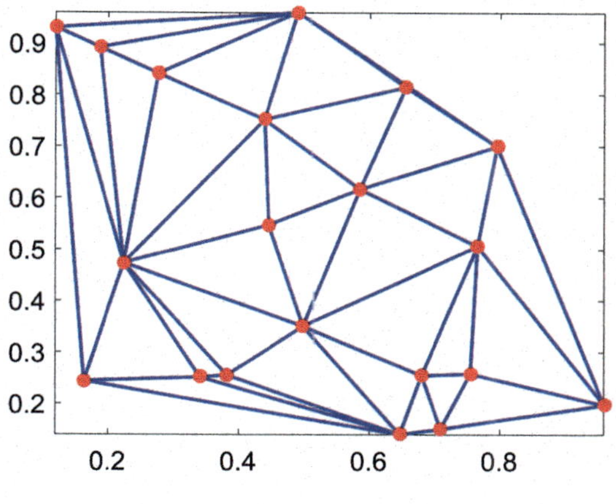

Fig. 13.1 Schematic of an irregular sampling network $\mathbb{S}_N$ (points denoted by circles) and its Delaunay triangulation. The straight lines depict the edges of the Delaunay triangles which are used for the estimation of the characteristic grid length $\hat{a}$ based on (13.7)

$$\hat{a} = \left(\frac{1}{N_0} \sum_{p=1}^{N_0} D_p^d \right)^{1/d}. \tag{13.6}$$

Self-consistent kernel bandwidth tuning For compactly supported (e.g., quadratic, triangular) kernel functions, the kernel terms in (13.4) and (13.5) imply that the respective summations involve only pairs of points such that $\|\mathbf{s}_n - \mathbf{s}_m\| \leq h_n$. If the kernel has infinite support (e.g., exponential of Gaussian kernels), the local bandwidth determines the characteristic distance beyond which the kernel's influence becomes negligible.

A simple recipe for the selection of *local kernel bandwidths* is proposed below. It uses an ad hoc *self-consistency principle* that requires the Nadaraya-Watson average of the squared distances between pairs to be equal to the square of the characteristic grid length.

Based on the self-consistency principle, the local kernel bandwidth can be obtained by solving the following nonlinear equations:

$$\frac{\sum_{m=1,\neq n}^{N} K\left(\frac{\|\mathbf{s}_n - \mathbf{s}_m\|}{h_n}\right) \|\mathbf{s}_n - \mathbf{s}_m\|^2}{\sum_{m=1,\neq n}^{N} K\left(\frac{\|\mathbf{s}_n - \mathbf{s}_m\|}{h_n}\right)} = \hat{a}^2, \text{ for } n = 1, \ldots, N. \tag{13.7}$$

Equations (13.6) and (13.7) imply that the local bandwidths are tuned so that the kernel-based average of the squared distances between the target point $\mathbf{s}_n$ and all other sampling points in $\mathbb{S}_N \setminus \mathbf{s}_n$ is equal to the characteristic length of the sampling grid.

More on kernel-based methods Different approaches are possible for determining the kernel bandwidths, including both local and global estimates [229, 368, 373]. If the sampling density is approximately uniform, a global estimate of the kernel bandwidth sounds reasonable. However, if the sampling density varies considerably, local estimates of the kernel bandwidth are more meaningful.

A different proposal for the local bandwidths is $h_n = \mu\, D_{n,[k]}(\mathbb{S}_N)$, where $D_{n,[k]}(\mathbb{S}_N)$ is the distance between $\mathbf{s}_n$ and its *k-nearest neighbor* in $\mathbb{S}_N$, and $\mu > 1$ is a free parameter. This approach has the advantage that a single parameter, i.e., μ, determines all the local bandwidths in connection with the neighborhood structure of the sampling network.

An analysis of the bias and variance of the kernel-based sample gradient estimators is currently unavailable. However, it should be accessible using the same concepts and tools as for non-parametric variogram estimators [322].

Squared curvature In contrast with the squared gradient, the definition of the *square curvature* sample-based constraint is not as straightforward. One possibility is to use linear combinations of squared gradient estimators with suitably selected kernel bandwidths motivated by (8.21). However, since one of the terms contains a minus sign, it is possible that the estimator becomes negative in cases of highly variable sampling density.

A better alternative for estimating the squared curvature is to employ the concept of the *graph Laplacian*. Taking the square of the graph Laplacian gives a non-negative number regardless of the variability of the sampling network. The kernel functions could then be used to assign weights to the sampling graph edges.

13.1.3 Parameter Inference with MoNC

As stated above, MoNC focuses on short-distance correlations. This choice has the advantage that the correlation model is not influenced by potential fluctuations of the sample function that commonly occur at large lags. The short-lag behavior is also more important for interpolation purposes. On the other hand, MoNC fails to capture information about the correlations at larger distances. In this sense, MoNC is similar to the *Yule-Walker* method used in time series parameter estimation [95].

Assumptions MoNC is based on the following assumptions: (i) The spatial averages $\overline{S_l}$, $l = 0, 1, 2$ are accurate and precise estimators of the stochastic expectations $\mathcal{E}_t[S_l]$ of the underlying random field, where $\mathcal{E}_t[\cdot]$ denotes the expectation with respect to the unknown probability density (ergodic assumption). (ii) The stochastic expectations $\mathcal{E}[S_l]$ of the parametric model approximate the expectations $\mathcal{E}_t[S_l]$ of the underlying field (model suitability assumption). (iii) The specified constraints suffice for practical estimation of the correlation structure (constraint sufficiency assumption).

Construction of objective functional The model parameters, $\boldsymbol{\theta}$, are obtained by solving numerically the system of nonlinear equations

$$\mathcal{E}[S_l] = \overline{S_l}, \; l = 0, 1, 2.$$

In practice, the solution is based on the minimization of a *convex objective functional* $\mathcal{U}(\vartheta_0^2, \vartheta_1^2, \vartheta_2^2)$, where $\vartheta_l := \mathcal{E}[S_l] - \overline{S_l}$ are the "contrasts" between sample and ensemble constraints. An obvious choice for $\mathcal{U}(\cdot)$ is some norm of the contrast vector $(\vartheta_0, \vartheta_1, \vartheta_2)^\top$.

Normalized correlations It is found empirically that optimization convergence improves if the objective functional is independent of the variance and the measurement units of $X(\mathbf{s}; \omega)$. Hence, in MoNC the dependence of the constraints on the field amplitude (which is essentially proportional to the variance) is eliminated from the optimization procedure by suitable normalization of the constraints. This is reminiscent of the variance elimination used in likelihood maximization (see Chap. 12).

Motivated by the above observations, the cost functional is expressed in terms of the *normalized correlations*

$$\rho_1(\hat{a}; \boldsymbol{\theta}') := \frac{\mathcal{E}[S_1(\hat{a})]}{\mathcal{E}[S_0]}, \quad \rho_2(\hat{a}; \boldsymbol{\theta}') = \frac{\mathcal{E}[S_2(\hat{a})]}{\mathcal{E}[S_0]}, \tag{13.8a}$$

where the *reduced parameter vector* $\boldsymbol{\theta}'$ comprises all the parameters in $\boldsymbol{\theta}$ except for the overall "amplitude" parameter (i.e., the variance in classical models or the scale parameter η_0 in SSRFs).

Since $\mathcal{E}[S_l]$, and $\overline{S_l}$ ($l = 0, 1, 2$) represent averages of non-negative terms they are positive quantities. Thus, the *normalized moments* ρ_1 and ρ_2 are finite and non-negative. In addition, they are independent of $\mathcal{E}[S_0]$.

Normalized sample correlations The sample counterparts of the normalized moments ρ_1 and ρ_2 are given by

$$\hat{\rho}_1(\mathbf{x}^*, \hat{a}) = \overline{S_1}/\overline{S_0} \propto \hat{a}^{-2}, \tag{13.8b}$$

$$\hat{\rho}_2(\mathbf{x}^*, \hat{a}) = \overline{S_2}/\overline{S_0} \propto \hat{a}^{-4}. \tag{13.8c}$$

Dimensionless moment ratios Taking advantage of the positive values of the normalized moments ρ_1 and ρ_2 we define the following *dimensionless moment ratios* that are independent of $\hat{a}$

$$\vartheta_l(\mathbf{x}^*, \hat{a}; \boldsymbol{\theta}') := 1 - \frac{\hat{\rho}_l(\mathbf{x}^*, \hat{a})}{\rho_l(\hat{a}, \boldsymbol{\theta}')}, \quad \text{for } l = 1, 2. \tag{13.8d}$$

In light of the above, the estimates $\hat{\boldsymbol{\theta}}'$ can be derived from the minimization of the following *MoNC objective functional* that measures the deviation between the respective sample and ensemble normalized moments:

$$\Phi_s(\mathbf{x}^*, \hat{a}; \boldsymbol{\theta}') = \vartheta_1^2(\mathbf{x}^*, \hat{a}; \boldsymbol{\theta}') + \vartheta_2^2(\mathbf{x}^*, \hat{a}; \boldsymbol{\theta}'). \tag{13.8e}$$

Then, the MoNC estimate of the parameter vector $\boldsymbol{\theta}'$ is given by

$$\hat{\boldsymbol{\theta}}' = \arg\min_{\boldsymbol{\theta}'} \Phi_s(\mathbf{x}^*, \hat{a}; \boldsymbol{\theta}').$$

Estimating the scale factor The above procedure leaves the amplitude of the random field undetermined. This means that the true covariance function is actually of the form $\lambda C_{xx}(\mathbf{r}; \hat{\boldsymbol{\theta}}')$, where λ is an unknown coefficient and $C_{xx}(\mathbf{r}; \hat{\boldsymbol{\theta}}')$ is the normalized covariance determined from $\hat{\boldsymbol{\theta}}'$. There are at least two distinct approaches for estimating λ.

The simplest approach is by means of the ratio of the sample variance over the estimated covariance at zero lag, i.e.,

$$\hat{\lambda} = \frac{\overline{S_0}}{C_{xx}(0; \hat{\boldsymbol{\theta}}')}.$$

This approach implies confidence in the variance estimate $\overline{S_0}$. However, the latter may not be accurate due to spatial correlations among the sampling sites and sample-to-sample (non-ergodic) fluctuations.

The second approach requires minimizing the following functional

$$\hat{\lambda} = \arg\min_{\lambda} \left\{ \left(\overline{S_1} - \lambda \mathcal{E}[S_1(\hat{a}; \hat{\boldsymbol{\theta}}')] \right)^2 + \left(\overline{S_2} - \lambda \mathcal{E}[S_2(\hat{a}; \hat{\boldsymbol{\theta}}')] \right)^2 \right\},$$

with respect to the unknown parameter λ. This approach assumes confidence in the estimates $\overline{S_1}$ and $\overline{S_2}$.

Computational efficiency The main advantage of MoNC is computational efficiency in terms of memory resources and computation speed. Since the method is based on local correlations, it does not require the evaluation, storage and inversion of large covariance matrices. Hence, it is a suitable candidate for big spatial data sets.

For uniformly spaced samples the estimation of the sample constraints is an $O(N)$ operation. For scattered samples, the complexity of constraint calculations is limited by the Delaunay triangulation, which is an $O(N \log_2 N)$ procedure, and the kernel summations. For compactly supported kernels, the complexity of kernel summations is $O(N\,M)$, where M is the average number of neighbors within the kernel bandwidth per sampling point.

The numerical complexity of the $\Phi_s(\mathbf{x}^*, \hat{a}; \boldsymbol{\theta}')$ optimization is in principle independent of N. In practice, it has been observed that the optimization time may slightly decrease with increasing N (for constant $\hat{a}$).

Remarks The MoNC has two potential weaknesses: (i) The complete dependence on short-range correlations and (ii) the difficulty of formulating a demonstrably non-negative square curvature sample estimator for irregularly sampled data. However, this can be addressed by means of the graph Laplacian (as mentioned above) or by using finite elements. The issue of short range can be remedied by including a number of "square gradient" terms that use different lags. However, MoNC then starts to resemble variogram estimation. Anisotropic correlations can also be handled by extending the method to include directional constraints.

13.1.4 MoNC Constraints for Time Series

As an example of the above general formulation, we present the stochastic and sample constraints in the one-dimensional case. Let $\mathrm{X}(s;\omega)$ be a $1D$ random process sampled uniformly with step α at the points s_n, $n = 1, \ldots, N$. Setting $\hat{a} = \alpha$, the local field functions (13.1) become

$$S_0(s_n;\omega) = (\mathrm{X}(s_n;\omega) - m_{\mathrm{x}})^2, \tag{13.9a}$$

$$S_1(s_n, \alpha; \omega) = \frac{1}{\alpha^2}[\mathrm{X}(s_n + \alpha; \omega) - \mathrm{X}(s_n; \omega)]^2, \tag{13.9b}$$

$$S_2(s_n, \alpha; \omega) = \frac{1}{\alpha^4}[\mathrm{X}(s_n + \alpha; \omega) + \mathrm{X}(s_n - \alpha; \omega) - 2\mathrm{X}(s_n; \omega)]^2. \tag{13.9c}$$

The sample constraints are given by the following averages

$$\overline{S_0(\mathbf{x}^*)} = \frac{1}{N}\sum_{n=1}^{N}\left(x_n^* - \overline{\mu}_x\right)^2, \tag{13.10a}$$

$$\overline{S_1(\mathbf{x}^*, \alpha)} = \frac{1}{N}\sum_{n=1}^{N}\left(\frac{x_{n+1}^* - x_n^*}{\alpha}\right)^2, \tag{13.10b}$$

$$\overline{S_2(\mathbf{x}^*, \alpha)} = \frac{1}{N}\sum_{n=1}^{N}\left(\frac{x_{n+1}^* - 2x_n^* + x_{n-1}^*}{\alpha^2}\right)^2. \tag{13.10c}$$

The stochastic constraints (13.2) are reduced to the simple expressions

$$\mathcal{E}[S_0] = C_{\mathrm{xx}}(0), \tag{13.11a}$$

$$\mathcal{E}[S_1] = \frac{2}{\alpha^2}\left[C_{\mathrm{xx}}(0) - C_{\mathrm{xx}}(\alpha)\right] = \frac{2\gamma_{\mathrm{xx}}(\alpha)}{\alpha^2}, \tag{13.11b}$$

$$\mathcal{E}[S_2] = \frac{2}{\alpha^4}\left[3C_{\mathrm{xx}}(0) + C_{\mathrm{xx}}(2\alpha) - 4C_{\mathrm{xx}}(\alpha)\right] = \frac{2\left[4\gamma_{\mathrm{xx}}(\alpha) - \gamma_{\mathrm{xx}}(2\alpha)\right]}{\alpha^4}. \tag{13.11c}$$

The normalized correlations (13.8a) used in the MoNC objective functional are modified accordingly as follows

$$\rho_1(\alpha; \boldsymbol{\theta}') = \frac{2\left[1 - \rho_{\mathrm{xx}}(\alpha; \boldsymbol{\theta}')\right]}{\alpha^2}, \tag{13.12a}$$

$$\rho_2(\alpha; \boldsymbol{\theta}') = \frac{2\left[3 + \rho_{\mathrm{xx}}(2\alpha; \boldsymbol{\theta}') - 4\rho_{\mathrm{xx}}(\alpha)\right]}{\alpha^4}. \tag{13.12b}$$

Upper bounds for sample constraints If the correlation function $\rho_{\mathrm{xx}}(\mathbf{r})$ decays monotonically to zero, it holds that $\rho_{\mathrm{xx}}(\alpha; \boldsymbol{\theta}') > 0$ for all $\alpha > 0$. Hence, it follows from (13.12a) that $\rho_1(\alpha; \boldsymbol{\theta}') \leq 2/\alpha^2$ and $\rho_2(\alpha; \boldsymbol{\theta}') \leq 6/\alpha^4$. These upper bounds can be used to test the accuracy of the sample estimators $\hat{\rho}_1(\mathbf{x}^*; \alpha)$ and $\hat{\rho}_2(\mathbf{x}^*; \alpha)$ that are derived from the one-dimensional version of (13.3).

13.1.5 Discussion

The basic premises of MoNC are (i) the ergodic condition and (ii) that knowledge of the entire covariance function is not necessary if our focus is the short-range behavior. For example, in the case of *spatial interpolation*, the short-range dependence of the covariance is more important than the covariance at large lags [774].

MoNC is motivated by Spartan spatial random fields, since the latter are defined by means of local geometric properties such as the gradient and curvature [229, 360, 373]. In addition, the ease with which the sample constraints can be calculated for data on spatial grids makes MoNC particularly useful in the reconstruction of missing data in remote sensing images.

The MoNC procedure is similar in spirit to the *Yule-Walker* method of moments estimator used for autoregressive processes, AR(p), which focuses on the correlation function at lags less than $p\delta t$ (δt being the sampling step) [95]. The connection between the SSRF model and AR(2) processes, as discussed in Chap. 9, further motivates the use of MoNC.

13.2 The Method of Maximum Entropy

This section focuses on the principle of maximum entropy and its applications in the estimation of spatial model parameters. The notion of entropy is important in physics, information theory and statistics.

13.2.1 Introduction to Entropy

Entropy in physics *Thermodynamic* entropy was introduced in the pioneering work of Ludwig Boltzmann and Willard Gibbs on statistical mechanics in the nineteenth century. Entropy is used in statistical mechanics to quantify the disorderly motion of ensembles of microscopic particles. It measures the number of distinct microscopic states that are compatible with a specific macroscopic state. In the thermodynamic sense, the entropy of a system is a measure of its disorder, so that higher temperatures imply higher entropy values than lower temperatures. The classical thermodynamical entropy S is given by the *Gibbs formula*

$$S = -k_B \sum_{n=1}^{N} p_n \ln p_n, \qquad (13.13)$$

where the summation is over the probabilities p_n of the microscopic states of the system and k_B is Boltzmann's constant. Entropy is a central concept in statistical physics [125, 245, 802]. Several generalizations of the Boltzmann-Gibbs expression (13.13) have been proposed that purport to generalize the concept of entropy for strongly interacting systems [410] (see also Chap. 14).

Entropy in information theory The concept of *information entropy* as a measure of the information content of probability distributions was introduced by Claude Shannon [739]. According to *Shannon's definition*, the entropy of a system that can occupy N discrete states is given by

$$S = -\sum_{n=1}^{N} p_n \ln p_n, \tag{13.14}$$

where $\{p_n\}_{n=1}^{N}$, are the probabilities of the discrete states. Information theory was founded on *Shannon's* definition of entropy [472]. Note that the information entropy can be defined with respect to the natural logarithm [as is done in (13.14)] or with respect to a logarithm in a different base. If the natural logarithm is used, the entropy is measured in *natural units of information (nats)*.

Why the negative sign The entropy definition (13.14) involves a negative sign in front of the summation over the microscopic states. To appreciate the origin of this sign, consider that since $p_n \in (0, 1]$ for all the accessible states, the logarithms $\ln p_n$ are negative numbers. Hence, the negative sign ensures that the entropy is a non-negative quantity.

The *Shannon entropy* provides a quantitative measure of uncertainty (unpredictability) in stochastic systems. High information entropy implies that the result of a measurement is uncertain, but also that new observations considerably improve our current knowledge of the system. In contrast, low information entropy indicates that sufficient knowledge about the system is already at hand, so that new observations do not add significant new information. Readers can find an enjoyable exposition of the relation between thermodynamic and information entropy in [517].

Joint entropy Above we reviewed the definition of entropy for individual random variables, which is based on marginal probability distributions. However, we often would like to define entropy for random variables that are distributed in space. In the simplest case of two discretely-valued random variables the joint entropy is defined as

$$S(\mathrm{X}, \mathrm{Y}) = \sum_{n=1}^{N} \sum_{m=1}^{M} p(x_n, y_m) \ln p(x_n, y_m),$$

where $p(x_n, y_m)$ is the joint probability for the random variables $\mathrm{X}(\omega)$ and $\mathrm{Y}(\omega)$ that take respectively N and M values. The inequality

$$S(\mathrm{X}, \mathrm{Y}) \leq S(\mathrm{X}) + S(\mathrm{Y}),$$

shows that the joint entropy is in general less than the sum of the respective entropies, reflecting the fact that joint dependence can reduce the uncertainty.

Next, let us consider the continuously-valued random vector

$$\mathbf{X}(\omega) = \big(X_1(\omega), \ldots, X_N(\omega)\big)^\top,$$

where the components $X_n(\omega)$, $n = 1, \ldots, N$, are random variables that take continuous values in $\mathbb{R}$. For example, $\mathbf{X}(\omega)$ could represent samples of a random field taken at the locations of the set $\mathbb{S}_N = \{\mathbf{s}_n\}_{n=1}^N$. Furthermore, let us assume that the joint pdf of the random vector $\mathbf{X}(\omega)$ is given by the function $f_X(\mathbf{x}; \mathbb{S}_N)$.

The *joint entropy*, $S(X)$, *of the random vector* $\mathbf{X}(\omega)$ is given by the expectation of the logarithm of $f_X(\mathbf{X}(\omega); \mathbb{S}_N)$, i.e.,

$$S(\mathbf{X}) = -\mathcal{E}\big[\ln f_X\big(\mathbf{X}(\omega); \mathbb{S}_N\big)\big].$$

Since the random variables $X_n(\omega)$ take values in $\mathbb{R}$, the expectation is expressed as

$$S(\mathbf{X}) = -\left(\prod_{n=1}^{N} \int_{-\infty}^{\infty} dx_n\right) f_X\,(\mathbf{x}; \mathbb{S}_N)\,\ln f_X\,(\mathbf{x}; \mathbb{S}_N)\,. \tag{13.15}$$

Conditional entropy For a set of random variables $\mathbf{X}(\omega)$ the conditional entropy given a set of measurements $\mathbf{y}^*$ can be defined based on the expectation with respect to the conditional pdf $f_X(\mathbf{x} \mid \mathbf{y}^*)$ as follows:

$$S(\mathbf{X} \mid \mathbf{y}^*) = \mathcal{E}_{\mathbf{y}^*}\left[\ln f_X\big(\mathbf{X}(\omega) \mid \mathbf{y}^*\big)\right].$$

The conditional entropy is constrained by the measured values (data) of the field at the sampling locations. It quantifies the uncertainty in $\mathbf{X}(\omega)$ given the information in the data $\mathbf{y}^*$.

Mutual information refers to the information that is shared by two sets of random variables. In the simplest case of two scalar random variables $X(\omega)$ and $Y(\omega)$ the mutual information is defined by

$$\begin{aligned} \mathrm{MI}(X, Y) &= S(X) + S(Y) - S(X, Y) \\ &= S(X) - S(X \mid Y) = S(Y) - S(Y \mid X). \end{aligned} \tag{13.16}$$

Hence, the mutual information of $X(\omega)$ and $Y(\omega)$ is obtained by subtracting the joint entropy from the sum of the marginal entropies. As evidenced in the second line of (13.16), mutual information also represents the average reduction of uncertainty regarding $X(\omega)$ if the value of $Y(\omega)$ is known. Equivalently, the mutual information represents the uncertainty reduction in $Y(\omega)$ if the value of $X(\omega)$ is known.

It can be shown that mutual information is *symmetric* and *non-negative*, i.e.,

$$\mathrm{MI(X, Y)} = \mathrm{MI(Y, X)}, \text{ and } \mathrm{MI(X, Y)} \geq 0.$$

Nonlinear dependence If two variables are independent, their mutual information is zero. Mutual information is positive if there is dependence between the variables. In contrast with the Pearson correlation which is designed to detect linear dependencies, mutual information measures nonlinear dependence. Thus, mutual information can detect relationships even in cases where the correlation coefficient is zero. On the other hand, mutual information does not distinguish between "positive" and "negative" relations. Such a distinction is meaningful in the case of linear dependence, where the sign is related to the slope of the linear function, but it is not obvious for nonlinear forms of dependence. For brief introductions to mutual information consider [673, Chap. 14] and [521].

Spatially extended variables Equation (13.16) defines mutual information for two random variables. It is straightforward to generalize this equation to random fields using the proper definitions of the marginal and joint entropy. Thus, it is possible to quantify the information added at a specific location given the values of the field at other locations. In the context of spatial data analysis, information theory has been used to analyze spatial uncertainty correlations, and to investigate the impact of additional information at one location on the uncertainty reduction at other locations [840].

Relative entropy This measure, also known as *Kullback-Leibler divergence*, quantifies the difference in entropy between two probability distributions using one of the two as the reference distribution. Hence, in contrast with the entropic measures above (joint and conditional entropy and mutual information) that focus on two (or more) random variables, relative entropy focuses on the comparison of two different probability distributions for the same random variable.

Consider a random variable $\mathrm{X}(\omega)$ which takes values in the set $\mathcal{X}$ and two different candidate probability distribution functions $p_1(x)$ and $p_2(x)$. The relative entropy of $p_1(\cdot)$ with respect to $p_2(\cdot)$ is given by

$$D_{KL}(p_1 \parallel p_2) = -\sum_{x \in \mathcal{X}} p_1(x) \ln \left(\frac{p_2(x)}{p_1(x)} \right). \tag{13.17}$$

In general, the relative entropy is not symmetric with respect to an interchange of the two distributions, i.e., it holds that $D_{KL}(p_1 \parallel p_2) \neq D_{KL}(p_2 \parallel p_1)$.

In the case of continuous random variables, the functions $p_i(\cdot)$ are replaced by the densities $f_i(\cdot)$, and the summation over states is replaced with an integral.

Note that equation (13.17) implies $D_{KL}(f_1 \parallel f_1) = 0$. The relative entropy $D_{KL}(f_1 \parallel f_2)$ measures the information gain if the density $f_1(\cdot)$ is used instead of $f_2(\cdot)$ [472]. Thus, in *Bayesian inference* the relative entropy is used to quantify the information gain achieved by the posterior distribution, $f_1(\cdot)$ in relation to the prior distribution, $f_2(\cdot)$ due to the integration of the data.

13.2.2 The Maximum Entropy Principle

The *principle of maximum entropy* maintains that the optimal probability distribution given any number of constraints (information) is the distribution that maximizes the entropy of the system conditionally on the constraints. Edwin Jaynes showed that the Boltzmann-Gibbs theory of statistical mechanics follows from the principle of maximum entropy without the additional assumptions (ergodicity, metric transmissivity and equidistribution) that are required in the traditional formulation of statistical mechanics [406, 407]. In the words of Jaynes [406]:

> Information theory provides a constructive criterion for setting up probability distributions on the basis of partial knowledge, and leads to a type of statistical inference which is called the maximum-entropy estimate. It is the least biased estimate possible on the given information; i.e., it is maximally noncommittal with regard to missing information.

The maximum-entropy method has been applied to image reconstruction [315, 754], in the construction of joint pdfs for random fields based on partial information [138, 141], in Bayesian data analysis [237, 752, 753] where it is used, among other things, to define prior distributions [120].

Is the maximum entropy principle universal? It should be mentioned that there are non-Hamiltonian (irreversible) physical systems for which entropy is not maximized at equilibrium. In addition, nonlinear systems that include stochastic terms may evolve towards configurations of reduced entropy. Such configurations can appear in processes of self-organization during which systems move from states of higher probability towards lower probability states, which correspond to reduced entropy values [500]. Recent research shows that the maximum entropy principle can be maintained for certain generalized forms of the entropy functional. Such generalizations might be more suitable for describing the entropy of complex dynamical systems [410].

13.2.3 Formulation of the Maximum Entropy Method

In this section we show how a *maximum entropy (MaxEnt)* probability density function can be obtained for the target random variable $X(\omega)$ or random field $X(\mathbf{s}; \omega)$ based on the optimization of the entropy under data-imposed constraints. For an introduction to the maximum entropy method see [156]. The application of the maximum entropy method to spatial data problems is discussed in [138].

The entropy of random vectors is defined in terms of the joint density $f_{\mathrm{x}}(\mathbf{x}; \mathbb{S}_N)$ by means of (13.15). The *maximum entropy solution* f_{x}^* that satisfies a set of data-based constraints $\{C\}$ is defined as follows

$$f_{\mathrm{x}}^* = \arg\max_{f_{\mathrm{x}}} S(\mathrm{X}), \text{ subject to } \{C\}.$$

The data constraints often involve knowledge of the support of $X(\mathbf{s};\omega)$ and various moments of the random variable (or field). In the following, we drop for simplicity the symbol "$*$" in the MaxEnt solution.

The Lagrangian function To maximize the entropy under the constraints, we can maximize the Lagrangian function $S(X \mid \{C\})$, which involves a weighted linear combination of K constraints, i.e.

$$S(X \mid \{C\}) = -\int_{\mathbb{R}^N} d\mathbf{x}\, f_x(\mathbf{x}; \mathbb{S}_N) \ln f_x(\mathbf{x}; \mathbb{S}_N) - \sum_{k=0}^{K} \mu_k\, g^{(k)}[f_x]. \tag{13.18}$$

In (13.18) the $\{\mu_k\}_{k=1}^{K}$ represent *Lagrange multipliers* that correspond to the data constraints, while the functions $g^{(k)}[f_x]$ represent the data-imposed constraints. The constraint for $k = 0$ enforces the normalization of the density function.[1] Maximizing the Lagrange function (13.18) with respect to the pdf leads to an exponential expression for the density. The maximization with respect to $\{\mu_k\}_{k=1}^{K}$ implies that the MaxEnt pdf will depend on the vector of the Lagrange multipliers $\boldsymbol{\mu} = (\mu_1, \ldots, \mu_K)^\top$. We do not explicitly show this dependence to avoid cumbersome expressions.

General constraints The first equation below (for $k = 0$) is used to ensure the proper normalization of the joint pdf. The remaining K constraints can be expressed in terms of the nonlinear—in general—functions $\Phi^{(k)}(\mathbf{x}^*)$ as follows:

$$\int_{\mathbb{R}^N} d\mathbf{x}\, f_x(\mathbf{x}; \mathbb{S}_N) = 1, \tag{13.19a}$$

$$\mathcal{E}\left[\Phi^{(k)}(\mathbf{x}; \mathbb{S}_N)\right] = \overline{\Phi^{(k)}(\mathbf{x}^*)},\; k = 1, \ldots, K. \tag{13.19b}$$

The last K equations enforce the equality of sample-based and stochastic constraints.

The expectation $\mathcal{E}\left[\Phi^{(k)}(\mathbf{x}; \mathbb{S}_N)\right]$ denotes the ensemble average of the function $\Phi^{(k)}(\cdot)$ with respect to the pdf, while $\overline{\Phi^{(k)}(\mathbf{x}^*)}$ the respective sample average. Note that the ensemble average depends on the Lagrange multipliers $\boldsymbol{\mu}$ via the pdf, while the sample average does not.

Based on the above, the functions $g^{(k)}[f_x]$ that appear in the Lagrangian expression (13.18) are given by

$$g^{(0)}[f_x] = \int_{\mathbb{R}^N} d\mathbf{x}\, f_x(\mathbf{x}; \mathbb{S}_N) - 1, \tag{13.20a}$$

[1] To review constrained optimization with Lagrange multipliers consider [673, Chap. 16.5] and [67].

$$g^{(k)}[f_{\mathrm{x}}] = \int_{\mathbb{R}^N} \mathbf{dx}\, \Phi^{(k)}(\mathbf{x})\, f_{\mathrm{x}}(\mathbf{x}; \mathbb{S}_{\mathrm{N}}) - \overline{\Phi^{(k)}(\mathbf{x}^*)}, \quad k = 1, \ldots, K. \tag{13.20b}$$

In terms of the functions $g^{(k)}[f_{\mathrm{x}}]$ the *maximum entropy constraints* (13.19) are expressed as

$$g^{(k)}[f_{\mathrm{x}}] = 0, \; k = 0, 1, \ldots, K. \tag{13.21}$$

MaxEnt pdf Maximizing the Lagrangian function (13.18) with respect to the pdf $f_{\mathrm{x}}(\mathbf{x}; \mathbb{S}_{\mathrm{N}})$ requires determining the *stationary point* of the Lagrangian expression with respect to the pdf and the Lagrange multipliers, i.e.,

$$\frac{\delta S(\mathrm{X} \mid \{C\})}{\delta f_{\mathrm{x}}} = 0, \qquad \frac{\partial S(\mathrm{X} \mid \{C\})}{\partial \mu_k} = 0, \; k = 1, \ldots, K. \tag{13.22}$$

The first of the above equations involves the functional derivative of the Lagrangian function; this can be accomplished using the tools described in Sect. 6.2.1.

The maximization leads to the following *exponential density* expression for the *maximum entropy* pdf:

$$f_{\mathrm{x}}(\mathbf{x}; \mathbb{S}_{\mathrm{N}}) = \exp\left[-1 - \mu_0 - \sum_{k=1}^{K} \mu_k\, \Phi^{(k)}(\mathbf{x})\right]. \tag{13.23}$$

The optimal values of the Lagrange multipliers are determined by solving the *constraint equations* (13.21). The constant μ_0 that normalizes the pdf is linked to the logarithm of the partition function, i.e.,

$$\mu_0 + 1 = \ln Z.$$

For the exponential density model (13.23), the moment equations (13.19) [equivalently (13.21)], are equivalent to the maximum likelihood solution [778]. The moment equations can be expressed as a nonlinear least squares problem. Hence, they can be solved by means of the *Levenberg-Marquardt (LM)* algorithm [611, 673, 800].

The Levenberg-Marquardt algorithm is an iterative, gradient-based procedure for solving nonlinear least squares problems. The LM algorithm for the MaxEnt problem iteratively updates the parameters $\{\mu_k\}_{k=0}^{K}$ as follows

$$\mu_k^{m+1} = \mu_k^m - \left[\mathbb{J}^\top \mathbb{J} + \lambda\, \mathrm{diag}\left(\mathbb{J}^\top \mathbb{J}\right)\right]^{-1} \mathbb{J}^\top \mathbf{g}\left(\boldsymbol{\mu}^m\right), \quad m = 0, 1, \ldots M_c.$$

In the above, the parameter $\lambda > 0$ is a damping factor used to control the convergence of the method. In spirit, λ is similar to the Tikhonov regularization parameter. The vector $\mathbf{g}(\cdot)$ contains the constraint functions [cf. (13.21)], and $\mathbb{J}$ is the Jacobian matrix of the constraints defined at each step m. The Jacobian is obtained by means of the partial derivatives of the constraints with respect to the Lagrange multipliers, i.e.,

$$\mathbb{J}_{k,l} = \left.\frac{\partial g_k}{\partial \mu_l}\right|_{\mu=\mu^m}, \quad \text{for } k, l = 1, \ldots, K.$$

The derivatives of the constraints make a non-zero contribution only for the first part of the constraint functions that involves the expectations over the MaxEnt pdf. The elements of the Jacobian matrix at the m-th step are evaluated based on the following covariance functions

$$\mathbb{J}_{k,l} = \mathbb{C}\mathrm{ov}\left\{\Phi^{(k)}(\mathbf{x}), \Phi^{(l)}(\mathbf{x})\right\}_m, \quad k, l = 1, \ldots, K.$$

The $\Phi^{(k)}(\mathbf{x})$ represent the constraint functions [cf. (13.19)], while the index "m" indicates that the expectations are calculated based on the MaxEnt pdf (13.23) with the Lagrange multipliers $\boldsymbol{\mu}^m$ obtained at the m-th step [778]. The covariance functions above can be explicitly evaluated for Gaussian MaxEnt pdfs. For non-Gaussian density functions, they can be calculated using Markov Chain Monte Carlo (MCMC) methods (see Chap. 16 for a general introduction to MCMC).

The performance of the LM algorithm depends on our ability to determine reasonable initial conditions for the parameters [673]. The method progresses iteratively until the constraint functions approach zero within a specified tolerance.

In order to illustrate the maximum entropy principle, we give two simple examples for univariate probability distributions. In the first case normalization is the only constraint, while in the second case the mean value is added as a constraint.

The uniform distribution Let us consider the discrete random variable $\mathrm{X}(\omega)$ that can take N different values $\{x_n\}_{n=1}^N$. What is the probability distribution $p_n = P(\mathrm{X}(\omega) = x_n)$ for $n = 1, \ldots, N$ that maximizes the entropy, if no other information is given?

Since the random variable is discrete the Lagrange function of the entropy is given by the following sum

$$S(\mathrm{X} \mid \{C\}) = -\sum_{n=1}^{N} p_n \ln p_n - \mu_0 \left(\sum_{n=1}^{N} p_n - 1\right).$$

To maximize the Lagrangian function we need to solve the equations

$$\frac{\partial S(\mathrm{X} \mid \{C\})}{\partial p_n} = 0 \Rightarrow \ln p_n = -(1 + \mu_0), \; n = 1, \ldots, N.$$

Hence, in light of the above equations and the normalization constraint, the maximum entropy solution is the *uniform distribution* with probability

$$p_n = p = \mathrm{e}^{-(1+\mu_0)} = \frac{1}{N}, \text{ for } n = 1, \ldots, N, \text{ and } \mu_0 = \ln N - 1.$$

The lack of specific information leads to the uniform distribution which does not distinguish between the states.

The exponential distribution As a second example, consider a non-negative random variable $X(\omega)$ with a known expectation equal to m_X. Our goal is to find the pdf $f_X(x)$ that maximizes the entropy given the expectation constraint. In contrast with the previous example, the random variable $X(\omega)$ takes continuous values in $[0, \infty)$.

In this case, the Lagrangian function incorporates in addition to the normalization a constraint for the expectation. Thus, the Lagrangian function is given by the following expression

$$S(X \mid \{C\}) = -\int_0^\infty dx\, f_X(x) \ln f_X(x) - \mu_0\, g^{(0)}[f_X] - \mu_1\, g^{(1)}[f_X],$$

where the constraints $g^{(0)}[f_X]$ and $g^{(1)}[f_X]$ are respectively given by

$$g^{(0)}[f_X] = \int_0^\infty dx\, f_X(x) - 1,$$

$$g^{(1)}[f_X] = \int_0^\infty dx\, x\, f_X(x) - m_X.$$

Then, by evaluating the functional derivative of $S(X \mid \{C\})$ with respect to $f_X(x)$ we obtain the equation

$$\frac{\delta S(X \mid \{C\})}{\delta f_X(x)} = 0 \Rightarrow \ln f_X(x) + 1 + \mu_0 + \mu_1 x = 0 \Rightarrow f_X(x) = e^{-\mu_0 - 1 - \mu_1 x}.$$

Enforcing the first constraint requires $g^{(0)}[f_X] = 0$ which leads to $e^{-\mu_0 - 1} = \mu_1$. The second constraint requires $g^{(1)}[f_X] = 0$ and leads to $e^{-\mu_0 - 1} = \mu_1^2\, m_X$, which in light of the first constraint implies that $1/\mu_1 = m_X$.

Thus, the maximum entropy pdf is given by the following exponential density expression

$$f_X(x) = \frac{1}{m_X}\, e^{-x/m_X}.$$

If you want a slightly more challenging example, consider deriving the maximum entropy pdf of a system with N centered (zero-mean) variables assuming that the covariance matrix and the average kurtosis coefficient for all the variables are known. The resulting maximum entropy pdf has the exponential form $f_X(\mathbf{x}) = \exp\left[-\mathcal{H}(\mathbf{x})\right]/Z$ with energy function $\mathcal{H}(\mathbf{x})$ given by (6.77).

13.2.4 Maximum Entropy Formulation of Spartan Random Fields

In this section we show that the joint probability density of Spartan random fields can be obtained by means of the maximum entropy principle. Centered (zero mean) Spartan spatial random fields follow an exponential Boltzmann-Gibbs joint density of the form $f_{\mathrm{x}}[x(\mathbf{s})] = Z^{-1}\mathrm{e}^{-\mathcal{H}_0}$ with the energy functional (7.4), which is repeated below for reasons of convenience

$$\mathcal{H}_0 = \frac{1}{2\eta_0 \xi^d} \int_{\mathcal{D}} \mathrm{d}\mathbf{s} \left\{ [x(\mathbf{s})]^2 + \eta_1\, \xi^2\, [\nabla x(\mathbf{s})]^2 + \xi^4 \left[\nabla^2 x(\mathbf{s})\right]^2 \right\}.$$

As we show below, the SSRF pdf can be derived from the principle of *maximum entropy*.

For a given realization $x(\mathbf{s})$, let us define the domain integrals Σ_i (for $i = 0, 1, 2$) as follows:

$$\Sigma_0[x(\mathbf{s})] := \int_{\mathcal{D}} \mathrm{d}\mathbf{s}\, x^2(\mathbf{s}), \tag{13.24a}$$

$$\Sigma_1[x(\mathbf{s})] := \int_{\mathcal{D}} \mathrm{d}\mathbf{s}\, [\nabla x(\mathbf{s})]^2, \tag{13.24b}$$

$$\Sigma_2[x(\mathbf{s})] := \int_{\mathcal{D}} \mathrm{d}\mathbf{s}\, [\nabla^2 x(\mathbf{s})]^2. \tag{13.24c}$$

The expectations $\mathcal{E}\{\Sigma_i[x(\mathbf{s})]\}$, where $i = 0, 1, 2$, represent *stochastic constraints*. In the case of random fields defined on continuum domains $\mathcal{D}$ these expectations are given respectively by (5.32) and (5.33). In the case of lattice random fields (on hypercubic grids) the respective discretized mean squared gradient and Laplacian are given by (8.20) and (8.21). In addition, we assume that *sample-based estimates*, $\overline{\Sigma_i(\mathbf{x}^*)}$, of the stochastic constraints can be estimated from the data.

We can now use the formulation outlined in the previous section. The Lagrangian function (13.18) becomes

$$S\,(\mathrm{X} \mid \{C\}) = -\int \mathcal{D}\{x(\mathbf{s})\}\, f_{\mathrm{x}}\,[x(\mathbf{s})]\, \ln f_{\mathrm{x}}\,[x(\mathbf{s})] - \sum_{k=0}^{3} \mu_k\, g^{(k)}[f_{\mathrm{x}}], \tag{13.25}$$

where $\mathcal{D}x(\mathbf{s})$ indicates functional integration, the functions $\{g^{(k)}[f_{\mathrm{x}}]\}_{k=0}^{3}$ represent the SSRF constraint-matching expressions, and the coefficients $\{\mu_k\}_{k=0}^{3}$ are Lagrange multipliers. The SSRF constraint-matching expressions are given by

$$g^{(0)}[f_{\mathrm{x}}] = \int \mathcal{D}\{x(\mathbf{s})\}\, f_{\mathrm{x}}\,[x(\mathbf{s})] - 1, \tag{13.26a}$$

$$g^{(k)}[f_x] = \int \mathcal{D}\{x(\mathbf{s})\}\, f_x\,[x(\mathbf{s})]\, \Sigma_{k-1}[x(\mathbf{s})] - \overline{\Sigma_{k-1}(\mathbf{x}^*)},\;\; k = 1, 2, 3. \quad (13.26\text{b})$$

Evaluating the stationary point of the Lagrangian function (13.25) involves calculating the functional derivative of the entropy with respect to the pdf [see (13.22)]. This leads to the *exponential density* expression for the *maximum entropy* pdf:

$$f_{\text{MaxEnt}}\,[x(\mathbf{s}); \boldsymbol{\theta}] = \mathrm{e}^{-1-\mu_0(\boldsymbol{\theta})-\mu_1\Sigma_0-\mu_2\Sigma_1-\mu_3\Sigma_2}, \quad (13.27)$$

where $\boldsymbol{\theta} = (\mu_1, \mu_2, \mu_3)^{\top}$ is the SSRF parameter vector and $\mu_0(\boldsymbol{\theta})$ normalizes the pdf through the partition function $Z = \exp(\mu_0 + 1)$.

The coefficients $\{\mu_k\}_{k=1}^{3}$ are in principle obtained by solving the system of the three SSRF constraint equations

$$\mathcal{E}\big\{\Sigma_k[x(\mathbf{s})]\big\} = \overline{\Sigma_k(\mathbf{x}^*)},\;\; \text{for } k = 0, 1, 2. \quad (13.28)$$

The pdf (13.27) is equivalent to the Gaussian Gibbs pdf with the SSRF energy functional (7.4), under the transformations $\mu_1 = 1/2\eta_0\xi^{a}$, $\mu_2 = \eta_1\xi^{2-d}/2\eta_0$, and $\mu_3 = \xi^{4-d}/2\eta_0$.

Discrete supports For *discrete supports*, the integral constraints (13.24) are replaced by discrete summations of suitably defined random field increments. For example, on *orthogonal grids* the derivatives are replaced by respective finite differences as in (13.1). The respective stochastic constraints involve the relations (8.20) and (8.21). On *irregular grids*, estimators of the constraints can be constructed using kernel functions [229, 368, 897]. Extensions of the squared gradient and squared curvature to irregular grids are discussed in Sect. 13.1.2.

In principle, SSRFs can be generalized to non-Gaussian pdfs by including higher-order moment constraints in the maximum entropy formulation [762, 763]. However, in such cases the derivation of explicit moment expressions requires the use of approximate or numerical approaches [365, 371]. Approximate approaches include the perturbation formalism and the variational approximation for non-Gaussian pdfs which are presented in Sect. 6.4.

13.3 Stochastic Local Interactions (SLI)

So far we have discussed the application of the Boltzmann-Gibbs probability density function $f \propto \mathrm{e}^{-\mathcal{H}}$ to random fields defined in the space continuum (Spartan random fields in Chap. 7) and on regular grids (Chap. 8). In the case of regular grids, we have focused on the connections between SSRFs and Markov random fields.

A practical question is whether a similar formulation to SSRFs can be developed for irregular grids starting from physically motivated energy functions and leading to sparse precision matrices. This question has been partly answered to the affirmative by the construction of *stochastic local interaction (SLI)* models that use kernel functions to implement local coupling of neighboring field values [368, 373].

13.3.1 Energy Function

In the following, we illustrate the main ideas using a simplified version of the SLI model presented in [368] that involves squared fluctuation and gradient terms. The motivation is to derive an energy function similar to the grid-based function (11.33) (except for the curvature term) which is valid for scattered data. Since the sampling point set $\mathbb{S}_N$ has an irregular geometry, it is not possible to use finite differences to approximate the square gradient. Hence, in this case we will use kernel functions to derive a generalized expression for the square gradient which involves differences between local neighbors around each point.

The energy function for this first-order model is given by

$$\mathcal{H}_0(\mathbf{x};\boldsymbol{\theta}) = \frac{1}{2\,\lambda}\left[S_0(\mathbf{x}) + \alpha_1\, S_1(\mathbf{x};\mathbf{h})\right], \tag{13.29a}$$

where $\boldsymbol{\theta} = (m_{\mathrm{x}}, \alpha_1, \lambda, \mu, k)^{\top}$ is the *SLI parameter vector*.

- m_{x} is for the mean of the field.
- The scale factor $\lambda > 0$ is proportional to the variance.
- The stiffness $\alpha_1 > 0$ determines the relative contribution of the gradient term.
- The neighbor order $k \in \mathbb{N}$ determines the near-neighbor distances used in the definition of the vector bandwidth $\mathbf{h}$ ($k = 1$ denotes the nearest neighbor of the target point in the sampling set $\mathbb{S}_N$, $k = 2$ denotes the second-nearest neighbor, etc.).
- The bandwidth scaling factor $\mu > 0$ determines the local bandwidths in combination with k.
- The vector bandwidth $\mathbf{h}$ is determined from k and μ (see below).

Non-stationarity Since the parameters of the Boltzmann-Gibbs pdf are uniform in space and do not discriminate between different directions, the SLI model presented above is *translation invariant and isotropic* with respect to the parameters. However, the stationarity of the model is broken due to the geometry of the sampling network that can lead to variations in the local neighborhood structure. This means that the variance changes at different locations and the covariance matrix depends on both the location and the lag.

Permissibility The first-order SLI model is *permissible* if the energy function is *non-negative*. In fact, if $\alpha_1 > 0$ the SLI function is positive for any configuration of the sampling set $\mathbb{S}_N$ and for any $\mathbf{x} \in \mathbb{R}^N$ that is not completely uniform.

The terms $S_0(\mathbf{x})$ and $S_1(\mathbf{x};\mathbf{h})$ correspond to sample averages of the squared fluctuations and the square gradient. $S_0(\mathbf{x})$ is given by the standard sample average

$$S_0(\mathbf{x}) = \frac{1}{N} \sum_{n=1}^{N} (x_n - m_{\mathbf{x}})^2. \tag{13.29b}$$

Square gradient on scattered grids There is no universally accepted definition of the gradient on scattered grids. Mimetic discretization approaches offer a possible alternative [390]. The SLI model does not require the full gradient, which is a vector quantity, but only its square which is a scalar quantity. Hence, it is not necessary to estimate a gradient direction. We thus estimate the square gradient by means of a kernel-weighted average of squared differences defined as follows

$$S_1(\mathbf{x}; \mathbf{h}) = \left\langle (x_n - x_m)^2 \right\rangle_{\mathbf{h}}, \tag{13.29c}$$

where $\langle \cdot \rangle_{\mathbf{h}}$ denotes the *Nadaraya-Watson two-point average* based on a vector of local bandwidths $\mathbf{h} = (h_1, \ldots, h_N)^\top$. To be dimensionally correct, we should divide the sample average by the square of a small length scale that represents the equivalent of the lattice step on regular grids. For example, we could use the characteristic grid length defined by (13.6). We might also be tempted to normalize $S_1(\mathbf{x}; \mathbf{h})$, so that its expectation converges to the expectation of the square of forward finite differences (13.2c) on a square (or cubic) grid. While this is possible, it is by no means necessary. We may as well assume that all the normalization factors are absorbed in the coefficient α_1.

13.3.2 Nadaraya-Watson Average

Next, we define the Nadaraya-Watson average of two-point functions $g(\cdot; \cdot)$ using kernel functions $K(\cdot)$ with locally adjustable bandwidth. We assume the following:

1. $\{x_n\}_{n=1}^N$ is a set of sample values at the locations $\{\mathbf{s}_n\}_{n=1}^N$.
2. $g(\cdot, \cdot)\colon \mathbb{R} \times \mathbb{R} \to \mathbb{R}$ is a symmetric, two-point scalar function of the field values, i.e., $g(x_n, x_m) = g(x_m, x_n)$, for $n, m = 1, \ldots, N$.
3. $K(\cdot)\colon \mathbb{R} \to \mathbb{R}_+$ is a kernel (weighting) function (see Definition 2.1).
4. The vector $\mathbf{h} = (h_1, \ldots, h_N)^\top$ contains the local bandwidths: $h_n > 0$ controls the neighborhood size around the location $\mathbf{s}_n$.

The two-point Nadaraya-Watson average is then defined as follows:

$$\langle g(x_n, x_m) \rangle_{\mathbf{h}} = \frac{\sum_{n,m=1}^N K\left(\frac{\|\mathbf{s}_n - \mathbf{s}_m\|}{h_n}\right) g(x_n, x_m)}{\sum_{n,m=1}^N K\left(\frac{\|\mathbf{s}_n - \mathbf{s}_m\|}{h_n}\right)}. \tag{13.29d}$$

Local bandwidths $k \in \mathbb{N}$ is an integer which determines the *order* of the near neighbors that will be used to construct the bandwidths. If $\mathbf{s}_n \in \mathbb{S}_N$, setting $k = 1$ yields zero distance, since the nearest neighbor of $\mathbf{s}_n$ in $\mathbb{S}_N$ is itself. If $\mathbf{s}_n \notin \mathbb{S}_N$, then $k = 1$ yields the finite distance between $\mathbf{s}_n$ and its first neighbor in $\mathbb{S}_N$. If $\mathbf{s}_n \notin \mathbb{S}_N$, setting $k = 2$ selects second-nearest neighbors. This choice performs well for compactly supported kernels, while $k = 1$ (nearest neighbors) is better for kernels with unbounded support (e.g., Gaussian, exponential). For compactly supported kernels $k = 2$ avoids zero-bandwidth problems that result from $k = 1$ if there are collocated sampling and prediction points. Typically we use $k \leq 5$ for the neighbor order.

For fixed k, $D_{n,[k]}(\mathbb{S}_N)$ denotes the distance between the target point $\mathbf{s}_n$ and its *k-nearest neighbor* in $\mathbb{S}_N$. The local bandwidth h_n associated with $\mathbf{s}_n$ is determined according to

$$h_n = \mu \, D_{n,[k]}(\mathbb{S}_N), \tag{13.30}$$

In the SLI model both μ and $k > 1$ are parameters. While k is set by the user, the value of μ is determined in the parameter estimation stage. The distance $D_{n,[k]}(\mathbb{S}_N)$ depends purely on the sampling point configuration, but μ also depends on the sample values.

Weight asymmetry In light of the above bandwidth definitions, the kernel average (13.29d) is asymmetric with respect to the contribution of two points $\mathbf{s}_n$ and $\mathbf{s}_m$, since the contribution of the point $\mathbf{s}_n$ (for all $m = 1, \ldots, N$) is

$$K\left(\frac{\|\mathbf{s}_n - \mathbf{s}_m\|}{h_n}\right)(x_n - x_m)^2,$$

while the contribution of the point $\mathbf{s}_m$ (for all $n = 1, \ldots, N$) is given by

$$K\left(\frac{\|\mathbf{s}_n - \mathbf{s}_m\|}{h_m}\right)(x_n - x_m)^2.$$

The bandwidths h_n and h_m are respectively determined from the neighborhoods of the points $\mathbf{s}_n$ and $\mathbf{s}_m$ and can thus be quite different, even for points that are near each other. An example that illustrates the fluctuations of the local bandwidth depending on the sampling point neighborhood is shown in Fig. 13.2. The relative size of the local bandwidth is indicated by the radius of the circle centered at $\mathbf{s}_n$ and the color scale. Larger bandwidths are typically observed near the boundary of the convex hull of the sampling set $\mathbb{S}_N$ where the sampling density is lower. Also note that neighboring points can have quite different bandwidths.

Non-Euclidean distances The SLI model is permissible even for kernel weights that are not based on the Euclidean distance $\|\mathbf{s}_n - \mathbf{s}_m\|$. In contrast, the permissibility of covariance functions depends on the definition of the distance metric. For example, the Gaussian function $\exp(-\|\mathbf{r}\|^2)$ is not a permissible covariance model

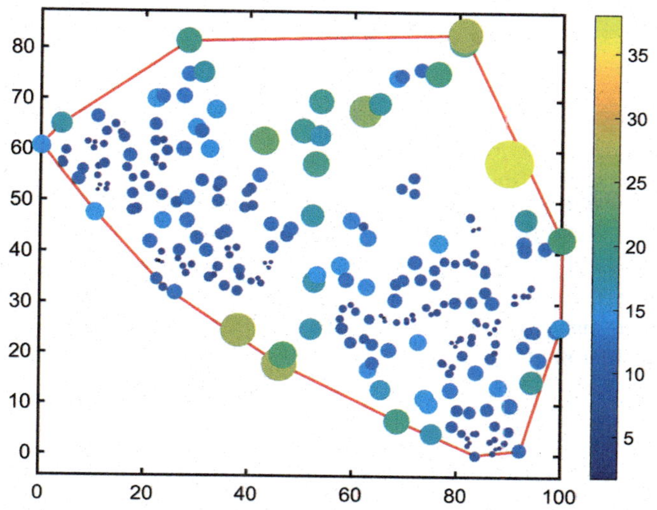

Fig. 13.2 Illustration of the relative size of local bandwidths based on the nearest neighbor distance (obtained for $k = 2$ since the distances are calculated between points in the same set). The radius of each circle is proportional to the bandwidth h_n that corresponds to the sampling point $\mathbf{s}_n$ (circle center)

for the *Manhattan distance metric* $\|\mathbf{r}\| = \sum_{i=1}^{d} |r_i|$ where $\mathbf{r} = (r_1, \ldots, r_d)^\top$. On the other hand, the Manhattan distance can be used in the kernel functions of the SLI model without breaking permissibility. This reflects the fact that the SLI model defined by (13.29) has a non-negative energy (13.29a) regardless of the distance metric used in the kernel functions. This advantageous property enables the application of SLI to spatial data that are supported on spaces (e.g., spherical surfaces) where the geodesic distance or some other distance metric is more suitable than the Euclidean distance.

13.3.3 Precision Matrix Formulation

In light of the equations (13.29) the SLI energy is a quadratic function of the field. In Sect. 11.6.2 we presented a precision matrix formulation of local interaction models on regular grids. It is straightforward to extend this formulation to irregular sampling grids using the energy function (13.29a). The latter is expressed as follows

$$\mathcal{H}_0(\mathbf{x}; \boldsymbol{\theta}) = \frac{1}{2}(\mathbf{x} - \mathbf{m}_x)^\top \mathbf{J}(\boldsymbol{\theta})\,(\mathbf{x} - \mathbf{m}_x). \tag{13.31a}$$

The *precision matrix* $[\mathbf{J}]_{n,m} = J_{n,m}(\boldsymbol{\theta})$ is given by means of

$$\mathbf{J}(\boldsymbol{\theta}) = \frac{1}{\lambda} \left[\frac{\mathbf{I}_N}{N} + \alpha_1 \, \mathbf{J}_1(\mathbf{h}) \right], \tag{13.31b}$$

where $\mathbf{I}_N$ is the $N \times N$ *identity matrix*, defined by $[\mathbf{I}_N]_{n,m} = \delta_{n,m}$, and $\delta_{n,m}$ is Kronecker's delta.

Gradient precision sub-matrix The gradient component, $\mathbf{J}_1(\mathbf{h})$, of the precision matrix is determined by the sampling pattern, the kernel function, and the vector bandwidth. We call $\mathbf{J}_1(\mathbf{h})$ a *network matrix* to emphasize its dependence on the network generated by the sampling points and all the links between these points allowed by the bandwidths. The elements of the gradient network matrix $\mathbf{J}_1(\mathbf{h})$ are given by the following equations which involve the normalized kernel weights $u_{n,m}(h_n)$:

$$[\mathbf{J}_1(\mathbf{h})]_{n,m} = -u_{n,m}(h_n) - u_{n,m}(h_m) + \delta_{n,m} \sum_{l=1}^{N} \left[u_{n,l}(h_n) + u_{l,n}(h_l) \right], \tag{13.32a}$$

$$u_{n,m}(h_n) = \frac{K\left(\frac{\|\mathbf{s}_n - \mathbf{s}_m\|}{h_n}\right)}{\sum_{n=1}^{N} \sum_{m=1}^{N} K\left(\frac{\|\mathbf{s}_n - \mathbf{s}_m\|}{h_n}\right)}, \quad \text{for } n, m = 1, \ldots, N. \tag{13.32b}$$

Properties of SLI precision matrix

1. The SLI precision matrix defined in (13.31b) is symmetric, real-valued, and *strictly positive definite*. The latter is guaranteed by the fact that the energy function (13.31a) defined by means of the precision matrix is equivalent to (13.29a). This is a convex function minimized at $\mathbf{x} = \mathbf{m}_x$ where $\mathcal{H}_0(\mathbf{m}_x; \boldsymbol{\theta}) = 0$. Hence, equation (13.31a) implies that $(\mathbf{x}')^\top \mathbf{J}(\boldsymbol{\theta})\, \mathbf{x}' > 0$ (where $\mathbf{x}' = \mathbf{x} - \mathbf{m}_x$) for all real-valued vectors $\mathbf{x}' \in \mathbb{R}^N$ that are not identically zero and for all $N \in \mathbb{N}$.
2. The SLI covariance matrix, $\mathbf{C} = \mathbf{J}^{-1}(\boldsymbol{\theta})$, is also strictly positive definite, being the inverse of a strictly positive definite matrix. The easiest way to understand this is to consider the eigenvalues $\{\lambda_n\}_{n=1}^N$ of the matrix $\mathbf{J}(\boldsymbol{\theta})$: since the latter is strictly positive definite, $\lambda_n > 0$ for all $n = 1, \ldots, N$. The eigenvalues of the matrix $\mathbf{C}$ are given by $\{\lambda_n^{-1}\}_{n=1}^N$ and thus are also positive.

   The properties 3–6 below refer to the *gradient network matrix* and are straightforward consequences of the definitions (13.31b) and (13.32):
3. *Symmetry:* The network matrices defined by (13.32) are *symmetric* by construction, i.e., $[\mathbf{J}_1(\mathbf{h})]_{n,m} = [\mathbf{J}_1(\mathbf{h})]_{m,n}$ for all $n, m = 1, \ldots, N$. Therefore, the SLI precision matrix $\mathbf{J}(\mathbf{h})$ given by (13.31b) is also symmetric.
4. *Vanishing row and column sums:* The row and column sums of the gradient network matrix (13.32) vanish, i.e.,

$$\sum_{m=1}^{N}[\mathbf{J}_1(\mathbf{h})]_{n,m} = \sum_{n=1}^{N}[\mathbf{J}_1(\mathbf{h})]_{n,m} = 0, \text{ for all } n, m = 1, \ldots, N. \tag{13.33}$$

5. *Diagonal elements:* Based on (13.32a), the diagonal elements of the gradient network matrix are given by the following expression

$$J_{1;n,n} \doteq [\mathbf{J}_1(\mathbf{h})]_{n,n} = \sum_{l=1,\neq n}^{N} \left[u_{n,l}(h_n) + u_{l,n}(h_l)\right]. \tag{13.34}$$

6. *Diagonal dominance:* Since the kernel weights are non-negative, it follows that the gradient network matrix $\mathbf{J}_1(\mathbf{h})$ is *diagonally dominant*, i.e.,

$$\left| J_{1;n,n}(\mathbf{h}) \right| \geq \sum_{m \neq n} \left| J_{1;n,m}(\mathbf{h}) \right|, \text{ for all } n = 1, \ldots, N.$$

The entire SLI precision matrix is diagonally dominant. To prove this consider that based on (13.31b) it holds

$$J_{n,n} = \frac{1}{\lambda N} + \frac{\alpha_1}{\lambda} J_{1;n,n}, \tag{13.35a}$$

$$J_{n,m} = \frac{\alpha_1}{\lambda} J_{1;n,m}, \quad m \neq n. \tag{13.35b}$$

Hence, the diagonal elements of the precision matrix take higher values than those of the scaled (by α_1/λ) gradient network matrix while the off-diagonal elements of $\mathbf{J}(\mathbf{h})$ are given by the respective gradient network matrix elements multiplied by α_1/λ. Hence, the diagonal dominance of the gradient network matrix suffices to prove the diagonal dominance of the precision matrix.

7. *Row and column sums of SLI precision matrix:* It also follows from (13.31b) and (13.33) that the sum of the SLI precision matrix elements over any row (or any column) is the same, i.e.,

$$\sum_{m=1}^{N}[\mathbf{J}(\boldsymbol{\theta})]_{n,m} = \sum_{n=1}^{N}[\mathbf{J}(\boldsymbol{\theta})]_{n,m} = \frac{1}{N\lambda}. \tag{13.36}$$

13.3.4 Mode Prediction

In this section we use the SLI model to estimate the unknown values of the process at the locations $\{\mathbf{z}_p\}_{p=1}^{P} \in \mathcal{G}$. The data vector $\mathbf{x}^*$ comprises measurements at the sampling point set $\mathbb{S}_N = \{\mathbf{s}_n\}_{n=1}^{N}$. The data are viewed as a realization from the random vector $\mathbf{X}(\omega)$ with joint pdf determined by the SLI energy function (13.31a). We also assume that the SLI parameters are known (see Sect. 13.3.5). Prediction points $\mathbf{z}_p$ are then inserted one by one in the network of sampling points. Since SLI is a local method, we assume that $\mathbf{z}_p$ is located inside the convex hull of $\mathbb{S}_N$.

To predict the unknown value of the field at $\mathbf{z}_p$, we modify the energy function (13.31a), allowing $\mathbf{z}_p$ to interact with the sampling points in $\mathbb{S}_N$. The energy function is modified due to these interactions as shown below in (13.39). The optimal value $\hat{x}_p$ at $\mathbf{z}_p$ minimizes the *SLI excess energy* generated by the inclusion of the prediction point. The minimum excess energy solution corresponds to the *mode* of the conditional (on $\boldsymbol{\theta}^*$ and $\mathbf{x}^*$) Boltzmann-Gibbs pdf which has the SLI energy (13.39) in the exponent.

Modified kernel weights Upon adding $\mathbf{z}_p$ to the set of sampling points, the weights (13.32b) in the gradient network matrix (13.32a) are modified as follows for points $\mathbf{s}_n, \mathbf{s}_m \in \mathbb{S}_N$:

$$u_{n,m}(h_n) = \frac{K\left(\frac{\|\mathbf{s}_n - \mathbf{s}_m\|}{h_n}\right)}{\sum_{k,l=1}^{N} K\left(\frac{\|\mathbf{s}_k - \mathbf{s}_l\|}{h_k}\right) + \sum_{k=1}^{N} K\left(\frac{\|\mathbf{s}_k - \mathbf{z}_p\|}{h_k}\right) + \sum_{k=1}^{N} K\left(\frac{\|\mathbf{s}_k - \mathbf{z}_p\|}{h_p}\right)}. \tag{13.37}$$

The first term in the denominator of (13.37) captures the interactions between sampling points and is identical to the denominator of (13.32b). The second term involves interactions between the sampling points and the prediction point, in which the bandwidth is controlled by the sampling points. Finally, the third term also involves interactions between the prediction point and the sampling points, but in this case the bandwidth is controlled by the prediction point. Figure 13.3 below illustrates the difference between the second and the third term.

The calculation of contributions from $\mathbf{z}_p$ in the denominator of (13.37) is an operation with computational complexity $O(N)$ compared to $O(N^2)$ for the interactions between sampling points. The latter term, however, is calculated once

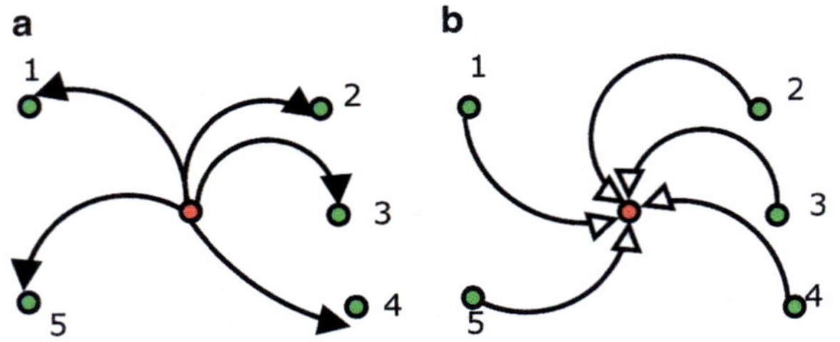

Fig. 13.3 Schematic diagrams of terms contributing to (13.39). They include the prediction point $\mathbf{z}_p$ (red center point) and five sampling points $\{\mathbf{s}_m\}_{m=1}^{5}$ (green circles) in its local neighborhood. The diagram (**a**) represents terms $u_{p,m}(h_p)$ and the diagram (**b**) represents terms $u_{m,p}(h_m)$. The point (circle marker) at the beginning of each arrow controls the bandwidth for the weight that involves the pair linked by the arrow

and then used for all the prediction points. The numerical complexity of the denominator calculation can be further improved for compactly supported kernel functions.

In addition to the weights that involve pairs of sampling points, the predictor includes weights for combinations of sampling and prediction points, i.e.,

$$u_{p,m}(h_p) = \frac{K\left(\frac{\|\mathbf{z}_p-\mathbf{s}_m\|}{h_p}\right)}{\sum_{n,m=1}^{N} K\left(\frac{\|\mathbf{s}_n-\mathbf{s}_m\|}{h_n}\right) + \sum_{n=1}^{N} K\left(\frac{\|\mathbf{s}_n-\mathbf{z}_p\|}{h_n}\right) + \sum_{n=1}^{N} K\left(\frac{\|\mathbf{s}_n-\mathbf{z}_p\|}{h_p}\right)}, \tag{13.38}$$

where $p = 1, \ldots, P$ and $m = 1, \ldots, N$. The denominator of (13.38) is identical to that of (13.37).

The weights $u_{p,m}(h_p)$ and $u_{m,p}(h_m)$ involve both $\mathbf{z}_p$, but they are in general different as illustrated schematically in Fig. 13.3. The left-hand-side diagram represents terms $u_{p,m}(h_p)$ with bandwidths determined by the prediction point, while the right-hand side diagram corresponds to terms $u_{m,p}(h_m)$ whose bandwidths are determined by the sampling points in the neighborhood of $\mathbf{z}_p$.

Modified SLI energy function Using the SLI precision matrix formulation (13.31a), the modified energy function is given by

$$\begin{aligned} 2\hat{\mathcal{H}}_0(\mathbf{x}^*, x_p; \boldsymbol{\theta}^*) &= 2\mathcal{H}_0(\mathbf{x}^*; \boldsymbol{\theta}^*) + J_{p,p}(\boldsymbol{\theta}^*)(x_p - m_{\mathrm{x}})^2 \\ &\quad + 2\sum_{n=1}^{N}(x_n^* - m_{\mathrm{x}})\, J_{i,p}(\boldsymbol{\theta}^*)\,(x_p - m_{\mathrm{x}}). \end{aligned} \tag{13.39}$$

The symmetry property of the SLI precision matrix justifies the prefactor 2 in the second line of (13.39), since $J_{i,p}(\boldsymbol{\theta}^*) = J_{p,i}(\boldsymbol{\theta}^*)$.

The first term in the energy (13.39) involves only interactions between sampling points. The second term is the self-interaction of the prediction point. Finally, the third term represents interactions between the prediction point and the sampling points. The second and third terms represent the excess SLI energy due to the "injection" of the prediction point.

The energy function (13.39) employs the "optimal" parameter vector $\boldsymbol{\theta}^*$ (see Sect. 13.3.5 below). By extending (13.32a), the elements of the gradient precision sub-matrix that involve the prediction point are

$$[\mathbf{J}_1(\mathbf{h})]_{p,p} = \sum_{n=1}^{N}\left[u_{n,p}(h_n) + u_{p,n}(h_p)\right], \tag{13.40a}$$

$$[\mathbf{J}_1(\mathbf{h})]_{p,n} = -\left[u_{n,p}(h_n) + u_{p,n}(h_p)\right], \text{ for } n = 1, \ldots, N \text{ and } \mathbf{s}_n \neq \mathbf{z}_p. \tag{13.40b}$$

SLI mode predictor The *SLI mode predictor* is the value that minimizes the modified SLI energy function (13.39), i.e.,

$$\hat{x}_p = \arg\min_{x_p} \hat{\mathcal{H}}_0(\mathbf{x}^*, x_p; \boldsymbol{\theta}^*). \tag{13.41}$$

Minimization of the quadratic energy function (13.39) with respect to x_p leads to the following SLI mode estimate conditional on the data $\mathbf{x}^*$:

$$\begin{aligned}\hat{x}_p &= m_{\mathrm{x}} - \frac{\sum_{n=1}^{N}\left[J_{n,p}(\boldsymbol{\theta}^*) + J_{p,n}(\boldsymbol{\theta}^*)\right](x_n^* - m_{\mathrm{x}})}{2\,J_{p,p}(\boldsymbol{\theta}^*)} \\ &= m_{\mathrm{x}} - \frac{\sum_{n=1}^{N} J_{p,n}(\boldsymbol{\theta}^*)\,(x_n^* - m_{\mathrm{x}})}{J_{p,p}(\boldsymbol{\theta}^*)},\end{aligned} \tag{13.42a}$$

where the precision matrix elements are given by (13.40) using the modified kernel weights (13.37) and (13.38). Note that (13.42a) is essentially the GMRF conditional expectation (8.7) (albeit with different precision matrix). This is not surprising since the SLI model generalizes grid-based GMRFs to scattered data.

Equation (13.42a) can also be expressed in matrix form as follows

$$\hat{x}_p = m_{\mathrm{x}} - \frac{1}{J_{p,p}(\boldsymbol{\theta}^*)}\,\mathbf{J}_{p,\mathbb{S}_N}(\boldsymbol{\theta}^*)\,\left(\mathbf{x}^* - \mathbf{m}_{\mathrm{x}}\right),\ \ p = 1, \ldots, P, \tag{13.42b}$$

where $\mathbf{J}_{p,\mathbb{S}_N} = (J_{p,1}, \ldots, J_{p,N})$.

SLI variance In analogy with the GMRF conditional variance (8.8), the *conditional variance* of the SLI predictor is given by

$$\sigma^2_{\mathrm{x}}(\mathbf{z}_p \mid \mathbf{x}^*) = \frac{1}{J_{p,p}}. \tag{13.43}$$

The SLI conditional variance is not spatially uniform, because it depends on the local variations of the sampling density which affect the weights in the precision matrix [cf. (13.38)]. Given the local nature of compactly supported kernel functions, the SLI variance is typically larger than respective kriging variances which are based on dense precision matrices.

The variance is reduced if the SLI method is implemented with non-compact kernels such as the exponential and the Gaussian. However, in this case the computational advantage of SLI is reduced since the summation over all the kernel weights is an $O(N^2)$ operation.

SLI PACF The *partial autocorrelation function* of the SLI model is given by the equation (8.9) for the GMRF PACF using the SLI precision matrix.

Multipoint SLI mode predictor The SLI mode predictor can be generalized to P prediction points as follows

$$\hat{\mathbf{x}} = \mathbf{m}_{\mathrm{x}} - \tilde{\mathbf{J}}_{\mathcal{G},\mathbb{S}_{\mathrm{N}}}(\boldsymbol{\theta}^*)\,(\mathbf{x} - \mathbf{m}_{\mathrm{x}}), \tag{13.44a}$$

where $\tilde{\mathbf{J}}_{\mathcal{G},\mathbb{S}_{\mathrm{N}}}(\boldsymbol{\theta}^*)$ is a $P \times N$ matrix whose elements are given by

$$[\tilde{\mathbf{J}}_{\mathcal{G},\mathbb{S}_{\mathrm{N}}}(\boldsymbol{\theta}^*)]_{p,n} = \frac{J_{p,n}(\boldsymbol{\theta}^*)}{J_{p,p}(\boldsymbol{\theta}^*)}, \quad p = 1, \ldots, P, \ \text{and}\ n = 1, \ldots, N. \tag{13.44b}$$

Properties of SLI predictor The SLI mode predictor satisfies certain useful properties that result from (13.42).

1. The SLI prediction (13.42b) is *unbiased* if $\mathcal{E}[x_n^*(\omega)] = m_{\mathrm{x}}$.
2. The SLI prediction is independent of the parameter λ that determines the amplitude of the fluctuations. The prediction (13.42) depends on the ratio of precision matrix elements which is independent of λ. This property is analogous to the independence of the kriging predictor from the random field's variance. The optimal value of λ is obtained from (13.46).
3. The SLI predictor is not constructed as an exact interpolator. It can become approximately exact only if the kernel function is compactly supported and if the kernel bandwidth is based on the nearest-neighbor distance.
4. The "naive" computational complexity of the SLI predictor is $O(N^2 + 2P\,N)$.

   The $O(N^2)$ term is due to the double summation of the kernel weights over the sampling points in (13.37). This term is calculated only once. The remaining operations per prediction point scale linearly with the sample size, hence the $O(2P\,N)$ dependence. The $O(N^2)$ dependence can be improved for compactly supported kernels using the sparsity of the precision matrix and KD-tree structures [742]. It is also worthwhile investigating analytical continuum approximations of the double summation of kernel weights.

13.3.5 Parameter Estimation

In the previous sections we assumed that the optimal SLI parameter vector $\boldsymbol{\theta}^* = (m_{\mathrm{x}}^*, \alpha_1^*, \lambda^*, \mu^*, k^*)^\top$ is known. In this section we discuss the estimation of the SLI parameters from the data.

Maximum likelihood estimation One possibility is to maximize the likelihood of the SLI model. The likelihood is defined as

$$\mathcal{L}(\boldsymbol{\theta}; \mathbf{x}^*) = \frac{\mathrm{e}^{-\mathcal{H}_0(\mathbf{x}^*;\boldsymbol{\theta})}}{Z(\boldsymbol{\theta})}, \tag{13.45}$$

where the energy function $\mathcal{H}_0(\cdot)$ is given by (13.31a). The partition function $Z(\boldsymbol{\theta})$ is given by the summation over all possible states. Since the SLI energy is a quadratic function of the field states, the partition function can be expressed—after performing the Gaussian integrals—in terms of the determinant of the precision matrix, i.e.,

$$Z = \sum_{\mathbf{x} \in \Omega} e^{-\mathcal{H}_0(\mathbf{x};\boldsymbol{\theta})} = (2\pi)^{N/2} \left[\det(\mathbf{J})\right]^{-1/2}.$$

Determinant calculation Maximum likelihood estimation can be pursued using standard procedures as outlined in Sect. 12.4. However, this approach is beset by the $O(N^3)$ computational complexity for the calculation of the determinant of the precision matrix. Newly developed, hierarchical factorization methods may reduce the computational complexity to $O(N \log^2 N)$ [19]. In addition, improved numerical complexity can be achieved for sparse positive definite matrices [680]. Methods for the efficient calculation of the log-determinant are also investigated for applications of *sparse Gaussian processes* [826].

The *scale factor* λ in the energy $\mathcal{H}_0(\mathbf{x}^*; \boldsymbol{\theta})$ can be explicitly obtained by partial maximization of the likelihood with respect to λ, in analogy with the variance estimation in Sect. 12.4.1.

Let $\boldsymbol{\theta}_{-\lambda|k} = (m_{\mathrm{x}}, \alpha_1, \mu)^\top$ represent the SLI parameter vector excluding the theoretically determine scale factor λ and the fixed neighbor order k. The latter is usually assigned a predefined low integer value between one and five. If $\tilde{H}(\mathbf{x}^*; \boldsymbol{\theta}_{-\lambda|k})$ is the energy given by (13.31) using $\lambda = 1$ for fixed k, the optimal vector $\boldsymbol{\theta}^*_{-\lambda|k}$ is defined as

$$\boldsymbol{\theta}^*_{-\lambda|k} = \arg\min_{\boldsymbol{\theta}_{-\lambda|k}} \mathrm{NLL}(\boldsymbol{\theta}_{-\lambda|k}; \mathbf{x}^*).$$

The optimal value λ_k^* of the scale factor is obtained by minimizing the negative log-likelihood with respect to λ. This calculation leads to [368, Appendix]

$$\lambda_k^* = \frac{2\tilde{H}(\mathbf{x}^*; \boldsymbol{\theta}^*_{-\lambda|k})}{N}. \tag{13.46}$$

Estimation based on cross validation Estimation of $\boldsymbol{\theta}^*$ by optimizing a specified cross validation metric is computationally more efficient than maximum likelihood estimation, because the computational complexity of the former is $O(N^2)$ in the worst case compared with $O(N^3)$ for the latter.

The following general *cross validation cost function* can be used

$$\Phi(\mathbf{x}^*; \boldsymbol{\theta}_{-\lambda|k}) = D\left(\hat{\mathbf{x}}_{\mathrm{loo}}(\boldsymbol{\theta}_{-\lambda|k}), \mathbf{x}^*\right). \tag{13.47}$$

In the cost function (13.47) $\hat{\mathbf{x}}_{\mathrm{loo}} = (\hat{x}_1, \ldots, \hat{x}_N)^\top$ is the prediction vector that comprises the *leave-one-out* SLI predictions $\hat{x}_n$ at the points $\mathbf{s}_n \in \mathbf{S}_{\mathrm{N}}$, based on the *reduced sampling set* $\mathbf{S}_{\mathrm{N}}^{(n)} = \mathbf{S}_{\mathrm{N}} \setminus \{\mathbf{s}_n\}$, for $n = 1, \ldots, N$. The SLI predictions are based on the interpolation equation (13.42) and do not involve λ.

The function $D(\mathbf{x}, \mathbf{y})$ represents a *statistical measure of distance* between the vectors $\mathbf{x}$ and $\mathbf{y}$. For example $D(\cdot, \cdot)$ could represent the root mean square error, the mean absolute error, the correlation coefficient, or some combination of these measures.

The optimal parameter vector $\boldsymbol{\theta}_{-\lambda|k}$ is determined by minimizing the cost functional (13.47), i.e.,

$$\boldsymbol{\theta}^*_{-\lambda|k} = \underset{\boldsymbol{\theta}_{-\lambda|k}}{\arg\min}\, \Phi(\mathbf{x}^*; \boldsymbol{\theta}_{-\lambda|k}). \tag{13.48}$$

The optimal scale factor λ^*_k is estimated from (13.46) using $\boldsymbol{\theta}^*_{-\lambda|k}$. The parameter estimation can be repeated for different values of k, in order to finally select the *optimal neighbor order* k^* according to

$$k^* = \underset{k\in\mathbb{N}}{\arg\min}\, \Phi(\mathbf{x}^*; \boldsymbol{\theta}^*_{-\lambda|k}). \tag{13.49}$$

Summary The main steps of the SLI parameter estimation based on cross validation are summarized below.

1. Select a kernel function.
2. Select the neighbor order $k \in \mathbb{N}$.
3. Choose the cross-validation function $D(\cdot, \cdot)$.
4. Determine the optimal parameter vector $\boldsymbol{\theta}^*_{-\lambda|k}$, conditional on the neighbor order k, by minimizing the cross-validation function based on (13.48).
5. Determine the optimal scale factor λ^*_k based on (13.46) using $\boldsymbol{\theta}^*_{-\lambda|k}$.
6. Repeat steps 2–5 for different values of k and select the optimal k^* that minimizes $\Phi(\mathbf{x}^*; \boldsymbol{\theta}^*_{-\lambda|k})$ according to (13.49).
7. Repeat the steps 1–6 for different types of kernel functions.

Estimation based on pseudo-likelihood The concept of pseudo-likelihood was introduced to address the computational challenges posed by maximum likelihood estimation in the case of Gaussian Markov random fields [70, 71, 682].

Conditional likelihood The conditional likelihood at the point $\mathbf{s}_n$ is defined as the likelihood of the datum x^*_n conditionally upon the values $\mathbf{x}^*_{-n}$ at all the points in $\mathbb{S}_N \setminus \{\mathbf{s}_n\}$, that is

$$\mathcal{L}(\boldsymbol{\theta}; x^*_n \mid \mathbf{x}^*_{-n}) = \mathcal{L}\left(\boldsymbol{\theta}; x^*_n \mid \mathcal{B}(\mathbf{x}^*_{-n})\right),$$

where $\mathcal{B}(\mathbf{x}^*_{-n})$ denotes the set of the process values inside the local neighborhood $B(\mathbf{s}_n)$ that are "connected" with the location $\mathbf{s}_n$ according to Definition 8.1.

The *pseudo-likelihood* $\mathcal{L}_{\text{pseudo}}(\boldsymbol{\theta}; \mathbf{x}^*)$ is then defined as the product of the conditional likelihoods over all the sites in $\mathbb{S}_N$, i.e.,

$$\mathcal{L}_{\text{pseudo}}(\boldsymbol{\theta}; \mathbf{x}^*) = \prod_{n=1}^{N} \mathcal{L}(\boldsymbol{\theta}; x^*_n \mid \mathbf{x}^*_{-n}). \tag{13.50}$$

Using equation (8.4) for the conditional pdf of GMRFs and equation (8.6) that links the precision matrix to the coupling parameters, the *SLI conditional likelihood*, can be expressed as follows

$$\mathcal{L}(\boldsymbol{\theta}; x_n^* \mid \mathbf{x}_{-n}^*) = \frac{J_{n,n}^{1/2}}{\sqrt{2\pi}} \exp\left[-\frac{J_{n,n}}{2} \left(x_n^* - m_x + \sum_{m=1}^{N} \frac{J_{n,m}(x_m^* - m_x)}{J_{n,n}} \right)^2 \right], \tag{13.51}$$

where the elements of the precision matrix are determined by (13.31b) and (13.32). Then, neglecting the irrelevant constant term $\sqrt{2\pi}$, the negative logarithm of the pseudo-likelihood becomes

$$-\ln \mathcal{L}_{\text{pseudo}}(\boldsymbol{\theta}; \mathbf{x}^*) = \sum_{n=1}^{N} \frac{J_{n,n}}{2} \left[x_n^* - m_x + \frac{\sum_{m=1}^{N} J_{n,m}(x_m^* - m_x)}{J_{n,n}} \right] - \frac{1}{2} \sum_{n=1}^{N} \ln J_{n,n}. \tag{13.52}$$

The pseudo-likelihood estimation approach bypasses the calculation of the determinant of the full precision matrix [391]. The function (13.52) can then be numerically minimized to determine the optimal parameter set $\boldsymbol{\theta}$ for any given kernel function and neighbor order k. The optimal values of k and λ can be determined using the same steps as in the cross validation approach.

13.4 Measuring Ergodicity

A main goal of spatial data analysis is to estimate the structure and parameters of the random field model from a single sample. The sample-based approach gives an accurate estimate of the mean and the covariance function if the random field satisfies ergodic conditions (see Sect. 4.1).

Even if ergodic conditions are in principle satisfied (i.e., at the limit of an infinite domain) for a given random field model, we would like to know if finite sample domains (that cannot expand at will) are sufficiently large to be considered as practically infinite; otherwise, significant *sample-to-sample fluctuations* can be expected between different simulated states of the random field. Below we discuss the definition of *ergodic indices* that can be used to quantify if the spatial domain is sufficiently large to allow assuming ergodic conditions [380].

Ergodicity is a property that cannot be falsified, since typically only a single realization of the field is possible. The reason is that disproving ergodicity requires showing that the ergodic conditions (4.2) or (4.3) are violated. These conditions involve the covariance function, which is not known a priori. If some information (e.g. asymptotic solutions of differential equations satisfied by the random field) is available about the large-lag behavior of the covariance function, this information can be used to investigate the validity of the ergodic conditions. If there is no other information than data, tests of ergodic criteria can only be based on the sample-based estimate of the covariance function. However, the latter is not a reliable estimate of the "true" covariance if the field is not ergodic, thus creating a chicken-and-egg problem. In practice, ergodic models are often used based on statistical moments that are estimated from spatial (instead of ensemble) averages. A thorough discussion of ergodicity is given in the book of Chilés and Delfiner [132].

An intuitive argument for the ergodic criterion (Slutsky's theorem), is presented in [92]. A sample function of a correlated random field over a domain $\mathcal{D}$ can be viewed as a collection of N_{eff} "statistically independent" random variables. Each of the independent random variables extends over a "correlated volume" v_{cor}, such that $N_{\text{eff}}\, v_{\text{cor}} = |\mathcal{D}|$. The correlated volume can be estimated by means of the integral $v_{\text{cor}} = \int d\mathbf{r}\, \rho_{\text{xx}}(\mathbf{r})$. Then, the number of independent units is approximately given by

$$N_{\text{eff}} = \left\lfloor \frac{|\mathcal{D}|}{\int d\mathbf{r}\, \rho_{\text{xx}}(\mathbf{r})} \right\rfloor,$$

where $\lfloor y \rfloor$ denotes the largest non-negative integer that is less than or equal to y. Slutsky's theorem focuses on the limit $\mathcal{D} \to \infty$. According to (4.2), ergodicity requires that $N_{\text{eff}} \to \infty$ as $\mathcal{D} \to \infty$. If this condition holds, it can be argued that the N_{eff} independent units adequately sample the entire probability distribution.

The integral of the correlation function, provided that it converges, is expressed in terms of the integral range ℓ_c according to (5.38). Hence, it follows that

$$N_{\text{eff}} = \left\lfloor \frac{|\mathcal{D}|}{\ell_c^d} \right\rfloor.$$

According to Slutsky's theorem, the random field is ergodic if the integral range is finite and independent of the domain size. This condition ensures that $N_{\text{eff}} \to \infty$ as the domain $\mathcal{D}$ increases. The connection between ergodicity and the integral range is investigated in Lantuéjoul [486].

13.4.1 Ergodic Index

A key practical question related to ergodicity is the following: Consider an ergodic random field sampled over a finite domain $\mathcal{D}$. Under what conditions can $\mathcal{D}$ be considered as practically infinite, in the sense that a single realization contains sufficient information about the field's correlations?

In particular, if $\mathcal{D}$ is large with respect to the correlation scale of the field we expect that the fluctuations of sample-based statistical estimates will not differ significantly between samples.

We first consider isotropic random fields $X(\mathbf{s}; \omega)$ with a radial covariance function, and we define the following *ergodic index*

$$I_{\text{erg}} = \frac{\ell_c^d}{|\mathcal{D}|}. \tag{13.53}$$

Slutsky's theorem (4.2) is satisfied for finite ℓ_c: in this case, $I_{\text{erg}} \to 0$ as $\mathcal{D} \to \infty$. The ergodic index takes finite values for large (but not infinite) domains. Hence, it allows us to measure the "deviation" of random field samples on finite-size domains from the ideal ergodic condition ($I_{erg} = 0$). If $I_{erg} \approx 0$, the domain is sufficiently large to contain adequate information about the field.

What is the cutoff value of I_{erg} above which the sample does not contain adequate information about the field? This question can only be answered properly in a probabilistic framework. However, if we arbitrarily choose $N_{eff} \geq 30$ (30 sometimes referred to as the "magic number in statistics") to ensure adequate sampling of the random field, a rough estimate of the *critical value of the ergodic index* is $I_{erg}^{crit} = 0.03$.

In Fig. 13.4 we illustrate the impact of the domain size and the random field parameters on the assessment of ergodicity based on the empirical ergodic index. The realizations shown in the nine plots are generated from two-dimensional SSRFs

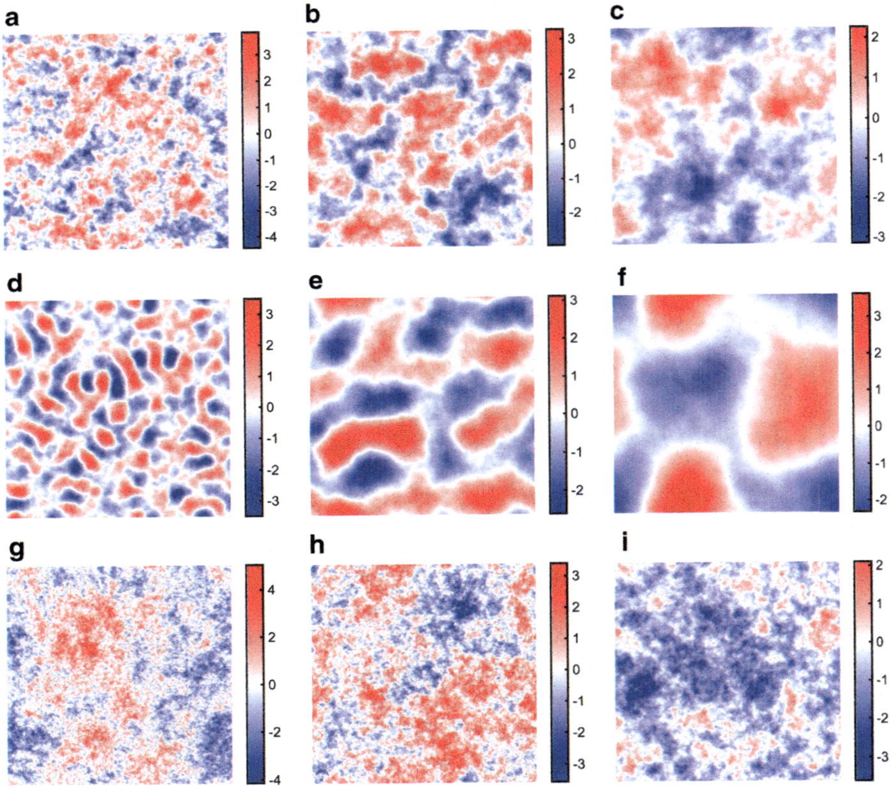

Fig. 13.4 Realizations of two-dimensional SSRFs with $\eta_1 = -1.95$ (middle row), $\eta_1 = 2$ (top row), and $\eta_1 = 100$ (bottom row) on a square domain with 512 points per side and a unit step. The characteristic length is $\xi = L/n$, where $n = 50$ (left column), $n = 20$ (middle column) and $n = 10$ (right column). The field is normalized by dividing with the corresponding standard deviation. (**a**) $\xi = L/50$, $\eta_1 = 2$. (**b**) $\xi = L/20$, $\eta_1 = 2$. (**c**) $\xi = L/10$, $\eta_1 = 2$. (**d**) $\xi = L/50$, $\eta_1 = -1.95$. (**e**) $\xi = L/20$, $\eta_1 = -1.95$. (**f**) $\xi = L/10$, $\eta_1 = -1.95$. (**g**) $\xi = L/50$, $\eta_1 = 100$. (**h**) $\xi = L/20$, $\eta_1 = 100$. (**i**) $\xi = L/10$, $\eta_1 = 100$

Table 13.1 Values of the ergodic index (13.53) for the SSRF realizations shown in Fig. 13.4. The shaded cells correspond to ergodic index values that exceed the critical value $I_{\mathrm{erg}}^{\mathrm{crit}} = 0.03$

η_1	-1.95	2	100
$L/\xi = 50$	0.0004	0.0050	0.0546
$L/\xi = 20$	0.0024	0.0314	0.3410
$L/\xi = 10$	0.0096	0.1257	1.3641

with different characteristic lengths ξ and rigidity parameters η_1. In all of the plots the same domain, a square with lengths of size $L = 512$ is used. The characteristic SSRF length ξ changes according to L/n where $n = 50, 20, 10$ as we move from the left to the right column. The rigidity parameter also changes from $\eta_1 = 2$ to $\eta_1 = -1.95$ and finally $\eta_1 = 100$ as we move from the top to the bottom row.

The values of the ergodic index for the realizations of Fig. 13.4 are shown in Table 13.1. The integral range of two-dimensional SSRFs is calculated based on (7.40). The main trends shown in the table are that the ergodic index increases with (i) the rigidity parameter and (ii) with the ratio ξ/L. The shaded cells correspond to ergodic index values that exceed the critical threshold $I_{\mathrm{erg}}^{\mathrm{crit}}$. Hence, they correspond to realizations that do not contain "complete" information about the field.

The realization shown in Fig. 13.4e meets the ergodicity criterion with a value of I_{erg} clearly below 0.03. A visual inspection of the plot, however, does not seem to corroborate this conclusion. The failure of the large-domain assumption is even more pronounced in Fig. 13.4f. The apparent failure of the index is due to the oscillations of the negative rigidity correlation function that reduce the integral range of the field, thus lowering I_{erg}. To illustrate the impact of negative values of the correlation function, consider the "extreme" case of a one-dimensional cosine function with wavelength λ. The integral range calculated over a length L which is a multiple of λ is equal to zero, in spite of the finite characteristic length.

13.4.2 Improved Ergodic Index

An improved ergodic index should be able to handle negative correlations and at the same time converge to the ergodic index (13.53) in the case of monotonically declining correlations. Such a new index can be constructed using the correlation spectrum (5.43). For monotonically declining covariance functions and $\alpha = 0$, it holds that $\ell_c = 2\pi\,\lambda_c^{(0)}$, where $\lambda_c^{(\alpha)}$ represents the correlation spectrum. Hence, the new ergodic index can be defined as follows

$$\tilde{I}_{\mathrm{erg}} = \frac{\left(2\pi\,\lambda_c^{(0)}\right)^d}{|\mathcal{D}|}. \tag{13.54}$$

Table 13.2 Ergodic index values for the SSRF realizations shown in Fig. 13.4 based on (13.55). The shaded cells correspond to $\tilde{I}_{\text{erg}}$ values that exceed the critical value $I_{\text{erg}}^{\text{crit}} = 0.03$

η_1	-1.95	2	100
$L/\xi = 50$	0.0017	0.0050	0.0546
$L/\xi = 20$	0.0108	0.0314	0.3410
$L/\xi = 10$	0.0431	0.1257	1.3641

The calculation of the zero-index correlation spectrum for different d and η_1 is presented in Sect. 7.3.4. Based on (5.44), in the case of a random field with monotonically declining spectral density the ergodic index (13.54) is equivalent to the ergodic index defined by (13.53). In the case of a spectral density with possibly negative η_1, the new index is given by the following

$$\tilde{I}_{\text{erg}} = \begin{cases} I_{\text{erg}}, & \text{if } \eta_1 \geq 0, \\ \frac{I_{\text{erg}}}{\sqrt{1-\eta_1{}^2/4}}, & \text{if } \eta_1 < 0. \end{cases} \tag{13.55}$$

The modified ergodic index (13.55) for the SSRF samples shown in Fig. 13.4 is shown in Table 13.2. The sample in Fig. 13.4f has a modified index equal to 0.0431, i.e., above the threshold. On the other hand, the index of the sample in Fig. 13.4e (i.e., 0.0108), remains below the threshold.

13.4.3 Directional Ergodic Index

In the case of anisotropic random fields, the domain $\mathcal{D}$ must contain many multiples of the respective characteristic length in each orthogonal direction. To simplify things, we consider a rectangular domain with sides $L_i, i = 1, \ldots, d$, and an ergodic random field such that the directions of its principal anisotropy axes coincide with the orthogonal domain sides.

Directional length scales $\lambda_{c,i}^{(0)}$ can be defined in each orthogonal direction $\hat{\mathbf{e}}_i$, for $i = 1, \ldots, d$, based on the following extension of the correlation spectrum (5.43)

$$\lambda_{c,i}^{(\alpha)} = \frac{\sup_{\mathbf{k}\in\mathbb{R}^d} \widetilde{C_{\text{xx}}}(\|\mathbf{k}\|)}{\int_{-\infty}^{\infty} \mathrm{d}k\, \widetilde{C_{\text{xx}}}(k\,\hat{\mathbf{e}}_i)}. \tag{13.56}$$

Then, we can define d *directional ergodic indices* by means of

$$\tilde{I}_{\text{erg},i} = \frac{2\pi\, \lambda_{c,i}^{(0)}}{L_i}, \quad \text{for } i = 1, \ldots, d. \tag{13.57}$$

An adequately large sample requires that all the directional indices be small. We can construct a *composite ergodic index*, which is based on the largest of the d indices as follows

$$\tilde{I}_{\text{erg}} = \max_{i=1,\ldots,d} \left\{ \tilde{I}_{\text{erg},1}, \ldots, \tilde{I}_{\text{erg},d} \right\}. \tag{13.58}$$

The above definition is conservative, since it is based on the largest value of the directional ergodic index, which corresponds to the direction most likely to deviate from favorable sampling conditions. Then, the criterion for a sample domain to be large enough for ergodic conditions to hold is that

$$\tilde{I}_{\text{erg}} \leq I_{\text{erg}}^{\text{crit}}. \tag{13.59}$$

The above indices provide practical, empirical guides for testing whether a given sample domain can be considered sufficiently large for ergodic conditions to approximately hold. We have shown that the integral range (in the case of monotonically declining correlation functions) and the correlation spectrum (which is better adapted to correlation functions that include negative values), allow quantifying the proximity to ergodic conditions. In the case of correlation functions that involve more than two parameters (i.e., more than the variance and the characteristic length ξ), both of these length scales are not uniquely determined by ξ. This has been demonstrated by means of the Spartan covariance function, where the proximity to ergodic conditions depends on the rigidity coefficient η_1 in addition to ξ.

Chapter 14
Beyond the Gaussian Models

Μανθάνων Μη Κάμνε
Do not get tired of learning
Delphic maxim

This chapter focuses on the modeling of non-Gaussian probability distributions. The main reason for discussing non-Gaussian models is the fact that spatial data often exhibit properties such as (i) strictly positive values (ii) asymmetric (skewed) probability distributions (iii) long positive tails (e.g., power-law decay of the pdf) and (iv) compact support.

In some cases, the departure from Gaussianity is mild and can be addressed with relatively simple fixes. In other cases, however, the differences are significant and require radically different thinking. In addition, the spatial and temporal scales of observation as well as the temporal resolution (for space-time data) play an important role in determining the probability distributions of the observables. Typical examples of environmental variables that exhibit non-Gaussian probability distributions involve precipitation [15, 16, 52, 785, 861], ocean waves [255, 584, 585, 872, 873], wind speed [341, 484, 583, 748], and spatial processes that follow skewed distributions with extreme values [182, 183].

A common approach for modeling non-Gaussian spatial data is based on *nonlinear transformations* of latent Gaussian random fields which generate either binary-valued (e.g., indicators) or continuous (e.g., lognormal) non-Gaussian random fields.

A different approach involves *intrinsically non-Gaussian* pdf models (e.g., Student's t-distribution) which admit closed-form expressions for the joint distribution. *Exponential Boltzmann-Gibbs pdfs* can also be used to define random field model based on suitable "energy" functionals (e.g., Ising model). This framework was employed to define (Gaussian) Spartan spatial random fields in Sect. 7.

D. T. Hristopulos, *Random Fields for Spatial Data Modeling*, Advances in Geographic Information Science, https://doi.org/10.1007/978-94-024-1918-4_14

Finally, a fourth approach (not common in spatial data analysis) involves *non-Gaussian perturbations* that are formulated in terms of Boltzmann-Gibbs densities with higher-order (than quadratic) terms in the energy functional. This perturbation approach was discussed in Chap. 6.

The following is an indicative list of non-Gaussian models available in the scientific literature.

- Extensions of Gaussian models to multivariate, non-Gaussian, symmetric probability distributions with explicit joint density expressions. These include the multivariate *Student's t-distribution and Wishart* (multivariate chi-square) models [26, 468]. Multivariate distributions are classified as spherical or elliptical if they have, respectively, spherical or ellipsoidal constant-density contours, such as the multivariate normal and Student's t-distributions [239].
- Asymmetric deformations of elliptical distributions that can be derived by means of multiplicative functions which modify the pdf and introduce skewness [284].
- Perturbation expansions of the non-Gaussian pdf around some variational Gaussian pdf as discussed in Chap. 6. This approach is often used in statistical physics [245, 474] and in machine learning [521, 561].
- Various transformations of Gaussian random fields, such as thresholding (level cuts, excursion sets), truncation, and continuous nonlinear transforms that lead to non-Gaussian probability densities [93, 487, 766, 804, 823].
- Intrinsically non-Gaussian models defined by Boltzmann-Gibbs exponential joint densities, such as the Ising, Potts, and clock models used in statistical physics [297, 593, 890].
- Non-Gaussian random fields can also be constructed as mixtures of Gaussian models. Random fields based on Gaussian mixtures are particularly suitable for modeling bimodal and multi-modal data distributions. This approach has mainly been pursued in pattern recognition and machine learning [18, 78, 681]. The maximization of the likelihood function of Gaussian mixtures is carried out using the *expectation-maximization* algorithm due to Dempster *et al.* [193].
- Copula models impose non-Gaussian dependence structure between random variables that follow arbitrary continuous marginal distributions. Various copula families have been developed for applications in epidemiology, mathematical finance and environmental spatial data [48, 147, 307, 445, 524, 525].
- Methods based on spin glasses and replica have been successful in the restoration of digital images [119]. Replica theory has also been used to calculate effective parameters in heterogeneous media [50, 369], and for the inference of statistical models in high-dimensional spaces [12]. It is an open question if it will have an impact in spatial data analysis as well.

Some of the aforementioned methods are well established (e.g., indicators, level sets, Box-Cox transforms, Hermite polynomials), others are gaining momentum (copulas, Tukey and Student random fields), while some other ones are promising but require further investigation (e.g., transforms based on κ-functions, replica theory, variational approach).

We first focus on non-Gaussian random fields generated by nonlinear transformations such as normal scores, Box-Cox, κ-exponential and κ-logarithm functions, Tukey random fields and Hermite polynomials. We then discuss intrinsically non-Gaussian models based on the multivariate Student's t-distribution and copulas. We close this chapter with a short discussion of the replica method. The following chapter deals with concepts and models pertaining to binary-valued random fields. These topics include the indicator random fields, the generalized linear model and the Ising model.

14.1 Trans-Gaussian Random Fields

Nonlinear transformations, such as the square root and the logarithm can be used to convert asymmetric marginal probability distributions to the Gaussian distribution. Conversely, random fields $X(\mathbf{s};\omega)$ with non-Gaussian marginal distributions can be generated by applying suitable inverse transforms to Gaussian random fields.

Definition 14.1 *Trans-Gaussian random field:* The random field $X(\mathbf{s};\omega)$ is called trans-Gaussian if there exists a nonlinear transformation $g(\cdot) : y \in \mathbb{R} \to x \in \mathbb{R}$ such that

$$g\left[X(\mathbf{s};\omega)\right] = m_{\mathrm{y}}(\mathbf{s}) + Y(\mathbf{s};\omega) + \varepsilon(\mathbf{s};\omega), \tag{14.1}$$

where (i) $m_{\mathrm{y}}(\mathbf{s})$ is a trend function, (ii) $Y(\mathbf{s};\omega)$ is a zero-mean, Gaussian random field, and (iii) $\varepsilon(\mathbf{s};\omega) \overset{d}{=} \mathcal{N}(0,\sigma_\varepsilon^2)$ is a Gaussian white noise process with variance σ_ε^2, which is independent of $Y(\mathbf{s};\omega)$. Equivalently, $m_{\mathrm{y}}(\mathbf{s})$ can be absorbed in $Y(\mathbf{s};\omega)$, which then becomes a Gaussian random field with mean $m_{\mathrm{y}}(\mathbf{s})$.

The transformation $g(\cdot)$, if it exists, maps the data values x_n^* to normally distributed values y_n^*, for $n = 1,\ldots,N$. The spatial model is then constructed using the transformed data. The latter can be interpolated on a regular grid using kriging methods, and the inverse transformation generates the respective state of the original (non-Gaussian) field. A similar approach can be used for simulation. However, the transformation method involves a "leap of faith" since it tacitly assumes that the joint probability distribution is also normalized along with the marginal distribution.

Trans-Gaussian Kriging For interpolation and simulation purposes, estimates $\{\hat{y}_p\}_{p=1}^P$ are generated based on the data at the prediction points $\{\mathbf{z}_p\}_{p=1}^P$. Predictions derived by means of kriging correspond to the mean of the conditional distribution at the prediction point. Since the conditional distribution of the predictions $\{\hat{y}_p\}_{p=1}^P$ is Gaussian, the mean and the median coincide.

Prediction of the original field $X(\mathbf{s};\omega)$ is then obtained by the inverse nonlinear transformation

$$\hat{x}_p = g^{-1}\left(\hat{y}_p\right) = \phi\left(\hat{y}_p\right), \quad p = 1,\ldots,P,$$

where $\phi(\cdot) = g^{-1}(\cdot)$ is the inverse of the normalizing transformation.

If the prediction $\hat{y}_p$ is obtained by means of kriging, the above method for estimating $\hat{x}_p$ is known as *trans-Gaussian kriging* [186, 667, 717].

Quantile invariance If the transformation $g(\cdot)$ is a *monotonically increasing function*, it preserves ordering, i.e., $y_1 \leq y_2$ is equivalent to $g(y_1) \leq g(y_2)$. Such transformations also preserve the quantiles of the distributions: if $\hat{y}_p$ corresponds to the q-th quantile of the marginal distribution $F_{\mathrm{y}}(y)$, then $F_{\mathrm{x}}(\hat{x}_p)$, where $\hat{x}_p = g^{-1}(\hat{y}_p)$ corresponds to the q-th quantile of the marginal distribution $F_{\mathrm{x}}(x)$. This property is known as *invariance of the quantiles* under monotone transformations.

The kriging prediction $\hat{y}_p$ of the transformed random field $\mathrm{Y}(\mathbf{z};\omega)$ corresponds to the *median*, which coincides with the mean, of the Gaussian conditional distribution $F_{\mathrm{y}}(y_{\mathbf{z}} \mid \mathbf{y}^*)$. Thus, under the nonlinear transformation $\hat{x}_p = \phi(\hat{y}_p)$, the Gaussian median $\hat{y}_p$ is mapped to the median of the conditional distribution $F_{\mathrm{x}}(x_{\mathbf{z}} \mid \mathbf{x}^*)$ by virtue of the quantile invariance.

The invariance of the median under the monotone transformation also implies that for fields $\mathrm{X}(\mathbf{s};\omega)$ with asymmetric marginal pdf, the prediction $\hat{x}_p$ is *biased*: The expectation of $\hat{x}_p$ (over the ensemble of sample realizations) does not coincide with the mean of $\mathrm{X}(\mathbf{s};\omega)$.

Bias correction If an unbiased estimate is desired, a *bias correction* should be applied. In the case of lognormal kriging this is shown in Chap. 11. Approximate analytical expressions for bias correction and for the mean-square error of trans-Gaussian kriging are given in [165, p. 137–138] and in [717, p. 270–271]. These expressions are based on Taylor series expansions of $\phi(y)$ around the mean m_{y} of the Gaussian field $\mathrm{Y}(\mathbf{s};\omega)$. We list the relevant results without proof below.

The following assumptions and notation are used.

- The transformed field $\mathrm{Y}(\mathbf{s};\omega)$ is a Gaussian, *intrinsic* random field with unknown (but constant) expectation m_{y} and variogram $\gamma_{\mathrm{yy}}(\mathbf{r})$.
- The ordinary kriging predictor is used to estimate the transformed field at the locations $\mathbf{z}_p$.
- The inverse transformation $\mathrm{Y} \mapsto \mathrm{X}$ is realized by means of the transfer function $\phi(\cdot)$, i.e., $\mathrm{X}(\mathbf{s};\omega) = \phi\left[\mathrm{Y}(\mathbf{s};\omega)\right]$.
- The OK prediction at the target point is denoted by $\hat{y}_{\mathrm{ok}}(\mathbf{z}_p)$ and the OK variance by $\sigma^2_{\mathrm{y;ok}}(\mathbf{z}_p)$.
- The trans-Gaussian kriging prediction is denoted by $\hat{x}_{\mathrm{tGk}}(\mathbf{z}_p)$.

Under the above assumptions, the trans-Gaussian kriging prediction with second-order bias correction is given by [717, p. 270–271]

$$\hat{x}_{\mathrm{tGk}}(\mathbf{z}_p) \approx \phi\left(\hat{y}_{\mathrm{ok}}(\mathbf{z}_p)\right) + \frac{\phi^{(2)}(m_{\mathrm{y}})}{2}\left[\sigma^2_{\mathrm{y;ok}}(\mathbf{z}_p) - 2m_{\mathrm{y}}\right], \tag{14.2}$$

where $\phi^{(2)}(m_{\mathrm{y}})$ denotes the second-order derivative of the transfer function $\phi(\cdot)$.

Similarly, the mean-squared prediction error of trans-Gaussian kriging, obtained by the first-order Taylor expansion of $\phi(\cdot)$ around the mean m_{y} is given by

$$\mathcal{E}\left[\left\{\hat{\mathrm{X}}_{\mathrm{tGk}}(\mathbf{z}_p;\omega) - \mathrm{X}(\mathbf{z}_p;\omega)\right\}^2\right] \approx \left[\phi^{(1)}(m_{\mathrm{y}})\right]^2 \sigma^2_{\mathrm{y;ok}}(\mathbf{z}_p), \tag{14.3}$$

where $\phi^{(1)}(m_{\mathrm{y}})$ is the first-order derivative of $\phi(\cdot)$.

The above approximations are valid only if higher-order corrections can be neglected. Hence, the analytical corrections only partially resolve the problem of bias. Numerical approaches based on the bootstrap idea have been proposed to overcome the limitations of the analytical approximations [323, 477, 683]. Bootstrap involves the generation of an ensemble of states (samples) with the desired (as estimated from the data) Gaussian characteristics which are then used for estimation and prediction. The ensemble of the nonlinear estimates of the field provides a predictive distribution at the target point.

Fields with zero values There exist natural processes that take non-negative real values including a number of zero values. These zeros cannot be ignored, since they might represent a crucial aspect of the process. For example, records of hourly precipitation are very likely to include significant number of zeros, especially in regions with a dry climate. In semi-arid areas, even monthly aggregated precipitation data include zero values during the dry season.

One possible approach for modeling this type of data is to use a product of two fields: a binary-valued *occurrence field* and a continuously-valued *intensity field*. This approach was discussed in reference to product composition in Sect. 3.6.5.

A different possibility is to use the transformation of a *latent Gaussian field*, conditionally on the exceedance of a specified threshold. For example, if the latent field takes values below the threshold, then it can be assumed that the precipitation field is equal to zero. On the other hand, if the latent field exceeds the threshold, the intensity of the precipitation is determined by a suitable nonlinear transformation [52]. Based on these ideas, the precipitation field $\mathrm{X}(\mathbf{s};\omega)$ can be expressed as

$$\mathrm{X}(\mathbf{s};\omega) = \begin{cases} 0, & \text{if } \mathrm{Y}(\mathbf{s};\omega) \leq \alpha_c \\ \phi\left[\mathrm{Y}(\mathbf{s};\omega)\right], & \text{if } \mathrm{Y}(\mathbf{s};\omega) > \alpha_c. \end{cases} \tag{14.4}$$

This approach has the advantage that both the occurrence and the intensity are controlled by the same latent field. In addition, assuming that the latent field is spatially correlated, a smooth spatial transition from locations of high intensity to locations with zero intensity is expected. This smooth transition is not guaranteed in the case of a product of independent occurrence and intensity fields.

14.1.1 Joint Density of Trans-Gaussian Random Fields

Let us now consider the multivariate distribution that corresponds to trans-Gaussian random fields. We assume the following:

1. The marginal pdf of $X(\mathbf{s};\omega)$ is given by the non-Gaussian density $f_{\mathrm{x}}(x)$.
2. The marginal distribution of the transformed random field $Y(\mathbf{s};\omega) = g\left[X(\mathbf{s};\omega)\right]$ is the standard normal distribution.
3. $Y(\mathbf{s};\omega)$ follows the joint normal pdf with covariance matrix $\mathbf{C}_{\mathrm{yy}}$. Then, the joint density for the N-dimensional random vector $\mathbf{Y}(\omega)$ is given by

$$f_{\mathbf{y}}(\mathbf{y}) = \frac{1}{(2\pi)^{N/2}\sqrt{\det(\mathbf{C}_{\mathrm{yy}})}}\,\mathrm{e}^{-\frac{1}{2}\mathbf{y}^{\top}\mathbf{C}_{\mathrm{yy}}^{-1}\mathbf{y}}.$$

In light of Jacobi's transformation Theorem A.2 and the univariate transformations $y_n = g(x_n)$ for $n = 1,\ldots,N$, the joint probability density of the random vector $\mathbf{X}(\omega)$ is given by

$$f_{\mathbf{x}}(\mathbf{x}) = \frac{\left|\det(\mathbb{J}_{\mathbf{y}\to\mathbf{x}})\right|}{(2\pi)^{N/2}\sqrt{\det(\mathbf{C}_{\mathrm{yy}})}}\,\mathrm{e}^{-\frac{1}{2}\mathbf{g}(\mathbf{x})^{\top}\mathbf{C}_{\mathrm{yy}}^{-1}\mathbf{g}(\mathbf{x})},$$

where

$$\mathbf{y} = \mathbf{g}(\mathbf{x}) = \left[g(x_1),\ldots,g(x_N)\right]^{\top}.$$

The matrix $\mathbb{J}_{\mathbf{y}\to\mathbf{x}}$ is the Jacobian of the transformation from $\mathbf{y}$ to $\mathbf{x}$. The Jacobian matrix elements are given by

$$[\mathbb{J}_{\mathbf{y}\to\mathbf{x}}]_{n,m} = \frac{\partial y_n}{\partial x_m} = g'(x_m)\,\delta_{n,m}, \quad \text{for } n,m = 1,\ldots,N,$$

where $\delta_{n,m}$ is the Kronecker delta and $g'(\cdot)$ is the first-order derivative of the univariate transformation function.

The conservation of probability under the transformation $x_n \mapsto y_n = g(x_n)$, for $n = 1,\ldots,N$, leads to the following expression for the derivative $g'(\cdot)$:

$$g'(x_n) = \frac{f_{\mathrm{x}}(x_n)}{\mathrm{e}^{-y_n^2/2}\big/\sqrt{2\pi}}.$$

Hence, we obtain the following expression for the elements of the Jacobian matrix

$$\frac{\partial y_n}{\partial x_m} = \sqrt{2\pi}\,\mathrm{e}^{y_n^2/2}\,f_{\mathrm{x}}(x_m)\,\delta_{n,m}, \quad \text{for } n,m = 1,\ldots,N.$$

In light of the above analysis, the joint pdf of the trans-Gaussian random field $X(\mathbf{s}; \omega)$ is given by the following expression (where $\mathbf{I}_N$ is the $N \times N$ identity matrix)

$$f_{\mathbf{x}}(\mathbf{x}) = \frac{1}{(2\pi)^{N/2}\sqrt{\det(\mathbf{C}_{\mathrm{yy}})}}\, e^{-\frac{1}{2}\mathbf{g}(\mathbf{x})^{\top}\left(\mathbf{C}_{\mathrm{yy}}^{-1}-\mathbf{I}_N\right)\mathbf{g}(\mathbf{x})} \prod_{n=1}^{N} f_{\mathrm{x}}(x_n). \tag{14.5a}$$

The above formula seems deceptively simple, but keep in mind that it depends on $g(\cdot)$ which needs to be determined from the data. Assuming that $g(\cdot)$ is a monotone transformation, we can use the invariance of the quantiles to obtain $\Phi(y) = F_{\mathrm{x}}(x)$ for $y = g(x)$. Here $\Phi(y)$ is the cdf of the *standard normal distribution* and $F_{\mathrm{x}}(x)$ is the cdf of the random variable $X(\omega)$. Thus, the transformed normal variable $y = g(x) = \Phi^{-1}\left[F_{\mathrm{x}}(x)\right]$ is expressed as

$$g(x) = \sqrt{2}\,\mathrm{erf}^{-1}\left[2F_{\mathrm{x}}(x) - 1\right], \tag{14.5b}$$

where $\mathrm{erf}^{-1}(\cdot)$ is the inverse error function, x represents the values of the original random variable, and y corresponds to the transformed Gaussian value.

14.2 Gaussian Anamorphosis

The transformations discussed in the preceding section map spatial data with some well defined, non-Gaussian marginal cdf to a transformed set, the values of which follow the normal distribution. These transformations are also known as *normal score transforms* and *Gaussian anamorphosis*. In the field of reliability analysis the name *Nataf iso-probabilistic transform* is used for Gaussian anamorphosis [597, 664].

Definition 14.2 The terms *normal scores transform* and *Gaussian anamorphosis* refer to the bijective (one to one and invertible) mapping $g : X(\mathbf{s}; \omega) \mapsto Y(\mathbf{s}; \omega)$ from a random field $X(\mathbf{s}; \omega)$ with non-Gaussian marginal distribution to a random field $Y(\mathbf{s}; \omega) = g\left[X(\mathbf{s}; \omega)\right]$ with a Gaussian marginal distribution [132].

Non-Gaussian to Gaussian If $F_{\mathrm{x}}(x)$ is the marginal cdf of $X(\mathbf{s}; \omega)$, the function $g(\cdot)$ is defined by $g : x \mapsto y$ such that $y = \Phi^{-1}\left[F_{\mathrm{x}}(x)\right]$.

(continued)

Gaussian to Non-Gaussian Conversely, assuming that the inverse cdf, $F_{\mathrm{x}}^{-1}(\cdot)$, exists, the *anamorphosis function* $\phi = g^{-1}$ defined by $\phi : y \mapsto x$ so that $x = F_{\mathrm{x}}^{-1}[\Phi(y)]$, is a bijective mapping from the Gaussian variable y to the target variable x.

The normal scores transform maps the sample values $\{x_n^*\}_{n=1}^N$ into a respective set of values $y_n^* = g(x_n^*)$ that follow the standard normal distribution. The term anamorphosis comes from a Greek word which means "to radically improve the system" or "to improve one's behavior." So, Gaussian anamorphosis could be viewed as an attempt to improve the "statistical behavior" of the data.

In the following, we consider the implementation of the normal scores transform by means of specific nonlinear functions. However, one needs to keep in mind that anamorphosis is not always feasible. For example, transforming a probability distribution with significant weight at a single value (e.g., if the observed process has a large number of zeros, a case commonly occurring in mineral reserves and precipitation data sets), transformation to a normal distribution does not work well [36].

14.2.1 Square Root Transformation

This transformation can be used for a random field $\mathrm{X}(\mathbf{s};\omega)$ that takes non-negative values and follows a skewed marginal probability distribution. Assuming that the random field $\mathrm{Y}(\mathbf{s};\omega) = \sqrt{\mathrm{X}(\mathbf{s};\omega)}$ follows the standard normal distribution, the marginal probability distribution of $\mathrm{X}(\mathbf{s};\omega)$ is given by

$$f_{\mathrm{x}}(x) = \frac{\mathrm{e}^{-x}}{\sqrt{2\pi\, x}}. \tag{14.6}$$

The above is the pdf of the *chi-squared distribution* with one degree of freedom. The pdf (14.6) has a weak (integrable) singularity at zero and decays predominantly as an exponential for $x \to \infty$.

14.2.2 Johnson's Hyperbolic Sine Transformation

A parametric transformation based on the hyperbolic sine function was proposed by Johnson [414]:

$$X(\mathbf{s};\omega) = \xi + \lambda \sinh\left(\frac{Y(\mathbf{s};\omega) - \gamma}{\delta}\right) \tag{14.7}$$

This transformation involves a location parameter ξ, a scale parameter λ and two shape parameters, γ, δ. Thus, the hyperbolic sine transformation is flexible enough to match any combination of mean, variance, asymmetry and skewness that can be inferred from the data.

The Johnson transformation is commonly used for risk and reliability analysis in geotechnical engineering [664].

14.2.3 *Box-Cox Transformation*

Asymmetric (skewed) random fields $X(\mathbf{s};\omega)$ that mildly deviate from the Gaussian pdf can be modeled by means of the so-called *Box-Cox transformation (BCT)* [93].

Assuming a *positive-valued* random field $X(\mathbf{s};\omega) : \mathbb{R}^d \rightarrow \mathbb{R}_{+}$, the Box-Cox transform is defined as follows:

$$Y(\mathbf{s};\omega) = \begin{cases} \frac{1}{\lambda}\left[X^{\lambda}(\mathbf{s};\omega) - 1\right], & \lambda \in \mathbb{R} \text{ and } \lambda \neq 0 \\ \ln\left[X(\mathbf{s};\omega)\right], & \lambda = 0. \end{cases} \tag{14.8}$$

The goal is to estimate the parameter λ from the data so that the marginal probability distribution of $Y(\mathbf{s};\omega)$ approximates the Gaussian distribution as closely as possible. This could be accomplished by using maximum likelihood estimation for λ in conjunction with a model selection criterion (e.g., AIC).

The *inverse Box-Cox* transformation is defined as follows

$$X(\mathbf{s};\omega) = \begin{cases} \left[\lambda\, Y(\mathbf{s};\omega) + 1\right]^{1/\lambda}, & \lambda \in \mathbb{R} \text{ and } \lambda \neq 0 \\ \exp\left[Y(\mathbf{s};\omega)\right], & \lambda = 0. \end{cases} \tag{14.9}$$

The case $\lambda = 0$ implies that $X(\mathbf{s};\omega)$ follows the lognormal distribution if $Y(\mathbf{s};\omega)$ is a normally distributed random field.

Properties of Box-Cox transformation

- BCT is well defined by means of (14.8) only for random fields $X(\mathbf{s};\omega)$ that take non-negative values. If $X(\mathbf{s};\omega)$ allows negative values which have a lower bound $x_{\min} < 0$, the BCT can be modified as follows:

$$Y(\mathbf{s};\omega) = \frac{\left[X(\mathbf{s};\omega) - x_{\min}\right]^{\lambda} - 1}{\lambda}, \quad \lambda \in \mathbb{R}. \tag{14.10}$$

- To apply the transformation (14.8), it is necessary that $X(\mathbf{s};\omega)$ be a *dimensionless* random field. If this is not the case, $X(\mathbf{s};\omega)$ can be turned into a non-dimensional field by dividing with a characteristic value $x^* \neq 0$ (e.g., the mean value, the median, or the geometric mean).

- For $\lambda = 0$ the BCT is the logarithmic function. The $\lambda \to 0$ limit of the power law in (14.8) can be evaluated using the Taylor series expansion of the numerator around $\lambda = 0$ and *l'Hospital's rule*, i.e.,

$$\frac{x^\lambda - 1}{\lambda} = \frac{e^{\lambda \ln x} - 1}{\lambda} = \frac{1 + \lambda \ln x + \frac{\lambda^2}{2} \ln^2 x + O(\lambda^3 \ln^3 x) - 1}{\lambda}.$$

Based on the above expansion, it follows that the limit $\lambda \to 0$ is given by

$$\lim_{\lambda \to 0} \frac{x^\lambda - 1}{\lambda} = \ln x. \tag{14.11}$$

Thus, the Taylor expansion confirms the logarithmic transformation used for $\lambda = 0$. The limit (14.11) is also used in the *replica trick* (see Sect. 14.8).

Example 14.1 Assume that $\mathrm{Y}(\mathbf{s}; \omega)$ is a Gaussian random field with mean m_{y}, standard deviation σ_{y}, and covariance function $C_{\mathrm{yy}}(\mathbf{r})$.

1. Find the expression for the marginal pdf, $f_{\mathrm{x}}(x)$, of the random field

$$\mathrm{X}(\mathbf{s}; \omega) = [1 + \lambda\, \mathrm{Y}(\mathbf{s}; \omega)]^{1/\lambda},$$

which is the BCT of $\mathrm{Y}(\mathbf{s}; \omega)$.

2. Find the mean and the covariance function of $\mathrm{X}(\mathbf{s}; \omega)$ for $\lambda = 0$.

Answer [1.] We use the change-of-variables Theorem A.1 to determine the pdf of $\mathrm{X}(\omega)$.[1] Let $x \doteq g(y)$, where $g(y) = (1+\lambda\, y)^{1/\lambda}$ and thus $y = g^{-1}(x) = (x^\lambda - 1)/\lambda$. Since the mapping $x \mapsto y$ is continuous and one-to-one, it follows that

$$f_{\mathrm{x}}(x) = f_{\mathrm{Y}}\left(y = g^{-1}(x)\right) \left| \frac{\mathrm{d}g^{-1}(x)}{\mathrm{d}x} \right|. \tag{14.12}$$

Combining the above with $y = (x^\lambda - 1)/\lambda$ we obtain the following pdf for the transformed variable:

$$f_{\mathrm{x}}(x) = \frac{x^{\lambda-1}}{\sqrt{2\pi}\, \sigma_{\mathrm{y}}} \exp\left[-\frac{\left\{(x^\lambda - 1)/\lambda - m_{\mathrm{y}}\right\}^2}{2\sigma_{\mathrm{y}}^2} \right]. \tag{14.13}$$

At the limit $\lambda \to 0$, the pdf (14.13) tends to the *lognormal pdf* given by (6.63), i.e.,

[1]Note that the role of $\mathrm{X}(\omega)$ and $\mathrm{Y}(\omega)$ is reversed herein with respect to the expression of Jacobi's theorem in the Appendix A, since the pdf of $\mathrm{Y}(\omega)$ is assumed to be known.

$$f_X(x) = \frac{1}{\sqrt{2\pi}\, x\, \sigma_y}\, e^{-(\ln x - m_y)^2/2\sigma_y^2}.$$

[2.] The mean and the covariance function of $X(\mathbf{s}; \omega)$ for $\lambda = 0$ are given by (6.64) and (6.65) respectively.

Examples of BCT probability density functions given by (14.13) are shown in the plots of Fig. 14.1. The right tails are longer for density functions corresponding to values of $|\lambda| \approx 0$ than for larger $|\lambda|$ (both for positive and negative λ). Hence, the *lognormal* pdf, obtained from (14.13) at the limit $\lambda \to 0$, has a heavier right tail (i.e., more density at large values) than any other BCT-derived density function. Thus, the BCT unfortunately cannot be used to transform a pdf that has a power-law tail into a Gaussian pdf.

The Box-Cox transform cannot guarantee that any probability distribution can be normalized. However, even in cases when an exact mapping to the normal distribution is not possible, BCT may still provide a useful approximation. Draper and Cox have investigated the problem of transforming probability distributions, including the exponential and Weibull models, to normality [211]. In addition, various BCT generalizations have been proposed to extend the scope of the approach [93, 868].

Intuitive understanding of BCT It is straightforward to understand BCT at the limit $\lambda = 0$, where it reduces to the logarithmic transform. We then obtain a pair of variables $X(\omega)$ and $Y(\omega)$ that are related through the transform pair $y = \ln x$, and $x = \exp(y)$. However, for $\lambda \neq 0$ the transforms $y = (x^\lambda - 1)/\lambda$ and $x = (1+\lambda y)^{1/\lambda}$ seem quite different from the logarithm and the exponential ... or are they not?

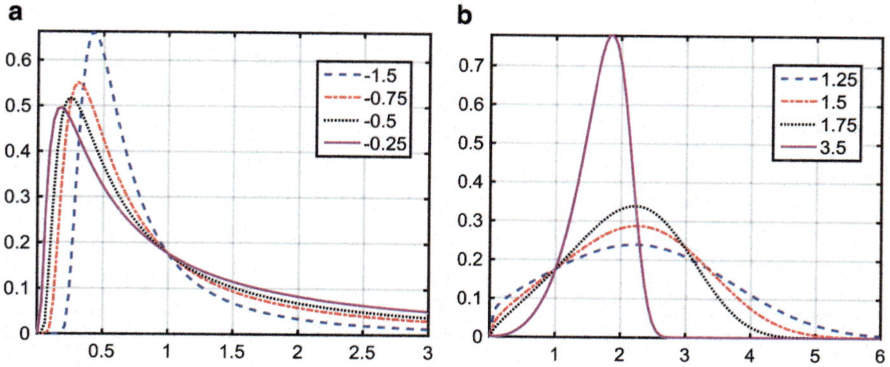

Fig. 14.1 Non-Gaussian probability density functions generated by the Box-Cox transform from a Gaussian pdf with $m = 1$ and $\sigma = 2$. The pdf curves are based on the equation (14.13). The curves in the left frame correspond to negative values of the Box-Cox exponent λ ($-1.5, -0.75, -0.5, -0.25$), while the curves in the right frame correspond to positive values of λ (1.25, 1.5, 1.75, 3.5). **(a)** $\lambda < 0$ **(b)** $\lambda > 0$

Let us consider *Euler's number* e for a moment: it is ubiquitous in science and engineering, and so we tend to forget that it is defined in terms of the following limit

$$\mathrm{e} = \lim_{n \to \infty} (1 + n^{-1})^n. \tag{14.14}$$

Thus, for $y \in \mathbb{R}$ the exponential $\exp(y)$ is defined by

$$\exp(y) = \lim_{n \to \infty} (1 + y/n)^n. \tag{14.15}$$

Let us now set in (14.15) $1/n = \lambda$. It now becomes obvious that the BCT pair of transforms (14.8) and (14.9) are generalizations of the logarithm and the exponential functions obtained for finite and possibly non-integer values of n, i.e., for $\lambda \neq 0$.

Practical matters The goal for data analysis is to determine the parameter λ so that the marginal distribution of $\mathrm{Y}(\mathbf{s}; \omega)$ is as close as possible to the normal. This can be accomplished by minimizing the log-likelihood of the data [93].

Alternatively, the parameter λ can be estimated using a least-squares approach. The latter minimizes the sum of the squared differences between the cumulative distribution function of the transformed data, i.e., the values $y_n = [(x_n^*)^\lambda - 1]/\lambda$, where $n = 1, \ldots, N$, and the cdf of the normal distribution, i.e.,

$$(\lambda^*, m_{\mathrm{y}}^*, \sigma_{\mathrm{y}}^*) = \underset{\lambda, m_{\mathrm{y}}, \sigma_{\mathrm{y}}}{\arg\min} \sum_{n=1}^{N} \left[\hat{F}_{Y_\lambda}(y_n) - F_0(y_n; m_{\mathrm{y}}, \sigma_{\mathrm{y}}) \right]^2 .$$

In the above, $\{\hat{F}_{Y_\lambda}(y_n)\}_{n=1}^N$ denotes the sample-based *empirical cdf of the transformed data*, and $F_0(y; m_{\mathrm{y}}, \sigma_{\mathrm{y}})$ the *normal cdf* with mean m_{y} and standard deviation σ_{y}. A different measure of dissimilarity between the two probability distributions can also be used, such as the Kullback-Leibler divergence.

Limitations of BCT BCT is a practical approach that allows building Gaussian models for non-Gaussian data. However, BCT is not without shortcomings, as we discuss below.

Existence The existence of a $\lambda \in \mathbb{R}$ such that the BCT data closely follow the normal distribution is not a priori guaranteed. Hence, it is not always possible to achieve even approximate normalization of the data by applying BCT.

Marginal action Linear geostatistical methods are based on the assumption that the joint distribution of the data is approximately Gaussian. On the other hand, BCT aims to normalize the marginal distribution of the data. Thus, it does not ensure that the joint distribution of the transformed data is close to the multivariate normal model. Since it is not easy to apply tests of multivariate normality [26] to scattered spatial data, multivariate normality is often assumed in practice but not rigorously established.

Prediction For the purpose of spatial prediction, the BCT is used to normalize the data and kriging is subsequently applied to the transformed values. To obtain the predictions for the original field, the inverse BCT is applied to the kriging predictions. If $\lambda = 0$ this approach is referred to as *lognormal kriging*.

The map resulting from the inversion of the kriging predictions needs to be corrected for bias. However, bias correction is not completely trouble free (see also Sect. 14.1). For example, in lognormal kriging the estimates of the original field, corrected for bias, depend crucially (exponentially) on the kriging variance of the spatial model inferred for the transformed data [823, p. 227], [168]. Thus, the kriging prediction involves considerable uncertainty and depends on the selected "optimal" variogram model. This can significantly influence the estimates of the target variable $X(\omega)$ and increase their dispersion.

14.3 Tukey g-h Random Fields

Tukey g-h (TGH) random fields are based on the Tukey g-h probability distributions [804]. The latter are obtained by means of the following nonlinear transform of the normal variate z

$$\tau_{g,h}(z) = \frac{1}{g} \left(e^{g\, z} - 1 \right) e^{h\, z^2/2}. \tag{14.16}$$

The above is a strictly monotone function of z if $h \geq 0$ and $g \in \mathbb{R}$. The transformation (14.16) has been used recently by Marc Genton and collaborators to generate random fields with flexible marginal distributions that can accommodate skewness and heavy tails [248, 860]. Hence, Tukey g-h random fields are suitable models for spatial extremes [861]. The Tukey g-h marginal distribution can also adequately approximate many standard probability distributions, including the Student's t-distribution, the exponential, as well as the Cauchy, Weibull, and logistic distributions [860].

Definition 14.3 Assume that z in (14.16) is replaced by the Gaussian random field $X(\mathbf{s};\omega)$. Then, $Y(\mathbf{s};\omega) = \tau_{g,h}\left[X(\mathbf{s};\omega)\right]$ is a Tukey g-h (TGH) random field. TGH random fields are thus trans-Gaussian random fields.

Skewness parameter The parameter g controls the skewness of the probability distribution of the TGH random field $Y(\mathbf{s};\omega)$: $g > 0$ yields a right-skewed distribution (the mean is higher than the median), while $g < 0$ makes the distribution left-skewed (the mean is lower than the median).

Tail parameter The parameter h governs the tail behavior of $Y(\mathbf{s};\omega)$. The tail becomes heavier, implying increased values of the marginal pdf at higher y, as h increases.[2]

[2]The parameter h of TGH random fields should not be confused with the normalized distance $\mathbf{h}$.

Expectation The expectation of Tukey g-h random fields is given by [860]

$$m_{\mathrm{y}} = \mathcal{E}[\mathrm{Y}(\mathbf{s};\omega)] = \frac{1}{g\sqrt{1-h}}\left[\mathrm{e}^{g^2/2(1-h)} - 1\right].$$

Covariance The covariance function for $h < 1/2$ is given by the following, somewhat complicated expression

$$C_{\mathrm{yy}}(\mathbf{s}_1,\mathbf{s}_2) = \frac{\exp\left[g^2\frac{1+\rho_{\mathrm{xx}}(\mathbf{s}_1,\mathbf{s}_2)}{1-h\,\{1+\rho_{\mathrm{xx}}(\mathbf{s}_1,\mathbf{s}_2)\}}\right] - 2\exp\left[\frac{g^2}{2}\,\frac{1-h\{1-\rho^2_{\mathrm{xx}}(\mathbf{s}_1,\mathbf{s}_2)\}}{(1-h)^2-h^2\,\rho^2_{\mathrm{xx}}(\mathbf{s}_1,\mathbf{s}_2)}\right] + 1}{g^2\sqrt{(1-h)^2 - h^2\,\rho^2_{\mathrm{xx}}(\mathbf{s}_1,\mathbf{s}_2)}} - m_{\mathrm{y}}^2.$$

The type of the covariance $C_{\mathrm{xx}}(\mathbf{r})$ of the latent Gaussian random field (i.e., stationary or non-stationary) determines the covariance function of $\mathrm{Y}(\mathbf{s};\omega)$ as well. In addition, if $h < 1/2$, properties of the latent Gaussian random field such as second-order stationarity, mean-square continuity and mean-square differentiability are transferred to the respective TGH random field.

The parameter estimation, spatial prediction, and simulation of TGH random fields are investigated in [860].

14.4 Transformations Based on Kaniadakis Exponential

As discussed above, the exponential $\exp(x)$ can be viewed as the asymptotic limit of the function $(1 + x/n)^n$ as $n \to \infty$. A generalization of the exponential is thus obtained by the function $(1 + x/n)^n$ for finite n. Other generalizations of the exponential function that are based on different asymptotic limits are also possible [426, 428, 429]. For example, Giorgio Kaniadakis has introduced expressions that extend the standard exponential and logarithm functions [427–431].

κ-exponential The so-called *κ-exponential* function is defined by

$$\exp_\kappa(y) = \left(\sqrt{1+\kappa^2y^2} + \kappa y\right)^{1/\kappa}, \text{ where } y \in \mathbb{R} \text{ and } \kappa \geq 0. \tag{14.17}$$

The κ-exponential is well defined even for $y < 0$, as it follows from the inequality $\sqrt{1+\kappa^2y^2} + \kappa y > |\kappa y| + \kappa y \geq 0$.

The above generalization emerges naturally in the framework of special relativity, where the parameter κ takes values $0 \leq \kappa < 1$ and is proportional to the reciprocal of the speed of light [428, 429].

The κ-exponential can also be introduced as the solution, i.e., $g(\cdot) = \exp_\kappa(\cdot)$, of the following functional equation

$$g\left(y_1 \overset{\kappa}{\oplus} y_2\right) = g(y_1)\, g(y_2), \tag{14.18a}$$

where

$$y_1 \overset{\kappa}{\oplus} y_2 = y_1\sqrt{1+\kappa^2 y_2^2} + y_2\sqrt{1+\kappa^2 y_1^2} \tag{14.18b}$$

is the relativistic summation formula [428]. The κ-exponential has been used in applications outside the scope of the relativistic regime [148, 149, 376]. Based on (14.17) the κ-exponential function is strictly positive for all $y \in \mathbb{R}$.

κ-logarithm The inverse function of the κ-exponential, called the *κ-logarithm*, is defined by means of the following function

$$\ln_\kappa(x) = \frac{x^\kappa - x^{-\kappa}}{2\kappa}, \quad \text{where } x \geq 0 \text{ and } \kappa \geq 0. \tag{14.19}$$

Based on the definition (14.19), it holds that for all $\kappa \geq 0$ the κ-logarithm satisfies the properties $\ln_\kappa(x) > 0$ for $x > 1$ and $\ln_\kappa(x) < 0$ for $x < 1$. In addition, based on the definitions of the κ-exponential and the κ-logarithm, it holds that

$$\ln_\kappa\left[\exp_\kappa(x)\right] = \exp_\kappa\left[\ln_\kappa(x)\right] = x.$$

The κ-logarithm can be used as a normalizing transformation for skewed data, similar to the BCT. For example, in Sect. 14.4.3 we derive a random field $X(\mathbf{s}; \omega)$ that follows a *κ-lognormal distribution* by taking the κ-exponential of a Gaussian random field $Y(\mathbf{s}; \omega)$. Conversely, the κ-logarithm function can be used to normalize a random field that follows the κ-lognormal distribution.

14.4.1 Properties of Kaniadakis Exponential and Logarithm Functions

The properties of the κ-exponential and κ-logarithm functions are thoroughly examined in [427]. Below we list the most important properties. We remind the reader that in the following it is assumed that $\kappa \geq 0$.

Properties of the κ-exponential The κ-exponential shares a number of properties with the ordinary exponential function.

1. The κ-exponential is a positive-valued function as it follows from (14.17).
2. The κ-exponential is a smooth function, it is defined for all real numbers, and it belongs in the class $\mathbb{C}^\infty$, i.e., it admits derivatives of all orders.
3. The derivative of the κ-exponential is positive everywhere, i.e., $\frac{\mathrm{d}\exp_\kappa(y)}{\mathrm{d}y} > 0$.
4. The value of $\exp_\kappa(y)$ at zero is equal to one.

5. The asymptotic limits of the κ-exponential at infinity are given by

$$\lim_{y\to\infty} \exp_\kappa(y) = \infty, \qquad \lim_{y\to\infty} \exp_\kappa(-y) = 0.$$

6. The inverse of the κ-exponential is defined via $\exp_\kappa(y)\,\exp_\kappa(-y) = 1$.
7. The ordinary exponential is obtained for $\kappa = 0$, i.e.,

$$\exp_{\kappa=0}(y) = \lim_{\kappa\to 0} \exp_\kappa(y) = \exp(y).$$

Most of the above properties are fairly obvious. Some of them are revisited and further explained below. Plots of the κ-exponential function for different values of κ are shown in Fig. 14.2. The κ-exponential increases slower than the standard exponential, both for $\kappa \geq 1$ and $0 \leq \kappa < 1$.

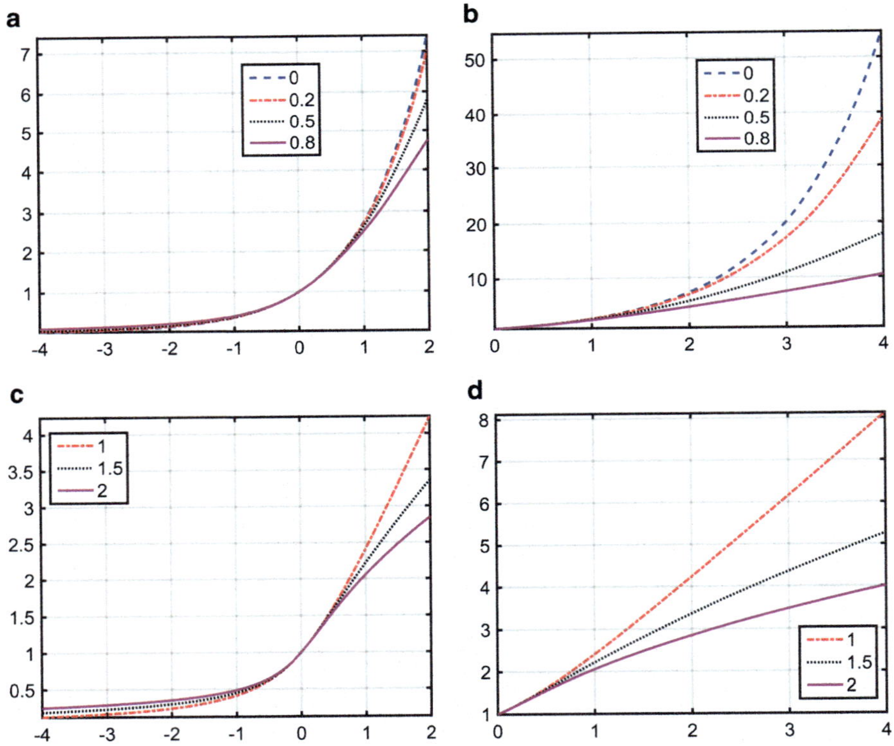

Fig. 14.2 Kaniadakis exponential functions for seven different values of κ. In the top row frames $\kappa = (0, 0.2, 0.5, 0.8)$ while in the bottom row frames $\kappa = (1, 1.5, 2)$. The standard exponential function ($\kappa = 0$) is also shown for comparison. **(a)** $0 \leq \kappa < 1$ **(b)** $0 \leq \kappa < 1$ **(c)** $\kappa \geq 1$ **(d)** $\kappa \geq 1$

κ-derivative Let the Lorentz factor be defined by $\gamma(y) = \sqrt{1+\kappa^2 y^2}$. Then, the Kaniadakis κ-derivative is defined by means of

$$\frac{\mathrm{d}}{\mathrm{d}_\kappa y} = \gamma(y)\,\frac{\mathrm{d}}{\mathrm{d}y}. \tag{14.20}$$

The κ-exponential is an eigenfunction of the κ-derivative, just as the ordinary exponential function is an eigenfunction of the ordinary derivative. Therefore,

$$\frac{\mathrm{d}}{\mathrm{d}_\kappa y}\,\exp_\kappa(y) = \exp_\kappa(y). \tag{14.21}$$

Taylor expansion The Taylor expansion of $\exp_\kappa(y)$ is given by [431]:

$$\exp_\kappa(y) = \sum_{n=0}^{\infty} \xi_n(\kappa)\,\frac{y^n}{n!}, \text{ for } (\kappa y)^2 < 1. \tag{14.22a}$$

where the functions $\{\xi_n(\kappa)\}_{n=0}^{\infty}$ are polynomials of κ defined by the following recurrence relations

$$\xi_0(\kappa) = \xi_1(\kappa) = 1, \tag{14.22b}$$

$$\xi_n(\kappa) = \prod_{j=1}^{n-1} \left[1-(2j-n)\kappa\right],\ n>1. \tag{14.22c}$$

The polynomials $\xi_1(\kappa)$ for the first seven orders are given by

$$\xi_0(\kappa) = \xi_1(\kappa) = \xi_2(\kappa) = 1, \tag{14.22d}$$

$$\xi_3(\kappa) = 1-\kappa^2, \tag{14.22e}$$

$$\xi_4(\kappa) = 1-4\kappa^2, \tag{14.22f}$$

$$\xi_5(\kappa) = (1-\kappa^2)(1-9\kappa^2), \tag{14.22g}$$

$$\xi_6(\kappa) = (1-4\kappa^2)(1-16\kappa^2). \tag{14.22h}$$

It is remarkable that the first three terms of the Taylor expansion (14.22) coincide with those of the ordinary exponential, i.e.,

$$\exp_\kappa(y) = 1 + y + \frac{y^2}{2} + (1-\kappa^2)\frac{y^3}{3!} + \mathcal{O}(\kappa^2 y^4).$$

The Taylor expansion implies that for $y \to 0$ and $\kappa \to 0$ the κ-exponential function $\exp_\kappa(y)$ tends to the ordinary exponential. More precisely,

$$\exp_\kappa(y) = \exp(y) + O(\kappa^2\, y^3). \tag{14.23}$$

In addition, based on (14.22) the sign of the leading corrections of $\exp_\kappa(y)$ with respect to $\exp(y)$ is negative.

Asymptotic behavior One of the most interesting properties of $\exp_\kappa(y)$ is its *power-law* asymptotic behavior as $y \to \infty$, that is [428, 429, 431]

$$\exp_\kappa(y) \sim |\, 2\kappa y|^{\,\mathrm{sign}(y)/|\kappa|}, \text{ as } y \to \pm\infty. \tag{14.24}$$

In light of the above, the κ-exponential for $\kappa > 0$ exhibits a right tail *power-law decay* for $y \to +\infty$, i.e.

$$\exp_\kappa(-y) \sim (2\kappa y)^{-1/\kappa}, \text{ as } y \to \infty. \tag{14.25}$$

Hence, $\exp_\kappa(-y)$ can be used to model distributions with *heavy tails*, i.e., power-law decay of the probability density function. Equation (14.24) also applies to negative values of y, i.e., to the left tail of the pdf.

Monotonicity and convexity The κ-exponential is a continuously differentiable function. Its derivatives are proportional to $\exp_\kappa(y)$, but they also involve a modifying factor that depends on κ and y. The modifying factor becomes equal to one at $\kappa = 0$. The first two derivatives are given by the following expressions

$$\frac{\mathrm{d}\exp_\kappa(y)}{\mathrm{d}y} = \frac{\exp_\kappa(y)}{\sqrt{1+\kappa^2\, y^2}}, \tag{14.26}$$

$$\frac{\mathrm{d}^2\exp_\kappa(y)}{\mathrm{d}y^2} = \exp_\kappa(y)\, \frac{\sqrt{1+\kappa^2\, y^2} - \kappa^2\, y}{\left(1+\kappa^2\, y^2\right)^{3/2}}. \tag{14.27}$$

It follows from (14.26) that the first derivative is non-negative for all $y \in \mathbb{R}$. Hence, the κ-exponential is a *monotonically increasing* function. In addition, the second derivative is positive for $\kappa \leq 1$ which implies that the κ-exponential is a *convex function*. For $\kappa > 1$ the κ-exponential has an *inflection point* at $y_* = \left(\kappa\sqrt{\kappa^2 - 1}\right)^{-1}$ where the second derivative changes sign.

Exponentiation Let $\phi(y)$ denote a real-valued function of y. The κ-exponential of $\phi(y)$ satisfies the following property:

$$\left\{\exp_\kappa\,[\phi(y)]\right\}^r = \exp_{\kappa/r}\,[r\phi(y)]\,,\; r \in \mathbb{R}. \tag{14.28a}$$

Survival function Let us use the notation $R_\kappa(y) = \exp_\kappa\,[\phi(y)]$. Furthermore, assume that $y \in [0, \infty)$ and that $\phi(y)$ is a monotonically decreasing function that takes values in $(-\infty, 0]$ so that $\phi(0) = 0$ and $\lim_{y\to\infty} \phi(y) = -\infty$.

Then, according to the definition above, $R_\kappa(y)$ is also a monotonically decreasing function that satisfies

$$0 \leq R_\kappa(y) \leq 1, \text{ with } R_\kappa(0) = 1 \text{ and } R_\kappa(\infty) = 0.$$

Thus, $R_\kappa(y)$ is a feasible *survival function*, i.e., $R_\kappa(y) = \text{Prob}[\mathrm{Y}(\omega) > y]$, for a random variable $\mathrm{Y}(\omega)$ that takes values in $\mathbb{R}_{+,0}$. The survival functions of a probability distribution is the complement of respective cumulative distribution function, i.e., $R_{\mathrm{x}}(x) = 1 - F_{\mathrm{x}}(x)$, where $x \in \mathbb{R}$. The name "survival function" [synonym: reliability function] has a simple interpretation in engineering: Let y represent the load applied to a system and the random variable $\mathrm{Y}(\omega)$ represents the strength of the system for the specific loading conditions. Assume that the load increases from zero towards a value given by y. The survival function represents the probability that the system will not have failed when the load becomes equal to y.

If we assume two different values κ and $\kappa' = \kappa/r$ where $r > 0$, equation (14.28a) connects the respective survival functions as follows

$$R_\kappa^r(y) = R_{\kappa/r}(y). \tag{14.28b}$$

Weakest link scaling In [376, 377] it was surmised that for interacting systems the parameter κ measures the size of the system in terms of elementary (independent) units. In particular, the system is assumed to contain N_{el} elementary units, where N_{el} is proportional to $1/\kappa$. In this perspective, the equation (14.28b) implies that the survival function for a system comprising $r\,N_{\mathrm{el}}$ elementary units is equal to the product of the survival functions of r systems that contain N_{el} elementary units each.

The product property (14.28b) of the survival function $R_\kappa(y)$ is very interesting, because it implies *weakest link scaling*: the entire system fails (the survival function becomes zero), if one (the weakest) of its components fails. This concept of weakest link scaling is central to the derivation of the Weibull distribution for a system that comprises a number of independent units (links) [838]. The Weibull survival function is given by

$$R(y) = \exp\left[-\left(y/y_s\right)^{m_s}\right], \tag{14.29}$$

where y_s is the scale parameter and m_s is the Weibull modulus.

The *Kaniadakis-Weibull* (synonym: κ-Weibull) distribution with survival function

$$R_\kappa(y) = \exp_\kappa\left[-(y/y_s)^{m_s}\right], \tag{14.30}$$

generalizes the Weibull distribution. One possible interpretation of (14.30) is that it provides an extension of the weakest link scaling principle to systems which comprise a number of dependent (interacting) links. The physical arguments that support this interpretation are presented in [377].

Mellin transform Let $f(y)$ be a real-valued function which is integrable over $\mathbb{R}_{0,+}$. The Mellin transform $\mathcal{M}[f(\cdot)](z)$ is defined by means of the integral

$$\mathcal{M}[f(\cdot)](z) = \int_0^\infty \mathrm{d}y y^{z-1} f(y),$$

for the values of z for which the integral exists. In the case of the κ-exponential, the Mellin transform is well defined for $0 < z < 1/\kappa$. When the exponent z is in this range, the κ-exponential Mellin transform is given by

$$\mathcal{M}[\exp_\kappa(\cdot)](z) = \int_0^\infty y^{z-1} \exp_\kappa(-y)\,\mathrm{d}y = \frac{(2\kappa)^{-z}}{1+\kappa z} \frac{\Gamma\left(\frac{1}{2\kappa} - \frac{z}{2}\right)}{\Gamma\left(\frac{1}{2\kappa} + \frac{z}{2}\right)} \Gamma(z)\,. \tag{14.31}$$

Properties of the κ-logarithm The κ-logarithm shares a number of properties with the ordinary logarithm.

- The κ-logarithm is defined for all positive real numbers $x \in \mathbb{R}_+$. It is a smooth function that belongs in the class $\mathbb{C}^\infty$, i.e., it admits derivatives of all orders.
- The first-order derivative of the κ-logarithm is positive, i.e., $\mathrm{d}\ln_\kappa(x)/\mathrm{d}x > 0$ everywhere.
- The limit of $\ln_\kappa(x)$ as x approaches zero from the right is equal to negative infinity, i.e.,

$$\lim_{x\to 0^+} \ln_\kappa(x) = -\infty.$$

- The asymptotic limit of the κ-logarithm at infinity is

$$\lim_{x\to\infty} \ln_\kappa(x) = \infty.$$

- At $x = 1$ the value of $\ln_\kappa(x)$ is $\ln_\kappa(1) = 0$.
- The ordinary logarithm is obtained at the limit $\kappa \to 0$, i.e.,

$$\ln_0(x) = \lim_{\kappa\to 0} \ln_\kappa(x) = \ln(x).$$

Plots of the κ-logarithm functions for different values of κ are shown in Fig. 14.3. The κ-logarithm function increases faster than the standard logarithm, at least for large x. The difference becomes more pronounced for higher κ.

κ-reflection symmetry Based on the definition of the κ-logarithm (14.19) it holds that $\ln_\kappa(x) = \ln_{-\kappa}(x)$.

κ-logarithm of inverse $\ln_\kappa(1/x) = -\ln_\kappa(x)$.

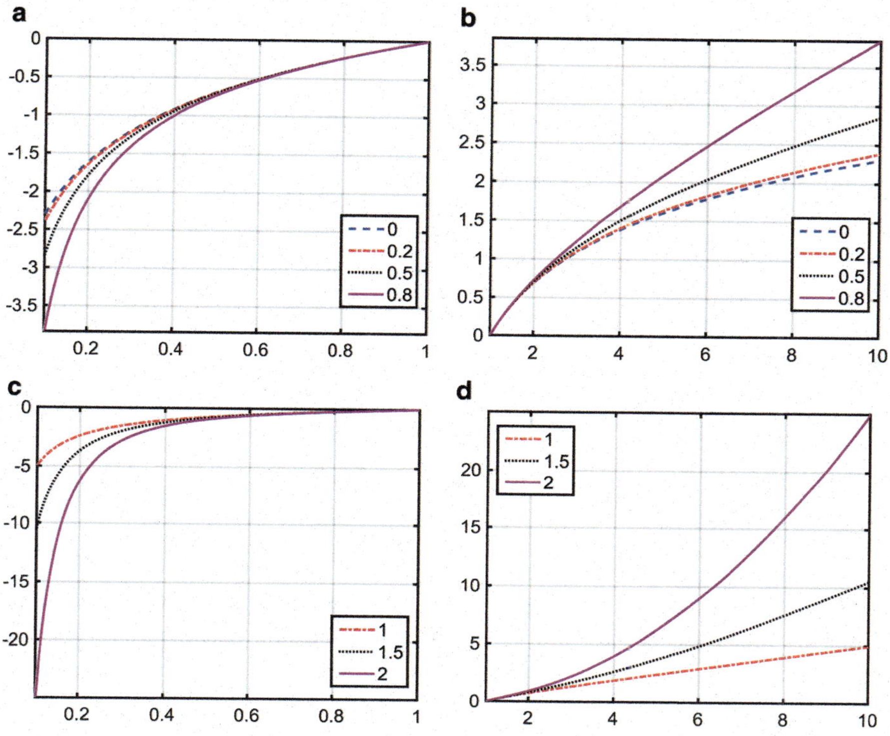

Fig. 14.3 Kaniadakis logarithm functions for seven different values of κ. In the top frames $\kappa = (0, 0.2, 0.5, 0.8)$ while in the bottom frames $\kappa = (1, 1.5, 2)$. The standard logarithm ($\kappa = 0$) is also shown for comparison. (**a**) $0 \leq \kappa < 1$ (**b**) $0 \leq \kappa < 1$ (**c**) $\kappa \geq 1$ (**d**) $\kappa \geq 1$

Small κ *expansion* The small κ expansion of (14.19) yields

$$\ln_\kappa(x) = \ln(x) + \frac{1}{6}\,\kappa^2 \ln(x)^3 + \frac{1}{120}\,\kappa^4 \ln(x)^5 + O(\kappa^6). \tag{14.32}$$

This recovers the natural logarithm for $\kappa = 0$ and shows that corrections of the κ-logarithm with respect to the ordinary logarithm are of third order in x.

Taylor expansion The Taylor expansion of $\ln_\kappa(1 + x)$ converges if $-1 < x < 1$ and it is expressed as

$$\ln_\kappa(1 + x) = \sum_{n=1}^{\infty} b_n(\kappa)(-1)^{n-1}\frac{x^n}{n}. \tag{14.33a}$$

where $b_1(\kappa) = 1$ and for $n > 1$ the $b_n(\kappa)$ coefficients are given by

$$b_n(\kappa) = \frac{1}{2}\left[(1-\kappa)\left(1-\frac{\kappa}{2}\right)\dots\left(1-\frac{\kappa}{n-1}\right)\right. \quad (14.33b)$$

$$\left. +(1+\kappa)\left(1+\frac{\kappa}{2}\right)\dots\left(1+\frac{\kappa}{n-1}\right)\right]. \quad (14.33c)$$

Based on the above, the first few terms of the Taylor expansion are given by

$$\ln_\kappa(1+x) = x - \frac{x^2}{2} + \left(1+\frac{\kappa^2}{3}\right)\frac{x^3}{3} - \dots .$$

Asymptotic limits For $\kappa > 0$, it follows that the κ-logarithm function has power-law tails at infinity and near $x = 0$, i.e,

$$\ln_\kappa(x) \sim \begin{cases} \dfrac{x^\kappa}{2\kappa}, & \text{as } x \to \infty, \\ -\dfrac{x^{-\kappa}}{2\kappa}, & \text{as } x \to 0^+. \end{cases} \quad (14.34)$$

Monotonicity and concavity The κ-logarithm function is continuously differentiable. Its first two derivatives are given by the following equations (for $x \geq 0$ and $\kappa > 0$):

$$\frac{\mathrm{d}\ln_\kappa(x)}{\mathrm{d}x} = \frac{1}{2x}\left(x^\kappa + x^{-\kappa}\right), \quad (14.35)$$

$$\frac{\mathrm{d}^2\ln_\kappa(x)}{\mathrm{d}x^2} = \frac{\kappa^2}{x^2}\ln_\kappa(x) - \frac{\mathrm{d}\ln_\kappa(x)}{\mathrm{d}x} = \frac{-(1-\kappa)}{2}x^{-2+\kappa} - \frac{1+\kappa}{2}x^{-2-\kappa}. \quad (14.36)$$

Note that $\mathrm{d}\ln_\kappa(x)/\mathrm{d}x > 0$ for all $x \geq 0$ and $\kappa \geq 0$, and thus the κ-logarithm function is monotonically increasing. In addition, the second derivative is negative for all x if $\kappa < 1$: hence, for $\kappa < 1$ the κ-logarithm is a *concave function*. In contrast, for $\kappa > 1$ the κ-logarithm has an *inflection point* at $x_* = \left(\frac{\kappa+1}{\kappa-1}\right)^{1/2\kappa}$.

Product rule The following two properties concern the κ-logarithm of products. They are reduced to properties of the ordinary logarithm for $\kappa = 0$.

$$\ln_\kappa(x^r) = r\,\ln_{r\kappa}(x), \quad (14.37)$$

$$\ln_\kappa(x\,y) = \ln_\kappa(x) \overset{\kappa}{\oplus} \ln_\kappa(y), \quad (14.38)$$

where $r \in \mathbb{R}$ and the summation symbol $\overset{\kappa}{\oplus}$ has been defined in (14.18).

Integral representation The κ-logarithm function can be expressed in terms of the following integral

$$\ln_\kappa(x) = \frac{1}{2}\int_{1/x}^{x} \frac{\mathrm{d}t}{t^{1+\kappa}}, \tag{14.39}$$

which in the limit $\kappa \to 0$ reduces to the integral representation of the ordinary logarithm.

14.4.2 *Kaniadakis Entropy*

The classical (Shannon-Boltzmann) definition of the entropy was discussed in Sect. 13.2. The Kaniadakis entropy is similarly defined using the κ-logarithm instead of the ordinary logarithm. In the case of a discrete probability distribution that admits N states, the Kaniadakis entropy is defined by the following sum over all states

$$S_\kappa = -\sum_{n=1}^{N} f_\kappa(x_n) \ln_\kappa f_\kappa(x_n) = -\sum_{n=1}^{N} \frac{f_\kappa(x_n)^{1+\kappa} - f_\kappa(x_n)^{1-\kappa}}{2\kappa}, \tag{14.40}$$

where $f_\kappa(x_n)$ is a distribution function that depends on κ and $\ln_\kappa$ is the κ-logarithm function.

The entropy S_κ represents a single-parameter deformation of the Shannon-Boltzmann entropy. The Kaniadakis entropy satisfies the Shannon-Khinchin [449] axioms of continuity, maximality, and expandability [20]. In addition, it satisfies a generalized additivity law.

As shown in Sect. 13.2.3, if the only constraints of the probability distribution are the normalization and the expectation, the *maximum entropy principle* leads to the exponential distribution. In the case of the Kaniadakis entropy, the maximum entropy principle under the *standard* normalization and expectation constraints leads to the following statistical distribution [430]

$$f_\kappa(x_n) = \frac{1}{\epsilon} \exp_\kappa\big(-\gamma\mu_1 x_n + \gamma\mu_2\big), \tag{14.41a}$$

where μ_1 and μ_2 are Lagrange multipliers, while the constants γ and ϵ are given by the following functions of κ:

$$\gamma = \frac{1}{\sqrt{1-\kappa^2}}, \quad \epsilon = \left(\frac{1+\kappa}{1-\kappa}\right)^{1/2\kappa}. \tag{14.41b}$$

The Kaniadakis probability distribution (14.41a) is a single-parameter deformation of the classical *Boltzmann-Maxwell distribution*. In the classical limit $\kappa \to 0$, the Kaniadakis distribution behaves as the Boltzmann-Maxwell distribution, i.e.,

$$f_0(x_n) \sim \exp\left(-\mu_1 x_n + \mu_2\right). \tag{14.42}$$

On the other hand, for $\kappa > 0$ the Kaniadakis distribution develops a power-law right tail according to (14.25), i.e.,

$$f_\kappa(x_n) \sim N\, x_n^{-1/\kappa}, \text{ as } x_n \to \infty. \tag{14.43}$$

14.4.3 κ-Lognormal Random Fields

It is possible to generalize the lognormal distribution using the Kaniadakis exponential and logarithm functions. Let us assume that $\mathrm{Y}(\mathbf{s};\omega)$ is a Gaussian random field with marginal probability distribution $\mathcal{N}(m,\sigma^2)$. We can define the *κ-lognormal random field* $\mathrm{X}(\mathbf{s};\omega)$ in terms of $\mathrm{Y}(\mathbf{s};\omega)$ and the κ-exponential as follows

$$\mathrm{X}(\mathbf{s};\omega) = \exp_\kappa\left[\mathrm{Y}(\mathbf{s};\omega)\right] = \left(\sqrt{1+\kappa^2\mathrm{Y}(\mathbf{s};\omega)^2} + \kappa\mathrm{Y}(\mathbf{s};\omega)\right)^{1/\kappa}. \tag{14.44}$$

The κ-lognormal random field takes only positive values. This property is inherited from the positiveness of the κ-exponential.

The inverse transformation from the κ-lognormal random field $\mathrm{X}(\mathbf{s};\omega)$ to the Gaussian random field $\mathrm{Y}(\mathbf{s};\omega)$ is given by the κ-logarithm, i.e.,

$$\mathrm{Y}(\mathbf{s};\omega) = \ln_\kappa\left[\mathrm{X}(\mathbf{s};\omega)\right] = \frac{\mathrm{X}(\mathbf{s};\omega)^\kappa - \mathrm{X}(\mathbf{s};\omega)^{-\kappa}}{2\kappa}.$$

We call the marginal distribution of the random field $\mathrm{X}(\mathbf{s};\omega)$ the *κ-lognormal distribution*. This distribution function can be obtained using the conservation of the probability under transformation of variables which is expressed by means of (14.12). We thus define the nonlinear transformations $x = g(y)$, where $g(\cdot) = \exp_\kappa(\cdot)$ and $y = g^{-1}(x)$ where $g^{-1}(\cdot) = \ln_\kappa(\cdot)$. Then, in light of (14.12) and using the first derivative of the κ-logarithm, given by (14.35), we obtain the *κ-lognormal* pdf

$$f_{\mathrm{x};\kappa}(x) = \frac{1}{\sqrt{2\pi}\,\sigma x}\, \mathrm{e}^{-(\ln_\kappa x - m)^2/2\sigma^2}\, \frac{x^\kappa + x^{-\kappa}}{2}, \quad x > 0. \tag{14.45}$$

The density function (14.45) is similar to the lognormal pdf. In fact, the product of the first two terms on the right-hand side is almost the lognormal pdf (6.63), the only difference being that $\ln x$ in the exponent is replaced by $\ln_\kappa x$. Finally, the third term takes values close to one as $\kappa \to 0$. Based on the limiting property of the κ-logarithm, $\ln_{\kappa=0}(x) = \ln x$, it follows that $f_{\mathrm{x};0}(x)$ is the lognormal pdf.

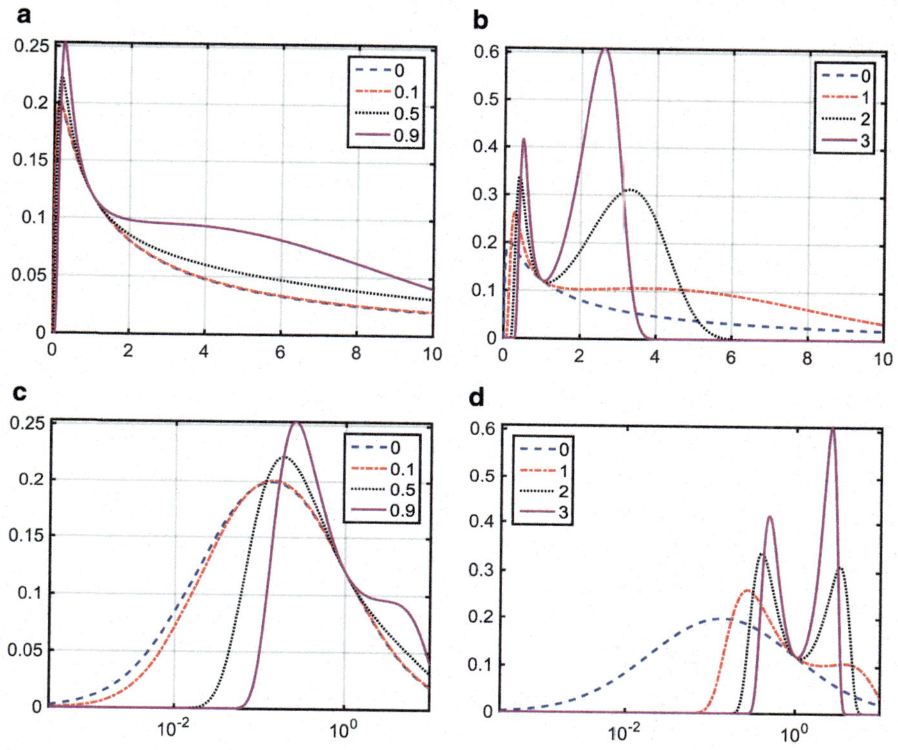

Fig. 14.4 κ-lognormal probability density functions given by (14.45) for different values of κ (left: $\kappa < 1$; right: $\kappa \geq 1$) and $m = 1$, $\sigma = 2$. The κ-lognormal density functions are compared with the lognormal ($\kappa = 0$) pdf. The pdf curves for $\kappa < 1$ are unimodal, while for $\kappa > 1$ they are bimodal. The plots in the bottom row use a logarithmic scale (in base 10) for the horizontal axis to allow better viewing of the left tail of the pdfs. (**a**) $\kappa < 1$ (**b**) $\kappa > 1$ (**c**) $\kappa < 1$ (**d**) $\kappa > 1$

A comparison of the κ-lognormal with the lognormal pdf is shown in Fig. 14.4. The plots in Fig. 14.4a, b display the κ-lognormal pdf for $\kappa < 1$ and $\kappa > 1$ respectively. The κ-lognormal is similar to the lognormal pdf for $\kappa < 1$. For $\kappa > 1$, however, the κ-lognormal density function becomes bimodal. For all the values of κ, the right tail of the κ-lognormal density is shorter than that of the lognormal.

The tail behavior is better illustrated in Fig. 14.4c, d which employ a logarithmic scale for the horizontal axis. On the other hand, the κ-lognormal density has more weight than the lognormal over an intermediate range of values that depends on κ, a property which could be useful in applications.

The plots in Fig. 14.5 display the ratio of the κ-lognormal density values over the respective values for the lognormal pdf. For $\kappa = 0.1$, the ratio is close to one over the displayed range of values $x \in [0, 100]$. As κ increases to 0.5, the ratio drops below one faster (i.e., at smaller x). This behavior is due to the faster than logarithmic increase of the κ-logarithm function as $x \to \infty$ (see Fig. 14.3).

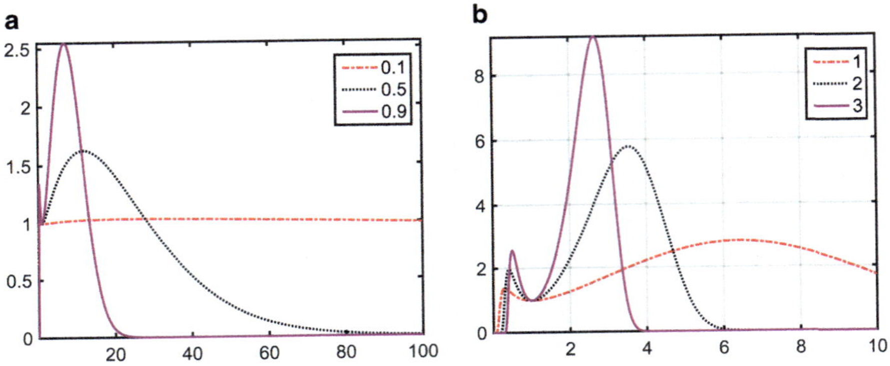

Fig. 14.5 Ratio of κ-logarithm probability density functions given by equation (14.45) over the lognormal pdf for different values of κ (left: $\kappa < 1$; right: $\kappa \geq 1$) and $m = 2$, $\sigma = 2$. Note the shorter range of the horizontal axis for $\kappa \geq 1$ due to the considerably shorter right tail of the pdfs. (**a**) $\kappa < 1$ (**b**) $\kappa > 1$

Since the κ-logarithm appears in the exponent of the pdf (14.45), it dominates the asymptotic regime and overshadows the differences between the rational functions x^{-1} (lognormal) and $x^{\kappa-1} + x^{-\kappa-1}$ (κ-logarithm). For $\kappa \geq 1$ the ratio is a non-monotonic (bimodal) function of x, which drops below one faster than for $\kappa < 1$.

Simulation Realizations of κ-lognormal random fields are shown in Fig. 14.6. They are generated by means of (14.44), based on a Gaussian random field $\mathrm{X}(\mathbf{s};\omega)$ with SSRF covariance. The realizations are generated on a square grid with $L = 512$ nodes per side. The SSRF parameters involve the characteristic length $\xi = L/50$, the rigidity parameter which takes values -1.9, 1.9, 10 and the scale parameter η_0 which is tuned to yield unit variance for all the rigidity parameters. The parameter κ takes the values $0.5, 1, 2$.

14.5 Hermite Polynomials

In this section we discuss Gaussian anamorphosis using expansions that are based on Hermite polynomials [485], [823, pp. 238–249], [132]. These mathematical functions are well known in physics due to their presence in the eigenfunctions of the Schrödinger equation for the harmonic quantum oscillator [478, 722]. Hermite polynomials have also been used in connection with the maximum entropy method to simulate non-Gaussian random processes [675]. In the latter approach, the maximum entropy method is used to determine the marginal pdf based on a number of available statistical moments. In this section we follow the presentation in [823].

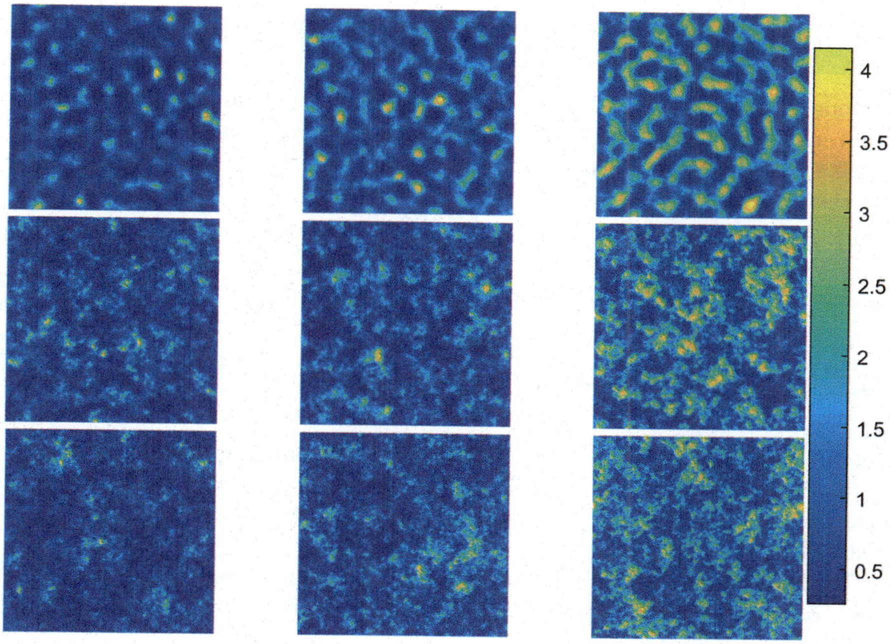

Fig. 14.6 Realizations of κ-lognormal random fields based on Gaussian SSRFs with Spartan covariance function. The columns correspond to $\kappa = 0.5$ (left), $\kappa = 1$ (middle) and $\kappa = 2$ (right). Rows correspond to $\eta_1 = -1.9$ (top), $\eta_1 = 1.9$ (middle) and $\eta_1 = 10$ (bottom)

The Hermite polynomials are defined by means of the derivatives of the Gaussian density function as follows[3]

$$H_m(y) = (-1)^m \frac{\mathrm{d}^m e^{-y^2/2}}{\mathrm{d}y^m}\, e^{y^2/2}, \quad \text{for } m \geq 0.$$

The factor $(-1)^m$ ensures that the highest-degree monomial of the Hermite polynomial has a positive sign. The first few polynomials are thus given by $H_0(y) = 1$, $H_1(y) = y$, $H_2(y) = y^2 - 1$, and $H_3(y) = y^3 - 3y$.

The Hermite polynomials can be constructed using the recursion relation

$$H_{m+1}(y) = -y\, H_m(y) - m\, H_{m-1}(y), \quad \text{for } m \geq 0.$$

[3]There are slight notational variations across different fields; for example, in physics the Gaussian exponent is y^2 instead of $y^2/2$, and in geostatistics the sign $(-1)^m$ is often dropped. Such differences lead to respective variations in the relations that involve the Hermite polynomials.

Remark In the following, $\mathcal{E}_0$ denotes the expectation with respect to the $\mathcal{N}(0, 1)$ standard normal distribution.

Orthogonality The crucial property of the Hermite polynomials which makes them suitable candidates for the Gaussian anamorphosis is their orthogonality with respect to the Gaussian density, i.e.,

$$\mathcal{E}_0 \left\{ H_m \left[\mathrm{Y}(\mathbf{s}; \omega) \right] H_k \left[\mathrm{Y}(\mathbf{s}; \omega) \right] \right\} = \frac{1}{\sqrt{2\pi}} \int_{-\infty}^{\infty} \mathrm{d}y \, \mathrm{e}^{-y^2/2} H_m(y) H_k(y) = m! \, \delta_{m,k}. \tag{14.46}$$

14.5.1 Hermite Polynomial Expansions

Let $\phi(\cdot)$ represent the inverse of the nonlinear transformation function $g(\cdot)$ used in the Gaussian anamorphosis Definition 14.2. Hence, we assume that for a non-Gaussian random field $\mathrm{X}(\mathbf{s}; \omega)$ there exists a Gaussian random field $\mathrm{Y}(\mathbf{s}; \omega)$ such that $\mathrm{X}(\mathbf{s}; \omega) = \phi \left[\mathrm{Y}(\mathbf{s}; \omega) \right]$.

The Hermite polynomials form an orthonormal base in Hilbert space for random fields $\mathrm{X}(\mathbf{s}; \omega) = \phi \left[\mathrm{Y}(\mathbf{s}; \omega) \right]$ that possess a finite second-order moment, i.e.,

$$\mathcal{E}_0 \left\{ \phi^2 \left[\mathrm{Y}(\mathbf{s}; \omega) \right] \right\} = \int_{-\infty}^{\infty} \mathrm{d}y \, \phi^2(y) \, \frac{\mathrm{e}^{-y^2/2}}{\sqrt{2\pi}} < \infty.$$

Functions $\phi \left[\mathrm{Y}(\mathbf{s}; \omega) \right]$ that admit finite second-order moments can be expanded in terms of Hermite polynomials of the Gaussian field $\mathrm{Y}(\mathbf{s}; \omega)$ as follows

$$\phi \left[\mathrm{Y}(\mathbf{s}; \omega) \right] = \sum_{m=1}^{\infty} \frac{c_m}{m!} H_m \left[\mathrm{Y}(\mathbf{s}; \omega) \right], \tag{14.47}$$

where the coefficients $\{c_m\}_{m=1}^{\infty}$ are given by the Gaussian expectations

$$c_m = \mathcal{E}_0 \left\{ \phi \left[\mathrm{Y}(\mathbf{s}; \omega) \right] \, H_m \left[\mathrm{Y}(\mathbf{s}; \omega) \right] \right\}. \tag{14.48}$$

Moments of Hermite polynomials The above equations may seem rather abstract. To give practical value to the Hermite expansion we need two ingredients. (i) The analytical expression of the function $\phi(\cdot)$, and (ii) a "recipe" for calculating expectations of terms that involve the Hermite polynomials over the Gaussian density.

Let us start with the second point. The orthogonality property (14.46) can be used to calculate the Gaussian expectations of Hermite polynomials of the Gaussian field $\mathrm{Y}(\mathbf{s}; \omega)$. For example, to calculate $\mathcal{E}_0 \left\{ H_m \left[\mathrm{Y}(\mathbf{s}; \omega) \right] \right\}$ we use (14.46) with $k = 0$. In light of $H_0(y) = 1$, the orthogonality relation leads to

$$\mathcal{E}_0 \{ H_m [\mathrm{Y}(\mathbf{s};\omega)] \} = \delta_{m,0}. \tag{14.49}$$

Similarly, we can obtain expressions for the variance of the function $H_m [\mathrm{Y}(\mathbf{s};\omega)]$ as well as the covariance between two terms $H_m [\mathrm{Y}(\mathbf{s};\omega)]$ and $H_k [\mathrm{Y}(\mathbf{s};\omega)]$, i.e.,

$$\mathbb{V}\mathrm{ar} \{ H_m [\mathrm{Y}(\mathbf{s};\omega)] \} = m! \left(1 - \delta_{m,0}\right), \tag{14.50}$$

$$\mathbb{C}\mathrm{ov} \{ H_m [\mathrm{Y}(\mathbf{s};\omega)], H_k [\mathrm{Y}(\mathbf{s};\omega)] \} = m!\, \delta_{m,k} \left(1 - \delta_{m,0}\right). \tag{14.51}$$

The above covariance equation implies that two Hermite polynomials of the Gaussian field $\mathrm{Y}(\mathbf{s};\omega)$ are uncorrelated if the Hermite polynomial degrees are different. In addition, the variance of the zero-order polynomial vanishes. This is expected since $H_0 [\mathrm{Y}(\mathbf{s};\omega)] = 1$ regardless of $\mathrm{Y}(\mathbf{s};\omega)$.

Moments of the Hermite expansion Using the equations (14.49)–(14.51) for the moments of the Hermite polynomials, we can evaluate the moments of the Hermite expansion (14.47) of the general function $\phi [\mathrm{Y}(\mathbf{s};\omega)]$ in terms of the coefficients $\{c_m\}_{m=1}^{\infty}$. More specifically

$$\mathcal{E}_0 \{ \phi [\mathrm{Y}(\mathbf{s};\omega)] \} = c_0. \tag{14.52}$$

$$\mathbb{V}\mathrm{ar} \{ \phi [\mathrm{Y}(\mathbf{s};\omega)] \} = \sum_{m=1}^{\infty} \frac{c_m^2}{m!}. \tag{14.53}$$

In addition, if $\phi_1 [\mathrm{Y}(\mathbf{s};\omega)]$ and $\phi_2 [\mathrm{Y}(\mathbf{s};\omega)]$ are two different functions of the random field $\mathrm{Y}(\mathbf{s};\omega)$, their covariance function is given by

$$\mathbb{C}\mathrm{ov} \{ \phi_1 [\mathrm{Y}(\mathbf{s};\omega)], \phi_2 [\mathrm{Y}(\mathbf{s};\omega)] \} = \sum_{m=1}^{\infty} \frac{1}{m!}\, c_{1,m}\, c_{2,m}, \tag{14.54}$$

where $c_{i,m}$ are respectively the expansion coefficients of the functions ϕ_i, $i = 1, 2$, respectively.

Truncated Hermite expansions In practice, the Hermite expansion is truncated at some high but finite order M, and equation (14.47) is approximated by

$$\phi^{(M)} [\mathrm{Y}(\mathbf{s};\omega)] = \sum_{m=1}^{M} \frac{c_m}{m!} H_m [\mathrm{Y}(\mathbf{s};\omega)]. \tag{14.55}$$

The order of the expansion can be determined by requiring that the variance of the truncated expression (14.55), given by $\sum_{m=1}^{M} \frac{c_m^2}{m!}$, approximates the variance of the data to a desired level of accuracy.

14.5.2 *Practical Use of Hermite Expansions*

If closed-form expressions exist for $F_{\mathrm{x}}(x)$, $\phi(\cdot)$ and $\phi^{-1}(\cdot)$, the Gaussian anamorphosis equations (14.5) can be used to transform the data and the joint pdf. In addition, estimates that are derived based on the normal field $\mathrm{Y}(\mathbf{s};\omega)$ can be back-transformed to obtain respective estimates for the original field $\mathrm{X}(\mathbf{s};\omega)$. For example, in the case of lognormal data, the transformation from the normal values y is given by $\phi(y) = \exp(y)$, while for data that follow the $\chi^2(1)$ distribution (chi squared distribution with one degree of freedom) the respective transformation is $\phi(y) \propto y^2$. The above transformations can easily be inverted by means of closed-form expressions.

The Hermite polynomial expansion is particularly useful for performing Gaussian anamorphosis if the probability distribution of the data $\{x_n^*\}_{n=1}^N$ is not known in closed form. In order to apply the Hermite polynomial expansion, the transformation function $\phi(\cdot)$ is needed. Hence, in the following it is assumed that an estimate of the transformation function can be obtained from the data.

Empirical Gaussian anamorphosis If a parametric expression of the marginal cdf cannot be deduced from the data, the alternative is to construct the empirical cdf, $\hat{F}_{\mathrm{x}}(x)$. The latter has the form of the well-known non-decreasing staircase function

$$\hat{F}_{\mathrm{x}}(x) = \frac{1}{N}\sum_{n=1}^{N} \mathbb{I}_{x^*_{[n]}\le x}(x), \tag{14.56}$$

where the values $x^*_{[n]}$ represent the ordered sample values:

$$x^*_{[1]} \le x^*_{[2]} \le \ldots \le x^*_{[N]},$$

and $\mathbb{I}_{x^*_{[n]}\le x}$ is the *indicator function* defined by

$$\mathbb{I}_{x^*_{[n]}\le x} = \begin{cases} 1, & \text{if } x^*_{[n]} \le x, \\ 0, & \text{otherwise.} \end{cases}$$

The empirical Gaussian anamorphosis produces a discontinuous (piecewise linear) transformation function which is constructed as follows

$$\hat{\phi}(y) = \frac{1}{N}\sum_{n=1}^{N} x^*_{[n]}\mathbb{I}_{y\in A_n}, \tag{14.57a}$$

where A_n, $n = 1, \ldots, N$ are the following half-open intervals

$$A_n = \left(\Phi^{-1}\left(\frac{n-1}{N}\right),\ \Phi^{-1}\left(\frac{n}{N}\right)\right], \tag{14.57b}$$

defined by means of the inverse cdf, $\Phi^{-1}(\cdot)$, of the standard normal probability distribution. Based on the form of the equation (14.57) the empirical transformation function $\hat{\phi}(\cdot)$ is discontinuous and resembles a staircase with steps of height $x^*_{[n]}$.

Smoothing the empirical transformation function The empirical Gaussian anamorphosis is non-invertible due to its step-function structure, which makes it unsatisfactory for modeling purposes. The truncated Hermite polynomial expansion (14.55) can be applied to smoothen the empirical transformation function. In order to calculate the Hermite expansion coefficients $\{c_m\}_{m=1}^M$ the estimated transformation function is inserted in (14.48). Then, the following expression is obtained for the Hermite coefficients:

$$c_m = \frac{1}{\sqrt{2\pi}} \int_{-\infty}^{\infty} \hat{\phi}(y) H_m(y) \mathrm{e}^{-y^2/2}, \; m = 1, \ldots, M.$$

In deriving the above we used the fact that the expectation $\mathcal{E}_0[\cdot]$ implies integration over the standard normal pdf given by $\frac{1}{\sqrt{2\pi}} \exp(-y^2/2)$. Taking into account the expression of the transformation function (14.57a) the following integral expression is obtained for the coefficients of the Hermite expansion

$$c_m = \frac{1}{\sqrt{2\pi} N} \sum_{n=1}^{N} x^*_{[n]} \int_{A_n} H_m(y) \mathrm{e}^{-y^2/2}, \; m = 1, \ldots, M.$$

However, even the truncated smoothed expansion produces a transformation function that does not increase monotonically outside the range of the sample values (for a discussion of this matter see [823, p. 247]).

Non-parametric kernel-based density estimation Non-parametric kernel density estimators allow continuous representations of the pdf that can improve the staircase estimate (14.56). The kernel density estimator of the marginal pdf of $\mathrm{X}(\mathbf{s}; \omega)$ based on the sample $\{x^*_n\}_{n=1}^N$ is given by

$$\hat{f}_{\mathrm{x}}(x) = \sum_{n=1}^{N} \frac{1}{N h} K\left(\frac{x - x^*_n}{h}\right), \tag{14.58}$$

where $K(\cdot)$ is a kernel function with the properties described in Definition 2.1 and h is the bandwidth parameter [734]. The latter can be estimated by means of various techniques that minimize specific risk functions. For example, the minimization of the mean integrated square error (MISE) of the estimator (14.58) can be used as a bandwidth selection criterion. There also exist empirical, approximate bandwidth formulas that are based on parametric assumptions [128].

The advantage of (14.58) is that it provides an explicit equation for the pdf at every $x \in \mathbb{R}$. An estimate of the empirical cdf is then obtained as follows

$$\hat{F}_{\mathrm{x}}(x) = \sum_{n=1}^{N} \frac{1}{N h} \widetilde{K}\left(\frac{x - x_n^*}{h}\right), \tag{14.59a}$$

where $\widetilde{K}(\cdot)$ represents the kernel's integral, i.e.,

$$\widetilde{K}\left(\frac{x - x_n^*}{h}\right) = \int_{-\infty}^{x} \mathrm{d}x' \, K\left(\frac{x' - x_n^*}{h}\right). \tag{14.59b}$$

Beyond marginal dependence Gaussian anamorphosis by means of the Hermite polynomial expansion focuses on the marginal distribution of the data. Hence, it does not guarantee that the bivariate distribution of the transformed variables $\mathrm{Y}(\mathbf{s}_1; \omega)$ and $\mathrm{Y}(\mathbf{s}_2; \omega)$, where $\mathbf{s}_1 \neq \mathbf{s}_2$, is jointly Gaussian. To remedy this shortcoming, Matheron developed *isofactorial models* which allow controlling the bivariate in addition to the marginal distribution [132, 823].

14.6 Multivariate Student's t-Distribution

The multivariate Student's t-distribution is a symmetric, heavy-tailed function which looks similar to the Gaussian distribution but its tails decay algebraically (sub-exponentially) instead of exponentially. The Student's t-distribution is foremost known for its applications in statistics and in particular in hypothesis testing. Due to the power-law dependence of the pdf for $|x| \gg 1$, the Student's t-distribution is also used to model heavy tails that appear in functions that describe complex systems and in various scientific measurements [41, 766].

14.6.1 Univariate Student's t-Distribution

The marginal Student's t-pdf is characterized by the parameter ν. The value of $\nu \geq 1$ controls the "weight" of the tail: higher values of ν imply less weight in the tails. The parameter ν is also known as the number of *degrees of freedom* of the probability distribution. As is explained below, this term reflects the fact that the Student's t-pdf applies to the standardized sample average statistic for a sample that involves $\nu + 1$ data. The Student's t-pdf is given by the following equation

$$f_{\mathrm{x}}(x) = \frac{\Gamma(\frac{\nu+1}{2})}{\sqrt{\pi \, \nu} \, \Gamma(\frac{\nu}{2})} \left(1 + \frac{x^2}{\nu}\right)^{-\frac{\nu+1}{2}}, \quad \text{for } x \in (-\infty, \infty). \tag{14.60}$$

The above can be used to model the marginal pdf of a random field $\mathrm{X}(\mathbf{s}; \omega)$.

The pdf (14.60) describes symmetrically distributed random variables $\mathrm{X}(\omega)$ with mean $m_{\mathrm{x}} = 0$ and variance which is equal to $\sigma_{\mathrm{x}}^2 = \nu/(\nu - 2)$ for $\nu > 2$ and infinite for $1 \leq \nu \leq 2$. The divergence of the variance for $\nu \leq 2$ is due to the increased dispersion caused by the the heavy tails. In a sense that will be specified below, the Student's t-distribution is a prototype for symmetric random variables with more weight in the pdf tails than normally distributed variables.

Origin of the heavy tails For $|x| \gg 1$ it follows from (14.60) that the pdf of the Student's t-distribution drops off asymptotically as a power law, i.e.,

$$\lim_{|x| \to \infty} f_{\mathrm{x}}(x) \propto |x|^{-\nu+1}.$$

The heavy tails of the Student's t-distribution can originate from the superposition of uncertainties. For example, consider the coarse-grained random variable

$$\overline{\mathrm{X}(\omega)} = \frac{1}{N} \sum_{n=1}^{N} \mathrm{X}_n(\omega),$$

where the random variables $\{\mathrm{X}_n(\omega)\}_{n=1}^N$ are independent and identically distributed with mean m_{x} and variance σ_{x}^2. If the $\mathrm{X}_n(\omega)$ represent independent samples, then $\overline{\mathrm{X}(\omega)}$ represents the *sample average*. The latter is also a random variable, since different realizations of $\{\mathrm{X}_n(\omega)\}_{n=1}^N$ produce different values of $\overline{\mathrm{X}(\omega)}$. The sample average is typically used to estimate the unknown population mean.

Next, let us assume that $\mathrm{X}_n(\omega) \overset{d}{=} \mathcal{N}(m_{\mathrm{x}}, \sigma_{\mathrm{x}}^2)$, for all $n = 1, \ldots, N$. Then, the standardized random variable $Z(\omega)$ defined by

$$Z(\omega) = \frac{\overline{\mathrm{X}(\omega)} - m_{\mathrm{x}}}{\sigma_{\mathrm{x}}/\sqrt{N}} \overset{d}{=} \mathcal{N}(0, 1),$$

follows the standard normal distribution. $Z(\omega)$ can be used to study the distribution of the sample average around the population mean. However, the above definition implies that the population variance σ_{x}^2 is known. If σ_{x}^2 is not known, it is approximated based on the unbiased sample variance estimator

$$\hat{\sigma}_{\mathrm{x}}^2 = \frac{1}{N-1} \sum_{n=1}^{N} (x_n - m_{\mathrm{x}})^2 .$$

Then, $Z(\omega)$ is replaced by the following standardized random variable

$$T(\omega) = \frac{\overline{\mathrm{X}(\omega)} - m_{\mathrm{x}}}{\sqrt{\hat{\sigma}_{\mathrm{x}}^2/N}}. \tag{14.61}$$

It can be shown that $T(\omega)$ follows the Student's t-distribution with $N - 1$ degrees of freedom [252]. The Student's t-distribution has a heavier tail than the Gaussian distribution that governs the population of the i.i.d. variables $X_n(\omega)$, reflecting the additional uncertainty caused by the unknown population variance.

The Gaussian limit For $\nu \gg 1$ the asymptotic formula for the gamma function [4, Eq. 6.1.46, p. 257] can be used, according to which

$$\lim_{\nu\to\infty} \frac{\Gamma(\frac{\nu+1}{2})}{\nu^{1/2}\Gamma(\frac{\nu}{2})} = 1.$$

In addition, at the limit $\nu \to \infty$ the power law can be approximated by the Gaussian exponential, thus leading to

$$\lim_{\nu\to\infty} \left(1 + \frac{x^2}{\nu}\right)^{-\frac{\nu+1}{2}} = \mathrm{e}^{-\nu \ln\left(1+\frac{x^2}{\nu}\right)/2} \approx \mathrm{e}^{-x^2/2}.$$

Hence, at the $\nu \gg 1$ limit the Student's t-distribution converges to the Gaussian.

***t* location-scale distribution** A location-scale transformation maps a random variable $X(\omega)$ into another variable $Y(\omega) = a + bX(\omega)$, where $a \in \mathbb{R}$ is the location and $b > 0$ the scale parameter. Then, both random variables follow the same probability distribution, i.e., $Y(\omega) \overset{d}{=} X(\omega)$. If $X(\omega)$ follows the Student's t-distribution, the random variable $Y(\omega)$ is said to follow the t location-scale distribution. The location parameter determines the mean of $Y(\omega)$ while the scale parameter is a multiplicative factor that determines (along with ν) the standard deviation of $Y(\omega)$.

14.6.2 Connection with the Tsallis Distribution

The q-Gaussian (synonym: Tsallis) distribution is used in statistical physics for heavy-tailed processes [802]. This distribution is closely related to the univariate Student's t-distribution despite some differences in parametrization. The q-Gaussian pdf is based on the q-exponential function which is defined by

$$\mathrm{e}_q(x) = \begin{cases} \mathrm{e}^x, & q = 1 \\ [1 + (1-q)x]^{1/1-q}, & q \neq 1, \text{ and } 1 + (1-q)x > 0 \\ 0^{1/1-q}, & q \neq 1, \text{ and } 1 + (1-q)x \leq 0. \end{cases} \tag{14.62}$$

The q-exponential coincides with the inverse Box-Cox transform for $q = 1 - \lambda$.

q-Gaussian distribution Based on the q-exponential, the *q-Gaussian* pdf is given by the expression

$$f(x) = \frac{\sqrt{\beta}}{C_q} e_q(-\beta x^2),$$

where $-\infty < q < 3$, and C_q is a normalization factor that depends on q.

The Student's t-distribution with $\nu > 0$ degrees of freedom corresponds to a q-Gaussian with $\beta = 1/(3-q)$ and $q = (\nu+3)/(\nu+1)$. This transformation implies that $3 > q > 1$ if $\nu > 0$. For $q = 1$ the q-Gaussian recovers the normal distribution. However, the q-Gaussian is also defined for $q < 1$. In this case the q-Gaussian has a compact support with boundaries at $\pm 1/\sqrt{\beta(1-q)}$.

Tsallis entropy The q-Gaussian distribution follows from the maximization of the non-extensive *Tsallis entropy* which is given by [801]

$$S_q[f] = \frac{1}{q-1}\left(1 - \int \mathrm{d}x \, [f(x)]^q\right),$$

where $f(x)$ is the pdf of the respective random variable $X(\omega)$. At the limit $q \to 1$ the above tends to the standard (Boltzmann-Gibbs) definition of entropy which is extensive. A physical observable A_n is called extensive if the limit $\lim_{n\to\infty} A_n/n$ exists, where n is the size of the system. A more detailed definition and differences between extensive and additive systems are given in [798].

The q-Gaussian distribution was proposed as a model for strongly correlated variables. Strong correlations are believed to generate the non-extensivity of the Tsallis entropy. Efforts to derive a generalized central limit theorem which would prove that the q-Gaussian is an attractor for some class of probability distributions have been inconclusive [349].

Different generalizations of the exponential (deformed exponentials) that also lead to pdfs with power-law tails and resemble the Student's t-distribution have been proposed by Kaniadakis [426, 428, 429, 824]. These general forms employ the q-Gaussian function which is derived from the κ-exponential function (14.17).

14.6.3 Multivariate Student's t-Distributions

The univariate Student's t-distribution can be generalized to higher dimensions by means of various extensions [468]. We focus on the most common generalization.

We consider the random vector $\mathbf{X}(\omega) = (X_1(\omega), \ldots, X_N(\omega))^\top$ with realizations $\mathbf{x} \in \mathbb{R}^N$. The vector components $X_n(\omega)$ are assumed to follow the Student's t-distribution with ν degrees of freedom. The random vector $\mathbf{X}(\omega)$ then follows the *multivariate Student's t-distribution (mSt-d)*, denoted by $\mathcal{T}_\nu(\mathbf{m}_x, \boldsymbol{\Sigma})$, if its joint pdf is given by

$$f_{\mathrm{X}}(\mathbf{x}) = \frac{\Gamma\left[(\nu+N)/2\right]}{(\nu\,\pi)^{N/2}\,\Gamma(\nu/2)\,\det(\boldsymbol{\Sigma})^{1/2}}\left[1+\frac{1}{\nu}(\mathbf{x}-\mathbf{m})^{\top}\boldsymbol{\Sigma}^{-1}(\mathbf{x}-\mathbf{m})\right]^{-(\nu+N)/2}, \tag{14.63}$$

where ν is the number of degrees of freedom of the marginal Student's t-distribution, N is the number of components (variates), $\mathbf{m}$ is the *location vector*, $\boldsymbol{\Sigma}$ is the $N \times N$ *scale matrix* (which is proportional to the covariance matrix), and $\det(\boldsymbol{\Sigma})$ is the determinant of the scale matrix. Any permissible covariance model can be used to generate $\boldsymbol{\Sigma}$, including the Spartan covariance family.

The mSt-d is a generalization of the multivariate Gaussian distribution with heavier tails. Compared with the Gaussian, the mSt-d includes the additional parameter ν which allows flexibility in modeling the asymptotic decay of the tails.

Based on the definition of the mSt-d, infinite-dimensional *Student's t-distributed random fields* have been proposed as models for *heavy-tailed spatial processes* including well-log data [690] and high-frequency (15-min interval) data of intermittent precipitation fields [785]. In the case of well-log data the long-tailed model can be justified due to the fact that each well log potentially integrates information from different geological layers which can have different variances.

Properties of mSt-d The multivariate Student's t-distribution satisfies a number of useful mathematical properties, which we present briefly below without proof following the exposition in [690]. In the following, it is assumed that $\mathbf{X}(\omega) \stackrel{d}{=} \mathcal{T}_{\nu}\,(\mathbf{m}_{\mathrm{X}}, \boldsymbol{\Sigma})$, where the joint pdf of Student's t-distribution is given by (14.63).

The Gaussian limit At the limit $\nu \to \infty$ the mSt-d tends to the multivariate Gaussian distribution.

The Cauchy limit For $\nu = 1$, the mSt-d becomes the multivariate Cauchy distribution.

Moments of first two orders The mean and the covariance of a random vector that follows the mSt-d are determined by the location vector and the scale matrix as follows

$$\mathcal{E}[\mathbf{X}(\omega)] = \mathbf{m}, \tag{14.64a}$$

$$\mathcal{E}[X_n(\omega)\,X_m(\omega)] = \frac{\nu}{\nu-2}\,\Sigma_{n,m}, \quad \text{for } n, m = 1, \ldots, N, \text{ and for } \nu \geq 3. \tag{14.64b}$$

For $\nu < 3$ the covariance matrix takes infinite values.

Invariance under linear transform If the random vector $\mathbf{X}(\omega)$ follows the mSt-d, its linear transforms $\mathbf{Y}(\omega)$ also follow the mSt-d with suitably scaled parameters.

More precisely, if $\boldsymbol{A}$ is an $p \times N$ matrix, then the p-dimensional random vector $\mathbf{Y}(\omega) = \boldsymbol{A}\,\mathbf{X}(\omega)$ follows the mSt-d:

$$\mathbf{Y}(\omega) \stackrel{d}{=} \mathcal{T}_{\nu}\left(\boldsymbol{A}\,\mathbf{m}_{\mathrm{X}},\, \boldsymbol{A}\,\boldsymbol{\Sigma}\,\boldsymbol{A}^{\top}\right).$$

Consistent marginalization The invariance under linear transformation implies that if the vector $\mathbf{X}(\omega)$ is partitioned into two sets $\mathbf{X}_1(\omega)$ and $\mathbf{X}_2(\omega)$ with dimensions N_1 and N_2 respectively, the joint pdfs of $\mathbf{X}_1(\omega)$ and $\mathbf{X}_2(\omega)$ are also mSt-d with ν degrees of freedom [468, p. 15–16]. This statement ensures that the mSt-d is consistent under marginalization. Consistency is a desired random field property as discussed in Sect. 5.9. In addition, consistency means that all the marginal pdfs follow the univariate Student's t-distribution with ν degrees of freedom. This is easily shown considering that the marginal distributions are obtained by setting $\boldsymbol{A}$ equal to a binary-valued $1 \times N$ matrix that contains a single non-zero element at the location of the specified marginal.

Zero correlation does not imply independence This maxim is well known from the theory of probability. Let us consider two random vectors $\mathbf{X}_1(\omega)$ and $\mathbf{X}_2(\omega)$ that follow the Student's t-distribution. In addition, assume that the respective cross-covariance matrices $\boldsymbol{\Sigma}_{1,2} = \mathbf{0}$, $\boldsymbol{\Sigma}_{2,1} = \mathbf{0}$ include only zeros. The lack of correlations, however, does not imply that $\mathbf{X}_1(\omega)$ and $\mathbf{X}_2(\omega)$ are statistically independent. Independence follows from the lack of correlations only at the Gaussian limit $\nu \to \infty$.

Conditional distributions The properties of conditional distributions are important, because prediction, as discussed in Chap. 10, and conditional simulation can be formulated using conditional distributions.

Let us assume that $\mathbf{Y}(\omega)$ is an $p \times 1$ random vector related to the $N \times 1$ random vector $\mathbf{X}(\omega)$, where $N > p \in \mathbb{N}$, via $\mathbf{Y}(\omega) = \boldsymbol{A}\,\mathbf{X}(\omega)$ where $\boldsymbol{A}$ is an $p \times N$ linear transformation matrix as discussed above. In addition, let us consider that the values of the random vector $\mathbf{Y}(\omega)$ are fixed to $(y_1, \ldots, y_p)^\top$.

The *conditional probability density function* of the vector $\mathbf{X}(\omega)$ given the values of $\mathbf{Y}(\omega)$ is expressed by means of the following multivariate Student's t-distribution

$$f_{\mathrm{x}|\mathrm{y}}(\mathbf{x} \mid \mathbf{y}) = \mathcal{T}_{\nu+p}(\mathbf{m}_{\mathrm{x}|\mathrm{y}}, \boldsymbol{\Sigma}_{\mathrm{x}|\mathrm{y}}), \tag{14.65}$$

where the conditional mean and scale matrices are given by means of

$$\mathbf{m}_{\mathrm{x}|\mathrm{y}} = \mathbf{m} + \boldsymbol{\Sigma}\,\boldsymbol{A}^\top \left(\boldsymbol{A}\,\boldsymbol{\Sigma}\,\boldsymbol{A}^\top\right)^{-1} (\mathbf{y} - \boldsymbol{A}\,\mathbf{m}), \tag{14.66a}$$

$$\boldsymbol{\Sigma}_{\mathrm{x}|\mathrm{y}} = \phi(\mathbf{y}) \left[\boldsymbol{\Sigma} - \boldsymbol{\Sigma}\,\boldsymbol{A}^\top \left(\boldsymbol{A}\,\boldsymbol{\Sigma}\,\boldsymbol{A}^\top\right)^{-1} \boldsymbol{A}\,\boldsymbol{\Sigma}\right], \tag{14.66b}$$

$$\phi(\mathbf{y}) = \frac{\nu}{\nu + p} \left[1 + \frac{1}{\nu} (\mathbf{y} - \boldsymbol{A}\,\mathbf{m})^\top \left(\boldsymbol{A}\,\boldsymbol{\Sigma}\,\boldsymbol{A}^\top\right)^{-1} (\mathbf{y} - \boldsymbol{A}\,\mathbf{m})\right]. \tag{14.66c}$$

Decomposition Let $V(\omega)$ be a scalar random variable such that $\nu\,V(\omega)$ follows the chi squared distribution with ν degrees of freedom, i.e., $\nu\,V(\omega) \overset{d}{=} \chi^2(\nu)$, and $\mathrm{U}(\omega) \overset{d}{=} \mathcal{N}(\mathbf{0}, \mathbf{I})$ an N-dimensional uncorrelated vector of Gaussian uncorrelated components ($\mathbf{I}$ is the $N \times N$ identity matrix). The random vector $\mathbf{X}(\omega)$ that follows the joint Student's t-distribution can be expressed in terms of $V(\omega)$ and $\mathbf{U}(\omega)$ as follows:

$$\mathbf{X}(\omega) = \mathbf{m} + V^{-1/2}(\omega)\, \boldsymbol{\Sigma}^{1/2}\, \mathbf{U}(\omega).$$

14.6.4 Student's t-Distributed Random Fields

The RF $\mathrm{X}(\mathbf{s};\omega)$ is called a Student's t-distributed or simply t-distributed random field if for all $N \in \mathbb{N}$ and all configurations of points $\{\mathbf{s}_1, \ldots, \mathbf{s}_N\}$, the random vector

$$[\mathrm{X}(\mathbf{s}_1;\omega), \ldots \mathrm{X}(\mathbf{s}_N;\omega)]^\top$$

follows the multivariate Student's t-distribution. The elements of the scale matrix $[\boldsymbol{\Sigma}]_{i,j}$ of the joint distribution are determined for any lag $\mathbf{r} \doteq \mathbf{s}_i - \mathbf{s}_j$ from a positive definite function $\Sigma(\mathbf{r})$. The notation

$$\mathrm{X}(\mathbf{s};\omega) \overset{d}{=} \mathcal{T}_\nu \left[m_\mathrm{x}(\mathbf{s}),\, \Sigma(\mathbf{s},\mathbf{s}')\right],$$

denotes that $\mathrm{X}(\mathbf{s};\omega)$ is a Student's t-distributed random field.

Conditional random field with Student's t-distribution Let us consider a conditional random field that is constrained by the observations $\mathbf{x}^* = (x_1^*, \ldots, x_N^*)^\top$ at the locations $\{\mathbf{s}_n\}_{n=1}^N$ of the sampling set $\mathbb{S}_\mathrm{N}$.

If the parent field $\mathrm{X}(\mathbf{s};\omega)$ follows the Student's t-distribution with unconditional mean $m_\mathrm{x}(\mathbf{s})$ and scale function $\Sigma(\mathbf{s},\mathbf{s}')$, the conditional random field also follows the Student's t-distribution with the following modified mean and scale functions:

$$\mathrm{X}(\mathbf{s};\omega) \mid \mathbf{x}^* \overset{d}{=} \mathcal{T}_{\nu+N} \left(m_{\mathrm{p|d}}(\mathbf{s}),\, \Sigma_{\mathrm{p|d}}(\mathbf{s},\mathbf{s}')\right). \tag{14.67}$$

1. The function $m_{\mathrm{p|d}}(\mathbf{s})$ is the *conditional mean* of the Student's t-distribution. It is given by the following function

$$m_{\mathrm{p|d}}(\mathbf{s}) = m_\mathrm{x}(\mathbf{s}) + \boldsymbol{\Sigma}_{\mathrm{s,d}}\, \boldsymbol{\Sigma}_{\mathrm{d,d}}^{-1} \left(\mathbf{x}^* - \mathbf{m}\right), \tag{14.68}$$

where

$$\boldsymbol{\Sigma}_{\mathrm{s,d}} = [\Sigma(\mathbf{s},\mathbf{s}_1), \ldots, \Sigma(\mathbf{s},\mathbf{s}_N)], \tag{14.69}$$

$$\left[\boldsymbol{\Sigma}_{\mathrm{d,d}}\right]_{n,m} = \Sigma(\mathbf{s}_n,\mathbf{s}_m)^\top, \quad , n, m = 1, \ldots, N, \tag{14.70}$$

and

$$\mathbf{m} = [m_{\mathrm{x}}(\mathbf{s}_1), \ldots, m_{\mathrm{x}}(\mathbf{s}_N)]^\top . \tag{14.71}$$

2. The function $\Sigma_{\mathrm{p|d}}(\mathbf{s}, \mathbf{s}')$ is the *conditional scale (dependence) matrix* entry for the locations $\mathbf{s}$ and $\mathbf{s}'$, which is defined by

$$\Sigma_{\mathrm{p|d}}(\mathbf{s}, \mathbf{s}') = \frac{\nu + N^*}{\nu + N} \left[\Sigma(\mathbf{s}, \mathbf{s}') - \boldsymbol{\Sigma}_{\mathrm{s,d}}\, \boldsymbol{\Sigma}_{\mathrm{d,d}}^{-1}\, \boldsymbol{\Sigma}_{\mathrm{d,s'}} \right], \tag{14.72a}$$

$$N^* = \left(\mathbf{x}^* - \mathbf{m}\right)^\top \boldsymbol{\Sigma}_{\mathrm{d,d}}^{-1} \left(\mathbf{x}^* - \mathbf{m}\right). \tag{14.72b}$$

Note that $N^* \in \mathbb{R}$ is an effective number of points that is not necessarily an integer number. In the above equations the following functions and matrices are used:

- $\Sigma(\mathbf{s}, \mathbf{s}')$ is the unconditional dependence function between the prediction points $\mathbf{s}$ and $\mathbf{s}'$.
- $\boldsymbol{\Sigma}_{\mathrm{d,d}}$ is the $N \times N$ scale matrix for all the points in $\mathbb{S}_{\mathrm{N}}$ [see (14.70)].
- $\boldsymbol{\Sigma}_{\mathrm{s,d}}$ is the $1 \times N$ dependence vector between the prediction point $\mathbf{s}$ and all the locations in $\mathbb{S}_{\mathrm{N}}$ [see (14.69)].
- $\boldsymbol{\Sigma}_{\mathrm{d,s'}}$ is the $N \times 1$ dependence vectors between all the locations in $\mathbb{S}_{\mathrm{N}}$ and the prediction point $\mathbf{s}'$.

3. The *conditional covariance function*, $C_{\mathrm{p|d}}(\mathbf{s}, \mathbf{s}')$, of the Student's t-distribution is obtained from the scale function (14.72) and the relation (14.64b) between the covariance and the scale function. Taking into account that the conditional random field has $N + \nu$ degrees of freedom, these equations lead to

$$C_{\mathrm{p|d}}(\mathbf{s}, \mathbf{s}') = \frac{\nu + N}{\nu + N - 2}\, \Sigma_{\mathrm{p|d}}(\mathbf{s}, \mathbf{s}'), \tag{14.73}$$

where $\Sigma_{\mathrm{p|d}}(\mathbf{s}, \mathbf{s}')$ is given by (14.72).

Estimation Parameter estimation for Student's t-distributed random fields by means of maximum likelihood is not easy. In particular, it is difficult to estimate accurately the value of ν that determines the tail behavior [690]. The spatial dependence function can be estimated using the geostatistical method of moments. Røislien and Omre propose a *hierarchical maximum likelihood* approach which is applicable if multiple samples are available [690]. A different approach assumes that the functional form and the range of the covariance function are known and uses Bayesian optimization methods [105, 761] to determine optimal values for the remaining parameters [738].

***T*-kriging** The conditional distribution of a Student's t-distributed random field can be used to formulate a spatial predictor. This extends the idea of optimal linear estimation (kriging) to non-Gaussian, heavy-tailed processes [738].

The *T-kriging prediction* is given by the conditional mean (14.68). The *prediction variance* is obtained from (14.73) and (14.72) which lead to

$$\sigma^2_{\rm tk}(\mathbf{s}) = \frac{\nu + N^*}{\nu + N - 2} \left[\Sigma(\mathbf{s}, \mathbf{s}) - \boldsymbol{\Sigma}_{\rm s,d}\, \boldsymbol{\Sigma}^{-1}_{\rm d,d}\, \boldsymbol{\Sigma}_{\rm d,s}\right], \quad \text{for } N + \nu \geq 3, \tag{14.74}$$

where N^* is given by (14.72b).

In fact, in light of (14.67), the *predictive distribution* is given by the multivariate Student's t-distribution with $N + \nu$ degrees of freedom. Hence, just as in the case of Gaussian kriging, in T-kriging the entire probability distribution at the prediction points is available. Some properties of the predictive distribution are listed below.

- As ν tends to infinity, the predictive distribution (14.67) tends to the Gaussian distribution, reflecting the convergence of the Student's t-distribution to the Gaussian.
- The predictive distribution also converges to the Gaussian if $N \to \infty$. This result suggests that for very large spatial data sets we cannot expect a gain in predictive performance by specifying a model with long tails.
- The predictive mean (14.68) has the same form as for simple kriging—provided that the same covariance function and respective parameters are used. In practice, however, the parameter inference procedure is expected to give different results for T-kriging than for simple or regression kriging (due to differences in the likelihood of the Gaussian versus the Student t-distributions). Hence, the T-kriging predictions are not necessarily the same as those of kriging with Gaussian assumptions.
- The conditional covariance of Student's t-distributed random fields differs significantly from the Gaussian conditional covariance. The former depends on the observations $\mathbf{x}^*$ through the parameter N^* defined in (14.72). Thus, in contrast with the (Gaussian) kriging variance which is independent of the data, the T-kriging variance (14.74) depends on the data via N^*. The dependence is reduced in the limits $\nu \to \infty$ and $N \to \infty$. In these limits the Student's t-distributed random field converges to the Gaussian random field with the respective mean and covariance function.

14.6.5 Hierarchical Representations

Student's t-distributed random fields can be generated from Gaussian random fields in a hierarchical representation. The parameters of the Gaussian random fields are considered to be distributed random variables that follow specific probability laws [690, 738, 874]. More precisely, the trend coefficients and the variance of the random field are modeled as random fields with their own probability distributions. This perspective is at the root of *multi-level or hierarchical stochastic modeling*.

Hierarchical representations have recently gained popularity in spatial statistics due to their flexibility and multiple-level representation of uncertainty [169, 271, 274]. Such representations usually involve three different levels:

1. The first level comprises a *model for the data*. The model involves some predictor variables, a random effect (random field) and spatial noise. At this stage it is assumed that the model parameters and the random effect are known. Thus, the data are *conditionally independent* given the parameters and the random field values.
2. At the second level, a *parametric model* is specified for the random effect while the model parameters are assumed to be known. For example, a zero-mean Gaussian model, conditional on the covariance function, is a typical choice for the random field.
3. Finally, at the third level *probability models* are defined for all the parameters. In the Bayesian framework, these models define the prior probability distributions both for the trend coefficients and the random effect parameters (e.g., the random field variance and characteristic length). In Bayesian language, the term *hyperparameters* is used to refer to parameters that are specified by prior distributions. In statistical physics, the term *disorder* is used to emphasize the probabilistic nature of the model parameters.

Let us now return to the hierarchical representation of Student's t-distributed random fields. We start at the second level of the three-level approach discussed above. Let $X(\mathbf{s};\omega)$ denote a Gaussian random field with a joint normal probability distribution that, conditionally on the trend and the variance, is given by

$$\text{(i)}\quad X(\mathbf{s};\omega)\mid \boldsymbol{\beta},\sigma_{\rm x}^2 \overset{d}{=} \mathcal{N}\left(\boldsymbol{\beta}\cdot\mathbf{f}(\mathbf{s}),\, \sigma_{\rm x}^2\,\rho_{\rm x}(\cdot)\right),$$

where the inner product of the K-dimensional coefficient vector $\boldsymbol{\beta}$ and the vector of basis functions $\mathbf{f}(\mathbf{s})$ determines the trend function (cf. Chap. 2), $\sigma_{\rm x}^2$ is the variance, and $\rho_{\rm x}(\cdot)$ is the correlation function of the random field.

Let us further assume that the conditional dependence of the trend coefficients on the variance is given by the multivariate normal distribution

$$\text{(ii)}\quad \boldsymbol{\beta}\mid\sigma_{\rm x}^2 \overset{d}{=} \mathcal{N}(\mathbf{m}_\beta,\,\sigma_{\rm x}^2\,\boldsymbol{\rho}_\beta),$$

where $\mathbf{m}_\beta$ is the expectation vector of the coefficients, and $\boldsymbol{\rho}_\beta$ is a $K\times K$ correlation matrix. Note that this construction assumes a proportional effect between the variance of the trend coefficients and the random field variance.

Finally, assume that the variance hyperparameter follows an *inverse gamma distribution*, i.e.,

$$\text{(iii)}\quad \sigma_{\rm x}^2 \overset{d}{=} \mathcal{IG}\left(\frac{\nu}{2},\,\frac{\nu\sigma_\upsilon^2}{2}\right),$$

where the pdf of the inverse gamma distribution is given by

$$\text{(iv)} \quad f(x;\kappa,\lambda) = \frac{\lambda^{\kappa}}{\Gamma(\kappa)} \left(\frac{1}{x}\right)^{\kappa+1}, \quad \kappa, \lambda, x > 0.$$

Then, the distribution of the random field $X(\mathbf{s};\omega)$ obtained after integrating over the hyperparameters is given by

$$X(\mathbf{s};\omega) \stackrel{d}{=} \mathcal{T}_{\nu}\left[m_{x}(\mathbf{s}),\, \Sigma(\mathbf{s},\mathbf{s}')\right],$$

where the mean function and the dependence matrix of the joint Student's t-distribution are given by

$$m_{x}(\mathbf{s}) = \mathbf{m}_{\beta} \cdot \mathbf{f}(\mathbf{s}), \tag{14.75}$$

$$\Sigma(\mathbf{s},\mathbf{s}') = \sigma_{\upsilon}^{2}\left[\rho_{x}(\mathbf{s},\mathbf{s}') + \mathbf{f}^{\top}(\mathbf{s})\,\rho_{\beta}(\mathbf{s},\mathbf{s}')\,\mathbf{f}(\mathbf{s}')\right]. \tag{14.76}$$

Superstatistics Hierarchical representations have been investigated in spatial statistics by various authors [169, 201, 203, 273]. In statistical physics, ideas similar to the hierarchical representation of random fields have been proposed in the study of nonlinear and non-equilibrium systems.

The theory of *superstatistics* describes the statistical mechanics of systems that can be construed to involve a superposition of different temperatures [55, 153, 328]. A connection between Bayesian statistics and superstatistics is established in [549]. The hierarchical representation has also been used to model the mechanical strength distribution of heterogeneous materials [377]. In particular, in systems that follow weakest-link scaling the possibility is investigated that the parameters of the independent link functions are random variables.

14.6.6 Log-Student's t-Random Field

If the random field $Y(\mathbf{s};\omega)$ follows the Student's t-distribution, then the exponentiated random field $X(\mathbf{s};\omega) = \exp\left[Y(\mathbf{s};\omega)\right]$ follows an asymmetric, heavy-tailed distribution that takes values in the interval $[0,\infty)$.

In particular, using the conservation of probability under transformation, it can be shown that $X(\mathbf{s};\omega)$ follows the log-Student's t-distribution with a univariate pdf given by

$$f_{x}(x) = \frac{\Gamma(\frac{\nu+1}{2})}{\sqrt{\pi\,\nu}\,\Gamma(\nu/2)}\, \frac{1}{x\left[1+\left(\ln x - m_{y}\right)^{2}/\nu\right]^{(\nu+1)/2}}.$$

The characteristic properties of the above pdf are (i) the divergence of $f_X(x)$ at $x \to 0$ because x tends to zero faster than $(\ln x)^2$ tends to infinity; (ii) the heavier than lognormal tail of the pdf for finite ν; and (iii) the convergence of the pdf to the lognormal density for $\nu \to \infty$.

The log-Student's t-distribution has been used to model data from heavy-tailed processes such as precipitation [536] and financial asset returns [79, 671].

14.6.7 Student's t-Distributed Process

Student's t-distributed processes were recently proposed in machine learning as an alternative to Gaussian processes by Shah *et al.* [738]. Student's t-distributed processes were derived using a Bayesian treatment of Gaussian regression that involves a nonparametric prior over the Gaussian process covariance kernel. The prior represents potential uncertainties about the covariance function and reflects the natural intuition that the covariance kernel does not have a simple parametric form.

In a characteristic pattern for two scientific disciplines that do not interact as much as they could, the results obtained in [738] are identical to those in [690], except for differences in parametrization. The main new insight in [738] is that the Student's t-distributed process can be derived from a Gaussian process by placing an inverse Wishart prior on the covariance kernel, thus establishing a direct connection between Gaussian and Student's t-distributed processes. In this setting, the heavy tails of the Student's t-distribution can be attributed to the broader distribution of process values that is caused by the uncertainty in the covariance function of the Gaussian process.

Inverse Wishart distribution In the Gaussian process perspective, the covariance matrix $\mathbf{C}$ (the covariance kernel in the machine learning jargon) is treated as a *random matrix*. The Wishart distribution is a reasonable probability model for real-valued, $N \times N$, symmetric, positive definite matrices [26, 78, 853].

However, Shah *et al.* argue that the Wishart distribution is not suitable for $N \gg 1$, because it requires that $\nu > N - 1$, where ν is the number of degrees of freedom of the Wishart distribution. Hence, large values of ν are required to model covariance matrices of large size, but then the Wishart distribution narrowly focuses on a single covariance matrix realization. Instead, they propose the *inverse Wishart distribution* with the following pdf

$$f_{\mathbf{C}}(\mathbf{C}) = \frac{\det(\boldsymbol{\Sigma})^{\nu/2}}{2^{\nu N/2}\,\det(\mathbf{C})^{(N+\nu+1)/2}\,\Gamma_N\left(\frac{\nu}{2}\right)}\, e^{-\frac{1}{2}\operatorname{Tr}(\boldsymbol{\Sigma}\,\mathbf{C}^{-1})}, \tag{14.77}$$

where $\mathbf{C}$ is the covariance matrix, $\boldsymbol{\Sigma}$ is an $N \times N$ positive definite scale matrix, $\operatorname{Tr}(\mathbf{A}) = \sum_{n=1}^{N} A_{n,n}$ is the trace operation, and $\Gamma_N(\cdot)$ is the *multivariate gamma function* defined as follows:

$$\Gamma_n(x) = \pi^{n(n-1)/4} \prod_{j=1}^{n} \Gamma\left[x + (1 - j)/2\right], \text{ for } z \in \mathbb{R},\ n \in \mathbb{N}.$$

Properties of the inverse Wishart distribution

1. If the $N \times N$ covariance matrix $\mathbf{C}$ follows the inverse Wishart distribution, i.e., $\mathbf{C} \stackrel{d}{=} \mathcal{IW}_N(\nu, \mathbf{\Sigma})$, the mean and the covariance of $\mathbf{C}$ exist only if $\nu > 2$; in particular $\mathcal{E}[\mathbf{C}] = \mathbf{\Sigma}/(\nu - 2)$.
2. The matrix $\mathbf{C}$ follows the inverse Wishart distribution, i.e., $\mathbf{C} \stackrel{d}{=} \mathcal{IW}_N(\nu, \mathbf{\Sigma})$, if and only if its inverse follows the Wishart distribution, i.e., $\mathbf{C}^{-1} \stackrel{d}{=} \mathcal{W}_N(N+\nu-1, \mathbf{\Sigma}^{-1})$.
3. The inverse Wishart distribution also satisfies the property of consistency under marginalization: if $\mathbf{C} \stackrel{d}{=} \mathcal{IW}_N(\nu, \mathbf{\Sigma})$, then any principal submatrix $\mathbf{C}_{11}$ of $\mathbf{C}$ follows the inverse Wishart distribution, $\mathbf{C}_{11} \stackrel{d}{=} \mathcal{IW}_N(\nu, \mathbf{\Sigma}_{11})$, where $\mathbf{\Sigma}_{11}$ is the respective principal submatrix of $\mathbf{\Sigma}$.

14.7 Copula Models

If the marginal distribution of the data is Gaussian, it is often assumed that the joint distribution of the spatial random field model is also Gaussian. A number of mathematical tools are available for modeling Gaussian joint distributions. If the marginal distribution is not Gaussian, different paths can be followed to model the data. One possibility, as discussed above, is to apply a nonlinear transformation that normalizes the marginal distribution. This is followed by the usual *leap of faith* that the joint distribution of the transformed data is also Gaussian.

Copulas provide an elegant solution to the problem of constructing a non-Gaussian joint pdf from an arbitrary marginal pdf. Copulas are multivariate probability distributions defined for transformed variables that follow the uniform marginal distribution. They can be used to impose non-Gaussian dependence structure between random variables with any continuous marginal distribution. Thus, copulas allow decoupling the marginal distribution from the spatial dependence.

The main idea underlying copula theory is that the joint probability distribution for a set of random variables is more simply manipulated by applying the probability integral transform theorem to each of the variables. The probability integral transform theorem is expressed as follows [27]:

Theorem 14.1 (Probability Integral Theorem) *If the cdf $F(x)$ of the random variable $\mathrm{X}(\omega)$ is a continuous function, then the random variable $\mathrm{Y}(\omega)$ defined by the nonlinear transformation $\mathrm{Y}(\omega) = F[\mathrm{X}(\omega)]$ follows the standard uniform distribution $U(0, 1)$.*

This transformation leads to a new set of variables that follow uniform marginal distributions, for which a joint distribution (copula) can be easily constructed. Let us first review some basic concepts of copulas [411, 599].

- If the function $C : [0, 1]^N \mapsto [0, 1]$ is an N-copula, then for all $u \in [0, 1]$ it holds that $C(1, \ldots, 1, u, \ldots, 1) = u$.
- If the function $C : [0, 1]^N \mapsto [0, 1]$ is an N-copula, then for all $u_n \in [0, 1]$ the identity $C(u_1, \ldots, u_N) = 0$ implies that at least one element of the set $\{u_n\}_{n=1}^N$ equals zero.
- Copulas and multivariate distributions are linked by means of *Sklar's theorem*:

  Theorem 14.2 *Given an N-dimensional cumulative distribution function (cdf) $F(x_1, \ldots, x_N)$ with continuous marginal distributions $F_1, \ldots, F_N$, there exists a unique N-copula $C : [0, 1]^N \mapsto [0, 1]$ such that*

  $$F(x_1, \ldots, x_N) = C\,[F_1(x_1), \ldots, F_N(x_N)]\,. \tag{14.78}$$

  In general, the joint cdf $F(x_1, \ldots, x_N)$ and the marginal cdfs $F_n(x_n)$ can be represented by different functional forms.
- Copulas can be constructed from probability distribution functions as follows

  $$C(u_1, \ldots, u_N) = F\left(F_1^{-1}(u_1), \ldots, F_N^{-1}(u_N)\right), \tag{14.79}$$

  where $0 \le u_n = F(x_n) \le 1$ $(n = 1, \ldots, N)$. The above follows directly from (14.78) in Sklar's theorem.
- Copulas are invariant under strictly increasing monotonic transformations of the variables. Hence, for copula-based models it is not necessary to transform the data (e.g., by taking logarithms). Copulas provide measures of statistical association that are invariant under such transformations.

A *copula density function* is defined by the N-order partial derivative of the copula with respect to each dimension, i.e.,

$$c(u_1, u_2, \ldots, u_N) = \frac{\partial^N C(u_1, u_2, \ldots, u_N)}{\partial u_1 \partial u_2 \ldots \partial u_N} = \frac{f(x_1, x_2, \ldots, x_N)}{\prod_{n=1}^N f_n(x_n)}, \tag{14.80}$$

where the $f_n(x_n)$ are the marginal pdfs obtained from $f_n(x) = \mathrm{d}F_n(x)/\mathrm{d}x$. The last equality in (14.80) is derived from (14.79) using the chain rule of differentiation $\partial/\partial u_n = (\partial x_n/\partial u_n)\,\partial/x_n$ and replacing u_n with $F(x_n)$.

The expression (14.80) can be recast in a form that makes explicit the dependence of the joint pdf on the copula density and the marginal pdfs. The copula density in the following equation imposes the spatial dependence.

$$f(x_1, x_2, \ldots, x_N) = c(u_1, u_2, \ldots, u_N) \prod_{n=1}^{N} f_n(x_n). \tag{14.81}$$

In the case of independent random variables $\{X_n(\omega)\}_{n=1}^N$ the copula density is equal to one, and the joint pdf becomes equal to the product of the marginal pdfs as expected.

The copula density in (14.81) enforces correlations (dependence) by linking together the independent marginal distributions. Thus, it is similar to the Jastrow factor used in variational wavefunctions of many-body quantum systems [259]. These wavefunctions involve a "free part" composed of anti-symmetric combinations of single-body wavefunction products. This part is generated by the Slater determinant which enforces the Pauli exclusion principle [23, 539]. The Jastrow factor is an empirical, many-body function that modifies the free component and introduces additional correlations.

14.7.1 Gaussian Copula

The Gaussian copula is obtained if the joint, N-dimensional cdf $F(x_1, \ldots, x_N)$ in (14.79) is the Gaussian joint cdf $\Phi(x_1, \ldots, x_N)$. Thus the Gaussian copula is given by

$$C(u_1, \ldots, u_N) = \Phi\left(F_1^{-1}(u_1), \ldots, F_N^{-1}(u_N)\right).$$

The joint Gaussian cdf is given by the following multivariate integral

$$\Phi(\mathbf{x}) = \prod_{n=1}^{N} \int_{-\infty}^{x_n} \mathrm{d}x'_n \frac{\exp\left[-\frac{1}{2}(\mathbf{x}' - \mathbf{m})^\top \mathbf{C}^{-1} (\mathbf{x}' - \mathbf{m})\right]}{(2\pi)^{N/2} [\det(\mathbf{C})]^{1/2}}, \tag{14.82}$$

where $\mathbf{x} = (x_1, \ldots x_N)^\top$, $\mathbf{x}' = (x'_1, \ldots x'_N)^\top$, $\mathbf{m} = (m_1, \ldots, m_N)^\top$ and $\mathbf{C}$ is a positive-definite covariance matrix.

Gaussian marginal densities If the marginal cdfs $\{F(x_n)\}_{n=1}^N$ are also Gaussian, it can be shown by means of (14.80) and (14.81) that the Gaussian copula leads to the well-known joint Gaussian pdf.

If the marginal distribution functions $F_n(x_n)$ ($n = 1, \ldots N$, $x_n \in \mathbb{R}$), are not Gaussian, the joint pdf based on the Gaussian copula is given by the following expression [715]

$$f(\mathbf{x}) = \frac{\exp\left[-\frac{1}{2}(\mathbf{y} - \mathbf{m})^\top \mathbf{C}^{-1} (\mathbf{y} - \mathbf{m})\right]}{(2\pi)^{N/2}[\det(\mathbf{C})]^{1/2}} \prod_{n=1}^{N} \frac{f_n(x_n)\sqrt{2\pi}\,\sigma_n}{\exp\left[-\frac{(y_n - m_n)^2}{2\sigma_n^2}\right]}, \tag{14.83}$$

where $y_n = \Phi^{-1}[F_n(x_n)]$ is the value of the normal quantile function at the probability level $F_n(x_n)$, and $\sigma_n^2 = C_{n,n}$ for all $n = 1, \ldots, N$.

Equivalence of Gaussian copula and trans-Gaussian random fields Joint pdfs obtained with the Gaussian copula are identical to joint pdfs of trans-Gaussian random fields [525]. Kazianka and Pilz expressed the equivalence as follows [444]:

> The trans-Gaussian kriging model using an almost surely strictly monotone transformation is equivalent to the Gaussian spatial copula model.

The equivalence is evidenced by comparing (14.5) with (14.83). The only difference between the two formulas is that in (14.5) the normalized vector $\mathbf{g}(\mathbf{x})$ is assumed to be standardized. In addition, the product of the marginal Gaussian densities in the denominator of the right-hand side of (14.83) is absorbed in the exponent of (14.5) in the term proportional to $\mathbf{I}_N$.

There is an increasing interest in applications of copulas to non-Gaussian spatial data. For example, the Gaussian copula was used in combination with χ^2 marginal pdfs to construct a likelihood function for cosmological data. This model was shown to perform better (in terms of AIC values) than the multivariate Gaussian likelihood [715]. The Gaussian copula and a transformed Gaussian copula were applied in geostatistical interpolation of groundwater quality data, leading to better performance than ordinary and indicator kriging methods [47, 48]. Geostatistical applications of copulas have been pursued in the Bayesian framework [442–445]. Vine copulas [412], i.e., tree-like graphical structures coupled with bivariate copulas, have been used to model skewed random fields [307].

14.7.2 Other Copula Families

Various families of copulas have been proposed in the scientific literature, often motivated by potential applications in risk analysis and finance [411]. A specific point of practical interest is the modeling of dependence between extreme values, i.e., values in the tails of the distribution. This interest is motivated by the high risks for property and life that can be caused by environmental spatial extremes (e.g., flooding, droughts, prolonged heat waves and cold spells). Since copulas allow modeling the tail behavior of the probability distribution, they are suitable spatial extremes models [183, 265, 712]. Different extreme-value copulas (e.g., the Gumbel copula) have thus been developed. A different approach uses the Gaussian copula to introduce dependence between marginal Weibull distributions [524].

The choice of the copula model is often based on mathematical convenience. Recently, however, a non-parametric approach was proposed for selecting a suitable structure for the copula model from the data [497]. Since the copula-based framework is a relative newcomer in spatial data analysis, it is prudent to carefully consider the assumptions used and their implications. The cautious reader should also consider the critical review of copulas' applications [573].

14.8 The Replica Method

The replica method was developed by physicists as a tool for calculating free energy expectations in disordered systems. First, recall that the free energy is proportional to the logarithm of the partition function. Disordered systems are described by joint probability density models with randomly distributed parameters. Hence, the expectation of the free energy is calculated over the probability distribution of the parameters. In the statistical physics jargon, this expectation operation is called *quenched average* over the disorder. The term "quenched" means that the partition function is evaluated first for each realization of the parameter vector. The logarithm of the partition function is subsequently averaged over the disorder ensemble [569]. In contrast, in the *annealed average* the partition function is averaged over the disorder before taking the logarithm. This is justified in dynamic systems if the disorder changes fast with time. The annealed average is easier to calculate than the quenched average, and thanks to the concavity of the logarithm it provides a lower bound for the quenched average [877]. Hence, it is often used as a first-order approximation of the quenched average.

To better understand the difference between the quenched and annealed averages consider that the first one involves the expectation $\mathcal{E}[\ln Z]$ while the latter involves $\ln \mathcal{E}[Z]$, where $\mathcal{E}[\cdot]$ implies the disorder average. The replica method involves the generation of multiple copies of the system, followed by an averaging over the probability distribution of the disorder over all the copies. Finally, the limit as the number of copies tends to zero is evaluated. The rationale for taking the limit of zero number of copies is elucidated below.

The replica method was introduced by Sam Edwards and Philip Anderson in their investigation of disordered magnetic materials known as *spin glasses* [223]. The disordered spatial arrangement of the magnetic moments in spin glasses resembles the irregular, non-crystalline atomic structure of conventional glasses. *Spin glasses* are treated as Ising models with randomly varying coupling strengths between spins [see (15.43) in Chap. 15.] Edwards and Anderson assumed that the spin coupling strengths $J_{n,m}$ follow the normal distribution. Then, they used replicas to calculate the average of the spin glass free energy over the disorder distribution of the coupling strengths. Giorgio Parisi extended the theory by formulating the concept of *replica symmetry breaking* [570, 571, 653], which applies to complex systems with many coexisting phases of similar energy (e.g., the infinite-range spin glass model) [654]. Such systems are non-ergodic and organize in different clusters, so that each cluster contains similar states. The emergence of distinct replica clusters breaks the identical replica symmetry (hence the term replica symmetry breaking).

14.8.1 Applications in Spatial Data Problems

To make the connection between replicas and spatial data analysis, recall that there is a connection between the free energy and the cumulant generating functional for exponential density models [495] (see also Sect. 6.4.2). Hence, if we know the disorder-averaged free energy we can also generate the disorder-averaged cumulants of the joint pdf. The replica method can thus be applied to calculations in hierarchical models that involve averages over different levels of uncertainty. Applications of the replica method are also being investigated with respect to problems of statistical inference [877]. Finally, replica analysis led to explicit expressions for *non-Gaussian maximum a posteriori* (MAP) estimation [677].

In order to appreciate the relevance of replicas for spatial data analysis, consider a Gibbs random field with joint pdf $f \propto \exp\left[-\mathcal{H}(\boldsymbol{\theta})\right]$. If such a random field is jointly Gaussian, i.e., if $\mathcal{H}(\boldsymbol{\theta})$ is a quadratic functional of the random field states (see Chap. 6), its cumulants can be generated from the Gaussian cumulant generating functional (CGF) (6.9) by calculating suitable derivatives according to (6.79).

In a *hierarchical model* setting, the energy functional $\mathcal{H}(\boldsymbol{\theta})$ involves a parameter vector $\boldsymbol{\theta}$ represented by a random vector or a random field. This type of *bi-level stochastic model* has been discussed in the framework of the hierarchical representation in Sect. 14.6.5. The replica method helps to determine the CGF of hierarchical models by enabling the calculation of the expectation over the probability distribution of the parameters.

The CGF can be expressed in terms of the logarithm of the partition function as shown in (6.80). For the moment we forget about the specific structure of the energy function that enters in the partition function. Let us assume, however, that the joint pdf depends on the random parameter vector $\boldsymbol{\theta}$ that is governed by a known joint distribution function.

> We will refer to expectations over the joint distribution of the parameter vector $\boldsymbol{\theta}$ by means of $\mathcal{E}_{\boldsymbol{\theta}}[\cdot]$. In statistical physics, the probability distribution of $\boldsymbol{\theta}$ represents the randomness due to *disorder*.

The logarithm as the limit of a power law The calculation of the CGF involves the calculation of the *quenched average of the free energy*, i.e., of the expectation $\mathcal{E}_{\boldsymbol{\theta}}[\ln Z(\boldsymbol{\theta})]$ according to (6.80). The first step of the replica method is that we can write the logarithm as the following limit:

$$\ln Z = \lim_{\lambda \to 0} \frac{Z^{\lambda} - 1}{\lambda}.$$

To see why this is true consider the following series expansion:

$$Z^{\lambda} = e^{\lambda \ln Z} = 1 + \lambda \ln Z + \sum_{m=2}^{\infty} \frac{(\lambda \ln Z)^m}{m!} \Rightarrow \ln Z = \frac{1}{\lambda}\left(Z^{\lambda} - 1\right) - O(\lambda).$$

The terms $O(\lambda)$ vanish at the limit $\lambda \to 0$. This is the same argument used to show that the Box-Cox transformation for $\lambda = 0$ is the logarithmic transform.

The replica trick The non-rigorous averaging procedure, which is responsible for the name *replica trick*, comprises the following steps:

1. Use an integer n instead of the real-valued exponent λ in the power-law transformation $(Z^n - 1)/n$.
2. Evaluate the average $\mathcal{E}_{\boldsymbol{\theta}}[Z^n(\boldsymbol{\theta})]$ assuming that $n \in \mathbb{N}$. The expectation operator trades places with the limit $n \to 0$ and n is treated as an integer in this step.
3. Finally, take the limit $n \to 0$. This implies that the *number of replicas* n is treated as a real number that can take values arbitrarily close to zero. Essentially, in this step we evaluate a function $G(n)$, where $n \in \mathbb{N}$, and then we assume that the function $G(n)$ is well defined for $n \in \mathbb{R}$. This procedure, which is known as *replica analytic continuation*, is not rigorously proved.

The replica method states that the expectation of $\ln Z(\boldsymbol{\theta})$ over the probability distribution of the parameter vector $\boldsymbol{\theta}$ is given by

$$\mathcal{E}_{\boldsymbol{\theta}}[\ln Z(\boldsymbol{\theta})] = \lim_{n \to 0} \frac{\mathcal{E}_{\boldsymbol{\theta}}\left[Z^n(\boldsymbol{\theta})\right] - 1}{n}. \tag{14.84}$$

The following are equivalent expressions for calculating the replica average

$$\mathcal{E}_{\boldsymbol{\theta}}[\ln Z(\boldsymbol{\theta})] = \begin{cases} \lim_{n \to 0} \frac{1}{n} \ln \mathcal{E}_{\boldsymbol{\theta}}\left[Z^n(\boldsymbol{\theta})\right], \\ \lim_{n \to 0} \frac{\partial}{\partial n} \ln \mathcal{E}_{\boldsymbol{\theta}}\left[Z^n(\boldsymbol{\theta})\right]. \end{cases} \tag{14.85}$$

Similarly, we can use the replica method to express the expectation of the inverse of the partition function as follows

$$\mathcal{E}_{\boldsymbol{\theta}}[Z^{-1}(\boldsymbol{\theta})] = \lim_{n \to 0} \mathcal{E}_{\boldsymbol{\theta}}\left[Z^{n-1}(\boldsymbol{\theta})\right]. \tag{14.86}$$

Replica fields In the following, for concreteness we focus on the partition function of a continuously-valued random field with realizations $x(\mathbf{s}) : \mathbb{R}^d \to \mathbb{R}$. In this case, the partition function is given by a functional integral, cf. (6.27).

The next step in the replica method involves the introduction of "replicated fields" (replicas) which are used to express the power of the partition function $Z^n(\boldsymbol{\theta})$. Using the index $\alpha \in \{1, 2, \ldots, n\}$ to denote different replicas of the field realizations $x(\mathbf{s})$, we obtain the following expression in terms of functional integrals, $\int \mathcal{D}x^{\alpha}(\mathbf{s})$, over the replica fields

$$Z^n(\boldsymbol{\theta}) = \prod_{\alpha=1}^{n} \left(\int \mathcal{D}x^{\alpha}(\mathbf{s})\, \mathrm{e}^{-\mathcal{H}[x^{\alpha}(\mathbf{s});\boldsymbol{\theta}]} \right)$$
$$= \left(\prod_{\alpha=1}^{n} \int \mathcal{D}x^{\alpha}(\mathbf{s}) \right) \mathrm{e}^{-\sum_{\alpha=1}^{n} \mathcal{H}[x^{\alpha}(\mathbf{s});\boldsymbol{\theta}]}. \tag{14.87}$$

Based on the above representation, the expectation of the n-th power of the partition function is given by

$$\mathcal{E}_{\boldsymbol{\theta}}[Z^n(\boldsymbol{\theta})] = \left(\prod_{\alpha=1}^{n} \int \mathcal{D}x^{\alpha}(\mathbf{s}) \right) \mathcal{E}_{\boldsymbol{\theta}} \left[\mathrm{e}^{-\sum_{\alpha=1}^{n} \mathcal{H}[x^{\alpha}(\mathbf{s});\boldsymbol{\theta}]} \right]. \tag{14.88}$$

At this stage, the expectation is applied to an exponential function instead of the logarithm. Thus, it can be calculated using the cumulant expansion (6.61). Note that the random field realizations $x(\mathbf{s})$ are not necessarily continuously valued. In the case of the Ising problem, for example, the permissible field values are ± 1. In addition, the field could be defined over a countable set of points instead of a continuum domain.

Averaging over the parameter distribution The next steps involve (i) calculating the expectation over the parameter distribution in (14.88), and (ii) evaluating the limit as $n \to 0$. The expectation depends on the functional form of $\mathcal{H}$ and the joint probability distribution of the parameter vector.

The calculations are simplified if the *energy $\mathcal{H}$ is a linear function* of *normally distributed parameters*. Then only the two non-zero cumulants enter the calculations. This is a straightforward extension of the cumulant expansion for Gaussian random fields as expressed in (6.62).

Below we illustrate the application of the replica method by means of a simple example.

Example 14.2 Let $\mathrm{Y}(\mathbf{s};\omega)$ be a normally distributed random field with mean m_{y} and covariance function $C_{\mathrm{yy}}(\mathbf{r})$ so that the random field

$$\mathrm{X}(\mathbf{s};\omega) = \exp\left[\mathrm{Y}(\mathbf{s};\omega)\right]$$

follows the lognormal probability distribution. Confirm that $\mathcal{E}[\ln \mathrm{X}(\mathbf{s};\omega)] = m_{\mathrm{y}}$ using the replica method.

Answer Using the replica method we express the expectation $\mathcal{E}[\ln \mathrm{X}(\mathbf{s};\omega)]$ as

$$\mathcal{E}[\ln \mathrm{X}(\mathbf{s};\omega)] = \lim_{n\to 0} \mathcal{E}\left[\frac{\mathrm{X}(\mathbf{s};\omega)^n - 1}{n} \right] = \lim_{n\to 0} \mathcal{E}\left[\frac{\mathrm{e}^{n\,\mathrm{Y}(\mathbf{s};\omega)} - 1}{n} \right].$$

We calculate the expectation of $e^{n\,\mathrm{Y}(\mathbf{s};\omega)}$ based on the cumulant expansion of Gaussian random fields (6.62), which leads to

$$\mathcal{E}\left[e^{n\,\mathrm{Y}(\mathbf{s};\omega)}\right] = e^{n\,m_{\mathrm{y}}+\frac{n^2\,\sigma_{\mathrm{y}}^2}{2}} = 1 + n\,m_{\mathrm{y}} + O(n^2).$$

From the two equations above it follows that $\mathcal{E}[\ln \mathrm{X}(\mathbf{s};\,\omega)] = m_{\mathrm{y}}$.

The application of the replica trick is rather trivial in this example which does not even involve the use of replicated fields.

14.8.2 *Relevant Literature and Applications*

The basic ideas of the replica method are introduced in [569]. Classical texts on replica theory and its applications in spin glasses include [209, 571]. A presentation that focuses on the concepts, interpretations and applications more than the mathematics of the replica theory is given in [772]. A simplified introduction to spin glass theory including the replica method is presented in [119], and validation of the replica trick in terms of simple (i.e., exactly solvable) models is discussed in [746].

The theory of replicas is beautiful but mathematically intricate, and the interpretation of some mathematical results has been a challenge for many years. A key issue is the validity of the analytic continuation of the replica average: the latter is calculated using an integer number of replicas, but the replica number is then assumed to take real or complex values as the limit of zero replica number is approached. Recent calculations for exactly solvable models with parametric disorder show that the results of replica analytic continuation agree with the respective closed-form solutions [746]. This study provided support for the performance of the replica analytic continuation. Studies in a similar spirit can provide useful tests for the performance of the replica method and help to determine the scope of its validity.

The lack of formal proof for analytic continuation has not prevented the successful application of the replica method by the statistical physics and machine learning communities to various problems in data modeling [12, 112, 321, 381, 569, 574, 610]. Contributions that focus on machine learning applications of the replica method are included in volumes of the *Neural Information Processing Systems* (NIPS) conference proceedings.[4]

Some applications of the replica method in problems of data analysis include the following:

---

[4]Available online at: https://papers.nips.cc/

- Analytical approximations of bootstrap averages [527].
- Applications to information processing, optimization, neural networks and image restoration [610].
- Unsupervised segmentation of multiscale images [381].
- Statistical inference in high-dimensional statistical problems, where the number of constraints is similar to the number of unknowns [12].
- Replica exchange Monte Carlo schemes for Bayesian data analysis [321] and replica-based analysis in Bayesian inference [877].
- Non-Gaussian maximum a posteriori (MAP) estimation [677].
- Learning curves for Gaussian processes on random graphs [806].

Further methodological advances driven by the replica method can be expected, for example in the area of data analysis in high-dimensional spaces [99, 112, 321, 574, 677, 806]. On the other hand, given the lack of a rigorous proof for replica analytic continuation, the validation of replica-based results using independent methods is recommended.

Chapter 15
Binary Random Fields

Remember that all models are wrong;
the practical question is how wrong do
they have to be to not be useful.
George E. P. Box

This chapter focuses on non-Gaussian random fields that admit two different values (levels). Such fields can be used to classify objects or variables in two different groups (categories). Binary-valued random fields are also known in geostatistics as *indicator random fields*. Indicator random fields generated by "cutting" Gaussian random fields at some specified threshold level are commonly used. The mathematical properties and the simulation of binary random fields are discussed below.

Magnetic spin systems also generate binary-valued random fields that represent spatial configurations of binary variables (spins). Spin variables represent intrinsic, quantized magnetic moments of electrons with two possible orientations, that can arbitrarily called "up" and "down". Spin models are widely studied in statistical physics to better understand the properties of magnetic materials. At the same time, these models have numerous applications in data science. The popular *Ising model* is presented in some depth below. We also briefly mention the classical rotator model that generalizes the Ising spin model to fields with continuous values.

Binary-valued random fields have applications in geostatistics, statistical mechanics, and in studies of porous media morphology. Spatial statistics and statistical physics have evolved without much interaction. A notable effort that promotes cross-disciplinary exchange is the volume edited by Mecke and Stoyan [560]. Nonetheless, some powerful methods from statistical physics that remain largely unknown to spatial statistics hold potential for data analysis applications. This chapter aims to further promote cross-disciplinary interaction.

Finally, we briefly discuss the modeling of non-Gaussian behavior by means of generalized linear models, model-based geostatistics, and autologistic models.

D. T. Hristopulos, *Random Fields for Spatial Data Modeling*, Advances in Geographic Information Science, https://doi.org/10.1007/978-94-024-1918-4_15

These approaches employ nonlinear transformations, such as the *logit transform*, to derive regression models for conditional probabilities.

15.1 Indicator Random Field

This section focuses on the *indicator random field* or *indicator function*. The indicator is a binary-valued random field that is often used to represent two distinct phases. Thus, the indicator function is similar to the *phase field* or phase function used in physical models of microstructure evolution [579]. The main difference is that the phases field is a continuous function in [0, 1] that leads to a diffuse interface between the phases, while the indicator function is restricted only to the values 0 and 1 and leads to a sharp interface between phases.

The indicator is employed to model different geological phases (e.g., porous versus solid matrix), excessive environmental pollution concentrations (the phases are defined with respect to a specific concentration threshold), and the occurrence of precipitation. The indicator function is also used in the analysis of non-Gaussian spatial data by means of indicator kriging methods. The indicator field has been applied in the physics literature to characterize porous media microstructure, to model random morphologies in composite heterogeneous materials, and to distinguish between different types of morphologies [8, 59, 60, 62, 189, 687, 688, 793, 797].

Thresholds and level cuts As stated above, the indicator function can be defined with respect to an underlying continuum random field $\mathrm{X}(\mathbf{s};\omega)$ that is known as the "parent" or "latent" field and a specified threshold x_{c}.

The realizations of the indicator field then take values according to the relation between the realization $x(\mathbf{s})$ of the latent field $\mathrm{X}(\mathbf{s};\omega)$ and the threshold x_{c}. Thus, the *indicator function* is a discrete, binary-valued random field that takes values equal to zero and one, i.e.,

$$\left\{ \mathrm{I}_{\mathrm{x}}(\mathbf{s}, x_{\mathrm{c}};\omega) \in \{0,\ 1\};\ \mathbf{s} \in \mathcal{D} \subset \mathbb{R}^d;\ \omega \in \Omega \right\}.$$

The realizations (states) of the indicator field will be denoted by the functions $\iota_{\mathrm{x}}(\mathbf{s}, x_{\mathrm{c}}) : \mathbb{R}^d \to \{0, 1\}$.

Definition 15.1 Let $\mathrm{X}(\mathbf{s};\omega) : \mathbb{R}^d \to \mathbb{A}_{\mathrm{x}} \subset \mathbb{R}$ denote a real-valued latent random field. Given a threshold value $x_{\mathrm{c}} \in \mathbb{A}_{\mathrm{x}}$, the indicator field $\mathrm{I}_{\mathrm{x}}(\mathbf{s}, x_{\mathrm{c}};\omega)$ with respect to x_{c} is defined as follows

$$\mathrm{I}_{\mathrm{x}}(\mathbf{s}, x_{\mathrm{c}};\omega) = \begin{cases} 1, \text{ if } & \mathrm{X}(\mathbf{s};\omega) \leq x_{\mathrm{c}}, \\ 0, \text{ if } & \mathrm{X}(\mathbf{s};\omega) > x_{\mathrm{c}}. \end{cases} \tag{15.1}$$

An equivalent definition of the indicator field based on the *unit step function* is as follows (where $\Theta(x) = 0$ if $x < 0$ and $\Theta(x) = 1$ if $x \geq 0$):

$$\mathrm{I_x}(\mathbf{s}, x_c; \omega) = \Theta\big(x_c - \mathrm{X}(\mathbf{s}; \omega)\big). \tag{15.2}$$

According to the above, we can generate an indicator function by "cutting" realizations of the parent random field at specified thresholds. This method of simulating indicator fields is known as the *level cut* [687].

15.1.1 Moments of the Indicator Random Field

In the following, we assume that the latent random field $\mathrm{X}(\mathbf{s}; \omega)$ is *second-order stationary.*

Mean indicator The expectation of the indicator function is given by the cdf of the latent field evaluated at the threshold x_c, i.e.,

$$\overline{\phi}(x_c) := \mathcal{E}[\mathrm{I_x}(\mathbf{s}, x_c; \omega)] = P[\mathrm{X}(\mathbf{s}; \omega) \leq x_c] = F_{\mathrm{x}}(x_c). \tag{15.3}$$

Non-centered indicator covariance The *non-centered* indicator covariance function is obtained by means of the bivariate cdf of the latent field evaluated at the threshold[1]

$$\begin{aligned} S_2(\mathbf{r}, x_c) :=& \mathcal{E}[\mathrm{I_x}(\mathbf{s}_1, x_c; \omega)\, \mathrm{I_x}(\mathbf{s}_2, x_c; \omega)] = P\big(\mathrm{X}(\mathbf{s}_1; \omega) \leq x_c, \mathrm{X}(\mathbf{s}_2; \omega) \leq x_c\big) \\ =& F_{\mathrm{x}}(x_c, x_c; \mathbf{s}_1, \mathbf{s}_2) = F_{\mathrm{x}}(x_c, x_c; \mathbf{r}), \quad \text{where} \quad \mathbf{r} = \mathbf{s}_1 - \mathbf{s}_2. \end{aligned} \tag{15.4}$$

The last equality follows from the fact that the bivariate $F_{\mathrm{x}}(x; \mathbf{s})$ depends only on the distance $\mathbf{r}$ due to the stationarity of the parent field.

Indicator variance The variance of the indicator function depends only on $\overline{\phi}$ and it is given by the following equation

$$\begin{aligned} \sigma_{\mathrm{I}}^2(x_c) =& \mathrm{Var}\,\{\mathrm{I_x}(\mathbf{s}, x_c; \omega)\} = \mathcal{E}[\mathrm{I_x}(\mathbf{s}, x_c; \omega)^2] - \mathcal{E}^2[\mathrm{I_x}(\mathbf{s}, x_c; \omega)] \\ =& \mathcal{E}[\mathrm{I_x}(\mathbf{s}, x_c; \omega)] - \mathcal{E}^2[\mathrm{I_x}(\mathbf{s}, x_c; \omega)] \\ =& \overline{\phi}(x_c) - \overline{\phi}^2(x_c) = \overline{\phi}(x_c)\left[1 - \overline{\phi}(x_c)\right]. \end{aligned} \tag{15.5}$$

[1] In the physics literature, the non-centered indicator covariance is often denoted by $S_2(\mathbf{r})$ and the centered covariance by $\chi(\mathbf{r})$.

In deriving the indicator variance, we used the identity $\mathrm{I}_{\mathrm{x}}^2(\cdot) = \mathrm{I}_{\mathrm{x}}(\cdot)$, which is based on the fact that the indicator function is either 0 or 1. The indicator variance is maximized at $\overline{\phi}(x_c) = 0.5$, i.e., when there are equal numbers of ones and zeros.

Indicator variogram According to the usual definition (3.44), the indicator variogram is given by

$$\gamma_{\mathrm{I}}(\mathbf{r}) = \frac{1}{2}\,\mathbb{V}\mathrm{ar}\left\{\mathrm{I}_{\mathrm{x}}(\mathbf{s}_1, x_c; \omega) - \mathrm{I}_{\mathrm{x}}(\mathbf{s}_2, x_c; \omega)\right\}. \tag{15.6}$$

The indicator increment field takes the values $-1, 0, 1$. Hence, it satisfies the inequalities

$$-1 \le \mathrm{I}_{\mathrm{x}}(\mathbf{s}_1, x_c; \omega) - \mathrm{I}_{\mathrm{x}}(\mathbf{s}_2, x_c; \omega) \le 1.$$

We can thus use *Popoviciu's variance inequality* [668]. This states that the variance of a random field which takes values in the bounded interval $[x_{\min}, x_{\max}]$ satisfies the inequality $\sigma_{\mathrm{x}}^2 \le (x_{\max} - x_{\min})^2/4$. Hence, in light of the inequality and the definition (15.6), the following upper bound is obtained for the variogram of the increment field

$$\gamma_{\mathrm{I}}(\mathbf{r}) \le \frac{1}{2}. \tag{15.7}$$

The upper bound is satisfied even if the latent field is statistically non-stationary (e.g., in the case of fBm fields).

"Higher" values of the indicator variogram at a given lag imply that the parent field's values are, on average, on opposite sides of the threshold, so that one of the indicator values is equal to zero and the other equal to one. Indicator variograms exhibit less variability if one of the phases dominates, i.e., if $\sigma_{\mathrm{I}}^2(x_c) \approx 0$.

15.1.2 *Level Cuts of Gaussian Random Fields*

Let us assume that the latent field $\mathrm{X}(\mathbf{s};\omega)$ is a zero-mean, statistically homogeneous, jointly Gaussian random field with the two-point correlation function $\rho_{\mathrm{xx}}(\mathbf{r})$.

The expectation of the indicator field defined by means of the equations (15.2) is then given by

$$\overline{\phi}(x_c) = \mathcal{E}[\mathrm{I}_{\mathrm{x}}(\mathbf{s}, x_c; \omega)] = \frac{1}{\sqrt{2\pi}} \int_{-\infty}^{x_c} \mathrm{d}x\, \mathrm{e}^{-x^2/2} = \frac{1}{2}\left[1 + \mathrm{erf}\left(\frac{x_c}{\sqrt{2}}\right)\right], \tag{15.8}$$

where $\mathrm{erf}(x)$ is the *error function* integral

$$\text{erf}(x) = \frac{2}{\sqrt{\pi}} \int_0^x \mathrm{d}t\, \mathrm{e}^{-t^2}. \tag{15.9}$$

The *indicator centered covariance function* is expressed as follows:

$$\begin{aligned} C_{\text{II}}(\mathbf{r}, x_c) =& \frac{1}{2\pi} \int_0^{\rho_{\text{xx}}(\mathbf{r})} \frac{du}{\sqrt{1-u^2}} \exp\left(-\frac{x_c^2}{1+u}\right) \\ =& \frac{1}{2\pi} \int_0^{\arcsin \rho_{\text{xx}}(\mathbf{r})} dz \exp\left(-\frac{x_c^2}{1+\sin z}\right). \end{aligned} \tag{15.10}$$

Sketch of the proof The indicator covariance function is defined by means of the standard equation

$$C_{\text{II}}(\mathbf{r}, x_c) = \mathcal{E}[\text{I}_{\text{x}}(\mathbf{s}, x_c; \omega)\, \text{I}_{\text{x}}(\mathbf{s}+\mathbf{r}, x_c; \omega)] - \mathcal{E}[\text{I}_{\text{x}}(\mathbf{s}, x_c; \omega)]\, \mathcal{E}[\text{I}_{\text{x}}(\mathbf{s}+\mathbf{r}, x_c; \omega)].$$

The first two-point expectation comprises a double integral over the field values at locations $\mathbf{s}$ and $\mathbf{s}+\mathbf{r}$. Hence, how is the univariate integral (15.10) justified?

The first step is to realize that the indicator covariance obeys the first-order ordinary differential equation

$$\frac{\partial C_{\text{II}}(\mathbf{r}, x_c)}{\partial \rho} = \int_{-\infty}^{x_c} \int_{-\infty}^{x_c} dx\, dy\, \frac{\partial^2 f_{\text{xx}}(x, y)}{\partial x \partial y} = f_{\text{xx}}(x_c, x_c),$$

where $f_{\text{xx}}(x, y)$ is the bivariate Gaussian density (6.10) [487], given by

$$f_{\text{xx}}(x, y) = \frac{1}{2\pi\, \sigma^2 \sqrt{1-\rho^2}} \exp\left(-\frac{(x^2 + y^2 - 2\rho\, x\, y)}{2\sigma^2\, (1-\rho^2)}\right).$$

The proof is based on a clever trick that uses the identity

$$\frac{\partial^2 f_{\text{xx}}(x, y)}{\partial x \partial y} = \frac{\partial f_{\text{xx}}(x, y)}{\partial \rho}.$$

The final result is obtained by integration of the ODE with respect to the correlation coefficient ρ.

Remark In the physics of scattering in porous media the non-centered covariance function is often referred to as the covariance or the two-point correlation function. The non-centered indicator covariance function is obtained by adding to (15.10) the square of the indicator expectation (15.8).

Indicator integral range The indicator integral range is defined by means of (5.38) as follows:

$$\ell_c(x_c) = \left[\int \mathrm{d}\mathbf{r}\, \rho_{\mathrm{II}}(\mathbf{r}; x_c) \right]^{1/d}, \tag{15.11}$$

where $\rho_{\mathrm{II}}(\mathbf{r}; x_c)$ is the indicator correlation function given by the ratio of the covariance over the variance, i.e.,

$$\rho_{\mathrm{II}}(\mathbf{r}; x_c) = \frac{C_{\mathrm{II}}(\mathbf{r}; x_c)}{\sigma_{\mathrm{I}}^2(x_c)}.$$

The stationary points of the integral range with respect to x_c occur where the derivative of $\ell_c(x_c)$ with respect to x_c vanishes. However, the derivative of $\ell_c^d(x_c)$ gives the same information and is easier to calculate. We can thus write

$$\frac{\mathrm{d}\ell_c^d(x_c)}{\mathrm{d}x_c} = \frac{\mathrm{d}}{\mathrm{d}x_c} \int \mathrm{d}\mathbf{r} \rho_{\mathrm{II}}(\mathbf{r}; x_c) = \int \mathrm{d}\mathbf{r}\, \frac{\partial \rho_{\mathrm{II}}(\mathbf{r}; x_c)}{\partial x_c}.$$

If x_c^* is a root of $\partial \rho_{\mathrm{II}}(\mathbf{r}; x_c)/\partial x_c$ for all $\mathbf{r} \in \mathbb{R}^d$, then x_c^* is also a root of $\mathrm{d}\ell_c^d(x_c)/\mathrm{d}x_c$ and thus a stationary point of the integral range x_c^*. The partial derivative of the correlation function with respect to the threshold x_c is given by

$$\frac{\partial \rho_{\mathrm{II}}(\mathbf{r}; x_c)}{\partial x_c} = \frac{1}{\sigma_{\mathrm{I}}^2(x_c)} \left[\frac{\partial C_{\mathrm{II}}(\mathbf{r}; x_c)}{\partial x_c} - \frac{C_{\mathrm{II}}(\mathbf{r}; x_c)}{\sigma_{\mathrm{I}}^2(x_c)} \frac{\partial \sigma_{\mathrm{I}}^2(x_c)}{\partial x_c} \right].$$

Since $\sigma_{\mathrm{I}}^2(x_c) > 0$, unless the indicator is uniformly equal to zero or one, we concentrate on the terms inside the brackets. Based on (15.10), the first term inside the brackets is given by

$$\frac{\partial C_{\mathrm{II}}(\mathbf{r}; x_c)}{\partial x_c} = -\frac{x_c}{\pi} \int_0^{\arcsin \rho_{\mathrm{xx}}(\mathbf{r})} \mathrm{d}z\, \mathrm{e}^{-x_c^2/(1+\sin z)} = -2x_c\, C_{\mathrm{II}}(\mathbf{r}; x_c).$$

Regarding the second term inside the brackets, in light of (15.5) the derivative of the indicator variance is given by

$$\frac{\partial \sigma_{\mathrm{I}}^2(x_c)}{\partial x_c} = \frac{\partial \overline{\phi}(x_c)}{\partial x_c} \left[1 - 2\overline{\phi}(x_c)\right].$$

Furthermore, the derivative of $\overline{\phi}(x_c)$ with respect to x_c is obtained from (15.8) as follows

$$\frac{\partial \overline{\phi}(x_c)}{\partial x_c} = \frac{1}{\sqrt{2\pi}}\, \mathrm{e}^{-x_c^2/2}.$$

Collecting all the terms and extracting the common factor $C_{\mathrm{II}}(\mathbf{r}; x_c)$, the derivative of the indicator correlation function with respect to x_c is given by

$$\frac{\partial \rho_{\mathrm{II}}(\mathbf{r}; x_{\mathrm{c}})}{\partial x_{\mathrm{c}}} = -\frac{C_{\mathrm{II}}(\mathbf{r}; x_{\mathrm{c}})}{\sigma_{\mathrm{I}}^2(x_{\mathrm{c}})} \left[2x_{\mathrm{c}} + \frac{\left(1 - 2\overline{\phi}(x_{\mathrm{c}})\right)}{\overline{\phi}(x_{\mathrm{c}}) \left(1 - \overline{\phi}(x_{\mathrm{c}})\right)} \frac{\mathrm{e}^{-x_{\mathrm{c}}^2/2}}{\sqrt{2\pi}} \right].$$

It is straightforward to see that the above function has a root at $x_{\mathrm{c}} = 0$, where the first term inside the square brackets is trivially zero, while the second term also vanishes because $\overline{\phi}(x_{\mathrm{c}}) = 1/2$ at $x_{\mathrm{c}} = 0$. Hence, $x_{\mathrm{c}} = 0$ is a stationary point for all $\mathbf{r}$.

We can also calculate the second derivative with respect to x_{c} (not shown here). This turns out to be negative at $x_{\mathrm{c}} = 0$, implying that $x_{\mathrm{c}} = 0$ is a maximum for the indicator covariance function at every $\mathbf{r}$. Moreover, a numerical evaluation of $\partial \rho_{\mathrm{II}}(\mathbf{r}; x_{\mathrm{c}})/\partial x_{\mathrm{c}}$ shows that $x_{\mathrm{c}} = 0$ is the only stationary point.

15.1.3 Connectivity of Extreme Values

The calculation above shows that the integral range of indicator fields generated by level cuts of zero-mean, stationary, Gaussian random fields is maximized at $x_{\mathrm{c}} = 0$. If the random field has a constant mean equal to m_{x}, the stationary point occurs at $x_{\mathrm{c}} = m_{\mathrm{x}}$. In contrast, the integral range is small for values $x_{\mathrm{c}} \to -\infty$ and $x_{\mathrm{c}} \to \infty$. This property is justified by the exponential decrease of the integrand in the indicator covariance integral (15.10) at the above limits.

Interestingly, these observations hold for all types of covariance functions. They also have been observed in numerical studies of indicator fields (see [889] and references therein). Hence, extreme values of the threshold tend to generate fluctuations that form isolated clusters instead of long-range connected paths. The short-range clustering property of Gaussian-based indicator fields is linked with reduced *connectivity of extreme values* of transport properties (e.g., fluid permeability) in porous media.

The connectivity of extreme values plays a significant role in subsurface flow and transport models. Typically, the fluid permeability (or equivalently, the conductivity) of the geological medium is modeled in terms of multivariate Gaussian or log-Gaussian random fields [173, 275, 300]. Such transport properties are determined by the structural properties of the medium and in particular by well-connected, long-range structures such as channels of high permeability, fractures, or low-permeability formations. Random fields with Gaussian or log-Gaussian joint distributions do not sufficiently capture the connectivity of extreme permeability values. Thus, the impact of connectivity in groundwater flow and solute transport is significant [300, 466, 889]. For example, Zinn and Harvey [889] opined:

> ... Thus, conductivity fields with the same conventional spatial statistics may produce very different groundwater fluxes, solute travel times, and solute spreading because of spatial patterns that are not characterized by these conventional statistics. In other words, the full marginal distribution of conductivity values and the spatial covariance function for these values may not provide sufficient information to estimate effective flow and transport parameters.

In the same study, an *extreme value rearrangement simulation algorithm* is proposed for generating random fields with high connectivity of extreme values from a generating Gaussian random field. This algorithm behaves according to the following general principles [889]:

- Values around the Gaussian mean are transformed to low values.
- Extreme values (either low or high) of the Gaussian field are transformed to high values.

The main steps of the extreme value rearrangement algorithm that aims to generate a well-connected low-value or high-value phase are as follows:

1. A realization of a Gaussian random field with specified marginal statistics and covariance function is generated.
2. The field values are standardized using a normal scores transform.
3. The absolute values of the standardized field are calculated, thus transforming extreme values into high values and values around the mean to low values.
4. A normal scores transform is applied again to generate a field with well-connected low values. This is due to the fact that low values are the transforms of around-the-mean values that exhibit higher connectivity.
5. If the goal is to generate a well-connected low-value phase, there is no need for this step. If the target is high connectivity of the high-value phase, the target field is obtained by "reflecting" low values to high and vice versa; given that the mean is zero, the reflection is equivalent to flipping the sign.

The above transformations maintain the classical (univariate) statistics and the covariance function, but they affect the connectivity of either the low or the high values of the field. The normal scores transform in step 2 can be reversed to obtain a Gaussian random field with the initial statistics (mean and variance). Finally, if the Gaussian random field has been derived by a nonlinear transformation of the data, the transformation also needs to be reversed. For example, in order to simulate permeability fields, the exponential function should be applied to the Gaussian log-permeability field generated by means of the previous steps.

Using the above algorithm and other similar methods [300, 466] it is possible to construct spatial patterns with the same marginal pdf and isotropic covariance function as Gaussian random fields, which however have higher connectivity of the extreme values than the Gaussian field. For example, using the extreme value rearrangement algorithm it is possible to obtain connected structures (of either high or low values) than span the entire domain, even though the integral range of the generating field is much smaller than the domain size [889].

Various studies suggest that standard stochastic models of flow and transport that are based on second-order statistics (i.e., Gaussian random fields) may not be adequate for problems of groundwater flow and transport. However, the importance of these observations regarding field-scale groundwater flow and solute transport is still debated. A recent publication that investigates three-dimensional numerical models of flow and transport shows that solute breakthrough curves[2] can be adequately approximated by means of the standard stochastic models [171, 174, 275]. This study suggests that (i) higher-order statistics (which are difficult to estimate from available geohydrological data anyhow) may not be necessary for the prediction of solute transport in three-dimensional aquifers, and (ii) the crucial parameters are the mean, variance, and integral range of the hydraulic conductivity [404].

Minkowski functionals Another possibility for modeling connectivity is the use of *Minkowski functionals*. The latter are global and additive measures that are based on concepts of integral geometry. Minkowski functionals are related to curvature integrals, which characterize connectivity and topology in addition to geometric properties of spatial patterns [558]. In three dimensional space the Minkowski functionals are related to familiar geometric quantities such as the volume, surface area, and integral mean curvature, as well as the Euler characteristic which is a topological property of the field.

The calculation of the Minkowski functionals for Gaussian random fields has been investigated in [10, 11]. Minkowski functionals have been used to characterize the structure of porous media [37, 820] and the large-scale structure of the universe [559, 725].

The motivation for Minkowski functionals is their ability to quantify topological properties. The idea that data represent shapes in high-dimensional spaces, and therefore their topology is important for characterization and prediction—especially of high-dimensional, noisy and incomplete data—has been pursued by mathematicians who coined the term *topological data analysis* [117, 288]. At least to my knowledge, however, topological approaches have not yet been investigated in the field of spatial data analysis.

15.1.4 Excursion Sets

The indicator random field can also be defined in terms of excursion sets. The latter comprise the spatial sub-domains in $\mathcal{D}$ over which the random field values exceed a certain threshold.

Definition 15.2 Let $\Phi(\mathbf{s})$ be an arbitrary, real-valued, scalar function defined over a domain $\mathcal{D} \subset \mathbb{R}^d$. Subsequently, for any threshold $x_c \in \mathbb{R}$, the excursion set of $\Phi(\mathbf{s})$ with respect to x_c is defined as [10]

$$A_{x_c}(\Phi) := \Big\{ \mathbf{s} \in \mathcal{D} \, \text{such that} \; \Phi(\mathbf{s}) \geq x_c \Big\}. \tag{15.12}$$

[2]A breakthrough curve $C(t)$ at a given location is the measured solute concentration $C(\cdot)$ as a function of time t.

We can thus define the excursion set for any realization $x(\mathbf{s})$ of a Gaussian random field $\mathrm{X}(\mathbf{s};\omega)$ with respect to the threshold x_c. Furthermore, we can extend the definition of $A_{x_c}(\cdot)$ so that the argument is a random field. Then, $A_{x_c}(\cdot)$ represents a *random excursion set* that comprises the set of points where the random field exceeds the threshold.

Then, we can define the indicator random field as follows

$$\mathrm{I}_{\mathrm{x}}(\mathbf{s}, x_c;\omega) = \begin{cases} 1, & \text{if} \quad \mathbf{s} \in A_{x_c}\big(\mathrm{X}(\mathbf{s};\omega)\big), \\ 0, & \text{otherwise.} \end{cases} \tag{15.13}$$

where $A_{x_c}\big(\mathrm{X}(\mathbf{s};\omega)\big)$ is defined in (15.12).

Remark Note that according to the definition (15.13) the indicator field takes values equal to one where the latent (Gaussian field) realization exceeds the threshold. This is in contrast with the indicator field defined above in (15.1)–(15.2). Readers should beware that both conventions are used in the literature.

Illustrations of indicator fields obtained from Gaussian SSRF latent fields are shown in Fig. 15.1. Three realizations of latent fields with increasing rigidity coefficient η_1 are shown. As evidenced in the indicator field realizations, increasing rigidity leads to rougher contours of the excursion set. This is expected, because as the rigidity increases the latent Spartan random field further departs from mean square differentiability (see discussion in Sect. 7.2.6).

Mean indicator The definition of the indicator according to (15.13), i.e., so that it equals one if the latent field exceeds the threshold, affects the indicator's expected value. In particular, equation (15.8) is replaced by

$$\mathcal{E}[\mathrm{I}_{\mathrm{x}}(\mathbf{s}, x_c;\omega)] = \frac{1}{\sqrt{2\pi}} \int_{x_c}^{\infty} \mathrm{d}x \, \mathrm{e}^{-x^2/2} = \frac{1}{2}\left[1 - \mathrm{erf}\left(\frac{x_c}{\sqrt{2}}\right)\right]. \tag{15.14}$$

Indicator covariance On the other hand, the centered indicator covariance function is still given by (15.4). We prove this property in the following example.

Example 15.1 Let $\mathrm{I}_{\mathrm{x}}(\mathbf{s}, x_c;\omega)$ represent the indicator random field defined by means of the level cut (15.2) and $\mathrm{I}'_{\mathrm{x}}(\mathbf{s}, x_c;\omega)$ the random field defined by means of the excursion set (15.13). Show that the centered covariance function is the same for both definitions.

Answer The above definitions of the indicator random field are related by means of the *reflection transformation* $\mathrm{I}_{\mathrm{x}}(\mathbf{s}, x_c;\omega) \mapsto \mathrm{I}'_{\mathrm{x}}(\mathbf{s}, x_c;\omega) = 1 - \mathrm{I}_{\mathrm{x}}(\mathbf{s}, x_c;\omega)$ which maps ones to zero and vice versa. The reflection implies that

$$\mathcal{E}[\mathrm{I}'_{\mathrm{x}}(\mathbf{s}, x_c;\omega)] = 1 - \mathcal{E}[\mathrm{I}_{\mathrm{x}}(\mathbf{s}, x_c;\omega)] = 1 - \overline{\phi}(x_c).$$

The centered covariance function of $\mathrm{I}'_{\mathrm{x}}(\mathbf{s}, x_c;\omega)$ is then given by

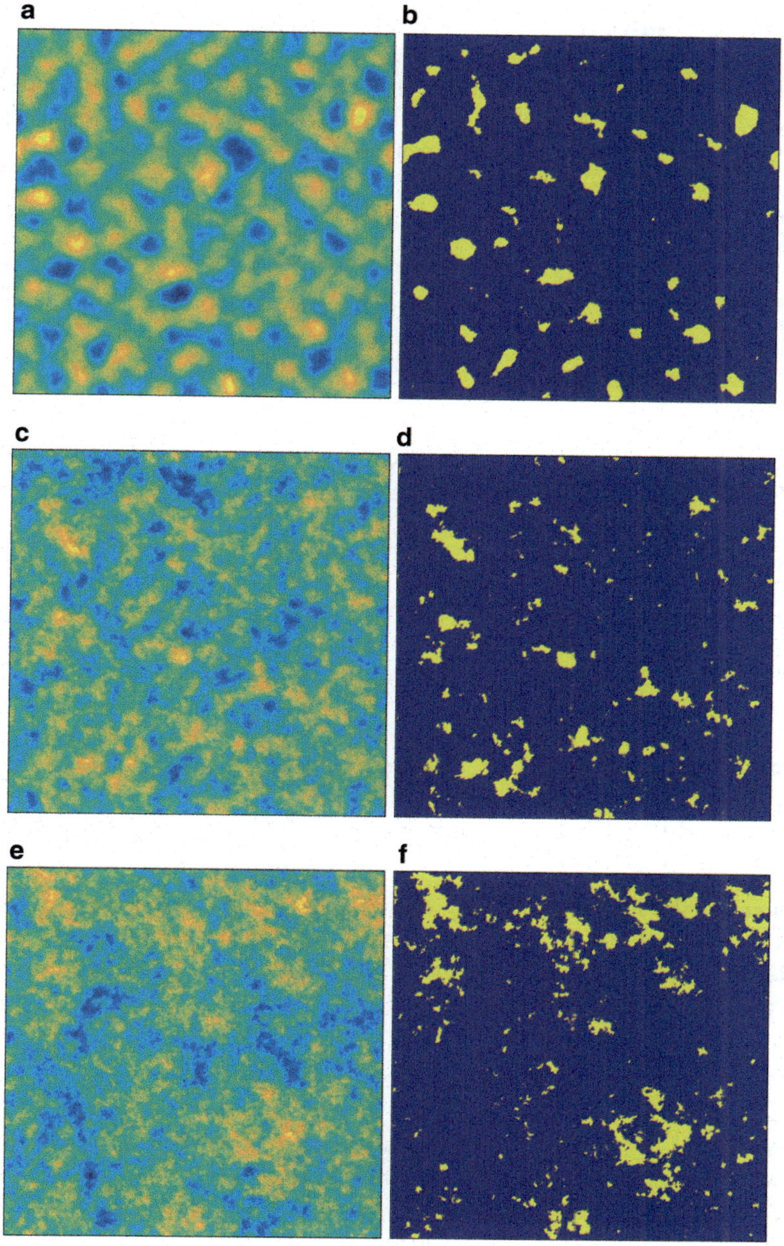

Fig. 15.1 SSRF latent field realizations for $\eta_1 = -1.9$ (top row), $\eta_1 = 1.9$ (middle row), and $\eta_1 = 10$ (bottom row) and corresponding excursion sets for $x_c = 1.5$. A square grid with 512 points per side and unit step is used. The latent fields (shown on the left column) have characteristic length $\xi = 10$ and η_0 such that the variance is normalized to $\sigma_x^2(\eta_1) = 1$. In the excursion set plots (right column), lighter areas correspond to $\iota_x(\mathbf{s}, x_c) = 1$, while darker areas to $\iota_x(\mathbf{s}, x_c) = 0$

$$C_{I'I'}(\mathbf{r}, x_c) = \mathcal{E}[\mathrm{I}'_x(\mathbf{s}_1, x_c; \omega)\, \mathrm{I}'_x(\mathbf{s}_2, x_c; \omega)] - \mathcal{E}[\mathrm{I}'_x(\mathbf{s}_1, x_c; \omega)]\, \mathcal{E}[\mathrm{I}'_x(\mathbf{s}_2, x_c; \omega)].$$

The second term on the right-hand side is equal to $\left(1 - \overline{\phi}(x_c)\right)^2$. The first term on the right-hand side is equal to the non-centered covariance function of $\mathrm{I}'_x(\mathbf{s}, x_c; \omega)$ which is expressed in terms of the moments of $\mathrm{I}_x(\mathbf{s}, x_c; \omega)$ as follows:

$$\begin{aligned}
\mathcal{E}[\mathrm{I}'_x(\mathbf{s}_1, x_c; \omega)\, \mathrm{I}'_x(\mathbf{s}_2, x_c; \omega)] =& \mathcal{E}\left[\{1 - \mathrm{I}_x(\mathbf{s}_1, x_c; \omega)\}\, \{1 - \mathrm{I}_x(\mathbf{s}_2, x_c; \omega)\}\right] \\
=& 1 - \mathcal{E}[\mathrm{I}_x(\mathbf{s}_1, x_c; \omega)] - \mathcal{E}\mathrm{I}_x(\mathbf{s}_2, x_c; \omega)] \\
& + \mathcal{E}[\mathrm{I}_x(\mathbf{s}_1, x_c; \omega)\, \mathrm{I}_x(\mathbf{s}_2, x_c; \omega)] \\
=& 1 - 2\overline{\phi}(x_c) + \overline{\phi}^2(x_c) + C_{\mathrm{II}}(\mathbf{r}, x_c).
\end{aligned}$$

By subtracting from the above the square of the indicator expectation, i.e., $(1 - \overline{\phi})^2$, it follows that the centered covariance functions of the indicator fields $\mathrm{I}_x(\cdot)$ and $\mathrm{I}'_x(\cdot)$ are identical, i.e.,

$$C_{I'I'}(\mathbf{r}, x_c) = C_{\mathrm{II}}(\mathbf{r}, x_c).$$

This proves the symmetry of the indicator covariance function with respect to the $\mathrm{I}_x(\mathbf{s}, x_c; \omega) \mapsto \mathrm{I}'_x(\mathbf{s}, x_c; \omega)$.

15.1.4.1 Special Limits of Indicator Covariance

We review certain explicit expressions for the indicator covariance function which have been derived at specific limits.

Level cut at the mean In this case we generate the indicator field by cutting the latent random field at the mean. At $x_c = 0$ the two phases are equally present, i.e., $\mathcal{E}[\mathrm{I}_x(\mathbf{s}; \omega)] = 0.5$. In addition, the indicator covariance admits the closed-form expression obtained by Berk [59]

$$C_{\mathrm{II}}(\mathbf{r}, x_c = 0) = \frac{1}{2\pi} \arcsin \rho_{xx}(\mathbf{r}). \tag{15.15}$$

Given that for $-1 \leq y \leq 1$ it holds that $\arcsin y \in [-\pi/2, \pi/2]$, from which it follows that the indicator covariance satisfies the bounds

$$-1/4 \leq C_{\mathrm{II}}(\mathbf{r}, x_c = 0) \leq 1/4. \tag{15.16}$$

The above result is in agreement with Popoviciu's variance inequality.

Asymptotic expressions Explicit expressions for the indicator covariance, in the form of convergent series, have been obtained by Roberts and Teubner for very small and very large values of the threshold x_c [688, Appendix A].

- *Cut at zero level:* The indicator covariance is given by the following series

$$C_{II}(\mathbf{r}, x_c) = \frac{1}{2\pi} \arcsin \rho_{xx}(\mathbf{r}) + \frac{e^{-x_c^2/2}}{2\pi} \sum_{n=1}^{\infty} \frac{(-1)^n \, x_c^{2n} \, a_n(\mathbf{r})}{2^n \, n!}, \tag{15.17a}$$

where the functions $a_n(\mathbf{r})$ depend on the two-point correlation function of the latent field as follows

$$a_n(\mathbf{r}) = \frac{2}{2n-1} \left[1 - \left(\frac{1-\rho_{xx}(\mathbf{r})}{1+\rho_{xx}(\mathbf{r})} \right)^{n-\frac{1}{2}} \right] - a_{n-1}(\mathbf{r}), \text{ and } a_0(\mathbf{r}) = 0. \tag{15.17b}$$

The series (15.17) converges fast for $|x_c| \approx 0$.
- *Cut at very large positive x_c:* Next, we consider the limit $x_c \gg 0$ which corresponds to $\overline{\phi} \approx 0$ according to the definition (15.13). Using the method of successive integration by parts [57, Chap. 6], the integral (15.14) is approximated by means of the following series

$$\overline{\phi}(x_c) := \mathcal{E}[\mathrm{I}_x(\mathbf{s}, x_c; \omega)] = \frac{e^{-x_c^2/2}}{\sqrt{2\pi}\, x_c} \left(1 + \sum_{n=1}^{\infty} \frac{(-1)^n 1 \times 3 \ldots (2n-1)}{x_c^{2n}} \right). \tag{15.18}$$

- At the same limit, an expansion of the covariance function is obtained in terms of the *parabolic cylinder functions* $U(\alpha, z)$; α is a real-valued or complex-valued parameter with real part $\Re(\alpha) > -3/2$ [4, p. 686]. Let us define the following functions

$$T_n(z) := \int_0^{\infty} dx \, e^{-zx - \frac{1}{2}x^2} \, x^{n-1} dx = \Gamma(n) \, e^{z^2/4} \, U\left(n - \frac{1}{2}, z\right) \tag{15.19}$$

and the two-point *non-centered correlation function* $p_2(y, x_c)$

$$p_2(y, x_c) = \frac{e^{-x_c^2/(1+y)}}{2\pi} \left[\frac{1+y}{x_c} T_1\big(\delta(y) x_c\big) - \frac{(1+y)^2}{x_c^2} \delta(y) \, T_2\big(\delta(y) x_c\big) + \frac{(1-2y)(1+y)^2}{2x_c^3} T_3\big(\delta(y) x_c\big) \right], \tag{15.20a}$$

where $\delta(y)$ is the following nonlinear function of y

$$\delta(y) = \sqrt{\frac{1-y}{1+y}}. \tag{15.20b}$$

Then, the *centered indicator covariance* function is given by

$$C_{\mathrm{II}}(\mathbf{r}, x_c) = p_2\big(\rho_{\mathrm{xx}}(\mathbf{r}), x_c\big) - p_2\left(0, x_c\right). \tag{15.21}$$

- *Cut at very large negative x_c:* In the opposite limit $x_c \ll 0$, that is, if $\overline{\phi} \approx 1$ according to the definition (15.13), the respective equations for the indicator moments of the first two orders are given by

$$\overline{\phi}(x_c) = 1 - \overline{\phi}(-x_c), \tag{15.22a}$$

$$p_2\big(\rho_{\mathrm{xx}}(\mathbf{r}), x_c\big) = 2\,\overline{\phi}(-x_c) - 1 + p_2\big(\rho_{\mathrm{xx}}(\mathbf{r}), -x_c\big), \tag{15.22b}$$

where $\overline{\phi}(-x_c)$ and $p_2\big(\rho_{\mathrm{xx}}(\mathbf{r}), -x_c\big)$ are given, respectively, by (15.18) and (15.20a). Finally, the indicator centered covariance function is given by (15.21).

Approximations of higher-order correlation functions can also be obtained by means of asymptotic methods [60, 688, 797].

15.1.5 Leveled-Wave Model

In spatial statistics the indicator function is defined in terms of level cuts or level sets of a Gaussian random field as discussed above. A more intuitive approach was introduced in physics by J. W. Cahn who uses the concept of the *stochastic standing wave* to represent the latent random field [114].

The standing wave is composed of N sinusoidal modes with fixed wavelength λ. The modes are further defined by sets of random directions $\{\hat{\mathbf{k}}_n(\omega)\}_{n=1}^{N}$, uniformly distributed random phases $\{\phi_n(\omega)\}_{n=1}^{N}$, and positive random variables $\{a_n(\omega)\}_{n=1}^{N}$ that correspond to realizations of the random amplitude $A(\omega)$. Putting everything together, the latent random field is generated by the following superposition

$$\mathrm{X}(\mathbf{s};\omega) = \frac{1}{\sqrt{N\,\mathcal{E}[A^2(\omega)]}} \sum_{n=1}^{N} a_n(\omega) \cos\left(\frac{2\pi}{\lambda}\hat{\mathbf{k}}_n(\omega)\cdot\mathbf{s} + \phi_n(\omega)\right).$$

In the above, the symbol ω emphasizes which variables are random. The expectation $\mathcal{E}[A^2(\omega)]$ represents the common mean square amplitude of the modes and is assumed to be finite. The normalization is chosen so that the root mean square of the wave amplitude is equal to one.

At the limit $N \to \infty$, the Central Limit Theorem guarantees that the random field $\mathrm{X}(\mathbf{s};\omega)$ follows the Gaussian distribution.[3] The indicator field is then produced by cutting the standing wave at a specified level x_c.

Berk [59, 60] generalized the leveled-wave method by allowing the magnitude of the wavevectors to vary as well, leading to the following standing-wave representation of the latent random field

$$\mathrm{X}(\mathbf{s};\omega) = \frac{1}{\sqrt{N\,\mathcal{E}[A^2(\omega)]}} \sum_{n=1}^{N} a_n(\omega) \cos\left(\mathbf{k}_n(\omega)\cdot\mathbf{s} + \phi_n(\omega)\right). \tag{15.23}$$

Statistical homogeneity All the random coefficients in the standing wave equation (15.23) are independent of the location. Hence, the standing wave $\mathrm{X}(\mathbf{s};\omega)$ is *statistically homogeneous*. In addition, there is no preference in the directions of the wavevectors are treated equivalently thus leading to an *isotropic* random field.

Radial symmetry of wavevector distribution Let us assume that $C_{\mathrm{xx}}(\mathbf{r})$ is the covariance function of $\mathrm{X}(\mathbf{s};\omega)$ with spectral density $\widetilde{C}_{\mathrm{xx}}(\mathbf{k})$. Then, the pdf of the wavevectors is given by

$$f_{\mathbf{k}}(\mathbf{k}) = \frac{\widetilde{C}_{\mathrm{xx}}(\mathbf{k})}{(2\pi)^d\,\sigma_{\mathrm{x}}^2}. \tag{15.24}$$

In (15.24), the variance of the standing wave is given by the spectral integral

$$\sigma_{\mathrm{x}}^2 = \frac{1}{(2\pi)^d} \int_{\mathbb{R}^d} \mathrm{d}\mathbf{k}\, \widetilde{C}_{\mathrm{xx}}(\mathbf{k}).$$

The above equations ensure that $f_{\mathbf{k}}(\mathbf{k})$ is a properly normalized pdf such that

$$\int_{\mathbb{R}^d} \mathrm{d}\mathbf{k}\, f_{\mathbf{k}}(\mathbf{k}) = 1.$$

The function $f_{\mathbf{k}}(\mathbf{k})$ represents the probability density of the mode wavevectors assuming that both the magnitudes and the wavevector direction are random. The pdf that corresponds to the magnitude of the wavevectors is obtained by integrating $f_{\mathbf{k}}(\mathbf{k})$ over all directions of $\hat{\mathbf{k}}$. This is equivalent to evaluating the following integral over the surface of the unit sphere $\mathcal{B}_d$ in d dimensions ($\mathrm{d}\Omega_d$ is the differential of the d-dimensional solid angle)

$$p_{\mathbf{k}}(\|\mathbf{k}\|) := \|\mathbf{k}\|^{d-1} \int_{\mathcal{B}_d} \mathrm{d}\Omega_{\hat{\mathbf{k}}}\, f_{\mathbf{k}}\left(\|\mathbf{k}\|\,\hat{\mathbf{k}}\right) = \|\mathbf{k}\|^{d-1} f_{\mathbf{k}}(\|\mathbf{k}\|)\, S_d. \tag{15.25}$$

[3] Provided that the probability distribution of the random variable $A(\omega)$ satisfies the CLT conditions.

The result follows from the radial symmetry of the density $f_{\mathbf{k}}(\mathbf{k})$ and the fact that the integral over the solid angle yields the surface of the unit sphere in d dimensions; the latter is obtained from (3.62b).

15.1.6 Random Porous Morphologies

Porous media involve synthetic materials, such as microemulsions, ceramics, and glasses. They also include natural media such as cell membranes, zeolites, sandstones, soils, and aquifers that involve pores with a broad range of sizes. Such media have a complex microstructure that is determined by a number of different morphological measures which include the volume fractions of the constituents or phases, the interfacial surface areas between them, as well as the statistical distribution of the constituents' geometric properties (e.g., orientation, shape, and size), and connectivity measures of the different components [526].

The indicator function is used to model the microstructure of porous media and to distinguish in a quantitative manner between different types of microstructure [348, 797]. In this context, the indicator field is viewed as a discontinuous *phase field* $I(\mathbf{s}; \omega)$ that takes the value of 1 inside the pore space and 0 inside the solid matrix.[4] The microstructure of porous media has been experimentally studied by means of scattering experiments [59, 60, 62, 189] and theoretical concepts of stochastic geometry [134].

The phase field $I(\mathbf{s}; \omega)$ represents the *porosity* $\phi(\mathbf{s})$ of the medium at each point in space. The morphology of the medium is determined by the realizations of the phase field. In the following, *stationary* phase fields will be considered.

In contrast with the indicator field $\mathrm{I}_x(\mathbf{s}, x_c; \omega)$, the phase field does not depend on a threshold x_c, since such dependence is not necessary if a latent field is not used. However, if the phase field is generated by level cuts of a Gaussian random field, the dependence on x_c reemerges.

The accurate determination of the microstructure of the medium from a set of correlation functions is a field of ongoing research. The initial investigations focused on the utilization of information encoded in low-order stochastic moments [9]. The research of Yeong and Torquato paved the way for incorporating higher-order correlation functions in the reconstruction process [869, 870]. Recent efforts focus on accurate and efficient estimation of higher-order correlation functions and their integration in efficient reconstruction algorithms [437, 526].

The results presented below have been obtained for binary random porous media. However, they are also useful in more general problems of spatial data analysis that can be expressed in terms of indicator random fields.

---

[4]The opposite assignment is also possible.

Second-order moments If the spatial average of the medium's porosity is equal to ϕ, and the medium is *second-order ergodic* (in the sense defined in Chap. 3), it follows that

$$\mathcal{E}[I(\mathbf{s};\omega)] = \phi. \tag{15.26}$$

Based on (15.5), the variance of the phase field is given by

$$\mathbb{V}\text{ar}\,\{I(\mathbf{s};\omega)\} = \phi\,(1-\phi)\,. \tag{15.27}$$

Hence, the phase field variance is completely determined from the mean porosity, and it is achieves its maximum value for $\phi = 0.5$.

Using the above equation for the variance, the *phase field correlation function* is given by the following ratio, where $C_{\mathrm{II}}(\|\mathbf{r}\|)$ is the phase field covariance

$$\rho_{\mathrm{II}}(\|\mathbf{r}\|) = \frac{C_{\mathrm{II}}(\|\mathbf{r}\|)}{\phi\,(1-\phi)}. \tag{15.28}$$

Specific interfacial area Porous media have complicated internal structure leading to various random morphologies. The interface between the pore space and the solid matrix is also a complex surface (a manifold) that is folded inside the three-dimensional space.

Let us denote the volume occupied by the porous medium by V and the area of the interface between the pore space and the solid matrix by $\mathcal{S}_{\mathrm{int}}$. The *specific interfacial area* between the pore space and the solid matrix is defined as the surface area of the interface divided the volume occupied by the medium, i.e., $S_V = \mathcal{S}_{\mathrm{int}}/V$. For simple interfaces this ratio tends to zero as the volume increases. For example, consider a cube with side length L, separated in two different domains (phases) by a plane that cuts through the middle. Then, the specific interfacial area is given by $S_V = L^2/L^3 = 1/L \to 0$ as $L \to \infty$. However, this inverse linear scaling of S_V with the domain size does not hold in the case of random porous media.

Isotropic random media For three-dimensional media that are modeled by means of a statistically isotropic phase field, the specific interfacial area is expressed in terms of the indicator correlation function as follows [189, 560]:

$$S_V = -4\,\phi\,(1-\phi)\lim_{\|\mathbf{r}\|\to 0}\rho'_{\mathrm{II}}(\|\mathbf{r}\|), \tag{15.29a}$$

where $\rho'_{\mathrm{II}}(\|\mathbf{r}\|)$ is the first derivative of the indicator correlation function with respect to the lag distance, i.e.,

$$\rho'_{\mathrm{II}}(\|\mathbf{r}\|) = \frac{\mathrm{d}\rho_{\mathrm{II}}(\|\mathbf{r}\|)}{\mathrm{d}\|\mathbf{r}\|}. \tag{15.29b}$$

Based on the definition of S_V in (15.29) and using the relation (15.28) for phase field covariance function, the specific interfacial area is given by

$$S_V = -4 \lim_{\|\mathbf{r}\| \to 0} \frac{\mathrm{d}C_{\mathrm{II}}(\|\mathbf{r}\|)}{\mathrm{d}\|\mathbf{r}\|}. \tag{15.30}$$

This expression is generalized to media of different dimensionality $d \neq 3$ by replacing the coefficient 4 with a dimension-dependent coefficient c_d (in $d = 1$, $c_1 = 2$, while in $d = 2$, $c_2 = \pi$) [797].

Anisotropic random media In the case of random porous media with anisotropic correlations, the indicator covariance $C_{\mathrm{II}}(\mathbf{r})$ is not a radial function. The expression (15.30) has been extended to anisotropic media by Berryman [63] as follows

$$S_V = -4 \lim_{\|\mathbf{r}\| \to 0} \frac{\mathrm{d}A_2(\|\mathbf{r}\|)}{\mathrm{d}\|\mathbf{r}\|}, \tag{15.31a}$$

where the function $A_2(\|\mathbf{r}\|)$ is the average of the anisotropic indicator covariance function $C_{\mathrm{II}}(\mathbf{r})$ evaluated over all directions of $\mathbf{r}$, i.e.,

$$A_2(\|\mathbf{r}\|) = \frac{1}{4\pi} \int_{\mathcal{B}_3} \mathrm{d}\theta \mathrm{d}\phi \, \sin\theta \, \rho_{\mathrm{II}}(\mathbf{r}). \tag{15.31b}$$

In the integral (15.31b), θ denotes the polar angle, ϕ the azimuthal angle, and $\mathcal{B}_3$ is the surface of the unit sphere. The product $\mathrm{d}\theta \mathrm{d}\phi \sin\theta$ represents the differential of the solid angle in three dimensions, and 4π is the solid angle that corresponds to the entire surface of the three-dimensional sphere.

Gaussian level cuts For a phase field that is generated by level cuts of a Gaussian latent random field, the specific interfacial area is still obtained from (15.30). The derivative of the indicator covariance function is determined using (15.10) for S_V and Leibniz's rule for differentiation. These lead to

$$S_V = -c_d \lim_{\|\mathbf{r}\| \to 0} \frac{\rho'_{\mathrm{xx}}(\|\mathbf{r}\|)}{\sqrt{1 - \rho^2_{\mathrm{xx}}(\|\mathbf{r}\|)}} \exp\left(-\frac{x_c^2}{1 + \rho_{\mathrm{xx}}(\|\mathbf{r}\|)} \right). \tag{15.32}$$

Note that $\lim_{\|\mathbf{r}\| \to 0} \rho_{\mathrm{xx}}(\|\mathbf{r}\|) = 1$, and thus the denominator of the above expression tends to zero. In order to properly evaluate the limit, we assume that $\rho_{\mathrm{xx}}(\|\mathbf{r}\|)$ is a radial function that is at least once differentiable at $\|\mathbf{r}\| = 0$ with respect to $\|\mathbf{r}\|$. Then, we use its Taylor expansion around $\|\mathbf{r}\| = 0$ as follows

$$1 - \rho^2_{\mathrm{xx}}(\|\mathbf{r}\|) = 1 - \left(1 + \rho'_{\mathrm{xx}}(0)\,\|\mathbf{r}\| + O(\|\mathbf{r}\|^2)\right)^2 = -2\rho'_{\mathrm{xx}}(0)\,\|\mathbf{r}\| + O(\|\mathbf{r}\|^2).$$

Based on the above, the specific interfacial area is given by the following expression

$$S_V = c_d \, \mathrm{e}^{-x_c^2/2} \lim_{\|\mathbf{r}\|\to 0} \frac{-\rho'_{xx}(\|\mathbf{r}\|)}{\sqrt{-2\rho'_{xx}(0)\,\|\mathbf{r}\|}} = \frac{c_d \, \mathrm{e}^{-x_c^2/2}}{\sqrt{2}} \lim_{\|\mathbf{r}\|\to 0} \sqrt{\frac{-\rho'_{xx}(\|\mathbf{r}\|)}{\|\mathbf{r}\|}}. \tag{15.33}$$

The above limit is finite for latent correlation functions that admit a first-order derivative proportional to $\|\mathbf{r}\|$ at the origin. This condition implies that the surface area $\mathcal{S}_{\text{int}}$ of the interface grows proportionally to the volume V occupied by the medium.

Example 15.2 Find the specific interfacial area of a phase field $\mathrm{I}_x(\mathbf{s}, x_c; \omega)$ generated by a level cut of a Gaussian random field with Gaussian correlation function, $\rho_{xx}(\|\mathbf{r}\|) = \exp\left(-\|\mathbf{r}\|^2/\xi^2\right)$ at threshold x_c.

Answer The Gaussian correlation function has a first-order derivative with respect to the distance $\|\mathbf{r}\|$ which is given by

$$\rho'_{xx}(\|\mathbf{r}\|) = \frac{-2\|\mathbf{r}\|}{\xi^2} \exp\left(-\|\mathbf{r}\|^2/\xi^2\right).$$

Hence, the zero-lag limit involved in the specific interfacial area (15.33) is

$$\lim_{\|\mathbf{r}\|\to 0} \sqrt{\frac{-\rho'_{xx}(\|\mathbf{r}\|)}{\|\mathbf{r}\|}} = \frac{\sqrt{2}}{\xi}.$$

Thus, based on the above limit and (15.33), the specific interfacial area is given by

$$S_V = \frac{c_d}{\xi} \, \mathrm{e}^{-x_c^2/2}. \tag{15.34}$$

Equation (15.34) shows that the maximum specific interfacial area (for fixed ξ) is obtained at $x_c = 0$, i.e., when there is an even mixture of both phases.

On the other hand, given a fixed x_c the specific interfacial area of the interface is reduced with increasing correlation length. This makes sense because the higher range of correlations implies fewer transitions between phases.

The exponential covariance function is non-differentiable at zero lag; yet, it is differentiable with respect to $\|\mathbf{r}\|$ around $\|\mathbf{r}\| = 0$, yielding the following Taylor expansion

$$\exp(-\|\mathbf{r}\|/\xi) = \sum_{n=0}^{\infty} \frac{(-1)^n \, \|\mathbf{r}\|^n}{n! \, \xi^n}.$$

Based on the above, the first-order derivative of the correlation function with respect to the distance $\|\mathbf{r}\|$ is given by $\rho'_{\mathrm{xx}}(\|\mathbf{r}\|) = -1/\xi + O(\|\mathbf{r}\|)$, and thus the limit of $\sqrt{\frac{-\rho'_{\mathrm{xx}}(\|\mathbf{r}\|)}{\|\mathbf{r}\|}}$ diverges at $\|\mathbf{r}\| \to 0$. This example demonstrates that latent fields with non-differentiable covariance functions have rougher interfaces that can lead to infinite S_V. Note that the divergence of the S_V results from the fact that the formulation used is based an ergodic medium, and the infinite-domain limit is assumed.

Specific interfacial area from scattering experiments Two useful relations for the phase field covariance have been obtained in scattering studies in porous media [797, p. 37]. The first equation is

$$C_{\mathrm{II}}(\mathbf{r}) = \phi\,(1-\phi) - \frac{S_V}{4}\|\mathbf{r}\| + O(\|\mathbf{r}\|^2). \tag{15.35}$$

This equation is valid for three-dimensional, isotropic, two-phase, heterogeneous media. Note that (15.35) yields the correct indicator variance at $\|\mathbf{r}\| = 0$, while the term proportional to $\|\mathbf{r}\|$ is in agreement with (15.30).

Remark The equation shown in [797] involves the non-centered covariance function and is given by $S_2(\mathbf{r}) = \phi - \frac{S_V}{4}\|\mathbf{r}\| + O(\|\mathbf{r}\|^2)$. This is in complete agreement with (15.35) in light of the identity $S_2(\mathbf{r}) = C_{\mathrm{II}}(\mathbf{r}) + \phi^2$.

The second equation applies to the specific interfacial area of simulated porous media that are obtained by means of levels cuts of Gaussian random fields [189, 792]

$$S_V = \frac{2}{\pi}\mathrm{e}^{-x_c^2}\sqrt{\frac{1}{3}\mathcal{E}[\|\mathbf{k}\|^2]}. \tag{15.36}$$

The expectation over the wavevector magnitudes in the above is based on the standing wave model presented in Sect. 15.1.5. As discussed therein, the latent field can be viewed as a superposition of harmonic modes with randomly distributed wavevectors. The pdf of the wavevector magnitudes is given by (15.25).

15.2 Ising Model

The original Ising model represents magnetic interactions between binary variables, called spins. The spins represent magnetic moments that are generated by the intrinsic angular momentum of elementary particles. However, the Ising model can be applied to other binary variables as well [766]. In statistics, it has attracted attention due to the work of Besag on lattice systems [69]. The essentials of the Ising model for statistical applications are reviewed in [165]. The Ising model has also found applications in the field of image analysis [852]. Finally, it is the topic of three studies in the collection [560] which focuses on connections between spatial statistics and statistical physics.

The Ising model occupies a central position as a non-Gaussian system in physics. Ising variables that interact with random strengths (couplings) provide a classical paradigm of collective phenomena in disordered systems, with applications that encompass glassy magnetism and frustration, combinatorial optimization, protein folding, stock market dynamics, and social dynamics [175]. The Ising model has also been successfully applied to neuroscience [729], image processing [279], the reconstruction of missing pixels in digital images [896], and physical models of social dynamics among other topics [768].

We believe that the Ising model can find new applications in spatial statistics, potentially leading to alternatives to indicator kriging that, unlike the latter, are not beset by Gaussian assumptions. The inference of the Ising model parameters from data, known as the *inverse Ising problem*, is currently a topic of intense research interest due to its potential applications in many areas of data science [607, 608].

15.2.1 Ising Model: Basic Facts

The Ising model was introduced by the physicist Ernst Ising in an effort to understand ferromagnetic materials. Ferromagnetism is the property of certain magnetic materials (like iron) to maintain their induced magnetization after the polarizing field is removed, provided that the temperature is below a critical value T_c.

The spins represent microscopic magnetic moments that can take the values $\sigma_n = \pm 1$, where n is the position index denoting the location $\mathbf{s}_n$. It is common to say that the spins point up if $\sigma_n = 1$ and down if $\sigma_n = -1$.

Magnetization Local magnetic moments (spins) measure the microscopic magnetic field at each site. Magnetization is a macroscopic variable that represents either the average value or the sum of the local magnetic moments in a given material sample. A non-zero magnetization can emerge as the result of the interactions between local magnetic moments (spins) in ferromagnetic materials and the effect of external magnetic fields. In spite of its simplicity, the Ising model exhibits complex behavior including phase transitions in $d \geq 2$ at the limit of an infinite system [218, 658].

A phase transition marks the passage of the system from a disordered state (above T_c) with zero magnetization to an ordered state with finite magnetization (below T_c).

The Ising system exhibits a phase transition in spatial dimensions higher than one. This is remarkable, since the model involves only local interactions, which, however, suffice to generate a long-range organization in the ordered phase of the system.

Spin-Indicator transformation Let us consider a system that contains N spins. The latter are binary random variables $S_n(\omega)$, and their values in any given realization are denoted by $\sigma_n = \pm 1$.

The spin values can be mapped to respective indicator values by means of the affine transformation

$$\mathrm{I_x}(\mathbf{s}_n;\omega) = \frac{1}{2}\left[1 + S_n(\omega)\right]. \tag{15.37}$$

The states of the Ising model are determined by an "energy function" which is a functional of the spin state (realization). The energy function embodies interactions between the spins and interactions of the spins with an external magnetic field defined by the vector $\mathbf{g} = (g_1, \ldots g_N)^\top$.

Ising model with nearest-neighbor coupling The simplest form of the Ising model is obtained for uniform interactions between *nearest-neighbor spins* and *constant external field*. This *translation invariant model* leads to a stationary spin field since its properties are independent of the location.

The energy of the translation invariant Ising system with nearest-neighbor interactions between N spins is expressed as follows[5]

$$\mathcal{H}(\boldsymbol{\sigma}) = -J \sum_{\langle n,m \rangle} \sigma_n \, \sigma_m - g \sum_{n=1}^{N} \sigma_n, \tag{15.38}$$

where $\boldsymbol{\sigma} = (\sigma_1, \ldots, \sigma_N)^\top$ is the vector of the spins at every site, J is the *spin coupling strength*, g is the *external field*, and

$$\langle n, m \rangle = \sum_{n=1}^{N} \sum_{m \in nnb(n)}$$

denotes a sum over all the *nearest-neighbor pairs* of points $\mathbf{s}_n$ and $\mathbf{s}_m$. For example, if the spins are evenly distributed on a 1D chain with periodic boundary conditions, $\sum_{m \in nnb(n)}$ translates into (i) $m = n \pm 1$ for $m \neq 1, N$, (ii) $m = \{N, 2\}$ for $n = 1$ and (iii) $m = \{1, N-1\}$ for $n = N$.

The probability distribution of the Ising states $\boldsymbol{\sigma}$ are determined by the *Gibbs exponential density*

[5] In the physics literature the symbol "h" is typically used to denote the external magnetic field. Herein we use "g" instead, since "h" denotes the normalized distance lag.

$$f_S(\sigma_1, \ldots, \sigma_N) = \frac{1}{Z} e^{-\mathcal{H}[\sigma]/k_B T}, \tag{15.39}$$

where k_B is Boltzmann's constant, T is the temperature, and Z is the *partition function*. The latter is evaluated by the sum of the exponential density over all possible configurations (states) of the spin system, i.e.,

$$Z = \sum_{\{\sigma_n\}_{n=1}^N} e^{-\mathcal{H}[\sigma]/k_B T}. \tag{15.40}$$

Connection with Markov random fields The restriction of the interactions to nearest neighbors (or to higher-order neighbors for this matter) implies that the energy function (15.38) and the distribution (15.39) define a *non-Gaussian Markov random field*. In fact, there are strong connections between the development of Markov random fields and the effort to provide a firm probabilistic setting for the Ising model [450].

Ferromagnetic model If $J > 0$ (ferromagnetic model), neighboring spins of the same sign give lower energy contributions than configurations with neighbors that have opposite spin sign. Hence, the configuration of the lowest energy is the one in which all the spins are aligned with the external field. This is the ground state of the system, and it is the only accessible state at $T = 0$.

At $T > 0$, according to (15.39) configurations with lower energy have a higher probability of occurrence than higher energy configurations. This means that the external field tends to align the spins in its direction, since this arrangement lowers the energy of the system.

Anti-ferromagnetic model If $J < 0$ (anti-ferromagnetic model), the spin-spin interaction favors neighboring spins with opposite signs over spins that have the same sign (see Fig. 15.2).

In this case, the configuration of lowest energy is not uniquely defined, since the external field tends to align all the spins in its direction, while the spin-spin interaction prefers an alternating spin configuration. Even in the absence of an

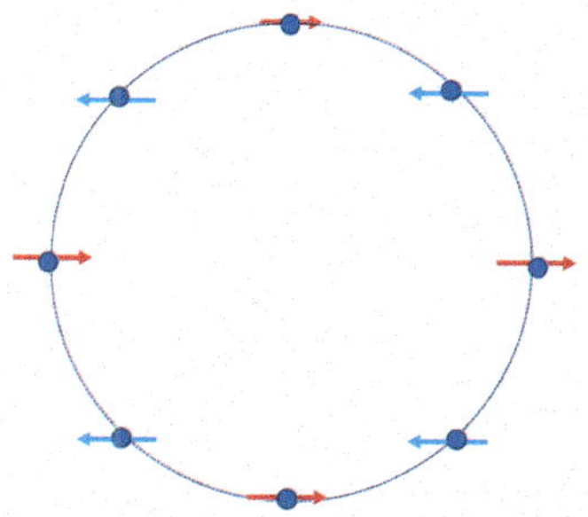

Fig. 15.2 Schematic of Ising spins with anti-ferromagnetic interactions leading to a configuration with alternating spin orientations that represents the lowest-energy state of the model

external field, the lowest energy state of the anti-ferromagnetic model is not always straightforward.

Geometrical frustration Let us consider N collinear spins with nearest-neighbor interactions. In the absence of an external field, the lowest energy state is obtained for an alternating, i.e., $1, -1, 1, -1, 1, \ldots$, spin configuration. Since there are $N-1$ nearest-neighbor pairs, the energy of the system is $\mathcal{H} = -(N-1)\,|J|$ (see Fig. 15.2). The mirror configuration $-1, 1, -1, 1, -1, \ldots$ shares the same energy. These two states are called *degenerate*, because their energies are indistinguishable.

For spins located at the nodes of a 2D square lattice, the lowest energy state is also obtained for an alternating arrangement of ± 1. If, however, the spins are placed at the vertices of a triangular lattice, it is not possible to find a configuration with the favorable (alternating sign) arrangement for all the spins pairs (see Fig. 15.3). Thus, anti-ferromagnetic spins on the triangular lattice are said to be *frustrated*, because it is impossible to find a state (spatial configuration) that is optimal for every spin. Hence, the geometry of the lattice is an important factor in determining the state with the highest probability.

Frustration is a key concept in the theory of spin glasses, which are generalizations of the Ising model with complex dependence of the local magnetic field and the interaction coupling [223]. However, unlike the simple Ising model, the frustration in spin glasses arises from variations of the coupling constant J and occurs regardless of the lattice structure.

Example 15.3 Consider a ring comprising N spins with nearest-neighbor, uniform, anti-ferromagnetic interactions. Think of the ring as a chain with periodic boundary conditions, so that the left-hand-side neighbor of the first spin is the last spin of the chain, and the right-hand side neighbor of the last spin is the first spin of the chain (as shown in Fig. 15.2). Determine the lowest-energy state of the ring.

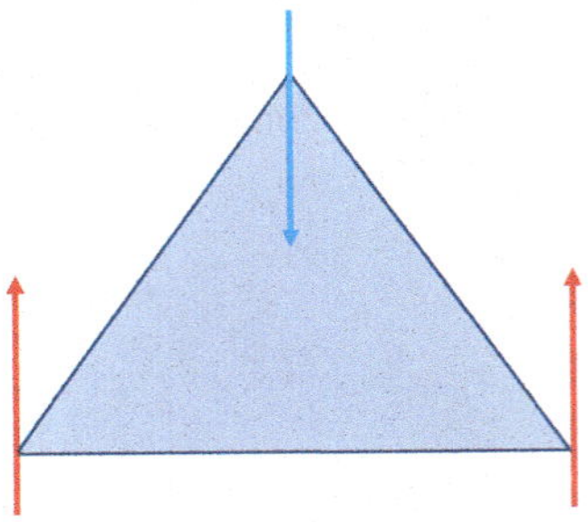

Fig. 15.3 Schematic of frustrated spin state of the anti-ferromagnetic Ising model at the vertices of a triangular cell. The resulting configuration does not satisfy the energetic requirements of all the spins (since they all would like to have an opposite-spin neighbor). Furthermore, there is no other spin configuration with lower energy

Answer The role of the boundary conditions is crucial in this problem. If the chain consists of an even number $N = 2L$ spins, then the two states with alternating spin values have the lowest possible energy, i.e., $E_0 = -N_p |J| = -2L |J|$, where N_p is the number of pairs. Note that due to the periodic boundary conditions, there is an additional spin pair compared with the linear chain. In this case all the nearest-neighbor spin pairs are in the lowest energy state.

On the other hand, if the chain contains an odd number of spins, $N = 2L + 1$, it is impossible for all the pairs to have the lowest energy: there will exist at least one frustrated pair consisting of neighbors with the same spin. The lowest-energy will thus be $E_0 = -2L |J| + |J| = -(2L - 1) |J|$. There are many more "lowest-energy" configurations than for $N = 2L$, because (i) the energetically unfavorable pair can be placed anywhere along the chain, and (ii) if all the spin values are flipped the energy of the chain remains unchanged.

Closed-form solutions The Ising model exhibits complex behavior in spite of its simplicity. Even the simple Ising model with uniform J and h has been solved explicitly only in $d = 1$.

In higher dimensions, a major breakthrough was the closed-form solution for the zero-field ($g = 0$), ferromagnetic, nearest-neighbor Ising model on square grids with periodic boundary conditions, derived by Lars Onsager [629]. The Onsager solution is presented and explained in the book by Huang [383]. Solutions have been derived for other planar geometries as well. However, so far closed-form solutions in the cases of non-zero external field and three dimensions remain elusive. This has been linked to the fact that both problems (i.e., finite external field and three dimensions are NP-complete) [144]. A recent study formulated various NP-complete computer science problems in terms of the Ising problem [515].

Magnetization The *magnetization* of the Ising model for a given realization $\{\sigma_n\}_{n=1}^{N}$ is given by the global average of the spin moments

$$M = \sum_{n=1}^{N} \sigma_n.$$

M can be viewed as a realization of the magnetization random variable $\mathcal{M}(\omega) = \sum_{n=1}^{N} S_n(\omega)$ which is equal to the sum of all the spins.

The existence of a *phase transition* in the ferromagnetic Ising model is marked by the presence of a finite magnetization at temperatures below the critical T_c. Mathematically, this is expressed as follows:

$$\lim_{N\to\infty} \mathcal{E}_+[\mathcal{M}(\omega)] = \begin{cases} M_0 > 0, & \text{if } T < T_c \\ 0, & \text{if } T \geq T_c. \end{cases}$$

The symbol $\mathcal{E}_{+}[\cdot]$ indicates a conditional expectation. The existence of a phase transition does not ensure the polarity of the spins, which can be either positive or negative (both polarities lead to the same value, equal to one, for the product of neighboring spins). Thus, it is necessary to polarize the system in one of the two possible states.

Mathematically, this can be done by requiring that all the spins at the boundary of the system be positive. A physical approach to breaking the symmetry between the positive and negative spin configurations is by using a magnetic field to align the spins, and then to slowly switch off the field, leaving the system in the polarized state.

Partition function and moments In the case of a uniform, non-zero external field g, the partition function (15.40) is analogous to the mMGF (4.70), while the logarithm of the partition function is analogous to the mCGF (4.71).[6] More precisely, the cumulant generating functional is $\mathcal{K}_{\mathrm{x}} = -k_{\mathrm{B}}\,T\,\ln Z$.

Hence, the *expected magnetization* of the Ising model is given by

$$\mathcal{E}[\mathcal{M}(\omega)] = -k_{\mathrm{B}}\,T\,\frac{\partial \ln Z}{\partial g}. \tag{15.41}$$

The expected magnetization per spin is given by $\mathcal{E}[\mathcal{M}(\omega)]/N$.

In comparison with Sect. 3.2.4, the auxiliary field u is replaced in (15.40) by $-g/k_{\mathrm{B}}\,T$. This explains the prefactor $-k_{\mathrm{B}}\,T$ in the calculation of the spin expectation. The above formulation works well if the external field g is nonzero; however, we could also use the mapping $u \mapsto -(g+\delta g)/k_{\mathrm{B}}\,T$ where δg is a small perturbation field, and then take the limit $\delta g \to 0$ at the end of the calculation. This modification is well-defined even at the limit $g \to 0$.

In addition, the magnetization variance is obtained from

$$\mathbb{V}\mathrm{ar}\,\{\mathcal{M}(\omega)\} = -k_{\mathrm{B}}\,T\,\frac{\partial^2 \ln Z}{\partial g^2}. \tag{15.42}$$

The Ising model partition function can be explicitly evaluated in one dimension. In two dimensions the partition function has only been explicitly calculated for zero external field. The tour de force calculation is known as the Onsager solution. It is described in pedagogical manner in [245]. Approximate solutions for the Ising partition function can be derived using mean-field theory, the Bethe-Peierls approximation, and the Gaussian model. Interested readers can find more details in [593].

Spin glasses If we allow for random local variations of the external field and the spin-spin interactions, so that the magnitude of the latter depends on the spatial lag, we obtain the following energy function

[6] The analogy is evident if we replace $\mathbf{u}$ with $\mathbf{g} = (g_1, \ldots, g_N)^{\top}$; for uniform external field it holds that $g_n = g$, for all $n = 1, \ldots, N$.

$$\mathcal{H}(\sigma) = -\sum_{n=1}^{N}\sum_{m=1}^{N} J_{n,m}\,\sigma_n\,\sigma_m - \sum_{n=1}^{N} g_n\,\sigma_n. \tag{15.43}$$

Models with the energy function (15.43) are known in statistical physics as "spin glasses", while in the neural network community as *Hopfield networks* and *Boltzmann machines* [350, 521]. Spin glasses were initially proposed by Sam Edwards and Philip Anderson to explain the observed phase transition in dilute magnetic alloys [223]. The disordered spatial arrangement of the magnetic moments in spin glasses resembles the irregular, non-crystalline atomic structure of conventional glasses.

Spin glasses are considerably more complicated than the simple Ising model, because the states that are energetically favorable depend on the local magnetic field and the strength of the inter-spin coupling. The two central features of spin glasses are *frustration*, i.e., the fact that is not possible to find spin configurations that are energetically favorable for all the spins, and *ergodicity breaking* which implies that the spin glass gets stuck in meta-stable configurations (local minima of the free energy). Solutions of spin-glass models have been obtained by means of the replica trick [223] and the replica symmetry breaking idea of Giorgio Parisi [571]. The inverse problem of inferring the coupling parameters and the external field values from the magnetization and the correlation functions is a current research topic, e.g. [463, 693].

Spin glasses have led to deep insights in other scientific fields such as *combinatorial optimization* [772] as well as graph, packing, coloring and partitioning problems [515]. To our knowledge, spin glasses and Boltzmann machines have not yet been used in the analysis of spatial data. Due to the complexity of these models parameter inference (learning) can be time consuming and "delicate". However, this is the price to be paid for the considerable flexibility offered by spin glasses.

15.2.2 One-Dimensional Ising Model

The nearest-neighbor Ising model can be solved explicitly in one dimension. We present the main features of the closed-form solution, setting $k_B T = 1$ for simplicity. The technical details can be found in statistical physics textbooks such as [122, 383, 593]. A characteristic property of the Ising model in $d = 1$ is the lack of a phase transition for all values of the spin coupling constant J.

We assume that the N spins are placed on a one-dimensional ring with periodic boundary conditions, so that the right-hand neighbor of the spin σ_N is the spin σ_1, while the former is the left-hand neighbor of the latter. This arrangement enforces a reflection symmetry, i.e. $\sigma_n = \sigma_{N-n}$.

Partition function The partition function of the model (15.38) is then expressed as follows

$$\ln Z = N \ln \lambda_+ + \ln \left[1 + (\lambda_- / \lambda_+)^N \right]. \tag{15.44}$$

The coefficients λ_- and λ_+ are the *eigenvalues of the Ising model's transfer matrix*. The latter is a 2×2 matrix with elements [383]

$$P_{n,m} = \exp \left[J \sigma_n \sigma_m + \frac{g}{2} (\sigma_n + \sigma_m) \right], \quad n, m = 1, \ldots, N,$$

where $\sigma_n, \sigma_m = \pm 1$. The transfer matrix is thus given by

$$\mathbf{P} = \begin{pmatrix} e^{J+g} & e^{-J} \\ e^{-J} & e^{J-g} \end{pmatrix}.$$

Finally, the eigenvalues of the Ising transfer matrix are given by the following equation in terms of J and the external field g:

$$\lambda_\pm = e^J \cosh(g) \pm \left[e^{2J} \sinh^2(g) + e^{-2J} \right]^{1/2}. \tag{15.45}$$

It follows from the above that $\lambda_- < \lambda_+$. Thus, for $N \gg 1$ it follows that $(\lambda_- / \lambda_+)^N \approx 0$, and the partition function can be approximated as

$$\ln Z \approx N \ln \lambda_+. \tag{15.46}$$

Expected spin value The expectation of the Ising spins is calculated from (15.41) (recalling that $k_B T = 1$). For $N \gg 1$ only the largest eigenvalue λ_+, plays a role. Thus, the approximate expression for the partition function (15.46) can be used.

The expected value of the Ising spin variables is given by the equation

$$\mathcal{E}[S_n(\omega)] = \frac{\sinh(g) + \frac{\sinh(g)\cosh(g)}{\sqrt{\sinh^2(g) + \exp(-4J)}}}{\cosh(g) + \sqrt{\sinh^2(g) + \exp(-4J)}}. \tag{15.47}$$

Since the denominator of (15.47) is a positive function for all J and g, the sign of $\mathcal{E}[S_n(\omega)]$ is determined by $\sinh(g)$, i.e., by the sign of g. If $g = 0$ and J is finite, it follows from (15.47) that $\lim_{g \to 0} \mathcal{E}[S_n(\omega)] = 0$. Hence, in the 1D model there is no spontaneous magnetization in the absence of zero external field, and thus no phase transition.

Minimum energy state Based on the expression of the energy (15.38), the state of minimum energy for the ferromagnetic Ising model ($J > 0$) is obtained if all the spins point in the same direction. The latter is determined from the sign of the uniform external field g. The expectation (15.47) represents the ensemble average over all the states of the spin field, and it is evidently different from the spatial spin average in the minimum energy state. Hence, in general the minimum energy state does not satisfy ergodic conditions.

For $g = 0$ the minimum energy state is degenerate, i.e., it corresponds to all the spins pointing in the same direction, regardless of their orientation. Since both of the degenerate states are equally probable, the spin expectation is zero.

Spin covariance The covariance function of the Ising model for two spins at distance r, i.e.,

$$C_{\sigma\sigma}(r) = \mathcal{E}[S_n(\omega)S_{n+r}(\omega)] - \mathcal{E}[S_n(\omega)]\mathcal{E}[S_{n+r}(\omega)],$$

can also be evaluated in closed form [53].

The centered covariance function of the Ising model is given by the following expression

$$C_{\sigma\sigma}(r) = \sin^2 2\theta \left(\frac{\lambda_-}{\lambda_+}\right)^{|r|}, \tag{15.48a}$$

where $\lambda_\pm$ are the transfer matrix eigevenvalues (15.45) and

$$\cot 2\theta = e^{2J} \sinh(g), \; 0 < \theta < \pi/2. \tag{15.48b}$$

The correlation length of the spin covariance follows straightforwardly from (15.48), by realizing that the correlations decay exponentially with distance. To clarify this statement, express the covariance (15.48) function as

$$C_{\sigma\sigma}(r) = \mathbb{V}\text{ar}\,\{S_n(\omega)\}\, e^{-|r|/\xi}.$$

Then, it follows from (15.48) that the variance and the correlation length of the Ising model are given by

$$\mathbb{V}\text{ar}\,\{S_n(\omega)\} = \sin^2 2\theta, \tag{15.49a}$$

$$\xi = \frac{1}{\ln\left(\frac{\lambda_+}{\lambda_-}\right)}. \tag{15.49b}$$

Ideally, we would have liked the external field g to control the magnetization and the interaction strength J to control the fluctuations. In contrast, both parameters have an impact on the spin expectation and the covariance of the Ising model.

15.2.3 Mean-Field Theory

In higher-dimensional spaces it is possible to obtain explicit but approximate solutions for the Ising model by linearizing the dependence of the energy on the spins. This can be accomplished by neglecting fluctuations. In this perspective, each spin random variable $S_n(\omega)$ interacts with an effective field that is generated by the magnetization of the remaining spins. The effective field is determined self-consistently, so that the spin expectation $\mathcal{E}[S_n(\omega)]$ agrees with the effective field. This approach is known as *mean-field theory* and is widely used in statistical physics. Mean-field theory can also be viewed as a *variational theory* which employs an independent-spin approximation for the joint density of the system [561]. In this perspective, the effective field is a variational parameter determined by minimizing the free energy (cf. Sect. 6.4.5).

There are various ways of generating mean-field theories that can lead to different approximations [297, 593]. We briefly describe below the main ideas and assumptions involved in the *Weiss mean-field theory*.

1. The Ising energy function is given by the nearest-neighbor interaction model (15.38), i.e.,

$$\mathcal{H}(\boldsymbol{\sigma}) = -J \sum_{\langle n,m \rangle} \sigma_n \, \sigma_m - g \sum_{n=1}^{N} \sigma_n .$$

2. The *local magnetization*, that is, the expectation of the spin variables is given by a constant value m:

$$m = \mathcal{E}[S_n(\omega)], \ \text{for all } n = 1, \ldots, N.$$

3. Each spin interacts with z nearest neighbors. For an Ising model on a lattice, z is the respective coordination number; the latter is an integer that depends on the lattice structure (e.g., $z = 4$ for the square lattice).

Mean-field energy approximation Based on the above, we replace the Ising energy function by means of the following expression

$$\mathcal{H}_0(\boldsymbol{\sigma}) = -J\, z\, m \sum_{n=1}^{N} \sigma_n - g \sum_{n=1}^{N} \sigma_n . \tag{15.50}$$

In the mean-field approximation, the interaction between the spin σ_n and its neighbors is replaced with the interaction between σ_n and the constant mean field m. Note that in the energy function (15.50) the spins are independent (they do not interact). Hence, it is possible to focus on the marginal energy of each spin which is given by

$$\widetilde{\mathcal{H}}_0(\sigma_n) = -J\,z\,m\,\sigma_n - g\,\sigma_n, \quad \text{for } n = 1, \ldots, N.$$

The *marginal pdf* for the local spin variables is then given by the simple expression

$$fs(\sigma_n) = \frac{\mathrm{e}^{-\widetilde{\mathcal{H}}_0(\sigma_n)}}{\sum_{\sigma_n=\pm 1} \mathrm{e}^{-\widetilde{\mathcal{H}}_0(\sigma_n)}} = \frac{\mathrm{e}^{-\widetilde{\mathcal{H}}_0(\sigma_n)}}{2\cosh\left(J\,z\,m + g\right)}. \tag{15.51}$$

Self-consistency The final step closes the loop in the hypothesis that the local magnetization is known (so far we have not specified how m is determined). The principle of self-consistency suggests that the value of m should be the same as the expectation of σ_n based on the marginal pdf (15.51), i.e.,

$$m = \sum_{\sigma_n=\pm 1} \sigma_n\, fs(\sigma_n) = \tanh\left(J\,z\,m + g\right). \tag{15.52}$$

The algebraic nonlinear equation (15.52) is solved numerically leading to the respective value of the local magnetization. Since $|\tanh(x)| \le 1$ for all $x \in \mathbb{R}$, it follows from (15.52) that $-1 \le m \le 1$. The solution of the self-consistency equation (15.52) is plotted in Fig. 15.4 as a function of J and g.

Properties of the mean-field solution

- *Symmetry:* It follows from (15.52) that $m(g, J) = -m(-g, J)$, i.e., flipping the sign of the external field changes the sign of the magnetization.
- *Critical point:* For $g = 0$ a possible solution of (15.52) is $m = 0$. However, if $J > J_c = 1/z$, the mean-field equation admits two non-zero solutions $\pm m_c$ (see Example 15.4 below).
- For $J > J_c$ there is a sudden transition of the solution from $m = -1$ to $m = 1$ as g changes from negative to positive sign (see Fig. 15.4).

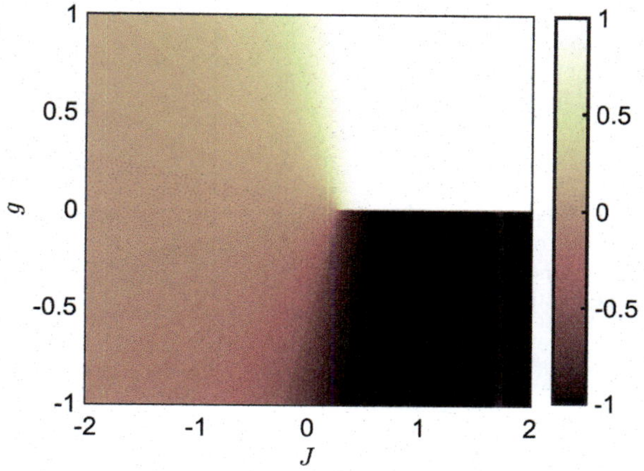

Fig. 15.4 Mean-field theory solutions for the Ising model magnetization with nearest neighbor interactions based on the equation (15.52). The horizontal axis measures the coupling strength J and the vertical axis the external field g. The coordination number is set to $z = 4$

- *Infinite-range model:* Notably, the mean-field solution becomes *exact* for an infinite-range Ising model in which all the spins interact with each other with the same coupling constant J [610, p. 8].
- The mean-field theory predicts an incorrect critical temperature for the two-dimensional Ising model. It also predicts the existence of a phase transition in the one-dimensional model, even though it is known that there is no phase transition in one dimension (cf. Sect. 15.2.2). However, the accuracy of the mean-field theory improves in higher dimensions where the role of the fluctuations is reduced.

Example 15.4 Determine the critical value of the coupling parameter J_c above which the mean-field equation (15.52) admits non-trivial (non-zero) solutions.

Answer For $g = 0$ the mean-field equation (15.52) becomes $m = \tanh(Jzm)$. This equation admits three solutions if the curves m and $\tanh(Jzm)$ intersect at points other than $m = 0$. Given that $\tanh(Jzm)$ is bounded in the interval $[-1, 1]$, this is only possible if the slope of $\tanh(Jzm)$ exceeds one (which is the slope of m) at $m = 0$ (Fig. 15.5). The slope of the hyperbolic tangent $\tanh(Jzm)$ is given by its first derivative with respect to m, i.e.,

$$u(m) = \frac{\mathrm{d}\tanh(Jzm)}{\mathrm{d}m} = \frac{Jz}{[\cosh(Jzm)]^2}.$$

For the slope to be larger than one, it is necessary that $Jz > [\cosh(Jzm)]^2$. The second term is minimized—it becomes equal to one—for $m = 0$. Hence, the *condition of criticality* is $J_c z = 1$. Values of $J > J_c$ lead to three mean-field solutions, namely 0 and $\pm m_c$, while $J < J_c$ implies that only the $m = 0$ solution is feasible.

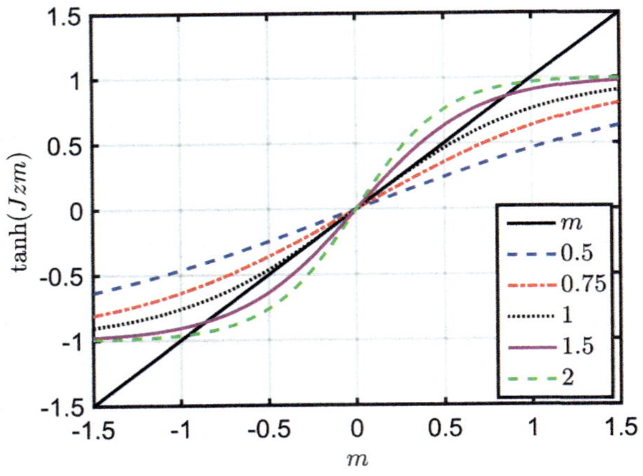

Fig. 15.5 Graphical solution of the mean field equation $m = \tanh(J\,z\,m)$: The left-hand side of the equation corresponds to the continuous straight line (black). The curved lines correspond to $\tanh(Jzm)$ for five different values of Jz equal to 0.5, 0.75, 1, 1.5, 2. For $J\,z \leq 1$ the equation admits only the solution $m = 0$. For $J\,z > 1$ the equation has three solutions: 0, $\pm m_0$

15.2.4 What Is the Connection with Spatial Data Analysis?

Equivalence of spin and indicator fields Let us assume that the magnetic moments σ_n take values ± 1 on a continuous spatial domain $\mathcal{D}$, thus defining the spin field $S(\mathbf{s}; \omega)$, where $\mathbf{s} \in \mathcal{D}$. The connection between the spin field and the associated indicator field is simply expressed as

$$\mathrm{I}_{\mathrm{x}}(\mathbf{s}; \omega) = \frac{1}{2}\left[1 + S(\mathbf{s}; \omega)\right].$$

Hence, in the case of a statistically homogeneous Ising model (i.e., a model with constant coupling coefficients and polarizing field), the expectation of the indicator is linked to the expected magnetization by means of

$$\mathcal{E}[\mathrm{I}_{\mathrm{x}}(\mathbf{s}; \omega)] = \frac{1}{2}\left(1 + m\right).$$

The stationary indicator covariance is related to the spin covariance by means of

$$C_{\mathrm{II}}(\mathbf{r}) = \frac{1}{4}\, C_{\sigma\sigma}(\mathbf{r}).$$

Hence, the indicator field expectation and covariance determine the spin expectation and covariance and vice versa.

Applications of Ising spin models In light of the above, the Ising model and other spin models (e.g., the classical Heisenberg, rotator, and Potts models) define Boltzmann-Gibbs probability distributions for indicator fields. These models are non-Gaussian by construction and therefore cannot be fully determined by purely Gaussian statistics.

Non-parametric (i.e., J- and g-free) spin models have been proposed for spatial data on regular lattices [896, 898, 901]. These models were successfully used to reconstruct missing data in digital images. The energy of the non-parametric models is based on correlation measures that include the squared gradient and curvature of each spin configuration. The data values are *discretized* with respect to several pre-defined levels that span the range between the minimum and maximum sample values. The spatial distribution at each level is described by means of a respective "spin" model. Inter-level correlations are incorporated by conditioning each level on the spatial configuration of the lower levels.

The parameter-free formulation of the spin-based models enables the accurate reconstruction of missing data, even if the underlying probability distribution is non-Gaussian. Extension of the spin-based models to scattered (irregular-grid) data can be accomplished by using kernel functions to implement the spin interactions between neighbors in the spirit described in [368]. Such an extension will add computational cost for tuning kernel bandwidth parameters. It presents, nonetheless, a promising approach for non-Gaussian data modeling.

The Ising model is a primitive example of a *non-Gaussian Markov* random field for binary-valued data [450]. In fact, the *auto-logistic model* for binary data introduced developed in the general framework of Markov random fields, is formally equivalent to the Ising model (see Bartlett's remarks in the discussion section of [69]).

Rotator models for continuous-value data Spin models can also be used to construct *non-Gaussian Markov random fields* for *continuously-valued* scalar variables X [901, 902]. An example is provided by the *planar rotator* (also known as the XY) model in which the spins $\boldsymbol{\sigma}_n$ on the lattice sites $\mathbf{s}_n \in \mathbb{Z}^2$, are represented by means of a rotation angle ϕ. The angle values are mapped by means of a nonlinear transformation to the values of the scalar field $X(\mathbf{s}; \omega)$ at the respective locations. The energy function for this model in the presence of an external field g is

$$\mathcal{H}(\boldsymbol{\sigma}) = -J \sum_{\langle n,m \rangle} \cos(\phi_n - \phi_m) - \sum_{n=1}^{N} g_n \cos\phi_n, \tag{15.53}$$

where the phase angles ϕ are connected to the original variables X by means of an invertible transformation $X \mapsto G(\phi)$ which implies that $\phi \mapsto G^{-1}(X)$.

The mapping $X \mapsto G(\phi)$ is reminiscent of the transformations in Gaussian anamorphosis. However, if the transformed variable follows the Gaussian law, the energy cost for "extreme" values explodes due to the quadratic dependence of the energy on the field values. On the other hand, the model (15.53) does not penalize extreme values as harshly, since the energy costs are determined by the bounded cosine functions. Hence, spin rotator models provide promising alternatives for non-Gaussian models.

15.2.5 Estimation of Missing Spins

The spin-based methods are currently developed only for gridded data. However, the extension to non-gridded data is in principle straightforward. Let us consider the Ising model with the energy function (15.38), repeated below for convenience:

$$\mathcal{H}(\boldsymbol{\sigma}) = -J \sum_{\langle n,m \rangle} \sigma_n \sigma_m - g \sum_{n=1}^{N} \sigma_n.$$

As usual, $\boldsymbol{\sigma} = (\sigma_1, \ldots, \sigma_N)^\top$ represents the spin field which takes values ± 1 at every site, ($\sigma = 1$ corresponds to an indicator value equal to one, while $\sigma = -1$ corresponds to an indicator value equal to zero), J is the interaction strength between spins, g is the external polarizing field, and $\sum_{\langle n,m \rangle}$ denotes summation over neighboring sites.

The pdf of the nearest-neighbor Ising model is given by the Boltzmann-Gibbs expression

$$f_{\sigma}(\boldsymbol{\sigma}) = \frac{\mathrm{e}^{-\mathcal{H}(\boldsymbol{\sigma})}}{Z},$$

where Z is the partition function.

If the spins are located at the sites of an irregular grid (scattered data), the Ising model's energy function can be generalized as follows

$$\mathcal{H}(\boldsymbol{\sigma};\boldsymbol{\theta}) = -J\sum_{n,m=1}^{N} K_h(\mathbf{s}_n - \mathbf{s}_m)\sigma_n\,\sigma_m - g\sum_{n=1}^{N}\sigma_n,$$

where $K_h(\mathbf{s}_n - \mathbf{s}_m)$ are *interaction weights* based on a *kernel function* (see Sect. 2.4.3) that decays with increasing distance $\mathbf{s}_n - \mathbf{s}_m$ and h is the *kernel bandwidth* that determines the range of influence of the kernel function. If the latter is compactly supported, the sum over m for a given n is restricted to a finite size neighborhood around the location $\mathbf{s}_n$.

Assuming that the functional form of the kernel function is known (selected by the user) and that a uniform bandwidth is used, the *parameter vector* $\boldsymbol{\theta} = (J, g, h)$ involves three parameters.[7] It is also possible to include additional parameters that allow the kernel bandwidth to adapt to the local neighborhood. Locally adaptive bandwidths allow "spreading" the interaction between spins to larger distances in areas that are sparsely sampled [368] (see also Sect. 13.3).

The estimation of the model parameters from the data can be based on the methods described in Sect. 13.3 for the stochastic local interaction models. Approaches based on cross validation and pseudo-likelihood are computationally more efficient than maximum likelihood which requires the calculation of the partition function. In order to use cross-validation based estimation (as well as for prediction at unmeasured locations), a predictive equation for missing spins is needed. This is briefly described below.

Provided that the inverse Ising problem can be solved so that the model can be inferred from the data, the "optimal" value of the discrete spin variable σ_n can be estimated by determining the *mode of the Ising conditional pdf*

$$f(\sigma_n \mid \boldsymbol{\sigma}_{-n}) \propto \mathrm{e}^{-\mathcal{H}(\sigma_n|\boldsymbol{\sigma}_{-n})}, \tag{15.54a}$$

where the conditional energy $\mathcal{H}(\sigma_n \mid \boldsymbol{\sigma}_{-n})$ is given by

[7] The parameter set can be further extended by including directionality in the spin coupling, a more complex distance measure in the kernel function, and spatial dependence of g, while the bandwidth can also be allowed to vary locally.

$$\mathcal{H}(\sigma_n \mid \boldsymbol{\sigma}_{-n}) = -\sum_{n=1}^{N}\left[J \sum_{m=1}^{N} K_h(\mathbf{s}_n - \mathbf{s}_m)\sigma_n\,\sigma_m + g\,\sigma_n \right]. \tag{15.54b}$$

Hence, the optimal spin indicator value (either one or minus one) is selected by minimizing the conditional energy function

$$\hat{\sigma}_n = \arg\min_{\sigma_n=\pm 1}\left[-J \sum_{m=1}^{N} K_h(\mathbf{s}_n - \mathbf{s}_m)\sigma_n\,\sigma_m - g\,\sigma_n \right]. \tag{15.55}$$

This step is analogous to the mode prediction of continuously-valued SLI models discussed in Sect. 11.6.

Comparison with indicator kriging Unlike indicator kriging, the mode estimate $\hat{\sigma}_n$ does not require the calculation of the indicator variogram. If the parameter vector $\boldsymbol{\theta}$ has been determined from the data (not necessarily an easy task), the mode estimate can be obtained by simple comparison of the two conditional energy values that result from $\sigma_n = \pm 1$. Finally, the mode estimate is exactly equal to either one or minus one (corresponding, respectively to indicator values of one and zero). Hence, indicator estimation based on the optimal spin value of the Ising model does not face the problem of non-integer or negative indicator values that occurs in kriging.

15.3 Generalized Linear Models

In classical statistics the *generalized linear model (GLM)* extends linear regression to problems that involve random variables with non-Gaussian distributions [197]. Typical examples include binary and count data. In addition to non-Gaussian distributions, GLMs can handle data that exhibit dependence of the response variance on the mean.

The main assumption in GLM is that the data vector $\{y_n^*\}_{n=1}^N$ involves identically distributed, independent measurements of a response variable $\mathrm{Y}(\omega)$. The latter typically follows a probability distribution that belongs in the exponential family, such as the binomial, Poisson, multinomial, Gamma and inverse Gaussian models. In addition, the response data are assumed to depend on a set of predictor variables $\mathbf{x}$.

Generalized linear models involve the following components:

1. A *linear model*, i.e., $u = \boldsymbol{\beta}^\top \mathbf{x}$ that involves the predictor (explanatory) variables in $\mathbf{x}$ and the vector of linear coefficients $\boldsymbol{\beta}$.
2. A nonlinear *link function*, $g(\cdot)$, which connects the linear model of the predictor variables with the expectation of the response variable.
3. A function that describes how the variance of the response variable depends on the expectation m_{y} of the response variable $\mathrm{Y}(\omega)$.

Using the formalism of Sect. 2.2, the expectation of the response variable is given by the following expression in terms of the predictor variables

$$m_{\mathrm{y}} \doteq \mathcal{E}[\mathrm{Y}(\omega)] = g^{-1}\left(\boldsymbol{\beta}^{\top}\mathbf{x}\right), \tag{15.56a}$$

$$u \doteq \boldsymbol{\beta}^{\top}\mathbf{x}, \tag{15.56b}$$

$$g(m_{\mathrm{y}}) = u, \tag{15.56c}$$

where $g(\cdot)$ is the link function that connects the linear model $\boldsymbol{\beta}^{\top}\mathbf{x}$ to the expectation, m_{y}, of the response.

GLM versus variable transformation The concept underlying GLMs is similar to that of normalizing transformations that operate on an original, non-Gaussian variable. Normalizing transformations are often used in time series analysis and in the Gaussian anamorphosis of spatial data. In the latter cases, nonlinear transforms of the original variable are modeled using linear and Gaussian assumptions.

However, the approach based on normalizing transforms is not always optimal. In particular, the applied transformation may not be valid for the entire range of values of the original variable. In contrast, the GLM approach constructs a linear model for a nonlinear function of the original variable's expectation. This approach has advantages in some cases. For example, in problems of groundwater flow the logarithm of the permeability is often modeled as a Gaussian random field. However, this modeling assumption excludes the presence of zero permeability values for which the logarithm is not a real number. On the other hand, if the GLM approach is followed, the assumption of a linear model for the logarithm of the expectation does not exclude zero permeability values.

15.3.1 Logistic Regression

Let us consider the random variable $\mathrm{Y}(\omega)$ that follows the *binomial distribution* $B(N, p)$ where N is the number of trials an p is the success probability for each trial. For example, this variable could represent the occurrence of exploitable ore at a given location or a rainfall event at a specified location and time.

The proportion of successful trials is modeled by means of the random variable $\Pi(\omega) = \mathrm{Y}(\omega)/N$. The expectation and the variance for the *proportion of success* are respectively given by $\mathcal{E}[\Pi(\omega)] = p$ and $\mathbb{V}\mathrm{ar}\{\Pi(\omega)\} = N^{-1}p(1-p)$.

Our goal is to construct a model for the probability of success (which is the mean of the binomial variable) in terms of some predictor variables specified by the feature vector $\mathbf{x}$. This is not always possible since the probability is bounded in $[0, 1]$, while a linear model of the form $u = \boldsymbol{\beta}^{\top}\mathbf{x}$ may take unbounded values. On the other hand, we could exploit the idea of a nonlinear link function.

Logit transform To overcome this potential problem, the *logit transform* is often used as the link function for the success probability. For any $p \in [0, 1]$ the *logit transform* is defined by

$$\text{logit}(p) = \ln \frac{p}{1-p}, \quad p \in [0, 1]. \tag{15.57}$$

Based on (15.57), the logit transform takes values in $\mathbb{R}$. The logit function is equal to the logarithm of the *odds ratio*, that is, the probability of success divided by the probability of failure.

Logistic function The *inverse-logit transform*, i.e., the function $\text{logit}^{-1}(\cdot)$ is given by the *sigmoidal function*

$$\sigma : u \in \mathbb{R} \mapsto p \in [0, 1], \quad p = \sigma(u) \doteq \frac{e^u}{1+e^u}, \quad u \in \mathbb{R}. \tag{15.58}$$

The function $\sigma(\cdot)$ is also known as the *logistic function.*

The logit transform (15.57) maps probabilities, which are defined over [0, 1], into unbounded variables defined in $\mathbb{R}$. The logistic function (15.58) inverts the logit transform and associates a probability with any given real value u. These transformations are quite handy since unbounded variables can be modeled using linear, Gaussian models.

The linear model In logistic regression, the goal is to predict the output classes of the discrete variable $\mathrm{Y}(\omega)$ based on a $\mathbb{R}^{N\times P}$ design matrix $\mathbf{X}$ which comprises N samples with P features each. A linear regression model can be constructed for the logit link function in terms of the feature vector as follows

$$\text{logit}(p_n) = \boldsymbol{\beta}^\top \mathbf{x}_n = u, \; n = 1, \ldots, N, \tag{15.59}$$

where $\mathbf{x}_n = (1, x_1, \ldots, x_{P-1})^\top$ is the feature vector and $\boldsymbol{\beta} = (b_0, b_1, \ldots, b_{P-1})^\top$ is the weight vector. The linear equation (15.59) is the core of *logistic regression.*

After estimating the weights $\boldsymbol{\beta}$ by solving the linear regression problem, the respective probabilities are obtained by means of the inverse logit function based on (15.58). Hence, the probability of success is given by

$$P\left(\mathrm{Y}_n = 1 \mid \mathbf{x}_n; \boldsymbol{\beta}\right) = \frac{1}{1+\exp\left(-\boldsymbol{\beta}^\top \mathbf{x}_n\right)} = \sigma\left(\boldsymbol{\beta}^\top \mathbf{x}_n\right), \tag{15.60}$$

$$P\left(\mathrm{Y}_n = 0 \mid \mathbf{x}_n; \boldsymbol{\beta}\right) = 1 - P\left(\mathrm{Y}_n = 1 \mid \mathbf{x}_n; \boldsymbol{\beta}\right). \tag{15.61}$$

Knowing the probability of success also determines the mean and subsequently the variance of the proportion of success.

Classification In addition, the probability values returned by the logistic function can be used for classification. Hence, logistic regression is often used for binary classification. Note that the logistic function (15.58) is a so-called *soft classifier*, because it does not uniquely specify if a certain variable $Y(\omega)$ is in a given category (i.e., if $y = 0$ or if $y = 1$); instead, it specifies the probabilities for the variable to belong in either category. The multinomial extension of logistic regression, called *SoftMax* regression, is used for classifying variables that can belong in multiple categories [561].

Estimation The optimal parameter vector $\hat{\boldsymbol{\beta}}$ is usually estimated by maximizing the likelihood. We assume that the data involves the set $\mathbb{D} = \{(y_n^*, \mathbf{x}_n)\}_{n=1}^N$ where $y_n^* \in \{0, 1\}$ are binary labels (classes) and $\mathbf{x}_n$ is the vector of features. The likelihood function is then given by

$$\mathcal{L}(\boldsymbol{\beta}; \mathbb{D}) = \prod_{n=1}^{N} \left[\sigma\left(\mathbf{x}_n^\top \boldsymbol{\beta}\right)\right]^{y_n^*} \left[1 - \sigma\left(\mathbf{x}_n^\top \boldsymbol{\beta}\right)\right]^{1-y_n^*}. \tag{15.62}$$

This expression for the likelihood assumes independence for the data while the probabilities of success and failure are determined from (15.60). The log-likelihood is then obtained from

$$\log \mathcal{L}(\boldsymbol{\beta}; \mathbb{D}) = \sum_{n=1}^{N} y_n^* \log \sigma\left(\mathbf{x}_n^\top \boldsymbol{\beta}\right) + (1 - y_n^*) \log \left[1 - \sigma\left(\mathbf{x}_n^\top \boldsymbol{\beta}\right)\right].$$

Finally, the optimal parameters are estimated by maximizing the log-likelihood or (equivalently) minimizing the negative log-likelihood, i.e.,

$$\hat{\boldsymbol{\beta}} = \arg\max_{\boldsymbol{\beta}} \log \mathcal{L}(\boldsymbol{\beta}; \mathbb{D}) = \arg\min_{\boldsymbol{\beta}} \left[-\log \mathcal{L}(\boldsymbol{\beta}; \mathbb{D})\right].$$

The negative log-likelihood for the logistic regression problem is known as the *cross entropy*. The latter is a convex function of the weights $\boldsymbol{\beta}$ and thus its minimization leads to a global minimum. However, unlike the Gaussian case, this minimum cannot be determined analytically and requires numerical methods.

Poisson regression Similarly to logistic regression, a GLM can be constructed for a random variable that follows the *Poisson distribution*. In this case, the rate parameter is connected to a linear model by means of the logarithmic link function. Both the expectation and the variance of the Poisson variable are equal to the rate parameter.

15.3.2 Model-Based Geostatistics

The original GLM formulation does not explicitly incorporate spatial dependence of the data. However, spatial correlations are important for most spatially distributed processes of interest. The framework of model-based geostatistics, introduced in [203], extends the GLM concept by incorporating spatial dependence. The latter is introduced by means of a *latent Gaussian field*. Hence, model-based geostatistics belongs to the wider class of *latent Gaussian models* [701]. A simplified introduction to model-based geostatistics is given in [110].

Below, we present the main ideas of model-based geostatistics for a response variable that follows the normal distribution. We will then show how the same framework can also be applied to non-Gaussian variables. For a detailed and more general presentation see [201].

Main steps of model-based geostatistics for Gaussian data

1. *Assumption:* The measurements of the response variable, $\{y_n^*\}_{n=1}^N$ are *mutually independent* random variables that follow the normal probability distribution, i.e.,

$$\mathrm{Y}_n(\omega) \mid x(\mathbf{s}_n) \overset{d}{=} \mathcal{N}\left[m_{\mathrm{y}}(\mathbf{s}_n), \tau^2\right].$$

   In the above, $m_{\mathrm{y}}(\mathbf{s}_n)$ represents the local mean of $\mathrm{Y}_n(\omega)$, and τ^2 is an *observation variance*. The latter is analogous to the nugget effect, since we have assumed that the measurements at different locations are mutually independent. Note that the distribution of $\mathrm{Y}_n(\omega)$ is conditioned on the value $x(\mathbf{s}_n)$ of a latent random field; the role of the latter is clarified below.
2. In the GLM approach, the local mean is assumed to vary in space according to

$$m_{\mathrm{y}}(\mathbf{s}) = \boldsymbol{\beta}^\top \mathbf{f}(\mathbf{s}) + \mathrm{X}(\mathbf{s}; \omega). \tag{15.63}$$

3. The $\mathrm{X}(\mathbf{s}; \omega)$ is a zero-mean, stationary *Gaussian random field*, i.e.,

$$\mathrm{X}(\mathbf{s}; \omega) \overset{d}{=} \mathcal{N}\left[0, C_{\mathrm{xx}}(\mathbf{r}; \boldsymbol{\theta}')\right].$$

   $\mathrm{X}(\mathbf{s}; \omega)$ is known as the *latent spatial process* (synonyms: spatial random effect or residual spatial variation). The function $C_{\mathrm{xx}}(\mathbf{r}; \boldsymbol{\theta}')$ is the covariance of the latent field, and $\boldsymbol{\theta}'$ is the parameter vector that determines $C_{\mathrm{xx}}(\cdot)$.
4. The vector $\mathbf{f}(\mathbf{s}) = \left(f_1(\mathbf{s}), \ldots, f_P(\mathbf{s})\right)^\top$ includes basis functions that model the explanatory variables (synonyms: fixed effects, auxiliary variables, predictors). We usually set $f_1(\mathbf{s}) = 1$, so that β_1 represents a constant offset. Commonly used explanatory variables include topographical parameters (e.g., longitude, latitude, elevation) and other application-specific predictors (e.g., orientation, local slope, distance from main water bodies or from geological faults, etc.).
5. *Parameter estimation:* The parameter vector $\boldsymbol{\theta}$ of the complete model includes the coefficient vector $\boldsymbol{\beta}$, the observation variance τ^2, and the covariance parameter vector $\boldsymbol{\theta}'$. The estimation of the model parameters can be accomplished by maximizing the likelihood $\mathcal{L}(\boldsymbol{\theta}; \mathbf{y}^*)$ of the model, i.e.,

$$\hat{\boldsymbol{\theta}} = \arg\min_{\boldsymbol{\theta}} \mathcal{L}(\boldsymbol{\theta}; \mathbf{y}^*),$$

where

$$\mathcal{L}(\boldsymbol{\theta}; \mathbf{y}^*) = P\left(\{y_n^*\}_{n=1}^N; \boldsymbol{\theta}\right).$$

The probability distribution $P(\cdot; \boldsymbol{\theta})$ is the normal distribution with a trend function $m_y(\mathbf{s}_n) = \boldsymbol{\beta}^\top \mathbf{f}(\mathbf{s}_n)$ and covariance matrix defined by

$$C_{n,m} = C_{\mathrm{xx}}(\mathbf{s}_n - \mathbf{s}_m; \boldsymbol{\theta}') + \tau^2 \delta_{n,m}, \quad n, m = 1, \ldots, N.$$

For more details see [201, Chap. 5.4].

Maximum likelihood estimation adopts the classical (frequentist) approach which is based on a single (optimal) estimate of the parameter vector. In the Bayesian approach, the posterior probability $P(\boldsymbol{\theta} \mid \mathbf{y}^*)$ is used for the parameters.

6. *Prediction:* The prediction of the response variable at a target point $\mathbf{z}$ where there are no observations can be formulated in the classical or the Bayesian framework. In both cases the prediction is based on the conditional local mean $m_y(\mathbf{z})$.

   In the *classical setting*, the optimal value $\hat{\boldsymbol{\theta}}$ of the parameter vector is used to formulate the prediction. Since the parameter vector is "frozen" at its optimal value, classical predictors are also known as *plug-in* spatial estimators. The plug-in prediction can be based on the conditional probability distribution $P\left[m_y(\mathbf{z}) \mid \mathbf{y}^*; \hat{\boldsymbol{\theta}}\right]$. The latter is Gaussian and can be analytically determined.

   If a single estimate is needed instead of the entire distribution, the expectation of the conditional local mean can be calculated, i.e.,

$$\hat{m}_y(\mathbf{z}) = \mathcal{E}\left[m_y(\mathbf{z})\right] = \int_{-\infty}^{\infty} \mathrm{d}x \, x \, \pi_{\mathbf{z}}(x \mid \mathbf{y}^*; \hat{\boldsymbol{\theta}}),$$

   where $\pi_{\mathbf{z}}(x \mid \mathbf{y}^*; \hat{\boldsymbol{\theta}})$ is the pdf of the conditional local mean at location $\mathbf{z}$.

   In the *Bayesian setting*, the prediction is formulated in terms of the marginal *posterior predictive distribution*. The latter is obtained by integrating over all the values of the parameter vector $\boldsymbol{\theta}$, where each value is weighted by means of the parameter posterior density. In analogy with (11.19) the posterior predictive density is given by the following integral

$$\pi_{\mathrm{pred};\mathbf{z}}\left(m_y \mid \mathbf{y}^*\right) = \int \mathrm{d}\boldsymbol{\theta} \, f_{\mathrm{post}}(\boldsymbol{\theta} \mid \mathbf{y}^*) \, \pi_{\mathbf{z}}\left(m_y \mid \mathbf{y}^*, \boldsymbol{\theta}\right). \tag{15.64}$$

   In this case m_y is treated as a variable and does not have a unique value.

If the data follow the Gaussian distribution, model-based geostatistics leads to formulations familiar in standard linear geostatistics and Gaussian process regression. Things become more interesting if the data follow a non-Gaussian distribution.

Non-Gaussian data The same framework can also be applied to data that do not follow the Gaussian distribution. In terms of the mathematical formalism there are two main differences. First, in Step 1 of the above, the $\{y_n^*\}_{n=1}^N$ are *mutually independent* random variables that follow a non-Gaussian probability distribution, i.e.,

$$\mathrm{Y}_n(\omega) \mid x(\mathbf{s}_n) \stackrel{d}{=} \pi \left[m_{\mathrm{y}}(\mathbf{s}_n), \boldsymbol{\theta}_d\right],$$

where $\boldsymbol{\theta}_d$ involves parameters other than the mean in the conditional marginal. Secondly, in Step 2 the linear model is connected to the expectation $m_{\mathrm{y}}(\mathbf{s})$ by means of a link function $g(\cdot)$, i.e.,

$$g\left[m_{\mathrm{y}}(\mathbf{s})\right] = \boldsymbol{\beta}^\top \mathbf{f}(\mathbf{s}) + \mathrm{X}(\mathbf{s}; \omega).$$

Due to the nonlinear link function, the estimation of the model parameters (i.e., the linear coefficient vector $\boldsymbol{\beta}$ and the covariance parameters $\boldsymbol{\theta}$ of the latent Gaussian field) is usually carried out in the Bayesian framework, using Markov Chain Monte Carlo methods to integrate the non-Gaussian posterior density for the parameters and the posterior predictive density [201, 203].

Binomial distribution Let us consider that the measurements of the response variable, $\{y_n^*\}_{n=1}^N$ take values in $\{0, 1\}$ and are drawn from the *binomial probability distribution*. The probability of K successful out of N trials in total is given by

$$P\,(K; N) = \frac{N!}{K!\,(N-K)!}\, p^K\,(1-p)^{N-K},$$

where p is the *success probability* per trial (i.e., the probability that the outcome is equal to one). The success probability is assumed to vary in space according to the following *logistic function*

$$p_{\mathbf{s}} = \frac{\exp\left[\mathrm{Z}(\mathbf{s}; \omega)\right]}{1 + \exp\left[\mathrm{Z}(\mathbf{s}; \omega)\right]} = \sigma\left[\mathrm{Z}(\mathbf{s}; \omega)\right], \tag{15.65}$$

where $\mathrm{Z}(\mathbf{s}; \omega)$ is a normally distributed (Gaussian) random field. Hence, the *link function* for the success probability is the logit function $\mathrm{logit}\,(p_{\mathbf{s}}) = \mathrm{Z}(\mathbf{s}; \omega)$.

The *normal random field* $\mathrm{Z}(\mathbf{s}; \omega)$ satisfies a linear model that includes both spatial explanatory variables (auxiliary variables) and a correlated spatial component, i.e.,

$$\mathrm{Z}(\mathbf{s}; \omega) = \boldsymbol{\beta}^\top \mathbf{f}(\mathbf{s}) + \mathrm{X}(\mathbf{s}; \omega). \tag{15.66}$$

For non-Gaussian problems there is no closed-form expression for the likelihood. This shortcoming complicates the application of MLE. In non-Gaussian cases, Bayesian inference is typically preferred to MLE. The calculation of the posterior and predictive probability density functions used in Bayesian inference is accomplished by means of either Markov Chain Monte Carlo (MCMC) methods or the integrated nested Laplace approximation (INLA) that was introduced by Rue Martino and Chopin [701]. INLA is a deterministic, iterative approach that allows faster computational times than MCMC.

15.3.3 Autologistic Models

The main motivation for the logit transform (and other link functions) is the construction of regression models for *conditional probabilities* $\pi_n = \pi(x_n \mid \mathbf{x}^*_{-n})$ (the π_n represent the probability of observing a value x_n at $\mathbf{s}_n$ given the values $\mathbf{x}^*_{-n}$ at the remaining sites except $\mathbf{s}_n$). The logit transformation thus opens the door to spatially indexed regression models for *conditional probabilities*. We discussed how this is implemented in terms of GLM for classical statistical models (in which the success probabilities lack spatial dependence), and for geostatistical applications (if the success probabilities are random fields) by means of model-based geostatistics.

Similar ideas can also be used for lattice models, where the sampling set $\{\mathbf{s}_n\}_{n=1}^{N}$ involves lattice sites. Simple linear regression models cannot ensure that the response variable, π_n, is bounded in [0, 1] as required. To remedy this situation, one defines the logit transformation of the probability, i.e.

$$\text{logit}(\pi_n) = \ln\left(\frac{\pi_n}{1-\pi_n}\right). \tag{15.67}$$

The logit transform takes values in $\mathbb{R}$. Hence, instead of constructing a model for the conditional probability $\pi_n(\cdot)$, a spatial regression model, known as *autologistic model*, is constructed for the respective logit function [70, 385, 699]. Known values of the logit field can be inverted using the inverse logit function (15.58) to obtain the respective values of the conditional probability, from which a respective value of the random field X(**s**; ω) can be determined.

For example, consider a binary random field that takes values $\sigma = 0, 1$ and satisfies the exponential pdf with Ising energy given by (15.38). In this case, the conditional probability $\pi(\sigma_n \mid \boldsymbol{\sigma}_{-n})$ is given by

$$\pi_n = \pi(\sigma_n \mid \boldsymbol{\sigma}_{-n}) = \frac{e^{g\,\sigma_n + J\sum_{\langle m\rangle}\sigma_n\sigma_m}}{1 + e^{g + J\sum_{\langle m\rangle}\sigma_m}},$$

where $\sum_{\langle m\rangle}$ indicates the summation over the nearest-neighbors of σ_n. Then, it is straightforward to show that

$$\text{logit}(\pi_n) = g\,\sigma_n + \sigma_n\,J\sum_{\langle m\rangle}\sigma_m,$$

which motivates the formulation of *autologistic models* [68, 69, 385, 682].

The auto-logistic formulation for lattice data is based on the concept of the *conditional probability*. Using conditional probabilities is computationally advantageous for model inference, because the likelihood function can be approximated by means of the *pseudo-likelihood function*. The latter is expressed as a product of univariate conditional probability functions and is thus computationally more tractable than the full likelihood.

Applications *Binary random fields* are used to model the presence or absence (i.e., the occurrence) of a spatial attribute of interest (e.g., rainfall). In the case of threshold-based indicator fields, the two possible values (0 versus 1) respectively represent whether the values of a certain attribute (e.g., pollutant concentration, mineral concentration, rainfall) are above a specific threshold.

In the paper introducing model-based geostatistics, Diggle, Tawn and Moyeed used the logit function to model the spatial distribution of the probability of bacterial infection [203]. The number of infections in the study area was assumed to follow a binomial distribution.

In the same study, the authors investigated the residual contamination from nuclear weapons testing on an island in the South Pacific. In this case, the intensity of the radioactivity was modeled as a latent random field. The observed counts (data), were assumed to follow the Poisson distribution conditionally on the latent intensity. The spatial variation of the Poisson counts followed from the variability of the latent intensity field. The respective link function for the Poisson distribution is the logarithmic transform. Comparing this model with conventional geostatistical analysis, the authors concluded that the latter tends to overly smooth the data and to underestimate the spatial extremes of the intensity field.

The logit transform has been used, among other applications, to model the spatiotemporal variation of precipitation fields [785]. In this framework, the precipitation field is considered as the product of a binary (0 or 1) occurrence field and an intensity field that takes values in $\mathbb{R}_+$. The indicator field that models precipitation occurrence is determined by a logistic regression model.

Logistic regression has also been recently used to model the occurrence of landslides using rainfall, topographic, geological, vegetation, and soil properties as independent predictor variables [494]. Furthermore, the logistic regression model can be extended to *categorical variables* that take values in an integer set which involves more than two values [346, 357, 561].

Chapter 16
Simulations

> Any one who considers arithmetical methods of producing random digits is, of course, in a state of sin.
>
> *John von Neumann*

Spatial interpolation generates "images" or "landscapes" of a physical process that are optimal in some specified sense as discussed in Chap. 10. This is often sufficient for a first assessment of spatial variability. However, interpolated fields tend to be overly smooth and to miss low-probability, extreme-value events. The latter are important for the assessment of environmental hazards, the estimation of mineral reserves, and the evaluation of distributed energy resources (solar, wave and aeolian). For such tasks that require a thorough analysis of the spatial variability, simulation is a more reliable tool (Fig. 16.1).

Simulation refers to the generation of synthetic realizations (states) of a random field by computational means. The generated realizations aim to *sample the joint pdf* of the field. Simulation involves the use of generators that produce pseudo-random numbers[1] and thus instill an element of randomness in the artificial realizations [673]. The generated states respect the univariate statistics (e.g., mean and variance) of the random field as well as statistical constraints of spatial dependence (e.g., the covariance function). *Conditional simulation* also respects the measurements of the field locally. *Unconditional simulation* is typically used in theoretical studies, while conditional simulation is used to model processes that are constrained by data (see also next section and Fig. 16.2 below).

Simulation is important for at least two reasons: (i) When we build new models of spatial dependence, we need to explore the spatial features of the respective

[1] The term "pseudo-random" denotes that the "random numbers" satisfy mathematical tests of randomness, although they are generated by deterministic, non-linear algorithms.

D. T. Hristopulos, *Random Fields for Spatial Data Modeling*, Advances in Geographic Information Science, https://doi.org/10.1007/978-94-024-1918-4_16

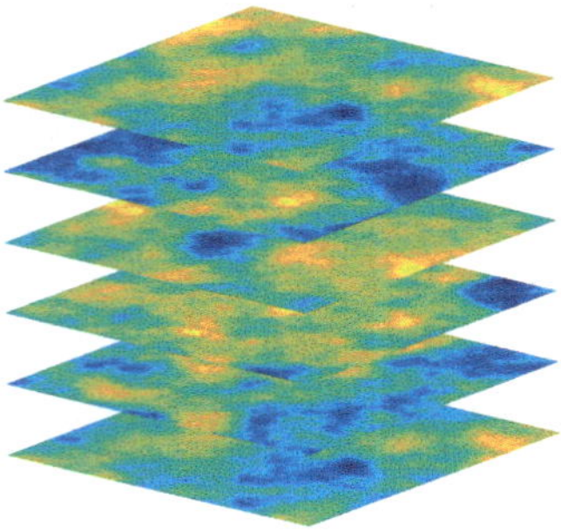

Fig. 16.1 Unconditional realizations of Gaussian SSRF with parameters $\eta_0 = 10^3$, $\xi = 50$, and $\eta_1 = 2$

realizations. We need both a visual appreciation of the landscape of fluctuations as well as quantitative measures of the spatial patterns that emerge. (ii) Often, only sparse spatial data are available. Based on the sparse information, we need to reconstruct probable scenarios and explore different possibilities, including even *low-probability* but *high-impact* cases (for example, zones of high-grade mineralization or pollution hot spots).

In spatial data analysis, simulation is often preferred to interpolation, because the former allows improved exploration of the *uncertainty* and *variability* (heterogeneity) of spatial patterns, which the interpolation methods tend to underestimate. There are two general categories of simulation: *point-based* and *object-based* methods. In the former, the goal is to simulate the values of the process of interest at the nodes of a specified grid. In the latter, the simulated objects represent patterns of geological interest (e.g., channels, layers). We will focus on point-based methods which have wider applicability in spatial data analysis. Object-based methods, however, are also important in the modeling of hydrocarbon reservoirs and ore exploitation studies.

The main disadvantages of simulation methods are the relative *complexity* compared to interpolation and the *computational cost* that could be considerable for large grids and non-Gaussian distributions. The advent of modern computers, however, has significantly reduced the computational time required for simulations and contributed to its acceptance as a practical tool for spatial data analysis.

What are Monte Carlo methods? The term *Monte Carlo* is often used in connection with statistical methods that employ random numbers. Monte Carlo methods exploit random sampling strategies to solve both deterministic and stochastic problems. For example, they can be used to evaluate multi-dimensional integrals that are not calculable by means of deterministic approaches [422]. A distinction between simulation and Monte Carlo is that simulation methods may use all the states (random numbers) generated, while Monte Carlo methods typically take into account only a subset of the generated states.

A common classification of Monte Carlo methods involves rejection sampling, importance sampling, and Markov Chain Monte Carlo (MCMC) methods [685].

Rejection sampling is based on the generation of independent identically distributed random numbers that are used to derive estimates of the target quantity.

A subset of these numbers is maintained for the estimation, based on a suitable criterion, while the remaining numbers are rejected. A common application for rejection sampling is the integration of high-dimensional functions.

Importance sampling employs random numbers generated from a convenient probability density function that differs from the target pdf. To correct for this discrepancy, the sampling function is multiplied by a compensating weight that is equal to the ratio of the target over the surrogate pdf.

MCMC methods, on the other hand, generate a sequence of correlated random states by means of suitable transition rules. The rules for the generation of new states ensure that the sampled distribution asymptotically tends to the target distribution.

We will consider importance sampling of the wavevector space in spectral simulation. We will also use MCMC for the simulation of random fields (e.g., by means of the sequential Gaussian algorithm). Nonetheless, other random field simulation methods, e.g., covariance matrix factorization and spectral approaches considered below, do not strictly belong to any of the three classes above: they do not reject any of the random numbers generated, and they do not belong to MCMC methods, since each realization is based on an independent set of identically distributed random numbers. Some spectral methods use importance sampling of the wavevector space, but factorization methods do not.

Kalos and Whitlock in the introduction of [422] state that simulation refers to computational algorithms that aim to solve natural stochastic processes. On the other hand, Monte Carlo refers to probabilistic methods that aim to solve problems which are not necessarily of a probabilistic nature. However, as pointed out by these authors, the distinction between simulation and Monte Carlo methods is not always clear. Since random field models are inherently stochastic and their simulation involves probabilistic methods, we could refer to spatial simulation methods as *Monte Carlo*.

An introduction to the simulation of stochastic models is given in the report published by Sandia laboratories [249]. Several geostatistics textbooks examine simulation at various levels of sophistication [35, 132, 138, 487, 543]. The statistical physics perspective is described in [479, 603]. In physics, Monte Carlo simulation is used to investigate systems that cannot be explicitly solved (e.g., interacting systems, materials with disorder and stochastic physical processes). More mathematical treatments of Monte Carlo simulation are given in [422, 505, 685]. Finally, a pragmatic presentation of Monte Carlo methods is given in Numerical Recipes [673].

16.1 Introduction

This chapter provides a first glimpse into spatial simulation methods. We focus on Gaussian random fields, non-Gaussian fields obtained from Gaussian ones by means of nonlinear transformations, binary-valued (e.g., Ising) random fields, and fields defined by Boltzmann-Gibbs probability distributions (Gaussian or otherwise).

It is known that Gaussian and Gaussian-transformed random fields do not accurately represent the spatial connectivity of extreme values. For example, binary random media generated by thresholding Gaussian random fields lack long-range features such as channels that extend throughout the entire medium (see also Chap. 15). Such features, however, are important in permeability models that are used in simulations of fluid flow in aquifers and oil reservoirs [326, 461, 523, 889]. Recent efforts to improve this shortcoming involve the use of geological training images for extracting higher-order statistics. This statistical information that describes patterns involving three or more points is then used in the simulation. We will not further discuss these *multipoint statistics* methods herein. The interested reader can find more information in [543, 619, 620, 781].

We will focus on *covariance matrix factorization* methods, *spectral* methods, and *Markov Chain Monte Carlo (MCMC)* methods including the Gibbs sampling scheme (exemplified by sequential simulation), the Metropolis-Hastings update algorithm (with emphasis on the Ising model), and simulated annealing. MCMC methods have a broad scope of applications, and specialized updating algorithms have been devised to overcome problems of slow convergence [479]. We close this chapter with a discussion of the *Karhunen-Loève (K-L)* expansion, which is a spectral method that provides an optimal eigenbase. To our knowledge, the K-L expansion has been explicitly solved only for one-dimensional random fields with specific covariance functions. Thus, in higher dimensions it is necessary to either use the covariance separability assumption (see Chap. 3), in order to construct the covariance as a product or one-dimensional component functions, or numerically calculate the eigenbase.

To condition or not? There is a distinction between *unconditional* and *conditional* simulations. In unconditional simulation the generated realizations respect global statistical constraints (e.g., the marginal probability law, the equations of two-point moments) without being locally restricted by observations. Simulations of this type can reproduce observed overall patterns and are often used for exploratory purposes.

On the other hand, *conditional simulation* generates states that respect both global constraints and the local constraints imposed by the observations. Hence, conditional simulations are more representative of reality if spatial data are available. The conceptual differences between the two approaches are explained with the schematic in Fig. 16.2. Most simulations used in theoretical studies in physics and engineering are unconditional. Conditional simulations are used in applied research fields (e.g., petroleum reservoir simulation, environmental pollution mapping) where the use of data is crucial in order to restrict the probability space generated by the natural heterogeneities and the inherent process uncertainties.

Definition 16.1 For *unconditional simulation* we use the following notation:

- $\mathbb{G}_P$: Set of prediction points. It usually contains the node coordinates of a regular grid $\mathcal{G} \in \mathbb{Z}^d$.
- $\mathbf{X}(\omega)$: Vector denoting the lattice random field defined at the point set $\mathbb{G}_P$.

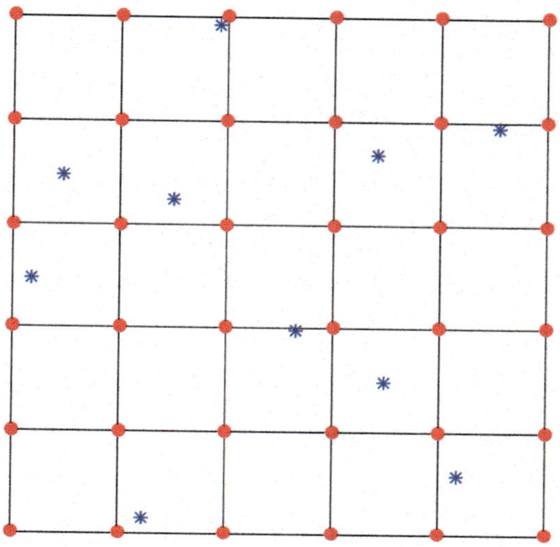

Fig. 16.2 Sampling point set $\mathbb{S}_N$ (stars) and simulation grid point set $\mathbb{G}_P$ (filled circles). In *unconditional simulation*, field values are generated at the grid nodes (points in $\mathbb{G}_P$) while the values at $\mathbb{S}_N$ are only used to estimate the spatial model. In *conditional simulation* the field is generated at the points $\mathbb{S}_N \cup \mathbb{G}_P$, but the sample values $\mathbf{x}^*$ are preserved in $\mathbb{S}_N$. The sample values thus impose constraints on the field values at the neighboring grid nodes due to the spatial continuity enforced by the covariance function

- $\mathbf{x}^{(m)}$, where $m = 1, 2, \ldots, M$: Realizations (states) of the random field $\mathbf{X}(\omega)$ indexed by m.[2] The vector $\mathbf{x}^{(m)}$ comprises $P \in \mathbb{N}$ values $x_p^{(m)}$ at the points $\mathbf{z}_p \in \mathbb{G}_P$ $(p = 1, \ldots, P)$.

Definition 16.2 For *conditional simulation*, in addition to the above we use the following notation:

- $\mathbb{S}_N$: Point set of measurement locations. It includes the sampling points $\mathbf{s}_n \in \mathbb{S}_N$ $(n = 1, \ldots, N)$.
- $\mathbf{x}^* = (x_1^*, \ldots x_N^*)^\top$: Vector of data values at the locations in $\mathbb{S}_N$.
- $\mathcal{G} \in \mathbb{Z}^d$: Simulation grid in d dimensions.
- $\mathbb{S}_N \cup \mathbb{G}_P$: Target point set that involves both the sampling and grid locations.
- $\mathbf{x}^{c(m)} = \left(x_1^*, \ldots, x_N^*, x_1^{(m)}, \ldots, x_P^{(m)}\right)^\top$, where $m = 1, 2, \ldots, M$: Conditional realization vector that comprises $N + P$ values at the points that belong in the target set $\mathbb{S}_N \cup \mathbb{G}_P$. The realization vector is constrained so that $x_n^{c(m)} = x_n^*$ for all $\mathbf{s}_n \in \mathbb{S}_N$.

Modus operandi In the following, it is assumed that the simulated random fields satisfy certain conditions.

1. We focus on the simulation of the field *fluctuations* (residuals) after potential trends have been identified and removed. Realizations of the entire field can be accomplished by adding the trend function to the simulated residuals.

[2]If there is no risk of confusion, the state index superscript (m) will be omitted.

2. The *parameters* of the spatial model are assumed to be known. For example, they can be estimated from the data using the methods of Chap. 12, or they can be specified by the user in exploratory analyses.
3. In the case of *non-Gaussian distributions*, unless otherwise noted, it will be assumed that some normalizing transform is applied (such as those described in Chap. 14), to enforce Gaussian behavior.
4. Certain simulation methods can only be applied to points that coincide with the nodes of a *regular grid* (e.g., FFT-based spectral methods). Other methods are *mesh-free* and can be used to simulate at any point in space (e.g., matrix covariance factorization and randomized spectral sampling).

Parameter uncertainty Geostatistical simulation is based on an "optimal estimate" of the spatial model parameters and a simulation algorithm that generates multiple realizations (conditional or non-conditional) that conform with the chosen parameter set. In this procedure, the uncertainty of the estimated parameters is ignored. In particular, the variogram uncertainty can be large if the size of the experimental sample is small. This is usually the case if data collection is difficult and expensive (e.g., if the data is collected from drill-hole samples).

Simulations that are based on the "optimal parameter set" may under-represent the natural variability of the modeled process. A recent proposal for addressing this issue calls for a *two-level simulation* approach. On the first level, an ensemble of parameter sets is generated based on the respective estimated confidence intervals for the parameters. On the second level, a number of realizations are generated for each parameter set using a simulation algorithm such as the ones described below. The number of the realizations generated for each particular parameter set is proportional to its a posteriori probability [649].

Sanity check The goal of simulation is to generate random field states that conform with a specific spatial model. Then, it should be tested if the produced realizations are consistent with the statistical constraints used to generate them. There are two approaches for testing the quality of the simulated realizations.

The *ensemble-based* approach calculates statistical averages over a finite but large number of realizations. The ensemble averages should accurately reproduce the theoretical values of quantities such as the expected value (mean), the variance, the probability distribution, and the two-point functions (e.g., the covariance function and the variogram). One may also consider *multipoint statistics* that involve higher-order information than that contained in two-point functions, e.g. [469].

The *sample-based* approach estimates the statistical averages based on a single realization. This approach is followed if the computational cost of the simulations is high. For example, studies of fluid flow in aquifers and hydrocarbon reservoirs may involve the simulation of the permeability field at millions of nodes. If the sample-based approach is used for reproduction testing, *ergodicity* is assumed in order to compare the sample-based averages with the theoretical values.

Statistical physics analogy Let us try to understand the difference between spatial interpolation and simulation using an analogy from physics. This is best illustrated if we keep in mind the Boltzmann-Gibbs (B-G) representation of random fields discussed in Sect. 6.1.1. In B-G, the joint pdf is an exponential density that is determined from the energy exponent $\mathcal{H}_0(\mathbf{x}; \boldsymbol{\theta})$. In physical systems, e.g., gases and semiconductors, *temperature* is the parameter that controls the random motion of the atoms and is thus responsible for any disorder present.

In the case of spatial data, the *variance* measures the configurational disorder (i.e., the amplitude of the random field fluctuations in space) of the studied variable. In the case of spatial interpolation—think of the kriging prediction discussed in Chap. 10—the optimal estimate of the field over a domain of interest is a relatively smooth state, represented in discrete form by the vector $\mathbf{x}^*$. The prediction is independent of the variance and represents a stationary state (minimum) of the energy, i.e.,

$$\mathbf{x}^* = \arg\min_{\mathbf{x}} \mathcal{H}_0(\mathbf{x}; \boldsymbol{\theta}).$$

In physical systems, this optimal state corresponds to the *ground state*, i.e., to the lowest energy state. The ground state is achieved at zero temperature (measured in degrees Kelvin). This explains the independence of the optimal state from temperature (i.e., from the variance of the field). The ground state in physical systems is often spatially uniform (translation invariant), although there are notable exceptions in strongly interacting systems.

In spatial data analysis, the "ground state" $\mathbf{x}^*$ is typically non-uniform, due to the data imposed constraints at the measurement locations. For example, consider the simple kriging prediction (10.35): if the target (prediction) point is far away from the sampling point, the kriging prediction is given by the uniform field expectation (ensemble mean).

In simulation studies, the computer generated probable states $\mathbf{x}^{(m)}$, $m = 1, \ldots, M$, are determined by means of random numbers that comply with the specified statistical properties of the random field. The simulated realizations are not necessarily optimal, in the sense of minimizing the energy $\mathcal{H}_0(\mathbf{x}^{(m)}; \boldsymbol{\theta})$. Instead, they are analogous to thermally excited states of a physical system that is determined by the same "energy" function $\mathcal{H}_0(\mathbf{x}; \boldsymbol{\theta})$ as the field. Each such state has a probability of occurrence that decays exponentially with the energy of the state according to (6.5a).

16.2 Covariance Matrix Factorization

The simulation method based on covariance matrix factorization is applicable to zero-mean Gaussian random fields $X(\mathbf{s}; \omega)$ with covariance function $C_{xx}(\mathbf{r})$. Actually, the covariance matrix factorization method does not require a translation invariant covariance matrix, and can thus be applied to random fields with non-stationary covariance function $C_{xx}(\mathbf{s}_1, \mathbf{s}_2)$ as well.

Let us denote by $\mathbf{C}$ the $P \times P$ covariance matrix of the prediction grid points with elements $[\mathbf{C}]_{p,q} = C_{xx}(\mathbf{z}_p - \mathbf{z}_q)$, where $p, q = 1, \ldots, P$. Furthermore, let us assume a *factorization* or *decomposition* of the covariance matrix $\mathbf{C}$. Since $\mathbf{C}$ is a real-valued, symmetric, positive definite matrix, the decompositions listed in Table 16.1 are valid.

The LU factorization applies to square matrices in general. The Cholesky factorization is applicable to square, positive definite matrices. The LU factorization is unique if the matrix $\mathbf{C}$ is *strictly positive definite*.

A positive-definite matrix has more than one square roots, i.e., matrices $\mathbf{A}$ such that $\mathbf{C} = \mathbf{A}\mathbf{A}$. However, the *principal square root* is the unique matrix $\mathbf{A}$ that satisfies the above equation and is also *positive definite* (i.e., its eigenvalues are non-negative). Since $\mathbf{A}$ is positive definite, it is also symmetric $\mathbf{A} = \mathbf{A}^\top$.

The *LDL' decomposition* has the advantage that it avoids the calculation of square roots and can thus be numerically more stable. Since $\mathbf{D}$ is a diagonal matrix, its square root, $\mathbf{D}^{1/2}$, is simply a diagonal matrix whose entries are the square roots of the entries in $\mathbf{D}$. Then, the Cholesky factorization of $\mathbf{C}$ can be obtained from the LDL' factorization by means of $\mathbf{C} = \mathbf{L}_c\mathbf{L}_c^\top$, where the Cholesky lower triangular matrix is given by $\mathbf{L}_c = \mathbf{L}\,\mathbf{D}^{1/2}$.

Unconditional simulations of the Gaussian random field $\mathbf{X}(\omega)$ can be generated using the following matrix product of the *covariance matrix factor* $\mathbf{A}$ and a vector $\boldsymbol{\epsilon}(\omega)$ of P independent, $\mathcal{N}(0, 1)$ random numbers

$$\mathbf{X}(\omega) = \mathbf{A}\,\boldsymbol{\epsilon}(\omega), \tag{16.1}$$

In (16.1) the matrix factor $\mathbf{A}$ represents any of the factorizations listed in Table 16.1. The factor $\mathbf{A}$ acts as a filter that removes "fast spatial variations" from the Gaussian white noise field $\boldsymbol{\epsilon}(\omega)$, thus enabling the emergence of spatial correlations (continuity).

Proof of covariance reproduction It is straightforward to show that the covariance of the simulated vector $\mathbf{X}(\omega)$ conforms with $\mathbf{C}$ as follows

Table 16.1 List of matrix factorizations that can be used for a real-valued, symmetric, positive definite covariance matrix $\mathbf{C}$. For positive definite matrices, the LU decomposition leads to $\mathbf{U} = \mathbf{L}^\top$

Method	Covariance factorization
LU	$\mathbf{C} = \mathbf{L}\mathbf{U}$ where $\mathbf{L}$ is lower triangular and $\mathbf{U}$ upper triangular matrices [299]
Cholesky	$\mathbf{C} = \mathbf{U}^\top\mathbf{U}$ where $\mathbf{U}$ is upper triangular and $\mathbf{U}^\top$ lower triangular matrices
Cholesky	$\mathbf{C} = \mathbf{L}\mathbf{L}^\top$ where $\mathbf{L}$ is lower triangular and $\mathbf{L}^\top$ its transpose
Square root	$\mathbf{C} = \mathbf{A}\mathbf{A}$ where $\mathbf{A}$ is the principal square root of $\mathbf{C}$ [343]
LDL'	$\mathbf{C} = \mathbf{L}\mathbf{D}\mathbf{L}^\top$ where $\mathbf{L}$ is lower triangular and $\mathbf{D}$ is a diagonal matrix

$$
\begin{aligned}
\mathcal{E}\left[X_i(\omega)\,X_j(\omega)\right] =& \mathcal{E}\left[\sum_{k=1}^{P}\sum_{l=1}^{P} A_{i,l}\epsilon_l(\omega) A_{j,k}\epsilon_k(\omega)\right] \\
=& \sum_{k=1}^{P}\sum_{l=1}^{P} A_{i,l}A_{j,k}\mathcal{E}\left[\epsilon_l(\omega)\epsilon_k(\omega)\right] \\
=& \sum_{k=1}^{P}\sum_{l=1}^{P} A_{i,l}A_{j,k}\,\delta_{k,l} = \sum_{k=1}^{P} A_{i,k}A_{j,k} \\
=& \left[\mathbf{A}\mathbf{A}^{\top}\right]_{i,j} = [\mathbf{A}\mathbf{A}]_{i,j} = C_{i,j},
\end{aligned}
$$

for all $i, j = 1, \ldots, P$. In the above we use the orthonormality of the random noise, i.e., $\mathcal{E}\left[\epsilon_l(\omega)\epsilon_k(\omega)\right] = \delta_{k,l}$, for $k, l = 1, \ldots, P$. The proof is shown for the principal square root decomposition, but its extension to the other decompositions is straightforward.

The main steps of unconditional random field simulation on a spatial grid $\mathcal{G}$ using covariance matrix factorization are given in the Algorithm 16.1.

Algorithm 16.1 Unconditional random field simulation using the method of covariance matrix factorization

1. Reshape the matrix of grid locations into a $P \times d$ matrix
2. Calculate the covariance matrix $\mathbf{C}$ between grid locations
3. Compute the factorization $\mathbf{A}^{\top}\,\mathbf{A}$ of $\mathbf{C}$

for $m \leftarrow 1, M$ **do**

  4.1 Generate a vector $\boldsymbol{\epsilon}^{(m)}$ of random numbers from $\mathcal{N}(0, 1)$ distribution

  4.2 Construct realization $\mathbf{x}^{(m)} \leftarrow \mathbf{A}\,\boldsymbol{\epsilon}^{(m)}$

  4.3 Reshape the $P \times 1$ vector $\mathbf{x}^{(m)}$ into a grid matrix

end for

Example We generate and visually compare realizations of random fields with different covariance dependence. The exponential, Gaussian, Spartan, and Matérn covariance will be used as representative models. In general, it is better to compare covariance models with the same integral range than with the same characteristic length ξ [380]. If the integral range is the same, the "islands" of high values and the low-value "valleys" will be statistically of similar size in realizations with different covariance models, thus facilitating the visual comparison. According to the discussion on ergodicity in Sect. 13.4, fields with equal integral ranges sampled over the same domain have the same ergodic index (13.53).

We arbitrarily select the characteristic length ξ of the Spartan covariance model as the reference scale. The integral range of an SSRF in two dimensions is given by (7.40). For the other three models, the integral range ℓ_c can be evaluated based on (5.38), which defines ℓ_c in terms of the spectral density, and from Table 4.4 which lists the spectral densities of common covariance models. These lead to $\ell_c = \sqrt{\pi}\,\xi_G$ for the Gaussian model, $\ell_c = \sqrt{2\pi}\,\xi_E$ for the exponential model, and $\ell_c =$

$2\sqrt{\pi\nu}\xi_M$ for the Matérn model. The expression for the integral range of the Spartan model depends on the value of η_1; assuming that $\eta_1 = 2$, we obtain $\ell_c = 2\sqrt{\pi}\xi_S$. For all the models to have the same integral range, the characteristic lengths ξ should obey the relations $\xi_G = 2\xi_S$, $\xi_E = \sqrt{2}\xi_S$, and $\xi_M = \sqrt{\nu}\xi_S$.

Four realizations obtained with the respective covariance models are shown in Fig. 16.3. The overall spatial patterns are similar—since the same random numbers are used; however, the realizations for the exponential and Spartan covariance (oriented along the main diagonal of the matrix plot) that are non-differentiable at the origin exhibit more small-scale roughness than those for the differentiable Matérn and Gaussian covariances (oriented along the orthogonal-to-the-main diagonal.

Computational issues According to Higham [343], the *principal square root factorization* of the covariance matrix is preferable to other square root factorizations for reasons of numerical stability. Regarding the computational cost, if the prediction grid $\mathcal{G}$ involves L nodes per side in d dimensions, the total number of nodes is $P = L^d$, and the size of the covariance matrix is $L^d \times L^d$.

The covariance factorization method is computationally intensive due to the decomposition of the covariance. The computation complexity of the factorization scales as $O(P^3) = O(L^{3d})$ for dense matrices. Better scaling with size can be obtained if the covariance matrix is sparse (see review in [882]). On the other hand, covariance matrix factorization methods avoid the approximations intrinsic in other (e.g., spectral, turning bands) methods. To alleviate the computational burden of covariance matrix factorization, a method based on *Chebychev matrix polynomial*

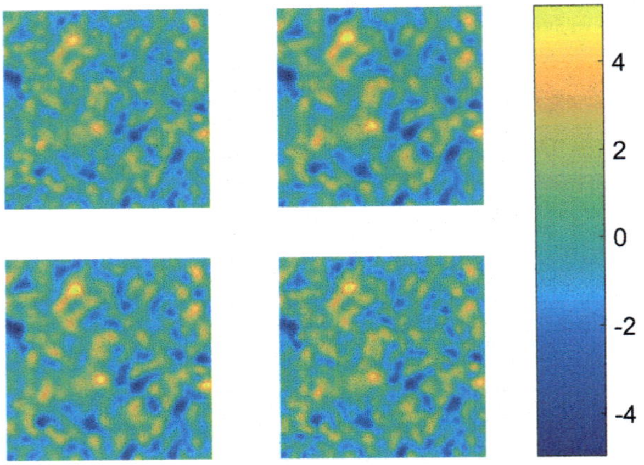

Fig. 16.3 Comparison of random field realizations obtained with the principal square root covariance matrix factorization. The realizations are generated on a square grid with 100 nodes per side using the following covariance functions: exponential (top left) with $\sigma_x^2 = 2$, $\xi = 2\sqrt{2}$, Gaussian (bottom left) with $\sigma_x^2 = 2$, $\xi = 4$, Matérn (top right) with $\sigma_x^2 = 2$, $\xi = 2\sqrt{1.5}$, $\nu = 1.5$, and Spartan (bottom right) with $\eta_0 = 25.15$, $\eta_1 = 2$, $\xi = 2$ models. The coefficient η_0 is selected so that $\sigma_x^2 \approx 2$ for the Spartan model. The same random numbers are used for all four realizations

approximations of the symmetric square root matrix has been proposed [200]. This method has been extended to conditional simulation as well.

Conditional simulations The covariance matrix factorization method can be extended to conditional simulations of random fields that respect the available sample data [624, Chap. 9]. We discuss conditional simulation based on covariance factorization in Sect. 16.7.

16.3 Spectral Simulation Methods

Spectral simulation methods are based on the sampling of the reciprocal space that involves the spectral content of random fields. For Gaussian random fields spectral methods employ the covariance spectral density. They are most suitable for the unconditional simulation of stationary random fields on regular grids, although extensions to more general cases (conditional simulation, non-stationary fields, irregular grids) are also possible. Spectral methods avoid the computationally costly operations that involve the covariance matrix. For simulations on regular grids, spectral methods capitalize on the speed of the *Fast Fourier Transform (FFT)* [673].

Different spectral methods and variations have been developed in engineering, mathematics, and physics. A review of the engineering literature is given in [771]. Random field simulation in theoretical studies of turbulence is reviewed in [469], while cosmological applications are reviewed in [66]. The latter two publications in particular address the issue of simulating *multiscale random fields*, i.e., random fields with a spectral density that retains significant weight over a large range of spatial frequencies (wavenumbers). Multiscale spectral densities are modeled by slowly decaying power laws.

The publication [66] introduces an algorithm that allows refining the resolution of the grid on specified parts of the simulation domain. The proposed algorithm is the equivalent of *adaptive mesh refinement* in numerical analysis extended to Gaussian random fields. Adaptive mesh refinement allows focusing on selected areas of the simulation domain where higher resolution is important. The multiscale refining method is actually based on the real-space convolution of Gaussian white noise with a transfer function which is efficiently computed by means of the FFT.

16.3.1 Fourier Transforms of Random Fields

First, we review the principles of spectral simulation for a real-valued, *continuum and stationary Gaussian random field* $X(\mathbf{s}; \omega)$. One could view this as a theoretical exercise, since simulated realizations are actually defined on discrete supports. However, the continuum formulation allows conceptual clarity by avoiding the intricate mathematical details imposed by the structure of discrete supports.

The Fourier transform of a single state (realization) of $\mathrm{X}(\mathbf{s};\omega)$ is given by (8.14). Generalizing this to every realization, we obtain the following equation for the Fourier transform of the random field $\mathrm{X}(\mathbf{s};\omega)$

$$\widetilde{\mathrm{X}}(\mathbf{k};\omega) = \mathrm{u}(\mathbf{k};\omega)\left[\widetilde{C}_{\mathrm{xx}}(\mathbf{k})\right]^{1/2}, \tag{16.2}$$

where $\mathrm{u}(\mathbf{k};\omega)$ is a complex-valued, zero-mean Gaussian random field with delta-like correlations in reciprocal space, i.e.,

$$\mathcal{E}[\mathrm{u}^{\dagger}(\mathbf{k}';\omega)\,\mathrm{u}(\mathbf{k};\omega)] = (2\pi)^d\delta(\mathbf{k}-\mathbf{k}'). \tag{16.3}$$

Comment Note that in real space the equation (16.2) is equivalent to the convolution of Gaussian white noise with a transfer function $T(\mathbf{s}-\mathbf{s}')$ that is obtained by the inverse Fourier transform of $\left[\widetilde{C}_{\mathrm{xx}}(\mathbf{k})\right]^{1/2}$. The white-noise convolution representation of Gaussian random fields is used in [66] to construct multiscale random fields with adaptive resolution.

The *orthogonality relation* (16.3) implies the following normalization condition

$$\int_{\mathbb{R}^d} d\mathbf{k}'\,\mathcal{E}[\mathrm{u}^{\dagger}(\mathbf{k}';\omega)\,\mathrm{u}(\mathbf{k};\omega)] = (2\pi)^d. \tag{16.4}$$

In addition, since the field $\mathrm{X}(\mathbf{s};\omega)$ is real-valued, the complex-valued Gaussian field $\mathrm{u}(\mathbf{k};\omega)$ must be *conjugate symmetric*, that is,

$$\mathrm{u}^{\dagger}(-\mathbf{k};\omega) = \mathrm{u}(\mathbf{k};\omega). \tag{16.5}$$

In light of the conjugation symmetry the orthogonality relation (16.3) becomes

$$\mathcal{E}[\mathrm{u}(\mathbf{k}';\omega)\,\mathrm{u}(\mathbf{k};\omega)] = (2\pi)^d\delta(\mathbf{k}+\mathbf{k}'). \tag{16.6}$$

The spectral decomposition (16.2) is the equivalent of (16.1) in continuum reciprocal space. As we show below, the construction (16.2) ensures that the field generated by means of the inverse Fourier transform, i.e.,

$$\mathrm{X}(\mathbf{s};\omega) = \mathrm{IFFT}\,[\widetilde{\mathrm{X}}(\mathbf{k};\omega)] \tag{16.7}$$

has a covariance function with spectral density $\widetilde{C}_{\mathrm{xx}}(\mathbf{k})$.

Example 16.1 Show that the random field $\mathrm{X}(\mathbf{s};\omega)$ generated by the inverse Fourier transform (16.7) has a covariance function $C_{\mathrm{xx}}(\mathbf{r})$ given by the inverse Fourier transform of $\widetilde{C}_{\mathrm{xx}}(\mathbf{k})$.

Answer We express the random field by means of the inverse Fourier transform

$$\mathrm{X}(\mathbf{s};\omega) = \int_{\mathbb{R}^d} d\mathbf{k}\, e^{i\mathbf{k}\cdot\mathbf{s}}\, \widetilde{\mathrm{X}}(\mathbf{k};\omega).$$

Based on this integral we obtain the following expression for the covariance function

$$\begin{aligned} C_{\mathrm{xx}}(\mathbf{r}) &= \frac{1}{(2\pi)^{2d}} \int_{\mathbb{R}^d} d\mathbf{k} \int_{\mathbb{R}^d} d\mathbf{k}'\, e^{i(\mathbf{k}\cdot\mathbf{s}+\mathbf{k}'\cdot\mathbf{s}')}\, \mathcal{E}[\widetilde{\mathrm{X}}(\mathbf{k};\omega)\,\widetilde{\mathrm{X}}(\mathbf{k}';\omega)] \\ &= \frac{1}{(2\pi)^{2d}} \int_{\mathbb{R}^d} d\mathbf{k} \int_{\mathbb{R}^d} d\mathbf{k}'\, e^{i(\mathbf{k}\cdot\mathbf{s}+\mathbf{k}'\cdot\mathbf{s}')} \left[\widetilde{C}_{\mathrm{xx}}(\mathbf{k})\,\widetilde{C}_{\mathrm{xx}}(\mathbf{k}')\right]^{1/2} \mathcal{E}[\mathrm{u}(\mathbf{k}';\omega)\,\mathrm{u}(\mathbf{k};\omega)] \\ &= \frac{1}{(2\pi)^{d}} \int_{\mathbb{R}^d} d\mathbf{k}\, e^{i[\mathbf{k}\cdot(\mathbf{s}-\mathbf{s}')]}\, \widetilde{C}_{\mathrm{xx}}(\mathbf{k}). \end{aligned}$$

The second line is obtained from the first by means of (16.2). The third line is obtained by means of the orthogonality relation (16.6).

16.4 Fast-Fourier-Transform Simulation

Spectral simulations on square/cubic grids can take advantage of the lattice structure regularity and the efficiency of the *Fast Fourier Transform (FFT)* [673]. In the following, we assume that $\mathrm{X}(\mathbf{s};\omega)$ is a second-order stationary, Gaussian random field with covariance function $C_{\mathrm{xx}}(\mathbf{r})$ and spectral density $\widetilde{C}_{\mathrm{xx}}(\mathbf{k})$. We focus on *square grids with L nodes per side* (for simplicity we assume that L is an even integer) and $P = L^2$ nodes in total. The results can be in a straightforward manner extended to higher dimensions, and with some mild bookkeeping modifications to odd-valued L. The main idea is quite simple and similar to the covariance matrix factorization method. However, the implementation involves some tedious mathematical details due to the discreteness of the grid support.

The filtering concept FFT simulation is based on the concept of filtering random fluctuations in the Fourier domain [523]. The main idea is to generate a vector of P Gaussian standard random numbers that is subsequently passed through a low-pass frequency filter which selectively removes higher wavenumbers. The Fourier filtering operation modifies the initially white spectrum of the random numbers thus enforcing spatial correlations in real space [317, 363, 482].

Computational complexity The random number generation and the filtering procedure have a computational cost that scales linearly with the lattice size. The random field in real space follows from the inverse Fast Fourier Transform (IFFT) of the filtered set. The *computational cost* of the latter scales as $O(P \log_2 P)$ operation. This accounts for the very fast performance of the FFT algorithm compared with methods based on the covariance matrix factorization that have computational complexity $O(P^3)$ and are memory intensive.

Implementation In light of equations (16.2) and (16.7) the FFT spectral method seems quite simple. The main implementation difficulties arise from the structure (geometry) of the reciprocal space and the normalization of the Fourier transform pair in the lattice field case. These issues are discussed below.

16.4.1 Discrete Fourier Transforms

The FFT-based simulation requires adapting the continuum models for use with *discrete Fourier transforms* (DFT). The direct and inverse DFT on a square lattice with *lattice step* equal to a, are given by the following Fourier series [673]

$$\widetilde{x}_{l_1,l_2} = a^2 \sum_{n_1=1}^{L} \sum_{n_2=1}^{L} x_{n_1,n_2}\, e^{-i\frac{2\pi}{L}[(n_1-1)(l_1-1)+(n_2-1)(l_2-1)]}, \quad l_1, l_2 = 1, \ldots, L, \tag{16.8a}$$

$$x_{n_1,n_2} = \frac{1}{Na^2} \sum_{l_1=1}^{L} \sum_{l_2=1}^{L} \widetilde{x}_{l_1,l_2}\, e^{i\frac{2\pi}{L}[(n_1-1)(l_1-1)+(n_2-1)(l_2-1)]}, \quad n_1, n_2 = 1, \ldots, L \tag{16.8b}$$

The lattice nodes are represented by the position vectors $\mathbf{s}_{n_1,n_2} = (n_1 a, n_2 a)$, where the integers n_1, n_2 are the column and row indices. The column (row) indices determine the x (y) coordinates of the vectors. The symbol x_{n_1,n_2} denotes the field value at $\mathbf{s}_{n_1,n_2}$ and $\widetilde{x}_{l_1,l_2}$ is the DFT at the wavevector $\mathbf{k}_{l_1,l_2}$.

Structure of reciprocal space The integers l_1, l_2 are respectively the column and row indices of the wavevector $\mathbf{k}_{l_1,l_2}$ in reciprocal space.

An index value l_1 or l_2 equal to one implies that the respective component of the spatial frequency is zero, i.e., $\mathbf{k}_{1,l_2}$ means that $k_x = 0$. Frequency indices from 2 to $L/2$ represent positive spatial frequencies and from $L/2 + 2$ to L negative frequencies. An index value equal to $L/2 + 1$ corresponds to the *Nyquist spatial frequency* π/a which is the highest wavenumber (in the given direction) that can be sampled on the present lattice. In the case of a square lattice, the organization of the wavevectors in reciprocal space is shown schematically in Fig. 16.4.

Based on the above, the *wavevectors* $\mathbf{k}_{l_1,l_2}$ are linked to the integer reciprocal indices as follows

$$\mathbf{k}_{l_1,l_2} = \frac{2\pi}{La}\left[\zeta(l_1), \zeta(l_2)\right], \quad \text{where} \tag{16.9a}$$

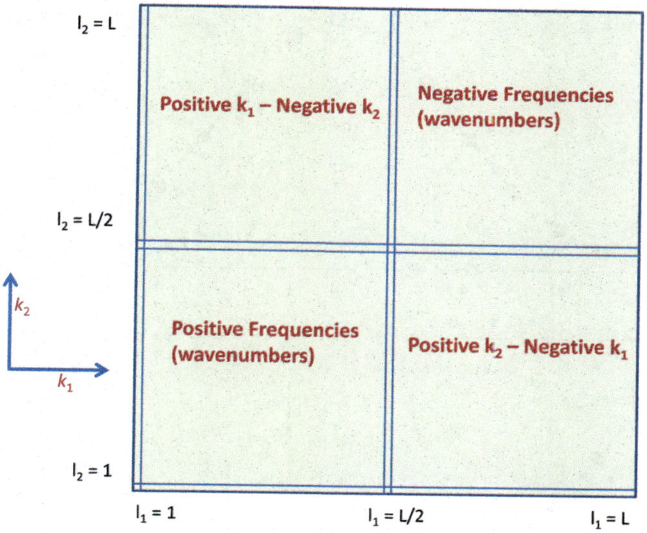

Fig. 16.4 Schematic that shows the partition of the reciprocal (Fourier) space for a square $L \times L$ lattice, where L is an even integer. This structure is used in Discrete Fourier transforms. The *Nyquist spatial frequency* in each of the orthogonal directions corresponds to a respective wavevector index value of $L/2 + 1$ (indicated by the second vertical line in the middle of the rectangle)

$$\zeta(l) = \begin{cases} l - 1, & \text{if } L/2 + 1 \geq l \geq 1, \\ -L - 1 + l, & \text{if } L \geq l > L/2 + 1. \end{cases} \tag{16.9b}$$

The discrete spectral density, $\widetilde{C}_{\rm xx}(\mathbf{k}_{lj})$ is an even function, obtained from the values of the continuum $\widetilde{C}_{\rm xx}(\mathbf{k})$ at the discrete frequencies $\mathbf{k}_{lj}$. Figure 16.5 illustrates the organization of the negative and positive wavevector sectors of the spectral density in reciprocal space using a one-dimensional example.

DFT normalization The DFT normalization factors in (16.8) ensure that the discrete series for the covariance function in real and reciprocal space converge to the respective Fourier transform integrals at the continuum limit $a \to 0$. This is because in the direct Fourier transform (16.8a)

$$a \sum_{n_1=1}^{L} \to \int_{-\infty}^{\infty} \mathrm{d}y,$$

while for the inverse Fourier transform (16.8b) it holds that

$$\frac{1}{La} \sum_{l_1=1}^{L} \to \frac{1}{2\pi} \int_{-\infty}^{\infty} \mathrm{d}k_y.$$

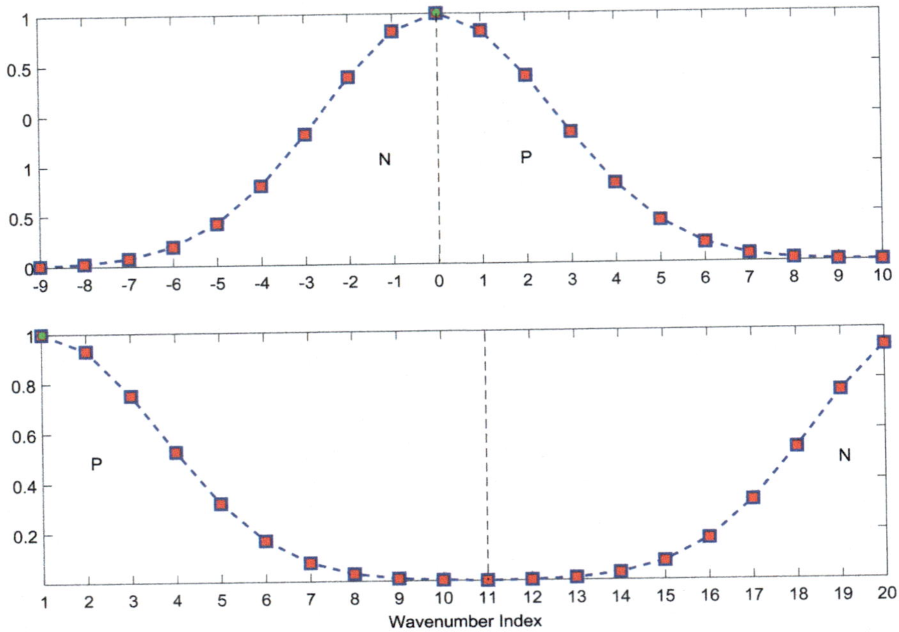

Fig. 16.5 Folding of the spectral density used in the Discrete Fourier transform for $L = 20$ points on a 1D lattice. The top diagram shows the intuitive plot of Gaussian spectral density (negative wavenumbers on the left, zero wavenumber at the center, and positive wavenumbers on the right side). The bottom diagram plots the same spectral density with the DFT wavenumber ordering: The zero wavenumber component is plotted first, followed by positive wavenumbers up to the Nyquist index value $L/2 + 1$, then followed by negative-wavenumber components, ordered from the most negative (index 12) to the least negative (index 20) wavenumber

Random field in reciprocal space The random field components in reciprocal (spectral) space are given by

$$\widetilde{x}(l_1, l_2) = u_{l_1,l_2} \sqrt{\widetilde{C}_{\mathrm{xx}}(\mathbf{k}_{l_1,l_2})}, \quad l_1, l_2 = 1, \ldots, L, \tag{16.10}$$

where the u_{l_1,l_2} $(l_1, l_2 = 1, \ldots, L)$ are entries of an $N \times N$ matrix $\mathbf{u}$ of *random complex-valued coefficients*, drawn from the normal distribution with zero mean and standard deviation c_0; the latter will be specified below.

Then, according to (16.8b) the inverse transform which generates the realization of the random field is given by

$$x_{n_1,n_2} = \frac{1}{Na^2} \sum_{l_1=1}^{L} \sum_{l_2=1}^{L} u_{l_1,l_2} \left[\widetilde{C}_{xx}(\mathbf{k}_{l_1,l_2}) \right]^{1/2} e^{i\frac{2\pi}{L}[(n_1-1)(l_1-1)+(n_2-1)(l_2-1)]}. \tag{16.11}$$

This equation is often referred to at the *spectral representation* of random fields [384]. It is rather straightforward to simulate random field states by implementing this equation. Notice also the close connection of the above with the spectral representation of random processes (9.56) (given in Theorem 9.1). In contrast with (9.56), the equation (16.11) employs a discrete sampling of the wavevector space.

The spatial dependence of the field in real space enters in $\widetilde{x}(l_1, l_2)$ only through the spectral density. This equation is the discrete analogue of the Fourier transform (16.2) for a specific realization of the random field. However, note that the replacement of the integral (16.2) with a discrete summation in (16.8b) implies that the DFT-based spectral method is only *approximate* [199].

The random coefficient matrix The matrix $[\mathbf{u}]_{l_1,l_2=1}^{L}$ should satisfy certain symmetry properties to guarantee that the inverse Fourier transform of $\widetilde{x}_{l_1,l_2}$ is real. Based on (16.8a) the DFT can be expressed as follows (below we use the scalar index notation for the wavevectors)

$$\widetilde{x}(\mathbf{k}) = a^2 \sum_{n=1}^{N} e^{-i\mathbf{k}\cdot\mathbf{s}_n}\, x(\mathbf{s}_n).$$

Then, the complex conjugate of $\widetilde{x}(\mathbf{k})$ is given by

$$\widetilde{x}^{\dagger}(\mathbf{k}) = a^2 \sum_{n=1}^{N} e^{-i\mathbf{k}\cdot\mathbf{s}_n}\, x(\mathbf{s}_n) \Rightarrow \widetilde{x}^{\dagger}(-\mathbf{k}) = a^2 \sum_{n=1}^{N} e^{i\mathbf{k}\cdot\mathbf{s}_n}\, x(\mathbf{s}_n) = \widetilde{x}(\mathbf{k}).$$

The above proves that $\widetilde{x}(\mathbf{k}) = \widetilde{x}^{\dagger}(-\mathbf{k})$. Based on (16.2) $\widetilde{x}(\mathbf{k}) = u_{\mathbf{k}} \sqrt{\widetilde{C}_{xx}(\mathbf{k})}$ and since $u_{\mathbf{k}}$ is the only complex-valued coefficient, it holds that $u_{-\mathbf{k}}^{\dagger} = u_{\mathbf{k}}$.

Conjugate symmetry In light of the organization of the reciprocal space (see Fig. 16.5), to ensure *conjugate symmetry* the matrix $\mathbf{u}$ should satisfy the following block-matrix structure

$$\mathbf{u} = \begin{bmatrix} u_{1,1} & \mathbf{b}_{1,R}^{\top} & u_{1,L/2+1} & (\mathcal{R}\mathbf{b}_{1,R}^{\top})^{\dagger} \\ \mathbf{b}_{C,1} & \mathbf{C} & \mathbf{b}_{C,L/2+1} & \mathcal{R}\mathbf{D}^{\dagger} \\ u_{L/2+1,1} & \mathbf{b}_{L/2+1,R}^{\top} & u_{L/2+1,L/2+1} & (\mathcal{R}\mathbf{b}_{L/2+1,R}^{\top})^{\dagger} \\ \mathcal{R}\mathbf{b}_{C,1}^{\dagger} & \mathbf{D} & \mathcal{R}(\mathbf{b}_{C,L/2+1}^{\dagger}) & \mathcal{R}(\mathbf{C}^{\dagger}) \end{bmatrix}. \tag{16.12}$$

- The elements $u_{1,1}$, $u_{1,L/2+1}$, $u_{L/2+1,1}$ and $u_{L/2+1,L/2+1}$ are real numbers. This ensures that the DFT values (16.8a), evaluated at the wavevectors $(0, 0)$, $(0, \pi)$, $(\pi, 0)$, (π, π), are real numbers.
- The blocks $\mathbf{b}_{1,R}^{\top} = (u_{1,2}, \ldots, u_{1,L/2})$ and $\mathbf{b}_{L/2+1,R}^{\top} = (u_{L/2+1,2}, \ldots, u_{L/2+1,L/2})$ are complex-valued row vectors with $L/2 - 1$ columns.
- The blocks $\mathbf{b}_{C,1}=(u_{2,1}, \ldots, u_{L/2,1})^{\top}$ and $\mathbf{b}_{C,L/2+1}=(u_{2,L/2+1}, \ldots, u_{L/2,L/2+1})^{\top}$ are complex-valued column vectors with $L/2 - 1$ rows.
- The blocks $\mathbf{C}$ and $\mathbf{D}$ are $(L/2-1)\times(L/2-1)$ square matrices of complex-valued coefficients.
- The operation $\mathcal{R}(\mathbf{A})$ denotes the point reflection of the array (vector or matrix) $\mathbf{A}$. If $\mathbf{A} = (a_1, \ldots, a_N)$, then $\mathcal{R}(\mathbf{A}) = (a_N, \ldots, a_1)$, while if $\mathbf{A}$ is a square $N \times N$ matrix, then $\mathcal{R}(\mathbf{A}) = \mathbf{A}'$ such that $a'_{i,j} = a_{N-i+1,N-j+1}$, for $i, j = 1, \ldots, N$. The reflection operation represents the arrangement of wavevectors in reciprocal space, in particular the fact that the zone of negative wavevector components begins with the most negative frequency to the right of the respective Nyquist limit.

Normalization of random coefficients The standard deviation c_0 of the random coefficients u_{l_1,l_2} is determined by the requirement that the correct field variance is obtained in the continuum limit. The random coefficients $u_{\mathbf{k}_n}$ can be expressed in terms of the real-valued random variables $\{v_n(\omega)\}_{n=1}^{N}$ and $\{w_n(\omega)\}_{n=1}^{N}$ as follows

$$u_{\mathbf{k}_n}(\omega) = \Re[u_{\mathbf{k}_n}(\omega)]+\mathrm{i}\Im[u_{\mathbf{k}_n}(\omega)] = v_n(\omega)+\mathrm{i}\, w_n(\omega), \;\; n = 1, \ldots, N. \tag{16.13a}$$

The equivalent of the orthogonality relation (16.6) in the discrete case is

$$\mathcal{E}[\mathrm{u}(\mathbf{k}'; \omega)\, \mathrm{u}(\mathbf{k}; \omega)] = \mathcal{E}[\mathrm{u}^{\dagger}(-\mathbf{k}'; \omega)\, \mathrm{u}(\mathbf{k}; \omega)] = c_0^2\, \delta_{\mathbf{k}+\mathbf{k}',\mathbf{0}},$$

where c_0^2 is the variance of the random coefficient $\mathrm{u}(\mathbf{k}; \omega)$. Using the above orthogonality relation, the normalization spectral integral (16.4) is discretized as follows

$$\frac{c_0^2}{L^2a^2} \sum_{\mathbf{k}'} \delta_{\mathbf{k}+\mathbf{k}',\mathbf{0}} = 1 \Rightarrow c_0 = \frac{1}{La}. \tag{16.13b}$$

The variance of a complex-valued normal random variable $\mathrm{u}(\mathbf{k}; \omega)$ with independent real, $v(\omega)$, and imaginary, $w(\omega)$, parts is equal to the sum of the variances of the real and imaginary components, i.e.,

$$\mathbb{V}\mathrm{ar}\,\{\mathrm{u}(\mathbf{k}; \omega)\} = \mathbb{V}\mathrm{ar}\,\{v(\omega)\} + \mathbb{V}\mathrm{ar}\,\{w(\omega)\}\,.$$

Hence, the *real*, $v(\omega)$, and *imaginary*, $w(\omega)$ parts of the random coefficients u_{l_1,l_2} follow the normal distributions with zero mean and standard deviations as follows

$$\text{If } l_1 \vee l_2 = 1, L/2+1 : v_{l_1,l_2} \overset{d}{=} \mathcal{N}(0, c_0), \; w_{l_1,l_2} = 0, \tag{16.13c}$$

$$\text{Otherwise: } v_{l_1,l_2} \overset{d}{=} \mathcal{N}\left(0, \frac{c_0}{\sqrt{2}}\right), \; w_{l_1,l_2} \overset{d}{=} \mathcal{N}\left(0, \frac{c_0}{\sqrt{2}}\right). \tag{16.13d}$$

Finally, the spectral representation of the lattice random field is fully determined by (16.10) and (16.13). The field in real space is obtained by numerically calculating the inverse DFT using the fast Fourier transform methodology.

The mathematical reason for which the imaginary part, $w(\omega)$ of the random coefficients is zero if either l_1 or $l_2 = 1, L/2+1$ is evidenced in (16.8a): For the above choice of wavenumbers, the phase of all the terms in the DFT (i.e., for all n_1 and n_2) is a multiple of π; hence, the respective imaginary part is zero. One may wonder whether certain coefficients should be purely imaginary. This would happen if all the terms in the series (16.8a) had phases equal to an odd multiple of $\pi/2$. However, this is not possible, since a series that contains terms with phase equal to an odd multiple of $\pi/2$ also includes terms with phases that are even multiples of $\pi/2$ (hence real numbers).

Algorithm 16.2 Spectral simulation of zero-mean, Gaussian, stationary random field with covariance function $C_{xx}(\mathbf{r})$ based on the Fourier transform method. The inverse FFT is used to obtain the real-space realization of the field from the DFT. The symbol "*" denotes multiplication of the respective matrix elements, i.e., if $\mathbf{A}$ and $\mathbf{B}$ are $P \times P$ matrices, then $[\mathbf{A} * \mathbf{B}]_{i,j} = A_{i,j} B_{i,j}$ for $i, j = 1, \ldots, P$. The computational efficiency of the algorithm is determined by the $O(P \log_2 P)$ complexity of the IFFT

1: Calculate the spectral density $\widetilde{C}_{xx}(\mathbf{k})$ that corresponds to the covariance $C_{xx}(\mathbf{r})$
2: Define the simulation lattice with $L = 2m$ nodes per side
3: Construct the spectral density matrix $\widetilde{\mathbf{C}}$ based on $\widetilde{C}_{xx}(\mathbf{k})$ at the wavevectors $\mathbf{k}_i$ defined by (16.9a)
4: Generate the random matrix $\mathbf{u}$ of normally distributed coefficients based on (16.13)
5: Calculate the DFT of the state: $\widetilde{\mathbf{x}} \leftarrow [\widetilde{\mathbf{C}}]^{1/2} * \mathbf{u}$ ▷ Fourier transform
6: Construct the simulated state by means of $\mathbf{x} \leftarrow \text{IFFT}(\widetilde{\mathbf{x}})$ ▷ Inverse Fast Fourier transform

A more rigorous mathematical treatment of DFT-based random field simulations in any dimension d and for general rectangular domains and lattice cells is given in [482]. Therein the following lemma is shown:

Fourier transform of white noise random field Assume that $\epsilon(\mathbf{s}; \omega)$ is a centered Gaussian random field with covariance function $\mathcal{E}[\epsilon(\mathbf{s}; \omega)\, \epsilon(\mathbf{s}'; \omega)] = \delta(\mathbf{s} - \mathbf{s}')$, i.e., a *Gaussian white noise*. Let the Fourier transform of the above field be defined by the following integral

$$\widetilde{\epsilon}(\mathbf{k}; \omega) = \int_{\mathbb{R}^d} d\mathbf{s}\, e^{i\mathbf{k}\cdot\mathbf{s}}\, \epsilon(\mathbf{s}; \omega).$$

The FT $\widetilde{\epsilon}(\mathbf{k};\omega)$ is a *centered, complex-valued Gaussian field* with *delta-function correlations*, i.e., $\mathcal{E}\left[\widetilde{\epsilon}(\mathbf{k};\omega)\widetilde{\epsilon}(\mathbf{k}';\omega)\right] = \delta(\mathbf{k}+\mathbf{k}')$.

The FT $\widetilde{\epsilon}(\mathbf{k};\omega)$ can be constructed from a real-valued, standard Gaussian white noise field $W(\mathbf{k};\omega)$. In particular, the random field

$$V(\mathbf{k};\omega) = \frac{1}{2}\left[W(\mathbf{k};\omega) + W(-\mathbf{k};\omega)\right] + \frac{\mathrm{i}}{2}\left[W(\mathbf{k};\omega) - W(-\mathbf{k};\omega)\right],$$

follows the same normal distribution as the FT $\widetilde{\epsilon}(\mathbf{k};\omega)$ [482, Lemma 1].

Circulant embedding The approach of simulation by means of circulant embedding combines the accuracy of the covariance matrix factorization with the computational efficiency of the spectral FFT-based method. Circulant embedding allows the simulation of the random field by embedding the $P \times P$ covariance matrix in a larger $M \times M$ matrix with block circulant structure. A *circulant matrix is a Toeplitz matrix* with the property that each column can be obtained from the preceding one by a circular shift of its elements. An $M \times M$ matrix is a Toeplitz matrix if its entries are constant on every diagonal. Covariance circulant embedding has been used to simulate stationary random fields in two dimensions [199, 511, 854].

16.5 Randomized Spectral Sampling

Most spectral simulation methods inherently assume a regular grid structure. Hence, they are not suitable for simulating random fields at the nodes $(\mathbf{s}_1, \ldots, \mathbf{s}_N)$ of irregular (unstructured) meshes. *Grid-free simulation* on the other hand can generate random field realizations at any point in space. A significant advantage of grid-free simulation is that new locations can be added, leading to progressive refinement, without reconstructing the entire random field model. In addition, the resolution at any given point can be increased at will, and is only limited by the accuracy bestowed by the sampling of the spectral domain [876].

The method of *mode superposition with randomized spectral sampling* supports grid-free simulation. This method has been used extensively in the scientific literature for the simulation of random fields [212, 363, 469]. In geostatistics, it is known as the *Fourier integral method* [132, 648]. The main idea is that a Gaussian, stationary random field with mean m_{x} and covariance function $C_{\mathrm{xx}}(\mathbf{r})$ can be approximately expressed in terms of the following superposition of N_M *modes*:

$$\mathrm{X}(\mathbf{s};\omega) \approx m_{\mathrm{x}} + \frac{\sigma_{\mathrm{x}}}{\sqrt{N_M}} \sum_{n=1}^{N_M} \left\{ c_n(\omega)\cos\left[\mathbf{k}_n(\omega)\cdot\mathbf{s}\right] + d_n(\omega)\sin\left[\mathbf{k}_n(\omega)\cdot\mathbf{s}\right] \right\}. \tag{16.14}$$

The coefficients $c_n(\omega)$ and $d_n(\omega)$ are zero-mean, unit-variance normal random variables, and the wavevectors $\mathbf{k}_n(\omega)$ are random samples from the reciprocal space. The wavevectors are selected so as to ensure a faithful reproduction of the spectral density $\widetilde{C}_{xx}(\mathbf{k})$. The number of modes is arbitrary, but in general $N_M \gg 1$ is necessary for accuracy.

The superposition (16.14) converges to the Gaussian random field $X(\mathbf{s}; \omega)$ with the specified covariance as the number of modes N_M tends to infinity. Approximations with a few thousand modes can accurately reproduce the statistics of $X(\mathbf{s}; \omega)$. The *numerical complexity* of this method is $O(2N_M P)$. Hence, it is computationally more intensive than the FFT-based simulation for the same number of points P if $N_M \gg \log_2 P$. However, the mode superposition does not rely on an underlying lattice symmetry and is thus applicable to any spatial distribution of points.

The elementary steps of spectral simulation with mode superposition are listed in the Algorithm 16.3. More details regarding the implementation of the algorithm's steps are given in the following sections.

Algorithm 16.3 Spectral simulation of zero-mean Gaussian random field $X(\mathbf{s}; \omega)$, where $\mathbf{s} \in \mathbb{R}^d$, with covariance function $C_{xx}(\mathbf{r})$ based on importance sampling of the wavenumbers and the inverse probability transformation

---

1: Calculate the spectral density $\widetilde{C}_{xx}(\mathbf{k})$ that corresponds to the covariance $C_{xx}(\mathbf{r})$.
2: Select a number of wavevector modes N_M ▷ Wavevector generation
3: **if** the spectral density is separable **then**
4:     Generate d wavevector component sequences using inverse transform method (16.19)
5: **else if** the spectral density is non-separable **then**
6:     **if** the spectral density is anisotropic **then** ▷ Isotropy restoration (if needed)
7:         Apply isotropy restoring transformations (see Sect. 3.3.)
8:     **end if**
9:     Generate N_M wavenumbers from the pdf (16.18) using the inverse transform method
10:     Generate N_M random unit wavevectors
11: **end if**
12: Generate $N_M \times 2$ random coefficients from $\mathcal{N}(0, 1)$ ▷ Coefficient generation
13: Construct the simulated state using the mode superposition (16.14) ▷ Synthesis

---

The core of the Algorithm 16.3 is the sampling of the wavevector space (steps 2–11) in order to select the wavevectors that correspond to the different modes in (16.14).

16.5.1 Simple Sampling Strategies

Various schemes are possible for sampling the d-dimensional mode wavevectors $\{\mathbf{k}_n(\omega)\}_{n=1}^{N_M}$. For example, the wavevectors can be generated by means of a *deterministic procedure*. This is straightforward in the case of lattice simulations where the wavevectors are generated by the reciprocal lattice primitive vectors [648].

More generally, however, the $\mathbf{k}_n(\omega)$ can be selected by *random sampling* of the wavevector space. In the case of isotropic random fields, this approach implies that $\mathbf{k}(\omega) = k(\omega)\,\hat{\mathbf{k}}(\omega)$, where the wavenumber random variable $k(\omega)$ is sampled uniformly in the interval $[0, \infty)$ and the unit vectors $\hat{\mathbf{k}}(\omega)$ from a uniform distribution on the surface of the d-dimensional hypersphere. However, purely random sampling is inefficient because it generates a disproportionately large number of wavenumbers for which the spectral density takes very small values.

16.5.2 Importance Sampling of Wavenumbers

A more efficient sampling strategy than completely random drawing is to preferentially sample wavevectors that correspond to higher spectral density values. This can be accomplished by drawing vectors $\mathbf{k}$ according to the following *wavevector mode pdf*

$$f_{\mathrm{k}}(\mathbf{k}) = \frac{\widetilde{C}_{\mathrm{xx}}(\mathbf{k})}{(2\pi)^d\,\sigma_{\mathrm{x}}^2}, \tag{16.15}$$

where $\widetilde{C}_{\mathrm{xx}}(\mathbf{k})$ is the spectral density of the random field. In light of the variance spectral integral (3.57)

$$\sigma_{\mathrm{x}}^2 = \frac{1}{(2\pi)^d} \int_{\mathbb{R}^d} \widetilde{C}_{\mathrm{xx}}(\|\mathbf{k}\|),$$

it follows that the wavevector mode pdf $f_{\mathrm{k}}(\mathbf{k})$ is indeed properly normalized, i.e.,

$$\int_{\mathbb{R}^d} \mathrm{d}\mathbf{k}\; f_{\mathrm{k}}(\mathbf{k}) = 1.$$

16.5.3 Importance Sampling for Radial Spectral Densities

If the random field $\mathrm{X}(\mathbf{s};\omega)$ is statistically isotropic, the spectral density is a radial function of $\|\mathbf{k}\|$ as discussed in Chap. 3. Hence, we can generate the random wavevectors $\mathbf{k}(\omega)$ using $\mathbf{k}(\omega) = \mathrm{k}(\omega)\,\hat{\mathbf{k}}(\omega)$, where $\hat{\mathbf{k}}(\omega)$ is a unit random vector and $\mathrm{k}(\omega)$ is a non-negative random variable that represents the magnitude of $\mathbf{k}(\omega)$. In this approach, the wavenumber and the direction unit vector are separately generated.

Wavenumber pdf The wavenumber pdf, $p_{\mathrm{k}}(\|\mathbf{k}\|)$, can be obtained by integrating $f_{\mathrm{k}}(\mathbf{k})$ over all possible directions, i.e.,

$$p_{\mathrm{k}}(\|\mathbf{k}\|) = \|\mathbf{k}\|^{d-1} \int_{\mathcal{B}_d} \mathrm{d}\Omega_d(\hat{\mathbf{k}})\; f_{\mathrm{k}}(\mathbf{k}), \tag{16.16}$$

where $\hat{\mathbf{k}}$ is the unit vector and $d\Omega_d(\hat{\mathbf{k}})$ is the solid angle differential subtended by the unit vector $\hat{\mathbf{k}}$ on the surface of the unit sphere. Given the radial spectral density $f_{\mathrm{k}}(\|\mathbf{k}\|)$, the wavenumber pdf is given by

$$\int_{\mathcal{B}_d} d\Omega_d(\hat{\mathbf{k}})\, f_{\mathrm{k}}(\mathbf{k}) = f_{\mathrm{k}}(\|\mathbf{k}\|) \int_{\mathcal{B}_d} d\Omega_d(\hat{\mathbf{k}}) = \mathcal{S}_d\, f_{\mathrm{k}}(\|\mathbf{k}\|),$$

where, according to (3.62b), the surface of the unit sphere in d dimensions is given by $\mathcal{S}_d = \frac{2\pi^{d/2}}{\Gamma(d/2)}$. Hence, in light of (16.16) the wavenumber pdf is given by

$$p_{\mathrm{k}}(\|\mathbf{k}\|) = \frac{2\pi^{d/2}}{\Gamma(d/2)}\, \|\mathbf{k}\|^{d-1}\, f_{\mathrm{k}}(\|\mathbf{k}\|). \tag{16.17}$$

Finally, combining (16.17) with (16.15) the *wavenumber pdf* is given by

$$p_{\mathrm{k}}(\|\mathbf{k}\|) = \frac{2^{1-d}}{\pi^{d/2}\Gamma(d/2)}\, \frac{\|\mathbf{k}\|^{d-1}\, \widetilde{C}_{\mathrm{xx}}(\|\mathbf{k}\|)}{\sigma_{\mathrm{x}}^2}. \tag{16.18}$$

Sampling the wavenumber pdf The sampling of arbitrary pdfs $p_k(\|\mathbf{k}\|)$ can be accomplished using the *inverse transform sampling method*, also known as the *transformation method* [673, Chap. 7].

The inverse transform method assumes that there is a *nonlinear, invertible transformation* $g(k) = u$, which maps the random variable $\mathrm{k}(\omega)$ into a random variable $\mathrm{U}(\omega)$ that can be simulated with existing random number generators; e.g., $\mathrm{U}(\omega)$ may follow the uniform or the Gaussian distribution. The existence of the inverse $g^{-1}(\cdot)$ ensures that a random number $k = g^{-1}(u)$ is obtained for each u. The *conservation of probability* under the transformation $g(\cdot)$ implies that the respective cumulative distribution functions are equal, i.e., $F_{\mathrm{k}}(k) = F_{\mathrm{U}}(u)$.

If $\mathrm{U}(\omega) \overset{d}{=} U(0, 1)$, then $F_{\mathrm{U}}(u) = u$. Consequently, in light of the conservation of probability, the value k that corresponds to a specific u is determined by the solution of the nonlinear equation $F_{\mathrm{k}}(k) = u$.

Wavenumber sampling based on inverse transform and uniform random variates

$$\|\mathbf{k}\| = k : \; F_{\mathrm{k}}(k) := \int_0^k dk'\, p_{\mathrm{k}}(k') = u, \;\text{ where } u \overset{d}{=} U(0, 1). \tag{16.19}$$

For an example of random numbers generated from the asymmetric, heavy-tailed Kaniadakis-Weibull (also κ-Weibull) distribution see [376]. Inverse transformations for some commonly used spectral densities are presented in Sect. 16.5.4 below.

16.5.4 Importance Sampling for Specific Spectral Densities

In this section, we illustrate the importance sampling of the wavevector space for three radial covariance functions (i.e., exponential, Spartan and Whittle-Matérn models) and for two separable anisotropic covariances (Gaussian and cardinal sine models). For non-separable covariance functions it may be preferable to use a transformed coordinate system in which the covariance becomes a radial function (see Sect. 4.3). After the isotropy-restoring transformation, importance sampling of the wavenumbers and uniform sampling of the unit wavevectors can be applied.

Exponential Covariance The spectral density of the isotropic exponential covariance $C_{xx}(\mathbf{r}) = \sigma_x^2 \exp(-\|\mathbf{r}\|/\xi)$ is given in Table 4.4. According to (16.18) the wavenumber pdf, $p_k(\|\mathbf{k}\|)$, in d dimensions is given by

$$p_k(\|\mathbf{k}\|) = \frac{2}{\sqrt{\pi}} \frac{\Gamma\left(\frac{d+1}{2}\right)}{\Gamma\left(\frac{d}{2}\right)} \frac{\|\mathbf{k}\|^{d-1}}{\left(1+\xi^2\|\mathbf{k}\|^2\right)^{(d+1)/2}}. \tag{16.20}$$

The pdf of the dimensionless wavenumber $\kappa = \xi\,\|\mathbf{k}\|$ is $p_\kappa(\kappa) = \xi\; p_k(\|\mathbf{k}\|)$. The cumulative wavenumber distribution function is derived by replacing $dk p_k(k)$ with $d(\kappa/\xi) p_\kappa(\kappa)$ which leads to

$$F_\kappa(\|\mathbf{k}\|\xi) = \int_0^{\|\mathbf{k}\|\xi} d\kappa p_\kappa(\kappa) = \frac{2}{\sqrt{\pi}} \frac{\Gamma\left(\frac{d+1}{2}\right)}{\Gamma\left(\frac{d}{2}\right)} \int_0^{\|\mathbf{k}\|\xi} d\kappa \, \frac{\kappa^{d-1}}{(1+\kappa^2)^{(d+1)/2}}.$$

Then, if u is a uniform random number from the distribution $U(0, 1)$, based on the inverse transform sampling (16.19), the wavenumber $\|\mathbf{k}\|$ is obtained from u by solving the following integral equation:

$$\int_0^{\|\mathbf{k}\|\xi} d\kappa \, \frac{\kappa^{d-1}}{(1+\kappa^2)^{(d+1)/2}} = u\, \frac{\sqrt{\pi}}{2} \frac{\Gamma\left(\frac{d}{2}\right)}{\Gamma\left(\frac{d+1}{2}\right)}. \tag{16.21}$$

Example 16.2 Derive the inverse transformation that determines the wavenumbers $\|\mathbf{k}\|$ for the *exponential covariance* based on the uniform variates u for $d = 2$.

Answer For $d = 2$ it holds that $\Gamma(3/2) = \sqrt{\pi}/2$. Then, the integral equation (16.21) that generates the modes of the exponential spectral density from uniform random numbers becomes

$$u = \int_0^{\|\mathbf{k}\|\xi} \mathrm{d}\kappa \, \frac{\kappa}{(1+\kappa^2)^{3/2}} = \frac{\sqrt{1+\|\mathbf{k}\|^2\xi^2} - 1}{\sqrt{1+\|\mathbf{k}\|^2\xi^2}}.$$

The solution of the above equation for $\|\mathbf{k}\|\xi$ leads to the following *wavenumber generator* for random fields with exponential covariance in $d = 2$:

$$\text{Exponential: } \|\mathbf{k}\|\xi = \frac{\sqrt{1-(1-u)^2}}{(1-u)}, \text{ for } u \overset{d}{=} U(0, 1). \tag{16.22}$$

Hence, the random variates $\|\mathbf{k}\|$ are obtained from uniform random variates u according to (16.22).

Whittle-Matérn covariance The Whittle-Matérn spectral density is given in Table 4.4. Based on (16.18) the wavenumber pdf $p_{\mathrm{k}}(\|\mathbf{k}\|)$, where $\mathbf{k} \in \mathbb{R}^d$, is given by

$$p_{\mathrm{k}}(\|\mathbf{k}\|) = \frac{\Gamma\left(\nu + \frac{d}{2}\right)}{\Gamma\left(\frac{d}{2}\right)\Gamma(\nu)} \frac{2\|\mathbf{k}\|^{d-1}}{\left(1+\xi^2\|\mathbf{k}\|^2\right)^{\nu+d/2}}. \tag{16.23}$$

The cumulative wavenumber distribution function is accordingly given by

$$F_\kappa(\|\mathbf{k}\|\xi) = \int_0^{\|\mathbf{k}\|\xi} \mathrm{d}\kappa p_\kappa(\kappa) = \frac{\Gamma\left(\nu + \frac{d}{2}\right)}{\Gamma\left(\frac{d}{2}\right)\Gamma(\nu)} \int_0^{\|\mathbf{k}\|\xi} \mathrm{d}\kappa \, \frac{2\kappa^{d-1}}{(1+\kappa^2)^{\nu+d/2}}.$$

Example 16.3 For the *Whittle-Matérn* covariance in $d = 2$, derive the inverse sampling transformation for the wavenumbers $\|\mathbf{k}\|$ based on uniform variates u.

Answer In $d = 2$ the wavenumber cdf is given by the following integral

$$F_\kappa(\|\mathbf{k}\|\xi) = \int_0^{\|\mathbf{k}\|\xi} \mathrm{d}\kappa \, \frac{2\kappa\,\nu}{(1+\kappa^2)^{\nu+1}}. \tag{16.24}$$

Using the variable changes $\kappa^2 \mapsto x$ and $1 + x \mapsto z$ the cdf integral becomes

$$F_\kappa(\|\mathbf{k}\|\xi) = \int_0^{\|\mathbf{k}\|^2\xi^2} \frac{\mathrm{d}x\,\nu}{(1+x)^{\nu+1}} = \int_1^{1+\|\mathbf{k}\|^2\xi^2} \frac{\mathrm{d}z\,\nu}{z^{\nu+1}} = 1 - \frac{1}{\left(1+\|\mathbf{k}\|^2\xi^2\right)^{\nu}}.$$

The inverse sampling transform (16.19) $F_\kappa(\|\mathbf{k}\|\xi) = u$ leads to the following *wavenumber generator* for random fields with Whittle-Matérn covariance in $d = 2$:

$$\text{Whittle-Matérn:} \quad \|\mathbf{k}\|\xi = 1 - \left(\frac{1}{1-u}\right)^{1/\nu}, \quad \text{for } u \stackrel{d}{=} U(0,1). \qquad (16.25)$$

SSRF Covariance Next, we focus on the mode wavenumber distribution for the Spartan spectral density given by (7.16). In light of (16.15), the pdf of the wavevector modes is given by

$$f_{\mathrm{k}}(\|\mathbf{k}\|) = \frac{\eta_0\, \xi^d}{(2\pi)^d\, \sigma_{\mathrm{x}}^2 \left[1 + \eta_1\, (\|\mathbf{k}\|\, \xi)^2 + (\|\mathbf{k}\|\xi)^4\right]}. \qquad (16.26)$$

Using (16.17) the following pdf is obtained for the *SSRF mode wavenumbers*:

$$\begin{aligned} p_{\mathrm{k}}(\|\mathbf{k}\|) =& \frac{A_d}{\sigma_{\mathrm{x}}^2} \frac{\eta_0\, \xi^d\, \|\mathbf{k}\|^{d-1}}{1 + \eta_1\, (\|\mathbf{k}\|\xi)^2 + (\|\mathbf{k}\|\xi)^4}, \\ A_d =& \frac{1}{2^{d-1} \pi^{d/2}\, \Gamma(d/2)}. \end{aligned} \qquad (16.27)$$

The conservation of probability under the transformation to uniform variates, i.e., $F_{\mathrm{U}}(u) = F_\kappa(\kappa)$, is expressed as follows

$$u = \frac{\eta_0\, A_d}{\sigma_{\mathrm{x}}^2} \int_0^{\|\mathbf{k}\|\xi} \mathrm{d}\kappa \frac{\kappa^{d-1}}{1 + \eta_1\, \kappa^2 + \kappa^4}. \qquad (16.28)$$

The above integral depends on the spatial dimension d. Below we present explicit expressions for the wavenumber variates in $d = 2$ based on [363].

Two spatial dimensions In $d = 2$, the integral (16.28) is expressed by means of the change of variable $\|\mathbf{k}\|^2 \mapsto z$ as follows

$$u = \frac{\eta_0}{4\pi\, \sigma_{\mathrm{x}}^2} \int_0^{\|\mathbf{k}\|^2 \xi^2} \mathrm{d}z \frac{1}{1 + \eta_1\, z + z^2}.$$

To evaluate explicitly the above integral we distinguish between three cases, depending on the value of η_1, as follows:

1. For $\boxed{\eta_1 = 2}$ the relation for the conservation of probability under transformation to the uniform law (16.28) leads to

$$u = \frac{\eta_0 \|\mathbf{k}\|^2\, \xi^2}{4\pi\, \sigma_{\mathrm{x}}^2 \left(1 + \|\mathbf{k}\|^2\, \xi^2\right)}.$$

Using (7.38b) for the SSRF variance this leads to the following wavenumber cdf

$$u = F_{\mathrm{k}}(\|\mathbf{k}\|) = \frac{\|\mathbf{k}\|^2\,\xi^2}{1 + \|\mathbf{k}\|^2\,\xi^2}. \tag{16.29}$$

The mode wavenumber variates are then obtained from uniform variates as follows

SSRF with $\eta_1 = 2$:

$$\|\mathbf{k}\|\,\xi = \sqrt{\frac{u}{1-u}}, \quad \text{where } u \overset{d}{=} U(0, 1). \tag{16.30}$$

2. For $\boxed{\eta_1 \neq 2}$ the conservation of probability integral (16.28) leads to

$$u = \frac{\eta_0}{4\pi\,\sigma_{\mathrm{x}}^2} \int_0^{\|\mathbf{k}\|^2\,\xi^2} \frac{\mathrm{d}z}{1 + \eta_1\,z + z^2} = \frac{\eta_0}{2\pi\,\sigma_{\mathrm{x}}^2\sqrt{\eta_1{}^2 - 4}} \tag{16.31}$$

$$\times \left[\tanh^{-1}\left(\frac{\eta_1}{\sqrt{\eta_1{}^2 - 4}}\right) - \tanh^{-1}\left(\frac{2\|\mathbf{k}\|^2\xi^2 + \eta_1}{\sqrt{\eta_1{}^2 - 4}}\right)\right]. \tag{16.32}$$

To solve this transcendental equation for the $\|\mathbf{k}\|$ variates, we distinguish two different regimes depending on the value of the discriminant $\Delta = \sqrt{\eta_1{}^2 - 4}$.

a. If Δ is real, i.e., for $\boxed{\eta_1 > 2}$, we use the *hyperbolic tangent identity*

$$\tanh^{-1}(\phi) = \frac{1}{2}\ln\left(\frac{1+\phi}{1-\phi}\right),$$

in light of which the transcendental equation becomes

$$u = \frac{\eta_0}{4\pi\,\sigma_{\mathrm{x}}^2\,\Delta}\ln\left[\left(\frac{\Delta + \eta_1}{\Delta - \eta_1}\right)\left(\frac{\Delta - \eta_1 - 2\|\mathbf{k}\|^2\,\xi^2}{\Delta + \eta_1 + 2\|\mathbf{k}\|^2\xi^2}\right)\right].$$

Based on (7.38c) for the SSRF variance we derive the following relation for the Spartan wavenumber cdf

$$u = F_{\mathrm{k}}(\|\mathbf{k}\|) = \ln\left[\left(\frac{\eta_1 + \Delta}{\eta_1 - \Delta}\right)\left(\frac{\eta_1 + 2\|\mathbf{k}\|^2\xi^2 - \Delta}{\eta_1 + 2\|\mathbf{k}\|^2\xi^2 + \Delta}\right)\right]\left[\ln\left(\frac{\eta_1 + \Delta}{\eta_1 - \Delta}\right)\right]^{-1}. \tag{16.33}$$

This equation is inverted by moving the logarithmic function of the denominator to the left-hand side, taking the exponential on both sides, and collecting all factors that depend on $(\eta_1 + \Delta)/(\eta_1 - \Delta)$ on the same side, side leading to

$$\frac{2 + (\eta_1 + \Delta)\|\mathbf{k}\|^2\xi^2}{2 + (\eta_1 - \Delta)\|\mathbf{k}\|^2\xi^2} = \left(\frac{\eta_1 + \Delta}{\eta_1 - \Delta}\right)^{1-u}.$$

Finally, the SSRF wavenumber variates are obtained from the respective uniform variates by means of the following equation

SSRF with $\eta_1 > 2$:

$$\|\mathbf{k}\|\,\xi = \left[\frac{2\,[\varphi(\eta_1; u) - 1]}{\Delta\,[\,\varphi(\eta_1; u) + 1\,] - \eta_1\,[\,\varphi(\eta_1; u) - 1\,]}\right]^{1/2}, \tag{16.34}$$

$$\varphi(\eta_1; u) = \left(\frac{\eta_1 + \Delta}{\eta_1 - \Delta}\right)^{1-u}, \quad \text{where } u \overset{d}{=} U(0, 1).$$

b. If Δ is imaginary, i.e., $\boxed{-2 < \eta_1 < 2}$, we define $\Delta = \mathrm{i}\,\Delta'$ where $\Delta' = \sqrt{4 - {\eta_1}^2} \in \mathbb{R}$. Using the identity $\tanh^{-1}(z) = -\mathrm{i}\tan^{-1}(\mathrm{i}z)$ which holds for complex numbers z, the transcendental equation (16.31) becomes

$$u = \frac{\eta_0}{2\pi\sigma_{\mathrm{x}}^2\,\Delta'}\left[\tan^{-1}\left(\frac{2\|\mathbf{k}\|^2\,\xi^2 + \eta_1}{\Delta'}\right) - \tan^{-1}\left(\frac{\eta_1}{\Delta'}\right)\right].$$

Using the equation (7.38a) for the SSRF variance, the transcendental equation expressing the conservation of probability simplifies into

$$u = F_{\mathrm{k}}(\|\mathbf{k}\|) = \frac{\tan^{-1}\left(\frac{2\,\|\mathbf{k}\|^2\,\xi^2 + \eta_1}{\Delta'}\right) - \left(\frac{\eta_1}{\Delta'}\right)}{\frac{\pi}{2} - \tan^{-1}\left(\frac{\eta_1}{\Delta'}\right)}. \tag{16.35}$$

Finally, by inverting the above equation with respect to $\|\mathbf{k}\|\,\xi$ we obtain the following equation for the Spartan wavenumber variates

SSRF with $|\eta_1| < 2$:

$$\|\mathbf{k}\|\,\xi = \left[\frac{\Delta' \tan \psi(\eta_1; u) - \eta_1}{2}\right]^{1/2}, \text{ where } \Delta' = \sqrt{4 - \eta_1{}^2},$$

$$\psi(\eta_1; u) = \frac{u\,\pi}{2} + (1-u)\,\tan^{-1}\left(\frac{\eta_1}{\Delta'}\right), \text{ where } u \stackrel{d}{=} U(0,1). \tag{16.36}$$

Reality check Does the inverse transformation method capture the SSRF spectral density? In order to test the ability of the inverse transformation method to generate wavenumbers from the Spartan wavenumber pdf, we compare the histograms of sampled wavenumbers with respective theoretical expressions in Fig. 16.6.

The sampled wavenumbers are obtained via the inverse transformation of $N_0 = 100{,}000$ uniform random numbers from the $U(0, 1)$ distribution. We use three different rigidity values: $\eta_1 = 2, 10, -1.5$ which correspond to the three different branches of the Spartan covariance function. The inverse transformation is respectively given by (16.30), (16.34), and (16.36).

We use histograms with $N_b = 15$ bins of uniform width distributed over the interval [0, 2] for $\eta_1 = 2, 10$ and [0, 1] for $\eta_1 = -1.5$. The bin centers are located at the wavenumbers $\{k_{b,i}\}_{i=1}^{15}$. The interval corresponding to each bin is $[k_{b,i} - \delta k, k_{b,i} + \delta k]$, where δk is the half width of the bins. The theoretical value for the bin height is derived by measuring the frequency of the wavenumbers that should be contained in each bin based on the pdf of the SSRF mode wavenumbers (16.27), i.e.,

$$N_{b,i} = N_0 \int_{k_{b,i}-\delta k}^{k_{b,i}+\delta k} \mathrm{d}k\; p_{\mathrm{k}}(\|\mathbf{k}\|) = N_0\left[F_{\mathrm{k}}(k_{b,i} + \delta k) - F_{\mathrm{k}}(k_{b,i} - \delta k)\right],$$

$$i = 1, \ldots, N_b. \tag{16.37}$$

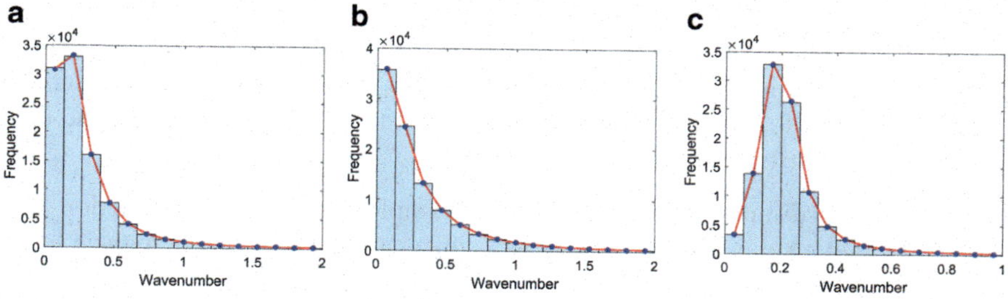

Fig. 16.6 Wavenumber importance sampling for Gaussian random field with Spartan covariance function in $d = 2$. Theoretical distribution of wavenumbers (filled circles) based on (16.37) and histograms of 100,000 sampled wavenumbers generated from $U(0, 1)$ random numbers by means of the inverse transformation method. Three rigidity values (one from each of the respective SSRF rigidity branches) are used. The same characteristic length, $\xi = 5$, is used for all three η_1 values. (**a**) $\eta_1 = 2$; (**b**) $\eta_1 = 10$; (**c**) $\eta_1 = -1.5$. The respective wavenumber cumulative distribution functions F_{k} used in (16.37) are given by (16.29) (**a**), (16.33) (**b**) and (16.35) (**c**)

The mode wavenumber cdf is given by the equations (16.29), (16.33), and (16.35) respectively for each of the three rigidity values. The resulting $N_{b,i}$, $i = 1, \ldots, 15$ are marked by filled circles in Fig. 16.6. Visually, the agreement between the sampled and the theoretically predicted wavenumbers per bin is excellent.

Anisotropic Gaussian covariance The geometrically anisotropic Gaussian covariance function in d dimensions is given by

$$C_{xx}(\mathbf{r}) = \sigma_x^2 \exp\left(-\sum_{i=1}^{d} \frac{r_i^2}{\xi_i^2}\right). \tag{16.38}$$

The spectral density is a simple modification of the isotropic expression given in Table 4.4 that accommodates distinct characteristic lengths ξ_i, $i = 1, \ldots, d$, i.e.,

$$\widetilde{C}_{xx}(\mathbf{k}) = \left(\prod_{i=1}^{d} \xi_i \sqrt{\pi}\right) \exp\left(-\sum_{i=1}^{d} \frac{k_i^2 \xi_i^2}{4}\right). \tag{16.39}$$

Based on the equation for the mode pdf (16.15) and the above spectral density the mode probability density function is given by

$$f_{\mathrm{k}}(\mathbf{k}) = \prod_{i=1}^{d} \left[\left(\frac{\xi_i}{2\sqrt{\pi}}\right) \exp\left(-\frac{k_i^2 \xi_i^2}{4}\right)\right]. \tag{16.40}$$

For the separable, anisotropic Gaussian spectral density each orthogonal wavevector component can be sampled from the zero-mean Gaussian distribution with variance equal to $2/\xi_i^2$, i.e., $k_i(\omega) \overset{d}{=} N(0, 2\xi_i^{-2})$ for $i = 1, \ldots, d$.

Hence, it is possible to simulate the orthogonal wavevector components with Gaussian random number generators. In this case, the separability of the covariance function implies that it is much simpler to sample the wavevector components in d directions than to sample the wavenumber $\|\mathbf{k}\|(\omega)$ and the direction of the unit wavevector $\hat{\mathbf{k}}(\omega)$ (for the latter see Sect. 16.5.5).

Anisotropic cardinal-sine covariance An example of a covariance function that exhibits negative holes is the anisotropic, product cardinal sine function given by

$$C_{xx}(\mathbf{r}) = \sigma_x^2 \prod_{i=1}^{d} \operatorname{sinc}\left(\frac{r_i}{\xi_i}\right).$$

A contour plot of the isotropic cardinal sine covariance is shown in Fig. 4.2.

The mode pdf $f_{\mathrm{k}}(\mathbf{k})$ is given by the following box-like uniform density

$$f_{\mathrm{k}}(\mathbf{k}) = \begin{cases} \frac{1}{2^d\, \xi_1\, \xi_2 \ldots \xi_d}, & \text{for } |k_i| < \xi_i^{-1},\ i = 1, \ldots, d, \\ 0, & \text{otherwise.} \end{cases} \tag{16.41}$$

The separable mode pdf (16.41) is expressed as $f_{\mathrm{k}}(\mathbf{k}) = \prod_{i=1}^{d} f_i(k_i)$, where $f_i(k_i) = 2/\xi_i$ for $|k_i\, \xi_i| \le 1$ and $f_i(k_i) = 0$ otherwise. The conservation of probability under the transformation $\mathbf{k}^\top \mapsto (x_1, \ldots, x_d)$ where $x_i \overset{d}{=} U(0,1)$ are uniform random variates, leads to

$$F_{\mathrm{x}}(x_i) := x_i = \int_0^{k_i} \mathrm{d}k'\, f_i(k'), \text{ for all } i = 1, \ldots, d.$$

The wavenumber distribution can thus be simulated from the d uniform random variates x_i and the equation

$$\text{Anisotropic cardinal sine: } k_i = -\frac{1}{\xi_i} + \frac{2\,x_i}{\xi_i}, \text{ for } i = 1, \ldots, d. \tag{16.42}$$

16.5.5 Sampling the Surface of the Unit Sphere

The simulation of the mode wavevectors for radial spectral densities involves the simulation of uniformly distributed unit random vectors $\hat{\mathbf{k}} \subset \mathbb{R}^d$, in addition to the simulation of wavenumbers discussed above. This section focuses on the simulation of unit random vectors.

The simulation of random unit vectors in d dimensions requires generating $d-1$ independent random orientation angles that define the respective vector according to (3.6.2). In $d = 2$ it suffices to generate the azimuthal angle ϕ while in $d = 3$ the polar angle θ and the azimuthal angle ϕ are necessary to define unit random vectors (see Fig. 16.7).

Seemingly, the simplest approach for simulating a unit random vector is to draw the orientation angles from respective uniform distributions. However, this choice does not lead to a uniform distribution of the unit vector on the d-dimensional hypersphere $\mathcal{B}_d$.

For example, in $d = 3$ the differential area subtended by the unit vector $\hat{\mathbf{k}}(\theta, \phi)$ on $\mathcal{B}_d$ is $\mathrm{d}A = \sin\theta\, \mathrm{d}\theta \mathrm{d}\phi$. The term $\sin\theta$ implies that the differential surface element $\mathrm{d}A$ tends to be larger near the equator and smaller near the poles. Hence, a uniform distribution of θ and ϕ does not lead to a uniform spacing of the unit vector endpoints on the surface of the sphere. Actually the "random points" will tend to accumulate more near the poles than around the equator (see Fig. 16.8a).

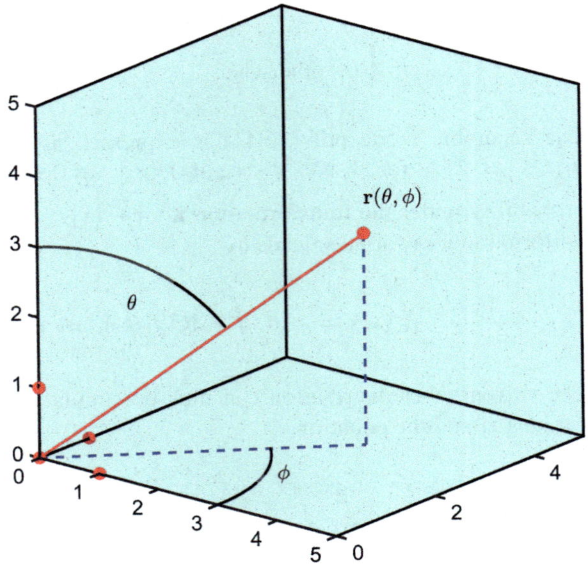

Fig. 16.7 Illustration of the polar (θ) and azimuthal (ϕ) angles in a three-dimensional spherical coordinate system using the standard physics notation

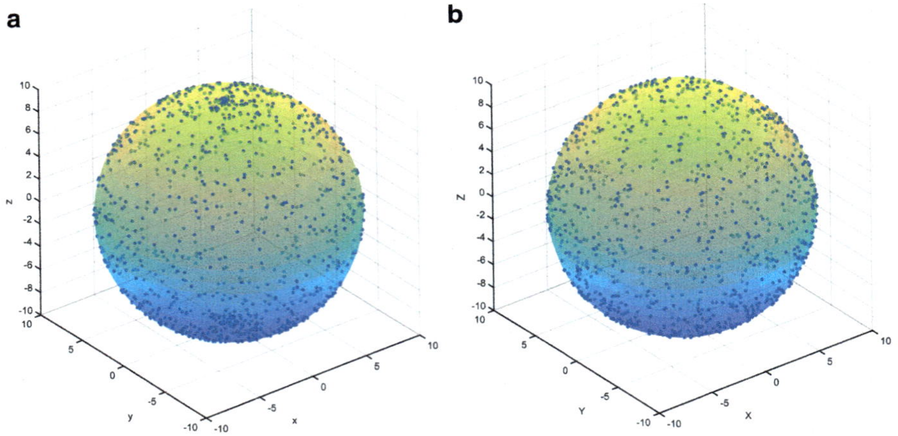

Fig. 16.8 Spatial distribution of 1000 random points on the surface of the unit sphere. (**a**) Points generated by uniform probability distribution of orientation angles. (**b**) Points generated by the Gaussian distribution according to (16.43). Note the concentration of points near the poles and the larger empty spaces near the equator in (**a**) which is in contrast with the more uniform point distribution in (**b**)

A general method for simulating random vectors the end-points of which are uniformly distributed on the random sphere (or the d-dimensional hypersphere, in general), is to simulate d independent random numbers $\{x_i\}_{i=1}^{d}$ from the *standard Gaussian distribution* and then to construct the unit random vector $\hat{\mathbf{k}}$ so that

$$\hat{\mathbf{k}}^{\top} = \left(\frac{x_1}{\sqrt{\sum_{i=1}^{d} x_i^2}}, \ldots, \frac{x_d}{\sqrt{\sum_{i=1}^{d} x_i^2}} \right), \tag{16.43}$$

where the x_i, $i = 1, \ldots, d$ are realizations of the random variables $X_i(\omega) \stackrel{d}{=} \mathcal{N}(0, 1)$.

Faster methods in three and four dimensions were designed by Marsaglia [281, 546, 662]. The difference between points generated by uniformly distributed orientation angles and those generated by normalized random vectors is shown in Fig. 16.8.

16.5.6 Stratified Sampling of Wavenumbers

Different sampling schemes, than the importance sampling methods presented in Sects. 16.5.2, 16.5.3, and 16.5.4, are possible for the wavenumbers $\|\mathbf{k}\|$. These schemes aim to take advantage of different aspects of the probability distribution. One possibility is *stratified sampling combined with Monte Carlo variance reduction techniques* [469]. The main idea in stratified sampling is to partition the wavenumber space into a number of bins of non-uniform width. The bin width is defined so that the integral of the spectral density per bin is approximately the same for all bins. A specified number of wavenumbers per bin are then generated according to a bin-specific pdf (importance sampling) that is based on the spectral density.

An alternative approach for bin design employs bins of the form $[k_n, k_{n+1})$, $n = 1, \ldots, n_{\max}$, where $k_1 = k_{\min}$ is a minimum wavenumber (corresponding to a maximal length scale) and $k_{n+1} = q\, k_n$, where $q > 1$. This sampling strategy is presumably suitable for *multiscale random fields* for which $\widetilde{C}_{\mathrm{xx}}(\|\mathbf{k}\|) = A/\|\mathbf{k}\|^{\alpha}$, where $A > 0$, $\alpha > 0$, for $\|\mathbf{k}\| \geq k_{\mathrm{c}}$ and $\widetilde{C}_{\mathrm{xx}}(\|\mathbf{k}\|) = 0$ for $\|\mathbf{k}\| < k_{\mathrm{c}}$.

16.5.7 Low-Discrepancy Sequences

All the sampling methods discussed so far involve the use of pseudo-random numbers. We have mentioned, however, that the use of uniform random numbers can result in non-uniform sampling of the target domain, as in the case of the surface of the unit sphere discussed above.

Consider the problem of evenly filling the square domain $[0, 1] \times [0, 1]$ in $\mathbb{R}^2$ with N random points (similar considerations hold for a cube in $\mathbb{R}^3$ or a hypercube in higher dimensions). Even filling of the domain implies that the distance between nearest neighbors is more or less uniform. Our first insight would probably be to generate N pairs of random deviates (u_i, v_i), $i = 1, \ldots, N$, from the uniform distribution $U(0, 1)$. However, it turns out that the random points thus generated do not evenly fill the square domain: certain areas contain point clusters, while other areas are sparsely covered (e.g., see Fig. 16.9a). This is perhaps surprising, but uniform random deviates do not evenly cover the square domain.

This shortcoming does not only affect uniform random numbers but also Gaussian deviates, if the latter are generated from the former by means of the *Box-Muller transformations*

$$\begin{aligned} x_1 &= \sqrt{-2\ln u_1}\,\cos 2\pi u_2, \\ x_2 &= \sqrt{-2\ln u_1}\,\sin 2\pi u_2, \end{aligned} \tag{16.44}$$

where u_1 and u_2 are uniform random deviates from the $U(0, 1)$ distribution and x_1, x_2 are the normal random deviates [673].

The deficiency of uniform random numbers to evenly fill domains in $\mathbb{R}^d$ is addressed by the development of *quasi-random numbers*. The latter are produced by means of so-called *low-discrepancy sequences* such as the Sobol, Faure, Niederreiter, and Halton sequences [673]. These sequences can be constructed sequentially, so that more points can be added as needed without necessitating the generation of the entire sequence from scratch. At the large sample limit, low-discrepancy sequences tend to uniformly fill the sampled space.

Low-discrepancy methods tend to evenly fill d-dimensional spaces, because they generate point sequences that avoid clustering [422, Chap. 4.5]. The ability of quasi-random numbers to fill space more evenly is illustrated in Fig. 16.9. The methods that use quasi-random numbers instead of pseudo-random numbers are known as

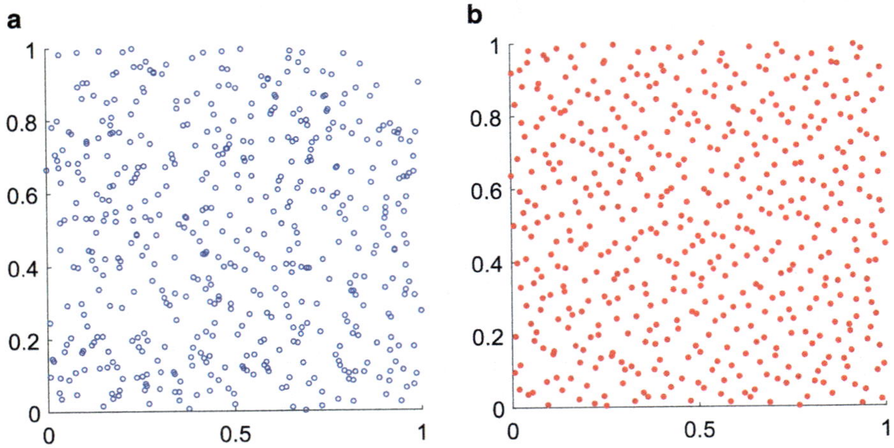

Fig. 16.9 Distribution of 500 random points on the unit square domain. (**a**) Points generated by means of the uniform random number generator. (**b**) Points generated by means of a Halton quasi-random sequence

quasi-Monte Carlo methods. Low-discrepancy sequences are successfully used in Monte Carlo integration of "difficult" multidimensional integrands [756]. Quasi-Monte Carlo methods converge faster than their classical Monte Carlo counterparts that are beset by the slow $O(N^{-1/2})$ convergence rate. The improved convergence rate is a signature of more efficient sampling of the domain of interest by means of quasi-random numbers than by means of pseudo-random numbers.

Quasi-Monte Carlo methods can also be applied to the sampling of the unit sphere. The generation of low-discrepancy point sequences for integration of functions on the unit sphere has been investigated in [98, 545]. Typically, random numbers for spherical integration are obtained by first generating low-discrepancy sequences on the d-dimensional cube $[0, 1]^d$ followed by mapping to the d-dimensional hyper-sphere. This approach, however, does not necessarily lead to optimal point distributions on the hyper-sphere. An improved approach for generating evenly spaced sampling patterns on the sphere by means of the *spherical Fibonacci point sets* was proposed in [545]. Several methods for generating uniformly spaced points on the sphere are compared in the review paper [329].

Quasi-Monte Carlo methods are promising for spatial data simulations. However, to our knowledge at least, with the exception of a few studies [358, 876] they have not been adequately investigated for this purpose.

16.5.8 Spectral Simulation of Spartan Random Fields

We employ the method of randomized spectral sampling described in the preceding sections to generate realizations of Spartan spatial random fields and to test the reproduction of two-point statistics by the generated states.

Sampling design The unit length square $\mathcal{D} = [0, 1] \times [0, 1] \subset \mathbb{R}^2$ is used as the study domain. A zero-mean Gaussian random field with Spartan covariance function is simulated on $\mathcal{D}$. We use importance sampling of the wavenumbers coupled with the inverse transformation method as described in Sect. 16.5.4. A total of $N_M = 10^4$ modes are used. For each mode a respective unit wavevector is generated from two Gaussian random numbers using (16.43) as shown in Sect. 16.5.5. The field is sampled at $N = 400$ randomly distributed locations inside $\mathcal{D}$.

We simulate SSRFs with three different rigidity values, $\eta_1 = -1.9, 2, 10$—one from each of the three rigidity branches. The respective covariance functions are given by (7.37). We use the scale factor $\eta_0 = 10$ for all three covariance functions. The characteristic length is taken as $\xi = 0.1$ for $\eta_1 > 0$, and $\xi = 0.03$ for $\eta_1 < 0$. The smaller length for $\eta_1 = -1.9$ is necessary in order to capture the oscillatory behavior of the Spartan covariance for negative rigidity. We generate $M = 100$ realizations from each random field.

Results For each realization we estimate the empirical variogram using the method of moments (12.10). A total of $N_c = 30$ variogram classes are used for lags between zero and two thirds of the domain length. This means that the variogram lag changes

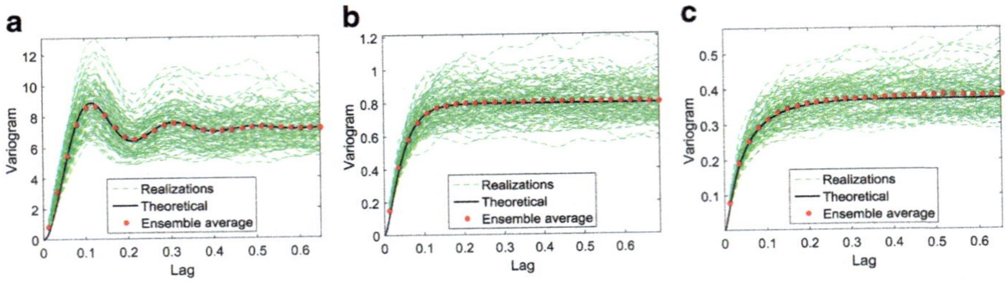

Fig. 16.10 Randomized spectral sampling simulation with wavenumber importance sampling: Variograms of Gaussian random field with Spartan covariance function in $d = 2$ for different values of η_1. Each plot displays the theoretical variogram based on the SSRF parameters used for the simulation (continuous line), the estimated variograms per realization (thin dashed lines) and the average of the estimated variograms (thicker dashed line) based on an ensemble of 100 realizations. Three rigidity values are used: (**a**) $\eta_1 = -1.9$ (**b**) $\eta_1 = 2$ (**c**) $\eta_1 = 10$

by ≈ 0.02 which is sufficient for resolving the SSRF correlations. The empirical variograms, $\hat{\gamma}_{xx}^{(m)}(\mathbf{r}_k)$, where $k = 1, \ldots, N_c$ and $m = 1, \ldots, M$ are plotted in Fig. 16.10 (thin dashed lines) along with the ensemble average (thicker dashed line) which is given by

$$\langle \hat{\gamma}_{xx}(\mathbf{r}_k) \rangle = \frac{1}{M} \sum_{m=1}^{M} \hat{\gamma}_{xx}^{(m)}(\mathbf{r}_k).$$

For comparison, the theoretical SSRF variogram functions that correspond to the input parameters η_0, η_1, ξ (thick continuous lines) are also shown. The theoretical variogram is given by $\gamma_{xx}(\mathbf{r}) = \sigma_x^2 - C_{xx}(\mathbf{r})$, where $C_{xx}(\mathbf{r})$ is defined by (7.37) and the variance by (7.38). As evidenced in the plots of Fig. 16.10, the ensemble averages provide excellent approximations of the theoretical variograms. At the same time, the empirical variograms of individual realizations show considerable scatter around the theoretical functions. The difference in the variogram sills among the three different η_1 is due to the dependence of the SSRF variance on both η_0 and η_1.

16.6 Conditional Simulation Based on Polarization Method

Conditional simulation generates random fields that are constrained by (i) the data values at the sampling points (ii) the general statistical constraints of the inferred probability distribution and (iii) the spatial continuity features (as embodied in the variogram function) inferred from the data.

The fact that conditional simulation fully respects the observations could be undesirable if the data are contaminated by noise or measurement errors. If such problems are expected, it is advisable to filter the data in order to restore a higher degree of smoothness and thus to improve the quality of the constraints.

One of the possible approaches for conditional simulation combines *unconditional realizations* of the random field with a *polarizing (correction) field* that enforces the constraints. The polarizing field is generated from the data and the unconditional realizations by means of an interpolation method such as kriging [200, 232, 419, 624, 626].

The polarization-based conditional simulation method involves the steps outlined below. Each simulated state (realization) is indexed by means of the simulation index $m = 1, \ldots, M$.

1. The data $\mathbf{x}^*$ sampled on $\mathbb{S}_N$ are interpolated using kriging at the nodes $\{\mathbf{z}_p\}_{p=1}^{P}$ of the prediction grid $\mathcal{G}$, thus generating the interpolated vector $\hat{\mathbf{x}}$. This interpolated vector remains fixed for different realizations generated by the conditional simulation.
2. An unconditional realization, $\mathbf{x}^{u,(m)}$, is generated at the set of points $\mathbb{S}_N \cup \mathcal{G}$ that includes both the prediction grid and the sample (data) points.
3. The values of the unconditional realization at *the sampling points* are interpolated on the *prediction grid nodes* thus generating the vector $\hat{\mathbf{x}}^{u,(m)}$.
4. Finally, the m-th (where $m = 1, \ldots, M$) conditional state $\mathbf{x}^{c,(m)}$ at the point set $\mathbb{S}_N \cup \mathcal{G}$ is given by

$$\mathbf{x}^{c,(m)} = \mathbf{x}^{u,(m)} + \left(\hat{\mathbf{x}} - \hat{\mathbf{x}}^{u,(m)}\right), \quad m = 1, \ldots, M, \tag{16.45}$$

   where $\mathbf{x}^{u,(m)}$ is the unconditional realization, $\hat{\mathbf{x}}$ is the vector comprising the data and their kriging interpolation on the grid, and $\hat{\mathbf{x}}^{u,(m)}$ is the vector that comprises the unconditional realization at the sampling points and their kriging interpolation on the grid points.
5. The *polarizing field* is represented by the difference $\hat{\mathbf{x}} - \hat{\mathbf{x}}^{u,(m)}$. At the locations of the sampling points this vector comprises the differences between the data and the unconditional realization values. At the locations of the prediction grid the polarizing field comprises the difference between the interpolated data and the interpolated unconditional values.

All the components in (16.45) are vectors of dimension $(N+P)\times 1$. In particular, they comprise the following elements

$$\mathbf{x}^{u,(m)} = \begin{bmatrix} x_1^{u,(m)} \\ \vdots \\ x_N^{u,(m)} \\ \hline x_{N+1}^{u,(m)} \\ \vdots \\ x_{N+P}^{u,(m)} \end{bmatrix}, \quad \hat{\mathbf{x}} = \begin{bmatrix} x_1^{*} \\ \vdots \\ x_N^{*} \\ \hline \hat{x}_{N+1} \\ \vdots \\ \hat{x}_{N+P}^{u,(m)} \end{bmatrix}, \quad \hat{\mathbf{x}}^{u,(m)} = \begin{bmatrix} x_1^{u,(m)} \\ \vdots \\ x_N^{u,(m)} \\ \hline \hat{x}_{N+1}^{u,(m)} \\ \vdots \\ \hat{x}_{N+P}^{u,(m)} \end{bmatrix}. \tag{16.46}$$

The following remarks hold with respect to the vectors defined in (16.46):

- The entries above the horizontal lines in $\mathbf{x}^{u,(m)}$, $\hat{\mathbf{x}}$, and $\hat{\mathbf{x}}^{u,(m)}$, represent values at the sampling locations, while the entries below the horizontal lines represent values at the nodes of the prediction grid.
- The first N elements of the kriged vector $\hat{\mathbf{x}}$ comprise the data vector $\mathbf{x}^{*}$. The vector $\hat{\mathbf{x}}$ is evaluated only once and is then used for all the simulated states.

Typically, the generation of the polarizing field that is used in the conditioning process is performed by means of simple kriging interpolation. It is straightforward to show, based on (16.45) and (16.46), that the values of the conditional states at the sampling locations coincide with the data. A recent study investigates conditioning with ordinary kriging (OK). The use of OK interpolation extends the scope of the polarization-based simulation method to *intrinsic random fields of order zero* (IRF-0) [232].

Unconditional realizations The unconditional realizations are generated at the data locations in addition to the nodes of the simulation lattice. Since this spatial configuration does not typically represent a regular grid structure, ideally a mesh-free simulation method, such as randomized spectral sampling (discussed in Sect. 16.5), covariance matrix factorization, or *turning bands* [233] should be used.

Alternatively, one can use a fast, grid-based simulation method, such as the spectral FFT method, to generate the unconditional realization at the lattice nodes. Then, each sampling point is assigned an approximate value which is equal to that of the nearest lattice node. This approximation allows taking advantage of the computational efficiency of the FFT method. The quality of the approximation improves as the simulation grid becomes denser, since nearest neighbors located very close to the sampling points can be found. This approach becomes exact if the data points coincide with nodes of the simulation grid.

Computational complexity The numerical bottleneck in the application of (16.45) is the kriging step, because the latter has a dominant computational complexity of $O(2\,N^3 + 2\,P\,N^2)$ *per simulated state*. This is due to the inversion of the covariance matrix and the solution of the kriging system. The inversion of the sample covariance matrix is carried out once, while the kriging system is solved for every prediction point.[3] In practice, the computational disadvantage of kriging can be mitigated by using a finite *search neighborhood* as discussed in Chap. 10.

Exampless Figure 16.11 shows 12 conditional realizations of a Gaussian random field $\mathbf{X}(\mathbf{s};\omega)$ with mean $m_x = 4$ and Spartan covariance function (7.37) with parameters $\eta_0 = 105, \eta_1 = 5, \xi = 3$. The field is simulated on a square lattice with 100 nodes per side. A total of 150 points are sampled from $\mathbf{X}(\mathbf{s};\omega)$ and used to condition the realizations by means of the polarization approach described above.

The unconditional realizations used to generate the conditional states are shown in Fig. 16.12. The smoothed interpolated map of the sample data is shown in Fig. 16.13. The map is generated by means of ordinary kriging using the theoretical covariance parameters of the Spartan function. Note that the unconditional realizations (Fig. 16.12) exhibit patterns of maxima and minima that move around in the domain from realization to realization, while the conditional states (Fig. 16.11) display persistent spatial patterns imparted by the influence of the conditioning points.

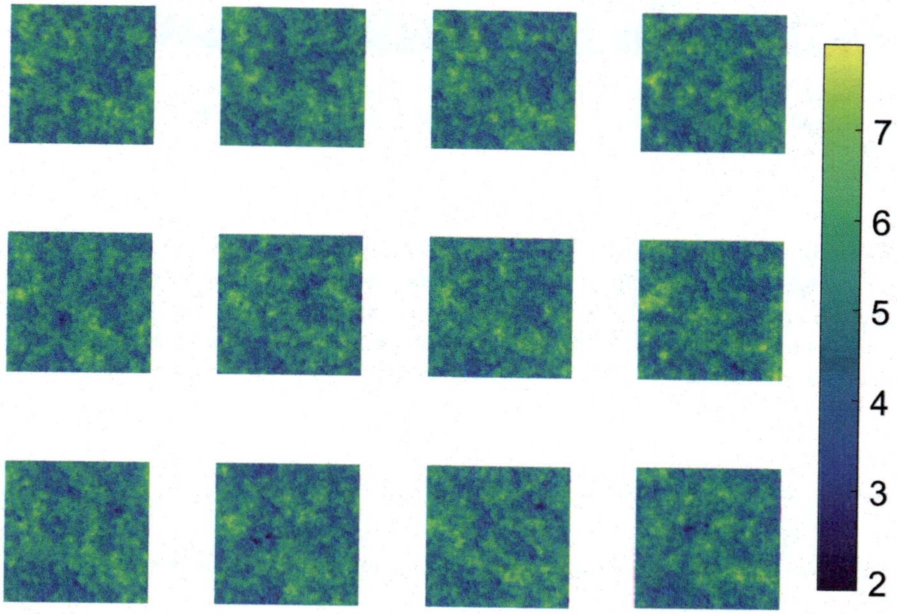

Fig. 16.11 Twelve conditional simulated realizations from a Gaussian random field with mean $m_x = 5$ and Spartan covariance function with $\eta_0 = 105, \eta_1 = 5, \xi = 3$. The field is simulated on a 100×100 square grid. We use 150 conditioning points (generated from same field) that are randomly distributed over the domain. The respective unconditional simulations are shown in Fig. 16.12. The kriged values of the unconditional simulations are generated using the theoretical covariance

[3]The factor 2 is due to the fact that kriging is applied twice, that is both to the data and the unconditional realization values at the sampling points.

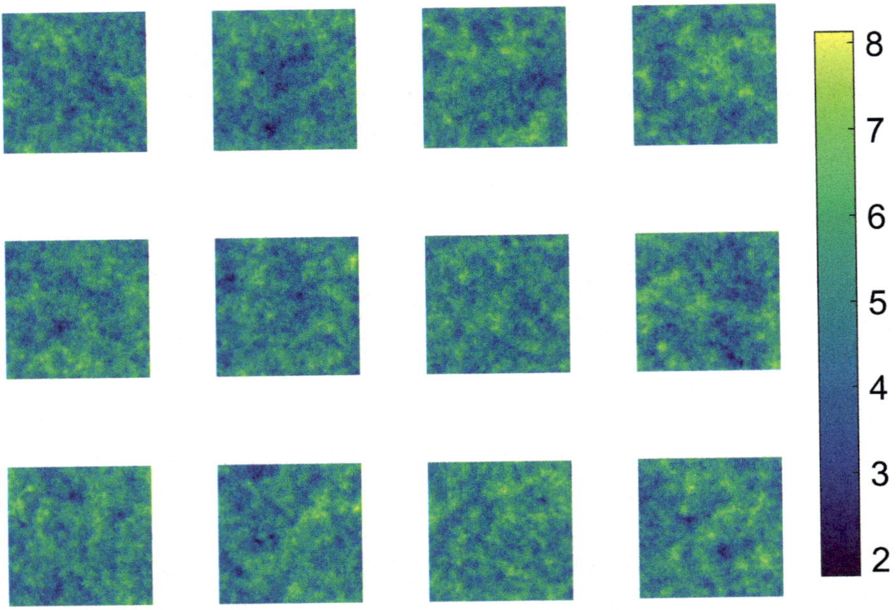

Fig. 16.12 Twelve unconditional simulated realizations from a Gaussian random field with mean $m_x = 5$ and Spartan covariance function with $\eta_0 = 105$, $\eta_1 = 5$, $\xi = 3$. The field is simulated on a 100×100 square grid. The simulated states are generated by means of the spectral FFT simulation method

Ensemble statistical properties Let us denote the conditionally simulated field by $X^{(c)}(\mathbf{s}; \omega)$, the kriged field by $\hat{X}(\mathbf{s}; \omega)$, the unconditional realization by $X^{(u)}(\mathbf{s}; \omega)$, and the kriged unconditional realization by $\hat{X}^{u}(\mathbf{s}; \omega)$. We view the kriged field as a random field, because different realizations of the sample data lead to different interpolated realizations. Then, the equation (16.45) for the realizations is expressed as follows for the conditional random field

$$X^{(c)}(\mathbf{s}; \omega) = X^{(u)}(\mathbf{s}; \omega) + \left[\hat{X}(\mathbf{s}; \omega) - \hat{X}^{u}(\mathbf{s}; \omega)\right]. \tag{16.47a}$$

The above can also be expressed by means of interpolation and simulation-based components as follows

$$X^{(c)}(\mathbf{s}; \omega) = \hat{X}(\mathbf{s}; \omega) + \left[X^{(u)}(\mathbf{s}; \omega) - \hat{X}^{u}(\mathbf{s}; \omega)\right] = \hat{X}(\mathbf{s}; \omega) + \epsilon(\mathbf{s}; \omega), \tag{16.47b}$$

where $\epsilon(\mathbf{s}; \omega)$ is the *estimation error* for the unconditional realizations. Equation (16.47a) is expressed in terms of the *polarizing field* $\hat{X}(\mathbf{s}; \omega) - \hat{X}^{u}(\mathbf{s}; \omega)$ while it counterpart (16.47b) in terms of the *estimation error field* $X^{(u)}(\mathbf{s}; \omega) - \hat{X}^{u}(\mathbf{s}; \omega)$.

Data conditioning Based on (16.45) and (16.46), it is straightforward to verify that at the sampling locations the simulated state respects the data values, i.e., $X^{(c)}(\mathbf{s}_n; \omega) = X^*_n(\omega)$, for $n = 1, \ldots, N$.

Note that we allow the *data vector* $\mathbf{X}^*(\omega) := \{X^*_n(\omega)\}_{n=1}^N$ to take different values between different realizations (denoted by the index ω).

Probability law By construction, if the unconditionally simulated states follow the data probability distribution, then so do the conditional simulations.

Covariance It can be shown that the covariance function of the conditional simulations is identical to the covariance function of the original field, e.g. [624, p. 161]. In the proof it is crucial to take into account the "freedom" of the data vector to vary respecting the probability distribution of the original field; that is, we do not restrict the calculation to a "frozen" data vector.

Conditional mean In calculating the conditional mean (and the variance below) the data vector is considered to be "frozen" at the observed values. It then follows from (16.47a) that

$$\mathcal{E}[X^{(c)}(\mathbf{s}; \omega) \mid \mathbf{X}^*(\omega) = \mathbf{x}^*] = \hat{x}(\mathbf{s}),$$

where $\hat{x}(\mathbf{s})$ is the function (response surface) generated by kriging the data $\mathbf{x}^*$. Figure 16.14 shows the average of the 12 conditional states realizations. Comparing this ensemble average with the kriged state shown in Fig. 16.13, we observe a close similarity between the two spatial patterns. However, the kriged state in Fig. 16.13 displays smoother variations. This is not surprising, however, since the ensemble

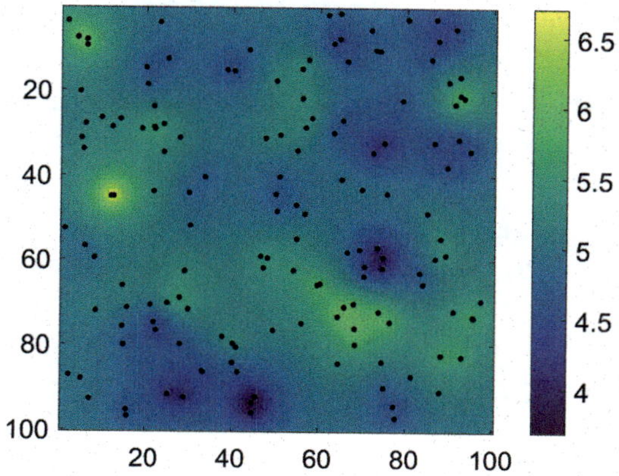

Fig. 16.13 Interpolated map of sample data generated from the Gaussian random field with mean $m_x = 5$ and Spartan covariance function with $\eta_0 = 105$, $\eta_1 = 5$, $\xi = 3$. One hundred and fifty conditioning points are randomly selected. Conditional data from the same field are simulated at these points using the method of covariance matrix factorization. Ordinary kriging with the theoretical Spartan field parameters is used for the spatial interpolation

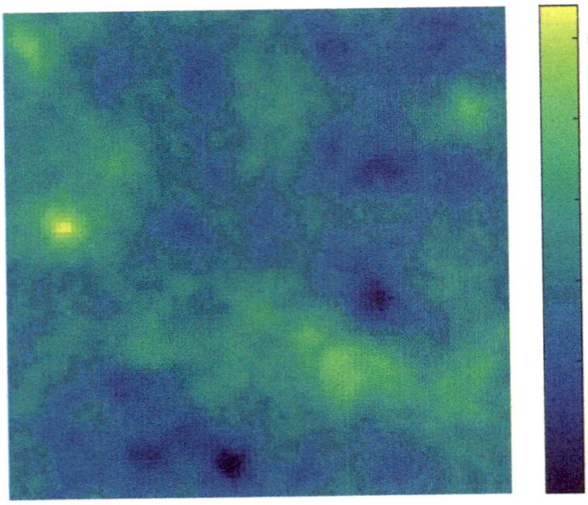

Fig. 16.14 Ensemble mean based on thirty conditional Gaussian random field realizations shown in Fig. 16.11

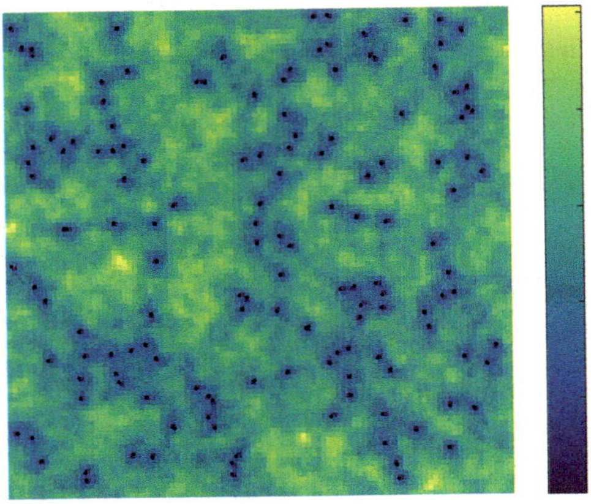

Fig. 16.15 Ensemble standard deviation based on thirty conditional Gaussian random field realizations shown in Fig. 16.11. The sampling sites are marked by black dots

average involves only 12 simulated states. Increasing the number of realizations will further smoothen the ensemble average and lead to better agreement with the kriged state.

Conditional variance It also follows from (16.47b) and the definition of the kriging variance, i.e., $\sigma^2_{\mathrm{kr}}(\mathbf{s}) = \mathcal{E}[\epsilon^2(\mathbf{s};\omega)]$, that the conditional variance constrained by the data is given by

$$\mathbb{V}\mathrm{ar}\left\{\mathrm{X}^{(c)}(\mathbf{s};\omega) \mid \mathbf{X}^*(\omega) = \mathbf{x}^*\right\} = \mathbb{V}\mathrm{ar}\left\{\mathrm{X}^{(u)}(\mathbf{s};\omega) - \hat{\mathrm{X}}^u(\mathbf{s};\omega)\right\} = \sigma^2_{\mathrm{kr}}(\mathbf{s}).$$

The *conditional variance* is equal to the kriging variance, and therefore it is lower than the (unconstrained) variance of the original field. A map of the standard deviation calculated at each prediction grid location over the 12 conditional realizations is shown in Fig. 16.15.

16.7 Conditional Simulation Based on Covariance Matrix Factorization

The matrix factorization method presented in Sect. 16.2 can be extended to conditional simulation as proposed by Davis [180]. The main difference between the unconditional and the conditional methods is that conditional simulation employs the *conditional covariance matrix*. Recall that the matrix factorization approach is applicable even if the covariance function is non-stationary. Hence, the fact that the conditional covariance matrix is by construction non-stationary does not present a problem.

For simplicity, in the following we assume that the simulated Gaussian random field has zero mean. This restriction is easily lifted by first subtracting from the data the estimated mean, which is added—at the end of the calculation—to the simulated zero-mean random field.

Let us define some relevant notation for the exposition of the covariance matrix factorization approach.

1. $\mathbf{x}^* = (x_1^*, \ldots, x_N^*)^\top$: Data vector at the sampling locations $\{\mathbf{s}_n\}_{n=1}^N$.
2. $\mathbf{X}_{\mathcal{D}}(\omega) = (\mathrm{X}(\mathbf{s}_1;\omega), \ldots, \mathrm{X}(\mathbf{s}_N;\omega))^\top$: Random field vector at the sampling locations.
3. $\mathbf{X}_{\mathcal{G}}(\omega) = (\mathrm{X}(\mathbf{z}_1;\omega), \ldots, \mathrm{X}(\mathbf{z}_P;\omega))^\top$: Random field vector at the locations of the simulation grid.
4. $\mathbf{C}_{\mathcal{D},\mathcal{D}} = \mathcal{E}[\mathbf{X}_{\mathcal{D}}(\omega)\,\mathbf{X}_{\mathcal{D}}^\top(\omega)]$: Covariance matrix between the data locations. The dimension of this matrix is $N \times N$.
5. $\mathbf{C}_{\mathcal{G},\mathcal{G}} = \mathcal{E}[\mathbf{X}_{\mathcal{G}}(\omega)\,\mathbf{X}_{\mathcal{G}}^\top(\omega)]$: Covariance matrix between the simulation grid locations. The dimension of this matrix is $P \times P$.
6. $\mathbf{C}_{\mathcal{G},\mathcal{D}} = \mathcal{E}[\mathbf{X}_{\mathcal{G}}(\omega)\,\mathbf{X}_{\mathcal{D}}^\top(\omega)]$: Covariance matrix between the data and the simulation grid locations. The dimension of this matrix is $P \times N$.

The $P \times P$ *conditional covariance matrix* is then given by

$$\mathbf{C}^c_{\mathcal{G},\mathcal{G}} = \mathbf{C}_{\mathcal{G},\mathcal{G}} - \mathbf{C}_{\mathcal{G},\mathcal{D}}\,\mathbf{C}^{-1}_{\mathcal{D},\mathcal{D}}\,\mathbf{C}^\top_{\mathcal{G},\mathcal{D}}. \tag{16.48}$$

The equation above is equivalent to (6.16) that defines the covariance matrix of the conditional normal pdf.

Covariance matrix factorization The $(P+N)\times(P+N)$ covariance matrix $\mathbf{C}$ that involves both the data locations and the prediction grid sites can be decomposed into an LU form. Since the full covariance is a symmetric and positive definite matrix with non-zero determinant, the transpose of the lower triangular matrix $\mathbf{L}$ is equal to the upper triangular matrix, i.e., $\mathbf{L}^\top = \mathbf{U}$. Hence, the LU factorization can be replaced by *Cholesky decomposition*. The factorization of $\mathbf{C}$ is thus expressed as follows:

$$\mathbf{C} = \mathbf{L}\mathbf{U} = \mathbf{L}\mathbf{L}^\top = \mathbf{U}^\top\mathbf{U}. \tag{16.49}$$

The Cholesky factorization of the covariance matrix $\mathbf{C}$ is equivalently expressed in terms of the following *block-matrix structure*

$$\mathbf{C} := \left[\begin{array}{c|c} \mathbf{C}_{\mathcal{D},\mathcal{D}} & \mathbf{C}_{\mathcal{D},\mathcal{G}} \\ \hline \mathbf{C}_{\mathcal{G},\mathcal{D}} & \mathbf{C}_{\mathcal{G},\mathcal{G}} \end{array}\right] = \left[\begin{array}{c|c} \mathbf{L}_{\mathcal{D},\mathcal{D}} & \mathbf{0}_{N\times P} \\ \hline \mathbf{L}_{\mathcal{G},\mathcal{D}} & \mathbf{L}_{\mathcal{G},\mathcal{G}} \end{array}\right] \left[\begin{array}{c|c} \mathbf{U}_{\mathcal{D},\mathcal{D}} & \mathbf{U}_{\mathcal{D},\mathcal{G}} \\ \hline \mathbf{0}_{P\times N} & \mathbf{U}_{\mathcal{G},\mathcal{G}} \end{array}\right], \tag{16.50}$$

where $\mathbf{0}_{P\times N}$ (respectively, $\mathbf{0}_{N\times P}$) is a $P \times N$ (respectively, $N \times P$) matrix with zero entries. Based on the above decomposition, the block covariance matrices are expressed in terms of their LU components by means of the following equations

$$\mathbf{C}_{\mathcal{D},\mathcal{D}} = \mathbf{L}_{\mathcal{D},\mathcal{D}}\,\mathbf{U}_{\mathcal{D},\mathcal{D}}, \tag{16.51a}$$

$$\mathbf{C}_{\mathcal{G},\mathcal{D}} = \mathbf{L}_{\mathcal{G},\mathcal{D}}\,\mathbf{U}_{\mathcal{D},\mathcal{D}}, \tag{16.51b}$$

$$\mathbf{C}_{\mathcal{D},\mathcal{G}} = \mathbf{L}_{\mathcal{D},\mathcal{D}}\,\mathbf{U}_{\mathcal{D},\mathcal{G}}, \tag{16.51c}$$

$$\mathbf{C}_{\mathcal{G},\mathcal{G}} = \mathbf{L}_{\mathcal{G},\mathcal{D}}\mathbf{U}_{\mathcal{D},\mathcal{G}} + \mathbf{L}_{\mathcal{G},\mathcal{G}}\,\mathbf{U}_{\mathcal{G},\mathcal{G}}. \tag{16.51d}$$

Note that in this decomposition of the covariance matrix the block $\mathbf{C}_{\mathcal{G},\mathcal{G}}$ does not involve only the LU matrices $\mathbf{L}_{\mathcal{G},\mathcal{G}}$ and $\mathbf{U}_{\mathcal{G},\mathcal{G}}$.

Simulated states Finally, the conditional states generated by covariance matrix factorization are given by the following equation

$$\mathbf{X}^{(c)}_{\mathcal{G}}(\omega) = \mathbf{L}_{\mathcal{G},\mathcal{G}}\,\boldsymbol{\epsilon}(\omega) + \mathbf{L}_{\mathcal{G},\mathcal{D}}\,\mathbf{L}^{-1}_{\mathcal{D},\mathcal{D}}\,\mathbf{x}^{*}, \tag{16.52}$$

where $\boldsymbol{\epsilon}(\omega)$ is a $P \times 1$ random noise vector drawn from the standard normal distribution $\mathcal{N}(\mathbf{0}, \mathbf{I}_P)$, where $\mathbf{I}_P$ is the $P \times P$ identity matrix.

As shown below the expectation and the covariance of the random vector $\mathbf{X}^{(c)}_{\mathcal{G}}(\omega)$ generated by (16.52) are given by the mean and the covariance, respectively, of the conditional Gaussian distribution.

1. Let us begin with the conditional expectation $\mathcal{E}[\mathbf{X}^{(c)}_{\mathcal{G}}(\omega)]$. Since $\mathcal{E}[\boldsymbol{\epsilon}(\omega)] = 0$, the expectation of the conditional states is given by

$$\mathcal{E}[\mathbf{X}^{(c)}_{\mathcal{G}}(\omega)] = \mathbf{L}_{\mathcal{G},\mathcal{D}}\,\mathbf{L}^{-1}_{\mathcal{D},\mathcal{D}}\,\mathbf{x}^{*}. \tag{16.53}$$

The conditional mean of the Gaussian distribution is given by (6.15). Taking into account that the unconditional expectation is zero, the conditional mean is given by

$$\mathbf{m}_{\mathcal{G}|\mathcal{D}} = \mathbf{C}_{\mathcal{G},\mathcal{D}}\,\mathbf{C}^{-1}_{\mathcal{D},\mathcal{D}}\,\mathbf{x}^{*}.$$

Based on (16.51a) the LU decomposition of the inverse data covariance matrix is given by

$$\mathbf{C}^{-1}_{\mathcal{D},\mathcal{D}} = \left[\mathbf{L}^{-1}_{\mathcal{D},\mathcal{D}}\right]^{\top}\mathbf{L}^{-1}_{\mathcal{D},\mathcal{D}},$$

while the decomposition of the data-grid covariance matrix $\mathbf{C}_{\mathcal{G},\mathcal{D}}$ based on (16.51b) becomes

$$\mathbf{C}_{\mathcal{G},\mathcal{D}} = \mathbf{L}_{\mathcal{G},\mathcal{D}}\mathbf{U}_{\mathcal{D},\mathcal{D}}.$$

Finally, using the matrix identity $\mathbf{A}^{\top}\mathbf{B}^{\top} = (\mathbf{BA})^{\top}$, we obtain

$$\begin{aligned}
\mathbf{C}_{\mathcal{G},\mathcal{D}}\,\mathbf{C}_{\mathcal{D},\mathcal{D}}^{-1} =& \mathbf{L}_{\mathcal{G},\mathcal{D}}\mathbf{U}_{\mathcal{D},\mathcal{D}}\left[\mathbf{L}_{\mathcal{D},\mathcal{D}}^{-1}\right]^{\top}\,\mathbf{L}_{\mathcal{D},\mathcal{D}}^{-1} \\
&= \mathbf{L}_{\mathcal{G},\mathcal{D}}[\mathbf{L}_{\mathcal{D},\mathcal{D}}]^{\top}\left[\mathbf{L}_{\mathcal{D},\mathcal{D}}^{-1}\right]^{\top}\,\mathbf{L}_{\mathcal{D},\mathcal{D}}^{-1} \\
&= \mathbf{L}_{\mathcal{G},\mathcal{D}}\left[\mathbf{L}_{\mathcal{D},\mathcal{D}}^{-1}\mathbf{L}_{\mathcal{D},\mathcal{D}}\right]^{\top}\,\mathbf{L}_{\mathcal{D},\mathcal{D}}^{-1} \\
&= \mathbf{L}_{\mathcal{G},\mathcal{D}}\,\mathbf{L}_{\mathcal{D},\mathcal{D}}^{-1}.
\end{aligned}$$

The above proves that the expectation $\mathcal{E}[\mathbf{X}_{\mathcal{G}}^{(c)}(\omega)]$ is identical to the conditional mean of the grid random vector, i.e.,

$$\mathcal{E}[\mathbf{X}_{\mathcal{G}}^{(c)}(\omega)] = \mathbf{m}_{\mathcal{G}|\mathcal{D}} = \mathbf{L}_{\mathcal{G},\mathcal{D}}\,\mathbf{L}_{\mathcal{D},\mathcal{D}}^{-1}\,\mathbf{x}^{*}.$$

2. To evaluate the covariance of the grid random vector $\mathbf{X}_{\mathcal{G}}^{(c)}(\omega)$, we first calculate the fluctuation of the grid vector, which is given by $\mathbf{X}_{\mathcal{G}}^{'(c)}(\omega) = \mathbf{X}_{\mathcal{G}}^{(c)}(\omega) - \mathbf{m}_{\mathcal{G}|\mathcal{D}}$. In light of (16.52) and the expression (16.53) for the conditional mean, we obtain

$$\mathbf{X}_{\mathcal{G}}^{'(c)}(\omega) = \mathbf{L}_{\mathcal{G},\mathcal{G}}\,\boldsymbol{\epsilon}(\omega).$$

The covariance matrix is then expressed as the expectation of the product of fluctuations, i.e.,

$$\begin{aligned}
\mathcal{E}\left[\mathbf{X}_{\mathcal{G}}^{'(c)}(\omega)\,\mathbf{X}_{\mathcal{G}}^{'(c)}(\omega)^{\top}\right] =& \mathcal{E}\left[\mathbf{L}_{\mathcal{G},\mathcal{G}}\,\boldsymbol{\epsilon}(\omega)\boldsymbol{\epsilon}^{\top}(\omega)\mathbf{L}_{\mathcal{G},\mathcal{G}}^{\top}\right] \\
=& \mathbf{L}_{\mathcal{G},\mathcal{G}}\,\mathbf{I}_{P}\,\mathbf{L}_{\mathcal{G},\mathcal{G}}^{\top}.
\end{aligned}$$

The above is derived using the orthogonality of the random noise vector $\boldsymbol{\epsilon}(\omega)$ and the LU factorization (16.49) of the grid covariance matrix.

Next, we show that $\mathbf{L}_{\mathcal{G},\mathcal{G}}\mathbf{L}_{\mathcal{G},\mathcal{G}}^{\top}$ is equal to the conditional covariance matrix. Based on (16.51d) it follows that

$$\mathbf{L}_{\mathcal{G},\mathcal{G}} = \left(\mathbf{C}_{\mathcal{G},\mathcal{G}} - \mathbf{L}_{\mathcal{G},\mathcal{D}}\mathbf{U}_{\mathcal{D},\mathcal{G}}\right)\mathbf{U}_{\mathcal{G},\mathcal{G}}^{-1}.$$

Then, the product $\mathbf{L}_{\mathcal{G},\mathcal{G}}\mathbf{L}_{\mathcal{G},\mathcal{G}}^{\top}$ is given by

$$\mathbf{L}_{\mathcal{G},\mathcal{G}}\mathbf{L}_{\mathcal{G},\mathcal{G}}^{\top} = \left(\mathbf{C}_{\mathcal{G},\mathcal{G}} - \mathbf{L}_{\mathcal{G},\mathcal{D}}\mathbf{U}_{\mathcal{D},\mathcal{G}}\right)\mathbf{U}_{\mathcal{G},\mathcal{G}}^{-1}\,\mathbf{L}_{\mathcal{G},\mathcal{G}}^{\top}. \tag{16.54}$$

We can further use $\mathbf{U}_{\mathcal{G},\mathcal{G}}^{-1}\,\mathbf{L}_{\mathcal{G},\mathcal{G}}^{\top} = \left(\mathbf{L}_{\mathcal{G},\mathcal{G}}^{\top}\right)^{-1}\,\mathbf{L}_{\mathcal{G},\mathcal{G}}^{\top} = \mathbf{I}_{P}$ to eliminate the last two terms in the above.

Using (16.51b) it follows that $\mathbf{L}_{\mathcal{G},\mathcal{D}} = \mathbf{C}_{\mathcal{G},\mathcal{D}}\,\mathbf{U}_{\mathcal{D},\mathcal{D}}^{-1}$, and based on (16.51c) it follows that $\mathbf{U}_{\mathcal{D},\mathcal{G}} = \mathbf{L}_{\mathcal{D},\mathcal{D}}^{-1}\,\mathbf{C}_{\mathcal{D},\mathcal{G}}$. Putting these together, we obtain

$$\mathbf{L}_{\mathcal{G},\mathcal{D}}\,\mathbf{U}_{\mathcal{D},\mathcal{G}} = \mathbf{C}_{\mathcal{G},\mathcal{D}}\,\mathbf{U}_{\mathcal{D},\mathcal{D}}^{-1}\mathbf{L}_{\mathcal{D},\mathcal{D}}^{-1}\,\mathbf{C}_{\mathcal{D},\mathcal{G}} = \mathbf{C}_{\mathcal{G},\mathcal{D}}\,\mathbf{C}_{\mathcal{D},\mathcal{D}}^{-1}\mathbf{C}_{\mathcal{D},\mathcal{G}}. \tag{16.55}$$

The equations (16.54) and (16.55) prove that the conditional covariance matrix $\mathbf{C}^c_{\mathcal{G},\mathcal{G}}$ is obtained from $\mathbf{L}_{\mathcal{G},\mathcal{G}}\mathbf{L}^\top_{\mathcal{G},\mathcal{G}}$.

The main steps of conditional random field simulation on a two-dimensional square grid using covariance matrix factorization are given in the algorithm 16.4.

Algorithm 16.4 Conditional simulation of Gaussian random field using covariance matrix factorization on a square grid with $P = L \times L$ nodes

1. Reshape the $L \times L \times d$ matrix of grid locations into a $P \times d$ location matrix
2. Calculate the conditional covariance matrix $\mathbf{C}^c_{\mathcal{G},\mathcal{G}}$ between grid locations according to (16.49)
3. Compute the Cholesky factorization $\mathbf{L}^c\,\mathbf{U}^c$ of $\mathbf{C}_{\mathcal{G},\mathcal{G}}$
4. Determine the matrices $\mathbf{L}_{\mathcal{G},\mathcal{G}}$, $\mathbf{L}_{\mathcal{G},\mathcal{D}}$, and $\mathbf{L}_{\mathcal{D},\mathcal{D}}$ from $\mathbf{L}^c$ and the block structure (16.50)

for $m \leftarrow 1, M$ **do**

  5.1 Generate a vector $\boldsymbol{\epsilon}^{(m)}$ of random numbers from the $\mathcal{N}(0, 1)$ distribution

  5.2 Construct realization $\mathbf{x}^{(m)} \leftarrow \mathbf{L}_{\mathcal{G},\mathcal{G}}\boldsymbol{\epsilon}^{(m)} + \mathbf{L}_{\mathcal{G},\mathcal{D}}\mathbf{L}^{-1}_{\mathcal{D},\mathcal{D}}\mathbf{x}^*$

  5.3 Reshape the $P \times 1$ vector $\mathbf{x}^{(m)}$ into an $L \times L$ matrix that represents the m-th realization

end for

16.8 Monte Carlo Methods (MCMC)

Monte Carlo is a famous resort along the French Riviera that is frequented by the rich and famous. Among its many charms, Monte Carlo's famous Casino helps money to effortlessly change hands, aided by the random events involved in gambling. Hence, the name of this minute principality became synonymous with games of chance and eventually (following the suggestion of Nicholas Metropolis at Los Alamos) with computational algorithms that exploit randomness.

Monte Carlo methods employ randomized sampling strategies in order to explore a given probability space. The goal of Monte Carlo methods is to evenly sample a multidimensional state space.

16.8.1 Ordinary Monte Carlo Methods

The main idea of ordinary Monte Carlo methods is to generate a sequence of random samples from the target distribution which are then used to estimate ensemble averages. Let us assume the following setting in order to concretely express this idea[4]:

[4]We use $\mathbf{X}(\omega)$ to denote a generic random vector that could represent Bayesian model parameters $\boldsymbol{\theta}$, samples from a continuous variable $X(\mathbf{s}; \omega)$, or a vector of binary variables $\boldsymbol{\sigma}(\omega)$; in the latter case, the integral in (16.56) over the values of the states is replaced by a summation.

1. $\{X_n(\omega)\}_{n=1}^N$ are continuously-valued random variables.
2. The random vector $\mathbf{X}(\omega) = (X_1(\omega), \ldots, X_N(\omega))^\top$ has joint pdf $f_X(\mathbf{x})$
3. The transformation $g(\cdot)\colon \mathbb{R}^N \to \mathbb{R}$ is a continuous function.

Then, the expectation of the random variable $g[\mathbf{X}(\omega)]$ can be estimated by means of the following *Monte Carlo average*

$$\mathcal{E}\{g[\mathbf{X}(\omega)]\} = \int d\mathbf{x}\, g(\mathbf{x})\, f_X(\mathbf{x}) \approx \frac{1}{M} \sum_{m=1}^{M} g\left[\mathbf{x}^{(m)}\right], \tag{16.56}$$

where the $\mathbf{x}^{(1)}, \ldots, \mathbf{x}^{(M)}$ represent realizations (samples) of the random vector $\mathbf{X}(\omega)$ drawn from the joint density $f_X(\mathbf{x})$.

The *Monte Carlo (MC) average*

$$\langle g(\mathbf{x}^{(m)})\rangle_{m=1}^{M} = \frac{1}{M} \sum_{m=1}^{M} g(\mathbf{x}^{(m)}), \tag{16.57}$$

is a random variable, since the outcome depends on M and the particular realizations involved. If the samples $\mathbf{x}^{(m)}$, where $m = 1, \ldots, M$ are independent and identically distributed and $\mathcal{E}\left\{|g[\mathbf{X}(\omega)]|^2\right\} < \infty$, it can be shown by means of the *Central Limit Theorem* that the MC average is asymptotically normally distributed around the true expectation, i.e.,

$$\lim_{M\to\infty} \langle g(\mathbf{x}^{(m)})\rangle_{m=1}^{M} \overset{d}{=} \mathcal{N}\left(\mathcal{E}\{g[\mathbf{X}(\omega)]\}, \frac{\mathbb{V}\mathrm{ar}\{g[\mathbf{X}(\omega)]\}}{M}\right). \tag{16.58}$$

Hence, the MC average represents a Gaussian random variable which in the limit $M \to \infty$ converges to the expectation $\mathcal{E}\{g[\mathbf{X}(\omega)]\}$. The fluctuations of the MC average have a variance equal to $\mathbb{V}\mathrm{ar}\{g[\mathbf{X}(\omega)]\}/M$. The variance dependence on M shows that the statistical error decreases slowly, i.e., with the square root of the number of samples. This slow convergence is characteristic of Monte Carlo methods [109].

The completely random sampling strategy of naive Monte Carlo is not suitable for probability distributions that carry more weight in specific regions of the multidimensional space compared to other, less probable regions.

16.8.2 Markov Chain Monte Carlo Methods

In ordinary Monte Carlo the states $\mathbf{x}^{(m)}$, $m = 1, \ldots, M$ used in the MC average (16.58) are independent. Markov Chain Monte Carlo (also known as MCMC) methods generate a sequence of correlated states. The goal of MCMC methods is to use strategies that allow focusing on the more important areas of the probability space. Hence, MCMC methods visit more often regions of the probability space that have a higher probability of occurrence.

Different algorithms have been developed for the implementation of MCMC methods. We will review *Gibbs sampling* and *Metropolis-Hastings*, two of the most commonly used algorithms. The former is used in sequential Gaussian simulation (see Sect. 16.9) while the latter is used in simulated annealing (see Sect. 16.10). More MCMC algorithms can be found in references such as [109, 470].

MCMC methods come particularly handy in the investigation of joint probability distributions that involve several variables. Various queries about such distributions can be expressed in terms of multidimensional integrals of the joint probability density function. In most cases, these integrals cannot be evaluated in closed form. Then, MCMC methods provide an efficient numerical integration framework.

Bayesian posteriors In Bayesian statistics, the posterior probability distributions of model parameters involve a normalizing factor that often cannot be explicitly evaluated because it represents a high-dimensional integral of some complicated function. This is better understood in terms of equation (11.18) for the posterior distribution of the parameters $\boldsymbol{\theta}$, i.e.,

$$f_{\text{post}}(\boldsymbol{\theta} \mid \mathbf{x}^*) = \frac{\mathcal{L}(\boldsymbol{\theta}; \mathbf{x}^*)\, f_{\text{prior}}(\boldsymbol{\theta})}{\int \mathrm{d}\boldsymbol{\theta}\, \mathcal{L}(\boldsymbol{\theta}; \mathbf{x}^*)\, f_{\text{prior}}(\boldsymbol{\theta})},$$

The normalization factor in the denominator involves integration of the likelihood function, $\mathcal{L}(\boldsymbol{\theta}; \mathbf{x}^*)$, and the prior pdf, $f_{\text{prior}}(\boldsymbol{\theta})$, product over the multidimensional space of the vector $\boldsymbol{\theta}$. This integral cannot in most cases be explicitly calculated.

Non-Gaussian distributions Evaluation of the normalizing factor is also challenging for non-Gaussian Boltzmann-Gibbs probability distributions. The calculation of the partition function (normalization constant) in such cases may require sampling of a high-dimensional space. A typical case is the partition function of the Ising model (discussed in Chap. 15) which is given by (15.40), i.e.,

$$Z = \sum_{\{\sigma_n\}_{n=1}^N} \mathrm{e}^{-\mathcal{H}(\boldsymbol{\sigma})} = \sum_{\{\sigma_n\}_{n=1}^N} \mathrm{e}^{J \sum_{\langle n,m \rangle} \sigma_n \sigma_m + g \sum_n \sigma_n},$$

where $\mathcal{H}(\boldsymbol{\sigma}) = -J \sum_{\langle n,m \rangle} \sigma_n \sigma_m - g \sum_n \sigma_n$ is the energy of a particular configuration $\boldsymbol{\sigma}$, $J \in \mathbb{R}$ is the coupling strength, and $g \in \mathbb{R}$ is the external field. The summation over the exponentials in the above equation is over all possible configurations of the set $\{\sigma_n\}_{n=1}^N$. This involves all the permutations with repetition of the binary variables $\sigma_n = \pm 1$, where $n = 1, \ldots, N$. The total number of states is thus 2^N which, even for a moderate system size $N = 10^3$, is a huge number $O(10^{301})$.

Not every state is worth visiting There are many travel destinations in the world, and our lives are not long enough to visit all of them. Fortunately, travellers can select based on their preferences the sites that will maximize some personal criterion of satisfaction, whether this involves watching Siberian tigers or exploring the ruins of ancient civilizations. Both problems discussed above (posterior normalization and Ising partition function) similarly involve the exploration of very large state spaces; however, while the space of feasible states is large, the states in this space are not equally probable. Often, the joint probability distribution is concentrated in a small region of the state space that provides the most "desirable destinations".

For example, consider the partition function of the Ising model given above for $J, g > 0$. A positive coupling constant, $J > 0$, implies that states $\{\sigma_n\}_{n=1}^N$ with uniform sign have lower energy than states with spins of different signs. The coupling constant does not specify whether the uniform sign should be positive or negative. However, $g > 0$ implies that states with a majority of positive values are more favorable (have lower energy) than states with a majority of negative values. Hence, states with a preponderance of positive values will make the dominant contributions to the partition function.

MCMC methods allow sampling complex probability spaces and enable the calculation of the desired expectations (e.g., mean, variance, covariance) without knowledge of the normalizing constant. The MCMC methods accomplish this by sampling a Markov chain which has the target distribution as its equilibrium limit [700].

Effective sample size In contrast with ordinary Monte Carlo methods, in Markov Chain Monte Carlo successive states are correlated. Hence, in MCMC the number of samples M used in the estimation of MC average variance (16.58) should be replaced by $M_{\text{eff}} = M/\tau_{\text{cor}}$, where τ_{cor} is the *autocorrelation time* of the correlations [438]. The autocorrelation time is measured in terms of Monte Carlo time (number of MC steps), e.g. [422, p. 109–110].

The autocorrelation time depends on the specific variable that is being simulated. For example, for a random variable $X(\omega)$ which could represent the random field value at a given location, or some average of the random field, or the magnetization of an Ising system, the autocorrelation time is defined as

$$\tau_{\text{cor}} = 1 + 2 \sum_{n=1}^{\infty} \rho_{\text{xx}}(n). \tag{16.59}$$

Various literature sources provide a more thorough introduction to MCMC methods than the brief account given herein. Particularly readable for physicists and engineers is [673, Chap. 15]. A synopsis focusing on the application of

MCMC to Gauss Markov Random fields is presented in [700]. More comprehensive mathematical treatments are given in [109, 568, 685].

16.8.3 Basic Concepts of MCMC Methods

In MCMC, the sampling of the *target distribution* $f_{\mathbf{x}}(\mathbf{x})$ is performed by means of a *Markov chain* which is a *memoryless random walk* over the space of probable states. The Markov chain is a stochastic process that proceeds sequentially by making transitions between feasible states. Schematically, the random walk corresponding to the Markov chain can be represented as follows:

$$\mathbf{x}^{(0)} \to \mathbf{x}^{(1)} \to \mathbf{x}^{(2)} \to \ldots \to \mathbf{x}^{(M)},$$

where the MC index (m) counts the states and determines the position of each state in the Markov chain. Hence, the MC index also measures the "simulation time".

State space This is the set $\mathbb{S}$ of all the feasible states that can be visited by the Markov chain. For example, in the case of the Ising model, the state space involves all the permutations with repetition of the binary variables $\sigma_i = \pm 1$, for $i = 1, \ldots, N$. Hence, for the binary Ising model the state space $\mathbb{S}$ comprises 2^N states.

Transition probability The *transition kernel* $P(\mathbf{X}^{(m)} = \mathbf{x} \mid \mathbf{X}^{(m-1)} = \mathbf{y})$, or $P(\mathbf{x}^{(m)} \mid \mathbf{x}^{(m-1)})$ for short,[5] defines the probability for making a transition from a state $\mathbf{x}^{(m-1)}$ to another state $\mathbf{x}^{(m)}$, for $m = 1, \ldots, M$. The transition probability should be defined for all initial and final states allowed by the state space. In addition, it should satisfy the following constraints for all $m = 1, \ldots, M$:

$$P(\mathbf{X}^{(m)} = \mathbf{x} \mid \mathbf{X}^{(m-1)} = \mathbf{y}) \geq 0,$$

This constraint simply expresses the *non-negativity* of the transition probability.

$$\sum_{\mathbf{x} \in \mathbb{S}} P(\mathbf{X}^{(m)} = \mathbf{x} \mid \mathbf{X}^{(m-1)} = \mathbf{y}) = 1.$$

The second constraint expresses the *closure property*, i.e., the fact that if a chain exits a state $\mathbf{y}$, it will (with probability one) make a transition to one of the available states in $\mathbb{S}$. Thus, the summation over $\mathbf{x}$ in the second equation is over the entire state space.

---

[5] Another common notation is $P(\mathbf{X}^{(m)} = \mathbf{x}_i \mid \mathbf{X}^{(m-1)} = \mathbf{x}_j) = p_{i,j}$.

In the case of continuously-valued fields, the summation symbol implies integration over all states. Then, the transition kernel for any given initial state $\mathbf{y}$ becomes a transition density function.

The Markov chain is called *homogeneous* if the transition probability is independent of the simulation time, that is, if

$$P(\mathbf{X}^{(m+1)} = \mathbf{x} \mid \mathbf{X}^{(m)} = \mathbf{y}) = P(\mathbf{X}^{(m)} = \mathbf{x} \mid \mathbf{X}^{(m-1)} = \mathbf{y}).$$

Lack of memory At each stage of the chain, the next state to be visited is independent of the past and depends only on the current state. This is mathematically expressed using transition probabilities as follows:

$$P(\mathbf{x}^{(m+1)} \mid \mathbf{x}^{(m)}, \mathbf{x}^{(m-1)}, \ldots, \mathbf{x}^{(1)}, \mathbf{x}^{(0)}) = P(\mathbf{x}^{(m+1)} \mid \mathbf{x}^{(m)}).$$

Equilibrium (stationary) distribution Homogeneous Markov chains tend to an *equilibrium (stationary) probability density* $\pi(\mathbf{x})$ such that

$$\sum_{\mathbf{x} \in \mathbb{S}} P(\mathbf{y} \mid \mathbf{x})\, \pi(\mathbf{x}) = \pi(\mathbf{y}), \;\text{ for all } \mathbf{y} \in \mathbb{S}. \tag{16.60}$$

Hence, the key property of the equilibrium distribution is its invariance under transitions. A desired property for any Markov chain is that its equilibrium distribution $\pi(\mathbf{x})$ should coincide with the target probability distribution $f_{\mathrm{x}}(\mathbf{x})$ that we aim to sample.

Example 16.4 Consider the AR(1) process $x_t = \phi\, x_{t-1} + \epsilon_t$, where $|\phi| < 1$ and $\epsilon_t \overset{d}{=} \mathcal{N}(0, 1)$. (i) Show that the above can be generated by means of a homogeneous MCMC. (ii) Determine the stationary distribution and the transition kernel.

Answer

(i) To show the equivalence of the AR(1) process with an MCMC consider that the AR(1) values can be generated by means of the following sequence of random variables

$$x_t(\omega) \mid x_{t-1} \overset{d}{=} \mathcal{N}(\phi\, x_{t-1}, 1), \;\text{ for } t = 1, 2, \ldots. \tag{16.61}$$

This equation generates a Markov chain, since the value at each step depends only on the previous value. The chain is homogeneous because the constant ϕ and the noise of the variance that determine the new states of the chain are constant (independent of time). Two sample AR(1) realizations are shown in Fig. 16.16.

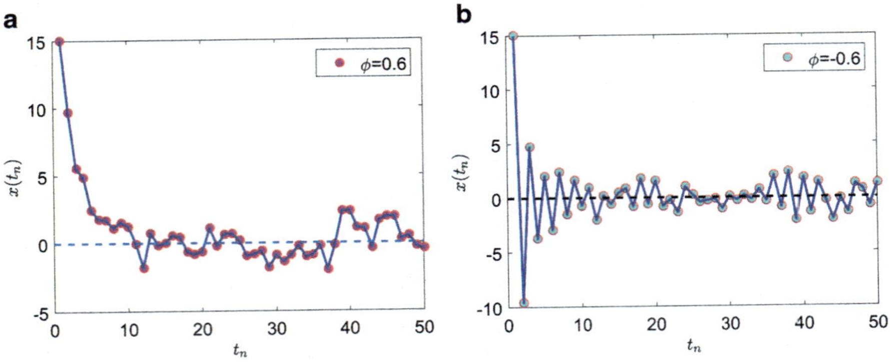

Fig. 16.16 Plots of the x_n, ($n = 1, \ldots, 50$) for two AR(1) processes with $\phi = 0.6$ (**a**) and $\phi = -0.6$ (**b**) respectively generated by means of (16.61). An initial value $x_0 = 15$ is used for both AR processes

(ii) The stationary distribution is Gaussian (due to the noise). The expectation of the stationary distribution is equal to zero, as can be seen by $\mathcal{E}[x_n(\omega)] = \phi^n x_0$, for $n = 1, 2, \ldots$. To calculate the variance, consider that

$$x_n(\omega) = \phi^n x_0 + \sum_{k=1}^{n} \epsilon_k(\omega)\phi^{n-k}.$$

In light of the orthogonality of the Gaussian white noise, i.e., $\mathcal{E}[\epsilon_k(\omega)\epsilon_l(\omega)] = \delta_{k,l}$, for $k, l = 1, 2, \ldots$, it follows that

$$\mathbb{V}\mathrm{ar}\,\{x_n(\omega)\} = \sum_{k=1}^{n} \phi^{2(n-k)} = \frac{1-\phi^{2n}}{1-\phi^2}.$$

Thus, the stationary distribution, obtained at the limit $n \to \infty$, is the following Gaussian

$$\pi(x) = \frac{\sqrt{1-\phi^2}}{\sqrt{2\pi}}\, \mathrm{e}^{-\frac{x^2\,(1-\phi^2)}{2}}.$$

The transition kernel density is determined from (16.61) which leads to the following time-independent Gaussian

$$P(y \mid x) = \frac{1}{\sqrt{2\pi}} \exp\left[-\frac{1}{2}(y-\phi x)^2\right].$$

The *transition kernel density* can then be shown to satisfy the continuum analogue of the equilibrium condition (16.60), which is expressed by means of the following convolution equation

$$\int_{-\infty}^{\infty} \mathrm{d}x\, P(y \mid x)\, \pi(x) = \pi(y).$$

Convergence of Markov chain This is a crucial property of Markov chains for practical purposes. Markov chains that are *irreducible* (every state can be reached from all other states after a finite number of transitions) and *aperiodic* (the chain can return to any given state at irregular times) have a *unique equilibrium distribution* to which the Markov chain converges [689].

While it is relatively easy to design Markov chains with the desired equilibrium distribution, it is also important to optimize convergence to the equilibrium distribution for reasons of computational efficiency. Practical issues involved in testing MCMC convergence are investigated in [109, 157] as well as in [685, Chap. 8–9].

MCMC around Europe Consider an adventurous perennial tourist who decides to sample European culture. She begins her journey in Berlin, choosing her next destination based on whim and the available train connections. She spends some time at each destination—the length of stay depending on the abundance of cultural attractions. Every time she decides to move, her destination is dictated by available means of transportation in the current city and the cultural potential of the probable destinations (transition probabilities) as well as her inner disposition at the time (random factor). Her travel plans are not linked to past visits (e.g., multiple visits to Paris are possible). Thus the tourist embarks on a memoryless Monte Carlo chain of travel that eventually (life span and resources permitting) samples the distribution of European cultural variability.

Burn-in phase Typically, the MCMC chain starts with an initial guess, $\mathbf{x}^{(0)}$, of the state variable. This initial state is considered successful if it lies in a high-probability region of the parameter space. However, it is more likely that the first steps of the Markov Chain produce low-probability states.

Note that in the Monte Carlo average (16.57) the contribution of the states $\mathbf{x}^{(m)}$ is not weighted according to their probability. Hence, the contribution of the initial states to the Monte Carlo average (16.57) can significantly bias the average, if the chain length M is not very large. Therefore, the evolution of Monte Carlo averages involves an initial *transient* (non-stationary) regime. After a number of initial steps the Markov Chain enters the stationary regime. In practice, it is customary to exclude the initial steps that sample the transient regime from the Monte Carlo average (16.57). This transient regime is known as the *burn-in phase* [109, 470]. If a transient part of the chain comprising K steps is discarded, the Monte Carlo average (16.57) becomes

$$\langle g(\mathbf{x}^{(m)}) \rangle_{m=K+1}^{M} = \frac{1}{M-K} \sum_{m=K+1}^{M} g(\mathbf{x}^{(m)}). \tag{16.62}$$

Determining the number of steps involved in the burn-in phase can be accomplished by tracking the evolution of the Monte Carlo average in order to determine when the chain enters the stationary regime. Then, the initial number of states preceding the stabilization is discarded from the average.

The term "burn-in phase" and its necessity are disputed [568]. The main counterargument is that in the asymptotic limit $M \to \infty$ the Monte Carlo average converges to the true expectation. In practice, the "burn-in" phase is useful if the initial (starting) guess represents a rather unlikely state and the length of the chain is not very long. In such cases, the transient regime may bias the Monte Carlo average, especially for chains of limited length.

Detailed balance The principle of detailed balance enforces *reversibility* in Markov chains. Reversibility expresses mathematically the physical *equilibrium condition*. Under equilibrium conditions, we expect that for any two states $\mathbf{x}$ and $\mathbf{y}$ of the Markov chain, the *transition rate* $\mathbf{x} \to \mathbf{y}$ is equal to the transition rate $\mathbf{y} \to \mathbf{x}$. The implementation of this principle requires determining a transition probability that makes the transition rates equal.

The transition rate from $\mathbf{x}$ to $\mathbf{y}$ is given by the probability of the chain being at $\mathbf{x}$ multiplied by the transition probability. Hence, the equal transition rate condition is expressed in terms of the following *detailed balance* equation

$$\pi(\mathbf{x})\, P(\mathbf{y} \mid \mathbf{x}) = \pi(\mathbf{y})\, P(\mathbf{x} \mid \mathbf{y}). \tag{16.63}$$

The left-hand side of (16.63) represents the transition rate from $\mathbf{x}$ to $\mathbf{y}$: it represents the *occupation probability* of the state $\mathbf{x}$, which is given by the equilibrium pdf $\pi(\mathbf{x})$, multiplied with the transition probability from $\mathbf{x}$ to $\mathbf{y}$. Similarly, the right-hand side of (16.63) represents the transition rate in the opposite direction, i.e., from $\mathbf{y}$ to $\mathbf{x}$.

In addition to ensuring reversibility, detailed balance provides a sufficient condition for equilibrium. In particular, if $\pi(\cdot)$ satisfies the detailed balance condition (16.63), then $\pi(\cdot)$ is an *equilibrium distribution*. To show this property, evaluate the summation over $\mathbf{x}$ on both sides of (16.63). Then, the equation (16.60) for the equilibrium distribution is obtained.

16.8.4 Metropolis-Hastings Sampling

The Metropolis-Hastings algorithm uses the concept of a *proposal distribution* in order to sample complex, potentially multimodal target distributions. The Metropolis updating step applies to transition probabilities that satisfy detailed balance, such as exponential densities. The Metropolis scheme is applied to exponential densities which allow defining transition probabilities that respect detailed balance.

The *Hastings updating scheme* extends the algorithm to more general probability distributions. It works by modifying the transition probabilities so that they satisfy detailed balance. The main ideas of the Hastings updating scheme are:

- A *convenient proposal distribution* (e.g., Gaussian), which we will denote by $Q(\mathbf{y} \mid \mathbf{x})$, is used. The choice of a suitable proposal distribution allows the efficient generation of proposal (trial) states of the Markov chain.
- The *acceptance probability* $\alpha(\mathbf{y} \mid \mathbf{x})$ is defined in terms of the proposal and the target distribution $f(\mathbf{x})$.
- The *transition probability* $P(\mathbf{y} \mid \mathbf{x})$ is defined so as to satisfy detailed balance.
- The transition probability is different from the acceptance probability (in the Metropolis case they coincide).

Proposal distribution The Metropolis-Hastings algorithm generates a chain of Monte Carlo states. The algorithm proposes at every iteration a new state which is drawn from the "proposal distribution". The proposed state is typically generated by means of a localized change from the current one (e.g., a simple flipping of spin values at two different locations). The "proposed" state is accepted or rejected according to an acceptance probability. The latter is defined so that the resulting transition rates satisfy the principle of detailed balance.

Acceptance probability The above insights can be expressed mathematically as follows: The *acceptance probability* $\alpha(\mathbf{y} \mid \mathbf{x})$ for the transition from a state $\mathbf{x}$ to a state $\mathbf{y}$ is defined as

$$\alpha(\mathbf{y} \mid \mathbf{x}) = \min\left(1, \frac{f(\mathbf{y})\, Q(\mathbf{x} \mid \mathbf{y})}{f(\mathbf{x})\, Q(\mathbf{y} \mid \mathbf{x})}\right), \tag{16.64}$$

where $Q(\mathbf{y} \mid \mathbf{x})$ is the *proposal distribution* for the transition $\mathbf{x} \to \mathbf{y}$. Note that the partition function Z (the normalizing constant of the target distribution) is not necessary to calculate the acceptance probability.

Transition probability The *transition probability* from state $\mathbf{x}$ to state $\mathbf{y}$ is respectively defined as follows:

$$P(\mathbf{y} \mid \mathbf{x}) = \begin{cases} Q(\mathbf{y} \mid \mathbf{x})\, \alpha(\mathbf{y} \mid \mathbf{x}), & \mathbf{x} \neq \mathbf{y} \\ Q(\mathbf{x} \mid \mathbf{x}) + \sum_{\mathbf{z} \neq \mathbf{x}} Q(\mathbf{z} \mid \mathbf{x})\, [1 - \alpha(\mathbf{z} \mid \mathbf{x})]\,, & \mathbf{x} = \mathbf{y}. \end{cases} \tag{16.65}$$

The second line corresponds to the probability that there is no transition out of the state $\mathbf{x}$. It is given by the proposal probability for the chain to remain in state $\mathbf{x}$ plus the probabilities that the transition to all other states that can be proposed fails.

To better understand the above definition of the transition probability consider the following metaphor: If the state $\mathbf{x}$ represents the country of your current residence, Greece for example, and $\mathbf{y}$ a country where you are considering to immigrate, e.g., Switzerland, the probability that you will make the transition from $\mathbf{x}$ to $\mathbf{y}$ is proportional to both the probability that you will receive a job offer in the destination country and the probability that you will accept the offer. The job offer probability is determined by the proposal distribution, while your decision regarding a specific offer is determined by the acceptance probability.

Detailed balance It is straightforward to show that the definitions (16.64) and (16.65) imply that the principle of *detailed balance* (16.63) is satisfied by $f(\mathbf{x})$, regardless of the form of the proposal distribution. Essentially, for any given $f(\cdot)$ and $Q(\cdot \mid \cdot)$ the Metropolis-Hastings updating selects the acceptance probability $\alpha(\cdot \mid \cdot)$ so that the probability density $f(\cdot)$ satisfies detailed balance. Since $f(\cdot)$ satisfies detailed balance, it is an equilibrium distribution of the Markov chain.

In order to ensure that $f(\cdot)$ is the *unique equilibrium distribution*, it is necessary to establish *irreducibility* and *aperiodicity* conditions for the Markov chain as discussed above. This can be accomplished by tailoring the proposal distribution accordingly, since the latter determines the transition probability according to (16.65).

Example 16.5 Show that the transition probabilities defined by the Metropolis-Hastings update scheme (16.64) satisfy the detailed balance equation (16.63) for the probability density $f(\mathbf{x})$.

Answer The detailed balance equation (16.63) is satisfied by $f(\mathbf{x})$ if the following relation holds for $\mathbf{x} \neq \mathbf{y}$:

$$f(\mathbf{x})\, P(\mathbf{y} \mid \mathbf{x}) = f(\mathbf{y})\, P(\mathbf{x} \mid \mathbf{y}).$$

Using the Metropolis-Hastings transition probability (16.65) the detailed balance equation is equivalent to

$$f(\mathbf{x})\, Q(\mathbf{y} \mid \mathbf{x})\, \alpha(\mathbf{y} \mid \mathbf{x}) = f(\mathbf{y})\, Q(\mathbf{x} \mid \mathbf{y})\, \alpha(\mathbf{x} \mid \mathbf{y}).$$

Based on the definition of the acceptance probability (16.64), it follows that either $\alpha(\mathbf{y} \mid \mathbf{x}) = 1$ or $\alpha(\mathbf{x} \mid \mathbf{y}) = 1$. Let us assume, without losing generality, that $\alpha(\mathbf{y} \mid \mathbf{x}) = 1$. Then, using (16.64) to calculate $\alpha(\mathbf{x} \mid \mathbf{y})$ the detailed balance condition becomes

$$f(\mathbf{x})\, Q(\mathbf{y} \mid \mathbf{x}) = f(\mathbf{y})\, Q(\mathbf{x} \mid \mathbf{y})\, \frac{f(\mathbf{x})\, Q(\mathbf{y} \mid \mathbf{x})}{f(\mathbf{y})\, Q(\mathbf{x} \mid \mathbf{y})},$$

which is identically true regardless of the form of the function $Q(\mathbf{x} \mid \mathbf{y})$.

Metropolis scheme In pure Metropolis updating a *uniform proposal distribution* is used, i.e., $Q(\mathbf{y} \mid \mathbf{x}) = c$. Assuming an exponential target distribution with density $f_{\mathrm{x}}(\mathbf{x}) \propto \exp\left[-\mathcal{H}(\mathbf{x})\right]$ the *Metropolis update acceptance probability* (16.64) becomes

$$\alpha(\mathbf{y} \mid \mathbf{x}) = \min\left(1,\ \frac{f(\mathbf{y})}{f(\mathbf{x})}\right) = \min\left(1,\ \mathrm{e}^{-[\mathcal{H}(\mathbf{y})-\mathcal{H}(\mathbf{x})]}\right). \qquad (16.66)$$

Acceptance probability The acceptance probability can be equivalently expressed as follows

$$\alpha(\mathbf{y} \mid \mathbf{x}) = \begin{cases} 1, & \text{if } \mathcal{H}(\mathbf{x}) > \mathcal{H}(\mathbf{y}), \\ \mathrm{e}^{-[\mathcal{H}(\mathbf{y})-\mathcal{H}(\mathbf{x})]}, & \text{if } \mathcal{H}(\mathbf{y}) \geq \mathcal{H}(\mathbf{x}). \end{cases} \tag{16.67}$$

Assume that $\mathbf{x}_1$ represents the current state in the chain. A proposal state $\mathbf{x}_2$ is drawn from the proposal distribution and is given an acceptance probability $\alpha(2 \mid 1)$. Since the simulated system follows an exponential pdf of the form $f(\mathbf{x}) \propto \mathrm{e}^{-\mathcal{H}(\mathbf{x})}$, we expect the acceptance probability to be equal to one if the energy of the proposed state is lower than the energy of the current state.

If the proposal (trial state) has higher energy than the current state, the proposal is not outright rejected but is given a second chance: a random number $r \overset{d}{=} U(0, 1)$ is drawn from the uniform distribution. The random number is then compared with the exponential "Boltzmann factor" $\exp(-\Delta\mathcal{H})$. If $\alpha(2 \mid 1) > r$ the proposal is accepted, while if $\alpha(2 \mid 1) < r$ the proposal is rejected (see lines 5–10 in Algorithm 16.5). This intuitive analysis is satisfied by the definition (16.67).

Transition probability Accordingly, the Metropolis transition probability is obtained from (16.65)

$$P(\mathbf{y} \mid \mathbf{x}) = \begin{cases} c\,\alpha(\mathbf{y} \mid \mathbf{x}), & \mathbf{x} \neq \mathbf{y} \\ c + c\,\sum_{\mathbf{z} \neq \mathbf{x}} [1 - \alpha(\mathbf{x}, \mathbf{z})], & \mathbf{x} = \mathbf{y}. \end{cases} \tag{16.68}$$

The *detailed balance* condition for transitions between the states $\mathbf{x}$ and $\mathbf{y} \neq \mathbf{x}$ can then be expressed in terms of the following ratio [125]

$$\frac{P(\mathbf{x} \mid \mathbf{y})}{P(\mathbf{y} \mid \mathbf{x})} = \frac{\pi(\mathbf{x})}{\pi(\mathbf{y})} = \frac{f(\mathbf{x})}{f(\mathbf{y})} = \mathrm{e}^{-[\mathcal{H}(\mathbf{x})-\mathcal{H}(\mathbf{y})]}. \tag{16.69}$$

Metropolis updating for Ising model The Metropolis updating procedure for an Ising spin model with fixed magnetization $M = \sum_{n=1}^{N} \sigma_n$ and energy given by (15.38) is summarized in the Algorithm 16.5. If the magnetization is not fixed, the updating step 3 is replaced by a single spin flip (change of sign) at a randomly selected site.

Metropolis updating for Ising model covariance The calculation of the Ising model covariance function in the case of *fixed magnetization M* is described in Algorithms 16.6–16.7. The initial state $\boldsymbol{\sigma}^{(0)}$ has K_1 positive spins and $K_2 = N - K_1$ negative spins, so that $K_1 - K_2 = N\,\overline{\sigma}$, where $\overline{\sigma} = M/N$ is the magnetization per spin. The Metropolis algorithm is used to update the spin states. The simplest updating move that can be proposed is the exchange of two spins at different locations. This leaves the overall magnetization (difference between positive and

Algorithm 16.5 Metropolis updating by means of spin exchange for the Ising model with N sites, nearest neighbor coupling, and fixed magnetization. The integer indices n_1, n_2 of the exchange sites are drawn randomly from the set $\{1, 2, \ldots, N\}$. The energy $\mathcal{H}$ of the Ising model is given by (15.38)

function METROPOLIS UPDATE($\boldsymbol{\sigma}^{(\mathrm{cur})}$)
  $\boldsymbol{\sigma}^{(\mathrm{temp})} \leftarrow \boldsymbol{\sigma}^{(\mathrm{cur})}$ ▷ Assign current state to trial state
  $i \leftarrow n_1 \stackrel{d}{=} \mathrm{Uniform}(1, N)$ ▷ Randomly select exchange site i
  $j \leftarrow n_2 \stackrel{d}{=} \mathrm{Uniform}(1, N)$, $(n_2 \neq n_1)$ ▷ Randomly select exchange site j
  $t \leftarrow \sigma_i^{(\mathrm{temp})}, \ \sigma_i^{(\mathrm{temp})} \leftarrow \sigma_j^{(\mathrm{temp})}, \ \sigma_j^{(\mathrm{temp})} \leftarrow t$ ▷ Propose update: $\sigma_i \leftrightharpoons \sigma_j$
  $P \leftarrow \min\left(\mathrm{e}^{-[\mathcal{H}(\boldsymbol{\sigma}^{(\mathrm{temp})}) - \mathcal{H}(\boldsymbol{\sigma}^{(\mathrm{cur})})]}, 1\right)$ ▷ Calculate transition probability
  $r \leftarrow u \stackrel{d}{=} \mathrm{Uniform}(0, 1)$ ▷ Generate random probability threshold
  if $r \leq P$ **then** ▷ Consider energetically unfavorable state
    return $\boldsymbol{\sigma}^{(\mathrm{new})} \leftarrow \boldsymbol{\sigma}^{(\mathrm{temp})}$ ▷ Accept trial state if transition is favorable
  else
    return $\boldsymbol{\sigma}^{(\mathrm{new})} \leftarrow \boldsymbol{\sigma}^{(\mathrm{cur})}$ ▷ Keep current state otherwise
  end if
end function

Algorithm 16.6 Metropolis MCMC method for evaluating the Ising model covariance function under the constraint of fixed magnetization M. The initial state $\boldsymbol{\sigma}^{(0)}$ has K_1 spins with value one and $K_2 = N - K_1$ spins with values equal to minus one so that $\mathcal{M} = (K_1 - K_2)$. The covariance is initialized with a zero vector of length $\lfloor N/2 \rfloor$. $N_{\min}$ is the sampling period used to obtain de-correlated MCMC samples

$\boldsymbol{\sigma}^{(\mathrm{cur})} \leftarrow \boldsymbol{\sigma}^{(0)}$ ▷ Initialize state
$N \leftarrow \mathrm{length}(\boldsymbol{\sigma}^{(0)})$ ▷ Determine length of spin vector
$N_c \leftarrow \lfloor N/2 \rfloor$ ▷ Set maximum covariance lag
$\mathbf{C} \leftarrow \mathrm{zeros}(N_c, 1)$ ▷ Initialize covariance vector
for $m \leftarrow 1, M_{\mathrm{ini}}$ **do** ▷ Burn-in phase loop
  $\boldsymbol{\sigma}^{(\mathrm{new})} \leftarrow$ METROPOLIS UPDATE($\boldsymbol{\sigma}^{(\mathrm{cur})}$) ▷ Update state
end for
$\boldsymbol{\sigma}^{(\mathrm{cur})} \leftarrow \boldsymbol{\sigma}^{(\mathrm{new})}$ ▷ Initialize spin state
$N_{\mathrm{sim}} \leftarrow 1$ ▷ Initialize sample index
for $m \leftarrow 1, M$ **do** ▷ Main MCMC loop
  $\boldsymbol{\sigma}^{(\mathrm{new})} \leftarrow$ METROPOLIS UPDATE($\boldsymbol{\sigma}^{(\mathrm{cur})}$)
  if $m \mod N_{\min} = 0$ **then**
    $\mathbf{c} \leftarrow$ CORREL($\boldsymbol{\sigma}^{(\mathrm{cur})}, \overline{\sigma}$) ▷ Sample correlations after $N_{\min}$ proposals
    $\mathbf{C} \leftarrow \mathbf{C} + \mathbf{c}$ ▷ Update the covariance cumulative sum
    $N_{\mathrm{sim}} \leftarrow N_{\mathrm{sim}} + 1$ ▷ Increase sample index
  end if
end for
$\mathbf{C} \leftarrow \mathbf{C}/N_{\mathrm{sim}}$ ▷ Calculate the MC covariance average

Algorithm 16.7 Sample estimate of Ising model covariance function

```
function CORREL(σ, σ̄)                          ▷ Inputs: spin state vector and magnetization
    N ← length(σ)                               ▷ Determine length of spin vector
    N_c ← ⌊N/2⌋                                  ▷ Set maximum lag
    c ← zeros(N_c, 1)                           ▷ Initialize covariance function
    for i ← 1, N_c do                           ▷ Loop over lags
        for j ← 1, N − i do                     ▷ Loop over sites
            c_i ← c_i + (σ_j − σ̄) (σ_{j+i−1} − σ̄)   ▷ Add pair correlations at lag i − 1
        end for
        c_i ← c_i/(N − i)                       ▷ Estimate covariance at lag i
    end for
    return c                                    ▷ Return covariance vector
end function
```

negative spins) constant. Other updating schemes are also possible. As a rule of thumb, proposed updates should lead to acceptance rates in the range 20%–50%.[6]

The covariance is typically calculated for lags between zero and $\lfloor N/2 \rfloor - 1$. The estimation of the sample covariance function for the Monte Carlo states is performed as shown in Algorithm 16.7.

The increment of the covariance estimate at each Monte Carlo step is evaluated by means of the function CORREL described in Algorithm 16.6. The simulation involves overall M Monte Carlo steps, but the spin state is sampled with a period of $N_{\min}$ steps, where $N_{\min} \propto N$. This means that the sample covariance is evaluated $N_{\text{sim}} = \lfloor N/N_{\min} \rfloor$ times. The intermittent sampling approach is used to avoid correlations between consecutive states and to achieve independent samples.

Other approaches In addition to the classical Metropolis algorithm there are *importance sampling* schemes such as the Swendsen-Wang and Wolff algorithms which propose moves that change simultaneously an entire cluster of spin values [470]. These algorithms are used to overcome the critical slowing down of the Metropolis updating scheme near phase transitions [479].

16.8.5 Gibbs Sampling

Unlike other Monte Carlo methods (e.g., rejection sampling or the Metropolis-Hastings algorithm), Gibbs sampling uses all the random numbers generated. Gibbs sampling is a special case of the Metropolis-Hastings scheme in which the acceptance probability is equal to one. Acceptance with probability one means that all the trial states are accepted. This feature of Gibbs sampling stems from the fact that it draws the proposed states from carefully constructed *univariate conditional*

[6] The acceptance rate is equal to the ratio of the accepted moves over the total number of proposals.

densities. Then, both the equilibrium and the proposal distributions are essentially given by the conditional distribution.

The method of Gibbs sampling is based on drawing random numbers that are locally determined from the conditional distribution of the target variable (e.g., specific parameter or random field value at some point) [279]. Explicit expressions for the conditional probability distributions are thus required. Hence, Gibbs sampling is less flexible than Metropolis-Hastings, because the former requires exact conditional distributions, while Metropolis-Hastings works with general proposal distributions.

If the target variable is defined over a P-dimensional space (e.g., a vector of P parameters or P values at the nodes of a simulation grid), a specific path is defined so that all the dimensions of the target space are "visited" in some random or deterministic order.

Posterior parameter distribution For the sake of example, we consider the simulation of the posterior distribution of parameters in a Bayesian setting. The posterior distribution $f_{\text{post}}(\boldsymbol{\theta} \mid \mathbf{x}^*)$ of the parameters $\boldsymbol{\theta}$ is given by (11.18).

The simulated parameter vector $\boldsymbol{\theta} = (\theta_1, \ldots, \theta_P)^\top$ at the m-th MC step is generated from the *conditional distribution* $f_{\theta_p}(\theta_p \mid \boldsymbol{\theta}_{-p}^{(m)})$, where

$$\boldsymbol{\theta}_{-p}^{(m)} = \left(\theta_1^{(m)}, \ldots, \theta_{p-1}^{(m)}, \theta_{p+1}^{(m-1)}, \ldots, \theta_P^{(m-1)}\right)^\top, \quad p = 1, \ldots, P,$$

is the reduced parameter vector that excludes the parameter θ_p.

The Gibbs sampler moves sequentially in the parameter space: at each step it samples from the conditional distribution that is based on the "current knowledge". This implies that updating the p-th parameter is conditioned on the values of the parameters $\theta_1^{(m)}, \ldots, \theta_{p-1}^{(m)}$ for the current MC step and the values $\theta_{p+1}^{(m-1)}, \ldots, \theta_P^{(m-1)}$ from the previous MC step.

The expectation of a function $g\left[\boldsymbol{\theta}(\omega)\right]$ of the parameters over the posterior distribution $f_{\text{post}}(\boldsymbol{\theta} \mid \mathbf{x}^*)$, i.e.,

$$\mathcal{E}\{g\left[\boldsymbol{\theta}(\omega)\right]\} = \prod_{p=1}^{P} \int \mathrm{d}\theta_p \, g(\boldsymbol{\theta}) \, f_{\text{post}}(\boldsymbol{\theta} \mid \mathbf{x}^*)$$

is then approximated in terms of the following Monte Carlo average

$$\mathcal{E}\{g\left[\boldsymbol{\theta}(\omega)\right]\} \approx \frac{1}{M} \sum_{m=1}^{M} g\left(\boldsymbol{\theta}^{(m)}\right).$$

The steps involved in the Gibbs sampler are outlined in the Algorithm 16.8. A pedagogical presentation of the Gibbs sampler is given in [118]. An application of Gibbs updating to the sequential simulation of random fields is given in Sect. 16.9.

Algorithm 16.8 Basic steps of Gibbs sampling

1: $\boldsymbol{\theta} \leftarrow \boldsymbol{\theta}^{(0)}$ ▷ Initialize parameter vector
2: **for** $m \leftarrow 1, M$ **do** ▷ Loop over the MC states
3: **for** $p \leftarrow 1, P$ **do** ▷ Loop over the parameters
4: $\theta_p^{(m)} \leftarrow r \overset{d}{=} f_{\theta_p}(\theta_p \mid \boldsymbol{\theta}_{-p}^{(m)})$ ▷ Sample from conditional distributions of P parameters
5: **end for**
6: **end for**

Markov Chain Monte Carlo methods are often criticized for their lack of computational efficiency and convergence problems [701]. These obstacles can be a problem for certain applications that require sampling complex, multidimensional spaces. However, recent research efforts focus on fertilizing MCMC with machine learning ideas. For example, reinforcement learning can be used to automatically tune the proposal distributions leading to efficient sampling methodologies [86].

16.9 Sequential Simulation of Random Fields

The sequential simulation algorithm is a *Markov Chain Monte Carlo (MCMC)* method based on the Gibbs sampler. The method is widely used to simulate Gaussian and non-Gaussian random fields. The latter are transformed to the Gaussian distribution by means of the normal scores transform (Gaussian anamorphosis) as explained in Chap. 14. Sequential simulation is described in detail in [132, 425, 692].

The sequential algorithm is a method of *conditional simulation* that owes its popularity to conceptual simplicity and ease of implementation. The term "sequential" denotes that the field values are generated in series, following a random path that visits all the lattice sites.

Let us assume that the simulation aims to generate field values at the nodes of the simulation lattice $\mathcal{G}$. Let us further assumed that the nodes of the lattice are visited in a certain random order $\mathbf{z}_{[1]}, \ldots, \mathbf{z}_{[P]}$, where the set $([1], [2], \ldots [P])$ denotes a random permutation of the set of integers $(1, 2, \ldots, P)$. Then, the values at each node grid node are randomly drawn from the *conditional distribution*

$$f_{\mathrm{x}}(x_{[p]} \mid \mathbf{x}_{[p-1]}; S_{p-1}),$$

where $x_{[p]}$ denotes the field value at the lattice node $\mathbf{z}_{[p]}$ that is visited at the p-th step.[7] On the other hand, $\mathbf{x}_{[p-1]}$ is a *sequentially updated sample set* that includes the data and the first $p-1$ simulated field values at the nodes $\mathbf{z}_{[1]}, \ldots, \mathbf{z}_{[p-1]}$. The point set S_{p-1} includes both the data locations and the grid nodes that have been visited until the previous step.

[7] The node $\mathbf{z}_{[\mathrm{p}]}$ is determined by the random order in which the lattice nodes are visited.

The set $\mathbf{x}_{[p]}$ is initially (for $p = 0$) assigned the values of the data set $\mathbf{x}^*$, and it is subsequently updated according to $\mathbf{x}_{[p]} = \mathbf{x}_{[p-1]} \cup \{x_{[p]}\}$. The procedure is repeated for $p = 1, \ldots, P$ until all the simulation lattice nodes have been visited.

In the case of Gaussian random fields that are fully determined by their mean and covariance function, the conditional distribution $f_{\mathrm{x}}(x_{[p]} \mid \mathbf{x}_{[p-1]}; S_{p-1})$ is evaluated in analogy to (6.14)–(6.16). Hence, the conditional distribution at the point $\mathbf{z}_p$ is given by the following equations for $p = 1, \ldots, P$:

$$f_{\mathrm{x}}(x_{[p]} \mid \mathbf{x}_{[p-1]}; S_{p-1}) = \mathcal{N}\left(m_{c;[p]}, \sigma^2_{c;[p]}\right), \tag{16.70a}$$

$$m_{c;[p]} = m(\mathbf{z}_{[p]}) + \mathbf{C}_{[p],[p-1]}\, \mathbf{C}^{-1}_{[p-1],[p-1]}\, (\mathbf{x}_{[p-1]} - \mathbf{m}_{[p-1]}), \tag{16.70b}$$

$$\sigma^2_{c;[p]} = \sigma^2_{\mathrm{x}}(\mathbf{z}_{[p]}) - \mathbf{C}_{[p],[p-1]}\, \mathbf{C}^{-1}_{[p-1],[p-1]}\, \mathbf{C}_{[p-1],[p]}, \tag{16.70c}$$

where $m_{c;[p]}$ is the conditional mean and $\sigma^2_{c;[p]}$ the conditional variance at the p-th step. The index $[p]$ indicates that the variables are evaluated at the point $\mathbf{z}_{[p]}$ conditionally upon the values (data and simulations) in the sample set S_{p-1}. The notation used in (16.70) is as follows:

- $\mathcal{N}\left(m_{c;[p]}, \sigma^2_{c;[p]}\right)$ is the marginal normal pdf with mean $m_{c;[p]}$ and variance $\sigma^2_{c;[p]}$.
- $m_{c;[p]}$ is the conditional mean at $\mathbf{z}_{[p]}$.
- $\sigma^2_{c;[p]}$ is the conditional variance at $\mathbf{z}_{[p]}$.
- $\mathbf{C}_{[p-1],[p-1]}$ is the covariance matrix for the point set S_{p-1}.
- $\mathbf{C}_{[p],[p-1]}$ is the covariance vector between $\mathbf{z}_{[p]}$ and the point set S_{p-1}.
- $\mathbf{m}_{[p-1]}$ is the vector of the unconditional expectations at the points in S_{p-1}.
- $m(\mathbf{z}_{[p]})$ is the unconditional expectation at $\mathbf{z}_p$.
- $\sigma^2_{\mathrm{x}}(\mathbf{z}_{[p]})$ is the unconditional variance at $\mathbf{z}_p$.

The calculation of the conditional mean and covariance requires the inversion of the covariance matrix $\mathbf{C}_{[p-1],[p-1]}$. In the case of GMRFs that are defined in terms of the precision matrix, the conditional distribution is readily obtained from equation (8.4).

The main steps of the sequential Gaussian simulation (SGS) method [692] are described in the Algorithm 16.9. Step 5.1 is equivalent to calculating the mean and variance of the conditional probability distribution according to (16.70). We have encountered similar calculations in Chap. 6 and in the context of the simple kriging equations (10.32) and (10.33). Step 5.2 implements the *Gibbs sampler*: random numbers are drawn from the full conditional Gaussian distribution with the respective conditional mean and variance at each stage.

Algorithm 16.9 Conditional simulation using the sequential simulation method

1. Preprocess the data (detrending, normal scores transform)
2. Estimate the spatial model that applies to the fluctuations

for $m \leftarrow 1, M$ **do** ▷ Loop over realizations
  3. Define the path $\Pi(\mathbf{z}_{[1]}, \ldots, \mathbf{z}_{[P]})$ ▷ Select a path that covers the grid
  4. $S \leftarrow \mathrm{S_N}$ and $\mathbf{x}_S \leftarrow \mathbf{x}^*$ ▷ Initialize the current data and sampling set
  for $p \leftarrow 1, P$ **do** ▷ Loop over grid sites
    5.1 Estimate $m_{c;[p]}$ and $\sigma^2_{c;[p]}$ ▷ Conditional mean and variance using (16.70)
    5.2 Generate r ▷ Random number from $\mathcal{N}\left(m_{c;[p]}; \sigma^2_{c;[p]}\right)$
    5.3 $x_{[p]} \leftarrow r$ ▷ Set the simulation value at $\mathbf{z}_p$
    5.4 $S \leftarrow S \cup \{\mathbf{z}_{[p]}\}$ and $\mathbf{x}_S \leftarrow \mathbf{x}_S \cup \{x_p\}$ ▷ Update the sampling and data sets
  end for
  6. $\mathbf{x}^{(m)} \leftarrow \mathbf{x}_S$ ▷ m-th realization of the field on the lattice
end for

7. Validation ▷ Confirm the probability distribution and variogram of simulations
9. Invert transformations ▷ Apply if data were transformed in Step 1

Path selection The sequential Gaussian simulation algorithm assumes that the grid nodes are visited by following a random path. At each point of the path, the conditional Gaussian distribution is determined, and a value is randomly drawn from it. Determining the conditional distribution requires solving a kriging system of increasing size. The escalating size results from the sequential addition of simulated points to the sample set S_p (the latter initially contains N points and at the end of the simulation path $N + P$ points).

Finding the conditional distribution is not too demanding computationally if GMRF or SLI models are used, since then the precision matrix is known and the conditional mean and variance can be estimated without solving a large linear system. However, for classical, covariance-based spatial models, kriging is used to estimate the conditional mean and variance. Then, the SGS computational cost increases proportionally with the third power of the sample set size, unless some simplifying approximations are applied.

Neighborhoods The standard approach for reducing the SGS computational cost uses a search neighborhood to restrict the number of neighbors of each simulation point and consequently the size of the kriging system. The neighborhood-based approach generates artifacts in the realizations, because the small errors introduced in the conditional distribution by the neighborhood cutoff progressively accumulate. The simulation path determines the locations and the size of these artifacts. The influence of the path on the simulation errors is comprehensively analyzed in a recent study [615].

The impact of using a finite search neighborhood on the quality of the simulated realizations is studied in [231]. The analysis therein is based on the reproduction quality of the histogram, the variogram, the indicator variograms, and the ergodic fluctuations of the first- and second-order statistics. It was established that, even if the ergodic index (13.53) is small (i.e., favorable conditions for ergodicity), the realizations derived by means of local neighborhoods may poorly reproduce the second-order statistics, while they may also be inconsistent with the stationarity and ergodicity assumptions. The largest impact was found for covariance functions that are differentiable at the origin and compactly supported.

The use of GMRFs with sparse precision matrices helps to overcome the computational complexity problem, since finite neighborhoods can be defined based on the sparsity pattern of the precision matrix [538].

Absence of burn-in phase Sequential Gaussian simulation is based on the Gibbs sampler which does not require a burn-in phase. Burn-in is unnecessary because the proposal distribution is the same as the conditional distribution of the simulated parameters. This leads to an acceptance probability equal to one as discussed above. This means that all the proposed moves in Gibbs sampling are good ones.

16.10 Simulated Annealing

Annealing is a metallurgical process used to improve the ductility and reduce the hardness of metals. It involves heating which allows the atoms in the crystal lattice to diffuse and thus reduces the number of lattice defects (e.g., dislocations). The heating helps to increase the ductility and reduce the hardness of materials thus making them more workable. The heating phase is followed by a slow cooling process during which the material gradually hardens and regains its solid state. The annealing process overcomes local energy minima (which correspond to states with crystal defects) and slow relaxation towards a globally optimal energetic configuration (a state without any crystal defects).

Simulated annealing (SA) is an MCMC algorithm that cleverly applies the principles underlying annealing to *constrained combinatorial optimization problems* [453, 454]. SA is an iterative optimization method based on the concept of *stochastic relaxation.* It can be used to optimize objective functions (possibly non-convex) over large spaces of probable configurations using a stochastic exploration algorithm. The objective function typically involves some form of "energy" and perhaps a number of additional constraints that represent structural features of the specific problem. A readable introduction to SA is given in [673].

The annealing temperature is controlled so as to allow transitions to configurations that are not energetically favorable. This allows the algorithm to escape from potential local minima. The initial temperature needs to be sufficiently high for the SA algorithm to escape local minima of the objective function. The transitions between states are determined by the *Metropolis step* that is typically based on the *Boltzmann distribution*, $f \propto \mathrm{e}^{-E/k_{\mathrm{B}}T}$, where E is the energy of the system and k_{B} is the Boltzmann constant. In non-thermodynamic applications of SA, $k_{\mathrm{B}}T$ is replaced by a fictitious temperature parameter. In optimization problems the energy E is replaced by the objective function.

Search for global minima In contrast with *gradient-based methods* and *greedy algorithms* that move downhill the energy landscape and can thus get stuck in local minima, SA is better equipped to search for the global minima of non-convex functions. This advantage stems from its ability to accept transitions that temporarily increase the energy, thus allowing the algorithm to escape local minima and explore other regions of the energy landscape. As the temperature is lowered, however, such jumps become less likely, and at zero temperature SA will converge to the first local minimum encountered. On the flip side, the flexibility afforded by simulated annealing comes at increased computational cost. In addition, there is no guarantee that a specific annealing schedule will lead to the optimal solution within a finite amount of time [392].

Conditional simulation of random fields SA has found wide-ranging applications in science and engineering. Mathematical reviews of SA are given in [685, 707]; the geostatistical perspective is described in [487]; a more physics-oriented presentation is given in [422].

In geostatistics, SA is used as a method of *conditional simulation*. The connection between SA and conditional simulation seems natural, since the latter can usually be recast into a constrained combinatorial optimization problem. In the cases of Gaussian random fields, this is accomplished as follows:

1. Generate a set of i.i.d. random numbers from the target probability distribution and randomly assign them to the sites of the simulation grid.
2. Iteratively perturb this initial configuration.
3. Use a stochastic update rule to generate a chain of updated configurations.
4. Find the configuration that minimizes a suitable objective function. The latter is selected so as to impose spatial continuity (e.g., specified covariance) and conditioning (e.g., fixed values at observation points) constraints. For example, one can use the following objective function

$$E\left(\mathbf{x}^{(n)}\right) = \sum_{k=1}^{K} \epsilon_k^2, \quad \epsilon_k^2 = \left[\frac{\gamma_{\mathrm{xx}}(\mathbf{r}_k) - \hat{\gamma}_{\mathrm{xx}}^{(n)}(\mathbf{r}_k)}{\gamma_{\mathrm{xx}}(\mathbf{r}_k)}\right]^2,$$

   where $\gamma_{\mathrm{xx}}(\mathbf{r})$ is the model variogram function, $\hat{\gamma}_{\mathrm{xx}}^{(n)}(\mathbf{r})$ is the estimated sample-based variogram from the current configuration, $\mathbf{x}^{(n)}$, during the n-th SA step, and ϵ_k^2 is the square of the relative variogram residual calculated for the lag $\{\mathbf{r}_k\}_{k=1}^{K}$.
5. The "final" (equilibrium) configuration obtained by means of the stochastic relaxation process represents one conditional realization of the random field.

The application of SA to geostatistical conditional simulation is studied in [196] and explained in more detail in [195]. Mining applications are reviewed in [210] and reservoir engineering applications are discussed in [446]. Finally, the use of SA in the restoration of digital images is reviewed in [852].

Implementation The main steps of a generic simulated annealing algorithm are presented in the pseudocode 16.10. In geostatistical simulation of Gaussian, zero-

Algorithm 16.10 Simulated annealing algorithm for the conditional simulation of a zero-mean, Gaussian random field. The spatial continuity properties of the field are defined by a quadratic energy function $E(\mathbf{x})$. The transition probabilities between configurations are evaluated by means of the Boltzmann distribution function. This implementation is known as Boltzmann annealing [392]

$T^{(1)}, \quad T_{\min}, \quad \rho_T, \quad i_{\max}, \quad j_{\max}$ ▷ Initialize parameters
Generate $N \times 1$ vector $\mathbf{x}^{(0)}$ from i.i.d. samples of $\mathcal{N}(m_{\mathrm{x}}, \sigma_{\mathrm{x}}^2)$ ▷ Initialize state
$k \leftarrow 1, \quad i \leftarrow 1, \quad j \leftarrow 0$ ▷ Initialize T, state, and accepted transition counters
$\mathbf{x}_{\mathrm{cur}} \leftarrow \mathbf{x}^{(0)}$ ▷ Initialize the current state
while $T^{(k)} > T_{\min}$ **do** ▷ Temperature loop
    while $i \leq i_{\max}$ **do** ▷ State loop
        while $j \leq j_{\max}$ **do** ▷ Accepted transitions loop
            Generate $\mathbf{x}_{\mathrm{new}}$ by perturbing $\mathbf{x}_{\mathrm{cur}}$ ▷ Propose new state
            Evaluate $E(\mathbf{x}_{\mathrm{new}})$ ▷ Update the energy
            $\mathrm{P}_{\mathrm{cur}\to\mathrm{new}} = \exp\left\{-[E(\mathbf{x}_{\mathrm{new}}) - E(\mathbf{x}_{\mathrm{cur}})]/T^{(k)}\right\}$ ▷ Metropolis step
            $r \leftarrow$ Uniform(0, 1) ▷ Generate random number for Metropolis step
            if $\mathrm{P}_{\mathrm{cur}\to\mathrm{new}} \geq r$ **then**
                $\mathbf{x}^{(i)} \leftarrow \mathbf{x}_{\mathrm{new}}, \ j = j + 1$ ▷ Update state & increase accepted transition counter
            else if $\mathrm{P}_{\mathrm{cur}\to\mathrm{new}} < r$ **then**
                $\mathbf{x}^{(i)} \leftarrow \mathbf{x}_{\mathrm{cur}}$ ▷ Maintain the current state
            end if
            $\mathbf{x}_{\mathrm{cur}} \leftarrow \mathbf{x}^{(i)}, \quad i \leftarrow i + 1$ ▷ Update state and increase state counter
        end while ▷ Close accepted transition loop
    end while ▷ Close states loop
    $k \leftarrow k + 1, \quad T^{(k)} \leftarrow \rho_T \, T^{(k-1)}$ ▷ Update temperature and T counter
end while ▷ Close temperature loop

mean random fields, the energy function $E(\mathbf{x})$ is usually given by the sum of square differences between the variogram of the current configuration and the model (theoretical) variogram. Different lags can be weighted differently in this energy function [196].

The successful implementation of simulated annealing requires defining various parameters used in the algorithm, the cooling schedule, the type of proposed transitions, and the termination criteria. We briefly discuss these topics below. More detailed accounts are given in the references listed above. A recent critical review of simulated annealing and its variants is presented in [392], where issues of convergence, methods to speed up the annealing process, and alternatives to the Boltzmann distribution are discussed.

Parameter selection A simulated annealing algorithm for the minimization of a specific energy function $E(\mathbf{x})$ where $\mathbf{x}$ represents the states of a random field $\mathrm{X}(\mathbf{s}; \omega)$, typically involves the following parameters:

1. *Initial temperature* $T^{(1)}$: This should be sufficiently high to allow escaping local energy minima and exploring different "energy valleys" in the opening stages.
2. *Annealing schedule*: This refers to the method used to "cool the system", aiming to converge to the optimal solution as the temperature evolves from the initial value $T^{(1)}$ towards a small final value (often near zero). The schedule can be a simple geometric progression, i.e.,

$$T^{(k+1)} = \rho_T \, T^{(k)}, \text{ for } k = 1, \ldots, K, \tag{16.71}$$

where $T^{(k)}$ is the temperature at the k-th annealing level and $\rho_T = 1 - \epsilon$, is a constant rate coefficient determined by small, non-negative constant $\epsilon = o(1)$. K is defined so that $T^{(K)} \geq T_{\min}$ and $T^{(K+1)} < T_{\min}$, where $T_{\min}$ is the minimum annealing temperature.

3. *Proposed moves:* This refers to the proposal of new configurations. The simplest proposal is to interchange the values of the current configuration at two randomly selected locations.
4. *Number of generated states per temperature level*, $i_{\max}$: This parameter defines the extent of the energy landscape exploration at each temperature. Typically, $i_{\max}$ is set equal to a multiple of the simulation grid size.
5. *Lower temperature limit*, $T_{\min}$: This limit defines the lowest temperature accessible by the algorithm, which is also used as a termination criterion. A practical choice is $T_{\min} = 0.01\, T^{(1)}$; however, the effectiveness of this choice depends on the initial temperature.
6. *Maximum number of accepted state transitions per temperature level*, $j_{\max}$: This parameter reduces the temperature if a large percentage of proposed transitions are accepted. A large acceptance ratio implies that the temperature is so high that almost every proposed transition is accepted. This number is also defined as a multiple of the simulation grid size.
7. *Energy level transition tolerance*, δE. This parameter defines the minimum change of the objective function that is considered meaningful (i.e., different than zero). It can optionally be used as a termination criterion. If the energy tolerance criterion is used, SA terminates if the energy is not lowered more than δE following a series of consecutive proposed transitions.

16.11 Karhunen-Loève (KL) Expansion

The Karhunen-Loève expansion also known as *Karhunen-Loève decomposition* is a *stochastic spectral method:* It decomposes random fields into an infinite series in terms of an orthogonal function basis. The main difference between other spectral methods and Karhunen-Loève expansion is that the latter employs an *optimal basis*. An optimal basis minimizes the mean square error of truncated approximations that involve a finite number of terms.

The Karhunen-Loève expansion is the generalization of *principal component analysis (PCA)* to continuum domains and is thus also known as *functional PCA* [416, 747]. Conversely, the Karhunen-Loève expansion of continuum systems that are numerically discretized for computational study can be derived by calculating the respective PCA. Some linear algebra fundamentals related to PCA are reviewed in the Appendix C.

There is a close link between PCA and the statistical physics renormalization group (RG) analysis. This relation is discussed in [97]. Both PCA and RG aim at the coarse-graining of complex systems by focusing on the most important modes. PCA assumes that the data can be successfully compressed by projecting the system onto a subspace spanned by a few dominant modes that capture most of the variability in the data. This approach is successful if the joint distribution is close to the Gaussian, in which case the dominant modes are determined by the eigenvalues and eigenvectors of the covariance matrix. RG, on the other hand, uses equations that describe the flow of the joint probability distribution under coarse-graining [417]. RG analysis can in principle be applied to systems with non-Gaussian probability distributions as well. If the joint pdf is available in the form of a Boltzmann-Gibbs expression, RG analysis can be theoretically applied to determine whether the probability distribution will flow towards the Gaussian or to a non-Gaussian fixed point upon coarse-graining.

Spectral methods that expand the random field in different (than Karhunen-Loève) orthogonal bases also exist. *Polynomial chaos*, also known as *Wiener chaos* expansion or *Wiener-Hermite* expansion after its inventor, Norbert Wiener [849], represents random variables as series of Hermite polynomials in a countable sequence of independent Gaussian random variables. Polynomial chaos was introduced in engineering applications by Roger Ghanem and Paul Spanos, who used truncated Wiener chaos expansions as trial functions in a Galerkin base, thus founding the *stochastic Galerkin method* [286].

Generalized polynomial chaos was developed later by Dongbin Xiu and George Karniadakis who obtained better approximations using polynomial expansions in terms of non-Gaussian random variables [238, 436, 859]. These spectral approaches have so far gained more acceptance in applied mathematics than in spatial statistics.

Polynomial chaos decompositions of random fields are expressed in terms of orthogonal Hermite polynomials of a Gaussian random variable. Hermite polynomial expansions are discussed herein in Sect. 14.5. Generalized polynomial chaos uses the general *Askey scheme* polynomial family that naturally incorporates general types of probability distributions (the polynomials from the Askey family are orthogonal with respect to specific probability distributions).

We only consider the Karhunen-Loève expansion, but interested readers can find information on other spectral methods in [286, 656, 858]. We first present the general theoretical frameworks of Karhunen-Loève expansions, followed by the calculation of the Karhunen-Loève expansion for the *Wiener process*, motivated by the latter's wide applicability as a model of Brownian motion. Then, we briefly comment on the numerical estimation of the Karhunen-Loève expansion. Finally, we calculate the Karhunen-Loève expansion for Spartan spatial random fields based on [803].

16.11.1 *Definition and Properties of Karhunen-Loève Expansion*

The Karhunen-Loève expansion is a very powerful tool in the analysis of random fields for the following reasons:

1. It enables separating the random (stochastic) components from the spatial components of the random field.
2. It provides an orthonormal basis for the covariance function. This basis can be used to construct a spectral series representation of the random field analogous to the Fourier expansion.
3. The terms (eigenfunctions) of the Karhunen-Loève expansion can be ordered according to the corresponding eigenvalues. Assuming a fast decay of the eigenvalues with increasing order, a small number of terms from the infinite series can be used to approximate the random field at the desired level of accuracy. Hence, the Karhunen-Loève expansion provides a controlled approach for *dimensionality reduction* in the simulation of spatial random fields.

The basic ingredients of the Karhunen-Loève expansion are contained in *Mercer's theorem* which allows the factorization of the covariance function in terms of an orthonormal set of functions that satisfy a characteristic Fredholm equation.

Theorem 16.1 (Mercer's theorem) *Let $K(\mathbf{s}, \mathbf{s}')$ represent a real, symmetric, continuous, and positive definite function from $\mathcal{D} \times \mathcal{D} \rightarrow \mathbb{R}$, where $\mathcal{D} \subset \mathbb{R}^d$ is a compact space. For simplicity we assume that $\mathbf{s} \in \mathbb{R}^d$ is a dimensionless position vector.*

According to Mercer's theorem, $K(\mathbf{s}, \mathbf{s}')$ can be expanded as the following series

$$K(\mathbf{s}, \mathbf{s}') = \sum_{m=1}^{\infty} \lambda_m \, \psi_m(\mathbf{s}) \, \psi_m(\mathbf{s}'), \tag{16.72a}$$

where the $\{\psi_m(\mathbf{s})\}_{m=1}^{\infty}$ are a countable sequence of orthonormal eigenfunctions that satisfy the integral equation

$$\int_{\mathcal{D}} \mathrm{d}\mathbf{s} \, \psi_m(\mathbf{s}) \, \psi_n(\mathbf{s}) = \delta_{n,m}, \quad m, n \in \mathbb{N}, \tag{16.72b}$$

and $\lambda_m > 0$, $m \in \mathbb{N}$, are the respective, non-negative eigenvalues. The eigenvalues λ_m and eigenfunctions $\psi_m(\mathbf{s})$ are solutions of the following homogeneous Fredholm integral equation of the second kind

$$\int_{\mathcal{D}} \mathrm{d}\mathbf{s}' \, K(\mathbf{s}, \mathbf{s}') \, \psi_m(\mathbf{s}') = \lambda_m \, \psi_m(\mathbf{s}), \; \textit{for } m \in \mathbb{N}. \tag{16.72c}$$

Let us now assume that the kernel function $K(\mathbf{s}, \mathbf{s}')$ represents a covariance function $C_{\mathrm{xx}}(\mathbf{s}, \mathbf{s}')$. Once the orthonormal expansion of the covariance function has been obtained, it is possible to construct a series expansion of a zero-mean random field $X(\mathbf{s};\omega)$ with that particular $C_{xx}(\mathbf{s},\mathbf{s}')$ as covariance function. This is formalized by means of the Karhunen-Loève expansion theorem [286, 508].

Theorem 16.2 *The Karhunen-Loève expansion of the zero-mean random field $X(\mathbf{s};\omega)$ with covariance $C_{xx}(\mathbf{s},\mathbf{s}')$ is given by the following superposition of eigenfunctions*

$$X(\mathbf{s};\omega) = \sum_{m=1}^{\infty} \sqrt{\lambda_m}\, c_m(\omega)\, \psi_m(\mathbf{s}), \tag{16.73}$$

where $\{c_m(\omega)\}_{m=1}^{\infty}$ is a countable set of respective projections of the random field on the Karhunen-Loève eigenfunctions, i.e.,

$$c_m(\omega) = \frac{1}{\sqrt{\lambda_m}} \int_{\mathcal{D}} d\mathbf{s}\, X(\mathbf{s};\omega)\, \psi_m(\mathbf{s}). \tag{16.74}$$

Hence, the $\{c_m(\omega)\}_{m=1}^{\infty}$ are zero-mean, uncorrelated random variables that satisfy the equations

$$\mathcal{E}[c_m(\omega)] = 0, \quad \textit{and} \quad \mathcal{E}[c_m(\omega)\, c_n(\omega)] = \delta_{n,m}, \ \forall n, m \in \mathbb{N}.$$

The series (16.73) converges uniformly on $\mathcal{D}$. In addition, for a Gaussian random field, the coefficients $c_m(\omega)$ are independent, Gaussian random variables.

The above construction does not make assumptions about the pdf of the random field. The type of the probability distribution that governs $X(\mathbf{s};\omega)$ affects the probability distribution of the coefficients.

Karhunen-Loève eigenvalues and variance Let us consider a stationary random field $X(\mathbf{s};\omega)$. Then, the variance is given by $\sigma_x^2 = C_{xx}(\mathbf{s},\mathbf{s})$. Using (16.72a) it follows that the field variance, which is uniform for all $\mathbf{s} \in \mathcal{D}$, is given by the following expansion

$$\sigma_x^2 = \sum_{m=1}^{\infty} \lambda_m\, \psi_m^2(\mathbf{s}). \tag{16.75a}$$

The above equation shows that the sum of the squared eigenfunctions weighted by the respective energy is constant and equal to the variance. If we integrate both sides of (16.75) using the orthonormality (16.72b) of the Karhunen-Loève eigenfunctions, we obtain the following relation between the field variance and the eigenvalues of the Karhunen-Loève expansion

$$|\mathcal{D}|\, \sigma_x^2 = \sum_{m=1}^{\infty} \lambda_m, \tag{16.75b}$$

where $|\mathcal{D}|$ is the volume of the bounded domain $\mathcal{D}$. This equation allows determining a number of Karhunen-Loève terms that can be used in order to achieve a desired approximation of the variance.

Orthogonality Based on Mercer's Theorem 16.1, the basis used to expand the covariance function is orthogonal (orthonormal, more precisely). According to the Karhunen-Loève Theorem 16.2 the respective random coefficients are also orthogonal. For this reason the Karhunen-Loève expansion is often called *bi-orthogonal*, since both the stochastic and the deterministic components are orthogonal.

Remark It is convenient to define the eigenfunctions $\psi_n(\cdot)$ as dimensionless quantities. Then, based on straightforward dimensional analysis of (16.72a), the units of the eigenvalues λ should be $[\mathrm{X}]^2$. In the Fredholm integral equation (16.72c), the kernel function (covariance) has units $[\mathrm{X}]^2$. Hence, in order for λ_m to also have units $[\mathrm{X}]^2$, the spatial integral in (16.72c) should be expressed in terms dimensionless coordinates. If the random field has a characteristic length ξ, the normalization can be accomplished by dividing all position vectors by ξ. The normalized form of the Karhunen-Loève expansion is demonstrated in Sect. 16.12.

Explicit solutions Determining the Karhunen-Loève orthonormal basis of the covariance function $C_{\mathrm{xx}}(\mathbf{s}, \mathbf{s}')$ requires solving the Fredholm equation (16.72c) of Mercer's theorem. Analytical solutions of this equation are not readily available. To our knowledge such solutions are limited to one-dimensional stochastic processes such as the *Wiener process*, the *Brownian bridge*, and the *Ornstein-Uhlenbeck process*.

An explicit solution is also available for the *exponential kernel* $\exp(-h)$, where $h = |s - s'|/\xi$ [286]. The lack of differentiability of the exponential function at the origin, which implies that the respective Gaussian realizations are also not differentiable, motivated the Karhunen-Loève expansion of the *modified exponential* model $(1 + h)\exp(-h)$, which is differentiable at the origin [767].

More recently, the Karhunen-Loève expansion of the flexible *Spartan random field* family was derived [803]. The latter includes the modified exponential as a separate case and also encompasses covariance functions with oscillatory dependence as well as damped, differentiable, exponential dependence. We believe that the Karhunen-Loève expansion of the 1D Spartan random field will find applications in the study of physical and engineered systems, since it essentially represents the optimal-basis expansion of the classical, damped harmonic oscillator in contact with a heat bath (i.e., driven by white noise).

16.11.2 Karhunen-Loève Expansion of the Wiener Process

The Wiener process, briefly discussed in Sect. 1.4.4, is a mathematical model of Brownian motion. The Karhunen-Loève expansion of the Wiener process can be explicitly calculated [656].

A random walk is a process defined in a discrete time domain. It can be represented as a non-stationary autoregressive process of order one, i.e., an AR(1), such that $W_t(\omega) - W_{t-1}(\omega) = \varepsilon_t(\omega)$, where the increments $\varepsilon_t(\omega)$ are Gaussian white noise (i.e., independent and identically distributed random variables).

The Wiener process is the limit of the random walk in continuous time domains. It is a zero-mean, non-stationary random process with covariance function $C_W(t, t') = \min(t, t')$.

Let us calculate the Karhunen-Loève expansion of the Wiener process over the interval $[0, T]$. The Fredholm equation (16.72c) is expressed for the covariance of the Wiener process as follows

$$\int_0^T \mathrm{d}t' \, C_W(t, t') \, \psi_m(t') = \lambda_m \, \psi_m(t), \quad \text{for all } t \in [0, T]. \tag{16.76a}$$

In light of the Wiener process covariance which is proportional to the absolute value of the time difference, the above equation is equivalently expressed as

$$\int_0^t \mathrm{d}t' \, t' \, \psi_m(t') + t \int_t^T \psi_m(t') = \lambda_m \, \psi_m(t). \tag{16.76b}$$

Differentiating with respect to t and using Leibnitz's rule to handle the derivatives of the integration limits we obtain the following integro-differential equation

$$\int_t^T \psi_m(t') = \lambda_m \frac{\mathrm{d}\psi_m(t)}{\mathrm{d}t}. \tag{16.76c}$$

A second differentiation leads to the following linear ordinary differential equation (ODE) with constant coefficients

$$\lambda_m \frac{\mathrm{d}^2\psi_m(t)}{\mathrm{d}t^2} + \psi_m(t) = 0. \tag{16.76d}$$

The solutions of the above ODE are the harmonic functions

$$\psi_m(t) = c_1 \cos\left(\frac{t}{\sqrt{\lambda_m}}\right) + c_2 \sin\left(\frac{t}{\sqrt{\lambda_m}}\right). \tag{16.76e}$$

The coefficients c_1 and c_2 are determined from the *boundary conditions* for the Karhunen-Loève eigenfunctions. These conditions are obtained from the Fredholm equation at the limits $t = 0$ and $t = T$.

1. Setting $t = 0$ in (16.76b) leads to $\psi_m(0) = 0$ which requires $c_1 = 0$.
2. Setting $t = T$ in the integro-differential equation (16.76c) leads to the boundary condition $\left.\frac{\mathrm{d}\psi_m(t)}{\mathrm{d}t}\right|_{t=T} = 0$. This condition determines the eigenvalues of the Karhunen-Loève expansion as follows:

$$\cos\left(\frac{T}{\sqrt{\lambda_m}}\right) = 0 \Rightarrow \lambda_m = \left(\frac{T}{\left(m - \frac{1}{2}\right)\pi}\right)^2, \text{ for } m \geq 1. \tag{16.76f}$$

Based on the expression for the eigenvalues, the respective expression for the Karhunen-Loève eigenfunctions is the following sinusoidal function

$$\psi_m(t) = \frac{\sqrt{2}}{\sqrt{T}} \sin\left(\left(m - \frac{1}{2}\right)\frac{\pi t}{T}\right), \tag{16.76g}$$

where $c_2 = \sqrt{2/T}$ to ensure *orthonormality* of the Karhunen-Loève eigenfunctions.

In light of the eigenfunction (16.76g) and the eigenvalue (16.76f) equations, the Karhunen-Loève expansion (16.73) of the Wiener process becomes

$$W(t) = \sum_{m=1}^{\infty} \frac{\sqrt{2T}}{(m - \frac{1}{2})\pi} \sin\left(\left(m - \frac{1}{2}\right)\frac{\pi t}{T}\right) c_m(\omega), \tag{16.77}$$

where $\{c_m(\omega)\}_{m=1}^{\infty}$ are random numbers drawn from the *standard normal distribution*.

The Wiener process (Brownian motion) $W(t;\omega)$ is a *self-affine process* with exponent $H = 1/2$. According to the definition of self-affinity (5.80), the Wiener process satisfies the scaling relation is $W(\lambda t;\omega) = \lambda^{1/2}\, W(t;\omega)$. The self-affinity also holds for the *truncated Karhunen-Loève approximation*

$$W^{(M)}(t) = \sum_{m=1}^{M} \frac{\sqrt{2T}}{(m - \frac{1}{2})\pi} \sin\left((m - \frac{1}{2})\frac{\pi t}{T}\right) c_m(\omega). \tag{16.78}$$

The above expansion satisfies $W^{(M)}(\lambda t) = \lambda^{1/2} W^{(M)}(t)$, because (i) $t \mapsto \lambda t$ does not affect the ratio t/T in the sinusoidal eigenfunction and (ii) the mapping $T \mapsto \lambda T$ implies that the prefactor $\sqrt{2\,T}$ in (16.78) is multiplied by $\lambda^{1/2}$.

Let us consider a process $W^{(M)}(t)$ sampled with a step Δt. Under the transformation $t \mapsto 2t$, the process $W(\lambda t)$ is obtained which is sampled with time step $2\Delta t$. In light of self-affinity, this process should be identical to the initial process multiplied with $\sqrt{2}$. This scaling behavior is illustrated in Fig. 16.17. The process $W_1(t)$ is sampled at 10^3 points in [0, 10], while $W_2(t)$ is sampled at 2×10^3 points in [0, 20]. The scaled process $\widetilde{W}_1(t)$ is generated from $W_1(t)$ by doubling the time step and multiplying with $2^{1/2}$. Although $\widetilde{W}_1(t)$ is defined at 10^3 points while $W_2(t)$ at 2×10^3 points, they both look exact copies of each other, thus confirming the self-affinity of the truncated Wiener process.

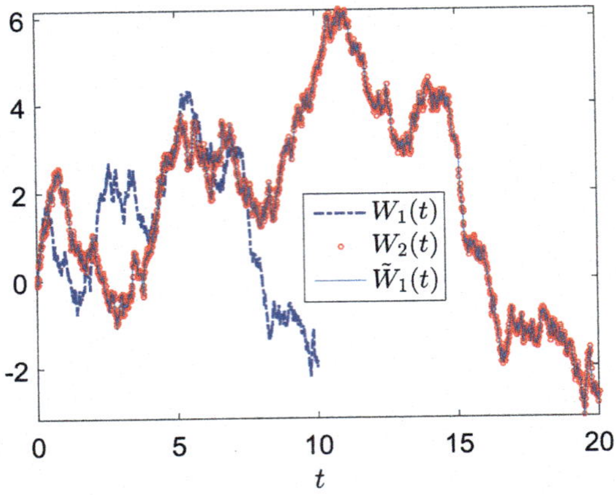

Fig. 16.17 Three graphs generated by means of the truncated Karhunen-Loève expansion (16.78). The same set of $M = 10^4$ Gaussian random numbers from $\mathcal{N}(0, 1)$ is used to generate $W_1(t)$ and $W_2(t)$. $W_1(t)$ is sampled at 10^3 points in $[0, 10]$ while $W_2(t)$ is sampled at 2×10^3 points in $[0, 20]$. The process $\tilde{W}_1(t)$ is generated from $W_1(t)$ by doubling the time step ($\lambda = 2$) and multiplying with $2^{1/2}$

16.11.3 Numerical Expansions

The Karhunen-Loève expansion can be explicitly derived only for certain low-dimensional covariance functions. In other cases, numerical methods are necessary for obtaining the Karhunen-Loève expansion. We discuss some issues that are relevant for the calculation of Karhunen-Loève expansions in cases of higher dimensionality.

Higher space dimensionality In spatial domains $\mathcal{D} \subset \mathbb{R}^d$, where $d > 1$ *separable covariance models* formed as the products of one-dimensional covariance functions—cf. (4.43)—can be used. Such models admit explicit Karhunen-Loève expansions [808]. In this case, however, the covariance model in 2D and 3D is not strictly isotropic, because it does not depend on the Euclidean norm of the lag vector $\mathbf{r}$, but rather on products of one-dimensional functions defined along the orthogonal axes. Numerical solutions of the Karhunen-Loève expansion are possible by means of various methods (see references in [803].)

Numerical K-L expansion In general, the K-L expansion for specific covariance functions can be obtained by solving numerically the respective Fredholm equation. It is also possible to derive an empirical Karhunen-Loève expansion, based on the covariance estimated from the data and subsequent numerical calculation of the Karhunen-Loève coefficients. The numerical solution of the Karhunen-Loève problem involves the discretization of the integral equation which implies replacing the random field by a random vector that represents the field values at the discretization points. The Karhunen-Loève expansion for discretized random fields is equivalent to *principal component analysis* and *proper orthogonal decomposition* [61, 127].

Truncated K-L expansion In principle, the Karhunen-Loève expansion contains an infinite sequence of eigenvalues and respective eigenfunctions, cf. (16.73). In practice, the expansion is truncated after a finite number of M terms. It is possible to control the accuracy of the approximation with respect to the reconstruction of the variance and the covariance function.

Assuming that the eigenvalues are numbered in the order of descending magnitude, i.e., $\lambda_1 \geq \lambda_2 \ldots \geq \lambda_M$, we define the truncated K-L expansion $\mathrm{X}^{(M)}(\mathbf{s}; \omega)$ as follows

$$\mathrm{X}^{(M)}(\mathbf{s}; \omega) = \sum_{m=1}^{M} \sqrt{\lambda_M}\, c_m(\omega)\, \psi_m(\mathbf{s}). \tag{16.79}$$

The *error of the Karhunen-Loève approximation* is defined as

$$\varepsilon^{(M)}(\mathbf{s}; \omega) = \mathrm{X}(\mathbf{s}; \omega) - \mathrm{X}^{(M)}(\mathbf{s}; \omega) = \sum_{M+1}^{\infty} \sqrt{\lambda_M}\, c_m(\omega)\, \psi_m(\mathbf{s}). \tag{16.80}$$

The truncated Karhunen-Loève expansion is *optimal* in the sense that its mean square error (MSE) is smaller than than the MSE of any other approximation (based on a different orthonormal basis set) that contains the same number of terms [286]. The optimality of the Karhunen-Loève expansion is proved by showing that the minimization of the mean square error is equivalent to the Karhunen-Loève Fredholm integral equation (16.72c).

Karhunen-Loève versus spectral superposition with randomized sampling Note the similarity between the truncated Karhunen-Loève expansion (16.80) and the expansion used in the spectral simulation (16.14). However, the number of modes involved in the expansions is quite different: The Karhunen-Loève expansion aims to reconstruct the field using an optimal, and thus small, number of modes. In contrast, the randomized spectral sampling used in (16.14) requires a high number of terms in order to achieve good accuracy.

On the other hand, explicit Karhunen-Loève expansions are difficult to obtain, while it is straightforward to obtain the series expansion used in randomized spectral simulation for most covariance functions. Hence, the optimality of the Karhunen-Loève expansion, and therefore its efficiency as a dimensionality reduction method, is not a free lunch. One has to pay the cost of either deriving an explicit solution of the Fredholm integral equation (16.72c)—which is feasible only in a few cases—or calculating the solution numerically.

Explained variance It is relatively easy to show, using the orthogonality of the eigenfunctions, that the variance of the truncated approximation is equal to the sum of the eigenvalues of the Fredholm integral equation (16.72c), i.e.,

$$\mathbb{V}\mathrm{ar}\left\{\mathrm{X}^{(M)}(\mathbf{s}; \omega)\right\} = \frac{1}{|\mathcal{D}|} \sum_{m=1}^{M} \lambda_m. \tag{16.81}$$

The variance of the truncated K-L expansion can thus be controlled if the eigenvalues are known. In fact, we can decide to truncate the series after M terms so that the ratio of the approximate (based on M Karhunen-Loève terms) over the true variance, $\sum_{m=1}^{M} \lambda_m / |\mathcal{D}| \sigma_{\mathrm{x}}^2$, achieves a desired level of accuracy such as 95%. The variance of the truncated Karhunen-Loève expansion is also known as the explained variance using the first M terms of the Karhunen-Loève expansion.

Error of covariance approximation Similarly to the variance, based on Mercer's covariance expansion (16.72a) the error of the Karhunen-Loève covariance approximation is defined as follows

$$\epsilon_C^{(M)}(\mathbf{s}, \mathbf{s}') = C_{\mathrm{xx}}(\mathbf{s}, \mathbf{s}') - \sum_{m=1}^{M} \lambda_m \, \psi_m(\mathbf{s}) \, \psi_m(\mathbf{s}'). \tag{16.82}$$

According to (16.82) the error of the covariance approximation is a function of the positions $\mathbf{s}$ and $\mathbf{s}'$.

16.12 Karhunen-Loève Expansion of Spartan Random Fields

This section focuses on the Karhunen-Loève expansion of Spartan spatial random fields [803]. We consider a 1D SSRF $\mathrm{X}(s; \omega)$, where $s \in [-L, L]$. To simplify the analysis we use normalized coordinates, i.e., $s/\xi \mapsto s$, and we express the equations of Mercer's Theorem 16.1 in terms of the normalized coordinates.

Theorem 16.3 *Let $C_{\mathrm{xx}}(h)$ where $h = |s - s'|$ represent the SSRF covariance* (7.27) *over the compact domain $s \in [-L/\xi, L/\xi]$. Then $C_{\mathrm{xx}}(h)$ can be expanded as follows:*

$$C_{\mathrm{xx}}(h) = \sum_{m=1}^{\infty} \lambda_m \, \psi_m(s) \, \psi_m(s'), \tag{16.83a}$$

where $\{\psi_m(s)\}_{m=1}^{\infty}$ is a countable set of orthonormal eigenfunctions which satisfy the integral equation

$$\int_{-L/\xi}^{L/\xi} \mathrm{d}s \, \psi_m(s) \, \psi_n(s) = \delta_{n,m}, \tag{16.83b}$$

and $\lambda_m > 0$, $m \in \mathbb{N}$, are the respective, non-negative eigenvalues. The eigenvalues λ_m and eigenfunctions $\psi_m(s)$ are solutions of the following homogeneous Fredholm integral equation of the second kind

$$\int_{-L/\xi}^{L/\xi} ds'\, C_{xx}(s,s')\, \psi_m(s') = \lambda_m\, \psi_m(s). \tag{16.83c}$$

Dimensional analysis of both sides of (16.83b) shows that the eigenfunctions $\phi_m(s)$ are dimensionless. Similarly, from (16.83) it follows that the eigenvalues, λ, have the same units as η_0, i.e., $[X]^2$ (the square brackets denote the dimensions of X) [49].

16.12.1 Main Properties of SSRF Karhunen-Loève Expansion

The SSRF Karhunen-Loève expansion has certain characteristic properties that are briefly explained below. These properties help to guide us through the steps involved in the calculation of the Karhunen-Loève expansion. They are clarified in the sections that will follow.

- The K-L function basis for Spartan spatial random fields is obtained by solving a fourth-order ordinary differential equation.
- The K-L eigenvalues are determined by solving (numerically) transcendental equations that are obtained from the respective boundary conditions.
- The K-L basis contains two distinct *eigenfunction branches*: the first branch involves a superposition of harmonic and hyperbolic eigenfunctions while the second branch involves only harmonic functions.
- Each branch contains eigenfunctions with both even, i.e., $\psi_m(s) = \psi_m(-s)$ and odd, i.e., $\psi_m(s) = -\psi_m(-s)$, symmetry. We will call them, respectively, *even-parity* and *odd-parity* eigenfunctions.
- The Karhunen-Loève expansion involves two *rigidity regimes* that are controlled by η_1: the first regime, denoted by R1, involves values $\eta_1 \geq 0$, whereas the second regime, denoted by R2, involves values $-2 < \eta_1 < 0$.
- The *first branch* contributes eigenfunctions in both rigidity regimes.
- The *second branch* contributes eigenfunctions only in R2, i.e., if $-2 < \eta_1 < 0$.
- The branch, the parity, and the rigidity regime are all factors that influence the transcendental equations for the K-L eigenvalues.

16.12.2 K-L ODE for Spartan Covariance

To obtain the Karhunen-Loève ODE from the respective Fredholm equation, we recall the inverse SSRF kernel defined by (9.30). The respective equations for the one-dimensional problem considered herein take the following form:

$$\int_{-L/\xi}^{L/\xi} ds''\, C_{xx}^{-1}(s,s'')\, C_{xx}(s'',s') = \delta(s-s'), \tag{16.84a}$$

$$C_{xx}^{-1}(s,s') = \frac{1}{\eta_0}\left(1 - \eta_1 \frac{d^2}{ds^2} + \frac{d^4}{ds^4}\right) \delta(s-s'). \tag{16.84b}$$

Remarks 1. The SSRF covariance function actually depends on the absolute values of the lag differences, i.e., on $|s - s''|$. 2. Since we use normalized coordinates—scaled by ξ—the Dirac delta function is also dimensionless. 3. In the following, we use $\widetilde{L} = L/\xi$ to denote the *scaled domain half-length*.

We rewrite the Fredholm equation (16.83c), changing the position labels, as follows

$$\int_{-L/\xi}^{L/\xi} ds''\, C_{xx}(s', s'')\, \psi_m(s'') = \lambda_m\, \psi_m(s').$$

We then operate on both sides of the above with the inverse SSRF kernel $C_{xx}^{-1}(s, s')$ and integrate over s' to obtain the following integral equation

$$\int_{-\widetilde{L}}^{\widetilde{L}} ds'\, C_{xx}^{-1}(s,s') \int_{-\widetilde{L}}^{\widetilde{L}} ds''\, C_{xx}(s',s'')\, \psi(s'') = \lambda \int_{-\widetilde{L}}^{\widetilde{L}} ds' C_{xx}^{-1}(s,s'')\, \psi(s').$$

Using the integral equation (16.84a) that defines the precision kernel on the left hand side and (16.84b) that defines the inverse SSRF kernel on the right hand side of the above equation, we obtain the following, *fourth-order, linear ODE*

$$\kappa_1(\lambda_m)\, \psi_m(s) + \eta_1\, \psi_m^{(2)}(s) - \psi_m^{(4)}(s) = 0, \text{ for } m \in \mathbb{N}, \tag{16.85}$$

where $\psi_m^{(n)}(s)$, $n = 2, 4$ represents the n-th order derivative of the eigenfunction $\psi_m(s)$, and the coefficient $\kappa_1(\lambda_m)$ is given by

$$\kappa_1(\lambda_m) = \frac{\eta_0}{\lambda_m} - 1, \; m \in \mathbb{N}. \tag{16.86}$$

16.12.3 K-L Eigenfunctions for Spartan Covariance

The general form of the *K-L eigenfunctions* that satisfy (16.85) is then given by

$$\psi_m(s) = a_1\, e^{\rho(\lambda_m)s} + a_2\, e^{-\rho(\lambda_m)s} + a_3\, e^{\mu(\lambda_m)s} + a_4\, e^{-\mu(\lambda_m)s}, \tag{16.87}$$

where the a_i $(i = 1, 2, 3, 4)$ are coefficients determined by respective boundary conditions obtained from the Karhunen-Loève Fredholm equation (see below).

The exponents $\pm\rho(\lambda_m)$, $\pm\mu(\lambda_m)$ are the roots of the *characteristic K-L SSRF polynomial* that is associated with the Karhunen-Loève ODE. The polynomial is obtained from (16.85) by means of the substitution $\psi_m^{(n)}(\mathbf{s}) \mapsto x^n$ which leads to

$$p_m(x) = \kappa_1(\lambda_m) + \eta_1 x^2 - x^4. \tag{16.88}$$

The roots of the above fourth degree *characteristic K-L SSRF polynomial* are given by the following radicals

$$\rho(\lambda_m) = \frac{1}{\sqrt{2}}\sqrt{\eta_1 - \sqrt{\eta_1{}^2 - 4 + \frac{4\eta_0}{\lambda_m}}} = \rho_{\text{re}}(\lambda_m) + i\rho_{\text{im}}(\lambda_m), \tag{16.89a}$$

$$\mu(\lambda_m) = \frac{1}{\sqrt{2}}\sqrt{\eta_1 + \sqrt{\eta_1{}^2 - 4 + \frac{4\eta_0}{\lambda_m}}} = \mu_{\text{re}}(\lambda_m) + i\mu_{\text{im}}(\lambda_m), \tag{16.89b}$$

where the subscripts "re" and "im" denote respectively the real and imaginary parts of the roots $\rho(\lambda_m)$ and $\mu(\lambda_m)$.

Eigenfunction branches The roots can be either real, imaginary or complex numbers for different values of λ_m and η_1. Two different eigenfunction branches are obtained that are differentiated by the type (real versus imaginary) of the roots. In the first eigenfunction branch, $\rho(\lambda_m)$ is an imaginary number while $\mu(\lambda_m)$ is a real number. The first branch is obtained for $\eta_1 > -2$—hence for both rigidity regimes R1 ($\eta_1 \geq 0$) and R2 ($-2 < \eta_1 < 0$)—and for eigenvalues $0 \leq \lambda_m < \eta_0$.

The second eigenfunction branch is obtained if both $\rho(\lambda_m)$ and $\mu(\lambda_m)$ are imaginary numbers. This condition is only materialized in the rigidity regime R2 and for eigenvalues such that $\eta_0 \leq \lambda \leq \eta_0/(4 - \eta_1{}^2)$.

Based on the nature (real or imaginary) of the roots in each branch, the first branch contains both harmonic and hyperbolic eigenfunctions, while the second branch only harmonic eigenfunctions. The eigenvalues that can be admitted in the two rigidity regimes are summarized in Table 16.2.

Inadmissible root combinations It is in principle possible to have two real roots or two complex-valued roots. These are accessible only for values of λ_m exceeding the

Table 16.2 Permissible range for the Karhunen-Loève SSRF expansion eigenvalues per eigenfunction branch and rigidity regime (R1 or R2)

	R1 ($\eta_1 \geq 0$)	R2 ($-2 < \eta_1 < 0$)
First branch	$0 \leq \lambda_m < \eta_0$	$0 \leq \lambda_m < \eta_0$
Second branch	—	$\eta_0 \leq \lambda_m \leq \frac{\eta_0}{1 - \eta_1{}^2/4}$

upper bound $\eta_0/(1-\eta_1{}^2/4)$ listed in Table 16.2. However, numerical investigations show that the transcendental K-L SSRF eigenvalue equations do not admit real-valued solutions that exceed this bound [803]. In addition, the first and second branch solutions accurately reconstruct the covariance function based on Mercer's expansion theorem (Theorem 16.1). Hence, we will not be further concerned with the possibility of both roots being real-valued or complex-valued.

16.12.4 K-L Eigenvalues for Spartan Covariance

The eigenvalues of the Karhunen-Loève SSRF expansion are determined by solving transcendental equations. These are determined by the respective boundary conditions of the SSRF Fredholm integral equation (16.83c) that are obtained in terms of the eigenfunctions and their first three derivatives at the endpoints of the domain (for details see [803]). The analytical expression of the transcendental equations depends on (i) the eigenfunction branch (ii) the parity of the eigenfunctions (see below) and (iii) the rigidity regime, i.e., whether $\eta_1 \geq 0$ or $-2 < \eta_1 < 0$.

To simplify the notation, we denote in (16.89) by $W(\lambda)$ the part of the roots $\rho(\lambda_m)$ and $\mu(\lambda_m)$ that depends on the eigenvalues. Hence, $W(\lambda)$ is given by

$$W(\lambda) = \frac{1}{2}\sqrt{\eta_1{}^2 - 4 + \frac{4\eta_0}{\lambda}}. \tag{16.90a}$$

In light of the lower and upper bounds satisfied by λ_m (see Table 16.2) the function $W(\lambda)$ defined above takes non-negative real value for all the λ values that are permitted in the two eigenfunction branches of the Karhunen-Loève SSRF expansion.

The expressions for the roots $\rho(\lambda_m)$ and $\mu(\lambda_m)$ are respectively modified as shown below.

First root, both branches The following expression applies to both rigidity regimes and eigenfunction branches, i.e., for $\eta_1 > -2$ and $0 \leq \lambda \leq \eta_0/(1-\eta_1{}^2/4)$

$$\rho(\lambda) = \mathrm{i}\sqrt{W(\lambda) - \frac{\eta_1}{2}}, \text{ for } \eta_1 > -2 \text{ and } 0 \leq \lambda \leq \frac{\eta_0}{1-\eta_1{}^2/4}. \tag{16.90b}$$

Second root, first branch This expression for $\mu(\lambda)$ applies in both rigidity regimes but only in the first eigenfunction branch

$$\mu(\lambda) = \sqrt{\frac{\eta_1}{2} + W(\lambda)}, \text{ for } \eta_1 > -2 \text{ and } 0 \leq \lambda < \eta_0. \tag{16.90c}$$

Second root, second branch The following expression for $\mu(\lambda)$ is applicable only in the second rigidity regime and for the second eigenfunction branch

$$\mu(\lambda) = \mathrm{i}\sqrt{\left|\frac{\eta_1}{2} + W(\lambda)\right|}, \text{ for } -2 < \eta_1 < 0 \text{ and } \eta_0 \leq \lambda < \frac{\eta_0}{1 - \frac{\eta_1{}^2}{4}}. \tag{16.90d}$$

The expressions of the transcendental equations depend on the eigenfunction branch and the eigenfunction parity. The roots of these equations with respect to $W(\lambda)$ will be denoted by w_m, $m \in \mathbb{N}$. The Karhunen-Loève SSRF eigenvalues are then determined from the roots $\{w_m\}_{m=1}^{\infty}$ as follows:

$$\lambda_m = \frac{\eta_0}{w_m^2 + 1 - \frac{\eta_1{}^2}{4}}, \quad m = 1, 2, \ldots. \tag{16.91}$$

Eigenvalue search intervals Given the inverse relation (16.90a) between $W(\lambda)$ and λ, the highest eigenvalue $\lambda_{\max} = \max(w_m)_{m=1}^{\infty}$ corresponds to the minimum $w_{\min} = \min(w_m)_{m=1}^{\infty}$. The eigenvalues $\{\lambda_m\}_{m=1}^{\infty}$ are obtained by solving transcendental equations that represent boundary conditions which result from the Fredholm integral equation (16.83c).

Different transcendental equations correspond to different parity sectors of each eigenfunction branch. The eigenvalues are found by numerically solving the respective equations. Hence, we need to specify search intervals for the roots $\{w_m\}_{m=1}^{\infty}$. In the *first branch*, since $\lambda < \eta_0$ it follows from (16.90a) that the roots $w_m \in (|\eta_1|/2, \infty)$ for all $m \in \mathbb{N}$ in R1 and R2. In the *second branch*, the bounds of λ_m specified in Table 16.2 imply that the $w_m \in (0, |\eta_1|/2]$ (for all $m \in \mathbb{N}$ in R2).

The eigenvalues obtained from (16.91) correspond to different parities and different branches depending on w_m. The index m orders the eigenvalues in descending order according to magnitude, i.e., $\lambda_1 > \lambda_2 > \lambda_3 \ldots$, but it does not distinguish between different eigenfunction parities and branches.

16.12.5 First Branch of K-L Eigenfunctions

In the first branch $\mu(\lambda)$ is real and $\rho(\lambda)$ is imaginary. The first branch is realized for all the permissible values of η_1, i.e., for $\eta_1 > -2$, and for $\lambda_m \in (0, \eta_0)$. In addition, for SSRFs with rigidity $\eta_1 > 0$ the Karhunen-Loève expansion is fully determined from eigenfunctions from the first branch.

The *first-branch K-L eigenfunctions* are given by the following superposition of harmonic and hyperbolic functions

$$\begin{aligned}\psi_m^{(1)}(s) &= c_1 \cos(\rho_{\text{im}}(\lambda_m)\, s) + c_2 \sin(\rho_{\text{im}}(\lambda_m)\, s) \\ &+ c_3 \cosh(\mu(\lambda_m)\, s) + c_4 \sinh(\mu(\lambda_m)\, s).\end{aligned} \tag{16.92}$$

To fully determine the eigenfunctions the coefficients $\{c_i\}_{i=1}^4$ should be specified. These are determined by the respective boundary conditions obtained in terms of the eigenfunctions and their first three derivatives at the endpoints of the domain (for details see [803]). There are two families of admissible solutions: one that contains *even functions* (even parity) and has $c_2 = c_4 = 0$ and one that contains *odd functions* (odd parity) and has $c_1 = c_3 = 0$.

First-branch, even-parity eigenfunctions The eigenfunctions in this sector satisfy the inversion symmetry $\psi_m^{(1,e)}(-s) = \psi_m^{(1,e)}(s)$. They are given by the following combination of cosine and hyperbolic cosine functions:

$$\psi_m^{(1,e)}(s) = \frac{1}{[Z_1(\lambda_m)]^{1/2}} \left[\alpha_1(\lambda_m)\cos(\rho_{\text{im}}(\lambda_m)\, s) + \cosh(\mu(\lambda_m)\, s)\right]. \tag{16.93a}$$

The coefficient $\alpha_1(\lambda_m)$ is the relative weight of the harmonic term with respect to the hyperbolic term, while $Z_1(\lambda_m)$ is a normalization constant that ensures the orthonormality of the eigenfunctions. The coefficients $\alpha_1(\lambda_m)$ and $Z_1(\lambda_m)$ are given by the following equations which involve dimensionless variables (below we drop the dependence of the characteristic polynomial roots μ and ρ on λ_m for reasons of brevity):

$$\alpha_1(\lambda_m) = \frac{\mu \sinh(\widetilde{L}\mu) g(\eta_1) + \left(\mu^2 + 1\right) \cosh(\widetilde{L}\mu)}{\rho_{\text{im}} \sin(\widetilde{L}\rho_{\text{im}}) g(\eta_1) - \left(1 - \rho_{\text{im}}^2\right) \cos(\widetilde{L}\rho_{\text{im}})}, \tag{16.93b}$$

$$Z_1(\lambda_m) = \widetilde{L} + \alpha_1^2 \left[\widetilde{L} + \frac{\sin\left(2\widetilde{L}\rho_{\text{im}}\right)}{2\rho_{\text{im}}}\right] + \frac{\sinh(2\widetilde{L}\mu)}{2\mu} + \frac{4\alpha_1}{\rho_{\text{im}}^2 + \mu^2}$$

$$\times \left[\rho_{\text{im}} \cosh\left(\widetilde{L}\mu\right) \sin\left(\widetilde{L}\rho_{\text{im}}\right) + \mu \cos\left(\widetilde{L}\rho_{\text{im}}\right) \sinh\left(\widetilde{L}\mu\right)\right], \qquad (16.93c)$$

where $g(\eta_1)$ is a function that depends on the rigidity coefficient as follows:

$$g(\eta_1) = \begin{cases} \frac{1}{\sqrt{2}}\left(\sqrt{\eta_1 - \sqrt{\eta_1{}^2 - 4}} + \sqrt{\eta_1 + \sqrt{\eta_1{}^2 - 4}}\right), & \eta_1 > 2 \\ \sqrt{\eta_1 + 2}, & -2 < \eta_1 \leq 2. \end{cases} \qquad (16.93d)$$

The two branches of $g(\eta_1)$ reflect the different forms of the SSRF covariance in the different rigidity regimes. With reference to the harmonic oscillator, they reflect the difference between the underdamped and critical regimes versus the overdamped oscillator regime.

Thus, the coefficients $\alpha_1(\lambda_m)$ and $Z_1(\lambda_m)$ are fully determined in terms of the eigenvalues, η_1, and the normalized domain half-length $\widetilde{L}$.

First-branch, even-parity eigenvalues The transcendental equation that determines the eigenvalues in this branch is

$$\rho_{\text{im}}(\lambda)\, I_2(\lambda)\, I_3(\lambda) + \mu(\lambda)\, I_1(\lambda)\, I_4(\lambda) = 0. \qquad (16.94a)$$

The coefficients $\{I_i(\lambda)\}_{i=1}^4$ are nonlinear functions of the eigenvalues. Their analytical expression depends on the rigidity parameter η_1 through $g(\eta_1)$—defined in (16.93d)—and the roots of the characteristic polynomial—defined in (16.90). The coefficients $\{I_i(\lambda)\}_{i=1}^4$ are given by the following functions

$$I_1(\lambda) = \left[1 - \rho_{\text{im}}^2(\lambda)\right] \cos\left[\widetilde{L}\,\rho_{\text{im}}(\lambda)\right] - g(\eta_1)\,\rho_{\text{im}}(\lambda)\,\sin\left[\widetilde{L}\,\rho_{\text{im}}(\lambda)\right], \qquad (16.94b)$$

$$I_2(\lambda) = g(\eta_1)\,\rho_{\text{im}}(\lambda)\cos\left[\widetilde{L}\,\rho_{\text{im}}(\lambda)\right] + \left[1 - \rho_{\text{im}}^2(\lambda)\right] \sin\left[\widetilde{L}\,\rho_{\text{im}}(\lambda)\right], \qquad (16.94c)$$

$$I_3(\lambda) = \mu^2(\lambda) + 1 + g(\eta_1)\,\mu(\lambda)\,\tanh\left[\widetilde{L}\,\mu(\lambda)\right], \qquad (16.94d)$$

$$I_4(\lambda) = g(\eta_1)\,\mu(\lambda) + \left[\mu^2(\lambda) + 1\right] \tanh\left[\widetilde{L}\,\mu(\lambda)\right]. \qquad (16.94e)$$

We rewrite (16.94) collecting all the terms (16.94b)–(16.94e). To keep it concise, we drop the dependence of g, μ, ρ_{im} on parameters. The eigenvalues of the *first branch, even sector* solution are given by the roots of the following transcendental equation:

Eigenvalue equation for the even sector of the first branch: $B_{1,e} = 0$

$$
\begin{aligned}
B_{1,e}(W) = & \left[g\rho_{\text{im}}^2 \left(\mu^2 + 1 + g\mu \tanh(\widetilde{L}\mu) \right) + \mu(1 - \rho_{\text{im}}^2) \right. \\
& \left. \times \left((\mu^2 + 1) \tanh(\widetilde{L}\mu) + g\mu \right) \right] \cos(\widetilde{L}\, \rho_{\text{im}}) \\
& + \left[\rho_{\text{im}} \left(1 - \rho_{\text{im}}^2 \right) \left(\mu^2 + 1 + g\mu \tanh(\widetilde{L}\mu) \right) - g\mu\rho_{\text{im}} \right. \\
& \left. \times \left((\mu^2 + 1) \tanh(\widetilde{L}\mu) + g\mu \right) \right] \sin(\widetilde{L}\, \rho_{\text{im}}).
\end{aligned}
\tag{16.94f}
$$

Numerical solution of transcendental eigenvalue equations The roots ρ and μ are defined in terms of (16.90), and $g(\eta_1)$ by means of (16.93d). The transcendental equation (16.94f) needs to be solved numerically. The dependence of (16.93d) on the Karhunen-Loève eigenvalues is through the roots of the characteristic polynomial (16.89). These roots depend nonlinearly on the eigenvalues $\{\lambda_m\}_{m=1}^{\infty}$, leading to a strongly non-uniform distribution of the latter over the search interval. The transcendental equation $B_{1,e}(W) = 0$ is thus solved numerically in terms of W, and the eigenvalues are obtained from the roots $\{w_m\}_{m=1}^{\infty}$ by inverting (16.90a).

First-branch, odd-parity eigenfunctions The odd-parity eigenfunctions satisfy the symmetry $\psi_m^{(1,o)}(-s) = -\psi_m^{(1,o)}(s)$. They are given by

$$
\psi_m^{(1,o)}(s) = \frac{1}{[Z_2(\lambda_m)]^{1/2}} \left\{ \alpha_2(\lambda_m) \sin\left[\rho_{\text{im}}(\lambda_m)\, s\right] + \sinh\left[\mu(\lambda_m)\, s\right] \right\},
\tag{16.95a}
$$

where

$$
\alpha_2(\lambda_m) = -\frac{\mu \cosh(\widetilde{L}\mu) g(\eta_1) + \left(\mu^2 + 1\right) \sinh(\widetilde{L}\mu)}{\rho_{\text{im}} \cos(\widetilde{L}\rho_{\text{im}}) g(\eta_1) - \left(1 - \rho_{\text{im}}^2\right) \sin(\widetilde{L}\rho_{\text{im}})},
\tag{16.95b}
$$

$$
\begin{aligned}
Z_2(\lambda_m) = & - \widetilde{L} + \alpha_2^2 \left[\widetilde{L} - \frac{\sin\left(2\widetilde{L}\rho_{\text{im}}\right)}{2\rho_{\text{im}}} \right] + \frac{\sinh(2\widetilde{L}\mu)}{2\mu} + \frac{4\alpha_2}{\rho_{\text{im}}^2 + \mu^2} \\
& \times \left[\mu \cosh\left(\widetilde{L}\mu\right) \sin\left(\widetilde{L}\rho_{\text{im}}\right) - \rho_{\text{im}} \sinh\left(\widetilde{L}\mu\right) \cos\left(\widetilde{L}\rho_{\text{im}}\right) \right].
\end{aligned}
\tag{16.95c}
$$

First-branch, odd-parity eigenvalues The eigenvalues in the odd sector are the roots of the following equation

$$\mu(\lambda)\, I_2(\lambda)\, I_3(\lambda) - \rho_{\mathrm{im}}(\lambda)\, I_1(\lambda)\, I_4(\lambda) = 0. \tag{16.96a}$$

Its compact form, in analogy with (16.94f) for the even sector, is given by
Eigenvalue equation for the odd sector of the first branch: $B_{1,o} = 0$.

$$\begin{aligned} B_{1,o}(W) = &\Big[g\rho_{\mathrm{im}}\mu \left(\mu^2 + 1 + g\mu \tanh(\widetilde{L}\mu)\right) - \rho_{\mathrm{im}}(1-\rho_{\mathrm{im}}^2) \\ &\times \left((\mu^2+1)\tanh(\widetilde{L}\mu) + g\mu\right)\Big] \cos(\widetilde{L}\,\rho_{\mathrm{im}}) \\ &+\Big[\mu\left(1-\rho_{\mathrm{im}}^2\right)\left(\mu^2 + 1 + g\mu \tanh(\widetilde{L}\mu)\right) + g\rho_{\mathrm{im}}^2 \\ &\times \left((\mu^2+1)\tanh(\widetilde{L}\mu) + g\mu\right)\Big] \sin(\widetilde{L}\,\rho_{\mathrm{im}}). \end{aligned} \tag{16.96b}$$

The equation $B_{1,o} = 0$ is also solved as a function of W where the latter is defined in (16.90a). The Karhunen-Loève eigenvalues are determined from the roots $\{w_m\}_{m=1}^{\infty}$ by means of (16.91).

16.12.6 *Second Branch of K-L Eigenfunctions*

In this case, both $\mu(\lambda)$ and $\rho(\lambda)$ are imaginary. This condition is obtained for $-2 < \eta_1 < 0$ and for $\eta_0/(1-\frac{1}{4}\eta_1{}^2) \geq \lambda \geq \eta_0$. The transcendental equation that determines the eigenvalues has a unique expression which corresponds to the underdamped oscillator regime.

The *K-L eigenfunction solutions from the second branch* are then given by

$$\begin{aligned} \psi_m^{(2)}(s) = &\, d_1 \cos(\rho_{\mathrm{im}}(\lambda_m)\, s) + d_2 \sin(\rho_{\mathrm{im}}(\lambda_m)\, s) \\ &+ d_3 \cos(\mu_{\mathrm{im}}(\lambda_m)\, s) + d_4 \sin(\mu_{\mathrm{im}}(\lambda_m)\, s). \end{aligned} \tag{16.97}$$

Second-branch, even-parity eigenfunctions These eigenfunctions satisfy the inversion symmetry $\psi_m^{(2,e)}(-s) = \psi_m^{(2,e)}(s)$. They are given by

$$\psi_m^{(2,e)}(s) = \frac{1}{[Z_3(\lambda_m)]^{1/2}} \left[\cos(\rho_{\mathrm{im}}(\lambda_m)\, s) + \alpha_3(\lambda_m)\cos(\mu_{\mathrm{im}}(\lambda_m)\, s)\right], \tag{16.98a}$$

where

$$\alpha_3(\lambda_m) = \frac{\rho_{\mathrm{im}} \sin(\widetilde{L}\rho_{\mathrm{im}}) g(\eta_1) + \left(1 - \rho_{\mathrm{im}}^2\right) \cos(\widetilde{L}\rho_{\mathrm{im}})}{-\mu_{\mathrm{im}} \sin(\widetilde{L}\mu_{\mathrm{im}}) g(\eta_1) + \left(1 - \mu_{\mathrm{im}}^2\right) \cos(\widetilde{L}\mu_{\mathrm{im}})}, \tag{16.98b}$$

$$Z_3(\lambda_m) = \widetilde{L} + \alpha_3^2 \left[\widetilde{L} + \frac{\sin\left(2\widetilde{L}\mu_{\mathrm{im}}\right)}{2\mu_{\mathrm{im}}} \right] + \frac{\sin(2\widetilde{L}\rho_{\mathrm{im}})}{2\rho_{\mathrm{im}}} + \frac{4\alpha_3}{\rho_{\mathrm{im}}^2 - \mu_{\mathrm{im}}^2} \times \left[\rho_{\mathrm{im}} \cos\left(\widetilde{L}\mu_{\mathrm{im}}\right) \sin\left(\widetilde{L}\rho_{\mathrm{im}}\right) - \mu_{\mathrm{im}} \cos\left(\widetilde{L}\rho_{\mathrm{im}}\right) \sin\left(\widetilde{L}\mu_{\mathrm{im}}\right) \right]. \tag{16.98c}$$

Second-branch, even-parity eigenvalues The eigenvalues λ_m are solutions of the following transcendental eigenvalue equation

$$\rho_{\mathrm{im}}(\lambda)\, I_2(\lambda)\, I_3^*(\lambda) - \mu_{\mathrm{im}}(\lambda)\, I_1(\lambda)\, I_4^*(\lambda) = 0, \tag{16.99a}$$

where the functions $I_1(\lambda)$, $I_2(\lambda)$ were defined in equations (16.94b)–(16.94c), while $I_3^*(\lambda)$ and $I_4^*(\lambda)$ are given by

$$I_3^*(\lambda) = \left[1 - \mu_{\mathrm{im}}^2(\lambda)\right] \cos\left[\widetilde{L}\, \mu_{\mathrm{im}}(\lambda)\right] - g(\eta_1)\mu_{\mathrm{im}}(\lambda) \sin\left[\widetilde{L}\, \mu_{\mathrm{im}}(\lambda)\right], \tag{16.99b}$$

$$I_4^*(\lambda) = g(\eta_1)\mu_{\mathrm{im}}(\lambda) \cos\left[\widetilde{L}\, \mu_{\mathrm{im}}(\lambda)\right] + \left[1 - \mu_{\mathrm{im}}^2(\lambda)\right] \sin\left[\widetilde{L}\, \mu_{\mathrm{im}}(\lambda)\right]. \tag{16.99c}$$

Remark Since the second branch plays a role only for $\eta_1 < 0$, it follows from (16.93d) that $g(\eta_1) = \sqrt{\eta_1 + 2}$ in this branch.

Note that in the second branch the terms $I_1(\lambda_m)$ and $I_2(\lambda_m)$ which depend only on $\rho_{\mathrm{im}}(\lambda)$ have the same form as in the first branch, reflecting the fact that in both branches the root $\rho_{\mathrm{im}}(\lambda_m)$ is imaginary. On the other hand, $I_3(\lambda_m)$ and $I_4(\lambda_m)$ are replaced by $I_3^*(\lambda)$ and $I_4^*(\lambda)$, because $\mu(\lambda_m)$ changes from real to imaginary in the second branch.

Eigenvalue equation for the even sector of the second branch: $B_{2,e} = 0$.

$$\begin{aligned}B_{2,e}(W) = & g\left[\rho_{\text{im}}^2\left(1-\mu_{\text{im}}^2\right)-\mu_{\text{im}}^2\left(1-\rho_{\text{im}}^2\right)\right]\cos(\widetilde{L}\mu_{\text{im}})\cos(\widetilde{L}\rho_{\text{im}}) \\ & +\left[\rho\left(1-\mu_{\text{im}}^2\right)(1-\rho_{\text{im}})+g^2\mu_{\text{im}}^2\,\rho_{\text{im}}\right]\cos(\widetilde{L}\mu_{\text{im}})\sin(\widetilde{L}\rho_{\text{im}}) \\ & -\left[g^2\mu_{\text{im}}\rho_{\text{im}}^2+\mu_{\text{im}}\left(1-\mu_{\text{im}}^2\right)(1-\rho_{\text{im}})\right]\sin(\widetilde{L}\mu_{\text{im}})\cos(\widetilde{L}\rho_{\text{im}}) \\ & +g\left[\mu_{\text{im}}\left(1-\mu_{\text{im}}^2\right)\rho_{\text{im}}-\mu_{\text{im}}\rho_{\text{im}}(1-\rho_{\text{im}})\right]\sin(\widetilde{L}\,\rho_{\text{im}})\sin(\widetilde{L}\,\mu_{\text{im}}).\end{aligned} \tag{16.99d}$$

Second-branch, odd-parity eigenfunctions Eigenfunctions in this sector are anti-symmetric, i.e., $\psi_m^{(2,o)}(-s) = -\psi_m^{(2,o)}(s)$. They are given by

$$\psi_m^{(2,o)}(s) = \frac{1}{[Z_4(\lambda_m)]^{1/2}}\left[\sin(\rho_{\text{im}}(\lambda_m)\,s)+\alpha_4(\lambda_m)\sin(\mu_{\text{im}}(\lambda_m)\,s)\right], \tag{16.100a}$$

where

$$\alpha_4(\lambda_m) = -\frac{\rho_{\text{im}}\cos(\widetilde{L}\rho_{\text{im}})g(\eta_1)+\left(1-\rho_{\text{im}}^2\right)\sin(\widetilde{L}\rho_{\text{im}})}{\mu_{\text{im}}\cos(\widetilde{L}\mu_{\text{im}})g(\eta_1)+\left(1-\mu_{\text{im}}^2\right)\sin(\widetilde{L}\mu_{\text{im}})}, \tag{16.100b}$$

$$\begin{aligned}Z_4(\lambda_m) = & \widetilde{L}+\alpha_4^2\left[\widetilde{L}-\frac{\sin\left(2\widetilde{L}\mu_{\text{im}}\right)}{2\mu_{\text{im}}}\right]-\frac{\sin(2\widetilde{L}\rho_{\text{im}})}{2\rho_{\text{im}}}+\frac{4\alpha_4}{\rho_{\text{im}}^2-\mu_{\text{im}}^2} \\ & \times\left[\mu_{\text{im}}\cos\left(\widetilde{L}\mu_{\text{im}}\right)\sin\left(\widetilde{L}\rho_{\text{im}}\right)-\rho_{\text{im}}\cos\left(\widetilde{L}\rho_{\text{im}}\right)\sin\left(\widetilde{L}\mu_{\text{im}}\right)\right].\end{aligned} \tag{16.100c}$$

Second-branch, odd-parity eigenvalues In this case, the eigenvalues are the roots of the following transcendental equation

$$\mu_{\text{im}}(\lambda)\,I_2(\lambda)\,I_3^*(\lambda)-\rho_{\text{im}}(\lambda)\,I_1(\lambda)\,I_4^*(\lambda)=0. \tag{16.101a}$$

Eigenvalue equation for the odd sector of the second branch: $B_{2,o}=0$.

$$\begin{aligned}B_{2,o}(W) =& g\rho_{\rm im}\mu_{\rm im}\left(1-\mu_{\rm im}^2-g\rho_{\rm im}\right)\cos(\widetilde{L}\mu_{\rm im})\cos(\widetilde{L}\rho_{\rm im})\\&+\mu_{\rm im}\left[\left(1-\mu_{\rm im}^2-g\rho_{\rm im}\right)\left(1-\rho_{\rm im}^2\right)\right]\cos(\widetilde{L}\mu_{\rm im})\sin(\widetilde{L}\rho_{\rm im})\\&-g\rho_{\rm im}\left(g\mu_{\rm im}^2+\rho_{\rm im}-\rho_{\rm im}\mu_{\rm im}^2\right)\sin(\widetilde{L}\mu_{\rm im})\cos(\widetilde{L}\rho_{\rm im})\\&+g\left[\mu_{\rm im}^2(\rho_{\rm im}-g)+\rho_{\rm im}^3\left(1-\mu_{\rm im}^2\right)-\rho_{\rm im}\right]\sin(\widetilde{L}\,\rho_{\rm im})\sin(\widetilde{L}\,\mu_{\rm im}).\end{aligned} \tag{16.101b}$$

16.12.7 Accuracy of Truncated Karhunen-Loève Expansions

Based on the relation (16.75) between the field variance and the eigenvalues of the Karhunen-Loève expansion, the variance is proportional to the superposition of the eigenvalues, i.e.,

$$2\widetilde{L}\sigma_{\rm x}^2=\sum_{m=1}^{\infty}\lambda_m, \tag{16.102}$$

where $2\widetilde{L}$ is the "volume" of the one-dimensional domain $[-L, L]$. If we use a Karhunen-Loève expansion that is truncated after M terms, the explained variance is given by (16.81). The accuracy of the approximation can be evaluated by means of the *relative variance error of the Karhunen-Loève approximation*, which is given as the relative difference between the true and the explained variance, i.e.,

$$\epsilon_{\sigma_{\rm x}^2}^{(M)}=1-\frac{1}{2\widetilde{L}\sigma_{\rm x}^2}\sum_{m=1}^{M}\lambda_m. \tag{16.103}$$

Similarly, the *covariance error* resulting from a truncated Karhunen-Loève expansion according to (16.82) is given by

$$\epsilon_C^{(M)}(s,s')=C_{\rm xx}(h)-\sum_{m=1}^{M}\lambda_m\,\psi_m(s)\,\psi_m(s'). \tag{16.104}$$

In the case of SSRFs, the dependence of the covariance error on the domain size, the rigidity coefficient η_1, and the truncation order M is investigated in [803].

16.12.8 Summary of SSRF Karhunen-Loève Expansion

The SSRF Karhunen-Loève expansion may seem mathematically complicated, but the main steps of the solution are straightforward. Below we emphasize certain properties of the Karhunen-Loève expansion and outline the main steps of the implementation.

- The SSRF Karhunen-Loève eigenfunctions are simple harmonic (sine and cosine) functions or combinations of harmonic and hyperbolic functions (hyperbolic sine and cosine).
- The periods of the harmonic functions and the characteristic lengths of the hyperbolic functions are determined by the roots (16.89) of the SSRF characteristic polynomial.
- The eigenbase (i.e., the eigenvalues and eigenfunctions) involves two different branches and each branch includes both even-parity and odd-parity eigenfunctions.
- The eigenbase depends on L and ξ only through the aspect ratio $\widetilde{L} = L/\xi$.
- The roots of the SSRF Karhunen-Loève characteristic polynomial depend on the K-L eigenvalues, and this dependence is transferred to the eigenfunctions. In particular, the roots are given by $\sqrt{\eta_1/2 \pm W(\lambda_m)}$, where $W(\lambda_m)$, defined by (16.90a), is the only term that depends on the eigenvalues λ_m.
- For $\eta_1 > 0$ the eigenbase comprises functions (both even-parity and odd-parity) from the first branch. The eigenvalues of the even-parity sector are given by the solution of the transcendental equation (16.94) and the eigenfunctions are determined by (16.93). In the odd-parity sector the eigenvalues are given by the solution of the transcendental equation (16.96) and the eigenfunctions are determined by (16.95).
- If $\eta_1 < 0$ a second branch of eigenfunctions develops with even- and odd-parity sectors as well. The eigenvalues of the even-parity sector are given by the roots of the transcendental equation (16.99), and the eigenfunctions are determined by (16.98). In the odd-parity sector the eigenvalues are given by the roots of the transcendental equation (16.101) and the eigenfunctions by (16.100).
- For $\eta_1 < 0$ the second branch contributes a finite number of the most significant eigenvalues. These are followed by an infinite sequence of less significant first-branch eigenvalues [803].
- If $\widetilde{L} \gg 1$ a large number of terms, e.g., $M = \mathcal{O}(100)$, is needed to achieve accurate representation of the SSRF (i.e., 95% of the total variance) by the truncated Karhunen-Loève expansion. The number of terms necessary to reach a specified level of accuracy depends on η_1. Large values of $\widetilde{L}$ imply that ergodic conditions approximately hold so that the full range of the random field variability can be "expressed" over a domain of length L. In such cases, a large number of Karhunen-Loève terms is necessary for convergence of the truncated expansion.
- As $\widetilde{L}$ decreases, a progressively smaller number of terms in the Karhunen-Loève expansion suffices to achieve a specified accuracy. In addition, the field

realizations tend to resemble deterministic functions, which can be described by fewer degrees of freedom.

The main steps of *SSRF simulation* using the Karhunen-Loève expansion are summarized in the Algorithm 16.11. The algorithm uses a stopping criterion based on the approximation of the field variance to a certain specified level p. Alternatively, it is possible to use a termination criterion based on a specified number of Karhunen-Loève terms.

Algorithm 16.11 Karhunen-Loève simulation of Spartan spatial random fields over the domain $[-L, L]$ using an approximation level equal to p for the SSRF variance

Assume that the SSRF parameters η_0, η_1, ξ are known
Calculate the SSRF variance σ_x^2 based on (7.29)
Set the approximation level $0 < p < 1$
Rescale positions and domain size: $s \leftarrow s/L$ and $L \leftarrow L/\xi$
if $\eta_1 < -2$ **then**
    Display *Non-permissible model*
    return
end if
if $-2 < \eta_1 < 0$ **then**
    Calculate the second-branch solution based on (16.98)–(16.101)
end if
Calculate the first branch solution based on (16.93)–(16.96)
Order the eigenvalues in descending order
$n \leftarrow 1$ and $E \leftarrow \lambda_1$
while $E \leq 2pL\sigma_x^2$ **do**
    Generate random number $r \leftarrow \mathcal{N}(0, 1)$
    Augment the eigenbase
    $n \leftarrow n + 1$
    $E \leftarrow E + \lambda_n$
    Based on (16.73) add new K-L component to field:
    $X \leftarrow X + r\sqrt{\lambda_n}\psi_n(s)$
end while

The Algorithm 16.11 focuses on the simulation of a one-dimensional SSRF. For simulations in higher spatial dimensions two approaches are possible: (i) A separable representation is assumed, i.e., $X(\mathbf{s}; \omega) = \prod_{i=1}^{d} X_i(s_i)$. The components $X_i(s_i)$, $i = 1, \ldots, d$ are separately simulated using the Algorithm 16.11. The resulting field in d dimensions is not a d-dimensional SSRF. (ii) The eigenvalues and eigenfunctions of the d-dimensional covariance matrix are numerically evaluated. It is straightforward to modify Algorithm 16.11 for the simulation of other covariance models.

16.12.9 Examples of K-L SSRF Expansion

We illustrate the application of the SSRF Karhunen-Loève expansion over the domain $\mathcal{D} = [-50, 50]$ using a SSRF with characteristic length $\xi = 5$. Hence, $\widetilde{L} = 10$ is the scaled domain half-size.

Positive-Rigidity An SSRF with rigidity coefficient $\eta_1 = 2$ has a modified exponential covariance function given by (7.27b). Since $\eta_1 > 0$, the expansion involves only the first branch.

The following results are obtained using the 25 leading terms of the Karhunen-Loève expansion. Figure 16.18 displays the transcendental functions $B_{1,e}(w)$ and $B_{1,o}(w)$ that determine the eigenvalues of the two parity sectors. These are oscillatory functions with a countable number of roots that determine the eigenfunctions of the Karhunen-Loève expansion.

Figure 16.19 shows eight eigenfunctions from each parity sector that correspond to the highest Karhunen-Loève eigenvalues. The eigenfunctions largely resemble harmonic functions, albeit they also contain hyperbolic components.

Figure 16.20 displays the Karhunen-Loève eigenvalues—note the regular alternation of even-parity and odd-parity eigenvalues—as well as a sample SSRF realization based on the leading $M = 25$ terms.

Finally, Fig. 16.20 shows the SSRF covariance function over the domain $\mathcal{D} \times \mathcal{D}$, as well as the Karhunen-Loève approximation and the approximation error. The latter exhibits an oscillatory structure with higher values near the edges of the $\mathcal{D} \times \mathcal{D}$ domain (Fig. 16.21).

Negative-Rigidity In this case the rigidity coefficient is $\eta_1 = -1.97$ (recall that the lower permissibility bound is $\eta_1 < -2$). The remaining parameters are identical to those used in the positive rigidity case. The negative rigidity coefficient implies that

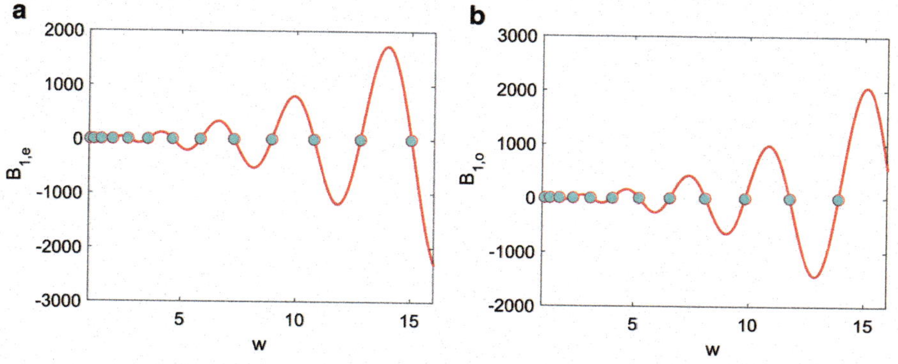

Fig. 16.18 First-branch eigenvalue functions $B_{1,e}(w)$, equation (16.94f) (even parity) and $B_{1,o}(w)$, equation (16.96b) (odd parity) for an SSRF with $\eta_1 = 2$. The roots of the equations $B_{1,e}(w) = 0$ and $B_{1,o}(w) = 0$, which correspond to the w_m in equation (16.91), are marked by the cyan circles. (**a**) Eigenvalues—even parity. (**b**) Eigenvalues—odd parity

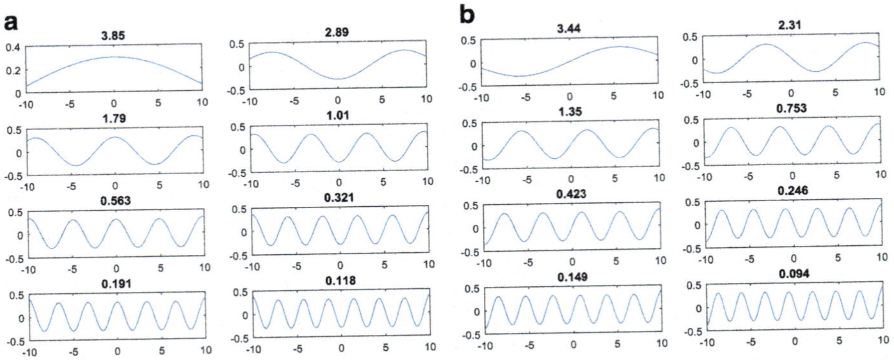

Fig. 16.19 First-branch eigenfunctions $\psi^{(1,e)}(s)$, equation (16.93) (even parity) and $\psi^{(1,o)}(s)$, equation (16.95) (odd parity) for an SSRF with $\eta_1 = 2$. The numbers at the top of the plots correspond to the eigenvalues λ_m. Eight functions are shown from each parity sector. (**a**) Eigenfunctions—even parity. (**b**) Eigenfunctions—odd parity

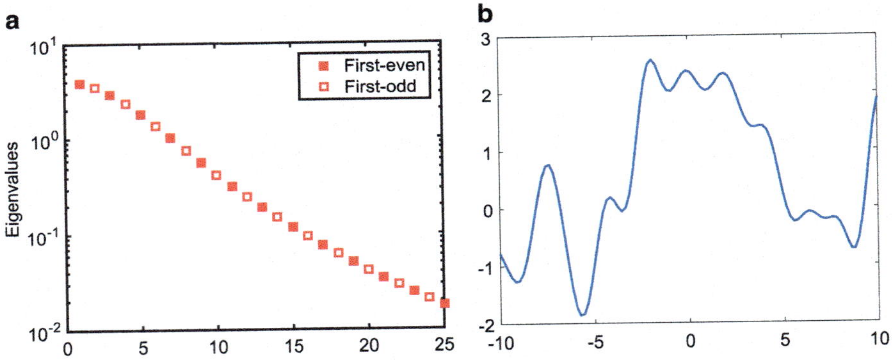

Fig. 16.20 (**a**) First-branch Karhunen-Loève eigenvalues (logarithmic vertical axis) with even parity (filled squares) and odd parity (open squares). (**b**) Sample realization based on 25 Karhunen-Loève terms for an SSRF with $\eta_1 = 2$

the covariance function is given by (7.27a); this expression exhibits an oscillatory dependence on the lag distance.

The following results are obtained using the 27 leading terms of the Karhunen-Loève expansion. Figure 16.22 shows the transcendental functions $B_{1,e}(w)$ and $B_{1,o}(w)$ which determine the eigenvalues in the two parity sectors.

Figure 16.23 displays the eight eigenfunctions from each parity sector with the highest eigenvalues. The same remarks apply as in the $\eta_1 > 0$ case. The respective plots for the second branch solutions are given in Figs. 16.24 and 16.25.

Figure 16.26 plots the Karhunen-Loève eigenvalues from both branches as well as a sample SSRF realization based on the leading $M = 27$ terms (i.e., the terms with the higher eigenvalues ranked from one to 27). The highest energy

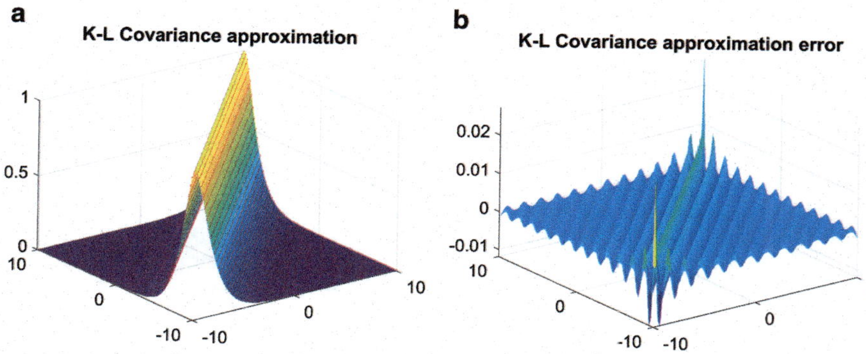

Fig. 16.21 (**a**) SSRF covariance function for $\eta_1 = 2$ (surface) and Karhunen-Loève approximation (red dots). (**b**) Error of Karhunen-Loève covariance approximation using equation (16.104) with $M = 25$ terms

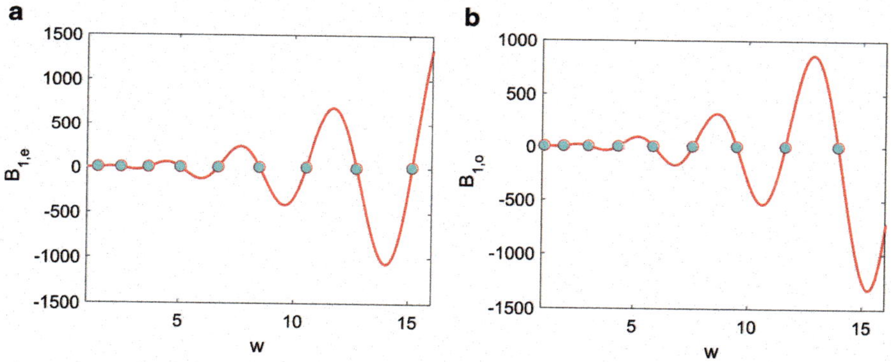

Fig. 16.22 First-branch eigenvalue functions $B_{1,e}(w)$, equation (16.94f) (even parity) and $B_{1,o}(w)$, equation (16.96b) (odd parity) for an SSRF with $\eta_1 = -1.97$. The roots of the equations $B_{1,e}(w) = 0$ and $B_{1,o}(w) = 0$, which correspond to the w_m in equation (16.91), are marked by the cyan circles. (**a**) Eigenvalues—even parity. (**b**) Eigenvalues—odd parity

eigenvalues come from the second branch that contributes a finite but important set of eigenvalues.

Finally, Fig. 16.27 displays the SSRF covariance function as well as the Karhunen-Loève approximation and the approximation error. A comparison of the two plots shows very good agreement between the covariance function and its Karhunen-Loève approximation.

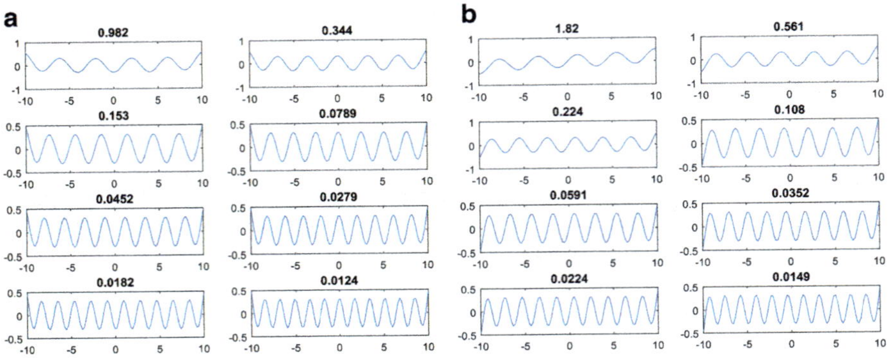

Fig. 16.23 First-branch eigenfunctions $\psi^{(1,e)}(s)$, equation (16.93) (even parity) and $\psi^{(1,o)}(s)$, equation (16.95) (odd parity) for an SSRF with $\eta_1 = -1.97$. The numbers at the top of the plots correspond to the eigenvalues λ_m. Eight functions are shown from each parity sector. (**a**) Eigenfunctions—even parity. (**b**) Eigenfunctions—odd parity

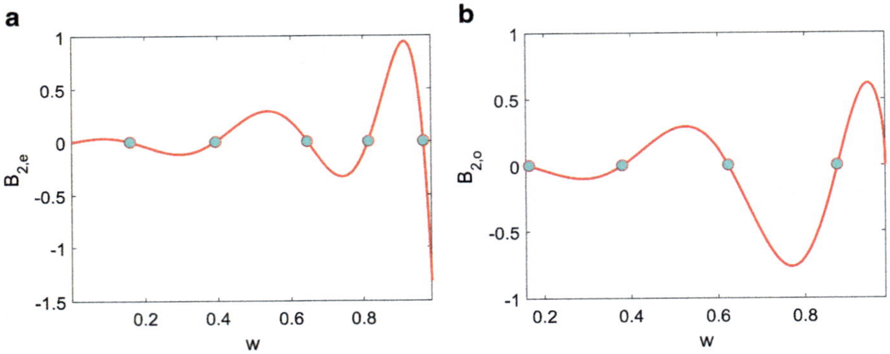

Fig. 16.24 Second-branch eigenvalue functions $B_{2,e}(w)$, equation (16.99d) (even parity) and $B_{2,o}(w)$, equation (16.101b) (odd parity) for an SSRF with $\eta_1 = -1.97$. The roots of the equations $B_{2,e}(w) = 0$ and $B_{2,o}(w) = 0$, which correspond to the w_m in equation (16.91), are marked by the cyan circles. (**a**) Eigenvalues—even parity. (**b**) Eigenvalues—odd parity

16.13 Convergence of Truncated K-L Expansion

The convergence of truncated Karhunen-Loève expansions of Spartan random fields was discussed in Sect. 16.12.8. More generally, the convergence of truncated Karhunen-Loève expansions was studied in [384]. In this paper the main factors affecting the convergence were identified as: (i) the ratio of the domain length over the correlation length of the SRF, (ii) the form of the covariance function, and (iii) the method (analytical or numerical) of solving the Karhunen-Loève Fredholm integral equation (16.83c). In general, the truncated expansion converges faster for lower values of the ratio $\widetilde{L}$ (i.e., the ratio of the domain length over the correlation

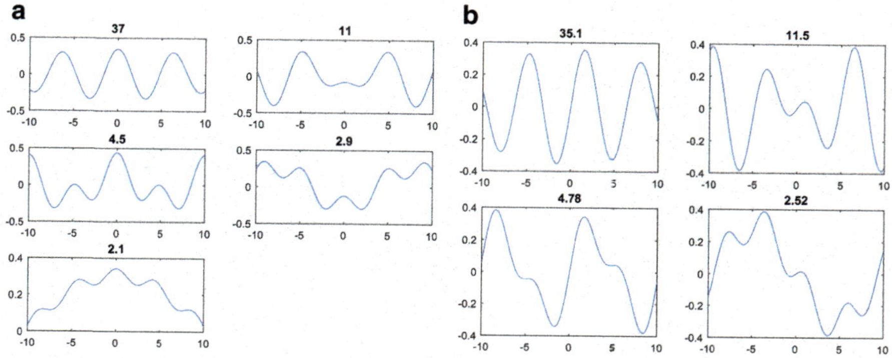

Fig. 16.25 Second-branch eigenfunctions $\psi^{(2,e)}(s)$, equation (16.98) (even parity) and $\psi^{(2,o)}(s)$, equation (16.100) (odd parity) for an SSRF with $\eta_1 = -1.97$. The numbers at the top of the plots correspond to the eigenvalues λ_m. There are five even-parity eigenfunctions and four odd-parity eigenfunctions in the second sector. (**a**) Eigenfunctions—even parity. (**b**) Eigenfunctions—odd parity

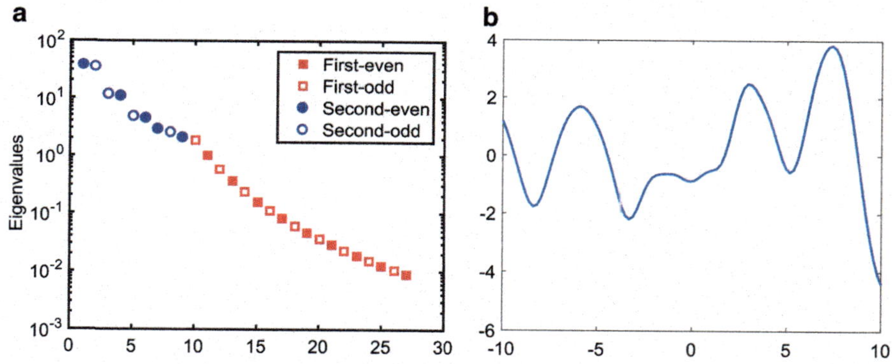

Fig. 16.26 (**a**) Karhunen-Loève eigenvalues from the first branch (squares) and from the second branch (circles) with even (filled marker) and odd (open marker) parity. The vertical axis uses a logarithmic scale. (**b**) Sample realization for an SSRF with $\eta_1 = -1.97$ based on $M = 27$ Karhunen-Loève terms

length). In addition, the Karhunen-Loève truncated expansion converges faster for smooth (differentiable) than for rough (non-differentiable) covariance functions. Finally, in the case of the exponential covariance model (which admits an explicit Karhunen-Loève solution), the convergence of the numerical truncated Karhunen-Loève expansion is slower than that of the explicit solution.

The comparison of the Karhunen-Loève expansion with the spectral representation (16.11) shows that the Karhunen-Loève expansion has an advantage over the spectral method if the random field is highly correlated (i.e., for long correlation lengths). In such cases the Karhunen-Loève expansion of the covariance function converges faster to the true model than the respective truncated spectral expansion

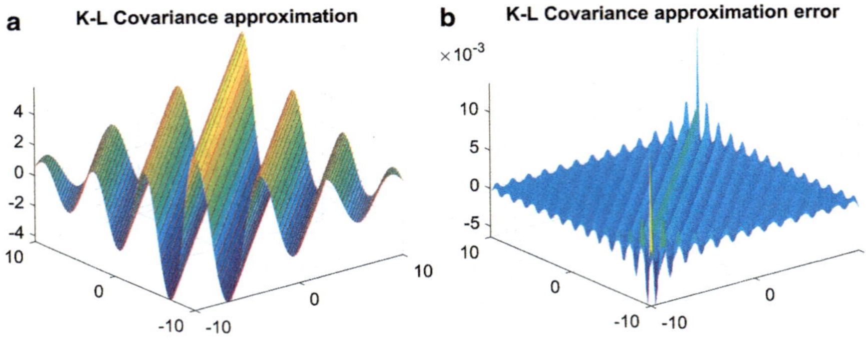

Fig. 16.27 (**a**) Covariance function (surface) and Karhunen-Loève approximation (red dots). (**b**) Error of Karhunen-Loève covariance approximation for an SSRF with $\eta_1 = -1.97$ using equation (16.104) with $M = 27$ terms

with the same number of terms [384]. If the domain length is large compared to the correlation length, both methods exhibit similar convergence. However, a large number of terms is necessary to obtain convergence of the truncated Karhunen-Loève expansion. While this is not a problem if the analytical Karhunen-Loève solution is available, the computational effort increases substantially if numerical solution of the expansion is needed.

An additional advantage of the Karhunen-Loève expansion is its applicability to non-stationary random fields. The expansion can be calculated, at least numerically, for non-stationary covariance models.

Chapter 17
Epilogue

> I had set out once to store, to codify, to annotate the past .... I had failed in it (perhaps it was hopeless?)—for no sooner had I embalmed one aspect of it in words than the intrusion of new knowledge disrupted the frame of reference, everything flew asunder, only to reassemble again in unforeseen, unpredictable patterns ....
> *Clea, The Alexandria Quartet*, by Lawrence Durrell

This book focuses on spatial data that can be represented by means of scalar spatial random fields defined in *continuum space*. Spatial data sets, on the other hand, consist of measurements collected over countable sets of points. Such measurements are treated as a sample of the underlying continuum random field.

This perspective excludes spatial processes that explicitly evolve on discrete structures such as *networks (or graphs)*. For example, the measurement of various quantities, such as passenger flow and epidemic spreading risks on the network created by airport hubs and flight routes is essentially different from the sampling of a continuum process. The network structure is unique, and it has a significant impact on contagion patterns during the spreading of epidemics [106]. For an introduction to networks from the perspective of physics consider [606]. Random fields on networks are studied in [320]. Research activity is also burgeoning in the related field of *graph signal processing* [633].

By adopting the random field framework, I implicitly assume that there is a statistical model that connects the predictor variables (in most cases discussed herein this means the space coordinates) with the response (i.e., the property modeled by the random field). The statistical model is hopefully rooted in some

D. T. Hristopulos, *Random Fields for Spatial Data Modeling*, Advances in Geographic Information Science, https://doi.org/10.1007/978-94-024-1918-4_17

deeper physical model of the underlying process, but in many cases it is just a convenient construction. The assumption that there exists an adequate mathematical model which admits explicit expression has been the norm in scientific inquiry and statistical modeling. This paradigm has recently been challenged by the proponents of the algorithmic approach, in which the statistical model is described in terms of rules. One has to follow the rules in a step-wise manner to their logical conclusion in order to generate different probable outcomes.

The *algorithmic approach* has been very successful in modeling big, multivariate data sets. Such data may contain a great number of predictor variables (e.g., think of predicting wind speeds based on several meteorological and topographic variables), that do not necessarily include spatial location information. A clear statement of the two-culture divide and an eloquent defence of the algorithmic approach is given in the paper by Leo Breiman, which is followed by a debate with eminent proponents of the model-based approach [100].

My perspective on the war of cultures is pragmatic: One should be open minded and avoid to discard opportunities that different viewpoints can provide. Instead, we should strive to compare and to understand the differences and capitalize on the advantages of each approach. This having been said, I should also mention for the benefit of the reader some interesting related topics which are not given the attention that they deserve herein.

List of notable omissions The first two items in the following list do not necessarily involve random fields, but they represent significant lines of research with applications in spatial data modeling. The remaining items on the list represent important areas of research that involves random fields.

- *Machine learning methods* developed by the computational science community can also be applied to spatial data. For example, *Gaussian Process Regression (GPR)* models are formulated in the Bayesian framework and have considerable similarities with geostatistics, as discussed briefly in Chap. 10. The GPR predictive equation is essentially the kriging equation. Various other approaches discussed herein (e.g., linear regression, logistic regression, mean-field theories, variational approximation, replicas, kernel functions, etc.) have strong links with machine learning and constitute essential ingredients in deep learning frameworks [561]. In addition, strong links are emerging between Gaussian processes and neural networks as discussed in Sect. 6.5.

  The algorithmic approach is tightly linked with machine learning [18, 78, 425, 521, 678]. Recently, algorithmic, model-free approaches such as *random forests* have also been applied to the analysis of environmental data [339]. Machine learning is closely linked with the algorithmic approach mentioned above.
- We only briefly mentioned *marked point processes*. Seismology provides a typical example of application of marked point processes: the locations of earthquakes can be viewed as a spatial point process, while the earthquake magnitudes represent the marks (intensities).

  Earthquakes may also be described as excursions of an underlying SRF that represents the Earth crust's stress field, above random, region-specific thresholds that represent the mechanical strength of the crust [375]. This perspective provides a physically motivated mechanism for generating marked point processes as a result of the competition between an excitatory SRF (e.g., stress) and a random resistance field (strength). However, not all marked point processes can be fitted in this framework. For an introduction to the theory and applications of point processes interested readers can consider [40, 177, 618].

- The theory of *Markov random fields* has a long history and is successfully used in image analysis and other spatial problems, but it is only briefly discussed in this book. Connections with the stochastic local interaction model are presented in Chaps. 7, 11 and 13. Interested readers will find more information on Markov random fields in [109, 320, 698, 852] and in the machine learning references above.
- *Vector, tensor, and multivariate random fields*, which are ongoing research topics, were only briefly mentioned (see Sect. 11.1.4). They are, however, important for the statistical description of turbulence [204, 557, 587], the properties of heterogeneous random microstructure [637], and in statistical analyses of multivariate data sets [26].

  One may think that such random fields can be studied as mere extensions of their scalar counterparts. However, the vector and tensor nature of the variables, in combination with possible physical constraints, such as continuity or zero circulation (curl), imply subtle mathematical properties that need to be carefully accounted for. A notable difference of the cross-covariance function and the autocovariance is that the cross-covariance of two vector components $X_i(\mathbf{s};\omega)$ and $X_j(\mathbf{s};\omega)$, where $i \neq j$, is asymmetric: $C_{i,j}(\mathbf{r}) \neq C_{i,j}(-\mathbf{r})$, and $C_{i,j}(\mathbf{r}) \neq C_{j,i}(\mathbf{r})$. On the other hand, symmetry is recovered if the spatial reflection is complemented by an interchange of the component indices, i.e., $C_{i,j}(\mathbf{r}) = C_{j,i}(-\mathbf{r})$. Another difference of the cross-covariance from the autocovariance is that the maximum of the cross covariance does not necessarily occur at zero lag [498].
- *Space-time random fields* are also left out of this volume. Their importance cannot be overstated, because they represent dynamic phenomena in spatially extended systems. Hence, they are quite relevant for studying dynamic properties of meteorological phenomena and transport in heterogeneous media. Their study is complicated by the fact that the time dimension is not a mere extension of the three-dimensional Cartesian space. In addition, the time dependence of space-time random fields is imparted by dynamic laws, and the physical constraints imposed by these laws have important consequences for the emergence of space-time correlations. Hence, mathematically permissible covariance models that fail to account for the dynamic laws may not accurately capture the space-time correlations. My personal perspective is that the study of space-time random fields should be based on the dynamic equations (either partial differential equations with random field coefficients or stochastic partial differential equations) of the underlying processes.

  More information on space-time fields can be found in [139–141, 169, 784]. The applied mathematics and fluid mechanics communities have developed space-time modeling frameworks that are based on orthogonal expansions, such as the proper orthogonal decomposition [61, 127] and similar approaches [714].

- *Complex-valued random fields* that have applications in modeling electromagnetic fields and propagating wavefronts are likewise not discussed.
- *Bayesian methods* are gaining popularity both in spatial statistics and in physics. One of the reasons for their appeal is that they naturally include expert and soft (uncertain) information in statistical models in the form of prior distributions. On the other hand, they often require significant computational resources, a concern that has been allayed by recent advances in computer technology and computational methods. Nevertheless, significant computational challenges remain for applications that involve big spatial data sets. Bayesian methods are briefly discussed in Chap. 10, and more information can be found in [154, 169, 271, 274, 278, 495, 678].

  Bayesian methods are not universally accepted as the golden standard in the scientific community. For example, in experimental physics the dependence of decisions that are based on limited information or prior assumptions can be considered as a weakness rather than a strength of Bayesian methods [403]. A collection of papers in [158] discuss the relative merits and disadvantages of the Bayesian and classical approaches from the perspective of physicists. On a more focused topic, Nobel Laureate Philip W. Anderson eloquently defended the use of Bayesian methods for *inductive reasoning* [24, 25].
- Given the emphasis of this book in macroscopic phenomena, applications of random fields in *quantum mechanics* are not discussed. Nonetheless, we referred to classical theories that originated in *statistical mechanics* [245, 471, 802, 903] and the *statistical theory of fields* [21, 399, 434, 467, 478, 495, 593, 890], since some of the principles and methods developed in these areas of physics are applicable in the statistical modeling of macroscopic systems.

Happy further reading!

Appendix A
Jacobi's Transformation Theorems

Often it is necessary to transform one random variable $X(\omega)$ into a different one by means of a *nonlinear transformation* $Y(\omega) = g\big(X(\omega)\big)$.

Given the pdf of the original variable $X(\omega)$, the question is how to express the pdf of the variable $Y(\omega)$ in terms of the pdf of $X(\omega)$. This can be accomplished by means of Jacobi's theorem [433, p. 102–104], [646].

We will discuss two forms of Jacobi's theorem, one that refers to the single-point (univariate) pdf, and a more general version that applies to multi-point (multivariate) pdf. This theorem is useful, for example, in determining the pdf of random variables obtained by means of *Box-Cox transforms*.

Theorem A.1 (Jabobi's univariate theorem) *Let $X(\omega)$ be a continuous random variable with pdf given by $f_X(x)$ and $Y(\omega) = g\big[X(\omega)\big]$ be a transformed variable.*

If the transformation $y = g(x)$, where $g(\cdot)$ is a continuous function, admits at most a countable number of real roots $x_q = g_q^{-1}(y)$, where $q = 1, \ldots, Q$, then $Y(\omega)$ has the following pdf

$$f_Y(y) = \sum_{q=1}^{Q} f_X\left(g_q^{-1}(y)\right) \left|\frac{\mathrm{d}g_q^{-1}(y)}{\mathrm{d}y}\right| .$$

The proof of the above theorem is straightforward and involves the use of the delta function. We assume for simplicity that $Q = 1$. Then, the pdf of $Y(\omega)$ is expressed as follows

$$f_Y(y) = \int_{-\infty}^{\infty} \mathrm{d}x\, f_X(x)\, \delta\left(y - g(x)\right) = \frac{f_X\left(g^{-1}(y)\right)}{\left|\frac{\mathrm{d}g\left(g^{-1}(y)\right)}{\mathrm{d}y}\right|} .$$

Using the chain rule of differentiation, it can be shown that

D. T. Hristopulos, *Random Fields for Spatial Data Modeling*, Advances in Geographic Information Science, https://doi.org/10.1007/978-94-024-1918-4

$$\left[\frac{\mathrm{d}g\left(g^{-1}(y)\right)}{\mathrm{d}y}\right]^{-1}=\frac{\mathrm{d}g^{-1}(y)}{\mathrm{d}y},$$

which completes the proof.

In the case of random fields, we would also like to know how the joint density of the random field transforms under nonlinear transformations. The multivariate version of Jacobi's theorem is given below.

Theorem A.2 (Jabobi's multivariate theorem) *Let* $\mathbf{X}(\omega)=(\mathrm{X}_1(\omega),\ldots,\mathrm{X}_n(\omega))^\top$ *and* $\mathbf{Y}(\omega)=(\mathrm{Y}_1(\omega),\ldots,\mathrm{Y}_n(\omega))^\top$ *denote two n-dimensional continuous random vectors.*

The transformed vector variable $\mathbf{Y}(\omega)$ *is obtained by means of the vector transformation* $\mathbf{y}=\mathbf{g}(\mathbf{x})$, *where* $\mathbf{g}(\cdot)$ *is a vector function with elements* $y_l=g_l(\mathbf{x})$, $l=1,\ldots,n$.

We assume that the functions $g_l(\cdot)$ *are continuous and possess continuous partial derivatives with respect to each of their arguments.*

A. *If the functions* $g_l(\cdot)$ *define one-to-one mappings, there exist unique inverse functions* $g_l^{-1}(\cdot)$ *so that*

$$\mathbf{X}(\omega)=\mathbf{g}^{-1}\left[\mathbf{Y}(\omega)\right].$$

Then, the joint pdf of the transformed vector, i.e., $f_{\mathbf{Y}}(\mathbf{y})$ *is given by means of the following equation*

$$f_{\mathbf{Y}}(\mathbf{y})=f_{\mathbf{X}}\left[\mathbf{g}^{-1}(\mathbf{y})\right]\,|\det(\mathbb{J})|, \tag{A.1}$$

where $\mathbb{J}$ *is the Jacobian of the transformation* $\mathbf{x}\mapsto\mathbf{y}$, *given by the following matrix of partial derivatives*

$$\mathbb{J}=\frac{\partial\left(g_1^{-1}(\mathbf{y}),\ldots,g_n^{-1}(\mathbf{y})\right)}{\partial(y_1,\ldots,y_n)}. \tag{A.2}$$

The above notation means that the elements of the Jacobian matrix are given by the following partial derivatives

$$\mathbb{J}_{i,j}=\frac{\partial g_i^{-1}(\mathbf{y})}{\partial(y_j)},\quad i,j=1,\ldots,n.$$

B. *If the vector function* $\mathbf{y}=\mathbf{g}(\mathbf{x})$ *admits at most a countable number of roots, i.e.,* $\mathbf{x}_q=\mathbf{g}_q^{-1}(\mathbf{y})$, *where* $q=1,\ldots,Q$, *then the joint pdf of the transformed vector,* *i.e., $f_{\mathbf{Y}}(\mathbf{y})$ is given by means of the following equation*

$$f_{\mathbf{Y}}(\mathbf{y}) = \sum_{q=1}^{Q} f_{\mathbf{X}}\left[\mathbf{g}_q^{-1}(\mathbf{y})\right] \left|\det(\mathbb{J}_q)\right|, \tag{A.3}$$

where $\mathbb{J}_q$ is the Jacobian corresponding to the q-th root, defined by

$$\mathbb{J}_j = \frac{\partial\left(g_{q,1}^{-1}(\mathbf{y}), \ldots, g_{q,n}^{-1}(\mathbf{y})\right)}{\partial(y_1, \ldots, y_n)}, \quad q = 1, \ldots, Q. \tag{A.4}$$

Remark The multivariate Jacobi's theorem also applies if $\dim[\mathbf{Y}(\omega)] = m < n$. In this case, the m-dimensional vector $\mathbf{Y}$ is augmented by an $(n-m)$-dimensional vector $\mathbf{X}'(\omega) = \mathbf{h}[\mathbf{X}(\omega)]$, where $\mathbf{h}(\cdot)$ is a simple function with continuous partial derivatives. The $n-m$ dummy variables in $\mathbf{X}'(\omega)$ are eliminated from the joint pdf of $\mathbf{Y}(\omega)$ by integration.

Appendix B
Tables of Spartan Random Field Properties

Table B.1 Variance of Spartan spatial random fields in one, two, and three dimensions for different values of the rigidity coefficient $\Delta = \sqrt{\eta_1{}^2 - 4}$ is the discriminant of the SSRF characteristic polynomial

σ_{x}^2	$-2 < \eta_1 < 2$	$\eta_1 = 2$	$\eta_1 > 2$
$d = 1$	$\frac{\eta_0}{2\sqrt{2+\eta_1}}$	$\frac{\eta_0}{4}$	$\frac{\eta_0}{2\Delta}\left(\frac{1}{w_1} - \frac{1}{w_2}\right)$
$d = 2$	$\frac{\eta_0}{2\pi\,\lvert\Delta\rvert}\left[\frac{\pi}{2} - \arctan\left(\frac{\eta_1}{\lvert\Delta\rvert}\right)\right]$	$\frac{\eta_0}{4\pi}$	$\frac{\eta_0}{4\pi\,\Delta}\ln\left(\frac{\eta_1+\Delta}{\eta_1-\Delta}\right)$
$d = 3$	$\frac{\eta_0}{4\pi\sqrt{2+\eta_1}}$	$\frac{\eta_0}{8\pi}$	$\frac{\eta_0}{4\pi\,\Delta}\,(w_2 - w_1)$

Table B.2 Integral range of Spartan spatial random fields in one, two, and three dimensions for different values of the rigidity coefficient. The lengths listed are normalized by ξ. In addition, $\Delta = \sqrt{\eta_1{}^2 - 4}$ is the discriminant of the SSRF characteristic polynomial

ℓ_c	$-2 < \eta_1 < 2$	$\eta_1 = 2$	$\eta_1 > 2$
$d = 1$	$2\sqrt{2+\eta_1}$	4	$\sqrt{2}\,\left(\sqrt{\eta_1+\Delta} + \sqrt{\eta_1-\Delta}\right)$
$d = 2$	$2\sqrt{\frac{\lvert\Delta\rvert}{1 - \frac{2}{\pi}\arctan\left(\frac{\eta_1}{\lvert\Delta\rvert}\right)}}$	$2\sqrt{\pi}$	$2\sqrt{\frac{\pi\,\Delta}{\ln\left(\frac{\eta_1+\Delta}{\eta_1-\Delta}\right)}}$
$d = 3$	$2\left(\frac{\pi}{2}\right)^{1/3}(2+\eta_1)^{1/6}$	$2\pi^{1/3}$	$\sqrt{2}\,\pi^{1/3}\left(\sqrt{\eta_1-\Delta} + \sqrt{\eta_1+\Delta}\right)^{1/3}$

D. T. Hristopulos, *Random Fields for Spatial Data Modeling*, Advances in Geographic Information Science, https://doi.org/10.1007/978-94-024-1918-4

Appendix C
Linear Algebra Facts

The following basic facts of linear algebra are drawn from the nicely written review of principal component analysis [747].

1. The inverse of an orthogonal matrix is its transpose. Hence, if $\mathbf{A}$ is *orthogonal*, then $\mathbf{A}^{-1} = \mathbf{A}^{\top}$.
2. For any $m \times n$ matrix $\mathbf{A}$, the products $\mathbf{A}^{\top}\mathbf{A}$ and $\mathbf{A}\mathbf{A}^{\top}$ are *symmetric* $n \times n$ and $m \times m$ matrices respectively.
3. A matrix $\mathbf{C}$ is *symmetric*, if and only if it *can be diagonalized by an orthogonal matrix.* A square matrix $\mathbf{C}$ can be diagonalized if it is similar to a diagonal matrix, i.e., if there is an invertible matrix $\mathbf{A}$ such that the matrix $\mathbf{A}^{-1}\mathbf{C}\mathbf{A}$ is diagonal.
4. *Principal value decomposition:* An $n \times n$ *symmetric* matrix $\mathbf{C}$ is diagonalized by the matrix of its *orthonormal eigenvectors*, i.e.,

$$\mathbf{C} = \mathbf{E}\mathbf{D}\mathbf{E}^{\top}, \tag{C.1}$$

   where $\mathbf{D}$ is the diagonal matrix (with the eigenvalues of $\mathbf{C}$ along the diagonal), and $\mathbf{E} = [\hat{\mathbf{v}}_1 \;\; \hat{\mathbf{v}}_2 \;\; \ldots \hat{\mathbf{v}}_n]$ is the orthogonal matrix built by the eigenvectors $\{\hat{\mathbf{v}}_i\}_{i=1}^{n}$ of $\mathbf{C}$ as columns.

   The *rank* of $\mathbf{C}$ is equal to $r \leq n$. If $r < n$ the matrix $\mathbf{C}$ is *degenerate* and has r non-zero eigenvalues with respective eigenvectors. Then, the matrix $\mathbf{E}$ is formed by arranging the r eigenvectors in respective columns and filling the remaining columns with any $n - r$ orthonormal vectors; similarly, the $n - r$ entries of the diagonal matrix $\mathbf{D}$ are filled with zeros.
5. For any arbitrary $m \times n$ matrix $\mathbf{A}$, the *symmetric* matrix $\mathbf{C} = \mathbf{A}^{\top}\mathbf{A}$ has a set of *orthonormal* $n \times 1$ *eigenvectors* $\{\hat{\mathbf{v}}_1, \hat{\mathbf{v}}_2, \ldots \hat{\mathbf{v}}_n\}$ and a set of associated *eigenvalues* $\{\lambda_1, \lambda_2, \ldots \lambda_n\}$.

   The set of vectors $\{\mathbf{A}\hat{\mathbf{v}}_1, \mathbf{A}\hat{\mathbf{v}}_2, \ldots \mathbf{A}\hat{\mathbf{v}}_n\}$ then form an *orthogonal basis*, and the length of the vector $\mathbf{A}\hat{\mathbf{v}}_i$ is $\sqrt{\lambda_i}$, where $i = 1, \ldots, n$.
6. *Singular value decomposition:* Let $\mathbf{A}$ be an $m \times n$ matrix and $\mathbf{C} = \mathbf{A}^{\top}\mathbf{A}$ the $n \times n$, rank-r symmetric matrix with non-zero eigenvalues λ_i and $n \times 1$ eigenvectors $\hat{\mathbf{v}}_i$,

D. T. Hristopulos, *Random Fields for Spatial Data Modeling*, Advances in Geographic Information Science, https://doi.org/10.1007/978-94-024-1918-4

where $i = 1, \ldots, r$, i.e.,

$$\mathbf{C}\,\hat{\mathbf{v}}_i = \lambda_i\,\hat{\mathbf{v}}_i, \;\text{ for } i = 1, \ldots, r.$$

Then, the following statements hold:

a. The $m \times 1$ vectors $\hat{\mathbf{u}}_i$ defined by $\mathbf{A}\,\hat{\mathbf{v}}_i = \sqrt{\lambda_i}\hat{\mathbf{u}}_i$, are *orthonormal.*
b. The $\sigma_i = \sqrt{\lambda_i}$ are called *singular values.*
c. Both the eigenvectors $\hat{\mathbf{v}}_i$ of $\mathbf{C}$ and the vectors $\hat{\mathbf{u}}_i$ form orthonormal bases in r-dimensional space.
d. The singular value decomposition (SVD) of the matrix $\mathbf{A}$ is given by

$$\mathbf{A} = \mathbf{U}\,\boldsymbol{\Sigma}\mathbf{V}^\top. \tag{C.2}$$

In the above decomposition, the $m \times n$ matrix of singular values $\boldsymbol{\Sigma}$ is given by

$$\boldsymbol{\Sigma} = \left[\begin{array}{cccc|c} \sigma_1 & 0 & \ldots & 0 & \\ 0 & \sigma_2 & \ldots & 0 & \\ \vdots & \vdots & \vdots & \vdots & \mathbf{0}_{r;n-r} \\ 0 & 0 & \ldots & \sigma_r & \\ \hline & & \mathbf{0}_{m-r;r} & & \mathbf{0}_{m-r;n-r} \end{array}\right],$$

where $\sigma_1 \geq \sigma_2 \ldots \geq \sigma_r$ are the rank-ordered singular values, and $\mathbf{0}_{k;l}$ is a matrix of zeros with k rows and l columns.

The $m \times n$ matrix $\mathbf{U}$ is given by

$$\mathbf{U} = \begin{bmatrix}\hat{\mathbf{u}}_1 & \hat{\mathbf{u}}_2 & \ldots & \hat{\mathbf{u}}_r & \hat{\mathbf{e}}_1 \ldots \hat{\mathbf{e}}_{m-r}\end{bmatrix},$$

and the $n \times n$ matrix $\mathbf{V}$ is given by

$$\mathbf{V} = \begin{bmatrix}\hat{\mathbf{v}}_1 & \hat{\mathbf{v}}_2 & \ldots & \hat{\mathbf{v}}_r & \hat{\mathbf{q}}_1 \ldots \hat{\mathbf{q}}_{n-r}\end{bmatrix},$$

where the $\{\hat{\mathbf{e}}_i\}_{i=1}^{m-r}$ and $\{\hat{\mathbf{q}}_i\}_{i=1}^{n-r}$ are orthonormal sets of vectors.
e. The $\hat{\mathbf{u}}_i$ are called *left singular vectors*, while the $\hat{\mathbf{v}}_i$ are called *right singular vectors* of the matrix $\mathbf{A}$, because $\mathbf{A}\hat{\mathbf{v}}_i = \sigma_i\hat{\mathbf{u}}_i$ and $\mathbf{A}^\top\hat{\mathbf{u}}_i = \sigma_i\hat{\mathbf{v}}_i$.

7. *Matrix inversion lemma:* This is also known as the Sherman-Morrison-Woodbury formula. Let $\mathbf{A}$ and $\mathbf{C}$ represent two $N \times N$ and $M \times M$, respectively, invertible matrices, and $\mathbf{U}$ and $\mathbf{V}$ represent $N \times M$, and $M \times N$ matrices. Then

$$(\mathbf{A} + \mathbf{U}\mathbf{C}\mathbf{V})^{-1} = \mathbf{A}^{-1} - \mathbf{A}^{-1}\mathbf{U}\left(\mathbf{C}^{-1} + \mathbf{V}\mathbf{A}^{-1}\mathbf{U}\right)^{-1}\mathbf{V}\mathbf{A}^{-1}. \tag{C.3}$$

The matrix inversion lemma is very useful in speeding up calculations if (i) the inversion needs to be calculated repeatedly with matrix $\mathbf{A}$ fixed while matrix $\mathbf{C}$ changes or (ii) $\mathbf{A} + \mathbf{UCV}$ is a low-rank update of $\mathbf{A}$, i.e., if $M \ll N$.

Appendix D
Kolmogorov-Smirnov Test

We can investigate different scenarios (trial distributions) for the univariate probability distribution of the data using the *Kolmogorov-Smirnov (K-S) test*. In contrast with the chi-square goodness-of-fit test which is applicable to discrete distributions, the K-S test is applicable to continuous distributions.

The *null hypothesis* of the K-S test is that the data are drawn from a specific model probability distribution (the trial distribution); we denote the cdf of this model by $F_{\mathrm{x}}(x; \boldsymbol{\theta})$, where $\boldsymbol{\theta}$ is the parameter vector.

The outcome of the test is based on the maximum distance D between the empirical cumulative distribution function and the respective trial cdf $F_{\mathrm{x}}(x; \boldsymbol{\theta})$.

The empirical cdf is estimated from the data, $\mathbf{x}^*$, by means of the following estimator:

$$\hat{F}_{\mathrm{x}}(x) = \frac{1}{N} \sum_{n=1}^{N} \mathbb{I}_{x^*_{[n]} \leq x}(x) \tag{D.1}$$

where $\mathbb{I}_A(x)$ is the *indicator function* of the set $A \subset \mathbb{R}$, i.e., $\mathbb{I}_A(x) = 1$, if $x \in A$ and $\mathbb{I}_A(x) = 0$, if $x \notin A$. The elements $x^*_{[n]}$ represent the ordered sample data:

$$x^*_{[1]} \leq x^*_{[2]} \leq \ldots \leq x^*_{[N]}.$$

The *Kolmogorov-Smirnov distance* D between the empirical (data) distribution, $\hat{F}_{\mathrm{x}}(x)$ and the trial distribution, $F_{\mathrm{x}}(x)$, is given by

$$D = \sup_{x \in \mathbb{R}} \left| \hat{F}_{\mathrm{x}}(x) - F_{\mathrm{x}}(x; \boldsymbol{\theta}) \right|, \tag{D.2}$$

where $\sup_A G(x)$ denotes the supremum (the least upper bound) of the function $G(\cdot)$ for $x \in A$. If the difference D is less than a certain *critical value* which depends on

D. T. Hristopulos, *Random Fields for Spatial Data Modeling*, Advances in Geographic Information Science, https://doi.org/10.1007/978-94-024-1918-4

the trial distribution $F_{\mathrm{x}}(x;\boldsymbol{\theta})$, the null hypothesis cannot be rejected; if on the other hand D exceeds the critical value, the null hypothesis is rejected.

In practice, the outcome of the K-S hypothesis test is determined by means of the *significance level* α, which is arbitrarily selected and reflects the critical value, and the p-value. The latter denotes the probability that the observed K-S distance occurs by chance. This means that the observed K-S distance can be justified as a fluctuation that is allowed by the trial distribution $F_{\mathrm{x}}(x;\boldsymbol{\theta})$.

What is a p-value? According to the recent statement of the American Statistical Association [833]:

> Informally, a p-value is the probability under a specified statistical model that a statistical summary of the data (e.g., the sample mean difference between two compared groups) would be equal to or more extreme than its observed value.

If $p > \alpha$ the null hypothesis that the data follow the trial distribution cannot be rejected. A typical value for the significance level is $\alpha = 5\%$. For problems related to the use, reporting in the literature, and interpretation of p-values we refer the readers to [393, 833].

The K-S test with the distance function (D.2) assumes that the parameter vector $\boldsymbol{\theta}$ of the trial distribution $F_{\mathrm{x}}(x;\boldsymbol{\theta})$ is known. For example, if the trial distribution $F_{\mathrm{x}}(x;\boldsymbol{\theta})$ is the normal distribution, it is assumed that both the mean, m_{x}, and the variance, σ_{x}^2 are known.

Unknown Distribution Parameters If the parameters of the trial distribution are unknown a priori, the straightforward K-S test as described above is inadequate. In particular, the p-value cannot be obtained by means of theoretical calculations. One could consider replacing the true (but unknown) parameter values with their sample estimates $\hat{\boldsymbol{\theta}}$. Then,

$$\hat{D} = \sup_{x\in\mathbb{R}} \left| \hat{F}_{\mathrm{x}}(x) - F_{\mathrm{x}}(x;\hat{\boldsymbol{\theta}}) \right|, \tag{D.3}$$

However, if the K-S test is based on $\hat{D}$, it neglects the fluctuations of sample-based parameter estimates between samples and therefore the fluctuations in $\hat{D}$.

In this case, the *parametric bootstrap* methodology described in [146, 244] should be followed. The parametric bootstrap is based on ideas proposed by the statistician Bradley Efron in [224]. The bootstrap uses Monte Carlo simulation to generate synthetic data from the trial distribution using the parameters that are estimated from the sample. The p-value is then determined as the percentage of the simulated states in which the K-S distance exceeds the K-S distance $\hat{D}$ obtained for the original sample.

An application of the parametric bootstrap to the investigation of the probability distribution of earthquake recurrence times is presented in [376].

The main steps of the parametric bootstrap are briefly described below.

1. The parameters $\boldsymbol{\theta}$ of the trial distribution are estimated from the sample $\{x_n^*\}_{n=1}^N$ using the method of maximum likelihood (ML) leading to an optimal parameter vector $\hat{\boldsymbol{\theta}}$.
2. The K-S distance $\hat{D}$ is evaluated based on (D.3). The latter uses the trial distribution $F_{\mathrm{x}}(x; \hat{\boldsymbol{\theta}})$ with the ML-estimated parameters.
3. A set of N_{sim} realizations $\{x_n^{(j)}\}_{n=1}^N$, where $j = 1, \ldots N_{\mathrm{sim}}$, with the same size N as the initial sample are generated from the trial distribution $F_{\mathrm{x}}(x; \hat{\boldsymbol{\theta}})$.
4. For each realization $j = 1, \ldots N_{\mathrm{sim}}$, the functional form of the trial distribution is fitted to the synthetic data $\{x_n^{(j)}\}_{n=1}^N$ and optimal parameters $\hat{\boldsymbol{\theta}}_j$ are estimated. The trial distribution with the new optimal parameters is denoted by $F_{\mathrm{x}}^{(j)}(x; \hat{\boldsymbol{\theta}}_j)$ for $j = 1, \ldots, N_{\mathrm{sim}}$.
5. The K-S distance between the *empirical distribution of the* j^{th} *realization*, $\hat{F}_{\mathrm{x}}^{(j)}(x)$, and the *optimal trial distribution* $F_{\mathrm{x}}^{(j)}(x; \hat{\boldsymbol{\theta}}_j)$ for the same realization is given by

$$d_j = \sup_{x \in \mathbb{R}} \left| F_{\mathrm{x}}^{(j)}(x; \hat{\boldsymbol{\theta}}_j) - \hat{F}_{\mathrm{x}}^{(j)}(x) \right|, \; j = 1, \ldots, N_{\mathrm{sim}}. \tag{D.4}$$

6. The p-value is then calculated as the percentage of realizations with K-S distance d_j that exceeds the distance $\hat{D}$ of the original sample, i.e.,

$$p = \frac{1}{N_{\mathrm{sim}}} \sum_{j=1}^{N_{\mathrm{sim}}} \mathbb{I}_{d_j > \hat{D}} \left(d_j \right). \tag{D.5}$$

7. The p-value reflects the probability that the K-S distance exceeds $\hat{D}$ purely by chance, if the null hypothesis is true. If the p-value exceeds the significance level, i.e.,

$$p > \alpha, \tag{D.6}$$

the null hypothesis cannot be rejected.

In applications of the bootstrapping approach outlined above, it is usually assumed that the sample $\{x_n^*\}_{n=1}^N$ comprises *statistically independent* observations. The bootstrapping approach can lead to erroneous results if the observations are correlated, which is often the case with data from complex systems. Such correlations are typically neglected in the ML estimation of the distribution parameters and the generation of the bootstrapped samples [285]. The presence of correlations essentially implies that the effective number of independent observations is smaller than N. Correlated sample data lead to higher rejection rates and stronger parameter fluctuations between bootstrapped samples [285].

There are two ways to address the problem of correlations. The first one is the *parametric* approach which assumes a statistical model for the correlations. This decorrelation approach is used in the modeling of spatial data. For example, in Chap. 12 the Gaussian likelihood function is based on the joint pdf of the data which involves the covariance matrix as a measure of dependence. The second approach is *non-parametric* and is based on estimating an effective sample size $N_{\rm eff}$ for the correlated data. This is determined as the ratio of N over a characteristic correlation scale. The bootstrap procedure is then applied to simulated samples of size $N_{\rm eff}$. For more details regarding the application of the non-parametric decorrelation approach consult [285].

Glossary

Adjacent nodes Two grid nodes are considered to be adjacent if they are located at opposite ends of the same edge of the grid.

Bayesian inference The term is used to denote a method of inference which is based on Bayes' theorem. An initial probability distribution is assumed (e.g., for the model parameters of interest), which is then progressively updated as more information becomes available, leading to a posterior probability distribution.

Bijective mapping A bijective mapping $X \mapsto Y$ defines a one to one correspondence between the sets X and Y. This means that each element of X is paired with one element of Y only and vice versa, and there are no elements that are left unpaired in either set. Bijective mappings are invertible.

Burn-in phase Initial sequence of transient states in a Markov Chain Monte Carlo during which the system is relaxed to the equilibrium distribution.

Cartesian product In set theory the Cartesian product of two sets $\mathcal{S}_1$ and $\mathcal{S}_2$ is the set $\mathcal{S} = \mathcal{S}_1 \times \mathcal{S}_2$ with elements s such $s = (s_1, s_2) \in \mathcal{S}$ where $s_1 \in \mathcal{S}_1$ and $s_2 \in \mathcal{S}_2$.

Confidence interval Confidence intervals are constructed around sample statistics (e.g., sample average, sample correlation coefficient). They are constructed so that they contain the respective "true" population parameter with a specified probability.

Compact space A space is called compact if it is closed (i.e., it contains all its limit points) and bounded (i.e., the maximum distance between any of its points is finite).

Convex function A function $f \colon \mathcal{D} \subset \mathbb{R}^d \to \mathbb{R}$ is a convex function if the following two conditions hold: (i) its support is a convex set, and (ii) for any $\mathbf{s}_1, \mathbf{s}_2 \in \mathcal{D}$ and $0 \le \theta \le 1$ it holds that $f(\theta \mathbf{s}_1 + (1-\theta)\mathbf{s}_2) \le \theta f(\mathbf{s}_1) + (1-\theta) f(\mathbf{s}_2)$.

Convex hull The convex hull of a set of points $\mathbf{s}_n \in \mathbb{R}^d$, $n = 1, \ldots, N$, is the smallest convex region in $\mathbb{R}^d$ that encloses all the points. A region is called

D. T. Hristopulos, *Random Fields for Spatial Data Modeling*, Advances in Geographic Information Science, https://doi.org/10.1007/978-94-024-1918-4

convex if for any two points that lie inside the region, the straight line that connects them lies entirely inside that region.

Convex set A set C is called convex if the line segment that connects any two points in C lies entirely in C. Hence, C is convex if for any set elements $a, b \in C$, and $0 \leq \theta \leq 1$, it holds that $\theta a + (1 - \theta)b \in C$.

Cross validation This refers to procedures by means of which the predictive accuracy of statistical models is tested. They involve the comparison between model predictions and a subset of the data that have not been used in model estimation.

Diagonal dominance A square $N \times N$ matrix $\mathbf{A}$ is called diagonally dominant if $A_{i,i} > \sum_{j=1,\neq i}^{N} |A_{i,j}|$ for all $i = 1, \ldots, N$.

Efficient estimator An unbiased estimator is called efficient if it has the lowest possible variance, which is determined by the lower Rao-Cramér bound.

Fixed point A fixed point for a given function or transformation is a point in the function's domain that remains invariant under the action of the function or transformation; that is, x^* is a fixed point for the function $f(x)$ if $f(x^*) = x^*$.

Greedy algorithm In optimization problems, greedy methods proceed by means of local moves that optimize the objective function relative to other possible local moves.

Hyperparameters In the Bayesian framework, a hyperparameter is any parameter that is "determined" by a prior distribution.

Ill-posed problems A problem is ill-posed if at least one of the following conditions hold: (i) A solution does not exist. (ii) The solution is not unique. (iii) The solution does not depend continuously on the data.

Kullback-Leibler (KL) divergence The KL divergence, also known as *relative entropy*, is a measure of the distance between two distributions. If the distributions have probability density functions $p(\cdot)$ and $q(\cdot)$, then the KL divergence is defined as $D(p \parallel q) = \mathcal{E}_p \{\ln [p(\cdot)/q(\cdot)]\}$, where $\mathcal{E}_p$ is the expectation with respect to the pdf $p(\cdot)$. (The following limits are used in applying the definition: $0 \ln(0/0) = 0$, $\ln(0/q) = 0$, $p \ln(1/0) = \infty$.)

Lattice We use the concept of a lattice as an ordered periodic pattern created by the repetition of an elementary unit cell. In statistics, the term lattice data often refers to measurements that are aggregated over certain regions [122]. Cells in this lattice may correspond to completely different areas in terms of size and shape. We do not follow the statistical definition of lattice data in this book. The term *regional data* which has been suggested seems more suitable [717].

Local random fields We use this term to refer to Boltzmann-Gibbs random fields the energy functional of which involves only local terms, such as terms proportional to the square of field derivatives. The spectral density of such random fields is a rational function the denominator of which is a polynomial of the square of the wavevector magnitude.

Matrix rank The rank of a matrix is the dimension of the vector space which is generated by its columns and is equal to the dimension of the matrix generated by its rows. The rank is thus the maximal number of linearly independent columns (rows) of the matrix. A matrix is called *full rank* if its rank is equal to $\min(C, R)$, where C is the number of columns and R the number of rows. Conversely, a matrix is called *rank deficient* if its rank is less than $\min(C, R)$.

Max-stable random fields A random field $X(\mathbf{s}; \omega)$, where $\mathbf{s} \in \mathbb{R}^d$ is a max-stable random field if there exist sequences of independent, identically distributed random fields $\{Y_1(\mathbf{s}; \omega), \ldots, Y_n(\mathbf{s}; \omega)\}$ and of two deterministic functions $\alpha_n(\mathbf{s}) > 0$, $\beta_n(\mathbf{s}) \in \mathbb{R}$ such that

$$X(\mathbf{s}; \omega) \stackrel{d}{=} \lim_{n\to\infty} \frac{\max_{i=1,\ldots n} \{Y_1(\mathbf{s}; \omega), \ldots, Y_n(\mathbf{s}; \omega)\} - \beta_n(\mathbf{s})}{\alpha_n(\mathbf{s})}$$

Multiscale random fields This term refers to random fields with spectral density that has significant weight over a large range of frequencies (or wavenumbers). Multiscale fields are used in studies of turbulence, porous media, and cosmology. Typically, the right tail of such spectral densities exhibits power-law dependence.

Non-extensive A quantity is called extensive if its magnitude is proportional to the size of the system. This is satisfied for additive quantities such as the classical entropy. However, not every quantity is extensive. An example of a non-extensive quantity is the Tsallis entropy.

Nyquist frequency In sampling theory, the Nyquist frequency is the maximum frequency that can be reconstructed from a given sample in terms of the Fourier series. For samples taken at regular intervals δt the sampling rate is $f_s = 1/\delta t$ and the Nyquist frequency is given by $f_c = f_s/2$.

Order notation Let $g(x)$ be a function whose behavior at the limits $x \to \infty$ is well understood. Then we say that (i) $f(x) = O[g(x)]$ as $x \to \infty$ if there exists a constant c such that $|f(x)| < c\ |g(x)|$ as $x \to \infty$ and (ii) $f(x) = \mathrm{o}[g(x)]$ as $x \to \infty$ if $\lim_{x\to\infty} f(x)/g(x) = 0$. Similar definitions hold for $x \to 0$.

Positive definite matrix A symmetric $N \times N$ real matrix $\mathbf{H}$ is said to be positive definite if $\mathbf{x}^\top \mathbf{H}\mathbf{x}$ is non-negative for every non-zero column vector $\mathbf{x}$ of dimension N. A diagonalizable matrix $\mathbf{H}$ is positive definite if and only if its eigenvalues are non-negative.

Precision matrix The precision matrix is defined as the inverse of the covariance matrix. Since the covariance matrix is positive definite, so is the precision matrix.

Principal irregular term The principal irregular term is the term with the lowest non-even power exponent in the series expansion of the variogram around zero lag. This term is typically proportional to $\|\mathbf{r}\|^\alpha$, where $0 < \alpha < 2$, or $\|\mathbf{r}\|^{2n} \ln \|\mathbf{r}\|$, where $n \in \mathbb{N}$.

Principal submatrix If $\boldsymbol{A}$ is an $n \times n$ matrix, a $m \times m$ submatrix of $\boldsymbol{A}$ is called principal submatrix of $\boldsymbol{A}$, if it is obtained by deleting the same $n - m$ rows and columns of $\boldsymbol{A}$. The determinant of a principal submatrix of $\boldsymbol{A}$ is called a principal minor of $\boldsymbol{A}$.

Regionalized variable Regionalized variables are functions of the spatial location within the domain of interest. They are modeled as realizations of spatial random fields.

Spartan spatial random fields These are local random fields with a specific structure of the Gibbs energy functional that involves the squares of the fluctuation, the gradient and the curvature of the field. In one dimension the realizations of the Spartan random field represent the position of a classical, damped harmonic oscillator driven by white noise.

Self-adjoint operator Let us consider a real-valued vector space $\mathbb{V}$ equipped with an inner product. Hence, if $\mathbf{v}, \mathbf{u} \in \mathbb{V}$ then $\langle \mathbf{v}, \mathbf{u} \rangle$ is a number in $\mathbb{R}$; Then, a linear operator $\mathbb{L}$ is called self-adjoint if $\langle \mathbf{v}, \mathbb{L}\,\mathbf{u} \rangle = \langle \mathbb{L}\,\mathbf{v}, \mathbf{u} \rangle$. If the vector space $\mathbb{V}$ has a finite-dimensional orthonormal base, then $\mathbb{L}$ is represented by a symmetric matrix. In quantum mechanics $\mathbb{V}$ usually refers to a complex vector space, and a self-adjoint operator is represented by means of a Hermitian matrix.

Self-affine A process $X(t; \omega)$ is self-affine if its probability law remains unchanged under simultaneous rescaling of both t and X. The definition is similarly extended to spatial random fields.

Self-similar A process $X(t; \omega)$ is self-similar if its probability law remains unchanged under a rescaling of t. The definition is similarly extended to spatial random fields.

Standard deviation The standard deviation is a measure of the dispersion of individual measurements around the mean. Is is equal to the square root of the variance, which is equal to the expectation of the square of the fluctuations around the mean.

Standard error The standard error measures the dispersion of a sample-based statistic (e.g., sample mean) around the true value (cf. confidence interval).

Statistic A statistic is a sample-based quantity (e.g., sample average) used to estimate a population parameter. Statistics represent functions of the sample values. In general, the sample values can be replaced by the respective random variables. This allows a wider perspective that accounts for possible many different samples. Thus, statistics (also known as sampling functions) are also random variables, since they are expressed as a function of random variables.

Stencil A stencil $\mathcal{S}$ is an $O(a^n)$ approximation for a differential operator D of order p if and only if $\mathcal{S}\left[p_q(\mathbf{s})\right] = D p_q(\mathbf{s})$ where $p_q(\mathbf{s})$ is any polynomial of degree $q = p + n - 1$ and a is any lattice step value.

Support The term support refers to the spatial domain over which a given measurement is defined. Measurements typically represent weighted integrals (averages) of the underlying physical process over a domain (support) that is defined by the characteristics of the measuring device or setup.

Supremum It denotes the smallest upper bound of a set. If the set contains a maximum, then the supremum coincides with the maximum.

Toeplitz matrix A matrix such that every descending diagonal is constant. An $N \times N$ matrix $\mathbf{A}$ is a Toeplitz matrix if $A_{i,j} = c_{|i-j|}$, for all $i, j = 1, \ldots, N$.

Uniform continuity A function $f(x)$ is continuous at point $x \in \mathbb{R}$ if for every $\epsilon > 0$, however small, there is a $\delta > 0$, independent of x, such that $|f(x) - f(y)| < \epsilon$

if $|x - y| < \delta$. In the case of local continuity, the value of δ depends on both ϵ and x. In the case of uniform (global) continuity, δ depends only on ϵ.

Voronoi tessellation Given a set of points on the plane, the Voronoi tessellation comprises a set of polygons, or Voronoi cells, that cover the plane so that (i) each cell has a sampling point at its center and (ii) the points contained in each cell are closer to the center of the polygon than to any other sampling point.

References

1. Ababou, R., Bagtzoglou, A.C., Wood, E.F.: On the condition number of covariance matrices in kriging, estimation, and simulation of random fields. Math. Geol. **26**(1), 99–133 (1994)
2. Abarbanel, H.: Analysis of Observed Chaotic Data. Springer, New York, NY, USA (1996)
3. Abrahamsen, P.: A Review of Gaussian Random Fields and Correlation Functions. Tech. Rep. TR 917, Norwegian Computing Center, Box 114, Blindern, N-0314, Oslo, Norway (1997)
4. Abramowitz, M., Stegun, I.A.: Handbook of Mathematical Functions with Formulas, Graphs, and Mathematical Tables. National Bureau of Standards, Washington, DC, USA (1972)
5. Abrikosov, A.A., Gorkov, L.P., Dzyaloshinski, I.E.: Methods of Quantum Field Theory in Statistical Physics. Courier Dover Publications, Mineola, NY, USA (2012)
6. Acker, J.G., Leptoukh, G.: Online analysis enhances use of NASA earth science data. EOS Trans. Am. Geophys. Union **88**(2), 14–17 (2007)
7. Addair, T.G., Dodge, D.A., Walter, W.R., Ruppert, S.D.: Large-scale seismic signal analysis with hadoop. Comput. Geosci. **66**(0), 145–154 (2014)
8. Adler, P.M.: Porous Media, Geometry and Transports. Butterworth and Heinemann, Stoneham, UK (1992)
9. Adler, P.M., Jacquin, C.G., Quiblier, J.A.: Flow in simulated porous media. Int. J. Multiphase Flow **16**(4), 691–712 (1990)
10. Adler, R.J.: The Geometry of Random Fields. John Wiley & Sons, New York, NY, USA (1981)
11. Adler, R.J., Taylor, J.E.: Random Fields and Geometry. Springer Science & Business Media, New York, NY, USA (2009)
12. Advani, M., Ganguli, S.: Statistical mechanics of optimal convex inference in high dimensions. Phys. Rev. X **6**(3), 031034 (2016)
13. Ahrens, J., Hendrickson, B., Long, G., Miller, S., Ross, R., Williams, D.: Data-intensive science in the US DOE: case studies and future challenges. Comput. Sci. Eng. **13**(6), 14–24 (2011)
14. Al-Gwaiz, M.A., Anandam, V.: On the representation of biharmonic functions with singularities in $\mathbb{R}^n$. Indian J. Pure Appl. Math. **44**(3), 263–276 (2013)
15. Allard, D.: Modeling spatial and spatio-temporal non Gaussian processes. In: Porcu, E., Montero, J., Schlather, M. (eds.) Advances and Challenges in Space-time Modelling of Natural Events. Lecture Notes in Statistics, vol. 207, pp. 141–164. Springer, Heidelberg, Germany (2012)
16. Allard, D., Naveau, P.: A new spatial skew-normal random field model. Comput. Stat. Theory Methods **36**(9), 1821–1834 (2007)

D. T. Hristopulos, *Random Fields for Spatial Data Modeling*, Advances in Geographic Information Science, https://doi.org/10.1007/978-94-024-1918-4

17. Allard, D., Senoussi, R., Porcu, E.: Anisotropy models for spatial data. Math. Geosci. **48**(3), 305–328 (2016)
18. Alpaydin, E.: Introduction to Machine Learning. MIT Press, Cambridge, MA, USA (2014)
19. Ambikasaran, S., Foreman-Mackey, D., Greengard, L., Hogg, D.W., O'Neil, M.: Fast direct methods for Gaussian processes. IEEE Trans. Pattern Anal. Mach. Intell. **38**(2), 252–265 (2016)
20. Amigó, J., Balogh, S., Hernández, S.: A brief review of generalized entropies. Entropy **20**(11), 813–833 (2018)
21. Amit, D.J.: Field Theory, the Renormalization Group, and Critical Phenomena, 2nd edn. World Scientific, New York, NY, USA (1984)
22. Anderson, E.R., Duvall Jr., T.L., Jefferies, S.M.: Modeling of solar oscillation power spectra. Astrophys. J. Part 1 **364**, 699–705 (1990)
23. Anderson, P.W.: Basic Notions of Condensed Matter Physics. Benjamin-Cummings, New York, NY, USA (1984)
24. Anderson, P.W.: The reverend Thomas Bayes, needles in haystacks, and the fifth force. Phys. Today **45**, 9 (1992)
25. Anderson, P.W.: More and Different: Notes from a Thoughtful Curmudgeon. World Scientific, Hackensack, NJ, USA (2011)
26. Anderson, T.W.: An Introduction to Multivariate Statistical Analysis, 3rd edn. John Wiley & Sons, New York, NY, USA (1984)
27. Angus, J.E.: The probability integral transform and related results. SIAM Rev. **36**(4), 652–654 (1994)
28. Anonymous: Hydrology Handbook, Management Group D, ASCE Manuals and Reports on Engineering Practice. Tech. Rep. No. 28, American Society of Civil Engineers, New York, NY, USA (1996)
29. Apanasovich, T.V., Genton, M.G., Sun, Y.: A valid Matérn class of cross-covariance functions for multivariate random fields with any number of components. J. Am. Stat. Assoc. **107**(497), 180–193 (2012)
30. Arfken, G.B., Weber, H.J.: Mathematical Methods for Physicists. Elsevier, Amsterdam, Netherlands (2013)
31. Arlot, S., Celisse, A., et al.: A survey of cross-validation procedures for model selection. Stat. Surv. **4**, 40–79 (2010)
32. Armstrong, M.: Problems with universal kriging. Math. Geol. **16**(1), 101–108 (1984)
33. Armstrong, M.: Basic Linear Geostatistics. Springer, Berlin, Germany (1998)
34. Armstrong, M., Diamond, P.: Testing variograms for positive-definiteness. Math. Geol. **16**(4), 407–421 (1984)
35. Armstrong, M., Galli, A., Beucher, H., Loc'h, G., Renard, D., Doligez, B., Eschard, R., Geffroy, F.: Plurigaussian Simulations in Geosciences. Springer Science & Business Media, Heidelberg, Germany (2011)
36. Armstrong, M., Matheron, G.: Disjunctive kriging revisited: Part I. Math. Geol. **18**(8), 711–728 (1986)
37. Arns, C.H., Knackstedt, M.A., Mecke, K.R.: Characterising the morphology of disordered materials. In: Mecke, K., Stoyan, D. (eds.) Morphology of Condensed Matter, pp. 37–74. Springer, Heidelberg, Germany (2002)
38. Atkenson, S.G., Moore, A.W., Schaal, S.: Locally weighted learning. Artif. Intell. Rev. **11**(1–5), 11–73 (1997)
39. Bachoc, F.: Asymptotic analysis of the role of spatial sampling for covariance parameter estimation of Gaussian processes. J. Multivar. Anal. **125**, 1–35 (2014)
40. Baddeley, A., Gregori, P., Mahiques, J.M., Stoica, R., Stoyan, D.: Case Studies in Spatial Point Process Modeling. Lecture Notes in Statistics. Springer, New York, NY, USA (2005)
41. Bailey, D.C.: Not normal: the uncertainties of scientific measurements. R. Soc. Open Sci. **4**(1), 160600 (2017)
42. Bak, P., Tang, C., Wiesenfeld, K.: Self-organized criticality – an explanation of 1/f noise. Phys. Rev. Lett. **59**(4), 381–384 (1987)

43. Bakr, A.A., Gelhar, L.W., Gutjahr, A.L., MacMillan, J.R.: Stochastic analysis of spatial variability in subsurface flows: 1. Comparison of one- and three-dimensional flows. Water Resour. Res. **14**(2), 263–271 (1978)
44. Banerjee, S., Carlin, B.P., Gelfand, A.E.: Hierarchical Modeling and Analysis for Spatial Data, 2nd edn. CRC Press, Boca Raton, FL, USA (2014)
45. Banerjee, S., Gelfand, A.E.: On smoothness properties of spatial processes. J. Multivar. Anal. **84**(1), 85–100 (2003)
46. Banerjee, S., Gelfand, A.E., Finley, A.O., Sang, H.: Gaussian predictive process models for large spatial data sets. J. R. Stat. Soc. Ser. B (Stat Methodol.) **70**(4), 825–848 (2008)
47. Bárdossy, A.: Copula-based geostatistical models for groundwater quality parameters. Water Resour. Res. **42**(11), W11416 (2006)
48. Bárdossy, A., Li, J.: Geostatistical interpolation using copulas. Water Resour. Res. **44**(7), W07412 (2008)
49. Barenblatt, G.I.: Dimensional Analysis. Gordon and Breach Science Publishers, New York, NY, USA (1987)
50. Barthelemy, M., Orland, H., Zerah, G.: Propagation in random media: calculation of the effective dispersive permittivity by use of the replica method. Phys. Rev. E **52**(1), 1123–1127 (1995)
51. Båth, B.M.: Spectral Analysis in Geophysics. Elsevier, Amsterdam, Netherlands (1974)
52. Baxevani, A., Lennartsson, J.: A spatiotemporal precipitation generator based on a censored latent Gaussian field. Water Resour. Res. **51**(6), 4338–4358 (2015)
53. Baxter, R.J.: Exactly Solved Models in Statistical Mechanics. Academic Press, San Diego, CA, USA (1982)
54. Beale, C.M., Lennon, J.J., Yearsley, J.M., Brewer, M.J., Elston, D.A.: Regression analysis of spatial data. Ecol. Lett. **13**(2), 246–264 (2010)
55. Beck, C., Cohen, E.: Superstatistics. Physica A: Stat. Mech. Appl. **322**, 267–275 (2003)
56. Ben-Avraham, D., Havlin, S.: Diffusion and reactions in fractals and disordered systems. Cambridge University Press, Cambridge, UK (2000)
57. Bender, C.M., Orszag, S.A.: Advanced Mathematical Methods for Scientists and Engineers I. Springer Science & Business Media, New York, NY, USA (1999)
58. Benson, D.A., Wheatcraft, S.W., Meerschaert, M.M.: The fractional-order governing equation of Lévy motion. Water Resour. Res. **36**(6), 1413–1423 (2000)
59. Berk, N.F.: Scattering properties of a model bicontinuous structure with a well defined length scale. Phys. Rev. Lett. **58**(25), 2718–2721 (1987)
60. Berk, N.F.: Scattering properties of the leveled-wave model of random morphologies. Phys. Rev. A **44**(8), 5069–5078 (1991)
61. Berkooz, G., Holmes, P., Lumley, J.L.: The proper orthogonal decomposition in the analysis of turbulent flows. Annu. Rev. Fluid Mech. **25**(1), 539–575 (1993)
62. Berryman, J.G.: Measurement of spatial correlation functions using image processing techniques. J. Appl. Phys. **57**(7), 2374–2384 (1985)
63. Berryman, J.G.: Relationship between specific surface area and spatial correlation functions for anisotropic porous media. J. Math. Phys. **28**(1), 244–245 (1987)
64. Berryman, J.G., Blair, S.C.: Use of digital image analysis to estimate fluid permeability of porous materials: application of two-point correlation functions. J. Appl. Phys. **60**(6), 1930–1938 (1986)
65. Bertschinger, E.: Path integral methods for primordial density perturbations-sampling of constrained Gaussian random fields. Astrophys. J. **323**, L103–L106 (1987)
66. Bertschinger, E.: Multiscale Gaussian random fields and their application to cosmological simulations. Astrophys. J. Suppl. Ser. **137**(1), 1–20 (2001)
67. Bertsekas, D.P., Nedi, A., Ozdaglar, A.E.: Convex Analysis and Optimization. Athena Scientific, Belmont, MA, USA (2003)
68. Besag, J.: Nearest-neighbour systems and the auto-logistic model for binary data. J. R. Stat. Soc. Ser. B Methodol. **34**(1), 75–83 (1972)

69. Besag, J.: Spatial interaction and the statistical analysis of lattice systems. J. R. Stat. Soc. Ser. B Methodol. **36**(2), 192–236 (1974)
70. Besag, J.: Statistical analysis of non-lattice data. The Statistician **24**(3), 179–195 (1975)
71. Besag, J.: Efficiency of pseudolikelihood estimation for simple Gaussian fields. Biometrika **64**(3), 616–618 (1977)
72. Besag, J., Kooperberg, C.: On conditional and intrinsic autoregressions. Biometrika **82**(4), 733–746 (1995)
73. Besag, J., Mondal, D.: First-order intrinsic autoregressions and the de Wijs process. Biometrika **92**(4), 909–920 (2005)
74. Beuman, T.H., Turner, A.M., Vitelli, V.: Extrema statistics in the dynamics of a non-Gaussian random field. Phys. Rev. E **87**(2), 022142 (2013)
75. Bhatia, R.: Positive Definite Matrices. Princeton University Press, Princeton, NJ, USA (2009)
76. Biermé, H., Richard, F.: Statistical tests of anisotropy for fractional Brownian texture. application to full-field digital mammography. J. Math. Imaging Vis. **36**(3), 227–240 (2010)
77. Birkholz, S., et al.: On the predictability of rogue events. Phys. Rev. Lett. **114**, 213901 (2015)
78. Bishop, C.M.: Pattern Recognition and Machine Learning. Springer, New York, NY, USA (2006)
79. Blattberg, R.C., Gonedes, N.J.: A comparison of the stable and Student distributions as statistical models for stock prices. J. Bus. **47**(2), 244–280 (1974)
80. Blei, D.M., Kucukelbir, A., McAuliffe, J.D.: Variational inference: a review for statisticians. J. Am. Stat. Assoc. **112**(518), 859–877 (2017)
81. Blocker, C.: Maximum likelihood primer. http://physics.bu.edu/neppsr/TALKS-2002/ (2002). New England Particle Physics Student Retreat Talks. [Online; accessed on 31 Oct 2018]
82. Blumer, A., Ehrenfeucht, A., Haussler, D., Warmuth, M.K.: Occam's razor. In: Shavlik, J.W., Dieterich, T.G. (eds.) Information Processing Letters, pp. 377–380. Morgan Kaufmann, San Mateo, CA, USA (1987)
83. Bochner, S.: Lectures on Fourier Integrals. Princeton University Press, Princeton, NJ, USA (1959)
84. Boissonnat, J.D., Cazals, F.: Natural neighbor coordinates of points on a surface. Comput. Geom. **19**(2–3), 155–173 (2001)
85. Boissonnat, J.D., Cazals, F.: Smooth surface reconstruction via natural neighbour interpolation of distance functions. Comput. Geom. **22**(1–3), 185–203 (2002)
86. Bojesen, T.A.: Policy-guided Monte Carlo: Reinforcement-learning Markov chain dynamics. Phys. Rev. E **98**(6), 063303 (2018)
87. Bolin, D., Lindgren, F.: Spatial models generated by nested stochastic partial differential equations, with an application to global ozone mapping. Ann. Appl. Stat. **5**(1), 523–550 (2011)
88. Bolin, D., Wallin, J.: Spatially adaptive covariance tapering. Spat. Stat. **18**(Part A), 163–178 (2016)
89. Bolthausen, E.: On the central limit theorem for stationary mixing random fields. Ann. Probab. **10**(4), 1047–1050 (1982)
90. Van den Bos, A.: Parameter Estimation for Scientists and Engineers. John Wiley & Sons, Hoboken, NJ, USA (2007)
91. Boubrima, A., Bechkit, W., Rivano, H.: Optimal WSN deployment models for air pollution monitoring. IEEE Trans. Wirel. Commun. **16**(5), 2723–2735 (2017)
92. Bouchaud, J., Georges, A.: Anomalous diffusion in disordered media: statistical mechanisms, models and physical applications. Phys. Rep. **195**, 127–293 (1990)
93. Box, G.E.P., Cox, D.R.: An analysis of transformations. J. R. Stat. Soc. Ser. B Methodol. **26**(2), 211–252 (1964)
94. Box, G.E.P., Jenkins, G.M., Reinsel, G.C., Ljung, C.M.: Time Series Analysis, 5th edn. Wiley, Hoboken, NJ, USA (2016)
95. Box, G.E.P., Jenkins, G.M., Reinsel, G.C., Ljung, G.M.: Time Series Analysis: Forecasting and Control, 4th edn. John Wiley & Sons, Hoboken, NJ, USA (2008)

96. Bracewell, R.N.: The Fourier Transform and its Applications, 3rd edn. McGraw-Hill, Boston, MA, USA (2000)
97. Bradde, S., Bialek, W.: PCA meets RG. J. Stat. Phys. **167**(3–4), 462–475 (2017)
98. Brauchart, J.S., Dick, J., Fang, L.: Spatial low-discrepancy sequences, spherical cone discrepancy, and applications in financial modeling. J. Comput. Appl. Math. **286**, 28–53 (2015)
99. Bray, A.J., Dean, D.S.: Statistics of critical points of Gaussian fields on large-dimensional spaces. Phys. Rev. Lett. **98**(15), 150201 (2007)
100. Breiman, L.: Statistical modeling: the two cultures. Qual. Control Appl. Stat. **48**(1), 81–82 (2003)
101. Briggs, I.C.: Machine contouring using minimum curvature. Geophysics **39**(1), 39–48 (1974)
102. Brillinger, D.: Modeling spatial trajectories. In: Gelfand, A.E., Diggle, P., Guttorp, P., Fuentes, M. (eds.) Handbook of Spatial Statistics, pp. 463–474. CRC Press, Boca Raton, FL, USA (2010)
103. Brillinger, D.R.: Trend analysis: time series and point process problems. Environmetrics **5**(1), 1–19 (1994)
104. Brits, L.: File:eulerangles.svg—wikimedia commons, the free media repository (2017). https://commons.wikimedia.org/w/index.php?title=File:Eulerangles.svg&oldid=230286188. [Online; accessed on July 5, 2018]
105. Brochu, E., Cora, V.M., De Freitas, N.: A tutorial on Bayesian optimization of expensive cost functions, with application to active user modeling and hierarchical reinforcement learning. arXiv preprint arXiv:1012.2599 (2010)
106. Brockmann, D., Helbing, D.: The hidden geometry of complex, network-driven contagion phenomena. Science **342**(6164), 1337–1342 (2013)
107. Brockwell, P.G., Davis, R.A.: Time Series: Theory and Methods, 2nd edn. Springer, New York, NY, USA (2006)
108. Brook, D.: On the distinction between the conditional probability and the joint probability approaches in the specification of nearest-neighbour systems. Biometrika **51**(3/4), 481–483 (1964)
109. Brooks, S., Gelman, A., Jones, G., Meng, X.L.: Handbook of Markov Chain Monte Carlo. CRC Press, Boca Raton, FL, USA (2011)
110. Brown, P.: Model-based geostatistics the easy way. J. Stat. Softw. **63**(12), 1–24 (2015)
111. Brown, R.: A brief account of microscopical observations made in the months of June, July and August 1827, on the particles contained in the pollen of plants; and on the general existence of active molecules in organic and inorganic bodies. Philos. Mag. **4**(21), 161–173 (1828)
112. Bun, J., Bouchaud, J.P., Potters, M.: Cleaning large correlation matrices: tools from random matrix theory. Phys. Rep. **666**, 1–109 (2016)
113. Bunde, A., Havlin, S.: Percolation I. In: Bunde, A., Havlin, S. (eds.) Fractals and Disordered Systems, pp. 59–113. Springer, London, UK (1991)
114. Cahn, J.W.: Phase separation by spinodal decomposition in isotropic systems. J. Chem. Phys. **42**(1), 93–99 (1965)
115. Cao, J., Worsley, K.J.: Applications of random fields in human brain mapping. In: Moore, M. (ed.) Spatial Statistics: Methodological Aspects and Applications. Lecture Notes in Statistics, vol. 159, pp. 169–182. Springer, Berlin, Germany (2001)
116. Carlson, J.M., Langer, J.S., Shaw, B.E., Tang, C.: Intrinsic properties of a Burridge-Knopoff model of an earthquake fault. Phys. Rev. A **44**(2), 884–897 (1991)
117. Carlsson, G.: Topology and data. Bull. Am. Math. Soc. **46**(2), 255–308 (2009)
118. Casella, G., George, E.I.: Explaining the Gibbs sampler. Am. Stat. **46**(3), 167–174 (1992)
119. Castellani, T., Cavagna, A.: Spin-glass theory for pedestrians. J. Stat. Mech: Theory Exp. **2005**(05), P05012 (2005)
120. Caticha, A., Preuss, R.: Maximum entropy and Bayesian data analysis: entropic prior distributions. Phys. Rev. E **70**(4), 046127 (2004)

121. Caughey, T.K., Stumpf, H.J.: Transient response of a dynamic system under random excitation. J. Appl. Mech. **28**(4), 563–566 (1961)
122. Chaikin, P.M., Lubensky, T.C.: Principles of Condensed Matter Physics, vol. 1. Cambridge University Press, Cambridge, UK (2000)
123. Chambers, R.L., Yarus, J.M., Hird, K.B.: Petroleum geostatistics for nongeostaticians: Part 2. Lead. Edge **19**(6), 592–599 (2000)
124. Chambers, R.L., Yarus, J.M., Hird, K.B.: Petroleum geostatistics for nongeostatisticians: Part 1. Lead. Edge **19**(5), 474–479 (2000)
125. Chandler, D.: Introduction to Modern Statistical Mechanics. Oxford University Press, Oxford, UK (1987)
126. Chang, J.C., Savage, V.M., Chou, T.: A path-integral approach to Bayesian inference for inverse problems using the semiclassical approximation. J. Stat. Phys. **157**(3), 582–602 (2014)
127. Chatterjee, A.: An introduction to the proper orthogonal decomposition. Curr. Sci. **78**(7), 808–817 (2000)
128. Chen, Y.C.: A tutorial on kernel density estimation and recent advances. Biostat. Epidemiol. **1**(1), 161–187 (2017)
129. Chen, Z., Ivanov, P.C., Hu, K., Stanley, H.E.: Effect of nonstationarities on detrended fluctuation analysis. Phys. Rev. E **65**(4), 041107 (2002)
130. Cherry, S., Banfield, J., Quimby, W.F.: An evaluation of a non-parametric method of estimating semi-variograms of isotropic spatial processes. J. Appl. Stat. **23**(4), 435–449 (1996)
131. Chilès, J.P.: The generalized variogram. Tech. Rep. réf. R121030JCHI, École des Mines de Paris, Centre de Géosciences, Géostatistique (2012). http://www.cg.ensmp.fr/bibliotheque/public/CHILES_Rapport_02272.pdf
132. Chilès, J.P., Delfiner, P.: Geostatistics: Modeling Spatial Uncertainty, 2nd edn. John Wiley & Sons, New York, NY, USA (2012)
133. Ching, J., Phoon, K.K.: Impact of autocorrelation function model on the probability of failure. J. Eng. Mech. **145**(1), 04018123 (2019)
134. Chiu, S.N., Stoyan, D., Kendall, W.S., Mecke, J.: Stochastic Geometry and its Applications. John Wiley & Sons, Chichester, West Sussex, UK (2013)
135. Chorti, A., Hristopulos, D.T.: Non-parametric identification of anisotropic (elliptic) correlations in spatially distributed data sets. IEEE Trans. Signal Process. **56**(10), 4738–4751 (2008)
136. Christakos, G.: On the problem of permissible covariance and variogram models. Water Resour. Res. **20**(2), 251–265 (1984)
137. Christakos, G.: A Bayesian/maximum-entropy view to the spatial estimation problem. Math. Geol. **22**(7), 763–777 (1990)
138. Christakos, G.: Random Field Models in Earth Sciences. Academic Press, San Diego (1992)
139. Christakos, G.: Modern Spatiotemporal Geostatistics. Oxford University Press, Oxford, UK (2000)
140. Christakos, G.: Spatiotemporal Random Fields, 2nd edn. Elsevier, Amsterdam, Netherlands (2017)
141. Christakos, G., Hristopulos, D.T.: Spatiotemporal Environmental Health Modelling. Kluwer, Boston (1998)
142. Christakos, G., Hristopulos, D.T., Bogaert, P.: On the physical geometry concept at the basis of space/time geostatistical hydrology. Adv. Water Resour. **23**(8), 799–810 (2000)
143. Christakos, G., Hristopulos, D.T., Miller, C.T.: Stochastic diagrammatic analysis of groundwater flow in heterogeneous porous media. Water Resour. Res. **31**(7), 1687–1703 (1995)
144. Cipra, B.A.: The Ising model is NP-complete. SIAM News **33**(6), 1–3 (2000)
145. Claeskens, G., Hjort, N.L.: Model Selection and Model Averaging, vol. 330. Cambridge University Press, Cambridge, UK (2008)
146. Clauset, A., Shalizi, C., Newman, M.: Power-law distributions in empirical data. SIAM Rev. **51**(4), 661–703 (2009)

147. Clayton, D.G.: A model for association in bivariate life tables and its application in epidemiological studies of familial tendency in chronic disease incidence. Biometrika **65**(1), 141–151 (1978)
148. Clementi, F., Di Matteo, T., Gallegati, M., Kaniadakis, G.: The κ-generalized distribution: a new descriptive model for the size distribution of incomes. Physica A **387**, 3201–3208 (2008)
149. Clementi, F., Gallegati, M., Kaniadakis, G.: A κ-generalized statistical mechanics approach to income analysis. J. Stat. Mech: Theory Exp. **2009**, P02037 (2009)
150. Cleveland, W.S., Devlin, S.J.: Locally weighted regression: an approach to regression analysis by local fitting. J. Am. Stat. Assoc. **83**(403), 596–610 (1988)
151. Cleveland, W.S., Loader, C.: Smoothing by local regression: principles and methods. In: Härdle, W., Shimek, G. (eds.) Statistical Theory and Computational Aspects of Smoothing, Proceedings of the COMPSTAT '94 Satellite Meeting, pp. 10–49. Springer (1996)
152. Clifford, P.: Markov random fields in statistics. In: Grimmett, G.R., Welsh, D.J.A. (eds.) Disorder in Physical Systems: A Volume in Honour of John M. Hammersley, vol. 19. Oxford University Press, Oxford, UK (1990)
153. Cohen, E.G.D.: Superstatistics. Physica D: Nonlinear Phenom. **193**(1–4), 35–52 (2004)
154. Congdon, P.: Bayesian Statistical Modelling, vol. 704. John Wiley & Sons, Chichester, West Sussex, England (2007)
155. Cotton, W.R., Bryan, G., Van den Heever, S.C.: Storm and Cloud Dynamics, vol. 99. Academic Press, Amsterdam, Netherlands (2010)
156. Cover, T.M., Thomas, J.A.: Elements of Information Theory, 2nd edn. John Wiley & Sons, Hoboken, NJ, USA (2006)
157. Cowles, M.K., Carlin, B.P.: Markov chain Monte Carlo convergence diagnostics: a comparative review. J. Am. Stat. Assoc. **91**(434), 883–904 (1996)
158. Cox, D.R.: Frequentist and Bayesian statistics: a critique. In: Lyons, L., Ünel, M.K. (eds.) Statistical Problems in Particle Physics, Astrophysics and Cosmology, Proceedings of PHYSTAT05, 12–15 Sept, pp. 3–6. World Scientific, Hackensack, NJ, USA (2006)
159. Cox, T.F., Cox, M.A.: Multidimensional Scaling. CRC Press, Boca Raton, FL, USA (2000)
160. Cramér, H.: On the theory of stationary random processes. Ann. Math. **41**(1), 215–230 (1940)
161. Cramér, H.: Mathematical Methods of Statistics (PMS-9), vol. 9, 1st edn. Princeton University Press, Princeton, NJ, USA (2016)
162. Cramér, H., Leadbetter, M.R.: Stationary and Related Stochastic Processes. John Wiley & Sons, New York, NY, USA (1967)
163. Crawford, F.S.: Berkeley Physics Course: Waves, vol. 3. McGraw-Hill, New York, NY, USA (1968)
164. Cressie, N.: The origins of kriging. Math. Geol. **22**(3), 239–253 (1990)
165. Cressie, N.: Spatial Statistics. John Wiley & Sons, New York, NY, USA (1993)
166. Cressie, N., Hawkins, D.M.: Robust estimation of the variogram: I. J. Int. Assoc. Math. Geol. **12**(2), 115–125 (1980)
167. Cressie, N., Johannesson, G.: Fixed rank kriging for very large spatial data sets. J. R. Stat. Soc. Ser. B (Stat Methodol.) **70**(1), 209–226 (2008)
168. Cressie, N., Pavlicová, M.: Lognormal kriging: bias adjustment and kriging variances. In: Leuangthong, O., Deutsch, C.V. (eds.) Geostatistics Banff 2004, Quantitative Geology and Geostatistics, pp. 1027–1036. Springer, Dordrecht, Netherlands (2005)
169. Cressie, N., Wikle, C.L.: Statistics for Spatio-temporal Data. John Wiley & Sons, New York, NY, USA (2011)
170. Creswick, R., Farach, H., Poole, C.: Introduction to Renormalization Group Methods in Physics. John Wiley & Sons, New York, NY, USA (1991)
171. Cushman, J.H.: The Physics of Fluids in Hierarchical Porous Media: Angstroms to Miles: Theory and Applications of Transport in Porous Media, 1st edn. Kluwer, Dordrecht, The Netherlands (1997)
172. Cushman, J.H., Bennethum, L.S., Hu, B.X.: A primer on upscaling tools for porous media. Adv. Water Resour. **25**(8–12), 1043–1067 (2002)
173. Dagan, G.: Flow and Transport in Porous Formations. Springer, Berlin, Germany (1989)

174. Dagan, G., Neuman, S.P.: Subsurface Flow and Transport: A Stochastic Approach. Cambridge University Press, Cambridge, UK (2005)
175. Dahmen, D., Bos, H., Helias, M.: Correlated fluctuations in strongly coupled binary networks beyond equilibrium. Phys. Rev. X **6**(3), 031024 (2016)
176. Dale, M.R.T., Fortin, M.J.: Spatial Analysis: A Guide for Ecologists. Cambridge University Press, Cambridge, UK (2014)
177. Daley, D.J., Vere-Jones, D.: An Introduction to the Theory of Point Processes. Probability and its Applications (New York), vol. I, 2nd edn. Springer, New York, NY, USA (2003)
178. Daley, R.: Atmospheric Data Analysis. Cambridge University Press, Cambridge, UK (1991)
179. Davies, S., Hall, P.: Fractal analysis of surface roughness by using spatial data. J. R. Stat. Soc. Ser. B (Stat Methodol.) **61**(1), 3–37 (1999)
180. Davis, M.W.: Production of conditional simulations via the LU triangular decomposition of the covariance matrix. Math. Geol. **19**(2), 91–98 (1987)
181. Davison, A.C., Gholamrezaee, M.M.: Geostatistics of extremes. Proc. R. Soc. A: Math. Phys. Eng. Sci. **468**(2138), 581–608 (2012)
182. Davison, A.C., Huser, R., Thibaud, E.: Geostatistics of dependent and asymptotically independent extremes. Math. Geosci. **45**(5), 511–529 (2013)
183. Davison, A.C., Padoan, S.A., Ribatet, M.: Statistical modeling of spatial extremes. Stat. Sci. **27**(2), 161–186 (2012)
184. De Iaco, S., Palma, M., Posa, D.: Geostatistics and the role of variogram in time series analysis: a critical review. In: Montrone, S., Perchinunno, P. (eds.) Statistical Methods for Spatial Planning and Monitoring, pp. 47–75. Springer Milan, Milano, Italy (2013)
185. De Micheaux, P.L., Liquet, B.: Understanding convergence concepts: a visual-minded and graphical simulation-based approach. Am. Stat. **63**(2), 73–78 (2012)
186. De Oliveira, V., Kedem, B., Short, D.A.: Bayesian prediction of transformed Gaussian random fields. J. Am. Stat. Assoc. **92**(440), 1422–1433 (1997)
187. Deans, S.R.: The Radon Transform and Some of its Applications. John Wiley & Sons, New York, NY, USA (1983)
188. Debnath, L., Mikusiński, P.: Hilbert Spaces with Applications. Academic Press, Amsterdam, Netherlands (2005)
189. Debye, P., Anderson, H.R., Brumberger, H.: Scattering by an inhomogeneous solid. II. The correlation function and its application. J. Appl. Phys. **28**(6), 679–683 (1957)
190. Dee, D.P., Da Silva, A.M.: Maximum-likelihood estimation of forecast and observation error covariance parameters. Part I: methodology. Mon. Weather Rev. **127**(8), 1822–1834 (1999)
191. Delicado, P., Giraldo, R., Comas, C., Mateu, J.: Statistics for spatial functional data: some recent contributions. Environmetrics **21**(3–4), 224–239 (2010)
192. Dempster, A.P.: Covariance selection. Biometrics **128**(1), 157–175 (1972)
193. Dempster, A.P., Laird, N.M., Rubin, D.B.: Maximum likelihood from incomplete data via the EM algorithm. J. R. Stat. Soc. Ser. B Methodol. **39**(1), 1–38 (1977)
194. Dennery, P., Krzywicki, A.: Mathematics for Physicists. Courier Dover Publications, Mineola, NY, USA (1996)
195. Deutsch, C.V.: Geostatistical Reservoir Modeling. Oxford, UK, New York, NY, USA (2002)
196. Deutsch, C.V., Cockerham, P.W.: Practical considerations in the application of simulated annealing to stochastic simulation. Math. Geol. **26**(1), 67–82 (1994)
197. Dey, D.K., Ghosh, S.K., Mallick, B.K.: Generalized Linear Models: a Bayesian Perspective. Marcel Dekker, New York, NY, USA (2000)
198. Di Federico, V., Neuman, S.P., Tartakovsky, D.M.: Anisotropy, lacunarity, and upscaled conductivity and its autocovariance in multiscale random fields with truncated power variograms. Water Resour. Res. **35**(10), 2891–2908 (1999)
199. Dietrich, C.R., Newsam, G.N.: A fast and exact method for multidimensional Gaussian stochastic simulations. Water Resour. Res. **29**(8), 2861–2869 (1993)
200. Dietrich, C.R., Newsam, G.N.: Efficient generation of conditional simulations by Chebyshev matrix polynomial approximations to the symmetric square root of the covariance matrix. Math. Geol. **27**(2), 207–228 (1995)

201. Diggle, P., Ribeiro, P.J.: Model-based Geostatistics. Springer Science & Business Media, New York, NY, USA (2007)
202. Diggle, P.J., Moraga, P., Rowlingson, B., Taylor, B.M., et al.: Spatial and spatio-temporal log-Gaussian Cox processes: extending the geostatistical paradigm. Stat. Sci. **28**(4), 542–563 (2013)
203. Diggle, P.J., Tawn, J.A., Moyeed, R.A.: Model-based geostatistics. J. R. Stat. Soc.: Ser. C: Appl. Stat. **47**(3), 299–350 (1998)
204. Ditlevsen, P.D.: Turbulence and Shell Models. Cambridge University Press, Cambridge, UK (2011)
205. Dolph, C.L., Woodbury, M.A.: On the relation between Green's functions and covariances of certain stochastic processes and its application to unbiased linear prediction. Trans. Am. Math. Soc. **72**(3), 519–550 (1952)
206. Donsker, M.D.: Justification and extension of Doob's heuristic approach to the Kolmogorov-Smirnov theorems. Ann. Math. Stat. **23**(2), 277–281 (1952)
207. Donsker, M.D.: On function space integrals. In: Martin, W.T., Segal, I. (eds.) Analysis in Function Space, pp. 17–30. MIT Press, Boston, MA, USA (1964)
208. Dorn, S., Enßlin, T.A.: Stochastic determination of matrix determinants. Phys. Rev. E **92**(1), 013302 (2015)
209. Dotsenko, V.: Introduction to the Replica Theory of Disordered Statistical Systems. Cambridge University Press, Cambridge, UK (2005)
210. Dowd, P.A., Dare-Bryan, P.C.: Planning, designing and optimising production using geostatistical simulation. In: Dimitrakopoulos, R. (ed.) Orebody Modelling and Strategic Mine Planning, 2nd edn., pp. 363–378. The Australasian Institute of Mining and Metallurgy, Spectrum Series, Carlton, Victoria, Australia (2007)
211. Draper, N.R., Cox, D.R.: On distributions and their transformation to normality. J. R. Stat. Soc. Ser. B Methodol. **31**(3), 472–476 (1969)
212. Drummond, I.T., Horgan, R.R.: The effective permeability of a random medium. J. Phys. A Math. Gen. **20**(14), 4661–4672 (1987)
213. Du, J., Zhang, H., Mandrekar, V.S.: Fixed-domain asymptotic properties of tapered maximum likelihood estimators. Ann. Stat. **37**(6A), 3330–3361 (2009)
214. Dubois, G.: Spatial interpolation comparison 97: foreword and introduction. J. Geogr. Inf. Decis. Anal. **2**(2), 1–10 (1998)
215. Dubois, G.: Report on the spatial interpolation comparison (SIC2004) exercise. In: Dubois, G. (ed.) Automatic Mapping Algorithms for Routine and Emergency Monitoring Data, vol. EUR 21595 EN, pp. 103–124. Office for Official Publications of the European Communities, Luxembourg, European Communities (2006)
216. Dubois, G., Galmarini, S.: Introduction to the spatial interpolation comparison (SIC)2004 exercise and presentation of the data sets. J. Geogr. Inf. Decis. Anal. **1**(2), 09–1–09–11 (2005)
217. Dubrule, O.: Indicator variogram models: do we have much choice? Math. Geosci. **49**(4), 441–465 (2017)
218. Dyson, F.J.: Existence of a phase-transition in a one-dimensional Ising ferromagnet. Commun. Math. Phys. **12**, 91–107 (1969)
219. E, W.: Principles of Multiscale Modeling. Cambridge University Press, Cambridge, UK (2011)
220. Ecker, M.D., Gelfand, A.E.: Bayesian modeling and inference for geometrically anisotropic spatial data. Math. Geol. **32**(1), 67–82 (1999)
221. Ecker, M.D., Gelfand, A.E.: Spatial modeling and prediction under stationary non-geometric range anisotropy. Environ. Ecol. Stat. **10**(2), 165–178 (2003)
222. Economou, E.N.: Green's functions in Quantum Physics, 3rd edn. Springer, Berlin, Germany (2006)
223. Edwards, S.F., Anderson, P.W.: Theory of spin glasses. J. Phys. F: Met. Phys. **5**(5), 965–974 (1975)
224. Efron, B., Tibshirani, R.J.: An Introduction to the Bootstrap. CRC Press, Boca Raton, FL, USA (1994)

225. Egozcue, J.J., Díaz-Barrero, J.L., Pawlowsky-Glahn, V.: Hilbert space of probability density functions based on Aitchison geometry. Acta Math. Sin. **22**(4), 1175–1182 (2006)
226. Einstein, A.: Über die von der molekularkinetischen theorie der wärme geforderte bewegung von in ruhenden flüssigkeiten suspendierten teilchen. Annalen der Physik **322**(8), 549–560 (1905). https://doi.org/10.1002/andp.19053220806
227. Eliazar, I., Klafter, J.: A probabilistic walk up power laws. Phys. Rep. **511**(3), 143–175 (2012)
228. Elogne, S.N., Hristopulos, D.T.: Geostatistical applications of Spartan spatial random fields. In: Soares, A., Pereira, M.J., Dimitrakopoulos, R. (eds.) geoENV VI-Geostatistics for Environmental Applications. Quantitative Geology and Geostatistics, vol. 15, pp. 477–488. Springer, Berlin, Germany (2008)
229. Elogne, S.N., Hristopulos, D.T., Varouchakis, E.: An application of Spartan spatial random fields in environmental mapping: focus on automatic mapping capabilities. Stochastic Environ. Res. Risk Assess. **22**(5), 633–646 (2008)
230. Embrechts, P., Maejima, M.: Selfsimilar Processes. Princeton University Press, Princeton, NJ, USA (2002)
231. Emery, X.: Testing the correctness of the sequential algorithm for simulating Gaussian random fields. Stochastic Environ. Res. Risk Assess. **18**(6), 401–413 (2004)
232. Emery, X.: Conditioning simulations of Gaussian random fields by ordinary kriging. Math. Geol. **39**(6), 607–623 (2007)
233. Emery, X., Lantuéjoul, C.: TBSIM: a computer program for conditional simulation of three-dimensional Gaussian random fields via the turning bands method. Comput. Geosci. **32**(10), 1615–1628 (2006)
234. Enßlin, T.A., Frommert, M.: Reconstruction of signals with unknown spectra in information field theory with parameter uncertainty. Phys. Rev. D **83**(10), 105014 (2011)
235. Enßlin, T.A., Frommert, M., Kitaura, F.S.: Information field theory for cosmological perturbation reconstruction and nonlinear signal analysis. Phys. Rev. D **80**(10), 105005 (2009)
236. Enßlin, T.A., Weig, C.: Inference with minimal Gibbs free energy in information field theory. Phys. Rev. E **82**(5), 051112 (2010)
237. Erickson, G., Smith, C.R. (eds.): Maximum-Entropy and Bayesian Methods in Science and Engineering: Foundations, vol. 1. Kluwer, Dordrecht, Netherlands (1988)
238. Ernst, O.G., Mugler, A., Starkloff, H.J., Ullmann, E.: On the convergence of generalized polynomial chaos expansions. ESAIM: Math. Model. Numer. Anal. **46**(2), 317–339 (2012)
239. Fang, K.T., Kotz, S., Ng, K.W.: Symmetric Multivariate and Related Distributions. Chapman and Hall, New Delhi, India (1990)
240. Farlow, S.J.: Partial Differential Equations for Scientists and Engineers. Courier Dover Publications, Mineola, NY, USA (2012)
241. Farmer, C.L.: Geological modelling and reservoir simulation. In: Iske, A., Randen, T. (eds.) Mathematical Methods and Modeling in Hydrocarbon Exploration and Production. Mathematics in Industry, vol. 7, pp. 119–212. Springer, Heidelberg, Germany (2005)
242. Farmer, C.L.: Bayesian field theory applied to scattered data interpolation and inverse problems. In: Iske, A., Levesley, J. (eds.) Algorithms for Approximation, pp. 147–166. Springer, Heidelberg, Germany (2007)
243. Feder, J.: Fractals. Plenum Press, New York, NY, USA (1988)
244. Feigelson, E.D., Babu, G.J.: Modern Statistical Methods for Astronomy. Cambridge University Press Textbooks, Cambridge, UK (2012)
245. Feynman, R.P.: Statistical Mechanics. Benjamin and Cummings, Reading, MA, USA (1982)
246. Feynman, R.P., Hibbs, A.R.: Quantum Mechanics and Path Integrals. Courier Dover Publications, Mineola, NY, USA (2012)
247. Feynman, R.P., Leighton, R.B., Sands, M.: The Feynman Lectures on Physics, vol. 1. Basic Books, New York, NY, USA (2010)
248. Field, C., Genton, M.G.: The multivariate g-and-h distribution. Technometrics **48**(1), 104–111 (2006)
249. Field, R.V.: Stochastic models: theory and simulation. Tech. Rep. SAND2008-1365, Sandia National Laboratories (2008)

250. Fiori, A., Jankovič, I., Dagan, G.: Flow and transport in highly heterogeneous formations: 2. Semianalytical results for isotropic media. Water Resour. Res. **39**(9), 1269 (2003)
251. Fisher, R.A.: On the mathematical foundations of theoretical statistics. Philos. Trans. R. Stat. Soc. A **222**, 309–368 (1922)
252. Fisher, R.A.: Applications of "Student's" distribution. Metron **5**(3), 90–104 (1925)
253. Foreman-Mackey, D., Agol, E., Ambikasaran, S., Angus, R.: Fast and scalable Gaussian process modeling with applications to astronomical time series. Astron. J. **154**(6), 220–240 (2017)
254. Forrester, A.I.J., Sóbester, A., Keane, A.J.: Multi-fidelity optimization via surrogate modeling. Proc. R. Soc. Lond. A: Math. Phys. Eng. Sci. **463**(2088), 3251–3269 (2007)
255. Forristall, G.Z.: On the statistical distribution of wave heights in a storm. J. Geophys. Res. Oceans **83**(C5), 2353–2358 (1978)
256. Fouedjio, F.: Second-order non-stationary modeling approaches for univariate geostatistical data. Stoch. Environ. Res. Risk Assess. **31**(8), 1887–1906 (2017)
257. Fouedjio, F.: A fully non-stationary linear coregionalization model for multivariate random fields. Stoch. Environ. Res. Risk Assess. **32**(6), 1699–1721 (2018)
258. Fouedjio, F., Desassis, N., Rivoirard, J.: A generalized convolution model and estimation for non-stationary random functions. Spat. Stat. **16**, 35–52 (2016)
259. Foulkes, W.M.C., Mitas, L., Needs, R.J., Rajagopal, G.: Quantum Monte Carlo simulations of solids. Rev. Mod. Phys. **73**(1), 33–83 (2001)
260. Franzke, C.L.E., O'Kane, T.J., Berner, J., Williams, P.D., Lucarini, V.: Stochastic climate theory and modeling. Wiley Interdiscip. Rev. Clim. Chang. **6**(1), 63–78 (2015)
261. Friedland, C.J., et al.: Isotropic and anisotropic kriging approaches for interpolating surface-level wind speeds across large, geographically diverse regions. Geomat. Nat. Haz. Risk **8**(2), 207–224 (2016)
262. Friedrich, R., Peinke, J., Sahimi, M., Tabar, M.R.R.: Approaching complexity by stochastic methods: from biological systems to turbulence. Phys. Rep. **506**(5), 87–162 (2011)
263. Frühwirth, R., Regler, M.: Data Analysis Techniques for High-energy Physics. Cambridge University Press, Cambridge, UK (2000)
264. Fuentes, M.: Approximate likelihood for large irregularly spaced spatial data. J. Am. Stat. Assoc. **102**(477), 321–331 (2007)
265. Fuentes, M., Henry, J., Reich, B.: Nonparametric spatial models for extremes: application to extreme temperature data. Extremes **16**(1), 75–101 (2013)
266. Fuglstad, G.A., Simpson, D., Lindgren, F., Rue, H.: Does non-stationary spatial data always require non-stationary random fields? Spat. Stat. **14**(Part C), 505–531 (2015)
267. Furrer, R., Genton, M.G., Nychka, D.: Covariance tapering for interpolation of large spatial datasets. J. Comput. Graph. Stat. **15**(3), 502–523 (2006)
268. Furutsu, K.: On the statistical theory of electromagnetic waves in a fluctuating medium. J. Res. Natl. Inst. Stand. Technol. **67D**(3), 303–323 (1963)
269. Gaetan, C., Guyon, X., Bleakley, K.: Spatial Statistics and Modeling, vol. 81. Springer, New York, NY, USA (2010)
270. Galetakis, M., Hristopuloss, D.T.: Prediction of long-term quality fluctuations in the South Field lignite mine of West Macedonia. In: Agioutantis, Z., Komnitsas, K. (eds.) Proceedings of the 1st International Conference on Advances in Mineral Resources Management and Environmental Geotechnology, pp. 133–138. Heliotopos Conferences, Athens, Greece (2004)
271. Gelfand, A.E.: Hierarchical modeling for spatial data problems. Spat. Stat. **1**, 30–39 (2012)
272. Gelfand, A.E., Banerjee, S.: Multivariate spatial process models. In: Gelfand, A.E., Diggle, P., Guttorp, P., Fuentes, M. (eds.) Handbook of Spatial Statistics, pp. 495–513. CRC Press, Boca Raton, FL, USA (2010)
273. Gelfand, A.E., Diggle, P., Guttorp, P., Fuentes, M.: Handbook of Spatial Statistics. CRC Press, Boca Raton, FL, USA (2010)
274. Gelfand, A.E., Schliep, E.M.: Spatial statistics and Gaussian processes: a beautiful marriage. Spat. Stat. **18**(Part A), 86–104 (2016)
275. Gelhar, L.W.: Stochastic Subsurface Hydrology. Prentice Hall, Englewood Cliffs, NJ (1993)

276. Gelhar, L.W., Axness, C.L.: Three-dimensional stochastic analysis of macrodispersion in aquifers. Water Resour. Res. **19**(1), 161–180 (1983)
277. Gell-Mann, M., Lloyd, S.: Information measures, effective complexity, and total information. Complexity **2**(1), 44–52 (1996)
278. Gelman, A., Carlin, J.B., Stern, H.S., Dunson, D.B., Vehtari, A., Rubin, D.B.: Bayesian Data Analysis, vol. 2. CRC Press, Boca Raton, FL, USA (2014)
279. Geman, S., Geman, D.: Stochastic relaxation, Gibbs distributions, and the Bayesian restoration of images. IEEE Trans. Pattern Anal. Mach. Intell. **6**(6), 721–741 (1984)
280. Gengler, S.: Spatial prediction of categorical variables in environmental sciences: a minimum divergence and Bayesian data fusion approach. Ph.D. thesis, Université Catholique de Louvain (2018). https://dial.uclouvain.be/pr/boreal/object/boreal:198388
281. Gentle, J.E.: Random Number Generation and Monte Carlo Methods. Springer, New York, NY, USA (2003)
282. Genton, M.G.: Highly robust variogram estimation. Math. Geol. **30**(2), 213–221 (1998)
283. Genton, M.G.: Classes of kernels for machine learning: a statistics perspective. J. Mach. Learn. Res. **2**, 299–312 (2002)
284. Genton, M.G.: Skew-elliptical Distributions and Their Applications: A Journey Beyond Normality. CRC Press, Boca Raton, FL, USA (2004)
285. Gerlach, M., Altmann, E.G.: Testing statistical laws in complex systems. Phys. Rev. Lett. **122**(16), 168301 (2019)
286. Ghanem, R., Spanos, P.D.: Stochastic Finite Elements: A Spectral Approach. Dover, Mineola, NY, USA (2003)
287. Ghil, M., Allen, M.R., Dettinger, M.D., Ide, K., Kondrashov, D., Mann, M.E., Robertson, A.W., Saunders, A., Tian, Y., Varadi, F., Yiou, P.: Advanced spectral methods for climatic time series. Rev. Geophys. **40**(1), 3.1–3.41 (2002)
288. Ghrist, R.: Barcodes: the persistent topology of data. Bull. Am. Math. Soc. **45**(1), 61–75 (2008)
289. Glimm, J., Jaffe, A.: Quantum Physics: A Functional Integral Point of View, 2nd edn. Springer, New York, NY, USA (1987)
290. Gneiting, T., Kleiber, W., Schlather, M.: Matérn cross-covariance functions for multivariate random fields. J. Am. Stat. Assoc. **105**(491), 1167–1176 (2010)
291. Gneiting, T., Sasvári, Z., Schlather, M.: Analogies and correspondences between variograms and covariance functions. Adv. Appl. Probab. **33**(3), 617–630 (2001)
292. Gneiting, T., Schlather, M.: Stochastic models that separate fractal dimension and the Hurst effect. SIAM Rev. **46**(2), 269–282 (2004)
293. Gneiting, T., Ševčíková, H., Percival, D.B., et al.: Estimators of fractal dimension: assessing the roughness of time series and spatial data. Stat. Sci. **27**(2), 247–277 (2012)
294. Gogolides, E., Constantoudis, V., Kokkoris, G., Kontziampasis, D., Tsougeni, K., Boulousis, G., Vlachopoulou, M., Tserepi, A.: Controlling roughness: from etching to nanotexturing and plasma-directed organization on organic and inorganic materials. J. Phys. D. Appl. Phys. **44**(17), 174021 (2011)
295. Gogolides, E., Constantoudis, V., Patsis, G.P., Tserepi, A.: A review of line edge roughness and surface nanotexture resulting from patterning processes. Microelectron. Eng. **83**(4), 1067–1072 (2006)
296. Goldberg, R.R.: Fourier Transforms. Cambridge University Press, London, UK (1970)
297. Goldenfeld, N.: Lectures on Phase Transitions and the Renormalization Group. Addison-Wesley, Reading, MA (1992)
298. Goldstein, H., Poole, C.P., Safko, J.S.: Classical Mechanics, 3rd edn. Addison-Wesley, San Francisco (2000)
299. Golub, G.H., Van Loan, C.F.: Matrix Computations, 4th edn. Johns Hopkins University Press, Baltimore, MD, USA (2013)
300. Gómez-Hernández, J.J., Wen, X.H.: To be or not to be multi-Gaussian? a reflection on stochastic hydrogeology. Adv. Water Resour. **21**(1), 47–61 (1998)

301. Gompper, G., Kraus, M.: Ginzburg-Landau theory of ternary amphiphilic systems. I. Gaussian interface fluctuations. Phys. Rev. E **47**(6), 4289 (1993)
302. Gompper, G., Kraus, M.: Ginzburg-Landau theory of ternary amphiphilic systems. II. Monte Carlo simulations. Phys. Rev. E **47**(6), 4301 (1993)
303. Goovaerts, P.: Geostatistics for Natural Resources Evaluation. Oxford University Press, Oxford (1997)
304. Goulard, M., Voltz, M.: Linear coregionalization model: tools for estimation and choice of cross-variogram matrix. Math. Geol. **24**(3), 269–286 (1992)
305. Gouyet, J.F., Rosso, M., Sapoval, B.: Fractal surfaces and interfaces. In: Bunde, A., Havlin, S. (eds.) Fractals and Disordered Systems, pp. 263–301. Springer, London, UK (1996)
306. Gradshteyn, I.S., Ryzhik, I.M.: Table of Integrals, Series, and Products, 7th edn. Academic Press, Boston (2007)
307. Graeler, B.: Modelling skewed spatial random fields through the spatial vine copula. Spat. Stat. **10**, 87–102 (2014)
308. Grassberger, P., Procaccia, I.: Measuring the strangeness of strange attractors. Physica D **9**(1–2), 189–208 (1983)
309. Gray, A.G., Moore, A.W.: N-body problems in statistical learning. In: Leen, T.K., Dietterich, T.G., Tresp, V. (eds.) Advances in Neural Information Processing Systems 13, Papers from Neural Information Processing Systems (NIPS) 2000, Denver, CO, USA, vol. 4, pp. 521–527. MIT Press, Boston, MA, USA (2001)
310. Grigoriu, M.: Stochastic Calculus: Applications in Science and Engineering. Springer Science & Business Media, New York, NY, USA (2013)
311. Gringarten, E., Deutsch, C.V.: Teacher's aide variogram interpretation and modeling. Math. Geol. **33**(4), 507–534 (2001)
312. Grohmann, C.H., Smith, M.J., Riccomini, C.: Multiscale analysis of topographic surface roughness in the Midland Valley, Scotland. IEEE Trans. Geosci. Remote Sens. **49**(4), 1200–1213 (2011)
313. Gu, G.F., Zhou, W.X.: Detrended fluctuation analysis for fractals and multifractals in higher dimensions. Phys. Rev. E **74**(6), 061104 (2006)
314. Guillot, D., Rajaratnam, B.: Functions preserving positive definiteness for sparse matrices. Trans. Am. Math. Soc. **367**(1), 627–649 (2015)
315. Gull, S.F., Skilling, J.: Maximum entropy method in image processing. IEE Proc. F (Commun. Radar Signal Process.) **131**(6), 646–659 (1984)
316. Gunning, J.: On the use of multivariate Lévy-stable random field models for geological heterogeneity. Math. Geol. **34**(1), 43–62 (2002)
317. Gutjahr, A., Bullard, B., Hatch, S.: General joint conditional simulations using a Fast Fourier transform method. Math. Geol. **29**(3), 361–389 (1997)
318. Guttmann, A.J.: Lattice green's functions in all dimensions. J. Phys. A Math. Theor. **43**(30), 305205 (2010)
319. Guttorp, P., Gneiting, T.: Studies in the history of probability and statistics: on the Matérn correlation family. Biometrika **93**(4), 989–995 (2006)
320. Guyon, X.: Random Fields on a Network: Modeling, Statistics and Applications. Springer, New York, NY, USA (1995)
321. Habeck, M., Nilges, M., Rieping, W.: Replica-exchange Monte Carlo scheme for Bayesian data analysis. Phys. Rev. Lett. **94**(1), 018105 (2005)
322. Hall, P., Fisher, N., Hoffman, B.: On the nonparametric estimation of covariance functions. Ann. Stat. **22**(4), 2115–2134 (1994)
323. Hall, P., Maiti, T.: On parametric bootstrap methods for small area prediction. J. R. Stat. Soc. Ser. B (Stat Methodol.) **68**(2), 221–238 (2006)
324. Halliday, D., Resnick, R., Walker, J.: Fundamentals of Physics Extended, vol. 1. John Wiley & Sons, Hoboken, NJ, USA (2010)
325. Hamilton, J.D.: Time Series Analysis, vol. 2. Princeton University Press, Princeton, NJ, USA (1994)

326. Hamzhepour, H., Sahimi, M.: Generation of long-range correlations in large systems as an optimization problem. Phys. Rev. E **73**(5), 056121 (2005)
327. Handcock, M.S., Stein, M.L.: A Bayesian analysis of kriging. Technometrics **35**, 403–410 (1993)
328. Hanel, R., Thurner, S., Gell-Mann, M.: Generalized entropies and the transformation group of superstatistics. Proc. Natl. Acad. Sci. **108**(16), 6390–6394 (2011)
329. Hardin, D.P., Michaels, T., Saff, E.B.: A comparison of popular point configurations on S^2. Dolomites Res. Notes Approx. **9**(1), 16–49 (2016)
330. Hastie, T., Tibshirani, R.J., Friedman, J.: The Elements of Statistical Learning, 2nd edn. Springer, New York, NY, USA (2008)
331. Hauer, B., Maciejko, J., Davis, J.: Nonlinear power spectral densities for the harmonic oscillator. Ann. Phys. **361**, 148–183 (2015)
332. Hawkins, D.M., Cressie, N.: Robust kriging: a proposal. J. Int. Assoc. Math. Geol. **16**(1), 3–18 (1984)
333. Heaton, M.J., Datta, A., Finley, A., Furrer, R., Guhaniyogi, R., Gerber, F., Gramacy, R.B., Hammerling, D., Katzfuss, M., Lindgren, F.: Methods for analyzing large spatial data: a review and comparison. arXiv preprint arXiv:1710.05013 (2017)
334. Helbing, D.: Globally networked risks and how to respond. Nature **497**(7447), 51–59 (2013)
335. Helgason, S.: Radon Transform. Birkhäuser, Boston, MA, USA (1980)
336. Helmig, R., Niessner, J., Flemisch, B., Wolff, M., Fritz, J.: Efficient modeling of flow and transport in porous media using multiphysics and multiscale approaches. In: Freeden, W., Nashed, M.Z., Zonar, T. (eds.) Handbook of Geomathematics, pp. 417–457. Springer, Berlin, Germany (2010)
337. Heneghan, C., McDarby, G.: Establishing the relation between detrended fluctuation analysis and power spectral density analysis for stochastic processes. Phys. Rev. E **62**(5), 6103 (2000)
338. Hengl, T.: A Practical Guide to Geostatistical Mapping of Environmental Variables, 2nd edn. University of Amsterdam, Amsterdam, Netherlands (2009). http://spatial-analyst.net/book/
339. Hengl, T., Heuvelink, G.B.M., Kempen, B., Leenaars, J.G.B., Walsh, M.G., Shepherd, K.D., Sila, A., MacMillan, R.A., de Jesus, J.M., Tamene, L., Tondoh, J.E.: Mapping soil properties of Africa at 250 m resolution: random forests significantly improve current predictions. PLoS One **10**(6), e0125814 (2015)
340. Hengl, T., Heuvelink, G.B.M., Rossiter, D.G.: About regression-kriging: from equations to case studies. Comput. Geosci. **33**(10), 1301–1315 (2007)
341. Hennessey Jr, J.P.: Some aspects of wind power statistics. J. Appl. Meteorol. **16**(2), 119–128 (1977)
342. Higdon, D., Swall, J., Kern, J.: Non-stationary spatial modeling. Bayesian Stat. **6**(1), 761–768 (1999)
343. Higham, N.J.: Computing real square roots of a real matrix. Linear Algebra Appl. **88–89**, 405–430 (1987)
344. Higham, N.J.: An algorithmic introduction to numerical simulation of stochastic differential equations. SIAM Rev. **43**(3), 525–546 (2001)
345. Higham, N.J.: Regularization. In: Higham, N.J., Dennis, M.R., Glendinning, P., Martin, P.A., Santosa, F., Tanner, J. (eds.) The Princeton Companion to Applied Mathematics, pp. 205–206. Princeton University Press, Princeton, NJ, USA (2015)
346. Hilbe, J.M.: Practical Guide to Logistic Regression. CRC Press, Boca Raton, FL, USA (2016)
347. Hildebrand, F.B.: Introduction to Numerical Analysis, 2nd edn. International Series in Pure and Applied Mathematics. McGraw-Hill, New York, NY, USA (1974)
348. Hilfer, R.: Local porosity theory and stochastic reconstruction for porous media. In: Mecke, K.R., Stoyan, D. (eds.) Statistical Physics and Spatial Statistics, pp. 203–241. Springer, Berlin, Germany (2000)
349. Hilhorst, H.J.: Note on a q-modified central limit theorem. J. Stat. Mech: Theory Exp. **2010**(10), P10023 (2010)
350. Hinton, G.E.: Boltzmann machine. Scholarpedia **2**(5), 1668 (2007). revision No. 91075

351. Hochberg, D., et al.: Effective action for stochastic partial differential equations. Phys. Rev. E **60**(6), 6343–6360 (1999)
352. Hodge, V.J., Austin, J.: A survey of outlier detection methodologies. Artif. Intell. Rev. **22**(2), 85–126 (2004)
353. Hoerl, A.E., Kennard, R.W.: Ridge regression: biased estimation for nonorthogonal problems. Technometrics **12**(1), 55–67 (1970)
354. Hoeting, J.A., Davis, R.A., Merton, A.A., Thompson, S.E.: Model selection for geostatistical models. Ecol. Appl. **16**(1), 87–98 (2006)
355. Hohn, M.E.: Geostastistics and Petroleum Geology. Kluwer, Dordrecht (1999)
356. Horváth, L., Kokoszka, P.: Inference for Functional Data with Applications, vol. 200. Springer Science & Business Media, New York, NY, USA (2012)
357. Hosmer Jr, D.W., Lemeshow, S., Sturdivant, R.X.: Applied Logistic Regression. John Wiley & Sons, Hoboken, NJ, USA (2013)
358. Hou, Z., Engel, D.W., Lin, G., Fang, Y., Fang, Z.: An uncertainty quantification framework for studying the effect of spatial heterogeneity in reservoir permeability on CO_2 sequestration. Math. Geosci. **45**(7), 799–817 (2013)
359. Hristopulos, D.T.: New anisotropic covariance models and estimation of anisotropic parameters based on the covariance tensor identity. Stoch. Environ. Res. Risk Assess. **16**(1), 43–62 (2002)
360. Hristopulos, D.T.: Permissibility of fractal exponents and models of band-limited two-point functions for fGn and fBm random fields. Stoch. Environ. Res. Risk Assess. **17**(3), 191–216 (2003)
361. Hristopulos, D.T.: Renormalization group methods in subsurface hydrology: overview and applications in hydraulic conductivity upscaling. Adv. Water Resour. **26**(12), 1279–1308 (2003)
362. Hristopulos, D.T.: Spartan Gibbs random field models for geostatistical applications. SIAM J. Sci. Comput. **24**(6), 2125–2162 (2003)
363. Hristopulos, D.T.: Spartan Gaussian random fields for geostatistical applications: non-constrained simulations on square lattices and irregular grids. J. Comput. Methods Sci. Eng. **5**(2), 149–164 (2005)
364. Hristopulos, D.T.: Approximate methods for explicit calculations of non-Gaussian moments. Stoch. Environ. Res. Risk Assess. **20**(4), 278–290 (2006)
365. Hristopulos, D.T.: Spatial random field models inspired from statistical physics with applications in the geosciences. Physica A: Stat. Mech. Appl. **365**(1), 211–216 (2006)
366. Hristopulos, D.T.: Spartan random fields and applications in spatial interpolation and conditional simulation. In: Proceedings of the 12th European Conference on the Mathematics of Oil Recovery, Oxford, UK. European Association of Geoscientists and Engineers (2010). Paper B004. Online at: http://www.earthdoc.org/detail.php?pubid=41284
367. Hristopulos, D.T.: Covariance functions motivated by spatial random field models with local interactions. Stoch. Environ. Res. Risk Assess. **29**(3), 739–754 (2015)
368. Hristopulos, D.T.: Stochastic local interaction (SLI) model: bridging machine learning and geostatistics. Comput. Geosci. **85**(Part B), 26–37 (2015)
369. Hristopulos, D.T., Christakos, G.: Variational calculation of the effective fluid permeability of heterogeneous media. Phys. Rev. E **55**(6), 7288–7298 (1997)
370. Hristopulos, D.T., Christakos, G.: Renormalization group analysis of permeability upscaling. Stoch. Environ. Res. Risk Assess. **13**(1–2), 131–160 (1999)
371. Hristopulos, D.T., Christakos, G.: Practical calculation of non-Gaussian multivariate moments in spatiotemporal Bayesian maximum entropy analysis. Math. Geol. **33**(5), 543–568 (2001)
372. Hristopulos, D.T., Elogne, S.: Analytic properties and covariance functions of a new class of generalized Gibbs random fields. IEEE Trans. Inf. Theory **53**(12), 4667–4679 (2007)
373. Hristopulos, D.T., Elogne, S.N.: Computationally efficient spatial interpolators based on Spartan spatial random fields. IEEE Trans. Signal Process. **57**(9), 3475–3487 (2009)

374. Hristopulos, D.T., Mertikas, S.P., Arhontakis, I., Brownjohn, J.M.W.: Using GPS for monitoring tall-building response to wind loading: filtering of abrupt changes and low-frequency noise, variography and spectral analysis of displacements. GPS Solutions **11**(2), 85–95 (2007)
375. Hristopulos, D.T., Mouslopoulou, V.: Strength statistics and the distribution of earthquake interevent times. Physica A **392**(3), 485–496 (2013)
376. Hristopulos, D.T., Petrakis, M., Kaniadakis, G.: Finite-size effects on return interval distributions for weakest-link-scaling Systems. Phys. Rev. E **89**(5), 052142 (2014)
377. Hristopulos, D.T., Petrakis, M.P., Kaniadakis, G.: Weakest-link scaling and extreme events in finite-sized systems. Entropy **17**(3), 1103–1122 (2015)
378. Hristopulos, D.T., Porcu, E.: Multivariate Spartan spatial random field models. Probab. Eng. Mech. **37**, 84–92 (2014)
379. Hristopulos, D.T., Tsantili, I.C.: Space-time models based on random fields with local interactions. Int. J. Mod. Phys. B **29**, 1541007 (2015)
380. Hristopulos, D.T., Žukovič, M.: Relationships between correlation lengths and integral scales for covariance models with more than two parameters. Stoch. Environ. Res. Risk Assess. **25**(1), 11–19 (2011)
381. Hu, D., Ronhovde, P., Nussinov, Z.: Replica inference approach to unsupervised multiscale image segmentation. Phys. Rev. E **85**(1), 016101 (2012)
382. Hu, K., Ivanov, P.C., Chen, Z., Carpena, P., Stanley, H.E.: Effect of trends on detrended fluctuation analysis. Phys. Rev. E **64**(1), 011114 (2001)
383. Huang, K.: Statistical Mechanics. John Wiley & Sons, New York, NY, USA (1987)
384. Huang, S.P., Quek, S.T., Phoon, K.K.: Convergence study of the truncated Karhunen–Loeve expansion for simulation of stochastic processes. Int. J. Numer. Methods Eng. **52**(9), 1029–1043 (2001)
385. Hughes, J., Haran, M., Caragea, P.C.: Autologistic models for binary data on a lattice. Environmetrics **22**(7), 857–871 (2011)
386. Hunt, A., Ewing, R.: Percolation Theory for Flow in Porous Media. Lecture Notes in Physics. Springer, Berlin, Germany (2009)
387. Hunt, A.G., Sahimi, M.: Flow, transport and reaction in porous media: percolation scaling, critical-path analysis, and effective-medium approximation. Rev. Geophys. **55**(4), 993–1078 (2017)
388. Hurst, H.E.: Long-term storage capacity of reservoirs. Trans. Am. Soc. Civ. Eng. **116**, 770–808 (1951)
389. Hurst, H.E., Black, R.P., Simaika, Y.: Long-term Storage: An Experimental Study. Constable, London, UK (1965)
390. Hyman, J.M., Steinberg, S.: The convergence of mimetic discretization for rough grids. Comput. Math. Appl. **47**(10–11), 1565–1610 (2004)
391. Hyvärinen, A.: Consistency of pseudolikelihood estimation of fully visible Boltzmann machines. Neural Comput. **18**(10), 2283–2292 (2006)
392. Ingber, L.: Adaptive simulated annealing. In: Hime, A., Ingber, L., Petraglia, A., Petraglia, M.R., Machado, M.A.S. (eds.) Stochastic Global Optimization and its Applications with Fuzzy adaptive Simulated Annealing, pp. 33–62. Springer, Heidelberg, Germany (2012)
393. Ioannidis, J.P.A.: Why most published research findings are false. PLoS Med. **2**(8), e124 (2005)
394. Isaaks, E.H., Srivastava, R.M.: Applied Geostatistics. Oxford University Press, New York, NY, USA (1989)
395. Iserles, A.: A First Course in the Numerical Analysis of Differential Equations, 2nd edn. Cambridge University Press, Cambridge, UK (2009)
396. Isichenko, M.B.: Percolation, statistical topography, and transport in random media. Rev. Mod. Phys. **64**(4), 961–1043 (1992)
397. Isihara, A.: The Gibbs-Bogoliubov inequality. J. Phys. A: Gen. Phys. **1**(5), 539–548 (1968)
398. Isserlis, L.: On a formula for the product-moment coefficient of any order of a normal frequency distribution in any number of variables. Biometrika **12**(1–2), 134–139 (1918)

399. Itzykson, C., Drouffe, J.M.: Statistical Field Theory, vol. 2. Cambridge University Press, Cambridge, UK (1991)
400. Jaakkola, T.S.: Tutorial on variational approximation methods. In: Opper, M., Saad, D. (eds.) Advanced Mean Field Methods: Theory and Practice, pp. 129–160. MIT Press, Cambridge, MA, USA (2001)
401. Jackson, J.D.: Classical Electrodynamics, 3rd edn. John Wiley & Sons, New York, NY, USA (1998)
402. Jacobs, K.: Stochastic Processes for Physicists: Understanding Noisy Systems. Cambridge University Press, Cambridge, UK (2010)
403. James, F.: Statistical Methods in Experimental Physics. World Scientific, Hackensack, NJ, USA (2006)
404. Jankovic, I., Maghrebi, M., Fiori, A., Dagan, G.: When good statistical models of aquifer heterogeneity go right: the impact of aquifer permeability structures on 3d flow and transport. Adv. Water Resour. **100**, 199–211 (2017)
405. Janson, S.: Gaussian Hilbert Spaces, vol. 129. Cambridge University Press, Cambridge, UK (1997)
406. Jaynes, E.T.: Information theory and statistical mechanics. Phys. Rev. **106**(4), 620–630 (1957)
407. Jaynes, E.T.: Information theory and statistical mechanics. Phys. Rev. **108**(2), 171–190 (1957)
408. Jensen, J.L.W.V.: Sur les fonctions convexes et les inégalités entre les valeurs moyennes. Acta Math. **30**(1), 175–193 (1906)
409. Jespersen, S., Metzler, R., Fogedby, H.C.: Lévy flights in external force fields: Langevin and fractional Fokker-Planck equations and their solutions. Phys. Rev. E **59**(3), 2736 (1999)
410. Jizba, P., Korbel, J.: Maximum Entropy Principle in statistical inference: case for non-Shannonian entropies. Phys. Rev. Lett. **122**(12), 120601 (2019)
411. Joe, H.: Dependence Modeling with Copulas. CRC Press, Boca Raton, FL, USA (2014)
412. Joe, H., Kurowicka, D.: Dependence Modeling: Vine Copula Handbook. World Scientific, Hackensack, NJ, USA (2011)
413. Johns, C.J., Nychka, D., Kittel, T.G.F., Daly, C.: Infilling sparse records of spatial fields. J. Am. Stat. Assoc. **98**(464), 796–806 (2003)
414. Johnson, N.L.: Systems of frequency curves generated by methods of translation. Biometrika **36**(1/2), 149–176 (1949)
415. Johnston, K., Ver Hoef, J.M., Krivoruchko, K., Lucas, N.: Using ArcGIS Geostatistical Analyst. http://dusk.geo.orst.edu/gis/geostat_analyst.pdf (2003). [Online; accessed on September 9, 2012]
416. Jolliffe, I.: Principal Component Analysis, 2nd edn. Springer, New York, NY, USA (2002)
417. Jona-Lasinio, G.: Renormalization group and probability theory. Phys. Rep. **352**(4–6), 439–458 (2001)
418. Journel, A.: Kriging in terms of projections. J. Int. Assoc. Math. Geol. **9**(6), 563–586 (1977)
419. Journel, A.G., Huijbregts, C.J.: Mining Geostatistics. Academic Press, London, UK (1978)
420. Joyce, S.G.: Exact evaluation of the simple cubic lattice Green function for a general lattice point. J. Phys. A **35**(46), 9811–9828 (2002)
421. Kaipio, J., Somersalo, E.: Statistical and Computational Inverse Problems, vol. 160. Springer Science & Business Media, New York, NY, USA (2006)
422. Kalos, M.H., Whitlock, P.A.: Monte Carlo Methods, vol. 1, 2nd edn. Wiley-Blackwell, Weinheim, Germany (2008)
423. Kamath, C.: Scientific Data Mining: A Practical Perspective. Society of Industrial and Applied Mathematics, Philadelphia, PA, USA (2009)
424. Kampen, N.G.V.: Stochastic Processes in Physics and Chemistry, 3rd edn. Elsevier, Amsterdam, Netherlands (2007)
425. Kanevski, M., Maignan, M.: Analysis and Modelling of Spatial Environmental Data. EPFL Press, Lausanne, Switzerland (2004)
426. Kaniadakis, G.: H-theorem and generalized entropies within the framework of nonlinear kinetics. Phys. Lett. **288**, 283–291 (2001)

427. Kaniadakis, G.: Non-linear kinetics underlying generalized statistics. Physica A **296**, 405–425 (2001)
428. Kaniadakis, G.: Statistical mechanics in the context of special relativity. Phys. Rev. E **66**(5), 056125 (2002)
429. Kaniadakis, G.: Statistical mechanics in the context of special relativity II. Phys. Rev. E **72**, 036108 (2005)
430. Kaniadakis, G.: Maximum entropy principle and power-law tailed distributions. Eur. Phys. J. B **70**, 3–13 (2009)
431. Kaniadakis, G.: Theoretical foundations and mathematical formalism of the power-law tailed statistical distributions. Entropy **15**, 3983–4010 (2013)
432. Kantz, H., Schreiber, T.: Nonlinear Time Series Analysis, vol. 7. Cambridge University Press, Cambridge, UK (2004)
433. Kaplan, W.: Advanced Calculus, 5th edn. Addison Wesley, Reading, MA, USA (1973)
434. Kardar, M.: Statistical Physics of Fields. Cambridge University Press, Cambridge, UK (2007)
435. Kardar, M., Parisi, G., Zhang, Y.C.: Dynamic scaling of growing interfaces. Phys. Rev. Lett. **56**(9), 889–892 (1986)
436. Karniadakis, G.E., Shu, C.H., Xiu, D., Lucor, D., Schwab, C., Todor, R.A.: Generalized polynomial chaos solution for differential equations with random inputs. Technical Report 2005-1, ETH Zürich, Zürich, Switzerland (2015)
437. Karsanina, M.V., Gerke, K.M.: Hierarchical optimization: fast and robust multiscale stochastic reconstructions with rescaled correlation functions. Phys. Rev. Lett. **121**(26), 265501 (2018)
438. Kass, R.E., Carlin, B.P., Gelman, A., Neal, R.M.: Markov chain Monte Carlo in practice: a roundtable discussion. Am. Stat. **52**(2), 93–100 (1998)
439. Katakami, S., Sakamoto, H., Murata, S., Okada, M.: Gaussian Markov random field model without boundary conditions. J. Phys. Soc. Jpn. **86**(6), 064801 (2017)
440. Katzfuss, M.: A multi-resolution approximation for massive spatial datasets. J. Am. Stat. Assoc. **112**(517), 201–214 (2017)
441. Kaufman, C.G., Schervish, M.J., Nychka, D.W.: Covariance tapering for likelihood-based estimation in large spatial data sets. J. Am. Stat. Assoc. **103**(484), 1545–1555 (2008)
442. Kazianka, H.: spatialcopula: a matlab toolbox for copula-based spatial analysis. Stoch. Environ. Res. Risk Assess. **27**(1), 121–135 (2013)
443. Kazianka, H., Pilz, J.: Copula-based geostatistical modeling of continuous and discrete data including covariates. Stoch. Environ. Res. Risk Assess. **24**(5), 661–673 (2010)
444. Kazianka, H., Pilz, J.: Spatial interpolation using copula-based geostatistical models. In: Atkinson, P.M., Lloyd, C.D. (eds.) geoENV VII – Geostatistics for Environmental Applications, pp. 307–319. Springer Netherlands, Dordrecht, Netherlands (2010)
445. Kazianka, H., Pilz, J.: Bayesian spatial modeling and interpolation using copulas. Comput. Geosci. **37**(3), 310–319 (2011)
446. Kelkar, M., Perez, G.: Applied Geostatistics for Reservoir Characterization. Society of Petroleum Engineers, Richardson, TX, USA (2002)
447. Kennedy, M.C., O'Hagan, A.: Predicting the output from a complex computer code when fast approximations are available. Biometrika **87**(1), 1–13 (2000)
448. Khachatryan, D., Bisgaard, S.: Some results on the variogram in time series analysis. Qual. Reliab. Eng. Int. **25**(8), 947–960 (2009)
449. Khinchin, A.Y.: Mathematical Foundations of Information Theory. Dover Publications, Mineola, NY, USA (1957)
450. Kindermann, R., Snell, J.L.: Markov Random Fields and their Applications, vol. 1. American Mathematical Society, Providence, RI, USA (1980)
451. King, G.: Ensuring the data-rich future of the social sciences. Science **331**(6018), 719–721 (2011)
452. King, P.R.: The use of renormalization for calculating effective permeability. Transp. Porous Media **4**(1), 37–58 (1989)
453. Kirkpatrick, S.: Optimization by simulated annealing. J. Stat. Phys. **34**(5), 975–998 (1984)

454. Kirkpatrick, S., Gelatt, C.D., Vecchi, M.P.: Optimization by simulated annealing. Science **220**(4598), 671–680 (1983)
455. Kitanidis, P.K.: Statistical estimation of polynomial generalized covariance functions and hydrologic applications. Water Resour. Res. **19**(2), 909–921 (1983)
456. Kitanidis, P.K.: Parametric estimation of covariances of regionalized variables. Water Resour. Res. **23**(4), 671–680 (1987)
457. Kitanidis, P.K.: Orthonormal residuals in geostatistics: model criticism and parameter estimation. Math. Geol. **23**(5), 741–758 (1991)
458. Kitanidis, P.K.: Generalized covariance functions in estimation. Math. Geol. **25**(5), 525–540 (1993)
459. Kitanidis, P.K.: Introduction to Geostatistics: Applications to Hydrogeology. Cambridge University Press, Cambridge, UK (1997)
460. Kitanidis, P.K.: Generalized covariance functions associated with the Laplace equation and their use in interpolation and inverse problems. Water Resour. Res. **35**(5), 1361–1367 (1999)
461. Kitanidis, P.K.: Persistent questions of heterogeneity, uncertainty, and scale in subsurface flow and transport. Water Resour. Res. **51**(8), 5888–5904 (2015)
462. Kitanidis, P.K., Lane, R.W.: Maximum likelihood parameter estimation of hydrologic spatial processes by the Gauss-Newton method. J. Hydrol. **79**(1–2), 53–71 (1985)
463. Kiwata, H.: Estimation of quenched random fields in the inverse ising problem using a diagonal matching method. Phys. Rev. E **89**(6), 062135 (2014)
464. Kleiber, W., Katz, R.W., Rajagopalan, B.: Daily spatiotemporal precipitation simulation using latent and transformed Gaussian processes. Water Resour. Res. **48**(1), W01523/17 (2012)
465. Kleinert, H.: Path Integrals in Quantum Mechanics, Statistics, Polymer Physics, and Financial Markets, 5th edn. World Scientific, Hackensack, NJ, USA (2009)
466. Knudby, C., Carrera, J.: On the use of apparent hydraulic diffusivity as an indicator of connectivity. J. Hydrol. **329**(3–4), 377–389 (2006)
467. Kopietz, P., Bartosch, L., Schütz, F.: Mean-Field Theory and the Gaussian Approximation. Lecture Notes in Physics, vol. 798. Springer, Berlin, Germany (2010)
468. Kotz, S., Nadarajah, S.: Multivariate t-distributions and their Applications. Cambridge University Press, New York, NY, USA (2004)
469. Kramer, P.R., Kurbanmuradov, O., Sabelfeld, K.: Comparative analysis of multiscale Gaussian random field simulation algorithms. J. Comput. Phys. **226**(1), 897–924 (2007)
470. Kroese, D.P., Taimre, T., Botev, Z.I.: Handbook of Monte Carlo Methods, vol. 706. John Wiley & Sons, Hoboken, NJ, USA (2013)
471. Kubo, R., Toda, M., Hashitsume, N.: Statistical Physics II: Nonequilibrium Statistical Mechanics, 2nd edn. Springer Science & Business Media, Berlin, Germany (1991)
472. Kullback, S.: Information Theory and Statistics. Dover Publications, Mineola, NY, USA (1997)
473. Kuwatani, T., Nagata, K., Okada, M., Toriumi, M.: Markov-random-field modeling for linear seismic tomography. Phys. Rev. E **90**(4), 042137 (2014)
474. Kuzemsky, A.L.: Variational principle of Bogoliubov and generalized mean fields in many-particle interacting systems. Int. J. Mod. Phys. B **29**, 1530010 (2015)
475. Kwaśnicki, M.: Ten equivalent definitions of the fractional Laplace operator. Fract. Calc. Appl. Anal. **20**(1), 7–51 (2017)
476. Lacasa, L., Toral, R.: Description of stochastic and chaotic series using visibility graphs. Phys. Rev. E **82**(3), 036120 (2010)
477. Lahiri, S.N., Maiti, T., Katzoff, M., Parsons, V.: Resampling-based empirical prediction: an application to small area estimation. Biometrika **94**(2), 469–485 (2007)
478. Lancaster, T., Blundell, S.J.: Quantum Field Theory for the Gifted Amateur. Oxford University Press, Oxford, UK (2014)
479. Landau, D.P., Binder, K.: A Guide to Monte Carlo Simulations in Statistical Physics, 3rd edn. Cambridge University Press, Cambridge, UK (2014)
480. Landau, L.D., Lifshitz, E.M.: Course of Theoretical Physics: Statistical Physics, vol. 5, 3rd edn. Butterworth-Heinemann, Oxford, UK (1980)

481. Lang, A.: Simulation of stochastic partial differential equations and stochastic active contours. Ph.D. thesis, Universität Mannheim, Germany (2007). https://ub-madoc.bib.uni-mannheim.de/1770/
482. Lang, A., Potthoff, J.: Fast simulation of Gaussian random fields. Monte Carlo Methods Appl. **17**(3), 195–214 (2011)
483. Lang, A., Schwab, C.: Isotropic Gaussian random fields on the sphere: regularity, fast simulation and stochastic partial differential equations. Ann. Appl. Probab. **25**(6), 3047–3094 (2015)
484. Lange, M.: On the uncertainty of wind power predictions – analysis of the forecast accuracy and statistical distribution of errors. J. Sol. Energy Eng. **127**(2), 177–184 (2005)
485. Lantuéjoul, C.: Cours de sélectivité. Tech. Rep. C-140, Centre de Géosciences/Géostatistique (1990). http://cg.ensmp.fr/bibliotheque/cgi-bin/public/bibli_index.cgi#1990. [Online; Accessed 31 Oct 2018]
486. Lantuéjoul, C.: Ergodicity and integral range. J. Microsc. **161**(3), 387–403 (1991)
487. Lantuéjoul, C.: Geostatistical Simulation: Models and Algorithms. Springer, Berlin, Germany (2002)
488. Lark, R.M.: Estimating variograms of soil properties by the method-of-moments and maximum likelihood. Eur. J. Soil Sci. **51**(4), 717–728 (2000)
489. Lauritzen, S.L.: Graphical Models, vol. 17. Clarendon Press, Oxford, UK (1996)
490. Lawler, G.F., Limic, V.: Random Walk: A Modern Introduction. Cambridge University Press, Cambridge, UK (2010)
491. Leach, R.: Characterisation of Areal Surface Texture. Springer, Berlin, Germany (2013)
492. Lee, J., Bahri, Y., Novak, R., Schoenholz, S.S., Pennington, J., Sohl-Dickstein, J.: Deep neural networks as Gaussian processes. arXiv preprint arXiv:1711.00165 (2017)
493. Lee, J., Bahri, Y., Novak, R., Schoenholz, S.S., Pennington, J., Sohl-Dickstein, J.: Deep neural networks as Gaussian processes. International Conference on Learning Representations (2018). https://openreview.net/forum?id=B1EA-M-0Z
494. Lee, S., Won, J.S., Jeon, S.W., Park, I., Lee, M.J.: Spatial landslide hazard prediction using rainfall probability and a logistic regression model. Math. Geosci. **47**(5), 565–589 (2015)
495. Lemm, J.C.: Bayesian Field Theory. Johns Hopkins University Press, Baltimore, MD, USA (2005)
496. Leow, A., et al.: Brain structural mapping using a novel hybrid implicit/explicit framework based on the level-set method. NeuroImage **24**(3), 910–927 (2004)
497. Li, B., Genton, M.G.: Nonparametric identification of copula structures. J. Am. Stat. Assoc. **108**(502), 666–675 (2013)
498. Li, B., Zhang, H.: An approach to modeling asymmetric multivariate spatial covariance structures. J. Multivar. Anal. **102**(10), 1445–1453 (2011)
499. Li, J., Heap, A.D.: Spatial interpolation methods applied in the environmental sciences: a review. Environ. Model. Softw. **53**, 173–189 (2014)
500. Liang, X.S.: Entropy evolution and uncertainty estimation with dynamical systems. Entropy **16**(7), 3605–3634 (2014)
501. Lim, S., Teo, L.: Generalized Whittle-Matérn random field as a model of correlated fluctuations. J. Phys. A Math. Theor. **42**(10), 105202 (2009)
502. Lim, S.C., Teo, L.P.: Gaussian fields and Gaussian sheets with generalized Cauchy covariance structure. Stoch. Process. Appl. **119**(4), 1325–1356 (2009)
503. Lindgren, F., Rue, H., Lindström, J.: An explicit link between Gaussian fields and Gaussian Markov random fields: the SPDE approach. J. R. Stat. Soc. Ser. B **73**(4), 423–498 (2011)
504. Lischke, A., Pang, G., Gulian, M., Song, F., Glusa, C., Zheng, X., Mao, Z., Cai, W., Meerschaert, M.M., Ainsworth, M., Karniadakis, G.E.: What is the fractional laplacian? arXiv preprint arXiv:1801.09767 (2018)
505. Liu, J.S.: Monte Carlo Strategies in Scientific Computing. Springer Science & Business Media, New York, NY, USA (2008)
506. Loader, C.: Local Regression and Likelihood. Springer, New York, NY, USA (1999)

507. Lodhia, A., Sheffield, S., Sun, X., Watson, S.S.: Fractional Gaussian fields: a survey. Probab. Surv. **13**, 1–56 (2016)
508. Loeve, M.: Probability theory. Springer, New York, NY, USA (1978)
509. Longley, P., Goodchild, M.F., Maguire, D.J., Rhind, D.W.: Geographic Information Systems and Science. John Wiley & Sons, Hoboken, NJ, USA (2005)
510. Longtin, A.: Stochastic dynamical systems. Scholarpedia **5**(4), 1619 (2010)
511. Lord, G.J., Powell, C.E., Shardlow, T.: An Introduction to Computational Stochastic PDEs. Cambridge University Press, New York, NY, USA (2014)
512. Lovejoy, S., Schertzer, D.: The Weather and Climate: Emergent Laws and Multifractal Cascades. Cambridge University Press, Cambridge, UK (2013)
513. Løvsletten, O.: Consistency of detrended fluctuation analysis. Phys. Rev. E **96**(1), 012141 (2017)
514. Lu, S., Molz, F.J., Liu, H.H.: An efficient, three-dimensional, anisotropic, fractional Brownian motion and truncated fractional Lévy motion simulation algorithm based on successive random additions. Comput. Geosci. **29**(1), 15–25 (2003)
515. Lucas, A.: Ising formulations of many NP problems. Front. Phys. **2**, 5 (2014)
516. Luo, J., Ying, K., He, P., Bai, J.: Properties of Savitzky Golay digital differentiators. Digital Signal Process. **15**(2), 122–136 (2005)
517. Lutz, E., Ciliberto, S.: Information: from Maxwell's demon to Landauer's eraser. Phys. Today **68**(9), 30–35 (2015)
518. Lyon, S.W., Sørensen, R., Stendahl, J., Seibert, J.: Using landscape characteristics to define an adjusted distance metric for improving kriging interpolations. Int. J. Geogr. Inf. Sci. **24**(5), 723–740 (2010)
519. Machta, B.B., Chachra, R., Transtrum, M.K., Sethna, J.P.: Parameter space compression underlies emergent theories and predictive models. Science **342**(6158), 604–607 (2013)
520. MacKay, D.J.C.: Introduction to Gaussian processes. NATO ASI Ser. F Comput. Syst. Sci. **168**, 133–166 (1998)
521. MacKay, D.J.C.: Information Theory, Inference, and Learning Algorithms. Cambridge University Press, Cambridge, UK (2003)
522. Makse, H.A., Davies, G.W., Havlin, S., Ivanov, P.C., King, P.R., Stanley, H.E.: Long-range correlations in permeability fluctuations in porous rock. Phys. Rev. E **54**(4), 3129–3134 (1996)
523. Makse, H.A., Havlin, S., Schwartz, M., Stanley, H.E.: Method for generating long-range correlations for large systems. Phys. Rev. E **53**(5), 5445–5449 (1996)
524. Malevergne, Y., Sornette, D.: High-order moments and cumulants of multivariate Weibull asset returns distributions: analytical theory and empirical tests: II. Financ. Lett. **3**(1), 54–63 (2005)
525. Malevergne, Y., Sornette, D.: Multivariate Weibull distributions for asset returns: I. Financ. Lett. **2**(6), 16–32 (2005)
526. Malmir, H., Sahimi, M., Jiao, Y.: Higher-order correlation functions in disordered media: computational algorithms and application to two-phase heterogeneous materials. Phys. Rev. E **98**(6), 063317 (2018)
527. Malzahn, D., Opper, M.: A statistical mechanics approach to approximate analytical bootstrap averages. In: Becker, S., Thrun, S., Obermayer, K. (eds.) Advances in Neural Information Processing Systems 15 [Neural Information Processing Systems, NIPS 2002, December 9–14, 2002, Vancouver, British Columbia, Canada], pp. 343–350 (2003)
528. Manchuk, J.G., Leuangthong, O., Deutsch, C.V.: The proportional effect. Math. Geosci. **41**(7), 799–816 (2009)
529. Mandelbrot, B.B.: Fractional Brownian motions, fractional noises and applications. SIAM Rev. **10**(4), 422–437 (1968)
530. Mandelbrot, B.B.: The fractal Geometry of Nature. W. H. Freeman and Company, New York, NY, USA (1983)
531. Mandelbrot, B.B.: Fractal analysis and synthesis of fracture surface roughness and related forms of complexity and disorder. Int. J. Fract. **138**, 13–17 (2006)

532. Mandelbrot, B.B., Passoja, D.E., Paullay, A.J.: Fractal character of fracture surfaces of metals. Nature **308**, 721–722 (1984)
533. Manovich, L.: Trending: the promises and the challenges of big social data. Debates Digital Humanities **2**, 460–475 (2011)
534. Mantoglou, A., Wilson, J.L.: The turning bands method for simulation of random fields using line generation by a spectral method. Water Resour. Res. **18**(5), 1379–1394 (1982)
535. Marchant, B.P., Lark, R.M.: The Matérn variogram model: implications for uncertainty propagation and sampling in geostatistical surveys. Geoderma **140**(4), 337–345 (2007)
536. Marchenko, Y.V., Genton, M.G.: Multivariate log-skew-elliptical distributions with applications to precipitation data. Environmetrics **21**(3–4), 318–340 (2010)
537. Marcotte, D.: Letter to the editor: comment on "understanding anisotropy computations" by M. Eriksson and P. P. Siska. Math. Geol. **35**(5), 643–646 (2003)
538. Marcotte, D., Allard, D.: Gibbs sampling on large lattice with GMRF. Comput. Geosci. **111**, 190–199 (2018)
539. Marder, M.P.: Condensed Matter Physics. John Wiley & Sons, Hoboken, NJ, USA (2010)
540. Mardia, K.V.: Should geostatistics be model-based. In: Proceedings of the 11th IAMG Conference, Beijing, China. International Association for Mathematical Geosciences, Houston, TX, USA (2007)
541. Mardia, K.V., Marshall, R.J.: Maximum likelihood estimation of models for residual covariance in spatial regression. Biometrika **71**(1), 135–146 (1984)
542. Mardia, K.V., Watkins, A.J.: On multimodality of the likelihood in the spatial linear model. Biometrika **76**(2), 289–295 (1989)
543. Mariethoz, G., Caers, J.: Multiple-point Geostatistics: Stochastic Modeling with Training Images. John Wiley & Sons, Chichester, West Sussex, UK (2015)
544. Markovič, D., Gros, C.: Power laws and self-organized criticality in theory and nature. Phys. Rep. **536**(2), 41–74 (2014)
545. Marques, R., Bouville, C., Ribardière, M., Santos, L.P., Bouatouch, K.: Spherical Fibonacci point sets for illumination integrals. Comput. Graphics Forum **32**(8), 134–143 (2013)
546. Marsaglia, G.: Choosing a point from the surface of a sphere. Ann. Math. Stat. **43**(2), 645–646 (1972)
547. Martin, P.C., Siggia, E.D., Rose, H.A.: Interpolation schemes for three dimensional velocity fields from scattered data using Taylor expansions. Phys. Rev. A **8**(1), 423–437 (1973)
548. Martin, W., Flandrin, P.: Wigner-Ville spectral analysis of nonstationary processes. IEEE Trans. Acoust. Speech Signal Process. **33**(6), 1461–1470 (1985)
549. Mathai, A.M., Haubold, H.J.: A pathway from Bayesian statistical analysis to superstatistics. Appl. Math. Comput. **218**(3), 799–804 (2011). Special Issue in Honour of Hari M. Srivastava on his 70th Birth Anniversary
550. Matheron, G.: Traité de Géostatistique Apliquée., vol. 1. Editions Technip (1962)
551. Matheron, G.: Principles of geostatistics. Econ. Geol. **58**(8), 1246–1266 (1963)
552. Matheron, G.: Le krigeage universel. Tech. rep., Les Cahiers du Centre de Morphologie Mathematique de Fontainebleau, École Nationale Superieure des Mines de Paris (1969). http://cg.ensmp.fr/bibliotheque/cgi-bin/public/bibli_index.cgi#1969. [Online; accessed on 31 Oct 2018]
553. Matheron, G.: The theory of regionalized variables and its application. Tech. rep., Les Cahiers du Centre de Morphologie Mathematique de Fontainebleau, École Nationale Superieure des Mines de Paris (1971). http://cg.ensmp.fr/bibliotheque/cgi-bin/public/bibli_index.cgi#1971. [Online; accessed on 31 Oct 2018]
554. Matheron, G.: The intrinsic random functions and their applications. J. Appl. Probab. **5**(3), 439–468 (1973)
555. Matheron, G.: Suffit-il, pour une covariance, d' être de type positif? Sciences de la Terre, Série Informatique Géologique **26**, 51–66 (1987)
556. Maybeck, P.S.: Stochastic Models, Estimation and Control, vol. I. Academic Press, New York, NY, USA (1979)

557. McComb, W.D.: The Physics of Fluid Turbulence. Oxford University Press, Oxford, UK (1990)
558. Mecke, K.R.: Additivity, convexity, and beyond: applications of Minkowski functionals in statistical physics. In: Mecke, K.R., Stoyan, D. (eds.) Statistical Physics and Spatial Statistics, pp. 111–184. Springer, Berlin, Germany (2000)
559. Mecke, K.R., Buchert, T., Wagner, H.: Robust morphological measures for large-scale structure in the Universe. Astron. Astrophys. **288**, 697–704 (1994)
560. Mecke, K.R., Stoyan, D.: Statistical Physics and Spatial Statistics. Springer, Berlin, Germany (2000)
561. Mehta, P., Bukov, M., Wang, C.H., Day, A.G.R., Richardson, C., Fisher, C.K., Schwab, D.J.: A high-bias, low-variance introduction to machine learning for physicists. Phys. Rep. **810**, 1–124 (2019)
562. Menafoglio, A., Guadagnini, A., Secchi, P.: A kriging approach based on Aitchison geometry for the characterization of particle-size curves in heterogeneous aquifers. Stoch. Environ. Res. Risk Assess. **28**(7), 1835–1851 (2014)
563. Menafoglio, A., Petris, G.: Kriging for Hilbert-space valued random fields: the operatorial point of view. J. Multivar. Anal. **146**, 84–94 (2016)
564. Menezes, R., Garcia-Soidán, P., Febrero-Bande, M.: A kernel variogram estimator for clustered data. Scand. J. Stat. **35**(1), 18–37 (2008)
565. Menezes, R., Garcìa-Soidán, P., Ferreira, C.: Nonparametric spatial prediction under stochastic sampling design. J. Nonparametric Stat. **22**(3), 363–377 (2010)
566. Metzler, R., Klafter, J.: The random walk's guide to anomalous diffusion: a fractional dynamics approach. Phys. Rep. **339**(1), 1–77 (2000)
567. Meurice, Y.: Simple method to make asymptotic series of Feynman diagrams converge. Phys. Rev. Lett. **88**(14), 141601 (2002)
568. Meyn, S.P., Tweedie, R.L.: Markov Chains and Stochastic Stability. Cambridge University Press, Cambridge, UK (2009)
569. Mezard, M., Montanari, A.: Information, Physics, and Computation. Oxford University Press, Oxford, UK (2009)
570. Mézard, M., Parisi, G., Sourlas, N., Toulouse, G., Virasoro, M.: Replica symmetry breaking and the nature of the spin glass phase. Journal de Physique **45**(5), 843–854 (1984)
571. Mézard, M., Parisi, G., Virasoro, M.A.: Spin Glass Theory and Beyond. World Scientific, Singapore, Singapore (1987)
572. Michalak, A.M., Hirsch, A., Bruhwiler, L., Gurney, K.R., Peters, W., Tans, P.P.: Maximum likelihood estimation of covariance parameters for bayesian atmospheric trace gas surface flux inversions. J. Geophys. Res. Atmos. **110**, D24107(1–16) (2005)
573. Mikosch, T.: Copulas: tales and facts. Extremes **9**(1), 3–20 (2006)
574. Mimura, K., Okada, M.: Statistical mechanics of lossy compression using multilayer perceptrons. Phys. Rev. E **74**(2), 026108 (2006)
575. Minasny, B., McBratney, A.B.: The matérn function as a general model for soil variograms. Geoderma **128**, 192–207 (2005)
576. Mirouze, I., Blockley, E., Lea, D., Martin, M., Bell, M.: A multiple length scale correlation operator for ocean data assimilation. Tellus A **68**, 29744 (2016)
577. Mishchenko, M.I., et al.: First-principles modeling of electromagnetic scattering by discrete and discretely heterogeneous random media. Phys. Rep. **632**, 1–75 (2016)
578. Mitchell, J.F.B., Lowe, J., Wood, R.A., Vellinga, M.: Extreme events due to human-induced climate change. Philos. Trans. R. Soc. A: Math. Phys. Eng. Sci. **364**(1845), 2117–2133 (2006)
579. Moelans, N., Blanpain, B., Wollants, P.: An introduction to phase-field modeling of microstructure evolution. Calphad **32**(2), 268–294 (2008)
580. Mohammadi, H., Le Riche, R., Durrande, N., Touboul, E., Bay, X.: An analytic comparison of regularization methods for Gaussian Processes. arXiv preprint arXiv:1602.00853 (2016)
581. Møller, J., Syversveen, A.R., Waagepetersen, R.P.: Log-Gaussian Cox processes. Scand. J. Stat. **25**(3), 451–482 (1998)

582. Molz, F.J., Liu, H.H., Szulga, J.: Fractional Brownian motion and fractional Gaussian noise in subsurface hydrology: a review, presentation of fundamental properties, and extensions. Water Resour. Res. **33**(10), 2273–2286 (1997)
583. Monahan, A.H., He, Y., McFarlane, N., Dai, A.: The probability distribution of land surface wind speeds. J. Clim. **24**(15), 3892–3909 (2011)
584. Monbet, V., Aillot, P., Prevosto, M.: Survey of stochastic models for wind and sea state time series. Probab. Eng. Mech. **22**(2), 113–126 (2007)
585. Monbet, V., Prevosto, M.: Bivariate simulation of non stationary and non Gaussian observed processes: application to sea state parameters. Appl. Ocean Res. **23**(3), 139–145 (2001)
586. Mondal, D.: Applying Dynkin's isomorphism: an alternative approach to understand the Markov property of the de Wijs process. Bernoulli **21**(3), 1289–1303 (2015)
587. Monin, A.S., Yaglom, A.M.: Statistical Fluid Mechanics: Mechanics of Turbulence. MIT Press, Cambridge, MA, USA (1971)
588. Montgomery, D.C., Jennings, C.L., Kulahci, M.: Introduction to Time Series Analysis and Forecasting. John Wiley & Sons, Hoboken, NJ, USA (2008)
589. Moore, C.J., Chua, A.J.K., Berry, C.P.L., Gair, J.R.: Fast methods for training Gaussian processes on large data sets. R. Soc. Open Sci. **3**(5), 160125 (2016)
590. Moro, E.: Network analysis. In: Higham, N.J., Dennis, M.R., Glendinning, P., Martin, P.A., Santosa, F., Tanner, J. (eds.) The Princeton Companion to Applied Mathematics, pp. 360–374. Princeton University Press, Princeton, NJ, USA (2015)
591. Moyeed, R.A., Papritz, A.: An empirical comparison of kriging methods for nonlinear spatial point prediction. Math. Geol. **34**(4), 365–386 (2002)
592. Müller, W.G.: Collecting Spatial Data: Optimum Design of Experiments for Random Fields, 3rd edn. Springer Science & Business Media, Berlin, Germany (2007)
593. Mussardo, G.: Statistical Field Theory. Oxford University Press, Oxford, UK (2010)
594. Nadaraya, E.A.: On estimating regression. Theory Probab. Appl. **9**(1), 141–142 (1964)
595. Nakanishi-Ohno, Y., Nagata, K., Shouno, H., Okada, M.: Distribution estimation of hyperparameters in Markov random field models. J. Phys. A Math. Theor. **47**(4), 045001 (2014)
596. Nan, T., Neuman, S.P., Riva, M., Guadagnini, A.: Analyzing randomly fluctuating hierarchical variables and extremes. In: Cushman, J.H., Tartakovsky, D.M. (eds.) The Handbook of Groundwater Engineering, 3rd edn., pp. 443–456. CRC Press, Taylor & Francis Group, Boca Raton, FL, USA (2016)
597. Nataf, A.: Determination des distributions dont les marges sont données. Comptes Rendus de l' Academie des Sciences **225**, 42–43 (1962)
598. Neal, R.M.: Bayesian Learning for Neural Networks, vol. 118. Springer Science & Business Media, New York, NY, USA (1996)
599. Nelsen, R.B.: An Introduction to Copulas, 2nd edn. Springer Science & Business Media, New York, NY, USA (2006)
600. Nelson, D., Piran, T., Weinberg, S.: Statistical Mechanics of Membranes and Surfaces. World Scientific, Hackensack, NJ, USA (2004)
601. Nerini, D., Monestiez, P., Manté, C.: Cokriging for spatial functional data. J. Multivar. Anal. **101**(2), 409–418 (2010)
602. Neumaier, A.: Solving ill-conditioned and singular linear systems: a tutorial on regularization. SIAM Rev. **40**(3), 636–666 (1998)
603. Neuman, S.P.: Generalized scaling of permeabilities: validation and effect of support scale. Geophys. Res. Lett. **21**(5), 349–352 (1994)
604. Neuman, S.P., Guadagnini, A., Riva, M., Siena, M.: Recent advances in statistical and scaling analysis of earth and environmental variables. In: Mishra, P.K., Kuhlman, K.L. (eds.) Advances in Hydrogeology, pp. 1–25. Springer, New York, NY, USA (2013)
605. Newman, M.E.J.: Power laws, pareto distributions and zipf's law. Contemp. Phys. **46**(5), 323–351 (2005)
606. Newman, M.E.J.: Networks: an Introduction. Oxford University Press, Oxford, UK (2010)
607. Nguyen, H.C., Berg, J.: Mean-field theory for the inverse Ising problem at low temperatures. Phys. Rev. Lett. **109**(5), 050602 (2012)

608. Nguyen, H.C., Zecchina, R., Berg, J.: Inverse statistical problems: from the inverse Ising problem to data science. Adv. Phys. **66**(3), 197–261 (2017)
609. Nisbet, R., Elder IV, J., Miner, G.: Handbook of Statistical Analysis and Data Mining Applications. Academic Press, Amsterdam, Netherlands (2009)
610. Nishimori, H.: Statistical Physics of Spin Glasses and Information Processing: An Introduction. International Series of Monographs on Physics, vol. 111. Clarendon Press, Oxford, UK (2001)
611. Nocedal, J., Wright, S.: Numerical Optimization, 2nd edn. Springer Science & Business Media, New York, NY, USA (2006)
612. Nørrelykke, S.F., Flyvbjerg, H.: Harmonic oscillator in heat bath: exact simulation of time-lapse-recorded data and exact analytical benchmark statistics. Phys. Rev. E **83**(4), 041103 (2011)
613. Novak, J., Sniady, P.: What is a free cumulant. Not. Am. Math. Soc. **58**(2), 300–301 (2011)
614. Novikov, E.A.: Functionals and the random-force method in turbulence theory. Sov. Phys. JETP **20**(5), 1290–1294 (1965)
615. Nussbaumer, R., Mariethoz, G., Gloaguen, E., Holliger, K.: Which path to choose in sequential Gaussian simulation. Math. Geosci. **50**(1), 97–120 (2018)
616. Nybø, R.: Fault detection and other time series opportunities in the petroleum industry. Neurocomputing **73**(10), 1987–1992 (2010)
617. Nychka, D., Bandyopadhyay, S., Hammerling, D., Lindgren, F., Sain, S.: A multi-resolution Gaussian process model for the analysis of large spatial data sets. J. Comput. Graph. Stat. **24**(2), 579–599 (2014)
618. Ogata, Y.: Seismicity analysis through point-process modeling: a review. Pure Appl. Geophys. **155**(2–4), 471–507 (1999)
619. Okabe, H., Blunt, M.J.: Prediction of permeability for porous media reconstructed using multiple-point statistics. Phys. Rev. E **70**(6), 066135 (2004)
620. Okabe, H., Blunt, M.J.: Pore space reconstruction using multiple-point statistics. J. Pet. Sci. Eng. **46**(1–2), 121–137 (2005)
621. Øksendal, B.: Stochastic Differential Equations: An Introduction with Applications. Springer, Berlin, Germany (2003)
622. Olea, R.A.: A six-step practical approach to semivariogram modeling. Stoch. Environ. Res. Risk Assess. **20**(5), 307–318 (2006)
623. Olea, R.A.: A practical primer on geostatistics. Tech. rep., US Geological Survey (2009). https://pubs.usgs.gov/of/2009/1103/. [Online; accessed on 31 Oct 2018]
624. Olea, R.A.: Geostatistics for Engineers and Earth Scientists. Springer Science & Business Media, New York, NY, USA (2012)
625. Oliver, D.: Calculation of the inverse of the covariance. Math. Geol. **30**(7), 911–933 (1998)
626. Oliver, D.S.: On conditional simulation to inaccurate data. Math. Geol. **28**(6), 811–817 (1996)
627. O'Malley, D., Cushman, J.H.: Anomalous dispersion. In: Cushman, J.H., Tartakovsky, D.M. (eds.) The Handbook of Groundwater Engineering, 3rd edn., pp. 497–505. CRC Press, Taylor & Francis Group, Boca Raton, FL, USA (2016)
628. Omre, H., Halvorsen, K.B.: The Bayesian bridge between simple and universal kriging. Math. Geol. **21**(7), 767–786 (1989)
629. Onsager, L.: Crystal statistics. I. A two-dimensional model with an order-disorder transition. Phys. Rev. **65**(3–4), 117–149 (1944)
630. Opper, M., Archambeau, C.: The variational Gaussian approximation revisited. Neural Comput. **21**(3), 786–792 (2009)
631. Oppermann, N., Robbers, G., Enßlin, T.A.: Reconstructing signals from noisy data with unknown signal and noise covariance. Phys. Rev. E **84**(4), 041118 (2011)
632. Oppermann, N., Selig, M., Bell, M.R., Enßlin, T.A.: Reconstruction of Gaussian and log-normal fields with spectral smoothness. Phys. Rev. E **87**(3), 032136 (2013)
633. Ortega, A., Frossard, P., Kovačević, J., Moura, J.M.F., Vandergheynst, P.: Graph signal processing: overview, challenges, and applications. Proc. IEEE **106**(5), 808–828 (2018)

634. Ortigueira, M.D., Machado, J.A.T.: What is a fractional derivative? J. Comput. Phys. **293**, 4–13 (2015)
635. Osborne, A., Provenzale, A.: Finite correlation dimension for stochastic systems with power-law spectra. Physica D: Nonlinear Phenom. **35**(3), 357–381 (1989)
636. Ostoja-Starzewski, M.: Random field models of heterogeneous materials. Int. J. Solids Struct. **35**(19), 2429–2455 (1998)
637. Ostoja-Starzewski, M.: Microstructural Randomness and Scaling in Mechanics of Materials. CRC Press, Boca Raton, FL, USA (2007)
638. Paciorek, C.: Technical vignette 3: Kriging, interpolation, and uncertainty. Tech. rep., Harvard School of Public Health (2008). https://www.stat.berkeley.edu/~paciorek/research/techVignettes/techVignette3.pdf
639. Paciorek, C.J., Schervish, M.J.: Nonstationary covariance functions for Gaussian process regression. In: Thrun, S., Saul, L.K., Schölkopf, B. (eds.) Advances in Neural Information Processing Systems 16 [Neural Information Processing Systems, NIPS 2003, 8–13 Dec 2003, Vancouver and Whistler, British Columbia, Canada], vol. 16, pp. 273–280. MIT Press, Cambridge, MA, USA (2004)
640. Paciorek, C.J., Schervish, M.J.: Spatial modeling using a new class of nonstationary covariance functions. Environmetrics **17**(5), 483–506 (2006)
641. Pain, J.C., Gilleron, F., Faussurier, G.: Jensen-Feynman approach to the statistics of interacting electrons. Phys. Rev. E **80**(2), 026703 (2009)
642. Painter, S.: Random fractal models of heterogeneity: the Lévy-stable approach. Math. Geol. **27**(7), 813–830 (1995)
643. Painter, S.: Evidence for non-Gaussian scaling behavior in heterogeneous sedimentary formations. Water Resour. Res. **32**(5), 1183–1195 (1996)
644. Painter, S.: Stochastic interpolation of aquifer properties using fractional Lévy motion. Water Resour. Res. **32**(5), 1323–1332 (1996)
645. Palma, W.: Long-memory Time Series: Theory and Methods, vol. 662. John Wiley & Sons, Hoboken, NJ, USA (2007)
646. Papoulis, A., Pillai, S.U.: Probability, Random Variables and Stochastic Processes. McGraw-Hill, New York, NY, USA (2002)
647. Pardo-Igúzquiza, E.: Bayesian inference of spatial covariance parameters. Math. Geol. **31**(1), 47–65 (1999)
648. Pardo-Igúzquiza, E., Chica-Olmo, M.: The Fourier integral method: an efficient spectral method for simulation of random fields. Math. Geol. **25**(2), 177–217 (1993)
649. Pardo-Iguzquiza, E., Chica-Olmo, M.: Geostatistical simulation when the number of experimental data is small: an alternative paradigm. Stoch. Environ. Res. Risk Assess. **22**(3), 325–337 (2008)
650. Pardo-Igúzquiza, E., Dowd, P.A.: Comparison of inference methods for estimating semivariogram model parameters and their uncertainty: the case of small data sets. Comput. Geosci. **50**, 154–164 (2013)
651. Pardo-Igúzquiza, E., Olea, R.A.: VARBOOT: a spatial bootstrap program for semivariogram uncertainty assessment. Comput. Geosci. **41**, 188–198 (2012)
652. Pardo-Igúzquiza, E.: Maximum likelihood estimation of spatial covariance parameters. Math. Geol. **30**(1), 95–108 (1998)
653. Parisi, G.: Infinite number of order parameters for spin-glasses. Phys. Rev. Lett. **43**(23), 1754–1756 (1979)
654. Parisi, G.: The physical meaning of replica symmetry breaking. arXiv preprint cond-mat/0205387 (2002)
655. Patra, M., Karttunen, M.: Stencils with isotropic discretization error for differential operators. Numer. Methods Partial Differ. Equ. **22**(4), 936–953 (2006)
656. Pavliotis, G.A.: Stochastic Processes and Applications. Springer, New York, NY, USA (2014)
657. Pearson, B., Fox-Kemper, B.: Log-normal turbulence dissipation in global ocean models. Phys. Rev. Lett. **120**(9), 094501 (2018)

658. Peierls, R.: On Ising's model of ferromagnetism. Math. Proc. Camb. Philos. Soc. **32**(3), 477–481 (1936)
659. Peng, C.K., Buldyrev, S.V., Havlin, S., Simons, M., Stanley, H.E., Goldberger, A.L.: Mosaic organization of DNA nucleotides. Phys. Rev. E **49**(2), 1685 (1994)
660. Perdikaris, P., Venturi, D., Royset, J.O., Karniadakis, G.E.: Multi-fidelity modeling via recursive co-kriging and Gaussian–Markov random fields. Proc. R. Soc. Lond. A **471**(2179), 20150018 (2015)
661. Pérez-Cruz, F., Vaerenbergh, S.V., Murillo-Fuentes, J.J., Lázaro-Gredilla, M., Santamaria, I.: Gaussian processes for nonlinear signal processing: an overview of recent advances. IEEE Signal Process. Mag. **30**(4), 40–50 (2013)
662. Petersen, W.P., Bernasconi, A.: Uniform sampling from an n-sphere. Tech. Rep. TR 97-06, Swiss Center for Scientific Computing, ETH, Zürich, Switzerland (1997)
663. Petrakis, M.P., Hristopulos, D.T.: Non-parametric approximations for anisotropy estimation in two-dimensional differentiable Gaussian random fields. Stoch. Environ. Res. Risk Assess. **31**(7), 1853–1870 (2017)
664. Phoon, K.K., Ching, J.: Risk and Reliability in Geotechnical Engineering. CRC Press, Boca Raton, FL, USA (2015)
665. Phythian, R.: The functional formalism of classical statistical dynamics. J. Phys. A Math. Gen. **10**(5), 777–789 (1977)
666. Pienaar, J.: Viewpoint: causality in the quantum world. Physics **10**, 86 (2017)
667. Pilz, J., Spöck, G.: Why do we need and how should we implement Bayesian kriging methods. Stoch. Environ. Res. Risk Assess. **22**(5), 621–632 (2008)
668. Popoviciu, T.: Sur les équations algébriques ayant toutes leurs racines réelles. Mathematica **9**, 129–145 (1935)
669. Porcu, E., Bevilacqua, M., Genton, M.G.: Spatio-temporal covariance and cross-covariance functions of the great circle distance on a sphere. J. Am. Stat. Assoc. **111**(514), 888–898 (2016)
670. Potthoff, J.: Sample properties of random fields III: Differentiability. Commun. Stoch. Anal. **4**(3), 335–353 (2010)
671. Praetz, P.D.: The distribution of share price changes. J. Bus. **45**(1), 49–55 (1972)
672. Press, S.J.: Multivariate stable distributions. J. Multivar. Anal. **2**(4), 444–462 (1972)
673. Press, W.H., Teukolsky, S.A., Vetterling, W.T., Flannery, B.P.: Numerical Recipes, The Art of Scientific Computing, 3rd edn. Cambridge University Press, Cambridge, UK (1997)
674. Pressé, S., Ghosh, K., Lee, J., Dill, K.A.: Principles of maximum entropy and maximum caliber in statistical physics. Rev. Mod. Phys. **85**(3), 1115–1141 (2013)
675. Puig, B., Akian, J.L.: Non-Gaussian simulation using Hermite polynomials expansion and maximum entropy principle. Probab. Eng. Mech. **19**(4), 293–305 (2004)
676. Ramsay, J.O., Silverman, B.W.: Functional Data Analysis, 2nd edn. Springer, New York, NY, USA (2005)
677. Rangan, S., Fletcher, A.K., Goyal, V.K.: Asymptotic analysis of MAP estimation via the replica method and applications to compressed sensing. IEEE Trans. Inf. Theory **58**(3), 1902–1923 (2012)
678. Rasmussen, C.E., Williams, C.K.I.: Gaussian Processes for Machine Learning. MIT Press, Cambridge, MA, USA (2006). www.GaussianProcess.org/gpml. [Online; accessed on 31 Oct 2018]
679. Reilly, R.C.: Mean curvature, the Laplacian, and soap bubbles. Am. Math. Mon. **89**(3), 180–198 (1982)
680. Reusken, A.: Approximation of the determinant of large sparse symmetric positive definite matrices. SIAM J. Matrix Anal. Appl. **23**(3), 799–818 (2002)
681. Ripley, B.D.: Pattern Recognition and Neural Networks. Cambridge University Press, Cambridge, UK (1996)
682. Ripley, B.D.: Spatial Statistics, vol. 575. John Wiley & Sons, Hoboken, NJ, USA (2005)
683. Rister, K., Lahiri, S.N.: Bootstrap based trans-gaussian kriging. Stat. Model. **13**(5–6), 509–539 (2013)

684. Rivoirard, J.: Introduction to disjunctive kriging and non-linear geostatistics. Oxford, Clarendon Press, New York, NY, USA (1994)
685. Robert, C.P., Casella, G.: Monte Carlo Statistical Methods. Springer, New York, NY, USA (1999)
686. Robert, C.Y.: Some new classes of stationary max-stable random fields. Statist. Probab. Lett. **83**(6), 1496–1503 (2013)
687. Roberts, A.P.: Statistical reconstruction of three-dimensional porous media from two-dimensional images. Phys. Rev. E **56**(3), 3203–3212 (1997)
688. Roberts, A.P., Teubner, M.: Transport properties of heterogeneous materials derived from Gaussian random fields: bounds and simulation. Phys. Rev. E **51**(5), 4141–4154 (1995)
689. Roberts, G.O.: Markov chain concepts related to sampling algorithms. In: Gilks, W.R., Richardson, S., Spiegelhalter, D.J. (eds.) Markov Chain Monte Carlo in Practice, vol. 57. Chapman and Hall, CRC, Boca Raton, FL, USA (1996)
690. Røislien, J., Omre, H.: T-distributed random fields: a parametric model for heavy-tailed well-log data. Math. Geol. **38**(7), 821–849 (2006)
691. Rosenburg, M.A., Aharonson, O., Head, J.W., Kreslavsky, M.A., Mazarico, E., Neumann, G.A., Smith, D.E., Torrence, M.H., Zuber, M.T.: Global surface slopes and roughness of the moon from the lunar orbiter laser altimeter. J. Geophys. Res. Planets **116**(E2), E02001 (2011)
692. Rossi, M.E., Deutsch, C.V.: Mineral Resource Estimation. Springer Science & Business Media, Dordrecht, Netherlands (2014)
693. Roudi, Y., Tyrcha, J., Hertz, J.: Ising model for neural data: model quality and approximate methods for extracting functional connectivity. Phys. Rev. E **79**(5), 051915 (2009)
694. Rozanov, J.A.: Markov random fields and stochastic partial differential equations. Math. USSR-Sbornik **32**(4), 515 (1977). http://stacks.iop.org/0025-5734/32/i=4/a=A08
695. Rubin, D.B.: The Bayesian bootstrap. Ann. Stat. **9**(1), 130–134 (1981)
696. Rubin, Y.: Applied Stochastic Hydrogeology. Oxford, UK, New York, NY, USA (2003)
697. Rue, H.: Fast sampling of Gaussian Markov random fields. J. R. Stat. Soc. Ser. B (Stat Methodol.) **63**(2), 325–338 (2001)
698. Rue, H., Held, L.: Gaussian Markov Random Fields: Theory and Applications. Chapman and Hall/CRC, Boca Raton, FL, USA (2005)
699. Rue, H., Held, L.: Conditional and intrinsic autoregressions. In: Gelfand, A.E., Diggle, P., Guttorp, P., Fuentes, M. (eds.) Handbook of Spatial Statistics, chap. 12, pp. 172–198. CRC Press, Boca Raton, FL, USA (2010)
700. Rue, H., Held, L.: Discrete spatial variation. In: Gelfand, A.E., Diggle, P., Guttorp, P., Fuentes, M. (eds.) Handbook of Spatial Statistics, chap. 12, pp. 172–198. CRC Press, Boca Raton, FL, USA (2010)
701. Rue, H., Martino, S., Chopin, N.: Approximate Bayesian inference for latent Gaussian models by using integrated nested Laplace approximations. J. R. Stat. Soc. Ser. B (Stat Methodol.) **71**(2), 319–392 (2009)
702. Ruelle, D.: Chance and Chaos. Princeton University Press, Princeton, NJ, USA (1991)
703. Ruiz-Alzola, J., et al.: Geostatistical medical image registration. In: Ellis, R.E., Peters, T.M. (eds.) Medical Image Computing and Computer-Assisted Intervention – MICCAI 2003, 6th International Conference, Montréal, Canada, November 15–18, 2003, Proceedings, Part II. Lecture Notes in Computer Science, vol. 2879, pp. 894–901. Springer (2003)
704. Runge, C.: Über empirische funktionen und die interpolation zwischen äquidistanten ordinaten. Zeitschrift für Mathematik und Physik **46**(20), 224–243 (1901)
705. Saibaba, A.K., Ambikasaran, S., Yue, L.J., Kitanidis, P.K., Darve, E.F.: Application of hierarchical matrices to linear inverse problems in geostatistics. Oil Gas Sci. Technol. Revue de l'IFP-Institut Francais du Petrole **67**(5), 857–875 (2012)
706. Sakamoto, H., Nakanishi-Ohno, Y., Okada, M.: Theory of distribution estimation of hyperparameters in Markov random field models. J. Phys. Soc. Jpn. **85**(6), 063801 (2016)
707. Salamon, P., Sibani, P., Frost, R.: Facts, Conjectures, and Improvements for Simulated Annealing. SIAM, Philadelphia, PA, USA (2002)

708. Samko, S.G., Kilbas, A.A., Marichev, O.I.: Fractional Integrals and Derivatives: Theory and Applications. Gordon and Breach, Amsterdam, Netherlands (1993)
709. Samorodnitsky, G., Taqqu, M.S.: Stable Non-Gaussian Random Processes: Stochastic Models with Infinite Variance. Chapman and Hall/CRC (2000)
710. Sampson, P.D., Guttorp, P.: Nonparametric estimation of nonstationary spatial covariance structure. J. Am. Stat. Assoc. **87**(417), 108–119 (1992)
711. Sandwell, D.T.: Interpolation of GEOS-3 and SEASAT altimeter data. Geophys. Res. Lett. **2**, 139–142 (1987)
712. Sang, H., Gelfand, A.E.: Continuous spatial process models for spatial extreme values. J. Agric. Biol. Environ. Stat. **15**(1), 49–65 (2010)
713. Sangoyomi, T.B., Lall, U., Abarbanel, H.D.I.: Nonlinear dynamics of the great salt lake: dimension estimation. Water Resour. Res. **32**(1) (1996)
714. Sapsis, T.P., Lermusiaux, P.F.J.: Dynamically orthogonal field equations for continuous stochastic dynamical systems. Physica D: Nonlinear Phenom. **238**(23–24), 2347–2360 (2009)
715. Sato, M., Ichiki, K., Takeuchi, T.T.: Copula cosmology: constructing a likelihood function. Phys. Rev. D **83**(2), 023501 (2011)
716. Savitzky, A., Golay, M.J.E.: Smoothing and differentiation of data by simplified least squares procedure. Anal. Chem. **36**(8), 1627–1639 (1964)
717. Schabenberger, O., Gotway, C.A.: Statistical Methods for Spatial Data Analysis. CRC Press, Boca Raton, FL, USA (2004)
718. Schäfer, F., Sullivan, T.J., Owhadi, H.: Compression, inversion, and approximate PCA of dense kernel matrices at near-linear computational complexity. arXiv preprint arXiv:1706.02205 (2017)
719. Schafer, R.W.: What is a Savitzky-Golay filter? IEEE Signal Process. Mag. **28**(4), 111–117 (2011)
720. Scheuerer, M.: Regularity of the sample paths of a general second order random field. Stoch. Process. Appl. **120**(10), 1879–1897 (2010)
721. Scheuerer, M., Schlather, M.: Covariance models for divergence-free and curl-free random vector fields. Stoch. Model. **28**(3), 433–451 (2012)
722. Schiff, L.: Quantum Mechanics, 3rd edn. McGraw-Hill, New York, NY, USA (1968)
723. Schlather, M.: Models for stationary max-stable random fields. Extremes **5**(1), 33–44 (2002)
724. Schlather, M.: Construction of covariance functions and unconditional simulation of random fields. In: Porcu, E., Montero, J., Schlather, M. (eds.) Advances and Challenges in Space-time Modelling of Natural Events. Lecture Notes in Statistics, vol. 207, pp. 25–54. Springer, Berlin, Germany (2012)
725. Schmalzing, J., Górski, K.M.: Minkowski functionals used in the morphological analysis of cosmic microwave background anisotropy maps. Mon. Not. R. Astron. Soc. **297**(2), 355–365 (1998)
726. Schmidt, A.M., O'Hagan, A.: Bayesian inference for non-stationary spatial covariance structure via spatial deformations. J. R. Stat. Soc. Ser. B (Stat Methodol.) **65**(3), 743–758 (2003)
727. Schmüdgen, K.: The Moment Problem, 1st edn. Graduate Texts in Mathematics 277. Springer, Cham, Switzerland (2017)
728. Schneider, C.L., Attinger, S.: Beyond Thiem: A new method for interpreting large scale pumping tests in heterogeneous aquifers. Water Resour. Res. **44**(4), W04427 (2008)
729. Schneidman, E., Berry, M.J., Segev, R., Bialek, W.: Weak pairwise correlations imply strongly correlated network states in a neural population. Nature **440**, 1007–1012 (2006)
730. Schoenberg, I.J.: Metric spaces and completely monotone functions. Ann. Math. **39**(4), 811–841 (1938)
731. Schölkopf, B., Smola, A.J.: Learning with Kernels: Support Vector Machines, Regularization, Optimization, and Beyond. MIT Press, Cambridge, MA, USA (2002)
732. Schröder, K.P., Connon Smith, R.: Distant future of the Sun and Earth revisited. Mon. Not. R. Astron. Soc. **386**(1), 155–163 (2008)
733. Schwartz, L.M.: Mathematics for the Physical Sciences. Dover, Mineola, NY, USA (2008)

734. Scott, D.W.: Multivariate Density Estimation: Theory, Practice, and Visualization, 2nd edn. John Wiley & Sons, Hoboken, NJ, USA (2015)
735. Scott, M.: Applied Stochastic Processes in Science and Engineering. University of Waterloo (2013). http://www.math.uwaterloo.ca/~mscott/Little_Notes.pdf. [Online; accessed on 31 Oct 2018]
736. Seber, G.A.F., Lee, A.J.: Linear Regression Analysis, 2nd edn. John Wiley & Sons, Hoboken, NJ, USA (2003)
737. Serinaldi, F., Kilsby, C.G.: Stationarity is undead: uncertainty dominates the distribution of extremes. Adv. Water Resour. **77**, 17–36 (2015)
738. Shah, A., Wilson, A.G., Ghahramani, Z.: Student-t processes as alternatives to Gaussian processes. In: Kaski, S., Corander, J. (eds.) Proceedings of the Seventeenth International Conference on Artificial Intelligence and Statistics, AISTATS 2014, Reykjavik, Iceland, 22–25 April 2014. JMLR Workshop and Conference Proceedings, vol. 33, pp. 877–885. JMLR.org (2014)
739. Shannon, C.E.: A mathematical theory of communication. Bell Syst. Tech. J. **27**(3), 379–423 (1948)
740. Shapiro, A., Botha, J.D.: Variogram fitting with a general class of conditionally nonnegative definite functions. Comput. Stat. Data Anal. **11**(1), 87–96 (1991)
741. Sheffield, S.: Gaussian free fields for mathematicians. Probab. Theory Relat. Fields **139**(3), 521–541 (2007)
742. Shen, Y., Ng, A.Y., Seeger, M.: Fast Gaussian process regression using KD-trees. In: Advances in Neural Information Processing Systems 18 [Neural Information Processing Systems, NIPS 2005, December 5–8, 2005, Vancouver, British Columbia, Canada], pp. 1225–1232. MIT Press, Cambridge, MA, USA (2005)
743. Shepard, D.: A two-dimensional interpolation function for irregularly-spaced data. In: Proceedings of the 1968 23rd ACM National Conference, pp. 517–524. ACM, New York, NY, USA (1968)
744. Sherman, M.: Spatial Statistics and Spatio-temporal Data: Covariance Functions and Directional Properties. John Wiley & Sons, Chichester, West Sussex, UK (2011)
745. Shilov, J.M., Gelfand, G.E.: Generalized Functions: Properties and Operations, vol. 2. Academic Press, Berlin, Germany (1968)
746. Shinzato, T.: Validation of the replica trick for simple models. J. Stat. Mech: Theory Exp. **2018**(4), 043306 (2018)
747. Shlens, J.: A tutorial on principal component analysis. arXiv preprint arXiv:1404.1100 (2014)
748. Shoji, T.: Statistical and geostatistical analysis of wind: a case study of direction statistics. Comput. Geosci. **32**(8), 1025–1039 (2006)
749. Shuman, D.I., Narang, S.K., Frossard, P., Ortega, A., Vandergheynst, P.: The emerging field of signal processing on graphs: extending high-dimensional data analysis to networks and other irregular domains. IEEE Signal Process. Mag. **30**(3), 83–98 (2013)
750. Shumway, R.H., Stoffer, D.S.: Time Series Analysis and its Applications. Springer Science & Business Media, New York, NY, USA (2000)
751. Siegmund, D.O., Worsley, K.J.: Testing for a signal with unknown location and scale in a stationary Gaussian random field. Ann. Stat. **23**(2), 608–639 (1995)
752. Sivia, D., Skilling, J.: Data Analysis: A Bayesian Tutorial. Oxford University Press, Oxford, UK (2006)
753. Skilling, J. (ed.): Maximum Entropy and Bayesian Methods, vol. 36. Springer Science & Business Media, Dordrecht, Netherlands (2013)
754. Skilling, J., Bryan, R.K.: Maximum entropy image reconstruction: general algorithm. Mon. Not. R. Astron. Soc. **211**(1), 111–124 (1984)
755. Skøien, J.O., Baume, O.P., Pebesma, E.J., Heuvelink, G.B.: Identifying and removing heterogeneities between monitoring networks. Environmetrics **21**(1), 66–84 (2010)
756. Sloan, I.H., Woźniakowski, H.: When are Quasi-Monte Carlo algorithms efficient for high dimensional integrals? J. Complex. **14**(1), 1–33 (1998)

757. Smith, R.C.: Uncertainty Quantification: Theory, Implementation, and Applications, vol. 12. SIAM, Philadelphia, PA, USA (2013)
758. Smith, R.L.: Spatial statistics in environmental science. In: Fitzgerald, W.J., Smith, R.L., Walden, A.T., Young, P.C. (eds.) Nonlinear and Nonstationary Signal Processing, pp. 152–183. Cambridge University Press, Cambridge, UK (2000)
759. Smith, R.S., O'Conell, M.D.: Interpolation and gridding of aliased geophysical data using constrained anisotropic diffusion to enhance trends. Geophysics **70**(5), V121–V127 (2005)
760. Smith, W.H.F., Wessel, P.: Gridding with continuous curvature splines in tension. Geophysics **55**(3), 293–305 (1990)
761. Snoek, J., Larochelle, H., Adams, R.P.: Practical Bayesian optimization of machine learning algorithms. In: Pereira, F., Burges, C.J.C., Bottou, L., Weinberger, K.Q. (eds.) Advances in Neural Information Processing Systems 25: 26th Annual Conference on Neural Information Processing Systems 2012. Proceedings of a meeting held December 3–6, 2012, Lake Tahoe, Nevada, United States, vol. 25, pp. 2951–2959 (2012)
762. Sobczyk, K., Trcebicki, J.: Approximate probability distributions for stochastic systems: maximum entropy method. Comput. Methods Appl. Mech. Eng. **168**(1), 91–111 (1999)
763. Sobczyk, K., Trebicki, J.: Analysis of stochastic systems via maximum entropy principle. In: Bellomo, N., Casciati, F. (eds.) Nonlinear Stochastic Mechanics, Proceedings of the IUTAM 1991 Symposium, Turin (Italy), pp. 485–497. Springer, Berlin, Germany (1992)
764. Solin, A., Sarkka, S.: Gaussian quadratures for state space approximation of scale mixtures of squared exponential covariance functions. In: Mboup, M., Adali, T., Moreau, E., Larsen, J. (eds.) 2014 IEEE International Workshop on Machine Learning for Signal Processing (MLSP), 21–24 Sept, Reims, France, pp. 1–6. IEEE (2014)
765. Song, H.R., Fuentes, M., Ghosh, S.: A comparative study of Gaussian geostatistical models and Gaussian Markov random field models. J. Multivar. Anal. **99**(8), 1681–1697 (2008)
766. Sornette, D.: Critical Phenomena in Natural Sciences. Springer, Berlin, Germany (2004)
767. Spanos, P.D., Beer, M., Red-Horse, J.: Karhunen–Loève expansion of stochastic processes with a modified exponential covariance kernel. J. Eng. Mech. **133**(7), 773–779 (2007)
768. Stauffer, D.: Social applications of two-dimensional Ising models. Am. J. Phys. **76**(4), 470–473 (2008)
769. Stauffer, D., Aharony, A.: Introduction to Percolation Theory. Taylor and Francis, London, UK (1994)
770. Steed, C.A., Ricciuto, D.M., Shipman, G., Smith, B., Thornton, P.E., Wang, D., Shi, X., Williams, D.N.: Big data visual analytics for exploratory earth system simulation analysis. Comput. Geosci. **61**, 71–82 (2013)
771. Stefanou, G., Papadrakakis, M.: Assessment of spectral representation and Karhunen–Loève expansion methods for the simulation of Gaussian stochastic fields. Comput. Methods Appl. Mech. Eng. **196**(21), 2465–2477 (2007)
772. Stein, D.L., Newman, C.M.: Spin Glasses and Complexity. Princeton University Press, Princeton, NJ, USA (2013)
773. Stein, M.: Asymptotics for spatial processes. In: Gelfand, A.E., Diggle, P., Guttorp, P., Fuentes, M. (eds.) Handbook of Spatial Statistics, pp. 79–87. CRC Press, Boca Raton, FL, USA (2010)
774. Stein, M.L.: Interpolation of Spatial Data: Some Theory for Kriging. Springer, New York, NY, USA (1999)
775. Stein, M.L.: Local stationarity and simulation of self-affine intrinsic random functions. IEEE Trans. Inf. Theory **47**(4), 1385–1390 (2001)
776. Stein, M.L.: The screening effect in kriging. Ann. Stat. **30**(1), 298–323 (2002)
777. Stein, M.L., Chi, Z., Welty, L.J.: Approximating likelihoods for large spatial data sets. J. R. Stat. Soc. Ser. B (Stat Methodol.) **66**(2), 275–296 (2004)
778. Stinis, P.: A maximum likelihood algorithm for the estimation and renormalization of exponential densities. J. Comput. Phys. **208**(2), 691–703 (2005)
779. Stone, M., Goldbart, P.: Mathematics for Physics: A Guided Tour for Graduate Students. Cambridge University Press, Cambridge, UK (2009)

780. Stratonovich, R.L., Silverman, R.A.: Topics in the Theory of Random Noise. General Theory of Random Processes, vol. I. Gordon and Breach Science Publishers, New York, NY, USA (1960)
781. Strebelle, S.: Conditional simulation of complex geological structures using multiple-point statistics. Math. Geol. **34**(1), 1–21 (2002)
782. Strogatz, S.H.: Nonlinear Dynamics and Chaos with Applications to Physics, Biology, Chemistry, and Engineering. CRC Press, Boca Raton, FL (2018)
783. Süli, E.: Numerical solution of partial differential equations. In: Higham, N.J., Dennis, M.R., Glendinning, P., Martin, P.A., Santosa, F., Tanner, J. (eds.) The Princeton Companion to Applied Mathematics, pp. 306–318. Princeton University Press, Princeton, NJ, USA (2015)
784. Sun, Y., Li, B., Genton, M.G.: Geostatistics for large datasets. In: Porcu, E., Montero, J., Schlather, M. (eds.) Advances and Challenges in Space-time Modelling of Natural Events, Lecture Notes in Statistics, pp. 55–77. Springer Berlin Heidelberg (2012)
785. Sun, Y., Stein, M.: A stochastic space-time model for intermittent precipitation occurrences. Ann. Stat. **9**(4), 2110–2132 (2016)
786. Swerling, P.: Statistical properties of the contours of random surfaces. IRE Trans. Inf. Theory **IT-8**(4), 315–321 (1962)
787. Swift, J., Hohenberg, P.C.: Hydrodynamic fluctuations at the convective instability. Phys. Rev. A **15**(1), 319–328 (1977)
788. Swift, J.B., Hohenberg, P.C.: Swift-Hohenberg equation. Scholarpedia **3**(9), 6395 (2008). https://doi.org/10.4249/scholarpedia.6395. Revision #91841
789. Tanizaki, H.: On regression models with autocorrelated error: small sample properties. Int. J. Pure Appl. Math. **5**, 161–175 (2003)
790. Teegavarapu, R.S.V., Meskele, T., Pathak, C.S.: Geo-spatial grid-based transformations of precipitation estimates using spatial interpolation methods. Comput. Geosci. **40**, 28–39 (2012)
791. Tessier, Y., Lovejoy, S., Schertzer, D.: Universal multifractals: theory and observations for rain and clouds. J. Appl. Meteorol. **32**(2), 223–250 (1993)
792. Teubner, M.: Level surfaces of Gaussian random fields and microemulsions. Europhys. Lett. **14**(5), 403–408 (1991)
793. Teubner, M., Strey, R.: Origin of the scattering peak in microemulsions. J. Chem. Phys. **87**(5), 3195–3200 (1987)
794. Tibshirani, R.: Regression shrinkage and selection via the Lasso. J. R. Stat. Soc. Ser. B Methodol. **58**(1), 267–288 (1996)
795. Tikhonov, A.N., Arsenin, V.Y.: Solutions of Ill-Posed Problems. V. H. Winston & Sons, Washington, DC, USA (1977)
796. Ton, J.F., Flaxman, S., Sejdinovic, D., Bhatt, S.: Spatial mapping with Gaussian processes and nonstationary Fourier features. Spat. Stat. **28**, 59–78 (2018)
797. Torquato, S.: Random Heterogeneous Materials: Microstructure and Macroscopic Properties. Springer, New York, NY, USA (2002)
798. Touchette, H.: When is a quantity additive, and when is it extensive? Physica A: Stat. Mech. Appl. **305**(1–2), 84–88 (2002)
799. Transtrum, M.K., Machta, B.B., Brown, S., Daniels, B.C., Myers, C.R., Sethna, J.P.: Perspective: sloppiness and emergent theories in physics, biology, and beyond. J. Chem. Phys. **143**(1), 010901 (2015)
800. Transtrum, M.K., Machta, B.B., Sethna, J.P.: Why are nonlinear fits to data so challenging? Phys. Rev. Lett. **104**(6), 060201 (2010)
801. Tsallis, C.: Possible generalization of Boltzmann-Gibbs statistics. J. Stat. Phys. **52**(1–2), 479–487 (1988)
802. Tsallis, C.: Introduction to Nonextensive Statistical Mechanics. Springer, New York, NY, USA (2009)
803. Tsantili, I.C., Hristopulos, D.T.: Karhunen-Loève expansion of Spartan spatial random fields. Probab. Eng. Mech. **43**, 132–147 (2016)
804. Tukey, J.W.: Exploratory Data Analysis, vol. 1. Addison-Wesley, Reading, MA, USA (1977)

805. Tzikas, D.G., Likas, A.C., Galatsanos, N.P.: The variational approximation for Bayesian inference. IEEE Signal Process. Mag. **25**(6), 131–146 (2008)
806. Urry, M.J., Sollich, P.: Replica theory for learning curves for Gaussian processes on random graphs. J. Phys. A Math. Theor. **45**(42), 425005 (2012)
807. Van Duijn, M.A., et al.: A framework for the comparison of maximum pseudo-likelihood and maximum likelihood estimation of exponential family random graph models. Soc. Networks **31**(1), 52 (2009)
808. Vanmarcke, E.: Random Fields: Analysis and Synthesis. World Scientific, Hackensack, NJ, USA (2010)
809. Vanmarcke, E.H.: Properties of spectral moments with applications to random vibration. J .Eng. Mech. Div. **98**(2), 425–446 (1972)
810. Vapnik, V.N.: The Nature of Statistical Learning, 2nd edn. Springer, New York, NY, USA (2000)
811. Vargas-Guzmán, J.A., Warrick, A.W., Myers, D.E.: Coregionalization by linear combination of nonorthogonal components. Math. Geol. **34**(4), 405–419 (2002)
812. Varin, C., Reid, N., Firth, D.: An overview of composite likelihood methods. Stat. Sin. **21**(1), 5–42 (2011)
813. Varouchakis, E.A., D. T. Hristopulos, Karatzas, G.: Improving kriging of groundwater level data using non-linear normalizing transformations-a field application. Hydrol. Sci. J. **57**(7), 1404–1419 (2012)
814. Varouchakis, E.A., Hristopulos, D.T.: Comparison of stochastic and deterministic methods for mapping groundwater level spatial variability in sparsely monitored basins. Environ. Monit. Assess. **185**(1), 1–19 (2013)
815. Varouchakis, E.A., Hristopulos, D.T.: Improvement of groundwater level prediction in sparsely gauged basins using physical laws and local geographic features as auxiliary variables. Adv. Water Resour. **52**(2), 34–49 (2013)
816. Vatsavai, R.R., Ganguly, A., Chandola, V., Stefanidis, A., Klasky, S., Shekhar, S.: Spatiotemporal data mining in the era of big spatial data: algorithms and applications. In: Chandola, V., Vatsavai, R.R., Gupta, C. (eds.) Proceedings of the 1st ACM SIGSPATIAL International Workshop on Analytics for Big Geospatial Data, Redondo Beach, CA, USA, November 6, 2012, pp. 1–10. ACM, New York, NY, USA (2012)
817. Venturi, D.: The numerical approximation of nonlinear functionals and functional differential equations. Phys. Rep. **732**, 1–102 (2018)
818. Vergara, R.C.: Development of geostatistical models using stochastic partial differential equations. Ph.D. thesis, MINES, Paris Tech (2018). http://cg.ensmp.fr/bibliotheque/public/CARRIZO_These_02513.pdf
819. Vergara, R.C., Allard, D., Desassis, N.: A general framework for SPDE-based stationary random fields. arXiv preprint arXiv:1806.04999 (2018)
820. Vogel, H.J., Weller, U., Schlüter, S.: Quantification of soil structure based on Minkowski functions. Comput. Geosci. **36**(10), 1236–1245 (2010)
821. Volfson, D., Vinals, J.: Flow induced by a randomly vibrating boundary. J. Fluid Mech. **432**, 387–408 (2001)
822. Vourdas, N., Tserepi, A., Gogolides, E.: Nanotextured super-hydrophobic transparent poly (methyl methacrylate) surfaces using high-density plasma processing. Nanotechnology **18**(12), 125304 (2007)
823. Wackernagel, H.: Multivariate Geostatistics. Springer, Berlin, Germany (2003)
824. Wada, T.: A nonlinear drift which leads to κ-generalized distributions. Eur. Phys. J. B **73**(2), 287–291 (2010)
825. Wahba, G.: Spline models for observational data. In: CNMS-NSF Regional Conference Series in Applied Mathematics, vol. 59, p. 35. SIAM, Philadelphia, Pennsylvania (1990)
826. Walder, C., Kim, K.I., Schölkopf, B.: Sparse multiscale Gaussian process regression. In: Cohen, W.W., McCallum, A., Roweis, S.T. (eds.) Machine Learning, Proceedings of the Twenty-Fifth International Conference (ICML 2008), Helsinki, Finland 5–9 June 2008. ACM International Conference Proceeding Series, vol. 307, pp. 1112–1119. ACM, New York, NY, USA (2008)

827. Wang, F., Fan, Q., Stanley, H.E.: Multiscale multifractal detrended-fluctuation analysis of two-dimensional surfaces. Phys. Rev. E **93**(4), 042213 (2016)
828. Wang, J.F., Stein, A., Gao, B.B., Ge, Y.: A review of spatial sampling. Spat. Stat. **2**(1), 1–14 (2012)
829. Wang, J.L., Chiou, J.M., Müller, H.G.: Functional data analysis. Ann. Rev. Stat. Appl. **3**, 257–295 (2016)
830. Wang, W.X., Lai, Y.C., Grebogi, C.: Data based identification and prediction of nonlinear and complex dynamical systems. Phys. Rep. **644**, 1–76 (2016)
831. Ward, L.M., Greenwood, P.E.: 1/f noise. Scholarpedia **2**(12), 1537 (2007). revision #90924
832. Warnes, J.J., Ripley, B.D.: Problems with likelihood estimation of covariance functions of spatial Gaussian processes. Biometrika **74**(3), 640–642 (1987)
833. Wasserstein, R.L., Lazar, N.A.: The ASA's statement on p-values: context, process, and purpose. Am. Stat. **70**(2), 129–133 (2016)
834. Watson, G.N.: A Treatise on the Theory of Bessel Functions, 2nd edn. Cambridge University Press, New York, NY, USA (1995)
835. Watson, G.S.: Smooth regression analysis. Sankhya Ser. A **26**(1), 359–372 (1964)
836. Weaver, A.T., Mirouze, I.: On the diffusion equation and its application to isotropic and anisotropic correlation modeling in variational assimilation. Q. J. R. Meteorol. Soc. **139**(670), 242–260 (2013)
837. Webster, R., Oliver, M.A.: Geostatistics for Environmental Scientists. John Wiley & Sons, Hoboken, NJ, USA (2007)
838. Weibull, W.: A statistical distribution function of wide applicability. J. Appl. Mech. – ASME **18**(1), 293–297 (1951)
839. Weller, Z.D., Hoeting, J.A.: A review of nonparametric hypothesis tests of isotropy properties in spatial data. Stat. Sci. **31**(3), 305–324 (2016)
840. Wellmann, J.F.: Information theory for correlation analysis and estimation of uncertainty reduction in maps and models. Entropy **15**(4), 1464–1485 (2013)
841. Wendland, H.: Piecewise polynomial, positive definite and compactly supported radial functions of minimal degree. Adv. Comput. Math. **4**(1), 389–396 (1995)
842. Wendland, H.: Scattered Data Approximation. Cambridge University Press, Cambridge, UK (2005)
843. Weron, R.: Estimating long-range dependence: finite sample properties and confidence intervals. Physica A: Stat. Mech. Appl. **312**(1), 285–299 (2002)
844. West, G.: Scale: The Universal Laws of Growth, Innovation, Sustainability, and the Pace of Life in Organisms, Cities, Economies, and Companies. Penguin, New York, NY, USA (2017)
845. Whitehouse, D.J.: Handbook of Surface and Nanometrology. CRC Press, Boca Raton, FL, USA (2010)
846. Whittle, P.: On stationary processes in the plane. Biometrika **41**(3/4), 434–449 (1954)
847. Whittle, P.: Stochastic processes in several dimensions. Bull. Int. Stat. Inst. **40**(2), 974–994 (1963)
848. Wick, G.C.: The evaluation of the collision matrix. Phys. Rev. **80**(2), 268–272 (1950)
849. Wiener, N.: The homogeneous chaos. Am. J. Math. **60**(4), 897–936 (1938)
850. Williams, C.K.I.: Computation with infinite neural networks. Neural Comput. **10**(5), 1203–1216 (1998)
851. Wilson, R., Li, C.T.: A class of discrete multiresolution random fields and its application to image segmentation. IEEE Trans. Pattern Anal. Mach. Intell. **25**(1), 42–56 (2002)
852. Winkler, G.: Image Analysis, Random Fields and Dynamic Monte Carlo Methods: A Mathematical Introduction. Springer, New York, NY, USA (1995)
853. Wishart, J.: The generalised product moment distribution in samples from a normal multivariate population. Biometrika **20A**(1–2), 32–52 (1928)
854. Wood, A.T.A., Chan, G.: Simulation of stationary Gaussian processes in $[0, 1]^d$. J. Comput. Graph. Stat. **3**(4), 409–432 (1994)
855. Wood, A.W., Leung, L.R., Sridhar, V., Lettenmaier, D.P.: Hydrologic implications of dynamical and statistical approaches to downscaling climate model outputs. Clim. Change **62**(1), 189–216 (2004)

856. Worsley, K.J., Cao, J., Paus, T., Petrides, M., Evans, A.: Applications of random field theory to functional connectivity. Hum. Brain Mapp. **6**(5–6), 364–367 (1998)
857. Wu, X., Zhu, X., Wu, G.Q., Ding, W.: Data mining with big data. IEEE Trans. Knowl. Data Eng. **26**(1), 97–107 (2014)
858. Xiu, D.: Numerical Methods for Stochastic Computations: A Spectral Method Approach. Princeton University Press, Princeton, NJ, USA (2010)
859. Xiu, D., Karniadakis, G.E.: The Wiener–Askey polynomial chaos for stochastic differential equations. SIAM J. Sci. Comput. **24**(2), 619–644 (2002)
860. Xu, G., Genton, M.: Tukey g-and-h random fields. J. Am. Stat. Assoc. **112**(519), 1236–1249 (2017)
861. Xu, G., Genton, M.G.: Tukey max-stable processes for spatial extremes. Spat. Stat. **18**(Part B), 431–443 (2016)
862. Yaglom, A.M.: Some classes of random fields in n-dimensional space, related to stationary random processes. Theory Probab. Appl. **2**(3), 273–320 (1957)
863. Yaglom, A.M.: Correlation Theory of Stationary and Related Random Functions, vol. I. Springer, New York, NY, USA (1987)
864. Yaglom, A.M.: Correlation Theory of Stationary and Related Random Functions. Supplementary Notes and References, vol. II. Springer, New York, NY, USA (1987)
865. Yaremchuk, M., Sentchev, A.: Multi-scale correlation functions associated with polynomials of the diffusion operator. Q. J. R. Meteorol. Soc. **138**(668), 1948–1953 (2012)
866. Yaremchuk, M., Smith, S.: On the correlation functions associated with polynomials of the diffusion operator. Q. J. R. Meteorol. Soc. **137**(660), 1927–1932 (2011)
867. Ye, M., Meyer, P.D., Neuman, S.P.: On model selection criteria in multimodel analysis. Water Resour. Res. **44**(3), W03428 (2008)
868. Yeo, I.K., Johnson, R.A.: A new family of power transformations to improve normality or symmetry. Biometrika **87**(4), 954–959 (2000)
869. Yeong, C.L.Y., Torquato, S.: Reconstructing random media. Phys. Rev. E **57**(1), 495–506 (1998)
870. Yeong, C.L.Y., Torquato, S.: Reconstructing random media. II. Three-dimensional media from two-dimensional cuts. Phys. Rev. E **58**(1), 224–233 (1998)
871. Yokota, Y., Gwinner, K., Oberst, J., Haruyama, J., Matsunaga, T., Morota, T., Noda, H., Araki, H., Ohtake, M., Yamamoto, S., Gläser, P., Ishihara, Y., Honda, C., Hirata, N., Demura, H.: Variation of the lunar highland surface roughness at baseline 0.15–100 km and the relationship to relative age. Geophys. Res. Lett. **41**(5), 1444–1451 (2014)
872. Young, I.R., Ribal, A.: Multiplatform evaluation of global trends in wind speed and wave height. Science **364**(6440), 548–552 (2019)
873. Young, I.R., Zieger, S., Babanin, A.V.: Global trends in wind speed and wave height. Science **332**(6028), 451–455 (2011)
874. Yu, S., Tresp, V., Yu, K.: Robust multi-task learning with t-processes. In: Proceedings of the 24th International Conference on Machine learning, ICML '07, pp. 1103–1110. ACM, New York, NY, USA (2007)
875. Yu, Y., Zhang, J., Jing, Y., Zhang, P.: Kriging interpolating cosmic velocity field. Phys. Rev. D **92**(8), 083527 (2015)
876. Zagayevskiy, Y., Deutsch, C.V.: Multivariate geostatistical grid-free simulation of natural phenomena. Math. Geosci. **48**(8), 891–920 (2016)
877. Zdeborová, L., Krzakala, F.: Statistical physics of inference: thresholds and algorithms. Adv. Phys. **65**(5), 453–552 (2016)
878. Zhang, H.: Inconsistent estimation and asymptotically equal interpolations in model-based geostatistics. J. Am. Stat. Assoc. **99**(465), 250–261 (2004)
879. Zhang, H., Wang, Y.: Kriging and cross-validation for massive spatial data. Environmetrics **21**(3–4), 290–304 (2010)
880. Zhang, J., Atkinson, P., Goodchild, M.F.: Scale in Spatial Information and Analysis. CRC Press, Boca Raton, FL, USA (2014)

881. Zhang, Z., Karniadakis, G.: Numerical Methods for Stochastic Partial Differential Equations with White Noise. Applied Mathematical Sciences, vol. 196. Springer, Cham, Switzerland (2017)
882. Zhong, X., Kealy, A., Duckham, M.: Stream kriging: incremental and recursive ordinary kriging over spatiotemporal data streams. Comput. Geosci. **90**, 134–143 (2016)
883. Zhu, Z., Zhang, H.: Spatial sampling design under the infill asymptotic framework. Environmetrics **17**(4), 323–337 (2006)
884. Zimmerman, D.L.: Another look at anisotropy in geostatistics. Math. Geol. **25**(4), 453–470 (1993)
885. Zimmerman, D.L.: Likelihood-based methods. In: Gelfand, A.E., Diggle, P., Guttorp, P., Fuentes, M. (eds.) Handbook of Spatial Statistics, pp. 45–55. CRC Press, Boca Raton, FL, USA (2010)
886. Zimmerman, D.L., Stein, M.: Classical geostatistical methods. In: Gelfand, A.E., Diggle, P., Guttorp, P., Fuentes, M. (eds.) Handbook of Spatial Statistics, pp. 29–44. CRC Press, Boca Raton, FL, USA (2010)
887. Zimmerman, D.L., Stein, M.: Constructions for nonstationary spatial processes. In: Gelfand, A.E., Diggle, P., Guttorp, P., Fuentes, M. (eds.) Handbook of Spatial Statistics, pp. 119–127. CRC Press, Boca Raton, FL, USA (2010)
888. Zimmerman, D.L., Zimmerman, M.B.: A comparison of spatial semivariogram estimators and corresponding ordinary kriging predictors. Technometrics **33**(1), 77–91 (1991)
889. Zinn, B., Harvey, C.F.: When good statistical models of aquifer heterogeneity go bad: a comparison of flow, dispersion, and mass transfer in connected and multivariate Gaussian hydraulic conductivity fields. Water Resour. Res. **39**(3), WR001146 (2003)
890. Zinn-Justin, J.: Quantum Field Theory and Critical Phenomena, 4th edn. Oxford University Press, Oxford, UK (2004)
891. Zinn-Justin, J.: Path integral. Scholarpedia **4**(2), 8674 (2009). revision #147600
892. Zinn-Justin, J.: Path Integrals in Quantum Mechanics. Oxford University Press, Oxford, UK (2010)
893. Zou, H., Hastie, T.: Regularization and variable selection via the elastic net. J. R. Stat. Soc. Ser. B (Stat Methodol.) **67**(2), 301–320 (2005)
894. Žukovič, M., Hristopulos, D.T.: Environmental time series interpolation based on Spartan random processes. Atmos. Environ. **42**(33), 7669–7678 (2008)
895. Žukovič, M., Hristopulos, D.T.: Spartan random processes in time series modeling. Physica A: Stat. Mech. Appl. **387**(15), 3995–4001 (2008)
896. Žukovič, M., Hristopulos, D.T.: Classification of missing values in spatial data using spin models. Phys. Rev. E **80**(1), 011116 (2009)
897. Žukovič, M., Hristopulos, D.T.: The method of normalized correlations: a fast parameter estimation method for random processes and isotropic random fields that focuses on short-range dependence. Technometrics **51**(2), 173–185 (2009)
898. Žukovič, M., Hristopulos, D.T.: Multilevel discretized random field models with spin correlations for the simulation of environmental spatial data. J. Stat. Mech: Theory Exp. **2009**(02), P02023 (2009)
899. Žukovič, M., Hristopulos, D.T.: A directional gradient-curvature method for gap filling of gridded environmental spatial data with potentially anisotropic correlations. Atmos. Environ. **77**, 901–909 (2013)
900. Žukovič, M., Hristopulos, D.T.: Reconstruction of missing data in remote sensing images using conditional stochastic optimization with global geometric constraints. Stoch. Environ. Res. Risk Assess. **27**(4), 785–806 (2013)
901. Žukovič, M., Hristopulos, D.T.: Short-range correlations in modified planar rotator model. J. Phys. Conf. Ser. **633**(1), 012105 (2015)
902. Žukovič, M., Hristopulos, D.T.: Gibbs markov random fields with continuous values based on the modified planar rotator model. Phys. Rev. E **98**, 062135 (2018)
903. Zwanzig, R.: Nonequilibrium Statistical Mechanics. Oxford University Press, New York, NY, USA (2001)

Index

D. T. Hristopulos, *Random Fields for Spatial Data Modeling*, Advances in Geographic Information Science, https://doi.org/10.1007/978-94-024-1918-4

L

Q